HANDBOOK OF TRANSPORTATION ENGINEERING

ABOUT THE EDITOR

MYER KUTZ is President of Myer Kutz Associates, Inc., a publishing and information services consulting firm. He is the editor of numerous books, including *Biomedical Engineering and Design Handbook*, Second Edition.

HANDBOOK OF TRANSPORTATION ENGINEERING

Volume II: Applications and Technologies

Myer Kutz   Editor

Second Edition

New York   Chicago   San Francisco   Lisbon   London   Madrid
Mexico City   Milan   New Delhi   San Juan   Seoul
Singapore   Sydney   Toronto

Library of Congress Cataloging-in-Publication Data

Handbook of transportation engineering / [edited by] Myer Kutz—2nd ed.
     p.   cm.
    Includes bibliographical references and index.
    ISBN 978-0-07-161492-4 (v. 1 : hardback)—ISBN 978-0-07-161477-1 (v. 2 : hardback)
    1. Transportation engineering—Handbooks, manuals, etc.   2. Traffic engineering—
Handbooks, manuals, etc.
   I. Kutz, Myer.
   TA1151.H34   2011
   629.04—dc22   2010050750

McGraw-Hill books are available at special quantity discounts to use as premiums and sales promotions, or for use in corporate training programs. To contact a representative please e-mail us at bulksales@mcgraw-hill.com.

Handbook of Transportation Engineering, Volume II: Applications and Technologies

1 2 3 4 5 6 7 8 9 0   DOC/DOC   1 9 8 7 6 5 4 3 2 1

ISBN 978-0-07-161477-1
MHID 0-07-161477-X

The pages within this book were printed on acid-free paper.

Sponsoring Editor
Larry S. Hager

Acquisitions Coordinator
Michael Mulcahy

Editorial Supervisor
David E. Fogarty

Project Manager
Vastavikta Sharma, Glyph International

Copy Editor
James K. Madru

Proofreader
Shivani Arora, Glyph International

Production Supervisor
Richard C. Ruzycka

Composition
Glyph International

Art Director, Cover
Jeff Weeks

For Carlos, Rafael, and Irena, starting on their life's journey

CONTENTS

Chapter 5. Traffic Control Systems: Freeway Management and Communications   *Richard W. Denney, Jr.*   5.1

Chapter 6. Traffic Signals   *Richard W. Denney, Jr.*   6.1

Chapter 7. Highway Sign Visibility   *Philip M. Garvey and Beverly T. Kuhn*   7.1

Chapter 8. Roadway Transportation Lighting   *John D. Bullough*   8.1

Chapter 9. Geometric Design of Streets and Highways   *Brian Wolshon*   9.1

Chapter 15. Tunnel Engineering   *Dimitrios Kolymbas*   15.1

Part IV   Non-Automobile Transportation

Chapter 16. Pedestrians   *Ronald W. Eck*   16.3

Chapter 17. Bicycle Transportation   *Lisa Aultman-Hall*   17.1

Chapter 18. The Spectrum of Automated Guideway Transit (AGT) and Its Applications   *Rongfang (Rachel) Liu*   18.1

Chapter 19. Railway Vehicle Engineering   *Keith L. Hawthorne and V. Terrey Hawthorne*   19.1

Chapter 20. Railway Track Design   *Ernest T. Selig*   20.1

Chapter 21. Improvement of Railroad Yard Operations   *Sudhir Kumar*   21.1

Chapter 28. Transportation Hazards   *Thomas J. Cova and Steven M. Conger*   28.1

Chapter 29. Hazardous Materials Transportation   *Linda R. Taylor*   29.1

Chapter 30. Incident Management   *Ahmed Abdel-Rahim*   30.1

Chapter 31. Security and Survivability of Surface Transportation Networks
Ahmed Abdel-Rahim and Paul W. Oman   31.1

CONTRIBUTORS

Hossam Abdelgawad   *Department of Civil Engineering, University of Toronto, Toronto, Ontario, Canada* (CHAP. 32)

Ahmed Abdel-Rahim   *Department of Civil Engineering, University of Idaho, Moscow, Idaho* (CHAPS. 30 AND 31)

Baher Abdulhai   *ITS Center, Department of Civil Engineering, University of Toronto, Toronto, Ontario, Canada* (CHAPS. 1, 2, AND 32)

Lisa Aultman-Hall   *Department of Civil and Environmental Engineering, University of Connecticut, Storrs, Connecticut* (CHAP. 17)

Ballard M. Barker   *School of Aeronautics, Florida Institute of Technology, Melbourne, Florida* (CHAP. 23)

Robert Britcher   *Montgomery Village, Maryland* (CHAP. 24)

John D. Bullough   *Lighting Research Center, Rensselaer Polytechnic Institute, Troy, New York* (CHAP. 8)

Steven M. Conger   *Center for Natural and Technological Hazards, Department of Geography, University of Utah, Salt Lake City, Utah* (CHAP. 28)

Thomas J. Cova   *Center for Natural and Technological Hazards, Department of Geography, University of Utah, Salt Lake City, Utah* (CHAP. 28)

Mohamed El Darieby   *PTRC Grid Computing Center, University of Regina, Regina, Saskatchewan, Canada* (CHAP. 1)

Richard W. Denney, Jr.   *Federal Highway Administration, Lorehsville, Virginia* (CHAPS. 5 AND 6)

Ronald W. Eck   *Department of Civil and Environmental Engineering, West Virginia University, Morgantown, West Virginia* (CHAP. 16)

Lily Elefteriadou   *Department of Civil and Coastal Engineering, University of Florida, Gainesville, Florida* (CHAP. 4)

Rune Elvik   *Institute of Transport Economics, Norwegian Center for Transportation Research, Oslo, Norway* (CHAP. 27)

Philip M. Garvey   *Pennsylvania Transportation Institute, Pennsylvania State University, University Park, Pennsylvania* (CHAP. 7)

William R. Graves   *School of Aeronautics, Florida Institute of Technology, Melbourne, Florida* (CHAP. 23)

Shauna L. Hallmark   *Department of Civil and Construction Engineering, Iowa State University, Ames, Iowa* (CHAP. 34)

Keith L. Hawthorne   *Transportation Technology Center, Inc., Pueblo, Colorado* (CHAP. 19)

V. Terrey Hawthorne   *Newtowne Square, Pennsylvania* (CHAP. 19)

Joseph E. Hummer   *Department of Civil, Construction, and Environmental Engineering, North Carolina State University, Raleigh, North Carolina* (CHAP. 10)

Lina Kattan   *Department of Civil Engineering, University of Calgary, Calgary, Alberta, Canada* (CHAPS. 1 AND 2)

M. Ali Khan   *Ali Khan & Associates, Moorestown, New Jersey* (CHAP. 14)

Kara Kockelman   *Department of Civil, Architectural, and Environmental Engineering, The University of Texas at Austin, Austin, Texas* (CHAP. 3)

Dimitrios Kolymbas   *University of Innsbruck, Faculty of Civil Engineering Sciences, Division of Geotechnical and Tunnel Engineering, Innsbruck, Austria* (CHAP. 15)

Beverly T. Kuhn   *System Management Division, Texas A&M, College Station, Texas* (CHAP. 7)

Sudhir Kumar   *Tranergy Corporation, Bensenville, Illinois* (CHAP. 21)

Yongqi Li   *Arizona Department of Transportation, Phoenix, Arizona* (CHAP. 13)

Rongfang (Rachel) Liu   *New Jersey Institute of Technology, Newark, New Jersey* (CHAP. 18)

Qingchang Lu   *Department of Urban and Regional Planning, University of Florida, Gainesville, Florida; School of Transportation Engineering, Tongji University, Shanghai, China* (CHAP. 35)

William H. Mason   *Department of Aerospace and Ocean Engineering, Virginia; Polytechnic Institute and State University, Blacksburg, Virginia* (CHAP. 22)

Thomas Miesner   *Pipeline Knowledge and Development, Katy, Texas* (CHAP. 26)

Julian Mills-Beale   *Department of Civil and Environmental Engineering, Michigan Technological University, Houghton, Michigan* (CHAPS. 11 AND 12)

Paul W. Oman   *Department of Computer Science, University of Idaho, Moscow, Idaho* (CHAP. 31)

Apostolos Papanikolaou   *National Technical University of Athens, Athens, Greece* (CHAP. 25)

Zhong-Ren Peng   *Department of Urban and Regional Planning, University of Florida, Gainesville, Florida; School of Transportation Engineering, Tongji University, Shanghai, China* (CHAP. 35)

Sarah Perch   *Department of Urban and Regional Planning, University of Florida, Gainesville, Florida* (CHAP. 35)

Judith L. Rochat   *U.S. Department of Transportation/RITA/Volpe Center, Cambridge, Massachusetts* (CHAP. 33)

Ernest T. Selig   *Department of Civil Engineering, University of Massachusetts; Ernest T. Selig, Inc., Hadley, Massachusetts* (CHAP. 20)

Suwan Shen   *Department of Urban and Regional Planning, University of Florida, Gainesville, Florida* (CHAP. 35)

Linda R. Taylor   *North Carolina State University, Raleigh, North Carolina* (CHAP. 29)

David Vanderpool   *Vanderpool Pipeline Engineers Inc., Littleton, Colorado* (CHAP. 26)

Brian Wolshon   *Department of Civil and Environmental Engineering, Louisiana State University, Baton Rouge, Louisiana* (CHAP. 9)

Zhanping You   *Department of Civil and Environmental Engineering, Michigan Technological University, Houghton, Michigan* (CHAPS. 11 AND 12)

Qian Zhang   *Department of Civil Engineering, Michigan Technological University, Houghton, Michigan; School of Civil Engineering, Xi'an University of Architecture and Technology, Xi'an, People's Republic of China* (CHAPS. 11 AND 12)

PREFACE

Volume II of the Second Edition of the *Handbook of Transportation Engineering* focuses on applications and technologies. It is divided into three parts:

Part I: Automobile Transportation: Traffic, Streets, and Highway, which contains 15 chapters

Part II: Non-Automobile Transportation, which consists of 11 chapters

Part III: Safety, Noise, and Air Quality, which contains 9 chapters

Of the 35 chapters in Volume II, 13 are entirely new to the handbook, 12 have been updated from the first edition, and 10 are unchanged. The purpose of these additions and updates is to expand the scope of the parts of the volume and provide greater depth in individual chapters.

The 13 new chapters in Volume II are

Chapter 2: Traffic Origin-Destination Estimation

Chapter 8: Roadway Transportation Lighting

Chapter 11: Pavement Engineering I: Flexible Pavements

Chapter 12: Pavement Engineering II: Rigid Pavements

Chapter 14: Modern Bridge Engineering

Chapter 15: Tunnel Engineering

Chapter 18: The Spectrum of Automated Guideway Transit (AGT) and Its Applications

Chapter 25: Maritime Transport and Ship Design

Chapter 26: Oil and Gas Pipeline Engineering

Chapter 29: Hazardous Materials Transportation

Chapter 31: Security and Survivability of Surface Transportation Networks

Chapter 32: Optimization of Emergency Evacuation Plans

Chapter 35: Climate Change and Transportation

The 12 chapters that contributors have updated are

Chapter 1: Traffic Engineering Analysis

Chapter 3: Traffic Congestion

Chapter 4: Highway Capacity

Chapter 5: Traffic Control Systems: Freeway Management and Communications

Chapter 6: Traffic Signals

Chapter 7: Highway Sign Visibility

Chapter 9: Geometric Design of Streets and Highways

Chapter 10: Intersection and Interchange Design

Chapter 16: Pedestrians

Chapter 30: Incident Management

Chapter 33: Transportation Noise Issues

Chapter 34: Transportation-Related Air Quality

Twenty-five chapters in Volume 2 were written by academics. Nineteen chapters emanate from the United States, three from Canada, and one each from Austria, Greece, and Norway. Of the remaining 10 chapters, 8 were written by experts working in private industry, mainly as consultants, and 2 emanate from contributors working for federal and state transportation departments. I would like to express my heartfelt thanks to all contributors for having taken the opportunity to work on this book. Their lives are terribly busy, and it is wonderful that they found the time to write thoughtful and complex chapters. I developed this handbook because I believed it could have a meaningful impact on the way many engineers and other transportation professionals approach their daily work, and I am gratified that the contributors thought enough of the idea that they were willing to participate in the project. I should add that most of the contributors to the first edition were willing to update their chapters, and it's interesting that even though I've not met most of them face to face, we have a warm relationship and are on a first-name basis. They responded quickly to queries during copyediting and proofreading. It was a pleasure to work with them—we've worked together on and off for nearly a decade. The quality of their work is apparent. Thanks also to my editors at McGraw-Hill for their faith in the project from the outset and to the personnel at Glyph International who guided the manuscript through production. And a special note of thanks to my wife, Arlene, whose constant support keeps me going.

MYER KUTZ
Delmar, New York

VISION STATEMENT

The first edition of the *Handbook of Transportation Engineering* was published in the fall of 2003. It was a substantial reference work, with 38 chapters addressing major areas of interest to transportation engineers and other professionals working in any phase of this civil engineering subdiscipline, including civil engineers, city and regional planners, public administrators, economists, social scientists, and urban geographers. The handbook was divided into five parts: transportation networks and systems, with five chapters covering topics where theory and practical application often intersect; traffic, streets, and highways, with ten chapters dealing with issues involved in moving automobiles, buses, and trucks efficiently; safety, noise, and air quality, with five chapters; non-automobile modes of transportation, with eight chapters covering movement by foot, bicycle, rail, and air; and a section on transportation operations and economic issues, with ten chapters covering issues of particular interest to planners and administrators. Despite the breadth of issues addressed in detail in the handbook, coverage was not as broad as I would have liked mainly because not all the assigned chapters could be delivered in time to meet the publication schedule, as is often the case with large contributed works (unless the editor keeps waiting for remaining chapters to stagger in while chapters already received threaten to become out of date). So, even as the first edition was being published, I looked forward to a second edition, when I could secure more chapters to fill in gaps in the coverage and allow contributors to add greater depth to chapters that had already been published.

The overall plan for the second edition of the *Handbook of Transportation Engineering* was to update or retain as-are most of the chapters that were in the first edition and add half as many new chapters, including chapters with topics that were assigned for the first edition but were not delivered, plus chapters with entirely new topics. Specifically, I was looking for new chapters on the relation of transportation to land use, sustainability, and climate change; hazardous materials transportation and emergency response; and engineering design aspects of pavement, bridges, tunnels, ships, pipelines, and people movers. Because of the size of the Second Edition, I recommended splitting it into two volumes, with 20 chapters in Volume 1 and 35 chapters in Volume 2. The split is uneven but natural: The first volume covers systems and operations, and the second volume covers applications and technologies.

The two volumes have been arranged as follows:

Volume I: Transportation Systems and Operations

    Part I: Transportation Networks and Systems

    Part II: Operations and Economics

Volume II: Transportation Engineering Applications and Technologies

    Part III: Automobile Transportation: Traffic, Streets, and Highways

    Part IV: Non-Automobile Transportation

    Part V: Safety, Noise, and Air Quality

Overall, two-thirds of the 55 chapters in the second edition are new or updated—17 chapters cover topics not included in the first edition and are entirely new, and 20 chapters have been updated. The Preface to each volume provides details about the parts of the handbook and individual chapters.

The intended audience for this handbook, as noted earlier, includes transportation engineers, civil engineers, city and regional planners, public administrators, economists, social scientists, and urban geographers, as well as upper-level students.

To meet the needs of this broad audience, I have designed a practical reference for anyone working directly, in close proximity to, or tangentially to the discipline of transportation engineering and who is seeking to answer a question, solve a problem, reduce cost, or improve the operation of a system or facility. The two volumes of this handbook are not research monographs. My purpose is much more practice-oriented: It is to show readers which options may be available in particular situations and which options they might choose to solve problems at hand. I want this handbook to serve as a source of practical advice to readers. I would like this handbook to be the first information resource a practitioner or researcher reaches for when faced with a new problem or opportunity—a place to turn to before consulting other print sources or even, as so many professionals and students do reflexively these days, going online to Google or Wikipedia. So the handbook volumes have to be more than references or collections of background readings. In each chapter, readers should feel that they are in the hands of an experienced and knowledgeable teacher or consultant who is providing sensible advice that can lead to beneficial action and results.

MYER KUTZ
Delmar, New York

HANDBOOK OF TRANSPORTATION ENGINEERING

P · A · R · T · III

AUTOMOBILE TRANSPORTATION— TRAFFIC, STREETS, AND HIGHWAYS

CHAPTER 1
TRAFFIC ENGINEERING ANALYSIS

Baher Abdulhai
ITS Center, Department of Civil Engineering, University of Toronto
Toronto, Ontario, Canada

Lina Kattan
Department of Civil Engineering, University of Calgary
Calgary, Alberta, Canada

Mohamed El Darieby
PTRC Grid Computing Center, University of Regina
Regina, Saskatchewan, Canada

1.1 TRAFFIC ENGINEERING PRIMER

1.1.1 Traffic Engineering

Traffic engineering, or arguably, traffic control and management in more modern terms, concerns itself with the provision of efficient mobility of people and goods while preserving safety and minimizing all harmful impacts on the environment. A broader look at traffic engineering might include a variety of engineering skills, including design, construction, operations, maintenance, and optimization of transportation systems. Practically speaking, however, traffic engineering focuses more on systems operations than on construction and maintenance activities.

1.1.2 Evolution of Current Transportation Systems and Problems

Automobile ownership and hence dependence and truck usage have been on the rise since the World War II. In the United States, the Federal Aid Highway Act of 1956 authorized the national system of interstate and defense highways. For a couple of decades thereafter, the prime focus was on the creation of this immense mesh of freeways, considered to be the largest public works project in the history of the planet. Very quickly, transportation professionals realized that the growth in automobile use and dependence was outpacing the growth in capacity building, not to mention other problems, such as lack of funds to maintain the giant infrastructure. Ever-rising congestion levels testify to this, and hence the sustainability of continued capacity creation came into the lime light, with fierce criticism by planners and environmentalists alike. Recognition quickly crystallized that we need to move "smarter" and make intelligent use of existing capacity before any attempt to add more. The Intermodal Surface Transportation Efficiency Act (ISTEA—reads "ice tea") of 1991 marked the formal birth of *intelligent transportation systems* (ITS). Heavy emphasis was placed on the use of technology to employ existing capacity efficiently instead of continued new construction, in addition to emphasizing intermodalism and modernization of public transport to curb automobile dependence. ISTEA dedicated $659 million U.S. dollars to research and development and experimental projects geared toward the intelligent use of the national transportation infrastructure. In 1998, the U.S. Congress passed the Transportation Equity Act for the 21st Century (TEA-21) that earmarked US $1.2 billion for mainstreaming ITS with emphasis on deployment. For the most part, similar initiatives took place all over the modern world, including Canada, Europe, Australia, and Japan. Hence the modern transportation engineering field focused on, in addition to the basics, a number of key direction, such as intermodalism, the use of technology to improve transportation provisions under ITS, managing ever-rising congestion through supply control and demand management, and protection of the environment.

1.1.3 Transportation Systems, Mobility, and Accessibility

Land transportation systems include all roadway and parking facilities dedicated to moving and storing private, public, and commercial vehicles. Such facilities serve two principal but contradicting functions: mobility and accessibility. Mobility is the commonsense objective of transportation, aiming at the fastest but safe movement of people or goods. Access to terminal points (homes, businesses) is also essential at trip ends. Mobility requires the least friction with terminal points, whereas accessibility requires slow speeds and hence contradicts mobility. Fortunately, roads systems evolved in a hierarchical manner to serve both without conflict. For urban areas, for instance, the American Association for State Highway and Transportation Officials (AASHTO) defines the hierarchy or roads as follows:

1. Urban principal arterial system including interstate highways, freeways, and other urban arterials, all have some level of access control to promote mobility; typified by high volumes and speeds.

2. Urban minor arterial street system, which augments the freeway system, emphasizes relatively high mobility while connecting freeways to collectors.

3. Urban collector street system, collecting traffic from local streets and streaming them onto arterials, with somewhat balanced emphasis on both mobility and accessibility.

4. Urban local street system, primarily providing access to terminal points, and hence deliberately discourages high mobility and emphasizes low volumes and speeds.

AASHTO has a somewhat similar classification for rural roads, defining the level of mobility versus accessibility provided by each class.

With the preceding hierarchical classification in mind, traffic control and management strategies must recognize and preserve the functional classification of the road at hand. For instance, improper provision of mobility on freeways and arterials might result in neighborhood infiltration by traffic, an undesirable and spreading phenomenon in today's congested urban areas.

1.1.4 Emerging Trends

Intelligent Transportation Systems (ITS).   Intelligent transportation systems (ITS), an emerging global phenomenon, is a broad range of diverse technologies applied to transportation in an attempt to save *lives, money,* and *time.* The range of technologies involved includes microelectronics, communications and computer informatics, and cuts across disciplines such as transportation engineering, telecommunications, computer science, financing, electronics, commerce, and automobile manufacturing. ITS and the underlying technologies will soon put a computer (that has the potential to eliminate human error) in each car to guide us to our destinations, away from congestion, to interact with the road, and even to drive itself. Although ITS sounds futuristic at best, it is becoming reality at a very fast pace. Already, real systems, products, and services are at work all over the world. However, a lot remains to be done; nevertheless, the future of ITS is promising. Many aspects of our lives have been more pleasant and productive as a result of the use of advanced technologies, and it is time for the transportation industry to catch up and benefit from technology. ITS can go beyond a transportation system whose primary controlling technology is the four-way traffic signal.

Some scattered computer-based solutions to transportation problems date back to the late 1950s and early 1960s. The first large-scale application of a computerized signal-control system in the world occurred in metropolitan Toronto, Canada, during the early 1960s. However, the ITS field as we know it today started to crystallize and mature only in the early 1990s. The field was first known as *intelligent vehicle and highway systems* (IVHS). It was further broadened to include all aspects of transportation, and the name was changed to *intelligent transportation systems* (ITS). Several driving forces gave rise to the ITS field. As mentioned earlier, transportation practitioners and researchers alike realized that road building can never keep pace with the increasing demand for travel. Some countries, such as the United States, invested billions of dollars in building road networks and infrastructure and are now faced with the challenge of revitalizing this huge network and making best use of its already existing capacity before expanding further. Another set of driving forces is environmental-related. Damage to the environment from traffic emissions rose to unprecedented alarming levels. In Canada, for instance, transportation represents the single largest source of greenhouse gas emissions, accounting for 27 percent of the total emissions, and this is estimated to increase to 42 percent by the year 2020. The problem is even greater in more car-dependant societies such as the United States. Road safety, or sometimes the lack thereof, and the escalating death tolls and injuries in traffic accidents each year are yet a third set of forces. For these reasons, collectively, more road building is not always viable or desirable. High-tech computer, electronic, and communications technologies offer one attractive and promising approach and hence the modern appeal for ITS. In addition, a healthy ITS industry also would have other non-traffic-related fringe societal benefits. Stimulation of new information technology (IT)–based industries and creation of new markets and jobs are some examples of such fringe benefits. Therefore, ITS is more than just intelligent solutions on the road, but rather a new strategic direction for national and international economies. The market share of ITS is projected to expand over the next decade from an annual world market of $25 billion in 2001 to $90 billion in 2011. A projected $209 billion will be invested in ITS between now and the year 2011 (ITS-America). Access to this sizable market is vital to the transportation and related technology sectors.

One important attribute of this emerging new face of the transportation industry—shaped by ITS—is that it is no longer restricted to civil engineers, nor is it restricted to a single department or agency. Given the broad range of technologies involved, the ITS field is multidepartmental, multiagency, and multijurisdictional, cutting across the public, private, and academic sectors. This broadness certainly will enhance potential, widen scope, and revolutionize the way we handle our transportation systems, but it also will pose institutional challenges that we must be aware of and prepared for.

ITS Subsystems.   Collectively, ITS aims to *enhance the utilization of existing roadway capacity,* as well as *increasing capacity* itself. Enhancing the use of existing capacity is achievable through improved distribution of traffic, dynamically sending traffic away from congested hot spots to underutilized segments of the network, and through the elimination of bottleneck-causing controls such as conventional toll plazas. Increasing the physical capacity itself is possible through the automation of driving and the elimination of the human behavior element altogether. This is the promise of automated highway systems that potentially could double or triple the number of vehicles a single lane can handle. From this perspective, one can divide ITS into two main categories of systems:

1. Advanced traffic management and traveler information systems (ATMS and ATIS or combined as ATMIS)

2. Advanced vehicle control and automated highway systems (AVCS and AHS)

ATMIS provide extensive traffic surveillance, assessment of recurring congestion owing to repetitive high demands and detection of nonrecurring congestion owing to incidents, traffic information and route guidance dissemination to drivers, and adaptive optimization of control systems (such as traffic signals and ramp meters). Current and near-future trends in ATMIS tend to rely on centralized management in traffic management centers (TMCs). TMCs gauge traffic conditions by receiving information from vehicle detectors throughout the network as well as the vehicles themselves as probes formulate control measures in the center and disseminate control to field devices, as well as information and guidance to drivers. Newer trends of "distributed" control are emerging but are not well crystallized yet. The main distinguishing characteristics of ATMIS are real-time operation and network-wide multijurisdictional implementation.

AVCS provides better control of the vehicle itself, either by assisting the driver or by automating the driving process in an autopilot-like fashion in order to increase capacity and enhance safety. Full highway system automation (AHS) can result in higher speeds at lesser headways and hence higher lane capacity. Automation can be applied to individual vehicles as free agents in a nonautomated mix of traffic or as fully automated lanes carrying platoons of electronically linked vehicles. Although AHS is technically promising, there remains an array of unsettled issues, including legal liabilities in the event of incident owing to any potential automatic controller failure, technical reliability issues, and social issues. Therefore, AHS remains to be "futuristic" at the current stage of ITS. The feasible alternative, however, is to use the technology to assist the driver who remains in control of the vehicle, that is, making the vehicle smarter. Such intelligent vehicles will detect obstacles on the road and in blind spots and warn the driver accordingly, maintain constant distance to the vehicle ahead, and alert a sleepy driver going off the road. As technology improves further, the role of the intelligent vehicle can move from a simple warning to full intervention and accident prevention by applying the breaks or overriding faulty steering decisions.

The prime distinction between ATMIS and AVCS is that ATMIS focus on smoothing out traffic flow in the network by helping the driver make best route-choice decisions and by optimizing the control systems in the network, whereas AVCS focus on the driver, the operation of the vehicle, and traffic maneuvers in the immediate vehicle vicinity. AVCS focus on enhancing the driver's awareness and perception, aiding decision making by providing early warning and potentially initiating action, and eventually using sensory inputs and computer control to substitute human sensory reactions and control. The need for bridging the gap between in-vehicle telematics and ITS traffic services leads to the emergence of the IntelliDrive initiative. Thus the IntelliDrive initiative is intended to provide a communications link between in-vehicle equipment and between vehicles and out-of-vehicle systems (commonly referred to as *infrastructure-based*). IntelliDrive combines leading-edge technologies—advanced wireless communications, on-board computer processing, advanced vehicle sensors, GPS navigation, smart infrastructure, and others—to provide the capability for vehicles to identify threats and hazards on the roadway and communicate this information over wireless networks to give drivers alerts and warnings (www.intellidrive.org).

Historically, research interest in Intellidrive [previously known as VII (vehicle infrastructure integration)] started with the automated highway system (AHS) initiative in the United States. In recent years, research thrust has been aimed at improving road safety. At present, research objectives have been broadened to include other areas of interest, such as the environment, security, and efficiency (AASHTO 2008; Shladover 2008; Misener 2008; Ball 2005; Fries et al. 2009; ITS International 2008; ITS America 2005).

A number of disciplines have to provide input for the success of IntelliDrive, particularly transportation engineering and computer, communications, and software engineering. Although there is an important role for automotive engineering, this discipline becomes essential when vehicle controls require modifications. The communication link that has been used most frequently is the dedicated short-range communication (DSRC) link.

Technology proof-of-concept tests took place in Michigan and California (U.S. Government 2008; Shladover 2008; Misener 2008; ITS World International Conference 2009). These tests have provided useful information, although of limited scope. The focus of tests was to examine an interoperable communications system that could support applications related to

1. Warning drivers of unsafe conditions or imminent collisions
2. Warning drivers to prevent a run-off-road accident or that speed around a curve is too high

3. Providing information to system operators on real-time congestion, weather conditions, road surface conditions, and traffic incidents

4. Providing information to operators on corridor capacity for real-time management, planning, and provision of corridor-wide alerts to drivers.

Thus no tests were carried out on supporting mobility and providing data for transportation management (ITS International 2008). Recently, it was recognized that the scope of IntelliDrive should be broadened to include a vehicle-to-vehicle capabilities as well as vehicle-to-infrastructure capabilities.

In addition to reporting progress, recent literature items have pointed a number facing IntelliDrive (Row et al. 2008; ITS International August 2008):

5. Public acceptance is far from certain owing to privacy concerns. However, this concern is not justified, given that according to design, the system has the capability to detect signal and speed violations, but there is no capability to identify and report violators. The system can alert the violators and/or approaching vehicles for the purpose of preventing collisions. There are other facets of public concerns. Advocates of recreational driving do not favor any kind of automation, even if it is the receipt of information. Also, critics of tolling are of the opinion that it will make driving very expensive for motorists in the lower-income bracket.

6. A number of technical and nontechnical issues have been reported.

   a. The perception is that costs are high.

   b. The auto manufacturers and agencies responsible for infrastructure-based systems have to work together in order for the IntelliDrive to function.

   c. The issue of how to update and maintain the in-vehicle units has to be addressed.

   d. Even if generic and easy-to-maintain in-vehicle terminals can be developed and updating can be done automatically via the Internet, the availability of the Internet in every vehicle is far from certain.

   e. Security of the in-vehicle and road-based units is a concern.

   f. Another issue is the requirement to digitize the inputs for the IntelliDrive system (e.g., existing traffic signs and roadway markings).

   g. Compatibility and formatting issues are a major concern. A projection of current developments suggests the likely outcome that initial products will be tailored to individual applications, and unless action is taken in the early stages of IntelliDrive research and development (R&D), it would be difficult to costly to overhaul the system.

   h. Regulations will be required to set in place access to the IntelliDrive data and communications between applicable agencies.

ITS User Services. Another way to look at the constituents of ITS is from the end-user perspective. In the United States, for instance, a collection of interrelated user services is defined and grouped into what could be called *user-service bundles*. As reported by the Intelligent Transportation Society of America (ITS-America, for short), 29 user services have been defined to date, as summarized in Table 1.1. This list of services and their definitions/descriptions are expected to evolve and undergo further refinements in time. User services are comprised of multiple technological elements or functions, which may be common with other services. For example, a single user service usually will require several technologies, such as advanced communications, mapping, and surveillance, which may be shared with other user services. This commonality of technological functions is one basis for the suggested bundling of services. In some other cases, the institutional perspectives of organizations that will deploy the services provided the rationale for the formation of a specific bundle. The users of this service or the ITS stakeholders include travelers using all modes of transportation, transportation management center operators, transit operators, metropolitan planning organizations (MPOs), commercial vehicle owners and operators, state and local governments, and many others who will benefit from deployment of ITS.

TABLE 1.1   ITS User Services

Bundle	User Services
1. Travel and traffic management	1. Pretrip travel information
	2. En-route driver information
	3. Route guidance
	4. Ride matching and reservation
	5. Traveler services information
	6. Traffic control
	7. Incident management
	8. Travel demand management
	9. Emissions testing and mitigation
	10. Highway rail intersection
2. Public transportation management	1. Public transportation management
	2. En-route transit information
	3. Personalized public transit
	4. Public travel security
3. Electronic payment	1. Electronic payment services
4. Commercial vehicle operations	1. Commercial vehicle electronic clearance
	2. Automated roadside safety inspection
	3. On-board safety monitoring
	4. Commercial vehicle administrative processes
	5. Hazardous materials incident response
	6. Commercial fleet management
5. Emergency management	1. Emergency notification and personal security
	2. Emergency vehicle management
6. Advanced vehicle control and safety systems	1. Longitudinal collision avoidance
	2. Lateral collision avoidance
	3. Intersection collision avoidance
	4. Vision enhancement for crash avoidance
	5. Safety readiness
	6. Precrash restraint deployment
	7. Automated vehicle operation
7. Information management	1. Archived data function
8. Maintenance and construction management	1. Maintenance and construction operation

ITS Architecture.   We deal with and benefit from *systems architectures* almost every day in our lives, although we might not know what an architecture is or what it is for. For instance, one day you purchase a television set, and later you purchase a videocassette recorder from a different retailer and by a different manufacturer, but you never worry about whether they will work together or not. Similarly, you could buy a low-end radio receiver and a high-end compact disc player and again assume that they will work together just fine. You travel with your FM radio receiver, and it works everywhere. This seamless operation of different systems or components of a system is not by chance, thanks to a mature industry and widely adopted architectures and related standards that ensure such interoperability. The ITS industry is evolving rapidly in different places all over the world, and different groups are pursuing its development. Users are at risk of investing in or adopting certain ITS equipment that works only locally. Similarly, if left without adequate guidance, stakeholders easily could develop systems solutions to their needs that are incompatible with those of their regional neighbors. For instance, an in-vehicle navigation system purchased in California might not work in Nevada, or a system purchased in Ontario, Canada, might not work once the user crosses the border to New York in the United States. Therefore, to ensure seamless ITS operation, some sort of global or at least national system architecture and related standards are needed. To fully maximize the potential of ITS technologies, system design solutions must be compatible at the system interface level in order to share data, provide coordinated, area-wide integrated operations, and support interoperable equipment and services where appropriate. An ITS architecture provides this overall guidance to ensure system, product, and service compatibility/interoperability without limiting the design options of the stakeholder. In this book, we once again use the U.S. ITS national system architecture only as an example. We are confident that similar efforts are underway in almost every ITS-active country, producing architectures that are more or less similar to the American architecture.

In the United States, the Congress directed the U.S. Department of Transportation (DOT) to promote nationwide compatibility of ITS. Spearheaded by the U.S. DOT and the Intelligent Transportation Society of America (ITS-America), four major teams were formed in 1993. The four teams proposed four different architectures. The most promising two architectures were selected, integrated, and refined in the form of the final architecture.

A rich set of documents describing every detail of the final architecture can be found on the ITS-America Web site (www.itsa.org), which we summarize in the following section.

It is important to understand that the architecture is neither a system design nor a design concept. It is framework around which multiple design approaches can be developed to meet the individual needs of users, while maintaining the benefits of a common architecture noted earlier. The architecture defines the *functions* (e.g., gather traffic information or request a route) that must be performed to implement a given user service, the *physical subsystems* where these functions reside (e.g., the roadside or the vehicle), the *interfaces/information flows* between the physical subsystems, and the *communications* requirements for the information flows (e.g., wire line or wireless). In addition, it identifies and specifies the requirements for the standards needed to support national and regional interoperability, as well as product standards needed to support economy-of-scale considerations in deployment. The function view of ITS is referred to as the *logical architecture,* as shown in Figure 1.1. Functions such as "Manage Traffic," for instance, can be further divided into finer processes, such as "Detect Pollution Levels" and "Process Pollution Data." The systems view is referred to as the *physical architecture,* as shown in Figure 1.1. The figure also shows communications requirements and information flows. The physical architecture partitions the functions defined by the logical architecture into systems and, at a lower level, subsystems based on the functional similarity of the process specifications and the location where the functions are being performed. The physical architecture defines four systems—traveler, center, roadside, and vehicle—and 19 subsystems. Subsystems are composed of equipment packages with specific functional attributes. Equipment packages are defined to support analyses and deployment. They represent the smallest units within a subsystem that might be purchased. In deployments, the character of a subsystem deployment is determined by the specific equipment packages chosen. For example, one municipal deployment of a traffic management subsystem may select "Collect Traffic Surveillance" and "Basic Signal Control" equipment packages, whereas a state traffic management center may select "Collect Traffic Surveillance" and "Freeway Control" packages. In addition, subsystems may be deployed individually or in aggregations or combinations that will vary by geography and time based on local deployment choices. A traffic management center may include a traffic management subsystem, an information provider subsystem, and an emergency management subsystem, all within one building, whereas another traffic management center may concentrate only on the management of traffic with the traffic management subsystem.

The architecture has identified four communication media types to support the communication requirements between the 19 subsystems. They are wire line (fixed to fixed), wide-area wireless (fixed to mobile), dedicated short-range communications (fixed to mobile), and vehicle to vehicle (mobile to mobile). Wire-line technology, such as leased or owned twisted wire pairs, coaxial cable, or fiberoptics, can be used by a traffic management center to gather information and to monitor and control roadway subsystem equipment packages (e.g., traffic surveillance sensors, traffic signals, and changeable message signs). Although wireless communications technologies also can be used in this case, they are used to provide fixed-to-fixed communications, and consequently, the architecture recognizes them as wire-line communications media. One- or two-way wide-area wireless (fixed-to-mobile) communications are suited for services and applications where information is disseminated to users who are not located near the source of transmission and who require seamless coverage,

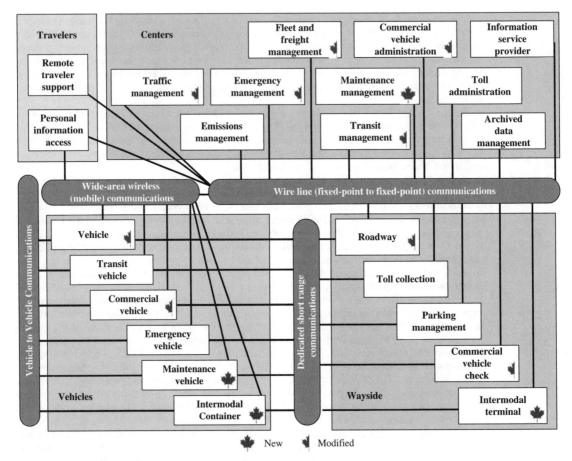

FIGURE 1.1 The Canadian physical architecture and communications connections.

such as the case for traveler information and route guidance dissemination. Short-range wireless communications are concerned with information transfer that is of a localized interest. There are two types of short-range wireless communications identified by the architecture. They are vehicle-to-vehicle communications that support AHS and dedicated short-range communications (DSRC) used in applications such as automated toll collection.

In conclusion, the basic benefit of the architecture is to provide a structure that supports the development of open standards. This results in numerous benefits: The architecture makes integration of complex systems easier, ensures compatibility, and supports multiple ranges of functionality and designs.

ITS and Potential Economic Stimulation.   As mentioned earlier, ITS is more than just solutions to traffic problems. Investments in the ITS industry are actually large-scale infrastructure investments that feature widespread application of high technology. The arguable question is whether ITS would promote national-level economic growth? IT related industries are increasingly becoming the heart of the economy of many industrialized nations worldwide. Directing IT investments toward large-scale infrastructure developments has the potential to promote new industries that would have an impact on long-term economic growth. This might take place for several reasons (Transportation Infostructures 1995). First, such investments would create large economies of scale for new computing and communications products even before they could attain such scale economies in the marketplace. This would offer more rapid returns on investment to supporters of such industries. Second, as a consequence of such success, capital markets, which are usually risk-averse, might be more inclined to support such industries and make larger funds available for their expansion. Third, this would speed the adoption of new generations of communications and computing technology.

Governments can play critical roles in accelerating the growth of the ITS industry. Public investments could shape the setting of system architectures and standards and influence the development of applications that would encourage private-sector investments by lowering the risk perceived by private investors. A national ITS mandate likely would reduce some of the risks and result in private companies taking a longer-range view of returns to their capital spending. As a matter of fact, the role of governmental leadership in the ITS industry can be appreciated easily if one contrasts the rapid growth of the ITS industry in a country such as the United States with that of its neighbor, Canada, for example. In the United States, the Intermodal Surface Transportation Efficiency Act of 1991 (ISTEA) promoted and accelerated ITS research and development using federal funds and resulted in large-scale involvement from the private sector. In Canada, on the other hand, the absence of a similar federal ITS mandate is severely crippling the growth rate of the ITS industry and the related job market, forcing Canadian talents and entrepreneurs in the ITS field to be export-oriented, shifting focus and effort toward the American and international markets in general.

Sustainability. A 1987 United Nations–sponsored report, entitled, "Our Common Future," defined *sustainable development* as a *"development that meets the needs of the present without compromising the ability of future generations to meet their own needs."* Thus any economic or social development should improve rather than harm the environment. Sustainability has recently begun to be applied to cities. With increasing environmental awareness, the urban transportation planning process has become more concerned with the air, land, water, and the likely ecological impact of transportation facilities.

As such, sustainable transportation planning states that cities can become more livable, more humane, and more healthy, but they must learn how to achieve this by using fewer natural resources, creating less waste, and decreasing their impact on their environment.

Cities are increasingly involved in pursuing this sustainability agenda. For this purpose, specific indicators are defined to guide cities to move toward more livable communities while reducing their impact on the earth and the ecosystem. Examples of such indicators are taken from Newman and Kenworthy (1998) and are listed below:

- Energy and air quality (e.g., the reduction of energy use per capita, air pollutants, and greenhouse gases)
- Water, materials, and waste (e.g., the reduction of total water use per capita, solid waste, consumption of building materials per capita)
- Land, green spaces, and biodiversity (e.g., preserve agricultural land and natural landscape and green space, increase the proportion of urban redevelopment to new developments, and increase the density of the population and employment in transit-oriented locations)
- Transportation (e.g., reduce auto use per capita, increase transit, walk, bike, and carpool use, and decrease parking spaces)
- Livability, human amenities, and health (e.g., decrease infant mortality, increase educational attainment, decrease transport fatalities, and increase proportion of cities allowing mixed-used and higher-density urban villages)

These indicators are a scaled-down version of the original 150 indicators suggested by the World Bank and the UN Center for Human Settlements. These indicators serve as evidence of how cities are contributing to global problems, namely, greenhouse gases and oil depletion. Every particular city has to define its own indicators that are applicable to its conditions and consequently manage both local and global issues.

In order to meet the sustainability indicators, cities have to develop sustainability plans, known also as *Local Agenda 21 plans*, as stated in Agenda 21:

> Each Local authority should enter into a dialogue with its citizens, local organizations and private enterprises and adopt a "local" Agenda 21. Local authorities should learn from citizens and local, civic, community, business, and industrial organizations the information needed for formulating the best strategies. This process will also increase household awareness of sustainable development issues [Sitarz 1994, p. 177].

Sustainability plans require two central approaches, namely, integrated planning and community participation. The integrated planning deals with the fusion of cities' physical and environmental planning with economic planning. On the other hand, community participation calls for public participation in planning. In other words, urban plans should be designed with local citizens to meet in harmony their local needs.

Automobile dependence is recognized as being a great threat to sustainability. In fact, one of the central arguments for sustainable development is concerned with the critical impacts of an automobile-based transportation system on a society. Thus, in essence, changes in travel behavior must occur in order to minimize transportation's impact on the environment (Newman and Kenworthy 1998). As such, three general approaches are to be implemented simultaneously to limit auto dependency and consequently change cities over time to become more sustainable. These three approaches are as follows:

- *Automobile technological improvements.* The technological improvement approach deals with the development of less polluting cars to reduce air pollutants and emissions.
- *Economic instruments.* This approach deals with setting the right road user charges to meet the real costs of auto usage, such as pollution cost, health costs, road and parking cost, etc.
- *Planning mechanisms.* This approach deals with the need of a non-automobile-dependent planning. The new urbanism trend encourages environmental friendly commuting modes, such as mass transit, cycling, and walking. This can be achieved by changing the urban fabric to become denser and involve mixed land use.

In addition, in order to reduce the automobile dependence in cities, five policies should be followed. These policies bring together the processes of traffic calming, state-of-the art mass transit, bicycle planning, and transit-oriented development, the neotraditional urban design of streets for pedestrians (in particular, the design of urban villages), growth management, and economic penalties on private transportation.

1.2 TRAFFIC STREAM PARAMETERS AND THEIR MEASUREMENT

1.2.1 Characteristics of Traffic Flow

Traffic flow can be divided into two primary types: *interrupted flow* and *uninterrupted flow. Uninterrupted flow* occurs when vehicles traversing a length of roadway are not required to stop by any cause external to the traffic stream, such as traffic control devices. Uninterrupted flow is regulated by vehicle-vehicle interactions on one side and by the interactions between vehicles and the roadway environment and geometry on the other side. An instance of uninterrupted flow includes vehicles traveling on an interstate highway or on other limited-access facilities where there are no traffic signals or signs to interrupt the traffic. Uninterrupted flow also can occur on long sections of rural surface highway between signalized intersections. Even when such facilities are experiencing congestion, breakdowns in the traffic stream are the result of internal rather than external interactions in the traffic stream.

Interrupted flow occurs when flow is interrupted periodically by external means, primarily traffic control devices such as STOP and YIELD signs and traffic signals. Under interrupted flow conditions, traffic control devices play a primary role in defining the traffic flow, whereas vehicle-vehicle inter-actions and vehicle-roadway interactions play only a secondary role. For instance, traffic signals allow designated movements to occur only part of the time. In addition, because of the repeated stop-ping and restarting of the traffic stream on such facilities, flow occurs in platoons.

1.2.2 Traffic Stream Parameters

Traffic stream parameters represent the engineer's quantitative measure for understanding and describing traffic flow. Traffic stream parameters fall into two broad categories: *macroscopic param-eters,* which characterize the traffic stream as a whole, and *microscopic parameters,* which character-ize the behavior of individual vehicles in the traffic stream with respect to each other.

The three macroscopic parameters that describe traffic stream are

- Volume or rate of flow
- Speed
- Density

Volume and Flow.   *Volume* is simply the number of vehicles that pass a given point on a roadway or a given lane or direction of a highway in a specified period of time. The unit of volume is simply *vehicles,* although it is often expressed as annual, daily, hourly, peak, and off-peak. The following sec-tions explain the range of commonly used daily volumes, hourly volumes, and subhourly volumes.

Daily Volumes.   Daily volumes are used frequently as the basis for highway planning, for gen-eral trend observations, and for traffic volume projections. Four daily volume parameters are used widely, namely, average annual daily traffic (AADT), average annual weekday traffic (AAWT), average daily traffic (ADT), and average weekday traffic (AWT).

- *AADT* is the average 24-hour traffic volume at a given location over a full year. Therefore, it is the total number of vehicles passing the site in a year divided by 365. AADT is normally obtained from permanent counting stations, typically bidirectional flow data rather than lane-specific flow data.
- *AAWT* is the average 24-hour traffic volume occurring on weekdays over a full year. AAWT is nor-mally obtained by dividing the total weekday traffic for the year by the annual weekdays (usually 260 days). This volume is of particular importance because weekend traffic is usually low; thus the average higher weekday volume over 365 days would hide the impact of the weekday traffic.
- *ADT* is the average 24-hour traffic volume at a given location for a period of time less than a year (e.g., summer, 6 months, a season, a month, a week). ADT is valid only for the period of time over which it was measured.
- *AWT* is the average 24-hour traffic volume occurring on weekdays at a given location for a period of time less than a year, such as a month or a season.

The units describing all these volumes are vehicles per day (vdp). Daily volumes are often not dif-ferentiated per lane or direction but rather are given as totals for an entire facility at a particular location.

Hourly Volumes.   As mentioned previously, daily volumes are used mainly for planning applica-tions. They cannot be used alone for design and operational analysis. Hourly volumes are designed to reflect the variation of traffic over the different time periods of a day. They are also used to identify single-hour or period of highest volume in a day occurring during the morning and evening commute, that is, *rush hours.* The single hour of the day corresponding to the highest hourly volume is referred to as the *peak hour.* The peak-hour traffic volume is a critical input in the design and operational analysis of transportation facilities. The peak-hour volume is usually a directional traffic, that is, the direction of flows is separated. Highway design as well as other operations analyses, such as signal design, must adequately serve the peak-hour flow corresponding to the peak direction.

Peak-hour-volumes sometimes can be estimated from AADT as follows:

$$DDHV = AADT \times K \times D$$

where DDHV = directional-design hourly volume (vph)

AADT = average annual daily traffic (24 hours) (vpd)

K = factor for proportion of daily traffic occurring at peak hour

D = factor for proportion of traffic in peak direction

K and D values vary depending on the regional characteristics of the design facilities, namely, rural versus urban versus suburban. K often represents the AADT proportions occurring during the thirtieth or fiftieth highest peak hour of the year. K factor is inversely proportional to the density of development surrounding the highway. As such, in design and analysis of rural areas, the thirtieth highest peak-hour volume is used, whereas in urbanized areas, the fiftieth highest is used. D factor depends on both the concentration of developments and the specific relationship between the design facility and the major traffic generators in the area. The *Highway Capacity Manual 2000* provides ranges for K and D factors depending on facility types and the corresponding regional characteristics of the area.

Subhourly Volumes.   Subhourly volumes represent traffic variation within the peak hour, that is, short-term fluctuations in traffic demand. In fact, a facility design may be adequate for design hour, but breakdown may occur owing to short-term fluctuations. Typical designs and operational analyses are based on 15-minute peak traffic within the peak hour (e.g., level of service analysis using the *Highway Capacity Manual*).

The peak-hour factor (PHF) is calculated to relate the peak flow rate to hourly volumes. This relationship is estimated as follows:

$$PHF = \frac{V}{4 \times V_{15}}$$

where PHF = peak-hour factor

V = peak-hour volume (vph)

V_{15} = volume for peak 15-minute period (veh)

The PHF describes trip generation characteristics. When PHF is known, it can be used to convert a peak-hour volume to an estimated peak rate of flow within an hour:

$$v = \frac{V}{PHF}$$

where v = peak rate of flow within hour (vph)

V = peak hourly volume (vph)

PHF = peak-hour factor

Speed.   The *speed* of a vehicle is defined as the distance it travels per unit of time. In other words, it is the inverse of the time taken by a vehicle to traverse a given distance. Most of the time, each vehicle on the roadway will have a speed that is somewhat different from the speed of the vehicles around it. In quantifying the traffic stream, the average speed of the traffic is the significant variable. The average speed, called the *space mean speed*, can be found by averaging the individual speeds of all the vehicles in the study area.

Space Mean Versus Time Mean Speed.   Two different ways of calculating the average speed of a set of vehicles are reported, namely, the space mean speed and the time mean speed. This difference in computing the average speed leads to two different values with different physical significance.

While the *time mean speed* (TMS) is defined as the average speed of all vehicles passing a point on a highway over a specified time period, the *space mean speed* (SMS) is defined as the average speed of all vehicles occupying a given section of a highway over a specified time period.

TMS is a point measure, and SMS is a measure relating to a length of highway or lane. TMS and SMS may be computed from a series of measured travel times over a measured distance as follows: TMS takes the arithmetic mean of the observation. TMS is computed as

$$\text{TMS} = \frac{\sum \frac{d}{t_i}}{n}$$

SMS is computed by dividing the distance of a given section of the highway by an average travel time as follows:

$$\text{SMS} = \frac{d}{\sum \frac{t_i}{n}} = \frac{nd}{\sum t_i}$$

where TMS = time mean speed (fps or mph)
      SMS = space mean speed (fps or mph)
      D = distance traversed (ft or mi)
      N = number of travel times observed
      t_i = travel time for the ith vehicles (sec or hr)

Wardrop (1952) derived the relationship between the time mean speed (TMS) and the space mean speed (SMS):

$$\text{TMS} = \text{SMS} + \frac{\sigma^2_{\text{SMS}}}{\text{SMS}}$$

where σ^2_{SMS} is the variance of the space-mean speed. Thus, as shown in this formula, TMS is usually larger than SMS, except in the case of a uniformly traveling flow.

Density. *Density* refers to the number of vehicles present on a given length of roadway or lane. Normally, density is reported in terms of vehicles per mile or vehicles per kilometer. High densities indicate that individual vehicles are very close to each other, whereas low densities imply greater distances between vehicles. Density is a difficult parameter to measure directly in the field. Direct measurements of density can be obtained through aerial photography, which is an expensive method. Hence density can be estimated from the density, flow, speed relationship, as explained in the following paragraphs.

Flow, Speed, and Density Relationship. Speed, flow, and density are all related to each other and are fundamental for measuring the operating performance and level of service of transportation facilities, such as freeway sections. Under uninterrupted flow conditions, speed, density, and flow are all related by the following equation:

$$\text{Flow} = \text{density} \times \text{speed} \qquad v = S \times D$$

where v = flow (vehicles/hour)
      S = space mean (average running) speed (miles/hour, kilometers/hour)
      k = density (vehicles/mile, vehicles/kilometer)

The general form of relationships among speed, density, and flow is illustrated in Figure 1.2, also known as the *fundamental diagram of traffic flow*. The relationship between speed and density is consistently decreasing. As density increases, speed decreases. This diagram and the preceding formula show that flow is zero under two different conditions:

• When density is zero. Thus there is no vehicle on the road.

• When speed is zero. Vehicles are at a complete stop because of traffic congestion.

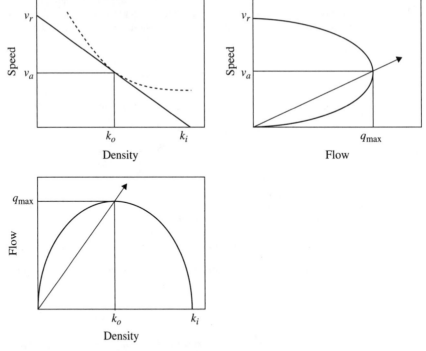

FIGURE 1.2   Fundamental flow-speed-density diagram.

In the first case, the speed corresponds to the theoretical maximum value, the *free-flow speed* v_0, whereas in the second case, the density assumes the theoretical maximum value, the *jam density* K_{jam}. The peak of the density-flow curve (and speed-flow curve) occurs at the theoretical maximum flow (i.e., capacity) of the facility. The corresponding speed v_c and density k_c are referred to as the *critical speed* and the *critical density* at which maximum capacity occurs.

Density is the most important of the three traffic-stream parameters because it is a measure most directly related to traffic demand and to congestion levels. In fact, traffic is generated from various land uses, bringing trips on a highway segment. Generated trips produce traffic density, which, in turn, produce flow rate and speeds. Density also gives an indication of the quality of flow on the facilities. It is a measure of proximity of vehicles, and it is also the basis for level-of-service (LOS) analysis on uninterrupted facilities. In addition, density readings in contrast to flow measurements clearly distinguish between congested and uncongested conditions.

1.2.3   Other: Gap, Headway, and Occupancy

Flow, speed, and density are macroscopic parameters characterizing the traffic stream as a whole. Headway, gap, and occupancy are microscopic measures for describing the space between individual vehicles. These parameters are discussed in the following paragraphs.

Headway.   *Headway* is a measure of the temporal space between two vehicles. More specifically, headway is the time that elapses between the arrival of the leading vehicle and the arrival of the following vehicle at the designated test point along the lane. Headway between two vehicles is measured by starting a chronograph when the front bumper of the first vehicle crosses the selected point and subsequently recording the time that the second vehicle's front bumper crosses over the designated point. Headway is usually reported in units of seconds.

Average values of headway are related to macroscopic parameters as follows:

$$\text{Average headway} = 1/\text{flow} \qquad \text{or} \qquad v = \frac{3600}{h_a}$$

where v = rate of flow (vehicle/hr)
h_a = average headway (sec)

Gap. *Gap* is very similar to headway, except that it is a measure of the time that elapses between the departure of the first vehicle and the arrival of the second at the designated test point. Gap is a measure of the time between the rear bumper of the first vehicle and the front bumper of the second vehicle, whereas headway focuses on front-to-front times. Gap is also reported in units of seconds. Figure 1.3 illustrates the difference between gap and headway.

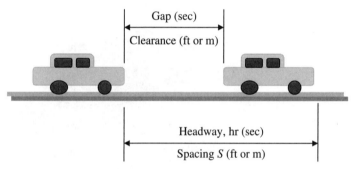

FIGURE 1.3   Illustration of gap and headway definition.

Occupancy. *Occupancy* denotes the proportion or percentage of time a point on the road is occupied by vehicles. Occupancy is measured using *loop detectors*. It is measured as the fraction of time that vehicles are on the detector.

Therefore, for a specific time interval T, occupancy is the sum of the time that vehicles cover the detector divided by T. For each individual vehicle, the time spent on the detector is determined as function of the vehicle's speed, headway, and length L plus the length C of the detector itself. That is, the detector is affected by the vehicle from the time the front bumper crosses the start of the detection zone until the time the rear bumper clears the end of the detection zone. Occupancy is computed as follows:

$$\text{LO} = \frac{(L+C)/\text{speed}}{\text{headway}} = (L+C) \times \text{density} = k \times (L+C) \qquad \text{Assuming flow} = \text{density} \times \text{speed}$$

where LO = lane occupancy, that is, percentage of time a lane is occupied with vehicles divided by total study time
K = density of flow
L = average vehicle length
C = length of detector

Therefore, if occupancy is measured as above, density can be estimated as

$$k = \frac{\text{LO}}{(L+C)}$$

1.2.4   Loop Detector as Measuring Device

The inductive loop detector is by far the most common form of detector used for both traffic counting and traffic management purposes. Loop detectors are used to measure traffic volume, flow rate, vehicle speed, and occupancy. Inductance loops are used widely in detectors systems and are known for their reliability in data measurements, flexibility in design, and relatively low cost (Figure 1.4).

Loop detector principal components include the following:

- One or more turns of insulated wire buried in a narrow, shallow saw cut in the roadway
- Lead-in cable that connects the loop to the detector via a roadside pull-out box
- Detector unit (or amplifier) that interprets changes in the electrical properties of the loop when a vehicle passes over it

Several data types could be determined from inductive loop detectors, namely, lane occupancy, traffic densities, traffic composition, average and instantaneous vehicle velocities, presence of congestion,

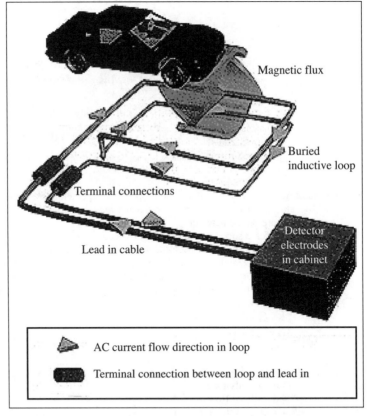

FIGURE 1.4   Car passing over inductive loop buried in pavement. The loop system becomes active when the detector unit sends an electric current through the cable, creating a magnetic field in the loop. When a vehicle passes over the loop, the metal of the vehicle disturbs the magnetic field over the loop, which causes a change in the loop's inductance. Inductance is an electrical property that is proportional to the magnetic field. The induced magnetic field increases the frequency of oscillation that is sensed by the detector unit. As such, the loop sensor detects a vehicle.

and length and duration of traffic jams. Depending on the technology used, these data can be determined directly or indirectly by the inductive loop detectors.

Loop detectors are also necessary to measure the data that will be used to construct a traffic model and to calibrate that model, as we will see in the next section (i.e., checking whether the behavior predicted by the model corresponds accurately enough with the real behavior of the system).

1.3 TRAFFIC FLOW THEORY

Knowledge of fundamental traffic flow characteristics, namely, speed, volume, and density and the related analytical techniques is an essential requirement in planning, design, and operation of transportation systems. Fundamental traffic flow characteristics have been studied at three levels: microscopic, mesoscopic, and macroscopic. Existing traffic flow models are based on time headway, flow, time-space trajectory, speed, distance headway, and density. These models lead to the development of a range of analytical techniques such as demand-supply analysis, capacity and level-of-service analysis, traffic stream modeling, shock-wave analysis, queuing analysis, and simulation modeling (May 1990).

In addition to traffic flow theory, traffic simulation models are also classified as microscopic, macroscopic, and mesoscopic. Microscopic simulation models are based on car-following principles and typically are computationally intensive but accurate in representing traffic evolution. On the other hand, macroscopic models are based on the movement of traffic as a whole by employing flow-rate variables and other general descriptors representing traffic at a high level of aggregation as a flow without distinguishing its parts. This aggregation improves computational performance but also reduces the detail of representation. Mesoscopic models lie between the other two approaches and balance accuracy of representation and computational performance. They represent average movement of a group of vehicles (packets) on a link. Microscopic analysis may be selected for moderate-sized systems, where there is a need to study the behavior of individual units in the system. Macroscopic analysis may be selected for higher-density, large-scale systems, in which a study of the behavior of groups of units is adequate. Knowledge of traffic situations and the ability to select the more appropriate modeling technique are required for specific problems. In addition, simulation models also differ in the effort needed for the calibration process. As such, microscopic models are the most difficult to calibrate, followed by mesoscopic models. Macroscopic models, on the other hand, are calibrated relatively easily.

1.3.1 Traffic Flow Models

Microscopic traffic flow modeling is concerned with individual time and space headway between vehicles, whereas macroscopic modeling is concerned with macroscopic flow characteristics. The latter are expressed as flow rates, with attention given to temporal, spatial, and modal flows (May 1990). This section describes the most known macroscopic, mesoscopic, and microscopic traffic flow models.

Macro Models.   In a macroscopic approach, the variables to be determined are

- The flow $q(x, t)$ (or volume) corresponding to the number of vehicles passing a specific location x in a time unit and at time period t
- The space mean speed $v(x, t)$ corresponding to the instantaneous average speed of vehicles in a length increment
- The traffic density $k(x, t)$ corresponding to the number of vehicles per length unit

These macroscopic variables are defined by the well-known equation

$$q(x, t) = k(x, t) \cdot v(x, t)$$

The static characteristics of the flow are completely defined by a fundamental diagram (as shown in Figure 1.2). The macroscopic approach considers traffic stream parameters and develops algorithms that relate flow to density and space mean speed. As such, various speed-density models have been developed and are shown to also fit experimental data. These models are explained below.

Greenshields Model.   The first steady-state speed-density model was introduced by Greenshields, who proposed a linear relationship between speed and density as follows:

$$u = u_f - \left(\frac{u_f}{k_j}\right) \times k$$

where u = velocity at any time
u_f = free-flow speed
k = density at that instant
k_j = maximum density

As mentioned previously, in these equations, as the flow increases, density increases and speed decreases. At optimal density, flow becomes maximum (q_m) at $u = u_f/2$ and $k = k_j/2$.

Greenberg Model.   A second early model was suggested by Greenberg (1959), showing a logarithmic relationship:

$$u = c \ \ln(k/k_j)$$

where u = velocity at any time
c = a constant (optimal speed)
k = density at that instant
k_j = maximum density

It is easy to show that in Greenberg model the maximum flow occurs at $q_m = ck_j/e$ and the corresponding optimal density $k_0 = k_j/e$.

Three-Dimensional Models.   The idea of considering all three fundamental variables (q, k, v) simultaneously first appeared in the transportation research board (TRB) SR-165. The notion of a three-dimensional model appeared in the form of Figure 1.5, where "$v = q/k$ represents the surface of admissible traffic stream models." The surface shown in Figure 1.5 is a continuous one; thus, by accepting that the $u = q/k$ relationship holds for the entire range of traffic operations, one can reasonably conclude that it suffices to study traffic modeling as a two-dimensional problem (Hall 1998).

Greenshields and Greenberg models are mainly applicable to uninterrupted flow. Unfortunately, these models are unable to cope with the added complexities that are generated by congestion. Other models such as shock-wave theory, the Lighthill and Whitham (LW) model, and queuing models can be used in the analysis of bottlenecks.

Lighthill and Whitham (LW) Model.   The continuous-flow approach proposed by Lighthill and Whitham (1955) represents the *aggregate behavior* of a large number of vehicles. This model is applicable to the distribution of traffic flow on long and crowded roads. The LW model reproduces qualitatively a remarkable amount of real traffic phenomena, such as decreasing speeds with increasing densities and shock-wave formation.

The LW model is derived from the physical law of incompressible fluids and is based on the following three fundamental principles (Cohen 1991):

1. *Continuous representation of variables.* This considers that at a given location x and time t, the traffic mean speed $v(x, t)$, flow $q(x, t)$, and density values $k(x, t)$ are continuous variables and satisfy the relation $v(x, t) = q(x, t)/k(x, t)$.

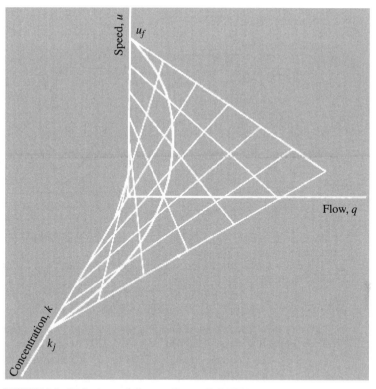

FIGURE 1.5   3D fundamental diagram. (*Source:* Hall 1998.)

2. *The law of conservation of mass.* This is a basic speculation of the simple continuum model that states that vehicles are not created or lost along the road. The law of the conservation of the number of vehicles leads to the continuity equation for the density $k(x, t)$:

$$\frac{\partial k(x, t)}{\partial t} + \frac{\partial q(x, t)}{\partial x} = 0$$

3. *The statement of fundamental diagrams.* The fundamental hypothesis of the theory is that at any point on the road, the speed u is a function of the density. In addition, speed is a decreasing function of concentration: $u = u(k)$.

Therefore, the law of traffic on a given section of the road during a given time period can be expressed in terms of an equation relating two of the three variables: flow, concentration, and speed (Cohen 1991).

For a macroscopic description of the theory, the flow q (veh/hr), density k (veh/km), and mean speed v (km/hr) are considered as differentiable functions of time t and space x (Papageorgiou 1998).

From the continuity equation with the flow-density relation $[q = q(k)]$ and the basic relation between traffic variables ($q = uk$), a differential equation of the density k is derived as follows:

$$\frac{\partial k(x, t)}{\partial t} + q'(k)\frac{\partial k(x, t)}{\partial x} = 0$$

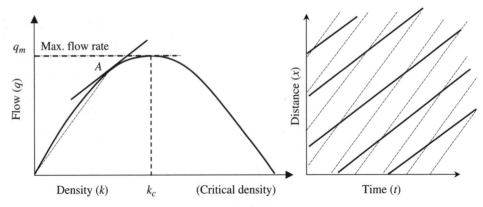

FIGURE 1.6   Speed and kinematic waves. (*Source:* Cohen 1991.)

The kinematic waves theory attempts to solve this partial differential equation to predict the concentration of flow at any point on the road at any time.

Figure 1.6 shows how the propagation speed of a shock wave corresponds to the slope of the tangent on the fundamental diagram.

This hypothesis implies that slight changes in traffic flow are propagated through the stream of vehicles along *kinematic waves*. As such, waves are propagated either

- forward, when density k is less than the critical density k_c, which corresponds to the uncongested region of the flow-density diagram, or
- backward, when k is greater than critical density k_c, which corresponds to the congested region of the flow-density diagram.

This property leads to a distinction between two types of flow, namely, uncongested flow (for $k < k_c$) and congested flow ($k > k_c$). In practice, once congestion occurs, disturbance propagates backward from downstream.

Under several assumptions and simplifications, the LW model is consistent with a class of car-following models. With regard to urban traffic flow in signalized networks, the LW model is more than sufficient because traffic flow dynamics are dominated by external events (red traffic lights) rather than by the inherent traffic flow dynamics (Papageorgiou 1998). Also, for freeway traffic flow, the LW model achieves a certain degree of qualitative accuracy and certainly is an improvement over purely static approaches. However, the LW model includes a number of simplifications and fails to reproduce some real dynamic phenomena observed on freeways.

Shock Waves. Flow-speed-density states changes over space and time (May 1990). With the prompt occurrence of such change, a boundary is established that marks a discontinuity of flow and density from one side of the boundary with respect to the other. This discrepancy is explained by the generation of shock waves. Basically, a shock wave exists whenever the traffic conditions change abruptly. As such, shock waves can be generated by collisions, by sudden increases in speed caused by entering free-flow conditions, or by a number of other means. A shock then represents a mathematical discontinuity (abrupt change) in k, q, or u.

Figure 1.7 shows two different densities, flows, and speeds of vehicles moving along a highway. The line separating these two flows represents the shock wave and is moving at a speed w_{AB}.

The propagation velocity of shock waves is given as

$$w_{AB} = q_B - q_A / k_B - k_A$$

where w_{AB} = propagation velocity of shock wave (mi/hr or km/hr)
q_B = flow prior to change in conditions (veh/hr)
q_A = flow after change in conditions (veh/hr)

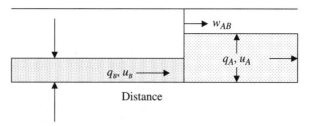

FIGURE 1.7   Shock-wave analysis fundamentals. (*Source:* May 1990.)

k_B = traffic density prior to change in conditions (veh/mi or veh/km)
k_A = traffic density after change in conditions (veh/mi or veh/km)

Thus the shock wave separating the two flows travels at an intermediate speed. Since the shock wave in Figure 1.7 is moving with the direction of traffic, it is a positive forward-moving shock wave. On the other hand, backward-moving shock waves or negative shock waves travel upstream or against the traffic stream.

Figure 1.8 demonstrates the use of traffic waves in identifying the occurrence of a shock wave and following its trajectory. The figure on the left represents a flow-concentration curve. The figure on the right represents the occurrence of a shock wave and follows its trajectory. On the qk curve, point A represents a situation where traffic travels at near capacity, implying that speed is well below the free-flow speed. Point B represents an uncongested condition where traffic travels at a higher speed because of the lower density. Tangents at points A and B represent the wave velocities of these two situations. The line connecting the two points on the qk curve represents the velocity of the shock wave. In the space-time diagram, the intersection of these two sets of waves has a slope equal to the slope of the line connecting the two points A and B on the qk curve. This intersection represents the velocity of the shock wave.

Second Order Model: Payne Model.   Payne (1971) proposed a method for relating macroscopic variables and car-following theories. Payne developed an extended-continuum model that takes into consideration drivers' reaction times and uses a dynamic speed equation, as shown below:

$$\frac{dV}{dt} = \frac{1}{\tau}[V_e(\rho) - V] - \frac{D(\rho)}{\rho\tau}\frac{\partial\rho}{\partial x}$$

where the term $[V_e(\rho) - V]/\tau$ is denoted by the *relaxation* term, and the term $[D(\rho)/\rho\tau](\delta\rho/\delta x)$ is denoted by the *anticipation* term.

The relaxation term allows for the delayed adjustment of the stream to a prespecified speed $V_e(\rho)$ as a result of reaction time τ and braking or acceleration procedures. On the other hand, the anticipation term allows for drivers to adjust their speeds in advance of changes in density lying ahead.

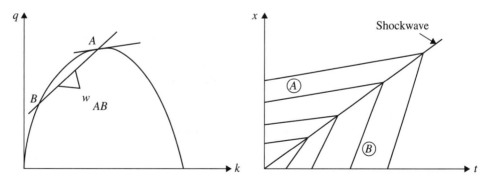

FIGURE 1.8   Shock-wave diagrams. (*Source:* Cohen 1991.)

The second-order model provides the possibility of a more realistic description of traffic flow (Kim 2002). As such, the shock-wave problem is alleviated through application of the diffusion terms. Moreover, unstable congested states are derived by the interplay between anticipation and relaxation effects in the model. The second-order model has a critical density above which uniform flow conditions are unstable and the wave is oscillating with ever-increasing amplitude. In addition, the presence of oscillating waves explains the stop-and-go traffic condition.

Microscopic Models.   Much research has been devoted to the concept that traffic stream behavior can be analyzed at the microscopic level. At this level, the behavior of individual drivers must be examined and modeled. Microscopic models use car-following laws to describe the behavior of each driver-vehicle system in the traffic stream as well as their interaction. Examples of microscopic models include car-following models, General Motors models, and cell transmission and cellular automata models.

Car Following.   These models are based on supposed mechanisms describing the process of one vehicle following another, the so-called follow-the-leader models (Lieberman and Rathi 1998). From the overall driving task, the subtask that is most relevant to traffic flow is the task of one vehicle following another on a single lane of roadway (car following). This particular driving subtask is relatively simple to describe by mathematical models compared with other driving tasks. Car-following models describe the process of the car-following task in such a way as to approximate the macroscopic behavior of a single lane of traffic. Hence car-following models form a bridge between individual car-following behavior and the macroscopic world of a line of vehicles and their corresponding flow and stability properties.

Pipes and Forbes Car-Following Models.   Car-following theories were developed in the 1950s and 1960s. Early models employed simple rules for determining the distance gap between vehicles. For example, Pipes (1953) argued that the rule that drivers actually follow is the following, as suggested by the California Motor Vehicle Code: "The gap that a driver should maintain should be at least one car length for every 10 mph of speed at which he is traveling."

Using the notation shown in Figure 1.9 for the gap and the vehicle speed, the resulting distance headway d can be written as

$$d_{min} = \left[x_n(t) - x_{n+1}(t)\right]_{min} = L_n \left[\frac{\dot{x}_{n+1}(t)}{(10)(1.47)}\right] + L_n$$

According to Pipes car-following theory, the minimum safe distance headways increase linearly with distance.

General Motors Models.   In reality, drivers conform to the behavior of the immediately leading vehicle. Under this notion, a stimulus-response relationship exists that describes the control process of the driver-vehicle system. Researchers at General Motors (GM) have developed car-following models

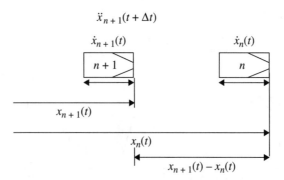

FIGURE 1.9   Car-following model.

and tested them using real-world data. The importance of these models lies in the discovery of the mathematical bridge linking the microscopic and macroscopic theories of traffic flow (May 1990).

The GM research team developed five generations of car-following models in terms of response-stimuli relationships. The general stimulus-response equation expresses the concept that a driver of a vehicle responds to a given stimulus according to the relation (May 1990)

$$\text{response} = \text{function }\{\text{sensitivity, stimuli}\}$$

where response = acceleration or deceleration of the vehicle, which depends on the sensitivity of the automobile and the driver himself or herself
sensitivity = ability of the driver to perceive and react to the stimuli
stimuli = relative velocity of the lead and following vehicles

The stimulus function may be composed of many factors, namely, speed, relative speed, intervehicle spacing, accelerations, vehicle performance, driver thresholds, etc. The relative velocity is the most used term. It is generally assumed in car-following modeling that a driver attempts to (1) keep up with the vehicle ahead and (2) avoid collisions. The response is the reaction of the driver to the motion of the vehicle immediately in front of him or her. The response of successive drivers is to react (i.e., accelerate or decelerate) proportionally to the stimulus.

From the notation of Figure 1.10, the model assumes that the driver of the following vehicle will space himself or herself from the leading vehicle at a distance such that in case the leading vehicle comes to an emergency stop, he or she will be able to come to a rest without crashing. Thus the spacing of the two vehicles at time t will be

$$d(t) = [x_n(t) - x_{n+1}(t)] = \Delta T \cdot x_{n+1}(t + \Delta t) + b_{n+1} + L - b_n$$

where b is the stopping distance of the vehicle.

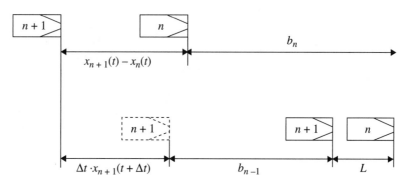

FIGURE 1.10 Simple car-following General Motors principle.

Assuming equal braking distances for the two vehicles and differentiating with respect to time t, we obtain

$$[\dot{x}_n(t) - \dot{x}_{n+1}(t + \Delta t)] = \Delta T \cdot \ddot{x}_{n+1}(t + \Delta t) \qquad \text{or} \qquad \ddot{x}_{n+1}(t + \Delta t) = \frac{1}{\Delta T}[\dot{x}_n(t) - \dot{x}_{n+1}(t + \Delta t)]$$

where b = is the stopping distance of the vehicle
Δt = reaction time (reciprocal of sensitivity)

The response of the following vehicle is to decelerate by an amount proportional to the difference in speeds. The measure of sensitivity is the reciprocal of the perception-reaction time of the driver. The response function is taken as the deceleration of the following vehicle. The response is lagged by the perception-reaction time of the following driver.

Another form of the simple car-following model is to distinguish the reaction time from the sensitivity by introducing a sensitivity term α as follows:

$$\ddot{x}_{n+1}(t+\Delta t) = \alpha[\dot{x}_n(t) - \dot{x}_{n+1}(t)]$$

The unit of the sensitivity term is sec^{-1}. The stimuli term $[\dot{x}_n(t) - \dot{x}_{n+1}(t)]$ could be positive, negative, or zero, causing the response to be, respectively, either an acceleration, deceleration, or constant speed.

The two parameters α (sensibility term) and Δt (reaction time) must be selected in such a way that traffic behaves realistically. The choice of these terms is associated with the concept of stability, which is explained below.

Traffic Stability. There are two important types of stability in the car-following system: local stability and asymptotic stability. *Local stability* is concerned with the response of a following vehicle to a fluctuation in the motion of the vehicle directly in front of it; that is, it is concerned with the localized behavior between pairs of vehicles. *Asymptotic stability* is concerned with the manner in which a fluctuation in the motion of any vehicle, say, the lead vehicle of a platoon, is propagated through a line of vehicles. The analysis of traffic stability determines the range of the model parameters over which the traffic stream is stable.

Improvements over the First Generation of the GM Model. The first GM model was derived using a functional value for acceleration with the assumption that driver sensitivity is constant for all vehicles (May 1990). In a revised version of the GM model, discrepancy from field values indicated that the sensitivity of the driver was higher whenever the headway was less. Accordingly, the GM model was adjusted to account for this error.

Further improvements in the sensitivity were introduced by the speed difference, that is, relative velocity, because as the speed difference increases, the sensitivity increases. Every system has a time lag to react to changes occurring ahead of it. This is accounted for by the term Δt, which represents the reaction time on the part of the following vehicle to accelerate and decelerate. Finally, the powers of the terms of speed and headway of the vehicle ahead were proposed, and these constants were called *speed component m* and *headway component l*. The resulting equation represents the fifth and final GM model and is stated as follows:

$$\ddot{x}_{n+1}(t+\Delta t) = \frac{\alpha_{l,m} \dot{x}_{n+1}^m(t+\Delta t)}{[x_n(t) - x_{n+1}(t)]^l} \cdot [\dot{x}_n(t) - \dot{x}_{n+1}(t)]$$

This is the generalized model, and all previous GM models can be considered a special case of this model (May 1990).

Macro-to-Micro Relationship. Gazis et al. (1959) studied the relationship between car-following models and macroscopic traffic stream models. They demonstrated that almost all macroscopic models were related to almost all car-following models (May 1990). As such, Gazis et al. derived a generalized macroscopic model from the car-following models. The generalized form of the derived macroscopic models is as follows:

$$v^{1-m} = v_f^{1-m}\left[1 - (k/k_j)^{l-1}\right]$$

For instance, the Greenshields model lies within the following feasible range: when $m = 0$ and $l = 2$. Figure 1.11 shows the speed-density relationship for a number of cases with $m = 1$ and $l = 2.0$ to 3.0.

New Trends in Microscopic Traffic Flow Modeling.

DAGANZO'S CELL-TRANSMISSION MODEL. Cell-transmission models of highway traffic, developed by Daganzo (1994), are discrete versions of the simple continuum (kinematic wave) model of

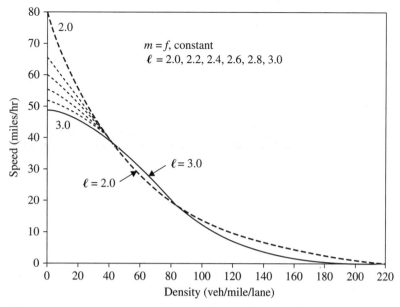

FIGURE 1.11 Speed-density relationships for various values of m and l. (*Source:* May 1990.)

traffic flow that are convenient for computer implementation. They are in the Godunov family of finite-difference approximation methods for partial differential equations. In cell-transmission models, the speed is calculated from the updated flow and density rather than being updated directly.

In the cell-transmission scheme, the highway is partitioned into small sections (cells). The analyst then keeps track of the cell contents (number of vehicles) as time passes. The record is updated at closely spaced instants (clock ticks) by calculating the number of vehicles that cross the boundary separating each pair of adjoining cells during the corresponding clock interval. This average flow is the result of a comparison between the maximum number of vehicles that can be sent by the cell directly upstream of the boundary and those which can be received by the downstream cell.

The sending (receiving) flow is a simple function of the current traffic density in the upstream (downstream) cell. The particular form of the sending and receiving functions depends on the shape of the highway's flow-density relation, the proximity of junctions, and whether the highway has special lanes (e.g., turning lanes) for certain vehicles (e.g., exiting vehicles). Although the discrete and continuum models are equivalent in the limit of "vanishing" small cells and clock ticks, the need for practically sized cells and clock intervals generates numerical errors in actual applications.

The cell-transmission representation can be used to predict traffic's evolution over time and space, including transient phenomena such as the building, propagation, and dissipation of queues.

CELLULAR AUTOMATA MODELS. Recently, there has been a growing interest in studying traffic flow with cellular automata (CA) models. CA models are conceptually simple rules that can be used to simulate a complex physical process by considering a description at the level of the basic components of the system (Jiang). Only the essential features of the real interactions are taken into account in the evolution rules. Through the use of powerful computers, these models can capture the complexity of real-world traffic behavior and produce clear physical patterns that are similar to real phenomena.

Nagel and Schreckenberg (1992) introduced the CA model for traffic. The rationale of CA is not to try to describe a complex system from a global point of view, as it is described using, for instance, differential equations, but rather modeling this system starting from the elementary dynamics of its interacting parts. In other words, CA does not describe a complex system with complex equations;

rather, it lets the complexity emerge through the interaction of simple individuals following simple rules. These simple models have been shown to reproduce, at least qualitatively, the features of real traffic flow.

TRANSIMS microsimulation has adapted the CA technique for representing driving dynamics and simulating traffic in entire cities. In these models, the basic idea is to formulate a model in space and time. The space is the road divided into grid points or cells (typically 7.5 m in length, which corresponds to the length that a car uses up in a jam). A cell is either empty or occupied by exactly one vehicle. In addition, car positions are updated synchronously in successive iterations (discrete time steps) (Dupuis and Chopard 2001). During motion, each car can be at rest or jump to the nearest neighbor site, along the direction of motion. The rule is simply that a car moves only if its destination cell is empty. In essence, drivers do not know whether the car in front will move or is stuck by another car. That is, the state of a cell $s(t)$ at a given time depends only on its own state one time step previously and the states of its nearby neighbors at the previous time step. This dynamic can be summarized by the following relation:

$$s_i(t+1) = s(t)_{i-1}[1 - s(t)_i] + s(t)_i s(t)_{i+1}$$

where t is the discrete time step.

All cells are updated together. Movement takes place by *hopping* from one cell to another using 1-sec time step, which agrees with the reaction-time arguments. Different vehicle speeds are represented by different hopping distances (Nagel and Rickert 2001). This implies, for example, that a hopping speed of five cells per time step corresponds to 135 km/h. Accordingly, the rules for car following in the CA model are

1. If no car is ahead, linear acceleration occurs up to maximum speed.

2. If a car is ahead, then velocity is adjusted so that it is proportional to the distance between the cars (constant-time headway).

3. Sometimes vehicles are randomly slower than what would result from (1) and (2).

Lane changing is done as pure sideways movement in a subtime step, before the forward movement of the vehicles; that is, each time-step is subdivided into two subtime steps (Nagel and Rickert 2001). The first subtime step is used for lane changing, whereas the second subtime step is used for forward motion. Lane-changing rules for TRANSIMS are symmetric and consist of two simple elements: Decide that you want to change lanes, and check if there is enough gap to "get in." A "reason to change lanes" is either that the other lane is faster or that the driver wants to make a turn at the end of the link and needs to get into the correct lane. In the latter case, the accepted gap decreases with decreasing distance to the intersection; that is, the driver becomes more and more desperate. In addition, details of the system, including lane changing, complex turns, and intersection configurations, are fully represented, and each driver is given a destination and a preferred path.

In more advanced work Nagel and Rickert (2001) proposed the parallel implementation of the TRANSIMS traffic microsimulation. In this parallelism, the road network is partitioned across many processors. This means that each CPU of the parallel computer is responsible for a different geographic area of the simulated region. The results show a significant speed-up in the computation efficiency.

Meso Models. The mesoscopic models fall in between the macroscopic and microscopic models. Mesoscopic traffic flow models describe the microscopic vehicle dynamics as a function of macroscopic fields. The gas-kinematic model, which is the most used mesoscopic traffic flow model, treats vehicles as a gas of interacting particles (Nagatani 2002). As such, when the number of vehicles is large, traffic flow is modeled in terms of one compressible gas. Prigogine and Herman have proposed the following Boltzmann equation for the traffic:

$$\frac{\partial f(x, v, t)}{\partial t} + v\frac{\partial f(x, v, t)}{\partial x} = -\frac{f(x, v, t) - \rho(x, t)F_{des}(v)}{\tau_{rel}} + \left(\frac{\partial f(x, v, t)}{\partial t}\right)_{int}$$

where the first term on the right-hand side represents the relaxation of the velocity distribution function $f(x, v, t)$ to the desired velocity distribution $\rho(x, t)F_{des}(v)$, with the relaxation time τ_{rel}, in the absence of the interactions of vehicles. The second term on the right-hand side takes into account the change arising from the interactions among vehicles.

1.3.2 Traffic Simulation Models

Computer simulation modeling has been a valuable tool for analyzing and designing complex transportation systems. Simulation models are designed to "mimic" the behavior of these systems and processes. These models predict system performance based on representations of the temporal and/ or spatial interactions between system components (normally vehicles, events, and control devices), often characterizing the stochastic nature of traffic flow. In general, the complex simultaneous interactions of large transportation system components cannot be described adequately in mathematical or logical forms. Properly designed models *integrate* these separate entity behaviors and interactions to produce a detailed quantitative description of system performance.

In addition, simulation models are mathematical/logical representations (or *abstractions*) of real-world systems that take the form of software executed on a digital computer in an experimental fashion (Lieberman and Rathi 1998). The inherent value of computer simulation is that it allows experimentation to take place off-line without having to go out in the real world to test or develop a solution. Specifically, simulation offers the benefits of being able to control input conditions, treat variables independently even though they may be coupled in real life, and most important, repeat the experiment many times to test multiple alternative performances (Middleton and Cooner 1999). The user of traffic simulation software specifies a "scenario" (e.g., highway network configuration and traffic demand) as model inputs. The simulation model results *describe* system operations in two formats: (1) statistical and (2) graphical. The numerical results provide the analyst with detailed quantitative descriptions of *what* is likely to happen. Traffic simulation models may be classified according to the *level of detail* with which they represent the transportation performance, as well as flow representation, namely (see Table 1.2):

- In *microscopic mo*dels, traffic is represented discretely (single vehicles); individual trajectories can be explicitly traced. Disaggregate performance measures are calculated based on explicit modeling of driver behavior.

- In *mesoscopic models*, traffic is represented discretely (vehicles or group of vehicles); individual trajectories can be explicitly traced as for microscopic models. However, aggregate performance measures are calculated as for macroscopic models.

- In *macroscopic models*, traffic is represented continuously following the fluid approximation. Individual trajectories are not explicitly traced. Aggregate performance measures are calculated using relations derived from fluid-approximation models (Cascetta 2001).

TABLE 1.2 Classification and Examples of Traffic Simulation Models

		Performance Functions	
		Aggregate	Disaggregate
Flow Representation	Continuous	Macroscopic For example: FREFLO, AUTOS, METANET	—

	Discrete	Mesoscopic
		For example: DYNASMART,
		    DYNAMIT, INTEGRATION
			Microscopic
			For example: INTRAS, CORSIM,
			    PARAMICS, CORSIM, AIMSUN2,
			    TRANSIMS, VISSIM, MITSIM

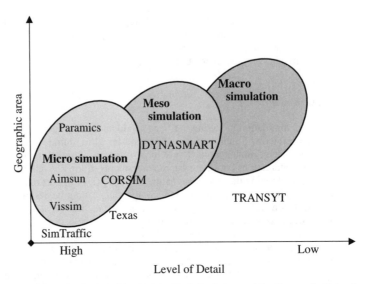

FIGURE 1.12   Scale and level of detail of simulation models. (*Source:* Institute of Transportation Studies, University of California, Irvine.)

Table 1.2 and Figure 1.12 illustrate the relationship between the type of simulation and the level of detail. In addition, Figure 1.12 illustrates the relation between the type of simulation and the size of the transportation network that needs to be analyzed.

Microscopic Simulation Models.   A *microscopic* model describes both the system entities and their interactions at a high level of detail. Microscopic models simulate the journey of each single vehicle through explicit driving behavior models of speed adjustment (Cascetta 2001). These models contain processing logic, which describes how vehicles behave, including acceleration, deceleration, lane changes, passing maneuvers, turning movement execution, and gap acceptance. Typically, vehicles are input into the project section using a statistical arrival distribution (a stochastic process) and are tracked through the section on a second-by-second basis (Hoogendoorn and Bovy 2001).

Microscopic models describe vehicles (and often drivers) individually with varying characteristics and multiple classes. In addition, these models describe the prevailing surrounding conditions in detail, such as traffic control logics (e.g., pretimed, actuated, adaptive). Their main trend has been to integrate as many traffic phenomena as possible. For instance, a lane-change maneuver at the micro level could use the car-following law for the subject vehicle with respect to its current leader. In addition, other detailed driving decision processes are reproduced, such as acceleration and braking and driving behavior (such as in-route choice, lane change, gap acceptance, and overtaking). The duration of the lane-change maneuver also can be computed. Such models can be solved only by event- or time-based simulation techniques.

Microscopic models provide very detailed traffic simulation on a small scale, yet they require a significant amount of data and effort for specification and calibration. Therefore, their application in *online traffic control* is limited owing to the large computation time. For these reasons, microscopic models are used primarily for *offline traffic* operations rather than transportation planning (Cascetta 2001). Microsimulators can prove quite useful to design intersections, to analyze accurate emission and fuel consumption modeling, to regulate traffic lights, and to study the impact of variable message signs or of ramp metering and any other form of traffic control (Marchal 2001). Moreover, because of the large computing resources needed and expensive calibration procedures, this highly detailed approach is practical only for small networks (i.e., a few hundred links). Nevertheless, recent advances in cellular automata models, such as TRANSIMS, precisely attempt to fill this gap with microsimulations of many thousands of link networks with the help of massive parallel computers.

Macroscopic Simulation Models.   Macroscopic models normally describe the movement of traffic as a whole by employing flow rate variables and other general descriptors representing movement at a high level of aggregation as a flow without distinguishing its parts. For instance, the traffic stream is represented in an aggregate manner using such characteristics as flow rate, density, and velocity. Macroscopic simulation takes place on a section-by-section basis rather than tracking individual vehicles. In addition, the choice of individual vehicle maneuvers, such as a lane change, is usually not explicitly represented (Schutter et al. 1999).

Macroscopic models are suited for large-scale, network-wide applications, where macroscopic characteristics of the flow are of prime interest. Generally, calibration of macroscopic models is relatively simple (compared with microscopic and mesoscopic models). However, macroscopic models generally are too coarse to correctly describe microscopic details and impacts caused by changes in roadway geometry (Hoogendoorn and Bovy 2001).

Mesoscroscopic Simulation Models.   Mesoscopic models fall in between macroscopic and microscopic models. Generally, they represent most entities at a high level of detail but describe their activities and interactions at a much lower level of detail than would a microscopic model. As such, a mesoscopic model does not distinguish nor trace individual vehicles; rather, it specifies the *behavior* of individuals in probabilistic terms. Individual vehicles with the same characteristics are represented by *packets* (small groups of vehicles). Mesoscopic models assume that packets of vehicles are moved together or that some "patterns" of decisions are modeled instead of individual decisions. Hence each vehicle within a packet has the same origin and destination, the same route, and the same driver characteristics. In addition, vehicles on a link have the same speed, which is generally time-dependent. While the traffic is represented discretely by tracing the trips of single packets as characterized by a departure time, overtaking, lane-changing, and car-following behavior are not modeled microscopically. Instead, the aggregate speed-volume interaction of traffic (*uq* relationship) is used on each link. Usually the smaller the size of the packet, the more realistic is the analysis. Mesoscopic models are able to simulate queue formation and spillbacks reasonably.

As such, the computation time needed for the simulation is reduced compared with micro models (Schutter et al. 1999) but is still high compared with macroscopic models. Mesoscopic models therefore can handle medium-sized transportation networks. They also have the advantage of enabling description behaviors of individual vehicles without the need to describe their individual time-space behavior.

Model Calibration and Validation.

Model Calibration.   In order to mimic real-world decisions, models must be calibrated. *Model calibration* is the process of quantifying model parameters using real-world data in the model logic so that the model can realistically represent the traffic environment being analyzed (Middleton and Cooner 1999).

Model calibration is conducted mainly by comparing user experimental conditions from simulation results with observed data from the real network (fields counts). The simulation is said to be accurate if the error between simulation results and the observed data is small enough. Thus model parameters should be optimized to match (possibly site-specific) observed settings. However, finding the model parameters requires a decision on a data set and a decision on an objective function that can *quantify* the closeness of the simulation to the observed data set. In general, calibration uses optimization techniques, such as generalized least squares, to minimize the deviation between observed and corresponding simulated measurements.

Sensitivity analysis should be conducted to test model robustness, as well as to study the impacts of changes in model parameters. Otherwise stated, the simulation is physically sound when a slight change in the experimental condition results in minimal oscillation in the simulated results. Sensitivity analysis is especially useful for complex microscopic models, in which effects of the parameters on the flow behavior are hard to analyze mathematically (Hoogendoorn and Bovy 2001). Sensitivity analysis is a time-consuming process because each parameter has to be individually analyzed. In fact, model transferability depends not only on an effect in the calibration process but also on an in-depth understanding of the sensitivity of model changes in parameters such as driver behavior (McDonald et al. 1994).

In general, vehicle characteristics and driver characteristics are the key parameters, which may be site-specific and require calibration (Taplin 1999). Polak and Axhausen (1990) identify three types of behavioral research needed to develop models, namely

- In-vehicle behavior in response to systems design
- Driving behavior (overtaking, gap acceptance, maneuver, signal behavior)
- Travel behavior, including route choice, compliance with ATIS advice, and responses to information from other sources

All three types of research try to find how various factors contribute to choices in order to simulate a stochastic model based on each driver determining his or her driving and travel decisions.

Calibration of Microscopic Models. Calibrating a microscopic traffic model is a difficult issue. Both traffic data and knowledge about the traffic behavior are needed. At the micro level, the traffic throughput is decided by driver behavior. The vehicle parameters are easily understood and possible to measure. Thus calibration of microscopic models is to be conducted at both the microscopic scale, with regard to vehicle-to-vehicle interactions (i.e., the calibration of the behavioral parameters), and the macroscopic scale, to ensure that the overall behavior is modeled correctly (McDonald et al. 1994). Therefore, reviewing the calibration of microscopic model also covers the calibration of macroscopic models.

Macroscopic models are relatively easy to calibrate using loop detector data (Cremer and Papageorgiou 1981). Data collection typically consists of minute-by-minute records of flows, average speeds/headways, and traffic composition, which can be extracted from loop data or video recordings of the road (McDonald 1994). Mostly, speed-density relations derived from observations are required. Kerner et al. (2000) show that traffic jam dynamics can be described and predicted using macroscopic models that feature only some characteristic variables that are to a large extent independent of roadway geometry, weather, and so forth. This implies that macroscopic models can describe jam propagation reliably without the need for in-depth model calibration.

Lind et al. (1999) suggest the collection of the following data to be used in the macroscopic calibration step:

- Flow and speed
- Travel time
- Headway
- Total queue time
- Maximum queue length in vehicle number
- Percentage stops
- Delay time

On the other hand, microscopic data are far more complex both to obtain and to calibrate. It is almost impossible to obtain data on all the behavioral parameters being modeled not only because of the huge number of parameters required (microscopic models have typically 20 or more parameters) but also because many behavioral parameters are not related directly to easily observable measurements. Brackstone and MacDonald (1998) recommend using suitable data sources (e.g., instrumented vehicles) to conduct such measurements. Examples include

- A *laser range-finder* capable of measuring the distances and relative speeds of immediately adjacent vehicles
- A *laser speedometer* capable of accurately determining vehicles' speeds and acceleration
- A *video-audio monitor* capable of providing permanent visual records

In addition, McDonald et al. (1994) propose other measurements relating driver behavior to road design, as well as observing driver eye position and duration.

Brackstone and MacDonald (1998) also suggest disassembling the model and testing it in a step-by-step fashion and adjustment of its distinctive entities individually. Whenever new behavioral rules are added to the model, they should be tested extensively and preferably in isolation.

In general, the lack of microscopic field data necessitates the analyst to be confined to use macroscopic or mesoscopic calibration data. Moreover, calibration generally attempts to reproduce macroscopic quantities, such as speed-density curves, by changing parameters describing driving behavior. Unfortunately, this cannot produce the optimal parameters because the number of degrees of freedom is far too large (Hoogendoorn and Bovy 2001). In fact, only few microscopic simulation models have been calibrated and validated extensively. Ben Akiva et al. (2002) propose the calibration framework outlined in Figure 1.13 for calibrating the MITSIMLab microsimulation package in Stockholm. The calibration process is divided into two steps, namely, aggregate and disaggregate calibration steps. First, using disaggregate data, individual models can be calibrated and estimated. Disaggregate data include information on detailed driver behavior such as vehicle trajectories of the subject and surrounding vehicles.

Aggregate data (e.g., time headways, speeds, and flows) are used to fine-tune parameters and estimate general parameters in the simulator. Aggregate calibration uses optimization techniques (such as generalized least squares techniques) to minimize the deviation between observed and corresponding simulated measurements.

Since aggregate calibration uses simulation output, which is a result of the interaction among all these components, it is impossible to identify the effect of individual models on traffic flow.

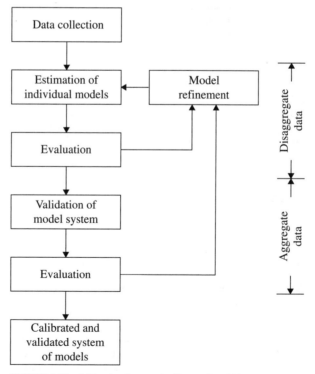

FIGURE 1.13   Calibration framework. (*Source:* Ben Akiva et al. 2002.)

In summary, microscopic model calibration is a tedious task, although the large number of sometimes unobservable parameters often plays a compromising role. Conversely, in macroscopic models, the number of parameters is relatively small and, more important, comparably easy to observe and measure.

Model Validation.   Verifying a calibrated model by running the model on data that are different from the calibration data set is commonly called *validation*. In fact, calibration is useless without the validation step. The validation process establishes the credibility of the model by demonstrating its ability to replicate actual traffic patterns. Similar to calibration, validation is done on two levels, namely, microscopic, and macroscopic. Microscopic validation deals with individual mechanisms, whereas macroscopic validation deals with aggregate performance measures. Macroscopic validation verifies data such as arterial travel time, average speeds, and delays as compared with real-world observations. Microscopic validation deals with car-following, lane-changing, and route-choice logic (McShane et al. 2004).

Similar to the calibration process, the validation of macroscopic models requires less effort than calibration of microscopic or mesoscopic models (Hoogendoorn and Bovy 2001). Accordingly, microscopic validation is carried less frequently than macroscopic and mesoscopic models.

1.4 CAPACITY ANALYSIS

Capacity and level of service (LOS) are fundamental concepts that are used repeatedly in professional practice. Determination of the capacity of transportation systems and facilities is a major issue in the analysis of transportation flow. *The Highway Capacity Manual* defines the capacity of a facility as "the maximum hourly rate at which persons or vehicles can be reasonably expected to traverse a point or uniform segment of a lane or roadway during a given time period under prevailing roadway, traffic, and roadway conditions" (TRB 2000). As such, capacity analysis estimates the maximum number of people or vehicles that can be accommodated by a given facility in reasonable safety within a specified time period. Capacity depends on physical and environmental conditions, such as the geometric design of the facilities and the weather. However, facilities are rarely planned to operate near capacity. Capacity analysis is only a means to estimate traffic that can be accommodated by a facility under specific operational qualities.

The Highway Capacity Manual (HCM 2000), produced by the Transportation Research Board (TRB), is a collection of standards that define all the parameters related to capacity studies for transport infrastructure. *HCM 2000* presents operational, design, and planning capacity analysis techniques for a broad range of transportation facilities, namely, streets and highways, bus and rail transit, and pedestrian and bicycle facilities.

1.4.1 Capacity and LOS

Capacity and *level of service* (LOS) are closely related and can be easily confused. While capacity is a measure of the demand that a highway potentially can service, level of service (LOS) is a qualitative measure of a highway's operating conditions under a given demand within a traffic stream and their perceptions by motorists and/or passengers. Thus LOS intends to relate the quality of traffic service to given volumes (or flow rates) of traffic. The parameters selected to define LOS for each facility type are called *measures of effectiveness* (MOEs). These parameters could be based on various criteria, such as travel times, speeds, total delay, probability of delay, comfort, and safety.

The Highway Capacity Manual defines six levels of service, designated by the letters A through F, with A being the highest level of service and F being the lowest. The definitions of these levels of service vary depending on the type of roadway or roadway element under consideration. For instance, in the case of basic freeway sections, the levels of service are based on density and are given in Table 1.3.

TABLE 1.3 Level of Service Definitions for Basic Freeway Segments

Level of service (LOS)	Density, pc/km/ln
A	0–7
B	7–11
C	11–16
D	16–22
E	22–28
F	>28

Source: Special Report 209: Highway Capacity Manual, 4th Edition, Copyright 2000 by the Transportation Research Board, National Research Council, Washington, DC.

It is important to understand the concept of *ideal condition* of a facility, a term often used in *HCM 2000*. In principle, an ideal condition of a facility is one for which further improvement will not achieve any increase in capacity. Ideal conditions assume good weather, good pavement conditions, and users familiar with the facility and no incidents impeding traffic flow. In most capacity analyses, prevailing conditions are not ideal, and computations of capacity, service flow rate, or LOS must include adequate adjustments to reflect this absence of ideal conditions.

The following sections describe the capacity and LOS analysis of three types of facilities, namely, a two-lane highway, a multilane highway, and weaving sections.

1.4.2 Two-Lane Highways

Two-lane highways differ from any other roadway in that it is the only road that requires the driver to share a lane with opposite traffic in order to pass a slow-moving vehicle. As traffic in one direction increases, passing maneuvers become more difficult. Hence directional analysis is the foundation of capacity studies of two-lane highways. Actually, both capacity analysis and LOS of two-lane highways address the two-way capacity of the facility.

The methodology for calculating LOS is to start with the maximum volume at ideal conditions and then adjust the maximum as less than ideal conditions become apparent.

Two-lane rural highways are divided into two classes, with differing LOS criteria:

• *Class I two-lane highways* are major intercity routes generally serving long-distance trips such as major intercity routes, primary arterials connecting major traffic generators, daily commuter routes, and primary links in state or national highway networks.

• *Class II two-lane highways* are usually access roads to class I facilities and serve shorter trips; examples include access routes to class I facilities, scenic or recreational routes that are not primary arterials, and roads through rugged terrain on which drivers do not expect high speeds.

HCM 2000 determines the LOS of class I two-lane highways as function of *percent time spent following* (PTSF) and average travel speed. PTSF is defined as the average percent of travel time that vehicles must travel in platoons behind slower vehicles owing to an inability to pass. Simulation studies indicated that PTSF could be estimated as the percentage of vehicles traveling at headways of *3* seconds or less (Harwood et al. 1999). PTSF was assumed to describe traffic conditions better than density because density is less evenly distributed on two-lane highways than on multilane highways and freeways (Luttinen 2001).

LOS for class I and class II two-lane highways is determined as a function of *percent time spent following*. However, the *average travel speed* (ATS) was selected as an auxiliary criterion for class I highways because ATS makes LOS sensitive to design speed. In addition, specific upgrades and downgrades can be analyzed by a directional-segment procedure.

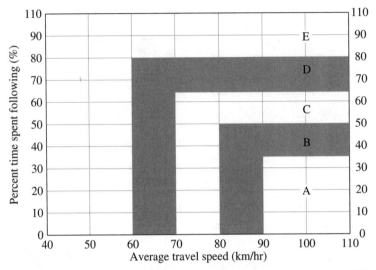

FIGURE 1.14   LOS graphical criteria for two-lane highways in class I. (*Source:* Exhibit 20-3 in *HCM 2000.*)

TABLE 1.4   LOS Criteria for Two-Lane Highways in Class II

LOS	Percent time spent following
A	≤40
B	>40–55
C	>55–70
D	>70–85
E	>85

Source: Exhibit 20-4 in *HCM 2000.*

Figure 1.14 and Table 1.4 illustrate the two-dimensional definition of LOS for, respectively, class I and class II two-lane highways (*HCM 2000*).

Analysis of Two-Way Segments.

Capacity and LOS.   Under ideal conditions, the total two-way capacity of a two-lane highway is up to 3,200 veh/hr. For short distances, a two-way capacity of between 3,200 and 3,400 veh/hr may be attained. Consequently, the capacity in a major direction is 1,700 pc/h until traffic flow in the minor direction reaches 1,500 veh/hr. After that, the two-way capacity becomes the determining factor.

The objective of operational analysis is to determine the LOS for a two-lane highway based on terrain, geometric, and traffic conditions. The traffic data needed include the two-hourly volume, the peak-hour factor, and the directional distribution of traffic flow. Typically, this methodology is applicable for highway sections of at least 3 km.

As such, the analysis procedure for two-lane highways is outlined in the following steps:

1. Determine free-flow speed
2. Determine demand flow rate
3. Estimate average travel speed
4. Estimate percent time spent following
5. Determine LOS
6. Estimate other traffic performance parameters

Determining Free-Flow Speed (FFS).   There are two options for determining the free-flow speed (FFS) of a two-lane highway. The first option is through conducting field measurements, and the other is to indirectly estimate FFS using *HCM* guidelines.

Field Measurement: FFS can be computed based on field data as follows:

$$FFS = S_{FM} + 0.0125 \frac{v_f}{f_{HV}}$$

where FFS = estimate free-flow speed (km/hr)
S_{FM} = mean speed of traffic measured in the field (km/hr)
v_f = observed flow rate for the period when field data were obtained (veh/hr)
f_{HV} = heavy-vehicle adjustment factor (obtained from *HCM 2000*)

Estimating FFS: FFS can be estimated indirectly if field data are not available. To estimate FFS, the analyst must characterize the operating conditions of the facility in terms of a base free-flow speed (BFFS) reflecting traffic characteristics and facility alignment. BFFS can be estimated based on design speed and posted design speed of the facility. Then FFS is estimated based on BFFS adjusted for lane and shoulder width and access points, as shown below:

$$FFS = BFFS - f_{LS} - f_A$$

where   FFS = estimated FFS (km/hr)
BFFS = base FFS (km/hr)
f_{LS} = adjustment for lane width and shoulder width
f_A = adjustment for access points

Note that f_{LS} and f_A adjustments are found in Exhibits 20-5 and 20-6 in *HCM 2000*.

Determining Demand Flow Rate. Demand flow rate is obtained by applying three adjustments to hourly demand volumes, namely, PHF, grade-adjustment factor, and heavy vehicle adjustment factors. Theses adjustments are applied as follows:

$$v_p = \frac{V}{PHF * f_G * f_{HV}}$$

where v_p = passenger car equivalent flow rate for 15-minute period (pc/hr)
V = demand volume for the full peak hour (veh/hr)
PHF = peak-hour factor
f_G = grade adjustment factor, which accounts for the effect of grade on passenger car operation (f_G depends on type of terrain and is obtained from Exhibits 20-7 and 20-8 in *HCM 2000*)
f_{HV} = heavy vehicle adjustment factor applied to trucks and recreational vehicles (RVs)

The heavy vehicle adjustment factor accounts for the effects of heavy vehicles. It is obtained as

$$f_{HV} = \frac{1}{1 + P_T(E_T - 1) + P_R(E_R - 1)}$$

where P_T = proportional of trucks in the traffic stream, expressed as decimal
P_R = proportional of RVs in the traffic stream, expressed as decimal
E_T and E_R = passenger car equivalent for respective trucks and RVs (E_T and E_R depend on type of terrain and are obtained from Exhibits 20-9 and 20-10 in *HCM 2000*)

Estimation of Average Travel Speed. The average travel speed is computed based on the FFS, the demand flow rate, and an adjustment factor for the percent no-passing zones. Average speed is determined from the following equation:

$$ATS = FFS - 0.0125v_p - f_{np}$$

where ATS = combined average travel speed for both directions of travel (km/hr)
f_{np} = adjustment for percentage of no-passing zones (Exhibit 20-11 in *HCM 2000*)
v_p = passenger car equivalent flow rate for peak 15-minute period (pc/hr)

Estimating Percent Time Spent Following.   The percent time spent following is computed based on the directional split and no-passing zones. It is computed as follows:

$$PTSF = BPTSF + f_{d/np}$$

where   PTSF = percent time spent following
       BPTSF = base percent time following for both directions of travel combined, obtained using the equation

$$BPTSF = 100(1 - e^{-0.000879 v_p})$$

where v_p = the passenger car equivalent flow rate for 15-minute period (pc/hr)
       $f_{d/np}$ = adjustment of the combined effect of the directional distribution of traffic and of the percentage of no-passing zones on percent time spent following (obtained from Exhibit 20-12 in *HCM 2000*)

Determining Level of Service (LOS).   LOS is obtained by comparing the passenger car equivalent flow rate v_p with the two-way capacity of 3,200 pc/hr. If v_p is greater than capacity, the LOS is F regardless of the speed or percent time spent following. Similarly, if the demand flow rate in either direction of travel is greater than 1,700 pc/hr, the level of service is also F. LOS of F corresponds to 100 percent time spent following and highly variable speeds. In the case where v_p is less than capacity, LOS criteria are applied for class I or class II. The LOS for a two-way segment of a class I facility is determined as a function of both ATS and PTSF that corresponds to locating a point in Exhibit 20-3 in *HCM 2000* (see Figure 1.14). Alternatively, LOS for a class II highway is determined from PTSF alone using the criteria in Table 1.4.

In addition to LOS analysis, *HCM 2000* provides procedures for computing volume/capacity ratio, total vehicle miles (or vehicle kilometers) of travel, and total vehicle hours of travel for two-lane highways. For instance, volume/capacity ratio is computed as follows:

$$v/c = \frac{v_p}{c}$$

where v/c = the volume/capacity ratio
       c = capacity taken as 3,200 pc/hr for a two-way segment and 1,700 pc/hr for a directional segment
       v_p = passenger car equivalent flow rate for peak 15-minute period

Analysis of Directional Segments.   *HCM 2000* presents an operational analysis procedure for directional segments. The methodology for directional segments is similar to the two-way segment methodology except that it analyzes traffic performance measures and LOS for one direction. However, the operational condition of one direction of travel depends on operational conditions not only for that direction of travel but also for the opposing direction of travel. *HCM 2000* methodology addresses three types of directional segments, namely, extended directional segments, specific upgrades, and specific downgrades. The directional-segment procedure incorporates the conceptual approach and revised factors discussed earlier in this chapter.

Hence the operational-analysis procedure for directional segments includes the following five steps:

1. Determination of free-flow speed
2. Determination of demand flow rate
3. Determination of ATS
4. Determination of PTSF
5. Determination of LOS

Determination of Free-Flow Speed.   The same procedure is followed as in the two-way segment analysis, taking each direction separately. As a result, the demand flow rate for the studied direction is determined as follows:

$$v_d = \frac{V}{\text{PHF} * f_G * f_{HV}} \quad \text{and} \quad v_o = \frac{V_o}{\text{PHF} * f_G * f_{HV}}$$

where v_d and v_o = are, respectively, the passenger car equivalent flow rate for the peak 15-minute period in, respectively, the direction analyzed and the opposing direction of travel

V = demand volume for the full peak hour (vph) for the direction analyzed in the left-hand-side formula and for the opposing direction in the right-hand-side formula

PHF = peak hour factor

f_G = grade adjustment factor obtained from Exhibits 20-7 and 20-8 in *HCM 2000*, as explained previously for the two-way segment analysis

f_{HV} = heavy vehicle adjustment factor applied to trucks and recreational vehicles (RVs) (f_{HV} is obtained as explained previously for the two-way segment analysis)

Determination of Demand Flow Rate.   Similar to two-way segment analysis, demand flow rate is obtained by applying three adjustments to hourly demand volumes, namely, PHF, grade adjustment factor, and heavy vehicle adjustment factors. These adjustments are applied similarly as in the two-way segment analysis. However, specific grades procedure, that is, specific upgrade or downgrade, is used if the downgrade exceeds 3 percent for a length of at least 1.0 km.

SPECIFIC UPGRADE.   In the specific upgrade procedure, the directional segment procedure is followed with different tables for f_G and f_{HV} (exhibits 20-13, 20-14, 20-15, 20-16, and 20-17 in *HCM 2000*).

SPECIFIC DOWNGRADE.   The specific downgrade procedure differs from the extended segment procedure in considering heavy vehicles effects. A special procedure is provided for locations where heavy trucks use crawl speeds on long and steep downgrades. f_{HV} for specific downgrade is given as

$$f_{HV} = \frac{1}{1 + P_{TC} * P_T(E_{TC} - 1) + (1 - P_{TC})P_T(E_T - 1) + P_R(E_R - 1)}$$

where P_{TC} = proportion of all trucks in the traffic stream using crawl speeds on specific downgrades

E_{TC} = passenger car equivalent for trucks using crawl speeds (obtained from Exhibit 20-18 in *HCM 2000*)

E_T and E_R = passenger car equivalent for trucks and RVs, respectively (E_T and E_R depend on type of terrain and are obtained from Exhibit 20-19 in *HCM 2000*)

Determination of Average Travel Speed.   Average travel speed (ATS) for directional analysis is computed from directional FFS using directional and opposing demand volumes, as well as adjustments for the percent no-passing zones. Average speed is determined from

$$\text{ATS}_d = \text{FFS}_d - 0.0125(v_d + v_o) - f_{np}$$

where ATS_d = average travel speed in the analysis direction (km/hr)

f_{np} = adjustment for percentage of no-passing zones in the analysis direction

v_p, v_o = passenger car equivalent flow rate for the peak 15-minute period (pc/hr) for the analysis direction and the opposing direction, respectively

Determination of PTSF.   The same procedure is followed as in two-way segment analysis with formula parameters considering directional PTSF_d rather than two-way PTSF:

$$\text{PTSF}_d = \text{BPTSF}_d + f_{np}$$

where PTSF_d = percent time spent following in the direction analyzed
BPTSF_d = base percent time following for the direction analyzed obtained using the following equation
f_{np} = adjustment of the percentage of no-passing zones on percent time spent following (obtained from Exhibit 20-20 in *HCM 2000*)

$$\text{BPTSF}_d = 100(1 - e^{av_p^b})$$

where v_p is the passenger car equivalent flow rate for the peak 15-minute period (pc/hr). The coefficients a and b are determined from Exhibit 20-21 in *HCM 2000*.

Determination of LOS.   LOS for directional analysis is obtained by comparing the passenger car equivalent flow rate v_p with the one-way capacity of 1,700 pc/hr. If v_p is greater than capacity, the level of service is *F*. In the case where v_p is less than capacity, then LOS criteria are applied to class I or class II, given by Table 1.4 and Figure 1.14, as in the two-way segment analysis.

Analysis of Directional Segments with Passing Lanes.   *HCM 2000* also presents an operational analysis procedure for directional segments containing passing lanes in level and rolling terrain. A *passing lane* is an added lane provided in one direction of travel on a two-lane highway in order to increase the availability of passing opportunities, which accordingly affects the LOS.

The procedure followed for analyzing this effect does not address added lanes in mountainous terrain or on specific upgrades known as *climbing lanes,* which are addressed separately.

The presence of passing lanes generally decreases PTSF and increases ATS on the roadway downstream of the passing lane; Figure 1.15 illustrates the effect of a passing lane on PTSF. The bold line shows the percent time spent following in the absence of a passing lane. The dotted line shows how the percent time spent following drops abruptly in the passing-lane section and slowly returns to the nonpassing-lane values downstream from the terminus of the passing lane. The figure also shows that the effective length of a passing lane actually is greater than its actual length.

The operational analysis procedure followed for directional segments containing passing lanes incorporates the following five steps:

1. Application of the directional segment procedure without the passing lane in place

2. Division of the segment into regions

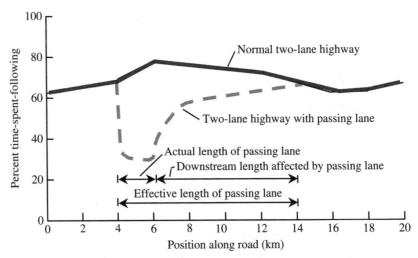

FIGURE 1.15 Operational effect of passing lane. (*Source:* Harwood and Hoban 1987.)

3. Determination of PTSF

4. Determination of ATS

5. Determination of LOS

Application of the Directional Segment Procedure Without Considering the Passing Lane in Place. The first step in a passing lane analysis is to apply the operational analysis procedure for the directional segment without considering the passing lane. The directional segment evaluated for a passing lane should be in level or rolling terrain. The results of the initial application of the directional segment procedure are estimates of PTSF and ATS for the normal two-lane cross section.

Dividing the Segment into Regions. The next step is to divide the analysis segments into four regions, namely

- Upstream of the passing lane
- Within the passing lane
- Downstream of the passing lane but within its effective length
- Downstream of a passing lane but beyond its effective length

These four lengths must, by definition, sum to the total length of the analysis segment. The analysis segments and their lengths will differ for estimation of PTSF and ATS because the downstream lengths for these measures differ. The length of the segment upstream of the passing lane (L_u) and the length of the passing lane itself (L_{pl}) are readily determined when the location (or proposed location) of the passing lane is known. The length of the downstream highway segment within the effective length of the passing lane (L_{de}) is determined from Exhibit 20-23 in *HCM 2000*. Once the lengths L_u, L_{pl}, and L_{de} are known, the length of the analysis segment downstream of the passing lane and beyond its effective length (L_d) can be determined as

$$L_d = L_t - (L_u + L_{pl} + L_{de})$$

where L_t = total length of analysis segment
L_u = length of two-lane highway upstream of the passing lane
L_{pl} = length of passing lanes including tapers
L_{de} = downstream length of two-lane highway within the effective length of the passing lane (L_{de} is obtained from Exhibit 20-23 in *HCM 2000*)

Determination of Percent Time Spent Following (PTSF). PTSF within lengths L_u and L_d is assumed to be equal to PTSF$_d$, as predicted by the directional segment procedure. Within the passing lane, PTSF is generally equal to 58 to 62 percent of its upstream value; this effect varies as a function of flow rate.

Within the downstream effective length of the passing lane, PTSF is assumed to increase linearly with distance from the within-passing-lane value to the normal upstream value. These assumptions result in the following equation for estimating PTSF for the analysis segment as a whole with the passing lane in place:

$$PTSF_{pl} = \frac{PTSF_d \left[L_u + L_d + f_{pl}L_{pl} + \left(\frac{1 + f_{pl}}{2} \right) L_{de} \right]}{L_t}$$

where $PTSF_{pl}$ = percent time spent following for the entire segment including the passing lane
$PTSF_d$ = percent time spent following for the entire segment without the passing lane
f_{pl} = factor for the effect of a passing lane on percent time spent following (f_{pl} is obtained from Exhibit 20-24 in *HCM 2000*)

If the analysis section is interrupted by a town or a major intersection before the end of full effective length of the passing lane, then L_d is not used, and the actual downstream length within the analysis segment L'_{de} is used. L'_{de} is less than the value of L_{de}, as tabulated in Exhibit 20-23 in *HCM 2000*. The preceding equation is replaced by

$$\text{PTSF}_{pl} = \frac{\text{PTSF}_d\left\{L_u + f_{pl}L_{pl} + f_{pl}L'_{de} + \left(\frac{1-f_{pl}}{2}\right)\left[\frac{(L'_{de})^2}{L_{de}}\right]\right\}}{L_t}$$

where L'_{de} is the actual distance from end of passing lane to end of analysis segment, $L'_{de} \leq L_{de}$ (L_{de} is obtained from Exhibit 20-23 in *HCM 2000*).

Determination of Average Travel Speed (ATS).   ATS with the passing lane in place is determined with a similar approach to PTSF except that ATS is increased rather than decreased by the presence of the passing lane. ATS within lengths L_u and L_d is assumed to be equal to ATS$_d$, as predicted by the directional segment procedure. Within the passing lane, ATS is generally 8 to 11 percent higher than its upstream value. Within the downstream effective length of the passing lane, ATS is assumed to increase linearly with distance from the within-passing-lane value to the normal upstream value. These assumptions result in the following equation for estimating ATS for the analysis segment as a whole with the passing lane in place:

$$\text{ATS}_{pl} = \frac{\text{ATS}_d * L_t}{L_u + L_d + \frac{L_{pl}}{f_{pl}} + \frac{2L_{de}}{1+f_{pl}}}$$

where ATS_{pl} = average travel speed for the entire segment including the passing lane
   ATS_d = average travel speed for the entire segment without using the passing lane
   f_{pl} = factor for the effect of a passing lane on average travel speed

Determination of LOS.   LOS for a directional segment containing a passing lane is determined in a manner identical to that for a directional segment without a passing lane except that PTSF_{pl} and ATS_{pl} are used in place of PTSF_d and ATS_d.

Operational Analysis Procedure for Directional Segments Containing Climbing Lanes on Upgrades.   A *climbing lane* is a passing lane added on an upgrade to allow traffic to pass heavy vehicles with reduced speed. The operational analysis procedure for a directional segment containing a climbing lane on an upgrade is the same as the procedure for passing lanes with the following modifications:

- In applying the directional segment procedure to the roadway without the added lane, the grade adjustment factor and the heavy vehicle adjustment factor should correspond to the specific upgrades. In cases where the added lane is not sufficiently long or steep, it should be analyzed as a passing lane rather than a climbing lane
- PTSF and ATS adjustment factors for a climbing lane are based on different factors than the factors for passing lanes (adjustments factors are obtained from Exhibit 20-27 in *HCM 2000*)

1.4.3   Multilane Highways

Multilane highways usually have four to six lanes, often with physical medians or two-way left turn lanes. Multilane highways do not usually provide uninterrupted flow because they provide partial or full access to adjacent facilities. In addition, traffic signals may be introduced with considerable distance in between. Owing to this long distance, flow between these interruptions may operate similarly to freeways (uninterrupted flow) between two fixed interruption points (two signalized intersections at least 2 miles apart). In general, the posted speed limit in multilane highways ranges between 60 and 90 km/hr.

This section discusses the methodologies for analyzing the capacity and LOS criteria for multilane highways based on the 2000 revision of the *Highway Capacity Manual (HCM 2000)*.

Multilane Highway Basic Characteristics.   Multilane highways have the following characteristics:

* They generally have posted speed limits between 60 and 90 km/hr.
* They usually have four or six lanes, often with physical medians or two-way left-turn-lane (TWLTL) medians, although they also may be undivided.
* Multilane highways are typically located in suburban communities leading to central cities or along high-volume rural corridors that connect two cities or significant activities generating a substantial number of daily trips.
* Traffic signals may be found along such highways, although traffic signals spaced at 3 km or less typically create urban arterial conditions.

Capacity and LOS analysis procedures for multilane uninterrupted flow begin with calibration of a characteristic set of speed-flow-density relationships for highways operating under ideal conditions. *Ideal conditions* imply no heavy vehicles and only the presence of familiar drivers in the facility. The main parameters describing the *HCM* speed-flow models are free-flow speed, capacity, and density at capacity. The *HCM 2000* has a standard family of curves for multilane highways under ideal conditions (Figure 1.16).

The general shapes of these curves are similar. When the flow rate exceeds a limit flow rate, speeds start decreasing below the free-flow speed with increasing flow rate toward the speed at capacity owing to an increasing level of interactions between vehicles. In addition, the figure shows that the capacity varies with the free-flow speed of the facility.

LOS for multilane highways is defined on the basis of the density-flow and speed-flow relationship. Thus LOS criteria reflect the shape of the speed-flow and density-flow curves. Speed remains relatively constant across LOS *A* to *D* but is reduced as capacity is approached.

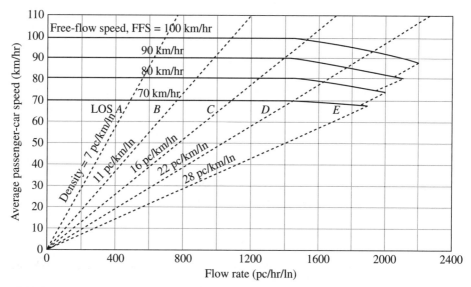

FIGURE 1.16   *HCM 2000* speed-flow models for four classes of multilane highways. (Based on functions given in *HCM 2000,* Chapter 21, Exhibit 21-3.)

Capacity and LOS Analysis.   This section discusses the methodologies for analyzing the capacity and LOS criteria for multilane highway based on *HCM 2000*. The methodology is based on determining the reduction in travel speed that occurs for less than base conditions.

The base, or ideal, conditions for multilane highways imply the following:

- 3.6-m minimum lane width
- A minimum of 3.6 m of total lateral clearance in the direction of travel
- No direct access point along the highway
- A divided highway
- Free-flow speed (FFS) > 95 km/hr
- Traffic stream composed of passenger cars only

The *HCM 2000* procedure is composed of the following steps:

Step 1: Calculation of free-flow speed (FFS)

Step 2: Determination of the flow rate

Step 3: Calculation of the LOS

Calculation of Free-Flow Speed (FFS).   FFS can be determined either from field measurement or by following *HCM* guidelines. If FFS is to be computed directly from field study, FFS is determined during periods of low to moderate flow, with an upper limit of 1400 pc/hr per lane.

Alternatively, FFS can be estimated indirectly from *HCM 2000* formula. The FFS decreases with decreasing lane width, decreasing lateral clearance and increasing access point (intersection and driveway) density, and is reduced for undivided highways. The *HCM 2000* formula is

$$\text{FFS} = \text{BFFS} - f_{LW} - f_{LC} - f_M - f_A$$

where BFFS = base free-flow speed

f_{LW} = adjustment factor for lane width (km/hr) (Exhibit 21-4 in *HCM 2000*)
f_{LC} = adjustment factor for lateral clearance (km/hr) (Exhibit 21-5 in *HCM 2000*)
f_M = adjustment factor for median type (km/hr) (Exhibit 21-6 in *HCM 2000*)
f_A = adjustment factor for access points (km/hr) (Exhibit 21-7 in *HCM 2000*)

DETERMINATION OF BASE FREE-FLOW SPEED (BFFS).   BFFS is based on the coded speed limit. As such, it is equal to

- Speed limit + 11 km/hr for posted speed limits between 65 and 75 km/hr
- Speed limit + 8 km/hr for posted speed limits between 80 and 90 km/hr

ADJUSTMENT FACTOR FOR LANE WIDTH.   The base condition for multilane highways requires lane widths of 3.6 m. FFS should be adjusted for narrower lanes. Adjustments factors for lane widths are computed from Exhibit 21-4 in *HCM 2000*. Note that lanes that are larger than 3.6 m are subject to no adjustment factors.

ADJUSTMENT FACTOR FOR LATERAL CLEARANCE (f_{LC}).   The *total lateral clearance* is defined as the sum of the total lateral clearance from the right edge added to the total lateral clearance from the left side as follows:

$$\text{TLC} = LC_R + LC_L$$

where LC_R = lateral clearance from the right edge of the travel lanes to roadside obstructions (maximum value of 1.8 m)
LC_L = lateral clearance from the left edge of the travel lanes to obstructions in the roadway median (maximum value of 1.8 m)

Lateral left clearance (LC_L) is computed for divided highways only. In all other cases, for instance, if a continuous two-way left turn lane exists or the facility is undivided, then LC_L is set to 1.8 m. Facilities with one-way traffic operation are considered divided highways because there is no opposing flow to interfere with traffic.

Once TLC is computed, the appropriate reductions in FFS are obtained from Exhibit 21-5 in *HCM 2000*. Linear interpolation is used for intermediate values.

ADJUSTMENT FACTOR FOR MEDIAN TYPE.   Exhibit 21-6 in *HCM 2000* states that the adjustment factor is taken as 2.6 if highway is undivided and 0.0 if highway is divided (including two-way left turning lanes).

ADJUSTMENT FACTOR FOR ACCESS POINTS (F_A).   Access points are intersections and driveways on the right side of the roadway in the direction of travel. The number of access points per kilometer is based directly on the values of the number of at-grade intersections with no traffic control devices and section length. A linear equation fits to the adjustment factor in Exhibit 21-7 in *HCM 2000*. The linear equation indicates that for each access point per kilometer, the estimated FFS decreases by approximately 0.4 km/hr.

Determination of Flow Rate.   The flow rate is obtained by applying two adjustments to hourly demand volumes, namely, PHF and heavy vehicle adjustment factors. Hence the flow rate is obtained as

$$v_p = \frac{V}{\text{PHF} * N * f_{HV} * f_p}$$

where v_p = passenger car equivalent flow rate for 15-minute period (pc/hr)
   V = demand volume for the full peak hour (veh/hr)
   PHF = peak hour factor
   N = number of lanes
   f_{HV} = heavy vehicle adjustment factor
   f_p = driver population factor

PHF.   As stated previously, PHF represents the variation in traffic flow within an hour.

ADJUSTMENT FACTOR FOR HEAVY VEHICLES (f_{HV}).   The adjustment factor for heavy vehicles applies to three types of vehicles, namely, trucks, recreational vehicles (RVs), and buses. Hence the heavy vehicle adjustment factor is based on calculating passenger car equivalents for trucks, buses, and RVs as follows:

$$f_{HV} = \frac{1}{1 + P_T(E_T - 1) + P_R(E_R - 1)}$$

where P_T, P_R = proportion of trucks, buses, and RVs, respectively, expressed as a decimal (e.g., 0.15 for 15 percent)
   E_T, E_R = passenger car equivalents for trucks, buses, and RVs, respectively
   f_{HV} = heavy vehicle adjustment factor for heavy vehicles

E_T AND E_R. E_T and E_R are obtained from Exhibit 21-8 in *HCM 2000* for general terrain based on the type of terrain (i.e., level, rolling, or mountainous). On the other hand, any grade of 3 percent or less that is longer than 1.6 km or a grade greater than 3 percent for a length longer than 0.8 km is treated as an isolated specific grade.

Additionally, downgrades and upgrades are treated differently because the impact of heavy vehicles differs significantly in each case. In both downgrade and upgrade cases, E_T and E_R factors depend on both the percent grade and the length of the related section. As such, Exhibits 21-9 and 21-10 in *HCM 2000* determine, respectively, the E_T and E_R factors for specific upgrades.

Similarly, Exhibit 21-11 in *HCM 2000* determines the E_T factor for specific downgrades greater than 4 percent. As for E_T factors for downgrades that are less than 4 percent and E_R factors for all downgrades, the same adjustment factors as for the level terrain are used (i.e., Exhibit 21-8 in *HCM 2000*).

DRIVER POPULATION FACTOR (f_p).   The driver population factor reflects the effect of familiarity and unfamiliarity of drivers with the road. Recreational traffic can decrease the capacity by 15 percent. f_p, the driver population factor, ranges from 0.85 to 1.0. For urban highways, the driver population factor is set to 1.0 to indicate that drivers are familiar with roadway and traffic conditions (assuming that most of the traffic is composed of commuters). Thus the analyst should consider f_p to be equal to 1.00, which corresponds to a weekly commuter, unless there is sufficient evidence that a lesser value reflecting more recreational trips should be used.

Calculation of LOS.   As stated previously, Figure 1.16 illustrates the relationships among LOS, flow, and speed. The LOS criteria reflect the shape of the speed-flow and density-flow curves, particularly because speed remains relatively constant across LOS *A* to *D* but is reduced as capacity is approached. LOS can be determined directly from this figure by entering the corresponding free-flow speed and density values.

The density of the flow is determined as follows:

$$D = \frac{v_p}{S}$$

where D = density (pc/km per lane)
v_p = flow rate (pc/km per lane)
S = average passenger car travel speed

LOS also can be determined by comparing the computed density with the density ranges provided in Figure 1.16.

1.4.4   Weaving Sections

Weaving is defined as the crossing of two or more traffic streams traveling in the same general direction along a significant length of the roadway without the aid of traffic control devices. Weaving areas typically are formed when a merging area is closely followed by a diverging area. Weaving sections are common design elements on freeway facilities such as near ramps and freeway-to-freeway connectors. The operation of freeway weaving areas is characterized by intense lane changing maneuvers and is influenced by several geometric and traffic characteristics.

One example of a freeway weaving section is a merge and a diverge in close proximity combined with a change in the number of freeway lanes that require either merging or diverging vehicles to execute a lane change. In such cases, traffic is subject to turbulence in excess of what is normally present on basic highway sections. Capacity is reduced in such weaving areas because drivers from two upstream lanes compete for space and merge into a single lane and then diverge into two different upstream lanes.

Since lane changing is a critical component of the weaving areas, lane configuration (i.e., the number of entry and exit lanes and relative placement) is one of the most vital geometric factors that needs to be considered in computing the capacity of a weaving section. Other factors, such as speed, LOS, and volume distribution, are also taken into consideration.

General Consideration of Geometric Parameters.   The *Highway Capacity Manual* identifies three geometric variables that influence weaving segment operations, namely, the configuration, length, and width.

Configuration.   The configuration of the weaving segment (i.e., the relative placement of entry and exit lanes) has a major effect on the number of lane changes required for weaving vehicles to complete their maneuver successfully. In traffic engineering, three types of weaving sections traditionally are distinguished based on the minimum number of lane changes required for completing the weaving maneuvers (TRB 1994, 1997, 2000):

- *Type A weaving sections.* Every weaving vehicle (i.e., vehicle merging or diverging) must execute one lane change. The most common type A configuration is a pair of on- and off-ramps connected by an auxiliary lane (Figure 1.17).

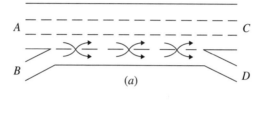

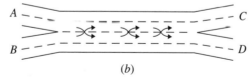

FIGURE 1.17 Type A weaving area. (*Source: HCM 2000.*)

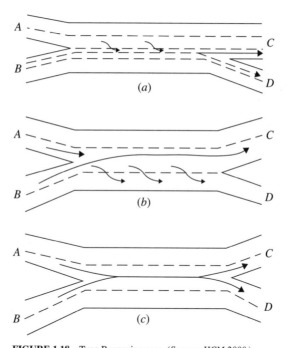

FIGURE 1.18 Type B weaving area. (*Source: HCM 2000.*)

- *Type B weaving sections.* One weaving movement can be made without making any lane change, whereas the other weaving movement requires at most one lane change. A common type B configuration has a lane added at an on-ramp. Merging traffic does not need to change lanes, but traffic diverging downstream must change onto this added lane to exit at the off-ramp (Figure 1.18).

- *Type C weaving sections.* One weaving movement can be made without making any lane change, whereas the other weaving movement requires at least two lane changes (Figure 1.19).

Weaving Length. The weaving length parameter is important because weaving vehicles must execute all the required lane changes for their maneuver within the weaving segment boundary from the entry gore to the exit gore. The length of the weaving segment constrains the time and space in

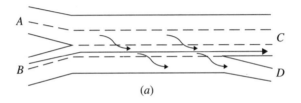

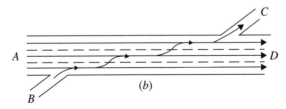

FIGURE 1.19 Type C weaving area. (*Source: HCM 2000.*)

which the driver must make all required lane-changing maneuvers. As the length of a weaving segment decreases (configuration and weaving flow being constant), the intensity of lane changing and the resulting turbulence increase. Similarly, by increasing the length of the weaving area, capacity is increased.

Weaving Width. The third geometric variable influencing operation of the weaving segment is its width, which is defined as the total number of lanes between the entry and exit gore areas, including the auxiliary lane, if present. As the number of lanes increases, the throughput capacity increases.

Four different movements may exist in weaving. The two flows crossing each other's path within the section are called *weaving flows,* whereas those that do not cross are called *nonweaving flows.* The proportional use of the lanes by weaving and nonweaving traffic is another important factor in the analysis of weaving sections.

The following two variables are defined as follows:

N_w = number of lanes weaving vehicles must occupy to achieve balanced equilibrium operation with nonweaving vehicle lanes

$N_w(max)$ = maximum number of lanes that can be occupied by weaving vehicles, based on geometric configuration

Accordingly, two distinguished types of operations are obtained:

- *Constrained operation [$N_w > N_w$ (max)].* When weaving vehicles use a significantly smaller proportion of the available lanes and nonweaving vehicles use more lanes. Constrained operation results in a larger difference in weaving and nonweaving vehicle speeds.

- *Unconstrained operation [$N_w \leq N_w$ (max)].* When configuration does not restrain weaving vehicles from occupying a balanced proportion of available lanes. Unconstrained operation results in small difference in weaving and nonweaving vehicle speeds.

Analysis of Weaving Area: HCM 2000 Computational Methodology. The latest *Highway Capacity Manual* (*HCM 2000*) procedures for weaving sections involve computing the speeds of weaving and nonweaving vehicles, calculating densities, and then performing a table lookup to assign LOS. LOS criteria for weaving segments are found in Exhibit 24-2 in *HCM 2000* (Table 1.5). The geometric characteristics of the weaving section, the characteristics of vehicles by type, and their distribution over the traffic stream are important issues to be considered in the analysis.

TABLE 1.5 LOS Criteria for Weaving Segments

| LOS | Density (pc/km/ln) | |
	Freeway weaving segment	Multilane and collector-distributor weaving segments
A	≤6.0	≤ 8.0
B	>6.0–12.0	>8.0–15.0
C	>12.0–17.0	>15.0–20.0
D	>17.0–22.0	>20.0–23.0
E	>22.0–27.0	>23.0–25.0
F	>27.0	>25.0

Source: Exhibit 24-2 in *HCM 2000*.

The procedure is outlined as follows:

1. Establish roadway and traffic condition types
2. Convert all traffic flows into equivalent peak-flow rates under ideal conditions (in pc/hr)
3. Construct a weaving segment diagram
4. Compute unconstrained weaving and nonweaving speed
5. Check for constrained operation
6. Compute average (space mean) speed of all vehicles in the weaving area
7. Compute average density of all vehicles in the weaving area
8. Determine LOS and weaving segment and capacity

Establish Roadway and Traffic Condition Types. This necessitates determining the following:

• Weaving length and number of lanes
• Type of configuration (i.e., type A, type B, or type C)
• Traffic composition and movement types

Equivalent Peak-Flow Rate Conversion. The first computational step in a weaving analysis is the conversion of all demand volumes to peak-flow rates under equivalent conditions:

$$v_i = \frac{V_i}{\text{PHF} \cdot f_{HV} \cdot f_p}$$

where v_i = peak-flow rate under equivalent ideal conditions for movement i, pc/hr
V_i = hourly volume, veh/hr
PHF = the peak-hour factor
f_{HV} = heavy vehicle factor (as in multilane highway methodology described previously)
f_p = driver population factor (as in multilane highway methodology described previously)

Weaving Segment Diagram. After volumes are converted to flow rates, a weaving diagram is constructed to identify the traffic flow rates by type of movement as follows:

• *Weaving movements.* Ramp-freeway, freeway-ramp
• *Nonweaving movements.* Ramp-ramp, freeway-freeway

Exhibit 24-4 in *HCM 2000* provides an example of the construction of such a diagram (Figure 1.20).

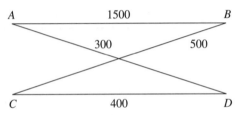

FIGURE 1.20   Weaving diagram. (*Source:* Exhibit 24-4 in *HCM 2000*.)

Unconstrained Weaving and Nonweaving Speeds.   Compute unconstrained weaving speed S_w and nonweaving speed S_{nw} using the equation

$$S_i = S_{min} + \frac{S_{max} - S_{min}}{1 + W_i}$$

where S_i = mean speed of a weaving or nonweaving movement (km/hr) (denoted as S_w and S_{nw} for weaving and nonweaving speeds, respectively)
  S_{max} = maximum speed expected in the section, assumed to be equal to SFF + 8 mph
  S_{min} = minimum speed expected in the section, normally assumed to be 24 km/hr
  W_i = weaving intensity factor ($i = w$ for weaving and $i = nw$ for nonweaving)

Setting S_{max} to free-flow speed plus 8 km/hr and S_{min} to 24 km/hr, the equation is simplified to

$$S_i = 24 + \frac{FFS - 16}{1 + W_i}$$

Determining Weaving Intensity.   The weaving intensity factors (for weaving and nonweaving flows) are measures of the influence of weaving activity on the average speeds of both weaving and nonweaving vehicles. These factors are obtained as follows:

$$W_i = \frac{a(1 + VR)^b \left(\dfrac{v}{N}\right)^c}{(3.28L)^d}$$

where L = length of weaving area (km)
  VR = volume ratio, given as weaving flow rate/total flow rate
  v = total adjusted rate of flow in the weaving section, pc/hr
  N = total number of lanes in the weaving section
  a, b, c, d = constants that depend on weaving configuration: constrained or unconstrained operation and weaving and nonweaving speeds ($a, b, c,$ and d factors are obtained from Exhibit 24-6 in *HCM 2000*)

Note that all predictions are conducted assuming that the type of operation is unconstrained.
  Determining the Type of Operation.   As mentioned previously, the type of operation depends on the proportional use of the lanes by weaving and nonweaving traffic. The type of operation is determined by determining the number of lanes N_w and N_w(max) required for weaving operation. N_w and N_w(max) are computed from Exhibit 24-7 in *HCM 2000*, depending on the configuration of the weaving segment (i.e., type *A, B,* or *C*).

- If $N_w < N_w$(max), unconstrained operation is valid, and speeds are as computed previously.
- Otherwise, if $N_w > N_w$(max), operation is constrained, so recompute speeds using $a, b, c,$ and d parameters corresponding to constrained operation.

Determining Average (Space Mean) Speed. Average speed is computed in *HCM 2000* as follows:

$$S = \frac{v_{nw} + v_w}{\dfrac{v_{nw}}{S_{nw}} + \dfrac{v_w}{S_w}}$$

where S_w = average speed of weaving vehicles (km/hr)
S_{nw} = average speed of nonweaving vehicles (km/hr)
v_w = weaving flow rate (pc/hr)
v_{nw} = nonweaving flow rate (pc/hr)

Determining Average Density. LOS criteria for uninterrupted weaving are based on density. Density is obtained from estimating average speed and flow for all vehicles in the weaving area. Accordingly, density is determined as

$$D = \frac{v/N}{S} = \frac{v}{NS}$$

where *D* is determined in pc/km/hr.

Determining LOS and Weaving Segment Capacity. LOS criteria for weaving areas based on average density of all vehicles in the section are as shown in Table 1.5. The capacity of weaving segments is represented by any set of conditions resulting in LOS at the E–F boundary condition, that is, an average density of 27 pc/km per lane for freeways or 25 pc/km per lane for multilane highways. Thus capacity depends on weaving configuration, length, number of lanes, free-flow speed, and volume ratio and should never exceed the capacity of similar basic freeway segments.

In addition, field studies suggest that capacities should not exceed 2,800 pc/hr for type A, 4,000 pc/hr for type B, and 3,500 pc/hr for type C. Furthermore, field studies indicate that there are other limitations related to the proportion of weaving flow *VR* as follows: *VR* = 1.00, 0.45, 0.35, or 0.20 for type A with two, three, four, or five lanes, respectively; *VR* = 0.80 for type B; and *VR* = 0.50 for type C. Although stable operations may occur beyond these limitations, operation will be worse than predicted by methodology, and failure is likely to occur.

Exhibit 24-8 in *HCM 2000* summarizes capacities under base conditions for a number of situations. These tabulated capacities represent maximum 15-minute. flow rates under equivalent base conditions and are rounded to the nearest 10 pc/hr. The capacity of weaving section under prevailing conditions is computed as follows:

$$c = c_b \times f_{HV} \times f_p$$

where c = capacity as an hourly flow rate for peak 15 minutes of the hour
c_b = capacity under base conditions (Exhibit 24-8 in *HCM 2000*)
f_{HV} = heavy vehicle factor
f_p = driver population factor

If capacity in peak hourly volume is desired, then it is computed as

$$c_h = c \times \text{PHF}$$

where c_h = capacity under prevailing conditions (veh/hr)
PHF = peak hour factor

1.5 CONTROL

1.5.1 Objectives

The task of traffic control is to specify the control parameters based on available measurements, estimations, or predictions in response to forces/disturbances acting on the system so as to achieve a prespecified goal or set of goals regarding the performance of the system. For instance, the task of ramp control is to specify the ramp metering rates based on measured volumes and occupancies on the main line and on the ramp in response to varying traffic demand so as to achieve minimum time spent in the network by travelers.

As you will see below, such control can come in many forms, varying from very simple to very advanced and hence complex. It can be *manual or automatic, regulatory or optimal, or open-loop or closed-loop*. Furthermore, it is important to distinguish, particularly in our field, between dynamic (i.e., traffic-responsive) control and adaptive control because the two types are often confused. The following is a brief control-theory primer for traffic applications. More elaborate treatments can be found in Papageorgiou (1999) and Dutton et al. (1997).

1.5.2 Control-Theory Primer

If the control task just described is undertaken by a human operator, we have a *manual* control system. If the control task is undertaken by a computer program (a control-strategy algorithm), we have an *automatic* control system. Hence *automatic control* refers to theories, applications, and systems for enabling a system to automatically perform certain prespecified tasks while achieving a set of goals in the presence of continuous action of external forces (disturbances) on the system. Figure 1.21 shows the basic elements of an automatic control system, and Figure 1.22 shows an example of a traffic control system. In both figures, the process is either the real physical system, such as an urban traffic network and flows, or a mathematical/computer model of the system. The control strategy is the "brain" that steers the system toward achieving its intended role by changing the control parameters or inputs such as traffic light timing plans. Disturbances are inputs to the systems that cannot be controlled directly, such as traffic demand or incidents. Process outputs are quantities that represent the behavior or performance of the system, such as travel times or delays, and can be inferred from direct measurements such as volumes and speeds. Those measurements can be fed directly into the control strategy or may have to be preprocessed for one reason or another, such as detecting incidents or forecasting near-future measurements based on current ones if a form of

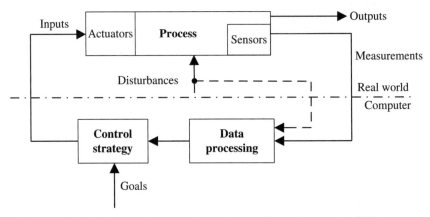

FIGURE 1.21 Basic elements of an automatic control system. (*Source:* Papageorgiou 1999.)

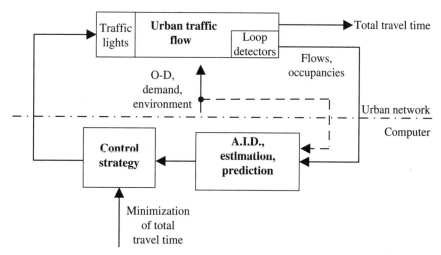

FIGURE 1.22   An example of a traffic control system. (*Source:* Papageorgiou 1999.)

proactive control is desired. In real-time applications, the control task is to specify in real time the process inputs, based on available measurements, to steer the process out toward achieving a goal despite the varying impacts of the disturbances. In the context of urban traffic control, the control strategy should calculate in real time the traffic light timing plans, the ramp metering rates, and traffic diversion plans, among other control actions, based on measured traffic conditions so as to achieve a goal such as minimizing travel times and maintaining densities at a certain desirable level despite the variations in demand and the occurrence of incidents that continuously disturb the state of the system.

One should not confuse the process and its model with the control strategy. The process model, for instance, a traffic simulator, is a mere replica of the real system that given the inputs and disturbance would accurately reproduce how the real system is likely to behave. The control strategy, on the other hand, is the algorithm that picks the proper control action given the system state and acting disturbances. Despite the continuous interaction between the process model and the control strategy, they are distinct and often separate components, the former to replicate the process behavior and the latter to make decisions on control actions that will influence the process behavior.

Regulator Versus Optimal and Open-Loop Versus Closed-Loop Control.   System control strategies can be regulatory or optimal. A regulator's goal is to maintain the system's performance and output near an exogenously prespecified desired value. In a traffic environment, an example is freeway ramp metering, where the controller attempts to maintain traffic occupancy downstream of the ramp at or below critical occupancy (corresponding to capacity). If y is the process output (e.g., downstream occupancy), d is the disturbance (e.g., upstream demand, ramp demand, incidents), and u is the control vector (e.g., ramp metering rate), then the system output is given by

$$y = R(u, d)$$

The regulator guarantees that $y = y_d$ despite the presence of disturbances. The regulator R can be designed using standard control theory (Papageorgiou 1999; Dutton et al. 1997).

One could argue that in the preceding equation, if a desired output is prespecified and y_d and the disturbance trajectory d are known, then one could solve for the control vector u that is required to yield the desired output, that is,

$$u = R^{-1}(y_d, d)$$

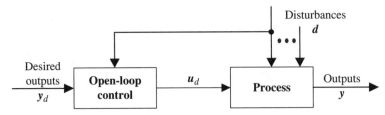

FIGURE 1.23 Open-loop regulator. (*Source:* Papageorgiou 1999.)

without having to measure the actual system output y, that is, without feedback. In this case, elimination of measurement of the actual system output turns the closed loop into an open loop, as shown in Figure 1.23. The main advantage of such open-loop control is rapidity, because the control system need not wait for any feedback from the process. However, the real process state is never known, and hence if the process model is inaccurate or nonmeasurable disturbances are present, the control would be off, and the process output would be far from y_d. For these reasons, closed-loop control would be preferable for most practical applications.

Recognizing that a desired output might not always be known and that, even when known, there is no guarantee that the process performance would be best, a better control strategy would be an *optimal* one. In such case, the control goal is to maximize or minimize quantity J, which is a function of the internal process (state) variables x and the applied control u; that is, the control problem is an optimization problem of the form

$$J(u, x) = y(u, x) \rightarrow \min$$

An urban traffic control example could be to minimize time spent in a freeway network using ramp metering. Note that in this case there is no need to specify desired values for the occupancies downstream of the ramps. Rather, the proper occupancy levels are obtained indirectly by applying the optimal control that minimizes time spent in the system by all travelers.

In the case of several competing optimization objectives (e.g., minimize travel time and reduce fuel consumption), the preceding equation would become

$$J(u, x) = \alpha_1 y_1(u, x) + \alpha_2 y_2(u, x) + \cdots$$

where α_i are weighting parameters such that $\Sigma \alpha_i = 1$.

Optimal control also can be open loop (aka P_1 problem) or closed loop (aka P_2 problem). In open-loop optimal control, the initial conditions and the disturbance trajectory in time are assumed to be known (forecasted) and the problem is to find the optimal control trajectory $u^*(t)$, as illustrated in Figure 1.24. This control trajectory u^* would be computed once at the beginning and applied afterwords without measuring any feedback from the process, that is, open loop. Hence any unexpected disturbances could throw the control off because it was optimal only for the initial disturbance trajectory.

A better closed-loop control looks for an optimal control law or policy that is applicable at all times to compute the optimal control vector, that is, finds the function R

$$u^*(t) = R[x(t), t]$$

that minimizes the objective function J. It is to be noted that a full control vector trajectory is not obtained once at the beginning, as in open-loop optimal control. Rather, the control law or policy R is obtained that can be applied in real time based on the latest system state, that is, closed loop, and hence independent of the initial forecasted disturbance trajectory. This type of control is capable of recovering from the impact of any unexpected disturbances, as shown in Figure 1.24.

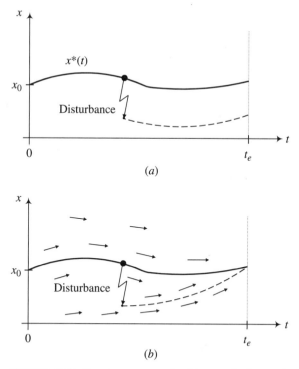

FIGURE 1.24 Open-loop versus closed-loop optimal control. (*Source:* Papageorgiou 1999.)

Both P_1 and P_2 problems can be solved using control theory (Papageorgiou 1999; Dutton et al. 1997). P_1 is much simpler to solve and therefore applicable to larger-size practical problems. However, it is not accurate, as explained earlier. P_2 is more accurate but suffers for the curse of dimensionality and is often infeasible for practical-size problems. Many compromises are possible, including hierarchical control using a combination of P_1 and P_2 formulations or repetitive optimizations with rolling horizons, which essentially solves a P_1 problem several times over the control time horizon. New approaches for solving P_2 problems using artificial intelligence are emerging that are less computationally intensive than standard control-theory approaches. A good example is the use of reinforcement learning (RL) to obtain optimal control policies (see Sutton and Barto 1998; Pringle and Abdulhai 2002; Kattan and Abdulhai 2003).

In closing this primer section, it is important to emphasize the difference between real-time control in general and adaptive control because the two are widely confused in the traffic control field. Any control system is by default dynamic or real time. When fed with varying disturbances over time, it responds with the proper (regulator or optimal) control as applicable. Most widely used traffic control systems are of this type; however, often they are erroneously labeled *adaptive* because they adapt to traffic conditions. From a control-theory perspective, the term *adaptive* means that the control-law parameters themselves are adjusted in real time in order to account for possible time-varying process behavior. For example, the control system is capable of "learning" in a sense, and it could respond to a given process state differently over time as the control policy evolves and adapts to time-varying process behavior. Driving and routing behavior, for instance, might vary widely from one city to another, and hence a generic signal control system might need to fine-tune itself according to local conditions. An attempt to develop such "true" adaptive control, applied to traffic signal control, can be found in Pringle and Abdulhai (2002).

1.5.3   Control Devices

Traffic control may be achieved by using various devices, namely, signs, traffic signals, ramp metering, islands, and pavement markings, to either regulate, guide, inform, and/or channel traffic. The *Manual on Uniform Traffic Control Devices* (*MUTCD*) defines all policies and guidelines pertaining to traffic control devices and for determining where and whether a particular control type is suitable for a given location and/or intersection (Garber and Hoel 2002). The *MUTCD*, published by the Federal Highway Administration, can be found online at the following Web site: mutcd.fwa.dot.gov.

The *MUTCD* defines the following five requirements for traffic control devices to be effective:

1. Fulfill a need
2. Command attention
3. Convey a clear, simple meaning
4. Command the respect of road users
5. Give adequate time for proper response

These five requirements are governed mainly by five major factors, namely, the design, placement, operation, maintenance, and uniformity of traffic devices (*MUTCD*). For instance, the traffic device should be designed with a combination of size, color, and shape that would be able to attract drivers' attention and clearly convey the intended message. In addition, the traffic device should be placed properly to fall within the field of vision of the driver. This would ensure that the driver has adequate time to respond while driving at normal speed.

1.5.4   Freeway Control: Ramp Metering

Freeway control is the application of control devices such as traffic signals, signing, and gates to regulate the number of vehicles entering or leaving a freeway to achieve some operational objective. Ramp metering has proven to be one of the most cost-effective techniques for improving traffic flow on freeways. Ramp meters can increase freeway speeds while providing increased safety in merging and reducing rear-end collisions on the ramps themselves. Although additional delays are incurred by the ramp traffic, mainline capacities are protected, and the overall operational efficiency, usually measured in terms of travel time or speed, is improved (Kachroo and Krishen 2000).

Warrants for Ramp Metering.   The *MUTCD* provides general guidelines for the successful application of ramp control. These guidelines are mainly qualitative. Makigami (1991) suggests the following warrants for installation of freeway ramp metering:

1. Expected reduction in delay to freeway traffic exceeds the summation of the expected delay to ramp users and the added travel time for diverted traffic on alternative routes
2. The presence of adequate storage space for the vehicles stopped at the ramp signal
3. The presence of alternative surface routes with adequate capacity to take the diverted traffic from the freeway ramps and either
   - there is recurring congestion on the freeway owing to traffic demand in excess of the capacity or
   - there is recurring congestion or a severe accident hazard at the freeway on-ramp because of inadequate ramp-merging area

Ramp Metering Objectives.   Typically, ramp meters are installed to address two primary objectives:

1. Control the number of vehicles that are allowed to enter the freeway
2. Break up the platoons of vehicles released from an upstream traffic signal

The purpose of the first objective is to balance demand and capacity of the freeway to maintain optimal freeway operation as well as to ensure that the total traffic entering a freeway section remains below its operational capacity. The purpose of the second objective is to provide a safe merge operation at the freeway entrance (Chaudhary and Messer 2000). The primary safety problem of the merging operation is incidents of rear-end collisions and lane-change collisions caused by platoons of vehicles on the ramp competing for gaps in the freeway traffic stream. As such, metering is used to break up these platoons and to enforce single-vehicle entry.

Advantages and Disadvantages of Ramp Metering. Implementation of ramp metering to control freeway traffic can bring both positive and negative impacts. Appropriate use of ramp meters can produce positive benefits, including increased freeway throughput, reduced travel times, improved safety, and reduced fuel consumption and emissions. Another important benefit is that short freeway trips may divert to adjacent underutilized arterial streets to avoid queues at the meters. If there is excess capacity on surface streets, it may be worthwhile to divert traffic from congested freeways to surface streets.

Inappropriate use of ramp meters can produce negative effects. Issues that have been getting more concern with the implementation of ramp metering include (Wu 2001)

- Freeway trips potentially may divert to adjacent arterial streets to avoid queues at the meters. If insufficient capacity exists on arterial streets, metering therefore can result in adverse effects.
- Queues that back up onto adjacent arterial streets from entrance ramps can adversely affect the surface network.
- Local emissions near the ramp may increase owing to stop-and-go conditions and vehicle queuing on the ramp.

As such, the greatest challenge of a ramp metering policy is its ability to prevent freeway congestion without creating large on-ramp queue overflows. Fortunately, the formation of excessively long queues may be avoided. By preventing freeway congestion, optimal ramp metering policies are able to service a much larger number of vehicles in the freeway than when no metering policies are used (Horowitz et al. 2002).

In fact, the positive impact of ramp metering on both the freeway and the adjacent road network traffic conditions was confirmed in a specially designed field evaluation in the Corridor Peripherique in Paris (Papageorgiou and Kotsialos 2000). As such, when an efficient control strategy is applied for ramp metering, the freeway throughput generally will increase. It is true that ramp metering at the beginning of the rush hour may lead to on-ramp queues in order to prevent congestion on the freeway. However, queue formation temporarily may lead to diversion toward the urban network. Nevertheless, owing to congestion avoidance or reduction, the freeway will be able to accommodate a higher throughput, thus attracting drivers from urban paths and leading to improved overall network performance.

Ramp Metering Components. Ramp meters (Figure 1.25) are traffic signals placed on freeway entrance ramps to control the rate of vehicles entering the freeway so that demand stays below capacity. Figure 1.25 shows the related ramp metering components, namely,

1. Ramp metering signal and controller
2. Upstream and downstream mainline loop detectors
3. Presence sloop detectors
4. Passage loop detectors
5. Queue loop detectors

Metering Rate and Control Strategies. Ramp metering involves determination of a metering rate according to some criteria, such as measured freeway flow rates, speeds, or occupancies upstream and downstream of the entrance ramp. The maximum practical single lane rate is generally at 900 vehicles per hour, with a practical minimum of 240 veh/hr.

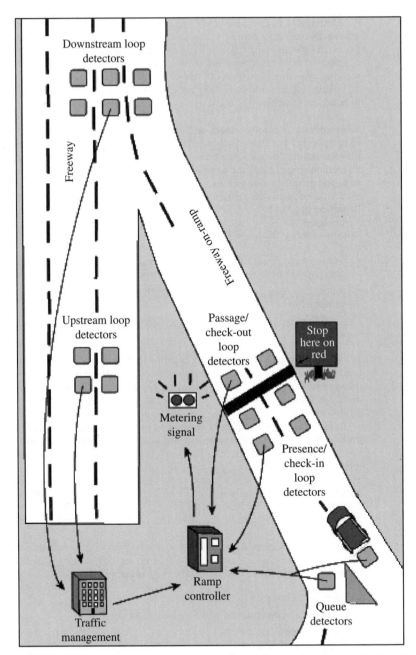

FIGURE 1.25 Ramp meter at freeway entrance. (*Source:* Partners for Advances Transit and Highways, www.path.berkeley.edu.)

When metering is to be used only as a means of improving the safety of the merging operation, then the metering rate is simply set at a maximum consistent with merging conditions at the particular ramp. The metering rate selected must ensure that each vehicle has time to merge before the following vehicle approaches the merge area.

Metering rates may be fixed for certain periods, based on historical data, or may be variable minute by minute (adaptive) based on measured traffic parameters. This latter metering scheme may be based on regulator or optimal control strategies. In addition, ramp metering may operate either in isolation or in coordinated fashion. The coordinated ramp meters control strategy establishes metering rates for various ramps on the basis of total freeway conditions.

Pretimed Ramp Metering Strategies. *Pretimed ramp metering* refers to a fixed metering rate, derived offline, based on historical data. Thus rates are not influenced by current real-time traffic conditions. Fixed-time ramp metering strategies are based on simple static models. They are derived for a freeway with several on- and off-ramps by subdividing sections, each containing one on-ramp (Papageorgiou 2000). Pretimed strategy typically is used where traffic conditions are predictable. Its benefits are mainly associated with accident reductions from merging conflicts, but it is less effective in regulating mainline conditions. The main drawback of preset strategies is that they may result in overly restrictive metering rates if congestion dissipates sooner than anticipated, resulting in unnecessary ramp queuing and delays. Owing to the absence of real-time observation and the reliance on historical observations, fixed ramp metering strategies are myopic. They lead either to overload of the mainline flow or to the underutilization of the freeway.

Coordinated Ramp Metering Strategies. *Coordinated metering* refers to the application of metering to a series of entrance ramps rather than dealing with individual ramps independently. As such, coordinated ramp metering strategy makes use of all available mainstream measurements on a freeway section to calculate simultaneously the ramp volume values for all controllable ramps included in the same section. This requires the presence of detectors upstream and downstream of each ramp, as well as a communication medium and central computer linked to the ramps.

Coordinated ramp metering can be either pretimed or actuated. In coordinated pretimed metering, the metering rate for each of these ramps is determined independently from historical data pertaining to each freeway section. In coordinated actuated metering, the metering rate is based on real-time readings. Freeway traffic conditions are analyzed at a central control system, and accordingly, metering rates at the designated ramps are computed. This centralized configuration allows the metering rate at any ramp to be affected by conditions at other locations within the network. This provides potential improvements over local ramp metering because of more comprehensive information provision and because of coordinated control actions (Papageorgiou and Kotsialos 2000). In addition to recurring congestion, system-wide ramp metering also can manage freeway incidents, with more restrictive metering upstream and less restrictive metering downstream of the incident. Based on capacity, queue length, and demand conditions, linear programming is used to determine the set of integrated metering rates for each ramp, as shown below.

Example of Computation of Pretimed Coordinated Ramp Metering Strategies. Fixed ramp metering strategies are based on simple static models. The freeway section with several on- and off-ramps is subdivided into sections, each containing one on-ramp. Mainline flow is given by (Papageorgiou and Kotsialos 2000)

$$q_j = \sum_{i=1}^{j} \alpha_{ij} r_i$$

where q_j = is the mainline flow section j, r_i is the on-ramp volume (ver/hr) of section i, and $\alpha_{ij} \in [0, 1]$ expresses the known portion of vehicles that enters the freeway section i and does not exit the freeway upstream of section j. To avoid congestion, the flow on freeway section j must be less than the capacity; thus $q_j \leq q_{\text{cap},j} \ \forall j$, where $q_{\text{cap},j}$ is the capacity of section j.

On the other hand, the ramp metering rates are subject to the following constraints:

$$r_{j,\min} \leq r_j \leq \min \{r_{j,\max}, d_j\}$$

where d_j is the demand on the ramp taken from historical data, and $r_{j,\max}$ is the ramp capacity at on-ramp j.

Optimal metering rates are found by optimizing a given criterion such as minimizing the total travel time in the network or minimizing the queue length on the ramps. As such, the problem might be formulated as a quadratic programming or linear programming problem. For instance, when the objective function is to minimize queue length at ramps, the problem is formulated as follows:

$$Z(r_i) = Arg\min_{r_i} \sum_j (d_j - r_j)^2$$

Subject to the following constraints:

$$q_j \le q_{cap,j} \; \forall j \qquad q_j = \sum_{i=1}^j \alpha_{ij} r_i \qquad r_{j,min} \le r_j \le \min \{r_{j,max}, d_j\}$$

This problem is solved easily by broadly available computer codes based on different algorithms such as gradient or conjugate gradient algorithms.

Actuated or Traffic-Responsive Ramp Metering Strategies. In contrast to the pretimed strategies, actuated ramp control strategies are based on real-time traffic measurements taken from main-line detectors placed in the immediate vicinity of the ramp to calculate suitable ramp metering values. Since occupancy is directly related to density, often occupancy data rather than flow is the commonly used parameter in actuated ramp metering. In fact, occupancy readings, in contrast to flow measurements, distinguish between congested and uncongested conditions. Furthermore, actuated traffic response strategies fall into two categories: regulator and optimal control. The difference in these two methodologies is detailed below.

Ramp Metering as a Regulator Problem. Regulator control strategies also fall into two types: reactive control and proactive control strategies. Under reactive control, metering rates are the difference between the upstream flow measured in the previous period, usually 1 minute earlier, and the downstream capacity. In the proactive control strategy, control-set metering rates are based on occupancy measurements taken upstream of the ramp during the previous period, usually 1 minute prior.

REACTIVE REGULATOR. Reactive ramp metering strategies are employed with the aim of keeping freeway traffic at prespecified set values based on real-time measurements (Papageorgiou and Kotsialos 2000). Metering rates are the difference between the upstream flow measured in the previous period, usually 1 minute earlier, and the downstream capacity. The algorithm determines the metering rate locally from input-output capacity considerations based on flow data as follows:

$$r(k) = \begin{cases} q_{cap} - q_{in}(k-1) & \text{if } o_{out}(k) \le o_{cr} \\ r_{min} & \text{else} \end{cases}$$

where $r(k)$ = number of vehicles allowed to enter in period t
q_{cap} = capacity of freeway section
$q_{in}(k-1)$ = upstream flow in period $k-1$

The upstream flow, $q_{in}(k-1)$ is measured by the loop detector, and the downstream capacity q_{cap} is a predetermined value.

The main criticism of these reactive algorithms is that they adjust their metering rates after the mainline congestion has already occurred. Traffic predictive algorithms, such as feedback regulators, have been developed to operate in a proactive fashion; that is, they anticipate operational problems before they occur.

FEEDBACK REGULATOR ALGORITHMS: PROACTIVE REGULATORS. The feedback regulator algorithms use the occupancy rather than the flow in the upstream to determine the ramp metering rate for subsequent periods. One example of such an algorithm is the Asservissement LINeaire d'Entree Autroutiere (ALINEA) (Papageorgiou et al. 1991). ALINEA is a local feedback control algorithm that adjusts the metering rate to keep the occupancy downstream of the on-ramp at a prespecified level, called the *occupancy set-point*. The feedback control algorithm determines the ramp metering rate as a function of the following:

- The desired downstream occupancy
- The current downstream occupancy
- The downstream occupancy
- The ramp metering rate from the previous period

The number of vehicles allowed to enter the motorway is based on the mainline occupancy downstream of the ramp and is given by

$$r(k) = r(k-1) + k_R[\hat{o} - o_{out}(k)]$$

where $r(k)$ = number vehicles allowed to enter in time period k
       k_R = a regulator parameter
       $\hat{o}$ = occupancy set-point
   $o_{out}(k)$ = downstream occupancy

While the reactive strategy reacts only to excessive occupancies o_{out}, proactive strategies such as ALINEA react smoothly even to slight differences $[\hat{o} - o_{out}(k)]$. As such, proactive strategies may prevent congestion by stabilizing the traffic flow at a high throughput level.

Ramp Metering as an Optimal Control Problem.   Optimal control strategies have a better coordination level than regulator strategies. They tend to calculate in real time optimal and fair set values from a more proactive and strategic point of view (Papageorgiou and Kotsialos 2000). Such an optimal approach takes into consideration the following:

- The current traffic condition on both the freeway and on the on-ramps
- Demand predictions over a sufficiently long time horizon
- The limited storage capacity of the on-ramps
- The ramp metering constraints such as minimum rate
- The nonlinear traffic flow dynamics, including the infrastructure's limited capacity
- Any incidents currently present in the freeway network

As such, the optimal control strategy is formulated as a nonlinear mathematical programming problem, with a parameter set to be adopted for the future time horizon, and the optimization is to be repeated after a specified time, with a new data set. For instance, the mathematical problem could be minimizing an objective criterion, such as minimizing the total time spent in the whole network (including the on-ramps), while satisfying the set of constraints (e.g., queue length, ramp capacity). This problem is solved online with moderate computation time by using suitable solution algorithms (Papageorgiou and Kotsialos 2000).

1.5.5   Surface Street Control

Warrants for Traffic Signals.   Traffic signals are one of the most effective methods of surface street control. The *MUTCD* provides specific warrants for the use of traffic control signals. These warrants are much more detailed than those for any other control device. This is justified by the significant cost and negative impact of misapplication of traffic signals compared with other control devices (McShane et al. 2004). The approach traffic volume represents the key factor in the *MUTCD* warrants. Other factors, such as pedestrian volume, accident data, and school crossings, also play a significant role (Garber and Hoel 2002).

Advantages and Disadvantages of Traffic Signals.   When placed properly, traffic signals offer the following advantages:

- Assign rights-of-way
- Increase capacity
- Eliminate conflicts, thus reducing the severity of accidents
- Allow for coordination plans at designated speeds
- Permit pedestrian movements
- Permit cross-street movements

Despite these advantages, when designed poorly, traffic signals or improperly operated traffic control signals can result in the following disadvantages:

- Volumes will increase
- Signals are no longer a safety device; in contrast, crashes often will increase
- Delays will increase
- Unjustified construction, operation, and maintenance costs will arise

Traffic Signal Design.   Traffic control systems are designed based on different control logics that can be divided into three categories. The simplest is the pretimed category, which operates in a prespecified set of cycles, phases, and interval durations. The second type is the actuated signal, which establishes cycle and phase duration based on actual traffic conditions in the intersection approaches. Pretimed and actuated signals can operate either in isolated or coordinated mode. Each intersection, in isolated signals, is treated as a separate entity with its timing plan independent of its neighboring intersections. On the other hand, coordinated signals operate mostly on the arterial or network level, where a series of intersections along arterial streets is treated as a single system, and the timing plans are developed together to provide good vehicle progression along the arterial or minimum travel time at the network level. Finally, the third type is the adaptive signal where the system responds to inputs, such as vehicle actuation, future traffic prediction, and pattern matching, that reflect current traffic conditions.

The main objective of signal timing at an intersection is to reduce the travel delay of all vehicles, as well as traffic movement conflicts. This can be achieved by minimizing the possible conflict in assigning an orderly movement of vehicles coming from different approaches at different times (phases) (Garber and Hoel 2002). While providing more phases reduces the probability of accident, it does not necessary reduce the delay of vehicles because each phase adds about 3 to 4 seconds of effective red. This means that the higher the number of phases, the more delay is incurred. The analyst therefore has to find a tradeoff between minimizing conflicting movements and vehicle delay.

In addition to the number of phases, designing signal timing plans requires a complete specification of the following (McShane et al. 2004):

1. Phasing sequence
2. Timing of "yellow" and "all red" intervals for each phase
3. Determination of cycle length
4. Allocation of available effective green time to the various phases, often referred to as *splitting* the green
5. Checking pedestrian crossing requirements

Moreover, the phase plan must be consistent with the intersection geometry, lane use assignments, and volumes and speeds at the designated location.

Isolated Pretimed Traffic Signals.   In pretimed traffic signals, the cycle length, phase, phase sequences, and intervals are predefined based on historical traffic data. However, the timing plan can change during the course of the day to respond to varying traffic demand patterns. Note that such intersections do not usually require the presence of loop detector sensors.

SIGNAL PHASING.   Signal phasing largely depends on the treatment of left-turn movements, which can fall in two different categories:

- *Protected left turns.* Where the left turns have an exclusive right-of-way (no conflict with through movement)

- *Permissive (permitted) left turns.* Where the left turns are made in the presence of conflicting traffic (finding gaps); need to find a sufficient gap to make a turn, potential conflict

The simplest phasing control for a four-leg intersection is a two-phase operation in which one phase allocates right-of-way to the north-south street traffic and the other phase allocates right-of-way to the east-west traffic. Such phasing provides separation in the major conflicting movements. However, the left turn must yield until an adequate gap is available.

Phases to protect left-turn movements are the most commonly added phases. However, adding phases may lead to longer cycle lengths, which lead to increased stop and delay to other approaches. Hence protected left turn is mainly justified for intersections with a high left-turn volume. Nevertheless, other consideration, such as the intersection geometry, accident experience, and high volume of opposing traffic, also might warrant left-turn protection.

YELLOW INTERVAL.   The objective of the yellow interval is to alert drivers to the fact that the green light is about to terminate and to allow vehicles that cannot stop to cross the intersection safely. A bad choice of yellow interval may lead to the creation of a dilemma zone. A *dilemma zone* is a distance in the vicinity of the intersection where the vehicle can neither stop safely before the intersection stop line nor continue through the intersection without accelerating before the light turns to red. To eliminate the dilemma zone, a sufficient yellow interval should be provided.

It can be shown that this yellow interval τ_{min} is obtained by applying the following formula:

$$\tau_{min} = \delta + \frac{u_0}{2(a+Gg)} + \frac{(W+L)}{u_0}$$

where δ = perception plus reaction time (sec)
   u_0 = speed limit (km/hr)
   W = width of the intersection (km)
   L = length of the intersection (km)
   a = comfortable deceleration rate (km/sec^2)
   g = the gravitational acceleration
   G = grade of the approach

For safety considerations, the yellow interval is not made less than 3 seconds and no more than 5 seconds. If the calculation requires more than 5 seconds, the yellow interval is supplemented with all-red phase.

CYCLE LENGTH OF PRETIMED SIGNALS.   Several methods have been developed to determine the optimal cycle length for an isolated intersection. The Webster method is described below.

In the Webster method, the minimum intersection delay is obtained when the cycle length C_0 is given by

$$C_0 = \frac{1.5L+5}{1-\sum\limits_{i=1}^{\varphi} Y_i}$$

where C_0 = optimal cycle length (sec)
   L = total lost time per cycle (usually you lose 3.5 seconds per distinct phase)
   Y_i = maximum value of the ratios of approach flows (called *critical movement*) to saturation flows for all traffic streams using phase i = V_{ij}/S_j
   ϕ = number of phases
   V_{ij} = flow on lane j having the right-of-way during phase i
   S_j = saturation flow in one lane j for a through lane

The computation of saturation flow rate is explained next.

SATURATION FLOW RATE.   Saturation flow rate is the maximum number of vehicles from a lane group that would pass through the intersection in 1 hour under the prevailing traffic and roadway conditions if the lane group was given a continuous green signal for that hour. The saturation flow rate depends on roadway and traffic conditions, which can vary substantially from one region to another. Thus the saturation flow rate is based on ideal saturation conditions, and then it is adjusted for the prevailing traffic and geometric conditions of the lane in consideration, as well as the area type. According to *HCM 2000*, the saturation flow rate is obtained as follows:

$$s = s_0 N f_w f_{HV} f_g f_p f_{bb} f_a f_{LU} f_{LT} f_{RT} f_{Lpb} f_{Rpb}$$

where s = saturation flow rate for the subject lane group
s_0 = base saturation flow rate per lane (usually taken as 1,900 or 2,000 veh/hr per lane)
N = number of lanes in the group
f_w = adjustment factor for lane width
f_{HV} = adjustment for heavy vehicles in the lane
f_g = adjustment for approach grade
f_p = adjustment factor for the existence of parking lane and parking activity adjacent to the lane group
f_{bb} = adjustment for the blocking effect of local buses that stop within the intersection area
f_a = adjustment factor for area type (C, B, D or suburban)
f_{LU} = adjustment factor for lane utilization
f_{LT} = adjustment factor for left turns in the lane group
f_{RT} = adjustment factor for right turns in the lane group
f_{Lpb} = pedestrian adjustment factor for left-turn movements
f_{Rpb} = pedestrian adjustment factor for right-turn movements

HCM 2000 also gives a procedure for computing the saturation rate from field measurements. If available, field measurements should be used instead of the analytical procedure given earlier.

Total Lost Time.   The lost time is critical in the design of signalized intersections because it represents time not used by any vehicle in any phase. During the green and yellow intervals, some time is lost at the very beginning when the light turns to green (i.e., startup lost time) and at the very end of the yellow interval (clearance lost time). The startup lost time represents the time lost when the traffic light turns green and before the vehicles start moving. The clearance lost time is the time between the last vehicle entering the intersection and the initiation of the green on the next phase.

Thus the lost time for a given phase i can be expressed as

$$l_i = G_{ai} + \tau_i - G_{ei}$$

where l_i = lost time for phase i
G_{ai} = actual green time for phase i (not including yellow time)
τ_i = yellow time for phase i
G_{ei} = effective green time for phase i

Accordingly, the total lost time for the cycle is given by

$$L = \sum_{i=1}^{\varphi} l_i + R$$

where R = total all-red during the cycle
l_i = lost time for phase i, where ϕ is the total number of phases

Allocation of Green Time. In general, the total effective green time available per cycle is given by

$$G_{te} = C - L = C - \left(\sum_{i=1}^{\varphi} l_i + R \right)$$

where G_{te} = total effective green time per cycle
C = actual cycle length

The total effective green time is distributed among the different phases in proportion to their Y_i values to obtain the effective green time for each phase:

$$G_{ei} = \frac{Y_i}{Y_1 + Y_2 + \cdots + Y_\varphi} G_{te}$$

Recall that $Y_i = V_{ij}/S_j$. Accordingly, the actual green time for each phase is obtained as

$$G_{ai} = G_{ei} + l_i - t_i$$

Pedestrian Crossing Requirements. Signals should be designed to provide an adequate time for pedestrians to cross the intersection safely. Thus the pedestrian crossing time serves as a constraint on the green time allocated to each phase of a cycle. Pedestrians can cross an intersection safely as long as there are not any conflicting movements occurring at the same time. Depending on the intersection geometry, in some cases the length of green time interval may not be sufficient for pedestrians to cross. As such, the minimum green time can be determined using the *HCM* expressions given as

$$G_p = 3.2 + \frac{L}{S_p} + \left(0.81 \frac{N_{ped}}{W_E} \right) \qquad \text{for } W_E > 3 \text{ meters}$$

$$G_p = 3.2 + \frac{L}{S_p} + \left(0.27 N_{ped} \right) \qquad \text{for } W_E \leq 3 \text{ meters}$$

where G_p = minimum green time (sec)
L = crosswalk length (m)
S_p = average speed of pedestrians, usually taken as 1.2 m/sec
W_E = effective crosswalk width
N_{ped} = number of pedestrian crossing during a given interval

Actuated and Semiactuated. In actuated signals the phase lengths are adjusted in response to traffic flow, as registered by the actuation of vehicle and/or pedestrian detectors. Signal actuation varies depending on the amount of vehicle detection used in the design. Fully actuated control requires the presence of detectors on all approaches. Semiactuated control operates with green time constantly allocated to major streets, except when there is a detector call from the minor street. Semi actuated signals usually are placed on minor streets with low traffic volumes and random peaks. Fully actuated signals adjust to traffic demand on all approaches. The green time allocation is based on calls from detector actuation as well as on minimum and maximum green time settings. In fully actuated signals, detectors are placed on all approaches. Fully actuated signals usually are placed in busy isolated intersections where demand fluctuations during the day make pretimed signals inefficient.

The main feature of actuated signals is the ability to adjust the signal's pretimed phase lengths in response to traffic flow. For instance, some pretimed phases could be skipped if the detector indicates no vehicles on the related approach. In addition, the green time for each approach is a function of the traffic flow, and can be varied between minimum and maximum duration lengths depending on flow. Moreover, cycle lengths and phases are adjusted at intervals set by vehicle actuation.

Actuated Signals: Signal Timing Parameters and Detector Placement.
Signal Timing Parameters.   Each actuated phase has the following features:

- *Minimum green time.* It intends to provide sufficient time for vehicles stored between detectors and stop line to enter the intersection.
- *Gap/Passage time.* It defines the maximum gap between vehicles arriving at the detector to retain a given green phase. It also allows a vehicle to travel from the detector to the stop line when a call is placed.
- *Maximum green time.* It is the maximum green time allowed for the green phase. A phase reaches this maximum when there is sufficient demand.
- *Recall switch.* Each actuated phase has one. When the switch is "on," the green time is recalled from a terminating phase. When the switch is "off," green is retained on the previous phase until a "call" is received.

MINIMUM GREEN TIME.   The minimum green time is usually set to be equal to an initial interval allowing all vehicles potentially stored between the detectors and the stop line to enter the intersection. A start up time of 4 seconds is also included, in addition to 2 seconds for each vehicle. Thus the minimum green time is given by the formula

$$G_{min} = 4 + \left[2 \times \text{Integer} \left(\frac{d}{20} \right) \right]$$

where G_{min} is the initial interval, and d is the distance between the detector and the stop line (ft). The denominator of 20 denotes the distance between stored vehicles in ft.

GAP/PASSAGE TIME.   The passage time is set as the time required for one vehicle to move from the point of detection to the stop line at its approach speed and is given by

$$P = \frac{d}{1.468S}$$

where P = passage time
d = distance between the detector and the stop line (ft)
S = approach vehicle speed (mph)

MAXIMUM GREEN TIME.   The maximum green time is generally set by working out an optimal cycle length and phase splits, as in the pretimed signal plan. The computed cycle length is usually increased by a factor of 0.23 or 1.5 to set the maximum green time.

DETECTOR LOCATION STRATEGIES.   The optimal detector location is computed based on the approach speed and the desired minimum green time. Two different approaches are presented. The first approach computes the detector location for an approach speed of 40 mph using the minimum green time G_{min} as follows:

$$d = 10G_{min} - 40$$

The second approach, places the detector back at a distance that is traveled by either 3 or 4 seconds, as follows:

$$d = 1.468S(3) = 4.404S \qquad \text{for a travel time of 3 seconds}$$

$$d = 1.468S(4) = 5.872S \qquad \text{for a travel time of 4 seconds}$$

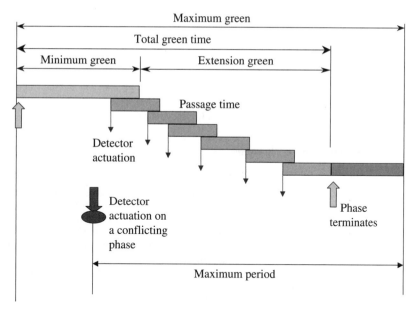

FIGURE 1.26   Operation of an actuated phase: maximum green.

PEDESTRIAN TIMES.   Pedestrian crossing time must be coordinated with vehicle signal phasing. The minimum green value should be set to accommodate safe pedestrian crossing time, as in the pretimed case. In addition, most actuated pedestrian signals must include a pedestrian pushbutton and, accordingly, an actuated pedestrian phase.

Figure 1.26 illustrates the operation of an actuated phase based on the actuated setting. When a green indication is initiated, it will be retained for at least the minimum green period. If an actuation occurs during the minimum green period, the green time is extended by an interval equal to the passage time starting from the time of actuation. If a subsequent actuation occurs within the passage time interval, another extension is added, and this process continues. The green is terminated in two cases: either a passage time elapses without an additional actuation or the maximum time is reached and there is a call on another phase.

Coordinated Signals.   In urban networks with relatively closely spaced intersections, signals must operate in a coordinated fashion. Coordinated signalization has the following objectives:

- Coordination of green times of relatively close signals to move vehicles efficiently through the set of signals without having vehicle queues blocking the adjacent intersections.

- Operation of vehicles in a platoon so that vehicles released from a signal maintain their grouping for more than 1,000 ft. This also would effectively regulate group speed.

Note that it is a common practice to coordinate signals less than a ½ mile apart on major streets and highways.

Advantages of Signal Coordination.   The primary benefit of signal coordination is the reduction in vehicle stops and delays. Reduced stops and delays also result in conservation of energy and protection of environment. Another benefit deals with speed regulation. Signals are set in such a way as to incur more stops for speeds faster than the design speed. In addition, grouping vehicles into platoons (shorter headways) leads to more efficient use of intersections. Finally, stopping fewer vehicles is especially important in short blocks with heavy flows, which otherwise may overflow the available storage.

However, a number of factors may restrain the preceding benefits and make the implementation of coordinated signals rather difficult. These factors are related to the following (McShane et al. 2004):

- Inadequate roadway capacity
- Existence of substantial side frictions, including parking, loading, double parking, and multiple driveways
- Complicated intersections, multiphase control
- Wide range of traffic speeds
- Very short signal spacing
- Heavy turn volumes, either into or out of the street

Moreover, easy coordination may not always be possible, and some intersections may make exceptions in the coordination system. As an example, a very busy intersection may be located in an uncongested area. The engineer may not want to use the cycle time of the busy intersection as the common cycle time. Another example is the existence of a critical intersection that causes queue spillback. Such an intersection may have low capacity and may not handle the delivered vehicle volume. This situation may be addressed either by detaching this intersection from the coordination system or by building the coordination around it and delivering it in the volume that would not create storage problems.

Signal Coordination Features: Time-Space Diagram and Ideal Offsets. The *time-space diagram* is simply a plot of signal indications as a function of time for two or more signals. The time-space diagram is usually scaled with respect to the distance. Figure 1.27 shows an example of a time-space diagram for two intersections. The *offset* is the difference between the green initiation times at two adjacent intersections. The offset is usually expressed as a positive number between zero and the cycle length.

Often, the ideal offset is defined as the offset that will cause the minimum delay. Frequently, the offset denotes the start of the green phase, that is, the delay initiation green phase for downstream signal so that leading vehicles from upstream intersections arrive at downstream intersections just as

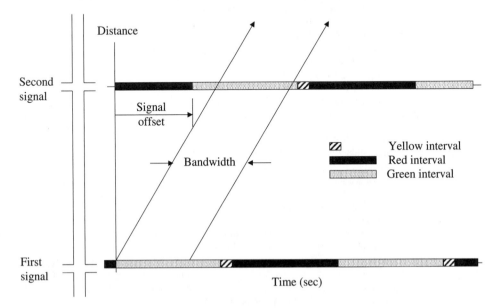

FIGURE 1.27 Time-space diagram for two intersections.

the light turns green (see Figure 1.27). Under ideal conditions, offset is equal to the travel time from the upstream intersection to the subject intersection, that is,

$$t(\text{ideal}) = \frac{L}{S}$$

where L is block length and S is vehicle speed (mps).

Another important definition related to coordinated signals is the bandwidth. The *bandwidth* is defined as a window of green through which platoons of vehicles can move. The time-space diagram shown in the figure illustrates the concept of offset and bandwidth.

The design of progression depends on the type of road. For instance, progression is easier to maintain on a one-way street than on a two-way street. Furthermore, a two-way street provides better progression than a network of intersections (grid).

One-way street progression is easier to achieve because traffic signal phasing is less of a constraint. In two-way streets, finding good progression becomes easy if an appropriate combination of cycle length, block length, and platoon speed exists. Therefore, whenever possible, in new developments, these "appropriate combinations" must be taken into account seriously.

Adaptive Signals.   The use of simulation in conjunction with field controllers to provide model optimization strategies is becoming common place. TRANSYT is perhaps the most used model for optimizing signals offline, and it has been quite successful in improving network performance when traffic demand is relatively constant (Stewart and Van Aerde 1998). However, such offline signal coordination methods, as well as signal progression, are gradually giving way in larger cities to adaptive traffic responsive schemes such as the Split, Cycle, Offset Optimization Technique (SCOOT), and the Sydney Coordinated Adaptive Traffic Control System (GATS).

Adaptive signal control systems help to optimize and improve intersection signal timings by using real-time traffic information to formulate and implement the appropriate signal timings. The adaptive system is able to respond to inputs, such as vehicle actuation, future traffic prediction, and recurrent as well as current congestion. The intelligence of traffic signal control systems is an approach that has shown potential to improve the efficiency of traffic flow.

Adaptive control strategies use real-time data from detectors to perform constant optimizations of the signal timing plan parameters (i.e., the split, cycle length, and offset timing parameters) for an arterial or a network. As such, traffic adaptive control requires a large number of sensors that monitor traffic in real time. As traffic changes, the splits, offsets, and cycle lengths are adjusted by small amounts, usually several seconds. The goal is always to have the traffic signals operating in a manner that optimizes the network-wide objective, such as minimizing queues or delays in the network.

REFERENCES

Abdulhai Baher and Kattan Lina. 2003. "Reinforcement Learning Crystallized: Introduction to Theory and Potential for Transport Applications," *Canadian Journal of Civil Engineering CSCE Journal* 30:981–92.

American Association of Highway and Transportation Officials (AASHTO). 2008. *VII Update*, Vol. 1, No. 3, November 2008.

Ball, W. 2005. "VII: A Proposed Conceptual Framework," General Motors, PowerPoint presentation, May 2005.

Ben-Akiva, M., Davol, A., Toledo, T., et al. 2002. "Calibration and Evaluation of MITSIMLab in Stockholm," submitted for presentation and publication at the 81st Transportation Research Board Meeting, January 2002.

Brackstone, M., and, McDonald, M. 1998. "Modeling of Motorway Operation," *Transportation Research Records* 1485.

Brundtland Commission (World Commission on Environment and Development). 1987. "Our Common Future," New York: Oxford University Press.

Cascetta, E. 2001. "Transportation Systems Engineering Theory and Methods," Lordecht, the Netherlands: Kluwer Academic Publishers.

Chaudhary, N. A., and Messer, C. J. 2000. "Ramp Metering Technology and Practice," Report No. FHWA/TX-00/2121-1. Texas Transportation Institute, The Texas A&M University System.

Cohen, S. 1991. "Kinematic Wave Theory." In *Concise Encyclopedia of Traffic and Transportation Systems*, M. Papageorgiou, ed., New York: Pergamon Press, pp. 231–34.

Cremer, M., and Papageorgiou, M. 1981. "Parameter Identification for a Traffic Flow Model," *Automatica* 17:837–43.

Daganzo F. C. 1994. "The Cell-Transmission Model: A Dynamic Representation of Highway Traffic Consistent with the Hydrodynamic Theory," *Transportation Research* B28:269–87.

De Schutter, B., Bellemans, T., Logghe, S., et al. (1999). "Advanced Traffic Control on Highways," *Journal A*, 40:42–51.

Dupuis, A. and Chopard, B. 2001. "Parallel Simulation of Traffic in Geneva Using Cellular Automata."

Federal Highway Administration. 1997. Freeway Management Handbook. Washington: U.S. Department of Transportation.

Fries, R., Chowdhury, M., and Brummond, J. 2009. *Transportation Security Utilizing Intelligent Transportation Systems*. New York: Wiley.

Gabart J. F. 1991. "Car Following Models." In *Concise Encyclopedia of Traffic and Transportation Systems*, M. Papageorgiou, ed., New York: Pergamon Press, pp. 231–34.

Garbacz, R. M. 2003. "Adaptive Signal Control: What To Expect." Available at http://209.68.41.108/itslib/AB02H473.pdf.

Garber, N., and Hoel, L. 2002. *Traffic and Highway Engineering*, 3d ed. Boston: Brooks/Cole Thomson Learning.

Gazis D. C., Herman R., and Potts, R. 1959. "Car-Following Theory of Steady-State Traffic Flow," *Operations Research* 9:545–95.

Greenberg H. 1959. "An analysis of traffic flow," *Operations Research* Vol. 7.

Hall, F. 1998. "Traffic Stream Characteristics" In *Traffic Flow Theory: A State-of the-Art Report,"* Revised monograph on traffic flow theory. Available at www.tfhrc.gov/its/tft/tft.htm.

Harwood, D. W., May, A. D., Anderson, I. B., et al. 1999. "Capacity and Quality of Service of Two-Lane Highways." NCHRP Final Report 3-55(3). Midwest Research Institute, University of California–Berkeley.

Harwood, D. W., and Hoban, C. J. 1987. "Low-Cost Methods for Improving Traffic Operations on Two-Lane Highways—Informational Guide," FHWA Report No. FHWA-IP-87-2. Federal Highway Administration, Washington.

Hoogendoorn, S. P., and Bovy P. H. L. 2001. State-of-the-Art of Vehicular Traffic Flow Modeling. Special Issue on Road Traffic Modeling and Control of the *Journal of Systems and Control Engineering*.

Horowitz, R., Skabardonis, A., Varaiya, P., and Papageorgiou, M. 2002. Design, Field Implementation and Evaluation of Adaptive Ramp Metering Algorithms, PATH RFP, 2001–2002.

ITS America. 2005. VII White Paper Series. Primer on Vehicle-Infrastructure Integration, October 2005.

ITS International. 2008. "Interview, Driven Man, Paul Brubaker Talks to ITS International about the Future of the US's Vehicle Infrastructure Integration Initiative," January–February 2008.

ITS International. 2008. "The Vehicle Infrastructure Integration Proof of Concept Test," August 2008.

Jiang, H. "Traffic Flow Simulation with Cellular Automata." Available at http://sjsu.rudyrucker.com/~han.jiang/paper/.

Jingcheng Wu. 2001. "Traffic Diversion Resulting from Ramp Metering," Master of Science Thesis, University of Wisconsin, Milwaukee.

Kachroo, P. and Krishen, K. 2000. "System Dynamics and Feedback Control Design Problem Formulations for Real-Time Ramp Metering," *Transactions of the Society for Design and Process Science (SDPS)* 4:37–54.

Lieberman, E. and Rathi, A. K. 1998. "Traffic Simulation" in *Traffic Flow Theory: A State-of-the-Art Report*. Revised Monograph on Traffic Flow Theory. Available at www.tfhrc.gov/its/tft/tft.htm.

Lighthill, M. J., and Whitham, G. B., 1955. "On Kinematic Waves II: A Theory of Traffic Flow in Long Crowded Roads," *Proceedings of the Royal Society* A229:317–45.

Lind, G., Schmidt, K, Andersson, H., et al. 1999. *Best Practice Manual*. SMARTEST Project Deliverable D8 Smartest: Simulation Modeling Applied to Road Transport European Scheme Tests. Available at www.its.leeds.ac.uk/smartest.

Marchal, F. 2001. "Contribution to Dynamic Transportation Models," Ph.D. dissertation, Department of Economics, University of Cergy, Pontoise, France.

Makigami, Y. 1991. "On-Ramp Control." In: *Concise Encyclopedia of Traffic and Transportation Systems*, M. Papageorgiou, ed., New York: Pergamon Press, pp. 285–89.

May, A. 1990. *Traffic Flow Fundamentals*. Englewood Cliffs, NJ: Prentice-Hall.

McShane, W. R., Roess, R. R. and Prassas, E. S. 2004. *Traffic Engineering* 3rd ed., Englewood Cliffs, NJ: Prentice-Hall, Pearson Education, Inc.

McDonald, M., Brackstone, M., and Jeffrey, D. 1994. "Data Requirements and Sources for the Calibration of Microscopic Motorway Simulation Models," 1994 IEEE Vehicle Navigation and Information Systems Conference Proceedings.

Middleton, M., and Cooner, S.A. 1999. "Simulation of Congested Dallas Freeways: Model Selection and Calibration," Texas Transportation Institute, Report 3943-1- Project Number 7-3943. Sponsored by the Texas Department of Transportation, November 1999.

Misener, J. 2008. "Spanning the Spectrum: Addressing a VII for Now and the Future," PATH Program, University of California, Berkley, presentation at Ottawa IEEE VTS Talk, March 4, 2008.

Nagatani Takashi. 2002. "The Physics of Traffic Jams," *Physica* 62:1331–86.

Nagel, K., and Schreckenberg, M. 1992. "Cellular Automaton Model for Freeway Traffic," *J. Physique I (Paris)* 2:2221.

Nagel, K., and Rickert, M. 2001. *Parallel Implementation of the TRANSIMS microsimulation.* New York: Elsevier Science.

Newman, P., and Kenworthy, J. 1998. *Sustainability and the Cities: Overcoming Automobile Dependence*, London: Island Press.

Papageorgiou, M. 1998 "Some Remarks on Macroscopic Traffic Flow Modeling," *Transportation Research Part A* 32:323–29.

Papageorgiou, M. 1999. "Automatic Control Methods in Traffic and Transportation." In *Operations Research and Decision Aid Methodologies in Traffic and Transportation Management*, P. Toint, M. Labbe, K. Tanczos, and G. Laporte, eds., New York: Springer-Verlag, pp. 46–83.

Papageorgiou, M., Blosseville, J.-M., and Hadj-Salem, H. 1989. "Macroscopic Modeling of Traffic Flow on the Boulevard Périphérique in Paris," *Transportation Research Part B*, 28:29–47.

Papageorgiou, M., Salem, H., and Blosseville, J. 1991. "ALINEA: A Local Feedback Control Law for On-Ramp Metering," *Transportation Research Record 1320.*

Papageorgiou, M., and Kotsialos, A. 2000. "Freeway Ramp Metering: An Overview." In: *2000 IEEE Intelligent Transportation Systems Conference Proceedings.* Dearborn, MI, October 1–3, 2000.

Payne, H. J. 1971. "Models of Freeway Traffic and Control, Simulation," *Council Proceedings* 1.

Polak, J., and Axhausen, K. 1990. "Driving Simulators and Behavioural Travel Research," TSU Ref. 559, University Transport Studies Unit, Oxford, U.K.

Row, S., Schagrin, M., and Briggs, V. 2008. "The Future of VII," U.S. DOT, published in *ITS International*, August 2008.

Schultz, J., and Larson, G. 2008. "VII Update: Designing a Better Transportation System Through Vehicle Infrastructure Integration, A Consortium of Transportation Organizations: VII Programs, USA," *American Association of State Highway and Transportation Officials* 1.

Shladover, S. E. 2008. "Vehicle-Infrastructure Cooperation and the Vehicle of the Future," PATH Program, University of California, PowerPoint presentation, IEEE VTS Talk, March 4, 2008.

Sitarz, D. 1994. *Agenda 21.* Boulder, CO: Earthpress.

Taplin, J. 1999. "Simulation Models of Traffic Flow." In: *34th Annual Conference of the Operational Research Society of New Zealand,* December 10–11, 1999, University of Waikato, Hamilton, New Zealand.

Tapio Luttinen. 2001. "Traffic Flow on Two-Lane Highways: An Overview," TL Research Report 1/2001, TL Consulting Engineers, Ltd.

TRB. 2000. *Highway Capacity Manual.* Special Report 209, Transportation Research Board, National Research Council, Washington.

Youngho Kim. 2002. "Online Traffic Flow Model Applying Dynamic Flow-Density Relations," Ph.D dissertation, University of Stuttgart, Germany.

U.S. Department of Transportation (US DOT). 2008. "VII: Vehicle Infrastructure Integration," Research and Innovative Technology Administration.

Wardrop, J. G. 1952. "Some Theoretical Aspects of Road Traffic Research," *Proceedings of the Institute of Civil Engineers* 1–2:325–78.

CHAPTER 2
TRAFFIC ORIGIN-DESTINATION ESTIMATION

Lina Kattan
Department of Civil Engineering, University of Calgary
Calgary, Alberta, Canada

Baher Abdulhai
ITS Center, Department of Civil Engineering, University of Toronto
Toronto, Ontario, Canada

2.1 INTRODUCTION

Origin-destination (O/D) matrices are of vital importance for transportation systems operation, design, analysis, and planning. These matrices contain information about the spatial and temporal distribution of activities between different traffic zones in an urban area. Each cell of the matrix represents the number of trips between an origin and a destination in the study area.

O/D matrices are used to estimate the present demand for transportation systems; then, based on anticipated future economic and population growth, land-use changes, and planning policies, these matrices are projected to identify and forecast future demand. O/D trips are assigned to a transportation network that is spatially defined to represent the actual transportation system (Meyer and Miller 2001). This is referred to as a *static assignment*.

Static assignment, however, is intended mainly for planning applications because it does not deal with congestion dynamics, as opposed to dynamic traffic assignment (DTA) (Jayakrishnan et al. 1995). DTA captures traffic dynamics by incorporating time dimension in the simulation of traffic flow. As such, the basic concept of DTA is to find the distribution of trip makers/vehicles over time and space in a given transportation network. Furthermore, DTA relies on the simulation of timely and accurate prevailing traffic conditions for the provision of reliable routing information to travelers. Thus DTA requires the estimation of dynamic O/D trip matrices and the modeling of changing network conditions in real time.

Therefore, time-dependent O/D is one of the essential components of DTA that, in turn, is vital for the success of intelligent transportation system (ITS) subsystems, such as advanced travel management systems (ATMS) and advanced travel information systems (ATIS). Dynamic O/D tables facilitate DTA schemes, which provide guidance, routing policies, and traffic information capable of achieving system-wide objectives (Hu 1997). Users may respond to guidance through trip rescheduling, mode change, and/or rerouting. Other applications, such as ramp metering strategies, also require O/D data to determine the optimal on-ramp diversion rates and optimal routing information for travelers between a given origin and destination (Khoshyaran 1995). Another benefit of dynamic O/D is also online control of traffic signals and optimization of adaptive control systems. Such systems are able to respond to varying O/D flows and predict variations in flow profiles and, accordingly, design traffic signal plans. It is to be noted that all these benefits will have the potential of reducing emissions by providing information relevant to both the supply-side (coordinating signals and ramp metering) and demand-side emission-reduction strategies [such as demand management via ATIS and advanced public transportation systems (ATPS)] (Horan 2000 chap. 27).

The most commonly used procedures for obtaining O/D demand tables are household surveys, roadside interviews, license plate studies, and returnable-postcard interviews. However, these procedures are rather expensive and time-consuming, and they are difficult to obtain. In the past two decades, there has been considerable interest in the estimation and/or updating of O/D flows using traffic counts, given the great economic advantages that such counts offer. Thus the objective of the static O/D estimation problem is to obtain an O/D matrix such that when assigned to the network, it reproduces the observed traffic flows. This problem is more widely researched than its dynamic counterpart. Recently, there has been growing interest in the problem of dynamic O/D estimation, which is the extension of the static problem to include time component.

From a certain point of view, the problem difficulty of estimating O/D flows by using traffic counts could be considered as the inverse assignment problem (Cascetta 2001). As shown in Figure 2.1, the traffic assignment problem deals with the estimation of flow patterns based on O/D flows and network and path-choice characteristics. Conversely, the O/D estimation problem is that of calculating O/D flows based on the network model, the path-choice characteristics, traffic, counts, and possibly the presence of historical information, that is, *a priori* demand matrix. The estimated O/D flows, once assigned to the network, should closely reproduce the observed traffic flows.

Similarly, the dynamic version of the O/D estimation problem consists of estimating O/D by time interval using time-varying traffic flow observations. Thus this problem is the reverse of the dynamic traffic assignment (DTA) process, and it uses real-time link observations to estimate dynamic versions of a seed time-dependent O/D matrix.

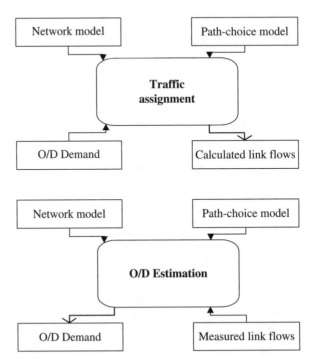

FIGURE 2.1 Dynamic O/D estimation and traffic assignment. (*Source:* Cascetta 2001.)

Moreover, the dynamic O/D estimation problem can be divided further into offline and online demand estimation. The offline problem is more relevant to planning applications as well as in the construction of a historical database (Ashok 1996). The online or real-time problem involves O/D estimation with regard to a real-time traffic management system.

2.2 APPROACHES TO O/D ESTIMATION

With or without incorporation of the time dimension, several techniques have been developed to solve the origin-destination (O/D) estimation problem. Most of these models deal with the *static* estimation problem; that is, O/D flows are assumed to be constant over a significant period of time and are estimated based on information about average traffic flow. Recently, increasing attention has been paid to the problem of *dynamic* O/D estimation. This latter is an extension of the static case whereby time-varying O/D flows are evaluated based on time-varying traffic observations (Cascetta 2001).

Two families of methods for O/D estimation exist in the literature: non-assignment-based and assignment-based techniques. In what follows, the non-assignment-based techniques are described briefly, followed by a more elaborate review of their assignment-based counterparts. Non-assignment-based approaches do not make use of a traffic assignment model to determine the route-choice patterns. This class of techniques is used often to estimate O/D flows in small networks where route choice is unimportant, for example, isolated intersections, interchanges, or small linear freeway segments.

On the other hand, methods in the assignment-based approach rely on an appropriate assignment model to develop route-choice patterns for drivers. These patterns, along with traffic flow measurements from the network links, are used to estimate the O/D flows. The assignment approach is applicable to large urban or freeway networks, where drivers have the choice of more than one route to reach their destination.

Furthermore, assignment-based techniques are formulated as a bilevel optimization problem for networks with cost functions having a symmetric Jacobian matrix (i.e., link cost interaction patterns are symmetric). Bilevel optimization problems that are inspired from game theory take two forms, the noncooperative Nash game and the cooperative Stackelberg game, both of which will be detailed in this chapter. The Stackelberg game condition is a closer representation of the O/D estimation problem. However, only a few researchers have examined the problem as a Stackelberg game solution. In fact, most of the existing solution approaches are more representative of the Nash game solution.

2.3 STATIC O/D ESTIMATION: A BRIEF OVERVIEW

In the static O/D estimation case, the computational complexity of the problem depends mainly on the techniques used. The proposed models are classified according to two categories of networks: uncongested networks and congested networks (Cascetta and Nguyen 1988). These two families of approaches are also known as *non-assignment-based* and *assignment-based techniques*. A traffic assignment is the process of distributing traffic demand in the form of an O/D matrix onto a simulated roadway network resulting in specific route choice and link volumes. In uncongested networks, non-assignment-based techniques are adopted. Thus the path-choice models are determined exogenously (i.e., a fixed proportion is allocated to every path) and therefore are not affected by the O/D demand flow across the network.

In the congested case, however, network congestion level is considered in the estimation process through a traffic assignment model (Yang et al. 1992). Cascetta and Nguyen (1988) demonstrated that the models proposed in the literature to solve the O/D estimation problem have the following general two-objective optimization form:

$$d^* = \arg \min_{d_{od}} [z_1(d_{od}, d') + z_2(f(d_{od}), f')] \tag{2.1}$$

where z_1 and z_2 are, in general, distance measures between paths 1 and 2, respectively, (1) the *a priori* matrix d' and the estimated demand vector d_{od} and (2) the observed link flows f' and the link flows resulting from the assignment of the estimated demand vector $F(d_{od})$.

As stated in equation (2.1), most assignment-based approaches rely on the availability of *a priori* information (e.g., a sample of O/D data, partial O/D information, or an outdated survey, etc.). This *a priori* information therefore is combined with newly obtained traffic counts in order to obtain an updated O/D matrix.

Most proposed formulations for the assignment-based O/D matrix estimation problems corresponding to congested networks have a bilevel structure, which was originally put forward by Nguyen (1977). In this bilevel structure, the upper-level problem estimates the O/D matrix assuming known O/D path flows, whereas the lower level solves a user-equilibrium assignment problem assuming that the O/D demand matrix, which is obtained from a previous step in the upper-level module, is fixed.

However, this bilevel programming structure is a particular case of the general fixed-point formulation and is valid only in cases where network assignment models are expressed by an optimization model satisfying certain mathematical properties such as continuous cost function with symmetric Jacobian matrix, meaning that link cost interaction patterns are symmetric (Cascetta 2001). For the more general case of a fixed-point formulation corresponding to the asymmetric Jacobian matrix, the network assignment model is formulated by a variational inequality (Cascetta 2001). Cascetta and Postorino (2001) considered this general case of O/D estimation for congested networks as a formulation of a fixed-point problem. The authors proposed different heuristic fixed-point solution algorithms, namely, functional iteration and the method of successive averages.

The reader may refer to Sheffi (1985), Nguyen (1977), Van Zuylen and Willumsen (1981), Cascetta (1984), Spiess (1990), Bell (1991), and Cascetta and Postorino (2001) for a further review of the literature of the static case.

2.4   DYNAMIC O/D ESTIMATION

Dynamic O/D estimation is an extension of its static counterpart with the incorporation of time dimension in the formulation. In the literature, the proposed procedures are classified into two broad families: dynamic traffic assignment (DTA)-based approaches and non-DTA-based approaches. A DTA model is equivalent to a traffic assignment model in the static case with the additional consideration of a possible variation of both O/D demand and network capacity over time. This incorporation of the time dimension in the assignment results in time-dependent (also known as *dynamic*) routes and flows.

2.4.1   Non-DTA-Based Approaches

Non-DTA-based approaches compute O/D parameters directly from the time series of input/output flows without the use of a DTA model. In fact, these approaches attempt to find the O/D split fractions in a closed network, wherein complete information is available on all the entry and exit points of the network at all time intervals. This class of techniques is often used in situations where route choice is unimportant.

Non-DTA models have the following equations as their core concept:

$$y_j(k) = \sum_{1}^{k}\sum_{i=1}^{M} b_{ij}(k) \times q_i(k) \tag{2.2}$$

Subject to

$$\sum_{j} b_{ij}(k) = 1 \ \forall \ k \qquad \text{and} \qquad b_{ij}(k) \geq 0 \ \forall \ k \tag{2.3}$$

where M is the number of origins, $y_j(k)$ is the number of vehicles counted at the destination during the observation interval k, $b_{ij}(k)$ is the split ratio (i.e., the proportion of demand heading from origin

i to destination j during interval time k), and $q_i(k)$ is the total number of vehicles generated from origin i during departure interval k.

A number of methods have been investigated in order to identify these turn proportions $b_{ij}(k)$ based on least square methods, constrained least square methods, Kalman filtering (Cremer and Keller 1987), recursive least square (Nihan and Davis 1987), a neural-network approach (Yang et al. 1992), bayesian updating (Van der Zijjp and Hamerslag 1994), and combined estimators (Bell 1991).

While non-DTA models offer computational advantages over their DTA-based counterparts, their application is limited to single intersections/interchanges or to small freeway segments (Chang and Tao 1999). This limitation is due to two major factors. First, the existence of node constraints at entry/exit flows in the problem formulation necessitates a closed network such as an intersection/interchange or freeway section so that complete information on incoming/outcoming flows is easily obtained. Second, the absence of a DTA model also plays a significant role in restricting the size of the network in these models because the importance of obtaining reasonable estimates of travel time and path choice without using a DTA model is possible only in small and simple networks where users do not have more than one path choice to reach their destinations.

2.4.2 DTA-Based Approaches

As the name indicates, methods in the DTA-based approaches rely on the existence of a reliable DTA model in order to conduct network assignment and consequently generate route-choice and link flow usage patterns (Chang and Tao 1999). In addition, most of these approaches require the existence of an *a priori* matrix in their computations. Unlike non-DTA models, DTA-based approaches do not require the location of traffic counts at exit and entrance points. Additionally, DTA-based approaches do not require any specific network topology. Consequently, in theory, these approaches are applicable to large urban or freeway networks where drivers may take one of several routes to reach a chosen destination.

Therefore, in the DTA-based approaches, the route-choice patterns obtained from a DTA model along with the *a priori* matrices and the flow measurements from the network links are used to estimate the O/D flows. In general, as in the static case, this problem is expressed through a fixed-point model. However, in the literature, only the specific case of bilevel programming optimization methods corresponding to cost functions with a symmetric Jacobian matrix is considered.

Problem Formulation of DTA-Based Dynamic OD Estimation. In the dynamic demand estimation problem, as an extension to the static case in the upper level of the bilevel structure, O/D vectors are estimated using a nonlinear optimization formulation with the objective of minimizing the discrepancies between (1) observed time-dependent link-level flows and the corresponding assigned values and (2) the time-dependent *a priori* demand and estimated demand. Similarly, in the lower level, the network equilibrium is satisfied using a DTA model, assuming known time-dependent O/D flows. The general solution algorithm consists of a series of iterations between these two optimization levels.

Therefore, assuming a fixed route choice (obtained from a DTA module at the lower level), the upper O/D level optimization estimates the likely demand flows as

$$d^* = \arg\min_{d_{od}[t]}\{z_1(d_{od}[t], d'[t]) + z_2(f_{lj}(d_{od}[t]), f'_{lj})\} \tag{2.4}$$

where $d_{od}[t]$ = the unknown O/D vector with departure interval t
  d^* = the estimated O/D
  $d'[t]$ = the *a priori* O/D vector
  f'_{lj} = the real observation counts at link l and observation interval j

$f_{lj}(d_{od}[t])$ is obtained from the previous DTA run (lower optimization level) and corresponds to the assigned traffic flow at link l and observation interval j. z_1 and z_2 can be considered the distance measures (or deviations) of (1) the unknown demand $d_{od}[t]$ from the *a priori* estimate $d'[t]$ and (2) the flows $f_{lj}(d_{od}[t])$ obtained by assigning $d_{od}[t]$ to the network and the observed traffic flows f'_{lj} (Cascetta 2001). These two types of deviations are usually combined into a weighted objective function, where the weighting value is determined by the corresponding variance-covariance matrices or by the analyst confidence in either information (i.e., either in the *a priori* information or the observation counts).

At the lower level, a DTA is conducted by assigning the O/D flow vector obtained from the upper estimation level to the network as follows:

$$f_{lj}(d_{od}[t]) = \text{assign}(d^* = d_{od}[t]) \tag{2.5}$$

In DTA-based approaches, the relationship between O/D flows and link volumes is concretized into the following equation (Cascetta 2001):

$$f_{lj}(d_{od}[t]) = \sum_{t=1}^{j}\sum_{od} m_{lj}^{od,t} d_{od}[t] + \varepsilon(l, t) \tag{2.6}$$

where f_{lj} is the flow at link l during time interval j and $m_{lj}^{od,t} \in [0, 1]$ is called the *link proportion matrix* or *assignment matrix*; $m_{lj}^{od,t}$ is written as M in vector form, and $\varepsilon(l, t)$ is the combined error terms in estimating traffic flows on link l during observation interval t. It is to be noted that, in the vector form, equation (2.6) is written as $f = Md + \varepsilon$.

Equation (2.6) reflects how the demand from all O/D pairs that departed from previous time steps t are mapped to link l at current observation interval j via the assignment matrix $m_{lj}^{od,t}$. This matrix represents the fraction of O/D flows d_{od} observed on link l at observation interval j and leaving their origin o at time slice t (with $t \in [l, j]$) and heading to destination d. Thus jt reflects the maximum time required for the slowest vehicle to reach link l from the farthest origin.

$m_{lj}^{od,t}$ is the most important and complicated component of equation (2.6). In fact, $m_{lj}^{od,t}$ captures the effects of departure time, route choice, modal choice, and travel time dispersion. Consequently, this assignment matrix is, as stated by Sun and Porwal (2000), "a complete description of the temporal-spatial characteristics of traffic in a network."

$m_{lj}^{od,t}$ is presented by vector form or as a two-dimensional matrix, as shown in Table 2.1. Each entry $m_{l,j}^{od,dep[t]}$ in this table represents the fraction of O/D flows $d_{od}[t]$ observed on link l at observation time j and leaving their origins at departure interval $[k]$ (i.e., at or before time slice j).

TABLE 2.1 The Assignment Matrix M

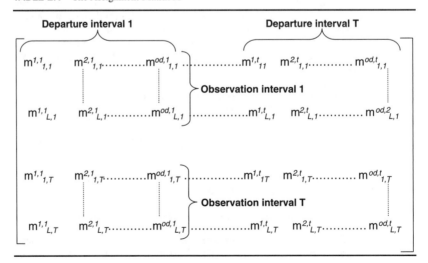

Therefore, for congested networks, M must be computed in order to determine the estimated demand. M is obtained either analytically or by using a DTA simulation package. However, during a DTA run, and particularly when dealing with congested networks, link-flow proportions are not constant. Link-flow proportions are themselves a function of route choice, which is also a function of unknown time-dependent demands. This complex relationship explains the circular dependency existing between O/D estimation on one hand and the DTA on the other (Figure 2.2).

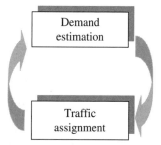

FIGURE 2.2 Circular dependency in O/D estimation.

Figure 2.2 indicates that for estimating the demand matrix, a traffic assignment is to be conducted in order to obtain the link-flow proportions. However, this traffic assignment run necessitates the presence of a time-dependent O/D demand.

Thus, for congested networks, the assignment matrix M is expressed as a function of the demand flow $d_{od}[t]$ (Cascetta 2001). Equation (2.4) is solved by imposing the condition that the flows $f_{lj}(d_{od}[t])$ correspond to the assignment of the demand vector $d_{od}[t]$. Therefore, if the relationship connecting the assignment matrix to link cost $M = M(c)$ is combined with the cost functions, which are themselves function of link flows $c = c(f)$, the relationship between link and demand flows is expressed through the assignment model: $f = v(d)$, also written as $f = \mathrm{assign}(d)$. Thus $M = M\{c[v(d)]\} = M\{c[\mathrm{assign}(d)]\}$. The dynamic O/D estimation problem then can be expressed through the general fixed-point model* obtained by

$$d^* = \arg\min_{d_{od}[t]}[z_1(d_{od}[t], d'[t]) + z_2(M\{c[v(d_{od}[t])]\}.d_{od}[t], f'_{lj})]$$
$$= \arg\min_{d_{od}[t]}[z_1(d_{od}[t], d'[t]) + z_2(M\{c[\mathrm{assign}(d_{od}[t])]\}.d_{od}[t], f'_{lj})]$$

(2.7)

The bilevel optimization problem is a specific case of the fixed-point formulation and requires that the assignment problem be expressed by an optimization model satisfying mathematical properties such as continuous cost functions with a symmetric Jacobian matrix (Cascetta 2001). Obviously, in the latter case, the two formulations become equivalent.

In vector notation, the estimated demand can be written as

$$d^* = x + \eta$$

(2.8)

with $E(\eta) = 0$ and $\mathrm{var}[\eta] = Z$.

where d^* is the estimated demand vector obtained from observation data, x is the true unknown demand vector, and η is the vector of sampling errors whose components are the deviations between the true unknown demand vector x and the estimated demand d^*. This error is due to the several assumptions underlying the assignment model, such as path-choice modeling, cost-function modeling, etc. (Cascetta and Postorino 2001).

Similarly, in vector form, equation (2.5) can be written as

$$\hat{f} = \hat{M}x + \varepsilon$$

(2.9)

with $E(\varepsilon) = 0$ and $\mathrm{var}[\varepsilon] = W$.

The error term ε in the preceding equation is explained by the fact that even if the actual demand vector were to be assigned to the network, the resulting assigned flows could be different from the actual one (Cascetta and Postorino 2001).

---

*If $\varphi(x)$ is an n-vectorial function of vector x, the point x^* is denoted as a fixed point if the function has a value equal to the argument $x^* = \varphi(x^*)$. In the problem at hand, the demand d is expressed as $d^* = \delta[M(d^*)]$.

The functional form of O/D estimation formulation, as stated in equation (2.4), depends on the type of information available (experimental or nonexperimental) and the probability laws associated with such information (Cascetta 2001). Current dynamic O/D DTA-based simulation methodologies use either the Kalman filter (Ashok and Ben Akiva 2002; Kang 1999; Zhou and Mahmassani 2004) or mathematical optimization techniques based on statistical inference approaches (Cascetta 2001), namely, bayesian inference estimation, maximum likelihood or entropy maximization (Wu 1996), generalized least squares (Cascetta 2001; Tavana 2001), and least relative error (LRE) (Hellinga 1994). Neural-network techniques also were applied to a long freeway corridor (Suzuki et al. 2000). Zhou et al. (2003) proposed a generalized least squares (GLS)–based dynamic O/D demand estimation model for planning applications using real-time link counts from multiple days. In this latter approach, the single-day formulation is extended to use link counts from multiple days to estimate the variation in traffic demand over multiple days. Sun and Porwal (2000) presented an approach for estimating dynamic O/D using observed section densities rather than observed link flow.

While all the previously mentioned literature relies on the presence of *a priori* demand tables, it is interesting to note that Chang and Tao (1999) developed an integrated DTA/non-DTA model using the Kalman filtering approach that does not need *a priori* information. The proposed methodology also takes advantage of the intersection turning fraction along with path-flow information from a DTA model.

Table 2.2 summarizes the functional forms of z_1 and z_2 terms of equation (2.1) for the static case. These formulations are easily extended to the dynamic case by incorporating time dimension in the formulation.

The terms Z and W values in Table 2.2 are the variance-covariance matrices representing the dispersion of collected data. In practice, these terms are represented by diagonal matrices. The covariances among all the components of Z^{-1} and W^{-1} are ignored, and only the variance term is computed. The variance term can be obtained from historical data. For instance, for the link term, assuming that traffic counts follow an independent Poisson distribution, and from time series of historical data, the variance of each link's flow can be calculated. Observed link-flow variance may reflect loop detector failure or any other problem related to the specific site location of detectors.

TABLE 2.2 Functional forms of the terms z_1 and z_2 in the static case (Cascetta 2001). x_{od} is the unknown O/D vector, d' is the *a priori* O/D vector, f'_l are the real observation counts at link l, $v_l(x)$ are the assigned traffic flow at link l, α_{od} is the sampling rate, n_{od} is the sampling size, Z and W are the variance-covariance matrices, η_{od} and ε_l are error terms, and M is the set of links with observation counts

	Distance from the initial estimate $z_1(x, d')$	Distance from the flow counts $z_2(v_l(x), f'_l)$
Generalized least squares (GLS)	$(d' - x)^T Z^{-1}(d' - x)$ or $\sum_{od}(x_{od} - d'_{od})^2 / \mathrm{Var}[\eta_{od}]$	$(f'_l - v_l(x))^T W^{-1}(f'_l - v_l(x))$ or $\sum_{l \in M}(f'_l - v_l(x))^2 / \mathrm{Var}[\varepsilon_l]$
Maximum likelihood	Poisson distribution $-\sum_{od}(n_{od} \ln(\alpha_{od} x_{od}) - \alpha_{od} x_{od})$ Multinomial distribution $-\sum_{od} n_{od} \ln x_{od}$	Poisson distribution $-\sum_{l \in M}(f'_l \ln v_l(x) - v_l(x))$ Multivariate normal $(f'_l - v(x))^T W^{-1}(f'_l - v_l(x))$ or $\sum_{l \in M}(f'_l - v_l(x))^2 / \mathrm{Var}[\varepsilon_l]$
Bayesian estimator	Poisson distribution $\sum_{od} x_{od} \ln[(x_{od}/d'_{od}) - 1]$ Multivariate normal $(d' - x)^T Z^{-1}(d' - x)$ or $\sum_{od}(x_{od} - d'_{od})^2 / \mathrm{Var}[\eta_{od}]$ Multinomial distribution $-\sum_{od} x_{od} \ln(x_{od}/d'_{od})$	Poisson distribution $-\sum_{l \in M}(f'_l \ln v_l(x) - v_l(x))$ Multivariate normal $(f'_l - v(x))^T W^{-1}(f'_l - v_l(x))$ or $\sum_{l \in M}(f'_l - v_l(x))^2 / \mathrm{Var}[\varepsilon_l]$

For the O/D term, computation of the variance depends on the type of sampling strategy followed (Cascetta 2001) in order to obtain the *a priori* demand table. For instance, in the case of simple random sampling, the variance can be calculated as (Cascetta 2001)

$$\mathrm{Var}[\hat{d}_{od}] = N^2 \hat{s}^2 (1 - \alpha)/n \tag{2.10}$$

where $\hat{s}^2$ is the sample estimate and is computed as

$$\frac{1}{(n-1)\sum_{i=1,\ldots,n}(n_{od}^i - \bar{n}_{od})^?}$$

where $\bar{n}_{od}$ is the average number of trips and α is the sampling rate in origin.

Thus the Z and W terms represent how much the individual observation link or O/D pair is weighted. In other words, the less an observed link or individual O/D pair weighs, the greater is the variance; that is, the "worse" is the estimate (Cascetta 2001). In some situations, it is reasonable to assume that these matrices are constant, thereby simplifying the process. In that latter case, the variances are a measure of analyst confidence in the validity of an *a priori* O/D estimate relative to the observation count.

2.5   IMPORTANT FACTORS IN THE O/D ESTIMATION PROBLEM

The quality of the estimated demand depends greatly on the accuracy of the input information, such as the quality of the *a priori* O/D data as well as traffic counts. In addition, O/D estimation places very high expectations on the quality of network assignment. This section explains the role of the *a priori* matrix in O/D estimation, as well as the factors to be considered when undertaking traffic assignment.

2.5.1   The *A Priori* Matrix

Unfortunately, because of the nonconvexity of both upper- and lower-level modules, the bilevel optimization problem does not have a unique solution. This nonuniqueness is also due to the fact that the problem is generally underspecified, because the number of unknowns (i.e., number of origins, number of destinations, and estimated departure intervals) is usually much smaller than the number of observations (i.e., number of observation count locations and observation intervals). As a result, the problem usually allows for an infinite number of optimal solutions that can reproduce a given set of observed counts (Cascetta 2001). Therefore, it is necessary to base the correct choice among the candidate set of matrices on other criteria. One important criterion is to limit the search to a smaller neighborhood, which is the closeness of the resulting matrix to historical matrices (i.e., *a priori* matrices).

This added condition is justified by the fact that trip making patterns are a function of the spatial and temporal distribution of activities in a given transportation network. Historical O/D tables already include valuable information about origin-destination movements, that is, the relations that affect trip making as well as their variations over space and time (Ashok 1996). Thus, specifying additional information in the form of the proximity to the *a priori* O/D would allow the incorporation of this important structural information into the real-time O/D estimation process. However, this high reliance on the *a priori* information underlines the importance of the accuracy of data in this matrix with regards to the estimation process.

In the dynamic demand estimation case, these *a priori* matrices most likely are obtained from a previous estimation step: either for the same period on a previous day or simply for an earlier departure time during the same day. These *a priori* tables are updated continuously and hence enclose the day-to-day variation in traffic patterns. Consequently, in a real-time estimation context, the estimation process would have indirectly taken into account all the experience gained over many prior estimations and therefore would be more robust in its structural content (Ashok 1996).

2.5.2   Choice of Count Locations

In the O/D estimation problem, the quality of the estimated O/D data is highly dependent on the accuracy of the input data, namely, the observation data and the *a priori* demand matrices. Both the number and the locations of observation counts affect the O/D estimation process.

Zhou et al. (2003) pointed out the importance of having the number of unknowns, that is, the time-dependent O/D entries (departure intervals, number of origin, number of destination), be less than the number of constraints embedded in the observation counts (i.e., number of links with flow observations in the network and observation intervals). If this inequality is satisfied, then the problem becomes overspecified, and the chances of obtaining a unique solution are increased. This might imply that the more observations we have, the better is the quality of the estimated results. However, traffic measurements at two neighboring sites are highly correlated and thus do not provide any extra information or equations. In addition, including observation readings from counting stations located on local streets would incorrectly force the estimated interzonal trips to take the demand that is normally carried by intrazonal trips.

Thus count locations should be designed according to their information content (Cascetta 2001). The problem of optimal counting locations can be formulated as a network design problem (i.e., an integer programming problem). Yang and Zhou (1998) introduced some basic rules for locating traffic counts and formulated the problem as a linear integer programming problem and solved it in a heuristic algorithm. These basic traffic location rules are

- *OD covering rule.* The traffic counting points on a road network should be located so that a certain portion of the trips between any O/D pair will be observed.
- *Maximal flow fraction rule.* For a particular O/D pair, the traffic counting points on a road network should be located at the links so that the O/D flow fraction between this O/D pair on these links is as large as possible.
- *Maximal flow intercepting rule.* Under a certain number of links to be observed, the chosen links should intercept as many flows as possible.
- *Link independence rule.* The traffic counting points should be located on the network so that the resulting traffic counts on all chosen links are not linearly dependent.

2.5.3   Factors to Be Considered When Running DTA

In addition to the *a priori* information, the O/D estimation process places very high expectations on the quality of network assignment. In other words, "any existing error in the route choice will superpose another error in the demand side, and even if the flows resemble well the counts after the adjustments, they will then do so for the wrong reason" (Spiess, personal communication, 2001). Consequently, it is vital to have an accurate assignment model that would realistically describe network flow distributions as well as route choices. However, in order to compute link-flow proportions for each iteration, the presumed demand should be assigned onto the network following a prespecified route-choice model. In fact, there are several time-dependent assignment rules that are either one or a combination of the following:

- Stochastic
- Deterministic
- User equilibrium
- System optimal

Furthermore, information received and route guidance obtained through ATIS may create more than one user class: informed travelers and uninformed travelers. As a result, in order to mimic observed traffic flows properly, the traffic simulation model must be calibrated properly to reflect systematically path-choice decisions, drivers' characteristics, classes, behavior, etc.

In summary, in the DTA-based approach, the quality of the estimated O/D data is highly dependent on the accuracy of the input data. Prior to undergoing the O/D estimation process, it is crucial to have (1) a well-calibrated traffic network able to mimic the real network, (2) accurate traffic counts, and (3) reliable *a priori* O/D demands. Otherwise, all the estimation processes are seriously questionable.

2.6 MEASURE OF PERFORMANCE OF THE QUALITY OF THE O/D ESTIMATION

The statistical performance of O/D estimation methods is usually measured by the divergence between the estimated demand d^* and the true demand vector d used in the simulation (Cascetta 2001). In general, two performance measurement indicators are used: The MSE_d and the RMSE_d, which are, respectively, the mean square error and the root mean square error between these two vectors, and they are expressed as follows:

$$\text{MSE}_d(d^*, d) = \frac{\sum_{\tau=1}^{\Gamma}\sum_{i=1}^{I}\sum_{j=1}^{J}(d^*_{\tau ij} - d_{\tau ij})^2}{\Gamma \times I \times J}$$

$$\text{RMSE}_d(d^*, d) = \sqrt{\frac{\sum_{\tau=1}^{\Gamma}\sum_{i=1}^{I}\sum_{j=1}^{J}(d^*_{\tau ij} - d_{\tau ij})^2}{\Gamma \times I \times J}}$$

Often, the quality of the performance of the estimation method is also measured in terms of the root mean square of traffic flows. Such a performance measure examines the consistency between the estimated flows of the traffic simulator and the observed flows and is expressed as

$$\text{RMSE}_v(v, \hat{v}) = \sqrt{\frac{\sum_{l=1}^{L}\sum_{t=1}^{T}(\hat{v}_{lt} - v_{lt})^2}{L \times T}}$$

where the notations used are as follows:

RMSE_d = the root mean square of errors of the estimated O/D flows

RMSE_v = the root mean square of errors of the estimated link flows

$d_{\tau ij}$ = the actual demand departing in the aggregate departure interval t going from zone i to zone j

$\hat{d}_{\tau ij}$ = the estimated demand departing in the aggregate departure interval t going from zone i to zone j

v_{lt} = the link flow in observation interval t on link l

$\hat{v}_{lt}$ = the estimated link flow in observation interval t on link l

I = the number of origin zones

J = the number of destination zones

L = the number of links with flow observations in the network

T = the observation intervals

Obviously, the estimated demand is better for lower RMSE_d, meaning that the solution is closer to the real demand. Additionally, lower RMSE_v results in a closer consistency between the simulated counts of the estimated demand and the observation counts.

It is important to emphasize the fact that the objection function focuses on minimizing the deviation between real observation flows and estimated link flows, whereas the performance function represented by RMSE_d expresses the deviation between the solution and the real demand. Therefore, since the objective function to be optimized is different from the measure of performance represented by the RMSE_d, some conflicting results are expected to occur. In other words, a smaller objective function value achieved by the O/D estimation process does not necessarily imply a better estimate of time-dependent O/D matrix. Obviously, since the true demand is unknown in reality, the measure of performance cannot be optimized through RMSE_d. The true demand can be revealed only indirectly in the link counts that witness the dynamic spatial and temporal behavior of the O/D demand while interacting within the traffic network.

2.7   SOLUTION ALGORITHMS FOR THE ORIGIN-DESTINATION PROBLEM

2.7.1   Conventional Iterative Solution Algorithm Methods for the Origin-Destination Problem

Generally, most solution algorithms adopted for the origin-destination problem focus on local hill-climbing approaches, where the solution adopted is the best local optimum in the vicinity of the *a priori* matrix. Existing solution algorithms for solving the O/D estimation problem for congested networks follow an iterative bilevel programming approach (Figure 2.3). As indicated in the figure, this approach starts by first assigning the *a priori* matrix to the network in order to obtain link-flow proportions and consequently the assignment matrix M. The demand then is estimated from the

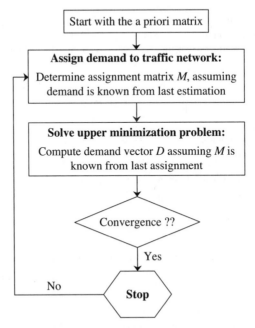

FIGURE 2.3   General framework of bilevel programming solution algorithms.

upper optimization level following a local search technique. In this step, M is assumed to be fixed and known from the previous traffic assignment run. Afterwards, M is reestimated in a subsequent iteration by assigning the newly estimated demand vector to the network. This process iterates until either a convergence or a stopping criterion is satisfied. The solution adopted is the best local optimum in the vicinity of the starting point, that is, the *a priori* matrix.

2.7.2   Advanced Approaches to the Solution of O/D Estimation

The previously described bilevel O/D estimation framework can be viewed as a game between two players, each trying to optimize his or her own objective function. The first player (i.e., the leader) is the O/D estimator, who tries to find the best O/D vector in equations (2.1) and (2.3), whereas the second player (i.e., the follower) is the DTA model, which tries to optimize route-choice patterns that satisfy the assignment rules in equations (2.2) and (2.4). In game theory, there is a distinction between two types of games: the noncooperative Nash game and the cooperative Stackelberg game. In the Nash game, the equilibrium state is characterized by the fact that neither player can improve his or her objective function by unilaterally changing his or her decision (Fisk 1984). Conversely, in the Stackelberg game, the player at the upper level (i.e., the leader) knows how the other player (i.e., the follower) will respond to any decision he or she may make.

Game Theory: Nash and Stackelberg Game Solution for the O/D Estimation Problem.   In the conventional iterative methods, the Nash game solution rather than the Stackelberg solution condition is achieved (Kim et al. 2001). This iterative bilevel formulation carries the approximate assignment results from a previous iteration to the following iteration (Figure 2.3). In fact, the Stackelberg solution is achieved only when the partial derivative of the link-flow proportions with respect to the demand is included in the formulation (Tavana 2001). Accordingly, the difference between solving the two games lies in the manner with which the derivative of the upper-level optimization is obtained. In a Nash game, the derivative of the assignment matrix with respect to the O/D flows is assumed to be equal to zero because the assignment is considered constant from a previous iteration, whereas in a Stackelberg game, it is not considered constant. However, this derivative denotes the change in route-choice proportions as a result of changes in demand flow and therefore should not be ignored in the O/D estimation problem (Tavana 2001). Consequently, a Stackelberg solution condition is more representative of the problem under consideration.

In the literature, to the best of our knowledge, only a few research efforts examined the O/D estimation problem with Stackelberg game conditions. Yang (1995) and Kim et al. (2001) examined application of the Stackelberg game conditions for the static O/D estimation problem. In addition, Tavana (2001) and Kattan and Abdulhai (2006) exploited the Stakelberg game conditions for the dynamic O/D estimation problem. All these approaches used the generalized least squares (GLS) estimation approach.

Advanced Approaches Based on Stakelberg Game Condition for the Static O/D Estimation Problem.   Yang (1995) proposed a bilevel programming method as an attempt to achieve a Stackelberg solution for the static O/D estimation problem. In this method, the anticipated change in the traffic flows, owing to the change in O/D flows, is explicitly included in the upper-level solution. Yang's method is known as a sensitivity analysis–based (SAB) approach. The SAB algorithm is an iterative algorithm that computes O/D flows in iteration n by using the sensitivity of iteration $n-1$. SAB also assumes linear approximation between previous iterations and current ones. However, because of this iterative linear approximation and the many-to-one mapping between link traffic and estimated O/D lows, SAB is still considered to be closer to a Nash game than to a Stackelberg game (Kim et al. 2001).

Kim et al. (2001) used an artificial intelligence technique, genetic algorithms (GAs), to achieve a Stackelberg solution for the static O/D estimation problem. GA methods are population-based search algorithms derived from Darwin's natural evolution and survival-of-the-fittest philosophies. In the

GA approach of Kim et al., the Stackelberg game is achieved by solving the problem in a one-level approach rather than by an iterative one. As such, the lower level is computed, as it is for all assignment-based approaches, from a traffic assignment model precisely reproducing the same assignment map used in the upper-level approach. While in the iterative approach the assignment matrix in the nth iteration is assumed to be the same as that of the $(n - 1)$th iteration, in the single-level approach, both the exact assignment matrix and simulated traffic counts are computed for each candidate O/D flow solution. Consequently, a Stackelberg solution is satisfied because the relationship between O/D flows and link flows is represented simultaneously and accurately. Kim et al. (2001) applied their proposed GA methodology to a sample network, and the results were compared with those of the SAB approach. The results of the GA approach were shown to yield a higher performance in generating the observed flow.

Advanced Approaches Based on Stakelberg Game Condition for the Dynamic O/D Estimation Problem. In an effort to achieve a Stackelberg solution for the dynamic O/D estimation problem, Tavana (2001) suggested an iterative Stackelberg solution whereby the link-flow and partial derivative values of the link-flow proportions, with respect to demand, are estimated numerically in the Dynasmart-P DTA package by tracing each vehicle temporally and spatially within the network. This process is recomputed for each iteration and for different demand levels and each O/D pair and departure intervals. In addition, a rolling-horizon framework is formulated to avoid error propagation in subsequent estimation steps.

Kattan and Abdulhai (2006) developed a dynamic O/D estimation framework called *DynODE* (*Dynamic Origin/Destination Estimation*) using advanced evolutionary algorithms (EAs). Additionally, these EA-based methods are augmented further with (1) a hybrid EA search mechanism also known as *memetic algorithms*, (2) EA parallelization, and (3) distributed computing to further improve the quality and efficiency of the solution. In contrast to more conventional O/D estimation methods, EAs are characterized by their global and multipoint searching capabilities, which decrease the dependency on the quality of the starting point. In addition, solving the O/D estimation problem using EAs eliminates the need for the iterative process. In other words, the EA approach computes the exact equilibrium flow corresponding to each candidate O/D vector, simultaneously satisfying both levels of the bilevel problem and the fixed-point formulation. This latter noniterative approach satisfies, in turn, a complete Stackelberg solution condition, meaning that the solution is optimal and superior to the Nash game solution condition. Moreover, owing to its fixed-point formulation, the DynODE framework approach is applicable to all networks irrespective of the assumptions underlying the modeling of the cost functions (i.e., for both the symmetric and asymmetric Jacobian matrices).

The results of intensive experimentations showed that the DynODE framework offers many advantages over more conventional approaches, most importantly, (1) the ability to estimate O/D tables with a gradient-free approach and (2) the reduce reliance on the initial (*a priori*) information. DynODE estimates the demand in a noniterative fashion. In fact, one of the biggest challenges of O/D estimation is computation of the assignment matrix. The EA-based framework presents a gradient-free approach that eliminates the cumbersome task of computing the assignment matrix in every iteration, as is the case with the more conventional DTA-based methods. Furthermore, conventional DTA-based approaches place high expectations on the quality of *a priori* information. Therefore, improvement over the starting point is usually limited because the adopted optimization techniques are likely to find the nearest local optimum. However, the EA techniques are multipoint searching methods, having global and probabilistic search capabilities. Thus this framework suggests that the solution of a DTA EA-based O/D estimation framework is independent of the initial starting point, and the role of the *a priori* demand lies mainly in giving a good starting search space by building the initial EA population. However, it does not have to be present necessarily in the objective function. In fact, the results of intensive experimentation demonstrate high performance irrespective of the historical information (Kattan 2005).

The other advantage of DynODE lies it its ability to combine any type of observation in the network (i.e., counts, densities, and travel time) with other sets of *a priori* information, such as outdated survey and probe data, automatic vehicle identification (AVI) data, and previous-step estimates or forecasts. Finally, although DynODE addresses mainly offline O/D estimation problems, online real-time O/D estimation is shown to be possible by applying advanced EA parallelization schemes. However, more multiprocessing and parallel computing systems are needed for real-time applications (Kattan and Abdulhai 2006).

2.8 SUMMARY

This chapter reviews the different approaches for estimating O/D flows from traffic counts, for the static as well as the dynamic case. Both assignment-based and non-assignment-based techniques are presented. O/D estimation problems are difficult to solve because of the nonconvexity of both upper- and lower-level optimization problems as well as the complex interaction between these two levels. This nonconvexity indicates the existence of many local solutions, and thus an optimal solution might be difficult to find. In addition, since the true O/D matrix is not known in practice, the objective function cannot be guaranteed to be close to the real matrix. The true demand can be revealed only indirectly in the link counts that witness the dynamic spatial and temporal behavior of the O/D demand while interacting within the traffic network. One possible remedy to this problem would be the possible incorporation of additional information such as link densities. Density reading in particular gives a better indication of the quality of flow on the facilities than link flows. Density readings, in contrast to flow measurements clearly distinguish between congested and uncongested conditions. This might improve the solution quality by decreasing the number of local point solutions. In addition, supplementing the estimation problem with observation data from other sources (e.g., cordon line counts, screenlines, vehicle probes fitted with GPS, AVI, partial ODs, etc.) in the form of additional constraints might more adequately represent the true demand in the mathematical problem and thus improve the estimation results.

To conclude, the estimation of dynamic demand is an intractable and challenging problem which will definitely occupy the research community in the years to come. Accounting for additional complexities, such as an integrated mode choice and an O/D estimation/prediction model, as well as its implementation of an online microscopic simulation model, would further complicate the problem. However, the deployment of intelligent transportation systems (ITS) is providing researchers and practitioners with an unprecedented amount of valuable online and archived data. Recently, GPS-enabled cellular phones are becoming one of the most promising vehicle-probe methods for the production of reliable traffic as well as O/D information. Today, more than 4 billion people carry mobile phones, and the number is continuing to rise (PA News 2009). By 2012, the number of new mobile phones that will be location-enabled via built-in GPS receivers will have doubled (Sacco 2007). The benefits from extracting knowledge from these databases will be immense. Therefore, these added complexities, coupled with the potential applications and fusion of various real-time data, may provide the opportunities needed to overcome some of the hurdles in estimating the time-dependent demand.

REFERENCES

Ashok, K. 1966. "Estimation and Prediction of Time-Dependent Origin-Destination Flows." Ph.D. dissertation, Massachusetts Institute of Technology. Department of Civil and environmental Engineering.

Ashok, K., and Akiva, B. 1996. "Estimation and Prediction of Time-Dependent Origin-Destination Flows: A Stochastic Mapping to Path Flows and Link Flows," *Transportation Science* 36:184–209.

Bell, M. 1991. "The Real Time Estimation of Origin-Destination Flows in the Presence of Platoon Dispersion," *Transportation Research Part B* 25:115–125.

Cascetta, E. 1984. "Estimation of Trip Matrix from Traffic Counts and Survey Data: A Generalized Least Squares Estimator," *Transportation Research Part B* 18:289–99.

Cascetta, E. 2001. *Transportation Systems Engineering Theory and Methods.* Dordre cht, The Netherlands: Kluwer Academic Publishers.

Cascetta, E., and Nguyen, S. 1988. "A Unified Framework for Estimating or Updating Origin/Destination Matrices from Traffic Counts," *Transportation Research Part B* 22:437–55.

Cascetta, E., and Postorino, M. N. 2001. "Fixed Point Approaches to the Estimation of O/D Matrices Using Traffic Counts on Congested Networks," *Transportation Science* 35:134–47.

Chan, W. T., Fwa, T. F., and Tan, C. Y. 1994. "Road-Maintenance Planning Using Genetic Algorithms," *Journal of Transportation Engineering* 120:693–709.

Chang, G., and Tao, X. 1999. "An Integrated Model for Estimating Time-Varying Network Origin-Destination Distributions," *Transportation Research Part A* 33:381–99.

Cheu, R. L., Sin, X., Ng, K. C, et al. "Calibration of FRESIM for Singapore Expressway using genetic algorithm," *Journal of Transportation Engineering* 124:526–35.

Cree, N. D., Maher, M. J. and Paechter, B. 1998. "The Continuous Equilibrium Optimal Network Design Problem: A Genetic Approach in Transportation Networks: Recent Methodological Advances," In: *Selected Proceedings of the 4th EURO Transportation Meeting*, Bell M. G. H., ed., 1998: 163–74.

Cremer, M., and Keller, H. 1987. "A New Class of Dynamic Methods for the Identification of Origin-Destination Flows," *Transportation Research Part B* 21:117–32.

Fisk, C. S. 1984. "Game Theory and Transportation Systems Modelings," *Transportation Research Part B* 18:301–13.

Hellinga, B. 1994. "Estimating Dynamic Origin-Destination Demands from Link and Probe Counts. Ph. D. dissertation, Queen's University, Kingston, Ontario, Canada.

Hu, S. 1997. "An Adaptive Kalman Filtering Algorithm for the Dynamic Estimation and Prediction of Freeway Origin-Destination Matrices. Ph.D. dissertation, Purdue, In.

Horan, T. 2000. "Environmental Issues." In: *Intelligent Transportation Primer*. Washington: Institute of Transportation Engineering.

Jayakrishnan, R., Tsai, W. K., and Chen, A. 1995. "A Dynamic Traffic Assignment Model with Traffic-Flow Relationships," *Transportation Research Part C* 3:51–72.

Kang, Y. 1999. "Estimation and Prediction of Dynamic Origin-Destination (O-D) Demand and System Consistency Control for Real-Time Dynamic Traffic Assignment Operation." Ph.D. Dissertation, University of Texas at Austin, 1999.

Kattan, L. 2005. "Dynamic Traffic Origin/Destination Estimation Using Evolutionary Based Algorithms." Ph.D. dissertation, Department of Civil Engineering, University of Toronto.

Kattan, L. and Abdulhai, B. 2006. "Non-iterative Approach to Dynamic Traffic Origin-Destination Estimation Using Parallel Evolutionary Algorithms," *Journal of Transportation Research Board (TRR)* 1964:201–10.

Khoshyaran, M. M. 1995. "Dynamic Estimation of Origin-Destination Matrices for Urban Network Using Traffic Counts. Ph.D. dissertation, Department of Civil and Environmental Engineering, Michigan State University, 1995.

Kim, H., Baek, S., and Lim, Y. 2001. "O-D Matrices Estimation Using Genetic Algorithms from Traffic Counts," *Journal of the Transportation Research Board* 1771:156–63.

Mahmassani, H. S. 2001. "Dynamic Network Traffic Assignment and Simulation Methodology for Advanced System Management Applications," *Networks and Spatial Economics* 1:267–92.

Mahmassani, H. S., and Sbayti, H. A. 2003. *Dynasmart-P Intelligent Transportation Networks Planning Tool, Version 0.930.4 User's Guide*. Maryland Transportation Initiative University of Maryland.

Memon, G. Q., and Bullen, A. G. R. 1996. "Multivariate Optimization Strategies for Real-Time Traffic Control Signals," *Journal of the Transportation Research Board* 1554:36–42.

Meyer, M. and Miller, E. 2001. *Urban Transportation Planning: A Decision-Oriented Approach*, 2d ed. New York: McGraw-Hill.

Nguyen, S. 1977. "Estimating an O/D Matrix from Network Data: A Network Equilibrium Approach." Publication No. 87, Centre de Recherches sur les Transports, Université de Montreal, Montreal, Quebec, Canada.

Nihan, N. L. and Davis, G. A. 1987. "Recursive Estimation of Origin-Destination Matrices from Input/Output Counts," *Transportation Research Part B* 21:149–63.

PA News. 2009. "Poor Drive Mobile Phone Growth." Retrieved April 9, 2009, from www.channel4.com/news/articles/science_technology/poor+drive+mobile+phones+growth/3007742.

Peeta, S., and Mahmassani, H. S. 995. "Multiple User Classes Real-Time Traffic Assignment for Online Operations: A Rolling Horizon Solution Framework," *Transportation Research Part C: Emerging Technologies* 3:83–98.

Peeta, S. and Ziliaskopoulos, A. K. 2001. Foundations of Dynamic Traffic Assignment: The Past, the Present, and the Future," *Networks and Spatial Economics* 1:233–65.

Renders, J.-M., and Bersini, H. 1994. "Hybridizing Genetic Algorithms with Hill-Climbing Methods for Global Optimization: Two Possible Ways." In: *Proceedings of the 1st IEEE Conference on Evolutionary Programming*, pp. 312–17.

Sacco, A. 2007. "GPS-Enabled Mobile Phone Shipments to More than Double Over Next Five Years. Retrieved April 9, 2009, from www.cio.com/article/160700/GPS_Enabled_Mobile_Phone_Shipments_to_More_Than_Double_Over_Next_Five_Years.

Sheffi, Y. 1985. *Urban Transportation Networks: Equilibrium Analysis with Mathematical Programming Methods.* Cambridge, MA: MIT Press.

Spiess, H. 1990. "A Gradient Approach for the OD Matrix Adjustment Problem, "Publication No. 693 at Centre de Recherches sur les Transports, Université de Montréal, Montréal, Canada.

Sun, C., and Porwal, H. 2000. "Dynamic Origin/Destination Estimation Using True Section Densities," California PATH Research Report UCB-ITS-PRR-2000-5, California PATH Program, Institute of Transportation Studies, University of California at Berkeley.

Suzuki, H., Nakatsuji, T., Tanaboriboon, Y., and Takahashi, K. 2000. "Dynamic Estimation of Origin-Destination Travel Time and Flow on a Long Freeway Corridor," *Journal of the Transportation Research Board* 1739:67–75.

Tavana, H. 2001. Internally Consistent Estimation of Dynamic Origin-Destination Flows from Intelligent Transportation Systems Data Using Bi-Level Optimizations. Ph.D. dissertation, University of Texas at Austin.

Van der Zijjp, N. J., and Hamerslag, R. "Improved Kalman Filtering Approach for Estimating Origin-Destination Matrices for Freeway Corridors," *Journal of the Transportation Research Board* 1443:54–64.

Van Zuylen, H. J. and Willumsen, L. G. 1981. "The Most Likely Trip Matrix Estimated from Traffic Counts," *Transportation Research Part B* 14:281–93.

Wu, J. A. 1996. "Real-Time Origin-Destination Matrix Updating Algorithm for On-Line Applications," *Transportation Research Part B* 31:381–96.

Yang, H., 1995. "Heuristic Algorithms for the Bilevel O-D Matrix Estimation Problem," *Transportation Research Part B* 29:231–42.

Yang, H., and Zhou, J. 1998. "Optimal Traffic Counting Locations for Origin-Destination Matrix Estimation," *Transportation Research Part B* 32:109–26.

Yang, H., Sasaki, T., and Iada, Y. 1992. "Estimation of Origin-Destination Matrices from Link Counts on Congested Networks," *Transportation Research Part B* 26:417–34.

Zhou, X., Qin, X., and Mahmassani, H. S. 2003. "Dynamic Origin-Destination Demand Estimation Using Multi-Day Link Traffic Counts for Planning Applications," CD-ROM, 82nd Annual Meeting of the Transportation Research Board, National Research Council, Washington.

Zhou, X., and Mahmassani, H. S. 2004. A Structural State Space Model for Real-Time Origin-Destination Demand Estimation and Prediction in a Day-to-Day Updating Framework," CD-ROM, 83rd Annual Meeting of the Transportation Research Board, National Research Council, Washington.

CHAPTER 3
TRAFFIC CONGESTION

Kara Kockelman

Department of Civil, Architectural, and Environmental Engineering
The University of Texas at Austin
Austin, Texas

3.1 INTRODUCTION

Congestion is everywhere. It arises in human activities of all kinds, and its consequences are usually negative. Peak demands for goods and services often exceed the rate at which that demand can be met, creating delay. That delay can take the form of supermarket checkout lines, long waits for a table at a popular restaurant, and after-work crowds at the gym. Yet the context in which we most often hear of congestion posing a serious problem to ourselves and to our economy is the movement of people and goods.

The average American reports traveling almost 80 minutes a day, over 80 percent of which is by automobile[*] (NHTS 2001). The Texas Transportation Institute (TTI) estimates that over 45 percent of peak-period travel or roughly one-third of total vehicle miles traveled occur under congested conditions in many U.S. metropolitan areas (Schrank and Lomax 2009). These include the predictable places such as Los Angeles (California), Washington (DC), and Atlanta (Georgia); they also include such places as San Diego (California), Tacoma (Washington), and Charlotte (North Carolina). Although crime, education, taxes, and the economy certainly are key issues for voters and legislators, polls regularly report congestion to be the number one local issue (see, e.g., Scheibal 2002; Knickerbocker 2000; Fimrite 2002).

Other types of transportation certainly are not immune to congestion either. Intercity trucking carries almost 30 percent of freight ton-miles shipped in the United States every year (BTS 2002) and 72 percent of the value shipped (CFS 1997). These trucks are subjected to the same roadway delays, resulting in higher priced goods, more idling emissions, and frayed nerves. The gates, runways, and traffic control systems of major airports are tested daily. Seaport berths, rail tracks, canals, and cables all have limitations as well. As soon as demand exceeds supply, goods, people, and information must wait in queues that can become painfully long. Although not stuck in queues, others find themselves waiting at the destinations for expected shipments, friends, family members, and colleagues that fail to arrive on time.

Engineers, economists, operations researchers, and others have considered the problem of congestion for many years. The confluence of growing traveler frustration, technological innovations, and inspirational traffic management policies from around the world provide added momentum for

---

[*]NPTS-reported trip making involves very distinct locations, such as home and work, school and shopping center. In reality, we are moving much more regularly, between bedrooms, around offices, and along supermarket aisles. Such travel, while substantial, is probably much less affected by congestion.

the modifications needed to moderate and, ideally, eliminate this recurring problem and loss. This chapter examines congestion's defining characteristics, its consequences, and possible solutions.

3.2   DEFINING CONGESTION

Notably, congestion is not always undesirable. Some "congested" experiences can be positive, and these tend to occur at one's destination—rather than en route. Myers and Dale (1992) argue that orchestrated congestion in public spaces, such as theater entry plazas, enhances public interaction, enables better land use mixing, and calms vehicular traffic. Taylor (2002) observes that congested city centers are often signs of vibrant city activity and prosperity. These forms of congestion may be, to some extent, desirable and are not the concern of this chapter.

This chapter stresses instead the undesirable form of congestion: the kind that impedes travel between two points, effectively adding access costs to a desired destination. The travel itself is not enjoyable; it is instead a necessary expenditure.[*] This form of congestion is a slowing of service. Queues (or lines) of travelers will form if demand exceeds capacity. But these are not necessary for congestion to occur. All that is necessary is that the service speed is less than the "free flow" or maximum speed, which exists when demand is light relative to capacity.

All transportation systems are limited by a capacity service rate. Operators at manual toll booths and transponder readers for electronic toll collection (ETC) cannot reliably serve more than a certain number of vehicles per hour. Commuter rail lines eventually fill up, along with their train cars. Port cranes exhibit functional limits, as do canals, runways, and pipelines. No system is immune; all physical pathways are constrained in some respect.

When systems slow down, delays arise. *Delay* generally is defined as the difference between actual travel time and travel time under uncongested or other acceptable conditions. The *Highway Capacity Manual 2000* (TRB 2000) defines *signalized intersection delay* as the sum of delay under uniform arrivals (adjusted for a progression factor), incremental delays (to account for randomness in arrival patterns), and any initial queue delays (to recognize spillovers from prior cycles). While vehicle detection and intersection automation can reduce signal delays dramatically, they cannot eliminate them. Any time two or more vehicles (users) wish to use the same space at the same time, delay will result—otherwise, there would be a crash! One must cede that space to the other. The mechanism may be a signal, a queue, pricing, rationing, or other policies.

In general, then, congestion is the presence of delays along a physical pathway owing to the presence of other users. Before discussing strategies to combat such delays, this chapter examines the general costs and consequences of congestion, its causes, and its quantification.

3.3   THE CONSEQUENCES OF CONGESTION

Automobile congestion has myriad impacts, from wasted fuel and added emissions to frayed nerves, more expensive goods, and elevated crash rates. Its clearest impact is delay, or lost time. Across the United States, this may average 20 hours per year per person. In Los Angeles, it is estimated to exceed 60 hours, which translates to 10 minutes a day, or one-sixth of one's average travel time (Schrank and Lomax 2002). Presented this way, even Los Angeles' numbers may seem acceptable. Why, then, does this issue so consistently top opinion polls as our communities' number one policy issue? There are many reasons. One is that dense car traffic is more difficult to navigate, even if speeds stay high. Such travel is unpleasant and tiring in many ways. Another is that congestion tends to be unpredictable, even when it is recurring. As a result, peak-period travelers—including trucks and buses—regularly arrive early or late at their destinations, creating frustration. These suboptimal arrival times carry a

---

[*]Mokhtarian and colleagues have examined the extent to which travel may be desirable in and of itself, resulting in "excess travel" (Mokhtarian and Salomon 2001; Redmond and Mokhtarian 2001; Salomon and Mokhtarian 1998). Richardson (2003) found reasonably high proportions of travelers in Singapore to exhibit a zero value of travel time, causing him to conclude that, among other things, travel on the air-conditioned transit system in that humid city can be a relatively enjoyable (or at least refreshing) experience.

cost: missed meetings and deliveries, loss of sleep, child care fees for late pickup, children waiting around for classes to begin, a supervisor's growing intolerance of missed work, and time spent unproductively, among other issues. Researchers have tried to quantify these additional costs.

Brownstone and Small (2005) reviewed a number of studies in this area, and Van Lint et al. (2008) characterized multiple methods for quantifying reliability. Bhat and Sardesai's (2006) and Small et al.'s (2005) analyses of departure time data (both stated and revealed) suggest that each minute of added uncertainty (measured as standard deviation or a difference in key quantiles of the distribution of travel times) in one's arrival time is valued on par with an added minute of (average) travel time. The review by Bates et al. (2001) of travel time reliability research found that every minute of *lateness* is regularly valued at two to five times a minute of travel time. And every minute *early* is valued at almost a minute of travel time (around 80 percent). In general, variation in travel time (as measured by standard deviation) is worth more—to the typical traveler—than the average travel time. Thus, even if one's commute trip *averages* 25 minutes, if it exhibits a 10-minute standard deviation,[*] it typically is not preferred to a guaranteed travel time of 35 minutes. Using loop detector data for samples of freeway sections in cities across the United States, Lomax et al. (2009) estimated early-departure "buffer times" in order to ensure on-time (or early) arrivals in 95 percent of one's trip making. For Austin, Texas, the required buffer was estimated to be 29 percent of the average travel time (using year 2007 data); in Los Angeles, it was 47 percent. In 90-region data set of Lomax et al., congestion was highly correlated with high buffer times (and thus low reliability). Clearly, congestion is costly—in a variety of ways.

Another reason for society's impatience with congestion may relate to equity and an ability to prioritize consumption or purchases. On a "free" system of public roadways, every trip is treated the same. In other words, more valuable trips experience the same travel times as less valued trips. Persons and goods with very low values of time pay the same time price as others, even though the monetary value of and/or willingness to pay for their trips can differ by orders of magnitude. Few options are available, and they can require significant adjustments: changes in one's home, work, school, or other locations; moved meeting times; and entirely forgone activities. In one exceptional circumstance, Silicon Valley pioneer Steve Jobs elected to purchase a helicopter to reduce his San Francisco Bay Area commute.

As mentioned earlier, some travelers find a moderate level of congestion acceptable, even desirable. Moreover, an evolution of in-vehicle amenities, from radios, air conditioning, and reclining seats to tape and CD players, stereo-quality sound systems, cellular phones, video players, heated seats, and Wi-Fi, plays a role in reducing the perceived costs of congestion. At the same time, the increasing presence of two-worker couples, rising incomes, just-in-time manufacturing and delivery processes, and complication of activity patterns for adults and children have resulted in a variety of travel needs and time constraints that can make delays more costly and stressful. The market for scarce roadspace breaks down during peak travel hours in many places. So what are the costs?

The TTI studies estimate Los Angeles' congestion costs to exceed $10 billion each year, or more than $800 per resident. This figure is based on speed and flow estimates across the region's network of roads, where every hour of passenger-vehicle delay is valued at nearly $20, every hour of delayed commercial truck travel is valued at just over $100, and each gallon of gasoline consumed by passenger vehicles (while delayed) is valued at $3.24 (Schrank and Lomax 2009). Together with costs from the nation's 438 other urban areas (using nearly the same unit-cost assumptions), the annual total reaches $87 billion. This is about 4 cents per vehicle mile traveled (VMT), or double U.S. gas taxes (which total roughly 40 cents/gal, or 2 cents/mi).[†] Essentially, the U.S. Highway Trust Fund's revenues could be doubled through the addition of these estimated costs. Of course, these neglect congestion in other travel modes and other U.S. locations, as well as environmental impacts, schedule delay, delivery difficulties, inventory effects, frustration, crash, and other costs. Carbon monoxide and hydrocarbon emissions are roughly proportional to vehicle *hours* of travel (Dahlgren 1994); oxides

---

[*]The standard deviation is the square root of the expected value of squared differences between actual and mean travel times: $\sigma_t = \sqrt{E[(t-\mu_t)^2]}$. For a normal distribution of travel times that averages 25 minutes, a standard deviation of 10 minutes implies a 16 percent probability that a trip will exceed 35 minutes.

[†]Passenger-vehicle occupancies were assumed to be 1.25 persons/vehicle, and 5 percent of congested-period VMT was assumed to be by trucks (Schrank and Lomax 2009).

of nitrogen also rise with slowed traffic, further worsening air quality (Beamon 1995). Congestion stymies supply chains and diminishes agglomeration economies (Weisbrod et al. 2001). It can increase crash risks dramatically (Edlin and Mandic 2006). Taken together, such indirect costs easily may exceed TTI's congestion cost estimates.

3.3.1   Congestion and Crashes

The consequences of congestion for crash frequency and severity are intriguing. The lowest levels of reported crash rates (per VMT) tend to occur at intermediate levels of flow (e.g., level of service C*). (see, e.g., Gwynn 1967; Brodsky and Hakkert 1983). Controlling for traffic density—rather than flow—also is key because low flows can occur under both uncongested (high speed) and congested (low speed) conditions. Garber and Subramanyan's recent work (2002) for weekday crashes on four highways indicates a steep reduction in police-reported crash rates (i.e., crashes per VMT) when densities are about half of critical density (where critical density corresponds to capacity flow rates).[†] Thus crash rates generally appear to rise as congestion sets in. However, speeds tend to fall under such conditions, especially when demand exceeds capacity. And lower collision speeds mean less severe traffic crashes (see, e.g., Evans 1991; Kockelman and Kweon 2002).

Overall, however, the negative safety implications of adding vehicles to even a moderately congested network may be quite sizable. Edlin and Mandic (2006) have estimated the annual marginal crash costs of congestion to be on the order of $1,700 to $3,200 per vehicle added to a congested state such as California. In places without congestion (such as Wyoming), their estimates are negligible.

Of course, crashes themselves also generate congestion, by distracting other drivers and blocking lanes and shoulders. And drivers frustrated by congestion may take risks that offset some of the benefits of reduced speeds, such as tailgating, using shoulders as traffic lanes, cutting across dense oncoming traffic, and speeding up excessively when permitted. Such behaviors are typical of "road rage" and may worsen congestion—and result in crashes. Little is formally known regarding the magnitude and nature of such indirect safety effects of congestion.

3.4   QUANTIFYING CONGESTION

The explicit measurement or quantification of congestion has many uses. Such measures help communities to identify and anticipate traffic problems by location, severity, and time of day. Their magnitude and ranking provide a basis for targeted investment and/or policy decisions. They also are useful as inputs for air quality models, which require travel speed and distance information. The following discussion provides a definition of congestion, along with methods for its estimation based on travel time formulas.

3.4.1   Congestion and the *Highway Capacity Manual*

For transmission of people, goods, or data, an important distinction exists between capacity (i.e., maximum flow) speeds and free-flow speeds. In the case of roadways, congestion can set in and speeds can fall well before capacity is reached. Chapter 23 of the *Highway Capacity Manual 2000* (TRB 2000) provides estimates of capacity and speeds for a variety of basic freeway segments. A traffic density of 45 passenger cars per lane-mile (pc/mi/ln) corresponds to (sustainable) capacity conditions, and it distinguishes levels of service E and F. On facilities with a 75 mi/hr free-flow speed

---

*Freeway level of service C implies conditions where speeds are near free-flow speeds, but maneuverability is "noticeably restricted". [*Highway Capacity Manual 2000* (TRB 2000), p. 13–10].

[†]Many slight crashes may occur under congested conditions, and go unreported to police. Thus it is possible that total (reported and unreported) crash rates stay stable or even rise under congested conditions.

(FFS), average speeds are expected to begin falling at relatively low densities (e.g., 18 pc/mi/ln). And at capacity conditions (2,400 pc/hr/ln and 45 pc/mi/ln), the predicted prevailing speed is just 53.3 mi/hr—well below the uncongested 75 mi/hr FFS. On freeways exhibiting lower free-flow speeds, congestion is expected to set in later, and speed reductions are less severe. For example, for FFS = 55 mi/hr, small drops in prevailing speed arise at 30 pc/mi/ln densities, and capacity speeds are 50 mi/hr (just 10 percent less than FFS). Travel on these lower speed facilities will take longer, however, even if their conditions do not qualify as "congested"—simply because their associated speeds are lower for every level of service.

Why do speeds fall as density increases? This is so because the smaller the spacing between vehicles, the more likely there is a conflict. Safety also sets an upper limit on how fast people want to travel. Human and vehicle response times are limited: We take time to perceive threats, and our vehicles take time to slow down. Thus drivers have maximum speed preferences, which govern when traffic is relatively light, and have minimum spacing preferences, which govern when traffic is relatively heavy.[*] Driver spacing requirements go up with speed, so there is a natural limitation on how many can traverse a given road section in any given time. The *Highway Capacity Manual* predictions for freeway lanes are just one illustration of these safety-response phenomena.

On uncontrolled (nonfreeway) multilane highways, *Highway Capacity Manual* estimates of speeds and capacity flows are lower than those found on freeway lanes. (Driveways and other access points, left turns in the face of oncoming traffic, and other permitted behaviors necessitate more cautious driving.) However, the density values defining levels of service are almost unchanged, and the magnitudes of speed reduction leading to capacity conditions are minor (on the order of 1 to 3 mi/hr).

On two-lane (undivided) highways, capacity flows are dramatically less (1700 pc/hr/ln), and levels of service are defined by the percentage of time that vehicles follow slower vehicles (PTSF) and, in the case of class I facilities, by average travel speed. If passing is not permitted along a section, average travel speeds are predicted to fall by as much as 4.5 mi/hr.

When travel is controlled by traffic signals, signal-related delay estimates define level of service. Speed calculations are not emphasized, although simulation software such as the FHWA's CORSIM (1999) can generate estimates of trip start times, end times, and distances, thereby predicting operating speeds.

Significantly, *Highway Capacity Manual* methods permit no traffic speed estimation beyond level of service E, and Chapter 23's speed-flow curves and tables disappear. Beyond level of service E, traffic conditions can be characterized by speeds as high as 50 mi/hr (on high-design freeways)—or by complete gridlock (0 mi/hr). When demand exceeds capacity, a queue develops, and straightforward speed models break down. At that point, it becomes more important to know when a traveler enters the queue than to know how many are entering it. And it is all characterized as level of service F.

Unfortunately, level of service F's oversaturated conditions are common in many regions and on many facilities. Local bottlenecks and incidents cause demand to exceed capacity, sending congestive shockwaves back upstream. Under these conditions, upstream speeds can fall well below those prevailing under capacity conditions, even to zero. While oversaturated traffic conditions are rather unstable and exhibit high variation, speeds can be approximated. The following two sections describe some applicable methods.

3.4.2 Roadway Conditions: The Case of Interstate Highway 880

The *Highway Capacity Manual* offers traffic predictions based on empirical evidence for a variety of roadway types, designs, and locations. Yet conditions on specific facilities can differ rather dramatically. Loop detectors embedded in highway pavements offer continuous data-collection opportunities for key traffic variables.

---

[*]Based on the traffic observations illustrated in Figures 3.1 and 3.2, Kockelman (2001) has estimated free-flow speed and spacing preferences for various freeway user classes.

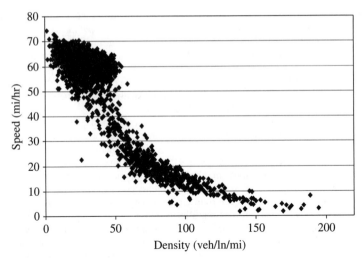

FIGURE 3.1 Speed versus Density (Lane 2 Observations from Northbound I.H. 880; Hayward, California).

Based on detector data from Interstate Highway (I.H.) 880's second-to-inside lane, observed speeds and densities are plotted as Figure 3.1. Speeds (measured in mi/hr) fall as density [measured in vehicles per lane mile (veh/ln/mi)] increases. Density is inversely related to vehicle spacing;[*] and as vehicles are added to a section of roadway, density rises and spacings fall. Drivers are inhibited by reduced spacings and the growing presence of others. For reasons of safety, speed choice, and reduced maneuverability, drivers choose lower speeds.

Speed multiplied by density is flow,[†] so Figure 3.1's information leads directly to Figure 3.2's speed versus flow values. Flow rates over these detectors may reach 3,000 vehicles per hour—or 1.2 seconds per vehicle for a brief 30-second interval. And average speeds appear to fall slightly throughout the range of flows from roughly 65 to 55 mi/hr. Travel times are affected as more and more vehicles enter the lane, densifying the traffic stream. At some point downstream of this detector station, demand exceeds capacity or an incident destabilizes traffic, and shock waves travel back upstream, forcing traffic into level-of-service F conditions. These oversaturated conditions correspond to speeds below 50 mi/hr in this lane on this facility.

Figure 3.3 is a photograph of congested traffic that could have come from this same section of I.H. 880. Since passenger vehicles average 15 to 18 feet in length (Ward's 1999), the image suggests an average spacing of roughly 40 feet per vehicle (per lane). Such spacing translates to a density of 132 vehicles per lane-mile. Assuming that Figure 3.1's relationships are predictive of Figure 3.3's traffic behaviors, this density corresponds to average speeds between 5 and 15 mi/hr. Since speed times density is flow, flows are likely to be around 1,300 veh/hr/ln, or about half of capacity.

As illustrated by Figure 3.2, a flow of just 1,300 veh/hr/ln also could correspond to a much higher speed (about 60 mi/hr)—and a much lower density (perhaps 25 veh/ln/mi—or a 211 ft/vehicle spacing). This contrast of two speeds (and two densities) for the same level of output (i.e., flow) is disturbing. Densities beyond critical (capacity level) density and speeds below critical speed identify a loss: The restricted roadspace could be more fully utilized and traveler delays could be avoided if only demand and supply were harmonized.

---

[*]Vehicles per unit distance (density) equal the inverse of distance per unit vehicle (spacing), where spacing is measured from the front of one vehicle to the front of the following vehicle (thereby including the vehicle bodies).

[†]Flow is vehicles per unit time, speed is distance per unit time, and density is vehicles per unit distance. Under stationary traffic conditions, vehicles maintain constant speeds, the mix of these vehicles (and their speeds) is unchanging, and density times space-mean speed (rather than the commonly measured time-mean speed or average of spot speeds) equals flow.

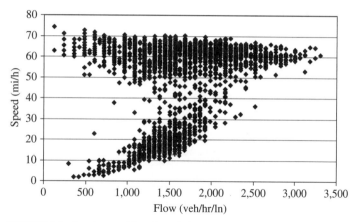

FIGURE 3.2 Speed versus Flow (Lane 2 Observations from Northbound I.H. 880; Hayward, California).

FIGURE 3.3 Image of Congestion.

This "tragedy of the commons" (Hardin 1968) plays itself out regularly in our networks: Heavy demand, downstream bottlenecks, and capacity-reducing incidents force upstream travelers into slow speeds. And facilities carry lower than capacity flows. In extreme cases, high-speed freeways as well as downtown networks become exasperating parking lots. To optimally avoid underutilization of scarce resources, one must have a strong understanding of demand versus supply. And bottlenecks are to be avoided if the benefits can be shown to exceed the costs.

A chain is only as strong as its weakest link, and downstream restrictions can have dramatic impacts on upstream roadway utilization. Bridges are common bottlenecks; they are relatively costly to build* and expand and thus often are constructed with fewer, narrower lanes and limited shoulders. Given a tradeoff in construction costs, travelers' willingness to pay, and traveler delays, theoretically, there is an optimal number of lanes to build in any section of roadway. Reliable information on demand (or willingness to pay) and associated link travel times (for estimates of delay) is necessary. Real-time roadway pricing, robust models of travel demand, and instrumentation of highways (for traffic detection) are key tools for communities aiming to make optimal investment—and pricing—decisions. This next section describes methods for estimating delays.

3.4.3   Link Performance Functions and Delay Estimation

Delay and the social value of this delay (such as the willingness of travelers to pay to avoid such delay) depend on the interplay of demand, supply, and willingness to pay. One must quantify congestion through link and network performance functions and transform the resulting delays into dollars.

Travel time (per unit distance) is the inverse of speed. Thus delays rise as speeds fall. And, as described earlier, speeds fall as density rises. Density rises as more and more users compete for limited road space, entering the facility and reducing intervehicle spacings, causing speeds to fall. Fortunately, densities can rise fast enough to offset reduced speeds, so their product (i.e., flow) rises. But flow can increase only so far: In general, it cannot exceed capacity. As soon as demand for travel across a section of roadway exceeds capacity, flow at the section "exit" will equal capacity, and a queue will form upstream, causing average travel times across the congested section to rise. Unfortunately, these travel times tend to rise exponentially as a function of demand for the scarce road space. Thus only moderate additions to demand can affect travel times dramatically. And if the resulting queuing impedes other system links, the delay impacts can be even more severe.

Referred to as the "BPR (Bureau of Public Roads) formula" (FHWA 1979), the following is a common travel-time assumption:

$$t(V) = t_f \left[1 + .15 \left(\frac{V}{C} \right)^4 \right] \tag{3.1}$$

where $t(V)$ is actual travel time as a function of demand volume V, t_f is free-flow travel time, and C is *practical capacity*, corresponding to approximately 80 percent of true capacity.[†] If one wishes to use true capacity (as is typical of most practice), parameter values of 0.84 and 5.5 [rather than the 0.15 and 4, respectively of equation (3.1)] can be used (Martin and McGuckin 1998).

Figure 3.4 illustrates the BPR relationship of equation (3.1) for a particular example where *true* capacity is 10,000 veh/hr, C is 8,000 veh/hr, and t_f is 1 minute (e.g., the time to traverse a 1-mile section at a free-flow speed 60 mi/hr). If demand exceeds capacity by 30 percent ($V = 1.3C$), travel times will be twice as high as those under free-flow/uncongested conditions. The resulting 2 min/mi

---

*Based on freeway construction-cost data from the Texas Department of Transportation, Kockelman et al. (2001) estimated bridge lanes to cost five to ten times as much as regular lanes.

†True capacity is understood to be the maximum service flow (MSF) under the *Highway Capacity Manual*'s (TRB 2000) level of service E. This practical capacity variable is a source of regular error and confusion in applications. Many (e.g., Garber and Hoel 2001) substitute a roadway's true capacity flow rate for C, resulting in an underprediction of travel times.

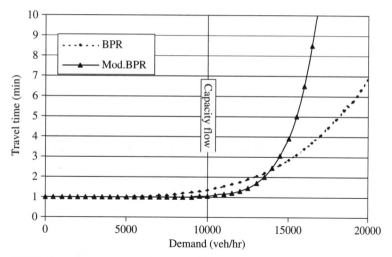

FIGURE 3.4 Travel Time versus Demand: BPR and Modified BPR Formulas.

pace implies speeds of just 30 mi/hr. It also implies 1 minute of delay for every mile of travel, with delay naturally defined as follows:

$$\text{Delay}(V) = t(V) - t_f = .15t_f \left(\frac{V}{C}\right)^4 \tag{3.2}$$

Is a one-to-one correspondence of delay to free-flow travel time common? Schrank and Lomax (2002) estimate 54 seconds of delay for every minute of peak-hour travel in the Los Angeles region. Kockelman and Kalmanje's (2003) survey results indicate that Austin, Texas, commuters perceive work-trip travel times that are almost double the free-flow time. Thus it may be in some regions that peak-period demand regularly exceeds capacity by 30 percent or more.

Of course, this 30 percent figure for a doubling of travel times is based on the BPR formula. Researchers have proposed modifications to this formula. Horowitz (1991) suggested replacing the two constants (0.15 and 4) with 0.88 and 9.8 (for use with 70 mi/hr design-speed freeways) and 0.56 and 3.6 (for 50 mi/hr freeways). Dowling et al. (1997) recommended 0.05 and 10 (for freeways) and 0.20 and 10 (for arterials). For comparison purposes, Dowling et al.'s freeway formula is included in Figure 3.4 as the "Modified BPR" curve. The two differ dramatically when demand exceeds capacity by more than 50 percent.

Beyond basic modifications to the BPR formula, there are other options. Arnott et al. (2005) rely on a piece-wise linear function, and Akçelik's (1991) formula recognizes demand duration. The duration of queuing has important consequences for total queue lengths and thus overall delays—which depend on when one can expect to enter the queue. In reality, traffic is dynamic, and vehicles occupy space. Links in the network can fill up, with traffic queues spilling back. These trigger delays on upstream links, cross streets, ramps, and even adjacent street systems. Simulation and dynamic traffic assignment techniques seek to address many of these realities and one day may replace static evaluations of traffic and associated congestion costs.

Modifications in BPR factors and the underlying formulas and/or use of newer modeling methods can have dramatic impacts on travel time estimates, travel—demand predictions, and policy implications.[*] Nevertheless, actual delay relationships remain poorly understood. The complexity of

---

[*]Nakamura and Kockelman's (2002) welfare estimates for selective pricing on the San Francisco Bay Bridge were very dependent on the bridge's travel-time function. Outputs of Krishnamurthy and Kockelman's (2003) integrated land-use–transportation models of Austin, Texas, were most affected by the exponential term in the BPR formula.

networks makes it difficult to measure travel times[*]—and even more difficult to ascertain real-time "demand" for congested-network links.

3.4.4 Delay Example: A Temporary Lane Loss

Relying again on Figure 3.4 and the BPR formula of equation (3.1), consider the impact of a loss of one lane. If capacity of 10,000 veh/hr corresponds to a four-lane high-design freeway, then the loss of one of these four lanes (by a crash or creation of a construction work zone, for example) results in an effective capacity of 7,500 veh/hr. At a demand of 13,000 veh/hr, travel times will jump by 115 percent from 2.0 to 4.3 min/mi. This is now 330 percent longer than travel time under free-flow conditions. Under this dramatic situation, speeds would be just 14 mi/hr—far lower than the free-flow speed of 60 mi/hr and well below the four-lane speed of 30 mi/hr.

The reason that travel times rise so dramatically once demand exceeds capacity is that a roadway (like an airport or any other constrained facility) can accommodate no more flow. It behaves much like a funnel or pipe that can release only so much fluid per unit of time. Any additional users will be forced to form a slow-moving queue, backing up and affecting the rest of the system (by blocking off-ramps and on-ramps or driveways and cross roads upstream of the limiting section). This is a classic bottleneck situation, where demand exceeds supply. Capacity-reducing incidents can affect supply instantly, leading to essentially the same low-speed, high-delay conditions for which recurring bottlenecks are responsible. Unfortunately, there is no guarantee that congestion can be avoided altogether; supply disruptions, through incidents and the like, can occur at most any time.

3.4.5 Recurring and Nonrecurring Congestion

The preceding example of a temporary lane loss may be recurring or nonrecurring, predictable or unpredictable. Recurring congestion arises regularly, at approximately the same time of day and in the same location. It results from demand exceeding supply at a system bottleneck, such as a bridge, tunnel, construction site, or traffic signal. Nonrecurring congestion results from unexpected, unpredictable incidents. These may be crashes, jackknifed trucks, packs of slow drivers, short-lived work zones, and/or foggy conditions.

Schrank and Lomax's (2009) extensive studies of regional data sets on travel, capacity, and speeds suggest that nonrecurring incidents account for roughly half of total delay across major U.S. regions. However, these percentages do vary. They depend on the levels of demand and supply, crash frequency, and incident response. For example, in the regularly congested San Francisco Bay Area, with its roving freeway service patrols, accidents and disabled vehicles account for 48 percent of total delay. In the New York–eastern New Jersey region, this estimate rises to 66 percent.

3.4.6 Evaluating the Marginal Costs of Travel

Whether a traveler opts to enter a facility that is congested for recurring or nonrecurring reasons, he or she pays a price (in travel time, schedule delay, and other costs) to use that facility. Because travel times rise with demand, his or her entry onto the facility (or at the back of the queue) also marginally increases the travel cost for others entering at the same time or just behind. This imposition of a cost, to be borne by others, is called a *negative externality*. Essentially, use of a congestible facility reduces the quality of service for others. This reduction in service quality is an external cost in the form of travel time penalties that others bear. Of course, all users (of the same congested corridor) bear it equally. So is it a problem?

---

[*]Lomax et al. (1997) recommend the use of probe vehicles to ascertain operating speeds. Loop detectors are presently only popular on freeway lanes and can assess only local conditions; frequent placement of loops is necessary to appreciate the extent of upstream queuing. Mobile GPS units offer hope for future traffic data collection and robust travel-time estimation.

Any time users "pay" a cost lower than society bears to permit the added consumption (in this case, use of a space-restricted facility), the good is overconsumed, and society bears more cost than it should. Economists have rigorously shown that in almost any market, goods should not be allocated beyond the point where marginal gain (or value to society) equals marginal cost to furnish the good. Marginal gains for most goods are well specified by consumers' willingness to pay. And marginal costs typically are absorbed by suppliers of those goods. In the case of road use, costs arise in many forms—and they are absorbed by many parties: Infrastructure provision and maintenance costs are absorbed by federal, state, and local agencies (and passed on through fuel and property taxes); travel-time costs are absorbed by travelers; environmental damages are absorbed by humans and other species (on, off, and far from the facilities themselves); and crash losses are felt by a variety of individuals (through pain, suffering, delay, and emergency medical services–related taxes).

The focus of this chapter is congestion and, therefore, travel times in excess of free-flow travel times. A road's available space is fixed, and under congested conditions, fellow travelers absorb the costs of delays arising from additional users. What are those marginal costs? Every link-performance function $t(V)$ implies these.

At a particular level of demand V, the marginal cost of an additional user $MC(V)$ is the change in total travel costs owing to that added user. Total costs $TC(V)$ are average cost per user $AC(V)$ multiplied by the number of users V. And travel time (per user) $t(V)$ is the average cost. Using this logic, the standard BPR travel-time function, and a little calculus (for continuous differentiation of the total-time formula), one has the following results:

$$AC(V) = t(V) = t_f \left[1 + .15 \left(\frac{V}{C} \right)^4 \right]$$

$$TC(V) = V \cdot t(V) \tag{3.3}$$

$$MC(V) = TC(V+1) - TC(V) \approx \frac{\partial TC}{\partial V} = AC(V) + 0.6 t_f \left(\frac{V}{C} \right)^4$$

The last of these three equations, the marginal cost of additional users, clearly includes the cost that the additional user experiences directly $AC(V)$. But it also includes a second term: the unpaid cost or negative externality that others endure in the form of higher travel times. In this BPR-based example, the externality also depends on the fourth power of the demand-to-practical-capacity ratio, but it is multiplied by a factor of 0.6 rather than 0.15. Thus, at certain levels of demand V, this second external cost will dominate the first average-cost term. It is this unpaid cost that is responsible for overconsumption of road space. Without assignment of ownership of the roads or some other method to ensure optimal use, excessive congestion results. If the roads are in heavy demand, the cost is more severe and the loss to society particularly striking.

Figure 3.5 plots the marginal cost curve above the standard BPR average cost curve. The difference between these two is the external costs that are associated with (and go unpaid by) additional users. At capacity levels of demand, this *difference* $(MC - AC)$ exceeds average cost by 7 percent. At demand of 13,000 veh/hr, it constitutes more than triple the average cost. Essentially, then, for optimal operations, perceived travel times or costs *should be* equivalent to 6.23 min/mi. Yet they are only 2.04 min/mi; the added travelers are enjoying an implicit subsidy of 4.18 min/mi at the expense of other travelers.

This situation may lead one to question: Who is the last "marginal" driver, and why (and how) should only he or she be penalized, when everyone should enjoy equal rights of access to the public right of way? The answer is that *all* drivers should weigh the true marginal cost of their trip before embarking—and then pay this cost (in the form of a toll) if their marginal gain from making the trip exceeds the total cost of time and toll. This requirement is placed on buyers of any good in regular markets, even for items as basic as clothing, shelter, food, and health care.[*] Private providers are not

---

[*]Education remains largely a public good. Regardless, subsidies targeting specific goods and consumer groups always can be provided (e.g., food stamps for low-income families).

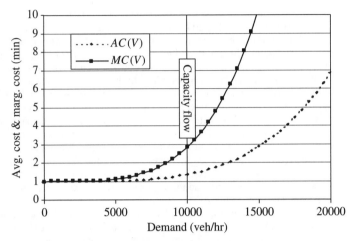

FIGURE 3.5   Average and Marginal Cost of Demand.

asked to provide goods below their marginal cost. Such excessive production is unwise. In considering whether and how to combat congestion, we should ask whether society should provide space on roadways at prices below marginal cost. This question brings us to this chapter's final section.

3.5 SOLUTIONS TO CONGESTION

Congestion may result from inadequate supply, imperfect information, flawed policies, or any combination of these three. To address these possibilities, a number of remedies have been proposed. Solutions may be supply- or demand-sided, long or short term, and/or best suited for recurring or nonrecurring congestion. They may be demographically, temporally, and spatially extensive—or limited; they may be costly or inexpensive, mode-specific or multimodal. They may involve mode subsidies, special lanes, pricing policies, access management, and/or land use controls. Virtually all produce both winners and losers.

In discussing various strategies for combating congestion, this section first details supply-side remedies, which generally aim to increase capacity by adding facilities or enhancing operations of existing facilities. On the demand side, the emphasis is on a moderation or modification of prices and preferences. In most regions, a combination of multiple solution types may be needed depending on the specifics of the problem, stakeholder support, and community aspirations.

3.5.1 Supply-Side Solutions

Traditionally, the solutions for congestion have been supply-sided. Engineers expected that roadway expansions and upgrades would relieve congestion. And as long as the latent demand for the facilities did not overtake the expansions [through, e.g., Downs' (1962) "triple-convergence principle" of route, time of day, and mode choice adjustments], peak-hour speeds rose and travel times fell.

The same phenomenon holds true at airports, where gates and runways may be added and aircraft may be made larger (to add seating). It holds true on railways, where engines and track may be added and headways reduced (through added trains and track as well as better coordination of train schedules), and at shipyards, where container cranes may be added or upgraded and berths may be extended. Traffic actuation and synchronization of signals, rationalization of freight networks, targeted enhancements of bottlenecks, headway-reducing vehicle-guidance technologies, and other remedies also are very helpful for specific applications.

Points of recurring congestion generally are often well suited for supply-side solutions. Returns on such investment will be more certain as long as similarly sized nearby bottlenecks do not negate the local expansion. (For example, a previously untested and thus undetected chokepoint may lie just downstream and lead to similar backups.) In cases of nonrecurring congestion, solutions providing rapid detection (e.g., paired loop detectors, video processing, or transponder tag reidentification), rapid response (e.g., roaming freeway service patrols), and real-time information provision (e.g., variable message signs) will have the greatest impact.

Capacity Expansion. There are a great many ways to expand the capacity of congestible systems. As long as increases in demand—through natural growth, changing preferences, and substitution (of origins, destinations, routes, modes, departure times, and other choices)—do not overtake these expansions, service times will fall. Yet, in many markets, capacity is so constrained and pent-up demand is so significant that travel times and speeds on expanded sections of the network may not fall in any perceptible way. Investigations by Hansen and Huang (1997), Noland and Cowart (1999), Fulton et al. (2000), Rodier et al. (2000), and Cervero (2002, 2003) have predicted relatively high (0.5 to 1.0) long-run elasticity of demand for road space (vehicle miles traveled), generally after controlling for population growth and income. Elasticity estimates of almost 1.0 (suggesting that new road space is almost precisely filled by new miles traveled) are not uncommon. However, several of these studies draw largely on California data, where congestion is relatively severe. In less congested regions, elasticities are expected to be lower. Even so, at roughly $5 million per added lane mile for freeway construction costs alone[*] (Kockelman et al. 2001; Klein 2001; Litman 2007), funding constraints regularly preclude major supply-side solutions. And in regions not in attainment with air quality standards and/or wishing to limit sprawl and other features of long-distance driving, building one's way out of congestion may not be a viable option.[†]

Alternative Modes and Land Use. To reduce roadway congestion, there are several supply-side enhancements of *alternative* modes that can cost less than new roadways while reducing driving and emissions. Improvements and expansions of bus, rail, ferry, and other services may qualify. However, transit already is heavily subsidized per trip in many countries. And even in downtown locations where its provision is reasonably extensive, transit ridership rates remain low in the United States. Thus it is unlikely to attract many travelers, particularly for long trips in a U.S. context.[‡]

Land-use solutions also have been proposed as a way to increase the use of alternative modes and diminish congestion. Transit use and walking are highest in high-density, mixed-used areas (see, e.g., Pushkarev and Zupan 1977; Kockelman 1997; Cervero and Kockelman 1997). Transit- and walking-oriented new urbanist designs strive to motivate mode shifts and reduce automobile reliance. But the resulting mode shifts are relatively weak, and neighborhood design—particularly in the form of higher development densities—is a poor instrument for combating roadway congestion (see, e.g., Boarnet and Crane 2001; Taylor 2002). Nevertheless, the shortening of overall trip lengths (and thus regional VMT), thanks to added density and mixing of activity sites, can be substantial (see, e.g., Ewing and Cervero 2001).

Managed Lanes. Addition of managed lanes, including high-occupancy vehicle (HOV) and high-occupancy toll (HOT) lanes, are an intermediate option (see, e.g., Kuhn et al. 2005; Obenberger 2004; Perez and Sciarra 2003). HOT lanes help to ensure against congestion for those whose trips

---

[*]Right-of-way acquisition, traffic diversion delays during construction, and other features of major highway projects in congested areas will add further to the overall expense of such projects.

[†]Even though VMT may rise in proportion to expanded capacity, with congestion remaining high, there may be sufficient benefits accruing to warrant such expansions. For example, if households can afford better homes and enjoy more choice in stores, schools, jobs, and other activities, thanks to expanded travel options, those benefits should be recognized.

[‡]There are a number of reasons for this. Land use patterns (including dispersed, low-density origins and destinations), parking provision, and relatively low gas prices are just a few.

are highly valued while facilitating full utilization of these special lanes (Peirce 2003). Fees can rise and fall (e.g., up to 40 cents/mi) to keep the HOT lanes flowing smoothly, whereas carpoolers (HOV users) and transit buses ride free—and fast. Thanks to revenues generated, agencies can float bonds to help cover some of the construction and other costs or spend the money on other services (such as increased transit service, roving freeway service patrols, and variable message signs with information on traffic conditions) (Dahlgren 2002). However, without pricing of substitute routes and services, it is difficult, if not impossible, to raise sufficient revenues from the private sector. Public financing is generally needed, at least in partnership with private investors. And HOV/HOT lane construction costs generally exceed those of standard freeway lanes owing to distinguishing features (such as longitudinal barriers and special access points).

Ramp Metering. Another supply-side strategy is ramp metering (May 1964; Newman et al. 1969). In contrast to expansion of existing systems and services, the objective may be *reduction* of ramp flows to keep main freeway lanes moving safely and swiftly. This form of supply restriction also tends to break up platoons of vehicles, effectively increasing the merge point's capacity while reducing travel times and improving safety (Chen et al. 2001; Klein 2001). It may penalize near-destination dwellers in favor of long-distance drivers, resulting in certain inequities (Levinson et al. 2002).

An extreme form of ramp metering is peak-hour ramp closure (see, e.g., Gervais and Roth 1966; Miesse 1967; Neurdorff et al. 1997; Pervedorous 2001). Of course, one also can pursue better incident management (as mentioned earlier) and information provision to system users. All these techniques help to moderate the use of key links in a system, thus affecting route choice and demand, the subject of the next section.

3.5.2 Demand-Side Solutions

Demand-side strategies seek to affect demand directly through policies and prices. Rather than expanding (or shrinking) existing services and facilities, one targets the relative prices of and/or access to such services and facilities. Reasonably common strategies include staggered/shifted work schedules, as well as working (and shopping) from home (via the Internet). Most trips are not work-related, however, and significant schedule flexibility already exists across many job sites. Real-time provision of network travel times (to warn potential motorists or corridor users of congestion), as well as bus arrival times (to enhance this competing mode's attractiveness), also can play a role in reducing congestion, albeit marginally.

Parking policies offer valuable examples of demand-side strategies. Most parking is provided "free" at offices, shopping centers, schools, and elsewhere. When space is plentiful, attendants are not needed, operations costs are zero, and maintenance costs may be minimal. Largely for the sake of cost-collection efficiency, parking costs are borne indirectly by users, through, for example, reduced salaries (to employees) and higher goods prices (for shoppers). Everyone bears these indirect prices, however, so there is no price-based incentive for not parking. Preferential parking and other perks for carpoolers and others who reduce total driving and parking demands provide a way to affect driving demand; however, there is a cost to these policies. Shoup's (1992, 1994) cash-out policy, now in place in California,* requires that the cash equivalent of parking expenses be given to employees who do not use the parking. This form of clear remuneration makes good sense to those who agree that markets naturally clear at optimal levels when pricing and other signals are unambiguous and consistent. This argument raises the case for congestion pricing.

Congestion Pricing. The objective of congestion pricing is efficient travel choices.† It is a market-based policy where selfish pursuit of individual objectives results in maximization of net social benefits. Such laissez-faire capitalism is the guiding light behind Adam Smith's (1776) "invisible hand."

---

*Under California Health and Safety Code Section 43845, this policy applies to businesses with 50 or more employees in air quality nonattainment areas.

†The FHWA (2009) offers a primer on the basics of congestion pricing, including descriptions of applications around the world.

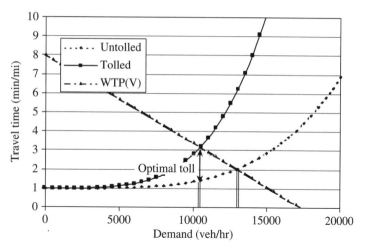

FIGURE 3.6 Demand versus Supply: Tolled and Untolled Cases.

When market imperfections are removed [through pricing of negative externalities (e.g., noise, emissions, and congestion), subsidy of positive externalities (e.g., public schools for educating all community members), and provision of adequate information], private pursuit of goods is optimal.[*]

In 1952, Nobel Laureate William Vickrey proposed congestion pricing for the New York City subway through the imposition of higher fares during congested times of day. This proposal was followed by theoretical support by Walters (1954, 1961) and Beckmann, McGuire, and Winsten (1956). If peak-period pricing recognizes the negative externality (i.e., time penalty) equivalent of each new, marginal user of a facility, optimal consumption and use decisions can result. In the case of Figure 3.5's standard BPR curve for travel time, the value of the negative externality, or the optimal toll, is the difference between the average and marginal cost curves:

$$\text{Optimal toll} = 0.6t_f \left(\frac{V}{C}\right)^4 \text{VOTT}_{\text{avg}} \tag{3.4}$$

where VOTT_{avg} is the monetary value of travel time of the average marginal traveler (roughly \$5 to \$15 per hour depending on the traveler and trip purpose).[†] Other variables are defined as for equation (3.1).

Figure 3.6 adds a downward-sloping demand or "willingness-to-pay" (WTP) curve to Figure 3.5's travel-time "supply" functions. The untolled supply curve represents the approach currently defining roadway provision, whereas the higher, tolled curve incorporates all marginal costs. The intersection of the demand curve with these two defines the untolled (excessively congested) and tolled

---

[*]Efficient travel choices will have effects on several related markets. For example, land-use choices and wage decisions will become more efficient by recognizing the true costs of accessing goods, services, and jobs.

[†]The toll should equal the value of the additional travel time the last vehicle adds to the facility. This last vehicle adds

$$\frac{dt(V)}{dV}V$$

where $V = \sum_i V_i(t, \text{toll})$ and i indexes the various classes of users demanding use of the road under that toll and travel time. The appropriate moneterarized value of travel time to interact with this added time is the demand-weighted average of VOTT_i values: $\text{VOTT}_{\text{avg}} = \sum_i \text{VOTT}_i \cdot V_i(t, \text{toll}) / \sum_i V_i(t, \text{toll})$. Given demand functions that are sensitive to time and toll and a link's performance function, it is not difficult to solve for the optimal toll. However, these inputs are tricky to estimate; they will be based on sample data and may involve significant error.

(less congested) levels of use on this roadway. The first results in 13,000 veh/hr, travel times of 2.0 min/mi, and delays of 1.0 min/mi. The latter results in 10,450 veh/hr, travel times of 1.44 min/mi, delays of 0.44 min/mi, and (the equivalent of) a toll of 1.75 min/mi (or 35 cents/mi, assuming a $12 per vehicle-hour value of travel time). In this example, neither situation is uncongested. Demand exceeds capacity in each instance, so queues will build for as long as the demand and supply curves remain at these levels.

Imposition of the net-benefit-maximizing "optimal toll" equates benefits (measured as willingness to pay, here in the form of time expenditures) and marginal costs. Under this 1.75 min/mi toll, realized demand is 10,450 veh/hr and toll revenues reach almost 18,300 minutes (1.75 × 10,450). Moreover, traveler benefits—measured as WTP in excess of travel time plus toll—exceed 25,100 minutes.[*] If time is converted to money at a rate of $12 per vehicle-hour (for the marginal traveler), the toll is worth 35 cents/mi, revenues are $3,600 per hour per mile, and the traveler benefits are worth $5,000/hr/mi.

Without the toll, traveler benefits exceed 39,000 minutes [(8 − 2) × 13,000/2], or $7800/hr/mi. But there are no toll revenues. Thus the $8,600 of value derived under the optimally tolled situation exceeds the laissez-faire approach by over 10 percent.

For a 10 percent addition in value during peak times of day on key roadway sections, should communities ask their public agencies to step in and start charging on the order of 35 cents per congested mile of roadway? There are many issues for consideration. First, the revenues do not go to those who bear the cost of congestion (i.e., the delayed travelers). If these revenues are not well spent, society loses. Second, if demand is highly inelastic (i.e., steeply sloped in Figure 3.6), realized demand will not change much under pricing; revenues simply will be transferred from traveler benefits to the collecting agent. Third, if the cost of implementing and enforcing congestion pricing is high, policy benefits may be wholly offset by implementation costs.

However, the benefits of pricing on highly congested links may well exceed 10 percent. While *average* travel times may rise by a factor of 2 during peak periods in congested areas such as Los Angeles, certain sections of roadway (such as key bridges) may experience much more severe delays. Imagine the same example in Figure 3.6 with a peak-period demand line that still begins at 8 min/mi but lies flatter and crosses the $AC(V)$ curve at 17,500 veh/hr, or 75 percent beyond capacity. In this case, untolled travel times are about 4.5 min/mi, and marginal costs lie 13.7 min/mi above that level (a significant externality and implicit subsidy). The optimal toll would be worth 3.6 min/mi (or 72 cents/mi), bringing demand down to 12,500 veh/hr. And tolled traveler benefits plus revenues would exceed untolled traveler benefits by a whopping 94 percent. Such a situation is probably a strong candidate for pricing.

Beyond increases in community benefits through removal of delay-related externalities, there are other advantages of congestion pricing. These include reduced emissions and gasoline consumption, because idling is reduced, closer destinations are chosen, and cleaner modes of transportation often are selected. They result in healthier lungs, less crop and property damage, and diminished threat of global warming. Such benefits are difficult to value, but Small and Kazimi's work (1995) puts the human health costs of emissions close to 5 cents/mi in air basins such as that of Los Angeles. In less polluted environments, health costs may be roughly 1 to 2 cents/mi (see, e.g., Ozbay and Berechman 2001).

Another key benefit is the allocation of road space to the "highest and best users."[†] Many users with high values of travel time (including, e.g., truck drivers delivering goods for just-in-time manufacturing processes or shift workers with constrained start times) presently are doing whatever they can to avoid congested roadways and avoid being late. Those with little or no value of travel time (e.g., high school students traveling to a shopping mall) are taking up that scarce space. By introducing monetary prices, the types of travelers using roadways during peak times of day would shift toward those who place higher value on their trip's time savings. Peak-period trip departures could be made later in the day, avoiding excessive addition of time to many travelers' journeys.

---

[*]Thanks to an assumed linear demand function, this computation is relatively simple. It involves the triangular area under the demand curve and above the 3.19-minute generalized cost: (8 − 3.19) × 10,450/2 = 25,132 minutes (per hour of flow on this mile-long section of roadway).

[†]*Highest and best use* is a normative economic term traditionally used to describe a market-derived (top-bid) use. A social or community-derived highest and best use may differ.

Of course, high values of time often correspond to higher wages and wealth, so communities and policymakers are understandably worried about the regressive impacts of congestion pricing—relative to the status quo. This is particularly true in the short term, when home, work, school, and other location choices are relatively fixed. Pricing will have land use effects that are difficult to predict; short-term activity location and timing inflexibilities suggest that restrained introduction of pricing will be necessary, and special cases of heavily affected low-income travelers (e.g., single parents with fixed work and child-care times that coincide with rush hour) may require credits and/or rebates. Toll revenues can be spent on alternative modes and other programs to benefit those most negatively affected by such policies. For example, revenues from London's downtown cordon toll of 8 pounds (roughly $12.50 in US dollars) go largely toward transit provision. This experiment may be considered a success: Speeds have doubled (from 9.5 to 20 mi/hr), bus ridership is up 14 percent, and only 5 percent of central London businesses claim to have experienced a negative impact (*Economist* 2003).

Equity Considerations: Minimum-Revenue Pricing, Rationing, and Reservations.   While various forms of congestion pricing already exist in southern California, Florida, Houston, Singapore, Milan, Rome, Trondheim, London, and now Stockholm, equity and revenue-distribution implications rouse public concern (Button and Verhoef 1998). These considerations also have spawned a number of creative policy proposals. They include Dial's (1999) "minimal revenue pricing" (where one route may be kept free between every origin-destination pair), DeCorla-Souza's (1995) "fast and intertwined regular (FAIR) lanes,"* and Daganzo's (1995) alternating tolled days across users.

Penchina (2004) has demonstrated that demand must be relatively inelastic for Dial's (1999) proposal to have important advantages over marginal-cost tolling (such as lower tolls, more stable tolls, and fewer tolled links). Given the variety of travel-choice substitutes people face for many activities (e.g., time of day, destination, and mode), relatively inelastic cases may be unusual.

Nakamura and Kockelman (2002) examined Daganzo's idea in the context of the San Francisco–Oakland Bay Bridge to assess winners and losers under a variety of alternate pricing and link-performance scenarios. Net benefits were significant, but as with almost any policy, they identified some losers—unless revenues are specially targeted and link-performance functions are favorable.

Another policy one might consider is reservation of scarce road capacity to allocate crossing of key network links (such as bridges) by time of day. This policy could involve a cap on free reservations and tolls for additional use. Airports grant (or sell) gate rights to airlines for exclusive use. And airlines sell specific seat assignments to passengers, effectively guaranteeing passage. Singapore rations vehicle ownership through a vehicle licensing system wherein only a certain number of 5-year licenses are auctioned off each year (online, through sealed bids). Coupled with the most extensive road-pricing system in the world, this city-state is a clear leader in congestion-fighting policies.

Owing largely to budget constraints, U.S. transportation agencies are looking more and more at toll roads as a congestion-fighting and roadway-provision option. These may be variably priced throughout the day to recognize the premium service they provide when demand is high (yet tolled travel times remain reasonable). In addition, smooth-flowing high-occupancy toll (HOT) lanes encourage peak-period travelers to shift to buses and carpooling while allowing better use of such lanes through tolling of single-occupancy vehicles (SOVs).† Toll roads and HOT lanes may be the intermediate policies that spark widespread congestion pricing in the United States. Yet the issue of equity and redistributive impacts remains.

Ideally, any selected solution will be cost-effective and relatively equitable. There is one possible solution that has the potential for near-optimal returns while ensuring substantial equity and efficiency. It is known as *credit-based congestion pricing* (Kockelman and Kalmanje 2005; Gulipalli and Kockelman 2008).

---

*FAIR lanes involve demarcating congested freeway lanes into tolled fast lanes and untolled regular lanes. Regular-lane drivers using electronic toll tags would receive credit for use of the facility; accumulated credits can be redeemed for use of the fast lanes.

†Toll collection can become an issue, however, because HOVs must be distinguished so that they are not charged.

Credit-Based Congestion Pricing.   Under a credit-based congestion-pricing plan, the tolling authority collects no net revenues (except, perhaps, to cover administrative costs), and travelers willing to reschedule their trips, share rides, or switch modes actually can receive money by not exhausting allocated cash credits. All collected tolls are returned in the form of per-driver rebates to licensed drivers in a region or regular users of a corridor.

A credit system for roadway use that provides for trading (and user rebates) addresses public concerns to a much greater extent than pricing alone—and thus could generate considerably more support. Credit banking and trading are becoming widespread in other domains. In 1999, the Organisation for Economic Cooperation and Development (OECD) found 9 programs involving tradable permit schemes in air pollution control, 5 in land-use control, 5 in water pollution control, 75 in fisheries management, and 3 in water resources (Tietenberg 2002). Today, many more applications exist, including greenhouse gas emissions trading across the European Union. As long as the cost of trading credits is not high, *any* distribution strategy should (in theory) result in an efficient use of the constrained resource (Tietenberg 2002).

In 2002, a National Academy of Science committee proposed a tradable credits system as an alternative to the current corporate average fuel economy standards (TRB 2002). For each gallon of fuel expected to be consumed by a new vehicle over its lifetime, the manufacturer would need a credit. The annual allocation of credits to a manufacturer would be based on the company's production level and the government's target for fuel consumption per vehicle.

Clearly, there is a distinction between industrial applications involving the trading of emissions, fuel economy credits, and the consumer-oriented credit-based congestion-pricing application considered here. In the case of roadway networks, where the temporal and spatial attributes of the congested resource are critical, appropriate link pricing also must exist. Roadway pricing, therefore, is somewhat more complex than emissions or fisheries regulation; one cannot simply cap total vehicle miles traveled (VMT) and allow VMT credit use at any time of day or at any point in the network. Continual monitoring of traffic conditions and recognition of key links and times of day are needed. However, just as companies achieved the imposed reductions in SO_2 emissions more easily than had been foreseen initially, travelers also may have greater flexibility in their demand for peak-period travel than some anticipate. The potential for and presence of such flexibility are key.

Congestion Pricing Policy Caveats.   While credit-based and other forms of pricing have the ability to reduce congestion to (and beyond) "optimal" levels, there are implementation issues and various other questions that require careful consideration. First, the costs of the technology can pose serious hurdles. Many commercial vehicles carry GPS systems, but these presently cost hundreds of dollars. If the price of congestion is less than $100 per vehicle in a region per year (as it is in most regions), GPS systems may not make economic sense. If much cheaper local radiofrequency systems are used, their roadside readers (presently costing thousands of dollars) are likely to be selectively distributed, leaving much of the network unpriced and some locations still congested. In either case, private third-party distribution of identification codes and/or formal legal protections probably are needed to ensure that privacy expectations are met. Second, policy administration is rife with special needs. If only a few regions per state merit pricing, visitors may need to buy or rent identification systems for travel through and around the priced systems.[*] For enforcement purposes, video-image processing of rear license plates is probably necessary at key network points in order to identify vehicles without transponders or system violators. Lists of license plate holders then would be necessary, and regions and states would have to share information (as they presently do for serious offenses).

In addition to technology costs and privacy questions, a pricing policy's distribution of benefits and costs should be evaluated for its equity implications as well as other considerations. Land markets will be affected, along with land-use patterns. Goods prices and wages may be noticeably affected in some markets. Of course, one can generally argue that such corrections to currently imperfect markets are welcome for a variety of reasons. And pricing may make many projects suddenly possible in an era of budget limitations. But such arguments do not make the

---

[*]Short-term visitors may be granted access without penalty depending on time of stay and use of the network.

transition to congestion-priced corridors and tolled networks easier for all affected stakeholders to swallow.

Finally, much travel-time is access time at trip ends. The drive from one's home to the first arterial may have little to no travel-time improvement available; similarly, the walk and elevator ride from one's parking space to one's office already may be minimized. These access times are not insignificant. Thus, even if congestion on all arterial roads is removed, total peak-period travel times may fall only 25 percent or less for most travelers, even in highly congested regions.[*] Congestion pricing is not a silver bullet, although it addresses some very vexing problems for residents of congested areas, particularly as populations grow and budgets and infrastructure investments stall. The world's motorization continues, and rationalization of network operations via thoughtful pricing merits close attention.

3.6 CONCLUSIONS

Congestion results from high demand for constrained systems and tends to manifest as delay. A relative scarcity of road space and other forms of transportation capacity has lead to substantial congestion losses in many travel corridors across many regions at many times of day. The delays may be severe or moderate; the congestion may be recurring and anticipated or nonrecurring and unpredictable.

Society pays for congestion not just through higher travel times and crash rates, uncertain and missed schedules, additional emissions, and personal frustration but also through higher costs for goods and services. After all, commercial delivery services must confront the same traffic delays that personal vehicle occupants face. These delays translate to lowered productivity and more expensive deliveries and commutes, resulting in higher prices for everyone.

The relationship between demand and delay is not clear-cut. As demand approaches capacity, travel times tend to rise. When demand exceeds capacity, queues form, and travel-time impacts are pronounced. Small additions to demand can generate significant delays—and minor reductions can result in significant time savings. Removal of bottlenecks, toll road provision, and subsidy of alternative modes are proving popular mechanisms to combat congestion. But more effective solutions are needed in most cases. Individual travel choices remain inefficient until travelers recognize and respond to the true costs of using constrained corridors and systems.

The good news is that new technologies are available to address the congestion issue. But robust estimates of individual demand functions and the marginal costs of additional users are needed to take congestion policies to the next level. Congestion pricing promises many benefits. And *credit-based* congestion pricing, as well as other strategies, may offset the burdens on automobile-dependent low-income populations substantially, particularly in the short term, when location and other requirements are relatively fixed for certain activity types.

Congestion is not just a roadway phenomenon. It affects nearly all pathways, including air, rail, water, and data ports. Fortunately, many of the strategies and technologies for resolving roadway congestion also apply to these other domains. Greater recognition of the negative externalities involved in oversaturation of our transportation systems will guide us to the most effective strategies for coping with congestion, with the promise of more efficient travel for everyone.

ACKNOWLEDGMENTS

I am grateful for the excellent suggestions of Tim Lomax, Steve Rosen, Sukumar Kalmanje, Jason Lemp, Steven Boyles, and Annette Perrone.

---

[*]Taylor (2002) makes this point in his *Access* article "Rethinking Traffic Congestion."

REFERENCES

Akçelik, Rahmi. 1991. Travel Time Functions for Transport Planning Purposes: Davidson's Function, Its Time-Dependent Form and An Alternative Travel Time Function. *Australian Road Research* 21:44–59.

Arnott, Richard, Rave Tilman, and Ronnie Schöb. 2005. *Alleviating Urban Traffic Congestion.* Cambridge, MA: MIT Press.

Bates, John, John Polak, Peter Jones, and Andrew Cook. 2001. The Valuation of Reliability for Personal Travel. *Transportation Research* 37E: 191–229.

Beamon, Benita. 1995. Quantifying the Effects of Road Pricing on Roadway Congestion and Automobile Emissions. Ph.D. Dissertation, Department of Civil and Environmental Engineering, Georgia Institute of Technology, Atlanta, GA.

Beckmann, M., McGuire, C. B., and Winsten, C. B. 1956. *Studies in the Economics of Transportation*, Chap. 4, New Haven, CT: Yale University Press.

Bhat, Chandra R., and Rupali Sardesai. 2006. The Impact of Stop-Making and Travel Time Reliability on Commute Mode Choice. *Transportation Research* 40B:709–730.

Boarnet, Marlon G., and Randall Crane. 2001. *Travel by Design: The Influence of Urban Form on Travel.* Oxford, England Oxford University Press.

Brodsky, H., and A.S. Hakkert, 1983. Highway Accident Rates and Rural Travel Densities. *Accident Analysis and Prevention* 151:73–84.

Brownstone, D., and K. A. Small. 2005. Valuing Time and Reliability: Assessing the Evidence from Road Pricing Demonstrations. *Transportation Research* 39A:279–293.

Bureau of Transportation Statistics (BTU). 2002. *National Transportation Statistics 2002.* Washington: Bureau of Transportation Statistics, U.S. Department of Transportation.

Button, Kenneth J., and Erik T. Verhoef (eds.). 1998. *Road Pricing, Traffic Congestion and the Environment.* Cheltenham, UK: Edward Elgar Publishing.

Cervero, Robert. 2002. Induced Travel Demand: Research Design, Empirical Evidence, and Normative Policies. *Journal of Planning Literature* 17:3–20.

Cervero, Rovert. 2003. Road Expansion, Urban Growth, and Induced Travel: A Path Analysis. *Journal of the American Planning Association* 69:145–64.

Cervero, Robert, and Kara Kockelman. 1997. Travel Demand and the Three D's: Density, Diversity, and Design. *Transportation Research* 2D:199–219.

Chen, Chao, Zhanfeng Jia, and Pravin Varaiya. 2001. Causes and Curves of Highway Congestion. *IEEE Control Systems Magazine* 21:26–33.

Commodity Flow Survey (CFS). 1997. U.S. Department of Transportation, Bureau of Transportation Statistics, and U.S. Department of Commerce, Economics and Statistics Administration, U.S. Census Bureau, Washington, DC.

Daganzo, Carlos. 1995. A Pareto Optimum Congestion Reduction Scheme. *Transportation Research* 29B: 139–54.

Dahlgren, Joy. 1994. An Analysis of the Effectiveness of High Occupancy Vehicle Lanes. Dissertation, Department of Civil Engineering, University of California at Berkeley.

Dahlgren, Joy. 2002. High-Occupancy/Toll Lanes: Where Should They be Implemented. *Transportation Research* 36A:239–55.

DeCorla-Souza, Patrick. 1995 Applying the Cashing Out Approach to Congestion Pricing. *Transportation Research Record No. 1450*:34–7.

Dial, Robert B. 1999. Minimal Revenue Congestion Pricing Part I. A Fast Algorithm for the Single-Origin Case. *Transportation Research* 33B:189–202.

Dowling, Richard D., W. Kittelson, J. Zegeer, and A. Skabardonis. 1997. NCHRP Report 387, Planning Techniques to Estimate Speeds and Service Volumes for Planning Applications. TRB, National Research Council, Washington.

Downs, Anthony. 1962. The Law of Peak-Hour Expressway Congestion. *Traffic Quarterly* 16:393–409.

Economist. 2003. Congestion Charge: Ken's Coup, March 22, p. 51.

Edlin, Aaron S., and Pinar Karaca Mandic. 2006. The Accident Externality from Driving. *Journal of Political Economy* 114:931–955.

Evans, Leonard. 1991. *Traffic Safety and the Driver.* New York: Van Nostrand and Reinhold.

Ewing, Reid, and Robert Cervero. 2001. Travel and the Built Environment: A Synthesis. *Transportation Research Record No. 1780*:87–113.

Federal Highway Administration (FHWA). 1979. *Urban Transportation Planning System* (UTPS). Washington: U.S. Department of Transportation.

Federal Highway Administration (FHWA). 1999. CORSIM, Version 4.2. Washington: Federal Highway Administration, U. S. Department of Transportation (Distributed through McTrans, University of Florida).

Federal Highway Administration (FHWA). 2009. Congestion Pricing, A Primer: Overview. Washington: U.S. Department of Transportation, Federal Highway Administration. Available at http://ops.fhwa.dot.gov/publications/fhwahop08039/fhwahop08039.pdf.

Fimrite, Peter. 2002. Traffic Tops List of Bay Area Banes, Weak Economy is Number 2 Bane, Survey Shows. *San Francisco Chronicle*. At http://sfgate.com/cgi-bin/article.cgi?file=/c/a/2002/12/05/MN51835.DTL#sections. (Accessed on December 5, 2002).

Fulton, Lewis M., Daniel J. Meszler, Robert B. Noland, and John V. Thomas. 2000. A Statistical Analysis of Induced Travel Effects in the U.S. Mid-Atlantic Region. *Journal of Transportation and Statistics* 3:1–14.

Garber, Nicholas, and Sankar Subramanyan. 2002. Feasibility of Incorporating Crash Risk in Developing Congestion Mitigation Measures for Interstate Highways: A Case Study of the Hampton Roads Area. Virginia Transportation Research Council, Final Report VTRC 02-R17. Charlottesville, VA.

Garber, Nicholas J., and Lester A. Hoel. 2002. *Traffic and Highway Engineering*, 3d ed. Pacific Grove, CA; Brooks-Cole.

Gervais, E. F., and W. Roth. 1966. An Evaluation of Freeway Ramp Closure. Traffic Division, Michigan State Highway Department.

Gulipalli, Pradeep, and Kara Kockelman. 2008. Credit-Based Congestion Pricing: A Dallas–Fort Worth Application. *Transport Policy* 15:23–32.

Gwynn, D.W. 1967. Relationship of Accident Rates and Accident Involvements with Hourly Volumes. *Traffic Quarterly* 21:407–418.

Hansen, Mark, and Yuanlin Huang. 1997. Road Supply and Traffic in California Urban Areas. *Transportation Research* 31A:205–218.

Hardin, Garrett. 1968. The Tragedy of the Commons. *Science* 162:1243–48.

Horowitz, Alan J. 1991. Delay Volume Relations for Travel Forecasting based on the 1985 Highway Capacity Manual. Report FHWA-PD-92-015. Washington: Federal Highway Administration.

Klein, Lawrence. 2001. *Sensor Technologies and Data Requirements for ITS*. Boston: Artech House.

Knickerbocker, Brad. 2000. "Forget Crime—But Please Fix the Traffic." *Christian Science Monitor*, February 16.

Kockelman, Kara. 1997. Travel Behavior as a Function of Accessibility, Land-Use Mixing, and Land-Use Balance: Evidence from the San Francisco Bay Area. *Transportation Research Record No. 1607*:116–125.

Kockelman, Kara. 2001. Modeling Traffic's Flow-Density Relation: Accommodation of Multiple Flow Regimes and Traveler Types. *Transportation* 28:363–374.

Kockelman, Kara, Randy Machemehl, Aaron Overman, et al. 2001. Frontage Roads in Texas: A Comprehensive Assessment. University of Texas at Austin, Center for Transportation Research, Report FHWA/TX-0-1873-2.

Kockelman, Kara, and Sukumar Kalmanje. 2005. Credit-Based Congestion Pricing: A Policy Proposal and the Public's Response. *Transportation Research* 39A:671–690.

Kockelman, Kara, and Young-Jun Kweon. 2002. Driver Injury Severity and Vehicle Type: An Application of Ordered Probit Models. *Accident Analysis and Prevention* 34:313–321.

Krishnamurthy, Sriram, and Kara Kockelman. 2003. Propagation of Uncertainty in Transportation–Land-Use Models: An Investigation of DRAM-EMPAL and UTPP Predictions in Austin, Texas. *Transportation Research Record No. 1831*:219–229.

Kuhn, Beverly, Ginger Goodin, Andrew Ballard, et al. 2005. *Managed Lanes Handbook*. Report No. 0-4160-24. Houston: Texas Transportation Institute.

Levinson, David, Lei Zhang, Shantanu Das, and Atif Sheikh. 2002. Ramp Meters on Trial: Evidence from the Twin Cities Ramp Meters Shut-Off. Transportation Research Board 81st Annual Meeting, Washington, DC.

Lomax, Tim, Shawn Turner, and Richard Margiotta. 2003. Monitoring Urban Roadways in 2001: Examining Reliability and Mobility with Archived Data. Washington: Federal Highway Administration Report FHWA-OP-02-029.

Lomax, Tim, Shawn Turner, Gordon Shunk, et al. 1997. NCHRP Report 398, Quantifying Congestion; Vol. 1: Final Report. Washington: TRB, National Research Council.

Martin, William A., and Nancy A. McGuckin. 1998. NCHRP Report 36, Travel Estimation Techniques for Urban Planning. Washington: National Research Council.

May, Adolf D. 1964. Experimentation with Manual and Automatic Ramp Control. *Highway Research Record No. 59*:9–38.

Miesse, C. C. 1967. Optimization of Single-Lane Freeway Traffic Flow by Elective Ramp Closure. *Transportation Research* 1:157–63.

Mokhtarian, Patricia, and Ilan Salomon. 2001. How Derived is the Demand for Travel? *Transportation Research* 35A:695–719.

Myers, Barton, and John Dale. 1992. Designing in Car-Oriented Cities: An Argument for Episodic Urban Congestion. In *The Car and the City: The Automobile, the Built Environment, and Daily Urban Life*, Martin Wachs and Margaret Crawford (eds.). Ann Arbor: University of Michigan Press.

Nakamura, Katsuhiko, and Kara Kockelman. 2002. Congestion Pricing and Roadspace Rationing: An Application to the San Francisco Bay Bridge Corridor. *Transportation Research* 36A:403–17.

National Household Travel Survey (NHTS). 2001. Washington: U.S. Department of Transportation, Federal Highway Administration.

Neudorff, L. G., J. E. Randall, R. Reiss, and R. Gordon. 1997. *Freeway Management and Operations Handbook*. Washington: Federal Highway Administration.

Newman, L., A. Dunnet and J. Meirs. 1969. Freeway Ramp Control: What It Can and Cannot Do. *Traffic Engineering,* June:14–25.

Noland, Robert.B., and William A. Cowart. 2000. Analysis of Metropolitan Highway Capacity and the Growth in Vehicle-Miles of Travel. Paper presented at the 79th Annual Meeting of the Transportation Research Board, Washington, DC.

Oak Ridge National Laboratory (ORNL). 2001. *1995 NPTS Databook*. Prepared for the U.S. DOT, Federal Highway Administration, ORNL/TM-2001/248.

Obenberger, Jon. 2004. Managed Lanes. *Public Roads* 68.

Ozbay, Kaan, Bekir Bartin, and Joseph Berechman. 2001. Estimation and Evaluation of Full Marginal Costs of Highway Transportation in New Jersey. *Journal of Transportation Statistics* 4:81–104.

Peirce, Neal. 2003. Congestion Insurance: "HOT" Lanes' Amazing Promise. *Washington Post*, March 5.

Penchina, Claude M. 2004. Minimal-Revenue Congestion Pricing: Some More Good-News and Bad-News. *Transportation Research* 38B:559–70.

Perez, Benjamin G., and Gian-Claudia Sciara. 2004. *A Guide for HOT Lane Development*. Washington: Federal Highway Administration.

Prevedouros, Panos D. 2001. Freeway Ramp Closure: Simulation, Experimentation, Evaluation and Preparation for Deployment. *ITE Journal* 71:40–44.

Pushkarev, Boris S., and Jeffrey M. Zupan. 1977. *Public Transportation and Land Use Policy*. Bloomington, IN: Indiana University Press.

Redmond, Lothlorien S., and Patricia L. Mokhtarian. 2001. The Positive Utility of the Commute: Modeling Ideal Commute Time and Relative Desired Commute Amount. *Transportation* 28:179–205.

Richardson, Anthony J. 2003. Some Evidence of Travelers with Zero Value of Time. *Transportation Record No. 1854*:107–113.

Rodier, Caroline, John Abraham, Robert Johnston, and John D. Hunt. 2000. Anatomy of Induced Travel Using an Integrated Land Use and Transportation Model in the Sacramento Region. Paper presented at the 79th Annual Meeting of the Transportation Research Board, Washington, DC.

Salomon, Ilan, and Patricia L. Mokhtarian. 1998. What Happens when Mobility-Inclined Market Segments Face Accessibility-Enhancing Policies? *Transportation Research* 3D:129–40.

Schrank, David, and Timothy Lomax. 2009. *The 2009 Urban Mobility Report*. Texas A&M University, Texas Transportation Institute.

Scheibal, Stephen. 2002. New Planning Group Kicks off Effort with Survey on What Area Residents Want. *Austin American Statesman*, August 26.

Shoup, Donald. 1992. Cashing Out Employer-Paid Parking. Report No. FTA-CA-11-0035-92-1. Washington: U.S. Department of Transportation.

Shoup, Donald. 1994. Cashing Out Employer-Paid Parking: A Precedent for Congestion Pricing? In *Curbing Gridlock, Peak-Period Fees to Relieve Traffic Congestion*, vol. 2. Washington: National Academy Press, pp. 152–200.

Small, Kenneth A., and Camilla Kazimi. 1995. On the Costs of Air Pollution from Motor Vehicles. *Journal of Transport Economics and Policy* 29:17–32.

Small, Kenneth.A., Clifford Winston, and Jia Yan. 2005. Uncovering the Distribution of Motorists' Preferences for Travel Time and Reliability. *Econometrica*, 73:1367–382.

Smith, Adam. 1776 (original text). *An Inquiry into the Nature and Causes of the Wealth of Nations*. London: Dent & Sons (1904 pub. date).

Taylor, Brian D. 2002. Rethinking Traffic Congestion. *Access* 21:8–16.

Tietenberg, Tom. 2002. The Tradable Permits Approach to Protecting the Commons: What Have We Learned? In *The Drama of the Commons*. Washington: National Research Council, National Academy Press, Chap. 6.

Transportation Research Board (TRB). 2000. *Highway Capacity Manual 2000*. Washington: Transportation Research Board, National Research Council.

Transportation Research Board (TRB). 2002, *Effectiveness and Impact of Corporate Average Economy (CAFÉ) Fuel Standards*. Washington: Transportation Research Board. National Academy Press.

Van Lint, J.W.C., H.J. Van Zuylen, and H. Tu. 2008. Travel Time Unreliability on Freeways: Why Measures Based on Variance Tell Only Half the Story. *Transportation Research Part A* 42:258–277.

Walters, A. A. 1954. Track Costs and Motor Taxation. *Journal of Industrial Economics. Ward's Automotive Yearbook 1998*. 1999. Ward's Communication, Intertec Publishing Corporation, 2(2):135–146.

Walters, A. A. 1961. The Theory and Measurement of Private and Social Costs of Highway Congestion, *Econometrica* 19.

Weisbrod, Glen, Donald Vary, and George Treyz. 2001. *Economic Implications of Congestion*. NCHRP Report 463. Washington: National Cooperative Highway Research Program, Transportation Research Board.

CHAPTER 4
HIGHWAY CAPACITY

Lily Elefteriadou

Department of Civil and Coastal Engineering
University of Florida
Gainesville, Florida

4.1 INTRODUCTION

How much traffic can a facility carry? This is one of the fundamental questions designers and traffic engineers have been asking since highways have been constructed. The term *capacity* has been used to quantify the traffic-carrying ability of transportation facilities. The value of capacity is used when designing or rehabilitating highway facilities to obtain design elements such as the required number of lanes. It is also used in evaluating whether an existing facility can handle the traffic demand expected in the future.

The definition and value for highway capacity have evolved over time. The *Highway Capacity Manual* (*HCM 2000*) is the publication used most often to estimate capacity. The current published version of the *HCM 2000* defines the capacity of a facility as ". . . the maximum hourly rate at which persons or vehicles reasonably can be expected to traverse a point or a uniform section of a lane or roadway during a given time period, under prevailing roadway, traffic and control conditions" (p. 2-2). Specific values for capacity are given for various types of facilities. For example, for freeway facilities, capacity values are given as 2,250 passenger cars per hour per lane (pc/hr/ln) for freeways with free-flow speeds of 55 mph and 2,400 pc/hr/ln when the free-flow speed is 75 mph (ideal geometric and traffic conditions).

For a long time, traffic engineers have recognized the inadequacy and impracticality of this capacity definition. First, the expression "maximum . . . that can reasonably be expected" is not specific enough for obtaining an estimate of capacity from field data. Second, field data-collection efforts have shown that the maximum flow at a given facility varies from day to day; therefore, a single value of capacity does not reflect real-world observations.

The main objective of this chapter is to provide transportation professionals with an understanding of highway capacity and the factors that affect it and to provide guidance on obtaining and using field values of capacity. The next part of this chapter discusses the history and evolution of capacity estimation. The third part discusses the factors that affect capacity. The fourth part presents uninterrupted flow capacity issues, whereas the fifth part presents interrupted flow capacity issues. The last part presents a vision for the future of defining and estimating capacity.

4.2 CAPACITY DEFINITION AND ESTIMATION METHODS

4.2.1 The *Highway Capacity Manual*: A Historical Perspective

The *Highway Capacity Manual* is the publication used most often to estimate capacity. The first edition of the *Highway Capacity Manual* (1950) defined three levels of roadway capacity: basic capacity, possible capacity, and practical capacity. *Basic capacity* was defined as "the maximum number of passenger cars that can pass a point on a lane or roadway during one hour under the most nearly ideal roadway and traffic conditions which can possibly be attained." *Possible capacity* was "the maximum number of vehicles that can pass a given point on a lane or roadway during one hour under prevailing roadway and traffic conditions." *Practical capacity* was a lower volume chosen "without the traffic density being so great as to cause unreasonable delay, hazard, or restriction to the drivers' freedom to maneuver under prevailing roadway and traffic conditions."

The second edition of the *Highway Capacity Manual* (1965) defined a single capacity, similarly to the possible capacity of *HCM 1950*. The term *basic capacity* was replaced by the term *capacity under ideal conditions*, whereas the term *practical capacity* was replaced by a series of *service volumes* to represent traffic operations at various levels of service (LOS). It is interesting to note that in the *HCM 1965*, the second chapter is titled, "Definitions," and begins as follows: "The confusion that has existed regarding the meaning and shades of meaning of many terms . . . has contributed . . . to the wide differences of opinion regarding the capacity of various highway facilities. . . . In fact, the term which is perhaps the most widely misunderstood and improperly used . . . is the word 'capacity' itself." Thus the definition of the term *capacity* allowed for various interpretations by different traffic analysts, and there was a desire to clarify the term. In *HCM 1965*, the definition of capacity was revised to read as follows: "Capacity is the maximum number of vehicles which has a reasonable expectation of passing over a given section of a lane or a roadway in one direction (or in both directions for a two-lane or three-lane highway) during a given time period under prevailing roadway and traffic conditions." This definition includes the term *reasonable expectation*, which indicates that there is variability in the numerical value of the maximum number of vehicles. Subsequent editions and updates of the *HCM* (1985, 1994, and 1997) define capacity in a similar manner, with the most recent definition (*HCM 2000*) as stated in the Introduction of this chapter. This most recent definition indicates that there is an expected variability in the maximum volumes, but it does not specify when, where, and how capacity should be measured, nor does it discuss the expected distribution, mean, and variance of capacity.

Capacity values provided in the *HCM* have increased over time. For example, *HCM 1950* indicated that the capacity of a basic freeway segment lane is 2,000 pc/hr/ln, whereas *HCM 2000* indicates that capacity may reach 2,400 pc/hr/ln for certain freeway facilities.

In addition to the definition of capacity, the *HCM* historically has provided (beginning with *HCM 1965*) relationships between the primary traffic characteristics (speed, flow, and density) that have been the basis of highway capacity analysis procedures, particularly for uninterrupted-flow facilities. Figure 4.1 presents a series of speed-flow curves that are provided in *HCM 2000* and illustrate the relationship between speed and flow for basic freeway segments and for various free-flow speeds (FFS) ranging from 55 to 75 mph. As shown in the figure, speed remains constant for low flows and begins to decrease as flow reaches 1,300 to 1750 pc/hr/ln. The capacity for facilities with FFS at or above 70 mph is 2,400 pc/hr/ln and decreases with decreasing FFS. For example, the capacity of a basic freeway segment with FFS of 55 mph is expected to be 2,250 pc/hr/ln.

Figure 4.2 provides the respective flow-density curves for basic freeway segments. Similarly to Figure 4.1, capacity values are shown to vary for varying free-flow speeds. This figure clearly illustrates the assumption used in the development of these curves that capacity is reached when density is 45 passenger cars per mile per lane (pc/mi/ln).

Both figures provide speed-flow-density relationships for undersaturated (i.e., noncongested) flow only. When demand exceeds the capacity of the facility, it will become oversaturated, with queues forming upstream of the bottleneck location. *HCM 2000* does not provide speed-flow-density relationships for oversaturated conditions at freeways because research has not been conclusive on this topic.

In summary, the definition of capacity within the *HCM* has evolved over time. There has been an implicit and, more recently, an explicit effort to include the expected variability of maximum

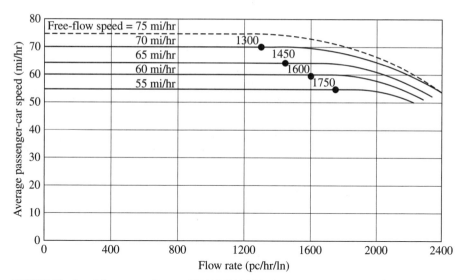

FIGURE 4.1   Speed-flow curves. (*Source: HCM 2000*, Exhibit 13-2.)

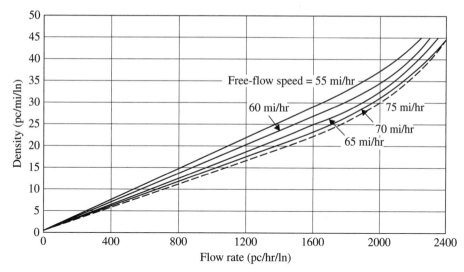

FIGURE 4.2   Flow-density curves. (*Source: HCM 2000*, Exhibit 13-3.)

volumes in the capacity definition, but there is no specific information in that document on where, when, and how capacity should be measured at a highway facility.

4.2.2 Other Publications

For a long time, researchers have recognized the inadequacy and impracticality of this definition for freeway facilities. Field data collection of capacity estimates has shown that there is wide variability in the numerical values of capacity at a given site. This section summarizes literature findings regarding capacity definition and estimation.

Persaud and Hurdle (1991) discussed various definitions and measurement issues for capacity, including maximum-flow definitions, mean-flow definitions, and expected-maximum-flow definitions. They collected data at a three-lane freeway site over 3 days. In concluding, they recommended that the mean queue discharge flow is the most appropriate partly owing to the consistency the researchers observed in its day-to-day measurement.

Agyemang-Duah and Hall (1991) collected data over 52 days on peak periods to investigate the possibility of a drop in capacity as a queue forms and to recommend a numerical value for capacity. They plotted prequeue peak flows and queue discharge flows at 15-minute intervals, which showed that the two distributions are similar, with the first one slightly more skewed toward higher flows. They recommended 2,300 pc/hr/ln as the capacity under stable flow and 2,200 pc/hr/ln for post-breakdown conditions, which corresponded to the mean value of the 15-minute maximum flows observed under the two conditions. The researchers recognized the difficulty in defining and measuring capacity given the variability observed.

Elefteriadou et al. (1995) developed a model for describing the process of breakdown at ramp-freeway junctions. Observation of field data showed that at ramp merge junctions, breakdown may occur at flows lower than the maximum observed, or capacity flows. Furthermore, it was observed that at the same site and for the same ramp and freeway flows, breakdown may or may not occur. The authors developed a probabilistic model for describing the process of breakdown at ramp-freeway junctions, which gives the probability that breakdown would occur at given ramp and freeway flows and is based on ramp-vehicle cluster occurrence. Similar to this research, Evans et al. (2001) also developed a model for predicting the probability of breakdown at ramp-freeway junctions that was based on Markov chains and considered operations on the entire freeway cross-section rather than on the merge influence area.

Minderhoud et al. (1997) discussed and compared empirical capacity estimation methods for uninterrupted flow facilities and recommended the product-limit method because of its sound theoretical framework. In this method, noncongested flow data are used to estimate the capacity distribution. The product-limit estimation method is based on the idea that each noncongested flow observation having a higher flow rate than the lowest observed capacity flow rate contributes to the capacity estimate because this observation gives additional information about the location of the capacity value. The article does not discuss transitions to congested flow nor discharge flow measurements.

Lorenz and Elefteriadou (2001) conducted an extensive analysis of speed and flow data collected at two freeway bottleneck locations in Toronto, Canada, to investigate whether the probabilistic models developed previously replicated reality. At each of the two sites, the freeway breakdown process was examined in detail for over 40 breakdown events occurring during the course of nearly 20 days. Examining the time-series speed plots for these two sites, the authors concluded that a speed *boundary* or *threshold* at approximately 90 km/hr existed between the noncongested and congested regions. When the freeway operated in a noncongested state, average speeds across all lanes generally remained above the 90 km/hr threshold at all times. Conversely, during congested conditions, average speeds rarely exceeded 90 km/hr, and even then they were not maintained for any substantial length of time. This 90 km/hr threshold was observed to exist at both study sites and in all the daily data samples evaluated as part of that research. Therefore, the 90 km/hr threshold was applied in the definition of breakdown for these sites. Since the traffic stream was observed to recover from small disturbances in most cases, only the disturbances that caused the average speed over all lanes to drop below 90 km/hr for a period of 5 minutes or more (15 consecutive 20-second intervals) were considered a true breakdown. The same criterion was used for "recovery periods." The authors recorded the frequency of breakdown events at various demand levels. As expected, the probability of breakdown increases with increasing flow rate. Breakdown, however, may occur at a wide range of demands (i.e., 1,500 to 2,300 veh/hr/ln). The authors confirmed that the existing freeway capacity definition does not accurately address the transition from stable to unstable flow, nor the traffic-carrying ability of freeways under various conditions. They suggest that freeway capacity may be described more adequately by incorporating a probability-of-breakdown component in the definition. A suggested definition reads "the rate of flow (expressed in pc/hr/ln and specified for a particular time interval) along a uniform freeway segment corresponding to the expected probability of breakdown deemed acceptable under prevailing traffic and roadway conditions in a specified direction." The value of the probability component should correspond to the maximum breakdown risk deemed acceptable for a particular time period. A target value for the acceptable probability of breakdown (or *acceptable breakdown risk*) for a freeway might be selected initially by the facility's design team and later revised by the operating agency or jurisdiction based on actual operating characteristics. With respect to the two-capacity phenomenon, the researchers observed that the magnitude of any flow drop following breakdown may be contingent on the particular flow rate at which the facility breaks down. Flow rates may remain constant or even increase following breakdown. This may explain the fact that some researchers have observed the two-capacity phenomenon and others haven't; it seems to depend on the specific combination of the breakdown flow and the queue discharge flow for the particular observation period. The article does not discuss maximum prebreakdown flow, however, nor does it directly compare breakdown flows with maximum discharge flows for each observation day.

Elefteriadou and Lertworawanich (2003) examined freeway traffic data at two sites over a period of several days, focusing on transitions from noncongested to congested state, and developed suggested definitions for these terms. Three flow parameters were defined and examined at two freeway bottleneck sites: the breakdown flow, the maximum prebreakdown flow, and the maximum discharge flow. Figure 4.3 illustrates these three values in a time series of flow-speed data at a given site.

It was concluded that

- The numerical value of each of these three parameters varies, and their range is relatively large, on the order of several hundred veh/hr/ln.
- The distributions of these parameters follow the normal distribution for both sites and both analysis intervals.

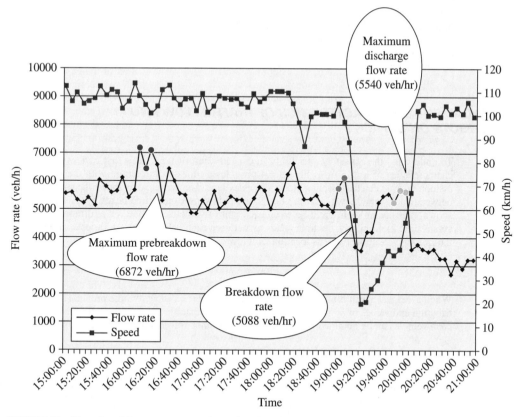

FIGURE 4.3   Illustration of three parameters on time-series plot of flow and speed.

- The numerical value of breakdown flows is almost always lower than both the maximum pre-breakdown flow and the maximum discharge flow.

- The maximum prebreakdown flow tends to be higher than the maximum discharge flow in one site, but the opposite is observed at the other site. A possible explanation for this difference may be that geometric characteristics and sight distance may result in different operations under high- and low-speed conditions.

- Brilon (2005) analyzed data collected over 3 years at uninterrupted flow facilities, focusing on cases of breakdown. *Breakdown* was defined as a reduction in average speed within one 5-minute interval to below a threshold of 70 km/hr. The researchers analyzed these data using the product-limit method, which is based on lifetime data analysis. They concluded that the concept of stochastic capacity is more realistic and useful than the traditional use of a deterministic capacity value.

Elefteriadou et al. (2006) discussed the definition and measurement methods for capacity. They concluded that there are three time periods of interest with regard to capacity on a freeway: prior to the breakdown of flow (drop in speeds), the interval immediately preceding breakdown, and the extended interval during the breakdown of flow. The authors indicated that regardless of which one of the three periods of interest is used to define and determine the capacity of a facility, the entire distribution should be obtained or a probability distribution function estimated over a large number of days.

In summary, several studies have shown that there is variability in the maximum sustained flows observed in the range of several hundred vehicles per hour per lane. Three different flow parameters have been defined (maximum prebreakdown flow, breakdown flow, and maximum discharge flow), any of which could be used to define capacity for a highway facility. The maximum values for each of these are random variables, possibly normally distributed. Prebreakdown flow is often higher than the discharge flow. The transition from noncongested to congested flow is probabilistic and may occur at various flow levels. The remainder of the chapter refers to these three as a group as *capacity*, with references to a specific one when appropriate.

4.3 FUNDAMENTAL CHARACTERISTICS OF TRAFFIC FLOW AND THEIR EFFECTS ON CAPACITY

To understand the causes of variability in capacity observations, let's first review the fundamental characteristics of traffic flow. Figure 4.4 provides a time-space diagram with the trajectories of a platoon of five vehicles traveling along a freeway lane. The vertical axis shows the spacing h_s (in feet) between each vehicle, whereas the horizontal axis shows their time headways h_t (in seconds). As shown, the spacing is different between each pair of vehicles if measured at different times (time 1 versus time 2). Similarly, the time headway between each pair of vehicles varies as they travel down the freeway (location 1 versus location 2). Flow can be expressed as

$$\text{Flow} = 3{,}600/\text{average}(h_t) \tag{4.1}$$

When average(h_t) is minimized, the flow is maximized (i.e., capacity is reached). Therefore, the distribution and values of h_t greatly affect the observed capacity of a facility. In the figure, the speed of each vehicle can be obtained graphically as the distance traveled divided by the respective time, or

$$\text{Speed} = \text{distance/time} = h_s/h_t \qquad \text{and} \qquad h_t = h_s/\text{speed} \tag{4.2}$$

Throughput is maximized when h_t is minimized or as spacing decreases and speed increases. Therefore, spacing and speeds also have an impact on the maximum throughput observed.

In summary, the microscopic characteristics of traffic, that is, the individual spacing, time headway, and speed of each vehicle in the traffic stream and their variability, result in variability in the

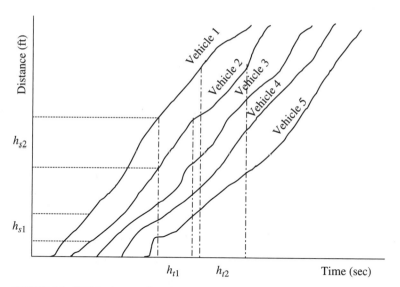

FIGURE 4.4   Vehicle platoon trajectories.

field capacity observations. The remainder of this section discusses factors that affect these microscopic characteristics of traffic and thus capacity.

The three components of the traffic system are the vehicle, the driver, and the highway environment. The vehicle characteristics, the driver characteristics, and the roadway infrastructure, as well as the manner in which the three components interact, affect the traffic operational quality and capacity of a highway facility. This section describes each of these three components and specific aspects that affect capacity.

4.3.1 Vehicle Characteristics

As discussed earlier, highway capacity is a function of speed, time headway, and spacing, which, in turn, are affected by the performance and size of the vehicles in the traffic stream. The variability of these characteristics contributes to the variability in capacity observations. Vehicle characteristics that affect capacity include

Wt/Hp (weight-to-horsepower ratio). The Wt/Hp provides a measure of the vehicle load to the engine power of the vehicle. It affects the maximum speed a vehicle can attain on steep upgrades (crawl speed), as well as its acceleration capabilities, both of which have an impact on capacity. Heavier and less powerful trucks generally operate at lower acceleration rates, particularly on steep upgrades. Slower-moving vehicles are particularly detrimental to capacity and traffic operational quality when there are minimal passing opportunities for other vehicles in the traffic stream.

Braking and deceleration capabilities. The deceleration capability of a vehicle decreases with increasing size and weight.

Frontal area cross section. Aerodynamic drag affects the acceleration of a vehicle.

Width, length, and trailer coupling. The width of a vehicle may affect traffic operations in adjacent lanes by forcing other vehicles to slow down when passing. In addition, the width, length, and trailer coupling affects the off-tracking characteristics of a vehicle and the required lane widths, particularly along horizontal curves. The encroachment of heavy vehicles in adjacent lanes affects their usability by other vehicles and thus has an impact on capacity.

Vehicle height. The vehicle height, even though not typically included in capacity analysis procedures, may affect the sight distance for following vehicles and thus may affect the resulting spacing and time headways and, ultimately, the capacity of a highway facility.

4.3.2 Driver Characteristics

Individual driver capabilities, personal preferences, and experience also affect highway capacity and contribute to the observed capacity variability. The driver characteristics that affect the capacity of a facility are

Perception and reaction times. These affect the car-following characteristics within the traffic stream. For example, these would affect the acceleration and deceleration patterns (and the trajectory) of a vehicle following another vehicle in a platoon. They also affect other driver actions such as lane changing and gap-acceptance characteristics.

Selection of desired speeds. The maximum speed at which each driver is comfortable driving at a given facility would affect the operation of the entire traffic stream. The effect of slower-moving vehicles in the traffic stream would be detrimental to capacity, particularly when high traffic demands are present.

Familiarity with the facility. Commuter traffic is typically more efficient in using a facility than those unfamiliar with the facility or those traveling for recreational purposes.

4.3.3 Roadway Design and Environment

The elements contained in this category include horizontal and vertical alignment, cross section, and traffic control devices.

Horizontal alignment and horizontal curves. Vehicles typically decelerate when negotiating sharp horizontal curves. In modeling speeds for two-lane highways (Fitzpatrick et al. 1999), it has been shown that drivers decelerate at a rate that is proportional to the radius of the curve.

Vertical alignment and vertical curves. Steep grades result in lower speeds, particularly for heavy trucks with low performance characteristics. Crawl speeds can be determined as a function of grade. Steep vertical-crest curves also would affect sight distances and may act as local bottlenecks.

Cross section. The number and width of lanes, as well as the shoulder width, have been shown to affect speeds and thus the capacity of a highway facility. Provision of appropriate superelevation increases the speed and thus enhances the efficiency of a highway facility.

Traffic control devices. The clarity and appropriateness of traffic control devices enhance the capacity of highway facilities.

Interactions between the three factors are also very important. For example, the effect of a steep upgrade on a heavy vehicle's performance is much more detrimental for capacity than generally level terrain. Also, the effect of a challenging alignment would be much more detrimental to an unfamiliar driver than to a commuter.

4.4 CAPACITY OF UNINTERRUPTED FLOW FACILITIES

Uninterrupted flow facilities are defined as those where traffic is not interrupted by traffic signals or signs. These include freeway segments, weaving segments, ramp junctions, multilane highways, and two-lane highways. This section provides procedures for obtaining maximum-throughput (i.e., capacity) estimates along uninterrupted flow facilities in field data collection or using microsimulation models. It provides guidance on observing and measuring (1) maximum prebreakdown throughput, (2) breakdown flow, and (3) maximum discharge flow. The last part of this section discusses capacity estimation for uninterrupted flow facilities using the *HCM 2000*.

4.4.1   Field Data Collection

The four important elements that should be considered when observing breakdown and maximum throughput are (1) site selection and measurement location, (2) definition of breakdown, (3) time interval, and (4) sample size.

Regarding site selection, the site should be regularly experiencing congestion and breakdown as a result of high demands and not as a result of a downstream bottleneck. For example, in Figure 4.5, which provides a sketch of a freeway facility with two consecutive bottlenecks (one merge and one lane drop), the bottleneck at location *B* will result in queue backup into location *A*. The downstream bottleneck location (i.e., location *B*) should be the data-collection point.

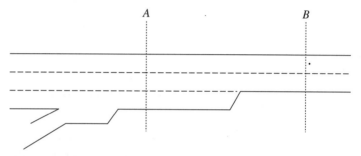

FIGURE 4.5   Freeway facility with two consecutive bottlenecks.

To graphically illustrate the importance of site and location selection, Figure 4.6 provides the speed-flow relationships and the time series of speed and flow for locations *A* and *B* across the freeway facility shown in Figure 4.5. Figure 4.6 illustrates that the maximum throughput at location *B* is equivalent to its potential capacity, that is, two-lane segment capacity. At location *A*, the potential capacity is that of a three-lane segment, which, however, cannot be attained owing to the downstream bottleneck. As soon as location *B* reaches capacity and breaks down, the queue created spills back into location *A*, which also becomes oversaturated (for additional discussion, see May 1990). The speed time series at location *B* shows the breakdown occurrence, which typically occurs with a relatively steep speed drop. At location *A*, speed drops gradually as a result of the downstream breakdown.

The second important element when measuring maximum throughput is the definition and identification of breakdown. In a previous study, Lorenz and Elefteriadou (2001) identified and defined breakdown at freeway merge areas as a speed drop below 90 km/hr with a duration of at least 15 minutes. Another study on weaving areas (Lertworawanich and Elefteriadou 2001) showed that the breakdown speed threshold exists at these sites as well but has a different value (80 km/hr). Brilon (2005) defined breakdown as a reduction of average speed within one 5-minute interval to below a threshold of 70 km/hr.

Given that speed drop and its respective duration can uniquely identify the presence of breakdown, it is recommended that breakdown be defined quantitatively using these two parameters. As illustrated in Figure 4.6, at the bottleneck (location *B*, part *b*), the speed drop is typically sharp, and the breakdown can be clearly identified.

The third element that is important in clearly defining maximum throughput is the selection of an appropriate time interval. Time intervals that are typical in traffic operational analysis range

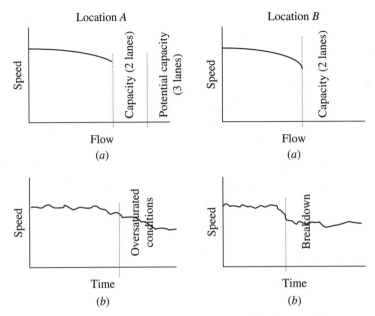

FIGURE 4.6   Speed-flow curves and time series for freeway locations *A* and *B*.

between 5 and 15 minutes. Previous research has demonstrated that maximum throughput increases for smaller intervals owing to general flow variability. Longer intervals result in averaging the flow over a longer time period, and thus the respective maximum throughput is lower.

The last element to be considered is the required sample size. Given the inherent variability of maximum throughput, it is important to observe an adequate number of breakdown events and the respective maximum prebreakdown and maximum discharge flows. Sample size determination equations should be used to establish the required number of observations for the desired precision in the maximum-throughput estimate.

Once these four elements are established, flow and speed data can be collected at the selected site(s), and time-series plots can be prepared (such as the one depicted in Figure 4.3) for each breakdown event. The breakdown can be identified based on the definition selected, and the respective breakdown flow can be obtained from the time series. Next, the maximum prebreakdown flows and the maximum discharge flows can be obtained for each breakdown event.

4.4.2 Additional Considerations for Obtaining Capacity from Microsimulation Models

The elements and procedure just outlined for field data collection can be followed when maximum-throughput information is obtained through a microsimulation model. Additional considerations include (1) simulation model selection and (2) development of simulated demand patterns, which are discussed below.

Microsimulation model selection is a very broad topic and is dealt with here only very broadly with regard to capacity and breakdown observations. The model selected should have the necessary stochastic elements and the capability to simulate breakdown as a random event. Stochastic elements include the vehicle and driver capabilities described earlier. Specifically, the acceleration and deceleration parameters for each vehicle, including the car-following models, should address and be calibrated for breakdown conditions. Similarly, the lane-changing algorithm of the model selected should consider and be calibrated for breakdown conditions.

The second consideration when using simulation modeling for capacity estimation is what demands to use and how to vary them so that breakdown is achieved. The most common technique is to run the model with incrementally higher demands starting at a sufficiently low below-capacity level. The analyst would need to "load" the network starting with relatively low demands and increasing them at constant intervals until breakdown is reached. Another, more complicated technique is to develop random patterns of demand to simulate the demand patterns in the field. For both techniques, the increments employed at each successive demand level would be a function of the desired interval in the capacity observations. Output data then can be collected on breakdown events and maximum throughput, similar to the field data collection.

4.4.3   Capacity Estimates in the *HCM 2000*

HCM 2000 provides capacity estimates for (1) basic freeway segments, (2) ramp merge segments, (3) weaving segments, (4) multilane highways, and (5) two-lane highways. *HCM 2000* provides, for each segment type and set of geometric conditions, a single value of capacity. For example, for freeway facilities, capacity values are given as 2,250 passenger cars per hour per lane (pc/hr/ln) for freeways with free-flow speeds of 55 mph, up to 2,400 pc/hr/ln when the free-flow speed is 75 mph (ideal geometric and traffic conditions). These values represent average conditions at similar sites around the United States, obtained based on general trends of maximum flows observed at various freeway locations. Note that the *HCM 2000* capacity definition is more closely aligned with the definition of maximum prebreakdown flow. The *HCM 2000* does not define breakdown flows and maximum-discharge flows, nor does it provide estimates for these at various facility types.

4.5   THE CAPACITY OF INTERRUPTED FLOW FACILITIES

Interrupted flow facilities are those where traffic flow experiences regular interruptions owing to traffic signs and signals. These include facilities such as signalized and unsignalized intersections and roundabouts, all of which are discussed in this section.

4.5.1   Signalized Intersections

The capacity of a signalized intersection very much depends on the phasing and timing plan. Traffic flow on a signalized intersection approach is regularly interrupted to serve conflicting traffic. Thus the capacity of the approach is a function of the amount of green given to the respective movements within a given time interval. For example, if the cycle length at a signalized intersection is 90 seconds and the eastbound traffic is given 45 seconds of green, then the total amount of time that the approach is given the right-of-way within an hour is

$$\text{Total green time} = \text{no. of cycles per hour} \times \text{green time}$$

$$= (3,600/90) \times 45$$

$$= 1,800 \text{ seconds}$$

This corresponds to $1,800/3,600 = 50$ percent of the full hour.

In addition to this time restriction of right-of-way-availability, the time headways observed at the stopline of a signalized intersection approach follow a different pattern as the traffic signal changes from green to yellow to red and then back to green.

Figure 4.7 illustrates a series of consecutive time headways (also called *discharge headways* when referring to queued vehicles at signalized intersection approaches) observed as vehicles depart from a single-lane approach. The horizontal axis in the figure represents time (in seconds), whereas

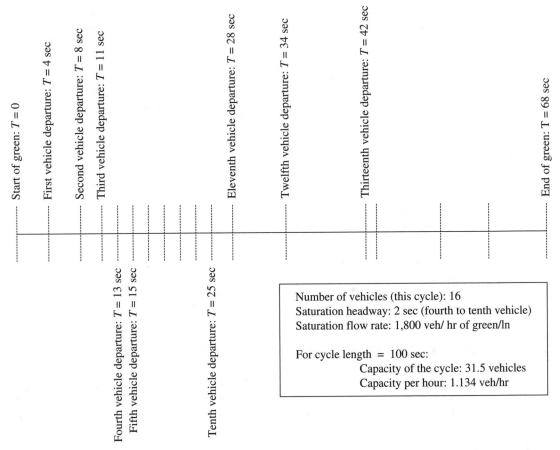

FIGURE 4.7 Time headways at a single-lane signalized approach.

the vertical dashed lines represent events (signal changes and vehicle departures). At the beginning of the green, there were 10 vehicles queued at the approach. The volume during this green interval is 16 vehicles. As shown, the first discharge headway, measured from the beginning of the green to the departure of the first vehicle in the queue, is also the largest among these first 10 queued vehicles. Subsequent discharge headways decrease gradually until they reach the *saturation headway* s_h level for the approach, which is the minimum-time headway observed under conditions of continuous queuing—in this example, 2 seconds. Next, *saturation flow* can be defined as the maximum through-put for the signalized intersection approach if the approach were given the green for a full hour. The saturation flow for a single lane can be calculated as

$$\text{Saturation flow (vehicles per hour of green)} = 3,600/\text{average}(s_h)$$

$$= 3,600/2 \tag{4.3}$$

$$= 1,800 \text{ veh/hr of green/ln}$$

Considering that signalized intersection approaches do not have the green for a full hour, the maximum throughput that can be achieved depends on the percent of time a given approach is given the green. Mathematically:

Maximum throughput per lane = % effective green × saturation flow

$$= g/C \times 3{,}600/s_h \qquad (4.4)$$

$$= (3{,}600 \times g)/(s_h \times C)$$

where g is the duration of effective green for the approach and C is the cycle length for the intersection.

Effective green is defined here as the time the approach is used effectively by this movement or the actual green time minus the lost time experienced owing to startup and acceleration of the first few vehicles (for discussion of lost time and its precise definition, consult the *HCM 2000*). In the example of Figure 4.7, the lost time is the extra time incurred by the first 10 vehicles after subtracting the saturation headways for each of these vehicles. Thus the effective green time is

Effective green time = actual green time − lost time

= 68 seconds − (25 seconds − 10 vehicles × 2 seconds)

= 63 seconds

In the example of Figure 4.7, the maximum throughput (i.e., capacity), assuming that the duration of the effective green remains constant through the entire hour and that the cycle length is 100 seconds, is estimated from equation (4.4) as

Capacity = (3,600 × 63 seconds)/(2 seconds × 100 seconds) = 1,134 veh/hr

An alternative method to estimate capacity within the hour is to estimate the capacity per cycle and multiply by the number of cycles within the hour. In the example of Figure 4.7, the capacity per cycle is

Capacity per cycle = effective green/saturation headway

= 63 seconds/2 seconds

= 31.5 vehicles per cycle

The total number of cycles in the hour is

Number of cycles = 3,600/100 seconds = 36 cycles per hour

Thus the capacity within an hour is

Capacity = capacity per cycle × number of cycles

= 31.5 vehicles per cycle × 36 cycles

= 1,134 veh/hr

4.5.2 Two-Way Stop-Controlled (TWSC) Intersections and Roundabouts

The operation of TWSC intersections and roundabouts (yield-controlled) is different from that of signalized intersections in that drivers approaching a stop or yield sign use their own judgment to proceed through the junction through conflicting traffic movements. Each driver of a stop- or yield-controlled approach must evaluate the size of gaps in the conflicting traffic streams and judge whether he or she can enter the intersection or roundabout safely. The capacity of a stop- or yield-controlled movement is a function of the following parameters:

- *The availability of gaps in the main (noncontrolled) traffic stream.* Note that *gap* is defined in *HCM 2000* as time headway; elsewhere in the literature, however, gap is defined as the time elapsing between the crossing of the lead vehicle's *rear bumper* and the crossing of the following vehicle's *front bumper*. In other words, the *HCM 2000* gap definition (which will be used in this chapter) includes the time corresponding to the crossing of each vehicle's length. The availability of gaps is a function of the arrival distribution of the main traffic stream.
- *The gap-acceptance characteristics and behavior of the drivers in the minor movements.* The same gap may be accepted by some drivers and rejected by others. Also, when a driver has been waiting for a long time, he or she may accept a shorter gap while have rejected longer ones. The parameter used most often in gap acceptance is the *critical gap*, defined as the minimum time headway between successive major street vehicles in which a minor street vehicle can complete a maneuver.
- *The follow-up time of the subject movement queued vehicles.* The *follow-up time* is the time headway between consecutive vehicles using the same gap under conditions of continuous queuing, and it is a function of the perception/reaction time of each driver.
- *The use of gaps in the main traffic stream by movements of higher priority.* This results in reduced opportunities for lower-priority movements (this is not applicable for roundabouts because there is only one minor movement).

The example provided below is a simplified illustration of the capacity estimation process for a stop- or yield-controlled movement. The capacity of the northbound (NB) minor through movement of Figure 4.8 is estimated based on field measurements at the intersection. There is only one major traffic stream at the intersection [eastbound (EB)] and one minor street movement (NB).

The critical gap was measured to be 4 seconds. It is assumed that any gap larger than 4 seconds will be accepted by every driver, whereas every gap smaller than 4 seconds will be rejected by every

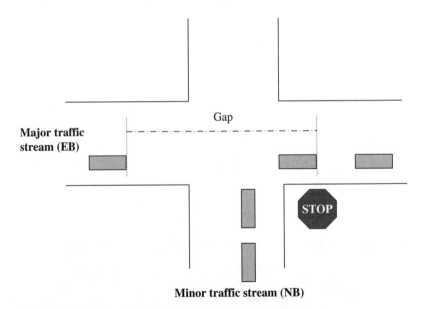

FIGURE 4.8 Capacity estimation for the NB through movement.

driver. The follow-up time was measured to be 3 seconds. Thus, for two vehicles to use a gap, it should be at least

$$4 \text{ seconds (first vehicle)} + 3 \text{ seconds (second vehicle)} = 7 \text{ seconds}$$

Conversely, the following equation can be used to estimate the maximum number of vehicles that can use a gap of size X:

$$\text{Number of vehicles} = 1 + (\text{gap size } X - \text{critical gap})/\text{follow-up time} \qquad (4.5)$$

Table 4.1 summarizes the field data and subsequent calculations for the intersection of Figure 4.8 and provides the respective capacity estimate. Column 1 of Table 4.1 provides the gaps measured in the field. Column 2 indicates whether the gap is usable, that is, whether it is larger than the critical gap. Column 3 uses equation (4.5) to provide the number of vehicles that can use each of the usable gaps. The total number of vehicles that can travel through the NB approach is the sum provided at the bottom of column 3 (26 vehicles over 103 seconds), or $(3,600/103) \times 26 = 909$ veh/hr, which is the capacity of the movement.

The *HCM 2000* methodology for estimating the capacity of TWSC intersections is based on the principles just outlined using mathematical expressions of gap distributions and probability theory for establishing the use of gaps by higher-priority movements.

TABLE 4.1 Capacity Estimation for a Stop-Controlled Movement

Gap size (sec) (1)	Usable gap (Y or N)? (2)	Vehicles in NB movement that can use the gap (veh) (3)
5	Y	1
2	N	0
7	Y	2
10	Y	3
4	Y	1
2	N	0
12	Y	3
17	Y	5
7	Y	2
3	N	0
2	N	0
3	N	0
13	Y	4
16	Y	5
103 seconds	9 usable gaps	26 vehicles

4.6 SUMMARY AND CLOSING REMARKS

This chapter provided an overview of the concept of highway capacity, discussed the factors that affect it, and provided guidance on obtaining and using field values of capacity. The definition of capacity within the *HCM* has evolved over time. Even in the earliest publications related to highway capacity, there is a general recognition that the *maximum throughput* of a highway facility is random, and thus there has been an effort to include the expected variability of maximum volumes in the capacity definition. Recently, several studies have used field data and have proven that there is variability in the maximum sustained flows observed in the range of several hundred vehicles per hour per lane. The cause of this variability lies in the microscopic characteristics of traffic, that is, the individual spacing, time headway, and speed of each vehicle in the traffic stream and their variability. This chapter also provided the fundamental principles of capacity estimation for uninterrupted and interrupted flow facilities and provided guidance on obtaining capacity estimates in the field and using simulation.

The discussion provided in this chapter is not exhaustive and is intended to provide only the fundamental principles of capacity estimation for various highway facilities. Detailed methodologies for estimating highway capacity for various facilities are provided in the *HCM 2000*, as well as elsewhere in the literature.

REFERENCES

Brilon, W., J. Geistefeldt, and M. Regler. 2005. Reliability of freeway traffic flow: A stochastic concept of capacity. In: *Proceedings of the 16th International Symposium on Transportation and Traffic Theory*, pp. 125–144. College Park, MD.

Elefteriadou, L., and P. Lertworawanich. 2002. Defining, measuring and estimating freeway capacity. Submitted to the Transportation Research Board, July.

Elefteriadou, L., F. Hall, W. Brilon, et al. 2006. Revisiting the definition and measurement of capacity. Presented at the 5th International Symposium on Highway Capacity and Quality of Service, Yokohama, Japan, July 25–29.

Elefteriadou, L., R. P. Roess, and W. R. McShane. 1995. The probabilistic nature of breakdown at freeway: Merge junctions. *Transportation Research Record No. 1484*:80–89.

Evans, J., L. Elefteriadou, and G. Natarajan. 2001. Determination of the probability of breakdown on a freeway based on zonal merging probabilities. *Transportation Research Part B* 35:237–54.

Fitzpatrick, K., L. Elefteriadou, D. Harwood, et al. Speed prediction for two-lane rural highways. Final report, FHWA-99-171, -172, -173, and -174, June. Available at www.tfhrc.gov/safety/ihsdm/pdfarea.htm.

Flannery, A., L. Elefteriadou, P. Koza, and J. McFadden. 1998. Safety, delay and capacity of single-lane roundabouts in the United States. *Transportation Research No. Record 1646*:63–70.

Highway Capacity Manual. 1950. Washington: U.S. Department of Commerce, Bureau of Public Roads.

Highway Capacity Manual. 1965. Washington: Highway Research Board, Special Report 87, National Academies of Science, National Research Council Publication 1328.

Highway Capacity Manual. 2000. Washington: Transportation Research Board, National Research Council.

Lertworawanich, P., and L. Elefteriadou. 2001. Capacity estimations for type B weaving areas using gap acceptance. *Transportation Research Record No. 1776*:24–34.

Lertworawanich, P., and L. Elefteriadou. 2003. A methodology for estimating capacity at ramp weaves based on gap acceptance and linear optimization. Transportation Research Part B: Methodological, 37B(5), 459–483.

Lorenz, M., and L. Elefteriadou. 2001. Defining highway capacity as a function of the breakdown probability. *Transportation Research Record No. 1776*:43–51.

May, A. D. 1990. *Traffic Flow Fundamentals*. Englewood Cliffs, NJ: Prentice Hall.

CHAPTER 5
TRAFFIC CONTROL SYSTEMS: FREEWAY MANAGEMENT AND COMMUNICATIONS

Richard W. Denney, Jr.
Federal Highway Administration
Lorehsville, Virginia

5.1 INTRODUCTION

The roadway system represents an asset, owned by the public or, occasionally, by private concerns, that provides a means of mobility for people, goods, and services in relatively small vehicles. Thus the roadway networks service a large number of disparate trips. Traffic is a fluid that comprises particles, each of which is controlled by a human operator seeking to travel from an origin to a destination. Each traveler in a road network makes decisions about how to use that network based on a limited understanding, through experience or through communication from network observers, of the conditions in that network. These decisions encompass three basic choices: when to depart, what route to take, and whether to use a car or some other transportation mode.

Roadways require huge physical construction projects and, therefore, are one of the costliest of public works activities. They consume land, require reshaping that land, impose difficult drainage challenges, displace landowners and residents, create boundaries, and demand extensive and expensive maintenance. Consequently, public agencies are often pulled between competing influences: building more roads to relieve congestion and building fewer roads to avoid a negative impact on quality of life. In this environment, agencies are strongly motivated to squeeze every possible bit of capacity a roadway network can offer. Doing so requires active traffic management using a variety of techniques to manage demand and capacity.

This chapter will present freeway management as a congestion-relief activity that is wholly tied to the demand and capacity relationship. Strategies will be presented for optimizing capacity and for managing demand, as will be the infrastructure required to perform this management. The chapter also will provide a description of freeway management activities in real time.

One of the most expensive and difficult components of real-time traffic management is establishing the ability to communicate with the management infrastructure in real time. Thus this chapter will present a discussion of communications methods and technology as they relate to traffic management to provide transportation engineers with a basic understanding that will help them to work effectively with communications engineers.

The general approach to this presentation loosely follows the *systems engineering approach*. In this approach, an engineer first defines user needs. These user needs are written down in a concept of operations that describes how those needs are addressed in practice (not how the system will address them but how humans address them). From those activity descriptions emerge a series of requirements that the system must fulfill to support those activities. Testing proves that the system components conform to the design and that the completed system fulfills the requirements. Finally, the system is used by the agency to perform the activities described in the concept of operations. A fundamental principle of this approach is *traceability*. At each step of this process, the relevant features are explicitly traced back to the relevant features at the previous step. Thus tests trace directly to requirements, and requirements trace directly to user needs. Based on this concept, the chapter will continue to refer back to the primary goal of freeway management to modify either demand or capacity to reduce congestion.

5.2   FREEWAY MANAGEMENT

5.2.1   What Is Freeway Management?

Simply put, freeway management is a collection of activities with the goal of minimizing congestion in a freeway network. Congestion is the result of demand exceeding capacity, so strategies to achieve the goal include increasing capacity and decreasing demand. While every freeway management activity traces back to this goal through these strategies, this simple statement is still perhaps a little too simple.

Freeway managers have another goal that is not so often stated, and that is to help travelers reconcile themselves to unavoidable congestion and to manage their lives around the resulting delay. Many freeway management activities present information about congestion to the motorist with no proposed solution in the hope that if the delayed traveler at least understands the nature of the delay, he or she will be calmer about it. The doctor may not be able to effect a cure, but the fully informed patient can at least make plans.

Freeway managers are concerned with congestion that results from routine network usage, such as commuter demand, and congestion from unexpected changes in capacity as a result of traffic crashes, stalled vehicles, construction, and other incidents. Thus freeway management includes both short-term strategies for incident management and long-term strategies for routine congestion.

5.2.2   Demand and Capacity: Congestion

Congestion is the natural result of demand exceeding capacity. A simple example is water pouring through a funnel. If more water pours into the funnel than the spout can pass, the excess water builds up in the funnel. Thus the capacity of water flow through the funnel is controlled only by the diameter of the hole at its tip. A bigger funnel may *store* more water, but it won't *move* more water unless that hole is larger. If the objective is to keep the water from filling up the funnel, we must either enlarge the hole or pour in less water.

Capacity.   The capacity of a freeway results from the operation of a fixed physical roadway. In the simplest case, a lane of a freeway carries at most a certain number of cars in an hour. In past years, it was assumed that, on average, drivers would not tolerate being closer together than 2-second headways. *Headway* is the time that passes between the front edge of one car passing a point and the front edge of the following car passing that point. The dynamics of the relationship between two cars is called *car-following theory*. Car-following rules are the subject of another chapter, but in this discussion, it is enough to be able to characterize headways. As drivers become more aggressive or more accustomed to congestion, their tolerance of close headways tightens. The traditional value of 2 seconds corresponds to a flow of 1,800 vehicles per hour (3,600 seconds/hour divided by 2 seconds/vehicle). Even in small cities, the tolerance is tighter than this, and the minimum capacity expected in most current situations is at least 2,000 vehicles per hour. The *Highway Capacity Manual* reveals that in certain circumstances, one might see as many as 2,400 vehicles per hour and higher flows than that for shorter periods.

Roadways, like rivers, are subject to constraints on capacity resulting from physical interruptions. For example, one of the busiest freeways in the United States, the Southwest Freeway in Houston, suffered a long-term capacity bottleneck where the freeway turned and rose over a bridge that spanned a railroad mainline. The combination of the upward slope (which only occurs at overpasses in flat Houston) and the horizontal curve created an obvious section of reduced capacity.

Most capacity constraints are less subtle, however. The typical capacity constraint occurs when a lane is dropped and the number of cars in each of the remaining lanes is higher than before the lane drop. This feature occurs in many places: in suburban areas where the freeway narrows at what was once the edge of urban development, in freeway interchanges where two-lane connectors merge into a single lane, at entrance ramps where two lanes merge into a single lane, and so on.

Most capacity problems occur downstream from merge points, where two roadways come together and are squeezed into a downstream roadway with less capacity than the total of the two entering roadways. For example, an entrance ramp of one lane merges into a four-lane freeway. The freeway still contains four lanes after the merge, but the capacity upstream from the merge was provided by five lanes.

The capacity constraints that most concern motorists and freeway managers alike are those which are not part of the physical features of the roadway. These include construction activities, traffic accidents, stalled vehicles, law-enforcement activities, and bad weather. Even a stalled vehicle parked on the shoulder can impose a slight reduction in capacity. These temporal reductions in capacity consume most of the attention of freeway managers partly because they are the easiest to manage and mostly because they have an unexpected impact on travelers and are the primary cause of reduced travel time reliability.

Demand.   Travelers have a need to move from one place to another. Their trips may be nondiscretionary and demand a specific arrival time (such as going to work) or discretionary with the option of going when it is most convenient (such as a shopping trip). Most traffic congestion results from nondiscretionary trips because motorists seek to avoid wasting time if they can help it and leave their departure to the last reasonable moment.

Travelers have other choices at their discretion, too. They may choose to travel by mass transit, for example. They also may choose to use a different route.

When we sum up the choices made by all travelers, we get a profile of demand that varies as the day progresses. At times, this profile may rise up over the limit of capacity, with the result that the traffic becomes congested. As with the funnel, the excess demand stacks up in a stream of slow-moving vehicles, or *queue.*

All traffic congestion is caused by excess demand. When the sum of the traffic on an entrance ramp and the traffic on the adjacent main lanes exceeds the capacity of the lanes downstream from the entrance ramp, a queue will form. The queue is quite sensitive to the amount of excess demand. For example, if the demand on a four-lane freeway is 8,400 vehicles per hour and the capacity in a bottleneck is also 8,400 vehicles per hour, we would expect unstable operation with very little queue. If the demand rose to 8,600 vehicles per hour, which is an increase of only a little over 2 percent, 200 cars would stack up into a queue in an hour. Each lane's 50 cars of queue, if we assume maximum, or *jam*, density to be about 250 cars per mile, would extend not quite a quarter of a mile. The speed in that queue would be a little less than 10 miles per hour on average (2,400 vehicles/hour at capacity divided by 250 cars/mile equals 9.6 mi/hr). A 10 percent increase in demand would create a mile of additional queue. Therefore, demand only has to exceed capacity by a few percentage points to create severe congestion.

But the problem is worse than just the simple arithmetic of demand and capacity because operational capacity is not constant. Queues are inefficient and reduce capacity somewhat just by their existence. The effect is subtle and not recognized by all experts, but only a small change can make a big difference.

In the case of a physical bottleneck, the only relief is a reduction in demand, but we can see that demand must be reduced to levels well below capacity to see much of a reduction in the queue. In our example with the flow of 8,600 and the capacity of 8,400 vehicles per hour, if the flow diminished to 8,400 vehicles per hour, the queue would remain where it is, unable to dissipate because it is already there and is being filled as fast as it is being emptied. Therefore, the demand must fall well below capacity. For example, if the demand diminished to 8000 vehicles per hour, the 200 cars in the queue would take a half hour to dissipate.

Bottlenecks caused by accidents are usually more severe. Let's say that an accident blocks a lane and reduces capacity in our example to 6400 vehicles per hour. When the accident is cleared and the capacity is restored, the front of the queue will be released. If the incident lasts an hour, the resulting queue will contain 2000 cars. At a capacity flow of 8400 vehicles per hour, those 2000 cars will take a little less than 15 minutes to clear, during which time the arriving flow will add another 2000 cars to the queue, which will take another 15 minutes to clear. Thus, the queue itself becomes a bottleneck that moves backward. Again, only a significant reduction in demand will allow the queue to clear completely.

From these examples, we can see that the smaller we can keep the queues, the easier it will be for them to dissipate when demand diminishes. Consequently, recognizing queues is one of the things we have to be able to do quickly to effectively manage freeways.

Reading a Freeway.   Freeway managers need to be able to understand what they are seeing when they observe freeway flow, and they need to gain that understanding almost instinctively. This section will discuss what a freeway bottleneck looks like.

Most drivers connect the queue with the capacity constraint. The queue forms upstream of the bottleneck, just as the water is stored in the funnel, but the bottleneck is the hole in its tip. In observing a queue, a freeway manager keeps moving downstream until the queue disappears and the traffic speeds up—this is the location of the bottleneck, within which traffic runs at capacity flow. Thus the bottleneck is always found just downstream of the queue. This is how freeway management systems identify bottlenecks automatically.

As we have seen, the speeds in the queue typically are at around 10 miles per hour or less. The speed of capacity flow is around 50 miles per hour or a little more. This value may be slightly lower in older freeways with tighter geometry.

Brake lights are a good indicator of queue formation, even at its inception. When an observer sees brake lights starting to come on from groups of drivers, he or she can expect the queue to start forming quickly. But drivers do not apply their brakes because of a capacity constraint; they apply their brakes because the car in front of them is too close. So brake lights are a secondary effect of a bottleneck, just as the queue is a secondary effect.

With this background, we can now restate our goal of minimizing congestion by the strategies of reducing demand or increasing capacity. What are these strategies? And what are the tactics we need to implement them? Before we can answer these questions, we must first understand how freeway systems monitor traffic.

5.2.3 Surveillance

Freeway management systems use two different approaches to monitoring traffic: automatic and those requiring human operators. The strategy is to know how many cars are on the facility and to know how fast they are going or some other measure of their performance. The tactics are based on available technologies, and this section will introduce the currently available options in light of their strategic value.

Electronic Surveillance.   The most basic form of surveillance in a freeway management system is point detection. Detection stations are constructed at regular intervals along a freeway, typically at half-mile or quarter-mile intervals. These stations use a detection technology to sense the presence of vehicles or to measure directly some aspect of their performance. These technologies can be broadly categorized into systems that measure the presence of a vehicle at a point, those which measure the movement of vehicles at a point, and those which record the passage of particular vehicles at various points.

Presence Detection.   The oldest and still most widely used detection technology is the inductive loop detector. On freeways, these are usually made by sawing a slot in the pavement in the form of a square 6 feet on a side in the middle of each lane. Wires are laid into the slot so that they loop the square around its perimeter three or four times, forming a coil. (The slot then is filled with a sealant to encapsulate the wires and to keep moisture out of the slot.) When electricity is applied to the wire, the current through the coil creates a weak electromagnetic field. When cars pass through the electrical field produced by the loop, they change the loop's *inductance*, which is an electrical property of coils. This change in inductance is detected by the attached electronic device. Loop detectors are relatively inexpensive to install, especially if they can be installed in pavement already closed to traffic for some other reason, such as during its initial construction or subsequent reconstruction.

Loop detectors are simple and accurate in freeway applications. They sense a vehicle when it is within about 3 feet of the edge of the loop and continue to sense it until it is 3 feet or less beyond the loop. Most freeway detection stations will place two loops in each lane, with about 10 feet of space between them. The small gap in time between detection at the first loop and detection at the second loop provides a direct measure of speed.

Loop detectors, therefore, can measure three things: the number of cars (volume), the percentage of time the loop is covered by cars (detector occupancy), and speed. *Occupancy* is a term used widely to describe the number of people in each vehicle, but in this application it means the percentage of time a detector is covered up. Occupancy is a surrogate for traffic density and speed. Of these three measures, occupancy is the most useful to freeway managers, and we will discuss why in a later section.

Other technologies are available that provide similar presence detection, such as magnetometers. Several new technologies also have emerged that allow presence detection without installing anything in the pavement. These are generally classed as *nonintrusive* detectors, and although their hardware usually costs more than loop detectors, they are far easier to install in pavement that is already carrying traffic. These technologies include acoustic detectors, infrared detectors, radiofrequency detectors, and certain examples of video-processing detectors.

Acoustic, infrared, and radio detectors work either by reflecting energy from vehicles being detected or by sensing the direct emissions of vehicles. These systems typically measure presence and some performance parameter such as speed, but some that use these technologies also can do simple classification of vehicles into trucks and cars. Of these, microwave detectors have become extremely popular, perhaps the most popular freeway detector for new installations.

The most glamorous of current detection technologies are the video-image-based systems. These systems digitize and evaluate video imagery by one of several means to find the vehicles and determine their performance. The key advantage to video-based detection systems is that the detection zone can be moved easily after installation by drawing them on a screen. This flexibility makes them ideal for maintaining detection capabilities within, for example, construction zones, where the lanes will be shifted from time to time. Even so, video-based detection has found more enthusiastic application for detection associated with traffic signals because they are at their worst when trying to count successive cars, especially in congested conditions as well as in conditions of fog or snow.

Vehicles as Probes. Presence detection has a basic limitation: It can only provide data for one physical location. A common need within freeway management is the ability to measure travel time, and point detectors do not provide this capability. Several technologies that use vehicles as probes measure travel times directly and overcome this limitation. Travel-time information has not been used in automated incident detection, but it has found popular use for letting motorists know the condition of freeways. In most applications, however, probe-vehicle data collection grew from another capability, and freeway managers have used the data as an opportunity.

For example, the Texas Department of Transportation developed the Houston area's entire automated surveillance capability in a remarkably short time by using existing electronic toll-collection systems. Even though toll facilities are not used widely in that part of the country, there were sufficient number of subscribers to the automatic toll-collection system for their two toll facilities to provide sufficient probe coverage throughout the network. The system in Houston uses automated toll readers on the freeways at 2-mile intervals in addition to their regular application at toll plazas. Studies have shown that as little as 3 or 4 percent of the traffic stream can provide reliable data about the traffic stream, so the system didn't need many equipped vehicles. Another example that uses toll-collection devices is the TRANSMIT system in New York.

An emerging technology, now used widely to populate speed and travel-time databases for state 511 motorist information systems and even for smart cell phone applications, makes use of cell phones as probes. They may either provide information to cell phone users in return for data, or they may obtain the information directly from the cellular network through cooperative agreements with cellular providers. In all cases, information based on cell phones is anonymous, making use solely of hardware identifiers that cannot be traced to individuals. Yet another technology measures travel times by reading the hardware addresses from passing Bluetooth devices.

Human Surveillance.   All freeway systems are built with an extensive capability for human surveillance using closed-circuit television. Humans have an ability to make judgments far beyond the skill of automated systems using electronic detection, but they are subject to loss of attention caused by boredom or distraction. Most systems combine electronic detection and human surveillance to optimize the strengths of each. The human operator might not be looking in the right place at the right time to see an incident occur, but once notified by an automated detection system, they can quickly determine what has happened and what should be done.

The major difficulty in establishing human surveillance is the infrastructure required to provide video to the location of the human operator. Because human operators are still more difficult to come by than infrastructure, however, agencies build control centers to bring all the surveillance information together. Control centers, in turn, require the surveillance information, including video, to be communicated from the field to the operator's workstation. The major cost of freeway management systems, therefore, is the communications infrastructure. With the advent of the public Internet and with agency-wide high-capacity data systems, traffic data, including video data, no longer require dedicated infrastructure, and the cost of communications with respect to the remainder of the system is falling rapidly.

Video monitoring technology is improving at a rapid pace, and this text would quickly become outdated if it attempted to document appropriate technology choices. Most cameras at the time of this writing use digital technologies for producing the image, convert them to a different digital standard for communication over networks that use the Internet protocols (though not necessarily part of the public Internet), and then distribute the video signals to workstations and video-display systems digitally. Many existing systems still use analog signals at various points in the process, although analog components of video monitoring systems are fast disappearing.

The typical arrangements of video cameras fall into three approaches: full redundant coverage, full coverage, and partial coverage. In the first category, cameras are placed frequently enough so that at least two cameras can see any given incident. This usually requires cameras at roughly half-mile spacing. Cameras can be placed at 1-mile intervals, but this requires cameras to be able to see half a mile, which stretches the capabilities of cameras and mounts unless the cameras include expensive image-stabilization technology. Typically, the traffic industry has avoided using such technologies in favor of simple cameras that are easier to maintain.

Partial coverage places monitoring cameras only at critical locations. Some traffic signal systems use partial coverage, but most freeway operators greatly prefer full coverage. The reason is that any section of freeway can become critical to the capacity of the facility should an incident occur there.

Cameras are usually mounted between 35 and 45 feet above the ground with due care given to terrain, field of view, and lens flare caused by being pointed into the sun. They are equipped with mounts that allow the cameras to be aimed from the control center, known as *pan-tilt-zoom mounts*. The sun is a major obstacle to clear viewing, and it will cause lens flare, blooming, and aperture errors when it is in the field of view. Most good designers, therefore, attempt to locate the cameras on the south side of east-west roadways and alternating on the east and west sides of north-south roadways. The key advantage to full redundant coverage is that an incident always can be viewed by a camera that is facing away from the sun. When designing camera locations in the field, most engineers will resolve troublesome locations by elevating themselves to camera position in a bucket truck and videotaping the available field of view using a camcorder. These tapes can be reviewed by agency engineers and operators to select the location that provides the most effective view.

Camera technology has exploded in the last few years, but engineers who write specifications for these cameras are still concerned with durability and maintainability. Many new installations are equipped with lowering devices so that the cameras can be lowered to ground level for maintenance. Some agencies use bucket trucks to reach cameras mounted at viewing height. A few agencies depend on ladder rungs and work platforms built into the poles, but these tend to be unusable for a variety of reasons by the normal maintenance technicians working for such agencies.

Most agencies build ground-level cabinets for their cameras, and most video manufacturers provide field equipment that allows a technician at the site to assume control of the camera for maintenance purposes.

5.2.4 Capacity Strategies

Freeway managers focus on two types of capacity constraints: queue formation and incidents. As we have seen, queues are themselves capacity constraints that linger after their cause has been removed. Minimizing queues resulting from physical bottlenecks is a major activity, but it will be discussed in more detail under demand strategies because improving the capacity within a physical bottleneck is not one of the capabilities of freeway management.

Capacity bottlenecks caused by incidents are another matter. Most freeway management systems are fully justified by the need to recognize, respond to, manage, and clear incidents as quickly as possible. There are many approaches for doing each of these.

Incident Recognition. Unlike the sea, rivers move while their waves do not. For a given traffic flow, routine bottlenecks will produce expected queues and congestion at expected places. When this congestion violates expectations, an incident might be at fault. A number of algorithms have been developed over the years to automatically warn of potential incidents. None are even close to perfect, but agencies that operate large systems are motivated to refine their approach as much as possible. For example, the California Department of Transportation reports managing as many as 60 incidents in a typical day in the Los Angeles area. If their incident-detection system flags 120 potential incidents, their operators will be overwhelmed. If their algorithm flags only 30 incidents, however, the operators will not discover the remaining incidents as quickly.

The literature reports four different categories of automated incident-detection technologies, including comparative, statistical, smoothing, and modeling methods. The comparative approach has been used most widely and is the basis for the algorithms used by the California Department of Transportation (Caltrans). Caltrans uses a collection of incident-detection algorithms that work by comparing volume and occupancy to thresholds that define normal operation. When the measured conditions cross the thresholds, the algorithm assumes abnormal operation and warns the operator. These algorithms have been refined over many years, and the best of them report accuracy rates of around 75 percent and false-alarm rates of around 1 percent. Agencies must spend a lot of time customizing the thresholds for normal operation, and they must aggressively maintain that customization as traffic conditions change over time.

The modeling approach has been used in Canada, based on successful research at McMaster University. The model predicts detector occupancy based on upstream data, and when measured occupancy and speed fall outside the predicted range, the algorithm warns of a possible incident. The modeling approach requires less tailoring to specific locations and is able to track trends in traffic over time, but it can be confused to some extent by the instability of the relationship between volume and congestion at the point of capacity. Nevertheless, the literature indicates that the McMaster method is at least as effective as the best of the Caltrans methods.

All these approaches share a common flaw: They detect incidents based on secondary effects. For example, if two cars collide, they will block at least one lane. The blocked lane will greatly reduce capacity in that freeway section, which will create a bottleneck. Traffic will queue up behind the bottleneck. Eventually, the queue will extend back and cover a freeway detector, causing the occupancy to shoot upward, and that spike in occupancy will require some time to appear in moving averages. The increase in that moving average, especially when coupled with the low occupancy at downstream locations, will flag the incident. The extending queue might not trigger the alarm for many minutes after the incident occurs. The best systems report a lag time of at least 3 minutes, but in many cases where the incident-induced congestion overlays existing congestion, it will be much longer.

Many of these incident-detection systems have become irrelevant in recent years with the explosion in the use of cellular telephones. Incidents often have been reported long before the queue builds back to the detector. Human operators also often will see the effects of incidents before the automated systems flag them, but many systems are operated with limited staff, and operator fatigue is a critical issue in watching for incidents. Fortunately, cellular technology has largely solved the problem of flagging incidents. It hasn't solved the problem of getting proper information about the incidents, which is why we see no trend away from providing a full surveillance capability.

Incident Response.   Incident response requires three elements: a notion that effects the incident at hand, a plan that addresses those effects, and a means of mobilizing that plan.

Earlier, I related reports of experience from Los Angeles suggesting that the number of daily incidents is in the range of dozens. As one would expect for such a metropolitan area, and from the point of view of the freeway managers, this is large number. From the perspective of all the traffic on the highway, though, incidents are rare events. Accident rates typically are measured in accidents per million miles of travel, which is an indication of how infrequent they are. The result of this relative rarity is that predicting incidents is nearly impossible, and measuring their effects by observing the freeway is even more difficult.

When an incident occurs, it often starts with a catastrophic effect on freeway capacity—often closing down the entire facility for a few minutes until following drivers get over their shock and pick their way around the wreckage. The one or two lanes will get through for a while until vehicles not completely disabled can be moved to the side. More lanes may be closed by emergency personnel and so on. The profile of closures for any one incident, therefore, is highly variable. This variability means that it is nearly impossible to develop sufficient data to understand the effects of incidents in enough detail to develop response plans.

Here is where the philosophy of freeway management has a large effect on how incident management is conducted. Some freeway managers do not attempt to prepare response plans in advance, and the systems they develop focus on the ability to develop responses on the fly. Other freeway managers want the responses to be stored in the system so that they can be implemented on command from a control center. Other systems combine these two philosophies, using an expert system to prepare elements of response plans on the fly from predesigned choices and decision trees.

For systems that are designed around stored response plans, the development of those plans is a substantial task. The development itself follows a basic outline:

- List all potential incident scenarios (categorized by location, time of day, number of lanes closed, and duration of closure).
- Analyze conditions for each scenario.
- Design a response plan based on the analysis.
- Program the response plan into the freeway system's database.

For example, the TransGuide system in San Antonio, Texas, initially covered 27 miles of freeway. The Texas Department of Transportation developed approximately 125,000 scenarios for the initial implementation of the system.

Some systems avoid building a massive flat database by following a decision-tree approach, which is sometimes called an *expert system*. In this approach, the possible responses are categorized into branches of decisions, where each of the categories listed earlier represents its own branches. Response elements common to an entire branch are applied when the decision to follow that branch is taken. The California Department of Transportation uses this approach in its District 12 freeway system that covers Orange County, south of Los Angeles.

The key advantage of these systems is that the response plans can be implemented immediately and in some cases automatically (although always with confirming approval from the system operator). The disadvantage is the massive task of defining the scenarios and designing responses for those scenarios and the maintenance required to keep them up to date. This approach may be overkill for incident management in a small city. In such cases, an experienced operator may be able to craft response plans on the fly.

Most agencies deploying incident-response scenarios will employ analysis tools to help understand the effects of incidents, usually involving one of several simulation models. Once incidents are modeled, freeway managers must determine how to respond to them. These responses fall into two categories: handling the emergency and managing traffic. Both are covered in the sections that follow. Agencies can't implement these plans, however, without the means to do so. Jurisdictional boundaries must be crossed between traffic management agencies and emergency-response agencies and often across geographic jurisdictions as well. For example, the typical freeway system in a major metropolitan area will involve dozens of agencies. The state transportation department usually will be the lead agency because it usually operates the freeway. But the emergency-response agencies, especially fire departments and emergency medical services, are usually attached to local governments, of which there may be many in a metropolitan region. State police may or may not exercise jurisdiction over freeways, and when they don't, local governmental police agencies will have to be involved. Freeway incidents often spill over into surrounding streets, and some systems are being designed with close coordination between state and local agencies because the latter usually operate the traffic signals. These jurisdictional issues easily can overwhelm the technical issues when implementing freeway management systems. Agreeing to incident-response plans before the fact is one way to resolve these issues by tying agencies together around a common understanding of the rules that will be used when incidents occur. Often, however, these common approaches remain elusive, even when a system is being constructed, and often the needed cooperation results from system construction rather than guiding it.

Traffic Management During an Incident.   A number of different techniques help to restore capacity during an incident, all of which are based on the principle of opening lanes as quickly as possible. Most states have passed laws that require motorists to move vehicles out of the travel lanes if possible and if there are no injuries, although many motorists are reluctant to do so for fear that the facts of the accident will become obscured, making it difficult to assign responsibility. To reinforce the law, most police agencies refuse to investigate accidents when the vehicles can be removed from the travel lanes and when there are no injuries.

One tactic to assist in clearing lanes after minor crashes involves creating protected sites for accident investigation outside the lanes of the freeway. This concept was popular among agencies because it could be implemented at a low cost, but it was not popular with motorists because the accident investigation sites were not easy to find during the emergency.

But most techniques employed during an incident attempt to manage demand by informing motorists of the accident so that they can take an alternate route, choose a different time to make their trip, or make use of a more efficient travel mode. These strategies are discussed in the next section on demand strategies.

5.2.5 Demand Strategies

Most freeway management strategies hope to modify the demand approaching an area of congestion. These strategies can be classified by the type of shift in demand they promote. At the beginning of this chapter, I stated that drivers have one of three choices in using the network: the route they take, the time they depart, and the mode of travel they use. Demand management seeks, therefore, to promote a spatial, temporal, or modal shift in demand.

Spatial Demand-Shift Methods.   All these approaches share the objective of encouraging (or directing) motorists to use a less-congested route in order to achieve a balanced overall use of network capacity. These methods primarily include motorist information systems but also include lane-use signals.

Motorist Information Systems.   When congestion on the freeway creates much longer delays than motorists would face on alternate routes, the freeway managers will try to inform them of the delay to encourage a diversion to those relatively uncongested routes. Freeway managers have to be careful, however, not to invoke too strong a reaction. A lane of a freeway carries between 2,000 and 2,600 vehicles per hour at capacity, whereas a lane on a surface street might carry only 500 vehicles per hour. Any large diversion from a freeway to a city street system will overwhelm the latter if the demand is heavy. Most freeway managers, therefore, are unwilling to direct motorists to an alternate route for fear of creating a worse problem. Most strategies hope to achieve as much as about 25 or 30 percent diversion, but no more, and a directive to follow a specific alternate route encourages too much compliance. Agencies usually, therefore, use general suggestions such as "Use Alternate Route" to allow motorists to pick their favorite and thus be responsible for their choice.

DYNAMIC MESSAGE SIGNS: When freeway managers first visualize a system, they typically think of video monitoring and overhead dynamic message signs. Therefore, fixed-location motorist information systems usually comprise the most important elements of freeway systems. Dynamic message signs are the most popular fixed-location information systems. These large overhead guide signs provide real-time information to a passing stream of traffic in the hope that drivers will be able to make more effective route choices.

Dynamic message signs can be divided into two broad categories: changeable message signs and variable message signs. These terms are often used interchangeably, but their meanings are distinct. *Changeable* message signs contain a fixed menu of messages that may be displayed, any one of which can be selected remotely by freeway system operators. *Variable* message signs provide the additional capability of editing and creating new messages remotely. Some display technologies fall into one category and not the other, but often this distinction is more imposed by the design of the system than by the hardware of the sign. A few of the major sign technologies that still may be in operation include

- Rotating drum
- Bulb matrix
- Flip disk
- Fiberoptic
- Light-emitting diodes

Light-emitting diodes (LEDs) are by far the most popular for new signs, to the point where signs using other technologies are now difficult to buy. LEDs are mounted in clusters that combine to form pixels. Typically, 8 to 14 LEDs will be clustered in a 2-inch pixel, and the clusters are illuminated directly by drawing electric current through the LEDs. Most dynamic message signs for freeway management in the western hemisphere are a single color against a black background, but LED technology already provides a full-color capability that can be used to provide graphic information to motorists rather than plain text. As of this writing, freeway managers have only scratched the surface of this potential.

Dynamic message signs typically are installed upstream from decision points to provide information about the relative choices. In most places, they are used to warn motorists of downstream incidents, but some agencies are using them for the display of useful nonincident real-time information.

One of the more heated debates within freeway management agencies is whether to display messages when there is no incident information to post. This debate has been resolved by Federal Highway Administration policy that requires agencies constructing signs with federal money to provide the capability to measure and display travel times on the signs in the absence of more pressing information.

For example, the Georgia NaviGAtor system displays travel-time information to known landmarks downstream from the sign. This sort of real-time messaging is not incident-related, but it has proved quite popular in cities where it has been implemented.

Many agencies have instituted strict policies on what may be displayed on dynamic message signs to control liability. Agencies usually require that messages include descriptions of what is happening and wherever possible avoid dictating what the drive should do about it. Freeway systems are intended to empower motorists to make better decisions about how to use the network, and an attempt to make those decisions for motorists bring the possibility of directing them to do something that will cause them injury. Thus most agencies are descriptive in their sign messages, not prescriptive.

HIGHWAY ADVISORY RADIO: Another well-established motorist information technology uses one of several protected bands of broadcast AM and FM radio to communicate to motorists. Because broadcast radio extends beyond specific facilities, these systems are used to provide wide-area information. Highway advisory radio (HAR) uses low-power transmissions (limited by the Federal Communication Commission to 10 watts with a range of about 5 miles) to provide messages using either looped recorded voice announcements or live announcers. The trend in major metropolitan areas is to use live announcers. For example, New Jersey Department of Transportation uses HAR with live announcers extensively along suburban freeways around New York City. Most of these systems use prerecorded voice loops, such as the systems along the New Jersey Turnpike. HAR facilities are provided at frequent intervals using alternate frequencies to provide continuous information along the facility.

An extension of HAR uses regular broadcast radio to provide traffic information over a metropolitan area. For example, the Minnesota Department of Transportation uses an FM broadcast frequency owned by the City of Minneapolis to provide continuous traffic information during the rush hour and music during off-peak periods. The live announcer's booth is located in the freeway management control center and includes full video traffic monitoring capabilities as well as a view of the freeway incident display.

511 CALL SYSTEMS: Many states are implementing systems that allow motorists to call into an information system by dialing 511 on cellular telephones. The system then provides information about operating conditions on the requested facilities, often using voice commands, based on information provided by freeway management systems and also by global travel-time monitoring systems that derive travel-time information from cellular telephone systems and other sources. These systems and their related providers may also provide this information on mobile data devices via the cellular network.

Temporal and Modal Demand-Shift Methods.   Once motorists are on the network, options for managing congestion are somewhat diminished because the car must go *somewhere*. Freeway control methods often seek to put motorists where they will do less harm to the demand-capacity relationship. A growing class of management techniques seeks to address *traveler* needs as opposed to *motorist* needs. The control methods will be presented first.

Ramp Metering.   Ramp control has been one of the most controversial techniques employed in freeway systems in many areas. In some locations, agencies attempted to implement ramp control widely only to be forced to retreat in the face of public outcry. In other places, established ramp-control systems have had to justify their existence in the face of political challenges. Many agencies use ramp control with no problem, and other agencies won't even consider the alternative.

Ramp metering forces a temporal shift in demand by restraining traffic from entering a freeway for a brief period to keep the entering flow below a certain threshold. The idea is that if enough ramps are controlled to a moderate extent, the overall entering flow on the freeway can be held to a total that keeps demand below capacity at key bottlenecks.

Thus ramp metering is best suited to freeways that have well-defined bottlenecks and a series of ramps feeding the bulk of traffic to that bottleneck.

Many motorists hate ramp meters. Minneapolis has established an aggressive use of ramp metering network-wide with careful refinement and maintenance of ramp-metering strategies over many years. Even so, removal of the ramp meters became a political issue with the election of a new governor, and the compromise was a detailed study to determine the effectiveness of the ramp-metering program. Considering the political environment at the time, it speaks highly for the effectiveness of the program that after the test, most ramp meters remained in operation. Other cities have not been able to demonstrate these benefits as clearly, and many had to remove most or all of their ramp meters after installation because of public pressure.

Thus most agencies that are willing to consider ramp metering at all cite two critical principles in its successful application, including

1. An aggressive, straightforward, and truthful public information campaign
2. Coupling ramp metering with popular freeway management services, such as effectively applied dynamic message signs and other motorist information systems

Traveler Information.   Another way to encourage both temporal and modal shifts in demand is to provide good information about traffic conditions to travelers before their departure time and mode choices are made. As of this writing, the most popular method for providing traveler information for freeway management is by use of an Internet Web site, which has replaced many older approaches. Most current freeway management systems provide open access to much of the information available to system operators. For example, the Georgia NaviGAtor Web site provides map displays showing currently congested sections of the freeway network, the messages being displayed on the various freeway dynamic message signs, locations of construction zones and incidents, and even snapshots from the video surveillance cameras. Many private providers of travel-time information supply information directly to the public. For example, as of this writing, the company Inrix provides a free iPhone application that shows freeway speed and incident data collected by its own system of cell phone probes.

If transit agencies cooperate with freeway management agencies and provide real-time transit information on their Web sites, then that information can help travelers to make effective mode choices. Many commuters use their personal cars because transit modes are less convenient and sometimes more time-consuming, but if unexpected congestion because of a major incident will cause an extensive increase in travel time, the commuter may decide to use the transit service if it is easy to identify and use.

5.2.6   Operations

The operation of freeway management systems includes manipulation of the various demand and capacity management strategies described earlier. However, agencies also must maintain information about their effectiveness so that they can evaluate their operations. Thus freeway management systems include software and hardware systems in the control centers to operate the field devices in the system and to maintain a database of activities, incidents, and responses. These systems are complex and expensive and often consume a large share of the budget for the entire system. Because software systems change so frequently, describing them in detail is beyond the scope of this book. In general terms, however, software systems are designed in accordance with the system engineering approach outlined at the beginning of this chapter. The more closely agencies follow this approach, the more predictable their software development and acquisition will be. Many agencies now employ qualified systems engineers to manage their software acquisition, maintenance, and operation, and this trend is growing. The reader interested in further study on software development and acquisition approaches should review the systems engineering literature, which is widely available at all technical bookstores.

Despite the fact that this book does not present software principles in detail, an understanding of good software management is critical to successful system deployments, and failures in the absence of good project management unfortunately are common. These failures repeatedly have resulted in cost overruns and blown schedules for system implementation.

5.3   COMMUNICATIONS

5.3.1   Why Learn Communications?

Communications infrastructure represents a huge portion of the cost of building traffic management systems. Most systems designers will say that at least two-thirds of the cost of systems will be used to establish communications with the field devices.

A review of systems projects around the country, particularly for traffic signal systems, reveals that the most common mistake in the design and implementation of these systems is a mismatch between the needs of the software systems in the control center and the services provided by the communications infrastructure. Managing implementation costs often point agencies to low-cost communications alternatives whose performance must be considered in the design of the software. Software providers, on the other hand, often do not see the requirements imposed by the communications system and design their software around unsupportable assumptions of communications performance. Only a good understanding of communications alternatives and architecture will prevent these mistakes from proliferating.

5.3.2 Architecture

The architecture of a communications network defines what kind of equipment will be used in the system. It does not necessarily define the medium; the same architecture often can be operated over different media, and different media frequently are used in the same system. In some cases, however, the selection of a particular architecture requires a specific medium or at the very least allows or precludes classes of media.

In general, architectural approaches to communications systems fall into two camps. The first camp accommodates systems that *require* reliable real-time transmission of messages. The second camp does not. The reason *require* is emphasized is that determination of the requirement usually hinges on the system implementation control messages over the communications link. Control messages often have real-time requirements, whereas surveillance messages usually do not. Thus the camps generally are demarked by systems that do remote control versus systems that delegate control to field devices. Freeway systems usually fall into the former camp, and most (but not all) traffic signal systems fall into the latter camp. The need to accommodate real-time control communications, therefore, becomes the single most important decision in the selection of communications architecture.

Real Time—Defined.   The term *real time* is often defined as messages being passed within a specified amount of time. To most traffic engineers, this time frame is usually 1 second. Other branches of the computer industry, however, have very different standards. In industrial control systems, for example, real-time operation often requires response to a stimulus within 100 microseconds. Some process-control systems boast of response times as short as 6 microseconds. Use of the term *real time* to describe a desired response horizon, therefore, is not very specific.

A much more rigorous and useful definition of real time is that the system must respond to external stimuli predictably—that is, deterministically. In a traffic system, for example, the very long 1-second event horizon is satisfactory, but the system must predictably provide that response horizon every time. In systems designed to accommodate unpredictable demand, including systems that use the Internet protocols over shared networks, the capacity of the system and its bottlenecks must be carefully designed not to push messages outside the performance required by the system.

Deterministic Architecture.   Previous editions of this book discussed deterministic architectures as the generally accepted means of dealing with real-time message-delivery requirements. This is no longer the case, and deterministic approaches using fixed multiplexing are still provided at the backbone level, but no longer for endpoint communications with field devices. All those systems now use shared channels carrying Internet-style packets of data in a shared stream using the Internet Protocol (IP). Some systems are still being installed with high-capacity private-channel backbone systems making use of fiberoptic networks. The private-channel scheme was developed by the telephone industry. When a caller places a call, a private channel is established between the caller and another party. That channel may traverse a huge variety of distribution lines, trunk lines, and switching equipment, but the size and capacity of that channel is predetermined. Virtually all technologies that are developed within the telephone industry use time-division multiplexing, including the T-Carrier standards for twisted-pair wire and the Synchronous Optical Network (SONET) standards for fiberoptic cable. These channels in modern systems are quite large and provide IP network connectivity for devices.

The Rise of the IP Cloud.   All current systems are being installed to provide IP-based communications between infrastructure components in traffic systems. The protocols are ubiquitous in computer operating systems and in-field devices, and intermediate routers and other networking devices are now available as commodities. The prevalence of IP-based communication encompasses all media, and IP-based routers, modems, and other devices are readily available using one of various wireless technologies, copper-wire technologies (including Ethernet), and fiberoptic technologies, often running over SONET or Asynchronous Transfer Mode (ATM) networks. The workstations and servers in the control centers are connected using standard Ethernet networking switches, routers, and cabling. Environmentally hardened field switches are generally available as well.

Reliability and Redundancy.   Many communications systems enhance reliability by providing redundant elements. IP-based networks generally provide redundancy by being able to navigate various routes as needed through a network that resembles a mesh or grid. Backbone systems, however, still may be implemented using self-healing rings, most often used for fiberoptic systems that employ the SONET standards. In a self-healing ring, signals are transmitted simultaneously clockwise and counterclockwise around a ring of hubs. Each hub is, therefore, connected to the other hubs in two ways. If the cable is broken between two hubs, then all the hubs are still accessible from one or the other direction of the ring. The equipment in a SONET system is designed to switch automatically between directions to correct faults.

Redundancy helps to improve reliability, but it does not reduce maintenance requirements. Actually, ring topology and other redundancy schemes increase maintenance requirements because the equipment is more complicated and, in some cases, duplicated (which means that more equipment must be maintained). Redundancy increases overall maintenance requirements but reduces the *urgency* of maintenance. When a link is damaged on a SONET, the system still functions, so operation of the system does not depend on immediate repairs. But the repairs must still be made quickly because the system no longer has the redundant backup capability. In many agencies, maintenance budgets are down to crisis levels, meaning that only repairs that are critical to system functionality will be performed. In such circumstances, ring redundancy may be a problem rather than a solution because the redundant elements will not be repaired as they fail. Once the redundant elements have failed, the system no longer benefits from the enhanced reliability.

The technology of the communication should not outstrip the ability of the maintaining agency.

5.3.3   Guidelines and Standards

Deterministic Architecture.   Private-channel architectures were developed initially by the telephone industry. Consequently, a host of standards is available for defining multiplexed digital communications. All these standards can be characterized in a single continuum known as the *North American Electrical Digital Hierarchy,* which is represented below:

Description	Total Data Transfer Rate	Composition
OC48	2.48832 Gb/s	48 DS3 (1344 DS0)
OC36	1.86624 Gb/s	36 DS3
OC24	1.24406 Gb/s	24 DS3
OC18	933.12 Mb/s	18 DS3
OC12	622.08 Mb/s	12 DS3
OC9	466.56 Mb/s	9 DS3
OC3	155.53 Mb/s	3 DS3
OC1	51.84 Mb/s	1 DS3 (plus SONET overhead)
DS3	44.736 Mb/s	28 DS1
DS2	6.312 Mb/s	4 DS1
DS1C	3.152 Mb/s	2 DS1
DS1 (T1)	1.544 Mb/s	24 DS0
DS0	64 Kb/s	Basic unit—one digital voice channel
DS0A	64 Kb/s	2.4, 4.8, 9.6, 19.2, or 56 Kb/s filling a single DS0
DS0B	64 Kb/s	20 2.4-, 10 4.8-, or 5 9.6-Kb/s subchannels on a single DS0

The hierarchy generally is divided into three sections. The slowest speed standards cover standard serial communications. Standard serial communications is becoming more and more rare in traffic system applications, but some traffic signal systems use them, and they are included here. These standards are defined by several agencies, including the Electronics Industries Association (for physical media) and the Consultative Committee for International Telephone and Telegraph (CCITT, for communications layers above the physical media). Communications at this level range in speed from 110 to 115,000 bits per second, but multiplexing equipment is only available to combine the speeds shown for DS0A and DS0B.

The middle level of the electrical digital hierarchy covers high-capacity communications over copper-wire links. These are commonly known as the *T1* or *T-Carrier standards* but are more specifically known as the *Digital Signal Types.* T1 is also used to describe DS1, or Digital Signal Type 1. The T1 standards emerged from the telephone industry, and the American National Standards Institute (ANSI) now maintains them. The basic unit of T1 is a digital voice channel, which is defined nominally to be 64 kilobits per second, which is known as DS0. A DS0 can be further subdivided, but any signal larger than DS0, all the way up through SONET speeds on fiber networks, first must be broken down into DS0 channels, sent across, however, many channels required, and reassembled on the other end. Higher standards are multiplexed combinations of multiple DS0 or unchannelized DS1 channels.

Nondeterministic Architectures. With the rise of Ethernet, new standards for packet-switched networks have emerged. The software protocol is now universally Transmission Control Protocol/Internet Protocol (TCP/IP), for the two main protocols used within the Internet. Messages (i.e., *packets*) that include TCP/IP management information are now typically being carried over protocol-driven media ranging from wireless digital packet protocols such as 802.11 (known as Wi-Fi) to conventional Ethernet. For hard-wired traffic systems, Ethernet is dominant.

Ethernet performance is divided into the following standards:

Standard	Copper Wire or Fiber Designation	Nominal Data Rate
Ethernet	10-BaseT (copper), 10-BaseF (fiber)	10 Mb/s
Fast Ethernet	100-BaseT (or F)	100 Mb/s
Gigabit Ethernet	1000-BaseX (or GigE)	1 Gb/s
10-Gigabit Ethernet	10-GigE	10 Gb/s

Ethernet networks are divided into sections that are connected using switches, routers, switch-routers, or bridges depending on the nature of the connection. To implement a wide-area Ethernet, all devices on the network must have an address that allows them to be identified by software-based protocols. This addressing capability is built into the TCP/IP. TCP/IP-equipped communications interfaces can handle the connection to plain serial devices that do not have this capability, bringing Ethernet possibilities even to traditional and existing traffic control equipment.

Polled and Nonpolled Protocols. Channel sharing creates the possibility of two devices attempting to use the channel at the same time. This potential conflict requires some discipline in how devices can access the channel. The simplest scheme precludes remote devices from speaking unless they are individually commanded to speak by a master device. This approach is known as a *polled protocol.* The master device sends out a poll request message to a specific field device. That field device then sends a response. Other devices will hear the transaction (or at least the master's side of the transaction) but will keep silent. The polled protocol can be characterized by the phrase *speak only when spoken to.* Most communications systems for transportation management applications use polled protocols, but this convention is giving way as field devices become more powerful and, therefore, more capable of processing complex protocols.

As of this writing, systems are emerging that allow field devices to report in to central systems without a previous poll request. The communications systems being designed for these systems have collision management built into them. For example, the 802.11 (Wi-Fi) standard for wide-area wireless communications has collision control as a basic service of systems supporting the standard.

Software Protocols. Most of the previous discussion involves the specific protocols of the communications medium. All these standards are applied at the bottom layer of a communications system. The bottom layer is the physical layer, the part that directly manages the signaling on the medium in question. Once the signals are translated into bits and bytes, they will be processed by computer programs within the communications software on the device. Each message is encapsulated with a series of wrappers, each of which defines some aspect of the packet of data. These wrappers are called *layers* and conform to a communications model developed by the International Standards Organization (ISO). The ISO seven-layer model has been called the Open Systems Interconnect model. Above the physical layer, the OSI model defines layers for interfacing with the physical layer, layers for handling transport issues such as breaking large messages apart and routing them through the network, layers for defining the protocol in use, and finally, layers that understand traffic systems.

A central principle of the multilayer model is that each layer need not know anything about the other layers. The application layer does not need to know anything about which protocol is used in the layers handling transport issues. Consequently, a profile of protocols can be defined, selecting the specific protocol for each layer as appropriate for the system in question.

In transportation systems, the profile of protocols used conforms to the National Transportation Communications for ITS Protocol (NTCIP). NTCIP will be discussed in a later section. NTCIP represents a special case of the multilayer approach using many of the Internet protocols but also borrowing some others.

5.3.4 Video Communications

The data-transmission requirements for full-motion video are so huge that they deserve a separate discussion. The addition of video surveillance to a system design fundamentally changes the requirements of the system. For data only, the major decision concerns selecting fixed-channel serial communications or IP-based packet communications. When video is added, the major decision affecting cost is whether or not to use analog video transmission or digital video transmission.

Analog Video. Full-motion video, in analog form, requires 6 MHz of bandwidth. This includes the 4.5 MHz required by the color-burst, or picture, and the remainder for audio. Analog video is a continuous scan of electrons moving horizontally in rows down the screen. In North America, video signaling is defined by the National Television Standards Committee (NTSC). NTSC video is defined as 525 lines (480 of which are visible) of vertical resolution, or scan rows, and all rows are scanned 30 times a second. For consumer television, the NTSC standard has been replaced by high-definition television, which is a digital protocol. Traffic monitoring cameras, however, still widely use the NTSC approach to creating the image, which is defined within the HDTV hierarchy as 480i, for 480 visible lines scanned in alternating pairs, or interlaced. All new traffic-monitoring video systems, however, use digital video streams employing compression and software-based switching.

Digital Video. If we divide each horizontal scan line into 700 picture elements, or *pixels*, then each scanned screen image contains 367,000 pixels. If each pixel defines a level of red, green, and blue (the additive primary colors used in constructing a color video image), and if each color can be defined at 256 levels, then each pixel needs 3 bytes to describe it. The size, in bytes, of a single frame of broadcast-quality video is, therefore, a little over 1 megabyte. If we transmit 30 frames a second, then the data stream is going by at 265 megabits per second. In practice, an analog video image can be stream digitized at less than broadcast quality at much lower rates. Typically, full-motion video is expected to consume 100 megabits per second. Even at this rate, the bandwidth requirements of digital video are vast compared with data requirements for most transportation management systems. Thus digital video is always compressed before being communicated.

The standards for digital video compression are in continuous development by the Motion Picture Expert Group, which is a working group of the ISO. Compressed images conforming to ISO standards are called by their acronym, *MPEG*. Most systems now can use MPEG compression to reduce the data rate for high-quality video down to about 1.5 megabytes per second and less if some loss of picture quality can be tolerated using MPEG4. Even using TCP/IP for managing streamed video, at least 50 channels of MPEG4 video can be transmitted at high quality over single Fast Ethernet network. With SONET, compressed video of high quality can be transmitted over one or two T1 links.

5.3.5 The National Transportation Communications for ITS Protocol

The NTCIP is an ongoing project resulting from a collaboration of the user community, represented by the Institute of Transportation Engineers (ITE) and the American Association of State Highway and Transportation Officials (AASHTO), and the National Electrical Manufacturers Association (NEMA). The general trend in systems toward more open architecture has generated the motivation to tackle the problem of noninterchangeability within systems. The NTCIP effort originally sought to resolve this problem. Over the course of the work, however, the NTCIP framers realized the need to expand the scope to cover different kinds of devices that must coexist on and share communications networks, and they expanded the scope of the NTCIP to include most communications links between control center and roadside systems components of ITS.

The NTCIP seeks to provide two fundamental capabilities: interchangeability and interoperability. *Interchangeability* has been defined as the ability to replace a device from one manufacturer with a similar device from another manufacturer without affecting the behavior of the system. *Interoperability* refers to the ability for dissimilar devices to share a single communications channel.

Interoperability.   In order to allow a variety of systems and devices to share the same communications channel, each message transmitted on that channel must be formatted in a way that the software in all the systems and devices understands. Achieving this understanding requires that messages be constructed in consistent ways with consistent protocol management information. It is not necessary for a traffic signal controller to understand the meaning of a message intended for a dynamic message sign, but it must understand the format of the message sufficiently to know that it *is* intended for a dynamic message sign.

Systems for traffic management are designed around different system architectures depending on the needs of the system. Thus NTCIP provides many options to allow efficient application in varied circumstances.

Each possible configuration requires specific protocols used at each layer of the multilayer protocol. This list is called a *profile*. Figure 5.1 shows the NTCIP Standards Framework. The profiles represent the paths that ascend and descend through this diagram.

Many of the protocols identified in the figure are beyond the scope of this book to discuss, and interested readers should learn about them in the current version of the *NTCIP Guide*, which is available for download at www.ntcip.org. The levels in the NTCIP framework approximately correspond to the layers in the ISO-OSI reference model. Most of these protocols are taken directly from the Internet, with the result that NTCIP messages with the appropriate profile can be communicated across the Internet.

For example, a simple freeway management system might include traffic sensors, dynamic message signs, and a control system with a simple single-channel multidrop on agency-owned copper wire connecting the system to the devices using serial communications. In this case, no routing is needed, and all messages for these devices will be small enough (<512 bytes) to be contained in a single packet. The profile, therefore, needs a protocol at the information, application, and subnetwork levels only. The protocols in the transport level are not needed by this simple network.

A complex system will need a different solution. For example, a freeway management system might communicate with field devices using a leased IP-cloud network, in which case the network between the system and the field devices is highly complex and largely unknown to the agency. Each message will need routing and end-to-end error checking and will require the User Datagram Protocol (UDP, for error checking) and the IP (for routing) contained in the transport level of the framework, in addition to the layers needed by the simple system. A message with UDP and IP information in it can even traverse the Internet.

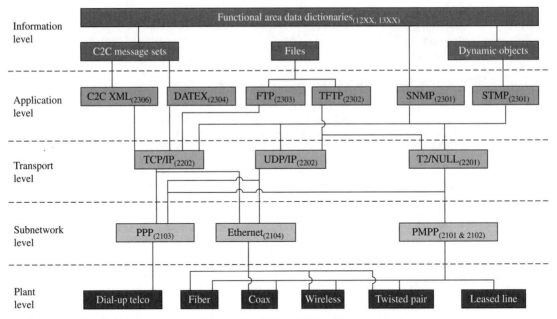

FIGURE 5.1   NTCIP standards framework (numbers refer to specific standards; see www.ntcip.org).

In addition to managing the network, interoperability requires that messages make use a standard way of defining themselves in terms of content. I am not talking about the content itself but rather how the content is identified. The Small Network Management Protocol (SNMP) in the application level of the NTCIP framework defines how message types are identified. Each of the messages supported by a particular device is stored in a management information base (MIB) that lists all the specific information needed by the protocol software to interpret the message. The SNMP software incorporates a device's MIB, and then uses that information to format messages as required by the system and device software. Each message object used by all devices in the transportation industry are defined within the standard, although manufacturers are allowed to create their own objects when the standard objects do not provide a specific feature they wish to support.

The content of each object determines the data it contains and how that data are defined. Thus the profile information for interoperability defines *syntax*, whereas the object definitions in the MIB define *semantics*. Standardized semantics are required for interchangeability.

Interchangeability.   Interchangeability allows similar devices from different makers to be interchanged on the network without any differences in the resulting operation. Interchangeability is more difficult than interoperability because it defines the functionality of the system. Interchangeability requires the common definition of message object content.

The status of the ongoing standards-development effort can be monitored on the Web at www.ntcip.org. This site includes the downloadable *NTCIP Guide*, which provides the best available fully detailed explanation of NTCIP and how it is applied.

Implementing NTCIP.   For each system, the designer has a series of determinations to make. The first is to determine the required profile, or what software protocols on the standards framework will be used in that system. The more protocols used, the bigger each packet will be because of the additional data added by each protocol. Consequently, the designer should use discretion in requiring protocols and use only those which will be needed.

The designer also must make sure that the data-flow requirements of the system, including the overhead added by the NTCIP, can be accommodated by the speed and capacity of the network being designed. Systems using complex protocols take necessary advantage of the much greater speeds available in IP-based networks compared with older serial communications links.

Once the profile is selected and an appropriate network is designed, then the designer must determine which objects in the standard MIBs will be needed. Some of the standards contain a table known as the *protocol requirements list* (PRL). The PRL identifies what users may need the device that is the subject of the standard to do. By selecting those needs, the PRL identifies a range of requirements that must be fulfilled. Those requirements then trace to *dialogs*, which identify sequences of communications processes and messages for fulfilling that requirement, and *objects*, as defined in the MIB. The connection between requirements, dialogs, and objects will be contained in a table called the *requirements traceability matrix*. Not all NTCIP standards provide these features, and many require evaluating the MIB directly to determine which objects and groups of objects must be included to provide the needed functionality for that system. Most of the objects in the MIB are optional, and they are grouped into mandatory and optional *conformance groups* that categorize the objects for a given device. All optional objects and conformance groups that are needed by the system must be specified during design of the system, either directly or through use of the PRL, if provided.

REFERENCES

The NTCIP Guide. NTCIP Joint Committee. Available at www.ntcip.org.

The Systems Engineering Guidebook for ITS. Web site sponsored jointly by Federal Highway Administration and the California Department of Transportation. Available at http://www.fhwa.dot.gov/cadiv/segb/.

Intelligent Transportation Systems. Web site by the Research and Innovative Technology Administration, Department of Transportation. Available at http://www.its.dot.gov/.

Traffic Control Systems Handbook. Federal Highway Administration, 2005. Available at http://ops.fhwa.dot.gov/publications/fhwahop06006/index.htm.

CHAPTER 6
TRAFFIC SIGNALS

Richard W. Denney, Jr.
Federal Highway Administration
Lorehsville, Virginia

6.1 INTRODUCTION

6.1.1 Background

At the intersection of two streets with no traffic control, drivers are expected to decide for themselves who is to go and who is to wait. Their only guidelines are the rules of the road, as defined in the *Uniform Vehicle Code* and the laws of the state, and their courtesy and cooperation. For intersections where the traffic demand is very light and the conditions favorable, drivers do not have trouble assigning their own right-of-way in this manner, and their delay is very small. As soon as traffic begins to build, drivers are faced with too many decisions to make in order to assign their own right-of-way safely, and some help from the governing body is required to assist them with these decisions. The simplest form of traffic control is a yield sign on one of the streets to let drivers know which movement has the right-of-way when a decision must be made. This, too, is only suitable for low traffic volumes. As traffic increases, drivers must be stopped so that they have time to make complex right-of-way decisions. A two-way stop, with stop signs on the minor street, serves this purpose.

Eventually, the traffic will build to a point where assignment of right-of-way is not the only problem, and moving the traffic efficiently also becomes a primary concern. Stop signs are effective in the former task, but they force every driver to stop and start, which is inefficient in that a large proportion of the time is spent just waiting for the drivers to get going. At this point, traffic signals can assign right-of-way much more efficiently because the delay associated with starting up is felt only by the first two or three cars in the *platoon*.

The ideal isolated traffic signal assigns just enough green time to clear the queue of traffic and then moves on to the next movement. When other signalized intersections are nearby, traffic engineers also try to arrange the timing so that the light will be green for platoons coming from the other signals when they arrive. These two goals sometimes compete. Current control equipment can provide a wide array of features, loosely organized into *fixed-time* and *traffic-actuated* capabilities. The former are usually used for calculating signal timings, especially in systems where those timings have to be consistent to maintain network flow, and actuated settings are used to allow the local controller to make adjustments based on the demand actually present at the intersection.

Before the actual discussion begins, the reader should be aware that although some very careful and scientific thought has gone into some of the principles, the traffic signal industry is very young, and much more is not known. Most experienced traffic professionals will happily depart from these methods when they do not seem to work on the street. But this freedom implies a warning:

Do not break the rules before learning them.

The reader also should be warned of the temptation to depend on computer programs without understanding their underlying methodologies. This strong temptation can be a downfall when traffic engineers are called on by the public and their representatives to explain their actions.

6.1.2 Definitions

Because of the complexity of traffic signals, a new jargon has evolved to help signal professionals communicate efficiently. These definitions are intended to make clear exposition in this chapter possible and thus are sometimes more precise or consistent than their usage within the industry might suggest. They are presented in logical order.

Traffic signal: Any power-operated device for warning or controlling traffic, except flashers, signs, and markings.

Legal Authority: Code refers to the *Uniform Vehicle Code*,[1] which is the basis for the motor vehicles laws in most states. It is assumed that the *Uniform Vehicle Code* is the controlling legal authority in this presentation. *MUTCD* refers to the *Manual on Uniform Traffic Control Devices*, as published by the Federal Highway Administration. Some states have their own versions of each of these documents. The federal *MUTCD* is available online at http://mutcd.fhwa.dot.gov.[2]

Traffic movement: A flow of vehicles or pedestrians executing a particular movement.

Approach: The roadway section adjacent to an intersection that allows cars access to the intersection. An approach may serve several movements.

Major street: The street in the intersection that has greater importance or priority, usually (but not always) because it carries the greater traffic volume.

Minor Street: The approaches that have the lesser importance, usually because they carry less traffic.

Right-of-way: The authority for a particular vehicle to complete its maneuver through an intersection.

Traffic signal system: A network of traffic signals coordinated to move traffic systematically. This system may be controlled at the individual intersections or by some central device or system of devices.

Signal indication: A particular message or symbol that is displayed by a traffic signal, such as a green ball or a WALK.

Green ball indication: Allows drivers to enter an intersection and go straight or turn right once it is safe to do so. If no left turn signal is present and no sign prohibits them, left turns may be executed in appropriate gaps in the opposing traffic.

Green arrow indication: Allows drivers to complete the movement indicated by the arrow without conflict. Right turn arrows may conflict with a WALK indication, although this practice may cause problems at high pedestrian locations.

Yellow indication: Drivers are notified that right-of-way is ceasing, and vehicles should stop, if practical.

Flashing yellow indication: Drivers should proceed through the intersection with more than usual caution.

Red indication: All vehicles must stop. Some turning movements are allowed after stopping, as provided in the *Code*.

Red arrow indication: A red signal applying to a particular movement.

Flashing red indication: All traffic must stop and then proceed when safe to do so. The meaning is the same as a stop sign.

WALK indication: Pedestrians may proceed to cross the intersection without conflict except from right-turning vehicles.

Flashing DONT WALK indication: A pedestrian clearance interval. Pedestrians in the crosswalk should complete their crossing, but no pedestrians should enter the crosswalk.

Steady DONT WALK indication: Pedestrians should wait at the curb; conflicting traffic has the right-of-way.

Signal section: A housing that contains a single indication.

Signal head: An assembly containing all the sections for a particular traffic movement or movements, using an arrangement as allowed in the *MUTCD*.

The following terms are purposely defined narrowly for clarity. The terms *phase* and *overlap* regularly cause confusion in traffic signal discussions, and these precise definitions help to keep things sorted out.

Traffic phase: The official definition is a particular traffic movement or combination of nonconflicting movements that receives a green indication simultaneously. But the term is based on the operation of the signal controller, and this definition obscures that fact. A signal head or group of signal heads is powered by a *load switch* (see Figure 6.2). The load switch provides the power necessary to illuminate the indications in each head, based on a command from the signal controller. A *phase,* therefore, is the logic in a signal controller that operates a load switch. Thus there is a direct connection between the phase, the load switch, the signal heads powered by that load switch, and the individual traffic movement served by those signal heads. The confusion comes when considering movements in plural. When multiple movements are served by signal heads wired to the same load switch, they are part of the same phase. When they are served by signal heads wired to separate load switches, they are served by separate phases, even if those phases start and stop at the same time.

A number of coding and numbering schemes have been proposed over the years, with no hope of a consistent standard, until the National Electrical Manufacturers Association (NEMA) developed a specification for a traffic-actuated controller. In that standard, a numbering system was used that now has become a de facto standard. The NEMA numbering method for movement in an intersection is as shown in Figure 6.1.

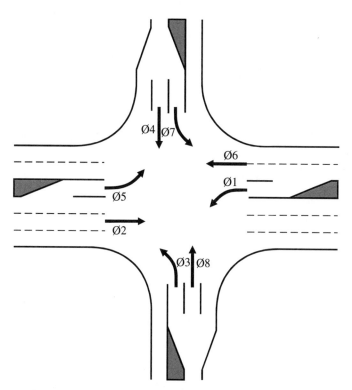

FIGURE 6.1   NEMA phase numbering scheme.

Sometimes, when no protected left turn movement is provided, the left-turners will be included in the straight-through phase number. For example, an intersection of two one-way streets may only use phases 2 and 4. A recent convention is to orient phase 2 in the primary direction on the major street. See section 6.3.8 for a discussion on how phasing works in a signal controller. There is no requirement for the phase numbering in the figure to run clockwise or for any particular phase to point in any particular direction as long as the resulting assignment is consistent with the rules underlying actuated controller operation.

Phase sequence: The order in which traffic phases are presented to drivers. The following terms apply to specific phase sequences:

Leading left: A protected left turn phase that displays green immediately before the opposing through movement is served. For example, if phase 5 comes up before phase 6, phase 5 is a leading left.

Lagging left: A protected left turn phase that displays green immediately following the opposing through movement is served. This may be accomplished by assigning the left turn to phase 6, for example, and the through movement to phase 5 or by declaring in the controller settings that the phase order serves phase 6 before phase 5.

Dual leading left: When the two opposing left turns are leading.

Dual lagging left: When the two opposing left turns are lagging.

Lead-lag lefts: When the left turn for one approach is leading, and the left turn for the opposing approach is lagging. For example, phases 2 + 5 (leading left), followed by phases 2 + 6 (through movements), followed by phases 1 + 6 (lagging left). The phase sequence 1 + 6, 2 + 6, 2 + 5 sometimes will be called *laglead*.

Split phase: When all the movements in one direction are served together, followed by all the movements in the opposite direction. With split phasing, the left turns and through movements in each direction are served simultaneously and often using the same load switch and phase. The opposing direction must be served by a different phase that is prohibited from operating at the same time to avoid conflicts. For example, split phasing on a side street is often operated by wiring all the signals in one direction to phase 3 and all the signals in the opposite direction to phase 4. Phases 3 and 4 cannot be served at the same time (see section 6.3.8 for a discussion of the rules that prevent this). Alternatively, they may be served by phases 4 and 8 in accordance with Figure 6.1 as long as those phases are set to operate exclusively in the controller.

Permitted left turns: When a green ball is displayed to left-turners, they may turn when it is safe to do so. If no left turn arrow is provided, the left turn is *permitted only*. If a green arrow is shown to the left-turners for part of the cycle and a green ball part of the time, then the operation is *protected-permitted*. When only a green arrow is presented to the left turners, or when so designated by a sign, then the operation is *protected only*. A recent development is to use a flashing yellow arrow to the left to denote the permitted portion of protected-permitted operation, with the green arrow used to denote the protected portion.

Overlap phase: A special feature of actuated controllers. An overlap phase is a special phase that displays green at the same time as two or more regular traffic phases. As with all phases, an overlap phase controls a separate load switch and the signal heads powered by that load switch. For example, an intersection may allow a movement during two other traffic phases (which cannot be run together). The overlap phase displays green at the same time as either of these two *parent phases* and stays green as long as either one of those phases is green. Do not confuse this hardware feature with the term *operational overlap*.

Operational overlap: Any interval when two normally distinct phases are green at the same time. One example is in the phase sequence 2 + 5, 2 + 6, 1 + 6. The 2 + 6 interval is called an *operational overlap*. Thus split phasing is lead-lag phasing with no operational overlap. The use of *overlap* in this way is declining in favor of using it only to describe overlap phases.

Cycle length: The time required to service all the phases at an intersection. This time is measured from the start of green of a particular phase to the start of green of that same phase again.

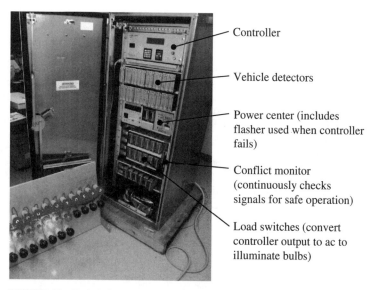

Controller

Vehicle detectors

Power center (includes flasher used when controller fails)

Conflict monitor (continuously checks signals for safe operation)

Load switches (convert controller output to ac to illuminate bulbs)

FIGURE 6.2   Typical signal controller cabinet being tested before installation.

Green split: Many times just called a *split.* The fraction of the cycle that is allotted to any one phase. This time may vary based on traffic actuation or be fixed or constrained by coordination timings. Depending on the controller and system software, the units may be either seconds or a percentage of the cycle.

Traffic signal controller: The device that directs the logic of signal timing. The traffic signal controller, or just *controller*, directs the phase order and interval length. The traffic signal controller is housed within a cabinet, as shown in Figure 6.2.

Master controller: A field device that manages the coordination of a small number of intersections and in some cases breaks communications between signal controllers and a signal system.

Local controller: The controllers in a signal system that receive coordination management either from a master controller or from a signal system or both.

Detectors: Equipment that detects the presence of traffic. A number of equipment and hardware technologies exist for this purpose.

These definitions are not intended to be comprehensive but rather to clear up the most commonly confused terms. Other terms will be described in the discussions that follow.

6.1.3 Advantages of Traffic Signals

Traffic signals are installed because they efficiently assign right-of-way to drivers approaching an intersection with crossing movements. Thus signals have the following advantages over other traffic control schemes. These advantages can be negated if the traffic signals are installed when not needed.

- They provide for orderly traffic movement.
- They allow higher capacity than other methods.
- They can promote systematic traffic flow when interconnected into a system.
- They can reduce the frequency of right-angle collisions (if such collisions are a problem).

6.1.4 Disadvantages of Traffic Signals

Traffic signals should never be considered a panacea for all intersection problems. This misconception is common, especially among members of the public, but hard to explain, given the large percentage of network delay that is caused by signals. In many urban areas, the majority of requests for signals concern locations where a lightly traveled minor street intersects a high-volume arterial highway. Because of volumes and speeds, the side-street drivers feel victimized, and their requests usually get political support. The political hierarchy also will complain about poor traffic flow on the major street and never realize the paradox they present. Traffic signals work because they occasionally turn red and delay the major street traffic. This interruption in major street flow is inherently detrimental; if it were not so, the signal would not be needed because adequate gaps already would exist. Traffic signals cannot add capacity to a free-flowing roadway; they can only reduce it.

Traffic signals do not always promote safety. The number of rear-end and sideswipe collisions actually may increase, particularly on high-speed roadways.

The important concept is that signal installation is a tradeoff between advantages and disadvantages. This tradeoff must be evaluated, even if the decision will be made in the political arena, because the consequences need to be communicated to the decision makers. The *MUTCD* requires that traffic signals be recommended by an engineering study before being installed.

6.1.5 Classification of Signals

Traffic signals include a large variety of traffic control devices, including pedestrian crosswalks, emergency signals to serve occasional use by emergency vehicles, lane-use signals for traffic management, freeway ramp meters, railroad crossings, and so on. This chapter will be limited to signals at roadway intersections for use by regular traffic and pedestrians.

6.1.6 When to Install a Traffic Signal

Generally, two levels of criteria are evaluated before an engineering study can recommend signal operation. The first level of evaluation is described by the minimum warranting conditions in the *MUTCD*, which are intended to suggest when a traffic signal is *not* appropriate. The *MUTCD* warrants from a threshold, not a conclusion, and are intended to establish a consistent and uniform floor below which signals will not be used.

Second and more important, an engineering analysis that analyzes the tradeoffs involved in signal installation must recommend a signal. The signal must solve more problems than it causes.

Both these levels of criteria are required by the *MUTCD*. The manual defines a traffic engineering study that includes the following data collection requirements:

- Sixteen-hour volume count on each approach in 15-minute intervals
- Classification study to determine the relative quantities of trucks and cars
- Pedestrian volume, where applicable
- Speed study on each approach
- Condition diagram to show the physical layout and features
- Accident history for the previous two years
- Operational data, such as delay, as necessary

Not all these data are collected at every location because agencies do not usually have the resources to go into this much detail. Any time any one of these factors is significant, however, some field data should be collected. By *significance*, the reader should understand that the critical factor is the one that motivated the request for the signal. If the request for the signal is based on perceived danger, crash statistics should be researched. If the request is motivated by excessive side-street delay, then that delay should be measured.

It is critically important to understand, however, that citizens who request a signal do not always explain their reasons in good faith. Most citizens requesting a signal will cite safety concerns, although in many cases the intersection in question shows no history of crashes related to right-of-way assignment. Often they will cite delay issues, and a delay study will show that a signal will increase the delay. These study results often will not persuade them because they address the stated problem rather than the true problem. In many cases, the true problem is that motorists are uncomfortable making tight gap-choice decisions, and that discomfort has motivated the call. Drivers would prefer to wait longer at a signal if it means they can enter the signal without discomfort. Traffic engineers must separate a sense of discomfort from true risk because accommodating the desire for comfort for side-street motorists may well cause considerable discomfort on the part of main-street drivers whose voices are not being heard.

The *MUTCD* defines eight general conditions that suggest that further consideration should be given. These can be grouped into several categories:

- Volume warrants are used to consider whether the general volume reaches a certain threshold or whether the flow on the main street is high enough to likely present too few opportunities to enter the street. They are evaluated on the basis of the 8 highest hours of the day, the 4 highest hours, or the peak hour, with the thresholds increasing for each. Pedestrian volumes also can be the basis for consideration. It should be noted that pedestrian volume in this context relates to whether there are sufficient numbers of pedestrians to warrant a signal just to serve pedestrians and is unrelated to the decision to use pedestrian signal displays at signals that will be built for other reasons.

- Safety warrants include school crossings as well as intersections that display a specific and measurable crash history.

- Network and system warrants include those intersections which may not (yet) meet the volume warrants but which rise to a specific standard of importance in the network or where a traffic signal might help to promote signal coordination.

Readers seeking detailed explanations of the warrants should consult the *MUTCD* directly.

6.2 OPERATIONAL OBJECTIVES

The expectations of motorists as individuals and the needs of motorists as a group vary as traffic demand increases. In light traffic conditions, motorists expect to be served on the next green signal. Therefore, traffic signals are timed to minimize the likelihood of requiring a motorist to wait through two red indications. A *cycle failure* occurs when a motorist who arrived during the red is not served on the next green. Typical traffic-actuated signal control seeks a gap in the traffic stream indicating that the waiting queue has cleared fully, which is consistent with this objective. Thus most agencies use traffic-actuated operation during light traffic conditions unless signals are so closely spaced that coordination is required even for very light conditions (a downtown grid is an example).

As traffic demand increases, the objective of minimizing cycle failures becomes unattainable. Cycles become excessively long, but moderate demand will cause long queues to form during these long cycles, and even the long green will not be adequate to clear the queue. Thus, in moderate traffic conditions short of congestion, the objectives of control shift to minimizing delay, stops, fuel consumption, emissions, or some combination thereof. Or the objective may be to maximize smooth flow such that motorists can drive through the network with the least interruptions. Driving through successive green signals is called *progression*. The methods presented in this chapter are aimed at moderate traffic conditions.

Eventually, excessive demand renders all such objectives also unattainable. When a traffic signal can no longer serve the demand presented to it, some of the arriving queue will remain unserved at the end of the green. This residual queue may grow from cycle to cycle if the demand fundamentally exceeds the capacity of the green interval. A growing residual queue, especially after every attempt has been made to eliminate it by adjusting signal operation, indicates that the intersection has entered a congested regime. Neither delay minimization nor progression is a useful objective in congested networks. Delay cannot be evaluated on the basis of one cycle, as it can during moderate conditions, and therefore, attempts at optimizing to minimize delay become mathematically intractable. And a residual queue blocks any attempt at progression.

The initial objective in responding to a congested traffic signal is to relieve the congestion by maximizing throughput. No existing tools attempt this optimization directly, and traffic engineers must undertake a trial-and-error process and apply a range of techniques to maximize intersection efficiency. If maximizing throughput becomes unattainable, then the final objective is to provide signal timing that manages queue formation so that it does the least damage throughout the network. Signal timing in congested conditions is beyond the scope of this chapter, but more information can be found in recent guidance produced by the Federal Highway Administration.[3]

6.3 INTERSECTION SIGNAL TIMING

6.3.1 Fixed-Time Signal Phasing and Timing

Even though nearly all signal controllers now used are traffic-actuated, understanding fixed-time signal operation provides the basis for understanding actuated control in coordinated systems.

6.3.2 Phasing

The first question to be answered in determining fixed-time operation is the order of phases. The primary question is the use of left turn phases. It has been said that if left turns did not exist, signal timing engineers would be out of work. Thus the decision to provide protected movements for left turners will have an effect on all other operational decisions.

Left Turn Phase Warrants. Unfortunately, no widely accepted standards for when left turns should receive their own phase have emerged, and practitioners have resorted to various limited research findings, agency policies based on tradition, and rules of thumb. In the 1980s, research at the University of Texas at Austin presented a methodology for deciding when to use left turn phases, and this method is presented here to give the reader at least a starting point. Other methods exist, and interested readers should review the literature and consider the methods used by their agencies.

The UT-Austin method assumes that signal timing is already known or has been determined provisionally. The next step is to compare the left turn volumes and green times against the formulas presented in the following table:

If the number of opposing lanes is . . .	. . . and the opposing traffic volume times the percentage of opposing green $Q_o(C/G)$ is . . .	. . . then the left turn volume must exceed
One	Less than 1,000	$770(G/C) - 0.634Q_o$
	Between 1,000 and 1,350	$480(G/C) - 0.348Q_o$
Two	Less than 1,000	$855(G/C) - 0.500Q_o$
	Between 1,000 and 1,350	$680(G/C) - 0.353Q_o$
	Between 1,350 and 2,000	$390(G/C) - 0.167Q_o$
Three	Less than 1,000	$900(G/C) - 0.448Q_o$
	Between 1,000 and 1,350	$735(G/C) - 0.297Q_o$
	Between 1,350 and 2,400	$390(G/C) - 0.112Q_o$

The formulas in this table were developed for intersections where adequate left turn lanes exist. A practitioner will filter this and all numerical methods through the judgment filter. Numerical methods may imply unrealistic precision.

Phase Sequence.   Traditional phasing uses two opposing leading left turn phases. This approach is the common default operation at uncoordinated traffic signals as long as the opposing left turns do not conflict physically. In coordinated systems, the phase sequence may be altered to improve flow through a succession of signals, as described in the section on coordination.

One situation that always should be avoided is a lagging protected left turn that opposes a permitted left turn. Left-turners in the permitted direction see the through-movement signal change to red and mistakenly assume that this change is occurring to the opposing through movement as well. But the opposing through movement is still green, coupled with its adjacent protected left turn. Drivers, in making a permitted left turn, will assume that they can clear during the yellow interval and will be surprised by oncoming traffic that still has a green.

6.3.3 Daily Patterns

Any intersection will experience a variety of different traffic demand patterns in any given day. Most new controllers provide many more pattern possibilities than any agency will use. One way to get a handle on the problem is to plot total intersection volume and major street volume on the same graph for a 24-hour period. Characteristics of the plot that would indicate the need for a new pattern would be when the total volume changes substantially (say, by greater than 30 percent) or when the fraction of total volume on the major street changes significantly. Without a tremendous amount of analysis work, these decisions must be made based on judgment; no cookbook methods will suffice.

Most agencies, however, do not attempt to design patterns for individual intersections based on daily demand patterns. Rather, they design coordination patterns for networks of signals and rely on local intersection actuation to accommodate variations that do not affect network operation. Coordination patterns will be covered later in this chapter.

6.3.4 Critical Lane Volumes

Once phasing has been determined, the designer must review the traffic using each phase to determine the demand for that phase that drives the decision of how much green time to provide. Most new traffic signals are traffic-actuated, but designs for traffic signals must be tested to determine that they can be operated with the lane configuration and phase assignments that have been assumed. The critical lane volume method allows the designer to test a design, and if it fails, the solution is to devise a phasing approach that allows critical movements to run at the same time rather than sequentially (this improves efficiency) or to provide additional lanes for the movements that fail (which improves effectiveness).

Correction factors have been formulated for capacity analysis to convert raw traffic volume to equivalent passenger cars. Factors also have been used to convert turning traffic to through-traffic equivalents. Finally, a representation of a multilane approach in terms of a single lane makes it possible to compare movements with differing numbers of lanes. These conversions make it possible to add sequential movements together meaningfully as a means of determining the demand for green time. But some of the conversion factors are presented with precision far in excess of accuracy. Field experience is essential, along with a healthy respect for variability. There is no sense in making signal timing decisions based on 5 percent effects when traffic routinely will vary by 20 percent or more from one cycle to the next or from one day to the next.

Here are some typical values to use as a starting point:

- One truck = 1.5 through passenger cars.
- One right-turner = 1.4 through passenger cars (1.25 if the curb return radius exceeds 25 feet).
- One left-turner = 1.6 for nonprotected movement or left turns from a through lane or 1.1 for protected movements from a left turn bay.

Once these equivalencies can be quantified, the designer can adjust the traffic volumes arriving at the intersection to equivalent through passenger cars and then assign them to lanes. All possible sequences are added to determine which series of phases represents the highest total demand, and this is the *total critical lane volume*. Refer to the example in a later section to see how this is done.

6.3.5 Cycle Length

The next question to be answered is the cycle length to be used. In many cases, this value is already determined because the signal will be part of a coordinated system that has a cycle length already defined, and the requirements of coordination often supersede the requirements of individual intersections. Even at traffic-actuated controllers, however, the cycle should be calculated and the results built into the maximum allowable green times under actuation, as described in later sections.

As other timing parameters are developed, the cycle length may need to be adjusted, but a common mistake is to define all the other timing parameters and add them up to see what cycle length to use. This usually results in cycle lengths that are too long. Cycle length calculation is not standardized, and many different techniques are used by different practitioners and agencies. Among practitioners who know how to calculate cycle length manually, the two most popular and widely used methods are the Poisson method and Webster's method.

Poisson Method.   In this method, one uses the Poisson statistical distribution to guess at the near-maximum size of the arriving platoon based on average demand. The details of this method are beyond the scope of this book, but it is mentioned because it exemplifies a method that seeks to minimize cycle failures and is appropriate only for light conditions. Most agencies use traffic actuation to control timings in light conditions, but the Poisson method might be useful for determining the maximum green for each phase under actuated control at intersections (or at times of the day) that do not experience heavy demand.

Webster's Method.   In the 1960s, F. V. Webster, of England's Road Research Laboratory, took a different approach by observing actual operation. He observed the relationship shown in Figure 6.3.

For each situation, Webster found an optimal cycle length that gave the least delay. As cycle length increases, the average wait time increases geometrically such that doubling the cycle length results in quadrupling the delay when the intersection is running at capacity. When cycle lengths are too short, too much of the cycle is wasted in queue startup lost time, and the number of cycle failures is so high that the queues build faster than they can be dissipated. Therefore, the optimal cycle length in uncongested conditions is the shortest cycle length that will provide the needed capacity.

Webster's formula is

$$C = \frac{1.5L + 5}{1 - Y}$$

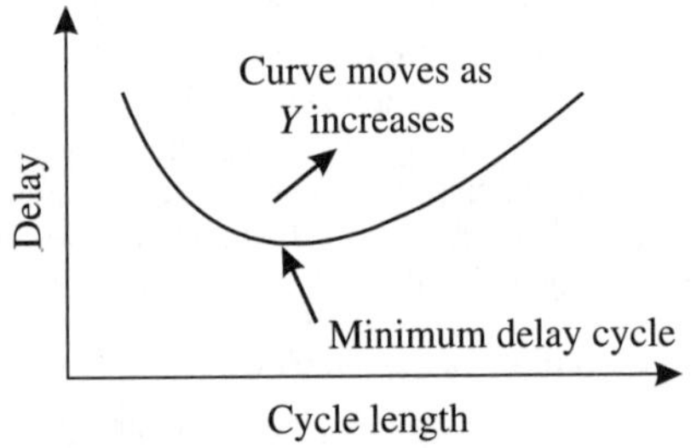

FIGURE 6.3 Webster's observed relationship between delay and cycle length.

where C = the optimal cycle length

L = total lost time per cycle (calculated by multiplying the number of critical phases by 4 seconds; the lost time is an effect of driver behavior and is only loosely related to clearance interval length)

Y = the total saturation flow, which is the sum of the critical phases divided by the maximum saturation flow

The maximum saturation flow is subject to some debate, with values ranging from below 1,800 to as much as 2,200 in aggressive urban conditions. This discussion will use 1,800 as the traditional value. This value is used only for cycle length calculation and should not be confused with total critical lane volume. Critical lane volume includes the effects of lost time during phase changes, whereas saturation flow describes only flow during the green.

Webster's cycle length goes to infinity as demand approaches saturation flow, and this is a flaw caused by the fact that Webster didn't study congested intersections.

Maximum cycle lengths are controversial in the traffic engineering community. The assumption is that longer cycles reduce the percentage of the cycle wasted in lost time and, therefore, provide more throughput. Recent research[3] has shown that excessively long cycles do not, in fact, increase throughput because the long queues are diluted by traffic that cannot reach turn lanes, and green time that serves long queues that lose efficiency because of the gaps left by those cars as they leave the through movement. The same effect occurs when a roadway adds a lane on the approach to the intersection—green times long enough to serve a queue in the narrow section will starve the intersection for demand, and throughput will decrease. Thus cycle lengths should be controlled to minimize the dilution of the traffic stream and the inefficiency of the departing queue. Most agencies develop a policy or standard practice, although many do not consider these effects and use cycle lengths that may be too long to meet their objective of maximizing throughput. I adopted a maximum cycle of 120 seconds except in rare special cases in the agencies for which I was responsible.

6.3.6 Splits and Minimum Phase Times

Green splits are calculated by dividing the cycle length in proportion to the critical lane volumes. Phases serving light traffic demand at otherwise heavily traveled intersections may require a disproportionately large portion of the cycle to avoid being too short. Signals should be long enough to provide the appearance of sane operation even if only one car is being served and also must provide adequate pedestrian crossing time.

Pedestrian Signal Timing. The *MUTCD* requires that pedestrians be given adequate time to cross an intersection, whether or not pedestrian signal indications are provided. This requirement does not have a pedestrian demand component, and even locations where no pedestrian traffic is expected are included. The principle conveyed in the *MUTCD* is that pedestrians can expect reasonable accommodation at every traffic signal unless pedestrian movement is specifically prohibited by the appropriate signaling.

When pedestrian signals are provided, two intervals are timed to accommodate pedestrians. The first is the WALK signal (or white walking person), which is used only to get waiting pedestrians started. The *MUTCD* suggests a minimum time for a WALK signal of 4 to 7 seconds, and different agencies will interpret this minimum differently, often with a stated policy that practitioners should follow. Pedestrian clearance intervals (flashing DONT WALK or orange hand indication) are calculated based on the length of the crosswalk. Sufficient time must be provided so that a pedestrian, walking at 3.5 feet per second, can cross the entire traveled width of the street before a conflicting signal is displayed. Some of this time can be provided during the clearance interval of the nonconflicting cars being served at the same time, even though many controllers will not use flashing DONT WALK during that period.

At intersections with a protective median of at least 6 feet in width, pedestrians can be provided sufficient time to cross only to and from the median. If this is the case, pedestrian signals (and pushbuttons, if actuated) must be provided in the median.

At locations without pedestrian signals, the green interval that serves pedestrians and adjacent cars still must be long enough to provide sufficient pedestrian clearance before conflicting traffic is released. Without pedestrian signals, the use of pushbuttons does not relieve the agency of this requirement because the pedestrian must know what to expect, using either pedestrian signals or sufficient green time.

Minimum Vehicular Movement Time.   Given that most intersections are controlled by actuated controllers, the minimum green time usually is controlled by the detection design. Most agencies have a policy for the minimum green time imposed by a coordination plan, and this usually ranges between 5 and 15 seconds to keep the intersection from looking as though it is malfunctioning even when only one car is being served.

Clearance Intervals.   Much discussion and research over the years have been devoted to clearance interval calculation. A traditional rule of thumb suggested 1 second for each 10 mi/hr of approach speed, although this approach no longer has a basis in research. An approach that does have a basis in research is to provide a constant 4-second clearance interval at all intersections, based on studies that have shown that drivers respond based on the assumption of that amount of clearance time no matter what is actually provided. By far the most common method, though, is the one that is based on providing sufficient time to cross the safe stopping distance. This method was developed by Robert Hermann of General Motors Labs in the 1960s and is now promulgated by the Institute of Transportation Engineers (ITE). Only the ITE method will be presented here.

The following equation derives from the physical laws of motion:

$$Clearance\,interval = t + \frac{v}{2d + gG} + \frac{l + w}{v}$$

where t = perception and reaction time of the driver, commonly considered 1 second at intersections, although some research suggests slightly longer periods

  v = approach speed, in feet per second (1.47 times speed in mi/hr)

  d = deceleration rate, in feet per second squared, normally taken to be 10 ft/sec^2

  g = acceleration due to gravity (32.2 ft/sec^2)

  G = grade of approach (in percent, positive for uphill approaches, negative for downhill)

  w = intersection crossing distance, in feet

  l = vehicle length, in feet, typically 20 feet

The first term is the time required for the driver to see the yellow signal and apply the brake, the second term is the time required to traverse the distance that would be required to stop, and the third term is the time required to drive through the intersection. The clearance interval includes yellow and red clearance (red clearance is the interval between the end of the yellow signal and the beginning of the conflicting green), although most methods (e.g., the publications of the ITE) suggest using the $(l + w)/v$ term for the red clearance interval.

6.3.7   Detailed Example of Phasing and Timing: Northwest Boulevard at Woody Hill Drive

Let's work through one typical example in detail. The example is a high-volume arterial street intersection with a relatively lightly traveled residential collector in the morning rush hour. It is perhaps a bit too realistic: The intersection is congested, and many of the usual techniques fail. But it presents a scenario that is seen all too commonly by practitioners.

Woody Hill Drive is a low- to moderate-volume residential collector street that feeds a neighborhood on one side of Northwest Boulevard and a shopping center and apartment community on the other side. Northwest Boulevard is the major arterial street connecting the city and its northwest suburbs. This roadway carries about 90,000 vehicles a day at this intersection. The current traffic demand is substantially higher than can be accommodated by the intersection geometry, and the resulting operation is highly congested during peak periods.

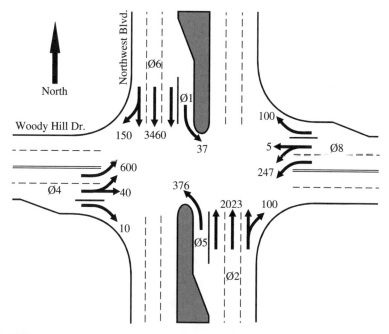

FIGURE 6.4   Example intersection layout.

This example concerns the morning peak period. During this period, this intersection creates a queue of over 1 mile, with typical delays of 20 to 25 minutes. This intersection is a good example of congested intersection timing. Base turning movement counts (in vehicles per hour) and layout for this intersection are shown below in Figure 6.4.

Phasing.   Let's assume in this case that the decision to use protected left turn north and south has already been made. A close look at the intersection will reveal that the width of the median opening is not enough to allow opposing two-lane left turns off of Woody Hill Drive to run simultaneously, and therefore, either lead-lag or split phasing must be used. The left turn volume on the side street is greater than the through volume, which is common with intersections such as this one, and split phasing, therefore, is the only remaining choice. Split phasing is preferred only when the left turn demand exceeds the through demand, especially when using split phasing allows the left turns to be served from more than one lane, as is the case here.

Adjusting Volumes.   The first step is to determine the truck correction factor. There are 10 percent trucks, and one truck is equal to 1.5 cars, so the correction factor looks like this:

$$(1 - 0.1) + (0.1)(1.5 \text{ cars/truck}) = 1.05$$

This value can be multiplied by the raw count to correct it to equivalent passenger cars. For movements served by the same phase, each volume is multiplied by the turning movement correction as follows:

Southbound Northwest Boulevard:

Right and through:

$$[(150)(1.25 \text{ right turn factor}) + 3,460](1.05) = 3,830 \text{ passenger car equivalents} \\ \text{per hour (pceph)}$$

Left turn:

$$(37)(1.1 \text{ left turn factor})(1.05) = 43 \text{ pceph}$$

Eastbound Woody Hill Drive:

Right turns have an exclusive lane, so its volume is adjusted separately:

$$(10)(1.25 \text{ right turn factor})(1.05) = 14 \text{ pceph}$$

Through volumes are much less than the average left turns per lane, so the exclusive through lane does not overflow, and the lane will contain both through and left turn traffic:

$$[40 + (600)(1.1)](1.05) = 735 \text{ pceph}$$

Northbound Northwest Boulevard:

Right and through:

$$[(100)(1.25) + 2{,}023](1.05) = 2{,}255 \text{ pceph}$$

Left turn:

$$(376)(1.1)(1.05) = 435 \text{ pceph}$$

Westbound Woody Hill Drive:

Right Turns:

$$(100)(1.25)(1.05) = 131 \text{ pceph}$$

Through volumes are much less than the average lane volume for the through and left lanes, and the exclusive through lane does not overflow.

$$[5 + (247)(1.1)](1.05) = 291 \text{ pceph}$$

We now must determine the average lane volume for the movements served by each phase. To do this, we just divide the total volume for that movement by the number of lanes. Sometimes we represent the heaviest lane on an approach by using a value slightly higher than the value obtained by dividing by the number of lanes.

When more than one movement uses a single phase (as is the case with split phasing, where the left turn and adjacent through movements are served by the same phase), the higher of the two average lane volumes is the critical lane volume for that phase.

Northwest Boulevard movements:

Movement served by phase 1:43/1 lane = 43

Movement served by phase 2:2255/3 lanes = 752

Movement served by phase 5:435/1 lane = 435

Movement served by phase 6:3830/3 lanes = 1,277

Woody Hill Drive movements (phases 4 and 8 are set in the signal controller to be exclusive, in order to provide split phasing):

Movement served by phase 4:735/2 lanes = 368

Movement served by phase 8:

Through and left: 291/2 lanes = 146

Right: 131/1 lane = 131

Because 146 > 131, 146 is the critical lane volume, and the through and left turn critical lane volume controls the green time requirements for these movements.

The critical movement combinations now must be evaluated.

Because the left turns from Northwest Boulevard can run simultaneously with each other or with nonconflicting through movements, their critical lane volumes must be combined. Of the four main street phases, only two combinations result in conflict. Movements served by phase 1 and phase 2 cannot run together without conflict, nor can movements served by phases 5 and 6. Consequently, the sequence 1–2 or 5–6 will be the critical sequence of movements on the main street. The critical lane volumes for each movement are added for these sequences to see which one must carry the higher traffic flow:

Phase 1 movement + phase 2 movement = 43 + 752 = 795 pceph per lane

Phase 5 movement + phase 6 movement = 435 + 1,277 = 1,712 pceph per lane

Because 1,712 > 795, the movements served by phases 5 and 6 are the critical movements for the main street. Because the minor street is using split phasing, the movements in opposite directions must run exclusively and, therefore, are both critical.

Cycle Length. We will use Webster's method for calculating the cycle length to demonstrate why it is not appropriate for congested conditions. First, we must determine the total critical lane volume:

Phases 5 + 6 + 4 + 8 movements = 435 + 1,277 + 146 + 368 = 2,226 pceph per lane

Given that the capacity of an intersection in terms of total critical lane volume is usually in the range of 1,300 to 1,500, we already know we have a problem at this intersection.

We can then use Webster's equation:

$$C = \frac{1.5L + 5}{1 - Y}$$

L is the total lost time, which is about 4 seconds for each separate phase serving a critical movement. There are four critical movements served by separate phases in the cycle at this intersection, so the total lost time is 16 seconds.

$$C = \frac{1.5(16) + 5}{1 - 2,226/1,900}$$
$$= -169$$

This bizarre result will always occur when the intersection is more heavily loaded than its capacity. The value of 1,900 is the *saturation flow*. As the total critical lane volume approaches this value, the resulting C increases to infinity. When 1,900 is exceeded, C becomes negative. Now we have seen the upper boundary of usefulness for Webster's method. Because the intersection is seriously congested, any cycle length will result in congestion, and the maximum cycle length allowed by policy should be used. In this case, we will assume that this maximum is 120 seconds. This situation is all too common at many locations in most cities, and it illustrates that traffic signals cannot add capacity; they can only distribute what capacity is provided by the available lanes. Traffic engineers are faced with situations repeatedly where they must rely on experience and judgment because reasonable design methods are overwhelmed.

Splits. Once the cycle length is determined, the green times for each phase are prorated based on the ratio of the critical lane volume for the movement served by that phase to the total critical lane volume. Clearance intervals are subtracted first because they are the same for all phases. To avoid making a complex example any more complex, we will assume that clearance times are set by policy in this jurisdiction at 4 seconds of yellow plus 1 second of red clearance.

Total available green time = 120 − 4 phases (5 seconds of clearance per phase) = 100 seconds

Phase 5 = (435/2,226) (100 seconds) = 20 seconds of green

Phase 6 = (1,277/2,226) (100 seconds) = 56 seconds

Phase 4 = (368/2,226) (100 seconds) = 17 seconds

Phase 8 = (146/2,226) (100 seconds) = <u>7 seconds</u>

Total   100 seconds

Minimum Intervals. These phase times must be checked to ensure that minimum green times are accommodated. Left turn minimum greens are assumed by policy to be 5 seconds, with 7-second minimums for through movements. These short minimums reflect timing policies in areas where the intersections are dramatically congested. All green times meet or exceed these values.

We also must check to make sure that we have acceptable pedestrian crossing times. Here we may have a problem. While the main-street times are more than sufficient to cross the narrow side street, the side-street times may not be adequate to cross the wide main street. There is a median wide enough to provide pedestrian refuge, although no pedestrian signals are installed. The crossing distance for pedestrians moving with phase 4 is 36 feet to the median and 48 feet from the median to the far side. At 3.5 feet per second, a minimum green time of 14 seconds is required, and the 17-second green time is sufficient. For phase 8, however, things are tight. Pedestrians moving with phase 8 also have a 48-foot crossing from the median to the far side, requiring 14 seconds of pedestrian clearance. The green time of 7 seconds is insufficient, even when including the clearance interval. Because of the congested nature of this intersection, and because of the relatively small number of observed pedestrians, the agency solves this problem by prohibiting pedestrian movement with appropriate signs. If this is not an option, then the longer green should be subtracted from the available green time, with the remaining movements apportioning what is left. To use the median for refuge, however, pedestrian signals must be installed there if used at the intersection.

Noncritical Phases. Now that we know the critical phase green times, we can proportion the time to the noncritical phases as follows:

Main street green time = phase 5 + phase 6 = 20 + 56 = 76 seconds

Total noncritical lane volumes for main street:

Phase 1 movement + phase 2 movement = 43 vehicles per hour per lane (vphpl)
+ 752 vphpl = 795 vphpl

The green times are determined as follows:

Phase 1 = (43/795)(76) = 4 seconds

Phase 2 = (752/795)(76) = <u>72 seconds</u>

Total   76 seconds

Check minimum green times: Phase 1 is 4 seconds, which is less than the minimum 5 seconds, so set phase 1 at the minimum and assign the remainder to phase 2.

The final interval timings, using the NEMA scheme for phase numbering, is as follows:

Phase			Length
1	+	5	5 seconds
1Y	+	5	4
1R	+	5	1
2	+	5	10
2	+	5Y	4
2	+	5R	1
2	+	6	56
2Y	+	6Y	4
2R	+	6R	1
4			17
4Y			4
4R			1
		8	7
		8Y	4
		8R	1
Total			120

This format explicitly spells out the entire operation at a glance and needs only to be accompanied by a diagram showing how phases are assigned to movements to fully describe the operation. Total phase length for any phase is obtained easily by adding the interval lengths for all intervals showing that phase number.

6.3.8 Traffic-Actuated Signals

Traffic-actuated signals vary their green time based on demand at the intersection, as measured on detectors installed on the approach. These detectors vary in technology, but the most common is the inductive loop detector. A loop of wire is installed in the street and is charged with a slight electric current, which creates an electrical field. As cars pass through the field, the inductance of the field changes, indicating the presence of a car. Other common technologies include video image detection and microwave radar detection. Detectors serve two purposes: They *call* phases to be served (when those phases are red), and they *extend* phases until the waiting cars are served (when those phases are green).

Actuated traffic controllers use the same numbering scheme used in the preceding section, and in reviewing the final signal timings of the example, we can see a pattern of operation emerge. This pattern is called *ring operation* and is the basis for intersection control in the United States. Figure 6.5 lays out the terms and the rules.

The section of the figure labeled "01" shows that all the phases to the left of the barrier serve main-street movements and phases to the right of the barrier serve minor-street movements. Also, any main-street phase in ring *A* can run simultaneously with any main-street phase in ring *B*, which is why the line that separates them is called the *compatibility line*. Ring rules state that phases in a ring are served sequentially and never simultaneously and that both rings must cross the barrier at the same time.

Within this context, each phase can either be served or skipped. The decision to skip a phase occurs when there is no car on the detector when the previous phase turns yellow. The previous phase will only turn yellow if there is demand on a competing phase, but that phase may not be the phase that immediately follows in the ring, which therefore, may be skipped. If a phase is served, it

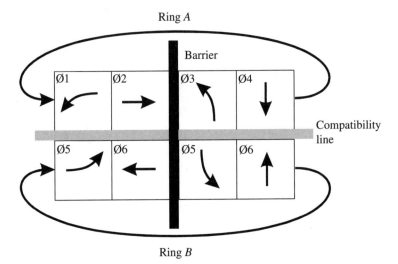

FIGURE 6.5   NEMA ring operation.

is served for a minimum period called the *minimum green* or *initial*. After the initial, the phase will rest in green until a car passes over a detector on a competing phase. At that point, the phase can be terminated by one of three processes.

- *Gap-out*. As soon as a competing phase has been called, the phase showing green will start a timer, called the *extension timer*, that counts down from the extension value to zero. The extension timer serves to extend the green for a short period to keep the signal green while tightly bunched cars are present. The extension is set to be long enough to allow closely spaced cars to reach the detector and reset the extension timer before it reaches zero. Once the gap in the traffic exceeds the extension time, the timer will reach zero, and the phase will turn yellow.

- *Max-out*. If traffic is heavy, a sufficient gap to allow gap-out may not occur in a reasonable period. Thus actuated control provides a maximum time to prevent excessive cycles lengths. When this time is reached, the phase will turn yellow. Some controllers won't start the max timer until there is competing demand, and others will time the maximum time and let the controller dwell in green, terminating immediately on a competing detection. The former method is more common. Maximum times are often set arbitrarily, but they should be set with some sensitivity to the demand. One useful approach for moderate traffic is to calculate the fixed-time operation as outlined in the preceding section and set the maximum greens to values perhaps 20 percent higher than that to allow some flexibility.

- *Force-off*. In coordinated signal systems, the network signal timing needs to impose discipline on the actuation process to keep the intersection in coordination with other intersections. Actuated controllers contain a cycle timer that starts with the end of the main-street green. When there are competing calls, the main-street green phase is held green until the time specified by the system, and then the controller can cycle through the other phases. Each of these phases may gap-out or max-out, but if they are extended beyond the time allowed by the system timings, they will be forced off. This force-off value is a setting in the controller that defines the point in the cycle timer at which this phase will be forced off. The relationship of this intersection with others is controlled by an *offset*, which is the time between the starting point of the cycle timer at this intersection and the starting point of an arbitrary system-wide cycle timer. All controllers calculate the master cycle timer in the same way, or it is provided by a signal system, either of which keeps them in coordination.

A whole chapter or indeed an entire book could be devoted to the individual settings and modifications that are possible to these basic timings, but they are beyond the scope of this book. Suffice to say that none of these values is constant. The minimum green (or initial), the extension, the maximum, and the force-off all can be modified by special features and processes based on demand conditions or design. In particular, the relationship of these values with the locations and dimensions of the detectors merit special study for those who are interested in deeper study.

6.4   TRAFFIC SIGNAL SYSTEMS

When traffic signals are close enough so that the platoons formed by one signal are still tightly formed at the next signal, traffic engineers can coordinate the intersections together so that the platoons avoid arriving at red lights as much as possible. This simply stated objective belies the tremendous complexity associated with this task, and indeed, the signal coordination topic is far more important and complicated than fixed-time or actuated control. The objectives of coordination are the same as for isolated control: minimizing some combination of negative performance (delay, stops, fuel consumption, emissions, etc.) or maximizing smooth flow.

The one objective that is a reality for nearly all agency practitioners is not stated in any of the research or computerized methods: minimize citizen complaint calls. Most practitioners seek to provide operation that promotes smooth, predictable flow and a sense of fairness.

Achieving signal coordination requires two branches of knowledge. The first is signal timing on a network scale, and the second is understanding how traffic signals can be tied together into functioning signal systems. Each of these topics will be discussed.

6.4.1   Signal Coordination

As was stated in the preceding section, signals are coordinated by means of an offset value that defines how much the cycle at one intersection is delayed from an arbitrary system zero point. These offset values are calculated to provide the best network flow based on one of the objectives mentioned earlier. No matter what the objective, however, traffic engineers use time-space diagrams to visualize these timing patterns. A typical time-space diagram is shown in Figure 6.6.

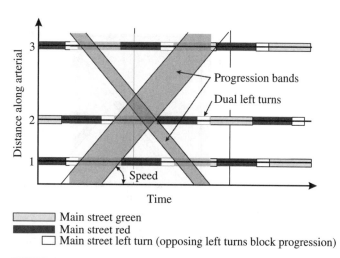

FIGURE 6.6   Time-space diagram.

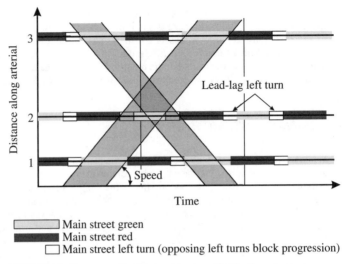

Main street green
Main street red
☐ Main street left turn (opposing left turns block progression)

FIGURE 6.7 Time-space diagram showing more effective phase sequence.

The sequence of phases is shown extending horizontally, and the distance along the road is shown vertically. Thus a constant speed can be represented by the sloped lines. If a driver can maintain a constant speed through all intersections, we say that we have achieved *progression,* and his or her course through the time-space diagram is a continuous line. A *progression band* is the collection of all such lines.

Phase sequence is also an issue in progression. The time-space diagram in Figure 6.7 is the same example as earlier, but with a different phase sequence at the middle intersection. Notice how the progression bands have been widened as a result.

Notice also the resonance between signal spacing, progression speed, and cycle length. Figure 6.8 shows the same system and the same splits, but with a cycle length that prevents good progression.

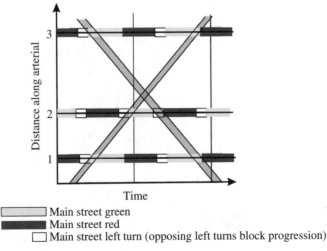

Main street green
Main street red
☐ Main street left turn (opposing left turns block progression)

FIGURE 6.8 Time-space diagram showing unresonant cycle length.

Calculating the correct cycle is the fundamental challenge for good coordinated signal timings. The time-space diagram illustrates the inflexible mathematical linkage among cycle length, signal spacing, and speed. The more regular the signal spacing, the more the correct cycle resonates, meaning that the pulses of traffic in one direction are reinforced by arriving at greens, which turn green at a time to send pulses in the opposite direction, which are also reinforced by arriving at a green. An unresonant cycle length causes platoons in one direction or the other (or both) to lose that progression.

A resonant cycle, therefore, is based on the travel time from intersection to intersection. A cycle length that is twice the travel time between intersections will provide the best possible progression, as shown in Figure 6.9. The *offset* is generally the amount of time the second intersection lags behind the first.

The following equation, therefore, quantifies the cycle length in terms of the distance between intersections.

$$C = 2(D/V)$$

Where D = the distance from one intersection to the next
    V = the speed of traffic

Many cities in the United States are laid out based on mile-spaced arterial streets, with signals at the arterial intersections and at intersections in between that serve collector streets. The travel time

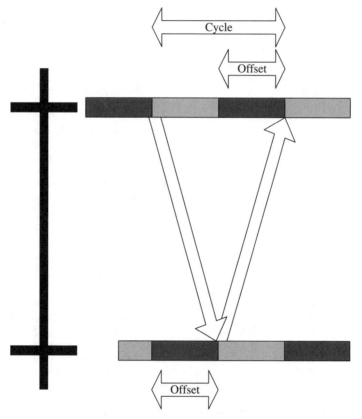

FIGURE 6.9 Cycle length is most resonant when there is twice the travel time between intersections..

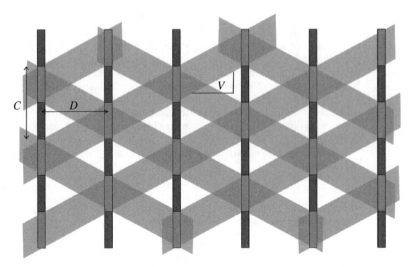

FIGURE 6.10 Single-alternate operation.

at 58.8 feet per second (40 mi/hr) is 45 seconds to go a half mile, and twice that travel time is 90 seconds. Therefore, a 90-second signal cycle works particularly well in mile-spaced arterial grids common to western American cities.

The preceding operation, known as *single-alternate operation,* requires that each intersection is offset from the previous intersection by half the cycle. Figure 6.10 shows the time-space diagram of a network running single-alternate operation.

Single-alternate operation provides full-bandwidth progression, meaning that all the green time at each intersection is within the progression bands.

For more closely spaced intersections, however, the resulting cycle from the preceding equation might prove too short. For example, an arterial street with signals spaced every quarter mile and speeds at 40 mi/hr will yield a 45-second cycle using the preceding equation. Few arterial intersections could provide for all the phases and pedestrian clearances required with a cycle length that short. For such networks, the first harmonic of the preceding equation still can provide some resonance and also a more usable cycle length. In that operation, the first two intersections have the same offset and, therefore, simultaneous green times, and the second two intersections are offset by half the cycle. This operation is known as *double-alternate operation,* and the equation for the cycle length is

$$C = 4(D/V)$$

For the arterial with traffic signals at quarter-mile intervals, the cycle would be 90 seconds. Figure 6.11 shows a time-space diagram with double-alternate operation.

As the figure shows, double-alternate operation provides progression bands only half as wide as the green times. This operation could be optimized for peak-flow traffic by making some adjustments to the timings without undermining the provision of two-way progression and basic resonance.

In tight grids of two-way streets, even double-alternate operation might require too short a cycle. For example, the typical downtown grid of streets with a 400-foot block spacing, a 30 mi/hr speed, and enough two-way streets to require two-way progression would yield a cycle of 18 seconds with single-alternate operation and 36 seconds with double-alternate operation, both of which are likely to be too short. In this case, the next harmonic, with a cycle length six times the travel time between intersections, will provide a usable progression. Thus the equation for *triple-alternate operation* is

$$C = 6(D/V)$$

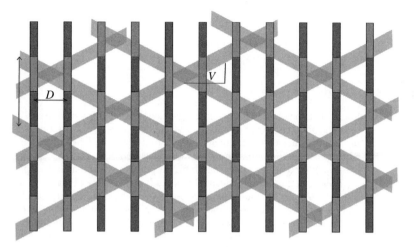

FIGURE 6.11   Double-alternate operation.

For the example downtown grid, the cycle would be 54 seconds. Most agencies would use a cycle of 60 seconds, which would lower the progression speed slightly to 27 mi/hr. Progressions speeds slightly slower than the normal speed of traffic, especially in tight, regular grids, trains and rewards repeat drivers for reducing their speed slightly. Figure 6.12 shows a time-space diagram for triple-alternate operation. As the figure shows, triple-alternate operation provides progression bandwidth that makes use of a third of the green time at each intersection.

The more regular the spacing in a network (even if some intersections are missing in the series, leaving gaps of some multiple of the regular spacing), the sharper is the resonance, and the better these equations work. As the spacing becomes more irregular, resonance becomes harder to find. It is

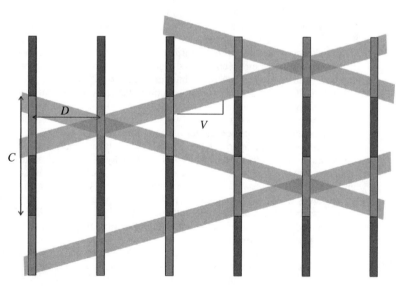

FIGURE 6.12   Triple-alternate operation.

possible to have signals spaced such that no progression is possible. In such cases, a range of cycles is attempted to make the best of the situation, often using computer-based tools for optimizing signal timing and for manipulating time-space diagrams.

6.4.2 System Design

Simply put, the objective of a traffic signal system is to implement signal timings in a way that can be operated and maintained efficiently. System designers devote a lot of energy, however, to determining which timings are implemented at what times and whether conditions provide an opportunity to adjust those timings. To summarize these considerations, three categories of signal systems will be presented: time-of-day operation, traffic-responsive operation, and adaptive control.

These systems are implemented physically in a variety of ways. At their simplest, the coordination settings may be installed in each controller, with accurate internal clocks synchronized to maintain the relationship between intersections. This is known as *time-based coordination* and is subject to clock drift and other problems. Most agencies prefer to use it only as a short-term solution or as a backup when communications are lost. Small systems may use a master controller on the street to maintain coordination that can be dialed into from a computer at the traffic engineer's office. These closed-loop systems also can be tied together into larger systems, which is known as a *distributed architecture*. Centralized systems maintain central control of all the signal controllers. These days, the only thing preventing all systems from such central operation is the cost of the required communications infrastructure.

Time-of-Day Operation.   Most signal systems in the United States use signal timings that are designed based on historical traffic data and then stored in the system to be implemented according to a fixed daily schedule. Time-of-day systems are still completely effective in many situations, especially considering the advanced features that traffic engineers can use to fine-tune them. These features include the ability to create exceptions for weekends, holidays, and other special-purpose plans. Some surveys show that time-of-day operation is used in as much as 95 percent of all signal systems. Some portray these findings as a waste of the traffic-responsive features in many of these same systems, but in fact, there is no evidence to prove that traffic-responsive operation is really more effective in common situations. Time-of-day operation works best when signal timings are designed to provide maximum versatility, which favors progression-based timings as opposed to computer optimizations for a narrow range of conditions based on delay and other measures of effectiveness.

Traffic-Responsive Operation.   Starting in the 1950s, some systems were able to use traffic detection to determine when preengineered signal timing plans would be brought into operation. These methods were first established in common practice with the advent of the Urban Traffic Control System (UTCS), a federal research program in the early 1970s that created a public-domain signal system that was the mainstay of U.S. large-scale systems for the following 15 years. UTCS traffic-responsive operation has grown into two general approaches for responding to measured demand by the implementation of a preengineering signal timing plan.

The more common approach is intended for arterial networks that have a significant inbound and outbound flow pattern. System detectors are usually regular intersection detectors that have been assigned also as system detectors. As with freeway systems, they measure volume (number of cars) and detector occupancy (percentage of time detector is covered by a car). These systems combine these values into a single parameter that they compare first with an *offset-level threshold*, which indicates whether it is an inbound, outbound, or average pattern. Within each offset level, they then determine which cycle to use and then which collection of splits to use. As with all traffic-responsive systems, they use existing timing patterns that were designed offline and stored in the system.

The other approach is better suited to grid networks where inbound and outbound influences might be harder to observe. It uses a pattern-recognition technology. Each signal timing plan is associated with a collection of system-detector volume and occupancy values. The system compares observed detector values against the stored values associated with each plan. The plan whose detector values show the least difference from measured values is the one selected.

With both these approaches, detector data are averaged over a control period, which may range from several minutes to as much as half an hour. When a plan is selected, the controllers are instructed to implement the plan, which may take up to several minutes. Each controller will shorten or lengthen its force-off values to nudge the cycle into the new plan incrementally, with the transition process controlled by a series of parameters set by the traffic engineer. If the control period is too short, a large percentage of the time might be spent in transition, and signal coordination is poor during those periods. If the control period is too long, the condition to which the system is responding might have changed yet again by the time the response is in place.

To make a traffic-responsive system work better than a time-of-day system, traffic engineers must spend considerable time with experimentation and fine-tuning. For this reason, most agencies are sufficiently constrained by budget and time to make traffic-responsive operation not worth the difficulty except in special cases.

Adaptive Control. Some newer systems (and some not so new) have the ability to measure traffic conditions and adjust signal timings on the fly. These systems have been grouped under the heading of *adaptive control*. A number of technologies for adaptive control exist in theory, one or two of which have received wide implementation throughout the world. None has yet been fully implemented based on compatibility with U.S.-style controllers that follow ring rules long enough to establish a market presence. Thus adaptive control systems have not been widely deployed in the United States.

Some adaptive control systems are based on incremental changes to the signal timing based on measured traffic flow on special system detectors, and others use those data to make predictions about traffic flow a few seconds or minutes into the future. All these systems require extensive fine-tuning during implementation—far more than with traffic-responsive operation. Unlike systems that use canned signal timing plans, however, adaptive systems should be able to track long-term trends in traffic such that signal timings do not have to be calculated again every few years. This is often reported as the key advantage of adaptive systems. Most adaptive systems, however, require an extensive implementation effort and a commitment on the part of the agency to maintain detectors. Those agencies unable to provide such support may find that they have to cease adaptive operation in favor of time-of-day control, so agencies should carefully consider their true operational objectives and the requirements those impose when deciding what sort of system to implement.

REFERENCES

1. *Uniform Vehicle Code.* Washington: National Committee on Uniform Traffic Laws and Ordinances, 2000.
2. *Manual on Uniform Traffic Control Devices.* Washington: Federal Highway Administration, 2003 (revised 2007).
3. Richard W. Denney, Jr, et al. *Signal Timing at Saturated Intersections.* FHWA-HOP-09-008. Washington: Federal Highway Administration, 2008. Available at http://ops.fhwa.dot.gov/publications/fhwahop09008/index.htm.

CHAPTER 7
HIGHWAY SIGN VISIBILITY

Philip M. Garvey
The Larson Institute
Pennsylvania State University
University Park, Pennsylvania

Beverly T. Kuhn
System Management Division
Texas A&M, College Station, Texas

7.1 MEASURES OF SIGN EFFECTIVENESS

Sign visibility is an imprecise term in that it encompasses both sign detectability and sign legibility. Sign detectability refers to the likelihood of a sign being found in the driving environment and is integrally associated with sign conspicuity. Sign conspicuity is a function of a sign's capacity to attract a driver's attention that depends on sign, environmental, and driver variables (Mace, Garvey, and Heckard 1994). Sign legibility describes the ease with which a sign's textual or symbolic content can be read. Sign legibility differs from sign recognizability in that the former refers to reading unfamiliar messages while the latter refers to identifying familiar sign copy. Sign legibility and recognizability in turn differ from sign comprehensibility in that the latter term implies understanding the message while the former merely involve the ability to discern critical visual elements.

While other measures of sign visibility such as blur tolerance (Kline et al. 1999) and comprehension speed (Ells and Dewar 1979) are sometimes used, sign visibility is most often assessed by determining threshold distance. The two thresholds often used are detection distance (the distance at which an observer can find a sign in the driving environment) and legibility distance (the distance at which an observer can read a sign's message). The intent of a sign designer is to provide the sign's observer with the maximum time to read the sign, and to do that the observer must find it prior to or at its maximum reading distance. Therefore, in designing a sign it is desirable to achieve a detection distance that is equal to or greater than that sign's legibility distance.

To describe legibility distance across signs with different letter heights, the term legibility index is often used. Legibility index (LI) refers to the legibility distance of a sign as a function of its text size. Measured in feet per inch of letter height (ft/in.), the LI is found by dividing the sign's legibility distance by its letter height. For instance, a sign with 10-inch letters legible at 300 feet has an LI of 30 ft/in. (300/10 = 30).

7.2 VISUAL PERCEPTION

In the initial stages of vision, light passes through the eye and is focused by the cornea and lens into an image on the retina. The retina is a complex network of nerve cells or light receptors composed of rods and cones. So named because of their basic shape, rods and cones perform different functions. Cones function best under high light intensities and are responsible for the perception of color and fine detail. In contrast, rods are more sensitive in low light but do not discriminate details or color. Rods and cones are distributed across the retina in varying densities.

The region of the central retina where a fixated image falls is called the fovea. The fovea has only cones for visual receptors and is about 1.5 to 2° in diameter. Beyond 2°, cone density rapidly declines reaching a stable low point at about 10°. Conversely, rod density rapidly increases beyond 2° and reaches a maximum at about 18° before dropping off (Boff, Kaufman, and Thomas 1986). From 18° outward toward the nose and ears, forehead, and chin, the number of rods decreases but still continues to be higher than the number of cones.

Functional detail vision extends to about 10°, worsening in the near periphery from about 10 to 18° and significantly deteriorating in the far periphery from about 18 to 100°. Occasionally the term *cone of vision* appears in the visibility literature. Although this expression is used freely, it is not well defined. In general, the cone of vision can be taken to refer to an area within the near periphery. The *Traffic Control Devices Handbook* (FHWA 2001) section on driver's legibility needs states:

> When the eye is in a fixed position it is acutely sensitive within a 5 or 6 degree cone, but is satisfactorily sensitive up to a maximum cone of 20 degrees. It is generally accepted that all of the letters, words, and symbols on a sign should fall within a visual cone of 10 degrees for proper viewing and comprehension.

7.2.1   Visual Acuity

Visual acuity is the ability to discern fine objects or the details of objects. In the United States, performance is expressed in terms of what the observer can see at 20 feet, referenced to the distance at which a "normal" observer can see the same object. Thus, we have the typical Snellen expressions 20/20, where the test subject can see objects at 20 feet that the standard observer can see at 20 feet, and 20/40, where the test subject can see objects at 20 feet and the standard observer can see at 40 feet. Visual acuity can be measured either statically and/or with target motion, resulting in static visual acuity (SVA) and dynamic visual acuity (DVA). Although there are questions as to its relationship to driving ability, static visual acuity is the only visual performance measure used with regularity in driver screening, with 20/40 visual acuity in one eye being the de facto threshold standard.

7.2.2   Contrast Sensitivity

Contrast threshold is the minimum detectable *luminance* difference between the dark and light portions of a target. (The term *luminance* is defined and discussed later in section 7.3.) Contrast sensitivity is the reciprocal of the contrast threshold; that is, contrast sensitivity = 1/contrast threshold. Therefore, a high contrast threshold represents low contrast sensitivity—in other words, a large difference between the darkest and lightest portions of a target is necessary for target visibility.

7.2.3   Visual Field

A driver's field of vision or visual field is composed of foveal and peripheral vision, or literally everything the driver can see. Visual acuity is highest in the central fovea. Beyond the fovea, vision deteriorates rapidly. In the near periphery, individuals can see objects but color and detail discriminations are weak. The same holds true for far peripheral vision. Aside from these general visual field break points, no specific designations of intermediate vision exist.

The related concept of useful field of view (UFOV) has gained widespread acceptance as a potentially useful tool to describe vision in natural settings (Ball and Owsley 1991). Proponents of UFOV assert that while an individual's visual field is physically defined by anatomical characteristics of the eye and facial structure, it can be further constricted by emotional or cognitive states, as with the common phenomenon of "tunnel vision" experienced under stressful conditions.

7.2.4 Glare

Another set of terms related directly to drivers' visual abilities concerns the adverse effects of light from external sources, which include, but are not limited to, signs, headlamps, and overhead lights. While signs are included in this listing, research has shown that they are not generally a source of glare or another measure of light pollution known as *light trespass* (Garvey 2005). Four types of glare exist: direct, blinding, disability, and discomfort. Direct glare comes from bright light sources or the reflectance from such sources in the driver's field of view (FHWA 1978a). Blinding glare and disability glare differ only in degree. Blinding glare results in complete loss of vision for a brief period of time, whereas disability glare causes a temporary reduction in visual performance ranging from complete to minor. Discomfort glare reduces viewing comfort for the observer.

7.3 PHOTOMETRY

Photometry is the measurement of radiant energy in the visual spectrum; that is, the measurement of light. Photometric equipment is designed and calibrated to match the human eye's differential sensitivity to color and to daytime-versus-nighttime lighting. Therefore, photometry is the measurement of light as people see it. The four most important photometric measurements used to describe sign visibility are luminous intensity, illuminance, luminance, and reflectance.

Luminous intensity, expressed in candelas (cd), is a description of a light source itself and is therefore independent of distance. That is, no matter how far away an observer is from a lamp, that lamp always has the same intensity. Luminous intensity is the photometric measurement most often specified by lamp and LED manufacturers.

Illuminance, or incident light, is expressed in units of lux (lx) and is a measure of the amount of light that reaches a surface from a light source. Illuminance is affected by distance and is equal to luminous intensity divided by the distance squared ($lx = cd/d^2$). For example, if a source emits a luminous intensity of 18 cd, the illuminance level measured 3 meters away would be $2 \, lx (18/3^2)$. If the source's intensity is unknown, illuminance is measured with an illuminance or lx meter placed on the surface of interest facing the light source.

Luminance is expressed in candelas per square meter (cd/m^2). Luminance is the photometric that most closely depicts the psychological experience of "brightness." Luminance can refer to either the light that is emitted by or reflected from a surface, and is an expression of luminous intensity (cd) over an extended area (m^2). Like luminous intensity, a source's luminance is constant regardless of distance. To measure luminance, a luminance meter is placed at the observer's position, aimed at the target of interest, and a reading is taken.

Reflectance is the ratio of illuminance to luminance and, as such, reflectance describes the proportion of incident light that is absorbed and the proportion that is reflected by a surface. If, for example, 100 lx hits an object's surface and that surface has a luminance level of 5 cd/m^2, that surface has a reflectance of 5 percent (5/100).

A related term often found in traffic sign literature is retroreflection. Retroreflection describes material that, unlike normal matte (i.e., diffuse) or specular (i.e., mirror-like) surfaces, reflects most incident light directly back to the light source. With regard to signs and other vertical surfaces, retroreflection is represented by a coefficient of retroreflection, known as R_A, although it is also commonly referred to as the "specific intensity per unit area," or SIA (ASTM 2001). A material's SIA is the ratio of reflected light to incident light. SIA values for signs are commonly expressed in candelas per lux per square meter ($cd/lx/m^2$).

7.4 FEDERAL TRAFFIC SIGN REGULATIONS

The *Manual on Uniform Traffic Control Devices* (*MUTCD*) (FHWA 2009) is the legal document that governs traffic control device design, placement, and use in the United States. All local, county, and state transportation departments must use this document as a minimum standard. According to the document, traffic control devices are "all signs, signals, markings, and other devices used to regulate, warn, or guide traffic, placed on, over, or adjacent to a street, highway, pedestrian facility, bikeway, or private road open to public travel . . . by authority of a public agency or official having jurisdiction, or, in the case of a private road, by authority of the project owner or private official having jurisdiction" (FHWA 2009).

7.4.1 Sign Height

The *MUTCD* requires that any sign placed at the side of a roadway have a minimum vertical height ranging from 5 to 7 feet, depending on the general location and the number of installed signs. For instance, a sign in a rural area or on freeways must have a vertical height of at least 5 feet, while a sign in a business, commercial, or residential area must have a vertical clearance of at least 7 feet (FHWA 2009). The reason behind this variation is that more vertical clearance is needed in areas where pedestrian traffic is heavier. To allow for the passage of the largest vehicles, overhead signs must have a vertical clearance of 17 feet.

7.4.2 Lateral Offset

As with sign height, lateral clearance requirements vary with sign location. For instance, the *MUTCD* states that traffic signs placed along a roadway should have a minimum lateral clearance of 6 feet from the edge of the shoulder. However in urban areas where lateral offset is often limited, offsets can be as small as 1 to 2 feet (FHWA 2009). These small lateral clearances reflect the lower vehicle speeds found in urban areas. The basic premise behind even the smallest minimum standards is a combination of visibility, safety, and practicality.

7.4.3 Longitudinal Sign Placement

The placement of directional signs must be far enough in advance of the location of the intended action so that the motorist can react and slow the vehicle or change lanes if necessary, after passing the sign and prior to reaching the appropriate crossroad or access road. The *MUTCD* (FHWA 2009) states, "When used in high-speed areas, Destination signs [i.e., conventional road guide signs] should be located 200 ft or more in advance of the intersection. . . . In urban areas, shorter advance distances may be used." The distance of 200 feet at 65 mph translates to approximately 2.0 seconds. For the more critical Warning Signs, such as Merge and Right Lane Ends, the *MUTCD* (FHWA 2009) states, ". . . [in] locations where the road user must use extra time to adjust speed and change lanes in heavy traffic because of a complex driving situation, . . . the distances are determined by providing the driver . . . [with] 14.0 to 14.5 seconds for vehicle maneuvers."

The *MUTCD* recommendations can best be thought of as absolute minimums. To establish more conservative recommended distances for longitudinal sign placement for the National Park Service, a formula developed by Woods and Rowan (1970) was combined with deceleration rates from the AASHTO green book. Table 7.1 contains the results of that formula for single lane approaches. Adding 4.0 seconds to the single lane approaches results in the multilane recommendations.

In some cases, such as high-speed highways, two signs may be necessary. In fact, the *MUTCD* (FHWA 2009) recommends, "major and intermediate interchanges Advance Guide signs should be placed at ½ mile and at 1 mile in advance of the exit with a third Advance Guide sign placed at 2 miles in advance of the exit if spacing permits."

7.4.4 Reflectorization and Illumination

Regulatory, warning, and guide signs must be illuminated in such a manner that they have the same appearance at night as in daytime. The *MUTCD* provides general regulations regarding

TABLE 7.1   Recommended Reading Time, Letter Height, and Longitudinal Sign Placement for Various Operating Speeds and Number of Words on a Sign

Operating speed (mph)	Number of words	Reading time (sec)	Letter height (in.)	Longitudinal sign placement distance (ft/sec)	
				Single-lane approach	Multilane approach
25–40	1–3	3.0–4.5	4–6		
	4–8	6.0	8	375/6.4	600/10.4
41–50	1–3	3.5–4.5	6–8		
	4–8	5.5–7.0	10 12	500/6.8	800/10.8
51–60	1–3	4.0–5.0	8–10		
	4–8	5.5–7.0	12–14	650/7.4	1000/11.4
61–70	1–3	4.0	10		
	4–8	5.5	14	725/7.1	1100/11.1

the illumination of traffic signs. For instance, signs may be illuminated in four ways: (1) by internal lighting (e.g., fluorescent tubes or neon lamps) that illuminates the main message or symbol and/or background through some type of translucent material, (2) by external lighting (including high-intensity discharge lamps or fluorescent lighting sources) that provides uniform illumination over the entire face of the sign, (3) by light-emitting diodes (LEDs) that illuminate the main message or symbol or portions of the sign border, and (4) by some other method such as luminous tubing, fiberoptics, incandescent panels, or the arrangement of incandescent lamps (FHWA 2009).

On freeway guide signs, all copy (legends, borders, and symbols) must be retroreflectorized with the sign background either retroreflectorized or otherwise illuminated. On overhead sign mounts, headlight illumination of retroreflective material may not be sufficient to provide visibility (FHWA 2009). In such situations, supplemental external illumination is necessary. Supplemental illumination should be reasonably uniform and sufficient to ensure sign visibility. Furthermore, the *MUTCD* provides standards for maintaining minimum retroreflectivity for signs and provides guidance on assessment or management methods for maintaining minimum levels to ensure nighttime sign visibility (FHWA 2009). The document also provides information on minimum maintained retroreflectivity levels according to sign color, sheeting type, placement, and text and symbol size (FHWA 2009).

7.4.5   Message Design

Both the *MUTCD* and *Standard Highway Signs* (FHWA 2002) specify guide sign content characteristics. Although they allow some flexibility for overall sign layout, these two federal documents provide detailed specifications for message design.

7.4.6   Font

With regard to font, the federal specifications are very specific. The letter style used must be one of the Standard Highway Alphabet series (e.g., Figure 7.1) (FHWA 2004). However, in 2004, the FHWA granted interim approval for use of the Clearview font for positive-contrast legends (i.e., lighter letters on a darker background) on guide signs, allowing any state to use this font in place of Standard Highway Alphabets on written request to the FHWA (e.g., Figure 7.2).

FIGURE 7.1

Bergaults

FIGURE 7.2

7.4.7   Case

With the exception of destination names, all signs must use only upper-case letters. For destination names (i.e., places, streets, highways) using mixed-case legends, the lower-case loop height should be 75 percent of the upper-case height.

7.4.8   Letter Height

As stated in the *MUTCD* (FHWA 2009), the general guidance for selecting letter height is based on a legibility index (LI) of 30 feet per inch of letter height (LI = 30). This is a 10 ft/in. reduction from the LI of 40 recommended in Standard Highway Signs (FHWA 2002) and the LI of 50 that was the standard for almost 50 years (FHWA 1988) prior to that. In addition to the general rule, the *MUTCD* (FHWA 2009) establishes 8 inches as the minimum letter height for freeway and expressway guide signs, although overhead signs and major guide signs require larger lettering. For example, on expressways, the minimum height ranges from 20 inches/15 inches (upper-/lower-case) for a major roadway to 10.6 inches/8 inches for a road classified as minor and 20 inches/15 inches to 13.3 inches/10 inches for freeways, respectively, for these two roadway types. Standard Highway Signs (FHWA 2004) states that the minimum letter height for the principal legend on conventional road guide signs on major routes is 6 inches; this drops to 4 inches for urban and less important rural roads. The long-held 4-inch minimum for street name signs found in previous versions of Standard Highway Signs (FHWA 1978b) was changed to 6 inches with the 2002 edition.

7.4.9   Border

All signs are required to have a border. The border must be the color of the legend and, in general, should not be wider than the stroke width of the largest letter used on that sign. Guide sign borders should extend to the edge of the sign, while regulatory and warning sign borders should be set in from the edge.

7.4.10   Spacing

Intercharacter spacings for the Standard Highway Alphabets must follow the spacing charts in the Standard Highway Signs book (FHWA 2004), whereas spacings for the Clearview font must follow the spacing tables for Clearview found under Interim Approvals in the *MUTCD* (FHWA 2009). The spacing between the sign copy and the border should be equal to the height of the upper-case letters. The spacing between lines of text should be 75 percent of the upper-case letter height. Spacing between unique copy elements (e.g., words and symbols) should be 1 to 1.5 times the upper-case letter height (FHWA 2004).

7.5 SIGN VISIBILITY RESEARCH

When reviewing the sign visibility literature, it is useful, if somewhat artificial, to divide the research by the type of visibility studied. The two main sign visibility research areas focus on sign detection and sign legibility. These areas are interdependent; one cannot read a sign that one cannot find, and there is little point in detecting a sign that cannot be read from the road. However, from a sign design and placement viewpoint, the characteristics that affect detection and legibility differ enough qualitatively to warrant separate consideration.

7.5.1 Detection

A sign's detectability is directly related to its conspicuity. However, while conspicuity is often thought of simply as target value, in that it describes how readily a sign is perceived, in reality, it is produced by a complex combination of sign factors, environmental qualities, and observer variables. Mace and Pollack (1983) wrote that "conspicuity . . . is not an observable characteristic of a sign, but a construct which relates measures of perceptual performance with measures of background, motivation, and driver uncertainty." This section addresses conspicuity and the characteristics of the environment, sign, and driver that directly affect both the likelihood and distance of sign detection.

Sign-Placement Variables.   One of the most important factors in sign detection has nothing to do with characteristics of the sign itself. That factor is sign placement. Because a sign that is not well placed cannot be seen and therefore cannot be read, a sign's positioning relative to its environment is key to its effectiveness. This section covers sign mounting height and offset and the immediate sign environment, or the sign's "surround."

Lateral and Vertical Offset.   Careful placement of signs along the roadway ensures that a driver has sufficient time to detect the sign and take necessary action. Pietrucha, Donnell, Lertworawanich, and Elefteriadou (2003) reported that for low-mounted storefront signs, the time a driver has to detect, read, and comprehend the sign depends on the position of the vehicle on the road's cross-section, the amount of traffic in both directions, and the sign's lateral offset. Upchurch and Armstrong (1992) found placement of signs with respect to restricting geometric features of the roadway, such as hills and curves, to be important in maximizing detection distance. Mace and Pollack (1983) stated that as the distance between a target sign and "noise" items increases, the sign becomes more conspicuous, although this conspicuity is eroded as the sign becomes located further from the center of the driver's visual field. Claus and Claus (1975) quantified this when they wrote that signs should be placed within 30° of the driver's line of sight. Matson (1955) suggested "that a sign should fall within a visual cone of 10° to 12° on the horizontal axis and 5° to 8° on the vertical axis" (in Hanson and Waltman 1967). Jenkins and Cole (1986) took this statement a step further when they wrote, "[I]f a sign is to be noticed . . . it will be within 10 degrees of his line of sight. When the eccentricity . . . becomes greater than this, the sign is most unlikely to be noticed at all." Jenkins and Cole's statement is supported by Zwahlen's (1989) study of nighttime traffic sign conspicuity in the peripheral visual field. Zwahlen found that retroreflective signs placed in the foveal region resulted in twice the detection distances of those located 10° outside this central visual area. A further result from Zwahlen's study was that signs located 20° and 30° outside the fovea resulted in one-third and one-quarter the detection distances, respectively, of centrally located signs.

As indicated by Zwahlen's study, sign placement is particularly important for retroreflective signs. This is because the angle between the vehicle and the sign strongly influences sign brightness and therefore nighttime sign detection (McNees and Jones 1987). A retroreflective material returns light back to its source as a function of entrance angle which describes the relative positions of the headlamps and sign. As a general rule, when this angle increases, the amount of reflected light seen by the driver decreases (King and Lunenfeld 1971). Thus, when placing retroreflective signs it is important to obtain entrance angles as close to the manufacturer's recommendations as is practically possible.

Surround.   Where a sign is placed in relation to other visual stimuli defines the visual complexity of the sign's surround. The factors that affect visual complexity include the number and overall density of noise in the driver's visual field and the density of noise items immediately adjacent to the sign (Mace and Pollack 1983). Research conducted on signs with various levels of retroreflectivity in different environments reveals that virtually any retroreflective sign can be seen at a reasonable distance in an environment that is not visually complex (Mace and Pollack 1983). In other words, if a sign does not have to compete with many other objects in a driver's cone of vision, it is conspicuous, even if its retroreflectivity is low. However, in an area that has more visual distractions, sign conspicuity becomes more a function of its retroreflectivity, size, color, and other variables (Mace and Pollack 1983; Mace, Garvey, and Heckard 1994). McNees and Jones (1987) supported Mace and his colleagues when they asserted that as the number of objects in the driver's cone of vision increases, the conspicuity of a sign decreases. However, when a sign is located in a visually complex environment, retroreflectivity may not be enough to ensure sign detection (Mace and Pollack 1983). Thus, in more complex environments, conspicuity boosters will be needed to achieve a desired detection distance. In such situations, additional lighting or sign redundancy may be necessary to provide adequate conspicuity to ensure timely sign detection. In such situations, additional lighting or sign redundancy may be necessary to provide adequate conspicuity to ensure timely sign detection. The *MUTCD* (FHWA 2009) discusses 11 methods for enhancing sign conspicuity, including adding LEDs to the legend, symbol, or border of a sign; increasing sign size; putting signs on the right- and left-hand sides of the road across from each other; adding red or orange flags to the signs; and reflectorizing the signpost.

Lighting Variables. The first step in visual perception is the detection of light. Differences in the quantity (e.g., luminance) and quality (e.g., color) of light are necessary to differentiate objects. A sign with the same luminance and color as its background is undetectable. Therefore, the term *lighting variables,* as used in this section, refers not only to illuminated nighttime sign display, but to all factors that fall within the category of photometric sign properties. While this category includes nighttime illumination techniques, it also covers daytime and nighttime sign luminance, sign color, and luminance and color contrast between the sign and surround.

External Luminance Contrast. Sign contrast affects both conspicuity and legibility distance. In sign visibility research, there are two types of luminance contrast: external and internal. Detection distance is affected by external sign contrast, which is the ratio of the sign's average luminance and the luminance of the area directly surrounding the sign. Legibility distance is affected by internal sign contrast, defined by the ratio of the luminance of a sign's content and its background.

As a sign's external contrast ratio increases, so does the sign's conspicuity (Forbes et al. 1968a; Mace and Pollack 1983). Mace, Perchonok, and Pollack (1982) concluded that in low visual complexity locations, external contrast and sign size are the major determinants of sign detection. Cooper (1988) goes a step further in stating that external contrast plays a far greater role in sign conspicuity than does sign size. While no research provides optimal and minimum values for external contrast, McNees and Jones (1987) found high-intensity background sheeting (i.e., ASTM type III, or encapsulated lens) with high-intensity copy, opaque background sheeting with button copy (i.e., letters embedded with "cat's eye" reflectors), and engineer-grade background sheeting (i.e., ASTM type I, or enclosed lens) with button copy to provide acceptable freeway guide sign-detection distances.

Sign Luminance. Mace and Pollack (1983) stated that sign conspicuity increases with higher sign luminance. Furthermore, Mace, Perchonok, and Pollack (1982) concluded that, with the exception of black-on-white signs, increasing sign luminance could even offset the detrimental effects of increased visual complexity. Pain (1969) stated, "a higher [overall] brightness enhances a high brightness ratio [external contrast] by roughly 10 percent." Zwahlen (1989) buttressed these findings when he concluded that increasing retroreflective sign SIA values can offset the negative effects of peripheral location. Research conducted on various types of commonly used retroreflective background sheeting combined with reflective copy concurs, indicating that conspicuity increases as sign retroreflectivity increases (McNees and Jones 1987). Research on white-on-green signs (Mace, Garvey, and Heckard 1994) goes further in reporting that sign brightness can actually compensate for sign size. Mace et al. found that small (24-inch) diamond-grade (i.e., ASTM type VII, or microprismatic) signs produced the same legibility distance as large (36-inch) engineer-grade signs.

Nighttime conspicuity research conducted by Mace, Garvey, and Heckard (1994) indicated that the relationship between sign brightness and detection distance is mediated by sign color. They found no difference in detection distance for either black-on-white or black-on-orange signs as a function of retroreflective material. However, higher-reflectance materials resulted in an improvement in detection distance for white-on-green signs at high and low visual complexity sites.

Color. Forbes et al. (1968b) concluded that "relative brightness is of most importance, but hue [color] contrast enhances the brightness effects in some cases." Of the background sign colors black, light gray, and yellow, Cooper (1988) found yellow to be the most effective for sign detection. Mace, Garvey, and Heckard (1994) reported that black-on-orange and white-on-green signs were detected at greater distances than black-on-white signs. This is consistent with the research of Jenkins and Cole (1986), which found black-on-white signs to provide particularly poor conspicuity. Mace, Garvey, and Heckard (1994) concluded that the reason for this was that white signs were being confused with other white light sources and that it was necessary to get close enough to the sign to determine its shape before recognizing it as a sign. Mace et al.'s research punctuates the interaction between various sign characteristics (such as shape and color) in determining sign conspicuity. Zwahlen and Yu (1991) furthered the understanding of the role color plays in sign detection when they reported their findings that sign color recognition distance was twice that of shape recognition and that the combination of a highly saturated color and specific shape of a sign could double a sign's average recognition distance.

Sign Variables. In addition to environmental and photometric variables, there are a number of characteristics related to sign structure and content that have been found to affect sign detection. These characteristics include the size and shape of the sign and the message design.

Size and Shape. The size and shape of a sign relative to other stimuli in the driver's field of vision play a role in determining the sign's conspicuity. Mace, Garvey, and Heckard (1994) found significant increases of around 20 percent in both nighttime and daytime detection distances with increases in sign size from 24 to 36 inches for black-on-white, black-on-orange, and white-on-green signs. In 1986, Jenkins and Cole conducted a study that provides corroborative evidence that size is a key factor in sign detection. Jenkins and Cole concluded that sign sizes between 15 and 35 inches are sufficient to ensure conspicuity, and that if signs of this size or bigger are not detected, the problem is with external contrast or surround complexity. In addition to the effects of sign size, Mace and Pollack (1983) concluded that conspicuity also increases if the shape of the sign is unique compared to other signs in the area.

Display. Forbes et al. (1968b) found green signs with high internal contrast to improve sign detection. In particular, these researchers found signs with bright characters on a dark background to have the highest conspicuity under light surround conditions and the reverse to be true for dark or nighttime environments. Hughes and Cole (1984) suggest that bold graphics and unique messages increase the likelihood of meaningful detection.

7.5.2 Legibility

Once a sign has been detected, the operator's task is to read its content; this is sign legibility. *Legibility* differs from *comprehensibility* in that *legibility* does not imply message understanding. Symbol signing provides a good example of this distinction. An observer could visually discern the various parts of a symbol and yet be unable to correctly report that symbol's meaning. The same is true for alphanumeric messages with confusing content. The problem with drawing a distinction between legibility and comprehension, however, is that familiar symbolic and textual messages are accurately reported at much greater distances than novel sign copy (Garvey, Pietrucha, and Meeker 1998). This well-documented phenomenon leads to the need to distinguish legibility from recognition.

Because recognition introduces cognitive factors, message recognition does not require the ability to discriminate all the copy elements—such as all the letters in a word, or all the strokes in a symbol—in order for correct identification to occur (Proffitt, Wade, and Lynn 1998). Familiar word or symbol recognition can be based on global features (Garvey, Pietrucha, and Meeker 1998). Sign copy recognition distances are, therefore, longer than would be predicted by either visual acuity or sign characteristics alone (Kuhn, Garvey, and Pietrucha 1998). In fact, one of the best ways to improve sign-reading distance is not through manipulation of sign characteristics, but rather by making the sign copy as familiar to the target audience as possible, a concept not lost on the advertising community.

However, modifications in sign design that improve sign legibility will enhance the reading distance for both novel and familiar content. The following sections provide an overview of more than 60 years of research on how to improve sign legibility. The research emphasizes the importance of sign characteristics such as photometric properties and symbol and textual size and shape.

Lighting Variables. The role of lighting variables in sign legibility is probably one of the best-researched areas in the sign visibility field. In this research, negative-contrast sign legibility (i.e., dark letters on a lighter background; e.g., regulatory and warning signs) is typically measured as a function of sign background luminance, and positive-contrast sign legibility (i.e., light letters on a darker background; e.g., guide signs) as a function of internal luminance contrast ratio.

Internal Contrast. Sivak and Olson (1985) derived perhaps the most well-accepted optimum contrast value for sign legibility. These researchers reviewed the sign legibility literature pertaining to sign contrast and came up with a contrast ratio of 12:1 for "fully reflectorized" or positive contrast signs using the average of the results of six separate research efforts. This 12:1 ratio would, for example, result in a sign with a 24 cd/m² legend and a 2 cd/m² background. This single, optimal ratio was expanded in a 1995 synthesis report by Staplin (1995) that gave a range of acceptable internal contrast levels between 4:1 and 50:1.

McNees and Jones (1987) found that the selection of retroreflective background material has a significant effect on sign legibility. These researchers found four combinations of sheeting and text to provide acceptable legibility distances for freeway guide signs: button copy on super engineer-grade (ASTM type II, or enclosed lens) background sheeting, high-intensity text on high-intensity background, high-intensity on super engineer, and high-intensity on engineer grade. Earlier research by Harmelink et al. (1985) concurs. These researchers found that observers favored high-intensity text on engineer-grade background, stating that this combination provided contrast ratios as good as those produced by high-intensity text on high-intensity background.

Sign Luminance. Khavanin and Schwab (1991) and Colomb and Michaut (1986) both concluded that only small increases in nighttime legibility distance occur with increases in sign retroreflectivity. More recently, however, Carlson and Hawkins (2002a) found that signs made of microprismatic sheeting resulted in longer legibility distances than those using encapsulated materials. The research of McNees and Jones (1987), Mace (1988), and Garvey and Mace (1996) supports that of Carlson and Hawkins.

Based on a review of the literature, Sivak and Olson (1983) suggested an optimal nighttime sign legend luminance of 75 cd/m² and a minimum of 2.4 cd/m² for negative contrast signs. With positive contrast signs, Garvey and Mace (1996) found 30 cd/m² to provide maximum nighttime legibility distance. Again using positive contrast signs, Garvey and Mace (1996) found that daytime legibility distance continued to improve with increases in luminance up to 850 cd/m², after which performance leveled off. In a study of on-premise sign visibility, Garvey, Pietrucha, and Cruzado (2008) found that nighttime legibility distance threshold for black-on-white (negative contrast) signs continued to increase until background sign luminance reached about 600 cd/m², above which legibility distance began to decrease.

Lighting Design. Overall, the literature indicates that a sign's luminance and contrast have a greater impact on legibility than does the specific means used to achieve these levels. Jones and Raska (1987) found no significant differences in legibility distance between lighted and unlighted overhead-mounted retroreflective signs for a variety of sign materials (McNees and Jones 1987). Other research extends this finding, indicating no significant difference in legibility distances for up to 10 different sign-lighting system types for freeway guide signs (McNees and Jones 1987; Upchurch and Bordin 1987). However, in a test track study that evaluated the effects of sign illumination type on storefront signs, Kuhn, Garvey, and Pietrucha (1999) found internally and neon-illuminated signs to perform 40 to 60 percent better at night than externally illuminated signs. A follow-up study (Garvey et al. in press) showed that even greater improvements (almost 70 percent on average and 240 percent in the best case) can be made when actual in-use, externally illuminated signs in the real world are upgraded to ones that use internal illumination.

In a study of changeable message sign (CMS) visibility, Garvey and Mace (1996) found retrore-flective and self-illuminated lighting design to provide equivalent legibility distances. Garvey and Mace did, however, find that the use of "black light" ultraviolet lighting severely reduced legibility. This was attributed to a reduction in internal luminance contrast and color contrast. Hussain, Arens, and Parsonson (1989) addressed this problem when they recommended the use of "white" fluorescent lamps for optimum color rendition and metal halide for overall performance (including color rendition) and cost-effectiveness.

Sign-Placement Variables.

Lateral Sign Placement.   Sign placement is as important to sign legibility as it is to detection. First, there is the obvious need to place signs so that traffic, pedestrians, buildings, and other signs do not block their messages. A less intuitive requirement for sign placement, however, involves the angle between the observer location and the sign (Prince 1958, in Claus and Claus 1974). Signs set off at large angles relative to the intended viewing location result in letter and symbol distortion. Prince recommended that the messages on signs at angles greater than 20° be manipulated (i.e., increased in height and/or width) to appear "normal" to the observer. Garvey (2006) took this a step further by developing a mathematical model to calculate letter heights for signs as a function of lateral offset.

Sign Variables.

Letter Height.   If a response to a guide sign is required, the typical behavior is speed reduction and a turning maneuver at the appropriate crossroad or interchange. On multilane roadways, the motorist may also have to change lanes. For warning and regulatory signs the motorist may have to reduce speed and will sometimes be required to change lanes or make a steering adjustment. Whether the sign is regulatory, warning, or for guidance, sign placement should allow sufficient time to comfortably react to the sign message after passing the sign. With the exception of corner-mounted street name signs, what occurs before the driver passes the sign should be limited to sign detection and reading for comprehension. Appropriate letter heights ensure sufficient time to accomplish the reading task.

READING SPEED.   Proffitt, Wade, and Lynn (1998) reported that the average normal reading speed for adults is about 250 words per minute (wpm), or 4.2 words per second. Research evaluating optimum acuity reserve (the ratio between threshold acuity and optimal print size) has demonstrated that optimal reading speeds result from print size that may be as much as four times size threshold (Bowers and Reid 1997; Yager, Aquilante, and Plass 1998; Lovie-Kitchin, Bowers, and Woods 2000). In fact, Yager, Acvilante, and Plass (1998) reported 0.0 wpm reading speed at size threshold. This explains some of the disparity between "normal" reading speed of above size threshold text and the time it takes to read a sign, which often begins at acuity threshold.

Research on highway sign reading provides evidence that it takes drivers approximately 0.5 to 2.0 seconds to read and process each sign word. Dudek (1991) recommended a minimum exposure time of "one second per short word . . . or two seconds per unit of information" for unfamiliar drivers to read changeable message signs. In a study conducted by Mast and Balias (1976), average advance guide sign reading was 3.12 seconds and average exit direction sign reading was 2.28 seconds. Smiley et al. (1998) found that 2.5 seconds was sufficient for 94 percent of their subjects to read signs accurately that contained three destination names; however, this dropped to 87.5 percent when the signs displayed four or five names.

McNees and Messer (1982) mentioned two equations to determine reading time: $t = (N/3) + 1$ and $t = 0.31N + 1.94$ (where t is time in seconds and N is the number of familiar words). In a literature survey on sign comprehension time, Holder (1971) concluded that the second equation was appropriate if the sign was located within an angular displacement of 10°. In their own research, McNees and Messer (1982) found that the time it takes to read a sign depends, among other things, on how much time the driver has to read it; in other words, signs are read faster when it is necessary to do so. However, they also found that as reading speed increases, so do errors; an example of the well-documented speed-accuracy tradeoff. McNees and Messer (1982) concluded that, "a cut-off of approximately 4.0 sec to read any sign was critical for safe handling of a vehicle along urban freeways." If the 4.0 seconds are plugged back into the second equation, the number of familiar words on the sign would be 6.7, or 1.7 words per second.

While it is impractical to specify a single minimum reading time that will allow all drivers to read and understand all signs, the research on sign-reading speed indicates that signs with four to eight words could be comfortably read and comprehended in approximately 4.0 seconds and signs with one to three words in about 2.5 seconds.

TASK LOADING. In addition to sign reading, the driver must also watch the road and perform other driving tasks. Considering overhead guide signs, McNees and Messer (1982) estimated that a 4.5-second sign-reading time would actually require an 11.0-second and sign-legibility distance. This results from adding 2.0 seconds for sign-clearance time (when the vehicle is too close to the sign for the driver to read it) and dividing the remaining 9.0 seconds equally between sign reading and other driving tasks. In looking at shoulder-mounted signs, Smiley et al. (1998) provide more practical estimates. These researchers allowed for 0.5 seconds clearance time and a 0.5-second glance back at the road for every 2.5 seconds of sign reading (based on eye movement research by Bhise and Rockwell 1973). This would require a 5.0-second legibility distance for 4.0 second of sign reading and 3.0 seconds legibility distance for 2.5 seconds of sign reading. This is assuming that the driver begins to read the sign as soon as it becomes legible. Allowing an additional 1.0 seconds for sign acquisition after it becomes legible, appropriate legibility distance for signs displaying four to eight words would be 6.0 seconds and for signs with one to three words would be 4.0 seconds. Based on these calculations and assuming a legibility index of approximately 40 ft/in., Table 7.1 provides reading times and recommended letter heights as a function of the number of words on the sign and travel speeds.

DIMINISHING RETURNS. While research indicates that legibility distance increases with letter height, a point of diminishing return exists (Allen et al. 1967; Khavanin and Schwab 1991). For example, doubling letter height will increase, but will not double, sign reading distance. Mace, Perchonok, and Pollack (1994) and Garvey and Mace (1996) found that increases in letter height above about 8 inches resulted in nonproportional increases in legibility distance. Garvey and Mace (1996) found that a sign with 42-inch characters produced only 80 percent of the legibility index of the same sign with 18-inch characters. That is, the 42-inch character produced a legibility distance of approximately 1,350 feet (LI = 32 ft/in.) while the 18-inch characters resulted in a legibility distance of about 800 ft (LI = 44 ft/in.).

Text versus Symbols. In a study of traffic sign comprehension speed, Ells and Dewar (1979) found symbolic signs to outperform those with textual messages. These researchers also discovered that symbolic signs were less susceptible than were text signs to visual degradation. In a 1975 visibility study, Jacobs, Johnston, and Cole assessed the legibility distance of almost 50 symbols and their textual counterparts. These researchers found that in the majority of cases the legibility distances for the symbols were twice that of the alphanumeric signs. This finding was replicated in Kline and Fuchs' (1993) research for a smaller set of symbols using young, middle-aged, and older observers. Kline and Fuchs' research also introduced a technique to optimize symbol legibility: recursive blurring, which results in symbols designed to "maximize contour size and contour separation." In other words, optimized symbols or logos will have elements that are large enough to be seen from a distance and spaces between the elements wide enough to reduce blurring between elements.

The literature clearly indicates that, from a visibility standpoint, symbols are superior to text. Symbols, however, require a different kind of comprehension than words. Symbol meaning is either understood intuitively or learned. Although traffic sign experts and traffic engineers agree that understandability is the most important factor in symbol design (Dewar 1988), other research has shown that what is intuitive to designers is not always intuitive to drivers, and that teaching observers the meaning of more abstract symbols is frequently unsuccessful. For example, in one study (Kline et al. 1990) even the relatively simple "HILL" symbol resulted in only 85 percent comprehension, while the "ROAD NARROWS" symbol accommodated only 52 percent of the respondents. In researching the Slow Moving Vehicle emblem, Garvey (2003) found correct symbol recognition to be approximately 30 percent for older and younger subjects under daytime and nighttime viewing conditions.

Upper-Case versus Mixed-Case.   Forbes, Moskowitz, and Morgan (1950) conducted perhaps the definitive study on the difference in sign legibility between text depicted in all upper-case letters and that shown in mixed-case. When upper- and mixed-case words subtended the same sign area these researchers found a significant improvement in legibility distance with the mixed-case words. Garvey, Pietrucha, and Meeker (1997) replicated this result with new sign materials, a different font, and older observers. They found a 12 to 15 percent increase in legibility distance with mixed-case text under daytime and nighttime conditions. It must be noted, however, that these results were obtained with a recognition task—that is, the observers knew what words they were looking for. In instances where the observer does not know the text, improvements with mixed-case are not evident (Forbes, Moskowitz, and Morgan 1950; Mace, Garvey, and Heckard 1994; and Garvey, Pietrucha, and Meeker 1997).

Font.   Assessing the effect of letter style on traffic signs has been limited by state and federal governments' desire to keep the font "clean," in other words, a sans serif alphabet that has a relatively constant stroke width. While sans serif letters are generally considered to provide greater legibility distance than serif letters (Prince 1957, in Claus and Claus 1974), a comparison of the sans serif Standard Highway Alphabet with Clarendon, the serif standard National Park Service font, revealed a slight improvement with the Clarendon font (Mace, Garvey, and Heckard 1994).

Currently, the only font allowed by FHWA on road signs is the Standard Highway Alphabet (FHWA 2004). However, research on highway font legibility (Garvey, Pietrucha, and Meeker 1997, 1998; Hawkins et al. 1999; and Carlson and Brinkmeyer 2002) has led the FHWA to allow any state that requests it to use Clearview on all their positive-contrast signs (Figure 7.3). Clearview was designed to reduce halation (Figure 7.4) resulting from the use of high-brightness retroreflective materials (i.e., microprismatic sheeting) and improve letter legibility for older drivers. Related research for the National Park Service has led that agency to accept NPS Roadway as an alternate to the NPS's current Clarendon font (Figure 7.5). NPS Roadway has been shown to increase sign legibility by 10.5 percent while reducing word length by 11.5 percent (Garvey et al. 2001).

Stroke Width.   Kuntz and Sleight (1950) concluded that the optimal stroke width-to-height ratio for both positive and negative-contrast letters was 1:5. Forbes et al. (1976) found increases in legibility distance of fully reflectorized, positive-contrast letters and decreases in legibility for negative-contrast letters when the stroke width-to-height ratio was reduced from 1:5 to 1:7. That is, light

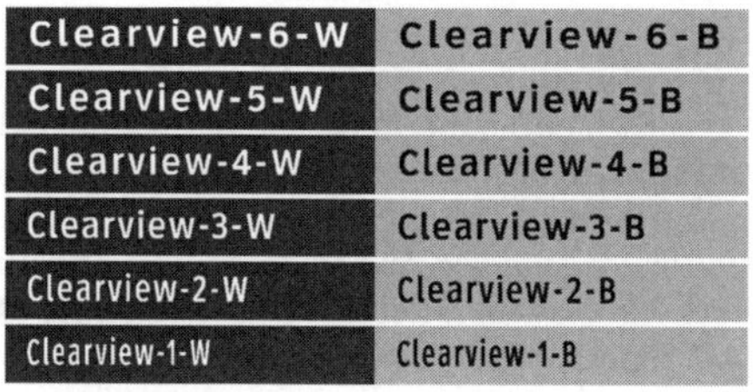

Clearview-6-W	Clearview-6-B
Clearview-5-W	Clearview-5-B
Clearview-4-W	Clearview-4-B
Clearview-3-W	Clearview-3-B
Clearview-2-W	Clearview-2-B
Clearview-1-W	Clearview-1-B

FIGURE 7.3

FIGURE 7.4

FIGURE 7.5

letters on a darker background should have a thinner stroke and dark letters on a lighter background should have a bolder stroke. Improved legibility for fully reflectorized, white-on-green signs with thinner stroke width was also found by Mace, Garvey, and Heckard (1994) for very high contrast signs, and for mixed case text by Garvey, Pietrucha, and Meeker (1998).

Abbreviations. In a study of changeable message sign comprehension, Huchingson and Dudek (1983) developed several abbreviation strategies. These researchers recommended the technique of using only the first syllable for words having nine letters or more; for example, *Cond* for *Condition*. This technique should not, however, be used if the first syllable is in itself a new word. A second method using the key consonants was suggested for five-to seven-letter words; for example, *Frwy* for *Freeway*. The *MUTCD* (FHWA 2009) provides lists of "acceptable" and "unacceptable" abbreviations for traffic control devices. Abbreviations, however, are to be used only as a last resort if limitations in sign size demand it, as they increase the possibility of incorrect sign interpretation. Alternative suggestions to deal with sign size limitations include selecting a synonym for the abbreviated word, reducing letter size, reducing message length, and increasing sign size.

Contrast Orientation. The research on this issue is clear; with the possible exception of tight inter-character spacing (Case et al. 1952), positive-contrast signs provide greater legibility distances than negative-contrast signs. As far back as 1955, laboratory research by Allen and Straub found that white-on-black signs (positive-contrast) provided longer legibility distances than black-on-white signs when the sign luminance was between 3 and 30 cd/m^2. Allen et al. (1967) replicated these results in the field. Garvey and Mace (1996) extended these results in their changeable message sign research with the addition of orange, yellow, and green signs. Positive-contrast signs resulted in improvements of about 30 percent over negative-contrast signs (Garvey and Mace 1996).

Color. Schnell et al. (2001) found small legibility improvements when comparing signs using fluorescent colors versus signs using matching nonfluorescent colors. However, in an evaluation of normal sign colors, Garvey and Mace (1996) found no difference in legibility distance that could not be accounted for by luminance, luminous contrast, or contrast orientation between signs using the following color combinations: white/green, black/white, black/orange, black/yellow, and black/red. This is also consistent with the findings of research on computer displays (Pastoor 1990). In general, the research indicates that if appropriate luminance contrast, color contrast, and luminance levels are maintained, the choice of specific colors for background and text does not affect legibility distance.

7.6 FINAL REMARKS

For any type of highway sign to be visually effective, it must be readable. While this seems like a fairly simple objective to achieve, the information presented herein indicates that the ability of a driver to detect and read a sign is a function of numerous human, environmental, and design factors with complex interrelationships. For example, visual acuity, contrast sensitivity, visual field, and glare can significantly impact the driver's ability to see the sign and read its message. Furthermore, the basic design and placement of the sign can greatly impact the detectibility and legibility of any highway sign. Features such as height, lateral offset, reflectorization, illumination, message design, font, case, letter height, border, and letter spacing are so critical that federal guidelines dictate these features to maximize the potential for a driver to see the sign within the highway environment. The photometric characteristics of the sign, including the internal contrast, luminance, and light design, can also directly impact how well a driver sees a sign. And if all of these factors are not enough, the location of a sign relative to the rest of the highway environment can either enhance its detectibility or force it to compete with other signs and objects for visibility. Thus, the task of designing detectible, legible, and understandable highway signs is a challenge that continues to be refined as we learn more about their role in the highway environment and their interaction with the driver. Ultimately, we hope to provide critical information to the traveler that they can use in a timely manner to navigate safely through the highway environment.

REFERENCES

Allen, T. M., and Straub, A. L. 1955. "Sign Brightness and Legibility." *Highway Research Board Bulletin* 127, 1–14.

Allen, T. M., F. N. Dyer, G. M. Smith, and M. H. Janson. 1967. "Luminance Requirements for Illuminated Signs." *Highway Research Record* 179, 16–37.

American Society for Testing and Materials (ASTM). 2001. *Standard Test Method for Coefficient of Retroreflection of Retroreflective Sheeting Utilizing the Coplanar Geometry.* Document Number ASTM E810-01, ASTM International.

Ball, K., and C. Owsley. 1991. "Identifying Correlates of Accident Involvement for the Older Driver." *Human Factors* 33(5):583–95.

Bhise, V. D., and T. H. Rockwell. 1973. *Development of a Methodology for Evaluating Road Signs.* Final Report, Ohio State University.

Boff, K. R., L. Kaufman, and J. P. Thomas. 1986. *Handbook of Perception and Human Performance,* vols. 1 and 2. New York: John Wiley & Sons.

Bowers, A. R., and V. M. Reid. 1997. "Eye Movement and Reading with Simulated Visual Impairment." *Ophthalmology and Physiological Optics* 17(5):492–02.

Carlson, P. J., and G. Brinkmeyer. 2002. "Evaluation of Clearview on Freeway Guide Signs with Microprismatic Sheeting." *Transportation Research Record* 1801, 27–38.

Carlson, P. J., and G. Hawkins. 2002a. "Legibility of Overhead Guide Signs Using Encapsulated versus Microprismatic Retroreflective Sheeting." Presented at the 16th Biennial Symposium on Visibility and Simulation, Iowa City, IA.

———. 2002b. "Minimum Retroreflectivity for Overhead Guide Signs and Street Name Signs." *Transportation Research Record* 1794, 38–48.

Case, H. W., J. L. Michael, G. E. Mount, and R. Brenner. 1952. "Analysis of Certain Variables Related to Sign Legibility." *Highway Research Board Bulletin* 60, 44–58.

Claus, K., and J. R. Claus. 1974. *Visual Communication through Signage,* vol. 1, *Perception of the Message.* Cincinnati: Signs of the Times.

Colomb, M., and G. Michaut. 1986. "Retroreflective Road Signs: Visibility at Night." *Transportation Research Record* 1093, 58–65.

Cooper, B. R. 1988. "A Comparison of Different Ways of Increasing Traffic Sign Conspicuity." *TRRL Report* 157.

Dewar, R. E. 1988. "Criteria for the Design and Evaluation of Traffic Sign Symbols." *Transportation Research Record* 1160, 1–6.

Dudek, C. L. 1991. *Guidelines on the Use of Changeable Message Signs.* Final Report, DTFH61-89-R00053, U.S. DOT, Federal Highway Administration, Washington, DC.

Ells, J. G., and R. E. Dewar. 1979. "Rapid Comprehension of Verbal and Symbolic Traffic Sign Messages." *Human Factors* 21, 161–68.

Federal Highway Administration (FHWA). 1978a. *Roadway Lighting Handbook.* U.S. Department of Transportation, FHWA, Washington, DC.

———. 2004. *Standard Highway Signs.* Washington: U.S. Department of Transportation, FHWA.

———. 2009. *Manual on Uniform Traffic Control Devices for Streets and Highways.* Washington: U.S. Department of Transportation, FHWA.

Forbes, T. W., J. P. Fry, R. P. Joyce, and R. F. Pain. 1968a. "Letter and Sign Contrast, Brightness, and Size Effects on Visibility." *Highway Research Record* 216, 48–54.

Forbes, T. W., K. Moskowitz, and G. Morgan. 1950. "A Comparison of Lower Case and Capital Letters for Highway Signs." *Proceedings, Highway Research Board* 30, 355–73.

Forbes, T. W., R. F. Pain, R. P. Joyce, and J. P. Fry. 1968b. "Color and Brightness Factors in Simulated and Full-Scale Traffic Sign Visibility." *Highway Research Record* 216, 55–65.

Forbes, T. W., B. B. Saari, W. H. Greenwood, J. G. Goldblatt, and T. E. Hill. 1976. "Luminance and Contrast Requirements for Legibility and Visibility of Highway Signs." *Transportation Research Record* 562, 59–72.

Garvey, P. M. 2003. "Motorist Comprehension of the Slow Moving Vehicle (SMV) Emblem." *Journal of Agricultural Safety and Health* 9(2):159–169.

Garvey, P. M. 2005. "On-Premise Commercial Sign Lighting and Light Pollution." *Leukos: The Journal of the Illuminating Society of North America* 1, 7–18.

Garvey, P. M. (2006). "Determination of Parallel Sign Legibility and Letter Heights." *United States Sign Council Foundation (USSCF) Research Project*, Final Report, Bristol, PA.

Garvey, P. M., M. T. Pietrucha, and I. Cruzado. 2008. "The Effect of Internally Illuminated On-Premise Sign Brightness on Nighttime Sign Visibility and Traffic Safety." *United States Sign Council Foundation (USSCF) Research Project*, Final Report, Bristol, PA.

Garvey, P. M., M. T. Pietrucha, S. J. Damin, and D. Deptuch. In press. "Relative Visibility of Internally vs. Externally Illuminated On-Premise Signs." *TRR. Journal of the Transportation Research Board.*

Garvey, P. M., and D. M. Mace. 1996. *Changeable Message Sign Visibility.* Publication No. FHWA-RD94-077.

Garvey, P. M., M. T. Pietrucha, and D. Meeker. 1997. "Effects of Font and Capitalization on Legibility of Guide Signs." *Transportation Research Record* 1605, 73–79.

———. 1998. "Development of a New Guide Sign Alphabet." *Ergonomics in Design* 6(3)7–11.

Garvey, P. M., A. Z. Zineddin, M. T. Pietrucha, D. T. Meeker, and J. Montalbano. 2001. *Development and Testing of a New font for National Park Service Signs.* U.S. Department of the Interior, National Park Service Final Report.

Hanson, D. R., and H. L. Waltman. 1967. "Sign Backgrounds and Angular Position." *Highway Research Record* 170, 82–96.

Harmelink, M. D., G. Hemsley, D. Duncan, R. W. Kuhk, and T. Titishov. 1985. "Evaluation of Reflectorized Sign Sheeting for Nonilluminated Freeway Overhead Guide Signs." *Transportation Research Record* 1010, 80–84.

Hawkins, H. G., D. L. Picha, M. D. Wooldridge, F. K. Greene, and G. Brinkmeyer. 1999. "Performance Comparison of Three Freeway Guide Sign Alphabets." *Transportation Research Record* 1692, 9–16.

Holder, R. W. 1971. "Consideration of Comprehension Time in Designing Highway Signs." *Texas Transportation Researcher* 7(3):8–9.

Huchingson, R. D., and C. L. Dudek. 1983. "How to Abbreviate on Highway Signs." *Transportation Research Record* 904, 1–3.

Hughes, P. K., and B. L. Cole. 1984. "Search and Attention Conspicuity of Road Traffic Control Devices." *Australian Road Research Board* 14(1):1–9.

Hussain, S. F., J. B. Arens, and P. S. Aparsonson. 1989. "Effects of Light Sources on Highway Sign Color Recognition." *Transportation Research Record* 1213, 27–34.

Jacobs, R. J., A. W. Johnston, and B. L. Cole. 1975. "The Visibility of Alphabetic and Symbolic Traffic Signs." *Australian Road Research* 5(7):68–86.

Jenkins, S. E., and B. L. Cole. 1986. "Daytime Conspicuity of Road Traffic Control." *Transportation Research Record* 1093, 74–80.

Khavanin, M. R., and R. N. Schwab. 1991. "Traffic Sign Legibility and Conspicuity for the Older Drivers." In *1991 Compendium of Technical Papers.* Washington, DC: Institute of Transportation Engineers, 11–14.

Kuhn, B. T., P. M. Garvey, and M. T. Pietrucha. 1998. "The Impact of Color on Typical On-Premise Sign Font Visibility." Presented at TRB's 14th Biennial Symposium on Visibility, Washington, DC, April.

———. 1999. "On Premise Sign Legibility and Illumination." In *1999 Compendium of Technical Papers.* Washington, DC: Institute of Transportation Engineers.

King, G. F., and H. Lunenfeld. 1971. *Development of Information Requirements and Transmission Techniques for Highway Users.* NCHRP Report 123, National Cooperative Highway Research Program, Washington, DC.

Kline, D. W., and P. Fuchs. 1993. "The Visibility of Symbolic Highway Signs Can Be Increased among Drivers of All Ages." *Human Factors* 35(1):25–34.

Kline, D. W., K. Buck, Y. Sell, T. L. Bolan, and R. E. Dewar. 1999. "Older Observers' Tolerance of Optical Blur: Age Differences in the Identification of Defocused Test Signs." *Human Factors* 41(3):356–64.

Kline, T. J. B., L. M. Ghali, D. W. Kline, and S. Brown. 1990. "Visibility Distance of Highway Signs among Young, Middle-Aged, and Older Observers: Icons Are Better than Text." *Human Factors* 32(5):609–19.

Kuntz, J. E., and R. B. Sleight. 1950. "Legibility of Numerals: The Optimal Ratio of Height to Width of Stroke." *American Journal of Psychology* 63:567–75.

Lovie-Kitchin, J. E., A. R. Bowers, and R. L. Woods. 2000. "Oral and Silent Reading Performance with Macular Degeneration." *Ophthalmology and Physiological Optics* 20(5):360–70.

Mace, D. J. 1988. "Sign Legibility and Conspicuity." In *Transportation in an Aging Society.* Special Report 218, vol. 2. Washington, DC: National Research Council, Transportation Research Board, 270–93

Mace, D. J., and L. Pollack. 1983. "Visual Complexity and Sign Brightness in Detection and Recognition of Traffic Signs." *Transportation Research Record* 904, 33–41.

Mace, D. J., P. M. Garvey, and R. F. Heckard. 1994. *Relative Visibility of Increased Legend Size vs. Brighter Materials for Traffic Signs.* Report FHWA-RD-94-035, U.S. Department of Transportation, Federal Highway Administration, Washington, DC.

Mace, D., K. Perchonok, and L. Pollack. 1982. *Traffic Signs in Complex Visual Environments.* Report FHWA-RD-82-102, U.S. Department of Transportation, Federal Highway Administration, Washington, DC.

Mast, T. M., and J. A. Balias. 1976. "Diversionary Signing Content and Driver Behavior." *Transportation Research Record* 600, 14–19.

McNees, R. W., and H. D. Jones. 1987. "Legibility of Freeway Guide Signs as Determined by Sign Materials." *Transportation Research Record* 1149, 22–31.

McNees, R. W., and C. J. Messer. 1982. "Reading Time and Accuracy of Response to Simulated Urban Freeway Guide Signs." *Transportation Research Record* 844, 41–50.

Pain, R. F. 1969. "Brightness and Brightness Ratio as Factors in the Attention Value of Highway Signs." *Highway Research Record* 275, 32–40.

Pastoor, S. 1990. "Legibility and Subjective Preference for Color Combinations in Text." *Human Factors* 32(2): 157–71.

Pietrucha, M. T., E. T. Donnell, P. Lertworawanich, and L. Elefteriadou. 2003. "Traffic Volume, Traffic Speed, Mounting Height, and Lateral Offset on the Visibility of On-Premise Commercial Signs." *United States Sign Council Foundation (USSCF) Research Project*, Final Report, Bristol, PA.

Proffitt, D. R., M. M. Wade, and C. Lynn. 1998. *Creating Effective Variable Message Signs: Human Factors Issues.* Final Contract Report, Proj No. 9816-040-940, VTRC 98-CR31, Virginia Department of Transportation, Richmond, VA.

Schnell, T., K. Bentley, E. Hayes, and M. Rick. 2001. "Legibility Distances of Fluorescent Traffic Signs and Their Normal Color Counterparts." *Transportation Research Record* 1754, 31–41.

Sivak, M., and P. L. Olson. 1983. *Optimal and Replacement Luminances of Traffic Signs: A Review of Applied Legibility Research.* UMTRI-83-43, University of Michigan Transportation Research Institute, Ann Arbor, MI.

———. 1985. "Optimal and Minimal Luminance Characteristics for Retroreflective Highway Signs. *Transportation Research Record* 1027, 53–56.

Smiley, A., C. MacGregor, R. E. Dewar, and C. Blamey. 1998. "Evaluation of Prototype Tourist Signs for Ontario." *Transportation Research Record* 1628, 34–40.

Staplin, L. 1995. *Older Driver and Highway Safety Literature Review and Synthesis.* Working Paper in Progress. U.S. Department of Transportation, FHWA, Washington, DC.

———. 1983. *Traffic Control Device Handbook.* U.S. Department of Transportation, FHWA, Washington, DC.

Upchurch, J. E., and J. D. Armstrong. 1992. "A Human Factors Evaluation of Alternative Variable Message Sign Technologies." In *Vehicle Navigation and Information Systems.* Oslo: Norwegian Society of Chartered Engineers, 262–67.

Upchurch, J. E., and J. T. Bordin. 1987. "Evaluation of Alternative Sign-Lighting Systems to Reduce Operating and Maintenance Costs." *Transportation Research Record* 1111, 79–91.

Woods, D. L., and N. J. Rowan. 1970. "Street Name Signs for Arterial Streets." *Highway Research Record* 325, 54.

Yager, D., K. Aquilante, and R. Plass. 1998. "High and Low Luminance Letters, Acuity Reserve, and Font Effects on Reading Speed." *Vision Research* 38:2527–31.

Zwahlen, H. T. 1989. "Conspicuity of Suprathreshold Reflective Targets in a Driver's Peripheral Visual Field at Night." *Transportation Research Record* 1213, 35–46.

Zwahlen, H. T., and J. Yu. 1991. "Color and Shape Recognition of Reflectorized Targets under Automobile Low-Beam Illumination at Night." *Transportation Research Record* 1327, 1–7.

CHAPTER 8
ROADWAY TRANSPORTATION LIGHTING

John D. Bullough
Lighting Research Center, Rensselaer Polytechnic Institute
Troy, New York

8.1 INTRODUCTION

Lighting and signaling systems are critical elements of the roadway visibility system. They include illumination systems, by which objects in and along the road and their surrounding backgrounds are made visible during nighttime driving conditions, and signal light systems that provide information to drivers regarding the status of traffic or of other vehicles. The latter systems are used during both daytime and nighttime. After brief introductions to the measurement and quantification of light and to lighting technologies, this chapter discusses the design and implementation of roadway lighting systems, requirements for vehicle lighting and signaling systems, and the visual characteristics of traffic signals. This chapter concludes with a short discussion of emerging issues in transportation lighting that are beginning to affect lighting practice. Information on sign visibility and roadway markings, two other critical elements of the roadway visibility system, is found elsewhere in this book.

8.2 CHARACTERIZATION AND MEASUREMENT OF LIGHT

8.2.1 Photometry

Light is visually perceived radiant energy, located between about 380 and 780 nanometers (1 nanometer = 10^{-9} meter) on the electromagnetic spectrum (Rea 2000). Light is only a small part of the entire electromagnetic spectrum, which also includes x-rays, ultraviolet (UV) radiation, infrared (IR) radiation, and radio and television waves. Electromagnetic radiation is measured in terms of its radiant power in watts. However, not all visible wavelengths are evaluated equally by the human visual system. A given amount of radiant power at 590 nanometers ("yellow" light) will not appear as bright as the same radiant power at 550 nanometers ("green" light). Furthermore, these relationships can change depending on the overall ambient light level. Under typical daytime light levels, the visual system is most sensitive to wavelengths near 555 nanometers, and sensitivity decreases for shorter or longer wavelengths. This is so because photoreceptors in the retina, called *cones*, are used in daytime or photopic vision, and the cones are maximally sensitive to light at 555 nanometers (Figure 8.1).

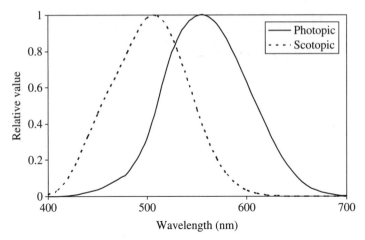

FIGURE 8.1 Photopic and scotopic luminous efficiency functions.

At very low levels, photoreceptors called *rods* are used for vision. This is called *scotopic vision*. Rods are maximally sensitive to light at 507 nanometers (see Figure 8.1). Scotopic vision is sometimes called *nighttime vision*, although most nighttime levels are actually in the range of levels called *mesopic vision*, where the transition from cones to rods occurs and the maximal sensitivity will lie somewhere between 555 and 507 nanometers depending on the light level. At high mesopic light levels, the maximum sensitivity will be closer to 555 nanometers, and at low mesopic levels, it will be near 507 nanometers. The implications of this fact on transportation lighting practice are discussed further in section 8.8.

Light is characterized in several different ways. A point source with a uniform *luminous intensity* of 1 candela (cd) will produce a total light output, or *luminous flux*, of 4π, or about 12.6 lumens (lm) on the interior surface of a sphere. Furthermore, if the source is placed at the center of the sphere, the luminous flux density, or *illuminance*, on the inner surface of the sphere will be 1 lm/ft^2 or 1 footcandle (fc) if the sphere radius is 1 foot and will be 1 lm/m^2 or 1 lux (lx) if the sphere radius is 1 meter (1 footcandle = 10.76 lux).

Light can also be measured in terms of *luminance*, which is the density of luminous intensity from a surface in a particular direction. For many practical purposes, luminance is analogous to the perceived brightness of a surface. It is measured in candelas per square meter; occasionally, this unit is referred to as a *nit*.

The measures of light that are used most commonly by illuminating engineers and in recommendations of organizations such as the Illuminating Engineering Society (IES) of North America or the Commission Internationale de l'Éclairage (CIE) are illuminance and luminance. For signal light applications, luminous intensity is the most commonly specified parameter. Portable meters for measuring illuminance and luminance in the lighted environment are available (Rea 2000). Measuring luminous intensity can be performed indirectly by measuring the illuminance from a signal light in a given direction and applying the following equation based on the *inverse-square law* (whereby the illuminance from a light source decreases as a function of the inverse of the square of the distance from the source) to calculate luminous intensity:

$$I = Ed^2 \tag{8.1}$$

where E is the illuminance from the light source (in lux), d is the distance between the light source and the measurement location (in meters), and I is the luminous intensity (in candelas). (Equation 8.1 also can be used if E is measured in units of footcandles and d in units of feet.)

8.2.2 Color Properties of Lighting

Color properties of light sources are characterized in several ways. Two commonly used ways are the *correlated color temperature* (CCT) and the *color rendering index* (CRI). The CCT of a light source is a measure of its color appearance. It is based on the temperature (in kelvins) of a blackbody radiator that produces light nearest in color to the light source being evaluated. Tungsten, used in incandescent lamp filaments, behaves much like a blackbody radiator. The temperature of a tungsten filament (about 2800 K in incandescent lamps) is approximately equal to its CCT. CCTs of other lamp types are based on how close their color is to a blackbody radiator (or tungsten filament) of a certain temperature. At higher temperatures, a heated tungsten filament will become "bluer" in appearance. At lower temperatures, it will become "redder."

The CRI of a light source is a rating that is designed to correlate with the light source's ability to render several standard test colors relative to a blackbody (incandescent type) or a daylight source. It has a maximum value of 100; a light source with a CRI approaching 100 is thought to have excellent color-rendering properties. Light sources with CRI values of 70 or higher generally are considered to have good color-rendering properties.

8.2.3 Optical Control

Light interacts with surfaces in several ways. It can be absorbed by materials. Dark materials have a low *reflectance* and absorb more light than bright materials. Light also can be reflected by materials, such as a mirror or a painted wall. In the case of a mirror, the reflections are called *specular*; in the case of a matte-painted surface, most of the reflected light is *diffuse*. Surfaces such as pavement reflect light with a combination of specular and diffuse reflections.

Light can be transmitted as through glass. Depending on the incident angle of the light, it also can undergo *refraction* ("bending"). Light fixtures (*luminaires*) can use glass or clear-plastic lenses to control the distribution of light through refraction or can redirect light (through reflection) from elements such as parabolic or ellipsoidal reflectors made of materials such as aluminum (Figure 8.2).

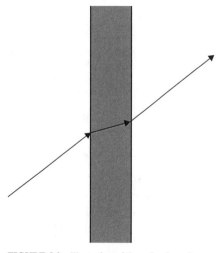

FIGURE 8.2 Illustration of the refraction of a ray of light by glass.

The spectral (color) properties of materials also affect the way they interact with light. For example, a red surface will reflect long visible wavelengths, which are perceived as red light, and will absorb other wavelengths. A blue gel (filter) transmits short visible wavelengths, which are perceived as blue light, and absorbs other wavelengths (Rea 2000).

8.3 VISION

8.3.1 Visual Performance

The primary purpose of lighting in the roadway environment is to aid vision. Because of concerns for energy conservation, there is a desire to provide enough light to allow people to safely and accurately perform visual tasks without providing excessive lighting that could, in fact, waste electrical energy. An understanding of visual tasks and the factors that affect visual performance can aid in providing the appropriate level of lighting for those tasks. *Visual performance* (in other words, the speed and accuracy with which a visual task can be performed) is affected by several factors (Rea and Ouellette 1991), including

- Contrast
- Size
- Duration

Contrast refers to the luminance difference between the critical detail of a task and its background, for example, the clothing worn by a pedestrian against the paved surface of the roadway behind the pedestrian. A commonly used equation for luminance contrast is

$$C = (L_t - L_b)/\max(L_t, L_b) \tag{8.2}$$

where L_t is the luminance of the target (in cd/m^2), L_b is the luminance of the background (in cd/m^2), and C is the luminance contrast in dimensionless units.

The absolute value of C in equation (8.2) will always be between 0 and 1, inclusive. When the target has a higher luminance than the background, C is positive. When the target has a lower luminance than the background, C is negative.

A reading task such as white letters on a black sign background would have a high contrast, approaching 1, whereas a dark-colored animal viewed against a dark background of foliage could have a low contrast, approaching 0. As the contrast of a task increases, visual performance also increases. Rea and Ouellette (1991) have published a comprehensive and validated model of visual performance based on speed and accuracy of visual processing. Once contrast is sufficiently high, further increases in contrast will have little effect on visual performance.

The *size* of the visual task is also important. A task with an infinitesimally small size will be invisible, with visibility increasing as size increases. As with contrast, once the size is sufficiently large, making it larger will not increase visual performance very much. For example, visual performance while reading 6-point type might be significantly better than visual performance reading 4-point type, but visual performance while reading 18-point type would be only marginally better than visual performance reading 12-point type.

The *duration* a visual task is presented also affects visual performance. For visual tasks that are visible for only very brief periods of time (<0.1 second), the intensity and duration are traded off such that a signal that appears for half the duration of a second signal would need to have twice the intensity to be as visible as the second signal. For presentation durations longer than 0.1 second, the duration of the visual task is much less important.

For a visual task with a specific contrast, size, and duration, lighting directly affects another factor that helps to determine the overall visual performance. The background luminance is affected by the illuminance falling on the surface surrounding the object and by the reflectance of that surface. Holding contrast, size, and duration constant, the higher the background luminance, the better visual performance will be. Luminance affects the lowest contrast that can be detected. As the luminance is increased, the minimum contrast decreases so that some objects that would be invisible at low luminances might become visible at higher luminances. Similarly, visual acuity (the ability to distinguish detail of the smallest objects that can be seen) improves with increasing light levels. In addition, both speed and accuracy of visual processing improve with higher light levels.

Dark surfaces, such as asphalt, will have a lower luminance than lighter surfaces, such as concrete, even for the same illuminance falling on them because the reflectance of asphalt is lower than that of concrete. For diffusely reflecting surfaces, the luminance L (in cd/m^2) of a surface can be estimated if its reflectance and the illuminance on the surface are known according to the formula

$$L = \rho E/\pi \tag{8.3}$$

where E is the illuminance on the surface (in lx), ρ is the reflectance of the surface (a dimensionless quantity ranging from 0 to 1, as described below), and L is the luminance (in cd/m^2).

Table 8.1 lists the approximate diffuse reflectances of a number of materials commonly used in the outdoor lighted environment (IES 1981). The reflectance of typical dark clothing is around 0.05.

Although the performance of a visual task improves with higher and higher light levels, the rate of improvement decreases when the light level is high. This *plateau effect* is inherent to most visual tasks encountered at work or at home when objects in the field of view are well above the visual threshold (Rea and Ouellette 1991). For very small, low-contrast visual tasks, however, the light level will have a relatively larger impact on visual performance.

The age of the occupant also should be considered when planning the lighting in a space. The amount of light reaching the retina of a 50-year-old person is approximately 50 percent of that of a 20-year-old (Rea 2000) because the lens of the eye yellows and thickens during aging, and because the pupil of the eye tends to decrease in size as a function of age.

TABLE 8.1 Reflectances of Common Materials Found Outdoors

Material	Reflectance
Asphalt (free from dirt)	0.07
Bluestone, sandstone	0.18
Brick, dark red, glazed	0.30
Cement	0.27
Concrete	0.40
Earth, moist, cultivated	0.07
Granolite pavement	0.17
Grass, dark green	0.06
Gravel	0.13
Macadam	0.18
Paint (white), new	0.75
Paint (white), old	0.55
Slate, dark clay	0.08
Snow, new	0.74
Snow, old	0.64
Vegetation (mean)	0.25

Source: IES 1981.

8.3.2 Glare

Although lighting is usually considered to be beneficial for visibility, excessive light can reduce visual performance by creating *disability glare*, which is the reduction in contrast caused by a bright light in the field of view that results in scattered light inside the eye. This scattered light creates a *veiling luminance* analogous to looking through a veil or haze. It has been demonstrated that the veiling luminance L_v from a light source in the field of view can be predicted using the following equation (Fry 1954):

$$L_v = 9.2E/[\theta(\theta + 1.5)] \tag{8.4}$$

where E is the illuminance (in lx) from the light source at the eyes of the observer, θ is the angular distance (in degrees) between the light source and the observer's line of sight, and L_v is the veiling luminance (in cd/m^2). The L_v term is added to both the target and the background luminances in equation (8.2) to quantify the equivalent reduction in luminance contrast for visual performance caused by the glare source.

Bright light sources also can cause *discomfort glare*, which is the annoying or even painful sensation created by the offending source. Mechanisms for discomfort glare are not well understood but differ from those for disability glare. Two lights that reduce visibility equally may be perceived as unequally uncomfortable; lights with substantial short-wavelength ("blue") spectral content tend to be rated uncomfortable compared with "yellowish" lights (Bullough 2009). Nonetheless, disability and discomfort glare often occur simultaneously in the nighttime lighted environment. See section 8.5 for a discussion of issues related to headlamp glare.

8.4 *LIGHT SOURCE TECHNOLOGIES*

Several types of light sources, or lamps, are used commonly in transportation lighting systems. Consult Rea (2000) and the National Lighting Product Information Program (www.lrc.rpi.edu/programs/nlpip) for detailed and regularly updated technical information on many different electric light sources. Important light source performance characteristics include luminous efficacy (lm/W), rated life, color-rendering characteristics, and *lumen maintenance* (defined as the percentage of light output that can be reasonably expected from a lamp after it has operated for approximately half its rated life).

8.4.1 Filament Lamps

Filament (incandescent and tungsten-halogen) lamps are available in many shapes and sizes. They are not used commonly in fixed roadway lighting systems but are the predominant light source in vehicle forward lighting. These lamps contain a tungsten filament inside a glass envelope (or bulb) and therefore have very good color-rendering characteristics. The small size of the filament makes optical control of light from these sources possible with relatively small optics. When a current is applied across the filament, it becomes so hot (around 2,800 K) as to radiate light, as well as significant heat (radiation in the infrared region of the spectrum). Typical incandescent lamps have luminous efficacies ranging from about 10 to 20 lm/W. Tungsten-halogen lamps, which use a halogen gas, are more efficient than standard incandescent lamps and can have luminous efficacies as high as 35 lm/W with nearly 100 percent lumen maintenance compared with around 80 percent for regular incandescent lamps. Reducing the operating voltage has several important effects on a filament lamp:

- Reduced light output (dimming)
- Reduced luminous efficacy (lm/W)
- Decreased CCT (resulting in a "yellower" appearance)
- Increased lamp life

A main advantage to filament sources in transportation systems is their low initial cost, but this is often offset by their relatively low efficacy and short operating lives (1,000 to 2,000 hours).

8.4.2 Fluorescent and Induction Lamps

Fluorescent lamps are low-pressure gas-discharge sources, where light is produced mainly by fluorescent powder coatings (phosphors) that are activated by UV energy generated by a mercury arc. Compared with other lamp types, traditional fluorescent sources are not used commonly in transportation applications. Fluorescent lamps have a typical rated life of between 8,000 and 20,000 hours, which is mainly limited by the life of the electrodes at the lamp ends that generate the electrical discharge. Fluorescent lamps require ballasts to operate, which provide the starting and operating voltages and currents that operate the fluorescent lamp properly. Some ballasts permit fluorescent lamps to be dimmed. The color of fluorescent lamps is determined by the phosphors used to coat the envelope. CCTs from 2,700 to 7,500 K, with good color-rendering properties, are readily available. Fluorescent lamps have substantially higher luminous efficacies (60 to 80 lm/W) than filament sources and relatively high lumen maintenance values (80 percent or higher).

A special type of fluorescent lamp known as the *electrodeless* or *induction fluorescent lamp* has found increasing acceptance for street lighting applications. These lamps can have more compact profiles than the tubular shapes required by conventional fluorescent lamps because they rely on a magnetic induction coil to provide the electric current that stimulates the mercury vapor inside the lamp envelope. Eliminating the electrodes results in lamp lifetimes of 60,000 hours with luminous efficacy comparable with that of conventional fluorescent lamps.

Very precise optical control of illumination from fluorescent and induction lamps is not possible without quite large optical systems that are impractical for many transportation applications, although induction lamps have been found to be suitable for illuminating streets and pedestrian walkways.

8.4.3 High-Intensity Discharge Lamps

High-intensity discharge (HID) lamps include mercury vapor (MV), metal halide (MH), and high-pressure sodium (HPS) lamps. Like filament lamps, they provide a relatively compact point of light; like fluorescent lamps, they are electrical discharge lamps, tend to have long lives, and require ballasts. Unlike incandescent and fluorescent lamps, starting and restrike times for many HID lamps can be as long as several minutes unless special additives are introduced into the arc tube to facilitate rapid starting.

MV lamps have substantially lower luminous efficacy than MH and HPS lamps and are being phased out by national energy legislation. MH lamps (Figure 8.3) are very similar in construction to mercury lamps, with an arc tube containing various metallic halide compounds in addition to mercury. These compounds emit light in different parts of the visible spectrum, and by using various mixtures of metallic halides, lamps with efficacies and CRI values much higher than mercury lamps are possible. Typical MH lamp efficacy ranges between 70 and 120 lm/W. CRI values range from 60 to over 90, and CCT ranges from 2,800 K (soft white) to over 6,000 K (a bluish white). Rated lamp life typically is from 10,000 to 20,000 hours, and lumen maintenance values are fair (between 60 and 80 percent is typical). MH lamps with ceramic arc tubes and improved starting gear have been developed recently that improve their life and lumen maintenance characteristics.

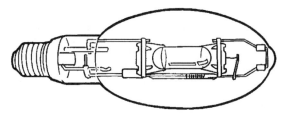

FIGURE 8.3 Drawing of a metal halide lamp. (*Courtesy of Lighting Research Center.*)

HPS lamps produce light by applying a current to sodium vapor. HPS lamps have two envelopes (Figure 8.4): an inner arc tube containing sodium and (usually) mercury and an outer glass bulb to absorb UV energy and stabilize the temperature of the arc tube. Typical HPS lamp efficacies range between 80 and 140 lm/W, with rated lamp lives of about 24,000 hours and excellent lumen maintenance (typically about 90 percent). HPS lamps radiate energy across the visible spectrum, although not uniformly. Light from standard HPS lamps is a yellowish color (CCT ranging from 1,900 to 2,200 K), and color rendering is quite poor (CRI of 22). Improved color-rendering HPS lamps can be created by increasing the sodium vapor pressure, which increases CRI to about 65 (while reducing efficacy and life) or by operating HPS lamps at higher frequencies (resulting in CCT between 2,700 and 2,800 K and CRI between 70 and 80). HPS lamps are the most common sources used in roadway lighting.

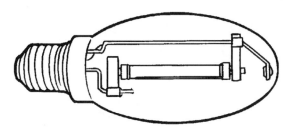

FIGURE 8.4 Drawing of a high-pressure sodium lamp. (*Courtesy of Lighting Research Center.*)

8.4.4   Low-Pressure Sodium Lamps

Low-pressure sodium (LPS) lamps generate light via a current applied to sodium vapor at a lower pressure than in an HPS lamp. The lamp shape is also different (more linear) because the arc itself is longer. These lamps emit a very narrow wavelength band of light near 589 nanometers and produce light that is pure yellow in appearance. Ballasts are required for operation of these lamps. They have typical rated lives of about 16,000 hours and nearly 100 percent lumen maintenance. LPS lamps have very high efficacy, near 180 lm/W. Because they produce monochromatic light, color rendering is essentially nonexistent with these lamps. LPS lamps are sometimes used for outdoor installations in the vicinity of observatories because it is relatively simple for the astronomical observers to filter out the narrow wavelength band emitted. LPS sources are used commonly in Europe for roadway lighting.

8.4.5   Light-Emitting Diodes

First invented in the 1960s, light-emitting diodes (LEDs) have increased in efficacy, in light output, and in the range of available colors so that they are being used for a number of transportation lighting applications. LED products presently make up a substantial proportion of the vehicle signal and traffic signal markets. The development of blue LEDs that emit short-wavelength light has permitted the subsequent creation of white-light LED systems in two ways: by combining red, green, and blue LEDs to create white light and by using phosphors together with blue LEDs that convert some of the blue light into yellow light, resulting in a mixture that is perceived as white light (Figure 8.5). Newer packages for LEDs have been developed in addition to the familiar 5-mm-diameter epoxy-capsule LED packages used for indicator lights, which produce greater light output and wider distributions than LEDs are often thought to exhibit. Some packages contain metal fins or slugs as heat sinks to help dissipate internal temperatures. Increased temperature inside the LED reduces its light output and over the long term can shorten the useful life of the source (Bullough 2003). Color rendering depends on the combination of phosphors used to generate white light, and luminous efficacies presently range from 30 to more than 60 lm/W.

While LEDs have very long operating lives, typically more than 50,000 hours without failing, they do exhibit gradual reductions in light output over time. White LED indicator-type packages with no heat sinking have been shown to reach half their initial light output within 10,000 hours. Improved packages have exhibited much improved lumen maintenance characteristics.

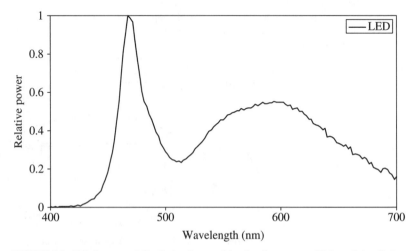

FIGURE 8.5   Relative spectral distribution from a white phosphor-converted light-emitting diode.

8.5 VEHICLE LIGHTING

8.5.1 Forward Lighting

Most of the roads in the United States are not lighted (NHTSA 2007); therefore, vehicle forward lighting systems are necessary for safe nighttime travel. Regulations for vehicle headlamps (in Federal Motor Vehicle Safety Standard 108) are based on standards published by the Society of Automotive Engineers (SAE) and require certain minimum or maximum luminous intensity values in various angular directions from the center of the headlamp. Two headlamps are required, and each one is required to meet the same photometric performance requirements. Additionally, there are two types of *beam patterns*, or distributions of luminous intensity, permissible for vehicle headlamps, known as the *high beam* (or *driving beam*) and the *low beam* (or *passing beam*).

Table 8.2 lists luminous intensity requirements for several angular locations for low-beam headlamps, and Table 8.3 lists them for high-beam headlamps. Two trends are evident from these tables: Requirements for high beams are less restrictive (i.e., have fewer maxima and generally higher luminous intensity values) than for low beams, and the high-beam pattern is symmetric, whereas the low-beam pattern is asymmetric, with more stringent maxima toward the left side (where oncoming traffic is more likely to be found). Figure 8.6 shows a graphic representation of the low-beam luminous intensity requirements, with locations of a straight, two-lane road overlaid for comparison. The luminous intensity values can be used in conjunction with equation (8.1) (cosine corrections for all the angles in Tables 8.2 and 8.3 are less than 3.5 percent) to estimate the corresponding vertical illuminances on objects located at various distances ahead of the headlamps. (Illuminances from each headlamp should be added together.)

TABLE 8.2 Selected Luminous Intensity Maxima and Minima for Low-Beam Headlamps

Angular location (degrees)	Maximum luminous intensity (cd)	Minimum luminous intensity (cd)
8° left, 0° up and 8° right, 0° up	—	64
8° left, 4° up and 8° right, 4° up	—	64
4° left, 0° up and 4° right, 0° up	—	125
4° left, 2° up and 4° right, 2° up	—	125
1.5° right, 0.5° down	20,000	8,000
6° left, 1° down	—	750
2° right, 1.5° down	—	15,000
9° left, 1.5° down and 9° right, 1.5° down	—	750
15° left, 2° down and 15° right, 2° down	—	700
1.5° left, 1° up	700	—
1.5° left, 0.5° up	1,000	—
1.5° left, 0.5° down	3,000	—
1° right, 1.5° up	1,400	—
1° right, 0.5° up, 2° right, 0.5° up, and 3° right, 0.5° up	2,700	—
4° right, 4° down	8,000	—

Source: Based on Federal Motor Vehicle Safety Standard 108.

Figure 8.7 illustrates a representative beam pattern for a low-beam headlamp projected onto a vertical wall. In order to control glare, many low-beam patterns exhibit relatively sharp vertical *cutoff* angles, above which there is relatively little light. The sharp cutoff also permits visual aiming of the headlamps by adjusting the vertical tilt; most headlamps in North America require the location of the right-side cutoff to be at 0° (Schoettle et al. 2008) or, in other words, at the same height as the

TABLE 8.3   Selected Luminous Intensity Maxima and Minima for High-Beam Headlamps

Angular location (degrees left/right, up/down)	Maximum luminous intensity (cd)	Minimum luminous intensity (cd)
0° right, 0° up	75,000	40,000
3° left, 1° up and 3° right, 1° up	—	5,000
0° right, 2° up	—	1,500
3° left, 0° up and 3° right, 0° up	—	15,000
6° left, 0° up and 6° right, 0° up	—	5,000
9° left, 0° up and 9° right, 0° up	—	3,000
12° left, 0° up and 12° right, 0° up	—	1,500
0° right, 1.5° down	—	5,000
0° right, 2.5° down	—	2,500
9° left, 1.5° down and 9° right, 1.5° down	—	2,000
12° left, 2.5° down and 12° right, 2.5° down	—	1,000
0° right, 4° down	12,000	—

Source: Based on Federal Motor Vehicle Safety Standard 108.

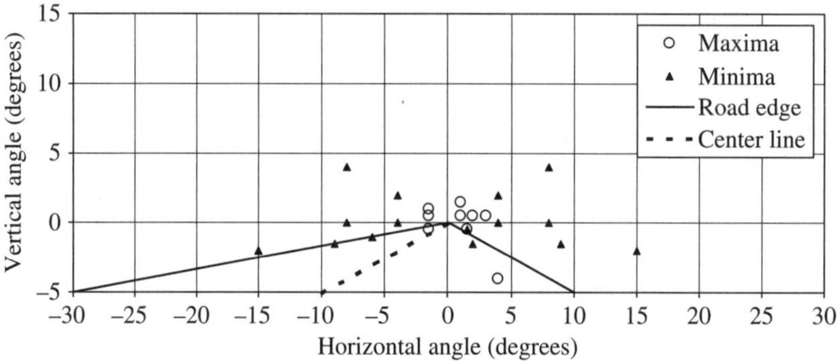

FIGURE 8.6   Angular locations of several low-beam luminous intensity minima and maxima superimposed onto a roadway scene.

FIGURE 8.7   Photograph of a low-beam headlamp pattern projected onto a wall.

headlamp. The cutoff on the left side, corresponding to likely locations of oncoming traffic, is often lower than the right-side cutoff to reduce glare.

The sharp cutoff of low-beam headlamps can restrict the forward visibility of drivers. Even on flat, straight roads, driving speeds in excess of 35 to 40 mph will make it impossible for a driver to detect and stop in time to see some potential hazards (Andre and Owens 1999). Use of high beams is warranted for these situations (except when approaching traffic is within several hundred feet), but drivers almost universally underuse their high-beam headlamps (Sullivan et al. 2004).

Most headlamps use tungsten-halogen (or more simply, halogen) lamps with reflector or projector optics to shape the beam pattern. A growing proportion of headlamps use HID sources, specifically MH lamps with a xenon fill to assist in rapid starting. HID headlamps must meet the same photometric requirements as other headlamps, but the higher luminous efficacy (see section 8.4) and greater light output from these headlamps often result in greater peripheral illumination than halogen headlamps, which can make detection of off-road objects such as pedestrians easier (Van Derlofske et al. 2002). HID headlamps tend to have higher CCTs than halogen headlamps; the higher short-wavelength ("blue") spectral content produces more discomfort glare for oncoming drivers than light from halogen headlamps, even if the illuminance at the eyes is equal. The U.S. National Highway Traffic Safety Administration (NHTSA) is evaluating public complaints about glare from HID headlamps (NHTSA 2007). Headlamps using LEDs are just beginning to be used on a few vehicle models. The relatively high cost of HID and LED headlamp systems will ensure that halogen headlamps remain a dominant light source for forward lighting in the coming decade.

The sharp vertical cutoff of low-beam headlamps makes vertical aim critical in the performance of these lighting systems. Few states in the United States require headlamps to be aimed properly as part of a safety inspection (NHTSA 2007). Recent measurements of vehicles (Skinner et al. 2010) have found that most of the vehicles measured had at least one misaimed headlamp. Upward misaim is probably the largest contributor to disability and discomfort glare from headlamps (Perel 1985; Sivak et al. 1998), and downward misaim can severely restrict forward visibility because of the sharp vertical cutoff.

Forward lighting systems that change dynamically in response to specific driving conditions are now beginning to show promise for changing the fixed, high-/low-beam pattern paradigm for headlamps (Wördenweber et al. 2007). These systems are commonly known as *adaptive forward lighting systems* (AFS). Already, some cornering and bending lights have begun to appear on certain vehicle models. Some bending lights use swiveling mechanisms that literally "turn" the entire beam pattern of one or both headlamps toward the direction of approaching curves. In Europe, some vehicles are equipped with a city/town beam with lower maximum luminous intensities but a broader distribution than typical low-beam patterns to facilitate detection of pedestrians while driving at the relatively lower speeds found in urban locations. The Economic Commission on Europe's Vehicle Regulation No. 123 describes requirements for AFS in many nations; U.S. Federal Motor Vehicle Safety Standard 108 is currently silent with respect to AFS.

8.5.2 Signal Lighting

Vehicles must have signal lights to allow the driver to communicate to others about braking, turning, and driving direction both during the daytime and at night. These commonly use filament sources and colored filters; an increasing proportion of vehicle signals use LEDs for signal light applications. Signals for different functions have different requirements regarding the color and luminous intensity. U.S. federal requirements for vehicle signal lights are based on SAE standards. Table 8.4 lists the color and allowable luminous intensity range for several signal functions.

Requirements for vehicle signal lights in Europe are not dramatically different from those in the United States regarding color and luminous intensity, with one major exception: Rear turn lights in the United States may be red or yellow (Table 8.4), and the luminous intensity requirements depend on the color. European regulations require rear turn signals to be yellow only. There is some evidence (Allen 2009) that yellow rear turn signals result in fewer crashes, perhaps because they tend to have

TABLE 8.4 Selected Color and Photometric Requirements for Vehicle Signals

Signal function	Required color	Minimum-maximum luminous intensity (cd)
Tail (presence) light	Red	2–18
Stop light	Red	80–300
Center high-mounted stop light	Red	25–130
Rear turn light	Red or yellow	80–300 (if red), 130–750 (if yellow)
Front turn light	Yellow	130–750
Backup light	White	80–300 (if two), 80–500 (if one)

Source: Based on Federal Motor Vehicle Safety Standard 108.

higher luminous intensities or because they are more readily distinguished from stop or tail lights. The NHTSA is presently considering whether yellow should be mandated for rear turn signals on vehicles in the United States.

8.6 ROADWAY LIGHTING

Two important purposes of roadway lighting are, according to the *American National Standard Practice for Roadway Lighting*, to "reveal the environment beyond the range of the vehicle headlights and ameliorate glare from oncoming vehicles" (IES 2000a). Crash rates along roadways generally are much higher during the night than during the daytime. Lighting is generally thought to reduce nighttime crash rates, and given the forward visibility limitations of low-beam headlamps described in section 8.5, such an expectation is not unwarranted. And increasing the luminances of objects along the roadway and their backgrounds will mitigate the effects of contrast-reducing veiling luminances (see equation 8.4) on visibility. In this section, the safety impacts of roadway lighting, the equipment used for roadway lighting, and some of the design criteria for roadway lighting are described.

8.6.1 Lighting and Safety

Almost all transportation agencies recognize that roadway lighting reduces nighttime crashes, and most empirical evidence is consistent with this notion (CIE 1992). Measuring the safety impacts of roadway lighting is difficult because crashes are (thankfully) relatively rare events, and it requires long periods of time to measure statistically reliable effects. Many factors may change along a roadway (e.g., daily traffic, road geometry, and number of lanes, to name a few), so before/after studies of roadway lighting can be confounded by different periods of time. Studies comparing different locations with and without roadway lighting require very careful selection of sites to avoid confounding (a location near a shopping center would be very different with respect to crashes than an isolated rural location). Some of these confounds can be reduced by considering daytime crashes as well as nighttime crashes, and a common metric used to assess the influence of roadway lighting (CIE 1992) is defined as follows:

$$r = (N_{with}/D_{with})/(N_{without}/D_{without}) \tag{8.5}$$

where N_{with} is the number of nighttime crashes with roadway lighting present, D_{with} is the number of daytime crashes with roadway lighting present, $N_{without}$ is the number of nighttime crashes without roadway lighting present, and $D_{without}$ is the number of daytime crashes without roadway lighting present. (When studying the impact of lighting after it has been installed, the subscript *without* can refer to the "before" conditions, and the subscript *with* can refer to the "after" conditions.) The parameter *r* is a dimensionless ratio that represents the proportion of nighttime crashes with roadway lighting relative to those without roadway lighting, so a value of r of less than 1 indicates an association between roadway lighting and a reduction in nighttime crashes. The quantity $1 - r$, expressed as a percentage, is often used to express the nighttime *crash-reduction factor* associated with roadway lighting.

Considering the daytime crashes in this way is supposed to help control for factors that might influence crashes between the "with" and "without" locations (or between the "before" and "after" periods). A nighttime crash-reduction factor of about 30 percent was found in a review of more than 60 studies of roadway lighting (CIE 1992). Although this single nighttime crash-reduction factor of 30 percent is often used by transportation agencies to predict the benefits of roadway lighting, it is only reasonable to suppose that roadway lighting does not carry the same benefit in all situations. Indeed, careful review of the literature has shown that in locations with complex roadway geometries, high potential for conflicts, or large pedestrian populations, the nighttime crash-reduction factor associated with lighting appears to be larger than along straight, access-controlled highways with little to no pedestrians (Rea et al. 2009).

Despite the control for many operational and geometric factors that is inherent in the use of daytime crashes to serve as a type of control condition, observational studies using with/without or before/after methodologies still can contain biases. One explanation is that roadway lighting is not assigned randomly to various roadway locations. Instead, if lighting is installed along an existing roadway, this might have been done in response to a higher than average observed number of crashes within a relatively short period of time of a few years (CIE 1992). If the increase was spurious, it is possible that the number of crashes will undergo *regression to the mean* and exhibit a relatively lower number of crashes simply because a roadway cannot experience an above-average number of crashes every year! Another explanation of bias might be that roadway lighting is often only one safety-related measure that might be installed along a roadway (IES 1989); others could include improved lane markings, medians, turn lanes, or traffic signals, and these might be partially responsible, along with lighting, for any safety improvement.

A recent study to attempt to reduce such potential sources of bias (Donnell et al. 2009) used multiple nonlinear regression modeling that included not only the presence of lighting but also the type of location (i.e., urban or rural), posted speed limit, percentage of heavy trucks, average daily traffic, number and width of lanes, presence of signalization, and other geometric and demographic variables in order to isolate the safety effect (if any) associated only with roadway lighting. The results were consistent with other studies (CIE 1992) in that there was generally a reduction in nighttime crashes associated with roadway lighting, but the magnitude of the effects was smaller, as might be expected. The nighttime crash-reduction factor associated with roadway lighting was closer to 10 percent for intersections and smaller for highway segments with no intersections or interchanges, not 30 percent, as commonly used by many transportation agencies.

Most studies of roadway lighting and safety effects treat lighting as either present or not. The discussion of visual performance from section 8.3 should reinforce the notion that improvements in visibility, if this is the mechanism accounting for safety improvements associated with lighting, will depend on the light level and other factors associated with a specific lighting configuration. A systematic examination of the visual performance benefits of roadway lighting of different light levels and spatial extents (i.e., one or two light poles at an intersection versus a continuously lighted roadway approaching an intersection) used in conjunction with vehicle headlamps and in locations varying in vehicle speeds, amount of ambient light from commercial properties, and for drivers of varying ages was conducted (Bullough et al. 2009a) to complement the statistical study of lighting and crashes by Donnell et al. (2009). The results of both studies yielded converging results. For example, the benefit of roadway lighting at rural intersections was found to be negligible (Donnell et al. 2009) even though such areas are inherently dark at night, and lighting would be expected to improve visual performance, particularly when driving with low-beam headlamps at speeds greater than 40 mph, which is common on many rural highways. The visibility analyses (Bullough et al. 2009a) found only modest visual performance improvements from "point" lighting consisting of only one or two poles at the intersection, especially for drivers along the secondary road of such an intersection. Continuous lighting of the rural highway, by comparison, would provide substantial improvement over no lighting or even "point" lighting but is rarely used in practice (Bullough et al. 2009a). Given these findings, drivers *might* be better served by lighting the approaching legs of a rural intersection and not only the immediate conflict area, but this hypothesis has yet to be verified.

Regarding impacts of roadway lighting on safety and security, the literature is very mixed. A review of studies conducted in the United States during the 1970s (Tien 1977) found little evidence to support the notion that roadway lighting reduces crime, and the picture has not gotten clearer since that time. Roadway lighting can, however, improve subjective impressions of the safety of a location (Leslie and Rodgers 1996). Surveys of exterior lighting in major urban areas, as well as suburban locations, found perceptions of safety and security to be positive when the average illuminance exceeded 10 lx; higher light levels were not judged as feeling particularly safer.

8.6.2 Roadway Lighting Equipment

For the most part, roadway lighting systems consist of luminaires mounted to poles, usually with individual controls consisting mainly of photocells. Poles are most commonly aluminum, steel, or wood. In many locations, existing utility poles are used to support roadway luminaires. If dedicated lighting poles are used, the utilities may be buried underground to improve the visual appearance. Cost may determine whether poles must be mounted along a single side of the road, or along both sides in an opposite or staggered arrangement.

The most common luminaires used for roadway lighting are so-called cobrahead luminaires (Figure 8.8), named for their distinctive shape. Cobrahead luminaires in the past have been equipped mainly with drop lenses (Figure 8.8*a*), but concerns for light pollution (see section 8.8) have made flat-lens cobrahead luminaires common (Figure 8.8*b*). Other luminaires used for functional roadway lighting include arm-mount or so-called shoebox luminaires (Figure 8.9) and post-top luminaires

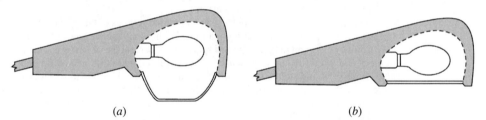

(a) $\hspace{9cm}$ (b)

FIGURE 8.8   Cobrahead luminaires with (*a*) drop lens and (*b*) flat lens. (*Courtesy of National Lighting Product Information Program.*)

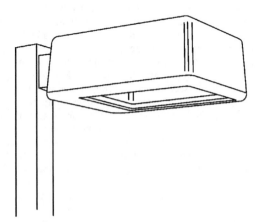

FIGURE 8.9   Arm-mounted luminaire. (*Courtesy of National Lighting Product Information Program.*)

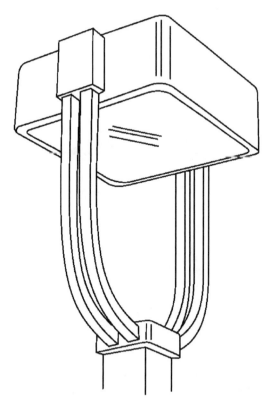

FIGURE 8.10   Post-top luminaire. (*Courtesy of National Lighting Product Information Program.*)

(Figure 8.10). Figure 8.11 shows the typical components of most roadway luminaires. In some locations, more decorative luminaires are used that may be historic or aesthetic in appearance, but the daytime appearance of such luminaires is often just as important as their optical performance (if not more so). Roadway lighting systems focused on maximum efficiency and functionality usually use the former types described earlier.

Most luminaires are designed for HID lamps, such as HPS or MH lamps (see section 8.4); those with induction lamps or using LEDs are much less common, and luminaire designs for these sources are more likely to be highly individualized. Technical information about many specific roadway luminaires can be obtained from the National Lighting Product Information Program (NLPIP 2004).

Roadway luminaires are classified in several important ways. The cutoff classification system, formerly used by the IES and still used by many lighting specifiers, classifies luminaires according to their luminous intensity distributions in various angular locations (using the nadir, the angle directly below the luminaire, as the reference for 0°), where light emitted by a luminaire might contribute to light pollution or glare (Figure 8.12). The luminous intensity limits (in candelas) for different cutoff classifications are made relative to the light output (in lumens) of the lamp inside the luminaire as follows:

- *Full cutoff.* The luminous intensity must be 0 for all angles equal to or greater than 90° above nadir and cannot exceed a value (in candelas) greater than 10 percent of the light output (in lumens) emitted by the lamp between 80° and 90° above nadir.

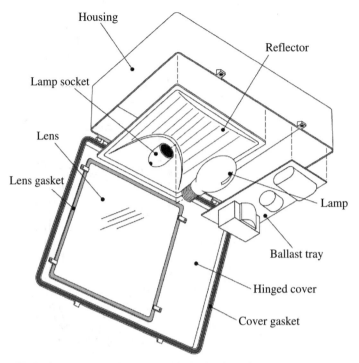

FIGURE 8.11   Drawing of a roadway luminaire showing main components. (*Courtesy of National Lighting Product Information Program.*)

• *Cutoff.* The luminous intensity cannot exceed a value (in candelas) greater than 2.5 percent of the lamp's light output (in lumens) for all angles equal to or greater than 90° above nadir and cannot exceed a value (in candelas) greater than 10 percent of the lamp's light output (in lumens) between 80° and 90° above nadir.

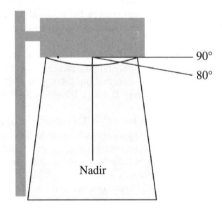

FIGURE 8.12   Luminaire angle references in the cutoff classification system. Nadir is defined as 0°, directly below the luminaire. (*Courtesy of National Lighting Product Information Program.*)

• *Semicutoff.* The luminous intensity cannot exceed a value (in candelas) greater than 5 percent of the lamp's light output (in lumens) for all angles equal to or greater than 90° above nadir and cannot exceed a value (in candelas) greater than 20 percent of the lamp's light output (in lumens) between 80° and 90° above nadir.

• *Noncutoff.* The luminous intensity exceeds a value (in candelas) greater than 5 percent of the lamp's light output (in lumens) for at least one angle greater than 90° above nadir or exceeds a value (in candelas) greater than 20 percent of the lamp's light output (in lumens) between 80° and 90° above nadir.

Referencing the luminous intensity requirements (in candelas) in the cutoff classification system to quantities having entirely different units (lumens) made the system potentially confusing, and the IES more recently adopted a system that simply uses the light output (in lumens) emitted in various angular regions to serve as the basis for classifying the distribution from a luminaire (IES 2007).

Roadway luminaires also can be classified by their *lateral distribution*. Put simply, the lateral distribution is an indication of how far *across* a roadway a luminaire might be expected to illuminate. The lateral distributions are classified into types arranged by Roman numerals (Figure 8.13). Type I is a narrow lateral distribution most suitable for two-lane roads without sidewalks. Type II, III, and IV distributions are increasingly broader, and type V is a symmetric distribution all around the luminaire (Rea 2000).

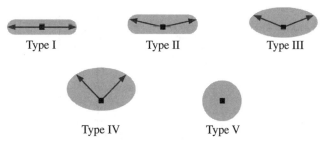

FIGURE 8.13 Comparison of type I, II, III, IV, and V lateral distributions. (*Courtesy of National Lighting Product Information Program.*)

The *vertical distribution* of a roadway luminaire (Rea 2000) relates to the position of the maximum luminous intensity emitted by a luminaire *along* the roadway, in units of luminaire mounting heights. If the maximum intensity value falls less than 2.25 mounting heights along the roadway from the luminaire, it is classified as having a short vertical distribution. If the maximum value falls between 2.25 and 3.75 mounting heights along the roadway from the luminaire, it is classified as having a medium vertical distribution, and if the maximum falls greater than 3.75 mounting heights along the roadway from the luminaire, it is classified as having a long vertical distribution.

Photometric reports are provided by manufacturers of most roadway luminaires that describe distributions of light from the luminaires for specific configurations such as different lamps, wattages, or lens types (drop lens versus flat lens). These often also include an isoilluminance plot such as that in Figure 8.14, which shows the illuminance on the ground surface produced by a luminaire along a grid demarcated in units of mounting height. The illuminance values in such plots are specific to a particular mounting height (such as 25 feet) but can be readily modified based on the inverse-square law (equation 8.1). For example, if Figure 8.14 is for a mounting height of 25 feet, but the luminaire were to be mounted at 30 feet, the illuminance values would have to be scaled by a factor of ($25^2/30^2 = 0.69$). If the luminaire were to be mounted at 20 feet, the scaling factor instead would be ($25^2/20^2 = 1.56$). Isoilluminance plots such as Figure 8.14 can be overlaid onto a scaled roadway plan to rapidly estimate illuminance distributions for different pole spacing configurations.

As mentioned previously, the most common control system for roadway lighting is a photocell mounted to an individual luminaire, which will switch the luminaire on and off at a specified ambient light level. Some systems will use centralized control via a single photocell or time clocks. Technological innovations are making centralized control systems attractive because these systems also can monitor performance of individual luminaires and alert the system operator when a lamp or ballast failure has occurred or is close to occurring (NLPIP 2004).

8.6.3 Roadway Lighting Practice

In North America, the *American National Standard Practice for Roadway Lighting* (IES 2000a) is the basis for the design of continuous roadway lighting systems by many agencies, including the members of the American Association of State Highway and Transportation Officials (AASHTO).

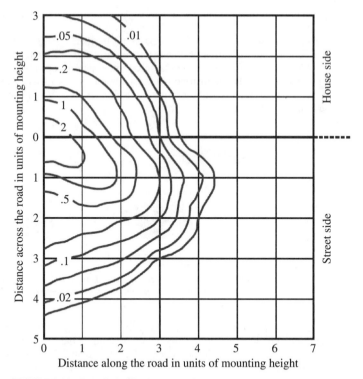

FIGURE 8.14   Example isoilluminance plot for a roadway luminaire. Illuminances in this example are in footcandles. 1 fc = 10.76 lx. (*Courtesy of National Lighting Product Information Program.*)

AASHTO (2005) provides guidance to the user regarding when roadway lighting is warranted based on daily traffic counts and nighttime crash histories. The IES (2000a) allows for three possible ways to design roadway lighting, on the basis of meeting illuminance criteria, luminance criteria, or visibility criteria, using a complex system known as *small-target visibility* (STV). In practice the STV methods is rarely used, and AASHTO (2005) recommends only the illuminance and luminance methods. In addition to the recommended light levels, all methods have minimum requirements for uniformity of lighting and limits on the maximum veiling luminance (see section 8.3.2) that can be created by the luminaires in a roadway lighting installation.

According to both the illuminance and luminance methods, the target light levels for different roadway situations differ based on the type of roadway (i.e., local, collector, major, or freeway) and the level of pedestrian conflict. The illuminance criteria differ for different pavement materials; they are higher for darker asphalt pavements than for lighter concrete pavements (IES 2000a). The luminance design method uses the pavement characteristics as an input to the design; both methods therefore attempt to provide similar roadway luminances regardless of the type of pavement.

As described earlier, most states in the United States base their roadway lighting practices on IES (2000a) and AASHTO (2005) guidelines. As an example, using the illuminance method, the IES (2000a) provides the following criteria for roadway lighting performance for local and major roadways, each with a medium pedestrian conflict level, on asphalt pavement:

- *Local roadway.* The average illuminance should be at least 7 lx, minimum illuminance should not be less than 1/6 the average, and the maximum veiling luminance from the installation should be no more than 40 percent of the average luminance of the roadway surface.

- *Major roadway.* The average illuminance should be at least 13 lx, the minimum illuminance should not be less than 1/3 the average, and the maximum veiling luminance from the installation should be no more than 30 percent of the average luminance of the roadway surface.

In practice, the calculations supporting roadway lighting design are performed via software that can accept a photometric file that contains tabulated luminous intensity data for a given luminaire. A user can input the geometric characteristics of the roadway and the locations and heights of the poles (all on the same side of the road or on both sides in an opposite or staggered configuration), and the software can calculate the resulting illuminances, uniformity values, and veiling luminances based on the equations provided by the IES (2000a). Meeting all the criteria using existing utility poles is difficult, and many residential street lighting applications do not meet IES (2000a) criteria for continuous roadway lighting. When light poles will be installed solely for the purpose of luminaire mounting, most software packages have a routine by which pole spacing starts very low, so that the resulting installation would be nearly certain to meet all light level, uniformity, and veiling luminance criteria, and then is increased iteratively until the longest pole spacing is identified for which all the criteria are met. This method results in a lighting configuration with the fewest number of poles and luminaires per unit distance and therefore the lowest operating costs. It is sometimes stated in the literature that luminaires with flat lenses require shorter pole spacing than luminaires with drop lenses because the flat lens may restrict high-angle light from the luminaire. An evaluation of many luminaires with different types of lenses (Zhang 2006) found significant overlap among luminaire types; many with flat lenses could achieve longer pole spacing than many with drop lenses.

It is important that such calculations incorporate appropriate factors for lumen maintenance because the light output from a new roadway system will decrease over time, as well as to account for the accumulation of dirt. A typical lumen maintenance factor for HPS lighting systems is 0.9 (NLPIP 2004), and typical *dirt depreciation factors* used by transportation agencies range from 0.8 to 0.9 (MNDOT 2006). Thus, a maintained light level from a lighting system after years of use could be about 20 to 30 percent lower than the initial light level when new and clean.

IES (2000a) and AASHTO (2005) criteria are primarily for continuous lighting along the length of roadways, excluding junctions such as intersections. Generally, intersection lighting criteria are based on adding the recommended light levels for the intersecting roadways so that the conflict area in an intersection will be lighted to a higher level than either of the individual roadways. When *point* (or partial) intersection or interchange lighting is to be provided instead of *continuous illumination*, the IES (2000a) and AASHTO (2005) provide guidance regarding placement of poles and luminaires.

8.7 TRAFFIC SIGNALS

This section discusses photometric requirements for traffic signals for providing adequate visibility according to the Institute of Transportation Engineers (ITE) and briefly touches on maintenance practices for signals using LEDs. Information about warrants for traffic signal systems and timing considerations are found elsewhere in this book.

8.7.1 Photometric Requirements

LED traffic signals have been gaining in market share since the late 1990s when they were first introduced on a widespread scale. In the United States, energy legislation enacted in 2006 mandated that all traffic signal modules meet the performance requirements of the U.S. ENERGY STAR Program, effectively mandating the use of LED signals for red and green indications. (LED signal modules use 80 to 90 percent less energy than their incandescent counterparts.) Yellow signal modules were excluded from the Energy Star requirements because the relatively high luminous intensity requirements for yellow LED traffic signals (Table 8.5) before 2005 (ITE 1998) made it impractical to construct yellow LED modules that were not difficult to manage thermally. As described in section 8.4, high internal temperatures can reduce the efficiency of some LED systems. The basis for the previous

TABLE 8.5 Photometric Requirements for Light-Emitting Diode Traffic Signals

LED traffic signal color	Diameter (mm)	ITE (1998) required luminous intensity (cd)	ITE (2005) required luminous intensity (cd)
Red		133	165
Yellow	200	617	410
Green		267	215
Red		339	365
Yellow	300	1571	910
Green		678	475

Source: ITE 1998, 2005.

luminous intensity requirements for different signal colors might have been the relative transmission of the filter glasses used to create these colors.

More recently, the ITE (2005) published a revised specification for LED traffic signals based on human factors research (Bullough et al. 2000; Freedman 2001) that suggested that the yellow signal did not need to have a luminous intensity nearly five times higher than that of the red signal in order to be equally visible. In the revised specification, the yellow signal needs to have only 2.5 times the luminous intensity of the red signal, which makes construction of yellow signal modules more practical. The viability of all-LED traffic signal systems has important implications for maintenance and energy use, especially for systems that convert partially to flashing-yellow modes during part of the night, and the lower power requirements can make flashing operation under battery backup during storms and power outages more practical as well.

8.7.2 Maintenance Issues

Aside from the reduced energy requirements, LED signals have been attractive because of their longer useful lives compared with incandescent signal systems. ITE's (2005) specification requires LED signal modules to meet performance requirements for a minimum of 5 years, and many agencies expect even longer performance. (In comparison, incandescent lamps for red and green indications typically last 1 to 2 years.) Unlike incandescent signals, LED modules rarely experience total "burnouts" in the field; they undergo gradual reductions in lumen maintenance, and failure of specific electrical components may cause partial failure, as evidenced by a portion of the indication failing to illuminate. While a partial or dimmed indication is better than none, it is challenging for agencies to identify at what point an LED signal module has fallen below the minimum performance requirements (ITE 2005). Limited experience from some agencies suggests that group replacement of LED modules every 6 to 8 years would be more cost-effective in the long term than individual replacement of modules on a piecemeal basis (Bullough et al. 2009b), especially given that it is time-consuming to assess module performance in the field.

8.8 SPECIAL TOPICS

8.8.1 Mesopic Vision

In section 8.2, the photopic and scotopic luminous efficiency functions were introduced (Figure 8.1). All photometric characterizations used in lighting for the built environment are based on the photopic function, which represents the spectral sensitivity of the cone photoreceptors in the eye to light at daytime light levels. At very low light levels corresponding to starlight with no moon present, the rod photoreceptors in the eye are used for seeing, and their spectral sensitivity is represented by the scotopic function in Figure 8.1. At many nighttime light levels experienced along roadways, a mixture of rods and cones contributes to vision. Since rods are more sensitive to shorter wavelengths in the "blue-green" portion of the visible spectrum, light sources with more power in this part of the spectrum will be more effective for vision at these so-called mesopic nighttime light levels. Different light sources can be quantified by a factor called the *scotopic/photopic (S/P) ratio* (IES 2006), which gives a relative indication of a light source's ability to stimulate the rod photoreceptors for an equivalent amount of cone stimulation. Table 8.6 lists the *S/P* ratios for several commonly used outdoor light sources (ASSIST 2009).

TABLE 8.6 *S/P* Ratios for Several Common Light Sources

Light source	*S/P* ratio
Low-pressure sodium (180 W)	0.25
High-pressure sodium, clear (250 W)	0.63
Mercury vapor, coated (175 W)	1.08
Incandescent (100 W)	1.36
Halogen headlamp (55 W)	1.43
Cool white fluorescent (34 W)	1.48
Metal halide, clear (400 W)	1.57
High-intensity discharge headlamp (35 W)	1.61
White light-emitting diode, 4,200 K (1 W)	2.04
Rare-earth fluorescent, 6,500 K (32 W)	2.19

Research to quantify visual performance from different light sources under mesopic conditions has been underway for more than a decade and ranges from basic laboratory studies to field studies involving driving actual vehicles along actual roadways (IES 2006). A unified system of photometry (Rea et al. 2004) bridging the photopic and scotopic systems has been developed to quantify the mesopic or unified luminance using the following equation:

$$L_u = 0.834L_p - 0.335L_s - 0.2 + (0.696L_p^2 - 0.333L_p - 0.56L_pL_s + 0.113L_s^2$$
$$+ 0.537L_s + 0.04)^{0.5} \tag{8.6}$$

where L_p is the photopic luminance (in cd/m^2), L_s is the scotopic luminance [in cd/m^2, where L_s can be defined by the quantity $L_p(S/P)$], and L_u is the unified luminance (in cd/m^2). Above a luminance of 0.6 cd/m^2, the unified luminance and the photopic luminance are equivalent, and below a luminance of 0.001 cd/m^2, the unified luminance and the scotopic luminance are equivalent.

For example, assuming a dark asphalt pavement with a diffuse reflectance of 0.1 and an HPS source ($S/P = 0.64$), use of equation (8.3) would yield a photopic pavement luminance of 0.3 cd/m^2 when the photopic illuminance on the pavement were 13.6 lx. Equation (8.6) would show this to be a unified luminance of 0.27 cd/m^2. Under an MH source ($S/P = 1.57$), the same unified luminance would be achieved when the photopic illuminance were only 10.0 lx rather than 13.6 lx.

The CIE is presently in the process of formally recommending a system of photometry having similar characteristics as the unified system of photometry (Rea et al. 2004).

8.8.2 Light Pollution

An increasing awareness of the potential negative impacts of roadway and other exterior lighting has developed among the public, particularly with regard to light pollution. The term *light pollution* encompasses different forms of nuisance or excess illumination: *sky glow*, the brightening of the nighttime sky that reduces visibility of the stars; *light trespass*, or unwanted light from nearby properties onto another property; and *glare*, the reduction in visibility and sensation of discomfort that bright lights can cause. The IES (2000b, 2000c) has published guidance for minimizing these negative impacts.

Recent trends in roadway lighting equipment (see section 8.6) have moved from luminaires having semicutoff distributions (often with drop lenses such as those illustrated in Figure 8.8*a*) to those with full cutoff distributions (often with flat lenses such as those in Figure 8.8*b*) to reduce the amount of light directed upward. While limiting the use of roadway luminaires that can direct light upward will tend to result in less light projecting into the sky (Zhang 2006), the majority of light leaving outdoor lighted locations is reflected from the ground and other surfaces (Waldram 1972). Comprehensive control of light pollution is often specific to the configuration of lighting and the entire site being lighted.

A quantitative system for assessing the different aspects of light pollution has been developed (Brons et al. 2008), called the *outdoor site-lighting performance (OSP) system*. OSP uses a virtual calculation box over the property (or right-of-way) boundaries including a ceiling 10 meters above any luminaire or architectural feature in the property being evaluated. For roadway installations with regularly spaced intervals of luminaires, the OSP calculation box can be simplified to include the repeated segment (Figure 8.15). OSP defines the *glow quantity* as the average illuminance on the surfaces of the calculation box divided by the average horizontal illuminance, representing the proportion of light used in the site that leaves the site either directly or through reflection. The OSP *trespass quantity* is defined as the maximum illuminance on any of the vertical surfaces of the box, representing light that could impinge an another property or into a window of an adjacent building. The OSP *glare quantity* is based on a model that takes into account the illuminance from each luminaire, the illuminance from the areas surrounding the luminaire, and the ambient illuminance from neighboring properties (Brons et al. 2008; Bullough et al. 2008). The model uses illuminance quantities because these can be readily calculated using many lighting software packages, and illuminance also can be measured in the field.

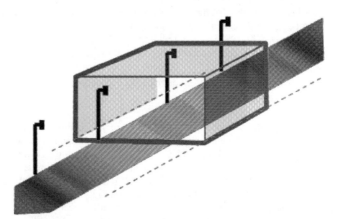

FIGURE 8.15   Illustration of the OSP calculation box placed over a roadway section.

REFERENCES

Allen K. 2009. *The Effectiveness of Amber Rear Turn Signals for Reducing Rear Impacts.* DOT HS 811 115. Washington: U.S. Department of Transportation.

Alliance for Solid State Illumination Systems and Technologies. 2009. *ASSIST Recommends: Outdoor Lighting—Visual Efficacy.* Troy, NY: Rensselaer Polytechnic Institute.

Andre J, Owens DA. 2001. The Twilight Envelope: A User-centered Approach to Describing Roadway Illumination at Night. *Human Factors* 43:620–30.

American Association of State Highway and Transportation Officials. 2005. *Roadway Lighting Design Guide.* Washington: American Association of State Highway and Transportation Officials.

Brons JA, Bullough JD, Rea MS. 2008. Outdoor Site-lighting Performance: A Comprehensive and Quantitative Framework for Assessing Light Pollution. *Lighting Research and Technology* 40:201–24.

Bullough JD. 2003. *Lighting Answers: LED Lighting Systems.* Troy, NY: National Lighting Product Information Program.

Bullough JD. 2009. Spectral Sensitivity for Extrafoveal Discomfort Glare. *Journal of Modern Optics* 56:1518–22.

Bullough JD, Boyce PR, Bierman A, et al. 2000. Response to Simulated Traffic Signals Using Light-emitting Diode and Incandescent Sources. *Transportation Research Record* 1724:39–46.

Bullough JD, Brons JA, Qi R, Rea MS. 2008. Predicting Discomfort Glare from Outdoor Lighting Installations. *Lighting Research and Technology* 40:225–42.

Bullough JD, Rea MS, Zhou Y. 2009a. *Analysis of Visual Performance Benefits from Roadway Lighting.* Washington: National Cooperative Highway Research Program.

Bullough JD, Snyder JD, Smith AM, Klein TR. 2009b. *Replacement Processes for Light Emitting Diode Traffic Signals.* Web-Only Document 146. Washington: National Cooperative Highway Research Program.

Commission Internationale de l'Éclairage. 1992. *Road Lighting as an Accident Countermeasure.* Report No. 93. Vienna: Commission Internationale de l'Éclairage.

Donnell ET, Shankar V, Porter RJ. 2009. *Analysis of the Safety Effects for the Presence of Roadway Lighting.* Washington: National Cooperative Highway Research Program.

Freedman M. 2001. *Visibility Performance Requirements for Vehicular Traffic Signals.* NCHRP Project 5-15 Report. Washington: National Cooperative Highway Research Program.

Fry GA. 1954. A Re-evaluation of the Scattering Theory of Glare. *Illuminating Engineering* 49:98–100.

He Y, Rea MS, Bierman A, Bullough JD. 1997. Evaluating Light Source Efficacy Under Mesopic Conditions Using Reaction Times. *Journal of the Illuminating Engineering Society* 26:125–38.

Illuminating Engineering Society. 1981. *IES Lighting Handbook: Reference Volume.* New York: Illuminating Engineering Society.

Illuminating Engineering Society. 1989. *Value of Public Lighting.* CP-31-1989. New York: Illuminating Engineering Society.

Illuminating Engineering Society. 2000a. *American National Standard Practice for Roadway Lighting.* RP-8-00. New York: Illuminating Engineering Society.

Illuminating Engineering Society. 2000b. *Addressing Obtrusive Light (Urban Sky Glow and Light Trespass) in Conjunction with Roadway Lighting.* TM-10-00. New York: Illuminating Engineering Society.

Illuminating Engineering Society. 2000c. *Technical Memorandum on Light Trespass: Research, Results, and Recommendations.* TM-11-00. New York: Illuminating Engineering Society.

Illuminating Engineering Society. 2006. *Spectral Effects of Lighting on Visual Performance at Mesopic Light Levels.* TM-12-06. New York: Illuminating Engineering Society.

Illuminating Engineering Society. 2007. *Luminaire Classification System for Outdoor Luminaires.* TM-15-07. New York: Illuminating Engineering Society.

Institute of Transportation Engineers. 1998. *Vehicle Traffic Control Signal Heads Part 2: Light-Emitting Diode (LED) Vehicle Signal Module (Interim).* Washington: Institute of Transportation Engineers.

Institute of Transportation Engineers. 2005. *Vehicle Traffic Control Signal Heads: Light-Emitting Diode Circular Signal Supplement.* Washington: Institute of Transportation Engineers.

Leslie RP, Rodgers PA. 1996. *Outdoor Lighting Pattern Book.* New York: McGraw-Hill.

Minnesota Department of Transportation. 2006. *Roadway Lighting Design Manual.* St. Paul: Minnesota Department of Transportation.

National Highway Traffic Safety Administration. 2007. *Nighttime Glare and Driving Performance: Report to Congress.* Washington: U.S. Department of Transportation.

National Lighting Product Information Program. 2004. *Specifier Reports: Parking Lot and Area Luminaires.* Troy, NY: National Lighting Product Information Program.

Perel M. 1985. Evaluation of Headlamp Beam Patterns Using the Ford CHESS Program. SAE Paper 856035. Warrendale, PA: Society of Automotive Engineers.

Rea MS (ed.). 2000. *IES Lighting Handbook*, 9th ed. New York: Illuminating Engineering Society.

Rea MS, Bullough JD, Fay CR, et al. 2009. *Review of the Safety Benefits and Other Effects of Roadway Lighting.* Washington: National Cooperative Highway Research Program.

Rea MS, Bullough JD, Freyssinier JP, Bierman A. 2004. A Proposed Unified System of Photometry. *Lighting Research and Technology* 36:85–111.

Rea MS, Ouellette MJ. 1991. Relative Visual Performance: A Basis for Application. *Lighting Research and Technology* 23:135–44.

Schoettle B, Sivak M, Takenobu N. 2008. Market-weighted Trends in the Design Attributes of Headlamps in the U.S. SAE Paper 2008-01-0670. Warrendale, PA: Society of Automotive Engineers.

Sivak M, Flannagan MJ, Miyokawa T. 1998. *Quantitative Comparisons of Factors Influencing the Performance of Low-Beam Headlamps*. UMTRI-98-42. Ann Arbor, MI: University of Michigan.

Skinner NP, Bullough JD, Smith AM. 2010. Survey of the Present State of Headlamp Aim. *Transportation Research Board 89th Annual Meeting*. Washington: Transportation Research Board.

Sullivan JM, Adachi G, Mefford ML, Flannagan MJ. 2004. High-beam Headlamp Usage on Unlighted Rural Roadways. *Lighting Research and Technology* 36:59–65.

Tien JM, O'Donnell VF, Barnett A, Mirchandani PB. 1979. *Street Lighting Projects National Evaluation Program: Phase 1 Report*. Washington: U.S. Department of Justice.

Van Derlofske J, Bullough JD, Hunter CM. 2002. Visual Benefits of High-intensity Discharge Forward Lighting. SAE Paper 2002-01-0259. Warrendale, PA: Society of Automotive Engineers.

Waldram JM. 1972. The Calculation of Sky Haze Luminance from Street Lighting. *Lighting Research and Technology* 4:21–6.

Wördenweber B, Wallaschek J, Boyce P, Hoffman DD. 2007. *Automotive Lighting and Human Vision*. New York: Springer.

Zhang C. 2006. Performance of Luminaire Metrics for Roadway Lighting Systems. M.S. thesis, Rensselaer Polytechnic Institute, Troy, NY.

CHAPTER 9
GEOMETRIC DESIGN OF STREETS AND HIGHWAYS

Brian Wolshon
Department of Civil and Environmental Engineering
Louisiana State University
Baton Rouge, Louisiana

9.1 INTRODUCTION

The geometric design of highways encompasses the design of features of the roadway associated with safe, efficient, and comfortable travel. The design process takes into account the range of different driver and vehicle types that are likely to use the facility, the topographic and land use characteristics of the area that surrounds it, and the costs involved in constructing and maintaining the highway. In addition, highway design also requires a balance between many overlapping and often competing functions and needs that will affect the safety, efficiency, and cost-effectiveness of the roadway.

The highway design process requires knowledge of most of the subdisciplines of civil engineering, including traffic and transportation engineering, geotechnical and materials engineering, structural engineering, hydraulic engineering, and surveying. It also requires an understanding of many aspects of the basic sciences, including physics, physiology, and psychology, among others. This chapter provides an overview of both the theory and practice of highway design. It summarizes the geometric highway design process by highlighting the fundamental engineering and scientific principles that underlie it and summarizing the ways in which they are applied to the development of a final design.

Since the highway design process encompasses such a wide range of design specialties, it is obviously not possible to cover all of them within a single book chapter. However, in recognition that the subject is considerably more complex and detailed than can be addressed adequately herein, I have included references for additional resources that provide a more detailed treatment of particular subjects. Readers are encouraged to review them as well as the related chapters of this book to gain a greater understanding of these specific topic areas.

The presentation of the material in this chapter approximates the coverage of similar subjects in the American Association of State Highway and Transportation Officials' *A Policy on Geometric Design of Highways and Streets.*[1] This manual, commonly referred to as the "Green Book," is the most widely referenced and applied highway design resource in the United States and serves as the foremost source of highway design theory and application. It has evolved over several decades based on a consensus of state and local highway agency design methods, standards, and assumptions that are themselves based on years of testing and experience.

Since the Green Book represents a compilation of design policies, it also allows ample latitude for the use of engineering judgment and experience that may be appropriate to address specific design issues. However, its wide scope and, in some cases, lack of specificity can leave some users overwhelmed with differing opinions, recommendations, and information when addressing specific design issues. Among the objectives of this chapter is an attempt to summarize some of the many details of the Green Book to make it more usable.

9.2 HIGHWAY FUNCTION AND DESIGN CONTROLS

The design of a highway needs to fit within a set of parameters that govern how it will be used, who will use it, and where it will be located. These parameters are a function of the traffic that it will serve and often the agency or entity that owns it. Thus the classification, whether functional or jurisdictional, is a fundamental component that drives final design. Conversely, the classification also can affect the makeup of the traffic that will be using the roadway, including their type (trucks, automobiles, buses, recreational vehicles, etc.), operating speeds, and to some extent, the type of trips they will make on it. The following sections summarize some of the differences between the functional and jurisdictional classifications of highways and describe some of the ways in which they affect their design.

Among the primary components that drive road design is its users. In addition to vehicles, users also include pedestrians and bicyclists. The composition of vehicles, commonly known as the *vehicle mix*, may itself include tractor semitrailer trucks, small cars, motorcycles, and nearly every vehicle type in between. What is important about these vehicle categories is that each of them has different heights, widths, lengths, and weights, as well as varying capabilities to accelerate, brake, turn, and climb hills. Similarly, individual drivers of these vehicles also differ in terms of their ability to see, hear, read, comprehend, and react to stimuli. Despite the differences in vehicle characteristics and user abilities, only a single road can be constructed. Thus it must be designed to safely and economically accommodate this spectrum vehicles and drivers. The following sections also will discuss some of the key aspects of vehicle and driver characteristics and the way they affect the design of the highway environment.

9.2.1 Roadway Functional Classification

Another key factor that guides the development of a roadway design is its *functional classification*. The functional classification influences both the type of traffic (automobiles, heavy trucks, etc.) the road will carry and the trip characteristics (high-speed through route, low-speed local access route, etc.) of these vehicles. Generally speaking, roads serve two functions, providing mobility to drivers who seek to move quickly and efficiently from one location to another and providing access to properties adjacent to the road. These functions, while complementary and often overlapping, also can be conflicting. To serve the range of mobility and access demands, a hierarchy of roadway functionality has evolved to serve traffic, access adjacent properties, and guide the development of their design.

The three primary road classifications within this functional hierarchy are *arterial, collector/ distributor*, and *local* roads and streets. There are also varying stages within each of these classifications. The basic function of arterial roadways is to accommodate through traffic over longer distances on routes that allow comparatively high speeds with a minimum of interruptions. An example of the highest level of arterial roadway is a freeway. Freeways tightly restrict access for entering and departing traffic, have no at-grade intersections, and allow traffic to operate at high speeds. At the opposite end of the functional classification spectrum are local roads. Local roadways are designed primarily to provide access to abutting lands. As a result, they often carry lower volumes of traffic than arterial roadways, are designed for low-speed traffic operation, and provide frequent points of access/egress. Residential neighborhood streets are the most common example of local roads. On them, posted speeds typically are in the range of 20 to 30 miles per hour, and access points (driveways) may be spaced as close as 50 to 100 feet apart. Collector/distributor roadways encompass the range of roads that are between the arterial and local categories. While they serve multiple purposes, including overlapping the arterial and local roles, their intended purpose is to serve as a transition between arterial and local streets by "collecting" traffic from local and arterial streets and "distributing" it up or down the functional hierarchy.

The distinction between the various roadway classifications and functions is not always a clear one and can become even more blurred as transportation needs and land-use requirements change over time. An example of how this can occur is often seen in the vicinity of residential neighborhoods when low-level collector distributor roads evolve (with or without conscious planning) to serve as minor arterial routes. An opposite case occurs when intensive land development takes place along arterial routes in areas that were once rural in character. In such cases, roads that once provided high mobility between distant locations now must adapt to provide access to new local developments. The functional conflicts illustrated in both these examples also lead to traffic congestion and safety problems as the traffic mix changes to combine both through traffic seeking to travel at higher speeds and low-speed local traffic turning in and out of local businesses. These situations, however, can be addressed (or avoided altogether) though a combination of effective land-use planning and travel-forecasting analyses with design rather than purely through design processes.

9.2.2 Design Parameters

In addition to operational considerations, a road's functional classification influences several other key design parameters, including the volume of traffic the road will carry, its composition (in terms on the percentage of trucks and other commercial and heavy vehicles), and its design speed. In turn, these parameters affect nearly all aspects of the geometric design as well many of the aspects of the pavement thickness requirements and culvert and bridge structural elements.

Design Vehicle Types.   To simplify the variety of vehicles that may use a roadway, the American Association of State Highway and Transportation Officials (AASHTO) has established a set of four general *design vehicle* groups based on representative sizes, weights, and operating characteristics. These vehicle classes include passenger automobiles, buses, trucks, and recreational vehicles and incorporate several specific subconfigurations within each. In addition to the specific dimension measurements that are shown in Table 9.1, each design vehicle also has a set of turning radii that also include off-tracking and vehicle overhang measurements for large vehicles.

Traffic Volume.   The amount of traffic that is expected to use a roadway typically increases over time. Forecasts of traffic volume usually come from planning studies that take into account future land-use

TABLE 9.1   Dimensions of AASHTO Design Vehicles

Design vehicle type	Symbol	Dimension (feet)							Typical kingpin to center of rear axle
		Overall			Overhang				
		Height (ft)	Width (ft)	Length (ft)	Front (in)	Rear (in)	WB_1	WB_2	
Passenger car	P	4.25	7.0	19.0	3.0	5.0	—	—	—
Single-unit truck	SU	11–13.5	8.0	30.0	4.0	6.0	20.0	—	—
Conventional school bus (65 pass)	S-BUS 36	10.5	8.0	35.8	2.5	12.0	21.3	—	—
Intermediate semi trailer	WB-50	13.5	8.5	55.0	3.0	2.0	14.6	35.4	37.5
Motor home	MH	12.0	8.0	30.0	4.0	6.0	20.0	—	—

Source: Ref. 1.

development, improvements to the surrounding road network, population demographics and behavioral characteristics, and future economic trends. The traffic-volume measures that are most useful in road design include the future *average daily traffic* (ADT) and the future *design hourly volume* (DHV).

DHV is closely related to the thirtieth highest hourly volume during the design year. This volume also can be used to assess the anticipated operational characteristics of planned roadways using the theories and computational procedures contained in the *Highway Capacity Manual (HCM).*[2] ADT measures typically are more useful for the design of minor and low-volume roads. It also should be noted that for areas that experience widely fluctuating seasonal or morning and evening peak-hour traffic volumes, additional analyses of demand may be necessary to provide adequate capacity for the anticipated demand. A more detailed discussion of the procedures used to acquire and compute both ADT and DHV can be found in Chap. 1.

Design Speed. Roads also must be designed under an assumed set of driver and vehicle operating characteristics. One of the most influential of these is *design speed.* Design speed affects most safety-related features of the design as well those associated with "rideability" and comfort. It also can affect the efficiency and capacity of a roadway. The selection of an appropriate design speed is based on a number of factors, including functional classification, the terrain and topography in the vicinity of the road, adjacent land use, and driver expectations.

In general, roads are engineered for design speeds between 15 and 75 miles per hour (mph). Design speed of 70 mph should be used for freeways, although depending on specific conditions (such as terrain, urban/rural conditions, etc.), actual design speeds may range from 60 to 80 mph. Most arterial roadways and streets are designed for between 30 and 60 mph. The design speed of collector/distributor roads is also in the lower area of this range and local streets, typically 30 mph and under.

Common practice dictates that the posted speed limit on a road is somewhat less than the design speed, although this is not universally true.[3] There are several different methods of posting speeds. Two popular methods include posting at 5 to 10 mph under the design speed, and the use of an 85th percentile running speed, which can be checked once the road is in operation.

Capacity and Level of Service. Similar to fluid flow in a pipe, there is a limit to the amount of traffic volume that a highway can accommodate. The maximum amount of flow is commonly referred to as the *capacity* of the roadway. Highway capacity is affected by a variety of factors that influence a driver's ability to maintain a desired operating speed and maneuver in traffic, including the number of lanes, the sharpness and steepness of curves, the number of heavy vehicles in the traffic stream, the quality of the signal coordination, and the location of lateral obstructions, among others. The *HCM* contains standardized procedures to estimate the capacity of highways in North America.[2] One of the fundamental aspects of highway design is the determination of how many lanes will be required to service the expected design traffic demand.

The *HCM* also contains procedures that allow alternative designs to be evaluated based on the anticipated quality of the operating conditions associated with each one. With them, the operational quality of service provided by a particular design can be designated on a *level-of-service* (LOS) rating scale of A to F, in which A is the highest quality of flow. On this scale, LOS level E describes the flow conditions at capacity and level F the conditions when the traffic demand exceeds the capacity of the roadway.

Typically, different roadways are designed to operate at different LOS values. In general, higher LOS values are sought for roads of higher functional classification. The AASHTO LOS guidelines for various functional classifications are shown in Table 9.2. As shown in the table, there are several factors that can influence LOS assessments of specific roadways, including the location of the roadway (rural/urban), the terrain (flat/mountainous), and the associated costs of constructing the desired facility.

9.2.3 Sight Distance

Another fundamental requirement of a highway design is the need to maintain adequate sight lines for drivers. This can be accomplished in a number of different ways, including manual checks on design profiles and through the application of theoretical stopping and passing sight-distance equations. At a

TABLE 9.2 Design Level of Service Based on Functional Classification

Roadway functional class	Appropriate level of service for combinations of area and terrain type			
	Rural level	Rural rolling	Rural mountainous	Urban and suburban
Freeway	B	B	C	C
Arterial	B	B	C	C
Collector/Distributor	C	C	D	D
Local	D	D	D	D

Source: Ref. 1.

minimum, all roads must be able to permit drivers operating vehicles at the design speed and within assumed conditions to bring a vehicle to a controlled stop in the event that an obstacle is blocking the lane of travel. This must be true longitudinally (i.e., straight ahead), around horizontal curves, and over and under vertical curves. Where economically and practically feasible, designers also may wish to provide enough advanced visible distance to drivers such that they are able to perform passing maneuvers in the lane of oncoming traffic.

The minimum lengths for sight distances are a function of a series of assumptions regarding the condition of the road and weather as well as the capabilities of a driver. The assumptions and resulting equations for determining minimum stopping and passing design requirements are described in the following sections.

Stopping Sight Distance. Design stopping sight distance lengths are computed as the sum of two separate distances, the distance traveled during the perception-reaction process and the distance covered during the deceleration from the design speed to a stop, as shown in equation (9.1).[1]

$$\text{SSD} = 1.468 V_d t_{p-r} + \frac{V_d^2}{30\left[\left(\dfrac{a}{32.2}\right) \pm G\right]} \tag{9.1}$$

where V_d = design speed, in miles per hour
t_{p-r} = perception reaction time, in seconds
a = deceleration rate, in feet per second squared
G = longitudinal grade of the road, in percent/100

In the perception-reaction distance portion of the equation, it is assumed that a driver will require 2.5 seconds to identify and react to a hazard stimulus. Thus 2.5 seconds is a standard value that is used despite prior research that has shown that actual reaction times may range from less than a second to more than 7 seconds depending on the level of expectation and complexity of the information given. The standard reaction time value does not account for all drivers who may be momentarily distracted from the driving task or who may be in a less than fully lucid state, such as those who may be drowsy, taking medication, or affected by other impairments.

The second part of the equation accounts for the distance traveled while the vehicle is decelerating. This distance is affected by several factors, including the initial speed of the vehicle, the friction conditions that exist between the vehicle tires and the road surface, and the grade of the road. The friction relationship assumes that the tires on the design vehicle will be well worn with less than full tread and that the road surface will be wet. These assumptions do not, however, take into account snow or icy conditions, in which the level of friction between the tires and the road surface would be effectively zero. Thus drivers are expected (and in most case required by law) to operate their vehicles in a manner "reasonable and prudent" for the apparent conditions, regardless of the design or posted speed limit. Road grade also can lengthen or shorten the required stopping distance depending on whether the road profile in the direction of travel is oriented uphill or downhill. The AASHTO stopping sight distances for flat grades are shown in Table 9.3.

TABLE 9.3   Recommended Stopping Sight Distances

Design speed (mph)	Brake reaction distance (ft)	Braking distance on a flat grade (ft)	Stopping sight distance	
			Calculated (ft)	Rounded for design (ft)
20	73.5	38.4	111.9	115
30	110.3	86.4	196.7	200
40	147.0	153.6	300.6	305
50	183.8	240.0	423.8	425
60	220.5	345.5	566.0	570
70	257.3	470.3	727.6	730
80	294.0	614.3	908.3	910

Note: Brake reaction distance assumes a 2.5-second driver perception reaction time and a vehicle deceleration rate of 11.2 ft/s^2
Source: Ref. 1.

Passing Sight Distance.   Unlike stopping sight distance, which is required for all designs, the provision for passing sight distance is an optional design feature. Typically, passing sight distance is provided where practicable and economically feasible. Since passing can increase the overall efficiency of a roadway, it is desirable to provide it on as high a proportion of the roadway as feasible, particularly on low-volume roads. Where it is not practically feasible to provide passing sight distance in a design, the incorporation of *passing lanes* is suggested.

Similar to the stopping sight distance equation, the computation of passing sight distance is made up of multiple parts that account for the various physical processes that occur during the maneuver. The equation is based on the passing need for a two-lane highway in which a driver would need to see far enough ahead into the opposing lane of traffic such that he or she can accelerate, drive around the passed vehicle, and return to his or her original lane while leaving an adequate safety distance between himself or herself and the oncoming vehicle.

The first component distance of a passing action is the *initial maneuver.* This part includes the distance covered as the driver of the passing vehicle checks the downstream clearance distance and begins accelerating to overtake the vehicle to be passed. The equation for computing this distance is as follows:[1]

$$d_1 = 1.468t_i + \left(v - m - \frac{at_i}{2}\right) \tag{9.2}$$

where t_i = time of initial maneuver, in seconds
a = average acceleration, in mph per second
v = average speed of the passing vehicle, in miles per hour
m = difference in speed of passed and passing vehicles, in miles per hour

One of the key assumptions in this equation is that the difference in speed between the passing and passed vehicles (the variable m in the equation) is 10 mph. In reality, this may or may not be the case.

The second distance component is that which is covered while the passing vehicle occupies the lane of oncoming traffic. This includes the distance in which any part of the passing vehicle encroaches into the lane of opposing traffic. The computation of this distance is based on the time that the passing vehicle is assumed to occupy the lane of opposing traffic, as shown below:[1]

$$d_2 = 1.468vt_2 \tag{9.3}$$

where t_2 = time passing vehicle occupies the left lane, in seconds
v = average speed of the passing vehicle, in miles per hour

TABLE 9.4 Recommended Passing Sight Distances

Design speed (mph)	Assumed vehicle speed (mph)		Stopping sight distance	
	Passed vehicle	Passing vehicle	Calculated (ft)	Rounded for design (ft)
20	18	28	706	710
30	26	36	1,088	1,090
40	34	44	1,470	1,470
50	41	51	1,832	1,835
60	47	57	2,133	2,135
70	54	64	2,479	2,480
80	58	68	2,677	2,680

Source: Ref. 1.

The third component accounts for the distance covered by a vehicle heading toward the passing vehicle in the opposing lane. For design purposes, it is assumed to be two-thirds of the d_2 distance. The fourth and final distance is safety separation distance between the passing and opposing vehicles. This distance varies and is based on the design speed of the road.

The AASHTO passing sight distance ranges for various operating speed ranges are shown in Table 9.4. Of course, it must be understood that the assumptions both for stopping and for passing conditions can vary significantly across any set of driver, vehicle, weather, and location conditions. However, an additional implied assumption is that it is unlikely that a set of near-worst-case conditions would occur simultaneously at a single time in a single place. Thus the available sight distances provided on most roads are more than adequate for the overwhelming majority of the time.

9.3 HORIZONTAL ALIGNMENT DESIGN

Although the construction of highways takes place in three-dimensional space, the procedures that govern their design have been developed within three separate sets of two-dimensional relationships that represent the *plan*, *profile*, and *cross-section* views. The plan view encompasses the *horizontal alignment* of the road in the *xy* plane. In it, the straight and curved sections of the road are described on a flat two-dimensional surface. The directions and lengths of straight or *tangent* segments of road are stated in terms of *bearings* and *distances* from one point to another and are referenced in a coordinate system of *northings* and *eastings* that are analogous to *y* and *x* rectangular coordinates, respectively. The profile view, or *vertical alignment*, represents the roadway in terms of uphill and downhill slopes, or *grades*, that connect points of vertical *elevation*. Finally, the cross-section is a "slice" of the roadway through the *yz* plane. The key components of the cross-section include the widths and cross-slopes of the lanes, shoulders, and embankment slopes, as well as the depths or thicknesses of the pavement and embankment material layers.

Horizontal alignments can be thought of a series of straight lines, known as *tangents*, that meet one another, end to beginning, at *points of intersection* (POIs). The goal of *horizontal curves* is to facilitate high-speed changes of travel direction between the various tangents. The sharpness of these curves dictates how abruptly a change in direction can be made at given design speed. The horizontal alignment also must be designed to provide adequate sight distances on these straight and curved sections of highway such that drivers will be able to operate their vehicles at the desire rate of speed without encountering any safety or efficiency obstacles.

The development of a horizontal alignment is based on a number of factors, the most important of which are safety, efficiency, access to adjacent properties, and cost. However, since both safety and efficiency can be designed into an alignment, the chief factors in most alignment selections are often cost and land access. The cost of an alignment may be affected by a number of factors, including right-of-way land acquisition, environmental considerations, and the cost of construction.

Construction cost can be increased significantly by special features such as bridges and tunnels and from added earthwork manipulation such as that necessitated by unsuitable soil conditions.

9.3.1 Horizontal Curve Design

The design of horizontal curves must allow a driver to maintain control of a vehicle, within a lane of travel, at desired rate of speed, while permitting him or her to see an adequate distance ahead such that he or she would be able to take evasive action (stop, change lanes, etc.) if a hazard is present in his or her path. To accomplish these objectives, two criteria have been developed: minimum-radius curve design and lateral sight distance within a curve.

Minimum Radius.   According to Newton's first law, a body in motion tends to stay in motion unless acted on by an external force. This theory is applied to horizontal curve design from the perspective of a vehicle rounding a curve. The natural tendency of a vehicle is to go straight rather than turn. As such, vehicles are prone to skid out of a curve that is too sharp for the speed at which they are traveling. To account for this fact, a minimum-curve-radius formula has been developed:[1]

$$R_{min} = \frac{V_d^2}{15\left(0.01e_{max} + f_{max}\right)} \tag{9.4}$$

where R_{min} = minimum radius of curvature, in feet
V_d = design speed, in miles per hour
e_{max} = maximum rate of superelevation, in feet per foot
f_{max} = coefficient of side friction

Equation (9.4) balances the constituent sideways, gravitational, and centripetal forces on a vehicle as it rounds a curve by effectively allowing all the external forces to sum to zero (i.e., the vehicle will neither skid out of the curve nor slide down the banking) and allowing a driver to maintain a travel-path lane position throughout the curve.

Curve Superelevation.   While equation (9.4) assumes that the curve will be superelevated throughout its length, in reality, this is rarely, if ever, the case. In actual practice, the lanes of tangent sections of road are designed with a crowed *cross-slope*. A crowned cross-slope facilitates the run-off of surface water across the lanes from a centerline high point to the outside edges of each lane. In a superelevated curve, the road cross-slope is sloped continuously from one edge to the other. Thus this roadway cross-slope change from a normal crown to a fully superelevated section must occur over a gradual transition distance. This transition distance can be accomplished in several different ways.

One method involves the design of a *spiral* or *transition curve* to transition the road alignment from a straight line to a curve. By definition, a spiral is a curve of constantly varying radius. When used to transition between a tangent and a curve, the spiral begins with an infinite radius (i.e., a straight line) and ends at the design radius of the curve. To transition from a normal crown to a fully superelevated section, the road surface cross section is transitioned gradually over this length.

Another common method of transition is the application of a *maximum relative gradient* (MRG). The MRG numerically designates the rate at which the road cross-slope, and correspondingly, the edge-of-pavement elevation, can change relative to the road centerline crown point. As shown in Table 9.5, the higher the design speed, the more gradual is the transition. For example, at a design speed of 50 mph, the MRG is 0.50 percent, or an elevation change of approximately 1 foot vertically for each 200 feet of longitudinal distance. This would mean that a two-lane highway of 24 foot width with a 2 percent normal crown cross-slope superelevated about the inside edge of pavement at 6 percent and designed with a 50 mph design speed would require a total longitudinal distance of approximately 120 feet to transition from a normal crown to a fully superelevated cross-slope.

TABLE 9.5 Maximum Relative Gradients

Design speed (mph)	Maximum relative gradient (%)	Equivalent maximum relative slope
20	0.74	1:135
30	0.66	1:152
40	0.58	1:172
50	0.50	1:200
60	0.45	1:222
70	0.40	1:250
80	0.35	1:286

Source: Ref. 1.

Each of the various methods of transition from a normal crown to a fully superelevated cross section (rotated about the centerline, inside edge of pavement or outside edge of pavement) has subtle advantages and disadvantages. In south Louisiana, for example, it is common practice to design superelevated transitions by rotating about the inside edge of pavement because of high groundwater conditions. While this type of design may require a small increase in the required quantity of embankment fill, it permits designers to maintain the elevation of the inside edge of pavement thereby and not force it any lower than its original elevation. It also should be noted that the required length of superelevation also increases with wider pavement widths (i.e., more lanes).

Horizontal Curve Sight Distance.   A key safety element of the design of horizontal curves is the provision of sight distance through the inside of the curve. Unlike a straight section of highway, the clear-vision area on a curve may need to be maintained beyond the edges of the roadway. Although stopping sight distance is measured along the alignment, the direct line of vision to the end of the stopping distance includes areas within the inside of the curve. This relationship is also illustrated schematically in Chapter 3 of the Green Book.

Using a special case of the circular horizontal curve calculation procedure in which the middle ordinate is measured perpendicularly from the travel path around the innermost lane to a sight obstruction inside the curve, an equation for calculating the minimum lateral clearance to provide adequate sight distance around a horizontal curve has been developed. The relationship, shown in equation (9.5), computes the middle ordinate distance from the center of innermost lane based on a corresponding curve radius and stopping sight distance requirement.

$$M = R\left(1 - \cos\frac{28.65S}{R}\right) \tag{9.5}$$

where S = stopping sight distance, in feet
   R = radius of the curve, in feet
   M = middle ordinate, in feet

Horizontal Alignment Layout.   The design of horizontal alignments also requires knowledge of several principles of curve layout that includes the calculation of various parts of these curves. While a discussion of this process is outside the scope of this chapter, readers are encouraged to review the included sources of information on this subject.[4–8]

9.4 VERTICAL ALIGNMENT DESIGN

Similar to horizontal alignments, profile design also can be regarded as a series of straight lines that meet one another at POIs, although in the case of profiles the tangents are uphill and downhill *grades* that meet at *points of vertical intersection* (PVIs) that are connected by *vertical curves*. Another difference is that the vertical curves are based on parabolic rather than circular curve relationships. Although much like horizontal design, the sharpness of these curves is dictated by how abruptly a change in direction can be made at given design speed. Thus the objective of a profile design is to provide a balance between vertical grade steepness and sight distance along the tangent and curved sections of highway so that drivers will be able to operate their vehicles at the desired rate of speed without encountering any significant efficiency or safety obstacles.

This section discusses the design considerations for the basic design and layout of vertical alignments. In recognition that vertical alignment design includes aspects that are considerably more complex than those presented here, readers are again urged to review the referenced resources for a more detailed treatment of the subject matter.

9.4.1   Maximum and Minimum Grades

Maximum road grades are based primarily on the ability of trucks and other heavy vehicles to maintain an efficient operating speed. A goal of any design is to permit uniform operating conditions throughout the length of the design segment. When assessing the impact of grade on operating conditions, both the steepness and the length of the grade should be considered. Most passenger cars are not affected by uphill or downhill grades equal to or less than 5 percent. However, heavy trucks have a comparatively lower weight-to-horsepower ratio that leads to a diminished ability to accelerate and maintain constant speeds on uphill grades.

In general, grades of 5 percent are considered the maximum for design speeds in the range of 70 mph. For lower-speed roads in the range of 30 mph, maximum grades in the range of 7 to 12 percent may be appropriate if the local terrain conditions are substantially rolling. These maximums also should be assessed in conjunction with the length of the sloped segment because it is also recommended that maximum grade be used infrequently and, where possible, for lengths of less than 500 feet.

Minimum grade requirements are based on the need to maintain surface runoff drainage. In general, a grade of 0.50 percent is considered to be the minimum to facilitate surface drainage on curbed roadway. However, grades of 0.30 percent also may be appropriate when pavement surfaces and cross-slopes are adequate to maintain surface flows. On uncurbed roadways, flat grades (0.0 percent slope) can be used when the pavement cross-slope is consistent and adequate to drain surface runoff laterally.

9.4.2   Vertical Curve Design

Vertical curves are used to transition vehicle direction between various uphill and downhill grades. Depending on the magnitude and type of grade (positive or negative), vertical curves may be either crest or sag, although curves that connect positive-to-positive and negative-to-negative tangent grades are monotonic in that they do not have a turning point. Vertical curve designs seek to provide a design that is safe, comfortable, and aesthetically pleasing. This is accomplished by providing proper sight distances and adequate drainage, and maintaining various vertical forces to within tolerable limits. The following sections present many of the key design equations and discuss both the theories that underlie them and their application for design situations.

9.4.3   Designing for Sight Distance

A vertical curve may be designed for several different criteria. One of the most critical is sight distance. When designing for sight distance, the goal is to design a curve that permits a driver to see far enough ahead along the road profile to permit a deceleration to a stop or to allow passing maneuvers. To achieve this, a curve must be flat enough at a given set of design speeds, driver eye height, object height, and road grades. However, the design of flatter curves also could require additional cutting and filling during construction. Thus when designing for sight distance, an attempt is made to minimize the length of the vertical curve to minimize construction costs while still providing adequate sight lines.

In the case of crest curves, the apex of the rise must not interfere with the stopping and passing sight-distance needs. This can be accomplished in several different ways, the most fundamental of which involves flattening the curve. To accomplish this, it is necessary to assume the height of the driver's eye in the design vehicle and the height of the target object. In the case of stopping sight-distance conditions the object is assumed to be the taillight of a preceding vehicle, and in the case of passing sight distance, the roof of an on-coming vehicle in the opposing lane.

In the case of sag curves, the ability to provide stopping sight distance is not limited by any geometric obstructions because a driver should be able to see across the curve when traveling on either of the approach grades. An exception to this would be at night when clear vision would be limited by darkness. Because of this, the limiting factor for sight distance on sag vertical curves is assumed to be the distance that vehicle headlights are able to cast light on the road ahead. And in general, this would equate to the stopping sight distance that would be computed in equation (9.1).

In both crest and sag vertical curve cases, two geometric relationships can occur between the curve length and the sight distance. The first is the case in which the stopping sight distance is contained within the length of the curve itself, known more commonly as the *S < L case*. This occurs under combinations of lengthy curves or in shorter curves with lower design speeds. The second is the *S > L case*, in which the stopping sight distance is longer than the curve itself and may extend prior to, after, or on both sides of the vertical curve. This is more common in shorter curves or when the *algebraic difference* between the right and left tangent grades is minimal.

The general equations for determining the minimum length of curve required to provide adequate sight distances on crest and sag curves for both the *S < L* and *S > L* cases are shown below:

Crest Curves.   When *S* is less than *L*,

$$L = \frac{AS^2}{100\left(\sqrt{2h_1} + \sqrt{2h_2}\right)^2} \tag{9.6}$$

When *S* is greater than *L*,

$$L = 2S - \frac{200\left(\sqrt{h_1} + \sqrt{h_2}\right)^2}{A} \tag{9.7}$$

where L = length of vertical curve, in feet
  S = stopping sight distance, in feet
  A = algebraic difference in grades, in percent
  h_1 = height of eye above the road surface, in feet
  h_2 = height of object above the road surface, in feet

Sag Curves.   When *S* is less than *L*,

$$L = \frac{AS^2}{200[2.0 + S(\tan 1°)]} \tag{9.8}$$

When *S* is greater than *L*,

$$L = 2S - \frac{200[2.0 + S(\tan 1°)]}{A} \tag{9.9}$$

where L = length of vertical curve, in feet
  S = light beam distance, in feet
  A = algebraic difference in grades, in percent

In all vertical curve design cases, the standard design assumption is that the driver' eye height will be 3.5 feet above the road surface. This represents the approximate height of a driver in a small sports car, for example. For stopping sight distance conditions, the object height is assumed to be 2.0 feet, or the approximate height of a rear taillight. When these standard values are substituted in the general equations (9.6) through (9.9), the specific-case equations shown below result.

Crest Curves. When S is less than L,

$$L = \frac{AS^2}{2,158} \tag{9.10}$$

When S is greater than L,

$$L = 2S - \frac{2,158}{A} \tag{9.11}$$

where L = length of vertical curve, in feet
 S = stopping sight distance, in feet
 A = algebraic difference in grades, in percent

Sag Curves. When S is less than L,

$$L = \frac{AS^2}{400 + 3.5S} \tag{9.12}$$

When S is greater than L,

$$L = 2S - \left(\frac{400 + 3.5S}{A}\right) \tag{9.13}$$

where L = length of vertical curve, in feet
 S = light beam distance, in feet
 A = algebraic difference in grades, in percent

The stopping sight distance curve length requirements for crest and sag are also shown graphically in Chapter 3 of the Green Book. In these figures, curve lengths are represented as a function of design speed and the algebraic differences between the left and right tangents. Also apparent in these figures are the two regions of $S < L$ and $S > L$ that are divided by a dashed curved line on both tables.

Another area of note on the AASHTO figures is the use of *K values*. *K* values are computed by dividing the length of a curve by the algebraic difference between the two tangent grades. Technically, *K* is a measure of the length of vertical curve per percent change in algebraic difference or, in effect, the sharpness of the curve. *K* values are also a convenient method of quickly relating design speeds and algebraic differences to required curve lengths. The figures also show a "drainage maximum" *K* of 167. This is to call attention to the fact that curves designed with *K* values greater than 167 will have relatively long, flat longitudinal grades in the vicinity of the curve turning point. Thus the need to maintain positive cross-slope drainage across the pavement surface will be of critical importance within the vicinity of the curve high or low points.

When computing crest curve length for the passing sight distance requirement, an object height of 3.5 feet is used to approximate the roof height of a small vehicle, and the resulting specific-case equations are shown below.

When S is less than L,

$$L = \frac{AS^2}{2,800} \tag{9.14}$$

When S is greater than L,

$$L = 2S - \frac{2,800}{A} \tag{9.15}$$

where L = length of vertical curve, in feet
 S = stopping sight distance, in feet
 A = algebraic difference in grades, in percent[1]

Although it is often difficult to design passing sight distance in hilly areas because of the added expense of earthwork, the addition of a passing/climbing lane on the uphill grade segments into and on the curve can make provisions for passing in these areas a viable option. Since it is generally assumed that vehicles operating at night will use headlamp illumination, no equations have been developed for passing sight distance conditions on sag vertical curves because a passing vehicle would be able to see the headlamps of an oncoming vehicles in the opposing travel lane.

While at a minimum vertical curve lengths must be designed to provide minimum sight distances, curves are also often designed to satisfy other needs. These can include matching the elevation of an intersection or drainage feature, providing adequate vertical clearance above or below obstructions, to limit the amount of vertical forces on the driver, and to coordinate with the associated horizontal alignment to provide safer and more distant perspectives of the highway ahead. Various procedures and equations have been developed specifically for these purposes and can be found in a number of related text references.[4–8] Similar to the design of horizontal alignments, vertical alignments also require knowledge of several basic principles of survey layout that are outside the scope of this chapter. However, readers are encouraged to also review the same references for detailed information on this subject.

9.5 CROSS-SECTION DESIGN

The third dimension of highway design is the cross-section. Roadway cross-section design encompasses elements such as the traveled way, shoulders, curbs, medians, embankments, slopes, ditches, and all other elements associated with the roadside area. While cross-section design deals primarily with the design of the widths and slopes of these elements, it also includes the design and analysis of safety-related aspects of roadside appurtenances such as sign supports, guardrails, and crash attenuators. As with horizontal and vertical alignment design, cross-section design is also affected by both the design speed and other operational considerations. The following sections present the elements and design aspects of roadway cross-section features and discuss many of the philosophies that are used in the design of roadside features.

9.5.1 Cross-Section Elements

A highway section is subdivided into separate regions based mainly on their likelihood to carry traffic. The portion that carries the highest-load traffic is the roadway itself. While this obviously includes the lanes of travel, it also can include the shoulders. And although shoulders are used less frequently by traffic, they still must be designed to support traffic stopping for emergencies or in some cases driving on the shoulder to avoid a blockage. As such, one of the key considerations in the development of road cross-section element design standards is the amount and travel speed of the traffic it will carry.

Traveled Way. The *traveled way* is the portion of the road used by vehicles under normal operation and at a minimum includes the travel lanes. The width of travel lanes varies by design speed, traffic mix, and functional classification. They can vary in width from 9 to 13 feet. Although 11-foot lanes are common in areas with low volume, low operating speeds, and/or restricted right-of-way, 12-foot lanes are recommended for most roadways.

The cross-slope of the travel lanes also may vary. While attempts should be made to minimize the road cross-slope for operational purposes, the main purpose of cross-slopes is to facilitate drainage. Recommended cross-slopes for travel lanes can vary from a little as 1 percent for some roads to as much as 6 to 8 percent for others. A specific slope design is primarily a function of the surface type and need to remove surface runoff. For smooth pavement surfaces such as asphalt and concrete, cross-slopes of 2 percent are recommended. Although in states of the Gulf Coast Region of the United States, where design rainfall intensities are significantly higher than those in other areas of the country, cross-slopes are commonly designed at 2.5 percent. For low-type surfaces and unpaved roadways, recommended cross-slope increases to around 6 percent.

Shoulders. Shoulders include the portion of the roadway immediately outside the traveled way. Although they are intended to serve traffic in special cases, they are neither designed nor meant for high-speed operation or high traffic volume. The benefits gained from the use of shoulders include

1. Aiding drivers in the recovery of temporary loss of control, or to provide room to perform emergency evasive action
2. Storing vehicles safely off the traveled way in emergency situations
3. Providing a safe means of accomplishing routine maintenance and navigational operations
4. Serving as a temporary traveled way during reconstruction or emergency operations
5. Serving as a primary clear area free of obstructions
6. Enabling greater horizontal sight distance in cut sections
7. Enhancing traffic flow and thereby capacity
8. Providing structural support to the pavement and traveled way[1]

Shoulder design also can vary significantly, although it is based primarily on the function of the facility. They may be paved or unpaved (grass and gravel), and their width may range from 2 to 12 feet. On high-speed, high-volume roadways, 12-foot shoulders are desirable, particularly in rural areas where wider right-of-ways are available. Since shoulders are not expected to be used for high-speed maneuvering, their cross-slopes are also higher than those of travel lanes, typically between 2 and 8 percent, although care must be taken to ensure that the algebraic difference in slope between the travel lane and shoulder is minimized so that severe slope breaks will not exist at the edges of the traveled way.

Medians. In areas where it is economically feasible, highways that are planned to carry high volumes of traffic at high operating speeds are often designed to separate opposing streams with the use of a center median. The benefits of roadway medians include

1. Physically separating high-speed opposing traffic, thereby minimizing the chances of serious head-on collisions
2. Providing a clear recovery area for inadvertent encroachments off the traveled way
3. Providing a means of safely storing stopped or decelerating left-turning vehicles out of the higher-speed through lanes
4. Providing a means of safely storing vehicles turning left out of a minor street or driveway as they wait for an available gap
5. Providing safe storage for pedestrians crossing a high-speed or wide divided highway
6. Restricting or regulating left turns to and from adjacent businesses except at designated locations[11]

As with shoulders, median design also varies in terms of widths, slopes, and surface. They may be as narrow as 2 to 4 feet but may be as wide as 100 feet. In urban designs, medians are typically 40 feet or less, although they are typically designed to accommodate shoulders and barriers (if warranted) and should allow adequate width for the storage of queued left-turning traffic and adequate turning radii for turning vehicles. On rural freeways, 60-foot median widths are common to remove the need for barriers to separate traffic on opposing roadways.

9.5.1 Roadside Design

Years of experience and research have shown that the proper design of the area immediately outside the traveled way is critical to reducing the potential for crash-related injuries and fatalities, particularly those involved in single-vehicle run-off-the-road types of accidents. The features that make up this area include embankment slopes (parallel and cross-slope), ditches, and recovery areas. The design of these elements needs to balance cost efficiency and functional needs with safety so that drivers will be able "to recover safely from loss-of-control situations or at least not suffer a serous injury or fatality should an accident occur."[9]

The most widely applied reference for the design and analysis of these areas is the AASHTO *Roadside Design Guide.*[9] In addition to a serving as reference on geometric design issues, the guide also offers guidance on the design and application of barriers, sign and lighting supports, and crash-attenuating devices.

Roadway Clear Zone.   To minimize the chances for injuries in run-off-the-road type accidents, the concept of a *clear recovery area* or *clear zone* was introduced to keep the area outside the traveled way clear of hazardous obstructions and to keep the cross-section slopes traversable for errant vehicles. A six-step hierarchy of design guidelines has been suggested to reduce the level of hazard in this area. These include

1. Removing obstructions from the roadside area
2. Relocating obstructions to less hazardous areas of the roadside
3. Redesigning obstructions to make them more traversable (in the case of a shoulder embankment) or less of a hazard
4. Reducing the hazard potential of obstructions by making them breakaway (such as sign supports)
5. Shielding obstacles from traffic by using barriers and crash attenuators
6. Delineating obstructions to make them more visible to drivers.[9]

Although the recommended clear zone depends on a number of factors such as traffic volume, roadway design speeds, and slope steepness, the use of a 30-foot clear-zone recovery area is recommended for most higher-classification facilities, including freeways. It also should be understood, however, that the required clear zone for any particular location will vary based on the amount of available right-of-way and terrain features.

Embankments and Slopes.   Among the most common hazards in the roadside vicinity are embankment slopes resulting from earthwork cuts and fills. Embankment slopes typically are categorized into *recoverable* and *nonrecoverable*. Recoverable slopes include those in which drivers are assumed to be able to recover control of an errant vehicle. The most desirable slopes are those which are flatter than 6:1, although steeper slopes up to 4:1 are recommended when limiting the lateral extent of more significant cut and fill heights. Nonrecoverable slopes are those steeper than 4:1, in which there is a likelihood of vehicle rollover and other serious accidents. The hazard potential of slopes is also related to the total height of the cut or fill section.[9]

In general, fill slopes are considered to be somewhat more hazardous than cut slopes because of the increased risk from accelerating down an embankment. Protective design measures, such as the use of guardrail, are usually recommended in the vicinity of high and steep slopes.

Barriers.   In areas where it is costly or not economically feasible to construct a suitable clear zone, roadside barriers such as guardrail and crash cushions may be appropriate. There are many different types of barrier designs that may be used based on the characteristics of a particular location. However, the decision as to whether to use a barrier has to be made with the understanding that a barrier is itself a potential hazard and that a judgment must be made in selecting and locating the barrier such that it both protects drivers from obstacles and minimizes the potential for damage and injury when hit.

In general, a barrier is warranted when the potential for damage and injuries that could be sustained from the hazard are deemed to be greater than those that would be sustained from a collision with the barrier itself. The general nature of this definition gives considerable latitude to the designer in making safety assessments where fixed hazards are present. Examples of obvious hazards are steep shoulder dropoffs, watercourses, and other substantial fixed objects located within the desired clear zone. The *Roadside Design Guide* offers more specific guidance for the use of guardrails in the vicinity of fill embankments based on a comparative risk associated with embankment fill slope steepness and height.

Another key element in barrier design is the length of need and the placement of the barrier relative to both the roadway edge and the hazard. Guidelines for placement are also specified in the *Roadside Design Guide*. This guide also details the design and use of other roadside reduction features, including crash attenuators and breakaway devices.

9.6 OTHER DESIGN CONSIDERATIONS

In addition to the general areas of geometric design principles that were presented in the preceding sections, there are also a range of specialized areas of design that are often required for the development of a final set of construction plans. These include the design of traffic control features, roundabouts, intersection channelization, railroad crossings, and roadway lighting, among many others. More specialized design principles are also required when attempting to modify existing designs to accommodate new development and/or reduce the occurrence or lessen the impact of cut-through traffic and high operating speeds. The following sections highlight some of the basic principles associated with the design of traffic control devices and intersections.

9.6.1 Traffic Control Devices

Traffic control is used to regulate and control the flow of traffic so that highway systems are able to achieve both safe and efficient operation. It is also a key element of highway design. Traffic control design elements are used to communicate information to drivers, including where and when to stop, when to go, how fast to travel, and which lane to turn from, as well as to give advanced warning of potentially hazardous conditions that may exist within the highway environment. Traffic control devices (TCDs) include road signs, traffic signals, pavement markings, and other warning and communication devices.

The most widely accepted reference for the design, placement, and maintenance of TCDs is the *Manual of Uniform Traffic Control Devices (MUTCD)*.[10] The *MUTCD* was developed to provide guidance and establish national uniformity for the use of TCDs. It includes guidelines for colors, shapes, symbols, sizes, patterns, and use of text on traffic signs, signals, and pavement markings. It also gives guidance on their appropriate use under routine conditions and under special conditions such as school zones and construction work areas.

Although TCDs are used to control traffic, they also must be used only within the limits of their intended purpose. For some of the most critical devices, such as stop signs and traffic signals, the *MUTCD* includes a series of *warrants* that should be followed to establish whether one of these devices is truly needed. TCDs never should be used indiscriminately and never should be used to alter or lessen the impacts of inadequate design and planning. One such example is the use of stop signs in residential areas as a speed control or traffic volume reduction measure. Inappropriate use of TCDs can lead to a general lack of importance placed on them or ignoring of them by drivers in these areas.

A more detailed description of the design and application of the *MUTCD* and various traffic control devices can be found in Chapters 5 through 7.

9.6.2 Intersections

When two highway alignments cross, an *intersection* is created. While some intersections can be avoided by creating a *grade-separated* intersection using a bridge or tunnel to achieve a vertical separation, the vast majority of intersections are *at grade*. Because of the obvious impact that intersections can have on both the safe and efficient operation of a highway, special consideration must be given to their design and control.

The most obvious difference between an intersection and a road segment is the numerous points of conflict. A typical four-legged intersection has 32 *conflict points*. These are points where traffic streams intersect (16 of the 32), traffic from one stream merges with another (8 of the 32), and vehicles depart a traffic stream (8 of the 32). To minimize the safety impacts while maintaining acceptable levels of operational efficiency, four design objectives should be followed when designing intersections, including

1. Minimizing the number of conflict points
2. Simplifying conflict areas
3. Limiting conflict frequency
4. Minimizing conflict severity[11]

The level of hazard of many of these conflict points can be moved, reduced, and/or eliminated with various design measures such as *traffic islands* and *exclusive turn lanes* and TCDs such as stop signs and traffic signals. These points are discussed further in the next chapter.

It is most desirable to have roadways intersect at right angles. A 90-degree intersection allows drivers a greater field of view of oncoming traffic and facilitates turning movements for large trucks. When perpendicular intersections are not possible, other design procedures for realigning one or more of the approaches can be used to mitigate or correct the effects of skewed intersections. Additional techniques, such as the use of channelization and the redesign of corner radii, can be used. Various techniques also can be used for intersecting roadways that cross at substantially varying grade profiles.

The provision of adequate sight distance is also a key element of intersection design. The evaluation and design of sight distances at intersections differ somewhat from those along roadway segments. As opposed to developing a design that provides stopping or passing sight distance along the mainline of a roadway, the design of sign distance at intersections seeks to provide an envelop of clear vision for drivers attempting a minor roadway to cross or turn onto a major roadway.

There are several different scenarios for which sight distance may be required at an intersection. These include the no-control, yield-control, stop-control, and signal-control cases. The no-control case requires clear vision areas on all approaches to the intersection so that vehicles approaching on any leg of the intersection can see approaching vehicles on the other approaching legs far enough in advance to adjust their speed to avoid a collision. The yield case is a special case of the stop-control condition in which additional clear sight areas are required because vehicles are permitted to enter or cross the intersecting roadway if no conflicting vehicles are present on the major highway. The stop-controlled case involves an analysis for the sight distance requirement of a vehicle on a stop-controlled minor street to make a left turn, right turn, or crossing maneuver. The required distance is affected by several factors, including the design vehicle, operating speed of the major road, grade of the minor street, and number of lanes/width of the median on the major road.

Required sight distances for right- and left-turning traffic at intersections are also graphically presented in Chapter 9 of the Green Book. The values represented within these figures, however, are for flat profile grades and passenger car vehicles; thus they do not reflect the effect of grade or vehicle type. The additional distances required for these factors, as well as more of the specifics for the location of driver eye position and multilane highways, also can be found in this reference.

9.6.3 Interchanges

Intersections that occur at freeways are based on the need to maintain access to and from these facilities. The specific location and configuration of freeway *interchanges* are primarily a function of the directional demand between the various intersecting freeways and roadways as well as other cost and right-of-way limitations. As with nonfreeway intersections, the objectives of freeway interchanges are to maximize the operational efficiency and safety aspects within the interchange vicinity. To achieve these objectives, eight interchange deign principals have been suggested, including

1. Avoiding the incorporation of left-hand entrances and exits
2. Striving for single exit designs from all approach directions

3. Placing exits in advance of (rather than after) the crossing roadway

4. Designing the interchange to permit the cross-road to pass over (rather than under) the freeway

5. Avoiding designs that require a weaving maneuver within the interchange area

6. Providing decision sight distance in advance of the interchange on all approaches

7. Using auxiliary lanes, two-lane ramps, and special ramp designs to provide *lane balance* at all interchange ramp terminals

8. Designing interchanges to fit the principle of *route continuity* rather than forecast traffic patterns[11]

Interchanges fall within two primary categories. These include *system interchanges* that occur at the intersection of two freeways or other controlled-access facilities and *service interchanges* that occur between a freeway and on other noncontrolled access roadway such as an arterial or collector roadway. Among the differences between these two categories is the need to maintain (without stopping) traffic on system interchanges. Since interchange design is also heavily influenced by the cost and availability of rights-of-way, various loop and ramp configurations also can be used to keep certain quadrants of the interchange clear of connecting roadways, such as those used in partial cloverleaf designs. These varied designs also can be used to facilitate left turn movements at interchanges and can eliminate the need for stops at the cross-road.

9.7   CONCLUSION AND SUMMARY

The design of highway facilities is a process that requires an understanding of the theories that describe human behavior and the interactions of the physical world and their application in a practical and cost-effective manner. Highway designs must accommodate a range of drivers and vehicle types that encompass a wide range of characteristics and capabilities as well as a number of other external influences such as weather, lighting, and even local driving customs. Because of this, there is no "one size fits all" design. Each design must be evaluated within the specific location, objective, and constraints of the particular location.

While certainly a key component, a proper geometric design on its own will not ensure the safe and proper use of a highway. Effective design must be integrated with adequate and continuous enforcement and education efforts. Another consideration is that conditions can change over time and/or the conditions assumed during the original design may turn out to be different once the highway is constructed and in operation. Thus both safety and operational conditions must be monitored over time to minimize accidents and delay problems that may occur.

A design must be developed within the context of many related topics discussed in the various chapters of this book, including pedestrian facilities, roadway lighting, parking, and safety countermeasures. Designers also need to stay current on evolving standards and practices of highway design, as well as the changing philosophies of design, including those associated with traffic calming, mixed land-use developments, environmental controls (including those for air and noise pollution), and other socioeconomic and environmental impacts. Information can be found both in published sources and through involvement in professional organizations.

REFERENCES

1. *A Policy on Geometric Design of Highways and Streets,* 5th ed. Washington: American Association of State Highway and Transportation Officials, 2004.

2. *Highway Capacity Manual.* Washington: Transportation Research Board, 2004.

3. Krammes, R. A., K. Fitzpatrick, J. D. Blaschke, and D. B. Fambro. "Speed—Understanding Design, Operating, and Posted Speed." Texas Transportation Institute Report 1465-1, College Station, TX, 1996.

4. Meyer, C., and D. Gibson. *Route Surveying and Design,* 5th ed. New York: HarperCollins, 1980.

5. Hickerson, T. *Route Location and Design,* 5th ed. New York: McGraw-Hill, 1964.

6. Crawford, W. *Construction Surveying and Layout,* 2nd ed. New York: Creative Construction Publishing, 1995.

7. Garber, N., and L. Hoel. *Traffic and Highway Engineering,* 3rd ed. California: The Wadsworth Group, 2002.

8. Schoon, J. *Geometric Design Projects for Highways,* 2nd ed. Washington: American Society of Civil Engineers Press, 2000.

9. *Roadside Design Guide,* 3rd ed. Washington: American Association of State Highway and Transportation Officials, 2002.

10. *Manual of Uniform Traffic Control Devices for Streets and Highways,* 2009 ed. Washington: Federal Highway Administration, 2009. Available at http://mutcd.fhwa.dot.gov/pdfs/2009/pdf_index.htm.

11. *Traffic Engineering Handbook,* 6th ed. Washington: Institute of Transportation Engineers, 2009.

12. Ewing, R. *Traffic Calming*: *State of the Practice*. Washington: Institute of Transportation Engineers, 1999.

CHAPTER 10
INTERSECTION AND INTERCHANGE DESIGN

Joseph E. Hummer
Department of Civil, Construction, and Environmental Engineering
North Carolina State University
Raleigh, North Carolina

10.1 INTRODUCTION

Intersections and interchanges are important parts of the highway system. They typically have much higher collision rates and cause much more delay than midblock segments. They are also particularly expensive parts of the highway system. The purpose of this chapter is to provide a summary of current intersection and interchange design concepts and to direct readers to sources of detailed information on those designs. The chapter discusses intersections first, and then interchanges. For each of these, basic design elements are presented first, followed by configurations.

10.2 BASIC INTERSECTION ELEMENTS

10.2.1 Spacing

For safe and efficient vehicular traffic, intersections should not be placed too close together. Drivers accelerating away from one intersection are not expecting to encounter traffic slowing for another intersection, for example. Also, a queue of vehicles from one intersection that blocks another intersection, called spill-back, will cause congestion to propagate and cause extra delay, as Figure 10.1 shows. Intersection spacing of at least 500 feet is typically desirable for vehicles. On the other hand, pedestrians and bicyclists enjoy shorter paths and greater mobility when intersections are more closely spaced, as in older central business districts of some U.S. cities. Intersection spacing as low as 300 feet works well for pedestrians and bicyclists.

Spacing between signalized intersections is critical. To ensure optimal progression in both directions on an arterial, signals should be spaced far enough apart that vehicles travel from one signal to the next in one-half the signal cycle length. For typical suburban speeds and cycle lengths, signal spacing around one-half mile provides for optimum two-way progression. Good two-way progression is often impossible with signal spacing from 500 to 2,000 feet.

FIGURE 10.1 Traffic queuing at one intersection spills back to block another intersection, adding to delay and causing potential safetyproblems. (*Source:* Daniel L. Carter.)

10.2.2 Location

Intersections are safer in some locations than others. One important guideline is that intersections should not be on a horizontal curve if possible. Horizontal curves could restrict sight distances to the intersection and to traffic signals or signs near the intersection. A horizontal curve requires some drivers turning at the intersection to make complex reverse turn maneuvers. Also, horizontal curves are often superelevated, which makes the profile of the intersecting street very difficult to negotiate for turning or crossing motorists. Relative to the vertical alignment of a road, intersections should not be near a crest vertical curve, again due to restrictions on sight distance.

10.2.3 Angle

The angle of intersection is important to operations. As the angle between the two streets departs further from 90°, vehicle and pedestrian time in the area conflicting with other traffic streams increases so delays and collisions increase. In addition, the paved area—and therefore construction and maintenance cost—will increase. Tradition is that these effects increase dramatically with angles less than 60° or greater than 120°.

Several options exist for treating existing or proposed intersections with unfavorable angles. One or both streets could be realigned with horizontal curves to create a more favorable angle. In some cases, one intersection can be made into two, creating an offset intersection as described below. Another option is to use islands to guide drivers and pedestrians better through the intersection and reduce the paved area (see below).

10.2.4 Number of Approaches

All else being equal, an intersection with fewer approaches will be safer and more efficient than an intersection with more approaches. The reason for this is the number of conflict points in the intersection—points where one traffic stream crosses, merges with, or diverges from, another traffic stream. A standard three-legged (T) intersection has 9 vehicle conflict points, a standard intersection between a two-way street and a one-way street has 13 vehicle conflict points, and a standard four-approach intersection has 32 vehicle conflict points. Among other effects, fewer conflict points mean fewer phases are needed at a traffic signal, which in turn means less lost time and lower delays.

There is a lively ongoing debate about whether it is safer and more efficient to create one four-legged intersection or two three-legged intersections separated by several hundred feet, an *offset intersection*. It appears that for some combinations of traffic volumes and spacing the offset intersection is a better choice than one standard four-legged intersection. Analysis tools like the *Highway Capacity Manual* (TRB 2000) and microscopic traffic simulation packages are good ways to examine the efficiency of the offset design compared to a standard four-legged design.

Intersections with five or more approaches are particularly inefficient and difficult for road users to negotiate. The treatments for such cases include terminating one or more legs before they reach the intersection, using horizontal curves to redirect one or more legs (likely creating another intersection), or making one or more legs one-way moving away from the intersection.

10.2.5 Turn Bays

Turn bays provide room for left-turning or right-turning vehicles to decelerate before their turns and/or to queue while waiting to turn. Left-turn bays are particularly effective at reducing delay and collisions by getting those vehicles out of the way of through vehicles. At busy signalized intersections, dual and triple left-turn lanes are used effectively to reduce the time that those vehicles need the right-of-way. Dual right-turn lanes are also used at some intersections. The drawbacks to using turn bays include higher right-of-way costs and longer crossing distances for pedestrians.

Through the years, many criteria have been published for left-turn and right-turn bays. The criteria are typically different for unsignalized and signalized intersections. For signalized intersections, one well-known set of turn bay criteria is provided in the *Highway Capacity Manual*. The Manual recommends:

- A single left-turn bay for peak hour left-turn volumes of 100 veh/hr or more
- A dual left-turn bay for peak hour left-turn volumes of 300 veh/hr or more
- A single right-turn bay for peak hour left-turn volumes of 300 veh/hr or more

The *Manual* also recommends additional through lanes for each 450 veh/hr of through volume. Many agencies use a triple left-turn bay for peak hour left-turn volumes of 500 veh/hr or more and a dual right-turn bay for peak hour right-turn volumes of 600 veh/hr or more.

Many other authors have provided turn bay criteria for signalized and unsignalized intersections. Pline (1996) synthesized information on left-turn treatments in the mid-1990s. Fitzpatrick and Wooldridge (2001) included this topic in their 2001 synthesis of research findings in geometric design.

Once a designer has decided to provide turn bays, he or she must decide on bay length. AASHTO (2004) recommends that overall turn bay length should be the sum of the taper length, deceleration length, and storage length. Taper rates into a turn bay are typically between 8:1 and 15:1, with lower rates for bays on urban roads with lower speeds. Taper rates into dual left-turn bays are sometimes even lower than 8:1 to maximize storage area. It is desirable to allow vehicles to decelerate fully after having departed a through lane, although this is sometimes impractical in urban areas. AASHTO (2004) states that typically lengths needed to decelerate from 45, 50, and 55 mph speeds to a full stop are 430, 550, and 680 feet, respectively, on grades of less than 3 percent. Storage lengths needed for unsignalized intersections are typically short, accommodating only a couple of vehicles. At signalized intersections, storage lengths typically must be much longer. AASHTO (2004) passes along traditional guidance that the turn bay should be able to store 1.5 to 2 times the average number of vehicles desiring storage per cycle. However, more sophisticated methods are available that consider more factors, including the possibility that a queue of through vehicles could block the entrance to the turn bay (Kickuchi, Chakroborty, and Vukadinovic 1993).

FIGURE 10.2   Positive offset left-turn bay. (*Source:* Daniel L. Carter.)

One recent innovation increasing the safety of turn bays that serve permissive left turns is providing a positive offset. A positive offset moves the intersection end of the left-turn bay as far to the left as possible, as shown in Figure 10.2, so that a left-turning driver's view of opposing through and right-turning traffic is not blocked by the opposing left-turn queue.

10.2.6 Islands

Providing islands to separate traffic streams, also called channelization, has a number of benefits and a few drawbacks. The benefits generally include (AASHTO 2004):

- Separation of conflicts
- Control of angle of conflict
- Reduction in excessive pavement areas
- Regulation of traffic and indication of proper use of intersection
- Arrangements to favor a predominant turning movement
- Protection of pedestrians
- Protection and storage of turning and crossing vehicles
- Location of traffic control devices
- Aesthetics

Separating and controlling the angle of conflicts are fundamental principles of intersection design that should lead to tangible collision reductions. However, designers should be cautious about applying islands everywhere, because in some places they do have drawbacks. Islands should help guide the motorist through the intersection using a natural path that meets their expectations, but islands that place unexpected obstacles in motorists' paths will likely cause collisions. In addition, very small islands are difficult for motorists to see and may pose a hazard. The AASHTO *Policy on Geometric*

Design of Highways and Streets (AASHTO 2004) recommends that the smallest size for a triangular turning island should be 50 square feet in an urban area and 75 square feet in a rural area.

One final note on choosing islands is that they tend to make the intersection look large and discouraging for pedestrians. Islands provide refuge for pedestrians, but the multiple-stage crossing can be slow and confusing for pedestrians. In places where the context calls for encouraging pedestrians, designers generally try to avoid wide channelized intersections.

Once a designer has decided to provide an island, he or she must decide on the type of surface, shape, and offset to travel lanes, among other details. Islands may have pavement flush with the surrounding travel lanes (separated only by pavement markings), may have a raised concrete surface, or may be landscaped with or without curbs. In deciding upon a surface, designers should consider the visual target value, the possibility of motorist violations, the protection afforded pedestrians using it as a refuge, maintainability, drainage, and other factors. Designers should also remember that any fixed objects placed on an island may be struck by errant motorists and may block drivers' or pedestrians' lines of sight. Designers should only use curbed islands at intersections with fixed-source lighting or where the curbs are delineated well. The shape of islands is typically dictated by the swept paths of the design vehicles chosen for the intersection and the widths of turning roadways (see below). The edge of an island is typically offset by 2 to 6 feet from the edges of the travel lanes or turning roadways surrounding the island, with a larger offset at points where driving paths diverge from each other. Island corners typically have one-foot to two-foot radii.

10.2.7 Curb Radii

The choice of curb radii is critical to optimal intersection function. Curb radii that are too small will lead to large vehicles slowing dramatically and/or encroaching over curbs and lane lines, causing delays and possibly collisions. Curb radii that are too large waste construction and right-of-way funds and cause longer intersection crossing times for pedestrians, which also causes delays and leads to collisions. Larger curb radii also decrease the distance from the curb to the right-of-way edge when the right-of-way edges are not curved (i.e., corner clearance), which could restrict pedestrian queuing area, among other effects. As noted above in the discussion of islands, choice of a design vehicle and the context of the intersection are critical in this aspect of design. In an exurban area with higher speeds and higher truck volumes larger curb radii (30 feet or more) are appropriate, while in a denser urban area with fewer large vehicles where pedestrians are encouraged smaller curb radii (15 to 25 feet) are appropriate. Parking lanes, with parking appropriately restricted near the intersection, help reduce the radius that would otherwise be needed.

One way a designer can compromise between the needs of large vehicles and pedestrians is by choosing the best type of intersection curb design. AASHTO (2004) provides for three types of intersection curb designs: a constant radius; a combination of taper, constant radius, and second taper; and a three-center compound curve. A constant radius is the simplest of the three choices to lay out and construct and is therefore the typical choice where neither large vehicle volumes nor pedestrian volumes are high, such as an intersection between two local residential streets. Since the tapered or compound curve designs more closely approximate modern vehicle turning paths, they use less space while accommodating larger vehicles at higher speeds with fewer encroachments. Tables in AASHTO (2004) recommend taper rates and curve radii given an intersection angle and a design vehicle.

10.2.8 Turning Roadway Widths

AASHTO (2004) recommends widths of turning roadways at intersections based on several factors, including the radius of the inner edge of the turn, the number of lanes, design vehicles, and the type of curb or shoulder. Recommended widths range from 12 feet for a 500-foot radius turn with one lane serving predominantly passenger cars to 45 feet for a 50-foot radius turn with two lanes serving large numbers of large semitrailer trucks.

10.2.9 Medians and Openings

Medians at intersections generally provide many of the same benefits as islands, described above, including separating opposing traffic directions (making head-on conflicts rarer), providing crossing pedestrians and vehicles refuge, providing a place for traffic control devices and light fixtures, and reducing pavement area. Since medians restrict left turns, median opening spacing can be a terrific access control mechanism.

The median width at an intersection is a major decision for a designer. Wider medians typically mean higher right-of-way costs, encourage vehicles turning left from the cross-street to "lock up" with each other, increase wrong-way movement potential, and mean higher minimum green times and lost times at signals, so engineers usually try not to design them wider than needed. Some of the key features of various median widths at intersections are:

- Narrower than 4 feet: difficult for drivers to see so the nose may be struck more often.
- Four feet or wider: provides a pedestrian refuge.
- Sixteen feet or wider: provides a pedestrian refuge and room for a left-turn bay.
- Twenty feet or wider: provides refuge for a crossing passenger car. According to the *Highway Capacity Manual,* at unsignalized intersections this refuge typically means a healthy improvement in level of service for the minor street.
- Twenty-eight feet or wider: provides a pedestrian refuge and room for a dual left-turn bay.
- Forty-nine to 59 feet: minimum median widths to allow U-turns by various trucks into the outer lane of a four-lane highway (AASHTO 2004). Allowing U-turning vehicles to encroach on the shoulder reduces these widths.

Several median nose designs are available. Designers typically design the median nose and the median opening width to conform to the turning paths of design vehicles, making sure that the paths of vehicles that are supposed to move simultaneously do not encroach on each other. AASHTO (2004) provides recommendations on minimum median opening widths based on design vehicle, median opening, nose shape, and intersection angle.

10.2.10 Access Points

To increase traffic safety and efficiency, driveways and minor streets should be separated from major intersections as far as possible. Some of the safety threats from driveways near major intersections include drivers accelerating away from an intersection being surprised by a driver ahead slowing to turn into a driveway, and a "double-threat" when a driver in one lane stops to let someone in or out of a driveway while a driver in an adjacent lane does not stop. Many agencies have programs to try to close existing driveways and minor street intersections near major street intersections or prohibit such driveways opening in the first place. Gluck, Levinson, and Stover reviewed corner clearance policies from many agencies across the United States and provided application guidelines. The midrange corner clearance called for in guidelines around the United States is 100 to 200 feet, with some agencies requiring up to 300 feet. One popular treatment to a problem of existing access points near a major intersection is installation of a narrow raised median to prohibit left turns into or out of the driveway, as Figure 10.3 shows.

10.2.11 Designing for Pedestrians and Bicycles

In recent years intersection designers have been taking the needs of nonmotorist road users, particularly pedestrians and bicyclists, into account more and more. In some dense urban areas, pedestrians and bicyclists are the dominant users, while even in suburban fringe areas a surprisingly high number of pedestrians and bicyclists need to be considered as they attempt to cross safely. Principles designers should adopt regarding pedestrians at intersections include:

FIGURE 10.3   A narrow, raised median helps treat a safety problem created by a driveway near a major intersection. (*Source:* Daniel L. Carter.)

- Provide as short a crossing as possible. Keep curb radii as short as possible and provide curb "bulb-outs" (sidewalk extensions into parking lanes) if possible.

- For wider crossings, consider medians and/or islands to provide pedestrian refuge.

- Provide a crossing path that is as direct as possible, since many pedestrians will attempt to walk on the shortest path anyway.

- Consider using a different or rougher pavement surface for the crosswalk to help alert and slow drivers. Figure 10.4 shows a "speed table" that raises the whole intersection surface and has been effective in some places by slowing vehicles appropriately in the presence of pedestrians.

- Provide appropriate curb ramps for wheelchairs, strollers, skaters, and other sidewalk users with wheels.

- Consult the *Highway Capacity Manual* for procedures to calculate the level of service for pedestrians at signalized or unsignalized intersections. Levels of service criteria are based upon delay and available queuing space for signalized intersections and upon delay for unsignalized intersections. Design treatments for intersections with poor pedestrian levels of service include medians, islands, larger corner queuing areas, and anything that would reduce the signal cycle length.

One of the main decisions facing designers regarding bicycles is how to route bicycle lanes through intersections. If there is an exclusive right-turn lane, the bicycle lane is typically placed between the turn lane and the through lanes, which means that bicycles must weave across right-turning vehicles at the beginning of the right-turn lane. If there is no exclusive turn lane, the bicycle lane will stay to the right of the through lanes and bicycles will conflict with right-turning vehicles at the intersection. In either case, clear sight lines and good traffic control devices will increase safety. AASHTO has published a *Guide for the Development of Bicycle Facilities* (AASHTO 1999) that provides many useful details on designing for bicycles at intersections.

FIGURE 10.4   A speed table can help slow vehicles, providing for a safer pedestrian crossing. (*Source:* Daniel L. Carter.)

10.2.12 Grades

Steep grades hamper traffic operations at intersections. Steep downgrades on an intersection approach increase stopping distances and make turning more difficult. Steep upgrades on an intersection approach make idling difficult for vehicles with manual transmissions and make acceleration slower for all vehicles, which in turn increases necessary gap sizes and sight distances for crossing and turning movements. Generally, grades under 2 percent do not cause many operational problems, grades from 2 to 4 percent begin to introduce noticeable problems, and grades over 4 percent should be avoided where practical.

10.2.13 Intersection Sight Distances

Intersection designers should ensure that approaching drivers can see all applicable traffic control devices and conflicting traffic soon enough to take appropriate actions. Sight may be blocked by the alignment of the road the driver is on, the alignment of the intersecting road, objects or terrain in the "sight triangle" near the corner, or objects or terrain in the median. Given the alignments and roadside obstacles, the designer needs to compute the intersection sight distance provided and compare that to the minimum sight distance recommended in AASHTO (2004). The AASHTO *Policy* helps the designer do this for six different situations, labeled case A through case F.

- Case A—intersections with no traffic control. Because vehicles typically slow down considerably on approach to this type of intersection, these intersection sight distances are lower than the stopping sight distance for any given design speed.
- Case B—intersections with stop control on the minor road. The sight distance of interest in this case is that for a vehicle that has stopped at the intersection and must be able to see the traffic on the major street far enough away to make a good decision about when to proceed. The calculation is different for vehicles turning left, turning right, and crossing the major road. The choice

of design vehicle and the presence of a median (making a left turn of a crossing maneuver into a two-stage process) affect the calculation greatly.

- Case C—intersections with yield control on the minor road. The computation in this case differs for vehicles that are crossing the major road from vehicles that are turning onto the major road. In both instances, intersection sight distances in case C exceed those for case B.

- Case D—intersections with traffic signal control. For this case, the AASHTO *Policy* states that the only minimal sight distances, allowing a vehicle at one stop bar to see vehicles at the other stop bars, are necessary, with three exceptions. First, permissive left-turning vehicles must be able to see gaps in oncoming through and right-turning traffic. Second, if flashing signal operation is allowed during off-peak times, the appropriate case B sight distances are recommended. Third, if right turns on red are allowed, the appropriate case B sight distances are recommended.

- Case E—intersections with all-way stop control. Like case D, the AASHTO *Policy* states that the only minimal sight distances, allowing a vehicle at one stop bar to see vehicles at the other stop bars, are necessary, but this time with no exceptions. All-way stop control is therefore an excellent treatment option for an intersection with sight distance problems.

- Case F—left turns from the major road. This should generally be provided at every point along a road where permissive left turns may be made now or in the future (i.e., designers should consider future intersection and driveway locations). If stopping sight distance has been provided along a road, and case B and C sight distances have been provided for each minor road intersecting the major road, the AASHTO (2004) states that no separate check for case F intersection sight distances may be needed.

Designers finding that an existing or proposed intersection does not provide the recommended intersection sight distance have a number of options to improve the situation, including realigning a road, installing a wide median, removing sight obstacles on the roadside, changing the type of traffic control, reducing speed limits, or installing warning signs.

10.3 INTERSECTION CONFIGURATIONS

10.3.1 Need for Alternatives

A basic configuration serves well at the vast majority of intersections. However, many intersections with a basic configuration are no longer serving today's higher traffic demand levels very well from a capacity or a safety point of view. Once good signal technology and some of the features discussed above, like turn bays and islands, have been installed, engineers have often exhausted the range of conventional improvements. Additional through lanes, bypasses, and structures are expensive and environmentally disruptive; alternative modes of travel and demand management remove few cars and trucks from the road; and intelligent transportation systems are years from making a major impact at most intersections. In this atmosphere, there is a great need for alternative intersection configurations as a practical way to improve safety and efficiency. The next few sections discuss some of those alternative configurations that designers should consider to improve intersection operations.

10.3.2 One-Way Approaches

An intersection involving one or two one-way streets operates much more efficiently and probably more safely than a similar conventional intersection between two two-way streets. Reasons for this greater efficiency and safety include:

- Signals at intersections involving one-way streets require fewer phases, reducing lost time. At an intersection between a two-way street and a one-way street, only one left-turning movement is opposed by vehicular traffic.

- It is possible to establish perfect signal progression along a one-way street with any signal spacing at any speed.
- The number of conflicts between traffic streams in the intersection is greatly reduced.
- Crossing pedestrians face fewer directions of conflicting vehicular traffic.

One-way streets also have some well-known drawbacks, including greater travel distances and potential for dangerous wrong-way movements. On the balance, however, converting conventional intersections into intersections involving one-way streets will improve traffic operations.

10.3.3 Traffic Calming

At some intersections, the primary goal is to discourage high vehicular volumes and/or speeds. *Traffic calming*, as it is termed, is typically appropriate on collector or local streets in residential areas and is often aimed at through traffic. Many measures have been used successfully at intersections to decrease vehicular volumes and speeds. Curb bulb-outs, textured crosswalks, and raised speed tables were mentioned above as design measures that assist crossing pedestrians, and these might be considered traffic-calming measures. Other design measures that engineers use to calm traffic at intersections include:

- Small traffic circles and roundabouts (see section below)
- Chokers (narrowing lanes with curbs and/or landscaping)
- Semidiverters (allowing one-way in or out of the intersection only on a particular approach)
- Forced-turn diverters (allowing no through movements for one street)
- Diagonal diverters (forcing two approaches to turn only left and the other two approaches to turn only right)
- Vehicular cul de sacs (usually allowing nonmotorized and emergency users to get through)

If done well, traffic calming at intersections can improve the overall quality of service and safety for road users and will improve the quality of the environment surrounding the intersections. If done poorly, though, traffic calming can cause collisions, increase travel times, increase frustrations, and harm the nearby environment. Some of the important issues designers need to examine when considering traffic calming at intersections include visibility to the design feature, forgiveness of the design feature if struck, driver expectations, aesthetics of the design feature, access for emergency vehicles, and whether problems mitigated at one intersection will simply migrate to another location. Ewing (1999) has written an excellent resource on traffic calming that will help designers negotiate some of these issues.

10.3.4 Roundabouts

A modern roundabout, as shown in Figure 10.5, offers greater safety and efficiency than a conventional intersection if designed and operated well. The idea of modern roundabouts came to the United States from Europe and Australia in the late 1980s, and modern roundabouts have since moved beyond the experimental stage to be a part of standard engineering practice in many U.S. agencies. Modern roundabouts differ from earlier, often unsuccessful, traffic circles because they have the following features:

- Yield control upon entry to the circle (traffic in the circle always has the right-of-way)
- Low design speeds for circulating traffic
- All traffic diverts from a straight-line path through the intersection
- No parking in or near the circle
- No pedestrians in the circle

FIGURE 10.5   Modern roundabout. (*Source:* Daniel L. Carter.)

Single-lane roundabouts have been proven in U.S. research to be safer than the conventional intersections they replaced (Persuad et al. 2001). Traffic-analysis software also typically shows that single-lane roundabouts that remain below capacity reduce delays compared to signalized intersections handling the same volumes. The capacity of single-lane roundabouts is typically about 30,000 vehicles per day, meaning that single-lane roundabouts are a good solution for intersections between two collector streets or minor arterials. Firm results on the safety and efficiency of roundabouts with two or more circulating lanes in the United States are not available to this point, however.

Roundabout designers face many decisions, including design speed, circle diameter, circle roadway width, approach flares, and splitter island size and shape. The FHWA has assembled an excellent information source for these and other roundabout design issues (FHWA 2000).

10.3.5 Auxiliary Through Lanes

Designers have sometimes successfully used short auxiliary through lanes—added before an intersection and dropped after the intersection—to increase capacity and reduce delay at the intersection. Frequently, such auxiliary lanes are an interim step in a general road-widening project. The main question designers have in considering auxiliary lanes is how long they have to be to allow safe diverges and merges and to entice enough motorists into the auxiliary lane to make a difference. If the lanes are too short, drivers will not choose to use them and the extra capacity provided on paper will not be realized in the field. Lee et al. (2005) have provided the best research available to this point guiding designers on auxiliary through lane lengths.

10.3.6 Median U-Turns

Reducing and separating conflicts between traffic streams, especially those involving left turns, often creates safer and more efficient intersections. In recent years designers have been able to choose from a menu of unconventional intersection designs that attempt, one way or another, to achieve that conflict reduction and separation. On top of that menu is the median U-turn, shown in

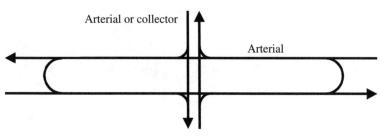

FIGURE 10.6   Median U-turns design.

Figure 10.6. Median U-turns have been in place for 40 years in Michigan, where over 1,000 miles are in service, as well as other states. They work by rerouting all left-turning drivers away from the main intersection to one of the U-turn roads typically located about 600 feet from the main intersection. The main intersection then requires only a simple two-phase signal.

Research has shown that the median U-turn typically produces lower overall travel times than a comparably sized conventional intersection or other unconventional intersection designs carrying the same traffic volumes (Reid and Hummer 2001). Arterials with median U-turn intersections also have lower average collision rates than arterials with conventional intersections (FHWA 2007). Other advantages of median U-turns over conventional intersections include easier progression for through traffic and fewer threats to crossing pedestrians. Disadvantages of the median U-turn design relative to a conventional intersection design include potential driver confusion, increased travel time for left-turning vehicles, wider rights-of-way along the arterial, and longer minimum green times for cross-streets to accommodate crossing pedestrians.

10.3.7   Other Unconventional Intersection Designs

Besides the median U-turn described above, there are at least 14 other unconventional intersection designs on the menu that offer advantages in certain situations. Figure 10.7 shows six of these other designs, with a note on how each works and the main advantage each offers. Most of the unconventional designs have been discussed in a peer-reviewed publication, and most have been placed in operation somewhere in the United States. Hummer and Reid (2002) provided a summary of the state-of-the-art for unconventional designs, the FHWA Signalized Intersections: *Informational Guide* (FHWA 2004) describes five unconventional designs, and another FHWA publication (2009a) provides recommended design details for six unconventional designs.

10.4   BASIC INTERCHANGE ELEMENTS

Interchanges are junctions between two or more roads where the conflicting traffic streams cross over or under each other using structures and connections between the roads are made using ramps. Interchanges serve two main purposes. First, one or more of the roads is a freeway that cannot have at-grade intersections. Second, an interchange is a safe way to add capacity and reduce delay at an intersection. The next few sections describe the basic design elements of an interchange.

10.4.1   Interchange Spacing

In general, interchanges should be spaced as far apart as possible. Interchanges that are too close force traffic streams to weave across each other, causing delays and collisions. Interchanges that

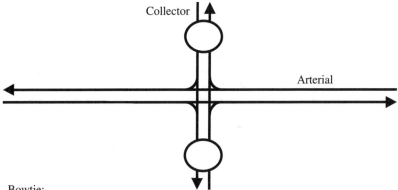

Bowtie:
- All left turns are routed to one of the roundabouts on the minor street
- Works like a median U-turn, but does not need a wide median on the major street

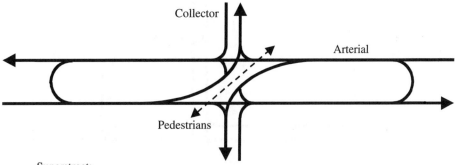

Superstreet:
- All minor street through and left turn traffic uses a median crossover
- Allows perfect progression with any signal spacing on the major street

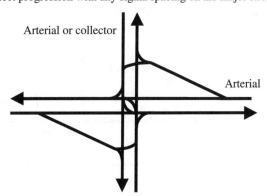

Jughandle:
- Left turns from the major street use a right-side ramp to the minor street
- Without left-turn bays or U-turns, the major street median can be narrow

FIGURE 10.7 A diagram and note on six other unconventional intersection designs.

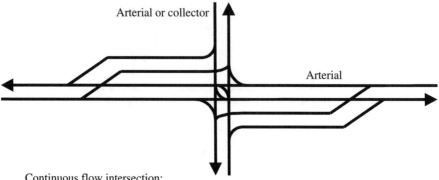

Continuous flow intersection:
- Left turns cross opposing traffic prior to the main intersection
- The main intersection has a two-phase signal with unopposed left turns

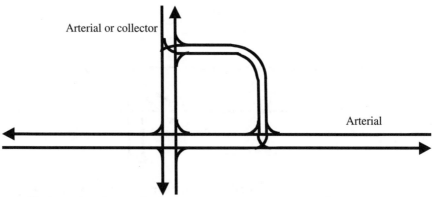

Single quadrant intersection:
- All left turns use the connecting roadway
- This generally vies with the median U-turn as most efficient unconventional design

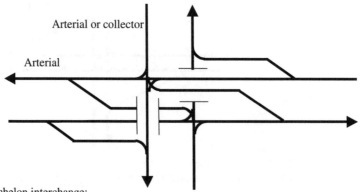

Echelon interchange:
- Half of each roadway meets on the top, while the other halves meet on the bottom
- Two-phase signals with unopposed left turns mean great efficiency

FIGURE 10.7   *(Continued)*

are too close also make signing difficult, forcing drivers to divert too much attention to reading too many signs and causing driver confusion. Closely spaced interchanges encourage short trips on the freeway and reduce the mobility function of the facility. AASHTO (2004) states that a general rule of thumb on minimum interchange spacing is 1 mile in urban areas and 2 miles in rural areas.

10.4.2 Route Continuity, Lane Continuity, and Lane Balance

Route continuity means drivers should not have to use a ramp to follow an important interstate route through an interchange, even if the route turns at the interchange. Figure 10.8 shows how a designer should employ route continuity at an interchange where the main interstate route turns. Route continuity is important to meet driver expectations, avoid driver confusion, and avoid sudden speed changes by large groups of drivers.

Lane continuity means that drivers should not have to change lanes to follow a certain main interstate route. For long stretches along a main interstate route, the freeway should contain the same basic number of through lanes. If auxiliary lanes are needed to boost capacity at some points along the route, they should be added (typically) to the right side and then terminated where they are no longer needed. Like route continuity, lane continuity helps meet driver expectations, avoids driver confusion, and ultimately eliminates collisions.

Lane balance is a related concept that helps designers decide how many freeway lanes are necessary before and after ramp junctions to satisfy driver expectations and increase efficiency. AASHTO (2004) states that:

1. At entrances, the number of lanes beyond the merging of two traffic streams should not be less than the sum of all traffic lanes on the merging roadways minus one, but may be equal to the sum of all traffic lanes on the merging roadway.

2. At exits, the number of approach lanes on the highway should be equal to the number of lanes on the highway beyond the exit, plus the number of lanes on the exit, minus one.

3. The traveled way of the highway should be reduced by not more that one traffic lane at a time.

There are exceptions to point 2 above for cloverleaf interchanges and where interchanges are closely spaced. Lane continuity and lane balance conflict at some interchanges, but the use of auxiliary lanes can remove the conflict.

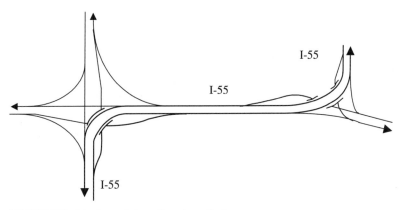

FIGURE 10.8 Interchange design with route continuity.

10.4.3 Overpass or Underpass

Sometimes the alignments of the freeway and crossroad dictate whether the freeway must be over or under the crossroad. Where the designer has a choice, though, it is an important one. Some of the positive aspects of placing the crossroad over the freeway include:

- A single structure, probably smaller than for the freeway going over the crossroad
- Off-ramps go uphill and on-ramps go downhill, allowing gravity to aid deceleration and acceleration
- Good visibility on the freeway to the crossroad, providing an early alert to exiting drivers
- Less noise impact from the freeway

Some of the positive aspects of placing the freeway over the crossroad include:

- Shorter structural spans
- No height restrictions on the freeway
- Less disturbance to the crossroad during construction
- Easier to drain the freeway

10.4.4 Types of Ramps

There are three basic types of ramps, as shown in Figure 10.9. Direct ramps are very common and accommodate most right-turn movements at interchanges. Direct ramps are inexpensive and can have higher design speeds—which drivers expect on freeway-to-freeway movements—but can accommodate left turns only if there is a stop sign or traffic signal at the crossroad. Indirect ramps are often used for left-turn movements. Indirect ramps can also allow higher design speeds, but they have high structural costs. Left turns not accommodated by indirect ramps are usually made with loop ramps. Loop ramps can accommodate left turns without additional structures, but they usually have relatively low design speeds, require large land areas, and cannot accommodate right turns without a stop sign or signal at the crossroad.

10.4.5 Consistent Ramp Pattern

Drivers expect a particular pattern of off-ramps and on-ramps as they negotiate an interchange, and the interchange will be safer if designers satisfy that expectation. Signing and visibility to decision points are also improved if the pattern is provided. In particular, the pattern is:

- All ramps merge into or diverge from the right side of the freeway.
- All off-ramps depart the freeway prior to the crossroad.
- All on-ramps join the freeway beyond the crossroad.

Left-side ramps obviously violate this pattern. Agencies rarely build new left-side ramps on freeway mainlines and often try to remove those that currently exist. Loop ramps often violate this pattern by merging into or diverging from the wrong side of the crossroad. Designers can make loop ramps fit the pattern by extending the ramp parallel to the freeway until the gore (the point where the ramp left edge line meets the freeway right edge line) is on the expected side of the crossroad.

10.4.6 Ramp Widths

AASHTO (2004) recommends ramp widths, like intersection turning roadway widths, at interchanges based on several factors, including the radius of the inner edge of the turn, design vehicles,

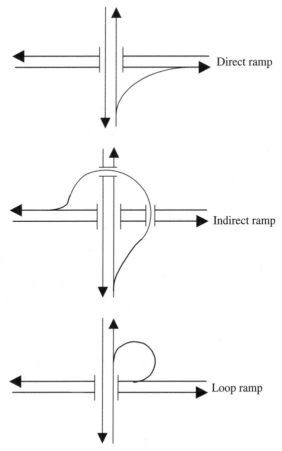

FIGURE 10.9 Three basic types of ramps.

whether passing a stalled vehicle is possible, and the type of curb or shoulder. Recommended widths for one-lane ramps range from 12 feet for a 500-foot radius turn serving predominantly passenger cars with no provision for passing a stalled vehicle, to 30 feet for a 50-foot radius turn with provision for passing a stalled vehicle serving large numbers of large semitrailer trucks. Recommended two-lane ramp widths range from 25 feet to 45 feet for those same conditions. For long ramps, designers sometimes choose two-lane ramps where a one-lane ramp would have been adequate because the difference in widths is not that great.

10.4.7 Ramp Acceleration and Deceleration Lengths

Ramp acceleration and deceleration lengths may be of either the tapered or the parallel type, as Figure 10.10 shows. Agencies across the United States are fairly evenly split on using tapered or parallel designs, and many believe that both designs provide safe and efficient merging and diverging if properly designed.

The acceleration or deceleration length provided between the end of the ramp curve and the end of the taper (i.e., where normal freeway lanes resume) is a critical design choice. Longer acceleration

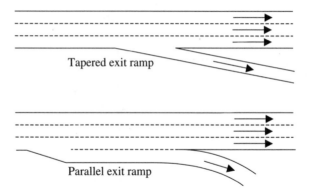

FIGURE 10.10   Tapered and parallel ramp terminals.

or deceleration areas mean better levels of service and fewer collisions, but if the areas are too long they may confuse drivers and they will waste pavement and right-of-way. AASHTO (2004) recommends minimum lengths for flat grades (less than 2 percent) ranging from 340 feet for deceleration from a 70-mph freeway design speed to a 50-mph ramp curve design to 1,620 feet for acceleration from a stop condition to a 70-mph freeway design speed. Grades have a substantial effect on acceleration or deceleration lengths needed. For example:

- Downgrades of 3 to 4 percent increase minimum deceleration lengths by a factor of 1.2.
- Downgrades of 5 to 6 percent increase minimum deceleration lengths by a factor of 1.35.
- Upgrades of 3 to 4 percent increase minimum acceleration lengths by a factor of 1.3 to 1.8.
- Upgrades of 5 to 6 percent increase minimum acceleration lengths by a factor of 1.5 to 3.0.

10.4.8 Weaving Areas

Weaving areas are created where an entrance ramp is followed within 2,500 feet by an exit ramp, where the ramps are connected by an auxiliary lane, and where the conflicting traffic streams are not controlled by stop signs or traffic signals. Weaving areas are usually found in the middle of cloverleaf interchanges or where two interchanges are close together. Weaving areas are relatively inefficient and unsafe because of the conflicting traffic streams. Currently, many highway agencies are not building new weaving areas on freeways and are, in fact, engaged in eliminating existing ones from their systems. Methods of eliminating weaving areas from freeways shown in Figure 10.11 include collector-distributor roads, braided ramps, and reversal of ramp direction (conversion to an X-interchange) where there are continuous one-way frontage roads.

10.4.9 Sign Placement

Guide sign placement is a critical decision for freeway designers because drivers moving at freeway speeds need long distances to read and respond properly to the sign messages. Five important human factors concepts that come into play in interchange sign placement are overloading, spreading, primacy, repetition, and redundancy. Overload is providing too much information to drivers at one time and can result in drivers making poor decisions or neglecting more important driving tasks. Spreading is a way to avoid overload by spacing signs for an interchange further apart. The *Manual on Uniform Traffic Control Devices* (*MUTCD*) (FHWA 2009) mandates that freeway guide signs not

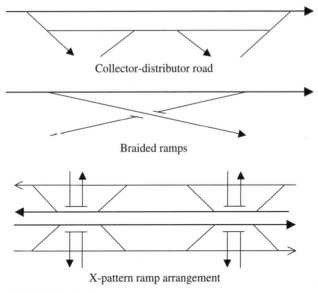

Collector-distributor road

Braided ramps

X-pattern ramp arrangement

FIGURE 10.11   Methods to eliminate weaving areas on freeways.

be closer than 800 feet apart, and ideally they should be spaced farther apart if possible. Primacy means providing the most important information, such as where precisely the gore is located, closest to the interchange. Designers place less important information, like guide signs for attractions and services, farther from the interchange. Repetition and redundancy mean providing important information more than once—repetition in the same form, and redundancy in different forms. This helps making sure drivers do not miss important information, especially in a complex environment with many sight restrictions. The *MUTCD* provides much more detailed suggestions and requirements for interchange signing.

10.4.10 Decision Sight Distances

When designing mainline and ramp alignments near interchanges, designers may use decision sight distance as the control instead of stopping sight distance. Decision sight distance allows drivers to take more time before deciding on a course of action when presented with a choice in a complex area. AASHTO (2004) provides decision sight distance values for five avoidance maneuvers:

 A: Stop on rural road

 B: Stop on urban road

 C: Speed, path, and/or direction change on rural road

 D: Speed, path, and/or direction change on suburban road

 E: Speed, path, and/or direction change on urban road

Decision sight distance values provided by AASHTO (2004) range from just slightly greater than stopping sight distance values at the same design speed for maneuver A to three times as high for maneuver E at low design speeds. Using minimum decision sight distances in alignment design will mean much flatter curves than using minimum stopping sight distances.

10.5 INTERCHANGE CONFIGURATION

Interchange designers have many factors to consider, but fortunately they have many configurations to choose from. The following paragraphs describe some of the basic interchange configurations and briefly describe their advantages and disadvantages.

10.5.1 Three-Legged

Many three-legged interchanges connect one freeway to another. There are two basic configurations of this type of interchange, the trumpet and the three-level, as Figure 10.12 shows. Trumpets have one left turn on a higher-speed indirect ramp and one left turn on a lower-speed loop ramp, so the lower volume left-turn movement or the left-turn movement from the road that has the lower functional class is typically relegated to the loop. Considering the relative advantages provided in Figure 10.12, designers typically use trumpets in lower-volume rural areas and three-level designs in higher-volume urban areas.

The trumpet is a very efficient and relatively low-cost interchange and thus designers often use it to connect freeways to other roads. Other common designs for three-legged freeway-to-surface street interchanges include the diamond (see below) and a modified diamond with a roundabout on the top of the "T."

10.5.2 Four-Legged Freeway-to-Freeway

There is a wide variety of four-legged freeway-to-freeway interchanges in service. Four common designs that are still considered for new interchanges include the cloverleaf, the single quadrant, the

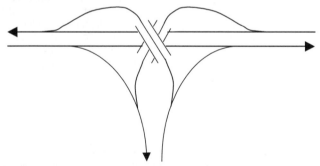

Three-level advantage: can fit in small right-of-way or can have high ramp speeds
Disadvantage: high bridge costs

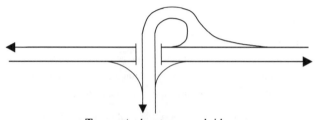

Trumpet advantage: one bridge
Disadvantages: high right-of-way, loop has low speed and capacity

FIGURE 10.12   Three-level and trumpet three-legged interchange designs.

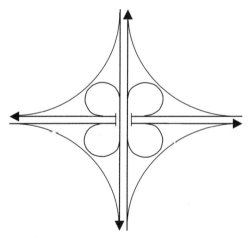

Cloverleaf advantages: one bridge, drivers expect this configuration
Disadvantages: large right-of-way, weaving areas break down easily

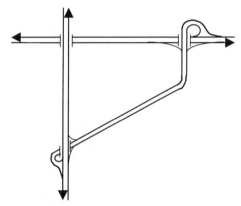

Single quadrant advantages: can use any quadrant, easy toll collection
Disadvantages: some long travel distances, violates expected ramp pattern

FIGURE 10.13   Common four-legged freeway-to-freeway designs.

pinwheel, and the four-level. Figure 10.13 shows these designs and the relative advantages of each. Considering those relative advantages, designers typically use cloverleaf designs in lower-volume rural areas, single-quadrant designs on a toll facility, and pinwheel and four-level designs in higher-volume urban areas.

10.5.3 Four-Legged Freeway-to-Surface Street

Most interchanges in the United States have four legs and connect freeways to surface streets. Figure 10.14 shows the most common configurations for this type of interchange. All of the four-legged freeway-to-freeway designs could also be employed in this category, but they are generally not, due to high costs and safety concerns with traffic coming off high-speed ramps onto surface

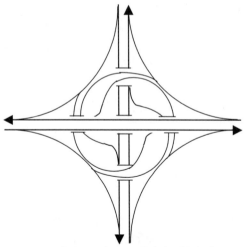

Pinwheel advantages: small right-of-way, all right-side indirect and direct ramps
Disadvantages: high bridge costs, low speed ramps

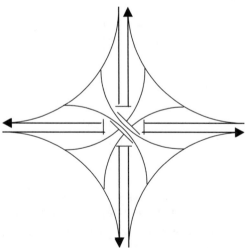

Four-level advantages: meets expected ramp pattern, efficient with high volumes
Disadvantages: very high bridge costs, steep ramp grades

FIGURE 10.13   (*Continued*)

streets. The diamond is the most common configuration of this type, but its drawbacks have become more exposed in recent years and other configurations are built now where diamonds would have been in the past. The single-point is probably the fastest growing type of interchange, especially in urban and suburban areas, because of its great efficiency in a compact area. Raindrops are common in Europe and are gaining acceptance in the United States due to great efficiency with a narrow bridge. Partial cloverleaf designs are common and quite efficient, particularly where the loops accommodate the left-turn movement from the freeway to the surface street as shown in Figure 10.14. Finally, the median U-turn interchange is a version of the median U-turn intersection described earlier in the chapter and is efficient because no left turns are made where the ramps meet the surface street.

A new interchange design that has gained great attention in the United States is the diverging diamond interchange (DDI), also known as the double-crossover diamond interchange (FHWA 2009c). The design includes two two-phase traffic signals. At each signal, the through movement in one direction of the surface street crosses the through movement in the other direction of the surface street. These crossovers allow all left turns to and from the surface street to be made with minimal conflicting traffic. The design looks well suited as a retrofit in places where an existing diamond interchange is failing operationally, especially owing to heavy left turn volumes, but has a

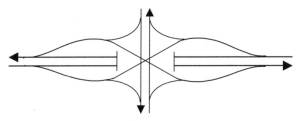

Diamond advantages: inexpensive, drivers expect this configuration
Disadvantages: high travel time, may need wide bridge

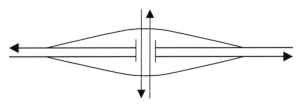

Single-point advantages: low travel time, small right-of-way
Disadvantages: long bridge, difficult pedestrian crossing

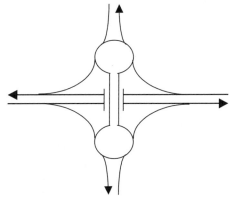

Raindrop advantages: narrow bridge, low travel times with moderate volumes
Disadvantages: large right-of-way at ramp terminals, driver confusion

FIGURE 10.14 Common four-legged freeway-to-surface street interchange designs.

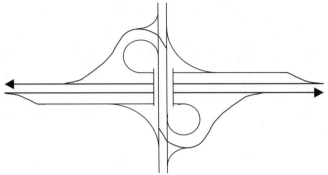

Partial cloverleaf advantages: low travel time, loops can be in any quadrant
Disadvantages: large right-of-way in loop quadrants, difficult pedestrian crossing

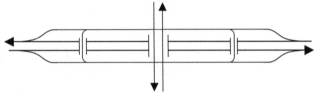

Median U-turn advantages: low travel time, narrow right-of-way
Disadvantages: driver confusion, three bridges needed

FIGURE 10.14   (*Continued*).

structurally sound bridge in place. The first DDI in the United States opened in 2009 in Springfield, Missouri, as Figure 10.15 shows, and provided great congestion relief at a reasonable cost with no evident safety or driver understanding issues.

10.5.4 Collector-Distributor Roads

Collector-distributor roads are one-way roads built to freeway standards running parallel to the freeway mainline that are meant to shield the mainline from multiple ramp merges and diverges. As noted earlier, collector-distributor roads are a good solution to at least move weaving areas off the mainline. Collector-distributor roads are also beneficial at other places where there is a high density of ramps.

10.5.5 Frontage Roads

Frontage roads are one-way or two-way roads built parallel to a freeway but not to freeway standards. Frontage roads typically serve land uses along the freeway and may keep some shorter trips off the freeway. Continuous one-way frontage roads are a common feature in Texas and other states and provide great efficiency, particularly with an X-pattern of ramps (shown earlier in Figure 10.11). Continuous one-way frontage roads are very expensive, however, particularly because they add another level at freeway-to-freeway interchanges. Two-way frontage roads sometimes pose problems for designers because they intersect the cross-street at an interchange very close to the ramp terminals.

FIGURE 10.15 The first diverging diamond interchange opened in the US, in Springfield, Missouri.

ACKNOWLEDGMENT

The photos and drawings for this chapter were provided by Daniel L. Carter.

REFERENCES

American Association of State Highway and Transportation Officials (AASHTO). 1999. *Guide for the Development of Bicycle Facilities,* 3rd ed. Washington: AASHTO.

——. 2004. *A Policy on Geometric Design of Highways and Streets*, 5th ed. Washington: AASHTO.

Castronovo, S., P. W. Dorothy, and T. L. Maleck. 1998. "Investigation of the Effectiveness of Boulevard Roadways." *Transportation Research Record* 1635.

Ewing, R. 1999. *Traffic Calming: State of the Practice,* Washington: Federal Highway Administration and Institute of Transportation Engineers.

Federal Highway Administration (FHWA). 2000. *Roundabouts: An Informational Guide.* FHWA-RD-00067. Washington: U.S. Department of Transportation.

——. 2004. *Signalized Intersections: Informational Guide.* FHWA-HRT-04-091. Washington: U.S. Department of Transportation.

——. 2007. *Synthesis of the Median U-Turn Intersection Treatment*, Safety, and Operational Benefits. FHWA-HRT-07-033. Washington: U.S. Department of Transportation.

——. 2009a. *Alternative Intersections/Interchanges: Informational Report.* FHWA-HRT-09-060. Washington: U.S. Department of Transportation.

——. 2009b. *Manual on Uniform Traffic Control Devices*. Washington: U.S. Department of Transportation.

——. 2009c. *Double Crossover Diamond Interchange*. FHWA-HRT-09-054. Washington: U.S. Department of Transportation.

Fitzpatrick, K., and M. Wooldridge. 2001. *Recent Geometric Design Research for Improved Safety and Operations*. NCHRP Synthesis 299. Washington: National Research Council, Transportation Research Board.

Gluck, J., H. S. Levinson, and V. Stover. 1999. *Impacts of Access Management Techniques*. NCHRP Report 420. Washington: National Research Council, Transportation Research Board.

Hummer, J. E., and J. D. Reid. 2002. "Access Management and Unconventional Arterial Designs: How Well Do the Various Design Accommodate Driveways?" In *Proceedings, Fifth National Conference on Access Management*. Austin, TX: Transportation Research Board, June 24.

Kickuchi, S., P. Chakroborty, and K. Vukadinovic. 1993. "Lengths of Left-Turn Lanes at Signalized Intersections." *Transportation Research Record 1385*.

Lee, J., N. M. Rouphail, and J. E. Hummer. 2005. "Lane Utilization Prediction Models for Lane-Drop Intersections." *Transportation Research Record 1912*.

Persaud, B. N., R. A. Retting, P. E. Garder, and D. Lord. 2001. "Safety Effects of Roundabout Conversions in the United States: Empirical Bayes Observational Before-After Study," *Transportation Research Record 1751*.

Pline, J. L. 1996. *Left-Turn Treatments at Intersections*. NCHRP Synthesis 225. Washington: National Research Council, Transportation Research Board.

Reid, J. D., and J. E. Hummer. 2001. "Travel Time Comparisons Between Seven Unconventional Arterial Intersection Designs." *Transportation Research Record 1751*.

Transportation Research Board (TRB). 2000. *Highway Capacity Manual*. Washington: National Research Council, TRB.

CHAPTER 11
PAVEMENT ENGINEERING I: FLEXIBLE PAVEMENTS

Qian Zhang

School of Civil Engineering, Xi'an University of Architecture and Technology, Xi'an, China
Department of Civil & Environmental Engineering, Michigan Technological University, Houghton, Michigan

Julian Mills-Beale

Department of Civil & Environmental Engineering, Michigan Technological University, Houghton, Michigan

Zhanping You

Department of Civil & Environmental Engineering, Michigan Technological University, Houghton, Michigan

11.1 INTRODUCTION

11.1.1 Pavement Types

Pavements can be sorted into three types: flexible pavements, rigid pavements, and composite pavements.

Flexible Pavements.   Flexible pavements appear in conventional flexible pavements and full-depth asphalt pavements. They are all layered systems. Based on the assumption that a layered system is infinite in a real extent, they all can be designed using Burmister's layered theory.

Conventional Flexible Pavements.   Conventional flexible pavements are structures composed of several layers. The top layers are usually constructed with materials having a better quality, whereas the bottom layers are constructed with materials having a lower quality than the top-layer materials. The advantage of this kind of structure is that some local materials with lower quality can be used in bottom layers of the pavement. A typical cross-section of a conventional flexible pavement is shown in Figure 11.1.

As shown in the figure, a conventional flexible pavement is made up of a surface course, binder course, base course, subbase course, compacted subgrade, and natural subgrade. The functions and materials usually used for the surface course, base course, and subbase course are further introduced in section 11.3. The binder course, also called the *asphalt base course*, is the layer directly below the surface course. There are two conditions based on which a binder course should be used. One is

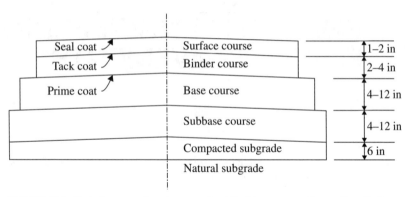

FIGURE 11.1   Typical cross-section of a conventional flexible pavement. (*Source*: Yang H. Huang. 2004. *Pavement Analysis and Design*, 2nd ed. Upper Saddle River, NJ: Pearson Education, p. 9.)

when the depth of the surface course is too thick to be well compacted as one layer. It then can be divided into two separate layers as a surface course and a binder course and compacted separately. The other is that binder course materials do not need to possess the same desirable qualities as those of the surface course materials. Therefore, mixtures with larger aggregate size and less asphalt content can be used in the binder course to reduce construction cost.

Usually three kinds of coats are needed in asphalt pavements: seal coat, tack coat, and prime coat. The *seal coat*, which is a thin asphalt surface treatment, is used to prevent water from penetrating into the pavement structure and to improve the skid resistance of the surface course. The *tack coat*, which usually consists of an asphalt emulsion diluted with water, contributes to the cohesion between the underlying course and the surface on top of it. The *prime coat*, which is made from low-viscosity cutback asphalt, is used to ensure a good bond between base course and the asphalt layer. Low-viscosity cutback asphalt is often used as a prime-coat material owing to its ease of absorption by the base material.

Full-Depth Asphalt Pavements.   Full-depth asphalt pavements are structures that are composed of one or several layers of hot-mix asphalt (HMA) paved directly above the subgrade or the improved subgrade. It was first proposed by the Asphalt Institute (AI) in 1960 to suit the requirements of heavy traffic and was intended to be the most economical a pavement. Figure 11.2 shows typical cross-section of a full-depth pavement. It is composed of asphalt surface, asphalt base, and prepared subgrade. The asphalt base has the same function as the binder course in conventional flexible pavement.

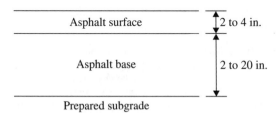

FIGURE 11.2   Typical cross-section of a full-depth asphalt pavement. (*Source*: Yang H. Huang. 2004. *Pavement Analysis and Design*, 2nd ed. Upper Saddle River, NJ: Pearson Education, p. 10.)

Rigid Pavements.   Rigid pavements refer to pavement constructed with portland cement concrete (PCC). This kind of pavement is often analyzed on the basis of plate theory. A typical cross-section of rigid pavements is shown in Figure 11.3. Rigid pavements can be further divided into four types: jointed plain-concrete pavement (JPCP), jointed reinforced-concrete pavement (JRCP), continuous reinforced-concrete pavement (CRCP), and prestressed-concrete pavement (PCP).

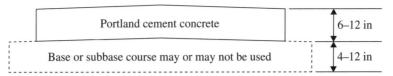

| Portland cement concrete | 6–12 in |
| Base or subbase course may or may not be used | 4–12 in |

FIGURE 11.3 Typical cross-section of a rigid pavement. (*Source*: Yang H. Huang. 2004. *Pavement Analysis and Design*, 2nd ed. Upper Saddle River, NJ: Pearson Education, p. 11.)

Composite Pavements. Composite pavements, which are usually composed of HMA on top of PCC, often can be found in rehabilitation construction instead of new pavement. With HMA as the top layer and PCC as the bottom layer, the pavement structure features a smoother and nonreflective surface and has a stronger base. Since the concrete slab is the major load-carrying layer in this pavement structure, the plate theory is applicable in the design of this kind of pavement. Another kind of pavement that is also categorized as composite pavement consists of asphalt surfaces and stabilized bases. In this kind of pavement, the most critical tensile stress or strain is located at the bottom of the asphalt layer. Two typical cross-sections for premium composite pavements are shown in Figure 11.4.

11.1.2 Pavement Materials

Aggregate, asphalt, portland cement, and soil are typical materials used in pavement construction. Fundamental properties of pavement material are analyzed based on their source, classification, and formation. For aggregates and soils, their particle or grain size, grading, shape, mineralogy, and chemical and physical properties are considered.

The asphalt cement or binder material has more complicated properties that are usually described in terms of rheologic properties such as aging, penetration, ductility, viscosity, temperature susceptibility, and stiffness. Portland cement, used in rigid pavements, is made from a mixture of limestone, marl, and clay or shale heated to a high temperature. Its properties are often discussed in terms of the percentage of silicon dioxide, aluminum oxide, ferric oxide, magnesium oxide, sulfur oxide, loss on ignition, and insoluble residue.

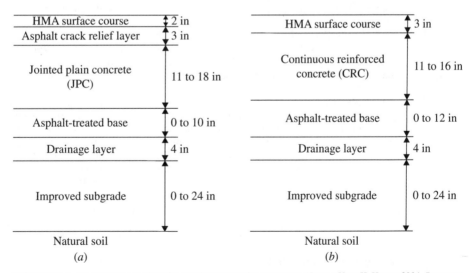

FIGURE 11.4 Typical cross-sections for premium composite pavements. (*Source*: Yang H. Huang. 2004. *Pavement Analysis and Design*, 2nd ed. Upper Saddle River, NJ: Pearson Education, p. 18.)

11.1.3 Traffic

Traffic flow on roads is composed of various kinds of vehicles. In pavement engineering, the design traffic loads are expressed in the form of the design life axle load repetitions by load increment and axle configuration (i.e., single axles and tandem axles). In fact, the traffic characteristics are described as the number of repetitions of an 18,000-lb (80-kilonewtons) single-axle load applied to the pavement on two sets of dual tires. This is usually referred to as the *equivalent single-axle load* (ESAL). It is a parameter presented by American Association of State Highway and Transportation Officials (AASHTO) in its 1986/1993 *Pavement Design Guide*. The conversion of other traffic axle loads to the 18,000-lb (80-kilonewtons) single-axle load is based on Miner's hypothesis.

11.1.4 Pavement Design Method

As mentioned earlier, theories used for the design of flexible pavements and rigid pavements are different. For flexible pavement, the layered theory is adopted on the assumption that its surface course, which is often a mixture of asphalt and aggregates, keeps an intimate contact with the underlying course. Under such an assumption, it is thought that traffic loads are transferred to the lower courses in flexible pavements. Therefore, the lower courses also play an important role in carrying the loads. However, for rigid pavement, it is assumed that its wearing surface, which is usually constructed with PCC, behaves as a slab on the top of underlying materials. Hence the plate theory is employed in the design of rigid pavements. Similarly, for composite pavements, the plate theory is also applied. Besides a comparatively detailed introduction of the 1993 AASHTO design method, a brief introduction of the development of the 2002 *Mechanistic-Empirical Pavement Design Guide* (*MEPDG*) led by the AASHTO is also presented in this chapter and in the next chapter.

11.2 PAVEMENT MATERIALS

11.2.1 Aggregates

Aggregates used in pavement engineering are mostly products of local natural rock supplies. Natural rocks are grouped into three categories according to their origin: igneous, sedimentary, and metamorphic. Usually, the overlying layers of natural rocks have to be removed or even blasted if they contain soil and decomposed rock. Then the stable rock is crushed, pulverized, and screened into the required pavement aggregate sizes. Lightweight aggregate, produced by heating clay to specific temperatures, and slag, a by-product of steel production, are also used in HMA pavement as aggregates.

Aggregate Mineralogy and Chemical Properties. Mineral composition and structure of aggregates determine their chemical and physical properties. Most natural aggregates are composed of a combination of minerals, with silica minerals, feldspars, ferromagnesian minerals, carbonate minerals, and clay minerals being the most common components found in aggregates. The rock and mineral constituents in aggregates are given in ASTM C294 (Table 11.1).

Aggregates' chemical properties, which are closely related to their chemical composition, can affect their adhesion on the asphalt binder and compatibility with antistripping additives, which may be added into the asphalt binder. The surface chemistry of aggregates plays an important part in this process. Aggregates are divided into three groups according to their affinity for water and for asphalt cement. Aggregates that appear to show a greater affinity for water than asphalt films are called *hydrophilic*, and they are easier to be detached or stripped in the presence of water. They basically tend to show acidic properties. Aggregates showing an affinity for asphalt are called *hydrophobic*, and they tend to be basic in nature. Usually, limestone and other calcareous aggregates are classified as hydrophobic, whereas most siliceous aggregates such as quartz, sandstone, and siliceous gravel are classified as hydrophilic. Some aggregates, such as trap rock, basalts, and

TABLE 11.1   Rock and Mineral Constituents in Aggregates

Minerals	Igneous rocks	Metamorphic rocks
Silica	Granite	Marble
Quartz	Syenite	Metaquartzite
Opal	Diorite	Slate
Chalcedony	Peridotite	Phyllite
Tridymite	Pegmatite	Schist
Cristobalite	Volcanic glass	Amphibolite
Silicates	Obsidian	Hornfels
Feldspars	Pumice	Gneiss
Ferromagnesian	Tuff	Serpentinite
Hornblende	Scoria	
Augite	Perlite	
Clay	Pitchstone	
Illites	Felsite	
Kaolins	Basalt	
Chlorites		
Montmorillonites	**Sedimentary rocks**	
Mica	Conglomerate	
Zeolite	Sandstone	
Carbonate	Quartzite	
Calcite	Greywacke	
Dolomite	Subgraywacke	
Sulfate	Arkose	
Gypsum	Claystone, siltstone	
Anhydrite	Argillite, and shale	
Iron sulfide	Carbonates	
Pyrite	Limestone	
Marcasite	Dolomite	
Pyrrhotite	Marl	
Iron oxide	Chalk	
Magnetite	Chert	
Hematite		
Goethite		
Ilmenite		
limonite		

Source: Freddy L. Roberts, Prithvi S. Kandhal, E. Ray Brown, Dah-Yinn Lee, and Thomas W. Kennedy 1966. *Hot Mix Asphalt Materials, Mixture Design, and Construction*, 2nd ed. Lanham, MD. National Asphalt Pavement Association Research and Education Foundation, p. 130.

siliceous limestone, show neutral properties because they bear both a positive and a negative charge after exposure to water.

Physical Properties of Aggregates.   It is worth noting that the physical properties of aggregates are more important than their chemical properties in the pavement engineering field. Aggregates used in HMA should be tough, strong, sound, hard, well graded, clean, and hydrophobic cubical particles of low porosity. Based on their sizes, they are often categorized into coarse aggregate, fine aggregate, and mineral fillers. In American Society for Testing and Materials (ASTM) definitions, coarse aggregates are particles retained on a No. 4 (4.75-mm) sieve. Fine aggregates are materials that pass a No. 4 (4.75-mm) sieve, and fillers are materials that can pass a No. 200 (75-μm) sieve with at least 70 percent. However, according to the AI standard specification, the No. 8 (2.36-mm) or the No. 10 (2.00-mm) sieve is suggested as the dividing line between coarse and fine aggregates. The property tests of aggregate required in pavement engineering are listed in Table 11.2. The individual specifications for them are available in ASTM D692, D1073, and D242.

TABLE 11.2 Properties and Tests of Aggregates for HMA

Properties	Significance	Test	Requirement
Hardness/toughness	Resistance to abrasion and degradation; quality of particles	ASTM C131 ASTM C535	Maximum percent weight loss; ASTM D692, D1073; AASHTO M283
Soundness	Resistance to freeze and thaw; Wet and dry	ASTM C88 AASHTO T103	Maximum percent weight loss; ASTM D692; D1073; AASHTO M283
Particle shape and surface texture	Friction; skid-resistance; compaction; mix stability	ASTM D3398 ASTM D4791 ASTM C1252	Minimum percent of crushed particles; ASTM D692; AASHTO M283
Resistance to polishing	Skid and wear resistance	ASTM D3319 ASTM E303 ASTM E660 ASTM D3042	Maximum polish value; minimum acid insolubles
Durability	Resistance to effect of weather and aging	ASTM D3744	Maximum durability index
Resistance to stripping	Resistance to effect of water	ASTM D1664 NCHRP 246, 274 AASHTO T283 ASTM D1075 AASHTO T182	Minimum percent of coated area; percent retained strength or modulus; number of freeze-thaw cycles to cause crack; AASHTO M283
Specific gravity and water absorption	Mix design calculations; porosity	ASTM C127 ASTM C128	—
Asphalt Absorption	Mix design calculations; durability	ASTM D2041 ASTM D4469	—
Gradation and size	Mix stiffness, stability, workability, fatigue resistance, fractural strength, durability, permeability, voids, compactibility	ASTM C117 ASTM C136, C702 AASHTO T30	Minimum and maximum percentages passing standard sieves; ASTM D3515
Mineral composition	Wettability and stripping potential; porosity	ASTM C294 ASTM C295	—
Cleanliness and deleterious substance	Resistance to weathering and effect of water	ASTM C117, D422 ASTM C123 ASTM C142 ASTM D2419 ASTM D4318	Maximum clay lumps and soft particles of 1 percent; minimum sand equivalent of 25–45 percent; P.I. not exceeding 4–6; percent fines (ASTM D242; D1073); AASHTO M382

Source: Freddy L. Roberts, Prithvi S. Kandhal, E. Ray Brown, Dah-Yinn Lee, and Thomas W. Kennedy. 1996. *Hot Mix Asphalt Materials, Mixture Design, and Construction*, 2nd ed. Lanham, MD. National Asphalt Pavement Association Research and Education Foundation. p. 130.

11.2.2 Bituminous Materials and Hot-Mix Asphalt Mixture

Source of Asphalt Materials. Asphalt materials are produced by distilling crude petroleum or obtained from natural deposits. Today, most asphalt materials used in pavement engineering are obtained from the processing of crude oils. The types of asphalt materials obtained from this distillation process can be asphalt cements, slow-curing liquid asphalt, medium-curing liquid asphalt, rapid-curing liquid asphalt, and asphalt emulsions. It is worthy to note that natural deposits of asphalt exist; they appear as native asphalts or as rock asphalts. Properties of native asphalts vary with their source. Softened with petroleum fluxes, native asphalts were once used widely as binders in pavement construction. Rock asphalts are seeps from sandstone or limestone filled with asphalt, and they can be used in road paving after the mined or quarried material has been suitably processed.

Properties of Asphalt Materials.   The rheologic and physical properties of asphalt materials will be described in this section. The physical properties and therefore physical testing of asphalt materials can be classified into five main aspects: consistency, durability, purity, safety, and others. The rheologic properties refer to such asphalt characteristics as age hardening, penetration, ductility, temperature susceptibility, viscosity, stiffness, and shear susceptibility.

Consistency represents the fluidity property of asphalt cement at any particular temperature. As a thermoplastic material, asphalt consistency changes with temperature. Tests suggested in testing asphalt consistency are the absolute viscosity test at 140°F (60°C), kinematic viscosity test at 275°F (135°C), penetration test, softening-point test, and ductility test.

Durability describes the aging degree of asphalt both in the short-term HMA mixing process and in the long-term pavement life cycle. The thin-film oven test (TFOT; ASTM D1754) and the rolling thin-film oven test (RTFOT; ASTM D2872) are used in measuring asphalt durability.

Since asphalt is a refined product from raw petroleum, there exist certain amounts of impurities in it. A solubility test (ASTM D2042) is conducted to determine its purity.

During the mixing and construction of asphalt pavements, the asphalt is heated to elevated temperatures. Some vapor will be generated during this heating process that may be ignited in the presence of a spark or a flame. Therefore, a flash point test needs to be conducted to ensure safety at the high temperatures (ASTM D92). The flash-point indicates the temperature to which asphalt may be safely heated without the danger of flash in the presence of an open flame.

Another important physical property of asphalt is its *specific gravity* (ASTM D70). Specific gravity is the ratio of the mass of asphalt at a given temperature to the mass of an equal volume of water at the same temperature. The value of the specific gravity of asphalt varies with temperature.

The rheologic properties of an asphalt binder have significant influences on the pavement performance. The following is a brief introduction of asphalt binder properties such as age hardening, penetration, ductility, viscosity, and temperature susceptibility.

Asphalt materials have a tendency to get harder during the construction and service life of the pavement. To test this tendency, the thin-film oven test and rolling thin-film oven test are used to simulate binder hardening during the heating, mixing, spreading, and compaction processes. The pressure aging vessel (PAV) is used to simulate binder hardening during the service life of the pavement.

Penetration and ductility are two traditional parameters that measure the consistency of the asphalt binder. The penetration test (ASTM D5) result is expressed and recorded as the depth of penetration (called the *penetration value*) in units of 0.1 mm. A greater penetration value indicates softer asphalt. Ductility test (ASTM D113) results give an idea of an asphalt binder's consistency under low temperatures. They are expressed as the distance to which the binder will elongate before breaking under certain temperatures. The greater the ductility value, the better the ductile property of the binder.

Asphalt is a thermoplastic material; its consistency is greatly influenced by temperature. Asphalt becomes softer at high temperature and becomes harder and more brittle at low temperature because its viscosity changes with the temperature distinctively. An optimal viscosity range exists for every particular type of asphalt. Figure 11.5 is the typical relationship between temperature and logarithm of viscosity. As shown in the figure, the viscosity of the asphalt increases when the temperature decreases. The steeper the slope, the higher the temperature susceptibility of the asphalt will be. This temperature-susceptibility nature of asphalt can affect the asphalt mixture's performance greatly during the pavement service life. Usually pavement is easier to crack if it suffers a sudden temperature drop when the viscosity of the asphalt is higher than the optimal range shown in Figure 11.5, whereas it is more likely to result in permanent

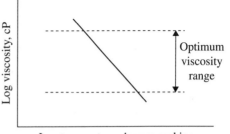

FIGURE 11.5 Typical relation between asphalt and temperature. (*Source*: Michael S. Mamlouk and John P. Zaniewski 2006. *Materials for Civil and Construction Engineering*. 2nd ed. Upper Saddle River, NJ: Pearson Prentice Hall, Pearson Education, p. 327.)

TABLE 11.3   Superpave Asphalt Binder Testing Equipments, Purposes, and Testing Methods

Equipment	Purpose	Performance parameters	Testing methods
Rolling thin-film oven (RTFO)	Simulate binder aging during HMA production and construction	Resistance to aging during construction	ASTM D2872 AASHTO T240
Pressure aging vessel (PAV)	Simulate binder aging during HMA service life	Resistance to aging during service life	ASTM D572
Rotational viscometer (RV)	Measure properties at high construction temperature	Handling and pumping	ASTM D4402 AASHTO TP48
Dynamic shear rheometer (DSR)	Measure binder properties at high and intermediate service temperatures	Resistance to permanent deformation and fatigue cracking	AASHTO TP5
Bending-beam rheometer (BBR)	Measure binder properties at low service temperatures	Resistance to thermal cracking	AASHTO TP1
Direct tension tester (DTT)	Measure binder properties at low service temperatures	Resistance to thermal cracking	AASHTO TP3

Source: Freddy L. Roberts, Prithvi S. Kandhal, E. Ray Brown, Dah-Yinn Lee, and Thomas W. Kennedy. 1996. *Hot Mix Asphalt Materials, Mixture Design, and Construction*, 2nd ed. Lanham, MD: National Asphalt Pavement Association Research and Education Foundation. p. 130.

deformation (rutting) if exposed to long-term high temperature and heavy traffic when the viscosity is lower than the optimal range.

Superpave Asphalt Binder Tests and Performance-Grade Binders.   Superpave asphalt binder tests and specifications are the research achievement of the 5-year Strategic Highway Research Program (SHRP) launched in 1987. The aim of this research program was to provide better performance-based asphalt binder and HMA mixtures test as well as a better specification system. Table 11.3 is a list of testing equipment used to conduct various superpave binder tests, the related purpose for testing, and the performance parameter of the pavement that is partly influenced by the asphalt binder.

Considering the temperature susceptibility of asphalt binders and the climate differences in different areas, the superpave binder specification (AASHTO MP1-93) suggests a binder grade selection process based on the tests listed in Table 11.4. It is worthy to note that these property indices of the binder remain the same for all performance grades, whereas the temperatures at which the properties are tested vary with the weather data collected for specified places where the pavement is going to be laid. For instance, if a binder is determined as PG 58-22 grade, this means that this binder can be used for a climate condition with the average seven-day maximum pavement temperature of 58°C and a minimum pavement design temperature of −22°C.

11.2.3   HMA Mixture Design Methodology

Asphalt mixture refers to such kinds of materials that are usually composed of asphalt binder, graded aggregate, and mineral filler mixed at a high temperature, spread, and compacted on the road as a pavement course. The pavement materials bear such properties as stability under high temperature, resistance to thermal cracking that might occur owing contraction at low temperature, resistance to fatigue cracking caused by repeated loadings, resistance to moisture damage, resistance to aging that would occur both in the mixture-producing process and during the service life, and skid resistance as a pavement surface. Two popular traditional asphalt mixture design methods that have been used for many years are the Marshall mixture design method and Hveem mixture design method. These two methods have their own design criteria. The Marshall mixture design method has the Marshall hammer and Marshall stability machine as its main design equipment. Marshall stability, flow measurement, voids ratio, density, voids in the mineral aggregate (VMA), and voids filled with asphalt (VFA)

TABLE 11.4 Performance-Grade Asphalt Binder Specification

Performance grade	PG 46			PG 52							PG 58					PG 64					
	-34	-40	-46	-10	-16	-22	-28	-34	-40	-46	-16	-22	-28	-34	-40	-10	-16	-22	-28	-34	-40
Average 7-day maximum pavement design temperature, °C[a]	<46			<52							<58					<64					
Minimum pavement design temperature, °C[a]	>-34	>-40	>-46	>-10	>-16	>-22	>-28	>-34	>-40	>-46	>-16	>-22	>-28	>-34	>-40	>-10	>-16	>-22	>-28	>-34	>-40
Original binder																					
Flash point temp, T48: minimum °C						230															
Viscosity, ASTM D 4402[b]: Maximum, 3 Pa·s (3,000 cP) Test temp., °C						135															
Dynamic shear, TP5[c]: $G^*/\sin\delta$, minimum, 1.00 kPa Test temp. @ 10 rad/sec, °C	46			52							58					64					
Rolling thin-film oven (T240) or thin-film oven (T179) residue																					
Mass loss, maximum, %						1.00															
Dynamic shear, TP5: $G^*/\sin\delta$, minimum, 2.20 kPa Test temp. @ 10 rad/sec, °C	46			52							58					64					
Pressure aging vessel residue (PP1)																					
PAV aging temperature, °C[d]	90			90							100					100					
Dynamic shear, TP5: $G^*\sin\delta$, maximum, 5,000 kPa Test temp. @ 10 rad/sec, °C	10	7	4	25	22	19	16	13	10	7	25	22	19	16	13	31	28	25	22	19	16
Physical hardening[e]						Report															
Creep stiffness, TP1[f]: S, maximum, 300 MPa m Value, minimum, 0.300 Test temp. @ 60 sec, °C	-24	-30	-36	0	-6	-12	-18	-24	-30	-36	-6	-12	-18	-24	-30	0	-6	-12	-18	-24	-30
Direct tension, TP3[f]: Failure strain, minimum, 1.0% Test temp. @ 1.0 mm/min, °C	-24	-30	-36	0	-6	-12	-18	-24	-30	-36	-6	-12	-18	-24	-30	0	-6	-12	-18	-24	-30

(*Continued*)

11.9

TABLE 11.4 Performance-Grade Asphalt Binder Specification (*Continued*)

	PG 70						PG 76					PG 82				
Performance grade	−10	−16	−22	−28	−34	−40	−10	−16	−22	−28	−34	−10	−16	−22	−28	−34
Average 7-day maximum pavement design temperature, °C[a]	<70						<76					<82				
Minimum pavement design temperature, °C[a]	>−10	>−16	>−22	>−28	>−34	>−40	>−10	>−16	>−22	>−28	>−34	>−10	>−16	>−22	>−28	>−34
Original binder																
Flash point temp. T48: Minimum °C	230															
Viscosity, ASTM D 4402[b]: Maximum, 3 Pa · s (3,000 cP): Test temp. °C	135															
Dynamic Shear, TP5[c]: $G^*/\sin\delta$, minimum, 1.00 kPa: Test temp. @ 10 rad/sec, °C	70						76					82				
Rolling thin-film oven (T240) or thin-film oven (T179) residue																
Mass loss, maximum, %	1.00															
Dynamic shear, TP5: $G^*/\sin\delta$, minimum, 2.20 kPa: Test temp. @ 10 rad/sec, °C	70						76					82				
Pressure aging vessel residue (PP1)																
PAV aging temperature, °C[d]	100(110)						100(110)					100(110)				
Dynamic shear, TP5: $G^*\sin\delta$, maximum, 5,000 kPa: Test temp. @ 10 rad/sec, °C	34	31	28	25	22	19	37	34	31	28	25	40	37	34	31	28
Physical hardening[e]	Report															
Creep stiffness, TP1[f]: S, maximum, 300 MPa: m-Value, minimum, 0.300: Test temp. @ 60 sec, °C	0	−6	−12	−18	−24	−30	0	−6	−12	−18	−24	0	−6	−12	−18	−24
Direct tension, TP3[f]: Failure strain, minimum, 1.0%: Test temp. @ 1.0 mm/min, °C	0	−6	−12	−18	−24	−30	0	−6	−12	−18	−24	0	−6	−12	−18	−24

Source: Freddy L. Roberts, Prithvi S. Kandhal, E. Ray Brown, Dah-Yinn Lee, and Thomas W. Kennedy. 1996. *Hot Mix Asphalt Materials, Mixture Design, and Construction,* 2nd ed. Lanham, MD: National Asphalt Pavement Association Research and Education Foundation, p. 94.

TABLE 11.5 Superpave Aggregate Properties

Design traffic level	Coarse aggregate angularity (% min)	Flat and elongated (% min)	Fine aggregate angularity (% min)	Sand equivalent (% min)
Light traffic	55/—	—	—	40
Medium traffic	75/—	10	40	40
Heavy traffic	85/80*	10	45	45

85/80* means minimum percentages of one fractured face/two or more fractured face.

Source: Michael S. Mamlouk, and John P. Zaniewski. 2006. *Materials for Civil and Construction Engineering.* 2nd ed. Upper Saddle River, NJ: Pearson Prentice-Hall, Pearson Education, p. 356.

are major parameters that are used in determination of the design asphalt content. In the Hveem method, the California kneading compactor is used to compact specimens, and the Hveem stabilometer is used to measure the Hveem stability. There is also a density and void analysis process in this method to obtain a reasonable design asphalt content. In the following section, a detailed description of the superpave mix design method is provided.

Superpave uses five steps in a complete mix design process: selection of aggregate, selection of binder, determination of the design aggregate structure, determination of the design binder content, and evaluation of moisture susceptibility.

Aggregate Selection.   Aggregate selection is based on the measurement of the following properties:

1. Coarse aggregate angularity measured by the percentage of fractured faces
2. Limitation on the amount of flat and elongated particles (ASTM D4791)
3. Fine aggregate angularity (AASHTO TP33)
4. Clay content that is measured and expressed with sand equivalents (ASTM D2419)

The concrete limitations of these characteristics are closely related to the design traffic level of the road. The required properties corresponding to each traffic level are shown in Table 11.5. Besides these properties, other important properties such as Los Angeles abrasion, soundness, and deleterious are also taken into account in the selection of an aggregate. Aggregates are also supposed to have a good gradation. The 0.45 power chart is recommended, and control points corresponding to the aggregate nominal size are specified by superpave to set certain limitations on aggregate gradation. In addition, a restricted zone is adopted in the superpave grading specification in order to make sure that too much natural sand is not used in a mixture and therefore the minimum VMA requirements are met. Figure 11.6 illustrates the grading requirements for 19-mm nominal size.

Binder Selection.   The asphalt binder is selected according to the temperature records and therefore the calculated maximum and minimum pavement temperature of the project location. The tests needed are shown in Table 11.3, which is stipulated by the performance-graded asphalt binder specification. Besides these properties, the specific gravity and rotational viscosity versus temperature relationship for the specified asphalt binder also should be determined. The specific gravity is used in the calculation of volumetric properties of the mix, whereas the viscosity is needed to determine the required mixing and compaction temperatures of the asphalt mixture.

Design Aggregate Structure.   Once the materials have been selected, three different aggregate gradations have to be worked out, and the corresponding optimal asphalt contents should be calculated using empirical equations suggested by the superpave specifications. Designers also have the latitude to estimate the asphalt content. Then trial specimens are compacted using the superpave gyratory compactor at a constant vertical pressure of 600 kPa and at an angle of 1.25°. Gyration numbers are determined on the basis of traffic level, as shown in Table 11.6.

As shown in Table 11.6, three compaction stages are specified in the superpave specification. They are the initial, design, and maximum stages. The initial compaction level N_{ini} is used for determining

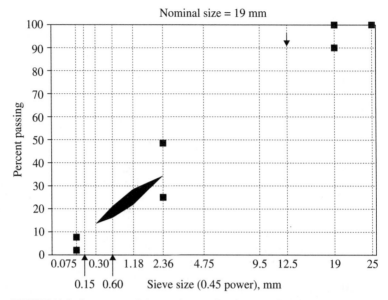

FIGURE 11.6 Superpave gradation requirements for 19-mm nominal size. (*Source*: Freddy L. Roberts, Prithvi S. Kandhal, E. Ray Brown, Dah-Yinn Lee, and Thomas W. Kennedy 1996. *Hot Mix Asphalt Materials, Mixture Design, and Construction*, 2nd ed. Lanham; MD: National Asphalt Pavement Association Research and Education Foundation, p. 240.)

TABLE 11.6 Number of Gyrations at Specific Design Traffic Level

	Traffic Level (10^6 ESAL*)			
	<0.3	0.3–3	3–30	>30
N_{ini}	6	7	8	9
N_{des}	50	75	115	125
N_{max}	75	100	160	205

Source: Michael S. Mamlouk, and John P. Zaniewski. 2006. *Materials for Civil and Construction Engineering*, 2nd ed. Upper Saddle River, NJ: Pearson Prentice Hall, Pearson Education, p. 357.

"tender" mixes. A tender mix at the initial number of gyrations may displace under rollers and may be unstable during service life. The design compaction level N_{des} corresponds to the compaction state that is expected at completion of the construction process. The maximum compaction N_{max} means the final density level of the pavement after a number of years of service. It is necessary to note that the maximum theoretical specific gravity G_{mm} of a loose mix with the same asphalt content and aggregate gradation should be obtained at the same time. The volumetric properties of the mix then are determined by measuring the bulk specific gravity G_{mb} of the compacted mix and G_{mm}. The volumetric parameters of voids in total mix (VTM), volume of voids in mineral aggregate (VMA), and void filled with asphalt (VFA) are calculated as

$$\text{VTM} = \left(1 - \frac{G_{mb}}{G_{mm}}\right)100 \tag{11.1}$$

$$VMA = 100\left(1 - \frac{G_{mb}(1 - P_b)}{G_{sb}}\right) \tag{11.2}$$

$$VFA = 100\left(\frac{VMA - VTM}{VMA}\right) \tag{11.3}$$

where G_{mb} = bulk specific gravity of compacted specimen
$\qquad G_{mm}$ = maximum theoretical specific gravity of mixture
$\qquad G_{sb}$ = bulk specific gravity of aggregate
$\qquad P_b$ = asphalt content

In addition, two more parameters, the percent of G_{mm} at N_{ini} and the dust-to-effective-binder-content ratio, are also used in the superpave mix design analysis. The percent of G_{mm} at N_{ini} can be calculated by using the following equation:

$$\%G_{mm}, N_{ini} = \%G_{mm, N_{des}} \frac{H_{des}}{H_{ini}} \tag{11.4}$$

where H_{ini} and H_{des} are the heights of the specimen at the initial and design number of gyrations, respectively. The percent of $G_{mm, N_{des}}$ is equal to $(100 - VTM)$.

The dust-to-effective-binder-content ratio is the quotient of the percent of aggregate passing the 0.075-mm sieve over the effective asphalt content and can be computed as

$$P_{ba} = 100\left(\frac{G_{ss} - G_{sb}}{G_{sb}G_{ss}}\right)G_b \tag{11.5}$$

$$P_{be} = P_b - \left(\frac{P_{ba}}{100}\right)P_s \tag{11.6}$$

$$\frac{D}{B} = \frac{P_D}{P_{bs}} \tag{11.7}$$

where P_{ba} = percent absorbed binder based on the mass of aggregate
$\qquad G_{se}$ = effective specific gravity of aggregate
$\qquad G_{sb}$ = bulk specific gravity of aggregate
$\qquad G_b$ = specific gravity of asphalt cement
$\qquad P_{be}$ = percent effective binder content
$\qquad P_b$ = percent binder content
$\qquad P_s$ = percent weight of aggregate
$\qquad D/B$ = dust-to-binder ratio
$\qquad P_D$ = percent dust or percent of aggregate passing the 0.075-mm sieve

When these results of compacted specimens are available, they must be checked against the criteria shown in Table 11.7.

Selection of Optimal Asphalt Content. Based on the results obtained from the preceding steps, the most qualified aggregate gradation can be decided. Then the optimal asphalt content can be determined. According to the superpave specification, the optimal asphalt content is the asphalt content that can make a mix of a 4 percent air void at N_{des}. In order to find the optimal asphalt content, eight specimens should be made at each of four binder contents: two specimens at the estimated optimal asphalt content, two at 0.5 percent less than the optimal, two at 0.5 percent more than the optimal, and

TABLE 11.7 Superpave Mix Design Criteria

Design air void	4 %					
Dust to effective asphalt [a]	0.6–1.2					
Tensile strength ratio	80% min					

	Nominal maximum size (mm)					
	37.5	25	19	12.5	9.5	4.75
Minimum VMA (%)	11	12	13	14	15	16

	G_{mm} & VFA requirements			
Design ESAL, in millions	Percent maximum theoretical specific gravity			Percent voids filled with asphalt [b,c,d]
	N_{ini}	N_{des}	N_{max}	
<0.3	≤91.5	96	≤98.0	70–80
0.3–3	≤90.5	96	≤98.0	65–78
3–10	≤89.0	96	≤98.0	65–75
10–30	≤89.0	96	≤98.0	65–75
≥30	≤89.0	96	≤98.0	65–75

[a] Dust-to-binder ratio range is 0.9 to 2.0 for 4.75-mm mixes.

[b] For 9.5-mm nominal maximum aggregate size mixes and design VFA ≥ 3 million, VFA range is 73–76 percent and for 4.75-mm mixes, the range is 75 to 78 percent.

[c] For 25-mm nominal maximum aggregate size mixes, the lower limit of the VFA range is 67 percent for design traffic levels < 0.3 million ESALs.

[d] For 37.5-mm nominal maximum aggregate size mixes, the lower limit of the VFA range is 64 percent for all design traffic levels.

Source: Michael S. Mamlouk and John P. Zaniewski. 2006. *Materials for Civil and Construction Engineering*, 2nd ed. Upper Saddle River, NJ: Pearson Prentice Hall, Pearson Education, p. 359.

another two at 1 percent more than the optimal. Note that specimens should be compacted using the N_{des} gyrations. Each specimen's volumetric properties are calculated, and the corresponding diagrams of volumetric properties versus asphalt content are plotted to find the optimal binder content. The content that corresponds to a 4 percent void ratio is the optimal binder content. The diagrams then are used for determination of the volumetric parameters at the selected binder content. If those parameters meet the criteria in Table 11.7, two specimens are prepared and compacted at N_{max} gyrations to see whether the G_{mb} data of these two specimens can also meet the criteria or not. If the data can meet the criteria, the moisture sensitivity of the mix then can be tested according to AASHTO T283.

Moisture Sensitivity Evaluation. The moisture sensitivity evaluation is conducted using the AASHTO T283 procedure. In the test, six specimens with a 7 percent air void at the optimal asphalt content are compacted. Three of them are conditioned by vacuum saturation, freezing, and thawing. The other three are not conditioned. The tensile strength of each specimen is tested on a Marshall stability machine having a modified loading head. The moisture sensitivity of the mix is measured by the ratio of the average tensile strength of conditioned specimens to that of the unconditioned ones. The required minimum ratio standard is 80 percent.

Characterization of Asphalt Mixtures. In addition to the physical properties of air voids, voids filled with asphalt cement (VFAs), and voids in mineral aggregate (VMAs), the characterization of asphalt mix is also expressed by mechanical tests and their associated parameters. As a nonlinear viscoelastic or viscoelastoplastic material, asphalt mixture performance under repeated dynamic loads of traffic and temperature changes varies with the rate of loading and the temperature. Since the asphalt mixture's resistance to permanent deformation in the wheel path (rutting), fatigue cracking, thermal cracking, aging, and moisture damage are of significant importance in its performance, these characteristics are tested in the laboratory to simulate the actual field condition.

Indirect Tensile Strength. Since the bottom of the asphalt pavement layer is subjected to tension under the action of applied traffic loads, the indirect tensile strength obtained by an indirect tensile test is used to evaluate the asphalt mixture's resistance to such a tension. The test result is important in the design of layer thickness. The test is conducted at a specified temperature, and a cylindrical sample of 102 mm in diameter and 64 mm in height is usually used. A compressive vertical load imposed by two curved loading strips moving at a rate of deformation of 51 mm/min is applied along the vertical diameter of the sample. Tensile stresses are developed in the horizontal direction. The sample fails in tension along the vertical diameter when the stresses reach its tensile strength. The indirect tensile strength can be calculated as

$$\sigma_t = \frac{2p}{\pi tD} \tag{11.8}$$

where σ_t = tensile strength, MPa (psi)
  P = load at failure, N (lb)
  t = thickness of sample, mm (inch)
  D = diameter of sample, mm (inch)

Resilient Modulus. The resilient modulus M_R measures the stiffness modulus of HMA. This index is used for evaluation of the structural response of an asphalt pavement system. To a certain extent, it has the same meaning and function as the modulus of elasticity, which applies to elastic materials, because it is impossible to characterize viscoelastic materials by using their modulus of elasticity. The index is obtained by the diametric tensile resilient modulus test (ASTM D4123). Equipment that is similar to that of the indirect tensile test is used to measure the resilient modulus. The difference is that the test equipment must be capable of applying a pulsating load along the vertical diameter of the specimen. A cylindrical specimen of 102 mm in diameter and 63.5 mm in height is used. A load of magnitude of between 5 and 20 percent of the indirect tensile strength is applied on the specimen with a duration of 0.1 second and a rest of period of 0.9 second. The recoverable horizontal deformation is measured by two linear variable differential transducers (LVDTs). Figure 11.7 shows typical load and horizontal deformation versus time relationships. The equation used for the calculation of the diametric tensile resilient modulus is

$$M_R = \frac{P(0.27 + v)}{t\Delta H} \tag{11.9}$$

where M_R = indirect tensile resilient modulus, MPa (psi)
  P = repeated load, N (lb)
  v = Poisson's ratio, typically 0.3, 0.35, and 0.4 at temperature of 5, 25, and 40°C, respectively
  t = thickness of specimen, mm (inch)
  ΔH = sum of recoverable horizontal deformation on both sides of specimen, mm (inch)

Dynamic Modulus. The dynamic modulus test is a triaxial compression test at designated temperatures and load frequencies. In this test, an axial sinusoidal compressive stress is applied to an unconfined or confined cylindrical specimen (ASTM D3497). A typical stress and strain pulse for the dynamic modulus test is shown in Figure 11.8. The so-called complex modulus E^* obtained from this test is used to illustrate the relationship between stress and strain of the asphalt mix. E^* can be expressed as the real and imagined parts with the equation $E' = E' + iE''$, where E' represents the elastic component of the complex modulus, and E'' is the viscous portion of it. The absolute value of the complex modulus $|E^*|$ is called the *dynamic modulus*. The definition of dynamic modulus is shown in the following equation

$$|E^*| = \frac{\sigma_0}{\varepsilon_0} \tag{11.10}$$

where σ_0 is the peak dynamic stress, and ε_0 is the peak recoverable axial strain. Another important parameter obtained from this test is the phase angle φ, which is a measurement of how much ε_0

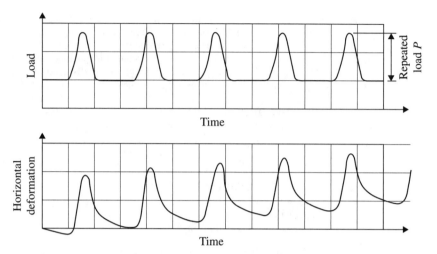

FIGURE 11.7   Typical load and horizontal deformation versus time relationship in the resilient modulus Test. (*Source*: Michael S. Mamlouk, and John P. Zaniewski. *Materials for Civil and Construction Engineering*, 2nd ed. Upper Saddle River, NJ: Pearson Prentice Hall, Pearson Education, p.380.)

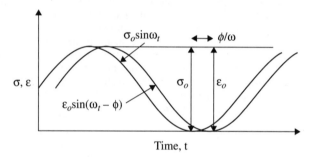

FIGURE 11.8   Stress and strain pulses for dynamic modulus test. (*Source*: Michael S. Mamlouk and John P. Zaniewski. 2006. *Materials for Civil and Construction Engineering*, 2nd ed. Upper Saddle River, NJ: Pearson Prentice Hall, Pearson Education, p. 365.)

lags behind σ_0 and an indication of the viscous properties of the mix. This is illustrated in equation form as

$$E^* = \left| E^* \right| \sin\varphi + i \left| E^* \right| \cos\varphi \qquad (11.11)$$

where t_i = time lag between stress and strain, seconds
t_p = time for a stress cycle, seconds

If $\varphi = 0$, the material is purely elastic, whereas a φ value of 90 refers to a purely viscous material.

Fatigue Characteristics.   The resistance to fatigue cracking is an important property of asphalt mixtures under repeated action of traffic load. Usually, the beam test with third-point loading is implemented to measure this property. The repeated tension-compression loads in the form of haversine waves are applied by an electrohydraulic or pneumatic testing machine at a 0.1-second duration with a 0.4-second rest period. The deflection of the beam at the midspan is measured through a linear variable differential transformer (LVDT). Typically, two types of controlled loading can be employed: constant stress and constant strain.

11.3 DESIGN OF FLEXIBLE PAVEMENTS

11.3.1 Structural Component of a Flexible Pavement

Typically, a flexible pavement is composed structurally of subgrade or prepared roadbed, subbase, base, and wearing surface. Figure 11.9 illustrates the structural composition of flexible pavement.

Subgrade (Prepared Road Bed). The subgrade is the foundation of the pavement. Usually, if the native soil along the horizontal alignment can be compacted to the required density, the upper layers of the pavement can be constructed above it. If it cannot reach the required density or stability after compaction, some selected materials need to be borrowed, and it must be ensured that the subgrade can attain a certain strength and stability status.

Subbase Course. Subbase course is the pavement course that is paved directly on top of the subgrade. A qualified subbase material should have certain gradation, plastic characteristics, and strength. If the available native subbase material is not suitable for use, it can be treated with other materials to gain the required properties. In highway engineering, this treatment practice of soils is called *stabilization*. The subbase may not be necessary when the subgrade material satisfies the standard of the subbase.

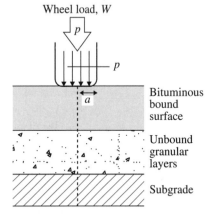

FIGURE 11.9 Structural composition of the flexible pavement. (*Source*: Nicholas J. Garber and Lester A. Hoel. *Traffic and Highway Engineering*, rev. 3rd ed. Pacific Grove, CA: Brooks/Cole Publishing Company, A Division of International Thomson Publishing, p. 963.)

Base Course. The base course is the course paved immediately above the subbase. Compared with the subbase, materials used for the base course should meet stricter standards. These standards are closely related to plasticity, gradations, and strength. Granular materials such as crushed stone, crushed or uncrushed slag, crushed or uncrushed gravel, and sand are used commonly in the construction of the base layer. Materials that fail to meet the required standards must be stabilized by portland cement, asphalt, or lime. Sometimes qualified base course materials also may be stabilized further with portland cement or asphalt to enhance their stiffness to ensure a better performance under heavy traffic.

Surface Course. The surface course is the course that is constructed directly on top of the base course. It is often composed of a mixture of mineral aggregate and asphalt materials in a flexible pavement. The surface course serves several functions. It needs to have enough strength to carry the traffic load. The material should have resistance to the abrasive effects of the vehicle tires and maintain a certain skid resistance to ensure safe driving. It also needs to have resistance to thermal and fatigue cracking in order to maintain the integrity of the surface and prevent moisture from entering the inner structure of the pavement. Typically, the thickness of the surface layer varies from 3 inches to more than 6 inches based on the expected traffic.

Stresses in Flexible Pavement. Flexible pavements are often considered as a multilayered elastic system in their structure design and analysis. Each layer's properties are expressed in terms of the modulus of elasticity, resilient modulus, and Poisson ratio. The subgrade layer typically is taken as infinite in both the horizontal and vertical directions, whereas the other layers are considered infinite in horizontal and finite in vertical directions. Stresses caused by applied traffic loads are taken into account in the design of flexible pavement structures. The distributions of the stresses and the temperature in a flexible pavement can be plotted as shown in Figure 11.10.

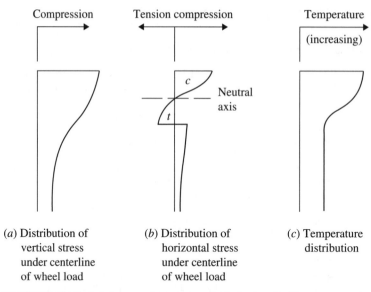

| (*a*) Distribution of vertical stress under centerline of wheel load | (*b*) Distribution of horizontal stress under centerline of wheel load | (*c*) Temperature distribution |

FIGURE 11.10   Typical stresses and temperature distribution in a flexible pavement under a wheel load. (*Source*: Nicholas J. Garber and Lester A. Hoel. 2004. *Traffic and Highway Engineering*, rev. 3rd ed. Pacific Grove, CA: Brooks/Cole Publishing Company, a Division of International Thomson Publishing, p. 963.)

As shown in the figure, the maximum vertical compression stress occurs immediately under the wheel. The magnitude of the vertical stresses diminishes with increasing depth from the pavement surface. In terms of the horizontal stress, the maximum horizontal compression stress also appears immediately under the applied wheel load. The compression stresses decrease with depth until, at a certain neutral axis, they reduce to zero. The horizontal tensile stresses occur below the neutral axis, as shown in Figure 11.10*b*. The temperature distribution in the pavement structure is similar to that of the vertical stresses and also has some effect on the magnitude of the stresses. Based on stress analysis, the pavement design, therefore, mainly focuses on the control of vertical and horizontal strains that can cause excessive cracking and excessive permanent deformation. It must be emphasized that the corresponding strain criteria are determined by taking into account the repeated load applications because they have profound influence on the generation of cracking and rutting in the pavement.

Since the flexible pavement is considered a multilayered system under repeated applications of load, the design process is quite complex. Fortunately, several computerized solutions for such a complicated system are available. Among them, the most commonly used are the AI method, the AASHTO method, and the California method. The AASHTO method will be introduced here.

11.3.2 AASHTO Design Method

The AASHTO flexible pavement design method was put forward on the basis of the AASHTO road test made in Ottawa, Illinois. The test was conducted on various kinds of road types, including flexible and rigid pavements paved on A-6 subgrade materials and short-span bridges. The pavement test sections were composed of pavement lengths of different designs with a length of at least 100 feet. The flexible pavements were made up of an asphalt concrete surface, a well-graded crushed limestone base, and a uniformly graded sand-gravel subbase. The pavement structures were diversified through a combination of surface thickness ranging from 1 to 6 inches, base thicknesses ranging from 0 to

9 inches, and subbase thicknesses ranging from 0 to 6 inches. In addition, a well-graded uncrushed gravel, bituminous plant mixture, or cement-treated aggregate also was used as a base type in some special sections.

Several thousands of load repetitions were applied by driving both single-axle and tandem-axle vehicles along the test sections. In order to guarantee the reliability of the test results, 10 different axle-arrangement and axle-load combinations were used, with single-axle loads ranging from 2,000 to 30,000 lb and tandem-axle loads ranging from 24,000 to 48,000 lb. Data regarding crack extension and patching work needed to keep certain levels of service were then acquired. Data on stresses transferred to the subgrade surface, temperature distribution in different layers, longitudinal and transverse profiles used to ascertain the development and extent of rutting, and surface deflections caused by slowly applied vehicle loads also were obtained. These data are the basis for the development of the AASHTO method, which was published in 1993.

Factors Considered in the 1993 AASHTO Method.   Factors such as pavement performance, traffic, roadbed soils (subgrade material), construction materials, local environment, drainage, and reliability are considered in the design of flexible pavement according to the 1993 AASHTO method.

Pavement Performance.   Both structural and functional performance is taken into account in terms of pavement performance. *Structural performance* refers to the pavement's capacity to carry the traffic load. Therefore, any distresses, including cracking, heaving, rutting, or raveling, can influence the pavement's structural performance. *Functional performance* demonstrates the efficiency with which the pavement serves the user. In the 1993 AASHTO design method, pavement performance is indicated by the index of *serviceability performance*. The *present serviceability index* (PSI) of the pavement is determined on the measurement of its roughness and distress, which is illustrated by cracking, patching, and rutting depth. The initial serviceability index p_i and the terminal serviceability index p_t are two indices used in the design. Here p_i refers to the serviceability index promptly after the initial construction of the pavement and p_t indicates the minimum acceptable value of the serviceability index before any resurfacing or reconstruction is needed. A value of 4.2 is recommended by AASHTO for p_i, whereas a value of 2.5 or 3.0 for major highways and 2.0 for lower classification highways is suggested for p_t.

TRAFFIC.   The traffic is expressed as the number of repetitions of an 18,000-lb (80-kN) single-axle load applied to the pavement on two sets of dual tires, that is, the so-called equivalent single-axle load (ESAL). The reason for choosing an 18,000-lb (80-kN) single-axle load is that experiments have shown that the number of single applications of an 18,000-lb single axle can represent the effect of any load applied on the pavement. In the AASHTO standards, the equivalent factors are the structural number (SN) and the terminal serviceability index, which are used in the design.

Roadbed Soils (Subgrade Material).   Resilient modulus M_r of the soil is used to show its strength property. The California bearing ratio (CBR) and the resilience R value of the soil can be converted to its resilient modulus. CBR is the ratio between the unit load for 0.1 piston penetration in the test specimen and the unit load for 0.1 piston penetration in standard crushed rock. The resilience R value shows the resistance of the soil to the horizontal pressure obtained by imposing a vertical stress of 160 lb/in on the soil sample. The equations used to convert the CBR or R value of the soil to an equivalent M_r are recommended by AASHTO as follows:

$$M_r \, (\text{lb/in}^2) = 1,500 \, \text{CBR} \qquad \text{(for fine-grain soils with soaked CBR of 10 or less)} \qquad (11.12)$$

$$M_r \, (\text{lb/in}^2) = 1,000 + 555 \times R \qquad \text{(for } R \leq 20) \qquad (11.13)$$

Construction Materials.   Construction materials that are used in a flexible pavement can be categorized as surface material, base material, and subbase material.

Subbase Material.   In the 1993 AASHTO method, a layer equivalent a_3 is introduced to describe the quality of the subbase material and also is used in the conversion from the actual thickness of the subbase to an equivalent *SN*. This index is often assigned a value according to the description of the material used. For example, a value of 0.11 was assigned to a sandy gravel

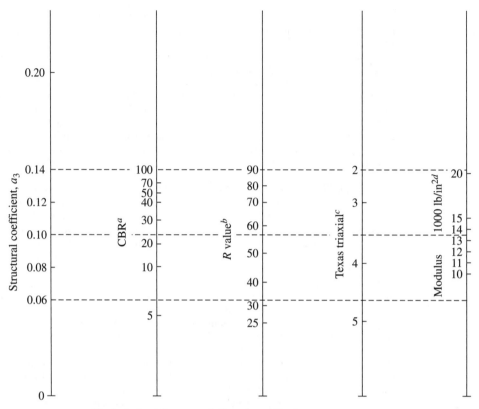

[a]Scale derived from correlations from illinois.

[b]Scale derived from correlations obtained from the Asphalt
Institute, California, New Mexico, and Wyoming.

[c]Scale derived from correlations obtained from Texas.

[d]Scale derived on NCHRP project 128, 1972.

FIGURE 11.11   Variation in granular subbase layer coefficient a_3 with various subbase strength parameters. (*Source*: Nicholas J. Garber and Lester A. Hoel. 2004. *Traffic and Highway Engineering*, rev. 3rd ed. Pacific Grove, CA: Brooks/ Cole Publishing Company, A Division of International Thomson Publishing, p. 996.)

subbase material by AASHTO. Charts that are related to different soil engineering properties have been developed. Figure 11.11 is one such chart for granular subbase materials. But a_3 values of subbase materials should be modified in accordance with the local environmental, traffic, and construction conditions.

Base Materials.   A structural layer coefficient a_2 must be decided for the selected base material and it can be determined by using Figure 11.12.

Surface Materials.   HMA is used most frequently as the surface material for flexible pavement. After determination of the composition of the asphalt binder and the graded aggregates, the structural layer coefficient for the surface material a_1 can be decided from Figure 11.13. Note that Figure 11.13 is available only for the dense-grade asphalt concrete with resilient modulus at 68°F.

Environment.   Temperature and precipitation are considered environmental factors that have an effect on pavement material properties. Stresses owing thermal action caused by temperature, variation in creep properties, and freezing and thawing of the subgrade soil are considered. One of the main influences of temperature is that the underlying material could be weakened

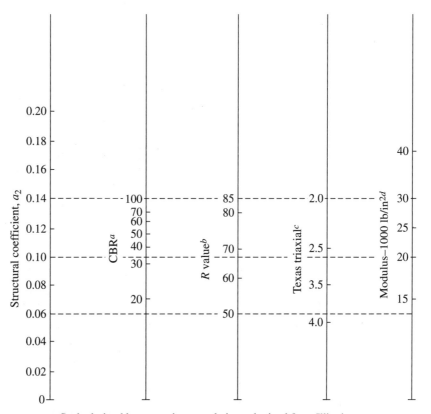

[a]Scale derived by averaging correlations obtained from Illinois.
[b]Scale derived by averaging correlations obtained from California,
   New Mexico, and Wyoming.
[c]Scale derived by averaging correlations obtained from Texas.
[d]Scale derived on NCHRP project 128, 1972.

FIGURE 11.12 Variation in granular base layer coefficient a_2 with subbase strength parameters. (*Source*: Nicholas J. Garber and Lester A. Hoel. 2004. *Traffic and Highway Engineering*, rev. 3rd ed. Pacific Grove, CA: Brooks/Cole Publishing Company, A Division of International Thomson Publishing, p. 997.)

in the thawing season. Therefore, the temperature has a significant effect on the strength of these materials. The modulus of some materials that are sensitive to frost action can be reduced by 50 to 80 percent of its normal value during the thawing period. Meanwhile, the resilient modulus of the subgrade soil also may vary during a year, even when there is no apparent thaw period in an area. Consequently, an effective annual resilient modulus is used in the design as a representation of the property of the subgrade material. This effective resilient modulus is an expression of the combined effect of the seasonal moduli in a year.

Two approaches are recommended by AASHTO to determine the effective resilient modulus. In the first method, a relationship between the resilient modulus and moisture content is established based on laboratory tests. Then, if the in situ moisture content of the subgrade soil is available, the corresponding resilient modulus can be obtained by using the relationship established. Note that the moisture content varies with the season and therefore the resilient modulus because a year is divided into several time intervals corresponding to different seasonal resilient moduli,

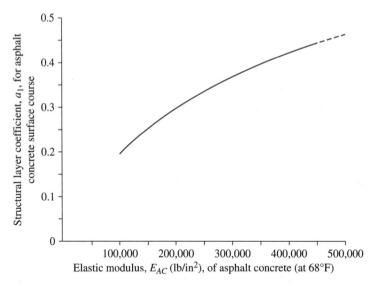

FIGURE 11.13   Chart for estimating structural layer coefficient of dense-graded/asphalt concrete based on the elastic (resilient) modulus. (*Source*: Nicholas J. Garber and Lester A. Hoel. 2004. *Traffic and Highway Engineering*, rev. 3rd ed. Pacific Grove, CA: Brooks/Cole Publishing Company, A Division of International Thomson Publishing, p. 998.)

as shown in Figure 11.14. It is also suggested by AASHTO that each time interval should be no less than one-half month. The chart, the vertical scale, or the equation in Figure 11.14 can be used to determine relative damage μ_f corresponding to each time period. An average μ_f then can be calculated. The effective resilient modulus then can be obtained by using the chart and μ_f.

Penetration of moisture from the pavement surface to the underlying material may be dangerous because the working condition of the underlying material can change greatly. Poor drainage may reduce the strength of both the base material and the subgrade material. To take this influence into account, a rapid-drainage layer is recommended in the pavement structure. The drainage layer can be the base layer or an additional layer under the subbase. The material selected must meet a filtering criterion. Meanwhile, a modification factor m_i for the base and subbase layer coefficients (a_2 and a_3) is used to modify the material property. The values of m_i are determined based on the quality of drainage, which is related to the time needed to drain the base layer to 50 percent of saturation and on the percentage of time when the pavement structure will be nearly saturated. Table 11.8 shows the definition of drainage quality levels suggested by AASHTO. Table 11.9 lists the corresponding m_i values.

Reliability.   The traffic data used in pavement design, which are expressed in terms of ESALs, are estimated data based on assumed traffic growth rates. Besides that, there are also certain uncertainties in pavement material performance prediction. This brings about uncertainties in pavement design. The 1993 AASHTO guide puts forward a reliability factor, known as the *reliability level R* percent, to consider these possible uncertainties. In pavement design, different reliability levels specify different assurance levels for different types of pavement designed. For instance, a 70 percent reliability design level means that the possibility of design performance success is 70 percent. Table 11.10 illustrates suggested reliability levels suggested by AASHTO.

Reliability factors are developed on the basis of chosen reliability levels. The overall variation S_0^2 represents the possible variation in both real pavement performance and traffic prediction during the service period. In AASHTO, the reliability factor F_R is given as

$$\log_{10} F_R = - Z_R S_0 \tag{11.14}$$

where Z_R = standard normal variate for a given reliability ($R\%$)
S_0 = estimate overall standard deviation

Month	Roadbed soil modulus M, (lb/in.²)	Relative damage u_f
Jan.	22000	0.01
Feb.	22000	0.01
Mar.	5500	0.25
Apr.	5000	0.30
May	5000	0.30
June	8000	0.11
July	8000	0.11
Aug.	8000	0.11
Sept.	8500	0.09
Oct.	8500	0.09
Nov.	6000	0.20
Dec.	22000	0.01
Summation: $\sum u_f =$		1.59

$$\text{Average: } \bar{u}_f = \frac{\sum u_f}{n} = \frac{1.59}{12} = 0.133$$

Effective roadbed soil resilient modulus, M_r (lb/in²) = 7250 (corresponds to $\bar{u}_f$)

FIGURE 11.14 Chart for estimating effective roadbed soil resilient modulus for flexible pavements design using the serviceability criteria. (*Source*: Nicholas J. Garber and Lester A. Hoel. 2004. *Traffic and Highway Engineering*, rev. 3rd ed. Pacific Grove, CA: Brooks/Cole Publishing Company, A Division of International Thomson Publishing, p. 999.)

TABLE 11.8 Definition of Drainage Quality

Quality of drainage	Water removed within[a]
Excellent	2 hours
Good	1 day
Fair	1 week
Poor	1 month
Very poor	(Water will not drain)

[a]Time needed to drain the base layer to 50 percent saturation.

Source: Adapted from AASHTO, 1993. *Guide for Design of Pavement Structures*, Washington: American Association of State Highway and Transportation Official.

TABLE 11.9   Recommended m_i Values

Quality of drainage	Percent of time pavement structure is exposed to moisture levels approaching saturation			
	1 Percent	1–5 Percent	5–25 Percent	25 Percent
Excellent	1.40–1.35	1.35–1.30	1.30–1.20	1.20
Good	1.35–1.25	1.25–1.15	1.15–1.00	1.00
Fair	1.25–1.15	1.15–1.05	1.00–0.80	0.80
Poor	1.15–1.05	1.05–0.80	0.80–0.60	0.60
Very poor	1.05–0.95	0.95–0.75	0.75–0.40	0.40

Source: Adapted from AASHTO. 1993. *Guide for Design of Pavement Structures*, Washington: American Association of State Highway and Transportation Official.

TABLE 11.10   Suggested Reliability Levels for Various Functional Classifications

Functional classification	Recommended level of reliability	
	Urban	Rural
Interstate and other freeways	85–99.9	80–99.9
Other principal arterials	80–99	75–95
Collectors	80–95	75–95
Local	50–80	50–80

Note: Results based on a survey of the AASHTO Pavement Design Task Force.
Source: Adapted from AASHTO. 1993. *Guide for Design of Pavement Structures*, Washington: American Association of State Highway and Transportation Official.

Table 11.11 shows Z_R values for different reliability levels *R*.

Structural Design.   The most commonly used flexible pavement surface materials are asphalt concrete, a single surface treatment, or a double surface treatment. Once the type of the surface is decided, the main task of the AASHTO flexible pavement design is to find a pavement *SN* sufficient to carry the predicted design ESAL. Note that this design process is eligible for ESALs greater than 50,000 in the pavement lifespan. According to the 1993 AASHTO guide, *SN* can be expressed as follows:

$$SN = a_1 D_1 + a_2 D_2 m_2 + a_3 d_3 m_3 \tag{11.15}$$

where
m_i = drainage coefficient for layer *i*
a_1, a_2, a_3 = layer coefficients representative of surface, base, and subbase course, respectively
D_1, D_2, D_3 = actual thickness in inches of surface, base, and subbase course, respectively

A fundamental design equation provided by the 1993 AASHTO guide is

$$\log_{10} W_{18} = Z_R S_0 + 9.36 \log_{10}(SN+1) - 0.20 + \frac{\log_{10}[\Delta PSI/(4.2-1.5)]}{0.40+[1094/(SN+1)^{5.19}]} + 2.32 \log_{10} M_r - 8.07$$

$$\tag{11.16}$$

TABLE 11.11   Standard Normal Deviation (Z_R) Values for Different Reliability Levels

Reliability (R %)	Standard normal deviation Z_R
50	–0.000
60	–0.253
70	–0.524
75	–0.674
80	–0.841
85	–1.037
90	–1.282
91	–1.340
92	–1.405
93	–1.476
94	–1.555
95	–1.645
96	–1.751
97	–1.881
98	–2.054
99	–2.327
99.9	–3.090
99.99	–3.750

Source: Adapted from AASHTO. 1993. *Guide for Design of Pavement Structures*. Washington: American Association of State Highway and Transportation Official.

where W_{18} = predicted number of 18,000-lb (80-kN) single-axle load applications
Z_R = standard normal deviation for a given reliability
S_0 = overall standard deviation
SN = structural number indicative of the total pavement thickness
$\Delta PSI = p_i - p_t$
p_i = initial serviceability index
p_t = terminal serviceability index
M_r = resilient modulus, in lb/in^2

To solve equation (11.16), a computer program and a chart as shown in Figure 11.15 are provided by AASHTO. The following is a simple demonstration of how to use the chart to determine the design *SN*.

Step 1. Choose the reliability level R, standard deviation S_0, initial serviceability index p_i, and terminal serviceability index p_t for the pavement under design.

Step 2. On the chart in Figure 11.15, draw a line connecting the selected reliability level value with the selected overall standard deviation value S_0. Then stretch this line to intersect the first T_L line on a point, say, A.

Step 3. Draw a line connecting A with the point corresponding to the estimated ESAL value, and extend this line to intersect the second T_L line at a point, say, B.

Step 4. Draw a line connecting point B with the resilient modulus M_r of the subgrade soil. Then extend this line to intersect the design serviceability loss chart at a point, say C.

Step 5. Draw a horizontal line from point C to intersect the design serviceability loss ΔPSI curve at point D. Note ΔPSI = initial serviceability index − terminal serviceability index.

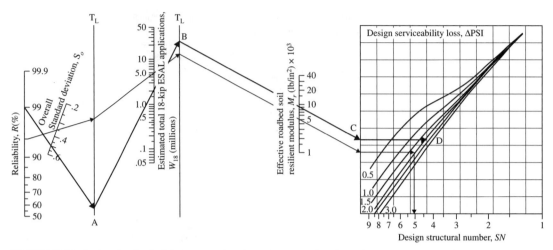

FIGURE 11.15   Design chart for flexible pavements based on using mean values for each input. (*Source*: Nicholas J. Garber and Lester A. Hoel. 2004. *Traffic and Highway Engineering*, rev. 3rd ed. Pacific Grove, CA: Brooks/Cole Publishing Company, A Division of International Thomson Publishing, p. 1004.)

Step 6. Draw a vertical line to intersect the design *SN*, and read the value on the intersection point.

Step 7. Determine the suitable structure layer coefficient for each selected material as follows:

- Find the a_1 value for asphalt cement from Figure 11.13. Note that its resilient value is known beforehand.
- Find the a_2 value for base course material from Figure 11.12 by using the CBR value.
- Find the a_3 value for subbase course material from Figure 11.11 by using the CBR value.

Step 8. Determine the suitable drainage coefficient m_i. The time period needed for the water to drain from within the pavement is known, and the drainage quality of the pavement can be obtained from Table 11.8. Then, according to the percent of time the pavement is exposed to moisture levels approaching saturation, the recommended m_i value can be obtained from Table 11.9.

Step 9. The appropriate layer thickness of the pavement structure can be decided by Equation (11.15).

In fact, several different combinations of D_1, D_2, and D_3 values may satisfy equation (11.15). The 1993 AASHTO guide has a limitation on minimum thickness of pavement, as illustrated in Table 11.12. The selection of layer thickness also may be related to maintenance and construction practices of the highway.

TABLE 11.12   AASHTO Recommended Minimum Thickness of Highway Layers

	Minimum thickness (inches)	
Traffic, ESALs	Asphalt concrete	Aggregate base
<50,000	1.0 (or surface treatment)	4
50,001–150,000	2.0	4
150,001–500,000	2.5	4
500,001–2,000,000	3.0	6
2,000,001–7,000,000	3.5	6
>7,000,000	4.0	6

Source: Adapted from AASHTO. 1993. *Guide for Design of Pavement Structures*. Washington: American Association of State Highway and Transportation Official.

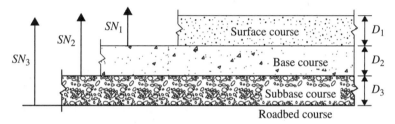

FIGURE 11.16 Procedure for determining thickness of layers using a layered analysis approach. (*Source*: Nicholas J. Garber and Lester A. Hoel. 2004. *Traffic and Highway Engineering*, rev. 3rd ed. Pacific Grove, CA: Brooks/Cole Publishing Company, A Division of International Thomson Publishing, p. 1007.)

In practice, the flexible pavement layer thicknesses can be calculated according to the following method based on Figure 11.16. In this method, the desired *SN* value above the subgrade is determined first. Then the corresponding *SN* values of the base and the subbase layers are determined separately by using the appropriate strength of each material. Using the differences of the computed *SN* values as shown in Figure 11.16, a minimum allowable thickness of each layer can be determined.

11.4 INTRODUCTION TO MECHANISTIC-EMPIRICAL PAVEMENT DESIGN

The AASHTO Joint Task Force on Pavements (JTFP) has been developing the 2002 *Mechanistic-Empirical Pavement Design Guide* (*MEPDG*) since 1996. Its interim edition, *A Manual of Practice*, is now available. Its corresponding software program is expected to be completed by mid-2011. The aim of the *MEPDG* is to provide significant revision to the present pavement design method. The *Guide* is applicable to designs for new, reconstructed, and rehabilitated flexible, rigid, and semirigid pavements. The *MEPDG* features a full integration of material characterization, climate condition, and traffic loading into pavement design. Performance and distress prediction models are used to aid the pavement designer in determining the desired pavement section.

As a whole design system, *MEPDG* has four input categories: traffic, material characterization, climate, and reliability. For traffic, the 2002 *Guide* adopts the full spectrum of axle load instead of the traditional ESAL. Two kinds of data are collected: weigh-in-motion (WIM) and automatic vehicle classification (AVC). The first indicates the number and configuration of axles in a series of load groups. The second shows the number and types of vehicles that use a specified highway in a period of time. Traffic data inputs are divided into three levels in *MEPDG,* with the most accurate being level 1 and the least accurate being level 3. Level 1 contains a site-specific measurement of traffic data. It requires traffic volume/classification and axle-load data associated with the highway under design, including number of trucks, axle loads of each truck class, average annual daily truck traffic (AADTT), truck traffic distribution factor (TTDF), monthly average daily truck traffic (MADTT) volume, hourly distribution factors, and lane distribution factors. Level 2 is based on a regional value or regression equation. The regional axle-load spectrum data are used in level 2. Level 3 traffic input is on the basis of default values or an educated guess. It needs regional or default classification and axle-load spectra data.

Material characterization considered in the *MEPDG* can be classified into two aspects: response properties and distress properties. *Response properties* refer to indices such as elastic modulus and Poisson ratio that are required in the prediction of stress, strain, and displacement states in the pavement. *Distress properties* can be used for the prediction of road performance indicators such as thermal cracking, permanent deformation, and fatigue cracking. In the *MEPDG*, the dynamic modulus of asphalt concrete is used as the elastic modulus. A master curve is presented, showing the relationship between the dynamic modulus, the frequency of loading, and the mixture temperature.

TABLE 11.13   Asphalt Materials Inputs Corresponding to Three Traffic Levels

Design type	Input level	Description
New	1	Conduct laboratory dynamic modulus test, perform tests on binder, simulate aging of mix, develop mix master curve
	2	Use predictive equation for dynamic modulus, perform tests on binders, develop mix master curve
	3	Use predictive equation for dynamic modulus, develop mix master curve
Rehabilitation	1	Backcalculate by FWD; develop mix master curve with aging; test cores in laboratory
	2	Determine mix properties from cores; develop mix master curve
	3	Develop undamaged mix master curve from typical mix properties; adjust for damage based on distress surveys

Source: Yang H. Huang 2004. *Pavement Analysis and Design*, 2nd ed. Upper Saddle River, NJ: Pearson Education, p. 719.

Used together with an asphalt aging model, the master curve is useful in determining the mixture modulus values in the design analysis period. The asphalt material inputs corresponding to three traffic levels are illustrated in Table 11.13.

For unbound materials, a nonlinear model is recommended to determine the resilient modulus of unbound bases, subbases, and subgrades:

$$M_R = K_1 P_a \left(\frac{\theta}{P_a} \right)^{K_2} \left(\frac{\tau_{\text{oct}}}{P_a} + 1 \right)^{K_3}$$ (11.17)

where M_R = resilient modulus
  p_a = atmospheric pressure to normalize stresses and modulus
  k_1, k_2, k_3 = regression constants that are functions of material types and properties
  θ = stress invariant, or the sum of the three principal stresses
  τ_{oct} = octahedral shear stress

The corresponding input requirements for determining the M_R of unbound materials suggested by the MEPDG are shown in Table 11.14.

Climate factors considered in the *MEPDG* are temperature and moisture. To establish site-specific variations, the *MEPDG* adopted the Federal Highway Administration's (FHWA) integrated climate model (ICM) into the 2002 *Guide*. The ICM is an environmental-effect model composed of

TABLE 11.14   Three Levels of Inputs for Determining M_R of Unbound Materials

Design type	Input level	Description
New	1	Use k_1, k_2, k_3 nonlinear coefficients, which can be determined by direct tests, predictive equations, or tabular summary based on soil classification group.
	2	Correlate M_R from predictive equations, CBR, or R values or plasticity/gradation properties.
	3	Correlate M_R from soil classification.
Rehabilitation	1	Backcalculate M_R using direct FWD measurements or tabular summary of backcalculated values by soil group.
	2	Correlate M_R from predictive equations, CBR, or R values, dynamic cone penetration test, plasticity/gradation properties, or layer coefficients.
	3	Select typical M_R as default based on soil classification and design conditions.

Source: Yang H. Huang 2004. *Pavement Analysis and Design*. 2nd ed. Upper Saddle River, NJ: Pearson Education, p. 721.

four parts: precipitation, infiltration and drainage, climate-material-structure, and frost heave and thaw settlement. The enhanced integrated climate model (EICM) will be an enhanced version of the ICM.

For reliability's sake, the research team wanted to use the Monte Carlo simulation method at the beginning of the research. But the large numbers of calculation prevented this method from practical usage. The residuals of distress models, namely, the differences between measured and predicted distresses, are adopted for pavement reliability based on the concept that residuals indicate the accuracy of the distress prediction model and that smaller residuals mean a more reliable design. However, the limitation of this method is that the residuals may not be so reliable because they are the results of calibrations. Under the assumption that the residuals are of normal distribution with a mean and a standard deviation, the distresses corresponding to specified reliability can be obtained.

Besides the input categories mentioned earlier, distress prediction models are also important constituent parts that are still under research for the 2002 *Guide*. These models can be classified into two categories. In the first category, stresses, strains, and displacements in the pavement structure can be calculated by using mechanistic structural models. The second category models focus on the prediction of the damage type and extent on the basis of mechanistic or empirical distress models. A revision of the new pavement under design or a rehabilitation of the existing pavement is recommended if either of the preceding damages go beyond its corresponding standard. These distress models are mostly regression equations with one or more regression constants. These constants play important roles in the accuracy of the predicted distress results. It is assumed that all the influencing factors that can possibly affect the prediction results are taken into account in these equations.

Compared with traditional flexible pavement design procedure, the 2002 *Guide* improves its design in two aspects. One is that it considers the longitudinal fatigue cracking that propagates from the top down. The other is that it also provides a prediction procedure for pavement roughness using the international roughness index (IRI). The following is a brief introduction to prediction models for pavement distresses such as fatigue, rutting, thermal cracking, and roughness.

Fatigue Cracking.   The traditional fatigue criterion equation shown in equation (11.18) is also applicable in the 2002 *Guide*.

$$N_f = f_1 (\varepsilon_t)^{-f_2} (E_1)^{-f_3} \tag{11.18}$$

where N_f = the allowable number of load repetitions to prevent fatigue cracking
E_1 = the elastic modulus of asphalt layer
f_1, f_2, f_3 = constants obtained from laboratory fatigue tests, with f_1 modified to correlate with field performance observation (The three coefficients can be determined by conducting fatigue beam tests with constant stress or constant strain control in laboratories.)

Permanent Deformation (Rutting).   Based on the assumption that the permanent strain of the asphalt mixture or unbound materials is proportional to its resilient strain, equations (11.18) and (11.19) can be used to determine the permanent strain of asphalt mixtures, respectively. The parameters in the equations can be obtained by conducting repeated load tests on cylindrical specimens.

$$\frac{\varepsilon_p}{\varepsilon_r} = 0.0007 \beta_{r1} T^{1.734 \beta_{r2}} N^{0.39937 \beta_{r3}} \tag{11.19}$$

where ε_p = permanent strain
ε_r = resilient strain
T = temperature of the specimen tested
N = number of load repetitions
$\beta_{r1}, \beta_{r2}, \beta_{r3}$ = all calibration factors

For unbound materials, equation (11.20), created by Tseng and Lytton (1998), can be used to calculate permanent deformation under repeated loads:

$$\delta_a = \beta_{S1} \varepsilon_v h \left(\frac{\varepsilon_0}{\varepsilon_r} \right) \left[e^{-\left(\frac{\rho}{N} \right)^\beta} \right] \tag{11.20}$$

where δ_a = permanent deformation of the unbound layer
β_{s1} = field calibration factor
e = base of natural logarithm with a value of 2.71828
N = number of load repetitions
ε_v = average vertical resilient strain obtained from the software program
h = thickness of the layer
$\beta, \rho, \varepsilon_0$ = regression constants representing material properties
ε_r = the resilient strain imposed in the laboratory test to acquire the three regression constants

Thermal Fracture.   The prediction model recommended by the 2002 *Guide* is fundamentally the same as the SHRP superpave performance models (Lytton et al. 1993), as shown in equation (11.21). Meanwhile, the indirect tensile test is suggested to predict thermal fracture of asphalt mixtures.

$$C_f = \beta_{t1} \beta_1 N_0 \left(\frac{\log^C / h_{ac}}{\sigma} \right) \tag{11.21}$$

where C_f = observed amount of thermal cracking in feet (m)
β_1 = calibration factor based on LTPP data
N_0 = standard normal distribution evaluated at 0
C = predicted crack depth by a crack propagation model
h_{ac} = thickness of AC layer
σ = standard deviation of the log crack depth
β_{t1} = field calibration factor

Roughness.   In the 2002 *Guide*, the increment international roughness index (ΔIRI) is recommended to evaluate pavement roughness. The corresponding equation is illustrated as follows:

$$\text{IRI} = \text{IRI}_0 + \Delta\text{IRI} \tag{11.22}$$

$$\Delta\text{IRI} = \text{function}(D_j, S_f) \tag{11.23}$$

where IRI$_0$ = newly constructed pavement smoothness
D_j = influence of surface distress
S_f = site factor or influence of nondistress variables

CHAPTER 12
PAVEMENT ENGINEERING II: RIGID PAVEMENTS

Julian Mills-Beale
Department of Civil and Environmental Engineering,
Michigan Technological University, Houghton, Michigan

Qian Zhang
Department of Civil Engineering, Michigan Technological University,
Houghton, Michigan
School of Civil Engineering, Xi'an University of Architecture and Technology,
Xi'an, People's Republic of China

Zhanping You
Department of Civil and Environmental Engineering,
Michigan Technological University, Houghton, Michigan

12.1 DESIGN OF RIGID PAVEMENTS

In the design of rigid pavements, the beam or plate theory is the underlying theory that governs the analysis. It is a simplified form of the layered theory in which it is assumed that the rigid pavement beam behaves like a medium thick plate with a plane before bending action and that still remains a plane after bending. Therefore, when traffic loads and temperature stresses are applied, the pavement's flexural strength allows only insignificant or minimal bending effect on the slab.

Even though rigid pavements have a relatively higher initial cost of construction than flexible pavements, they are inexpensive to maintain and have a longer service life. A typical rigid pavement thickness ranges from about 5 to 12 inches.

Rigid concrete pavements are either designed as simple plain, reinforced, or unreinforced pavements. Simple plain pavements are also known as *jointed plain concrete pavements* (JPCPs). Under the reinforced pavement class, jointed reinforced (JRCPs) and continuously reinforced (CRCPs) concrete pavements are the two main pavements of interest. Another pavement type that is currently used in the industry is the prestressed concrete pavement.

In the design and construction of rigid pavements, Portland cement, coarse aggregates, fine aggregates, and water are the main constituent materials, with or without steel reinforcement in the concrete. In special cases of environmental and climatic conditions, the use of admixtures in the preparation of the Portland cement concrete may be necessary.

This chapter deals with the main types of rigid pavements, the imposed stresses on the pavements, and the methods used for rigid pavement design.

12.2   *RIGID PAVEMENT TYPES*

Rigid concrete pavements are designed either as plain, simple reinforced or as continuously reinforced pavements. Another pavement type that is currently used in the industry is the prestressed concrete pavement. The following section describes in detail the characteristics and qualities of these types of pavements and the instances where they are used by highway agencies.

12.2.1   Plain Concrete Pavements

Plain concrete pavements are used by highway agencies on highways with low traffic volume or when cement-stabilized subbases are in place under the pavements. They are also known as *jointed plain concrete pavements* (JPCPs). In plain concrete pavements, no dowel bars for load transfer are applied in the pavement system. Furthermore, no temperature steel is found in plain concrete pavements. The plain concrete pavements have joints placed at distances ranging from about 8 to 20 feet in order to reduce or prevent the occurrence of cracking. Figure 12.1*a* shows this type of rigid pavement.

12.2.2   Simply Reinforced Concrete Pavements

A combination of dowel bars and temperature steel is used in simply reinforced concrete pavements to facilitate load transfer across joints and negate the effect of temperature stresses, respectively. The simply reinforced concrete pavements are also known as the *jointed reinforced concrete pavements* (JRCPs) and this type of pavement is shown in Figure 12.1*b*. It is common to find tie bars used at

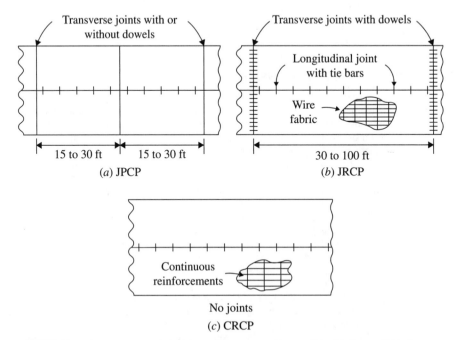

FIGURE 12.1   The three types of rigid concrete pavements. (*Source:* Yang H. Huang. 2004. *Pavement Analysis and Design*, 2nd ed. Upper Saddle River, NJ: Pearson Education.)

longitudinal joints in simply reinforced concrete pavements. The joints in these pavements normally range from 25 to 100 feet. The extended lengths of the joints in simply reinforced concrete pavements call for the use of the dowel bars. The slab length dictates the amount of temperature steel needed throughout the entire slab.

12.2.3 Continuously Reinforced Concrete Pavements

In this rigid pavement system, transverse joints are eliminated, and as such, continuously reinforced concrete pavements (CRCPs) are also known as *joint-free pavements*. Figure 12.1*c* shows continuously reinforced concrete. The advent of continuously reinforced concrete pavements was based on the hypothesis by engineers that a joint-free design would reduce the slab thickness drastically by 1 to a maximum 3 inches. In lieu of the absent transverse joints, construction joints and expansion joints are employed where necessary. A case in point is when continuously reinforced pavement is used on bridges. Highway agencies have found the frequent occurrence of transverse cracks to be characteristic of this type of pavement. To address the situation, the reinforcements are designed at convenient spacings to hold the cracks in place.

12.3 RIGID PAVEMENT MATERIALS

Rigid concrete pavements are basically made up of Portland cement concrete with or without steel reinforcement. The Portland cement concrete consists of Portland cement, water, and fine and coarse aggregates. This section details the specification requirements for the rigid pavement materials used by highway agencies.

12.3.1 Fine Aggregates

Fine aggregate materials, described as the aggregates passing through the 4.75-mm or No. 4 gradation sieve, are an important component in the Portland cement concrete. Sand is used most commonly, with proper gradation requirements, soundness, and cleanliness as key specifications. American Association of State and Highway and Transportation Officials (AASHTO) Standard M6 details the grading requirements used by highway agencies in selecting the most appropriate fine aggregate. The gradation requirement of AASHTO M6 is showed in Table 12.1. The silt (fines passing a No. 200 sieve) content specification in the fine aggregate is between 2 and 5 percent by weight.

TABLE 12.1 The Gradation Requirement for Fine Aggregates Used in Portland Cement Concrete (AASHTO Standard M6)

Sieve (M92)	Mass passing (%)
3/8 inch (9.5 mm)	100
No. 4 (2.36 mm)	95–100
No. 8 (2.36 mm)	80–100
No. 16 (1.18 mm)	50–85
No. 30 (600 μm)	25–60
No. 50 (300 μm)	10–30
No. 100 (150 μm)	2–10

Source: American Association of State and Highway and Transportation Officials. *Standard Specifications for Transportation Materials and Methods of Sampling and Testing*, 16th ed., Washington: AASHTO.

The fine aggregates used also should be free of deleterious materials in order not to compromise the hardening strength of the concrete. AASHTO Standard T21 is used to evaluate the cleanliness of the fine aggregates. The change to dark-color behavior of a sodium hydroxide solution when the fine aggregates are mixed in it depicts the presence of organic materials in the fine aggregates. Once organic material presence is confirmed, the sand is used to make about 2 inch cubes, and their strength is tested. This strength is compared with cubes from the same sand that has been washed in a 3 percent hydroxide solution. The sand in the former case passes specification only when its strength is at least 95 percent of that of the sand in the latter case.

12.3.2 Coarse Aggregates

Similar to the requirements for the fine aggregates, coarse aggregates used for Portland cement concrete should pass a gradation requirement. This specification is the American Society for Testing and Materials (ASTM) Standard C33 and is shown in Table 12.2.

TABLE 12.2 The Gradation Requirement for Coarse Aggregates used in Portland Cement Concrete (ASTM Standard C33)

| | % Passing by weight | | |
| | Aggregate designation | | |
Sieve designation	2 inches to No. 4 (357)	1½ inches to No. 4 (467)	1 inch to No. 4 (57)
2½ inches (63 mm)	100	—	—
2 inches (50 mm)	95–100	100	—
1½ inches (37.5 mm)	—	95–100	100
1 inches (25.0 mm)	35–70	—	95–100
¾ inches (19.0 mm)	—	35–70	—
½ inch (12.5 μm)	10–30	—	25–60
3/8 inch (12.5 μm)	—	10–30	—
No. 4 (4.75 mm)	0–5	0–5	0–10
No. 8 (2.36 mm)	—	—	0–5

Source: American Society for Testing and Materials (ASTM). 1992. *ASTM Standards: Concrete and Mineral Aggregates*, Vol. 04.02. Philadelphia: ASTM.

Typical coarse aggregates used are inert gravel, steel slag, and limestone. The inertness of the coarse aggregates ensures that the chemical hydration process is unaffected by the reactive nature of the aggregates. A lowly absorptive coarse aggregate is also desired to keep the water in the concrete matrix for a progressive hydration or hardening process. Coarse aggregates with absorptive capacities under 2 percent make good workable Portland cement concrete.

The maximum percent of foreign or deleterious materials in the coarse aggregates must be limited in addition to the aggregate's ability to resist abrasion and its soundness quality. In terms of the abrasion-resistant behavior, AASHTO Standard T96 (known as the *Los Angeles rattler test*) is used. The specification permits maximum allowable loss in weight ranges from 30 to 60 percent. Under the action of steel spheres, coarse aggregates retained on the 2.360-mm or No. 8 sieve are subjected to abrasion for 500 times. After the test, the coarse aggregate is sieved on a No. 12 sieve, and the weight retained is determined. The loss in weight is calculated as the retained weight subtracted from the original weight, expressed as a percentage.

12.3.3 Water

Water used for Portland concrete cement should satisfy drinking requirements. The water therefore should be free of organic matter, oil, acids, and alkalis, or in instances where they occur, they should be below permissible maximum limits.

12.3.4 Portland Cement

Portland cement consists of dicalcium silicate, tricalcium silicate (C_3S), and tetracalcium alumino-ferrite (C_4AF). There are five main types of Portland concrete cement, as specified by AASHTO Standard M85. ASTM Standard C150 may be an alternative in place of AASHTO M85.

The five Portland cement types specified for rigid pavement use by AASHTO M85 are

- *Type I.* This is the most common type supplied by manufacturers for general-purpose construction works. Type I cement comes with no special qualities or characteristics.

- *Type II.* When moderate heat of hydration is needed or the concrete will be exposed to moderate sulfate action, type II cement is specified.

- *Type III.* Type III is known as the high-early-strength cement and finds use in construction where the earliest possible concrete strength is desired.

- *Types IA, IIA, and IIIA.* These three types of concrete are relatively more durable and have special resistant properties against the effect of calcium chloride and deicing salts. The striking difference between the traditional types I, II, and III and types IA, IIA, and IIIA is that the latter contain about 4 to 8 percent of entrapped air (generated by air-entraining agents ground with the total mix.

- *Type IV.* This type of Portland cement concrete is used in conditions where low heats of hydration occur.

- *Type V.* Type V Portland cement is used in situations where the concrete will be subjected to high sulfate action.

12.3.5 Reinforcement Steel

Reinforcement steel is used either as dowel bars, tie bars, or temperature steel. Where reinforcement steel is used as a load-transfer mechanism across transverse joints, it is known as *dowel bars,* whereas in rigid pavements where steel reinforcement is used to tie two slabs together, the name given to it is *tie bars.*

Dowel Bars. In the application of a load-transfer mechanism across transverse joints, dowel bars are used. Typical dowel bar diameters range from 1 to 2 inches with lengths of 2 to 4 feet. Dowel bars normally are spaced at 1 to 1½ feet on center spanning across the width of the slab. When using dowels bars, engineers lubricate one end of the bar to allow for unrestricted expansion.

Tie Bars. Tie bars are used in tying two longitudinal joint sections together. Comparatively, tie bars have smaller diameters and larger center-to-center spacings than dowel bars. To ensure a durable bonding between the joint sections, tie bars usually are deformed or have hooks.

Temperature Steel. To reduce the effect of temperature stresses in rigid pavements, *temperature steel* is used. It is a complete mesh of both longitudinal and transverse steel wires held together at predetermined regular intervals. Temperature stresses are due to the progressive expansion and contraction of the slab during hot and cold weather conditions. Temperature steel controls the crack width in pavements by holding edges of developed cracks tight together. It must be noted, however, that temperature steel does not prevent the occurrence of cracks.

The cross-sectional area of the steel is vital in addressing the temperature-stress conditions. The size and spacing of the steel wires dictate the cross-sectional area per foot width of the slab. In designing the temperature steel required, the following factors are considered:

- Maximum anticipated stress in the concrete pavement

- Thickness of the pavement

- Elastic moduli of the cement concrete and the steel used

- Length of the pavement between the expansion joints

The cross-sectional area required for a reinforced steel mesh in a pavement can be estimated using the following equation:

$$A_s = 1/n(Lh\gamma_c f - 288 P_c h) \tag{12.1}$$

where A_s = the cross sectional area
$\quad\quad n$ = ratio of steel modulus to concrete modulus (E_s/E_c)
$\quad\quad L$ = length of slab (ft)
$\quad\quad h$ = concrete thickness (in.)
$\quad\quad \gamma_c$ = unit weight of concrete (lb/ft^3)
$\quad\quad f$ = coefficient of friction
$\quad\quad P_c$ = maximum desired stress in concrete (lb/in^2)

12.4   JOINTS IN RIGID PAVEMENTS

Joints are used in rigid pavements to serve two primary purposes:

1. Increase the bonding behavior of two adjacent pavements sections constructed at different times, that is, between one closing day's work and the beginning of the next
2. Reduce to the barest minimum the inherent temperature stresses in the pavement

The types of joints known in rigid pavements are (1) expansion joints, (2) contraction joints, (3) hinge joints, and (4) construction joints. Figure 12.2 shows all the typical joint types used in rigid pavements.

12.4.1   Expansion Joints

Highway agencies use expansion joints at regular spaced intervals to provide adequate allowance for the increasing slab-length dimensions. Under rising temperatures, the pavement slab tends to buckle or blow up as the length increases and the weight of the slab fails to provide a resisting downward stress. Typically, the expansion joints are between ¾ and about 1.5 inches wide and are designed in the longitudinal direction. Expansion joints are constructed in combination with smooth, load-transfer dowel bars and filler materials. The filler materials may be either cork, rubber, bituminous materials, or fabrics. Currently, expansion joints are phasing out owing to the inability of the load-transfer mechanism to transfer the loads.

12.5   RIGID PAVEMENT STRESSES

The main causes of stresses in rigid pavements are

- Expansion and contraction of the pavement slab resulting from repeated temperature changes
- Dynamic and repeated action of traffic wheel loads
- Loss of structural integrity of the subbase and subgrade underneath the rigid concrete slab
- Volumetric changes in the rigid concrete slab

12.5.1   Temperature-Induced Stresses

Owing to the effect of temperature differentials between the bottom and top of the slab, a rigid concrete slab could experience curling downward (daytime) and upward (nighttime). These curling stresses occur around the region of the slab edges. Highway agencies do not factor temperature curling stresses into pavement thickness design because joints and reinforcement steel reduce the effects of temperature stresses.

It is possible for a slab to experience daytime temperature differentials of between 2 and 4°F for every inch of the slab and between 1 and 2°F during nighttime. Rigid pavement research has shown that the temperature differential had no direct relationship with slab thickness.

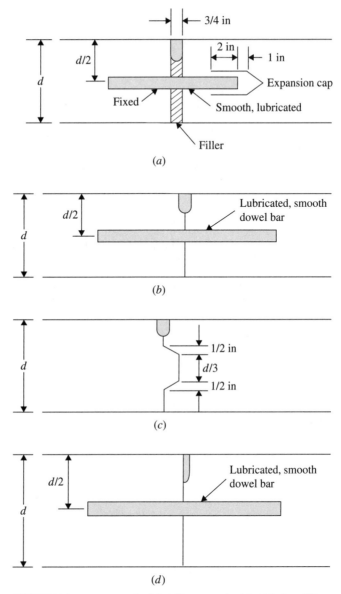

FIGURE 12.2 (*a*) An expansion joint; (*b*) a contraction joint; (*c*) a keyed hinge joint; (*d*) butt joint. (*Source:* Nicholas J. Garber and Lester A. Hoel. 2002. *Traffic and Highway Engineering*, 3rd ed.)

The weight of the slab acts as an opposite resistive stress to keep the slab in its original position and shape. During the day, tensile and compressive stresses occur at the bottom and top of the slab, respectively. On the contrary, during the night, tensile and compressive stresses occur at the top and bottom, respectively.

Temperature differentials in rigid pavements are largely influenced by the latitude location of the slab and seasonal variation conditions. For example, in locations in close proximity to the equator, the pavement slab has extremely high surface temperatures owing to the direct action of the sun's rays.

In calculating the maximum curling stresses at the edge and interior of the slab, the following equations are used:

$$\sigma_{xe} = \frac{C_x E_c et}{2} \tag{12.2}$$

$$\sigma_{xi} = \frac{E_c et}{2}\left(\frac{C_x + \mu C_y}{1 - \mu^2}\right) \tag{12.3}$$

where
σ_{xe} = the maximum curling stress (lb/in²) at the edge of the slab in the direction of the slab length

σ_{xi} = the maximum curling stress (lb/in²) at the interior of the slab in the direction of the slab length

E_c = the modulus of elasticity of the concrete (lb/in²)

μ = the Poisson ratio of the concrete

e = the concrete coefficient of thermal expansion and contraction for every degree Fahrenheit

C_x, C_y = are coefficients that are related to relative stiffness of the concrete

t = temperature difference between the top and bottom of the slab expressed in degrees Fahrenheit

12.5.2 Traffic Load-Induced Stresses

The Westergaard equations were developed by Westergaard in 1926 to provide the first analytical solutions to traffic-induced stresses on concrete pavements slabs. Enhanced versions of the Westergaard equations were published in 1927, 1933, 1939, 1943, and 1948. In his analysis, three critical locations of the traffic loads on the concrete pavement were considered. Westegaard's critical load locations are

- *Location 1:* When the load is located at the corner of a rectangular slab. Such a scenario is uncommon because pavements are wider (Westergaard provided no equations to solve this loading case)
- *Location 2:* When the load is located at the interior of the slab and at a considerable distance away from the slab edges
- *Location 3:* When the load is at the edge of the slab and at a considerable distance from the slab corners

12.5.3 The Modified Westegaard Equations

In the development of the Westergaard equations, the underlisted day and night slab behaviors were considered:

1. During the daytime, the temperature at the slab surface is higher than the temperature at the bottom. This allows for the slab to curl downward.
2. During the nighttime, the reverse condition occurs, where the temperature at the bottom of the slab surface is higher than the temperature at the surface. This allows for the slab to curl upward.

In 1936, the Bureau of Public Roads undertook an extensive research effort directed at the structural design of concrete pavements. The findings of their investigations led to the development of modified forms of the Westergaard equations. Based on the critical locations of the traffic wheel loads, the Bureau of Public Roads formulated the following equations in calculating the stress:

1. Condition of edge loading when the concrete slab curls downward in the daytime:

$$\sigma_e = \frac{0.572P}{h^2}\left[4\log_{10}\left(\frac{\ell}{b}\right)+0.359\right] \tag{12.4}$$

2. Condition of edge loading when the concrete slab curls upward at night:

$$\sigma_e = \frac{0.572P}{h^2}\left[4\log_{10}\left(\frac{\ell}{b}\right)+\log_{10}b\right] \tag{12.5}$$

3. Condition of interior loading:

$$\sigma_i = \frac{0.316P}{h^2}\left[4\log_{10}\left(\frac{\ell}{b}\right)+1.069\right] \tag{12.6}$$

where σ_e = the maximum stress (lb/in^2) induced in the bottom of the slab, directly under the load P and applied at the edge and in the direction parallel to the edge

σ_i = the maximum tensile stress (lb/in^2) induced at the bottom of the slab directly under the load P applied load in pounds, including the allowance for impact

H = the thickness of the slab (in)

ℓ = the radius of relative stiffness, calculated from,

$$\left[\frac{\sqrt{E_c h^3}}{12(1-\mu^2)k}\right]^{(1/4)}$$

E_c = the modulus of elasticity of concrete (lb/in^2)

μ = the Poisson ratio for concrete, given as 0.15 in this case

k = the modulus of the subgrade (lb/in^3)

b = the radius of equivalent distribution of pressure (inches), given by $\sqrt{(16a^2+h^2)}-0.675h$ for ($a < 1.724h$) and $b = a$ for ($a < 1.724h$)

a = the radius of contact area of load (inches)

12.5.4 Revised Edge Loading Equations

In further expanding the scope of knowledge on traffic wheel loads on concrete pavements, Ioannides et al. revised the edge-load equations. They postulated that

1. For a semicircular loaded area condition, the edge-loaded stress can be calculated using

$$\sigma_e = \frac{0.803P}{h^2}\left[4\log_{10}\left(\frac{\ell}{a}\right)+0.282\frac{a}{\ell}+0.650\right] \tag{12.7}$$

2. For a circular loaded area condition, the edge-loaded stress be calculated using

$$\sigma_e = \frac{0.803P}{h^2}\left[4\log_{10}\left(\frac{\ell}{a}\right)+0.666\frac{a}{\ell}-0.034\right] \tag{12.8}$$

12.6   *FAILURE MECHANISMS IN RIGID PAVEMENTS*

In the design of rigid pavements, fatigue cracking has been the major consideration until the issue of pumping and erosion came up for analysis by engineers. This section describes in detail the fatigue cracking and pumping mechanisms that influence the performance of rigid pavements.

12.6.1   Fatigue Cracking

The critical parameter in evaluating the allowable number of traffic load repetitions to initiate fatigue cracking in rigid pavements is the stress ratio between the flexural tensile stress and the concrete modulus of rupture. Designers have considered that fatigue cracking in rigid pavements is commonly caused by edge stresses at the midslab. In reality, a less significant percentage of the traffic load is applied at the pavement edge. Therefore, an equivalent number of edge loads is used in place of the total number of load repetitions on a rigid pavement in order to estimate the same amount of anticipated fatigue damage.

12.6.2   Pumping under Rigid Pavements

Pumping under rigid pavements is the phenomenon that occurs when a combination of water and fine aggregate materials (i.e., gravel, sand, clay, or silt) are ejected under the action of dynamic traffic loads. The pumped materials are transported through the joints or cracks in the pavement system. Highway agencies suspect pumping when base or subgrade material accumulates close to joints and cracks and on the surface of the rigid pavement. Pumping leads to the loss of structural integrity of the pavement support system.

12.6.3   Faulting of Joints

Faulting is caused largely by the direct consequence of pumping action under rigid pavement slabs. Faulting is more often associated with transverse joints than longitudinal joints. When loose materials are pumped under the slab, the trailing or leading slab close to the joint becomes depressed. This differential in slab height between the trailing and leading slabs is known as *faulting*. Faulting is also caused by the absence of a load-transfer mechanism. In the field, faulting is measured by the change in elevation between the trailing and leading slabs.

12.6.4   Blowups

In extremely hot weather conditions, blowups of the transverse joint or crack in rigid pavements are a common feature. Owing to inadequate width of a joint to permit expansion, the inbuilt compressive expansive stresses cause an upward movement or shattering action around the slab zone. In order to prevent blowups, maintenance pavement engineers remove incompressible materials such as plastics, organics, and coarse aggregates that have found their way into the joints between slabs. Typically, a highway engineer will estimate blowups by count number.

12.7   *PORTLAND CEMENT CONCRETE OVERLAYS*

This section briefly describes rigid concrete overlays [portland cement concrete (PCC) overlays] over either a flexible pavement or another PCC pavement. Overlays extend the design life of an existing pavement considerably while enabling it to perform its core function of serviceability and safety.

12.7.1   PCC Overlays on Flexile Pavements

PCC overlays on flexible pavements become the most prudent choice of overlay when the underlying asphalt pavement has lost its material and structural integrity beyond repair. Although PCC on flexible pavements is rare, its use has achieved considerable success the world over. The major design consideration is to use the existing asphalt pavement as the foundation. One of the low points in the design of PCC overlays over asphalt pavements is the frequent occurrence of reflective cracking resulting from the upward propagation of existing fatigue cracks. This is remedied by ensuring that all cracks of intense severity are either fully repaired or sealed.

12.7.2   PCC Overlays on PCC Pavement

Unbonded, semibounded, and fully bonded overlays are the three principal types of PCC overlays on PCC pavement commonly used in the concrete pavement industry.

Unbonded Overlay.   *Unbonded overlay*, also known as a *separated overly*, is the type in which an HMA separation layer is placed between the old pavement and the newly constructed overlay to avoid the occurrence of reflective cracking. The badly cracked existing pavement therefore will not be able to transmit cracks to the new overlay. The separation layer also functions as an excellent working platform for laying a continuously uniform overlay. Unbonded overlays could be either plain, reinforced, or continuously reinforced concrete.

Semibonded Overlay.   In this type of PCC overlay over PCC pavement, engineers assume a certain degree of bonding between the two pavements. During the construction of semibonded or partially bonded overlays, fresh concrete is poured on existing slabs that have not lost much of their structural integrity. Temperature stress buildup from the underlying layers has been known to cause problems for such partially bonded overlays. To address this setback, joint spacing in the partially bonded overlay is shortened as much as practicable.

Fully Bonded Overlay.   Bonded overlays are used when the existing pavement has an insignificant level of distress or all distresses have been repaired sufficiently. Fully bonded overlays are also known as *monolithic layers*. A number of steps are taken for the complete preparation of the existing pavement surface before the overlay is laid. These are

- Removal of all foreign materials or contaminants (solid and liquid) from the surface of the pavements, for example, grease, oil, paint, organics, and sand
- Application of a cement grout or sometimes liquid epoxy resin as the bonding material

Plain concrete is the most common type of bonded overlay over PCC pavements. Steel-reinforced concrete is used occasionally when thicker overlays are required.

12.8   RIGID PAVEMENT THICKNESS DESIGN

In the design of rigid pavement thicknesses, AASHTO and the Portland Cement Association (PCA) design methods are used extensively. This section describes in detail the AASHTO and PCA methods for designing rigid pavement thicknesses in the transportation industry. In both design approaches, the fundamental objective is to determine a slab thickness capable of withstanding the anticipated traffic wheel loads over the design life.

12.8.1   The AASHTO Design Approach

The AASHTO design method for rigid pavements was developed and published in the early 1960s and further updated a decade later (1970s) and two decades later (1980s). It is used by highway agencies for design and analysis of plain concrete, simply reinforced concrete, and continuously reinforced concrete pavements. The latest update features many aspects of research work, field tests, and empirical findings on rigid pavements. The first publication in the 1960s was based primarily on results of AASHTO road tests.

The major aspects of the AASHTO design method are (1) pavement thickness determination, (2) amount of reinforcing steel required, and (3) joint design. The 1993 AASHTO procedure for rigid pavement design considers many structural, environmental, field performance, and statistical reliability factors in completing the design. The main factors considered are

- Subgrade strength
- Subbase strength
- Concrete properties
- Pavement performance
- Traffic
- Drainage
- Statistical reliability

12.8.2   Subgrade Strength

The modulus of subgrade reaction k is used as the representative value for the subgrade strength and is obtained by dividing the load (lb/in^2) on a loaded area by the deformation (in inches). AASHTO Test T222 specifies the 30-inch-diameter plate-bearing test used in finding the k value. The k value also can be estimated using empirical charts between k and known engineering soil parameters. Pavement engineers also can provide close estimates of k values based on experience. Figure 12.4 provides the approximate empirical interrelationships between the bearing values and soil classifications.

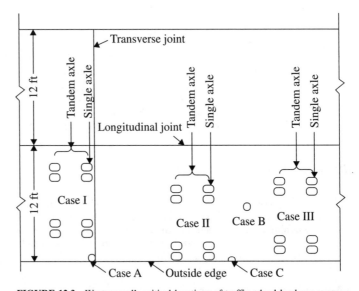

FIGURE 12.3   Westergaard's critical locations of traffic wheel loads on a pavement section. (*Source:* Nicholas J. Garber and Lester A. Hoel. 2002. *Traffic and Highway Engineering*, 3rd ed.)

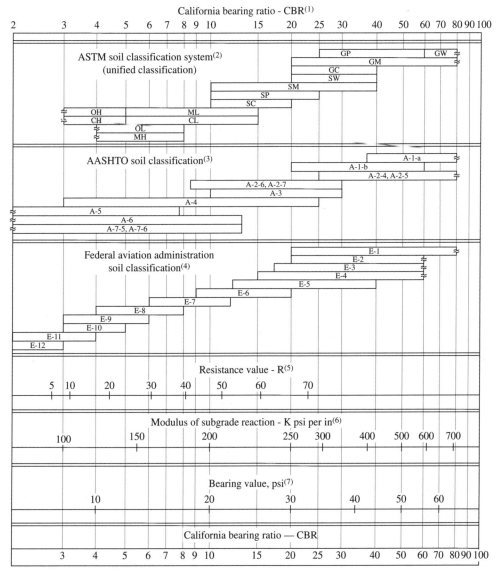

FIGURE 12.4 Empirical interrelationships between soil classification groups and bearing values. (*Source: R. G. Pacard. 1984. Thickness Design for Concrete Highway and Street Pavements. Skokie, IL: Portland Cement Association.*)

(1) For the basic idea, see O.J. Porter, "Foundations for Flexible Pavements," Highway Research Board *Proceedings of the Twenty-second Annual Meeting*, 1942, Vol. 22, pages 100–136.
(2) ASTM Designation D2487.
(3) "Classification of Highway Subgrade Materials," Highway Research Board *Proceedings of the Twenty-fifth Annual Meeting*, 1945. Vol. 25, pa 376–392.
(4) *Airport Paving*, U.S. Department of Commerce, Federal Aviation Agency, May 1948, pages 11-16. Estimated using values given in FAA *Design Manual* for *Airport Pavements*. (Formerly used FAA Classification: Unified Classification now used.)
(5) C. E.Warnes, "Correlation Between *R* Value and *k* Value," unpublished report, Portland Cement Association, Rocky Mountain-Northwest Region, October 1971 (best-fit correlation with correction for saturation).
(6) See T.A. *Middlebrooks* and G.E. Bertram. "Soil Tests for Design of Runway Pavements," Highway Research Board *Proceedings of the Twenty second Annual meeting*, 1942, Vol. 22, page 152.

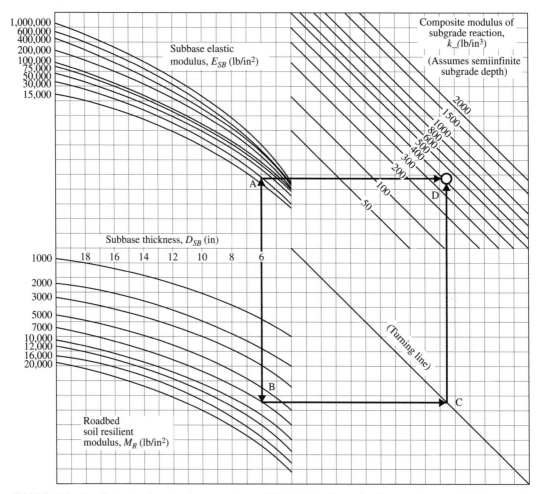

FIGURE 12.5 Empirical chart for estimating composite modulus of subgrade reaction K_∞ assuming a semiinfinite subgrade depth greater than 10-feet below the surface of the subgrade. (*Source:* AASHTO. 1993. *Guide for Design of Pavement Structures.* Washington: American Society of State Highway and Transportation Officials.)

The k value undergoes seasonal variation within each calendar year. A number of factors are known to affect the modulus of subgrade reaction. These factors lead to a new k value known as the *effective modulus of subgrade reaction*. They are (1) the seasonal effect on the resilient modulus of the subgrade, (2) the type and thickness of subbase material above the subgrade, (3) the effect of potential subbase erosion, and (4) the location of bedrock within the region of 10 feet of the subgrade.

Figure 12.5 is used to determine the composite modulus of subgrade reaction for types of subbase material, their elastic and resilient moduli, and the subbase thickness. The subbase elastic modulus is used as the representative value for the strength of the subbase. Furthermore, the material type and its thickness influence the economics of design.

The design also accounts for the loss of support under the slab, which may result in future erosion of the subbase. Table 12.3 provides the typical loss-of-support factors for different materials. Using these factors, the effective k value is reduced, as indicated in Figure 12.6.

TABLE 12.3 Range of Values for Loss-of-Support Factors for Different Materials

Material type	Loss of support (LS)
Cement-treated granular base $(E = 1,000,000$ to $2,000,000$ lb/in^2)	0.0 to 1.0
Cement aggregate mixtures $(E = 500,000$ to $1,000,000$ lb/in^2)	0.0 to 1.0
Asphalt-treated base $(E = 500,000$ to $1,000,000$ lb/in^2)	0.0 to 1.0
Bituminous stabilized mixtures $(E = 40,000$ to $300,000$ lb/in^2)	0.0 to 1.0
Lime-stabilized mixtures $(E = 20,000$ to $70,000$ lb/in^2)	1.0 to 3.0
Unbound granular materials $(E = 15,000$ to $45,000$ lb/in^2)	1.0 to 3.0
Fine-grained or natural subgrade materials $(E = 15,000$ to $45,000$ lb/in^2)	2.0 to 3.0

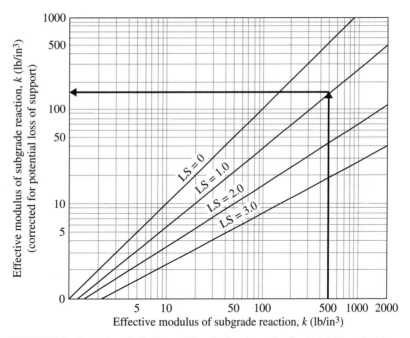

FIGURE 12.6 Correlation of effective modulus of subgrade reaction for potential loss of subbase support. (*Source:* AASHTO. 1993. *Guide for Design of Pavement Structures*. Washington: American Society of State Highway and Transportation Officials.)

When bedrock is located within 10 feet deep of the subgrade surface and spanning over a large portion of the road longitudinal and transverse alignment, the effective k value is also increased. Figure 12.7 is used in this design approach to account for the presence of subgrade.

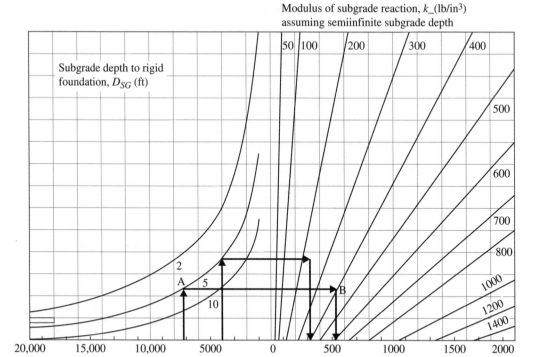

FIGURE 12.7 Empirical chart to modify modulus of subgrade reaction considering the effects of rigid foundation near surface (within 10 feet). (*Source:* AASHTO. 1993. *Guide for Design of Pavement Structures*. Washington: American Society of State Highway and Transportation Officials.)

12.8.3 Subbase Strength

The specification requires the use of graded granular materials and stabilized materials as the subbase for the rigid pavement. In Table 12.4, specifications of six types of subbase materials are given. In using this table, the following conditions and precautions apply:

1. Subbase thickness is normally greater than 6 inches and is designed and constructed to cover 1 to 3 feet outside the region of the pavement system.
2. Subbase types A through E typically are used in the upper 4-inch layer of the subbase.
3. Type F is used in the upper 4-inch layer.
4. Special project location conditions such as the presence of frost-susceptible soils allow for the use of types A, B and F, which contain the least amount of percent fines.

12.8.4 Traffic

In analyzing the traffic load application on rigid pavements, the 18,000-lb equivalent standard axle loads (ESALs) are used. ESAL factors are considered in the AASHTO design approach, and they are related directly to the terminal serviceability index P_t and pavement slab thickness. The *terminal serviceability index* P_t is defined according to the 1993 AASHTO *Guide for Design for Pavement Structures* as the lowest acceptable serviceability of the pavement before the highway agency can

TABLE 12.4 Recommended Particle Size Distributions for Different Types of Subbase Materials

Sieve designation	Type A	Type B	Type C (cement treated)	Type D (lime treated)	Type E (bituminous treated)	Type F (granular)
Sieve analysis % passing						
2 in	100	100	—	—	—	—
1 in	—	75–95	100	100	100	100
3/8 in	30–65	40–75	50–85	60–100	—	—
No. 4	25–55	30–60	35–65	50–85	55–100	70–100
No. 10	15–40	20–45	25–50	40–70	40–100	55–100
No. 40	8–20	15–30	15–30	25–45	20–50	30–70
No. 200	2–8	5–20	5–15	5–20	6–20	8–25
		(The minus Nos. 200 material should be held to a practical minimum.)				
Compressive strength lb/in^2 at 28 days			400–750	100		
Stability						
Hveem stabilometer					20 min	
Hubbard field					1000 min	
Marshall stability					500 min	
Marshall flow					20 min	
Soil constants						
Liquid limit	25 max	25 max				25 max
Plasticity index	N.P.	6 max	10 max		6 max	6 max

Source: Standard Specifications for Transportation Materials and Methods of Sampling and Testing, 20th ed. Washington: American Society of State Highway and Transportation Officials, 2000.

commence resurfacing or reconstruction. Tables 12.5 and 12.6 give the ESAL factors used when P_t is assumed to be 2.5. A P_t of 2.5 is typical for use in the design of major highways.

A slab thickness, in inches, is first assumed, and then it is used for calculation of the number of accumulated ESALs. Finally, the accumulated ESALs are used to establish the required thickness. When the calculated thickness is significantly different from the assumed initial thickness, the accumulated ESALs are recomputed. The procedure is conducted on a repetitive basis until the assumed and computed thicknesses are approximately the same.

12.8.5 Concrete Properties

The 28-day flexural strength or modulus of rupture of the concrete is used in the design, and this is determined from the three-point loading test (AASHTO Standard T97).

12.8.6 Drainage

The AASHTO design approach specifies the use of a drainage factor C_d in the design performance analysis. Typical C_d values are given in Table 12.7. Factors influencing the C_d value include the quality of drainage and the percent of time the pavement is exposed to moisture levels approaching the saturation state.

12.8.7 Pavement Performance

The AASHTO design approach assigns an initial serviceability index P_i of 4.5, whereas a terminal serviceability index P_t is selected as 2.5. The P_t of 2.5 is usually used by highway agencies and engineers for major highways.

TABLE 12.5 ESAL Factors for Rigid Pavements, Single Axles, and P_t of 2.5

Axle load (kip)	Slab thickness, D (in)								
	6	7	8	9	10	11	12	13	14
2	.0002	.0002	.0002	.0002	.0002	.0002	.0002	.0002	.0002
4	.003	.002	.002	.002	.002	.002	.002	.002	.002
6	.012	.011	.010	.010	.010	.010	.010	.010	.010
8	.039	.035	.033	.032	.032	.032	.032	.032	.032
10	.097	.089	.084	.082	.081	.080	.080	.080	.080
12	.203	.189	.181	.176	.175	.174	.174	.173	.173
14	.376	.360	.347	.341	.338	.337	.336	.336	.336
16	.634	.623	.610	.604	.601	.599	.599	.599	.598
18	1.00	1.00	1.00	1.00	1.00	1.00	1.00	1.00	1.00
20	1.51	1.52	1.55	1.57	1.58	1.58	1.59	1.59	1.59
22	2.21	2.20	2.28	2.34	2.38	2.40	2.41	2.41	2.41
24	3.16	3.10	3.22	3.36	3.45	3.50	3.53	3.54	3.55
26	4.41	4.26	4.42	4.67	4.85	4.95	5.01	5.04	5.05
28	6.05	5.76	5.92	6.29	6.61	6.81	6.92	6.98	7.01
30	8.16	7.67	7.79	8.28	8.79	9.14	9.35	9.46	9.52
32	10.8	10.1	10.1	10.7	11.4	12.0	12.3	12.6	12.7
34	14.1	13.0	12.9	13.6	14.6	15.4	16.0	16.4	16.5
36	18.2	16.7	16.4	17.1	18.3	19.5	20.4	21.0	21.3
38	23.1	21.1	20.6	21.3	22.7	24.3	25.6	26.4	27.0
40	29.1	26.5	25.7	26.3	27.9	29.9	31.6	32.9	33.7
42	36.2	32.9	31.7	32.2	34.0	36.3	38.7	40.4	41.6
44	44.6	40.4	38.8	39.2	41.0	43.8	46.7	49.1	50.8
46	54.5	49.3	47.1	47.3	49.2	52.3	55.9	59.0	61.4
48	66.1	59.7	56.9	56.8	58.7	62.1	66.3	70.3	73.4
50	79.4	71.7	68.2	67.8	69.6	73.3	78.1	83.0	87.1

Source: AASHTO. 1993. *Guide for Design of Pavement Structures.* Washington: American Society of State Highway and Transportation Officials.

12.8.8 Statistical Reliability

The reliability levels $R\%$ and the overall standard deviation S_o are incorporated in the AASHTO design method charts. These are similar to the statistical reliability used in flexible pavement design.

12.8.9 Design Procedure

In the 1986 AASHTO design guide, an equation to establish the required thickness of slab that can satisfactorily resist the project design ESALs was developed. This equation is given as

$$\log_{10}W_{18} = Z_R + 7.35 \log_{10}(D+1) - 0.066 + \left\{ \frac{\log_{10} + [\Delta PSI/(4.5-1.5)}{1 + \left[(1.624 \times 10^7)/(D+1)^{8.46} \right]} \right\}$$

$$+ (4.22 + 0.32 P_t)\log_{10}\left(\frac{S'_c C_d}{215.63J} \right) \left\{ \frac{D^{0.75} - 1.132}{D^{0.75} - \left[18.42/(E_c/k)^{0.25} \right]} \right\} \tag{12.9}$$

TABLE 12.6 ESAL Factors for Rigid Pavements, Tandem Axles, and P_t of 2.5

Axle load (kip)	Slab thickness, D (in)								
	6	7	8	9	10	11	12	13	14
2	.0001	.0001	.0001	.0001	.0001	.0001	.0001	.0001	.0001
4	.0006	.0006	.0005	.0005	.0005	.0005	.0005	.0005	.0005
6	.002	.002	.002	.002	.002	.002	.002	.002	.002
8	.007	.006	.006	.005	.005	.005	.005	.005	.005
10	.015	.014	.013	.013	.012	.012	.012	.012	.012
12	.031	.028	.026	.026	.025	.025	.025	.025	.025
14	.057	.052	.049	.048	.047	.047	.047	.047	.047
16	.097	.089	.084	.082	.081	.081	.080	.080	.080
18	.155	.143	.136	.133	.132	.131	.131	.131	.131
20	.234	.220	.211	.206	.204	.203	.203	.203	.203
22	.340	.325	.313	.308	.305	.304	.303	.303	.303
24	.475	.462	.450	.444	.441	.440	.439	.439	.439
26	.644	.637	.627	.622	.620	.619	.618	.618	.618
28	.855	.854	.852	.850	.850	.850	.849	.849	.849
30	1.11	1.12	1.13	1.14	1.14	1.14	1.14	1.14	1.14
32	1.43	1.44	1.47	1.49	1.50	1.51	1.51	1.51	1.51
34	1.82	1.82	1.87	1.92	1.95	1.96	1.97	1.97	1.97
36	2.29	2.27	2.35	2.43	2.48	2.51	2.52	2.52	2.53
38	2.85	2.80	2.91	3.03	3.12	3.16	3.18	3.20	3.20
40	3.52	3.42	3.55	3.74	3.87	3.94	3.98	4.00	4.01
42	4.32	4.16	4.30	4.55	4.74	4.86	4.91	4.95	4.96
44	5.26	5.01	5.16	5.48	5.75	5.92	6.01	6.06	6.09
46	6.36	6.01	6.14	6.53	6.90	7.14	7.28	7.36	7.40
48	7.64	7.16	7.27	7.73	8.21	8.55	8.75	8.86	8.92
50	9.11	8.50	8.55	9.07	9.68	10.14	10.42	10.58	10.66
52	10.8	10.0	10.0	10.6	11.3	11.9	12.3	12.5	12.7
54	12.8	11.8	11.7	12.3	13.2	13.9	14.5	14.8	14.9
56	15.0	13.8	13.6	14.2	15.2	16.2	16.8	17.3	17.5
58	17.5	16.0	15.7	16.3	17.5	18.6	19.5	20.1	20.4
60	20.3	18.5	18.1	18.7	20.0	21.4	22.5	23.2	23.6
62	23.5	21.4	20.8	21.4	22.8	24.4	25.7	26.7	27.3
64	27.0	24.6	23.8	24.4	25.8	27.7	29.3	30.5	31.3
66	31.0	28.1	27.1	27.6	29.2	31.3	33.2	34.7	35.7
68	35.4	32.1	30.9	31.3	32.9	35.2	37.5	39.3	40.5
70	40.3	36.5	35.0	35.3	37.0	39.5	42.1	44.3	45.9
72	45.7	41.4	39.6	39.8	41.5	44.2	47.2	49.8	51.7
74	51.7	46.7	44.6	44.7	46.4	49.3	52.7	55.7	58.0
76	58.3	52.6	50.2	50.1	51.8	54.9	58.6	62.1	64.8
78	65.5	59.1	56.3	56.1	57.7	60.9	65.0	69.0	72.3
80	73.4	66.2	62.9	62.5	64.2	67.5	71.9	76.4	80.2
82	82.0	73.9	70.2	69.6	71.2	74.7	79.4	84.4	88.8
84	91.4	82.4	78.1	77.3	78.9	82.4	87.4	93.0	98.1
86	102.0	92.0	87.0	86.0	87.0	91.0	96.0	102.0	108.0
88	113.0	102.0	96.0	95.0	96.0	100.0	105.0	112.0	119.0
90	125.0	112.0	106.0	105.0	106.0	110.0	115.0	123.0	130.0

Source: AASHTO. 1993. *Guide for Design of Pavement Structures.* Washington: American Society of State Highway and Transportation Officials.

TABLE 12.7 Recommended Drainage Coefficient C_d for Rigid Pavements

Quality of drainage	Percent of time pavement structure is exposed to moisture levels approaching saturation			
	Less than 1 percent	1–5 percent	5–25 percent	Greater than 25 percent
Excellent	1.25–1.20	1.20–1.15	1.15–1.10	1.10
Good	1.20–1.15	1.15–1.10	1.10–1.00	1.00
Fair	1.15–1.10	1.10–1.00	1.00–0.90	0.90
Poor	1.10–1.00	1.00–0.90	0.90–0.80	0.80
Very poor	1.00–0.90	0.90–0.80	0.80–0.70	0.70

Source: AASHTO. 1986. *Guide for Design of Pavement Structures.* Washington: American Society of State Highway and Transportation Officials.

where
Z_R = standard normal variant corresponding to the selected level of reliability
S_o = overall standard deviation
W_{18} = predicted number of 18-kip ESAL applications that can be carried by the pavement structure after construction
D = concrete pavement thickness to the nearest ½ inch
ΔPSI = change in serviceability index, that is, the design serviceability loss given by $p_i - p_t$; p_i and p_t being the initial serviceability and terminal serviceability index, respectively
E_c = elastic modulus of the concrete to be used in construction (lb/in^2)
S_c = modulus of rupture of concrete used (lb/in^2)
J = load-transfer coefficient, assumed to be 3.2
C_d = drainage coefficient

Engineers opt for one of two choices in the determination of pavement thickness using the equation. The first is the use of the two empirical charts, whereas the second alternative is the use of computer iteration techniques in design of the thickness. The iterative computer approach is preferred owing to its flexibility in allowing for a repetitive assumption of thickness D to calculate the effective modulus of subgrade reaction and ESAL factors.

12.9 THE PCA DESIGN APPROACH

The PCA rigid pavement thickness design approach was developed in 1961 and revised in 1984. It is used by highway agencies for pavement thickness design of plain concrete, simply reinforced concrete, and continuously reinforced concrete pavements. In using the PCA design method, the anticipated traffic load, concrete flexural strength, and subbase and subgrade support are significant.
This design approach is based on

- Theoretical development of stresses on layered structural systems
- Results of model and full-scale tests
- Empirical results of concrete pavements under typical traffic loads

12.9.1 Design Traffic Wheel Loads

The design traffic is the accumulated number of single and tandem axles of different loads estimated for the entire life of the concrete. Three principal traffic-volume parameters essential in calculating the estimated traffic loads are (1) the average daily traffic (ADT), (2) the average daily truck traffic (ADTT) in opposite directions, and (3) the axle load distribution of the truck traffic. It is

TABLE 12.8a   Truck Distribution for Multiple-Lane highways

One-way ADT	Two lanes in each direction		Three or more lanes in each direction		
	Inner	Outer	Inner[a]	Center	Outer
2,000	6	94	6	12	82
4,000	12	88	6	18	76
6,000	15	85	7	21	72
8,000	18	82	7	23	70
10,000	19	81	7	25	68
15,000	23	77	7	28	65
20,000	25	75	7	30	63
25,000	27	73	7	32	61
30,000	28	72	8	33	59
35,000	30	70	8	34	58
40,000	31	69	8	35	57
50,000	33	67	8	37	55
60,000	34	66	8	39	53
70,000	—	—	8	40	52
80,000	—	—	8	41	51
100,000	—	—	9	42	49

[a]Combined inner one or more lanes.

Source: Darter, M.I., B.F. McCullough, and J.L. Brown, 1973b. "Reliability Concepts Applied to the Texas Flexible Pavement System," Highway Research Record 407, 180–190; Highway Research Board.

TABLE 12.8b   Percentage of Total Truck in Design Lane

Number of traffic lanes in two direction	Percentage of trucks in design lane
2	50
4	45 (35–48)[a]
6 or more	40 (25–48)

[a]Probable range.

Source: AI, 1981a. Thickness Design—Asphalt Pavements for Highways and Streets, Manual Series No. 1; Asphalt Institute.

TABLE 12.8c   Lane Distribution Factor

No. of lanes in each direction	Percentage of 18-kip ESAL in design lane
1	100
2	80–100
3	60–80
4	50–75

Source: AASHTO, 1986. Guide for Design of Pavement Structures; American Association of State Highway and Transportation Officials.

particularly important to note that the ADTT considers only trucks with a minimum of six tires. The truck volume in each direction of travel either can be assumed to be the same or can have significant variation. When the design engineer concludes that the truck volume in each direction is significantly different, a directional adjustment factor is assigned. The truck distribution factors are shown in Table 12.8.

Load Safety Factors (LSFs).   The design axle loads must be multiplied by load safety factors (LSFs) per the conditions and values shown below:

1. When low truck volumes are expected on the pavement, for example, for county roads and residential streets, an LSF of 1.0 is used.

2. When moderate or medium truck volume is expected, for example, for highways and arteries, an LSF of 1.1 is used.

3. When high truck volumes are expected, for example, for interstate and multilane projects, an LSF of 1.2 is used.

4. When the pavement serviceability level is to be increased throughout the design life, the LSF is increased to 1.3.

5. A factor of safety of 1.1 or 1.2 is added to the LSF criteria given to make allowance for unpredicted truck traffic volume.

12.9.2 PCA Design Procedure

The PCA design procedure involves two key parts. The first is the fatigue analysis, and the second is the erosion analysis. This section describes the objectives, relevant information, and steps required for each analysis.

12.9.3 Concrete Flexural Strength

Using ASTM C78, *Annual Book of ASTM Standards*, highway agencies use the average 28-day flexural strength of concrete as determined by the third-point method. The flexural strength is indicated as the modulus of rupture of the concrete. In establishing the design tables and charts, the inherent point-to-point variation in strength of the concrete slab is considered. Furthermore, since it is a known engineering fact that concrete gains strength with age, this strength variation with age is also factored in the design charts and tables.

12.9.4 Subbase and Subgrade Support

In terms of subgrade or subbase support, the subgrade reaction value k is used in the design and analysis. This is known as the *Westergaard modulus of subgrade reaction*. The plate-bearing test is used to determine the k value. Another approach employed by highway agencies is to use other soil classification tests to relate and define the estimated k value. This approach is shown in Figure 12.4. Typically, the summer or fall k value is used.

In projects where unusually high traffic loads are anticipated on the road pavement and the engineer deems the subgrade to have low k value or weak subbase material, the soil can be stabilized with cement. The AASHTO Soil Classification System specifies that the A-1, A-2-4, A-2-5, and A-3 soil types can be used as cement stabilization of subbase or subgrade. Furthermore, ASTM Standards D560 and D559 require that laboratory freeze-thaw and wet-dry tests be conducted to define the quantity of cement needed for the stabilization. Table 12.9 provides the design k values for both untreated and cement-treated subbases.

12.9.5 Fatigue Analysis

In this procedure, the minimum thickness of the fatigue pavement required to limit the effect of fatigue cracking is determined. Two types of load repetitions, the expected axle repetitions and the allowable axle repetitions, are needed for this analysis. The specification criterion is that the expected axle repetitions must be lower than the allowable axle repetitions. In calculating the allowable number of load repetitions on the pavement, the equivalent-stress-ratio factor needs to be known. This is the ratio of the equivalent stress of the pavement to the concrete's modulus of rupture. Tables 12.10 and 12.11 show the equivalent stress values for single and tandem axles. Table 12.10 is used for pavements without any concrete shoulders, whereas Table 12.11 is used in cases where concrete shoulders are integrated into the design. The subgrade or subbase k value and the slab thickness influence the equivalent-stress-ratio factor. Figure 12.8 provides an empirical chart for determining the allowable load repetitions for any concrete slab. Highway agencies typically use the fatigue

TABLE 12.9 Design k Values for Untreated and Cement Treated Subbases

(a) Untreated granulated subbases				
Subgrade k value (lb/in^3)	Subbase k value (lb/in^3)			
	4 in	6 in	9 in	12 in
50	65	75	85	110
100	130	140	160	190
200	220	230	270	320
300	320	330	370	430
(b) Cement treated subbases				
Subgrade k value (lb/in^3)	Subbase k value (lb/in^3)			
	4 in	6 in	9 in	12 in
50	170	230	310	390
100	280	400	520	640
200	470	640	830	—

Source: Robert G. Packard 1984. *Thickness Design for Concrete Highway and Street Pavements.* Skokie, IL: Portland Cement Association.

TABLE 12.10 Equivalent Stress Values for Single and Tandem Axles (Without Concrete Shoulder)

Slab thickness (in)	k of Subgrade-subbase (lb/in^3) (single-axle tandem axle)						
	50	100	150	200	300	500	700
4	825/679	726/585	671/542	634/516	584/486	523/457	484/443
4.5	699/586	616/500	571/460	540/435	498/406	448/378	417/363
5	602/516	531/436	493/399	467/376	432/349	390/321	363/307
5.5	526/461	464/387	431/353	409/331	379/305	343/278	320/264
6	465/416	411/348	382/316	362/296	336/271	304/246	285/232
6.5	417/380	367/317	341/286	324/267	300/244	273/220	256/207
7	375/349	331/290	307/262	292/244	271/222	246/199	231/186
7.5	340/323	300/268	279/241	265/224	246/203	224/181	210/169
8	311/300	274/249	255/223	242/208	225/188	205/167	192/155
8.5	285/281	252/232	234/208	222/193	206/174	188/154	177/143
9	264/264	232/218	216/195	205/181	190/163	174/144	163/133
9.5	245/248	215/205	200/183	190/170	176/153	161/134	151/124
10	228/235	200/193	186/173	177/160	164/144	150/126	141/117
10.5	213/222	187/183	174/164	165/151	153/136	140/119	132/110
11	200/211	175/174	163/155	154/143	144/129	131/113	123/104
11.5	188/201	165/165	153/148	145/136	135/122	123/107	116/98
12	177/192	155/158	144/141	137/130	127/116	116/102	109/93
12.5	168/183	147/151	136/135	129/124	120/111	109/97	103/89
13	159/176	139/144	129/129	122/119	113/106	103/93	97/85
13.5	152/168	132/138	122/123	116/114	107/102	98/89	92/81
14	144/162	125/133	116/118	110/109	102/98	93/85	88/78

Source: R. G. Packard. 1984. *Thickness Design for Concrete Highway and Street Pavements.* Skokie, IL: Portland Cement Association.

TABLE 12.11 Equivalent Stress Values for Single and Tandem Axles for Pavements with Concrete Shoulders

Slab thickness (in)	k of Subgrade-subbase (lb/in³) (single-axle tandem axle)						
	50	100	150	200	300	500	700
4	640/534	559/468	517/439	489/422	452/403	409/388	383/384
4.5	547/461	479/400	444/372	421/356	390/338	355/322	333/316
5	475/404	417/349	387/323	367/308	341/290	311/274	294/267
5.5	418/360	368/309	342/285	324/271	302/254	276/238	261/231
6	372/325	327/277	304/255	289/241	270/225	247/210	234/203
6.5	334/295	294/251	274/230	260/218	243/203	223/188	212/180
7	302/270	266/230	248/210	236/198	220/184	203/170	192/162
7.5	275/250	243/211	226/193	215/182	201/168	185/155	176/148
8	252/232	222/196	207/179	197/168	185/155	170/142	162/135
8.5	332/216	205/182	191/166	182/156	170/144	157/131	150/125
9	215/202	190/171	177/155	169/146	158/134	146/122	139/116
9.5	200/190	176/160	164/146	157/137	147/126	136/114	129/108
10	186/179	164/151	153/137	146/129	137/118	127/107	121/101
10.5	174/170	154/143	144/130	137/121	128/111	119/101	113/95
11	164/161	144/135	135/123	129/115	120/105	112/95	106/90
11.5	154/153	136/128	127/117	121/109	113/100	105/90	100/85
12	145/146	128/122	120/111	114/104	107/95	99/86	95/81
12.5	137/139	121/117	113/106	108/99	101/91	94/82	90/77
13	130/133	115/112	107/101	102/95	96/86	89/78	85/73
13.5	124/127	109/107	102/97	97/91	91/83	85/74	81/70
14	118/122	104/103	97/93	93/87	87/79	81/71	77/67

Source: R. G. Packard. 1984. *Thickness Design for Concrete Highway and Street Pavements.* Skokie, IL: Portland Cement Association.

analysis for pavements expected to carry light traffic, as well as in cases where the doweled-jointed pavements will carry medium traffic loads.

Erosion Analysis. While the fatigue analysis considers the use of a stress-ratio factor, the erosion analysis involves the use of an erosion factor. The erosion factor, like the stress-ratio factor, depends on the thickness of the slab and the subbase-subgrade k value.

During the life of the concrete pavement, failure mechanisms such as pumping, faulting, and foundation and shoulder erosion arise owing to deflection within the slab. This phase of the PCA design approach, known as the *erosion-analysis procedure*, seeks to find the minimum slab thickness possible to limit the effect of pumping, faulting, and foundation and shoulder erosion. This erosion-analysis criterion serves as a guiding framework for engineers in determining the effort an axle load takes to deflect a slab. Modification of the erosion analysis is permissible according to the project-specific climatic and environmental conditions.

The erosion analysis finds use predominantly in concrete pavement projects with medium or heavy traffic loads and undoweled joints. It is also used for heavy-traffic pavements with doweled joints. Different erosion factors used for the various types of pavement construction are found in Tables 12.12 through to 12.15. Furthermore, using Figures 12.9 and 12.10, the allowable load repetitions can be found.

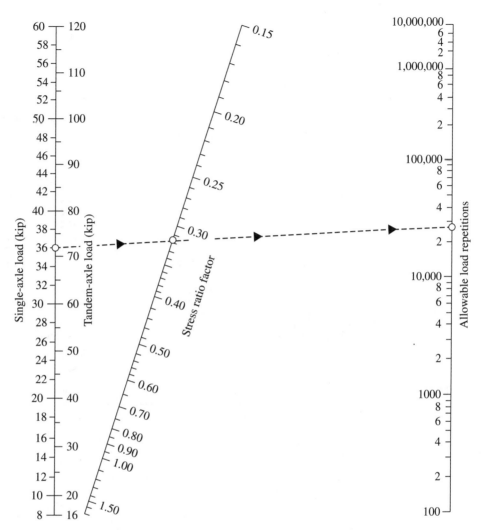

FIGURE 12.8   Allowable load repetitions for fatigue analysis based on the stress ratio. (*Source:* R. G. Packard. 1984. *Thickness Design for Concrete Highway and Street Pavements*. Skokie, IL: Portland Cement Association.)

TABLE 12.12 Erosion Factors for Single and Tandem Axles (Doweled Joints, without Concrete Shoulder)

Slab thickness (in)	k of Subgrade-subbase (lb/in³) (single-axle tandem axle)					
	50	100	200	300	500	700
4	3.28/3.30	3.24/3.20	3.21/3.13	3.19/3.10	3.15/3.09	3.12/3.08
4.5	3.13/3.19	3.09/3.08	3.06/3.00	3.04/2.96	3.01/2.93	2.98/2.91
5	3.01/3.09	2.97/2.98	2.93/2.89	2.90/2.84	2.87/2.79	2.85/2.77
5.5	2.90/3.01	2.85/2.89	2.81/2.79	2.79/2.74	2.76/2.68	2.73/2.65
6	2.79/2.93	2.75/2.82	2.70/2.71	2.68/2.65	2.65/2.58	2.62/2.54
6.5	2.70/2.86	2.65/2.75	2.61/2.63	2.58/2.57	2.55/2.50	2.52/2.45
7	2.61/2.79	2.56/2.68	2.52/2.56	2.49/2.50	2.46/2.42	2.43/2.38
7.5	2.53/2.73	2.48/2.62	2.44/2.50	2.41/2.44	2.38/2.36	2.35/2.31
8	2.46/2.68	2.41/2.56	2.36/2.44	2.33/2.38	2.30/2.30	2.27/2.24
8.5	2.39/2.62	2.34/2.51	2.29/2.39	2.26/2.32	2.22/2.24	2.20/2.18
9	2.32/2.57	2.27/2.46	2.22/2.34	2.19/2.27	2.16/2.19	2.13/2.13
9.5	2.26/2.52	2.21/2.41	2.16/2.29	2.13/2.22	2.09/2.14	2.07/2.08
10	2.20/2.47	2.15/2.36	2.10/2.25	2.07/2.18	2.03/2.09	2.01/2.03
10.5	2.15/2.43	2.09/2.32	2.04/2.20	2.01/2.14	1.97/2.05	1.95/1.99
11	2.10/2.39	2.04/2.28	1.99/2.16	1.95/2.09	1.92/2.01	1.89/1.95
11.5	2.05/2.35	1.99/2.24	1.93/2.12	1.90/2.05	1.87/1.97	1.84/1.91
12	2.00/2.31	1.94/2.20	1.88/2.09	1.85/2.02	1.82/1.93	1.79/1.87
12.5	1.95/2.27	1.89/2.16	1.84/2.05	1.81/1.98	1.77/1.89	1.74/1.84
13	1.91/2.23	1.85/2.13	1.79/2.01	1.76/1.95	1.72/1.86	1.70/1.80
13.5	1.86/2.20	1.81/2.09	1.75/1.98	1.72/1.91	1.68/1.83	1.65/1.77
14	1.82/2.17	1.76/2.06	1.71/1.95	1.67/1.88	1.64/1.80	1.61/1.74

Source: R. G. Packard. 1984. *Thickness Design for Concrete Highway and Street Pavements.* Skokie, IL: Portland Cement Association.

TABLE 12.13 Erosion Factors for Single and Tandem Axles (Aggregate Interlock Joints, Without Concrete Shoulder)

Slab thickness (in)	k of Subgrade-subbase (lb/in³) (single-axle tandem axle)					
	50	100	200	300	500	700
4	3.46/3.49	3.42/3.39	3.38/3.32	3.36/3.29	3.32/3.26	3.28/3.24
4.5	3.32/3.39	3.28/3.28	3.24/3.19	3.22/3.16	3.19/3.12	3.15/3.09
5	3.20/3.30	3.16/3.18	3.12/3.09	3.10/3.05	3.07/3.00	3.04/2.97
5.5	3.10/3.22	3.05/3.10	3.01/3.00	2.99/2.95	2.96/2.90	2.93/2.86
6	3.00/3.15	2.95/3.02	2.90/2.92	2.88/2.87	2.86/2.81	2.83/2.77
6.5	2.91/3.08	2.86/2.96	2.81/2.85	2.79/2.79	2.76/2.73	2.74/2.68
7	2.83/3.02	2.77/2.90	2.73/2.78	2.70/2.72	2.68/2.66	2.65/2.61
7.5	2.76/2.97	2.70/2.84	2.65/2.72	2.62/2.66	2.60/2.59	2.57/2.54
8	2.69/2.92	2.63/2.79	2.57/2.67	2.55/2.61	2.52/2.53	2.50/2.48
8.5	2.63/2.88	2.56/2.74	2.51/2.62	2.48/2.55	2.45/2.48	2.43/2.43
9	2.57/2.83	2.50/2.70	2.44/2.57	2.42/2.51	2.39/2.43	2.36/2.38
9.5	2.51/2.79	2.44/2.65	2.38/2.53	2.36/2.46	2.33/2.38	2.30/2.33
10	2.46/2.75	2.39/2.61	2.33/2.49	2.30/2.42	2.27/2.34	2.24/2.28
10.5	2.41/2.72	2.33/2.58	2.27/2.45	2.24/2.38	2.21/2.30	2.19/2.24
11	2.36/2.68	2.28/2.54	2.22/2.41	2.19/2.34	2.16/2.26	2.14/2.20
11.5	2.32/2.65	2.24/2.51	2.17/2.38	2.14/2.31	2.11/2.22	2.09/2.16
12	2.28/2.62	2.19/2.48	2.13/2.34	2.10/2.27	2.06/2.19	2.04/2.13
12.5	2.24/2.59	2.15/2.45	2.09/2.31	2.05/2.24	2.02/2.15	1.99/2.10
13	2.20/2.56	2.11/2.42	2.04/2.28	2.01/2.21	1.98/2.12	1.95/2.06
13.5	2.16/2.53	2.08/2.39	2.00/2.25	1.97/2.18	1.93/2.09	1.91/2.03
14	2.13/2.51	2.04/2.36	1.97/2.23	1.93/2.15	1.89/2.06	1.87/2.00

Source: R. G. Packard. 1984. *Thickness Design for Concrete Highway and Street Pavements.* Skokie, IL: Portland Cement Association.

TABLE 12.14 Erosion Factors for Single and Tandem Axles (Doweled Joints, Concrete Shoulder)

Slab thickness (in)	k of Subgrade-subbase (lb/in³) (single-axle tandem axle)					
	50	100	200	300	500	700
4	3.28/3.30	3.24/3.20	3.21/3.13	3.19/3.10	3.15/3.09	3.12/3.08
4.5	3.13/3.19	3.09/3.08	3.06/3.00	3.04/2.96	3.01/2.93	2.98/2.91
5	3.01/3.09	2.97/2.98	2.93/2.89	2.90/2.84	2.87/2.79	2.85/2.77
5.5	2.90/3.01	2.85/2.89	2.81/2.79	2.79/2.74	2.76/2.68	2.73/2.65
6	2.79/2.93	2.75/2.82	2.70/2.71	2.68/2.65	2.65/2.58	2.62/2.54
6.5	2.70/2.86	2.65/2.75	2.61/2.63	2.58/2.57	2.55/2.50	2.52/2.45
7	2.61/2.79	2.56/2.68	2.52/2.56	2.49/2.50	2.46/2.42	2.43/2.38
7.5	2.53/2.73	2.48/2.62	2.44/2.50	2.41/2.44	2.38/2.36	2.35/2.31
8	2.46/2.68	2.41/2.56	2.36/2.44	2.33/2.38	2.30/2.30	2.27/2.24
8.5	2.39/2.62	2.34/2.51	2.29/2.39	2.26/2.32	2.22/2.24	2.20/2.18
9	2.32/2.57	2.27/2.46	2.22/2.34	2.19/2.27	2.16/2.19	2.13/2.13
9.5	2.26/2.52	2.21/2.41	2.16/2.29	2.43/2.22	2.09/2.14	2.07/2.08
10	2.20/2.47	2.15/2.36	2.10/2.25	2.07/2.18	2.03/2.09	2.01/2.03
10.5	2.15/2.43	2.09/2.32	2.04/2.20	2.01/2.14	1.97/2.05	1.95/1.99
11	2.10/2.39	2.04/2.28	1.99/2.16	1.95/2.09	1.92/2.01	1.89/1.95
11.5	2.05/2.35	1.99/2.24	1.93/2.12	1.90/2.05	1.87/1.97	1.84/1.91
12	2.00/2.31	1.94/2.20	1.88/2.09	1.85/2.02	1.82/1.93	1.79/1.87
12.5	1.95/2.27	1.89/2.16	1.84/2.05	1.81/1.98	1.77/1.89	1.74/1.84
13	1.91/2.23	1.85/2.13	1.79/2.01	1.76/1.95	1.72/1.86	1.70/1.80
13.5	1.86/2.20	1.81/2.09	1.75/1.98	1.72/1.91	1.68/1.83	1.65/1.77
14	1.82/2.17	1.76/2.06	1.71/1.95	1.67/1.88	1.64/1.80	1.61/1.74

Source: R. G. Packard. 1984. *Thickness Design for Concrete Highway and Street Pavements.* Skokie, IL: Portland Cement Association.

TABLE 12.15 Erosion Factors for Single and Tandem Axles (Aggregate Interlock Joints, Concrete Shoulder)

Slab thickness (in)	k of Subgrade-subbase (lb/in³) (single-axle tandem axle)					
	50	100	200	300	500	700
4	3.46/3.49	3.42/3.39	3.38/3.32	3.36/3.29	3.32/3.26	3.28/3.24
4.5	3.32/3.39	3.28/3.28	3.24/3.19	3.22/3.16	3.19/3.12	3.15/3.09
5	3.20/3.30	3.16/3.18	3.12/3.09	3.10/3.05	3.07/3.00	3.04/2.97
5.5	3.10/3.22	3.05/3.10	3.01/3.00	2.99/2.95	2.96/2.90	2.93/2.86
6	3.00/3.15	2.95/3.02	2.90/2.92	2.88/2.87	2.86/2.81	2.83/2.77
6.5	2.91/3.08	2.86/2.96	2.81/2.85	2.79/2.79	2.76/2.73	2.74/2.68
7	2.83/3.02	2.77/2.90	2.73/2.78	2.70/2.72	2.68/2.66	2.65/2.61
7.5	2.76/2.97	2.70/2.84	2.65/2.72	2.62/2.66	2.60/2.59	2.57/2.54
8	2.69/2.92	2.63/2.79	2.57/2.67	2.55/2.61	2.52/2.53	2.50/2.48
8.5	2.63/2.88	2.56/2.74	2.51/2.62	2.48/2.55	2.45/2.48	2.43/2.43
9	2.57/2.83	2.50/2.70	2.44/2.57	2.42/2.51	2.39/2.43	2.36/2.38
9.5	2.51/2.79	2.44/2.65	2.38/2.53	2.36/2.46	2.33/2.38	2.30/2.33
10	2.46/2.75	2.39/2.61	2.33/2.49	2.30/2.42	2.27/2.34	2.24/2.28
10.5	2.41/2.72	2.33/2.58	2.27/2.45	2.24/2.38	2.21/2.30	2.19/2.24
11	2.36/2.68	2.28/2.54	2.22/2.41	2.19/2.34	2.16/2.26	2.14/2.20
11.5	2.32/2.65	2.24/2.51	2.17/2.38	2.14/2.31	2.11/2.22	2.09/2.16
12	2.28/2.62	2.19/2.48	2.13/2.34	2.10/2.27	2.06/2.19	2.04/2.13
12.5	2.24/2.59	2.15/2.45	2.09/2.31	2.05/2.24	2.02/2.15	1.99/2.10
13	2.20/2.56	2.11/2.42	2.04/2.28	2.01/2.21	1.98/2.12	1.95/2.06
13.5	2.16/2.53	2.08/2.39	2.00/2.25	1.97/2.18	1.93/2.09	1.91/2.03
14	2.13/2.51	2.04/2.36	1.97/2.23	1.93/2.15	1.89/2.06	1.87/2.00

Source: R. G. Packard. 1984. *Thickness Design for Concrete Highway and Street Pavements.* Skokie, IL: Portland Cement Association.

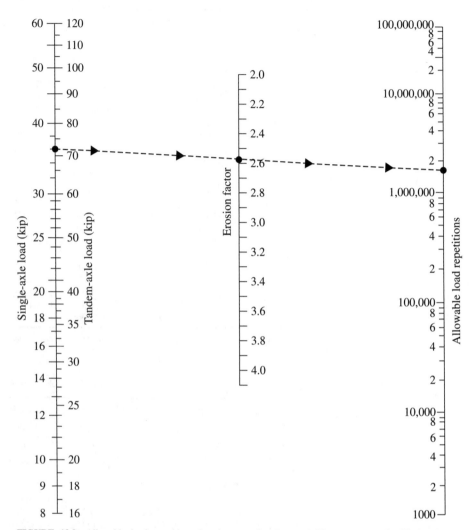

FIGURE 12.9 Allowable load repetitions based on erosion factors (without concrete shoulders). (*Source:* R. G. Packard. 1984. *Thickness Design for Concrete Highway and Street Pavements*. Skokie, IL: Portland Cement Association.

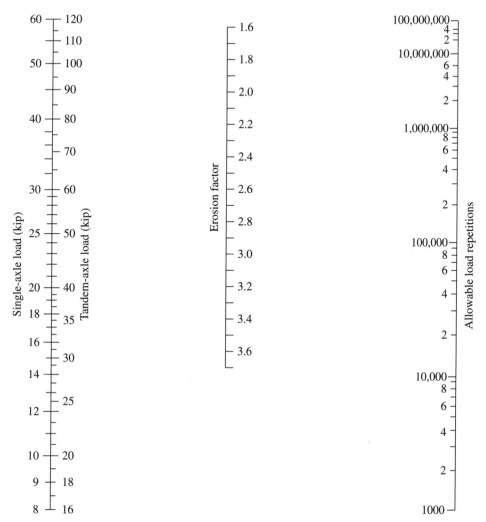

FIGURE 12.10   Allowable load repetitions based on erosion factors (with concrete shoulders). (*Source:* R. G. Packard. 1984. *Thickness Design for Concrete Highway and Street Pavements*. Skokie, IL: Portland Cement Association.)

CHAPTER 13
PAVEMENT TESTING AND EVALUATION*

Yongqi Li
Arizona Department of Transportation
Phoenix, Arizona

13.1 INTRODUCTION

A "good" pavement provides satisfactory riding comfort, structural integrity, and safe skid resistance. This chapter describes the major tests and evaluation procedures commonly used to assist pavement engineers to maintain pavements on their highway systems at such a level. These procedures evaluate pavement structural capacity, roughness, surface friction, and distress conditions.

Pavement structural capacity is an engineering concept to indicate the ability of pavement to carry the designed traffic loads adequately. It can be defined in terms of either the mechanical response (displacement, stress, and strain) of the pavement under simulated wheel-loading conditions or an index that characterizes the load-bearing abilities of pavement material layers such as structural number (AASHTO 1993). The evaluation of existing pavement structural capacity is required for the determination of pavement maintenance and rehabilitation (M&R) strategy and design. Pavement roughness reflects the traveling public's perception of the quality of the highways and is related to vehicle operating costs. It is defined as the distortion of the road surface, which contributes to an undesirable or uncomfortable and unsafe ride. Uses of roughness information include determining the need for pavement improvements from the user's perspective, identifying severely rough sections as M&R projects, and implementing a smoothness-based specification to measure the initial quality of the contractor's work. Skid resistance is the pavement surface's ability to resist sliding when braking forces are applied to the vehicle tires. Without adequate skid friction, the driver may not be able to retain directional control and stopping ability. The major reason for collecting skid resistance data is to identify and repair pavement sections with low level of friction and thus to prevent or reduce accidents. The friction data are a major factor affecting the prioritization of pavement M&R projects and the determination of the appropriate M&R strategies.

Pavement distress surveys are another important part of pavement evaluation. They record visible distresses, such as cracking, flushing, and patching. Distress surveys are needed for assessing the maintenance measures needed to prevent accelerated, future distress, or the rehabilitation strategies needed to improve the pavement. The distresses survey data, together with the test results of pavement structural capacity, roughness, and skid resistance, constitute the basic database of a pavement management system, which assists decision-makers of a highway agency in finding optimum strategies for providing and maintaining pavements in a satisfactory condition.

---

*Reprinted from the First Edition.

There have been many pavement-testing devices and evaluation methods, the majority of which are in the forms of ASTM or other organization's standards. This chapter describes only the basic principles and applications of those most widely used by highway agencies or major research facilities. For the detailed testing procedures readers should consult the published standards (ASTM 1999; AASHTO 1990; ISO 1998).

13.2   EVALUATION OF PAVEMENT STRUCTURAL CAPACITY

The testing techniques for pavement structural capacity include nondestructive deflection testing (NDT), full-scale accelerated pavement testing (APT), and destructive methods. In practice, NDT is more widely used than the other two approaches for many reasons, including convenience, efficiency, low cost, fewer traffic interruptions, less pavement damage, high measurement accuracy, and high reliability. NDT results are usually used to formulate pavement M&R strategies; determine overlay design thickness, load limits, and load transfer across joints (concrete pavements); and detect void and remaining structural life. Destructive tests are restricted to pavements showing severe evidence of distress (Hudson and Zaniewski 1994). APT is mainly used in academic circles (Highway Research Board 1962; Hass and Metcalf 1996).

13.2.1   Nondestructive Deflection Testing and Evaluation

Based on loading mode, deflection-measuring devices are categorized as static, steady-state dynamic, and impulse; impulse devices are the most widely used. Fewer highway agencies use steady-state dynamic devices. Automated static deflection-measurement equipment is used only in Europe (Croney and Croney 1991). The well-known Benkelman beam, a static device that played an important role worldwide in pavement research, evaluation, and design in the last four decades, now is mainly used for pavement research.

Static Equipment.   The Benkelman beam is typical of this type. It is a simple hand-operated deflection measurement device (Figure 13.1). It consists of a lever arm supported by an aluminum frame. It is used by placing the tip of the lever arm between the dual tires of a loaded truck at the point

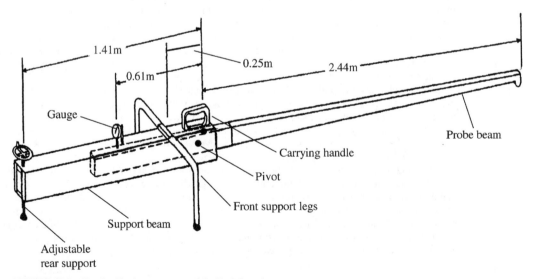

FIGURE 13.1   Sketch of basic components of the Benkelman beam.

where deflection is to be measured. As the loaded vehicle moves away from the beam, the rebound or upward displacement of the pavement is recorded by the dial gauge or an LVDT displacement transducer, which is installed at the rear end of the beam. When it is used with a pavement accelerated loading facility (ALF), the beam tip can be positioned at one side of the dual tires of the ALF. Thus, the beam will not block the movement of the wheels and the pavement deflections can be measured at any loading speeds. This is particularly useful for the investigation of the effect of loading speeds on the pavement response.

Because Benkelman beam testing is very slow and labor intensive some special vehicles mounted with automated deflection beam devices were developed based on the same principles as the Benkelman beam. The most common one of this type is the La Croix deflectograph, which is manufactured in France and has been widely used in Europe. The testing speed is approximately 2 to 4 km/hr. Neither the Benkelman beam nor the deflectograph measures the absolute deflection of the pavement because the beam supports in both cases are to some extent within the influence of the truck axles during testing.

Steady-State Dynamic Deflection Equipment.   This type of equipment induces a steady-state vibration to the pavement with a dynamic force generator, which can be either electromechanical or electrohydraulic. Pavement deflections are measured with velocity transducers. Figure 13.2 shows a typical force output. The dynaflect was one of the first devices of this kind available in the market (Smith and Lytton 1984). It is trailer-mounted and can be towed by a standard vehicle. Its electromechanical force generator has a pair of counter-eccentric masses, rotating in opposite directions at 8 Hz, to produce a cyclic force in the vertical direction. The static weight of the unbalanced mass is normally 907 kg and the produced peak-to-peak dynamic force of 4536 kN at 8 Hz is distributed between two rigid load wheels. The resulting deflection basin is measured by five geophones that are mounted on a placing bar at 0.3-m intervals. Other devices in this category include Road Rater and the U.S. Army Engineer Waterways Experiment Station vibrator. The steady-state dynamic deflection equipment is highly reliable. Their shortcoming is that they require a relatively large static preload and pavement resonance may affect deflection measurements.

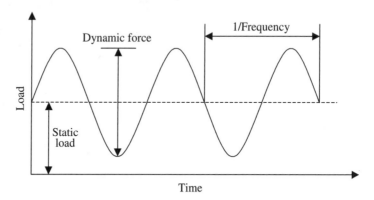

FIGURE 13.2   Typical vibrating steady-state force by Dynaflex.

Impulse Deflection Equipment.   Impulse deflection devices deliver a transient impulse load (Figure 13.3) to the pavement and the resulting pavement deflections are measured using either geophones or seismometer. The force is generated by a one-mass or two-mass (falling weight) system (Figure 13.4), which is raised to one or more predetermined heights and dropped with a guide system. These devices are typically referred to as falling weight deflectometers (FWD). The force magnitudes are mainly determined by the drop height and the weight of the mass and can vary from 13,620 kN to over 227,000 kN depending on the model used. FWD devices have

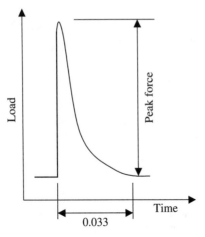

FIGURE 13.3 Typical load pulse produced by FWD.

relatively low static preloads, which eliminates the negative effects of a high preload usually associated with steady-state dynamic deflection equipment. The loading plate, mass system, type, and precision of deflection and load sensors, the location where the falling weight system is mounted at the vehicle, and other features of FWD vary from one manufacturer to another. The sensor spacing is adjustable. Typically, seven deflection sensors are located to measure the deflections at the center of the loading plate, 0.3, 0.6, 0.9, 1.2, 1.5, and 1.8 m from the center. The primary impulse-deflection equipment commercially available includes the Dynatest FWD, the JILS FWD, the Phonix FWD, and the KUAB FWD (Smith and Lytton 1984). To reduce the individual equipment error and the between-equipment error of the measurement, four regional FWD calibration centers have been established for the load cell and displacement sensors of a FWD to be calibrated regularly using the Strategic Highway Research Program SHRP procedure as the reference (FHWA 1993).

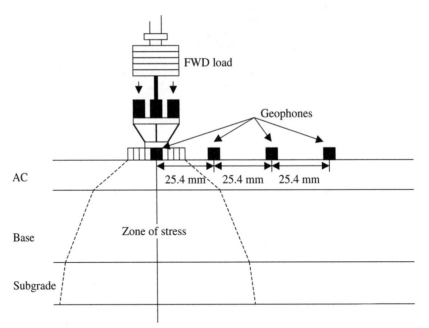

FIGURE 13.4 Schematic of stress zone within pavement structure under the FWD load. (*Source:* AASHTO 1993.)

Comparison between FWD and Other Deflection Devices. Theoretically, it is impossible to compare the deflections measured from different devices due to the difference of loading modes between FWD and other deflection devices and the intrinsic complexity of pavement systems' response to loading of different modes. From a practical standpoint, however, there may be situations where there are no alternatives but to develop the statistical correlation between different devices. In these cases, such correlations should be used cautiously. It should be kept in mind that the correlation equation is a function of pavement type, time of testing, material properties, and many other variables under which it was developed and it is impossible to estimate its accuracy on any given pavement. It has been suggested that the literature that documented the correlation equation should be tracked down to check the R^2 and the conditions under which it was derived to make the judgment whether the equation is suitable to the pavement and the conditions under which it will be used (FHWA 1994).

Influencing Factors of Measured Deflection. Load is the primary factor affecting measured deflection. The deflections measured from different types of devices inevitably differ when the peak load magnitude applied is equal. The same type of equipment also provides significantly different results at the same load magnitude if the load pulse shape and duration are different. The peak values of the center deflection can vary as much as 10 to 20 percent for an FWD (Royal Institute of Technology 1980).

Deflection measures the overall mechanical response of pavement system to load. Any factor affecting the mechanical properties of pavement layers has a direct impact on the measured deflection. For example, water content and high temperature can cause a dramatic reduction of the moduli of subgrade soil and asphalt layer, respectively, resulting in high measured deflection. Due to the limited pavement dimensions and variation of pavement distress, different testing locations in the same section can cause a significant difference in deflection measurement. For flexible pavement, deflections measured near cracks are normally much higher than the measurements in nondistressed areas (FHWA 1998). Similarly, deflection measurements near longitudinal joints, transverse joints, or corners are higher than those measured at mid-slab for concrete pavements. Thermal and moisture gradient in the vertical direction of concrete slabs cause curling and warping and have a significant influence on deflection measurements. Measurements taken at night or in the early morning are considerably different from those obtained in the afternoon.

In cold areas that experience freeze-thaw, the influence of season on deflection measurements shows a clear pattern (Figure 13.5; Scrivner et al. 1969). In the period of deep frost, the measured deflection is the lowest due to the frozen subgrade. The pavement is the weakest and the deflection

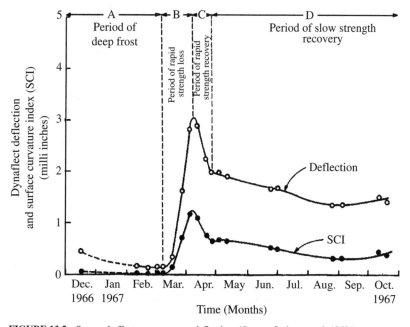

FIGURE 13.5   Seasonal effects on pavement deflection. (*Source:* Scrivner et al. 1969.)

increases rapidly in the thaw period, during which the frost begins to disappear and the subgrade becomes soft. When the excess of free water from the melting frost leaves the pavement, the soil begins to gain strength, rapidly at first and slowly afterwards.

Because deflection measurement is influenced by so many factors, it is important for an agency to develop a standard deflection test procedure. Ideally, deflection measurements should be conducted at approximately the same temperature and in the same season. If possible, adjustment factors should be applied to account for temperature and moisture variations. In cold areas, it is desirable to measure deflection during the spring thaw period, when the deflection measurements are at the maximum. It is particularly important to make sure that the measured deflections at different conditions are adjusted to the standard condition especially when pavement overlay design is based on measured deflections. In any case along with deflection measurements, the measurement locations, pavement condition, and temperature should be recorded.

Use of Measured Deflection.   The measured deflections are mainly used to determine overlay thickness design, material properties of pavement layers, load transfer across joints (concrete pavement), void detection, and layer bonding.

Overlay Thickness Design.   The commonly used methodology is to establish a regression relationship between the actual traffic loads and the pavement structural capability indices, which are represented by deflection basin parameters, and other factors under a defined criterion of the final condition of the pavement. The procedure used to design overlays in Arizona is typical of this methodology. Through a statistical analysis of the pavement performance data of several historical overlay projects, the following regression equation was established (Way et al. 1984):

$$\text{Log } L = 0.0587T(2.6 + 32.0 \ D5)^{0.333} - 0.104 \ \text{SVF} + 0.000578P_0 + 0.0653 \ \text{SIB} + 3.255 \qquad (13.1)$$

where L = design 18 kip ESALs
$\quad\quad T$ = overlay thickness, in.
$\quad\quad$ SVF = seasonal variation factor
$\quad\quad P_0$ = roughness before overlay, in./mi
$\quad\quad$ SIB = spreadability index before overlay. For the FWD, SIB = 2.7 (FWD SI)$^{0.82}$
$\quad\quad D5$ = #5 Dynaflect sensor reading, mils. For the FWD, $D5 = 0.16$ (FWD $D7$)$^{1.115}$
$\quad\quad$ SI = $(D_0 + D_1 + D_2 + D_3 + D_4 + D_5 + D_6) * 100/D_0$

The thickness is determined from the above regression equation as follows:

$$T = \frac{(\log L - 3.255) + 0.104 \ \text{SVF} + 0.000578P_0 - 0.0653 \ \text{SIB}}{0.0587(2.6 + 32.0 \ D5)^{0.333}} \qquad (13.2)$$

Like any other regression relationships, the validity of a design equation of this type is limited to the specific conditions and pavement types under which the equation was established. Usually the R^2 of the regression equation is very low.

Pavement Layer Moduli Back-Calculation.   Back-calculation is the process of estimating the fundamental engineering properties (elastic moduli or Poisson's ratio) of pavement layers and underlying subgrade soil from measured pavement surface deflections. This process can be illustrated using the point load case of Boussinesq's problem. Assuming that a semi-infinite medium is homogeneous, isotropic and linear elastic, the surface deflection D_0 at the center line of the circular uniformly distributed load p, is calculated from:

$$D_0 = \frac{2ap(1 - \mu^2)}{E} \qquad (13.3)$$

where a = radius of the circular area
$\quad\quad E$ = elastic modulus
$\quad\quad \mu$ = Poisson's ratio

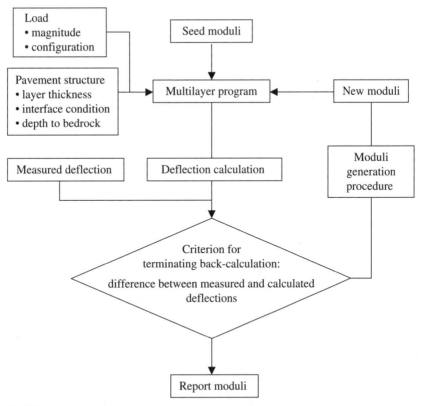

FIGURE 13.6 Common procedure of back-calculation programs.

Given the values of a, p, μ and measured D_0, E can be back-calculated by substituting the known values into equation (13.3).

For a typical pavement, which is usually a multilayer system (Figure 13.4), the equation for calculating pavement surface deflections is closed form. However, there is no closed-form solution to the back-calculation of the layer moduli from the measured deflections and other parameters as in Boussinesq's problem. The existing back-calculation programs determine pavement layer moduli by an iterative procedure (Figure 13.6; Lytton 1989). They search for the set of layer moduli from which the calculated deflections agree with the measured within a given tolerance limit. Different programs may use different moduli interactive procedures and tolerance limits. Examples of these programs include MODULUS, BISDEF, ELSDEF, and CHEVDEV. Elastic layered programs BISAR, ELSYM5, CIRCLY, and CHEVRON are generally used to calculate deflections. ILLI_ BACK is a back-calculation program for rigid pavements that uses Westergard's equation to calculate deflections.

It is very important to note that the back-calculated moduli do not necessarily reflect the fundamental property of the layer materials as the moduli determined in the laboratory do (Ullidtz and Coetzee 1995). A comparison study showed that not only are laboratory-determined moduli not identical in values to back-calculated moduli statistically but there is literally no correlation between the two at all with the R^2 of 0.013 (Figure 13.7; Mamlouk et al. 1988). The significant difference between the two moduli may be partly caused by the factors that affect the accuracy of laboratory-determined moduli, such as disturbance of the samples and nonrepresentive stress condition. However, the major source of this discrepancy is related to the back-calculation procedure itself. First, the elastic layered model used in back-calculation to calculate deflections is based on the

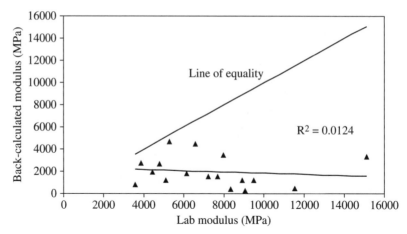

FIGURE 13.7 Correlation between lab and back-calculated moduli. (Produced from the data in Mamlouk et al. 1988.)

assumption that the pavement materials conform to an idealized condition of being linearly elastic, uniform, and continuous. This assumption ignores discontinuities (cracks, voids), nonlinearity and plasticity of materials, material variability, the effects of temperature gradients, and many other factors. This inevitably causes errors of the back-calculated moduli. Furthermore, the analytical solution of deflection itself does not guarantee that there is a unique solution of layer moduli for a given set of deflections. In other words, there is no solid theoretical base for the back-calculation of moduli from measured deflections for a typical pavement.

Therefore, it is critically important to note that back-calculation procedure can by no means prove its accuracy itself. A set of back-calculated moduli that provide a deflection basin perfectly matching the measured one does not necessarily mean that they are correct or accurate. In fact, the only way to evaluate the reliability of back-calculated moduli is to use the directly measured moduli as the reference.

Joint Load Transfer. When a load is applied near a transverse joint, both the loaded and the unloaded sides deflect because a portion of the load that is applied to the loaded side is carried by the other side through load transfer at the joint. Efficient load transfer can reduce the edge stresses and deflections, thus reducing fatigue damage and minimizing pumping. Therefore, load transfer at the joint or crack is essential for satisfactory performance of rigid pavement. Deflection testing can be used to measure the load transfer efficiency. The test can be conducted by using the FWD, with the loading plate positioned adjacent to the joint and a deflection sensor located across the joint (Figure 13.8). The load transfer efficiency (LTE) can be calculated as follows (FHWA 1998):

$$LTE = (\delta_u / \delta_l) * 100 \tag{13.4}$$

where δ_u is deflection on unloaded side of joint and δ_l is deflection on loaded side of joint. The deflection load transfer of the joint can be evaluated approximately by the following scale (FHWA 1998):

Good: greater than 75 percent

Fair: 50–75 percent

Poor: less than 50 percent

The test result of load transfer is greatly affected by temperature. When the temperature increases, the adjacent concrete slabs expand and more contact between them occurs at the joint, causing the load transfer to increase. When the temperature decreases, the slabs contract and the joint opens,

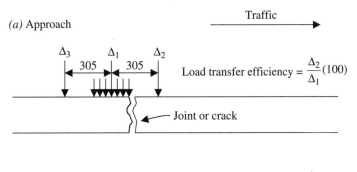

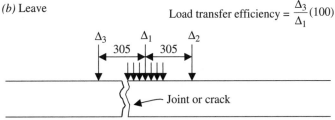

FIGURE 13.8 Arrangement of deflection sensors for determining load transfer efficiency. (After FHWA 1994, 5-29).

causing the load transfer between slabs to decrease. Studies show that on a summer day when the pavement temperature rises substantially, load transfer at the same joint may increase from 50 percent in the morning to 90 percent in the afternoon. Testing is usually recommended to be performed when the joint is open. It should be noticed that the LTE calculated from equation (13.4) is less than 100 percent in most cases but can be greater than 100 percent in some conditions.

Detection of Voids. Pumping and the subsequent voids beneath the slab are one of the major causes of deterioration of concrete pavements. Deflection testing can be conducted to detect voids below slab corners. The common techniques include corner deflection profile and variable load corner deflection analysis (AASHTO 1993).

In the first method, the deflection at the approach and leave corners are measured and plotted. The corners that exhibit the lowest deflections are expected to have full support value. Therefore, if the deflection at a corner is significantly greater than the lowest value, it can be concluded that voids exist at the corner. In the second method, the corner deflections are measured at three load levels and the load-deflection curve is plotted (Figure 13.9). Studies show that for the locations with no voids, the intercept at the deflection axis is very near the origin (less than 50 μm). Thus, if the intercept at the deflection axis is significantly greater than 50 μm it can be suspected that there are voids at the corner. Because slab curling caused by temperature gradient and moisture substantially affect the deflection measurement, the deflection testing to detect voids should be conducted at the appropriate time to eliminate or reduce the influence of slab curling.

Detection of Loss of Bonding. Consider a composite pavement system as illustrated in Figure 13.10. A wheel load $p(r)$ acted upon this system. It is assumed that the two Poisson ratios are equal in magnitude. Using the plate theory on composite section, the equivalent flexural rigidity can be computed as follows (Li and Li):

$$D_e = \frac{E_1 h_e^3}{12(1-\mu^2)}$$

(13.5)

where h_e is the equivalent thickness with respect to the type of interface.

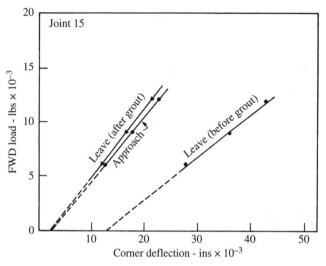

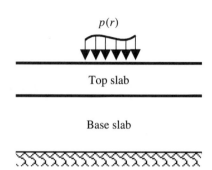

FIGURE 13.9   Variable load corner deflection analysis for void detection. (*Source:* AASHTO 1993.)

FIGURE 13.10   Composite pavement structure. (*Source:* Li and Li.)

For an unbonded interface, the equivalent thickness is expressed as

$$h_e = h_1 \left[1 + \frac{E_2}{E_1} \left(\frac{h_2}{h_1} \right)^3 \right]^{1/3} \tag{13.6}$$

where h and E are the thickness and elastic modulus of the corresponding layer material, respectively, as shown in Figure 13.11*a*.

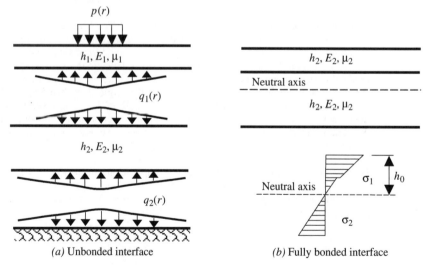

(a) Unbonded interface   (b) Fully bonded interface

FIGURE 13.11   Analytical approaches to unobonded and fully bonded slabs. (*Source:* Li and Li.)

For a fully bonded interface, the equivalent thickness is computed below:

$$h_e = h_1 \left[1 + \frac{E_2}{E_1}\left(\frac{h_2}{h_1}\right)^3 + \frac{3\dfrac{E_2}{E_1}\dfrac{h_2}{h_1}\left(1+\dfrac{h_2}{h_1}\right)^2}{\left(1+\dfrac{E_2}{E_1}\dfrac{h_2}{h_1}\right)} \right]^{1/3} \tag{13.7}$$

where all variables are as defined earlier.

Examination of the two equations indicates that the determination of equivalent thickness includes two same terms, that is, $[1 + (E_2/E_1)(h_2/h_1)^3]$, for both the unbonded and fully bonded interfaces. The only difference is the third term on the right side in equation (13.7), which accounts for the effect of the bonding across the interface. The contribution of a fully bonded interface is an increase in the equivalent thickness in terms of the third term in equation (13.7). Based on this observation, the coefficient of bonding K is introduced as follows:

$$h_e = h_1 \left[1 + \frac{E_2}{E_1}\left(\frac{h_2}{h_1}\right)^3 + K\frac{3\dfrac{E_2}{E_1}\dfrac{h_2}{h_1}\left(1+\dfrac{h_2}{h_1}\right)^2}{\left(1+\dfrac{E_2}{E_1}\dfrac{h_2}{h_1}\right)} \right]^{1/3} \tag{13.8}$$

For an unbonded interface, $K = 0$. For a fully bonded interface, $K = 1.0$. K is extended to characterize a partially bonded interface by assuming that K is a number varying from zero to one. By substituting equation (13.8) into equation (13.5) and subsequently substituting equation (13.5) into Westergard's equation, the deflection can be related to the coefficient of bonding K.

13.2.2 Destructive Structural Evaluation

The techniques used for destructive pavement evaluation generally involve taking cores or cutting trenches transversely across pavements. The objectives are usually to remove samples for inspection and testing and to diagnose the causes and mechanisms of pavement failures. Because of the extreme complexity and variability of pavement failures, sometimes postmortem is the only way to review what really happens inside pavements. Such evaluation procedures are used mostly on test roads and occasionally on in-service pavements.

Trenches were cut at the AASHO test road and other APT test sites to investigate the permanent deformation accumulated at the top of each of the structural layers (Highway Research Board 1962; Sharp 1991; Li et al. 1999). At the AASHO test road, it was found that rutting was mainly due to decrease in thickness of the pavement layers, which was caused by lateral movement of the materials. Postmortem was conducted on the asphalt pavements with cement-treated base (CTB) at Australia's Accelerated Loading Facility (ALF). It was found that the failure mechanism of the CTB pavements was debonding of the CTB layers under the repeated loading, followed by the penetration of water and subsequent erosion at the layer interfaces, eventually causing the break-up of the top CTB layer (Figure 13.12; Sharp 1991).

13.2.3 Full-Scale Accelerated Pavement Testing (APT)

"Full-scale accelerated pavement testing is defined as the controlled applications of a prototype wheel loading, at or above the appropriate legal load limit to a prototype or actual, layered, structural pavement system to determine pavement response and performance under a controlled, accelerated,

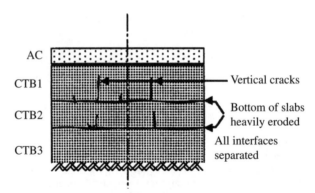

FIGURE 13.12 Typical failure mode of CTB pavements. (*Source:* Sharp 1991.)

accumulation of damage in a compressed time period" (Metcalf 1996). The acceleration of pavement damage process is achieved by means of increased load magnitude and frequency, imposed climatic conditions, thinner than standard pavements, or a combination of these factors. Based on the length of tested pavements and loading facilities, full-scale APT can be broadly classified into two groups: test roads and test tracks (Metcalf 1996). Test roads usually have a long loop of tested pavement (several miles) with several sections; loading of traffic is accomplished by either directed actual traffic or calibrated and controlled test trucks. An example of this group includes the well-known Road Test conducted between 1958 and 1960 by the AASHO (Highway Research Board 1962). Other more recent ones include MnRoad (Minnesota), WesTrack (Nevada), and NCAT Track (Alabama) (Epps et al. 1999; Harris, Buth, and Van Deusen 1994; Brown and Powell 2001). Test tracks have a shorter test pavement and a specially designed loading facility, which is of circular, linear, or free-form layout. Linear tracks have been widely used in the United States, including the Australian-designed Accelerated Loading Facility (ALF) (Figure 13.13) and the South African-designed and manufactured Heavy Vehicle Simulator (HVS) (Sharp 1991; Harvey, Prozzi, and Long 1991).

APT is an essential part of pavement research strategy. It also has a specific application to modifying existing design for conventional materials to heavier traffic, evaluating new pavement materials and structural designs, investigating pavement failure mechanisms, estimating remaining pavement life, and evaluating environmental effects. A successful application of APT was the AASHO test

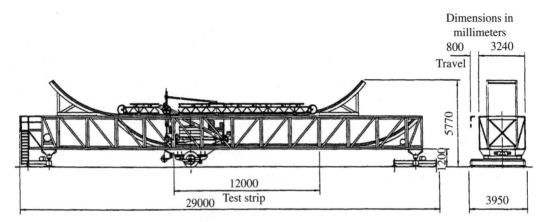

FIGURE 13.13 Schematic diagram of ALF. (*Source:* Sharp 1991.)

road, which resulted in the AASHTO Pavement Design Guide (AASHTO 1993). Most of the recent APT programs in the United States have focused on evaluating the SHRP Superpave mix design, innovative mix design, and the relationship between pavement roughness and fuel consumption.

APT facilities usually have instrumented test pavements to measure pavement responses under accelerated loading, such as the horizontal strain at layer interfaces, vertical pressures at the surface of subgrade, and displacement at multiple depths. Climate data are also collected. The survival rate and reliability of the instruments (strain gauges, pressure cells, displacement gauges, and moisture gauges) are usually very low due to the debonding between the instruments and measured pavement materials. It is also a challenging task to analyze and interpret the data from these gauges.

13.3 EVALUATION OF PAVEMENT ROUGHNESS

Roughness is an indicator of a car driver's perception of pavement riding comfort and safety. The measurement of roughness depends primarily on the vertical acceleration experienced by a driver. Vertical acceleration in turn depends on three factors: the pavement profile, the vehicle mass and suspension parameters, and the travel speed. Based on the ways to record and characterize vertical acceleration, roughness measurement methods are classified as response-type and profile-type. In a response-type system, a transducer installed on a passenger vehicle or a special trailer measures and accumulates deflections of the vehicle suspension as it travels on the road. The vertical movement of accumulated suspension stroke, normalized by the distance traveled, is used to quantify roughness. On the other hand, instead of recording the actual vertical movements of a real vehicle's suspension directly, a profile-type roughness measurement system records the pavement longitudinal profile. It then runs a mathematical model, which simulates the suspension response of a real vehicle moving on the road, to calculate the simulated suspension motion and normalize it by the traveled distance to give profile index. This profile index is used to quantify pavement roughness. Profile-type roughness measurement systems have many advantages over response-type systems and are the most widely used nowadays.

13.3.1 Response-Type Road Roughness Measuring Systems (RTRRMS)

The Bureau of Public Roads (BPR) Roughometer (Figure 13.14), was the earliest mechanical RTRRMS (Hass, Hudson, and Zaniewski 1994). It is a trailer-mounted system. The trailer, with a heavy rectangular chassis, is supported on a central wheel. The wheel is supported by two single-leaf springs positioned one on each side of the wheel. Two dashpot-damping assemblies are installed between the chassis and the wheel axle. With the appropriately selected parameters of the mass,

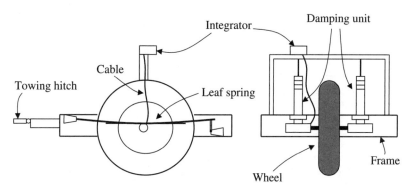

FIGURE 13.14 Sketch of basic components of BPR Roughometer. (After Sayers and Karamihas 1998.)

dashpot damping unit, and spring, the trailer approximately represents one-quarter of a passenger car. In operation, the differential movement between the wheel axle and the chassis is summed by a mechanical integrator unit installed on the chassis. The roughness index is given as the integrated vertical movements divided by the traveled distance, with a unit of m/km. Other RTRRMS systems following the concept of the BPR Roughometer include the TRRL bump integrator, Australia's NASRA meter, and the Mays Ride Meter. The last was the most popular in the United States before profilers were widely adopted.

The other type of RTRRMS uses an accelerometer as the primary motion sensor. The accelerometer can be mounted on either a special trailer or the vehicle body or axle. The output of the accelerometer can be integrated twice to obtain the vertical movement. The integration process can magnify the effect of noise signal. The axle-mounted accelerometers are not as sensitive to the vehicle parameters as the mechanical RTRRMS. Generally, the accelerometer-based RTRRMS uses the summarized root mean square value of acceleration (RMSA) to quantify pavement roughness (Hass, Hudson, and Zaniewski 1994).

Because each and every factor that influences vehicle response to the pavement surface impacts measures from response-type systems, these systems are host vehicle-dependent and inevitably have two problems. First, they are not stable with time. Measures made at different times cannot be compared with confidence. Second, they are not transportable, meaning that measures from different equipments on the same pavement are seldom reproducible. In practice, response-type systems have to be regularly calibrated by the measures from a valid profiler (Gillespie, Sayers, and Segel 1980).

13.3.2 Profile-Type Road Roughness Measuring Systems

Using a profiler to measure pavement roughness involves two steps. The first step is to measure the pavement longitudinal profile. The second step is to run a mathematical model against the profile to simulate a vehicle's suspension response to the actual pavement. Mathematical models simulating different roughness measurement devices result in different roughness indices. The most widely used one is International Roughness Index (IRI), which comes from a mathematical model simulating BPR-like devices (Sayers, Gillespie, and Paterson 1986; Sayers 1995).

Profile Measurement.   The rod and level, familiar to most civil engineers, can be used to measure pavement profile. However, the accuracy requirements for measuring a profile valid for computing roughness are much more stringent than is normal for road surveying. The elevations must be taken at close intervals of 0.3 meters or less, and the individual height measures must be accurate to 0.5 millimeters or less. Therefore, using the rod and level method to measure a profile for roughness is extremely slow and labor-intensive. One alternative to road and level equipment is Dipstick, which is faster than road and level for measuring profiles suited for roughness analysis. When it is "walked" along the line being profiled, a precision inclinometer installed in the device measures the differential height between the two supports, normally spaced 0.3 meters apart. The elevation at the second support is calculated from that at the first support.

The rod-and-level method and Dipstick measure the real profile and are usually used as a reference to check the accuracy of other profiler. The most widely used equipment for routine pavement profiling is the inertial profiler (Figure 13.15; Karamihas and Gillespie 1999). As the vehicle moves at a certain speed, an accelerometer mounted in the vehicle to represent the vertical axis of the vehicle measures vertical acceleration of the vehicle body. The accelerations are integrated twice to obtain the vertical movement d_1 of the vehicle body. A transducer is used for measuring the distance between the pavement surface and the vehicle d_2. The elevations of the pavement surface are obtained by adding d_2 to d_1. A distance odometer measures the distance along the pavement. The on-board data acquisition system records the data to obtain the profile files for real time or postroughness analysis. The inertial profilers not only work at normal highway speeds, but they require a certain speed (usually >15 km/hr) to function.

The profile from an inertial profiler does not look like one measured statically from either the road and level or Dipstick. For example, Figure 13.16 shows profiles measured from the Dipstick

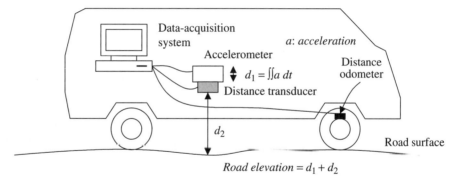

FIGURE 13.15   Schematic of inertial profiler. (After Karamihas and Gillespie 1999.)

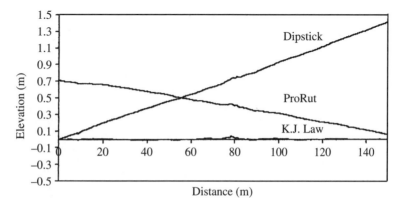

FIGURE 13.16   Unfiltered profiles.

and two inertial profilers. The different appearances are mainly due to the fact that neither of these devices accurately measures the grade and long undulations, which, for roughness analysis, we are not interested in. In order to make practical use of a profile, we have to remove the components of the grade and long undulations from the measured profile. In fact, after the grade and long undulations have been removed, the profiles from different devices match very closely to each other. They all present the pavement unevenness in such a scale that it causes severe disturbance to the public driving over it. Figure 13.17 shows the same profiles in Figure 13.16 after the road grade and long waves have been removed.

Removal of the grade and long undulations involves filtering techniques. A moving average is a simple filter commonly used in roughness analysis of profiles. It replaces each profile point with the average of several adjacent points. For a profile that has been sampled as interval ΔX, the filter is defined as

$$p_{fL}(i) = \frac{1}{2n+1} \sum_{j=i-n}^{i+n} p_{(i)} \qquad (13.9)$$

where $p_{(i)}$ and $p_{fL(i)}$ are the values of the original and filtered profile point at ith location, respectively, B is the base length of the moving average, and $B = 2n\,\Delta X$. The effect of a moving average filter is to smooth the profile by averaging out the point-by-point fluctuations. More importantly, it can also

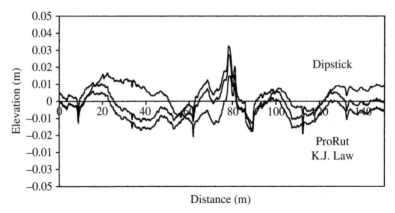

FIGURE 13.17   Profiles filtered with a 91-meter high-pass filter.

be used to filter out the smoothed components from the original profile to obtain the deviations from the smoothed profile. The filter to achieve this is to subtract the smoothed profile from the original:

$$p_{fH(i)} = p_{(i)} - p_{fL(i)} \qquad (13.10)$$

The filter $p_{fH(i)}$ is usually called a high-pass filter, while the filter $p_{fL(i)}$, which is used to smooth a profile, is usually called a low-pass filter. When the sampling interval ΔX is very small (<25.4 millimeters) a low-pass filter with a small B (about 0.3 meters) is usually applied to the original profile to average out the fluctuation within the small range in which we are not interested. Then a high-pass filter with a relatively large B (30 to 90 meters) is applied to the original or smoothed profile. Figure 13.17 is the resulting profiles after a high-pass filter with a base length of 90 meters is applied to the profiles shown in Figure 13.16. A high-pass filter is mandatory for roughness analysis, but a low-pass filter is not. It is the high-pass filtered profile that is used to calculate profile indices.

IRI.   The IRI is a profile index based on a mathematical model called a quarter-car (Figure 13.18; Karamihas and Gillespie 1999). The quarter-car model calculates the suspension deflection of a simulated mechanical system with a response-to-pavement profile similar to that of a passenger car. The IRI is defined as the accumulated suspension motion from the quarter-car with a set of specified parameters moving on an appropriately filtered profile divided by the distance traveled. It is also expressed in unit of m/km. The computer program to calculate the IRI from a profile can be found elsewhere (Sayers, Gillespie, and Querioz 1986).

    The IRI originally came out of the effort of the World Bank in 1982 to search for a calibration standard for roughness measurements, especially from response-type devices (Sayers, Gillespie, and Querioz 1986). Since the IRI is essentially a computer-based virtual response-type system, it has a good correlation with the roughness indices from response-type systems. In addition, the IRI is defined as a property of a profile and independent of profilers as long as they provide valid profiles. It is reproducible, portable, and stable with time. The IRI is an ideal pavement condition indicator. It captures the roughness qualities of a pavement profile that impact vehicle response. Figure 13.19 shows IRI ranges represented by different classes of road (Sayers and Karamihas 1998). Currently most states in the United States use inertial profiler as the profile measurement equipment and IRI as roughness index. Other profile indices are not in broad use because there is little reason to use them when IRI can be conveniently calculated.

    It should be noted that the accuracy of an IRI measurement depends solely on the accuracy of profile measurement. A profile measurement used for IRI calculation has to be particularly accurate in a range of wavelengths from 1.3 to 30 m. The guidelines for measuring a longitudinal pavement profile to use in computing IRI can be found elsewhere (Karamihas and Gillespie 1999).

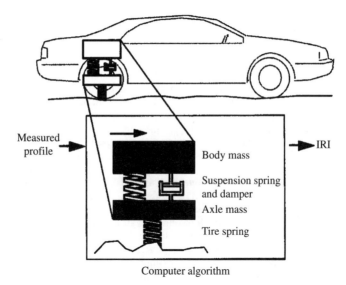

FIGURE 13.18 Quarter-car model for IRI. (*Source:* Karamihas and Gillespie 1999.)

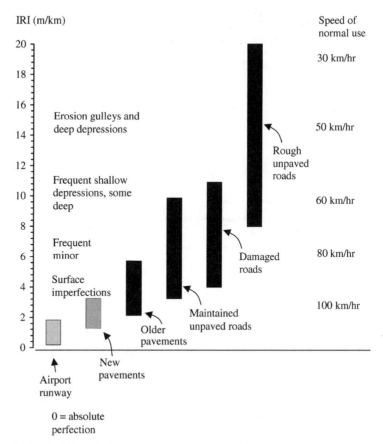

FIGURE 13.19 IRI of different classes of road. (After Sayers and Karamihas 1998.)

13.3.3 Profilograph

Profilographs have been widely used for evaluating the smoothness of concrete pavements during construction. The design of these devices is all based on the same principle, as shown in Figure 13.20. These devices consist of bogie wheel sets at front and rear, a recording wheel at the center, and a recorder to record the vertical movement of the recording wheel. A profilograph actually performs the similar function as a high-pass moving average filter does. The average is established by the bogie wheels, and deviations are measured relative to the average, which is not for points on the same path as in the case of profiles. The profilogram from a profilograph is usually recorded on a scale of 1:300 longitudinally and 1:1 vertically. In the following, the Arizona method to determine the profile index (PrI) and locate individual high areas that need to be grinded from the profilogram is described (Arizona DOT 1999).

The device for determining the PrI is a plastic scale 1.7 inches wide and 21.12 inches long, representing a pavement length of one tenth of a mile. Near the center of the scale is an opaque band 0.2 inch wide extending the entire length. On either side of this band are scribed lines 0.1 inch apart parallel to the opaque band. These lines, called scallops, serve as a scale to measure deviations of the graph above or below the blanking band. The plastic scale should be placed over the profile in such a way as to blank out as much of the profile as possible. When it is impossible to blank out the central portion of the trace, the profile should be broken in short sections (Figure 13.21). The height of all the scallops appearing both above and below the blanking band should be measured and totaled to the nearest 0.05 inch. Only the scallops ≥0.03 inch high and ≥0.08 inch long (2 feet pavement length) are included in the count. The PrI is defined as the total count of tenths of an inch divided by 10 times the miles of profile.

The device to determine scallops in excess of 0.3 inch is a plastic template having a line 1 inch long scribed on one face with a small hole at either end and a slot 0.3 inch from the parallel to the scribed line (Figure 13.22). The 1-inch line corresponds to a pavement distance of 25 feet. At each prominent peak on the profile trace, the template should be placed so that the small holes intersect the profile trace to form a chord across the base of the peak or indicated bump. The line on the

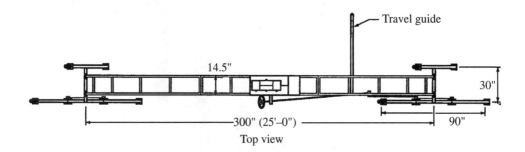

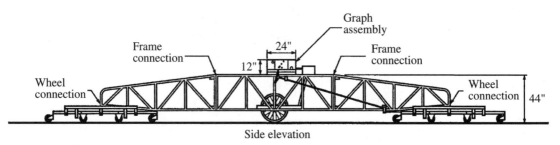

FIGURE 13.20 Schematic of a profilograph.

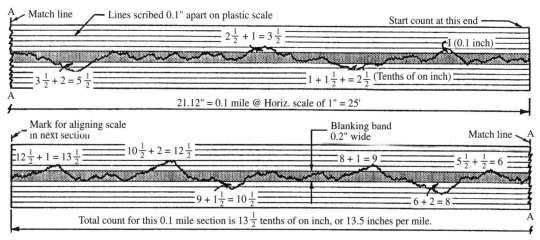

FIGURE 13.21 Example of showing method of deriving profile index from profilograms. (*Source:* Arizona Department of Transportation 1999.)

template need not be horizontal. A line should be drawn using the narrow slot as a guide. Any portion of the trace extending above this line indicates the length and height of the deviation in excess of 0.3 inch. When the distance between easily recognizable low points is less than one inch, a shorter chord length should be used in making the scribed line on the template tangent to the trace at the low points. The baseline for measuring the height of bumps should be as near 25 feet (1 inch) as possible, but in no case should exceed this value.

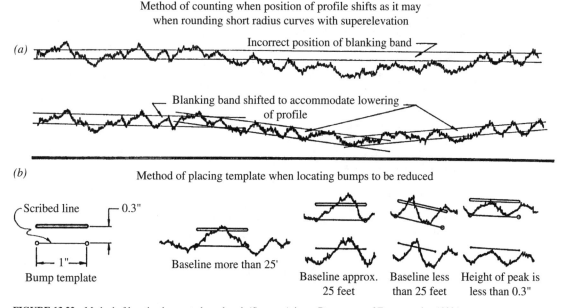

FIGURE 13.22 Method of locating bumps to be reduced. (*Source:* Arizona Department of Transportation 1999.)

13.4   *EVALUATION OF SKID RESISTANCE*

The coefficient of friction is obtained by dividing the frictional force in the plane of interface by the force normal to the plane. For pavements, the coefficient of friction measured when the tire is sliding on the wet pavement is usually called skid resistance. Skid resistance is very important to driving safety. Without adequate skid resistance the driver may not be able to retain directional control and stopping ability and an accident may occur. Skid resistance measurement is a major pavement test.

It has been demonstrated that the skid resistance decreases as the velocity of the tire surface relative to pavement surface increases (Moyer 1934). This relative velocity is called the slip speed. There are four types of pavement skid resistance measuring devices: locked wheel, side force, fixed slip, and variable slip. They measure the skid resistances at different slip speeds. The locked-wheel method simulates emergency braking without an antilocking system, the side-force method measures the ability to maintain control in curves, and the fixed-slip and variable-slip methods concern braking with antilock systems. The locked-wheel method is most widely used in North America. Outside the United States, side-force and fixed-slip methods are commonly used.

The locked-wheel system measures friction at a 100 percent slip condition in which the slip speed is equal to the vehicle speed. After the test wheel is locked and has been sliding for a suitable distance, the frictional force is measured and the skid number, SN, is calculated by multiplying the friction value by 100. Fixed-slip devices measure skid resistance at a constant slip, usually between 10 and 20 percent. The test wheel is driven at a lower angular velocity than its free-rolling velocity. These devices measure low-speed skid resistance. Variable-slip devices measure skid resistance at a predetermined set of slip ratios. Some locked-wheel testers can operate in a mode in which they measure the friction as the test wheel changes from free-rolling to the fully locked condition (0 to 100 percent slip). Side-force devices measure skid resistance in the yaw mode, where the wheels are turned at some angle to the direction of travel. The sideways frictional force varies with the magnitude of yaw angle, and it is desirable to use an angle that is relatively insensitive to small variations. The yaw angle is relatively small and the slip speed for these devices, approximately $V \sin \alpha$ (where V is the vehicle speed and α is yaw angle), is therefore also small even though the vehicle speed is high. A fairly simple and widely used version of a yaw mode device is the Mu-Meter.

Another machine widely used to assess skid resistance is the Pendulum Tester, which was developed by the British Transport and Road Research Laboratory (Biles, Sabey, and Cardew 1962). It involves dropping a spring-loaded rubber shoe attached to a pendulum. The results are reported as British Pendulum Numbers (BPN). It is mainly used for laboratory testing and areas such as intersections. Also, because the shoe contacts the pavement at a relatively low speed, the results do not correlate well with locked-wheel trailer test results conducted at 40 mph.

Correlations among various full-scale skid measurement devices are usually very poor, largely because each device measures a different aspect of the frictional interaction between vehicle and pavement (Hass, Hudson, and Zaniewski 1994). For example, side-force devices and fixed-slip devices are low-slip-speed systems and therefore are sensitive to microtexture but insensitive to macrotexture, whereas locked-wheel testers are high-slip systems and mostly reflect the effect of macrotexture (Leu and Henry 1983).

Having recognized the dependence of friction measurement on slip speed researchers have recommended that macrotexture depth be measured with skid resistance. In 1992, the World Road Association (formerly PIARC) conducted extensive pavement friction and texture tests with various types of devices. As a result of these tests, the International Friction Index (IFI) was proposed (PIARC 1995). The IFI is composed of two numbers that jointly describe the full feature of the skid resistance of a pavement: the speed constant (S_p) and the friction number (F60). The speed constant reflects the macrotexture characteristics and is linearly related to a macrotexture measurement (TX). The friction number (F60) is determined from a measurement of friction by:

$$\text{F60} = A + B \text{ FRS } e^{S-60/S_p} + C\text{TX} \tag{13.11}$$

where FRS = measurement of friction by a device operating at a slip speed (S)
   A, B, C = constants determined for a specific friction device.

The value of C is zero for a smooth-tread tire because the measured friction already reflects the effect of macrotexture. However, the term of C TX is needed for ribbed or patterned test tires because they are relatively insensitive to macrotexture. The values of these constants for the devices in the experiment are reported in ASTM Standard Practice E-1960. One advantage of the IFI is that F60 and S_p are sufficient to describe the friction as a function of slip speed. Another advantage is that F60 will be the same regardless of the slip speed. This allows any device operate at its safe speeds in different conditions and reported on a common scale.

The traditional device to measure pavement macrotexture is the sandpatch method (ASTM Standard Practice E-965). It requires spreading a specified volume of Ottawa sand or glass spheres of a specific sieve range in a circular motion. The mean texture depth (MTD) is obtained by dividing the volume by the area of the circular patch. Another device for characterizing macrotexture is the outflow meter (Henry and Hegmon 1975). This is a transparent cylinder that rests on a rubber annulus placed on the pavement. When the measurement is conducted, the cylinder is filled with water and the valve at the bottom of the cylinder is then opened. The time for the water level to fall by a fixed amount is measured. The time is reported in seconds as the outflow time (OFT). The OFT reportedly has a good correlation with MTD (Henry 2000). The most recent method is to evaluate pavement macrotexture from pavement profile measured by laser-based profilers. The measured pavement profile is used to calculate the mean profile depth (MPD) (ASTM Standard Practice E-1845; ISO Standard 13472). It was found that the MPD is the best macrotexture measure for the prediction of skid resistance (PIARC 1995).

In practice, skid-resistance evaluation, especially for the purpose of assessing pavement future rehabilitation needs, should consider changes on a time and/or traffic basis, as well as on a climatic effect basis. The latter can involve both short and longer periods of time (i.e., rainfall or icing of a short duration versus seasonal climatic changes). On a short-term basis, skid-resistance changes can occur rapidly, usually because of rainfall. On a somewhat longer, seasonal basis, skid resistance may fluctuate. On a still longer basis of, say, several years or several million vehicle passes, most pavements show a continual decrease of skid resistance.

13.5   PAVEMENT DISTRESS SURVEY

Distress surveys measure the type, severity, and density, or extent and location, of various pavement surface distresses. For asphalt pavements, the survey usually includes cracking, patching, potholes, shoving, bleeding, rutting and raveling, etc. The survey for concrete pavements measures cracking, joint deficiencies, various surface defects, and miscellaneous distresses. A pavement distress survey must be conducted following a well-developed survey manual to ensure the consistency and uniformity of the survey data. A manual includes a clear definition of the distress type and a definite procedure to determine or calculate the severity and density of each distress (Shabin and Kohn 1979; Strategic Highway Research Program 1993). Currently, most pavement distress surveys are performed walking along the road or from a moving vehicle. Walking surveys provide the most precise data of the distresses and are generally used for project review. Driving along the road, usually on the shoulder, at slow speed is suitable for pavement network survey. In recent years, several types of automated distress survey equipment based on film, video, or laser technology have been developed. Some agencies have started using these kinds of automated survey equipment (Hass, Hudson, and Zaniewski 1994).

13.6   SUMMARY

Almost all pavement tests are empirically based, and the results produced by the various devices and test methods are affected by many factors. Thus, it is important to follow the standard methods for the calibration and use of the devices. However, the discussion in this chapter focuses primarily on the principles of the test methods; it is not intended to be used as a substitute for reading and following the specifications of the standard test methods. Readers should consult officially published standards, such as the *Book of ASTM Standards,* for detailed information.

REFERENCES

American Association of State Highway and Transportation Officials (AASHTO). 1990. *Standard Specifications for Transportation Materials and Methods of Sampling and Testing.* Washington, DC: AASHTO.

———. 1993. *AASHTO Guide for Design of Pavement Structures.* Washington, DC: AASHTO.

American Society for Testing and Materials (ASTM). 1999. *Annual Book of ASTM Standards.* West Conshohocken, PA: ASTM.

Arizona Department of Transportation. 1999. *ADOT Materials Testing Manual.* Materials group, Intermodal Transportation Division, Arizona Department of Transportation, Phoenix.

ASTM Standard Test Method E-965. "Measuring Pavement Macrotexture Depth Using a Volumetric Technique." *Book of ASTM Standards,* vol. 04.03, ASTM, West Conshohocken, PA.

ASTM Standard Practice E-1845. "Calculation Pavement Macrotexture Profile Depth." *Book of ASTM Standards,* vol. 04.03. ASTM, West Conshohocken, PA.

ASTM Standard Practice E-1960. "Calculating International Friction Index of a Pavement Surface." *Book of ASTM Standards,* vol. 0403, ASTM, West Conshohocken, PA.

Biles, C. G., B. E. Sabey, and K. H. Cardew. 1962. *Development and Performance of the Portable Skid Resistance Tester.* ASTM Special Technical Publication No. 326.

Brown, E. R., and R. B. Powell. 2001. "A General Overview of Research Efforts at the NCAT Pavement Test Track." Paper presented at International Symposium on Maintenance and Rehabilitation of Pavements and Technological Control, Auburn, AL, July 29–August 1.

Croney D., and P. Croney. 1991. *The Design and Performance of Road Pavements.* New York: McGraw-Hill.

Epps, J. A., R. B. Leahy, T. Mitchell, C. Ashmorem, S. Seeds, S. Alavi, and C. L. Monismith. 1999. "Westrack—the Road to Performance-Related Specifications." Paper presented at First International Conference on Accelerated Pavement Testing, Reno, NV, October 18–20.

Federal Highway Administration (FHWA). 1993. *SHRP FWD Calibration Protocol.* Long-Term Pavement Performance Group, FHWA, McLean, VA, April.

———. 1994. *Pavement Deflection Analysis.* National Highway Institute Course No. 13127, FHWA-HI94-021, U.S. Department of Transportation, FHWA, Washington, DC, February.

Gillespie, T. D., M. W. Sayers, and L. Segel. 1980. *Calibration of Response-Type Road Roughness Measuring Systems."* NCHRP Report 228, National Research Council, Washington, DC.

Harris, B., M. Buth, and D. Van Deusen. 1994. *Minnesota Road Research Project: Load Response Instrumentation Installation and Testing Procedures.* Mn/Pr-94/01, Minnesota Department of Transportation, Office of Materials Research and Engineering, Maplewood, MN.

Harvey, J. T., J. A. Prozzi, and F. Long. 1999. "Application of CAL/APT Results to Long Life Flexible Pavement Reconstruction." Paper Presented at First International Conference on Accelerated Pavement Testing, Reno, NV, October.

Hass, R., W. R. Hudson, and J. Zaniewski. 1994. *Modern Pavement Management.* Malabar, FL: Malabar.

Henry, J. J. 2000. *Evaluation of the Italgrip Systems.* Highway Innovative Technology Evaluation Center, Civil Engineering Research Foundation, Washington, DC.

Henry, J. J., and R. R. Hegmon, 1975. Pavement Texture Measurement and Evaluation. ASTM Special Technical Publication No. 583. West Conshohocken, PA: American Society for Testing and Materials.

Highway Research Board (HRB). 1962. *The AASHO Road Test: Report 5—Pavement Research.* HRB Special Report 61-E.

International Standards Organization (ISO). 1998. ISO Standards. Geneva, ISO.

ISO Standard 13473. 1998. "Characterization of Pavement Texture by Use of Surface Profiles—Part 1: Determination of Mean Profile Depth." ISO, Geneva.

Karamihas, S. M., and T. D. Gillespie. 1999. *Guideline for Longitudinal Pavement Profile Measurement.* NCHRP Report 434, National Research Council, Washington, DC.

Leu, M. C., and J. J. Henry. 1983. "Prediction of Skid Resistance as a Function of Speed from Pavement Texture." *Transportation Research Record* 946.

Li, Y., J. B. Metcalf, S. A. Romanoschhi, and M. Rasoulian. 1999. "The Performance and Failure Modes of Louisiana Asphalt Pavements with Soil Cement Bases under Full-Scale Accelerated Loading." *Transportation Research Record* 1673, 9–15.

Li, S., and Y. Li. "Mechanistic-Empirical Characterization of Bonding between Ultra-Thin White-Topping and Asphalt Pavement." Paper submitted to ASTM.

Lytton, R. L. 1989. *Backcalculation of Pavement Layer Pavements and Backcalculation of Moduli.* ASTM STP 1026, American Society for Testing and Materials, West Conshohocken, PA, 7–38.

Mamlouk, M. S., W. N. Houston, S. L. Houston, and J. P. Zaniewski. 1988. *Rational Characterization of Pavement Structures Using Deflection Analysis,* vol. 1. report FHWA-A788-254, Arizona Department of Transportation, Phoenix.

Metcalf, J. B. 1996. *Application of Full-Scale Accelerated Pavement Testing.* NCHRP Report 235, National Research Council, Washington, DC.

Moyer, R. A. 1934. *Skidding Characteristics of Automobile Tires on Roadway Surfaces and Their Relation to Highway Safety.* Bulletin 120, Iowa Engineering Experiment Station, Ames.

PIARC. 1995. *International PIARC Experiment to Compare and Harmonize Texture and Skid Resistance Measurements.* PIARC Report 01.04.T, The World Road Association, Paris.

Royal Institute of Technology. 1980. *Bulletin 1980: 8 Testing Different FWD Loading Times.* Royal Institute of Technology, Department of Highway Engineering, Stockholm, Sweden.

———. 1998. *Techniques for Pavement Rehabilitation—Reference Manual (Sixth Edition).* National Highway Institute Course No. 13108, FHWA-HI-98-033, U.S. Department of Transportation, Federal Highway Administration, Washington, DC, August.

Sayers, M. W. 1995. "On the Calculation of International Roughness Index from Longitudinal Road Profile." *Transportation Research Record* 1501, 1–12.

Sayers, M. S., and S. M. Karamihas. 1998. *The Little Book of Profiling—Basic Information about Measuring and Interpreting Road Profiles.* Ann Arbor: University of Michigan Transportation Research Institute.

Sayers, M. W., T. O. Gillespie, and W. D. O. Paterson. 1986. *Guideline for Conducting and Calibrating Road Roughness Measurements.* World Bank Technical Paper Number 46, World Bank, Washington, DC.

Sayers, M. W., T. D. Gillespie, and C. A. Querioz. 1986. *The International Road Roughness Experiment—Establishing Correlation and Calibration Standard for Measurements.* World Bank Technical Paper Number 45, World Bank, Washington, DC.

Scrivner, F. H., R. Peohl, W. M. Moore, and M. B. Phillips. 1969. *Detecting Seasonal Changes in Load-Carrying Capabilities of Flexible Pavements.* National Cooperative Highway Research Program Report 76, National Research Council, Highway Research Board, Washington, DC.

Shahin, M. Y., and S. D. Kohn. 1979. *Development of Pavement Condition Rating Procedures for Roads, Streets, and Parking Lots—Volume I Condition Rating Procedure.* Technical Report M-268, Construction Engineering Research Laboratory, U.S. Corps of Engineers.

Sharp, K. G. 1991. "Australian Experience in Full-Scale Pavement Testing Using the Accelerated Loading Facility." *Journal of the Australian Road Research Board* 21(3):23–32.

Smith, R. E., and R. L. Lytton. 1984. Synthesis Study of Nondestructive Testing Devices for Use in Overlay Thickness Design of Flexible Pavements. Report No. FHWA/RD-83/097, U.S. Department of Transportation, Federal Highway Administration, Washington, DC, April.

Strategic Highway Research Program. 1993. *Distress Identification Manual for the Long-Term Pavement Performance Project.* SHRP-P-338, National Research Council, The Strategic Highway Research Program, Washington, DC.

Ullidtz, P., and N. F. Coetzee. 1995. "Analytical Procedures in Nondestructive Testing Pavement Evaluation." *Transportation Research Record* 1482, 61–66.

Way, G. B., J. F. Eisenberg, J. Delton, and J. Lawson. 1984. Structural Overlay Design Method for Arizona. Paper presented at AAPT annual conference, AAPT, Scottsdale, AZ, April 9–11.

CHAPTER 14
MODERN BRIDGE ENGINEERING

Mohiuddin Ali Khan
Ali Khan & Associates
Moorestown, New Jersey

14.1 BASIC CONCEPTS

14.1.1 Definitions

The old English word *brycg* means a "bridge," as defined by the *Oxford English Dictionary*. The word has Germanic root from the word *brugjo*. The structural concepts in the first bridges made by humans were simple. Trees, bamboo poles, or wooden logs or planks were placed on either side of a stream. This was followed by stone arches.

14.1.2 Benefits to Society and Civilization

The surface of earth is not smooth but has valleys and depressions in the form of contours. A bridge allows a highway or railway to span over sharp changes in elevations and serves as an essential part of the highway or railway. Transportation of vehicles and trains over an uneven surface is also possible by means of allowing traffic in one direction by an earth embankment or a culvert.

As an essential part of a highway, bridges help to distribute foodstuff and other goods to shops, even in the remotest of places. Simply expressed, without bridges, no food or milk would be readily available for breakfast and travel to school or hospital restricted.

14.1.3 Implementing Quality-Assurance/Quality-Control (QA/QC) Procedures in Planning, Design, Construction, and Maintenance

A project manager's role is important in the class and quality of finished products, whether a new design, deck widening or rehabilitation, retrofit, and repair. This chapter presents a summary of design procedures only. AASHTO and State Codes are applicable.

14.1.4 An Overview of Recent Development in Bridge Engineering

Bridges come in wide variety, forms, and shapes. Engineering is governed by procedures laid down in design and construction codes. The design codes have been changing every few years. Some of the major changes in the past two decades are as follows:

Accelerated bridge construction (ABC). The constant need for repair, retrofit, widening, and replacement need not shut down the traffic route for several months. These essential operations require quick and accelerated construction so that the user is not put to a disadvantage and resulting losses in commerce and trade are prevented. More efficient and powerful cranes, long trucks for hauling, and quality control have led to ABC.

New Construction Materials. Use of high-performance steel, high performance concrete, fiber reinforced polymer concrete, isolation bearings, composites, and so on has made longer spans, shallower superstructure depths, and lighter structures possible.

Remote-Sensor Techniques for Monitoring. Inspections have been made more informative and reliable.

Improvements in Design Methods and Codes. Old methods of design such as load-factor design (LFD) have been replaced with load-resistance-factor design (LRFD), which improved live loads, load combinations, methods of analysis, and detailed design and placed them on a more reliable probability theory. Design and rating codes are in much greater detail now.

Developments in Computer Applications and Software. Analysis and design software is readily available with optimal design options. Graphics and computer-aided design (CAD) such as Micro-Station and Auto-CAD have led to quick turnout of planning and construction drawings.

Improvements in Training Methods. University training and continuing-education requirements for licensure have resulted in well-equipped engineers who are able to tackle complex problems efficiently. Seismic analysis, seismic retrofit, and scour analysis and countermeasures design are better understood now than previously.

The following topics will present a description and summarize the scope of engineering effort. Owing to research in the United States and abroad, each topic can be a subject in itself:

- Introduction to highways and railway bridges: average daily traffic and operations, superstructure and substructure components, truck vehicles, Highway Loading HL-93 and Cooper loads, right-of-way, utilities, deck drainage, bridge security, and funding issues
- Highway structures: culverts, bridges, wing walls, retaining walls, sign structures, noise walls, causes of failures and preventive methods, footbridges, movable bridges, covered and historical bridges, and floating bridges
- Steel, concrete, timber, and aluminum bridges
- Problems of old and new bridges
- Codes of practice, AASHTO, American Railway Engineering and Maintenance of Way Association (AREMA) and state codes
- Bridges on rivers, hydrology, hydraulics, riverine and coastal hydraulics, erosion and scour analysis
- Hydraulic Engineering Circular HEC-18, HEC-23, and HEC-25 procedures for river bridges
- Planning of short-span and medium and long-span bridges, through-bridges, slab-beam bridges, cable-stayed and segmental bridges, and skew and curved bridges
- Geo-technical considerations, shallow and deep foundations
  - Diagnostic design, cost-benefit analysis and life-cycle costs, stress analysis, arching action and load resistance factor design, load combinations, seismic analysis, deflection criteria, vibrations
- Constructability issues, computer applications and software for old and new bridges
- New construction materials, high-performance and ultra-HPC, high-performance steel 70W and 100W
- Precast, fiber reinforced polymer (FRP) concrete and orthotropic decks; integral and semi-integral abutments and piers; solved examples of design problems
- Asset-management methods, structural health monitoring, remote sensors

- Structural evaluation, load resistance factor rating (LRFR) methods of rating analysis, safety and forensic engineering
- Maintenance methods for cracking, rupture, corrosion, and fatigue; recent developments in repair techniques, retrofits, and widening; repair methods for concrete and steel

Bridge design is governed by application of AASHTO Code as modified by State Code.

14.1.5 Summary of Topics and Overview of Methods Used

Sec. No. Discipline/Topic	Description of Topic in Bridge Engineering
1. Bridge design (load and resistance factor)	Summary of preliminary and final design, safety, economy, optimization, design for constructability, uniform probability of failure for all members AASHTO limit states, strength, serviceability, extreme events, fatigue, deflection, and constructability; safety load factors, dynamic impact factor, lane load-reduction factor, distribution factor, resistance factor, performance index, HL-93, legal (rating) and permit loads, fatigue and deflection trucks
2. Bridge geometry	Vertical curve, horizontal curve, spiral curve, profile grade elevation
3. Columns: Cap frame pier	Load combinations for analysis and detailed design
4. Deck drainage design	AASHTO method of deck drainage calculations, scupper
5. Deck slab and parapet	Reinforced-concrete deck slab, orthotropic grid, laminated timber deck, New Jersey barrier
6. Elastomeric and Multirotational bearings	Elastomeric laminated bearing pad, neoprene pad, plain and laminated bearings
7. Glulam timber beam	Load combinations, moving load analysis, strength, deflection
8. Hammerhead pier	Load combinations, analysis, and detailed design
9. Integral abutment	Moment connection of back wall with girder ends
10. Mechanically Stabilized Earthwork (MSE) wall	Evaluation of geometry, proprietary calculations
11. Noise barrier wall design	Determination of loads, stability analysis, footing and stem
12. Prestressed concrete beam	Single- and two-span continuous prestressed concrete I-beam, load combinations, moving load analysis, strength, deflection, rating analysis
13. Scour countermeasures	Hydrologic and hydraulic analysis, scour analysis, selection of countermeasures
14. Seismic retrofit and Isolation Bearing	Maximum reactions for load combinations in longitudinal and transverse directions
15. Single and continuous spans	Load combinations, moving load analysis, strength
16. Single- and two-cell box Culverts	Dead load and earth loads, moments and shear force diagrams, crack control check, fatigue stress check
17. Steel composite girders	Load combinations for analysis and detailed design and fatigue checks, rating analysis
18. Stub and full height abutment on spread footings	Determination of loads, stability analysis, spread footing and stem, wingwall; load combinations for analysis and detailed design
19. Stub and full height abutment on piles	Determination of loads, stability analysis, pile cap and stem design, pile design; load combinations for analysis and detailed design
20. Three-sided precast culverts	Dead and earth loads, moments and shear force diagrams, crack control check, fatigue stress check; load combinations for analysis and detailed design
21. Temporary support during construction	Determination of loads, underpinning, prop design for staged construction
22. Tie-back wall	Determination of loads, tie-back forces, stability analysis, footing and stem design
23. Wall pier	Load combinations, analysis, and detailed design

14.1.6   Multidisciplinary and Coordination Tasks

Bridge engineering requires a host of multidisciplinary tasks other than engineering computations and structural design for completion of the bridge. These include the following:

- Highway, surveying, geotechnical, water resources, estimating, CADD and construction management
- Bridge engineer catering to owner requirements, constructability, construction staging, cost considerations, coordination with outside agencies, budgets, plan preparation, resolution of comments, and project specifications

Design projects also should address right-of-way acquisition, community consensus, and construction permit requirements. Rehabilitation projects are discussed to address as-built plans, existing conditions, material properties, and physical conditions.

Finally, training of engineers should identify what should be taught and by whom. Case studies that show practical applications for successful completion of infrastructure projects, including lessons learned and client feedback need to be studied.

14.1.7   Users and Beneficiaries of a Bridge

The information presented here was developed originally by me for the textbook, *Bridge and Highway Structure Rehabilitation and Repair* (McGraw-Hill, 2010).The users of a given bridge are more varied than you would find for any other structure and include pedestrians, joggers, and marathon runners as well as

- Automobiles
- Bicycles
- Buses
- Ferry ramps
- High Occupancy Vehicles (HOVs)
- Permit loads
- Rapid transit
- Taxis
- Trucks

14.1.8   Miscellaneous Uses

Floating bridges are temporary bridges that float over water and are used for crossing waterways by lightweight traffic.

- Bridges also can carry a pipeline or waterway for water transport. An aqueduct is a bridge that carries water.
- A viaduct is a bridge that connects points of equal height.
- A combined road/rail bridge carries both road and rail traffic.
- Military bridges are temporary steel bridges used for military transport.
- An auxiliary bridge is also used for staged construction. Bailey-type or Mebee- type proprietary bridges are quick to erect.

14.1.9   Locations

A bridge is not required at a level crossing that is used for train traffic in one direction and truck traffic in the other direction.

- Guarded level crossings serve as traffic signals. Unguarded level crossings are for low railway traffic such as in a rural area but are not preferred because of the hazard of unexpected accidents.
- A highway bridge is located over an intersection to allow flow of traffic in two directions at right angles, and it is either over a highway or over a railroad.
- Bridges over rivers are either movable or fixed.
- Pedestrian or foot bridges are required to cross urban streets with high-volume traffic, over highways, or in parks over streams.
- Drawbridges, which used to serve as an entrance to castle in the olden days, are not seen nowadays.

14.1.10 Number of Spans

Most bridges in practice are single or double spans. Three or more span bridges are less common because of the limited width of the river underneath or because the number of traffic lanes for the lower level does not exceed six lanes with shoulders.

14.1.11 Minimum Span Length

A short span length of 20 feet generally is considered the minimum length to be classified as a bridge. For less than 20 feet, a three-sided precast culvert or a twin-cell box culvert may be used.

14.1.12 Long-Span Bridges

The longest span permissible depends on the type of bridge. Bridges such as the Golden Gate and the Verrazano Narrows in the United States have been pioneers in long-span construction. The Akashi-kaikyo Bridge in Japan until recently was the longest span in the world. However, the recent construction of a suspension bridge over Yangtse River in China has surpassed the Japanese span length. Segmental arch bridges supported on piers also can be classified as long spans. Single- and very-long-span bridges over waterways require special security arrangements and do not have considerable structural redundancy in design.

14.1.13 New Bridge Design versus Maintenance Design

The importance of continued availability of a bridge on a highway cannot be underestimated. A failure or a shutdown will

- Result in casualties
- Cause discomfort to thousands of users each day by longer detours
- Result in loss of trade and commerce

Bridges on military routes serve as an essential defense requirement for the country. The distinct aspects of a new design may include

- Widening
- Maintenance design
- Retrofit and repair
- Procedures and specifications laid down in the AASHTO *LRFD Specifications*. These methods are repetitive and are fairly straightforward. On the other hand, maintenance of an existing bridge requires site-specific procedures that are not always covered in the design specifications.

- The *LRFR Manual of Bridge Evaluation* addresses inspection, load rating, fatigue evaluation, materials, and nondestructive load testing. Based on rating, either the bridge can be posted for a lower live load or retrofitted for the actual live load or for extreme events of earthquake or provided with scour countermeasures against a flood event.

Maintenance procedures would require field inspections, structural evaluation, rating and rehabilitation, retrofit, or even replacement. There is an old saying that a stitch in time saves nine. Unnecessary expense involved in saving thousands of bridges at a late stage can be avoided by timely inspection and repair and by using modern Nondestructive testing (NDT) techniques or remote-sensor technology.

By means of proper and regular maintenance, the life of a bridge can be increased indefinitely. The most common maintenance tasks include

1. Replacing the reinforced-concrete decks because of cracks generated by the use of deicing salts, corrosion of reinforcing steel, creep and shrinkage, extreme temperature, impacts from heavy trucks, etc.
2. Painting of steel girders because of corrosion from a river environment
3. Repair and retrofit of malfunctioning bearings

A bridge engineer's main tasks include being fully familiar with

1. LRFD design specifications
2. Applications of computer software
3. Preparing construction specifications
4. Translating detailed design into CAD drawings
5. Understanding of constructability issues

14.2 HISTORY AND AESTHETICS

14.2.1 Introduction

The Romans are known as the greatest bridge builders. They supplied water from long distance by building aqueducts.

Ancient Roman Bridges.   Arch bridges such as the historical Alcántara Bridge over the River Tagus in Spain stand even today. The Romans used an earlier composition of cement to bind natural stone. Pozzolana which consisted of water, lime, sand, and volcanic rock, formed a strong cement substitute. It appears that the technology for cement manufacture was lost and then rediscovered later using a different approach.

European segmental arch bridges date back to 2nd century AD. Open-spandrel wooden segmental arches were used in the massive Roman era Trajan Bridge built in 105 AD.

Greek Bridges.   The Greeks designed arch bridges to accommodate chariots. The ancient Arkadiko Bridge is a corbel arch bridge dating to the Greek Bronze Age (13th century bc) and is one of the oldest arch bridges still in existence and use. Several intact arched stone bridges can be found in the Peloponnese in southern Greece.

Indian Bridges.   Brick-and-mortar bridges were built after the Roman era. Examples include the early bridges in ancient India. In India, the use of stronger bridges employing plaited bamboo and iron chain was seen around the 4th century. Puspa Gupta, the chief architect of Emperor Chandragupta I, is known to have built masonry bridges. Mughal Dynasty rulers in India between sixteenth and seventeen centuries built a number of bridges both for military and commercial purposes.

Chinese Bridges. The Chinese built the earliest timber and stone bridges. The oldest surviving stone bridge in China is the Zhaozhou Bridge, built during the Sui Dynasty from 595 to 605 A.D. This bridge is the world's oldest open-spandrel stone segmental arch bridge.

German Bridges. The Germans introduced many innovations in the 18th century to the design of timber bridges. The first book on bridge engineering was written by Hubert Gautier in 1716.

English Bridges. During the Industrial Revolution, in the erection of the Iron Arch Bridge over River Severn in England, cast iron was used for the first time. In the 19th century, for longer spans, truss systems of wrought iron were developed.

French Bridges. Gustave Eiffel of France used more modern steel that was manufactured by using a process similar to the Bessemer process. Steel has a higher tensile strength than wrought iron, and longer bridges were built.

Welded Bridges. In 1927, welding pioneer Stefan Bryla designed the very first welded road bridge, which was used in Poland. The American Welding Society (AWS) has developed comprehensive specifications for the design of fillet and butt welds that are used widely throughout the world.

14.2.2 Historic Bridges

Historic bridges are mainly associated with the nineteenth century. Covered bridges often serve as prominent local landmarks. Historic preservation or heritage conservation is a professional endeavor that seeks to preserve, conserve, and protect aesthetic bridges as artifacts of historic significance. In each state, the SHPO (state historic preservation organization) is assigned the task of maintaining such landmarks, which are sometimes shown on postcards.

A covered bridge is a bridge with enclosed sides and a roof, often accommodating only a single lane of traffic. Typically, they have been wooden, although some newer ones are of concrete or metal with glass sides.

Other names for the discipline of historic preservation include *urban conservation, landscape preservation,* and *heritage conservation.* In the United States, Pennsylvania has more covered bridges (over 200) than any other state, many of which can be seen in Bedford, Somerset, Washington, Chester, and Lancaster counties. Vermont has more covered bridges per square mile than any other place in the world, with 107 bridges located throughout the state. Oregon has the largest number of historical covered bridges in the western United States.

14.2.3 Aesthetic Planning

Appearance of the bridge should be pleasant to look at—"A thing of beauty is a joy forever." A bridge should have visual relationship with the surrounding area and visual impact.

1. The qualities will give the structure an appearance of a carefully thought-out design process is
   - Unity
   - Consistency
   - Continuity
2. Sound planning also leads to safety and effective operation at intersections.
3. Aesthetics is a requirement for mass production. New bridge facades should preferably blend with the appearance of existing bridges in the vicinity.
4. Bridges should have an open appearance and avoid abrupt changes in elevation or curvature. Abrupt changes in beam depth should be avoided when possible. Whenever sudden changes in

the depth of the beams in adjacent spans are required, care should be taken in the development of details at the pier.

- Avoid mixing structural elements, for example, concrete slab and steel beam superstructures or cap-and-column piers with wall-type piers.

Continuous superstructures should be provided for multiple-span bridges. Where construction joints cannot be avoided, the depths of spans adjacent to the joints preferably should be the same. Use of very slender superstructures over massive piers needs to be avoided.

- Bridge lighting can make a big difference in terms of aesthetics.
- Use of precast mechanically stabilized earthwork (MSE) walls: For abutments, wing walls, retaining walls, and MSE walls are gaining popularity owing to their elegant styles, low cost, and quick construction.

5. Careful attention to the details of the structure lines and forms will result in a pleasing structure appearance.

- No patchwork in concrete
- No dissimilar steel painting
- Use of innovative ideas and new technologies: Lightweight and weather-resistant transparent noise barrier sheets allow in-depth vision. By incorporating polyamide filaments, broken sheets can be held in place in the event of impact by a vehicle.

14.2.4 More Aesthetic Requirements for Pedestrian Bridges

Aesthetics are more of an issue in pedestrian and foot bridges located in parks and along highways than for other bridge types over which vehicles travel at a fast speed. Planning of rural bridges for aesthetics will ensure better acceptance than for those located in urban areas.

14.2.5 Bridge Planning for Very Long Spans

The architecture of long-span bridges has always attracted attention. Such bridges are pleasant to look at and also serve as cultural icons. Like movable bridges, they are a subject in themselves requiring specialized analysis and design. They include

- Arch bridges
- Cable-stayed bridges
- Segmental bridges
- Suspension-cable bridges
- Truss bridges

14.3 TYPICAL STRUCTURAL COMPONENTS

In bridges, the use of nonstructural elements is much less common than in a building, for example. The bridge designer has to deal with a wide variety of components with unique structural concepts.

14.3.1 Variety of Superstructure Elements (in Alphabetical Order)

- Approach slabs
- Bearings

- Deck slabs, reinforced or orthotropic steel
- Diaphragms
- Drainage system, gratings, and scuppers
- Longitudinal girders
- Parapets and median barriers
- Sidewalks

14.3.2 Variety of Substructure Elements (in Alphabetical Order)

- Cantilevered MSE walls or integral abutments
- Drilled shafts, short or long piles, spread footings
- Full-height or stub abutments
- Pier types such as solid walls, columns and caps, hammerhead-shaped
- Retaining walls such as cantilevered, cable-tied and buttressed
- Scour countermeasures
- Wingwalls, such as cantilevered, splayed, or with normal returns

14.3.3 Other Highway Structures Besides Bridges

1. Culverts serving as small bridges. Spans generally are less than 20 feet and can be planned as single- or two-cell square or rectangular reinforced-concrete culverts. For much smaller gaps in earthwork, steel or cast-iron pipes can be used.
2. For regular culverts, the modern trend is to use precast concrete with a three-sided culvert, doing away with bottom slab.
3. Electronic sign structures (displaying changeable and variable message).
4. Mechanically stabilized and cable-tied walls.
5. Noise walls, which are free-standing cantilever walls without subjected to earth pressure.
6. Plate sign structures: Common types are overhead, cantilever, and butterfly or wall-mounted.

14.3.4 Deck and Overlay Construction Materials

Continuous structural deck slabs vary in thickness, which is not likely to exceed 9 inches. For durability, it needs to be protected by one or more sacrificial overlays. Given the recent advancements in materials engineering, older and conventional materials are being discarded in favor of stronger composite materials. Use of structural deck slab material depends on span lengths, which are governed by girder spacing (between 5 and 15 feet) and the magnitude of truck live load with dynamic impact.

Castin-place reinforced concrete (with stay-in-place metal formwork) using HPC fiber-reinforced polymer (FRP) concrete deck slab

1. Lightweight aggregate concrete for higher seismic zones
2. Orthotropic decks using welded-steel angles in two directions within concrete
3. Precast prestressed concrete slab panels assembled in a factory away from the site
4. Timber deck panels using sawn-lumber planks

There are several overlay materials in use such as corrosion inhibitor aggregate concrete and latex modified concrete. They serve as wearing surface and are resistant to wear and tear besides offering a shield to the structural deck. They are one to two inches thick and can last twenty years or more depending upon Average Daily Truck Traffic (ADTT). The wearing surface thickness is neglected in structural slab calculations due to deterioration in strength with age. The roughness coefficient and frictional force with the tires may affect girder vibrations.

14.3.5 Selection of Modern Girder and Components Construction Materials

Due to research and development there is a much wider choice now than it used to be in the past:

1. Aluminum beams for smaller spans and lighter loads
2. Beams using ASTM Grade 50 steel and weathering steel (ASTM 50W)
3. For smaller beam depths, use of high-performance steel (HPS 70W and HPS 100W)
4. Mild steel diaphragms or web stiffeners (ASTM Grade 36 steel)
5. Timber beams (sawn lumber or glulam)
6. Arches using masonry blocks (burnt bricks or plain concrete)

14.3.6 Materials for Girder Bearings

The types of materials have changed from steel rocker and roller bearings due to maintenance problems. Elastomeric bearing pads are frequently being used. The following types are more common:

1. Plain elastomeric pads without steel and masonry base plates
2. Sandwich elastomeric pads with steel plates
3. Bronze isolation bearings

14.3.7 Steel Deck Joints

Deck joints have maintenance issues. These are being replaced by ends of beams made integral with the abutment wall. Joints are placed in the deck in the transverse direction near each girder support (at an abutment or pier) and can have 2 to 4 inches wide gaps to allow for thermal movement. They serve as expansion joints in the deck slab. Usually, two parallel steel angles are used to shield the sharp concrete edges of slab cutout. The gap width is set as per required coefficient of thermal expansion and contraction. Undesirable gaps are filled up with expansion material or polymer pads to prevent damage to tires or dust accumulation.

14.4 STRUCTURAL SYSTEMS

14.4.1 Mathematical Models

Without analysis for deformation or stress, the sizing of structural elements would be mere guess-work. The AASHTO *LRFD Specifications 2007* devote a full section (section 4) to analytical tools. The use of applied mathematics is considered a real engineering task. Computer software is based on recommended formulae and methodology, structural concepts, idealized diagrams of the model, and load paths. For example, the analysis of a plate surface (deck slab or retaining wall) is based on differential field equations which were converted to algebraic equations by means of finite-difference method. Relevant boundary conditions were applied. However, with the availability of fast computing power and use of algorithms, large sized matrices can be solved without difficulty by iterative numerical algebra and by partitioning operations. Stiffness matrix concepts are being utilized in finite-element methods, resulting in greater accuracy in computing deflections, moments, and forces.

In general bridge systems, the basic configuration and load paths of primary members would fall under the following structural systems:

1. *Common beam action.* The slab or girder is subjected to bending, torsion, and shear.

2. *Cantilever beam or slab action.* Small spans are subjected to bending and shear.

3. *Arch action.* Arched spans are subjected to primary axial compression.

4. *Truss action.* Truss spans are subjected to axial tension in ties and compression in struts, neglecting bending.

5. *Suspension-cable or cable-stayed system.* Load transfer occurs through very high axial tension in long-span bridges.

A simplified approach is assumed, in which the forces and moments cause deformations and stresses corresponding to the actions. Single dimension, two- and three-dimensional geometry and resulting mathematical models are considered. Bridge structural components can be idealized as subjected to a maximum of three moments acting at right angles, with primary and secondary bending moments (including torsional moments) accompanied by a set of three forces, which can be axial tension, compression, or shear forces also acting in x, y, and z directions. The theoretical resultants of moments and forces will act along principal planes, which are planes subjected to cumulative or peak stresses.

14.4.2 Redundancy Requirements

Through bridges are upstanding fascia girders with transverse floor beams. They have the least structural redundancy because if one of the two girders is damaged, collapse of the bridge becomes imminent.

Various physical parameters which represent theoretical concepts govern analytical results. There should be no ambiguity about their definitions, which are as follows:

Redundant and Nonredundant bridges.   As defined by the theory of indeterminate structures, use of a two-girder system or through trusses will have the least redundancy or least indeterminacy. The disadvantage is that the vulnerability to collapse or the probability of failure is increased against environmental loads. The formation of a plastic hinge in a single girder in an indeterminate multigirder system will only cause failure local to that girder. Other girders will be able to distribute moments, thereby offering combined resistance and avoiding failure. Hence the greater the number of girders, the higher is the degree of redundancy. A minimum of four girders is currently permitted by most highway agencies.

Similarly, a singlespan system will theoretically require only one plastic hinge to form at the location of maximum positive moment (midspan) for failure and has lesser redundancy than continuous spans. A continuous girder system will require two or more plastic hinges to be formed at the locations of maximum positive and negative moments to induce failure. Hence there is a need to adopt higher redundant systems to minimize or prevent failure.

Composite and Noncomposite Girder Sections.   Use of shear connectors makes it possible for composite action between the deck slab and the top flange of the girder. This provides a tremendous advantage in preventing global buckling of the top flange. During construction stages and until the full strength of concrete is achieved, the girder section will behave as noncomposite and will be designed as such. Composite action will prevent separation of slab from the beam. It indirectly contributes to redundancy since moment redistribution between slab acting as flange and girder acting as rib will be possible and unified resistance should be preferred in planning of bridge.

Behavior of Compact and Noncompact Sections.   To prevent local buckling, the top flange needs to be braced laterally by the composite slab. Flange shapes and sizes which meet long-column effect and local buckling requirements of slenderness and lateral support are called *compact sections.*

The advantages of compact sections are that they are capable of generating full plastic behavior leading to improved strength with local buckling resistance capability compared to noncompact sections. Noncompact sections can develop yield moment but only partially develop a plastic response. Noncompact sections are generally safe but are uneconomical compared to compact sections.

Compact or stocky sections will also indirectly contribute to increased redundancy. Hence it is important to try and achieve a compact section while sizing a plate girder. One method of reducing local buckling in the compression flange of a noncompact section is to provide transverse diaphragms. The long-column effect still needs to be checked for global buckling for total compression flange length between bearings.

14.4.3 Developing Bridge Geometry

Some knowledge of three-dimensional coordinate geometry is required for the analysis and efficient layout of a bridge in plan and elevation. The common plan geometry is as follows:

- Rectangular deck in plan with normal boundaries
- Rectangular deck in plan with skew boundaries
- Curved deck in plan with normal boundaries at ends
- Curved deck in plan with skew boundaries at ends

As part of connecting-highway planning, in the vertical plane, the bridge profile grade line (PGL) can be on a transitional curve, which is usually made convex for effective deck drainage. Double-decker bridges have two levels of traffic flow. Accordingly, the framework supporting the deck will be much higher. Examples are the San Francisco–Oakland Bay Bridge and the George Washington Bridge in North Jersey, where commuter traffic is extraordinarily heavy.

14.4.4 Typical Bridge Cross-Section

An essential section of the bridge in transverse direction will show the width of bridge with traffic lanes, shoulders and sidewalks the location of median barriers, parapets, and railings, if used. Girder depth, web stiffeners, girder spacing, and thickness of deck slab are detailed.

Diaphragm or cross-frame details are also shown, although located over twenty feet. Upwards camber is provided to the girders to offset dead-load deflection. Top flange of girders is made composite by connecting to the structural deck slab, by providing two or more longitudinal rows of shear connectors.

14.4.5 Use of Span-Deflection Criteria in the Design of Bridge Beams

The primary reasons for deflection control are as follows:

1. To ensure small deflection under live load for using small deflection slab/beam bending theory (without introducing thin-plate theory). Nonlinearity of load-deflection or stress-strain relationships will not be used for routine analysis.

2. To prevent cracking of concrete in deck slabs under repeated changes in vertical movements of steel girders.

The AASHTO *LRFD* has recommended optional maximum live load deflection of a girder (usually occurring at midspan) to L/800 for no-sidewalk bridges and L/1000 for sidewalk bridges.

Computation of liveload deflection includes 30% impact (dynamic) factor. Currently, liveload deflection is calculated for single line interior and fascia girders separately rather than for the whole composite system with diaphragms.

Computation of maximum static liveload deflection is necessary for the following:

1. Stationary or slow-moving traffic condition with no impact
2. Fast-moving traffic with impact

Parametric Study.   True live-load deflections based on composite action need to be computed using finite-element idealization for several three-dimensional systems. Optimization of girder sizes based on combined stress and deflection criteria would greatly help designers in not making mistakes in selecting correct sizes, leading to both safety and economy.

True deflections depend upon

1. Values of the distribution coefficients being used
2. Cross-frame stiffness and spacing
3. Girder spacing: Spacing of 8 to 12 feet
4. Four-, five-, and six-girder systems
5. Spans between 100 and 240 feet
6. Inclusion of parapets with deck slab stiffness
7. Grade of steel: Grade 50W, 70W, and 100W
8. Shallow beams and deep beams
9. Single and multiple spans

Vibration Control.   New types of expansion bearings and dampers are being used for minimizing vibrations arising out of moving loads, impact, wind, and earthquakes. Emphasis may be made to replace bearings on bridges have significant vibrations.

a. Use of Orthotropic Slabs

Use of orthotropic and exothermic deck slabs is on the rise, for example, in New Jersey. Load distribution in two directions of equal strength may minimize live-load deflection and influence vibration.

b. Use of Integral Abutment Bridges

Partial or full fixity at the ends of girders will give lower deflections, especially when there are pin-jointed connections with approach slab. Effects of vibration will be minimized.

14.4.6   General Dimension and Detail Requirements

Diaphragms and Cross-Frames.   The arbitrary requirement for diaphragms spaced at not more than 25.0 feet in the AASHTO Standard Specifications has been replaced by a requirement for rational analysis that often will result in the elimination of fatigue-prone attachment details.

Diaphragms or cross-frames may be placed at the end of the structure, across interior supports, and intermittently along the span. The need for diaphragms or cross-frames should be investigated for all stages of construction and the final condition.

This investigation should include, but is not be limited to, the following:

• Transfer of lateral wind loads from the bottoms of the girders to the deck and from the deck to the bearings
• Stability of the bottom flange for all loads when it is in compression
• Stability of the top flange in compression prior to curing of the deck
• Distribution of vertical dead and live loads applied to the structure

Diaphragms or cross-frames required for conditions other than the final condition may be specified to be temporary bracing. If permanent cross-frames or diaphragms are included in the structural model used to determine force effects, they should be designed for all applicable limit states for the calculated force effects.

At a minimum, diaphragms and cross-frames should be designed to transfer wind loads according to the provisions of the AASHTO *LRFD Specifications*, Article 4.6.2.7, and should meet all applicable slenderness requirements in Article 6.8.4 or Article 6.9.3. Connection plates for diaphragms and cross-frames should satisfy the requirements specified in Article 6.6.1.3.1. At the end of the bridge and intermediate points where the continuity of the slab is broken, the edges of the slab should be supported by diaphragms or other suitable means, as specified in Article 9.4.4. The AASHTO *LRFD Specifications* have eliminated the mandatory use of 25-foot spacing, but not for use with the girders.

Article 6.7.4.2 Straight I-Sections.   Diaphragms or cross-frames for rolled beams should be at least one-half the beam depth. Diaphragms or cross-frames for plate girders should be as deep as practicable. End cross-frames or diaphragms should be proportioned to transmit all the lateral forces to the bearings. End diaphragms should be designed for forces and distortion transmitted by the deck and deck joint. End moments of the diaphragms should be considered in the design of the connection between the longitudinal component and diaphragms.

If a bottom lateral bracing system is not present and the bottom flange is not adequate to carry the superimposed force effects, the first procedure should be used, and the deck should be designed and detailed to provide the necessary horizontal diaphragm action.

Limiting Slenderness Ratio.   Tension members other than rods, eyebars, cables, and plates should satisfy the slenderness requirements specified below:

- For main members subject to stress reversals,

$$\frac{\ell}{r} \le 140$$

- For main members not subject to stress reversals,

$$\frac{\ell}{r} \le 200$$

- For bracing members,

$$\frac{\ell}{r} \le 240$$

where ℓ = unbraced length (inches), and r = minimum radius of gyration (inches).

Compression members shall satisfy the slenderness requirements specified herein.

- For main members,

$$\frac{K\ell}{r} \le 120$$

- For bracing members,

$$\frac{K\ell}{r} \le 140$$

where K = effective length factor specified in Article 4.6.2.5, ℓ = unbraced length (inches), and r = minimum radius of gyration (inches).

Conclusions

- Vertical beam deflections (under symmetric and nonsymmetric truck loads) are not changed owing to use of bracing members or their stiffness.
- The pinned ends do not contribute to grid action.
- Also, the bracing members are not composite with deck slab for grid action to occur.
- However, X-frames are relatively more effective than straight diaphragms owing to their increased depth.

14.4.7 Bridge Drainage Issues

Deck Drainage.   To maintain traffic flow, the deck surface need not be submerged. The discharge volume and time will depend upon the intensity of rainfall. A small cross-slope and longitudinal slope (between 1 and 4%) on the deck slab needs to be provided using standard procedures. Hence slopes are either in one direction or in both directions of the PGL. Spacing of scuppers and inlets should not exceed 300 feet on deck or approach slabs in normal rainfall intensity areas.

Need for Underground Soil Drainage.   Perforated metal pipes of 6 to 12 inches in diameter allow any subsurface water in the soil behind the abutment and in wingwall retaining soil having elevated water table, to enter the pipes. These pipes are located behind abutments and wingwalls and under the approach slabs above the footing elevation. A longitudinal pipe slope of 1:20 or more is usually provided. The purpose is to keep the soil dry as moist soil can settle and exert greater unit pressure on the walls. Submerged soil may increase the liquefaction potential under the footings during an earthquake, leading to differential settlement. By proper soil drainage, any fluctuations in water table are avoided. Subsoil water in the pipe assembly is drained off into public sewers in urban areas or to streams in rural areas.

14.4.8 Role of Forensic Engineering

Bridges may fail unexpectedly or due to continued neglect. Conducting forensic engineering involves legal aspects of the damage caused to affected parties or victims. Bridges get damaged from sabotage, fatigue and fracture, old age, and inherent defects, or they may be demolished and replaced from lack of compliance to normal functions such as increase in traffic.

The basic procedures of conducting forensic engineering investigation are applied following a failure. They include the following:

- Expert witnessing
- Technical knowledge
- Detective skills

Forensic engineers explain how and why failures occur. The forensic process gets to the root of the problem. It gives a clear insight into structural misbehavior or lack of maintenance. However, more than one party may share the blame, and the extent of contribution by each party may not be easy to prove, especially in the case of older bridges that failed.

A forensic engineer gets to know that if design and reconstruction criteria were correctly implemented, the failure could be prevented. The expected life of 75 years or more for modern bridges and their components may not be achieved without regular inspection, structural evaluation, preventive action, and timely rehabilitation. Collapse due to earthquake, for example, could be avoided when retrofits are applied or standard details such as ductile moment resistance connections were used in the original design.

14.5   *LIFE-CYCLE COSTS*

Life-cycle costs are linked to the quality of planning. If initial cost does not cover all the structural requirements, the life-cycle cost for repair and rehabilitation will be much higher.

1. A vast majority of bridges in practice are small single-span structures. When skew angles over 45° are involved, adjacent prestressed concrete beam design should be chosen only after careful review, since conventional joint details and reinforcement become quite complicated, as do the size of the bearings and bridge seats. Bulb-tees or I-beams should be preferred.

2. For curved spans with prestressed concrete adjacent box beams or slab units are seldom chosen because of the increased cost of the wider chord alignment and the complications that arise with regard to bridge railing anchorage and end transitions. Prestressed concrete bulb-tees, I-beams, or spread boxes are alternates worth considering.

3. Concrete bulb-tees, I-beams, or spread box beams should be considered if vertical clearance requirements can be satisfied.

4. At locations where either long piles or poor bearing capacity is anticipated, prestressed adjacent box or adjacent slab design has the disadvantage of having a heavier superstructure. Under these conditions a spread box, bulb-tee, or concrete I-beam with deck slab configuration might be considered to reduce the loads.

5. Prestressed concrete adjacent beam design is often chosen over steel beams when a structure must be opened to traffic quickly. This type of construction eliminates the need for deck slab forming. It can also accommodate a temporary asphalt wearing surface if the time of the year prohibits placement of the concrete deck.

6. Where significant space must be provided for utilities, a spread system using steel girders, concrete I-beams, or bulb-tees is the preferred choice. Spread concrete box units can also accommodate some utilities.

7. Adjacent prestressed concrete boxes or slabs are preferred over streams where ice and/or debris is a problem. The smooth underside of adjacent units reduces the potential for snagging.

14.5.1   Superstructure Replacement Process

The collapse of I-35 bridge over Mississippi River on August 1, 2007, and many others has confirmed the need for regular maintenance. Otherwise, replacement is inevitable. The replacement process includes

1. Decision making to replace the bridge based on inspection and structural evaluation
2. Getting funding approved
3. Advertising or solicitation for design
4. Selection of consultants
5. Feasibility studies
6. Holding public meetings
7. Resolving right of way and utilities relocation issues
8. Approving preliminary and detailed design, contract, and documents
9. Selection of contractors
10. Monitoring construction and implementing construction schedule

From start to finish, the planning and design duration may extend well over one year. Even construction of a bridge smaller than 20 feet requires the same legal and financial approval procedures to be followed as required for a much longer bridge.

14.5.2 Case Study of Planning a Bridge

Based on a recent planning study of a 240-feet span length curved-girder bridge, the following considerations were adopted by the author:

1. *Maintenance and protection of traffic.* A detour during construction may not always be feasible. Staged construction or an accelerated schedule for construction may be necessary. There will be no impacts on traffic or on wetlands by using top to down construction and a temporary bridge.

2. Resolving difficult issues such as right-of-way, environmental permits, and utilities relocation may increase design or construction time.

3. *Public involvement.* A feedback from the users and taxpayers helps in selection of type of bridge and in planning for a minimum traffic disruption.

4. *Aesthetics:*
   - Open appearance.
   - Avoiding abrupt changes
   - Pier geometry made pleasant to look at
   - Use of MSE walls with patterns.

5. *Context Sensitive Design (CSD)* community or social groups involved. Historic or sensitive features. If existing bridge is registered on SHPO as historic, preservation features are applicable.

14.5.3 Superstructure Planning

1. Full composite action between slab and beam is assumed due to adequate use of shear connectors. Composite action helps in distributing dead loads from parapets, median barrier, and sidewalks to longitudinal girders equally.

2. *Planning for minimum dead weight.* In seismic zones, it is important to minimize inertia forces. Deck slab weight contributes the most to total dead weight and can be made lighter by using
   - Lightweight aggregate concrete
   - An orthotropic deck slab in place of solid slab

3. For the required distance between abutments of 240 feet, the possibilities are
   - Use of one single span of 240 feet length
   - Use of two equal spans of 120 feet
   - Use of two unequal spans of, say, 100 feet and 140 feet length
   - Use of three equal spans of 80 feet
   - Use of three unequal spans

   Use of unequal spans in urban areas is more frequent when required horizontal clearance from the new pier to a skew railway track is considered mandatory. However, by providing a guiderail in place of 30 feet minimum horizontal clearance, abutments and piers can be protected and the span length reduced.

4. *Number of girders and spacing.* Equal spacing (preferably between 6 and 12 feet) is most desirable. Fewer but deeper girders may reduce minimum vertical clearance of 16 feet, 6 inches. It is desirable to use an odd number of girders (with the middle girder logistically placed to carry a median barrier). For future staging, a longitudinal construction joint at the top flange of the middle girder location helps.

5. Fascia girder needs to be designed independently of the interior girder. However, fascia girder is generally lighter than interior girder and the latter is used in place of fascia. For future widening it is important to have equal strength girders.

6. Through girders are not generally used due to lack of redundancy. Risk of failure is greater. Minimum number of repeated girders is four.

7. *Effect of deck overhang.* Overhang dead weight negative moment at fascia girder reduces deck positive midspan moment in the adjacent panel. Without an overhang, fascia girder web is subjected to out-of-plane bending and torsion. For arriving at an optimum cantilever span length:

> maximum negative dead load moment at fascia = maximum positive dead load moment at midspan of deck panel

8. Generally overhang is made thicker than the deck. Additional weight of parapet at edge of overhang compensates for one truck wheel on the deck panel. The other wheel is assumed to act at or near fascia girder. Maximum overhang span is restricted to 5 feet due to expected large deflection at free end of cantilever which supports a parapet weight of 600 lb/ft.

14.5.4 Substructure Planning

This involves cost considerations and control of construction schedule. The alternate studies include

1. Use of integral abutment (no deck joints)
2. Wall abutments, such as
   - Full height
   - Medium height
   - Stub abutment (must fit into existing topography)
3. Extensive cut and fill needs to be avoided
4. Type of piers include
   - Column bent
   - Pile bent
   - Wall
   - Hammerhead type

14.6 SELECTION OF BRIDGE TYPE FOR SPAN LENGTHS

14.6.1 Girder Selection

Span Lengths Less than 40 Feet.

1. There are options to use deep fills, culverts, underpasses, tunnels, and/or masonry and concrete arches.
2. *Precast or cast-in-place reinforced-concrete structures.* Reinforced-concrete structures for culverts and short-span bridges consist of four-sided boxes, three-sided frames, and arch shapes.
3. *Deck slabs or deck/girder designs.* Prestressed slab units, stress-laminated timber decks, and concrete or timber decks with steel or timber girders cover this entire span range.

Conventional reinforced-concrete slabs, however, are inefficient for spans greater than 25 feet owing to their excessive depth and heavy reinforcement.

Spans Between 40 and 100 Feet.

1. Adjacent prestressed concrete slab units, prestressed concrete box units, concrete I-beams.

2. Bulb-tees are usually preferred over concrete I-beams.

3. Conventional composite design systems employing concrete decks and steel stringers can be used for the entire span range.

4. Special prefabricated bridge panels with concrete decks and steel beams can reach spans approaching 100 feet. They have the advantage of reduced field construction time.

Span Lengths Between 100 and 200 Feet.   Special modified prestressed concrete box beam units up to 60 inches deep can span up to 120 feet. Prestressed concrete I-beams and bulb-tee beams can span up to approximately 140 feet. Composite steel-plate girder systems can easily and economically span this range. Single spans up to 200 feet have been used.

Span Lengths Between 200 and 300 Feet.   A through or deck truss should be considered. Plate girders with webs splayed at supports can be used, especially at the lower end of this span range. Arches, slant-leg rigid frames, and concrete or steel box girders are also viable options.

Multiple-Span Arrangements.   For multiple-span bridges, a continuous design should be used whenever possible to eliminate deck joints. In the case of multiple simple-span prestressed unit bridges, the deck slab should be made continuous for live load over the intermediate supports.

Spans over 300 Feet.   Long multiple-span structures can utilize a variety of construction types and materials.

1. Steel bridges:
   - Through or deck trusses with girder approach spans
   - Trapezoidal box beams
   - Variable-depth girders (I-shaped beams and box girders)
   - Hybrid girders using conventional steel for web and high-performance steel for flanges
   - Cable-stayed girders or box beams
   - Deck or through arches
   - Cable-stayed bridges
   - Suspension bridges

2. Concrete bridges:
   - Segmental box designs
   - Cable-stayed trapezoidal boxes
   - Deck arches
   - Floating bridges/Pontoons
   - Post-tensioned, spliced bulb-tees
   - Segmental viaducts with variable-depth units

14.7   SUBSTRUCTURE RETROFIT MEASURES

1. Bearing strengthening—use of restrainers
2. Foundation improvement
3. Bearing replacement
4. Elastomeric bearings
5. Use of isolation bearings

14.7.1   Seismic Retrofit

1. Seismic retrofit goals:
   - Minimize the risk of unacceptable damage
   - Unacceptable damage:
     - Loss of life
     - Collapse of all or part of bridge
     - Loss of use of vital transportation route (essential route)
2. Seismic retrofit process: To upgrade bridges based on new seismic design criteria
   - Evaluate and upgrade the seismic resistance of existing bridges
   - Preliminary screening—inventory
   - Detailed evaluation:
     - Vulnerability rating
     - Seismic bridge ranking
   - Design retrofit measures
3. Retrofit measures:
   - Strengthening members—bearing strengthening by restrainers
   - Seat width improvement
   - Column strengthening—FRP wrapping or jacketing of column
   - Foundation improvement
   - Bearing replacement:
     - Elastomeric bearing
     - Isolation bearing
     - Dampers
   - Retrofitting for continuity

14.7.3   Project Scoping Process

Scoping is the first major stage of a project, where most important decisions are made. The outcome of scoping process is to develop

1. Project objectives
2. Design criteria
3. Feasible alternatives
4. A reasonable cost estimate
5. Historic preservation

The scope of work applies to the following aspects:

1. Life-cycle cost evaluation
2. Maintenance and protection of traffic
3. Hydraulic and scour studies
4. Seismic retrofits
5. Environmental considerations and acquiring permits
6. Performing value engineering for optimum project cost

To prevent fatigue, repainting of steel bridges is required at regular intervals (usually fifteen to twenty years) especially for structures located over rivers. Because of the need to remove and dispose of existing lead-based paints, repainting costs have escalated in recent years and in some cases have become nearly comparable with the cost of the bridge itself. The cost-effectiveness of fresh paint, which may last less than 20 years, needs to be investigated.

14.7.4 Examples of Introducing Economy in Design

1. Optimizing substructure design
2. Using the LRFD method
3. Using HPC and HPS
4. Prestressing the deck slab and longitudinal beams
5. Reducing superstructure depth
6. Widening girder spacing
7. Use of jointless (integral abutment) bridges
8. Eliminating bearings
9. Reducing field splices

14.7.5 Examples of Introducing Economy in Construction

1. Accelerated bridge construction
2. Quick erection using subassemblies
3. Increased shipping lengths

14.8 CONSTRUCTION BUDGET

14.8.1 Introduction

Most traffic congestions occur in urban areas where transportation needs are intense. Hence, with time, as a bridge gets older, maintenance costs increase, and the cost per vehicle usage is likely to increase. The higher the occupancy of vehicles, the lower will be the cost per person.

From records of past expenditures, a computer program can be developed for estimating daily cost per vehicle usage. For selection of major repairs, retrofit, rehabilitation, or replacement, a discount rate may be applied. Variable interest and inflation rates need to be included. Use may be made of software developed for present worth and annual worth. Table 14.1 lists the approximate unit costs of bridges.

Unit costs may vary for each location and project up to about 50% depending upon availability of labor and materials. Specialized contractors and special cranes will be needed for long-span bridges, and competitive unit rates may not be possible.

14.8.2 Life-Cycle Cost Analysis

Structural engineering projects can cost hundreds of millions of dollars. Highway agencies and owners have to develop a projected planning of yearly returns from their huge investment in constructing a bridge. A normal life of 75 years is assumed, although by regular maintenance, bridge life usually can be prolonged. Yearly maintenance expenses in terms of bridge inspection and rating,

TABLE 14.1   Approximate Unit Cost of Bridges (Prevalent in 2011)

Bridge Type	Unit Cost	Range of Span Length
Steel girder	$275–$300/ft^2	Small
Steel girder	$300–$350/ft^2	Medium
Steel box girder	$350–$400/ft^2	Medium
Steel arch	$450–$500/ft^2	Long
Concrete PC girder	$200–$225/ft^2	Small
Concrete PC girder	$225–$250/ft^2	Medium
Concrete box girder	$215–$230/ft^2	Medium
Segmental concrete span-by-span	$450–$475/ft^2	Long
Segmental concrete, Balanced	$475–$500/ft^2	Long
Cable-stayed bridge	$550–$600/ft^2	Long
Suspension bridge	$650–$700/ft^2	Very long

repairs, painting, occasional deck resurfacing, and any damage from accidents need to be allocated in advance.

Life-cycle cost analysis is an asset-management and economic analysis tool. It is useful in selecting the preferred alternatives by assigning values to their relative merits. It considers both short- and long-term costs. Computer software is available to compute life-cycle costs (see FHWA, *Life-Cycle Cost Analysis Primer*).

14.9 ANALYSIS AND DESIGN OF SUPERSTRUCTURE AND SUBSTRUCTURE

14.9.1 Design Methods and Details

The three criteria are

1. Strength control
2. Deflection control
3. Crack control

There are several considerations in design, of which avoiding overstress is the more important. Magnitude of stress depends upon response of the material and its age. Response of a material is based on the physical properties of that material, such as elastic modulus and modulus of rupture, and on size and shape of the member. Response is usually expressed as

1. Elastic stress
2. Yield stress
3. Plastic stress

Failure or Collapse Stress.   Generally, the extreme limit of elastic stress is yield stress, the extreme limit of yield stress is plastic stress, and extreme limit of plastic stress is failure or collapse stress.

1. Glass, ceramic, and plastics display fragile behavior.
2. Concrete displays brittle behavior.
3. Steel displays ductile behavior.

Hence glass needs to be wire-reinforced, plastics need to be fiber-reinforced, and concrete needs to be made composite with reinforcing rods.

Stress history is based on applied loads. Evaluation of various stages in the life of bridge components and chronological assessment of their performance are required so that limits can be placed either in design or in practice to prevent failure.

1. Fabrication stress
2. Transportation stress
3. Erection stress
4. Stresses resulting from maintenance loads

Stresses at Demolition or at Failure.   Many failures can be avoided by paying attention to changes in design technology and details. For example, new concrete materials such as lightweight and heavy-weight aggregates, structural plastics, and glass composites have different unit weights than conventional wet concrete. Such materials display nonhomogeneous and nonisotropic behavior. Current load factors and resistance factors need to be modified in light of experimental results.

14.9.2 Analytical and Computer Modeling Methods

Every physical parameter should be included in any analysis, namely

1. Geometry and curvature
2. Deck width, number of lanes, overhang
3. Girder spacing
4. Materials—reinforced concrete, steel, aluminum, prestressed concrete, timber, or masonry
5. Plan aspect ratio of deck slab and thickness
6. Skew angle

Computing method will be based on AASHTO Section 4.

14.9.3 Analytical Approach to Composite Slab-Beam Bridges

Idealization of composite deck and girder action can be obtained by the *simplified beam theory* using a transverse distribution factor. The bridge deck is idealized as a series of T-beams (line girders composite with deck slab) in the longitudinal direction. This approach is used commonly in the United States. The primary mode of deformation is by bending and torsion.

In older AASHTO codes, the distribution of loads in a transverse direction was based on a simpler equation for the more common distribution factor (*DF*), namely, *DF* = girder spacing/5.5. Spacing varies between 3.5 and 16 feet, and the distribution factor variation so computed differs widely for each spacing. With laboratory studies on the effect of girder spacing and calibration in existing bridges, it was possible to arrive at alternate methods to LFD method.

Simplified beam theory uses distribution formulas for the slab-beam system given below that take into account spacing, span, and beam stiffness, such as for moments in inferior beams (Table 4.6.2.2.2b-1):

$$DF = 0.15 + (S/3)^{0.6}(S/L)^{0.2}(Kg/12Lts^3)^{0.1}$$

where S = girder spacing $(0.075 + S/9.5)$
L = span length $(20'$ to $240')$
ts = slab thickness $(4.5''$ to $16'')$
Kg = stiffness $(10,000$ to $7,000,000)$

Owing to composite action between girders and the deck slab, arching action may result, thereby increasing the compressive forces in the top flanges of girders. The *Canadian Bridge Code* is taking advantage of this inherent behavior of a slab restrained at the boundaries acting as a dome and deck reinforcement computed to minimize the reinforcement.

It is seen that compact rolled-steel joists (compactness defined by Icy/Iyy) perform better transverse stress distribution than fabricated-plate girders. Use of noncompact plate girder sections will allow only partial moment and shear distributions compared with girders with compact sections.

Basic Assumptions for Analysis.

- Diaphragms can be modeled as a coexisting beam or truss in the transverse direction. Truss action by X-frame and K-frame configuration of diaphragm needs to be considered.
- A concrete deck may be assumed to be isotropic, that is, having equal strength in all directions. Certain types of grid decks may be orthotropic. Poisson's ratio for a slab-beam system for one-way action is neglected in longitudinal beam theory.
- Vertical shear deformation in the deck slab is considered negligible.
- Wheel loads may be approximated as an equivalent patch load acting under the wheel width. A 45° distribution in both directions in plan may be assumed and located away from the wheel load and distributed across depth of slab.

For monolithic slab-beam bridges, AASHTO section 4.6.1 shows a simplified line model of superstructure, neglecting modeling in the transverse distribution if span length exceeds 2.5 times deck width. Primary flexural members may be idealized as

1. *Line girder.* Stiffness matrix methods, moment distribution, strain energy, and area-moment methods
2. *Finite strip.* AASHTO section 4.6.1 describes limitations of the strip method for skew slabs. Suitable for slab-beam bridges spanning in a direction parallel with traffic. The widths of equivalent strips of concrete, steel, and timber decks are specified in Table 4.6.2.1.3.1-1. Empirical equations for deck bending moments and deflections are given in detail in section 4.6.2.1.8.
3. Grillage
4. Continuum (finite-element modeling)
5. Frames

An alternative to using a computer program is to develop equations for maximum bending moment, shear forces, and reactions. Based on different upper bounds, corresponding values can be generated using Excel spreadsheets or MathCAD software. I have developed such an approach.

14.9.4 Internal and External Effects

Factored resistance at any location of structure is greater than the sum of factored load effects acting at that location.

$$\text{Factored resistance} = \phi R_n$$

$$\phi R_n > \eta_i \Sigma \gamma_i Q_i \qquad \text{when the } \gamma_i \text{ selected is maximum}$$

$$\phi R_n > \Sigma \gamma_i Q_i / \eta_i \qquad \text{when the } \gamma_i \text{ selected is minimum}$$

$$\text{Factored resistance} = \phi R_n$$

where ϕ = resistance factor
  γ_i = load factor (AASHTO Tables 3.4.1-1 and 3.4.1-2)
  Q_i = nominal force effect
  R_n = nominal resistance (It is the mean or an identified level of strength.)
  η_i = load modifier (when (η_i) = 1, both equations are the same; see Table 14.2)

For an example of the load modifier (η) value used for reinforced concrete bridge design, see section 1.3.2.1.

TABLE 14.2.   AASHTO-Recommended Values for Load Modifier η:

	Strength	Service	Fatigue	Deflection	Construction (Wet Concrete)
Ductility η_D	0.95	1.0	1.0	1.0	1.0
Redundancy η_R	0.95	1.0	1.0	1.0	1.0
Importance η_I	1.05	N/A	N/A	1.0	1.0
$\eta_i = (\eta_D\,\eta_R\,\eta_I)^*$	0.95	1.0	1.0	1.0	1.0

*LRFD maximum value of $\eta_i = 1.0$; Table 5.5.

For load factor minimum value, $\eta_i = 1/(\eta_D\,\eta_R\,\eta_I) = 0.95$.

14.9.5 Resistance Factors

Resistance factors play an important role in the LRFD method and are based on observations. Within the life span of the bridge, LRFD method balances the probability of a reduced ultimate capacity of the section in bending and shear with the elastic moments and shears magnified by specified load factors. Strength reduction factor has a minimum value of 0.85 and maximum value of 0.90.

14.10 DEFLECTION CONTROL

When deflection limits are implemented for moving loads, it leads to a conservative approach in designing bridges. The deflection limit does not take advantage of modern materials for girders such as HPS 70W. The issue of deflection limits has greater visibility today than in the past. Refer to AASHTO 2.6.2.1, the Federal Highway Administration (FHWA), the New Jersey Department of Transportation (NJDOT), and New Jersey Turnpike regulations. The turnpike is experiencing some of the heaviest truck loads ever.

The 2007 AASHTO provisions are a reemphasis that deflection limits are not required, and some wording was added to the commentary on this issue. However, AASHTO is not attempting to introduce a deflection limit except for optional L/d criteria. Deflection limits were established with serviceability and long-term durability in mind. This approach in design of girders has prevented using shallow girders. It fitted in well with the use of I-girders because only the flanges are expected to resist bending, and the web plate resists shear. A deeper girder increases the lever arm between flanges and is economical.

In particular, the purpose of restrictions is

- To eliminate damage to structural components through resulting overstress
- To eliminate damage to nonstructural components
- To avoid loss of aesthetic appearance
- To ease vehicle rideability during sudden deflection from rigid pavement to a deflecting deck
- To minimize vibrations and to avoid any discomfort in meeting human response
- To provide maximum vertical underclearance by minimizing deflections
- To increase redundancy in the system by increasing the number of girders required to lower the deflection

The present study is aimed at knowing

- Is a mandatory deflection limit required?
- Is an optional deflection limit required?

- What is an optimal deflection limit?
- What is the resulting vibration limit?
- Should the span-to-deflection ratio be used?
- Should the span-to-depth ratio be used?
- Should the same criteria be used for all materials?
- Will the limits developed for Grade 36 steel be applicable for higher grades such as Grade 50 or Grade 70?

14.10.1 Literature Review on Deflection Limits

AASHTO LRFD Maximum Live-Load Deflection Guidelines. A literature search of the current state of the practice was conducted. Presentation of the results given below will better determine the exact nature of the research approach.

The origin of deformation requirements in bridges dates back to 1871 with a set of specifications established by the Phoenix Bridge Company. The 1958 American Society of Civil Engineers' (ASCE) report has been cited widely for the evolution of deflection-limit requirements. The criteria are empirical and based on arbitrary limits:

1. Span-to-depth ratio (L/d) limits were used as indirect measures of deflection limits.
2. Wu (2003) reported that for rolled-steel beams, pony trusses, pin-connected through trusses, and wood-plank decks, both deflection limits and L/d ratios were derived empirically.
3. AREA established L/d ratios for railway bridges prior those specified by AASHO for highway bridges.
4. Debate on the necessity of deformation requirements in current bridge design specifications centers on two main points:
   a. Do excessive deformations cause structural damage?
   b. Do deflection limits provide effective control of vibrations under normal traffic flow?
5. The ASCE report of 1958 has not come up with any evidence of serious structural damage owing to excessive vertical deflections. However, unfavorable psychological reactions to vibrations caused more concerns than the bridge safety itself.
6. Burke in 2001 explained that if deflection limits were not mandated, the effective service life of reinforced-concrete decks rather than steel girders could become considerably less than their normal replacement interval of 30 years.
7. In 2002, National Cooperative Highway Research Program (NCHRP) Project 20-7 by Roeder et al showed that bridges suffered structural damage due to excessive deformation.
8. In 2004, Azizinamini et al showed that deflection limits were not considered as a good method of controlling vibrations.
9. The optional AASHTO deflection limits may be summarized as follows (section 2.5.2.6.2):
   a. Minimum single-span length = 20 feet.
   b. Maximum single-span length = 300 feet.
   c. $L/800$ for bridges without sidewalks.
   d. $L/1000$ for bridges with sidewalks.
   e. $L/300$ for vehicular load on cantilever arms.
   f. $L/375$ for pedestrian loads on cantilever arms.
   g. $L/425$ for timber girders.
   h. The *NJDOT Bridge Design Manual* (2002) has adopted these criteria.

10. In a problem-free building floor, the allowable live-load deflection value is nearly three times higher; for example, the maximum live-load deflection limit is $L/325$.

However, span lengths in bridges are 5 to 10 times higher, and the magnitude and frequency of live loads also are much higher. Hence allowable deflections in bridge girders can be as low as $L/1000$, although some states such as New Hampshire use an $L/1600$ limit.

14.10.2 AASHTO Optional LRFD Deflection Criteria

Optional criteria use

1. Span/girder depth ratio.
2. Span/superstructure depth ratio.
3. Superstructure depth = total depth of structural deck slab + depth of haunch + depth of girder.
4. Thickness of a steel deck pan plate connected to the top of top flange is neglected.
5. Composite action between slab and beam is assumed to be due to shear connectors. Refer AASHTO Table 1.
6. Timber bridge beams have severe live-load deflection limitations and behave differently from steel or concrete beams.

14.10.3 Deflection Limits by Other States

See Table 14.3. Most states follow the 2007 Optional Deflection Criteria given in section 2.5.2.6.2 and summarized above.

14.10.4 Method of Computing Deflections

1. The current method prescribed by AASHTO is to use a line girder and apply a live-load distribution factor and multiple-lane reduction factor.
2. There are approximations in the method of applying live-load distribution from each lane for load sharing by the single girder under consideration. Code deflection calculations are based on use of distribution coefficients (DF), which may not give a true deflection value in all cases owing to the complexity of bridge geometry.
3. The line-girder code method can be calibrated against the three-dimensional model using the stiffness method. Software such as SAP2000 will be used.
4. In addition to HL-93, a number of truck loads such as permit loads and maximum legal loads will be considered. It is likely that deflection owing to the New Jersey permit load of 200 kips (compared with a 72-kip HL-93 truck) will be exceeded.

14.10.5 Effect of Boundary Conditions

1. Deflections of both interior girders and fascia girders will be evaluated separately for bridge decks with or without sidewalks.
2. *Skew slabs.* Most bridge decks have skew edges because roads do not usually run at right angles. Skew correction factors in the AASHTO code apply only to moments but not to deflections. Hence finite-element models (FEMs) for skew slabs are used for checking deflections in the line-girder method, especially for decks with sharp skew angles.
3. Fascia girder deflections are also affected by increased thickness of the sidewalk on the cantilever side. The code deflection method does not consider variations in deck slab thickness.

TABLE 14.3   Maximum Live-Load Deflection by State (Based on a recent review)

		Maximum Live-Load Deflection	
DOT	State	Steel	Concrete
CALTRANS	CA	Simple-continuous spans: 1/800, 1/1000 (pedestrian), 1/300 (cantilever), 1/375 (cantilever, pedestrian)	
DelDOT	DE	See AASHTO 2.5.2.6.2	See AASHTO 2.5.2.6.2
FDOT	FL	See AASHTO 2.5.2.6.2	See AASHTO 2.5.2.6.2
IDOT	IL	See AASHTO 2.5.2.6.2	See AASHTO 2.5.2.6.2
INDOT	IN	Same as AASHTO 2.5.2.6.2 Vehicular load, general . . . Span/800, Vehicular and/or pedestrian loads . . . Span/1000	Same as AASHTO 2.5.2.6.2 Vehicular load, general . . . Span/800 Vehicular and/or pedestrian loads . . . Span/1000,
KSDOT	KS	See AASHTO 2.5.2.6.2	See AASHTO 2.5.2.6.2
MHD	MA	Similar to AASHTO 2.5.2.6.2 Preferably vehicular load, general . . . 1/1000 But not more than 1/800 Vehicular and/or pedestrian loads . . . 1/1000	Similar to AASHTO 2.5.2.6.2 Preferably vehicular load, general . . . 1/1000 But not more than 1/800 Vehicular and/or pedestrian loads . . . 1/1000,
MoDOT	MO	See AASHTO 2.5.2.6.2	See AASHTO 2.5.2.6.2
NJDOT	NJ	Span/1000 (vehicular, general) Span/1000 (pedestrian) Span/400 (cantilever arm) Span/400 (pedestrian and cantilever arm)	
NYSDOT	NY	Same as AASHTO 2.5.2.6.2 Vehicular load, general . . . Span/800 Vehicular and/or pedestrian loads . . . Span/1000	Same as AASHTO 2.5.2.6.2 Vehicular load, general . . . Span/800 Vehicular and/or pedestrian loads . . . Span/1000
ODOT	OH	See AASHTO 2.5.2.6.2	See AASHTO 2.5.2.6.2
PennDOT	PA	Span/800 (vehicular, general) Span/1000 (pedestrian) Span/300 (cantilever arm) Span/375 (pedestrian and cantilever arm)	Span/800 (vehicular, general) Span/1000 (pedestrian) Span/300 (cantilever arm) Span/375 (pedestrian and cantilever arm)
TxDOT	TX	See AASHTO 2.5.2.6.2	See AASHTO 2.5.2.6.2
VDOT	VA	See AASHTO 2.5.2.6.2	See AASHTO 2.5.2.6.2
WVDOT	WV	See AASHTO 2.5.2.6.2	See AASHTO 2.5.2.6.2
WDOT	WA	See AASHTO 2.5.2.6.2	See AASHTO 2.5.2.6.2
AASHTO (optional)		Vehicular load, general . . . Span/800 Vehicular and/or pedestrian loads . . . Span/1000 Vehicular load on cantilever arms . . . Span/300, Vehicular and/or pedestrian loads on cantilever . . . Span/375.	Vehicular load, general . . . Span/800 Vehicular and/or pedestrian loads . . . Span/1000, Vehicular load on cantilever arms . . . Span/300, Vehicular and/or pedestrian loads on cantilever . . . Span/375

4. Special boundary conditions such as jointless decks (integral abutments) will have partial fixity. Deflections in such cases will deviate from the code method of evaluation, which neglects fixity boundary conditions in single spans.

14.10.6 Variety of Bridge Live Loads

Engineers are more concerned with serviceability and long-term durability. The original concept was that deflection limits provide a great benefit to bridge service life by restricting vibrations.

Factors related to live-load conditions are

- Truck weight
- Axle spacing
- Speed
- Tire properties
- Deck overlay roughness
- Multiple truck configurations

The *Journal of Nondestructive Evaluations* reported variations in loads and load combinations used by different highway agencies in the United States.

On major highways, the following truck and distributed loads are applicable:

- *HL-93 design live load:* Roadside weigh-in-motion (WIM) sites tend to record any overload. The results of overloads for the past several years show that AASHTO HL-93 loads are assumed to be blanket loads, which may not always be the maximum truck load. The difference in live load used by each state is as much as 1000 percent. European truck loads are of different intensities from HL-93.

- *Permit truck load.* We need to deal with real live loads, for example, permit loads are many times higher in magnitude and are expected to govern for causing maximum deflections. Motor vehicle agency (MVA) records will have an accurate inventory from their permit applications to ply on New Jersey highways.

- *Maximum legal load.* There is a requirement for a maximum legal load to restrict the heavier truck loads from out-of-state trucks from plying on New Jersey roads. In general, permit loads are heavier than legal loads, but for small spans, legal loads may control.

- Not every bridge, such as those on local roads, is designed to carry permit or HL-93 loads. Many county bridges are posted for smaller truck loads. Deflection limits there may not be applicable.

14.10.7 Conclusions on Deflection Limits

The general consensus is that deformation requirements are desirable though optional for both prevention of structural damage and control of vibrations. The validity of deflection limits and span-to-depth ratios as a means of vibration control for higher-strength steel materials need to be investigated.

14.11 COMPARISONS OF LRFD DESIGN METHOD WITH ALLOWABLE STRESS DESIGN (ASD) AND LOAD FACTOR DESIGN METHODS (LFD) METHODS

14.11.1 Load Combinations

Using appropriate load factors, strength load combinations for typical dead and live loads may be expressed only as

$$ASD - 1.0DL + 1.0(LL + \text{I}) \leq \text{resistance safety factor}$$

$$LFD - 1.3DL + 2.17(LL + I) \leq \phi R_n$$

$$LRFD - 1.25DC_1 + 1.25DC_2 + 1.50DW + 1.75(LL + I) \leq \phi R_n$$

The three sources in analysis and design are textbooks: Khan, Barker, Puckett etc., AASHTO LRFD Specifications (Formulae), and Detailed Procedures in Supporting Manuals and Computer Softwares (STLRFD, CONSPAN, SAP2000 etc.)

1. *HL-93 truck-load effect.* The influence line is divided into regions of positive and negative ordinates. By placing the load at various locations on the influence line of each region, the location of the maximum (peak) ordinate is found. For each peak of the influence line, the first axle of the truck is placed over the peak, and the other axles that follow are placed in their respective positions. The effect of this load position is computed by multiplying the axle load with the influence line ordinate under the load.

2. *Lane-load effect.* The positive regions of the influence line are loaded with the uniform lane load. The sum of the positive areas of the influence line is multiplied by the value of the uniform-lane load, and the result is stored as the positive lane-load effect. The negative lane-load effect is calculated similarly using the negative areas of the influence line.

3. *Dynamic load allowance.* The dynamic load allowance (impact) is computed in accordance with *LRFD Specifications*, Article 3.6.2. For the fatigue limit state, the dynamic load allowance *IM* is 15 percent. For all other limit states, the dynamic load allowance *IM* is 33 percent.

To compute the combined effects of live load and dynamic load allowance, multiply the live load by $1 + (IM/100)$. The dynamic load allowance is not applicable to pedestrian loads or to the HL-93 loads. It is applicable to the H and HS lane loadings.

14.12 USE OF SOFTWARE FOR GIRDER DESIGN

There are a large number of software products available for both analysis and design. These include PennDOT and SAP2000 programs and University of Maryland Merlin-dash and Leap software. Penn DOT's STLRFD program checks all applicable specifications for steel girder flexure, shear, fatigue, deflection, stiffeners, and shear connectors. Both positive and negative flexure is considered at each analysis point, as appropriate.

The program performs the following steps at each analysis point:

1. Check the flexure type for a given load combination. The flexure type can be positive flexure, negative flexure, or positive and negative flexure.

2. Obtain the combined load effects (factored moments, shears, stresses, etc.) and appropriate section properties based on the flexure type.

3. Obtain the variables required to check the specifications, such as width and thickness of the compression flange, depth of the web in compression, and lower moment at the bracing points for checking the unbraced length.

4. Check the *LRFD Specifications* for each limit state, as well as for uncured slab, fatigue, construction, and deflection, as summarized in LRFD Table 1.

There are in-house software developed for internal use. Permission for software use can be obtained from the owner and results can be checked by more than one software. All software should conform to latest provisions of AASHTO and State Code.

14.13 PROGRAM OUTPUT

- Unfactored moments including impact
- Unfactored shears and deflections including impact
- Reactions including impact
- Distribution factors

The live load can be one of the following:

1. Tandem + lane governs
2. Truck + lane governs

3. 90 percent tandem pair + lane governs

4. 90 percent truck pair + lane governs

5. Truck alone governs

6. 25 percent truck + lane governs

Selected software application is required and alternate solutions need to be compared for economical and optimum design.

14.14   DIAPHRAGMS AND X-FRAMES

14.14.1   Role of Diaphragms as Structural Elements

Cross-frames are conventional bracing systems that are used to prevent twist of a bridge's girder cross-section and are used on a bridge specifically to maintain stability of the bridge's girders during construction and pouring of the deck slab. Diaphragms consist of either steel angles or channels or wide flange I-girders. They serve as primary structural members for horizontally curved members under live load. They serve the following important functions:

1. During construction, they serve as temporary members for bracing adjacent girders and maintaining stability of girder flanges in compression during deck construction.

2. They support utility pipes under the deck between girders.

3. They resist wind forces acting on the superstructure.

4. They prevent twisting of the girder web and torsion under live load.

5. They minimize girder vibrations.

6. They distribute dead loads and truck loads both to near and far girders.

7. They reduce girder deflection and moments under the truck.

14.14.2   External and Intermediate Diaphragms

External diaphragms are located at the ends of girders to support the slab edge at bearing stiffeners. They are heavier than intermediate girders owing to greater shear force and reaction.

14.14.3   Types of Configurations Used by Different States

In a survey of most U.S. states, the common types of diaphragms are

1. X-bracing

2. K-bracing

3. Single-cross deep girder

4. Small top girder with X-bracing

5. Small top girder with K-bracing

6. Two small parallel girders

The tasks for cross-frame and diaphragm analysis and design include

1. Connection details

2. Empirical and truss method of analysis

3. Stress evaluation using proposed two-dimensional models

4. Stress distribution using proposed three-dimensional models

5. AASHTO *LRFD Specifications* for diaphragms and cross bracings

14.14.4   Literature Survey

Only a few studies examining the influence of cross-frame members on straight and curved bridge responses under seismic loads have been completed. The influence of cross-frames on the seismic performance of straight steel I-girder bridges was investigated by Azizinamini and colleagues. In their study, a two-span continuous composite bridge consisting of five haunched girders with two different types of cross-frames, X frames and K frames, was analyzed using software SAP90. Cross-frame influence on maximum bottom flange lateral displacements, maximum moments developed in the webs, and maximum total base shears was examined. *The studies showed that the differences in behavior between X and K cross-frames were negligible.* In a curved girder system, however, both during construction and while in service, cross-frames between the girders must be designed as primary load-resisting members and must be distributed adequately along the girder span.

14.14.5   Conclusions Drawn from the Literature Survey on Cross-Frames

1. The combination of results from natural frequencies, stresses, and displacements indicated that, for this structure, although certain parameters for X-type cross-frames were higher than those for K-type frames, the behavior of the two systems could be considered nearly identical.

2. It appears that, for a structure of similar curvature, when vertical displacement is of concern, an increase in the number of cross-frames may prove to be uneconomical because there is no corresponding increase in the efficiency of the system. However, when lateral displacement is of concern, an increased number of cross-frames would lead to a reduction in lateral displacements.

3. When dynamic response is a concern, upper lateral bracing appeared to provide the most benefit for this structure, and its use should be considered over part of the bridge length, especially when the curvature is sharp and the use of temporary supports is not practical.

4. "Highway Improvements Drive UH Engineering Efforts in Bridge Design," by Todd Helwig and Reagan Herman, Department of Civil and Environmental Engineering, University of Houston, Cullen College of Engineering (2005).

According to Helwig and Herman, this type of framing is expensive, and there can be long-term problems associated with its use, such as fatigue, added cost, and increased length of inspections throughout the life of the bridge. Cross-frames also are expensive to fabricate and require heavy machinery to install.

The University of Houston research proposes that permanent metal deck forms that are already used in bridge construction to hold the wet concrete in place as it cures can be used for stability bracing, thereby eliminating a number of cross-frames from the bridge. By using an element of the bridge that is already in place, the cost of these bridges is lowered significantly without reducing their stability. While cross-frames still will be required at all support locations and some intermediate locations, using the permanent metal deck forms for bracing will allow elimination of approximately half of all intermediate cross-frames. The connection details developed at the University of Houston allow the metal deck forms to be used to provide bracing for bridge girders, thus lessening the demand for cross-frames on the bridge.

14.14.6   Conclusions on Cross-Frame versus Diaphragms

Only locations of bracing points are required to restrict the length of a compression flange. There is no input requirement in straight girder programs for the sizes of cross-frames. The LRFD method and program treats cross-frames as a secondary member because their dead weight is neglected in DC_1 load. Three-dimensional analysis has shown that stresses in cross-frame or diaphragm members are small compared with the girder flanges. The type of bracing, whether cross-frame or diaphragm, to be used is optional. Only when computed distribution factors are used is there a requirement to input the type. However, both cross-frames and diaphragms provide identical results for girder analysis, including for maximum live load deflections.

The requirement for 25-foot spacing for diaphragms is replaced by calculated distances. Only in curved girders is it a maximum of 30 feet. Cross-frames should be nearly as deep as the girders. They can be connected directly to web stiffeners by bolts or welded to gusset plates. Diaphragms should be at least half the depth of girder. Either a single cross-girder or two small girders are used.

Cross-frames at supports (abutments and piers) are more important than intermediate cross-frames because they support the edge of concrete deck slab. Live-load deflections are not controlled by the $L/800$ or $L/1000$ requirement. Most European countries also do not have live-load deflection limits. The AASHTO minimum depth-to-span ratios are optional and a guidance for selecting a stiff bridge.

14.15   USE OF HIGH-PERFORMANCE STEEL

High-performance steel (HPS) design follows the same design criteria and good practice as provided in section 6, "Steel Structures," of the AASHTO *LRFD Bridge Design Specifications.* Use of HPS 70W generally results in smaller members and lighter structures. The designers should pay attention to deformations, global buckling of members, and local buckling of components. The live-load deflection criteria are considered optional, as stated in section 2, Article 2.5.2.6.2, of the AASHTO *LRFD Specifications.* The reason for this is that past experience with bridges designed under the previous editions of the AASHTO specifications has not shown any need to compute and control live-load deflections based on the heavier live load required by the AASHTO *LRFD Specifications.*

The AASHTO *HPS Guide* encourages the use of hybrid girders, that is, combining the use of HPS 70W and Grade 50W steels. A hybrid combination of HPS 70W in the negative moment regions and Grade 50W or HPS 50W in other areas results in the optimal use of HPS and attains the most economy.

14.15.1   First HPS 70W Bridge

The Nebraska DOT was the first to use HPS 70W in the design and construction of the Snyder Bridge, a welded plate girder steel bridge. The bridge was opened to traffic in October 1997. It is a 150-foot simple-span bridge with five lines of plate girders of 4 feet, 6 inches deep. The original design used conventional Grade 50W. Modern bridges have come a long way from the earlier covered bridges which are no longer built (Figure 14.1).

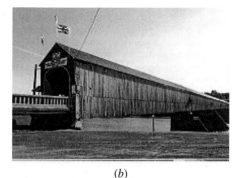

(a)   *(b)*

FIGURE 14.1   (*a*) A small span covered bridge (*b*) A long span covered bridge.

When HPS first became available, the Nebraska DOT replaced the Grade 50W steel with HPS 70W steel of equal size. The intent was to use this first HPS 70W bridge to gain experience with the HPS fabrication process. The fabricators concluded that there were no significant changes needed in the HPS fabrication process.

14.15.2 HPS Cost Study

Material cost and fabricated cost for HPS hybrid girders is about 20 percent higher than for Grade 50 steel. This is more than offset by avoiding painting costs with weathering steel and savings in weight. For ultimate load behavior, HPS girders are much stronger.

The study concluded that

1. HPS 70W results in weight and depth savings for all span lengths and girder spacing.
2. Hybrid designs are more economical for all the spans and girder spacing. The most economical hybrid combination is Grade 50 for all webs and positive-moment top flanges, with HPS 70W for negative-moment top flanges and all bottom flanges.
3. The LRFD treats deflection as an optional criterion with different live-load configurations. If a deflection limit of $L/800$ is imposed, deflection may control HPS 70W designs for shallow web depth.

Some optimization techniques include

- Use uncoated HPS steels.
- Use HPS 70W steel for flanges and webs over interior supports, where moments and shears are high.
- Use hybrid girder sections for composite sections in positive bending, where moments are high but shears are low.
- Use undermatching fillet welds with HPS 70W to reduce the cost of consumables.
- Use of constant-width plates to the greatest extent possible (consider plate-width changes at field splices wherever practical).
- Use the new AASHTO *Guide for Highway Bridge Fabrication with HPS 70W Steel.* Recommendations in the guide should be followed, with no more stringent requirements added.

14.16 DESIGN OF CONTINUOUS DECK SLAB ON BEAMS

The following AASHTO *LRFD Specifications* sections are applicable to the design of conventionally reinforced-concrete decks:

1. Deck slabs (4.6.2.1)
2. Minimum depth (9.7.1.1)
3. Empirical design (9.7.2)
4. Traditional design (9.7.3)
5. Strip method (4.6.2.1)
6. Live-load application (3.6.1.3.3, 4.6.2.1.5)
7. Distribution reinforcement (9.7.3.2)
8. Overhang design (A13.4, 3.6.1.3.4)
9. Minimum negative flexure concrete deck reinforcement (6.10.1.7)

14.16.1   Deck Replacement of Slab Composite with Repeated Beams

Empirical Design: Alternative to Traditional Method (9.7.2.4).   Flexure in longitudinal direction: Design as a series of longitudinal strips. Idealize width of strip = width of wheel + 2 × effective depth of slab:

- Maximum B.M. owing to dead load = $w_d(L^2/8)$.
- Maximum B.M. owing to lane load = $w_L(L^2/8)$.
- Maximum B.M. owing to truck load is calculated from influence lines. For deck slabs, Puncher's influence line diagrams were used.

AASHTO Analytical Method.   Analyze as a strip of unit width continuous over beam flanges. Many States in the United States have ready-made simplified structural solutions for deck thickness and rebars based on AASHTO empirical methods. These are used frequently in practice and were developed by examination of repeated drawing details and calculations performed for a large number of spans. Some of the approximations in the current AASHTO code or state code design methods include the following:

1. Boundary effects of skew and curved decks are not considered.
2. For arching action at supports (9.7.2.1) arising from reverse bending curvature, planar or membrane forces will be generated in addition to bending. Three-dimensional modeling and analysis will be required.
3. Conventional methods do not consider additional thickness for transverse deck drainage, camber thickness, thickness of concrete haunches on top of flanges, or any groove formations. In addition,
   a. Added stiffness owing to stay-in-place folded steel or aluminum formwork
   b. Secondary stresses such as those resulting from creep and shrinkage stresses contributing to cracking
   c. Daily thermal stress variation and during summer and winter months
   d. Composite behavior of wearing-surface thickness using special concrete (such as latex-modified or corrosion-inhibitor aggregates for forming defense against tire friction and braking forces)
4. Effects of shear deflection are neglected.
5. Fracture mechanics formulas are applied for deck cracking.
6. For approach slab analysis, the approach slab behind integral abutments is itemized as a structural member during construction. For analysis, it needs to be idealized as slab on grade and acts as a plate on an elastic foundation. The geotechnical properties of subgrade material will be required.

14.16.2   More Accurate Analytical Method

The deck slab is idealized using FEM. The ultimate load behavior of R.C. elements needs nonlinear analysis. For concrete, the stress-strain curve is nonlinear during the cracking stage. The tangential stiffness method is used.

LRFD Design Methods for Deck Slab Design.   Use MathCAD or spreadsheets to program equations and load and resistance factors to comply with the AASHTO *LRFD Specifications.*

Analysis. Main reinforcement perpendicular to traffic. Analyze as a strip of unit width continuous over beam flanges.

Deck thickness. This is based on live-load deflection control, which is a function of girder spacing. Thus $s/h < 20$, where s is girder spacing, and $h_{min} = (s + 3000)/30 < 175$ (AASHTO Table A2.5.2.6.3-1).

Structural depth. Structural depth >7 inches (for thinner slabs and wide girder spacing, transverse prestressing may be used to prevent crack formation).

Dead-load moments. Dead-load bending moments can be evaluated by the moment distribution method, the stiffness matrix method, or by coefficients given in the handbook.

The highest positive and negative moments are in end spans of a continuous deck and least in middle spans. The greater the number of beams, the higher is the redundancy. Equal spacing of girders is assumed. For transverse dead-load bending moments for repeated beams, $M_{max} = +wL^2/12$ and $-wL^2/8$. For four beams, $M_{max} = +wL^2/10$ and $-wL^2/10$. For five or more beams (use influence lines at $0.4L$ of the first span), $M = 0.0744wL^2$. Adjust for overhang moment: $-wL^2/2$. For the second span, $M = +0.0471wL^2$ and $-0.1141wL^2$.

For analysis, external actions owing to load must balance internal actions: $\Sigma M = 0$. Hence

$$M_u = \phi M_n = A_s f_y (d - a/2) \qquad a = A_s f_y/0.85f'_c b \qquad A_s = (M_u/\phi)/f_y jd$$

For design, $\phi M_n \geq M_u$. *Note:* The sign ($\geq$) is to ensure minimum requirements and results in a higher overall factor of safety for the bridge. Check for shear $\Sigma V = 0$ is not required.

- Check ductility: $a \leq 0.35d$.
- Minimum reinforcement: $\rho \geq 0.03f'_c/f_y$.

$$1.2M_{cr} < M_u \quad \text{or} \quad 1.33\Sigma M \text{ (factored moments)}$$

$$M_{cr} = f_r I/yt$$

$$M_u = M_{DC} + M_{DW} + M_{LL+I}$$

Rupture stress in slab concrete: $f_r = 2.4\sqrt{f'_c}$

- Maximum spacing, $s_{max} = 18$ inch or 1.5 h (5.10.3.2).
- For bottom distribution reinforcement, refer to (9.7.3.2): $220/\sqrt{S} < 67\%$ of primary reinforcement (placed perpendicular to traffic), where S is the effective span length.

For top distribution reinforcement, $A_s > 0.11 A_g/f_y$

- Maximum reinforcement:

$$c/d_e < 0.42 \text{ (5.7.3.3.1-1)}$$

- Shrinkage and temperature reinforcement:

$$\text{Temp } A_s \geq 0.11A_g/f_y \text{ (5.10.8.2)}$$

- Minimum cover is 1.0 inch or more based on Table 5.12.3.1 or state code requirements.
- Crack control: $f_s \leq Z/(d_c A)^{0.33} \leq 0.6f_y$ (5.7.3.4).

- Unfactored moment to calculate tensile stress in reinforcement:

$$M = M_{DC} + M_{DW} + M_{LL+I}$$
$$n = E_s/E_c \quad E_s = 29{,}000 \text{ ksi MPa (5.4.3.2)}$$
$$E_c = 33{,}000 \ W_c^{1.5} (5.4.2.4)$$

Overhang Design.

- Design as a nonredundant member.
- Cantilever span < 6 feet (3.6.1.3.4).
- Distance to face of barrier < 3 feet (4.6.2.2.1).
- Minimum edge depth = 8 inches for overhang supporting parapets or deck-mounted posts and 12 inches for side-mounted posts.

Live Loads.

- Equivalent live load = 1.0 kip/ft located 1.0 foot from face of railing.
- Strength I limit state for HL-93 loads
- Extreme event II for collision from vehicles

Negative moments at fascia girders from parapet loads have the beneficial effect of reducing positive moments in deck continuous spans. Since extreme state collision moment is distributed over 5 to 10 feet of deck, it is generally smaller than deck negative moment. Owing to small cantilever span, strength I overhang moment is less than deck negative moments. Cantilever design does not control the deck design.

The overhang slab is usually cast composite with the parapet by placing U- or L-shaped rebars from the overhang slab inside the barrier. Overhang thickness is increased for the extreme case of collision with the parapet or railing. In such cases, the overhang will not crack, and the parapet or railing can be made sacrificial (A13.4.2).

$$\text{Overhang } 1140 + 0.833 \times +ve \text{ moment } 660 + 0.55S - ve \text{ moment } 1220 + 0.25S$$

$$IM = 33\% \text{ of } LL \ (3.6.2.1)$$

- *Number of lanes:*

$$N = \text{Int(roadway width/12.0) (3.6.1.1.1)}$$

- *Multiple presence factors:*

$$m = 1.2 \text{ for one loaded lane}$$
$$= 1.0 \text{ for two}$$
$$= 0.85 \text{ for three}$$

- Tire contact area = 2D inch × 10 inch (3.6.1.2.5). Wheel load is applied as a distributed load.
- *Fatigue limit state:* Not required for multigirder applications (5.5.3).
- *Empirical design:* Alternate method to traditional method (9.7.2.4).

14.16.3   Simplified LRFD Calculations for Replacement of Deck

For solved example, see *Rehabilitation and Repair of Bridge and Highway Structure*, by Mohiuddin Ali Khan (McGraw-Hill, 2010).

14.17 BASIC STEPS FOR LRFD DESIGN OF CONCRETE BRIDGE GIRDERS

The following procedure is intended to be a generic overview of the design process using simplified methods. Relevant latest AASHTO *LRFD 2007* with *2008 Interims* are applicable. Sections are listed in parentheses.

1. Limit states (1.3.2) are used. Load combinations are based on strength, serviceability, extreme events, and fatigue. Deflection checks are optional.
2. Develop general plan, elevation, and section based on average daily traffic, number of lanes, and length of bridge.
3. Bridge type (prestressed planks, I-girders, box beams, T-bulbs, etc.):
   - Structure depth (2.5.2.6.3)
   - Reinforcement (5.14.1.3.2)
   - Minimum reinforcement (5.7.3.3.2 and 5.7.3.4)
   - Temperature and shrinkage reinforcement (5.10.8)
4. Deflections:
   - Optional live-load deflection control (2.5.2.6.2)
   - Optional criteria for span-to-depth ratios (2.5.2.6.3)
5. AASHTO-prescribed load factors and resistance factors:
   - Select resistance factors
   - Strength limit state (conventional) (5.5.4.2.1)
   - Select applicable load combinations and load factors (3.4.1, Table 3.4.1-1)
6. Select load modifiers:
   - Ductility (1.3.3)
   - Redundancy (1.3.4)
   - Operational importance (1.3.5)
7. Calculate live-load force effects:
   - Live loads (3.6.1) and number of lanes (3.6.1.1.1)
   - Multiple presence (3.6.1.1.2)
   - Dynamic load allowance (3.6.2)
8. Apply distribution factor for moment (4.6.2.2.2):
   - Interior beams with concrete decks (4.6.2.2.2b)
   - Exterior beams (4.6.2.2.2d)
   - Skewed bridges (4.6.2.2.2e)
9. Apply distribution factor for shear (4.6.2.2.3):
   - Interior beams (4.6.2.2.3a)
   - Exterior beams (4.6.2.2.3b)
   - Skewed bridges (4.6.2.2.3c, Table 4.5.3.2.2c-1)
10. Compute reactions to substructure (3.6).

Refer to solved example given in R. Barker and J. *Puckett's Design of Highway Bridges*, Chapter 7, John Wiley & Sons.

14.17.1 Precast Prestressed Beams

Typical precast prestressed (P/S) beams will conform to section 5.14.1.2.2, and joints will conform to 5.14.1.2.6.

14.17.2 Cast-in-Place T-Beams and Multiweb Box Girders

Typical sections will conform to 5.14.1.3:

- Effective flange widths (4.6.2.6)
- Strut-and-tie areas (5.6.3)
- Calculate force effects from other loads as required.
- Investigate service limit state.
    1. P/S losses (5.9.5)
    2. Stress limitations for P/S tendons (5.9.3)
    3. Stress limitations for P/S concrete (5.9.4)
        a. Before losses (5.9.4.1)
        b. After losses (5.9.4.2)
    4. Durability (5.12)
    5. Crack control (5.7.3.4)
    6. Fatigue, if applicable (5.5.3)
    7. Deflection and camber (2.5.2.6.2, 3.6.1.3.2, and 5.7.3.6.2)
- Investigate strength limit state:
    1. Flexure:
        a. Stress in P/S steel—bonded tendons (5.7.3.1.1)
        b. Stress in P/S steel—unbonded tendons (5.7.3.1.2)
        c. Flexural resistance (5.7.3.2)
        d. Limits for reinforcement (5.7.3.3)
    2. Shear (assuming no torsional moment):
        a. General requirements (5.8.2)
        b. Sectional design model (5.8.3)
            (1) Nominal Shear Resistance (5.8.3.3)
            (2) Determination of β and θ (5.8.3.4)
            (3) Longitudinal Reinforcement (5.8.3.5)
            (4) Transverse Reinforcement (5.8.2.4) (5.8.2.5) (5.8.2.6) (5.8.2.7)
            (5) Horizontal Shear (5.8.4)

Check details:

1. Cover requirements (5.12.3)
2. Development length—reinforcing steel (5.11.1 and 5.11.2)
3. Development length—prestressing steel (5.11.4)
4. Splices (5.11.5 and 5.11.6)
5. Anchorage zones:
    a. Posttensioned (5.10.9)
    b. Pretensioned (5.10.10)
6. Ducts (5.4.6)
7. Tendon profile limitation:
    a. Tendon confinement (5.10.4)
    b. Curved tendons (5.10.4)
    c. Spacing limits (5.10.3.3)

8. Reinforcement spacing limits (5.10.3)
9. Transverse reinforcement (5.8.2.6, 5.8.2.7, and 5.8.2.8)
10. Beam ledges (5.13.2.5)

14.18 DESIGN OF SOLID-SLAB BRIDGES

14.18.1 General Procedures for Design

The design approach for slab bridges (A5.4) is similar to that for beam and girder bridges with some exceptions:

1. Check minimum slab depth (2.5.2.6.3)
2. Determine live-load strip width (4.6.2.3)
3. Determine live load for decks and deck systems (3.6.1.3.3)
4. Design edge beam (9.7.1.4)
5. Check shear
6. Check distribution reinforcement

14.18.2 Voided or Cellular Slab Bridges

1. Check whether voided slab or cellular construction (5.14.4.2.1)
2. Check minimum and maximum dimensions (5.14.4.2.1)
3. Design diaphragms (5.14.4.2.3)
4. Check design requirements (5.14.4.2.4)

14.19 BASIC STEPS FOR DESIGN OF STEEL BRIDGE GIRDERS

14.19.1 Design Procedure

Refer to solved example given in R. Barker and J. *Puckett's Design of Highway Bridges*, Chapter 8, John Wiley & Sons. The following procedure is intended to be a generic overview of the design process using simplified methods. Relevant latest AASHTO *LRFD 2007* with *2008 Interims* are applicable. Sections are listed in parentheses. Limit states (1.3.2) are used. Load combinations are based on strength, serviceability, extreme events, and fatigue. Deflection checks are optional. Develop the general plan, elevation, and section based on average daily traffic, number of lanes, length of the bridge, and bridge type (I-girders, box beams).

1. I-girder:
    a. Composite (6.10.1.1) or noncomposite (6.10.1.2)
    b. Hybrid or nonhybrid (6.10.1.3)
    c. Variable web depth (6.10.1.4)
    d. Cross-section proportion limits (6.10.2)
2. Box girder:
    a. Multiple boxes or single box (6.11.1.1 and 6.11.2.3)
    b. Hybrid or nonhybrid (6.10.1.3)

    c. Variable web depth (6.10.1.4)

    d. Cross-section proportion limits (6.11.2)

3. Select resistance factors: Strength limit state (6.5.4.2)

4. Select load modifiers:

    a. Ductility (1.3.3)

    b. Redundancy (1.3.4)

    c. Operational importance (1.3.5)

5. Select load combinations and load factors (3.4.1):

    a. Strength limit state (6.5.4.1, 6.10.6.1, and 6.11.6.1)

    b. Service limit state (6.10.4.2.1)

    c. Fatigue and fracture limit state (6.5.3)

6. Calculate live-load force effects:

    a. Select live loads (3.6.1) and number of lanes (3.6.1.1.1)

    b. Multiple presence (3.6.1.1.2)

    c. Dynamic load allowance (3.6.2)

    d. Distribution factor for moment (4.6.2.2.2)

        (1) Interior beams with concrete decks (4.6.2.2.2b)

        (2) Exterior beams (4.6.2.2.2d)

        (3) Skewed bridges (4.6.2.2.2e)

    e. Distribution factor for shear (4.6.2.2.3)

        (1) Interior beams (4.6.2.2.3a)

        (2) Exterior beams (4.6.2.2.3b)

        (3) Skewed bridges (4.6.2.2.3c)

    f. Stiffness (6.10.1.5)

    g. Wind effects (4.6.2.7)

    h. Reactions to substructure (3.6)

7. Calculate force effects from other loads identified earlier.

8. Design required sections—flexural design:

    a. Composite section stresses (6.10.1.1.1)

    b. Flange stresses and member bending moments (6.10.1.6)

    c. Fundamental section properties (D6.1, D6.2, and D6.3)

    d. Constructability (6.10.3)

        (1) General (2.5.3 and 6.10.3.1)

        (2) Flexure (6.10.3.2, 6.10.1.8, 6.10.1.9, 6.10.1.10.1, 6.10.8.2, and A6.3.3)

        (3) Shear (6.10.3.3)

        (4) Deck placement (6.10.3.4)

        (5) Dead-load deflections (6.10.3.5)

    e. Service limit state (6.5.2 and 6.10.4)

        (1) Elastic deformations (6.10.4.1)

        (2) Permanent deformations (6.10.4.2)

            (i) General (6.10.4.2.1)

            (ii) Flexure (6.10.4.2.2, Appendix B, 6.10.1.9, and 6.10.1.10.1)

    f. Fatigue and fracture limit state (6.5.3 and 6.10.5)

        (1) Fracture (6.10.5.2 and 6.6.2)

        (2) Special fatigue requirement for webs (6.10.5.3)

    g. Strength limit state (6.5.4 and 6.10.6)

        (1) Composite sections in positive flexure (6.10.6.2.2 and 6.10.7)

        (2) Composite sections in negative flexure and noncomposite sections (6.10.6.2.3, 6.10.8, Appendix A, Appendix B, and D6.4)

        (3) Net section fracture (6.10.1.8)

        (4) Flange-strength reduction factors (6.10.1.10)

9. Dimension and connection detail requirements

    a. Material thickness (6.7.3)

    b. Bolted connections (6.13.2)

        (1) Minimum design capacity (6.13.1)

        (2) Net sections (6.8.3)

        (3) Bolt spacing limits (6.13.2.6)

        (4) Slip-critical bolt resistance (6.13.2.2 and 6.13.2.8)

        (5) Shear resistance (6.13.2.7)

        (6) Bearing resistance (6.13.2.9)

        (7) Tensile resistance (6.13.2.10)

    c. Welded connections (6.13.3)

    d. Block shear rupture resistance (6.13.4)

    e. Connection elements (6.13.5)

    f. Splices (6.13.6)

        (1) Bolted splices (6.13.6.1)

        (2) Welded splices (6.13.6.2)

    g. Cover plates (6.10.12)

    h. Diaphragms and cross-frames (6.7.4)

    i. Lateral bracing (6.7.5)

14.19.3 AASHTO LRFD Shear Capacity Evaluation of Steel Girders

All section numbers are based on AASHTO Specifications 2007.

1. Compact and noncompact sections

2. Strength limit states I to V:

    a. Construction limit state and uncured slab

    b. For stiffened webs, refer to 6.10.7.3.

    c. For unstiffened webs, refer to 6.10.7.2.

3. Interior panels of compact sections:

    a. Strength limit states I to V

  4. Interior panels of noncompact sections:
   a. Strength limit states I to V
  5. Noncompact section compression flange buckling:
   a. Strength Limit States I to V
  6. Noncomposite noncompact section lateral torsional buckling:
   a. Positive and negative flexure
   b. Strength limit states I to V
   c. $L_b < 1.76r_1(E/F_{yc})^{1/2}$

14.19.4 Evaluation of Fatigue Resistance

  1. Fatigue limit state
  2. Category C fatigue details and at sections:
   a. Refer to 6.6.1.2.5.
   b. Refer to 6.10.4.3 for web fatigue stress limit for moment.
   c. Refer to 6.10.4.4 for web fatigue stress limit for shear.
   d. Refer to 6.10.7.4 for stud and channel shear connectors fatigue.

14.19.5 Evaluation of Deflection

  1. Service limit states I to IV
  2. Refer to 6.10.3.2.
  3. Control of permanent deflection

14.19.6 Transverse and Bearing Stiffeners

   Strength limit states I to V:
  1. Refer to 6.10.8.
  2. Refer to 6.10.8.2.1 for bearing stiffener location.
  3. Refer to 6.10.8.2.4 for bearing stiffener geometry.

14.19.7 Shear Connectors

   Strength limit states I to V: Refer to 6.10.7.4.

14.19.8 Evaluation of Wind Effects M_w

  1. Strength limit states III and V only
  2. Construction limit state: Refer to 6.10.5.3.2 and 6.10.5.3.3.
  3. Refer to Sec. 6.10.5.3.2b for noncompact web slenderness.

14.20   *BEARINGS DESIGN*

14.20.1   Design Procedure

The reactions for each load combination should be applied to the substructure through bearings. Establish a minimum seat width. Bearings should conform to section 6.11.1.2 and bearing (5.7.5).

14.20.2   Bearing Retrofit

Types of Bearings.   Bridges may be upgraded for seismic resistance using isolation bearings to replace both fixed and expansion bearings. The need for multirotational or isolation bearings may be investigated for higher seismic zones.

Elastomeric bearings are relatively shallow, and differential height built-up is required. Pot bearings will be guided, nonguided, or fixed types. They offer better compatibility with existing bearings in terms of stiffness characteristics than elastomeric bearings.

Corrective Measures.   Expansion bearings can cause a malfunction owing to lack of maintenance. If any deficiencies exist, corrective measures must be incorporated into the contract plans. Expansion rocker bearings or pin/hinge-type bearings located on bridge abutments and piers have been known to perform poorly owing to bearing toppling and out-of-plane rotation. Temporary bearing repairs will be good for 5 to 10 years.

The weak links in bearing assemblies are the girder-to-sole-plate bolts or welds, pin heads, dowels, and the anchor bolts, all of which are capable of shear failure during a seismic event. At least one constructible scheme for bearing replacement must be shown in the contract documents. All related analyses, including the effects of jacking on all connections and elements, must be performed. Bearing design will be based on the LRFD method.

Typical modifications to steel bearings to withstand loadings should include increasing size, number, or embedment of anchor bolts. The options available for upgrading the support system for seismic resistance are bearing retrofit or bearing replacement, including the following:

1. Prevent toppling of existing bearings by installing longitudinal displacement stoppers.

2. Modify the existing bearings to resist seismic loads.

14.20.3   Alternate Schemes

Owing to the large number of rocker type bearings available for abutments and piers, it may be expensive to replace all the bearings. It is proposed to replace only the damaged, frozen, or contracted bearings. Selective replacements will be considered based on the seismic evaluation. The following schemes will be considered as alternates:

Scheme 1: Replace corroded expansion (high rocker-type) bearings and fixed bearings with appropriately guided, nonguided, or fixed pot bearings.

Scheme 2: Replace fixed bearings with elastomeric bearings, and replace all expansion bearings with an elastomeric bearing with or without a sliding plate.

Scheme 3: Maintain and retrofit bearings as required, and provide a keeper and/or catcher system to prevent toppling of rocker bearings and to limit movement of all bearings in the event of a shear failure.

Scheme 4: Maintain and retrofit bearings as required, and provide longitudinal restraints/ties as appropriate to reduce the risk of toppling of the rocker bearings.

14.20.4 Jacking of Girders

Bearings are usually replaced by jacking with hydraulic jacks placed on abutment seats where expansion bearings are located. Jacking beams need to be connected to webs by welding stiffener plates. Bolted connections at the ends of jacking beams need to be designed for the dead load. Jacking is usually not done under live load, and the traffic lanes need to be closed.

Consideration should be given to any required jacking procedure and constraints. Girders should be raised uniformly in a transverse direction in order to avoid inducing stresses into the superstructure. The following jacking design guidelines need to be considered:

1. Deck should be closed to traffic until jacking is done and bearing replacement is completed.
2. Do not include impact load to design jacking force requirements.
3. If shims and blocks are used for temporary support under traffic, design must include impact.
4. If strengthening during jacking is required, all details must be shown in contract documents.
5. A special provision or notes on the contract drawings will be added for the contractor to submit alternate schemes for seismic retrofit.
6. Minimum support widths for bridge seat: For minimum width requirements, a spreadsheet will be developed based on the AASHTO formula, and available bearing widths at each abutment and pier will be checked.

14.21 SUBSTRUCTURE DESIGN

14.21.1 Basic Procedures for Design

Compile force effects for substructure members based on the following effects and lateral loads:

1. Wind (3.8)
2. Water (3.7)
3. Scour (2.6.4.4.2)
4. Ice (3.9)
5. Earthquake (3.10 and 4.7.4)
6. Temperature (3.12.2, 3.12.3, and 4.6.6)
7. Superimposed deformation (3.12)
8. Ship collision (3.14 and 4.7.5)
9. Vehicular collision (3.6.5)
10. Braking force (3.6.4)
11. Centrifugal force (3.6.3)
12. Earth pressure (3.11)

14.21.2 Compile Load Combinations and Analyze Structure

1. Table 3.4.1-1
2. Special earthquake load combinations (3.10.8)

14.21.3   Design Compression Members (5.7.4)

1. Factored axial resistance (5.7.4.4)
2. Biaxial flexure (5.7.4.5)
3. Slenderness effects (4.5.3.2.2 and 5.7.4.3)
4. Transverse reinforcement (5.7.4.6)
5. Shear (usually Earthquake and ship collision induced) (3.10.9.4.3)
6. Reinforcement limits (5.7.4.2)
7. Durability (5.12)
8. Detailing and seismic (5.10.11)

14.21.4   Design Shallow or Deep Foundations (Geotechnical Considerations)

1. Scour
2. Footings (5.13.3)
3. Abutments, piers, and walls (11)
4. Pile detailing (5.13.4)

14.22   COMPARATIVE STUDY OF AASHTO AND SIMPLIFIED FORMULA

Simplified equations for moving load moments and shears (Table 14.4).

14.22.1   Construction Scheme

While structural analysis and design are based on theoretical assumptions, it is important that the constructed model match as closely as possible the theoretical model. Also, designing for construction loads should ensure trouble-free fabrication, erection, and construction.

Constructability issues are related to the following and can be site-specific:

- Transportation of girders
- Crane capacity
- Materials storage and construction duration

Failure during construction is the biggest single source of bridge collapse. Even though bridge members are designed for long-term loads, short-term design of temporary works is required.

1. There is a need to define a feasible method of construction in the contract documents. However, an alternate proposal from the contractor for a detailed workable procedure needs to be considered. It will be reviewed for safety and quality assurance.
2. The contractor's selection should be based on available resources, ingenuity, and experience.
3. Fabrication procedures should be described in the technical specifications. Vertical camber of a girder can be achieved by cutting the web to required contour and by heat curving the flanges to the required contour prior to welding with the web.
4. An erection plan showing location and capacity of the crane and sizes and elevations of temporary supports should be developed. It will be unique for each project for access to site, construction easement, and utilities relocation.

TABLE 14.4 Maximum Live Load Moments and Shear Forces for H-20 Truck (Alternate Lane Loads Apply)

Comparative Study of AASHTO and Simplified Formula				
	H20 Truck Moments		H20 Truck Shear	
Span Length (feet)	AASHTO LL Moment (kip-ft)	$M_{max} =$ $10/L(L-2.8)^{2}$* (kip-ft)	AASHTO Reaction/Shear Force (kips)	$V_{max} =$ $8(5L-14)/L$* (kips)
Small spans				
20	160	147.92	34.4	
30	246.6	246.61	36.3	49.60
40	346.0	345.96	36.8	55.20
Medium spans				
50	445.6	445.57	58.5	58.56
60	558.0	545.31	60.8	60.80
70	707.0	645.12	62.4	62.40
80	872.0	744.98	63.6	63.60
90	1053.0	844.87	64.5	64.53
100	1250.0	944.78	65.3	65.28
110	1463.0	1044.72	65.9	65.89
120	1692.0	1144.65	66.4	66.40
Long spans				
130	2063.1	2063.10	67.6	66.83
140	2242.8	2242.89	70.8	67.20
150		2422.704		67.52
160		2602.54		67.8
170		2782.40		68.06
180	Lane load with 18 kips concentrated load for bending governs	2962.27	Lane load with 26 kips concentrated load for shear governs	68.27
190		3142.16		68.46
200		3322.05		68.64
210		3501.96		68.8
220		3681.88		68.95
230		3861.80		69.08
240		4041.73		69.2

*Developed by the author.

5. Temporary supports and falsework should be designed for horizontal and vertical reactions from construction loads.

6. Precautions during construction: To prevent debris from falling below during deck slab construction, "containment methods," such as installing wire nets below the deck, will be evaluated.

7. Anticipated construction loads should be specified for each stage of construction.

14.22.3 Constructability Review

The constructability issues for proposed widening should be considered carefully. The subassemblies fabricated should be able to fit properly in the field. Constraints and field verification of conditions will help in planning. The major constraint is that the widened bridge should conform to existing conditions, including traffic count of truck live loads. For all widening, the available existing bridge plans must depict the actual field conditions.

1. The constructability review should include a simulation of construction sequencing requirements on the project. An on-site review may be necessary to ensure that all project elements are considered in the constructability review. Designers should become familiar with current local bridge construction practices so that most designs can be built with conventional equipment or local materials.

2. Overseeing construction and quality control:
   - Identifying constructability issues at the design stage
   - Preparing detailed contract documents
   - Developing a realistic construction schedule
   - Application of critical path method (CPM)
   - Shop drawings review
   - Request for information (RFI)
   - Design change notice (DCN) procedures
   - Developing quality-assurance (QA) and quality-control (QC) procedures for the project

14.23 ACCELERATED BRIDGE CONSTRUCTION (ABC)

14.23.1 Introduction

As bridges get older, highway agencies have on their agendas hundreds of bridges to reconstruct and open to the public in any given year. During lengthy construction periods, traffic problems are compounded at multiple bridge sites. This results in loss of useful personnel-hours.

14.23.2 Scope and Limitations of ABC Approach

ABC is useful for emergency replacement of bridges damaged by vehicular accidents, ship collisions, floods, or earthquakes, for which accelerated planning and design also will be necessary. Using ABC, formwork or much of concrete placement and curing are not required. However, activities such as bore-hole tests, pile driving, shop drawing review, and closure pour are unchanged. An ABC design code based on Accelerated Bridge Planning (ABP) for alternatives and connections will be required. While there is saving in construction time, design effort is increased owing to numerous precast joints and components.

Span lengths in concrete bridges are restricted to about 100 feet owing to transport restrictions of heavy components. While cast-in-place construction is time-tested, ABC applications are of recent origin. Compared with a unified integral abutment bridge, a precast bridge with numerous deck joints is likely to be weaker during earthquakes. Full-scale testing is required to develop confidence, especially for curved bridges. Maintenance and Protection of Traffic (MPT), approach-slab construction, permits, and utilities relocation are unavoidable constraints.

Contractors, in general, have trained technicians in formwork and cast-in-place construction, but new training in ABC is required. Overemphasis on incentives/disincentives pressurizes the contractor into adopting unrealistic schedules at the expense of quality control. Also, the manufacturing nature of precast products creates a proprietary system and monopolistic environment that may lead to unemployment of some number of construction workers.

14.23.3 Objectives of ABC/ABP

The primary objectives are constructability, erection, serviceability, durability, maintainability, inspectability, economy, and aesthetics. Detailed objectives are as follows:

1. *Develop modern codes and construction specifications.* Construction mistakes, ship collisions, scour, design deficiency, overload, fatigue, and earthquakes are the main causes of failure.

2. *Introduce specialized training of designers and field staff.* Use of applied mathematical methods and fracture mechanics in design need to be developed and promoted.

3. Monitor construction activities on the "critical path" for saving time and long-term rehabilitation costs.

4. Ensure safety during construction and of the bridge users after completion.

5. Obtain traffic counts for selecting full or partial detour, and apply temporary construction staging.

6. Optimize the size and number of girders, and use modern material and equipment.

7. Selecting pleasant bridge colors and aesthetics would keep "America the Beautiful."

Superstructure. Crews can cut the old bridge spans into segments and remove them, prepare the gaps for the new composite unit, and then set the new fabricated unit in place in an overnight operation. The quicker installation minimizes huge daily delay-related costs and daily traffic-control costs.

Construction usually is scheduled for the fall months, when weather is more predictable. A single-course deck will save a minimum of 6 weeks in construction time compared with a two-course deck.

Use of HPS: I designed bridges with HPS 70W hybrid girders in New Jersey recently. The hybrid allows longer span and lighter girders. Shallower girders improve vertical underclearance, reduce the number of girders to be constructed, eliminate painting, and weathering steel provides enhanced resistance to fracture.

Parapets. A number of parapets besides Jersey barriers are used. The NJDOT allows its contractors to use slip forms for increasing speed of construction, as done successfully on the Route I-195 and Route I-295 interchange.

Substructure. Integral abutments with fewer piles have been used successfully in New Jersey. They can be constructed more quickly than conventional bridges. An example of an integral abutment bridge using prestressed concrete box beams on Route 46 over Peckman's River was designed by me.

Precast post-tensioned pier caps were used recently on the Route 9 over the Raritan River and by me on the interchange of US Route 322 and NJ Route 50. Drilled shaft foundations and concrete cylinder piles of 36 to 66 inches in diameter are in use. Precast sheeting has been used for retaining walls and abutments. MSE abutments have performed extremely well.

Scour Countermeasures. Minimal marine life disruption and quick construction are achieved by using gabion baskets, articulated concrete, or cable-tied blocks in lieu of traditional sheet piles. I prepared a *Handbook for Scour Countermeasures* for the NJDOT jointly with CUNY, and it was approved by the FHWA, and in addition, I helped develop sections 45 and 46 of the NJDOT *Bridge Design Manual.*

14.23.4   Understanding Rapid Construction and Associated Needs

Requirements. Many innovative concepts are presented here. Each concept is a subject in itself. Modern prefabricated construction materials and methods are vastly different from traditional methods and require innovative ideas for making the system safe and efficient. ABC can be promoted by understanding and analysis.

The owner's requirements are clear: reduction in schedule, wider decks, reduced seismic effects, increased bridge ratings, longer service life, cost savings, and lower maintenance. Hence ABP should promote durability and compliance with environmental/preservation laws.

Accelerated Planning and Design.   An evaluation matrix for construction cost, life-cycle cost, environmental and social impact, constructability, future maintenance, inspection, and aesthetics needs to be prepared at the planning stage. Planning must address differences in ABC of small, medium, and long spans.

Efficient structural planning is necessary to minimize the constraints in meeting accelerated construction goals such as erection during extreme weather. Following factors may be considered:

1. *Improving aesthetics.* To improve appearance, an exterior arched beam can be provided to hide the fascia girder.
2. Installing interactive touch screens that feature bridge information for motorists.
3. Increasing rider comfort by using a durable deck overlay protective system to prevent deck cracking and post-tensioning to prevent cracks in concrete and noncorrosive reinforcement, thus improving crashworthiness of barriers and parapets.
4. Improving bearing performance—use of multirotational and isolation bearings in seismic zones.
5. Improving bridge lighting and drainage methods, and installing solar roof panels at approaches for bridge lighting and signage.
6. Using grouting methods for improving foundation soil, including liquefaction mitigation to protect foundations in saltwater against corrosion, and increasing resistance to earthquakes, liquefaction, and erosion of soil under footings; preventing fender damage from vessel collision.
7. Constructing bioretention ponds that collect and filter runoff from the bridge deck.
8. Introducing ground improvement techniques, modifying soil.

Modern Concrete Technology and Weathering Steel.   Concrete bridges are used more commonly for the smaller spans because rust in steel members increases corrosion and maintenance costs.

1. *Ultra HP FRC.* Compressive strength reaching 30 ksi is possible along with flexural strength of 7 ksi.
2. *Using HP lightweight aggregates.* Lightweight HPC reduces deadweight, enables longer span lengths, reduces the number of piers in a river, and allows fish travel to have the least obstruction.
3. Spliced girders of varying depth enable lightweight concrete to achieve spans of over 200 feet.
4. *Overlays.* Use of silica fume and high early strength Latex Modified Concrete (LMC) will open the deck to traffic within 3 hours of curing. Silica fume, pozzolans, fly ash, and slag may be used to reduce concrete permeability and heat of hydration. Fly ash and cenospheres are preferred for high-performance concrete (HPC) in bridge decks, piers, and footings. By-products of coal fuel such as fly ash, flue gas desulfurization materials, and boiler slag provide extraordinary technical, commercial, and sustainable advantages.
5. *Self-consolidating concrete (SCC).* Since vibration time is saved, SCC helps ABC; this is a more workable concrete with lower permeability than conventional concretes.
6. Develop preferred alternative structural solutions and optimization of girders using HPS 70W, 100W, and hybrid steel girders. Use of weathering steel minimizes painting costs.

Prefabrication of Components and Connection Details.

1. Contractor must be knowledgeable about latest technology and availability of new bridge components. Develop typical connection details for precast deck panels, piers, and abutments. Improve the quality of the superstructure by fabricating in a more controlled factory environment.
2. To facilitate one-time shopping for construction products, large supply stores on the pattern of Home Depot and Walmart need to be set up. Greater technical service be provided on bridge products such as precast deck units, girders with welded shear connectors, diaphragms, bearings, parapets, precast pier units, etc. Quick assembly is possible by systems such as Conspan, Inverset, Effideck, and SpaanSpan drop-in deck panels and post-tensioning in both directions.

3. Apply segmental construction for long spans to use ABC. The balanced cantilever method eliminates the need for formwork.

Design Code and Specifications Development.

1. Codes are being revised every few years. Future codes for ABP should incorporate principles of sustainability and context-sensitive design. *Practice needs to be backed with sound theory.* It is not correct to use Rankine or Coulomb empirical formulas in seismic zones. Also, most pier caps act as deep beams. Their shear-reinforcing details should use the strut and tie model.

2. Fabricators and erectors should follow AASHTO construction specifications, and fit-up should be assumed to be performed under the no-load condition, for which evaluation of erection stress is required.

3. Construction load combinations not covered by the AASHTO LRFD *Bridge Design Code* are summarized:

   a. For strength I conditions, use all dead load of bridge components (*DC*), utilities (*DW*), construction equipment load (*CEL*) such as screed, and construction live load (*CLL*). $\gamma_p = 0.9$ to 1.25. Using AASHTO notations, the construction live load combination is $\gamma_p(DC + DW) + 1.5(CEL + CLL)$.

   b. For strength III conditions, use wind on the superstructure, including forming. Construction wind load combination is $\gamma_p(DC + DW) + 1.5(CEL) + 1_{0.4}(WS)$.

   c. For strength V conditions, use construction wind load on equipment (*WCEL*); the load combination is $\gamma_p(DC + DW) + 1.5(CEL) + 1.35(CLL) + 0.4(WS) + 1.0(WCEL)$; $\gamma_p = 0.9$ to 1.25.

4. Specification developments will ensure quality assurance with ABC.

Training in New Technology.   It is necessary that construction personnel be given on-the-job training in the use of new technology.

1. Training aspects should cover tasks such as understanding erection procedures, erection drawings, modern concrete technology, and use of new steel, processed timber, etc. in bridge construction.

2. Continuing education in design and construction and rigorous certification requirements need to be introduced for certain levels of construction personnel.

Greater Scope of Postdesign Activities.   A workshop should be conducted to evaluate constructability. Topics would include equipment locations, construction duration, access, right-of-way, and material availability. To resolve constructability issues and any misinterpretations of contract drawings, designers should be consulted during preparation.

14.23.5 Promoting Rapid Construction

Alternative Design-Build Construction Procedure.   This approach still has led to faster turnout. For large projects, a design-build-finance-operate-maintain approach is most likely to be a complete situation. Simpler design-build is more common by placing the builder and the designer on one team.

Modern Construction Equipment.   Success of ABC is due to powerful equipment. Different erection equipment is required for girders, box beams, trusses, arches, and cable-stayed and suspension-cable bridges. Timely availability, leasing facility, and an experienced erection team will be necessary.

The erection contractor should use robotics, cranes such as tower cranes (for maximum lightweight pickup of 20 tons and heights greater than 400 feet), lattice-boom crawler cranes, mobile lattice-boom cranes, mobile hydraulic cranes, and lattice ringer cranes for varying heavy pickups and accessories. In addition, the specialized technology for the superstructure roll-in and roll-out method using self-propelled modular transport (SPMT) is illustrated in recently published FHWA material.

Improved Traffic Control and Maintenance and Protection of Traffic.   Improve contractor and motorist safety by reducing the time that a work zone is in use. Work on approaches and the bridge deck will require a detour. Investigate alternatives, either to detour one or both directions of traffic, and provide for pedestrians. An 8- to 10-hour night window is required. A traffic count needs to be performed to assess the impact on traffic flow during construction.

14.23.6   Erection Methods and Precautions to Prevent Construction Failures

A study of bridge failures that I carried out concluded that most failures occur during construction or erection. An ABC system must avoid future failures.

1. Failure of connections owing to overstress from bolt tightening, failure of formwork, local buckling of scaffolding, crane collapse, and overload were some of the causes.
2. Stability of girders during stage construction and the deck placement sequence needs to be investigated and temporary bracing provided. Expansion bearings need to be temporarily restrained during erection.
3. Some flexibility in selecting bolt splice locations may be permitted with approval of the designer.
4. Curved and skew bridges require special considerations such as uplift at supports, achieving cambers, and reducing differential deflections between girders during erection.

14.23.7   Complying with Permitting Regulations and Insurance Against Liabilities

The purpose of permits is to maintain air and water quality, health, and noise abatement. Stream encroachment, navigation permits, grants (tideland, riparian, etc.), environmental assessments (EAs), environmental impact statements (EISs), categorical exclusions (CE), and department of environmental protection (DEP) regulations would apply. The contractor will expedite acquiring the applicable construction permits. Each type of insurance will be covered by state law and should be project-specific (e.g., workers' compensation and employers' liability).

14.23.8   Resolving Utility Relocation Issues

The utility task usually is on a critical path; the project manager will begin coordination with the utility owners right after the notice to proceed. Consultants must have state-of-the-art equipment and custom rigs to provide a full array of field services for subsurface utility assignment.

14.23.9   Continued Development and Research in ABC Methods

Extending the Service Life of a Bridge.   ABC methods were evolved well ahead of design codes. Research is required in many areas:

1. Develop strengthening methods and corrosion-mitigation techniques; fabricate stronger girders by eliminating the need for shear stiffeners with the use of a folded web plate in steel girders.
2. Develop new methods to monitor foundations and to strengthen against scour, earthquake, and impact damage.
3. Revive and develop the concept of a full canopy on bridges and approach slabs to facilitate mobility, improved drainage, skid prevention, and the use of deicing agents.

Optimal Use of Construction Materials.   Conduct research into the use of FRP composite materials, geomaterials, geosynthetic products, and lightweight, high-performance, and ultra-high-performance concretes and steels. Develop appropriate limit-state criteria for the use of these materials, details, components, and optimizing structures for adoption into the *LRFD Specifications.*

Duration of Bridge Construction.   Develop contracting strategies such as realistic incentives/disincentives to encourage speed and quality. The mass-production management techniques adopted by the automobile/aircraft industry may be considered.

Mathematical Methods, ABC Codes, and Technical Specifications.   Learn and disseminate knowledge of emerging technology and methods.

1. A strong backup of mathematical technique, closed-form solutions, and formulas from mathematicians is needed.
2. Identify and calibrate the service-limit states for unusual construction conditions.
3. Continued development of LRFR provisions is required. Integrate information from maintenance and operations into code development and vice versa. Promote automated data collection and reporting damage models using the data collected.
4. Begin transition to a performance-based specification with an accompanying design manual.
5. Develop and incorporate security performance standards, especially for long spans.

14.23.10 Conclusions

1. Initiatives taken by the FHWA have led to considerable progress in implementing ABC concepts. ABC-related design needs to be made part of AASHTO and state bridge design codes and specifications.
2. To gain confidence in structural behavior, full-scale testing of joints in a precast curved deck is required.
3. Mathematical methods of analysis applicable to discontinuities of components need to be developed.
4. Application of the latest techniques in concrete manufacture, composites, HPS, and hybrid materials is feasible. Integrated software covering all aspects of ABC design and drawings should be developed.
5. Deterrents and bottlenecks such as Maintenance and Protection of Traffic (MPT), construction easement, right-of-way, permit approvals, and utilities relocation need to be resolved and administrative procedures simplified to facilitate ABC.
6. Certification and training of construction personnel, continuing education of engineers in rapid construction techniques, and construction management ABC courses at universities are recommended.

14.24 DESIGN REVIEW BASED ON CONSTRUCTABILITY

The design should be reviewed to identify any of the following potential construction problems:

1. Is the sequence of construction practical?
2. How will each component be constructed?
3. Can the structure actually be built in accordance with the plans?
4. Is the design economical, or can it be constructed with conventional equipment by experienced contractors who typically bid bridge projects?

5. Transportation and storage of materials at site

6. MPT: How will traffic be maintained at each stage of construction?

7. Right-of-way during construction: How will the contractor access the site?

8. MSF: Maintaining stream flow during construction for a bridge over a stream

9. Dewatering the work area

10. Removal of forms and temporary sheet piles

11. Utility relocations

12. Delays in acquiring construction permits and affect on the schedule

14.24.1 Allowances to Be Made in Design and Construction

1. Subzero and freezing temperatures

2. Hot weather

3. Low and high wind

4. Construction during all traffic lanes open

5. Construction during partial lane closure

6. Nighttime construction

14.24.2 Safety Considerations

1. Working adjacent to fast, unpredictable currents and rapidly rising water levels can be extremely dangerous. Safety of construction workers is an important aspect of emergency work.

2. Floating or subsurface debris contributes to hazard during emergency work.

3. Weather conditions (i.e., rain, snow, or darkness) may endanger safety further.

4. OSHA recommendations for slopes of excavations in soils. The following maximum values of slopes should be used for excavation of sloping structures:

   - Solid rock—90°
   - Compacted angular gravels—0.5:1 (63°, 26 feet)
   - Average soil—1:1 (45°)
   - Compacted sand—1.5:1 (33°, 41 feet)
   - Loose sand—2:1 (26°, 34 feet)

   The angle of repose should be flattened when an excavation has water conditions.

14.24.3 Traffic and Utilities Issues

The following steps are required:

1. *Site access.* Adequate access to the site should be provided for trucks to deliver materials.

2. *Right-of-way.* Construction easements and rights-of-way may have to be purchased for the duration of construction.

3. *Possible detours.* Detours, lane closures, or nighttime work may be necessary. Coordination with traffic control would be required.

4. Emergency vehicles and school bus services should not be affected by lane closures.

5. *Utilities.* Relocation of any utilities at the sides of abutments or piers may be necessary for the duration of construction. Coordination with utility companies would be required.

6. Four weeks before the start of construction, traffic police needs to be informed of shutdowns or detours of one or more lanes.

7. Warning signs showing dates and times of shutdowns are required to be posted well in advance for the information of users.

8. Construction time must be kept to a minimum or performed at night.

14.25 MAINTAINING THE ENVIRONMENT

14.25.1 Environmental Concerns

1. The purpose of reconstruction is to benefit transport facility users. The process of short-term construction and long-term impacts should not be adverse to road users or local residents.

2. The National Environmental Policy Act (NEPA) articulated national environmental policy, established federal agency responsibility, and created a basis for the federal decision-making process.

3. A large number of environmental concerns must be addressed related to water:
   - Maintaining water quality
   - Providing fish passage
   - Preventing wetlands contamination
   - Preventing stream encroachment
   - Improving drainage
   - Mitigating construction impact on a floodplain
   - Soil erosion and sediment transport: Minimizing the erosion of native substrate owing to sediment transport after installation

4. Issues related to ecology:
   - *Preservation of vegetative species:* Ecology (flora and fauna), minimizing impacts to natural vegetation by controlling construction access points
   - Revegetation of disturbed areas with species may be required
   - Landscape preservation
   - Preservation of endangered species

5. Maintaining air and water quality:
   - Maintaining air quality and preventing air contamination, pollution
   - Noise mitigation from construction vehicles, pile driving, concreting, excavation, welding, etc.
   - Relocation hazards of underground and bridge supported utilities
   - Reactions with acid-producing soils

14.25.2 Permit Requirements

Engineering data and documentation are required for permit approval. As per regulations, the following reports/proforma need to be submitted:

- An environmental assessment (EA) is required when the significance of impact on the environment is not clearly established.

- An environmental impact statement (EIS) needs to be prepared when a replacement or new bridge (usually with four or more lanes) has a significant impact on natural, ecological, or cultural resources, including endangered species, wetlands, floodplains, groundwater, fauna, and flora. An EIS is required when there are impacts on properties protected by the DOT Act or on those protected under the Historic Preservation Act. Significant impacts on noise and air quality need to be avoided.

- An action that does not have significant effect on the environment falls under Categorical Exclusion CE. Examples provided by the FHWA are reconstruction or modification of two-lane bridges, adding pedestrian or bicycle lanes, widening for shoulders, installation of signs, etc.

14.26 SEISMIC EFFECTS

14.26.1 Introduction

Bridges should be designed to have a low probability of collapse. The seismic hazard at a bridge site should be characterized by an acceleration response spectrum for the site and site factors. A site-specific procedure should be used if the bridge is located within 6 miles of an active fault or when a long-duration earthquake is expected. For peak ground acceleration (PGA) coefficients, maps given in AASHTO Interim Specifications 2008 should be referred to.

Bridge movements will be influenced by soil type and profile. Site class A (hard rock) to site class F (peat or highly organic clays) are the extreme conditions defined by AASHTO Table 3.10.3.1-1.

1. *Site coefficients and site effects.* Given the large variations in the values of site coefficients, soil profiles will be based on soil composition at the site. A geotechnical investigation should be performed to determine the soil conditions, whether cohesive or cohesion-less, the type of rock, sand, or gravel, and stiff clay, soft clay, or silt.

2. *Liquefaction.* The potential for soil liquefaction and liquefaction-related ground instability should be investigated at relevant locations along project alignments. The effects of settlement of footings, loss in bearing capacity, and increased lateral earth pressures should be considered in the design of abutments, walls, and footings.

3. *Seismic slope instability and landslides.* The potential for seismic-induced slope movements and landslides along the proposed alignment should be investigated. Mitigation measures should be incorporated in the design of abutments, walls, and footings.

14.26.2 Computation of Seismic Forces

The magnitude of seismic forces will depend on

1. The dead weight of the structure
2. Ground motion (acceleration coefficient)
3. The type of soil
4. The fundamental period of vibration
5. Importance classification
    1. The seismic design of new highway structures should follow the requirements of the AASHTO *LRFD Specifications* for design of bridges, with current interims.
    2. Suspension cable, cable-stayed, arch, and movable-type bridges are not covered by the AASHTO *LRFD Bridge Design Specifications*.

3. A seismic design is usually not required for buried structures or culvert structures. The guidance provided in 3.10.1 of the AASHTO *LRFD Bridge Design Specifications* should be referred to.

4. Provide seismic ductility design at the locations where plastic hinges will form on all new structures. Ductility is an important characteristic of structures because a ductile structure can absorb much more force than a nonductile structure before it fails. Conversely, nonductile structures such as unreinforced masonry or inadequately reinforced concrete are very dangerous because of the possibility of brittle failure.

14.26.3 Seismic Design of Abutments and Walls

1. Abutments do not need to be designed for seismic forces from the superstructure except for bridges with integral abutments. For bridges with semi-integral abutments, only the width of the bridge seat should be checked.

2. Abutments should be designed for static earth pressure and additional seismic-induced forces using the Mononobe-Okabe method, which is an extension of Coulomb's method for soil pressure on retaining walls.

3. Backfill is assumed to be unsaturated so that liquefaction effects are negligible. The backfill is assumed to be cohesionless.

4. Seismically induced active and passive pressures will be considered.

5. Foundation types for integral abutments. The abutment and pile design should assume that the girders transfer horizontal forces from seismic loads, in addition to moments and vertical forces.

6. *Retaining walls.* Seismic loads as specified in 3.11.4 of the AASHTO *LRFD Bridge Design Specifications* and FHWA Publication No. FHWA HI-99-012 should be referred to. Seismic forces on wingwall/retaining wall may be neglected for an economical design, except for design of wingwalls/retaining walls belonging to bridges that are classified as critical.

7. *Mechanically stabilized earthwalls (MSEs).* Seismic design of MSEs should be in accordance with FHWA-SA-96-071, "Mechanically Stabilized Earth Walls and Reinforced Soil Slopes—Design and Construction Guidelines."

14.26.4 Bearings

1. The current AASHTO *LRFD Specifications for Highway Bridges* also should be referred to for guidance in providing designs for pot, disk, and elastomeric-type bearings.

2. The AASHTO *Guide Specifications for Seismic Isolation Design* should be used for designing isolation bearings when they have been deemed necessary for accommodating seismic loads. These bearings have special performance characteristics that will alter the dynamic response of a bridge.

3. Superstructure forces can be reduced by factors of 2 to 5 in the lower seismic zones, and there are corresponding reductions in the forces transferred to piers and abutments.

14.26.5 Load Combinations

Extreme event I load combinations should be applicable. The guidance provided in section 3.4 of the AASHTO *LRFD Bridge Design Specifications* should be referred to. Forces from dead-load analysis and analysis for other applicable loads for extreme events will be combined, with the forces from single- or multimode analysis as follows:

- 100 percent of longitudinal seismic forces + 30 percent of transverse seismic forces
- 100 percent of transverse seismic forces + 30 percent of longitudinal seismic forces

14.26.6   Resistance Factors

Design of reinforced-concrete sections of abutments and piers will be based on ϕ factors. The guidance provided in 5.5.4.1 of the AASHTO *LRFD Bridge Design Specifications* should be referred to.

14.27   SEISMIC DESIGN OF NEW HIGHWAY STRUCTURES

1. Reinforcing bars provided in foundation and column/wall should be adequate to resist the design moments and shear forces.

2. Standard details for ductility will be followed. Whenever appropriate, the design detailing requirements recommended by Applied Technology Council (1996) Publication ATC-32—"Improved Seismic Design Criteria for California Bridges" should be considered.

3. Seismic detailing of reinforced-concrete footings, abutments, and piers should conform to the examples of standard details shown here.

14.28   SEISMIC RETROFIT OF EXISTING HIGHWAY BRIDGES

14.28.1   Scope of Retrofit

The seismic retrofit design of existing highway structures should follow the guidelines of the FHWA publication, *Seismic Retrofitting Manual for Highway Bridges*. Inspection procedures discussed in the *Handbook of Bridge Inspection*, by Sung H. Park, also may be consulted. FHWA Research Reports FHWA-IP-87-6, FHWA-RD-83-007, and FHWA-RD-94-052 also contain acceptable references for retrofit details.

1. Steel or concrete box girders

2. Continuous steel or concrete girders

3. Bridges that do not use conventional elastomeric bearing pads or those that rely on a system of fixed or sliding bearings for transmitting elongation changes

4. Longer spans generally in excess of 150 feet

5. Curved bridges with a sharp radius generally less than 1000 feet

6. Concrete segmental, steel cable–supported or bridges with complex or unusual geometry

A seismic retrofit report should be prepared to provide a determination as to a bridge structure's eligibility for a seismic retrofit.

1. A flowchart to provide guidance in determining if a bridge structure qualifies as a seismic retrofit candidate can be found in section 1.45.12. The results of the analysis, performed in accordance with the flowchart, should be provided in the seismic retrofit report.

2. In preparing the seismic retrofit report, the following guidance should be followed. Initially, seismic retrofitting of a bridge structure should be considered only under the following conditions:

   - The planned work will involve widening of a deck by more than 30 percent of its area.

   - The planned work will involve an entire deck replacement.

   - The planned work will involve superstructure rehabilitation or replacement, major abutment or pier repairs to bearing seat areas, or bearing repairs or replacement.

3. Several methods of seismic retrofit are outlined for bearings and expansion joints within the FHWA *Retrofit Manual*.

14.29 SEISMIC ANALYSIS OF INTEGRAL ABUTMENTS

14.29.1 Introduction

In recent years, integral abutments have been used widely. Their advantages may be summarized as follows:

1. They minimize maintenance of the deck slab by eliminating deck joints or expansion dams.

2. Integral abutments, in the form of stub or spill-through abutments and supported on a single row of piles, result in lower foundation costs

3. For integral abutments located over waterways and streams, up to a 50 percent reduction in local scour is attainable.

4. Owing to the special rotation mechanism provided on top of piles, beams expand and contract more efficiently than those in conventional bridges.

There are no significant disadvantages, except that a single-span bridge becomes statically indeterminate when beam ends are cast integral with the concrete abutment. Seismic forces for a single-span bridge are no longer negligible. The AASHTO LRFD load combinations of extreme events such as earthquakes become applicable.

14.29.2 Definition of Problem

The FHWA has laid down an analysis method of integral abutments for thermal forces only. Recently published NCHRP Draft Code 12-49 suggests a different approach for designing integral abutments for seismic moments. It is observed that design considerations for integral abutments deal mainly with thermal analysis only. End conditions of girders generally are assumed as pinned connections and seldom as partially fixed connections.

The tops of piles are free to rotate and translate. Hence full fixity, which requires zero rotation at the tops of piles, is not likely to occur. For displacement compatibility between tops of piles and beam ends, beams therefore will not have full fixity. However, during a seismic event, the ends of beams will lock at the interface with the relief slab boundary, giving rise to partial beam fixity and frame action in the longitudinal direction. In the transverse (stronger) direction, abutments are continuous and cast monolithic over piles, and multiple frame action would result.

The effect of underestimating fixity moments may result in cracking of concrete in the deck slab located directly over abutments. Seismic modeling must be considered, and an adequate thickness of slab must be provided.

14.29.3 Structural Planning

The parameters to be considered in the theoretical model include

- *Pile spacing.* Maximum pile spacing will be determined by frame analysis. Use a minimum pile spacing of 2.5 times the diameter of the pile, as allowed by AASHTO. For a composite frame action between beams and piles, place a pile preferably at every beam end.

- *Beam spacing.* By increasing the spacing of beams, the mass transferred and the seismic forces acting on the piles are reduced significantly. Avoid adjacent box beams because larger seismic forces would result owing to an increase in mass from the close spacing of beams. Also, the thinner deck slab used with adjacent box beams will increase tensile stresses at the tops of beams.

- *Type of beam.* For small spans and for beams spanning over rivers, where corrosion is a problem, spread concrete beams may be used. In other cases, steel beams are preferable to concrete beams owing to the smaller dead weight of steel beams.

- *Pile length and column-soil interaction.* An equivalent pile height can be determined from analysis of a single pile with a program such as L-Pile or COMP624P. The height of the pile can be obtained from computer output for first zero displacement of pile.

- *Pile orientation.* Since freedom from seismic and thermal displacements, rather than strength, is the main consideration, place the minor axis of the pile perpendicular to the direction of the beam.

14.29.4   Computer Model

Integral abutments will be idealized as a single pile bent in the longitudinal direction for a single span and a pile bent continuously over piers for multiple spans. For seismic analysis in the transverse direction, continuous frames need to be considered. Given the larger number of piles, transverse displacements during a seismic event will be smaller than displacements in the longitudinal direction.

14.29.5   Assumptions for Analysis

Frictional forces during a seismic event will be neglected. Unlike thermal movements, which are gradual and have well-defined directions, seismic forces take place in random directions, including upward. Vertical components are negligible. Given the small heights of abutments under the deck slab, passive pressure is negligible. Mononobe-Okabe seismic pressure is neglected.

Case Study of a Route 46 Bridge in New Jersey.   Simplified frame model: For a region that does not have a high degree of seismicity, and all the counties are classified as zone 2. Hence a quasi-static method was used rather than a dynamic analysis such as the response spectrum method. The uniform load method is applicable to a low-seismicity regular bridge with an acceleration coefficient of 0.18. Elastic seismic response coefficient Cs was computed. A frame analysis program, STAAD-Pro, was used. Seismic load combinations of (DL + 100% longitudinal + 30% transverse) and (DL + 30% longitudinal + 100% transverse) were used.

14.29.6   Conclusions

Modeling of integral abutments for thermal forces is a statically determinate problem. The same model that allows for thermal expansion and contraction cannot be used for the modeling of seismic forces. During a seismic event, displacements are high, and locking of the beam ends takes place. Owing to rotational compatibility between beams and the pile, moment distribution will take place according to the relative stiffness of the beams and pile.

The case study for seismic analysis of the Route 46 bridge shows that the pile design is based on pile behavior for an embedded pile and for an unembedded top portion. The unembedded pile length needs to be checked for slenderness ratio (kL/r) for seismic moments in the two directions, in addition to axial forces from dead and live loads. The effective height of the pile can be evaluated from a pile analysis program. With a seismic analysis, the reinforcement requirements at the top of the deck slab can be calculated. An interactive approach is required to include soil interactions.

14.30   CONSTRUCTION OVER RIVERS

Before starting a design, it is important to know the ground and river realities and construction factors that are likely to affect the selection of countermeasures. The following constructability issues need to be evaluated:

1. *Duration of construction.* The available flow width may be reduced owing to construction of cofferdams, embankments, and dams.

2. Flow velocities through the reduced channel opening will increase, thereby scour in the channel and around the structure is increased. Hence construction of the preceding items should be done during off-flood season.

3. *Maintenance and protection of traffic.* During installation, small cranes or pile-driving equipment may be parked on a lane or shoulders. A lane closure then would be required.

4. Coordination with traffic police and local officials would be necessary.

5. *Underwater work.* The health and safety of construction personnel may be of concern if the water depth is high. Trained divers will be required.

6. *Access to site.* A temporary road for transportation of materials and equipment adjacent to the channel bank may be difficult to construct. Wooden mats are used when lane width is restricted.

7. *Temporary works.* Temporary construction works may be required. More economical alternatives implementing quick construction and safety needs to be carefully evaluated.

8. *Safety of personnel.* Given the instability of the banks because of recent floods (for banks with slopes steeper than 1:1), sudden collapse of the bank may occur. Occupational Safety and Health Administration (OSHA) safety standards must be followed.

9. *Environmental risks.* Pollution of the river from construction materials may occur. Turbidity dams on both sides of the channel are needed. The channel needs to be cleaned. Approvals for stream encroachment permits would be necessary.

10. *Impact on existing utilities.* The effect of driving sheeting or bed armoring on existing utilities needs to be evaluated. Utilities may be relocated in such cases. Coordination and approval from utility companies would be required.

11. *Impact on right-of-way.* Countermeasures may extend into adjacent property limits. Rights-of-way need to be purchased in such cases. Similarly, encroachment on adjacent property during construction may occur. A construction easement needs to be determined and permits obtained.

12. *Specialized work.* Modern countermeasures require new construction techniques. The contractor performing such tasks needs to train his or her construction crew for such techniques.

13. *Availability of labor and plant.* Some types such as gabions, interlocking blocks, and stone pitching require experienced labor. Since local labor may not be familiar with the work, bringing labor from long distances may be an added expense.

14. *Limited vertical clearance under the bridge.* It may be difficult to construct a cofferdam or drive sheeting under a bridge if restricted vertical clearance exists. Placement of countermeasures also will be difficult.

14.30.1 Erosion and Scour at Bridge Sites

Scour is the result of erosive action of running water excavating and carrying away material from the bed and banks of a stream. According to the AASHTO *LRFD Specifications* (section C3.7.5), "Scour is the most common reason for the failure of highway bridges in the United States."

A countermeasure is defined by HEC-23 as a measure incorporated at a stream/bridge crossing system to monitor, control, inhibit, change, delay, or minimize stream and bridge stability problems and scour. The method of scour analysis can be made a part of the design code for protecting foundations.

Erosion is an old subject. It is eventually being placed on a scientific basis using scientific disciplines such as

1. Hydrology

2. Bridge hydraulics

3. Soil cohesion

4. Scour analysis

Existing design guidelines presented in hydraulic engineering circulars are applied differently to old and new bridges. Scour-critical bridges across the United States are currently being retrofitted using different standards for countermeasures. Their design procedures depend on individual bridge owners, who happen to be in a large number, representing hundreds of city, county, and state government agencies.

14.30.2 Codes and Design Guidelines

The following FHWA and AASHTO publications serve as major resources for scour analysis and design:

- HEC-18, "Evaluating Scour at Bridges"
- HEC-20, "Stream Stability at Highway Structures"
- HEC-23, "Countermeasures"
- HEC-25, "Tidal Scour Will Be Used"
- AASHTO *LRFD Specifications for Design of Bridges*
- AASHTO *Model Drainage Manual*
- NCHRP publications

14.30.3 Design Floods for Bridge Scour

The aim should be to design bridges for all times and for all events. Local scour is removal of material from around piers, abutments, spurs, and embankments caused by an acceleration of flow and resulting vortices induced by obstructions to the flow.

Local features at a bridge, such as abutments, piers, weirs, cofferdams, and dikes, may obstruct and deflect flow. The substructures increase the local flow velocities and turbulence levels, giving rise to vortices that may increase erosion of the river bed. At the piers, local scour is computed using the CSU equation. It depends on many factors, including length of the pier, width of the pier, and the angle of attack.

Abutment scour is computed from Froehlich's and Hire's equations. It depends on many factors, including the length of embankment. The flow of water is on both sides of a pier, generating vortices and eddy currents, whereas the flow is on one side only for abutments. This results in higher scour depths at piers than local scour at abutments.

AASHTO (LRFD) load combinations for extreme conditions are applicable. The extreme event limit states relate to flood events with peak return periods (usually 100 years) in excess of the design life of the bridge (usually 75 years).

Foundations of new bridges and bridges to be widened or to be replaced should be designed to resist scour for 100- or 500-year flood criteria or less but whichever event may create the deepest scour at foundations.

14.30.4 AASHTO (LRFD) Load Combinations for Extreme Conditions

There are generally two limit states that may involve consideration of the effects of scour:

1. A 100-year flood (the check flood for scour or superflood is a 500-year flood)

2. A vessel collision with the bridge

In addition to the preceding, there are ice loads or debris logging operations, etc. causing scour that the designer may determine as significant for a specified watershed.

Use AASHTO's "Load Combination Extreme Event II" as follows:

$$\text{Permanent loads} + WA + FR + CV$$

With all load factors equal to 1.0, nonlinear structural effects must be included and can be significant. It is anticipated that the superstructure must not collapse and the entire substructure (including piles) may be replaced.

14.30.5 Review of HEC-18 Scour Analysis Equations

Five Scour Conditions.

1. Laursen's equation for live-bed contraction scour
2. Laursen's equation for clear water contraction scour
3. Hire live-bed abutment scour equation
4. Froehlich's live-bed abutment scour equation
5. Local pier scour equation by CSU

The main parameters influencing local scour are hydrology, kinematic viscosity, acceleration, depth of approach flow, attack angle, bed material D_0, density of sediment and water, pier width, and shape coefficient. In addition, there are interactions between some of these variables that reduce the number of independent variables.

For years, bridge designers in the United States have used HEC-18 as the principal tool to determine scour depths. While the relationships in the HEC-18 equations may serve as a good starting point for scour analysis, practitioners and professionals now recognize that the equations for contraction and/or local scour often will overpredict scour depth in many hydraulic and geologic conditions.

The project impact of overpredicting scour depth can be significant because designers have only two options: (1) extend and/or stiffen the substructure or (2) provide countermeasures. Either option can increase construction costs substantially.

The equations in HEC-18 were developed originally using laboratory flume tests with select soil types, mostly sand and fine gravel. Subsequent field scour studies have shown the relationships to be reasonably reliable for this kind of soil.

However, the HEC-18 relationships are limited in two important aspects. First, some of the equations contain built-in factors of safety that are not easily recognizable or adjustable. This can lead to overly conservative design values for scour in low-risk or noncritical hydrologic conditions. A second limiting aspect of the HEC-18 relationships is insensitivity to a broad range of soil and rock textures. Sand and fine gravel are the most easily eroded bed material, but streams frequently contain much more scour-resistant materials such as compact till, stiff clay, and shale. The consequences of using design methods based on a single soil type are especially significant for many major physiographic provinces with distinctly different geologic conditions and foundation materials.

14.30.6 Recording and Coding Guide

In the absence of scour analysis, the *Coding Guide* of National Bridge Inspection Standards (NBIS) should be used to classify a bridge if it is scour-critical. Items 61, 71, and 113 are of concern during underwater inspections. Item 113 is used for scour rating.

14.31   COMPUTER SOFTWARE AND DATA

For analysis, an approved commercial software may be used or may be developed in-house. A spreadsheet based on HEC-18 equations may be developed and used in lieu of software. The following software may be used for detailed analysis and design of scour and scour counter-measures.

14.31.1   Hydrology

- TR55
- PENNSTATE Program
- USGS Stream Stat
- FEMA flood records
- Long-term gauge data

14.31.2   Hydraulics

- HEC-RAS
- WSPRO (for riverine flow)
- UNET (for unidirectional tidal flow)
- DYNET (for three-dimensional tidal flow)

14.31.3   Scour Analysis

- HEC-RAS (developed by the Army Corps of Engineers)
- Excel Spreadsheets based on HEC-18 equations

HEC-18 scour equations for contraction and local and long-term scours are used widely. Scour analysis methods are semiempirical and appear more conservative in their approach. For contraction and local scour, only a few equations for computing scour depth have been recommended by HEC-18. Melville and Coleman have listed and compared a number of equations developed by various researchers. Given the difficulties in modeling soil media, the FHWA and state codes recommend design based only on noncohesive soil conditions.

14.31.4   Countermeasures Design

A few computer methods and software products are currently available for detailed design. Vendors supply armoring methods and products that are proprietary in nature. They also provide technical specifications that were developed by the vendor. Design techniques for countermeasures are more empirical than the analytical methods and are based on engineering judgment and experience. However, Excel spreadsheets can be developed for well-known riprap, gabion, and articulated concrete blocks design based on formulas listed in HEC-23.

14.32 CONCEPTUAL TS&L QUANTITY ESTIMATE: STEEL BRIDGE (SAMPLE COSTS ONLY)

Item	Unit	Unit Cost	Quantity				Item Cost
			Length (m)	Width (m)	Depth (m)	Vol (m³)/ Area (m²)	
Mobilization	L.S.	$50,000.00					$50,000.00
Removal of Existing Bridge	C.M.	$200.00	19.583634	8.864698	0.711209	123.467977	$24,693.60
Removal of Existing Substruture	C.M.	$200.00	35.96684	5.435665	0.762009	297.951234	$59,590.25
Class 3 Excavation	C.M.	$60.00	42.88	4	1.5	514.56	$30,873.60
Temporary Earth Support System	S.M.	$45.00	104		2	416	$18,720.00
Class A Cement Concrete (Substructure)	C.M.	$1,310.00	39	4.561	1.2	426.9096	$559,251.58
Class A Cement Concrete (Footing)	C.M.	$1,310.00	39	1.6	1	124.8	$163,488.00
Class AA Cement Concrete	C.M.	$1,500.00	30.88	15.43	0.2	95.29568	$142,943.52
Latex Modified Concrete	S.M.	$95.00	30.88	15.43	0.03	14.294352	$1,357.96
Reinforcement Bars	Kg	$2.90				26579.5272	$77,080.63
Steel Piles	L.M.	$290.00	0	0	0	0	$———
			Mass (Kg/m)	Length (m)	No. of Beams	Weight (Kg)	
Structural Steel	Kg	$3.15	152	16.43	22	54,941.92	$173,067.05
						Subtotal	$1,301,066.18
						20% Contingency =	$ 260,213.24
						Total =	$1,561,279.41

BIBLIOGRAPHY

1. AASHTO. *Manual for Condition Evaluation and Load and Resistance Factor Rating of Highway Bridges.* American Association of State Highway and Transportation Officials, Inc., Washington, DC, 2003.

2. AASHTO. *LRFD Bridge Design Specifications*, 3d ed. American Association of State Highway and Transportation Officials, Inc., Washington, DC, 2004.

3. AASHTO. *Construction Handbook for Bridge Temporary Works.* American Association of State Highway and Transportation Officials, Inc., Washington, DC, 1995.

4. AASHTO. *LRFD Bridge Construction Specifications*, 2nd ed. American Association of State Highway and Transportation Officials, Inc., Washington, DC, 2004.

5. AASHTO. *Guide Specification for Seismic Design of Highway Bridges.* American Association of State Highway and Transportation Officials, Inc., Washington, DC.

6. AASHTO. *Standard Specifications for Structural Supports for Highway Signs, Luminaries, and Traffic Signals.* American Association of State Highway and Transportation Officials, Inc., Washington, DC.

7. AASHTO. *Guide Design Specifications for Bridge Temporary Works*. American Association of State Highway and Transportation Officials, Inc., Washington, DC, 1995.

8. AASHTO. *Manual for Condition Evaluation and Load and Resistance Factor Rating of Highway Bridges*. American Association of State Highway and Transportation Officials, Inc., Washington, DC, 2003.

9. AASHTO. *LRFD Bridge Design Specifications*, 3d ed. American Association of State Highway and Transportation Officials, Inc., Washington, DC, 2004.

10. AASHTO. *Construction Handbook for Bridge Temporary Works*. American Association of State Highway and Transportation Officials, Inc., Washington, DC, 1995.

11. AASHTO. *LRFD Bridge Construction Specifications*, 2nd ed. American Association of State Highway and Transportation Officials, Inc., Washington, DC, 2004.

12. AASHTO. *Guide Specification and Commentary for Vessel Collision Design of Highway Bridges*. American Association of State Highway and Transportation Officials, Inc., Washington, DC, 1991.

13. Aftab Mufti, Baidar Bakht, and Leslie Jaeger. "Bridge Superstructures: New Developments." National Book Foundation, Islamabad, 1997.

14. AISC. "Load and Resistance Factor Design." In *Specification for Structural Steel Buildings and Commentary*, 2nd ed. American Institute of Steel Construction, Chicago, IL, 1993.

15. AISI. *High Performance Steel Bridge Concepts*. J. Muller International, November 1996.

16. American Iron and Steel Institute. Steel Bridge Forum, ASCE Structures Congress, Seattle, WA, 2003.

17. "Transverse Stiffener Requirements in Straight and Horizontal Curved Steel I-Girders." *ASCE Journal of Bridge Eng.*, March–April 2007

18. Benedict, S. T., N. Deshpande, N. M. Aziz, and P.A. Conrads. "Trends of Abutment: Scour Prediction Equations Applied to 144 Field Sites in South Carolina." USGS, OFR 2003–295, 2006.

19. Biezma, María Victoria, and Frank Schanack. "Collapse of Steel Bridges." *Journal of Performance of Constructed Facilities* 21:398–405, 2007.

20. Brockenbrough, R. L., and B. G. Johnston. *Steel Design Manual*. US Steel Corporation, Pittsburgh, PA, 1968.

21. Brockenbrough, R. L., and F. S. Merritt. *Structural Steel Designer's Handbook*, 4th ed. McGraw-Hill, New York, 2006.

22. Connor, Robert J. "Extending Service Life of Steel Bridges Through Field Instrumentation and Preemptive Retrofits." ATLSS Engineering Research Center, Lehigh University, Bethlehem, PA.

23. FHWA. *Manual on Use of Self-Propelled Modular Transporters to Remove and Replace Bridges*. Federal Highway Administration, Washington, DC.

24. FHWA. *Bridge Seismic Retrofit Manual and Its Applications in Missouri Highway Bridges*, ed. by W. Phillip Yen, John D. O'Fallon, James D. Cooper, and Jeffrey J. F. Ger. Federal Highway Administration, Washington, DC, and Missouri Department of Transportation, St. Louis, MO.

25. Khan, Mohiuddin Ali, and Richard W. Dunne. "Recent Developments in the Design of New Jersey Bridges Using Accelerated Bridge Construction Concepts." FHWA Conference on Accelerated Bridge Construction, Baltimore, MD.

26. Jakovich and Vasant Mistry. "National Perspective on Accelerated Bridge Construction." 24th International Bridge Conference, Pittsburgh, PA.

27. Khan, Mohiuddin Ali. *Bridge and Highway Structure Rehabilitation and Repair*. McGraw-Hill, New York, 2010.

28. Khan, Mohiuddin Ali. *Handbook of Scour Countermeasures*, prepared by A. Agrawal, Y. Zhihua. New Jersey Department of Transportation, Bureau of Research, Trenton, NJ 2007.

29. Khan, Mohiuddin Ali. *Simplified Seismic Design*.", (Textbook under preparation) Elsevier, New York.

30. Khan, Mohiuddin A., Edwin Rossow, and Surendra Shah. "Shear Design of High Strength Concrete Beams." In ASCE-SEI Congress Proceedings, Philadelphia, PA, 2000.

31. Khan, Mohiuddin Ali. "Elastic Composite Action in Slab-Beam Systems." M.Phil. thesis, University of London, 1968.

32. Khan Mohiuddin Ali. "Long-Span Prestressed Cable Nets." Ph.D. thesis, University of Southampton, London, 1972.

33. Morice, P. B., and G. Little. "The Analysis of Right Bridge Decks Subjected to Abnormal Loading." Cement and Concrete Association, Wrexham Springs, UK.

34. NSBA. *Steel Bridge Design Handbook.* NSBA, Chapter 23.

35. NCHRP 12-33. "Detailed Planning Research on Accelerating the Renewal of America's Highway." NCHRP, 2003.

36. NCHRP Products for AASHTO Committees. "Procedures for Testing Bridge Decks." NCHRP, 2009.

37. NCHRP Products for AASHTO Committees. "Procedures and Software for Estimating Bridge Life Cycle Costs." NCHRP Report 483. NCHRP, 2002.

38. NCHRP. "Methodology for Considering Redundancy in Highway Bridge Substructures." NCHRP Report 458. NCHRP.

39. NCHRP. "Procedure for Validating Bridge Design and Rating Software." NCHRP Repost 485. NCHRP, 2002.

40. NCHRP. "Design Specifications and Details for Steel Bridge Integral Connections with Substructures." NCHRP Report 527, NCHRP, 2004.

41. NCHRP. "Rational Design and Construction Guidelines for Geosynthetic Reinforced Soil Bridge Abutments and Approaches with Flexible Facing Elements." NCHRP Report 556. NCHRP, 2004.

42. NCHRP. "Detailed Planning Research on Accelerating the Renewal of America's Highways." NCHRP, 2003.

43. NSBA. "Design for Constructability." In *Steel Bridge Design Handbook.* NSBA, 2007, Chapter 13.

44. Ontario Ministry of Transportation and Communication. *Ontario Highway Bridge Design Code.* Toronto, Canada.

45. *PCI Bridge Design Manual.*

46. Pradeep Kumar, T. V., and D. K. Paul. "Force-Deformation Behavior of Isolation Bearings." *ASCE J. Bridge Eng.*, July–August 2007.

47. Rahrig, Philip, American Galvanizing Association, "Galvanizing Plate Girders." *Modern Steel Construction*, August 2002.

48. Timoshenko, S. P., and S. Woinowsky-Kreiger. *Theory of Plates and Shells*, 2nd ed. McGraw-Hill, New York, 1959.

49. USDOT. NHI Course No. 13061.

50. USDOT, National Highway Institute, Participant Notebook, Course No. 13061, 1999.

51. USDOT, FHWA. *Manual on Use of Self-Propelled Modular Transporters to Remove and Replace Bridges.* Washington, DC, June 2007.

52. Wardhana, Kumalasari, and Fabian Hadipriono. "Analysis of Recent Bridge Failures in the United States." *Journal of Performance of Constructed Facilities* 17:144–150, 2003.

53. Wasserman, E. P., and J. H. Walker. "Integral Abutment for Steel Bridges." Tennessee DOT, Knoxville, TN.

References on Cross-Frames

1. AASHTO. *Guide Specifications for Horizontally Curved Highway Bridges.* American Association of State Highway and Transportation Officials, Washington, DC, 2003; as revised by Interim Specification for Bridges, Washington DC, 1993.

2. Maneetes, H., and D. G. Linzell. "Cross-Frame and Lateral Bracing Influence on Curved Steel Bridge Free Vibration Response." *Journal of Constructional Steel Research* 59, 2003.

3. AISC. *Manual of Steel Construction, Load and Resistance Factor Design.* American Institute of Steel Construction, Chicago, IL, 1994.

4. Azizinamini, A., R. Pavel, and H. R. Lotfi. "Effect of Cross Bracing on Seismic Performance of Steel I-Girder Bridges." In *Proceedings of Structures Congress XV: Building to Last*, SEI-ASCE, 1996.

5. Davidson, J. S., M. A. Keller, and C. H. Yoo. "Cross-Frame Spacing and Parametric Effects in Horizontally Curved I-Girder Bridges." *J Struct Engng, ASCE*, 122(9), 1996.

6. Meyer, C. *Finite Element Idealization for Linear Elastic Static and Dynamic Analysis of Structures in Engineering Practice.* Task Committee on Finite Element Idealization, ASCE, New York, 1987.

7. Schelling, D., A. H. Namini, and C. C. Fu. "Construction Effects on Bracing on Curved I-Girders." *J Struct Engng, ASCE*,115(9):2145–2165, 1989.

8. Yoo, C. H., and P. C. Littrell. "Cross-Bracing Effects in Curved Stringer Bridges." *J Struct. Eng, ASCE*, 112(9):2127–2140, 1986.

9. Yoon K, and Y. Kang. "Effects of Cross Beams on Free Vibration of Horizontally Curved I-Girder Bridges." In *Proceedings of the 1998 Annual Technical Session and Meeting, Structural Stability Research Council*, 1998. p. 165–174.

CHAPTER 15
TUNNEL ENGINEERING

Dimitrios Kolymbas

University of Innsbruck, Faculty of Civil Engineering Sciences
Division of Geotechnical and Tunnel Engineering
Innsbruck, Austria

15.1 INTRODUCTION

There is a continuously increasing number of reasons to move construction activities underground. Consequently, tunneling is nowadays a booming technology and comprises, besides financial and contractual aspects, many technological considerations for site investigation and description, planning, fire protection, excavation, mucking, and support. The design of the lining poses requirements on the prediction of the rock behavior, and there are still open questions referring to the behavior of squeezing and swelling rocks. Also, the management of groundwater poses severe tasks for the involved engineers, as well as the prediction of possible surface settlement. It is impossible to tackle all these questions in this chapter, and therefore, the book *Tunnelling and Tunnel Mechanics: A Rational Approach to Tunnelling,*[*] by the author, is recommended for further reading.

15.1.1 Notations in Tunneling

Considering the cross and longitudinal sections of tunnels shown in Figure 15.1, the various locations are denoted by the indicated names. The word *chainage* is used to identify a point along the axis of a tunnel defined by its distance from a fixed reference point.

15.1.2 Cross-Sections

The shape of a tunnel cross-section is also called its *profile*. Various profiles are conceivable, for example, rectangular profiles. The most widespread profiles, however, are circular and (mostly oblate) mouth profiles. The choice of profile aims at accommodating the performance requirements of the tunnel. Moreover, it tries to minimize bending moments in the lining (which is often academic because the loads cannot be assessed exactly), as well as costs for excavation and lining. Further aspects for the choice of profile are ventilation, maintenance, risk management, and avoidance of claustrophobia of users.

A mouth profile is composed of circular sections (Figure 15.2). The ratio of adjacent curvature radii should not exceed 5 ($r_1/r_2 < 5$). The minimum radius should not be smaller than 1.5 meters.

---

[*]D. Kolymbas, *Tunnelling and Tunnel Mechanics: A Rational Approach to Tunnelling*, New York: Springer, 2008.

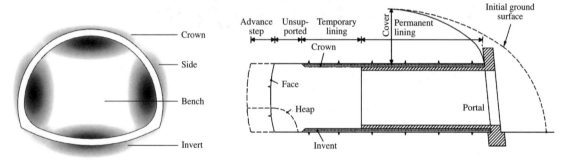

FIGURE 15.1 Parts of a tunnel cross-section (*left*); longitudinal sections of heading (*right*).

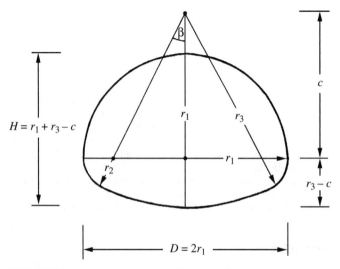

FIGURE 15.2 Geometry of a mouth profile (example).

Note that in the case of weak rock, the lower part of the lining also receives load from the adjacent ground. Therefore, a curved profile also is advisable from a statical point of view in the invert.

Typical Tunnel Cross-Sections	Area (m^2)
Sewer	10
Hydropower tunnels	10–30
Motorway (one lane)	75
Rail (one track)	50
Metro (one track)	35
High-speed rail (one track)	50
High-speed rail (two tracks)	80–100

Figures 15.3 and 15.4 provide some examples of tunnel cross-sections.

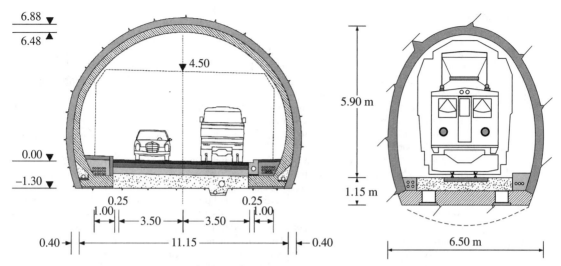

FIGURE 15.3 Hochrheinautobahn=Bürgerwaldtunnel, A 98 Waldshut-Tiengen (*left*); Vereina Tunnel South, Rätische Bahn; one track (*right*).

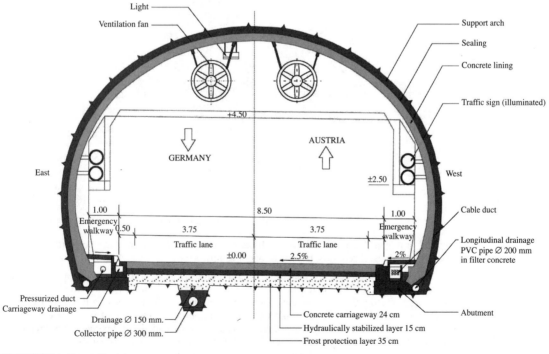

FIGURE 15.4 Füssen Tunnel.

15.1.3 Alignment

For choice of the proper alignment, several aspects must be taken into account: geotechnical, traffic, hydrologic, and risk-management issues. The main geotechnical aspect is to avoid bad rock and adverse groundwater. The choice of alignment also depends on the excavation method (drill and blast or tunnel boring machines). To minimize disturbance of the environment, aspects of vibration (e.g., owing to blasting), noise, and ventilation (the disposal of polluted air should not impair abutters, and ventilation shafts should not be too long) should be considered. In road tunnels, straight alignments longer than 1,500 meters should be avoided because they could distract the driver. Furthermore, to avoid excessive concentration on one point, the last few meters of a tunnel should have a gentle curve in plan view.

A dilemma is the choice between a high-level tunnel and a base tunnel (Figure 15.5). On the one hand a high-level tunnel is much shorter and has reduced geologic risk (because of the reduced cover). On the other hand, operation is more expensive because of increased power consumption and increased wear of the waggons. Velocity is reduced, and traffic interruptions or delays during winter must be factored in. A base tunnel is much longer and therefore much more expensive and difficult to construct. But it offers many operational advantages. Consider, for example, the Engelberg tunnel on the Stuttgart-Heilbronn Motorway. The old high-level tunnel had a length of approximately 300 meters and access ramps with a 6 percent slope. As a result, heavy trucks had to slow down, which led to traffic jams. The new base tunnel (completed in 1999) has a length of 2×2.5 kilometers and a maximum slope of 0.9 percent. Each tube has three drive lanes and one emergency lane. Thus traffic capacity has increased dramatically.

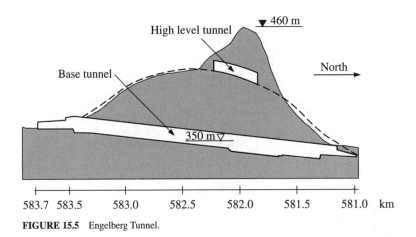

FIGURE 15.5   Engelberg Tunnel.

Further aspects for the choice of alignment include the following:

- *Depth.* For tunneled undercrossings, a sufficient cover is needed to avoid surface settlements and daylight collapses. It requires, however, longer ramps.

- *Longitudinal slopes.* There are the following limitations: $s > 0.2$ to 0.5 percent for water drainage, $s < 2$ percent for rail tunnels, and $s < 15$ percent for road tunnels (owing to exhaust gases). Inclinations of more than 3.5 percent make ventilation more difficult because it has to overcome the chimney effect. The latter becomes particularly important in the case of fire. Therefore, considerably lower limits for longitudinal inclination are recommended: $s < 4$ percent in bidirectional road tunnels. If the tunnel length exceeds 400 meters, then the inclinations should not exceed 2 percent. With an increasing slope, the production of exhaust gases also increases, and therefore ventilation costs rise.

15.2 FIRE PROTECTION

Because of the confined space, fires in tunnels can be disastrous.[*,†] In 1995, a fire in the metro tunnel of Baku caused 289 casualties. Other disastrous fires in metros occurred in 1903 in Paris (84 casualties) and in 1987 in London (Kings Cross Station; the fire cost 31 lifes). Between 1978 and 1999, 97 casualties resulted from accidents in tunnels.[‡] In 1999, 12 people died in the Tauern Tunnel, and the Montblanc Tunnel disaster claimed 41 lives. For the Montblanc disaster, the following reasons have been reported: obsolete ventilation, inefficient warning systems, insufficient communication between the French and the Italian sides, and when the fire broke out, only one fireman was on duty.

The United States has not seen many fire disasters, probably thanks to the rigorous fire protection regulations and frequent application of double tubes. In 1982, a fire in the Caldecott Tunnel (Oakland, CA) claimed 7 victims.[§]

In peak-traffic times, a road tunnel will hold up to 100 persons per kilometer and lane. In rail tunnels, the figure is, on average, 800 persons per train. These people are exposed to fire hazard, which can be due to motor conflagration or conflagration of truck freight, accident, overheated axes, arson, etc. During heading, fire can be released by leakage of methane from rock or oil from machines.[¶]

People are at danger from smoke, the toxic products of combustion, lack of oxygen, heat, and panic. Fire smoke is very dense and can reduce visibility below 1 meter, leading to loss of orientation.[**] The extraction of smoke must occur as close as possible to the fire source. Fume hoods (dumpers) should be provided with a sufficiently small spacing. It is important to open only the hoods nearest the fire source. This can be achieved either with remote control or with melting wires.[††] The localized extraction of smoke can be supported if the air is forced to blow toward the smoke source. For two-lane road tunnels with a longitudinal inclination, $i < 3$ percent (for $i > 3$ percent, special investigations are needed). The Austrian recommendation RVS 9.261 requires to extract at least 120 m^3/s of smoke over a length of 150 meters through large dumpers.

For design of the remedy measures, it is important to assume a realistic "design fire" including temperature and smoke production. Based on recent disasters and field tests, fire scenarios have been modified. Now one has to assume smoke production of 240 m^3/s, fire power up to 100 MW, durations between 30 minutes and several hours, and increases in temperature above 1,000°C within 5 minutes. The rise in temperature with time is specified in several standards:[‡‡]

- RABT curve in Germany (T_{max} = 1,200°C within 5 minutes)
- Eurocode 1-2-2 (hydrocarbon curve, T_{max} = 1,100°C)
- Rijkswaterstaat-curve (T_{max} = 1,350°C within 60 minutes; such high temperatures can appear when trucks carrying gasoline are involved)

To estimate the speed of events, the following facts should be taken into account:

- Usual gasoil tanks resist fire for about 3 minutes.
- Thermal convection winds attain speeds of up to 3 m/s. Thus smoke can spread out over a length of 180 meters before being detected. The propagation speed of smoke equals the speed of escape of individuals.

---

[*]This is, however, not the case for all tunnel fires. Before the disaster of 1999, there have been in the Montblanc road tunnel 17 fires of trucks, and in the Gotthard road tunnel, there have been 42 fires of vehicles from 1992 to 1998.

[†]A list of fire disasters in tunnels can be found in U. Schneider et al., "Versuche zum Brandverhalten von Tunnelinnenschalenbeton mit Faserzusatz," *Bautechnik* 78 (2001), Heft 11, 795–804. See also "Fire Protection in Tunnels," *Tunnel*, 2/2002, 58–63; K. Kordina, "Planning Underground Transport Facilities to Cope with Fire Incidents," *Tunnel* 5/2004, 9–20.

[‡]J. Day, "Road Tunnel Design and Fire Life Safety," *Tunnels and Tunnelling International*, October 1999, 29–31.

[§]The U.S. fire protection regulations are compiled in the *National Fire Protection Association (NFPA)* 502.

[¶]Therefore, oil with a of low combustibility is now used.

[**]Illuminated hand rails indicate the route to the nearest emergency exit; see www.nils.nl.

[††]C. Steinert, "Dimensioning Semicross Ventilation System for Cases of Emergency," *Tunnel* 1, 1999, *Tunnel* 2, 1999, 36–52.

[‡‡]K. Kordina and R. Meyer-Ottens, *Beton Brandschutz Handbuch*, 2. Auflage, Verlag Bau + Technik, Düsseldorf.

The transition of smoldering to open fire (so-called flash-over) with the accompanied sharp rise of temperature can set on within 7 to 10 minutes.

A layering of heated gases over the cool air helps the efficiency of ventilation and facilitates escape. This layering can be perturbed by fast movements of air, for example, owing to moving cars, sprinklers, and fire-induced convection. Therefore, longitudinal air velocity should be reduced to 1 m/s in the case of a fire. The air velocity in normal tunnel operation has a strong influence on the convection and layering of smoke in case of fire and therefore should not exceed 2.5 m/s (although in naturally ventilated mountain tunnels values up to 14 m/s have been observed).* Consequently, the fire ventilation has to achieve the two following objectives: massive smoke extraction in a limited section around a fire and control of the longitudinal air velocity. When designing ventilators, it has to be taken into account that their power is reduced by up to 50 percent in heated air.

Also, the safety training for tunnel operator personnel is important.† The safety and rescue plan should be discussed in advance among the designer, contractor, owner, and rescue services. Such an integrated procedure is advantageous even from a financial point of view because it helps to avoid expensive modifications. Given the remote locations of most tunnels, rescue plans should not rely primarily on the arrival of the rescue services but rather on enabling the endangered people to rescue themselves. This so-called self-rescue phase lasts only a few minutes and is crucial. The escape speed ranges from 2.5 to 5 m/s but is reduced to 0.5 to 1 m/s for older or infirm people. The best means of rescue is a second tube with a sufficient number of ventilated crosswalks. Alternatively, a rescue adit parallel to the main tunnel can be provided.

Fire-protected niches are problematic in view of ventilation and also for psychological reasons. Experience derived from the fire disasters in the Mont-Blanc and Tauern tunnels is that the semi-transverse and transverse ventilations were unable to extract the smoke sufficiently fast. Safety equipment should be redundant and should work according to the fail-safe principle, that is, on failure of a safety-relevant component, the system enters a safe state. An outfall of electric power supply should release a substitute supply within 10 seconds.

Fire combat comprises active and passive measures. Active measures aim at the extinction of the fire by means of fire detectors, fire extinguishers, sprinklers, emergency ventilation and telecommunication devices. Passive measures aim at minimizing the damage, for example, by means of fire-resistant concrete, synthetic materials that do not produce toxic gases when burning, safe electric cables placed below the carriageway, transverse drain pipes that collect leaking fuel, use of materials of low porosity (that do not fill with fuel), and clear signals indicating the escape routes.

Fire protection measures should by verified with tests. Cold smoke tests are comparatively easy and inexpensive, but only real fire tests can provide a genuine verification of the safety level achieved.

15.3   GEOTECHNICAL INVESTIGATIONS

The underground is a vast unknown that can hide many unpleasant surprises (e.g., weak zones, water inrushes, etc.). Therefore, a detailed site investigation is necessary not only for technical purposes but also for the contractual regulations of all involved parties. Geotechnical investigations aim at collecting all ground properties that are relevant to the heading.‡ Usually, site investigation, design, and construction are executed by different specialists, companies, and authorities, as well as at different times. Thus it may happen that necessary data are not available when needed, that is, when no more money or time are available for further investigations. The expenses for site investigation can make up

---

*K. Pucher and P. Sturm, "Fire Response Management bei einem Brand im Tunnel," *Oesterreichische Ingenieur- und Architektenzeitschrift* 146, Heft 4/2001, 134–138.

†The company DMT operates a training center for fire fighting in Dortmund.

‡See, that is, ETB (Empfehlungen des Arbeitskreises 'Tunnelbau'), RVS 9.241 and RVS 9.242, DS 853 (Eisenbahntunnel Planen, Bauen und Instandhalten), DIN 4020 (Geotechnische Untersuchung für bautechnische Zwecke), SIA199 (Erfassen des Gebirge sim Untertagebau), AFTES guidelines for caracterization of rock masses useful for the design and the construction of unterground structures (www.aftes.asso.fr).

to 3 percent of the construction costs.[*] Usually, the site investigation is released by the owner, who should provide the bidders with data that are as complete as possible. Otherwise, the owner will have to face the possibility of increased claims. According to experience from the United States[†] over 55 percent of claims relate to unforeseen ground conditions, and they decrease with increasing exploration, as shown in Table 151.

TABLE 15.1   Influence of Exploration on Claims

Meter of exploration boring per meter of tunnel alignment	Claims relative to the bid price
0.5	30–40%
1	<20%
1.5	<10%

Information on the underground is attained step by step. Each step reveals the proper subsequent investigations, (Glossop[‡] said: "If you do not know what you should be looking for in a site investigation, you are not likely to find much of value."). In general, at least three steps or phases are applied:

Phase 1: Preliminary investigation

Phase 2: Main ground investigation

Phase 3: Further investigation undertaken as part of the project itself

All investigations must be well documented. The geotechnical investigations should be done in close collaboration with engineering geology, a discipline that deals with the site investigation for civil engineering projects.[§]

15.3.1   Preliminary investigation

The preliminary investigation consists of

- Preliminary appreciation of the site, that is, an examination of existing information (also called *desk study*), and it comprises topographic, geologic, hydrologic, and other maps (e.g., aerial photos to find faults); evidence and experience from nearby construction sites; a walk-over survey to detect jointing, weakness zones, etc.; and engineering geologic mapping along tunnel alignment. The desk study aims at determining whether the envisioned tunnel is feasible, to what extent geology affects the excavation, and support and further investigations. It is the basis for the provisional design and investigation of variants (alternative alignments).

- Preliminary ground investigation, that is, site work to confirm feasibility and to enable plans for the next phase of the investigation to be formulated.

15.3.2   Main Site Investigation

The main site investigation comprises information that must be acquired ad hoc and serves the design, tendering, assessment of environmental impact, and execution, as well as damage analysis. In the frame of site investigation, a geotechnical report has to be worked out.

---

[*]According to Norwegian experience, the relative costs of geotechnical investigations (related to construction costs) are 1 percent for simple tunnels, 3 to 4 percent for storage caverns, and 5 to 6 percent for special projects.
[†]R. A. Robinson, M. A. Kucker, and J. P. Gildner, "Levels of Geotechnical Input for Design-Build Contracts." In: *Rapid Excavation and Tunneling Conference, 2001 Proceedings*, pp. 829–839. Society for Mining, Metallurgy, and Exploration, Inc., Littleton, Colorado.
[‡]R. Glossop, 1968 Rankine Lecture.
[§]P. N. W. Verhoef, *Wear of Rock Cutting Tools*. Balkema, 1997.

15.3.3 Investigation During and After Construction

It is never possible before hand to obtain complete information on the strata that will be encountered by a tunnel and to anticipate their behavior. For this reason, provision must be made for observation and any other necessary investigation during construction. Thus this investigation aims at continuously updating the acquired information and to check the validity of the prognoses. It comprises mappings of the tunnel face and wall and measurements of deformations, settlements, stresses, vibrations, and groundwater (Figure 15.6).

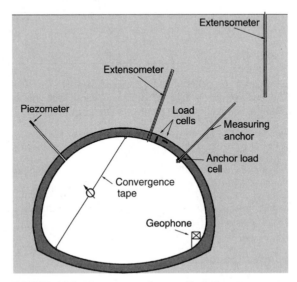

FIGURE 15.6 Measurements for tunneling. Convergence tapes are rather obsolete. Nowadays, convergence is mostly measured via remote sensing.

15.4 DRIVING

The driving of a tunnel comprises the following actions: excavation, support of the cavity, and removal of the excavated earth (mucking). It is distinguished between conventional (also called *incremental* or *cyclic*) driving and continuous driving. This section introduces the several methods applied. A rigorous classification is difficult because these methods are often combined. It has become usual to distinguish between conventional (or incremental) driving, on the one hand, and continuous (or tunnel boring machine driving), on the other hand. This is, however, not reasonable: tunnel boring machine (TBM) driving consists of several steps and thus is incremental.

The main characteristic of conventional driving is that it proceeds in small advance steps, whose lengths range between 0.5 and 1.0 meters in soft ground. This length is an important design parameter because the freshly excavated space has to remain stable for a while (at least 90 minutes) until the support has been installed. The length of advance steps also influences the settlement of the ground surface: Reducing the advance length decreases the surface settlement considerably.

15.4.1 Full-Face and Partial-Face Excavation

Large cavities are less stable than small ones. Therefore, in many cases, the tunnel cross-section is not excavated at once but in parts. For cross-sections > 30 to 50 m weak rock, the face must be supported (with a heap of soil, shotcrete, or nails) or excavated in parts. In the early era of tunneling, many

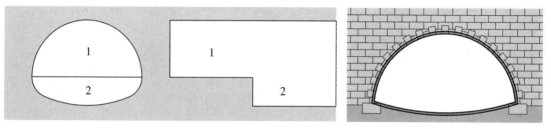

FIGURE 15.7   Top driving, cross, and longitudinal sections. (1) calotte, (2) bench (*left*). Concept to explain the support of crown (*right*). The excavation support acts in a similar way as an arch of a masonry bridge. The weight of the overburden is concentrated in the tow abutments.

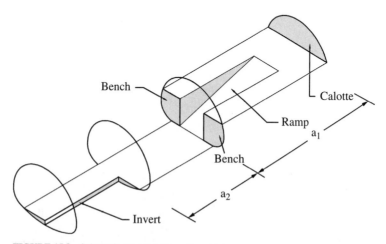

FIGURE 15.8   Schematic representation of top driving.

different schemes for partial-face excavation were developed. The terminology is neither unique nor systematic. The various types of contemporary partial-face excavations include

Top driving. The crown is excavated before the bench (Figures 15.7 and 15.8). The temporary support of the crown with shotcrete can be conceived as a sort of arch bridge. This explains why the abutments are prone to settlement, which induces settlement of the ground surface. Countermeasures are to enlarge the abutments (so-called elephant feet) or to strengthen them with micropiles or the construction of a temporary invert. The latter must be constructed soon after driving the crown, that is, not more than two to five advance steps beyond it.[*] A quick construction of a temporary invert of the crown section or, better, the speedy excavation and support of the bench and invert helps to avoid large settlements of the abutments of the crown arch. This means that the length $a = a_1 + a_2$ (Figure 15.8) should be kept as small as possible. On the other hand, a_1 should be sufficiently large to enable efficient excavation and support works in the crown.

If the crown and bench are excavated simultaneously, then the ramp must be moved "continuously" forward (i.e., every now and then). Alternatively, the ramp is not placed at the center (as shown in Figure 15.8) but on the side of the bench. Then the other side of the bench can be excavated over a longer distance. If the excavation of a ramp may cause instability, then the ramp must be heaped

---

[*]Richtlinien und Vorschriften für den Straßenbau (RVS 9.32); K. Kovári and F. Descoeudres, eds., *Tunnelling Switzerland.* Swiss Tunnelling Society, 2001, ISBN 3-9803390-6-8.

up after excavation and support of the bench. In 1985, a collapse occurred during the driving of the Kaiserau Tunnel in Germany. It was caused by a slit that was made to construct the ramp and which rendered the temporary invert of the calotte ineffective.

Sidewall drift. The side galleries are excavated and supported first. They serve as abutment for the support of the crown, which is subsequently excavated (Figures 15.9 and 15.10). This type of driving is approximately 50 percent more expensive and slower than top driving. Therefore, it is preferred in soils/rocks of low strength. Note that a change from top driving to sidewall drift is difficult to accomplish.

In all types of partial excavations, attention should be paid to connections of the lining segments constructed at different steps. The final lining (including its reinforcement) should be continuous without any weak points (interruptions or slits).

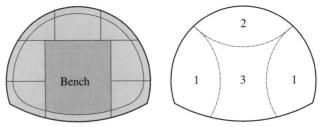

FIGURE 15.9   (*Left*) Core driving. (*Right*) Sidewall drift.

FIGURE 15.10   Niedernhausen Tunnel. (*Left*) Sidewall drift. (*Tunnel* 9, 2/2000, 19.) (*Right*) Hydraulic hammer. (Krapp Beroo.)

15.4.2 Excavation

Excavation is the process of detaching the rock using the following methods and tools:

Hammer. Pneumatic and hydraulic hammers (Figure 15.10) can be applied in weak rock and achieve performances comparable with drill and blast (see later in this section). In addition, they avoid the vibrations caused by drill and blast. For example, the hydraulic hammer HM 2500 of Krupp Berco has the following excavation performance:

Unconfined strength of the rock (MPa)	Excavation capacity (m³/h)
40–50	40
70–80	26
80–100	20

FIGURE 15.11   (*Left*) Excavator. (Liebherr 932.) (*Right*) Roadheader. (Voest Alpine Am 100.)

These performances can be improved by loosening blasts. Hydraulic hammers are carried by wheel or track vehicles. Dust is treated (not completely successfully) with spraying water from nozzles or water hoses.

Excavators. Boomed backhoe buckets excavate weak rock, whereas thin rippers and hydraulic chisels are applied whenever hard rock inclusions are encountered (Figure 15.1). In order to follow the prescribed tunnel profile exactly, these tools must be sufficiently free to rotate. Excavators can exhibit a high performance if the strength of the rock is moderate, that is, if the rock is either soft or jointed (RMR < 30).[*] Ripping is applied for an RMR between 30 and 60. Another criterion for the applicability of ripping is a propagation velocity of *P* waves of between 1 and 2 km/s.

Roadheaders (boom cutters). These tools are used for moderate rock strengths and for laminated or joined rock. The cutter is mounted on an extension arm (boom) of the excavator and fragments the rock into small pieces (Figure 15.11). Thus, overprofiling can be limited, and also, the loosening of the surrounding rock is widely avoided. One has to provide for measures against dust (suction or water spraying). The required power of the motors increases with rock strength.

Tunnel boring machines (TBMs). TBMs are applicable to rock of medium to high strength ($50 < q_u < 300$ MN/m^2) if its abrasivity is not too high.[†] TBMs excavate circular cross-sections with a rotating cutterhead equipped with disk cutters (Figure 15.12). To press the cutterhead against the rock, the TBM is propped at the tunnel wall by means of extendable grippers. Therefore, the rock must have a sufficient strength.

The support can be installed soon after the excavation. The classification of TBM as "continuous" driving instead of "incremental" (also called *cyclic*) is misleading because a TBM advances in strokes. Regular stops are needed mainly for maintenance of the excavation tools.

The design of a TBM consists of determination of thrust, torque, size, and spacing of disks. The advance rate (m/h) is given by the product penetration rate (m/h) × TBM utilization (%).

Since the TBM fills up the excavated space more or less completely, the systematic support can only be installed beyond it (i.e., beyond a working space of approximately 10 to 15 meters length). In weak rock, however, support measures such as rock bolts and wire meshes have to be applied adjacent to the cutterhead.

The minimum curvature radius is 40 to 80 meters; with backup equipment, the minimum radius is 150 to 450 meters. TBMs are often protected against cave-ins by cylindrical steel shields.

Drill and blast. Drill and blast was first applied in 1627 by the Tyrolean Kaspar Weindl in a silver mine in Banská Stiavnica (former Schemmnitz, Slovakia). It is suitable for hard rock (e.g., granite, gneis, basalt, and quartz), as well as for soft rock (e.g., marl, loam, clay, and chalk). Thus it is applicable to rocks with varying properties. Moreover, drill and blast is advantageous

---

[*]RMR is *rock mass rating*, according to Bieniawski.
[†]q_u is the unconfined strength of rock.

FIGURE 15.12   Gripper TBM: (1) shield, (2) arch segments, (3) annular errector, (4) drilling equipment for rock bolting, (5) protection canopy, (6) protection girder, (7) grippers. (Brochure Herrenknecht TBM.)

for relatively short tunnels, where a TBM does not pay, and for very hard rock and noncircular cross-sections.

To keep drill and blast economical, the steps involved (i.e., drilling, charging, tamping, igniting, ventilating, and support) must be coordinated in such a way that downtimes are avoided. The most time-consuming operations are drilling and charging. More details on drill and blast excavation are presented in the next section.

15.4.3   Drill and Blast

The drill and blast method consists of several subsequent steps (i.e., drilling, charging, tamping, ignition, extraction of fumes by ventilation, mucking, and support), which are described subsequently.

Drilling, Charging, and Tamping of Blastholes.   Rotary and percussion drilling is applied to drive blastholes within a diameter range from 17 to 127 millimeters (mostly being ca 40 millimeters) with drilling rates of up to 3 m/min into the rock (Figure 15.13). The prescribed positions, orientations, and lengths of blastholes must be kept precisely; therefore, the drilling equipment is mounted on

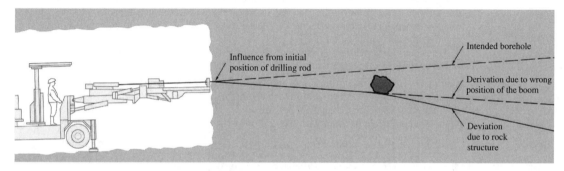

FIGURE 15.13   Advance drilling: intended and real borehole positions.

tire carriages, called *jumbos*, with two to six booms. The length of the blastholes corresponds to the advance step (usually 1 to 3 meters). To achieve good blast results, the advance step should not exceed the minimum curvature radius of the tunnel cross-section. For parallel cut (see section 15.4.3), the advance step can be longer.

Lengthy charges are applied in tunneling. Charging depends on the type of explosive. Cartridges are pushed with the help of rods, and ANFO, a mixture of Ammonium Nitrate and Fuel Oil, and emulsions are cast or pumped into the boreholes.

Explosion is the "instantaneous" transformation of the solid (or fluid) explosive into a mixture of gases called *fume*. To achieve pounding, the fume must be contained, that is, its expansion must be hindered. This is why blastholes are tamped, that is, plugged. Since the impact of the fume is supersonic, the strength of the plug is immaterial. Thus a sufficient tamping is obtained with sand or water cartridges. In long blastholes, even the inertia of the air column provides a sufficient tamping. Tamping increases pounding of the rock and reduces the amount of toxic blast fumes by improving the chemical transformation of the charge. Tamping, in particular, with water, is also a countermeasure against dust production.

Detonation is oxidation, where oxygen is present in compound form within the explosive. The reaction front propagates within the explosive with a detonation velocity that amounts up to 8 km/s and depends on the chemical composition, size, containment, and age of the explosive. The detonation front leaves behind the fume, which is a highly compressed gas mixture. One kilogram of explosive produces a gas volume of nearly 1 m^3 under atmospheric pressure. The highly compressed fume exerts a large pressure on its containment. The energy content of an explosive is not overly high, but the rate at which this energy is released corresponds to a tremendous power.

Ignition.   Modern explosives are inert to hits, friction, and heat. They can be ignited only with a (smaller) initial explosion. Therefore, ignition occurs through

> *Electric detonators.* These consist of a primary charge, which is susceptible to heat, and a less susceptible secondary one. The primary charge is ignited by means of an electric glow wire. A retarding agent can be added in such a way that the explosion is released some milliseconds after closing the circuit. The detonators are placed in the bottoms of the blastholes. Electronic detonators have a higher retardation accuracy (which is important for smooth blasting) and can be ignited with a coded signal.

> *Detonating cords.* These cords (∅ 5 to 14 millimeters) have a core made of explosive and are ignited with an electric detonator. The detonation propagates along the cord with a velocity of approximately 6.8 km/s. Modern variants (Nonel, Shockstar) are synthetic flexible tubes whose inner walls are coated with 10 to 100 g/m of explosive (Nitropenta). Detonating cords allow bunched ignitions with only one electric detonator.

The power of detonators is reduced with time, but appropriate storage ensures a long life.

Distribution of Charges and Consecution of Ignition.   The explosion aims at (1) breaking the rock into pieces manageable for haulage, (2) avoiding overbreak or an insufficient excavation profile (so-called smooth blasting), and (3) not disturbing the surrounding rock. To this end, several schemes (drilling and ignition patterns) have been developed empirically for the distribution of charges and the order of ignition. It is distinguished between production and contour drillholes. The most efficient excavation is obtained if the fume pushes the rock against a free surface. This can be achieved, for example, with a V-cut (edge or fan cut)[*]: The blastholes in the central part of the face are conically arranged and ignited first (Figure 15.14). The surrounding blastholes are ignited consecutively with a delay of some milliseconds. Thus the rock is pared progressively from the cut to the contour.

Parallel blastholes (parallel cut) are easier to drill precisely and enable longer advance steps but require more explosive than conically arranged ones (V-cut). Several unloaded drillholes are provided in the parallel cut, thus creating a cavity against which the detonation pushes the rock. Thus, the

---

[*]J. Johansen, *Modern Trends in Tunneling and Blast Design*. Rotterdam, Balkema, 2000.

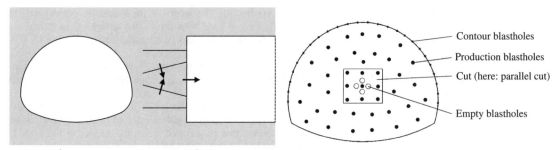

FIGURE 15.14   Wedge cut (*left*), distribution of blast holes (*right*).

efficiency is increased. For smooth blasting, the contour holes have a small spacing (e.g., 40 to 50 centimeters) and are charged with detonating cords. Smooth blasting helps to minimize the costly aftertreatment (postprofiling).

In the case of anisotropic, jointed, or stratified rock, the distribution of blastholes and the ignition pattern have to be adapted to the structure of the rock. The various work steps for drill and blast are repeated cyclically. Adjusting and optimization of the blasting scheme with regard to the individual rock properties are therefore recommended.[*]

Safety provisions.   Only licensed persons are allowed to handle explosives. Receipt, consumption, and return of explosives must be registered in a tractable way. Transportation must conform to the regulations for transportation of dangerous materials.[†] The electric circuit for ignition should be checked. The blasted rock can be spread over large distances. Therefore, a safety distance of 200 to 300 meters off the face should be maintained. Electromagnetic fields (e.g., radiofrequency radiation from mobile telephones) may cause in advertent initiation of electroexplosive devices. Therefore, a safe distance must be kept from radiation sources.[‡]

Contemporary explosives have a positive oxygen balance but still do contain toxic gases such as CO_2, CO, and nitrogen oxides. Also, dust is dangerous, in particular, quartz dust. Therefore, following an explosion, work should be resumed only after a ventilation time of at least 15 minutes with an air velocity (averaged over the largest cross-section) of at least 0.3 m/s. During ventilation, personnel should stay either in the open air or within a fume-protection container. This is also advisable because parts of the shotcrete lining can be detached (owing to the explosion shock) and fall down.

Vibrations.   Vibrations owing to drill and blast (as well as owing to TBMs)[§] propagate in the underground as elastic waves and can affect constructions and human well-being. The disturbance is controlled (for frequencies > 10 Hz)[¶] by the maximum vibration velocity, which can be measured by means of so-called geophones. Rating criteria for disturbance and hazard owing to vibrations can be found in standards such as DIN 4150 and ÖNORM S 9020. Transient vibrations (e.g., owing to drill and blast) can be noticed from $v_{max} = 0.5$ mm/s and cause complaints from $v_{max} = 5$ mm/s. Humans are particularly susceptible to and less tolerant of continued vibrations, especially during the night. Vibration hazard is usually overrated, and this may cause preexisting fissures in buildings to be attributed to drill and blast. Therefore, residents in the neighborhood should be informed about possible hazards. In addition, a perpetuation of evidence should be carried out. Vibrations owing to drill and

---

[*]"Computerised Drill and Blast Tunneling," *Tunnels & Tunnelling*, July 1995, 29–30, and G. Girmscheid, "Tunnel-bau im Sprengvortrieb—Rationalisierung durch Teilrobotisierung und Innovation," *Bautechnik* 75 (1998), 11–27.

[†]www.sprenginfo.com.

[‡]P. Röh, "Beeinflussung elektrischer Zündanlagen durch mobile und stationäre Funkeinrichtungen," *Nobelhefte* 2002, 5–14.

[§]R. F. Flanagan, "Ground Vibration from TBMs and Shields," *Tunnels & Tunnelling*, 25, 10,1993, 30–33; J. Verspohl, "Vibrations on Buildings Caused by Tunnelling, *Tunnels & Tunnelling*, 1995, BAUMA Special Issue, 81–85.

[¶]Values of the peak velocity for domestic structures exposed to transient vibration can be found in *Tunnels & Tunnelling International*, June 2002, 31–33.

FIGURE 15.15 (*Left*) Modern shield driving. *Tunnelling Switzerland*, K. Kovári & F. Descoeudres (eds.), Swiss Tunnelling Society, 2001. (*Right*) Historic shield tunneling beneath the St. Clair River.

blast can be reduced by splitting the explosion into several smaller explosions with a sequence of some milliseconds. For example, Nonel detonators enable 25 explosions within 6 seconds.[*] TBM vibrations are usually not noticeable at distances greater than 45 meters.

15.5 SHIELD DRIVING

Shields are used for driving in weak rock and soil.[†] A shield is a steel tube with a (usually) circular cross-section (Figure 15.15). Its front is equipped with cutters. The shield is pushed forward into the ground by means of jacks. A so-called blade shield is equipped with blades that can be protruded separately (Figure 15.16).

Jacks. The jacks have a stroke between 0.8 and 1.5 meters, and they operate with pressures up to 400 bar and apply forces up to 3 MN. Their abutment is the already installed lining. The shield can be steered by applying different pressures to the jacks distributed along its circumference. On this, shields should be not too lengthy ($L < 0.8\,D$), and the cavity diameter should be slightly larger than the outer diameter of the shield; that is, a shield-to-rock clearance of approximately 25 millimeters should be provided for.

After each advance stroke, the hydraulic pistons of the jacks are retracted in such a way that an additional slice of lining can be added under the protection of the shield tail (Figure 15.17). During the subsequent advance stroke, the tail gap is filled with grout. Coordination of the individual steps is important. Typical advance rates are 0.5 to 2 m/h.

Excavation. The excavation is either full face with rotary cutterheads (Figure 15.18), or selective with boom headers or roadcutters (roadheaders) that excavate the face in a series of sweeps. The cutterhead can be driven either electrically or hydraulically. Electric drive has a higher efficiency but is more difficult to control. Hydraulic drive is more flexible. The necessary thrust and torque are determined empirically. Note, however, that in general the cutterhead is not expected to support the face. At standstill, the openings should be closed with panels. In sticky soils, the wheel should be as open as possible. Adhesion of sticky soil (e.g., plastic clay) is a considerable handicap and can be combated with flushing.

---

[*]*Tunnels & Tunnelling*, July 1995, 8.

[†]See also B. Maidl, M. Herrenknecht, and L. Anheuser, *Maschineller Tunnelbau im Schildvortrieb*, Ernst & Sohn Verlag, Berlin, 1995; M. Kretschmer and E. Fliegner, *Unterwassertunnel*, Ernst & Sohn Verlag, Berlin, 1987.

FIGURE 15.16 Shield with blades and roadheader. (Herrenknecht.)

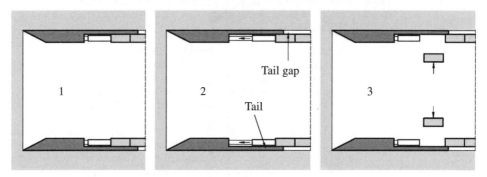

FIGURE 15.17 Sequences of shield driving.

To excavate the ground, the cutterhead is equipped with disk cutters, drag bits (for rock and embedded blocks), chisels, scrapers for sand and gravel, and scrapers for cohesive soil. Some maintenance work to the cutterhead can be undertaken only from ahead. To this end, the adjacent soil can be stabilized by grouting or freezing. Then the shield is retracted so that a space is created from which the inspection can proceed.

Face support. An unstable face can cause surface settlements and therefore should be supported, for example, with panels (Figure 15.19, *right*) or with the cutterhead (if $5 < c_u < 30$ kN/m^2). Alternatively, the soil is allowed to run out on to platforms to form supporting heaps (platform shield; Figure 15.19, *left*). Very soft soil can be displaced laterally. Unsupported faces are very problematic or impossible, especially below water level.

FIGURE 15.18 Cutterheads with scrapers and discs

FIGURE 15.19 Platform shield. (Les Vignes Tunnel, Herrenkrecht.) Shield with face breasting flaps. (Wagss and Freytag.)

If the cutterhead has to support the face, it must cover a part of the face, which should be as large as possible. The muck enters through slots. Alternatively, panels can be elastically mounted on the spokes of the cutterhead.

Lining segments. These are made of precast reinforced concrete or cast iron and have secant lengths up to 2.2 meters and widths between 0.6 and 2.0 meters (Figure 15.20). Several segments form a ring. Each ring is closed with a so-called key segment[*] (Figure 15.21).

Wider segments speed up the rate of lining but are disadvantageous for curved tunnels. The segments are moved with vacuum erectors and fixed temporarily with straight or curved bolts. Alternatively, plugs and studs can be used.

Usually, segments are made of concrete C35/45. A higher concrete strength is not necessary and renders the segments more brittle (edges can flake). The previously used hollow or ribbed segments are no longer in use. The segments also must be designed to carry the loads during transport and installation. The use of steel fiber reinforced concrete (e.g., with 30 kg of fibers

---

[*]See also *Sachstandsbericht Tübbinge*. Öesterreichische Vereinigung für Beton- and Bautechnik, Wien 2005.

FIGURE 15.20   Segmental tunnel lining. (Metro Madrid). Placing of small disks of wood in the joint.

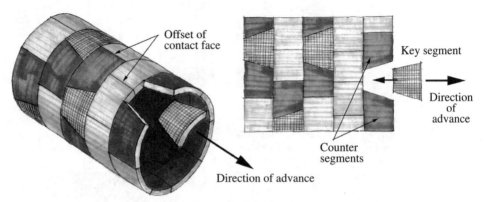

FIGURE 15.21   Array of segments: (*left*) perspectively, (*right*) unfurled.

per cubic meter of concrete) simplifies the manufacturing process and reduces the hazard of edge flaking. Steel fibers are often used together with bar reinforcement. With the introduction of spheroidal graphite iron (SGI), cast-iron segments are now used increasingly.[*] Their tensile strength allows them to carry bending moments, and they are up to 30 to 40 percent lighter than reinforced-concrete segments. In addition, they occupy a smaller part of the tunnel cross-section, and their joints are sufficiently waterproof. Corrosion is small (0.4 mm/year).

Planar joints with roughness less than 1 millimeter have been applied successfully, whereas groove and tongue (tenon and mortise) joints can be damaged easily. Since the lining segments are compressed against each other, they do not need to be connected. However, they have to be bolted together for installation. If the transverse forces are high, the individual rings can slip relative to each other, thereby reducing these forces. To preserve this mobility, stuffings (e.g., thin plates of wood) are placed in the joints (see Figure 15.20).

The lining is waterproofed by means of gaskets placed between the segments. The gaskets are compressed and thus become watertight. There are also water-expansive gaskets. In most cases, the lining is much more expensive (approximately 80 percent of the tunnel construction costs) than the shield machine and therefore should be designed carefully.

---

[*]*Tunnel & Tunnelling*, January 1998.

FIGURE 15.22 Double shield.

Tail void closure. The outer diameter of the shield is larger than the outer diameter of the lining so that the moving shield leaves behind a tail void, whose thickness is up to 20 centimeters. Such a large void can cause the lining rings to shift and/or large surface settlements and, therefore must be filled (closed) with mortar (tail gap grouting or back grouting). The mortar should set as fast as possible but not too fast (otherwise, it cannot be pumped in). To avoid settlements, the ring tail closure should be done as soon as possible after excavation, and the grouting pressure should be equal to the primary normal stress. Of course, there is no exact way to fulfill this requirement because the primary stress varies from point to point and is, at that, hardly known. In addition, the grouting pressure field cannot be controlled precisely. The slot between lining and shield tail must be plugged; otherwise, the grouted mortar can escape. The plugging is achieved with steel brushes whose bristles are filled with grease. A new method of keeping the grouting pressure constant within the tail void is to provide a compliant sealing lip that yields only if a threshold pressure is reached so that the void can be filled with a constant pressure.[*] It turns out that tail gap grouting cannot completely reverse settlements, even if the volume of the grout considerably exceeds the volume of the gap. This can be explained if one considers the mechanical behavior of ground at loading-unloading cycles.[†]

Double shield. This shield consists of two parts (Figure 15.22). At the front shield, the cutterhead is installed; at the rear shield (gripper shield), the grippers for lateral bracing and the device for placement of the segments are installed. Front and rear parts are connected via a telescopic section. Thus the excavation can continue while lining segments are placed. The double shield is a combination of TBM and shield and is intended to operate in varying rock.

In good rock, the cutterhead is buttressed against the rear part, which is connected to the adjacent rock via the grippers. Thus the lining segments can be installed while the cutterhead works. In weak rock, the two parts of the shield are jointed, and the shield operates as a conventional one. That is, the excavation has to stop while the lining segments are installed.

Tunnel driving machines. Many shields are equipped with a cutterhead, which is the main feature of a TBM. This is why the notions *shield* and *TBM* are often confounded, that is, taken as

---

[*]S. Babendererde, "Grouting the Shield Tail Gap," *Tunnels & Tunnelling International*, November 1999, 48–49.
[†]M. Mähr, "Settlements from Tail Gap Grouting Due to Contractancy of Soil," *Felsbau* 22 (2004), no. 6, 42–48.

synonyms. This is, however, wrong: Shields are used in loose ground, whereas (unshielded or open) TBMs are used in hard rock. A combination of both methods is applied in rocks with varying properties. As a generic term for *shield* and *TBM,* the word *tunnel driving machine* has been launched. A generally accepted classification is

$$
\text{TBM}
\begin{cases}
\text{open (for rock tunnels)} \\[2ex]
\text{shielded (for weak or jointed rock/soil)}
\begin{cases}
\text{open face} \\[1ex]
\text{closed face}
\begin{cases}
\text{slurry} \\
\text{EPB}
\end{cases}
\end{cases}
\end{cases}
$$

A TBM needs to be protected with a shield if the rock is caving in.[*]

15.5.1 Shield Driving in Groundwater

If a shield operates below groundwater level, a sufficient safety against uplift must be ensured for all situations to be encountered. In addition, the shield must be protected against the inrush of water and soil. This can be achieved by supporting the face with compressed air or a pressurized fluid.

Slurry shield. Some of the disadvantages of the support by compressed air are avoided if the support of the face is accomplished by a pressurized slurry, which in most cases is a bentonite suspension. There is no danger of blowouts, and all work can be done under normal atmospheric pressure. The support of the ground is achieved by a seepage force that presupposes the formation of a mud cake (made of bentonite) on the soil surface.

The soil is excavated with a cutterhead. If the ground is very soft, excavation even can be accomplished with a water jet. The muck is mixed with the slurry at a ratio of 1:10 and pumped away to a separation plant. Therefore, stones and blocks must be crushed first. The costs for separation rise at increased content of silt and clay.

The slurry pressure needed to support the ground must be estimated or calculated. To maintain the prescribed pressure and to replace any losses owing to mucking, a reliable control of the pressure must be guaranteed. An air cushion is provided in a part of the pressurized space. Owing to the high compressibility of air, the pressure of this air cushion is much less susceptible to small volume changes. By tuning the air pressure and balancing the removed and added slurry, the pressure can be controlled with an accuracy of between 0.05 and 0.1 bar.

For maintenance reasons and in order to remove blocks, the excavation chamber can be entered via an air lock. During maintenance, the slurry is replaced by compressed air. Note, however, that pressurized air is a risky method of support, because it can escape easily (blowouts). For this reason, daylight collapses occurred, for example, during the drivings of the Grauholz and the Westerschelde tunnels. An alternative is to freeze the soil ahead of the face and carry out the maintenance work under the protection of frozen soil.

Earth-pressure-balance (EPB) shield. Instead of a slurry, the face is supported with a mud formed of the excavated soil. The soil enters the excavation chamber through openings in the cutterhead. In most cases, water and some other additives (e.g., polymer foams) are added to render the excavated soil supple. Otherwise, heat will be developed owing to friction with the rotating cutterhead. The thick consistence of the mud (compared with a slurry) calls for a higher cutterhead torque (approximately 2.5 times higher than for slurry shields). On the other hand, the torque is limited by the available friction between lining and rock. Thus the diameters of EPB shields are limited to approximately 12 meters. Control of the mud pressure needed to support the face is

---

[*]Recommendations for selecting and evaluating tunnel boring machines. "Deutscher Ausschuß für unterirdisches Bauen, Öesterreichische Gesellschaft für Geomechanik, SIA-Fachgruppe für Untertagebauten," *Tunnel* 5/1997, 20–35. See also *Taschenbuch für den Tunnelbau,* Glückauf Verlag, 2001.

achieved by tuning the rotation speeds of the cutterhead (approximately 2 to 3 rpm) and the screw conveyor, which removes the muck from the front chamber (approximately 4 to 5 rpm). The mud should be sufficiently thick to plug the conveyor screw; otherwise, the pressure in the pressure chamber will drop. Advancing the shield by the jacks also helps to control the pressure. Given the compressibility of the mud and the inhomogeneous nonhydrostatic pressure distribution, the pressure control is not precise; it fluctuates by ±0.5 bar. The mud contains 50 to 70 percent solids and thus can be mucked with trucks or conveyor belts. In general, its dumping capability can be ensured easily if the bentonite content is not too high. The shield cannot start working until the pressure chamber is filled. This is much simpler with slurry shields than with EPB shields. The appropriate conditioning of the muck with foam etc., makes the application of EPB shields possible in a large variety of grounds (including gravel). Attention should be paid to heterogeneous ground, where the various soil layers need different support pressures.

15.5.2 Tunneling with Box or Pipe Jacking

The tunnel driving proceeds with jacking of precast support elements (pipes or boxes/frames) while the ground is excavated or pushed away at the face (Figure 15.23). Box jacking is usually applied to build subways under existing roads or rail tracks without interrupting the traffic.

The only difference between pipe jacking and shield driving is the position of the jacks: For pipe jacking, they are situated in the start shaft and/or at intermediate jack stations, whereas for shield driving, they are placed directly behind the shield. The excavated diameter is slightly larger than the pipe diameter so that the resulting gap can be grouted with bentonite to reduce friction.

FIGURE 15.23 Pipe jacking.

15.6 ROCK EXCAVATION

The words *drilling*, *boring,* and *cutting* are more or less synonymous in denoting rock excavation.

15.6.1 Drilling of Boreholes

In tunneling, boreholes are drilled for the purposes of exploration (site investigation), drill and blast, grouting, and installation of bolts, spiles, and other types of reinforcement. Drilling consists of break-out, removal of the rock, and cooling of the core bit.[*] Removal of the drill dust and chips and cooling are accomplished by flushing. Exploration drillings are flushed with water, whereas percussion

[*]See also J. A. Franklin and M. B. Dusseault, *Rock Engineering*, McGraw-Hill, 1989.

drillings are flushed with compressed air. Viscous fluids are used if the wall of the borehole has to be supported. Drilling of blastholes requires high speed, low wear of the core bit and high precision (an accuracy of 0.1° is required for accurate blasts), whereas exploration drillings aim at good core recovery and stable borehole walls.

15.6.2 Rock Excavation with Disk Cutters

The advance rate of a TBM is a very important quantity for planning and bidding tunnel projects. It therefore should be predicted as precisely as possible. So-called TBM performance prediction models have been developed, among them the models of the Norwegian Institute of Technology (NTH) and the Colorado School of Mines (CSM).

The excavation of rock is a very complex process that depends on factors that are hardly controllable. Thus most of the TBM prediction models are based on empirical correlations of the several controlling parameters.

Disk cutters (also called *disks*) exert a high pressure and thus break the rock (Figures 15.24 and 15.25). The disks exert forces up to $F = 250$ kN that fragment the rock into flat chips (fragments).

15.6.3 Abrasion

Abrasion of the excavation tools is an important issue. Several methods have been proposed to rate the abrasivity of rock. Among them are

CERCHAR (Laboratoire du Centre d'Etudes et Recherches des Charbonnages). The tip of a steel conus is loaded with 7 kilograms and scratched six times over a length of 1 centimeter along a fresh fracture surface of the investigated rock. The flattening of the tip owing to abrasion then is measured. CERCHAR indices, varying from 1 to 6, then are assigned to diameters of the truncated tip (varying from 0.1 to 0.6 millimeters).

LCPC (Laboratoire Central des Ponts et Chaussees). First the rock is broken down (a sample of 500 grams with grains sizing from 4 to 6.3 millimeters is whirled with a standardized steel propeller at 4,500 rpm). The index *ABR* is defined as the weight loss of the propeller (owing to abrasion) per 1 ton of rock.

FIGURE 15.24   (*Left*) Undercutting at the Uetliberg Tunnel. S. Mauerhofer, M. Glättli, J. Bolliger, and O. Schnelli, "Uetliberg Tunnel: Stage Reached by Work and Findings with the Enlargement Tunnel Boring Machine TBE," *Tunnel* 4/2004. (*Right*) Disk cutters from Lötschberg base tunnel. (*Left part*) Worn disk cutters.

Disk cutting   Undercutting

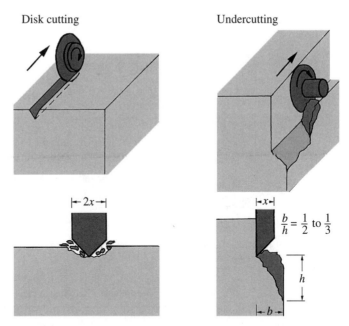

FIGURE 15.25 Principles of rock cutting: (*left*) Usual technique; (*right*) undercutting. L. Baumann and U. Zischinsky, "Neue Löse and Ausbautechniken zur maschinellen Fertigung von Tunneln in druckhaftem Fels: In: *Innovationen im unterirdischen Bauen*, STUVA Tagung 1993, 64–69.

Schimazek's coefficient of abrasivity. $F = Vd\sigma_z/100$; with V, volumetric percentage of quartz; d, mean size (in millimeters) of quartz grains; σ_z, tensile strength (MN/m²) of the rock. Examples of rating: $F < 0.05$, not very abrasive; $F > 2$, extremely abrasive.

15.7 PROFILING

For practical reasons, the excavated profile does not coincide with the intended one. Underprofile (i.e., deficient excavation) can be detected by means of templates or geodetic devices. Subsequently, the remaining rock has to be removed (scaling), for example, with excavators or cautious blasting.

Overprofile (also called *overbreak*), that is, surplus excavation, can be due to bad geologic conditions and thus is inevitable and/or can be due to improper excavation. It causes additional costs for the contractor because it has to be filled with shotcrete. Thus the distinction between both types of overbreak is of economic importance. A certain amount of overbreak, represented by a strip of the width d (Figure 15.26), should be allowed for, as required by the technical equipment. d should be specified by the contractor. A strip of the width D (Table 15.2) is to specify the *geologic* overbreak, that is, the overbreak imposed by the geologic conditions, and is to be payed to the contractor.[*]

If not otherwise specified, SIA 198 recommends the following expressions for D, with A being the theoretical tunnel cross-section.

---

[*]Swiss Code SIA 198.

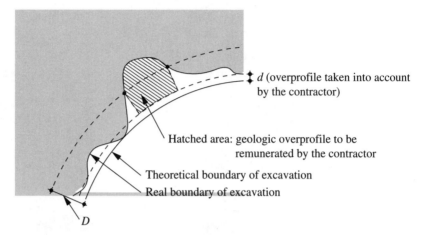

d (overprofile taken into account by the contractor)

Hatched area: geologic overprofile to be remunerated by the contractor

Theoretical boundary of excavation

Real boundary of excavation

D

FIGURE 15.26 Refundable overbreak (according to SIA 198) for tunnels excavated conventionally or with a roadheader.

TABLE 15.2 Strip Width *D* Referring to Figure 15.26

Excavation method	*D*
Drill and blast	Max($0.07\sqrt{A}$, 0.40 m)
Roadheader	Max($0.05\sqrt{A}$, 0.40 m)
Soil, without shield	Max($0.05\sqrt{A}$, 0.40 m)
Shield	Max($0.03\sqrt{A}$, 0.25 m)
TBM	Max($0.03\sqrt{A}$, 0.20 m)

Source: SIA 198.

15.8 MUCKING

The removal of excavated rock/soil is known as *mucking* and consists of loading up, transporting, and unloading of the muck (also called *spoil*). For transport (haulage), the following variants are available:

Trackless transport. Usual earthwork trucks are used. Of course, provision should be taken for the reduced production of harmful combustion products. A good carriageway is important to enable a speed of about 50 km/h. The inclination is limited to approximately 7 percent, and the necessary width of the tunnel is 7 to 8 meters. The requirements for a fresh-air supply are very high: A dumper of 300 PS (220 kW) needs 2,000 m^3 of fresh air per minute.

Railbound mucking (track transport). This is applicable in crowded spaces, that is, for spans < 6 meters. Usual track gauges are between 600 and 1,435 millimeters, the maximum inclination is 3 percent, the minimum curvature radius is about 7 meters, and the velocity is limited to about 20 km/h. The locomotive is powered by a diesel or electric motor. In the latter case, the power is delivered by accumulators. The power consumption is lower than with trucks.

Continuous conveyors. Conveyors have a very large transport capacity (upto 200 t/h).[*] A separate transport for personnel is needed, as well as a stone crusher and a dosing device. Apart

---

[*]S. Wallis, "Continuous Conveyors Optimise TBM Excavation in the Blue Mountains and Midmar," *Tunnel* 3/95, 10–20.

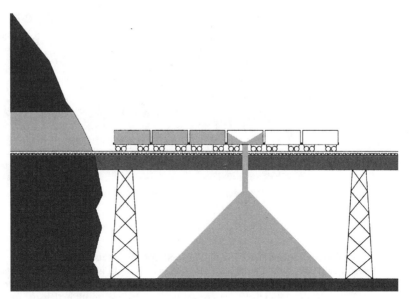

FIGURE 15.27 Unloading from bottom outlets.

from the high transport capacity, their main advantage is safe, clean, and silent transport. However, intensive maintenance is needed because their outfall implies downtime of all other work. A minimum curvature radius of 500 meters is required. Conveyors can be mounted on rail waggons.

Tunnel muck is utilized, if possible, as aggregate for shotcrete and cast concrete and for fills. The use as aggregate requires an appropriate mineralogic composition and, possibly, crushing and/or washing to remove the fines. In general, muck from drill and blast in igneous rock is more appropriate as aggregate than muck from TBMs.

If muck cannot be integrated immediately after driving, it has to be stored in a disposal dump. While the unloading of trucks is straightforward, rail tracks need special unloading facilities (an example is given in Figure 15.27).

15.9 SUPPORT

15.9.1 Basic Idea of Support

Usually the lining of a tunnel is not loaded by the stress that initially prevailed in the ground. Luckily, the initial (or primary) stress is reduced by deformation of the ground that occurs during excavation but also after installation of the lining (here, *lining* is understood as the shell of shotcrete, which is placed as soon as possible after excavation). Deformation of the ground (soil or rock) implies a reduction of the primary stress. This is a manifestation of *arching*. Since the deformation of the ground is connected with the deformation of the lining, it follows that the load acting on the lining depends on its own deformation. This is always the case with soil-structure interaction and constitutes an inherent difficulty for design because the load is not an independent variable. Thus the question is not "What is the pressure acting upon the lining?" but rather "What is the relation between pressure and deformation?"

FIGURE 15.28   Pull-out test of an anchor. (Dywidag DSI, Info 10.)

15.9.2   Steel Meshes

Steel meshes (mesh size ≥100 millimeters, $\varnothing < 10$ mm; concrete cover ≥ 2 cm) are mounted manually and therefore, should be not too heavy. A usual weight is 5 kg/m². Mesh installation is labor-intensive and relatively hazardous because personnel are exposed to small rock falls. For drill and blast, mounting the mesh adjacent to the face (proximity < 1 meter) can result in damage from the subsequent blast.

15.9.3   Rock Reinforcement

The mechanical properties of rock (be it hard rock or soft rock and soil) in terms of stiffness and strength can be improved by the installation of various types of reinforcement.[*] Steel bars[†] can be fixed at their ends and pretensioned against the rock. In this way, the surrounding rock is compressed, and as a consequence, its stiffness and its strength increase. Such reinforcing bars are called *anchors* or *bolts*.[‡] An alternative type of reinforcement consists of bars that are connected with the surrounding rock over their entire length, for example, by grout. Such bars are not pretensioned and are called *nails*.[§] Rock with nails is a composite material whose stiffness is increased compared with the original rock. A third action of reinforcement occurs when a steel bar (dowel) inhibits the relative slip of two adjacent rock blocks. In this case, the bar is loaded by transverse forces and acts as a plug. The use of names stated here (*anchor*, *bolt*, *dowel*) is, however, not unique, and they are often interchanged.

Connection with the Adjacent Rock.   The connection, that is, the force transfer between reinforcement and surrounding rock, is achieved either mechanically or by means of grout (cement mortar or synthetic resin). It is checked by means of pull-out tests (Figure 15.28). Mechanical connections can be loaded immediately after installation. They consist of

Wedges.   Conical wedges are placed at the end of the borehole. They can be moved in a longitudinal direction either by hammering (*slot and wedge anchors*, Figure 15.29) or by rotating a thread (*expanding-shell anchors*, Figure 15.30) in such a way that they force their containment (which is either a slotted bar or a shell) to grip into the rock. The transmission of a concentrated force is only possible in sufficiently hard rock (compression strength > 100 MPa). Slot and wedge anchors can be loosened by shocks and vibrations (e.g., owing to blasting).

---

[*]A good introduction to rock reinforcement is given in the book *Rock Engineering,* by J. A. Franklin and M. B. Dusseault, McGraw-Hill, 1989, which is, however, out of print.

[†]Anchors from synthetic material (e.g., fiberglass) or wood are used as temporary reinforcement of the face. They can be easily demolished during the subsequent excavation.

[‡]The two words are synonymous. Some authors, however, use *bolt* for forces < 200 kN, typically achieved with bars with diameter ≤ 25 millimeters and length less than 2 to 3 meters.

[§]In ground engineering, this type of reinforcement is called *nailing*.

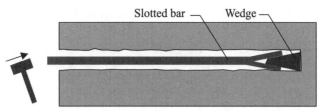

FIGURE 15.29   Slot and wedge anchor.

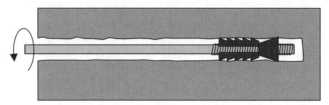

FIGURE 15.30   Expanding-shell anchor.

Tubular steel rockbolts.   A contact over the entire bolt length is achieved by the expansion of a tubular steel (also called *hollow anchor*) against the borehole wall. The expansion is either elastic (Figure 15.31) or is achieved by means of water pressure (Swellex rockbolt by Atlas Copco, Figure 15.32). The shear force is transferred to the rock by friction.

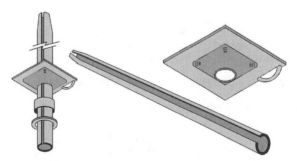

FIGURE 15.31   Hollow anchor systems.

FIGURE 15.32   Swellex anchor.

Alternatively, the connection with the rock can be achieved by means of cement mortar or resin. The resulting support is not immediate because setting and hardening need some time.

Grouted rockbolts.[*] The annular gap between rebar and drillhole wall is filled with cement or resin grout. Before grouting, the drillhole must be thoroughly flushed with water or air to ensure a clean rock surface. It also should be ensured that the rock does not contain wide-open joints into which the grout may disappear. This can be avoided by using geotextile containments of the grout. Cement grout (mortar) consists of well-graded sand and cement in ratios between 50:50 and 60:40. To obtain a sufficient strength, the water-to-cement ratio should be 40 percent or less by weight. The mortar should set (i.e., obtain the required strength) within 6 hours. To increase plasticity, a bentonite fraction of up to 2 percent of the cement weight can be added. Note that cement grout can be damaged by vibrations owing to blasting. In some cases, grouting is not allowed until the heading has advanced by 40 to 50 meters.

Synthetic resins harden very quickly (2 to 30 min) by polymerization when mixed with a catalyst. The two components are either injected or introduced into the drillhole within cartridges that are subsequently burst by introducing the rebar.

Cement mortar can be introduced in several ways:

- Grouting into the annular gap between rebar and drillhole wall.
- Perforated tubes filled with mortar are placed into the borehole. The subsequent introduction of the anchor (by means of hammer blows) squeezes the mortar into the remaining free space (*perfobolts*). However, perfobolts are obsolete.
- Self-boring or self-drilling anchors (SDAs).[†] The rod is a steel tube of 42 to 130 millimeters diameter driven into the rock with rotary-percussion drilling equipment, flushing, and a sacrificial drill bit. Standard delivery lengths vary between 1 and 6 meters. Two rods can be connected with couplers. Grouting of mortar occurs through the tube with pressures up to 70 bar.

At the head of the anchor, the tendon (steel rod) is fixed against a bearing plate or faceplate in such a way that the anchor tension is converted into a compressive force at the rock face. Spherical washers allow workers to fix the tendon against the faceplate as well in cases where the tendon is not perpendicular to the rock surface. The faceplate also helps to fix the wire mesh.

Application. There are two types of rockbolt applications in tunneling:

Spot bolting. Individual rockbolts are placed to stabilize isolated blocks.

Pattern bolting. Systematic installation of a more or less regular array of rockbolts. Despite some rational approaches, the design of pattern bolting is empirical. For coal mines, Bieniawski suggested the following approach:[‡] Based on the RMR value of the rock mass and the span s, one obtains the *rock load height* h_t as $h_t = s(100 - RMR)/100$. The bolt length l then is determined as $l = \min (s/3; h_t/2)$, and the spacing of the bolts is taken as 0.65 to 0.85l.

In unstable rock, bolts should be installed immediately after drilling of the hole. In so doing, the long-established safe practice of working forward from solid or secured ground toward unsupported or suspect ground must be adhered to. Apart from tunneling, anchors are also applied in other fields of civil engineering and mining, for example, to secure tied-back retaining walls and slopes or to prevent uplift by hydrostatic pressure.

15.9.4 Timbering

In the early days of tunneling, timbering was the only means for temporary support. Nowadays, it is used mainly for the support of small and/or irregular cavities (e.g., resulting from inrushes).

---

[*]So-called SN anchors. This name originates from their first application in Store Norfors in Sweden. An alternative etymology attributes SN to soil and nail.

[†]So-called IBO anchors, with the brands TITAN by Ischebek and MAI by Atlas Copco.

[‡]J. A. Franklin, and M. B. Dusseault, *Rock Engineering*, McGraw-Hill, 1989, Sec. 16.3.2.

FIGURE 15.33   Timbering in the metro of Madrid (2000).

Timbering has been used systematically (according to the old Belgian tunneling method) during the recent construction of the Madrid metro (Figure 15.33).

Wood is easy to handle and transport and indicates imminent collapse by cracking. On the other hand, the discontinuous contact with the rock is problematic. The spacing of the timber frames is usually 1 to 1.5 meters. Care must be taken for a sufficient longitudinal bracing.

15.9.5 Support Arches

Support arches are composed of segments of rolled-steel profiles or lattice girders (Figures 15.34 and 15.35). The arch segments are placed and mounted together with fixed or compliant joints (to accommodate large convergences). Contact with the adjacent rock is achieved with wooden wedges or with bagged packing, that is, bags filled with (initially) soft mortar. Usually the arches are covered subsequently with shotcrete. This leads to a garland-shaped shotcrete surface that protrudes into the cavity at the locations of the arches. To achieve good contact between the shotcrete surface and a geosynthetic sealing membrane, the sag between two adjacent arches should not exceed 1/20 of their spacing. Together with their contribution to support, arches also help to check the excavated profile. They also can serve to mount forepoling spiles in a longitudinal direction. Clearly, U-shaped rolled-steel profiles have a much higher bearing capacity than lattice girders.

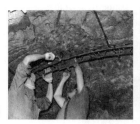

FIGURE 15.34   Rolled steel profiles, lattice girder (pantex-3-arch), connecting the segments, mounting with rock bolts.

FIGURE 15.35   Mounting of girder arches, Zürich-Thalwil Tunnel. (*Source:* Tunnelrg Switzerland, 2.)

15.9.6   Support Recommendations Based on RMR

Bieniawski recommends the support measures shown in Table 15.3. They are based on RMR[*] and refer to a tunnel of 10 meters diameter.

TABLE 15.3   Support Measures Based on RMR

RMR	Heading	Anchoring Ø 20 mm, fully bonded	Shotcrete	Ribs
81–100	Full face, advance 3 m	—	—	—
61–80	Full face, advance 1 to 1.5 m, complete support 20 m from face	Locally bolts in crown, 3 m long, spaced 2.5 m, with occasional wire mesh	5 cm in crown where required	—
41–60	Top heading and bench: 1.5 to 3 m advance in top heading, commence support after each blast, complete support 10 m from face	Systematic bolts 4 m long, spaced 1.5 to 2 m in crown and walls with wire mesh in crown	5 to 10 cm in crown, 3 cm in sides	—
21–40	Top heading and bench: 1 to 1.5 m advance in top heading, install support concurrently with excavation 10 m from face	Systematic bolts 4 to 5 m long, spaced 1 to 1.5 m in crown and walls with wire mesh	10 to 15 cm in crown and 10 cm in sides	Light ribs spaced 1.5 m where required
≤20	Multiple drifts: 0.5 to 1.5 m advance in top heading, install support concurrently with excavation	Systematic bolts 5 to 6 m long, spaced 1-1.5 m in crown and walls with wire mesh, bolt invert	15 to 20 cm in crown, 15 cm in sides, and 5 cm in face	Medium to heavy ribs spaced 0.75 m with steel lagging and forepoling if required, close invert

Source: Bieniawski, 1984.

[*]Z. T. Bieniawski, *Rock Mechanics Design in Mining and Tunnelling,* Balkema, 1984.

FIGURE 15.36   Special forepoling rig, *Rotex*; forepoling, schematically.

15.9.7   Forepoling

If the strength of the ground is so low that the excavated space is unstable even for a short time, a predriven support is applied in such a way that an excavation increment occurs under the protection of a previously driven canopy. The traditional method of forepoling was to drive 5- to 7-millimeter-thick steel sheets up to 4 meters beyond the face into the ground or 1.5- to 6-meter-long steel rods (called *spiles*) with a spacing of 30 to 50 centimeters. Nowadays, forepoling is achieved by spiling, pipe roof, grouting, and freezing.

Spiling. This method consists of drilling a canopy of spiles, that is, steel rods or pipes, into the face (Figure 15.36). A typical length is 4 meters. To give an idea, 40 to 45 tubes, ∅ 80 to 200 mm, each 14 meters long, enable a total advance of 11 to 12 meters (the last 2 to 3 meters serve as abutment of the canopy). In order for the spiles to act not only as beams (i.e. in longitudinal direction) but also to form a protective arch over the excavated space, the surrounding soil is grouted through the steel pipes or sealed with shotcrete. Thus a connected canopy is formed that consists of grouted soil reinforced with spiles. Drilling 40 tubes takes about 10 to 12 hours and grouting another 10 to 12 hours. Spile rods also can be placed into drillholes. The remaining annular gap is filled with mortar, whose setting, however, may prove to be too slow. Alternatively, *self-drilling* rods are used.

Pipe roof. This method is similar to spiling, with the only difference being that large-diameter (> 200-millimeter) steel or concrete tubes are jacked into the soil above the space to be excavated. The larger diameter provides a larger bearing capacity. Sometimes the tubes are filled with concrete. The steel tubes act only as beams and do not form an arch. Pipe roofs do not protect the overburden soil from considerable settlements.

15.9.8   Face Support

Unstable faces can be backed (buttressed) with a heap of muck or reinforced with fiberglass rods. Both methods are temporary and have to be removed before or during the next excavation step. In soft underground and for tunnel diameters larger than 4 meters, the face should not be vertical but inclined about 60 to 70°.

15.9.9   Sealing

Surface support, such as shotcrete, may act in two distinct ways. With uniform convergence of the rock, the support responds with arching, that is, mobilization of axial thrust within the lining. At local spots of weakness or at keystones (in the case of jointed rock), shear and tensile stresses are mobilized within the lining. At an initial stage of these deformations, a thin lining is sufficient to resist loosening of the rock and the related increase in loads. This supporting action is called *sealing*

FIGURE 15.37   Rolling formwork (Engelberg base tunnel); rolling formwork (Nebenwegtunnel, Vaihingen/Enz). (*Source: Tunnel* 3/2001, 30–31.)

and can be obtained with thin layers of shotcrete. A recent development is to seal with 3- to 6-mm-thick spray-on polymer liners. They have good adhesive bond when applied to clean rock, and they develop a good performance in tension and shear. It should be mentioned, though, that creep is still an open question. In contrast to shotcrete, the compliant nature of synthetic liners allows them to continue to function over a wide displacement range.[*]

15.9.10 Permanent Lining

The usual thickness of a permanent lining is at least 25 centimeters. For reinforced and watertight linings, a minimum thickness of 35 centimeters is recommended.[†] Blocks of 8 to 12 meters in length are separated with extension joints. Usually concrete C20/25 is used. Concretes of higher strengths develop higher temperatures during setting (fissures!) and are more brittle.

The concrete is poured into rolling formworks (Figure 15.37) and compacted with vibrators in the invert and with external vibrators in the crown (one vibrator for 3 to 4 m^2). It is difficult to achieve complete filling of the crown space with concrete: The pumping pressure should be limited; otherwise, the rolling formwork can be destroyed. Possibly unfilled parts should be regrouted with pressures of ≤2 bar, 56 days after concreting. Use of self-compacting concrete[‡] potentially can help to avoid incomplete filling of the formwork. Within 8 hours, the concrete should attain a sufficient strength that the formwork can be removed. However, there are cases reported where the setting was insufficient, and the lining collapsed after early removal of the formwork.

15.10 STRESS AND DEFORMATION FIELDS AROUND A DEEP CIRCULAR TUNNEL

The analytical representation of stress and deformation fields in the ground surrounding a tunnel succeeds only in some extremely simplified special cases, which are rather academic. Nevertheless, analytical solutions offer the following benefits:

- They are exact solutions, and they provide insight into the basic mechanisms (i.e., displacements, deformation, and stress fields) of the problem under consideration.

---

[*]D. D. Tannant, "Development of Thin Spray-on Liners for Underground Rock Support: An Alternative to Shotcrete?" In: *Spritzbeton Technologie* 2002, by W. Kusterle, University of Innsbruck, Institut für Betonbau, Baustoffe und Bauphysik, 2002, 141–153.

[†]"Concrete Linings for Mined Tunnels: Recommendations by DAUB," Dec. 2000, *Tunnel*, 3/2001, 27–43.

[‡]This is a concrete of high flowability (spread > 70 cm).

- They provide insight into the role and importance of the parameters involved.
- They can serve as benchmarks to check numerical solutions.

In this section, some solutions are introduced that are based on Hooke's law, the simplest material law for solids. The underground is regarded here as a linear elastic, isotropic semi-infinite space, that is bound by a horizontal surface, the ground surface. The tunnel is idealized as a tubular cavity with circular cross-section. Before its construction, the so-called primary stress state prevails. This stress state prevails also after construction of the tunnel over a sufficiently large distance (so-called far field).

15.10.1 Some Fundamentals

The equilibrium equation of continuum mechanics written in cylindrical coordinates reveals the mechanism of arching in terms of a differential equation. For axisymmetric problems, as they appear in tunnels with circular cross-section, the use of cylindrical coordinates is advantageous. In axisymmetric deformation, the displacement vector has no component in the θ direction: $u_\theta \equiv 0$. The nonvanishing components of the strain tensor reduce in case of axial symmetry to

$$\varepsilon_r = \frac{\partial u_r}{\partial r} \qquad \varepsilon_\theta = \frac{u_r}{r} \qquad \varepsilon_z = \frac{\partial u_z}{\partial z}$$

where u_r and u_z are the displacements in the radial and axial directions, respectively.

The stress components σ_r, σ_θ, and σ_z are principal stresses. The equation of equilibrium in the r direction reads

$$\frac{\partial \sigma_r}{\partial r} + \frac{\sigma_r - \sigma_\theta}{r} + \varrho g \cdot e_r = 0 \tag{15.1}$$

and in the z direction

$$\frac{\partial \sigma_r}{\partial z} + \varrho \boldsymbol{g} \cdot \boldsymbol{e}_z = 0 \tag{15.2}$$

where ϱ is the density, $\varrho \boldsymbol{g}$ is the unit weight, and $\boldsymbol{e}_r$ and $\boldsymbol{e}_z$ are unit vectors in the r and z directions. The second term in equation (15.1) describes arching. This can be seen as follows: If r points to the vertical direction z, then equation (15.1) reads

$$\frac{d\sigma_z}{dz} = \gamma - \frac{\sigma_x - \sigma_z}{r} \tag{15.3}$$

where the term $(\sigma_x - \sigma_z)/r$ is responsible for that fact that σ_z does not increase linearly with depth (i.e., $\sigma_z = \gamma z$). In the case of arching, that is, for $(\sigma_x - \sigma_z)/r > 0$, σ_z increases underproportionally with z. Note that this term, and thus arching, exists only for $\sigma_r \neq \sigma_\theta$. This means that arching is due to the ability of a material to sustain deviatoric stress, that is, shear stress. No arching is possible in fluids. This is why soil/rock often "forgives" shortages of support, whereas (ground) water is merciless. The equilibrium equation in the θ-direction, $(1/r) \cdot (\partial \sigma_\theta / \partial \theta) = 0$, is satisfied identically because all derivatives in the θ direction vanish in axisymmetric stress fields.

With reference to the arching term $(\sigma_\theta - \sigma_r)/r$, attention should be paid to r. At the tunnel crown, r is often set equal to the curvature radius of the crown. However, this is not always true. If we consider the distribution of σ_z and σ_x above the crown at decreasing support pressures p, we notice that for $K = \sigma_x/\sigma_z < 1$, the horizontal stress trajectory has the opposite curvature than the tunnel crown (Figure 15.38).

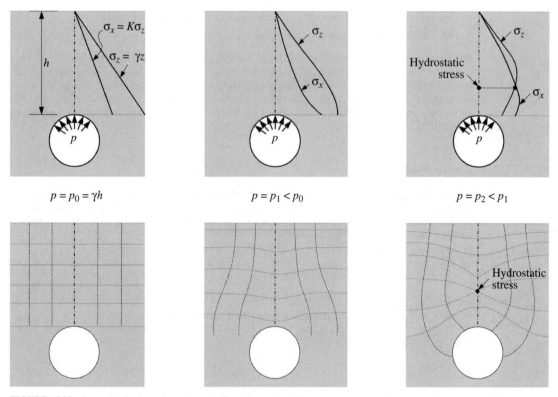

FIGURE 15.38 Stress distributions above the crown for various values of support pressure p and corresponding stress trajectories. At the hydrostatic point (*right*), the curvature radius of the stress trajectory vanishes, $r = 0$.

15.10.2 Geostatic Primary Stress

An often encountered primary stress (for horizontal ground surface) is $\sigma_{zz} = \gamma z$, $\sigma_{xx} = \sigma_{yy} = K\sigma_{zz}$, where z is the cartesian coordinate pointing downward, γ is the specific weight of the rock, and K is the so-called lateral stress coefficient. For noncohesive materials, K has a value between active and passive earth pressure coefficients $K_a \leq K \leq K_p$, and we often can set $K = K_0 = 1 - \sin \varphi$. The stress field around a tunnel has to fulfill the equations of equilibrium $\frac{\partial \sigma_{xz}}{\partial z} + \frac{\partial \sigma_{xz}}{\partial x} = \gamma$, $\frac{\partial \sigma_{xz}}{\partial z} + \frac{\partial \sigma_{xz}}{\partial x} = 0$, as well as the boundary conditions at the ground surface ($z = 0$) and at the tunnel wall. We assume that the tunnel is unsupported, so the normal and shear stress at the tunnel wall must disappear. The analytical solution to this problem is extremely complicated[*] and consequently offers no advantages over numerical solutions (e.g., according to the method of finite elements; Figure 15.39). The analytical solution can be simplified if one assumes the primary stress in the neighbourhood of the tunnel as constant (Figure 15.40) and not as linearly increasing: $\sigma_{zz} \approx \gamma H$, $\sigma_{xx} \approx K\gamma H$. This approximation is meaningful for deep tunnels ($H \gg r$). The stress field around the tunnel then can be represented in polar coordinates as follows (Kirsch 1898):

---

[*]R. D. Mindlin, "Stress Distribution Around a Tunnel," *ASCE Proceedings*, April 1939, 619–649.

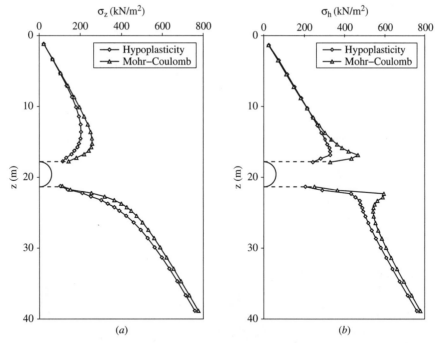

FIGURE 15.39 Distributions of vertical and horizontal stress along the vertical symmetry axis. Circular tunnel cross section ($r = 1.0$ m). Numerically obtained with hypoplasticity and Mohr-Coulomb elastoplasticity. (P. Tanseng, "Implementations of Hypoplasticity and Simulations of Geotechnical Problems: Including Shield Tunnelling in Bangkok Clay," PhD thesis University of Innsbruck, 2004.)

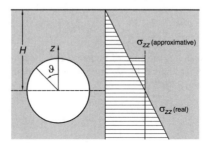

FIGURE 15.40 Distribution of vertical stress in the environment of a tunnel.

$$\sigma_{rr} = \gamma H \left[\frac{1+K}{2} \left(1 - \frac{r_0^2}{r^2} \right) \right] + \gamma H \left[\frac{1-K}{2} \left(1 + 3\frac{r_0^4}{r^4} - 4\frac{r_0^2}{r^2} \right) \cos 2\vartheta \right]$$

$$\sigma_{\vartheta\vartheta} = \gamma H \left[\frac{1+K}{2} \left(1 + \frac{r_0^2}{r^2} \right) \right] - \gamma H \left[\frac{1-K}{2} \left(1 + 3\frac{r_0^4}{r^4} \right) \cos 2\vartheta \right] \qquad (15.4)$$

$$\sigma_{r\vartheta} = -\gamma H \frac{1-K}{2} \left(1 - 3\frac{r_0^4}{r^4} + 2\frac{r_0^2}{r^2} \right) \sin 2\vartheta$$

From equation (15.4) it follows that at the invert and at the crown (for $r = r_0$ and $\vartheta = 0$ or π), the tangential stress reads $\sigma_{\vartheta\vartheta} = \gamma H (3K - 1)$. Thus, for $K < \frac{1}{3}$, one obtains tension, that is, $\sigma_{\vartheta\vartheta} < 0$.

15.10.3 Hydrostatic Primary Stress

The special case of equation (15.4) for $K = 1$ (i.e., for $\sigma_{xx} = \sigma_{\vartheta y} = \gamma H$) reads with $\sigma_\infty := \gamma H^*$

$$\sigma_r = \sigma_\infty \left(1 - \frac{r_0^2}{r^2} \right)$$

$$\sigma_\vartheta = \sigma_\infty \left(1 + \frac{r_0^2}{r^2} \right) \tag{15.5}$$

$$\sigma_{r\theta} = 0$$

This solution exhibits axial symmetry because the radius r is the only independent variable and ϑ does not appear.

Equation 15.5 can be generalized regarding the case that the tunnel wall is subjected to a constant pressure p (so-called support pressure):[†]

$$\sigma_r = \sigma_\infty \left(1 - \frac{r_0^2}{r^2} \right) + p \frac{r_0^2}{r^2} = \sigma_\infty - (\sigma_\infty - p) \frac{r_0^2}{r^2}$$

$$\sigma_\vartheta = \sigma_\infty \left(1 + \frac{r_0^2}{r^2} \right) - p \frac{r_0^2}{r^2} = \sigma_\infty + (\sigma_\infty - p) \frac{r_0^2}{r^2} \tag{15.6}$$

$$\sigma_{r\vartheta} = 0$$

With the application of pressure p, the wall of the tunnel yields by the amount $u(r_0)$. The displacement of the tunnel wall $u(r_0)$ can be computed as a function of p with the help of the solution by Lamé.[‡] One obtains the following linear relationship:

$$u(r_0) = r_0 \frac{\sigma_\infty}{2G} \left(1 - \frac{p}{\sigma_\infty} \right) \tag{15.7}$$

Figure 15.41 shows the plot of equation (15.7). If the support pressure p is smaller than σ_∞, then one obtains from Figure 15.42 a radial stress σ_r that increases with r and a tangential stress σ_θ that decreases with r.

15.10.4 Plastification

According to equation (15.6), the principal stress difference $\sigma_\vartheta - \sigma_r = 2(\sigma_\infty - p)(r_0/r)^2$ increases when p decreases. Now we must take into account that rock is not elastic; rather, it may be regarded as elastic only as long as the principal stress difference $\sigma_\vartheta - \sigma_r$ does not exceed a threshold that is given by the so-called limit condition. Only such stress states are feasible, to which applies:

$$\sigma_\vartheta - \sigma_r \le (\sigma_\vartheta + \sigma_r) \sin \varphi + 2c \cos \varphi \tag{15.8}$$

If p is sufficiently small, the requirement expressed by equation (15.6) is violated in the range $r_0 < r < r_e$ (r_e is still to be determined). Consequently, the elastic solution (15.6) cannot apply within this range.

---

[*]In denoting principal stresses, double indices can be replaced by single ones, that is, $\sigma_{rr} \equiv \sigma_r$ etc.
[†]One obtains the stress field as a special case of the solution that Lamé found in 1852 for a thick tube made of linear elastic material. See L. Malvern, *Introduction to the Mechanics of a Continuous Medium*, Prentice-Hall, 1969, p. 532.
[‡]Thereby u is presupposed as small so that the tunnel radius r_0 may be regarded as constant.

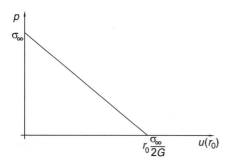

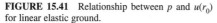

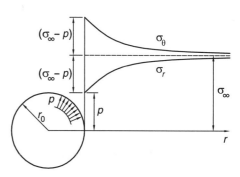

FIGURE 15.41 Relationship between p and $u(r_0)$ for linear elastic ground.

FIGURE 15.42 Stress field in linear elastic ground.

Rather, another relation applies that is to be deduced as follows: We regard equation (15.3) in the radial direction for $\gamma = 0$ and $c = 0$ and set into it the equation $\sigma_\vartheta = K_p \sigma_r$ with $K_p := (1 + \sin\varphi)/(1 - \sin\varphi)$. In soil mechanics, K_p is called the *coefficient of passive earth pressure*. With the requirement $\sigma_r = p$, for $r = r_0$, so we finally obtain:

$$\sigma_r = p\left(\frac{r}{r_0}\right)^{K_p - 1}$$

$$\sigma_\vartheta = K_p p\left(\frac{r}{r_0}\right)^{K_p - 1}$$

The range $r_0 < r < r_e$ where the above solution applies is called the *plastified zone*. The word *plastic* highlights the fact that with stress states that fulfil conditions (15.8), deformations occur without stress change (so-called plastic deformations or flow).

If the range $r_0 \le r \le r_e$ is plastified, then the elastic solution (15.6) must be slightly modified: σ_e is the value of σ_r for $r = r_e$. In place of equation (15.6), now applies for $r_e \le r < \infty$:

$$\sigma_r = \sigma_\infty - (\sigma_\infty - \sigma_e)\frac{r_e^2}{r^2}$$

$$\sigma_\vartheta = \sigma_\infty - (\sigma_\infty - \sigma_e)\frac{r_e^2}{r^2}$$

$$\sigma_{r\vartheta} = 0$$

For $r = r_e$, the stresses read $\sigma_r = \sigma_e$, and $\sigma_\vartheta = 2\sigma_\infty - \sigma_e$. They must fulfill the limit condition, that is, $\sigma_\vartheta = K_p \sigma_r$ or $K_p \sigma_e = 2\sigma_\infty - \sigma_e$. It then follows that

$$\sigma_n = \frac{2}{K_p + 1}\sigma_\infty \tag{15.9}$$

At the boundary $r = r_e$, the radial stresses of the elastic and plastic ranges must coincide: $p(r_e/r_0)^{Kp-1} = \sigma_e$. The radius r_e of the plastic range reads

$$r_e = r_0\left(\frac{2}{K_p + 1}\frac{\sigma_\infty}{p}\right)^{\frac{1}{K_p - 1}} \tag{15.10}$$

If we evaluate equation (15.10) for the case $r_e = r_0$, we obtain the support pressure p^* at which plastification sets in:

$$r_0 = r_0 \left(\frac{2}{K_p+1} \frac{\sigma_\infty}{p} \right)^{\frac{1}{K_p-1}}$$

$$p = p^* = \frac{2}{K_p+1} \sigma_\infty = (1 - \sin \varphi)\sigma_\infty$$

The distributions of σ_r and σ_ϑ in the case of plastification are represented in Figure 15.43.

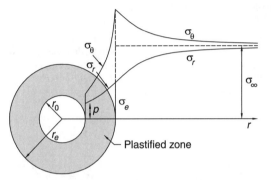

FIGURE 15.43   Distributions of σ_r and σ_ϑ within the plastic and elastic ranges. Note that σ_ϑ is continuous at $r = r_e$, and therefore, $\sigma_r(r)$ is smooth at $r = r_e$.

If the rock exhibits friction and cohesion, one obtains the following solution for the plastified zone:

$$\sigma_r = (p + c \cot \varphi) \left(\frac{r}{r_0} \right)^{K_p-1} - c \cot \varphi$$

$$\sigma_\vartheta = K_p (p + c \cot \varphi) \left(\frac{r}{r_0} \right)^{K_p-1} - c \cot \varphi$$

(15.11)

For $r = r_e$, the elastic stresses $\sigma_r = \sigma_e$ and $\sigma_\vartheta = 2\sigma_\infty - \sigma_e$ also must fulfill the limit condition. Thus

$$\sigma_e = \sigma_\infty (1 - \sin \varphi) - c \cos \varphi$$

(15.12)

At the boundary $r = r_e$, the radial stresses of the elastic and plastic zones must coincide

$$(p + c \cot \varphi) \left(\frac{r_e}{r_0} \right)^{K_p-1} - c \cot \varphi = \sigma_\infty (1 - \sin \varphi) - c \cos \varphi$$

Thus the radius r_e of the plastic zone is obtained as

$$r_e = r_0 \left(\frac{\sigma_\infty (1 - \sin \varphi) - c(\cos \varphi - \cot \varphi)}{p + c \cot \varphi} \right)^{\frac{1}{K_p-1}}$$

(15.13)

Setting $r_e = r_0$, we now obtain $p^* = \sigma_\infty (1 - \sin \varphi) - c\cos \varphi$. Again, $\sigma_\theta(r)$ is continuous at $r = r_e$, and consequently, $\sigma_r(r)$ is smooth at $r = r_e$.

15.10.5 Ground Reaction Line

The linear relationship (15.7) shows how the tunnel wall moves into the cavity if the support pressure is reduced from σ_∞ to p. This relationship applies to linear elastic ground. Now we want to see how the relationship between p and $u(r_0)$ reads if the ground is plastified in the range $r_0 \leq r < r_e$. Within, this range, plastic flow takes place, (i.e., the deformations increase without stress change). In order to specify plastic flow, one needs an additional constitutive relationship, the so-called flow rule. This is a relationship between the strains ε_r and ε_θ (ε_z vanishes per definition for the plane deformation we are considering). With the volumetric strain $\varepsilon_v := \varepsilon_r + \varepsilon_\vartheta$, the flow rule reads in a simplified and idealized form: $\varepsilon_v = b\varepsilon_r$. b is a material constant that describes the dilatancy (loosening) of the material.[*] Note, however, that several definitions of dilatancy exist. For $b = 0$, isochoric (i.e., volume-preserving, $\varepsilon_v = 0$) flow occurs. We express the strains with the help of the radial displacement u: $\varepsilon_r = d_u/d_r$, $\varepsilon_\vartheta = u/r$, and obtain thus $du/dr + u/r = b(du/dr)$, from which follows $u = \left(C / \frac{1}{r^{1-b}} \right)$ The integration constant C follows from the displacement $u = u_e$ at $r = r_e$ according to the elastic solution (equation 15.7):

$$u_e = r_e \frac{\sigma_\infty}{2G} \left(1 - \frac{\sigma_e}{\sigma_\infty} \right) \frac{C}{r_e^{\frac{1}{1-b}}} \tag{15.14}$$

From the two last equations, we obtain

$$u = r_e \frac{\sigma_\infty}{2G} \left(1 - \frac{\sigma_e}{\sigma_\infty} \right) \left(\frac{r_e}{r} \right)^{\frac{1}{1-b}} \tag{15.15}$$

If we introduce here the relations (15.9) and (15.10) for σ_e and r_e, we obtain for $r = r_0$, $c = 0$, and $p < p^*$:

$$u(r_0) = r_0 \sin \varphi \frac{\sigma_\infty}{2G} \left(\frac{2}{K_p + 1} \frac{\sigma_\infty}{p} \right)^{\frac{2-b}{(K_p - 1)(1-b)}} \tag{15.16}$$

This relationship applies to $p < p^*$, whereas for $p \geq p^*$; the elastic relationship (15.7) applies. Figure 15.44 shows the relationship between p and $u(r_0)$, which is called the *characteristic* of the ground (also called *Fenner-Pacher curve*, or *ground reaction curve* or *ground line*).

For the case $\varphi > 0$, $c > 0$, we can obtain the relationship between the cavity wall displacement $u(r_0)$ and p, if we use equation (15.15) with equations (15.12) and (15.13):

$$u(r_0) = r_0 \left[\frac{\sigma_\infty (1 - \sin \varphi) - c(\cos \varphi - \cot \varphi)}{p + c \cot \varphi} \right]^{\frac{2-b}{(K_p - 1)(1-b)}} \times \frac{\sigma_\infty}{2G} \left(\sin \sigma + \frac{c}{\sigma_\infty} \cos \varphi \right) \tag{15.17}$$

Contrary to equation (15.16), equation (15.17) supplies a finite displacement $u(r_0)$ for $p = 0$ (Figure 15.45), that is, in a cohesive material the cavity can persist also without support, whereas it

---

[*]The angle ψ: $= \arctan b$ can be called the *angle of dilatancy*.

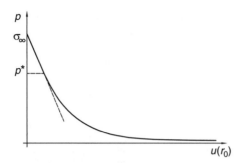
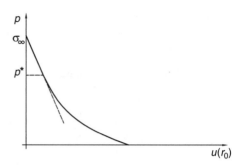

FIGURE 15.44   Ground reaction line with plastification (noncohesive ground).

FIGURE 15.45   Ground reaction line with plastification (cohesive ground).

closes up in noncohesive material (if unsupported). The tunnel wall displacement $u(r_0)$ is called in tunneling *convergence*. The equations deduced in this section cannot be evaluated easily for the case $\varphi = 0$ and $c > 0$. In this case, a solution can be obtained from equation (15.8). In the plastic range $r_0 < r \le r_e$,

$$\sigma_\infty = 2c \ln \frac{r}{r_0} + p, \, \sigma_\vartheta = \sigma_r + 2c$$

Furthermore, $r_e = r_0 \exp(\sigma_\infty - c - p)/2c$ and $\sigma_e = \sigma_\infty - c$. With equation (15.15) we obtain for $p < p^* = \sigma_\infty - c$:

$$u(r_0) = r_0 \frac{c}{2G} \left[\exp\left(\frac{\sigma_\infty - c - p}{2c} \right) \right]^{\frac{2-b}{1-b}} \tag{15.18}$$

The ground reaction line, that is, the dependence of p on $u(r_0)$ clearly shows that the pressure exerted by the rock on the lining is not a fixed quantity but depends on the rock deformation and thus on the rigidity of the lining. This is a completely different perception of load and causes difficulties for many civil engineers, who are used to considering the loads acting on, say, a bridge as given quantities.

15.10.6 Support Reaction Line

Now we want to see how the resistance p of the support changes with increasing displacement u. Considering equilibrium (Figure 15.46), the compressive stress in the support is easily obtained as $\sigma_a = pr_0/d$. This stress causes the support to compress by $\varepsilon = \sigma_a/E$ (E is the Young's modulus of the support).

The circumference of the support shortens by the amount $\varepsilon 2\pi r_0$, that is, the radius shortens by the amount $u = \varepsilon r_0$. From here follows a linear relationship between u and p (*characteristic of the support*, *support reaction line*, or *support line*):

$$p = \frac{Ed}{r_0^2} u \quad \text{or} \quad u = \frac{r_0^2}{Ed} p$$

We assume, for simplicity, that this linear relationship applies up to the collapse of the support, where $p = p_l$ (Figure 15.47).

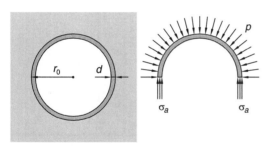

FIGURE 15.46 Forces acting on and within the support.

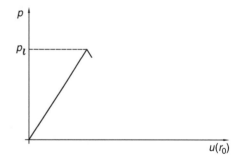

FIGURE 15.47 Support reaction line.

The reaction lines of the ground and the support serve to analyze the interaction between ground and support. To this purpose, they are plotted together in a p-u diagram. The characteristic of the support is represented by

$$u(p) = u_0 + \frac{r_0^2}{Ed} p$$

Here, the part u_0 takes into account that the installation of the support cannot take place immediately after excavation. Tunneling proceeds at a finite speed, and before the support can be installed, the tunnel remains temporarily unsupported in the proximity of the excavation face. Therefore, at the time of construction of the support, the cavity wall already has moved by the amount u_0 into the cavity. The influence of u_0 is shown in Figure 15.48: If u_0 is small (case 1), then the support cannot take up the ground pressure and collapses. If u_0 is large (case 2), then the ground pressure decreases with deformation so that it can be carried by the support. According to the terminology of the *New Austrian Tunneling Method* (NATM):

> . . . ground deformations must be allowed to such an extent that around the tunnel deformation resistances are waked and a carrying ring in the ground is created, which protects the cavity. . . .

The influence of the stiffness (Young's modulus E) on the support can be illustrated similarly in a diagram (Figure 15.48). A stiff support (case 1) cannot carry the ground pressure and collapses, whereas a flexible support (case 2) possesses sufficient carrying reserves.

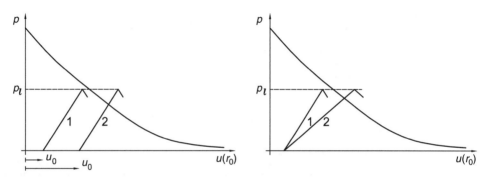

FIGURE 15.48 To the influence of u_0 (*left*); influence of the support stiffness (*right*).

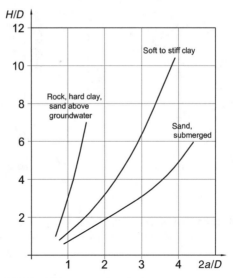

FIGURE 15.49   Estimation of A by Peck.

15.11 SETTLEMENT OF THE SURFACE

Apart from the assessment of stability, the determination of settlements at the surface is very important in tunneling. However, in geotechnical engineering, deformations can be forecast with less accuracy than stability. This is mainly because the ground has a nonlinear stress-strain relationship so that one hardly knows the distribution of the stiffnesses. We consider here some rough estimations of the settlement of the ground surface owing to the excavation of a tunnel. One should be aware of their limited accuracy.

A realistic description of the measured settlement is obtained according to Peck by the Gauss-distribution

$$u_v = u_{v,\text{max}} \cdot e^{-x^2/2a^2}$$

The parameter a (standard deviation) is to be determined by adjustment to measurements. It equals the x coordinate at the inflection point of the Gauss curve. It can be estimated with the diagram of Peck (Figure 15.49)[*] or according to the empirical formula[†]

$$2a/D = (H/D)^{0.8} \tag{15.19}$$

where D is the diameter of the tunnel, and H is the depth of the tunnel axis (Figure 15.50). For clay soils, $a \approx (0, 4, \ldots, 0, 6)\ H$; for non-cohesive soils, $a \approx (0, 25, \ldots, 0, 45)\ H$. Another estimation of a is given in Table 15.4.[‡]

[*]R. B. Peck, "Deep Excavations and Tunneling in Soft Ground: State-of-the-Art Report. In: *Proceedings of the 7th International Conference on Soil Mechanics and Foundation Engineering*, Mexico City, 1969, 225–290.

[†]M. J. Gunn, "The Prediction of Surface Settlement Profiles Due to Tunneling." In *Predictive Soil Mechanics*, Proceedings Wroth Memorial Symposium, Oxford, 1992.

[‡]J. B. Burland et al., "Assessing the Risk of Building Damage Due to Tunneling: Lessons from the Jubilee Line Extension, London. In: *Proceed. 2nd Int. Conf. on Soil Structure Interaction in Urban Civil Engineering*, Zürich 2002, ETH Zürich, Vol. 1, 11–38.

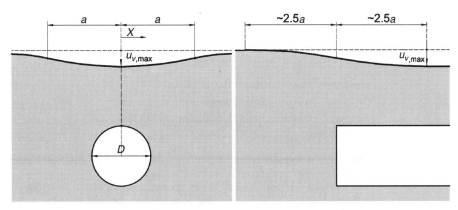

FIGURE 15.50 Settlement trough over a tunnel (*left*); approximate distribution of the surface settlements in tunnel longitudinal direction. The curve shown coincides reasonably with the function $y = \text{erf } x = \frac{1}{\sqrt{2\pi}} \int_0^x e^{-y^2/2} dy$ (*right*).

TABLE 15.4 Estimation of a

Soil	a/H
Granular	0.2–0.3
Stiff clay	0.4–0.5
Soft silty clay	0.7

The horizontal displacements u_h of the ground surface follow from the observation that the resulting displacement vectors are directed toward the tunnel axis (as shown in Figure 15.51) that is,

$$u_h = \frac{x}{H} u_v$$

The distribution of the settlements in the longitudinal direction of a tunnel under construction is represented in Figure 15.50. The volume of the settlement trough (per current tunnel meter) results from the Gauss distribution as

$$V_u = \sqrt{2\pi} \cdot a \cdot u_{v,\max} \tag{15.20}$$

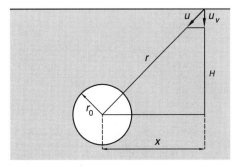

FIGURE 15.51 Vertical displacement at the ground surface.

and is usually designated as volume loss[*] (ground loss). The volume loss amounts to some percent of the tunnel cross-section area per current meter. If this ratio is known by experience for a given soil type, then the maximum settlement $u_{v,max}$ can be estimated with equations (15.19) and (15.20). Mair and Taylor[†] give the following estimated values for V_u/A:

Unsupported excavation face in stiff clay	1–2 percent
Supported excavation face (slurry or earth mash), sand	0.5 percent
Supported excavation face (slurry or earth mash), soft clay	1–2 percent
Conventional excavation with sprayed concrete in London clay	0.5–1.5 percent

The volume loss depends on the skill of tunneling. Given the improved technology now available, the volume loss has been halved over the last few years.

The evaluation of numerous field surveys and lab tests with a centrifuge leads to an empirical relationship[‡] between the volume loss V_u related to the area A of the tunnel cross-section and the stability number $N := (\sigma_v - \sigma_t)/c_u$. Here, σ_v is the vertical stress at depth of the tunnel axis, σ_t is the supporting pressure (if any) at the excavation face, and c_u is the undrained cohesion. If N_L is the value of N at collapse, then:

$$V_u/A \approx 0.23 \, e^{\,4.4N/N_L}$$

The estimations represented here refer to the so-called greenfield. If the surface is covered by a stiff building, then the settlements are smaller.[§]

[*]This designation is based on the conception that the soil volume V_u is dug additionally to the theoretical tunnel volume.

[†]R. J. Mair and R. N. Taylor, "Bored Tunneling in the Urban Environment," *14th Int. Conf. SMFE*, Hamburg, 1997.

[‡]S. R. Macklin, "The Prediction of Volume Loss Due to Tunneling in Over Consolidated Clay Based on Heading Geometry and Stability Number," *Ground Engineering*, April 1999.

[§]"Recent Advances into the Modeling of Ground Movements Due to Tunneling," *Ground Engineering*, September 1995, 40–43.

P · A · R · T · IV

NON-AUTOMOBILE TRANSPORTATION

CHAPTER 16
PEDESTRIANS

Ronald W. Eck

Department of Civil and Environmental Engineering
West Virginia University
Morgantown, West Virginia

16.1 INTRODUCTION AND SCOPE

A *pedestrian* is defined as any person on foot. While everyone is a pedestrian at one time or another, until recently, in the United States, walking has been viewed primarily as a recreational activity. However, for relatively short trips, walking can be an efficient and inexpensive mode of transportation.

There are many factors that influence choice of travel mode and, specifically, the decision to walk. The National Bicycling and Walking Study (Federal Highway Administration 1992) showed that there is a three-tiered hierarchy of factors.

1. *Initial considerations.* Many people rely on their automobile to go virtually anywhere and never seriously consider the option of walking. An individual's attitudes and values also play a role (e.g., walking may be considered as "not cool"). Perceptions are also important in the decision to walk (e.g., safety concerns about traveling at night). Finally, there are situational constraints that if they do not preclude the decision to walk, they do require additional planning and effort. Examples include needing a car at work or having to pick up children from soccer practice.

2. *Trip barriers.* Concern for safety in traffic is a frequently cited reason for not walking. This is particularly true where there are no alternatives to walking along high-speed, high-volume roadways. There may be problems with access and linkage (e.g., lack of connections between neighborhoods and shopping areas or parks). Environmental factors such as rugged topography or extremes in weather also can be considered as barriers.

3. *Destination barriers.* Lack of support from employers or coworkers can act as a barrier (e.g., relaxing the dress code or establishing a policy of flex time).

Increased levels of walking can result in benefits in terms of health and physical fitness, the environment, and transportation. Studies have demonstrated that even low to moderate levels of exercise, such as regular walking, can reduce the risk of coronary heart disease, stroke, and other chronic diseases; help to reduce health care costs and contribute to greater functional independence in later years; and improve the quality of life at every stage. Replacing automobile trips with walking trips could result in significant economic benefits. Since it is nonmotorized and nonpolluting, walking would result in reduced emissions, energy consumption, and congestion. Facilitating walking means additional travel options for those unable to drive or who choose not to drive for some trips. Where there is a truly intermodal transportation system in which walking is an important component, the livability of communities is enhanced.

Passage of the Intermodal Surface Transportation Efficiency Act of 1991 (ISTEA) helped to focus new attention on walking as an important nonmotorized mode of transportation. ISTEA requires each state and metropolitan planning organization (MPO) to include nonmotorized elements in their transportation plans.

ISTEA has opened up new sources of potential funding for nonmotorized transportation improvements, planning, and programs, as well as for recreational trails and intermodal linkages. With these funding programs and the planning requirements established by ISTEA, there is increased interest in pedestrian transportation.

This chapter takes a comprehensive look at incorporating pedestrians into the transportation system. While the chapter provides an overview, it also directs readers to sources of more detailed technical information.

16.2   CHARACTERISTICS OF PEDESTRIANS

16.2.1   Ambulation

In normal walking, the lead foot swings forward, and as it approaches the end of the stride, the heel comes down on the surface gently as the ankle allows the foot to rotate forward at a controlled rate until the sole also makes contact with the floor. Little horizontal force is applied during the heel-strike phase, but if the heel edge slips at this point, the lead foot slides forward. Therefore, the characteristic slip pattern is that the victim falls backwards. This type of fall usually results in the most severe injury. Forward slips may occur when the sole of the foot slides in the retreating portion of the step. This tends to produce a forward fall, where the injury is often less severe. Many slips result from the more complex activities of turning and changing direction. In these instances, the fall pattern is less predictable.

A trip occurs when a pedestrian's foot is impeded in the walking process by striking against some obstacle in the path of travel. If the interruption of motion is great enough, the pedestrian will fall forward. The person's reflex action is to extend the upper arms to absorb the energy of impact. Falling onto the extremities in this fashion can cause injury to wrist, forearm, elbow, upper arm, and shoulder. One or both of the knees also may strike the ground.

Pedestrians tend to look ahead to their objective; the normal line of sight is about 15° below horizontal relative to the eyes. Pedestrians do not usually look down deliberately unless something attracts their attention. Consequently, even small changes in surface elevation or characteristics are not always seen. However, even if someone is looking down at the surface, a problem may not be perceived. Color, texture, low light levels, and glare can obscure changes in walking surfaces.

Good design takes this human factor into account in two ways. First, many problems can be avoided by giving thoughtful attention to safety in facility planning, design, construction, and maintenance. Second, where something cannot be avoided, for example, a single step, the designer must build in visual and tactile cues that will cause the pedestrian to look down and see the hazard before tripping over it or slipping on it.

Another characteristic of pedestrians is that they always will take the least-energy route (i.e., the shortest distance and flattest path) between two points. They tend to cross streets at the most convenient locations rather than at designated crossings. If it is easier to cross at midblock than to walk a long distance to a corner crosswalk, pedestrians will take the most direct route and cut across the street between intersections. Where a choice of routes is available, substantial barriers may be needed to prevent pedestrians from taking the direct route, where that route includes hazards (e.g., at-grade crossing of a busy arterial roadway).

According to the American Association of State Highway and Transportation Officials (AASHTO) *Guide for the Planning, Design, and Operation of Pedestrian Facilities* (2004), most pedestrian trips are 0.25 mile or less. The guide indicates that 1 mile is generally the limit that most people are willing to travel on foot.

Effective pedestrian facilities are designed to accommodate the full range of users. There is no such thing as a standard pedestrian. The stature, travel speeds, endurance limits, physical strength, and judgment abilities of pedestrians vary greatly. Users include children, older adults, families, and people with and without disabilities. Pedestrians may be carrying packages or luggage, walking a dog, pushing children in strollers, or pulling delivery dollies. Pedestrians transporting items cannot react as quickly to potential hazards because they are more physically taxed and distracted.

16.2.2   Seeing and Being Seen

Visibility of a pedestrian in a traffic stream is influenced by environmental conditions, behavior, and attire. The key factor is the degree of contrast between the pedestrian and his or her environment.

Environmental factors can affect the visibility of pedestrians to motorists. Rain, snow, fog, shadows, and glare all reduce visual range and acuity. The National Highway Institute's (NHI's) participant workbook (1996) notes that vehicles themselves are equally important. Dirty or cracked windshields not only reduce vision but magnify the effects of glare.

There are also *visual screens* in the driving environment. Moving vehicles, particularly buses and commercial vehicles, can block pedestrians' and motorists' views of one another. Stationary features such as parked vehicles, shrubs, structures, and traffic signal controller boxes can have the same effect.

About one-half of fatal pedestrian crashes occur in low-light or dark conditions. At night, pedestrians frequently are difficult to see because they lack conspicuity. All the factors affecting conspicuity become increasingly critical during times of reduced light or darkness. For example, according to Federal Highway Administration (FHWA) data, at night, the average driver of a vehicle operating with low-beam headlights will see a pedestrian in dark clothing in the roadway about 80 feet in front of the vehicle. At speeds greater than about 20 to 25 mph, drivers do not have enough time to perceive the pedestrian, identify that it is a pedestrian, and make a decision to stop or swerve in time to avoid striking the pedestrian.

Pedestrians tend to overestimate their visibility to motorists. Motorists who have been involved in nighttime crashes with pedestrians often remark, "I don't know what I hit. I thought I struck an animal." This is an indication that they did not see the pedestrian until it was too late to react. While a pedestrian can see the headlamps of an approaching vehicle when the vehicle is several blocks away, this does not mean that the driver of that vehicle can see the pedestrian. Pedestrians should be encouraged to wear retroreflective material when they walk at night to enhance their visibility to motorists.

16.2.3   Groups of Particular Concern

Older Adults.   By 2050, it is estimated that 20 percent of the U.S. population will be older than age 65. Although aging itself is not a disability, most persons aged 75 years or older have a disability. Many of the characteristics commonly associated with aging can limit mobility. The aging process often causes a general deterioration of physical, cognitive, and sensory abilities. Characteristics may include

- Vision problems such as degraded acuity
- Reduced range of joint motion
- Reduced ability to detect, localize, and differentiate sounds
- Limited attention span, memory, and cognitive abilities
- Reduced endurance
- Decreased agility, balance, and stability
- Slower reflexes
- Impaired judgment, confidence, and decision-making abilities

Older adults generally need more frequent resting places and prefer more sheltered environments. The FHWA publication (1999) on designing pedestrian facilities for access notes that many older adults have increased fears for personal safety. Statistics confirm these fears, indicating that older pedestrians appear to be at increased risk for crime and crashes at places with no sidewalks, places with sidewalks on only one side, and places with no street lights. Older pedestrians would benefit from accessible paths that are well lit and well-policed.

Ambulation of older adults is affected by their reduced strength. Travel over changes in levels, such as high curbs, can be difficult or impossible for older adults.

Because older people tend to move more slowly than younger pedestrians, they require more time to get across streets than many other sidewalk users. For many years [including the 2003 edition of the *Manual on Uniform Traffic Control Devices (MUTCD)*], engineers designed traffic signals based on an average pedestrian walking speed of 4 feet per second. The 2009 edition of the *MUTCD* includes slower walking speeds for calculating pedestrian clearance time. The pedestrian clearance time is based on 3.5 feet per second and the sum of the walk time and pedestrian clearance time is based on 3.0 feet per second.

The reduced visual acuity of older people can make it difficult for them to read signs or detect curbs. Older people are more dependent on high contrast between sign backgrounds and lettering. Contrast-resolution losses can cause them to have difficulty seeing small changes in level, causing trips and falls on irregular surfaces.

Children.   Children have fewer capabilities than adults owing to their developmental immaturity and lack of experience. Compared with adults, children tend to exhibit the following characteristics:

- One-third less peripheral vision
- Less accuracy in judging speeds and distances
- Difficulty in localizing the direction of sounds
- Overconfidence
- Inability to read or comprehend warning signs and traffic signals
- Unpredictable or impulsive actions
- Trust that others will protect them
- Inability to understand complex situations

Disabled Persons.   According to the 2000 Census, one in every five Americans over the age of 15 has a disability. Between 10 and 12 million Americans have vision disabilities. In fact, 85 percent of Americans living to their full life expectancy will suffer a permanent disability. People with disabilities are also more likely to be pedestrians than other adults because some physical limitations can make driving difficult. According to the FHWA publication on providing access (1999), disabilities can be divided into three categories: mobility, sensory, and cognitive.

People with mobility impairments include those who use wheelchairs, crutches, canes, walkers, orthotics, and prosthetic limbs. However, there are many people with mobility impairments who do not use assistive devices. Characteristics common to people with mobility limitations include substantially altered space requirements to accommodate assistive device use, difficulty in negotiating soft surfaces, and difficulty negotiating surfaces that are not level.

Although sensory disabilities are more commonly thought of as total blindness or deafness, partial hearing or vision loss is much more common. Other types of sensory disabilities can affect touch, balance, and the ability to detect the position of one's own body in space. Color blindness is considered a sensory defect.

Cognition is the ability to perceive, recognize, understand, interpret, and respond to information. It relies on complex processes such as thinking, knowing, memory, learning, and recognition. Cognitive disabilities can hinder the ability to think, learn, respond, and perform coordinated motor skills. Such individuals might have difficulty navigating through complex environments such as city streets and might become lost more easily than other people.

The aforementioned FHWA publication (1999) on designing for access presents an excellent overview of the characteristics associated with different types of disabilities. The publication also reviews design approaches for accommodating specific categories of disabilities.

16.3 PLANNING

16.3.1 Conceptual Planning

Conceptual planning is relatively simple, consisting of determining the general direction that walkways should take. Focus should be on pedestrian generators such as schools, shops, cultural attractions, and work and play places. First, look at routes that presently exist before establishing new ones. Privacy, views, access, and local character must be understood and incorporated into the planning.

Initially, planners should determine where people want or need to travel, the routes they might travel, and who those people are. The most likely users of improved pedestrian facilities are

- Children who must be driven to school, play, and other activities
- Parents who have to drive children and would appreciate safe walking routes so their children can move around the community by themselves
- Older people who may not drive but who have time to walk and who may be able to carry out some of their daily chores, enjoy the outdoors, and exercise all on the same trip
- Commuters who may be able to walk to bus or carpool stops
- Recreational users, especially those who jog or walk regularly, who would benefit from improved routes and separation from fast-moving traffic

Simple pedestrian volume counts usually do not yield enough information about where people are going or coming from, trip purpose, and any special pedestrian needs that should be met. Such data are best obtained through an origin/destination survey that should include the following information:

1. Locations of major pedestrian generators such as parking facilities, transit stations, and major residential developments
2. Locations of significant pedestrian attractions such as shopping centers, office and public buildings, theaters, colleges, hospitals, and sports arenas
3. Existing and potential pedestrian routes between major destinations
4. Time periods in which major pedestrian flow occurs

Some questions to consider include

1. Do existing routes satisfy the heaviest travel demand? Can a need for new routes be clearly identified?
2. Do existing routes require improvement to resolve circulation problems?
3. Which areas seem to be preferred locations for development of new activities to generate pedestrian movement?

The AASHTO *Guide for the Planning, Design and Operation of Pedestrian Facilities* (2004) reviews issues that should be considered when integrating pedestrians in a variety of types of transportation planning studies.

16.3.2 Access and Linkages

There is no question that transportation modes have influenced the ways cities have grown and the forms they have taken. Before the advent of the automobile, cities were more compact in terms of area and population. However, in the United States, clearly the automobile is the dominant transportation mode. One manifestation of this is the phenomenon of suburbanization that occurred after World War II. Characteristics of suburbs include (1) suburban land-use planning encourages low density and separation of land-use types, (2) street design standards typically require wide streets that encourage high-speed traffic and sometimes do not require sidewalks, (3) it is not easy to use public transportation in suburban locations, and (4) barriers to walking are created unintentionally. Consequently, suburban activities essentially require the use of a car and generate large amounts of vehicular traffic.

Pedestrian travel is often an afterthought in the development process. The results are impassable barriers to walking both within and between developments. For example, early suburban communities had no sidewalks. Later, some communities required developers to install sidewalks. Consequently, in most suburbs, there is a patchwork of sidewalks that start and stop but often are not linked.

Suburban neighborhood design can be modified to encourage walking. A pedestrian-oriented neighborhood should include the following characteristics:

- Streets that are laid out in well-connected patterns on a pedestrian scale so that there are alternative automobile and pedestrian routes to every destination.
- A well-designed street environment that encourages intermodal transportation. These streets should include pedestrian-scale lighting, trees, sidewalks, and buildings that are within close walking distance to the sidewalk.
- Residential and internal commercial streets should be relatively narrow to discourage high-speed automobile traffic.
- On-street parallel parking is recommended, where it can be used as a buffer between pedestrians and motor vehicle traffic. Parked cars also serve to slow down passing traffic.
- Building uses are often interspersed, that is, small homes, large homes, outbuildings, small apartment buildings, corner stores, restaurants, and offices.
- In addition to streets, there are public open spaces, around which are larger shops and offices as well as apartments.

Local zoning ordinances can be revised to require more attention to the needs of pedestrians. Some examples are discussed below.

Residential subdivision layout should provide safe, convenient, and direct pedestrian access to nearby (within 0.5 mile) and adjacent residential areas, bus stops, and neighborhood activity centers such as schools, parks, commercial and industrial areas, and office parks. Cul-de-sacs have proven to be effective in restricting automobile through traffic. However, they also can have the effect of restricting pedestrian mobility unless public accessways are provided to connect the cul-de-sacs with adjacent streets. Nonmotorized path connections between cul-de-sacs and adjacent streets (shown in Figure 16.1) should be provided wherever possible to improve access for pedestrians. Pedestrian facilities should be designed to meet local and statewide design standards.

In some high-density residential areas, regulations require off-street parking and reduced lot frontage. This results in home fronts that consist largely of garage doors. Ordinances should be modified to allow for rear-lot access (alleys) or other innovative solutions in these areas. Parking codes can be modified to allow for a "reduced parking option" for developments located on bus routes and that provide facilities that encourage biking and walking.

One of the most important factors in a person's decision to walk is the proximity of goods and services to homes and workplaces. The most conducive land use for pedestrian activity is one with a higher-density mix of housing, offices, and retail spaces. Major pedestrian improvements will occur as land-use changes reduce the distances between daily activities. Such land-use changes include increasing density and mixing land uses. While converting suburban locations to accommodate pedestrians is more difficult than downtown, such low-density development offers opportunities not possible

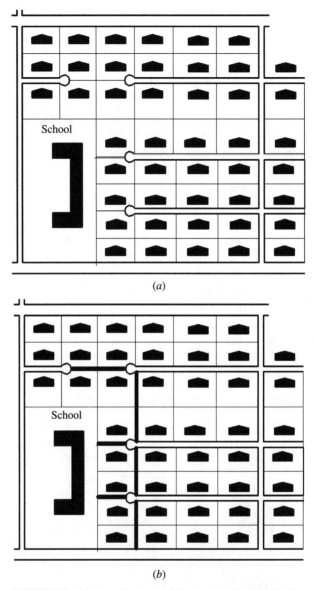

FIGURE 16.1 (*a*) Conventional subdivision layout showing lack of pedestrian connectivity between streets and other land uses. (*b*) Subdivision layout that provides pedestrian connections (*indicated by the heavy lines*) between streets and other land uses.

in built-up neighborhoods. A strategy of (1) linking internal spaces where possible and (2) making the street usable for pedestrians will enhance suburban living for many people.

Extending and/or improving pedestrian access within a suburban community may eliminate some need for a car, allowing increased flexibility for those who have to drive. Most walkways should be planned in conjunction with roads so that pedestrians can reach all developments that are located along the road.

Sidewalk width should vary to adjust to physical conditions and pedestrian volumes. Sidewalks near stores and schools should be wider to accommodate more people. Where there is a view, the sidewalk should be widened and a bench and landscaping added. Locations of commuter bus stops should be noted. Opportunities for shortcuts that make access easier should be identified. The success of suburban transit depends in part on the adequacy of the sidewalks and ease with which people can walk to bus stops.

Entrances to many commercial and retail centers are oriented toward automobile travel. Pedestrian access to storefronts is not only difficult and awkward but often also unsafe. A typical shopping center or strip mall is separated from the roadway by a wide parking lot. There are often no pathways linking store entrances to the sidewalks along the street, and in fact, there may be no sidewalks along the street to be linked. Parking lots with multiple access points allow traffic circulation in different directions, creating hazards and confusion for walkers. Figure 16.2 illustrates a location that is not pedestrian friendly.

Such locations can be redeveloped to better serve pedestrians. As older commercial/retail locations undergo renovations, they should be redesigned to serve customers who arrive via automobile, transit, bicycle, and on foot. Specific methods include

- Maximize pedestrian and transit access to the site from adjacent land uses.
- Provide comfortable transit stops and shelters with pedestrian connections to the main buildings.
- Transit stops and pedestrian drop-offs should be located with reasonable proximity to building entrances.
- Provide attractive pedestrian walkways between the stores and adjacent sites.
- Ensure that fencing and landscaping do not create barriers to pedestrian mobility.
- Rework entrances and orient buildings toward pedestrians and transit facilities instead of parking lots.
- Connect all buildings on site to each other via attractive pedestrian walkways, with landscaping and pedestrian-scale lighting. Provide covered walkways between buildings, if possible.
- Minimize pedestrian-automobile conflicts by consolidating auto entrances into parking lots.

While many transit agencies in the United States have expended significant planning and design efforts to meet the needs of pedestrians in transit stations, relatively little attention has been devoted to the pedestrian environment to and from the stations. This hinders consideration of the door-to-door

FIGURE 16.2   Lack of pedestrian connections to and within a retail center.

experience of using public transportation. It is not unusual for several different entities to maintain independent control over the various facilities that are used by someone walking to/from a single transit stop. State and local governments with responsibility for constructing and maintaining pedestrian facilities should cooperate with transit agencies, the private sector, and interested citizens in developing action programs to reduce barriers to pedestrian access to transit.

When given sidewalks and traffic-calmed streets to walk along, safe and convenient ways to cross streets, and a comfortable and attractive environment, most people are willing to walk farther to reach public transportation. Unfortunately, matters beyond the boundaries of the transit station typically have not received much attention in the United States. For example, park-and-ride lots are often located near freeways and/or shopping areas where residential housing is quite far away and there are no paths or sidewalks located near the park-and-ride lots.

However, the experience of some transit agencies shows that there is growing awareness of the need to look at the larger environment that surrounds and leads to transit stations and bus stops. Examples of enhancing pedestrian access include placing pedestrian signals and detectors at key intersections and providing wider sidewalks, pedestrian shortcuts to key destinations, and trees.

16.3.3 Pedestrian Level of Service

The *Highway Capacity Manual* (Transportation Research Board 2000) suggests that the quality of a pedestrian's experience is similar to an automobile-based measure; that is, vehicles that are moving at slow speeds or are stopped owing to congestion have low levels of service. The manual acknowledges that environmental factors also influence pedestrian activity. While speed, space, and delay are one set of measures of quality, there are other indicators of a good walking environment. Burden (1999) indicates that high pedestrian densities may indicate success. He notes that there are a number of attractions throughout the world where large crowds give excitement and security to a place.

Unlike high-speed motorists, whose travel is facilitated when the environment lacks detail, pedestrian travel is slow and interactive with the block, street, or neighborhood. The slower pace requires attention to detail. In fact, as Berkovitz (2001) points out, pedestrians are often intimidated by empty sidewalks and long travel distances, particularly along high-speed roads.

Landis et al. (2002) point out that evaluating the performance of a roadway section for walking is much more complex in comparison with that of the motor vehicle mode. While motor vehicle operators are largely insulated in their travel environment and are influenced by relatively few factors, pedestrians are relatively unprotected and are subjected to a variety of environmental conditions. Landis et al. (2002) note that there is not yet consensus among planners and engineers on which features of a roadway environment have statistically reliable significance to pedestrians. Several *walkability audits* have been developed that include a large number of features of the entire roadway corridor environment. Landis et al. (2002) present a list of factors on which it is generally agreed that pedestrians' sense of comfort and safety within a roadway corridor is based. The list includes personal safety (i.e., threat of crashes), personal security (i.e., threat of assault), architectural interest, pathway or sidewalk shade, pedestrian-scale lighting and amenities, presence of other pedestrians, and conditions at intersections.

16.4 PEDESTRIAN SAFETY

The engineering literature on pedestrian safety tends to focus on pedestrian crash types, that is, incidents where a pedestrian comes into contact with a motor vehicle. This is logical because many of these "crashes" are reported through the uniform crash reporting systems in each state. However, reliance on such data significantly understates the magnitude of the pedestrian safety problem because it typically ignores crashes that occur on private property, such as parking lots or driveways. More important, incidents where pedestrian slip, trip, or fall on a sidewalk, stair, ramp, or other facility are not included. As described below, available data suggest that these non-motor-vehicle-involved incidents occur more frequently than motor-vehicle-involved crashes. This section examines both types of incidents by discussing common crash types and/or injury mechanisms. Tort liability issues are outlined, along with risk-management strategies for enhancing pedestrian safety.

16.4.1 Pedestrian Crash Types

According to FHWA data, approximately 5,000 pedestrians are killed each year and about 70,000 are injured as a result of collisions with motor vehicles. In some large urban areas, pedestrians account for as much as 40 to 50 percent of traffic fatalities. Many more injuries probably are not reported to recordkeeping agencies.

An FHWA study (Hunter, Stutts, and Pein 1997) found that compared with their representation in the overall U.S. population, young persons (<25 years of age) were overrepresented in pedestrian–motor vehicle crashes. Older adults (ages 25 to 44 years) and the elderly (age 65+ years) were under-represented. However, elderly pedestrians in crashes were more than twice as likely to be killed (15 versus 6 percent) than young persons.

In general, pedestrian crashes occurred most frequently during the late afternoon and early evening hours. These are times when exposure is probably highest and visibility may be a problem. Alcohol or drug use was noted in about 15 percent of the pedestrian crashes overall. This increased to 31 percent for pedestrians in the 25 to 44 age group. Alcohol/drug crashes also were more frequent on weekends and during hours of darkness.

Forty-one percent of crashes occurred at roadway intersections, and an additional 8 percent occurred in driveways or alley intersections. About two-thirds of crashes were categorized as urban. Fifteen percent of the pedestrian crashes reported occurred on private property, primarily in commercial or other parking lots. The elderly were overrepresented in commercial parking lot crashes, young adults in non–commercial parking lot crashes, and children under age 10 in collisions involving driveways, alleys, or yards.

National Cooperative Highway Research Program (NCHRP) Report 500, Volume 10: "A Guide for Reducing Collisions Involving Pedestrians" (Transportation Research Board 2004), indicates that 12 specific pedestrian crash types have been identified for use with crash data to identify safety problems and corresponding countermeasures. They also can be used to help educate safety professionals, as well as the general public, about the types of situations that pose danger to pedestrians. These 12 crash-type groupings are outlined below.

- *Midblock dart/dash.* The pedestrian walked or ran into the roadway and was struck by a vehicle.
- *Multiple threat.* The pedestrian entered the traffic lane in front of stopped traffic and was struck by a vehicle traveling in the same direction as the stopped vehicle. The stopped vehicle may have blocked the sight distance between the pedestrian and the striking vehicle and/or the motorist may have been speeding.
- *Mailbox or other midblock destination.* The pedestrian was struck while getting into or out of a stopped vehicle or while crossing the road to/from a mailbox, newspaper box, ice cream truck, etc.
- *Failure to yield at an unsignalized intersection.* At an unsignalized intersection or midblock location, a pedestrian stepped into the roadway and was struck by a vehicle.
- *Bus-related.* The pedestrian was struck by a vehicle either (1) by crossing in front of a commercial bus stopped at a bus stop, (2) going to or from a school bus stop, or (3) going to or from or waiting near a commercial bus stop.
- *Turning vehicle at intersection.* The pedestrian was attempting to cross at an intersection and was struck by a vehicle that was turning right or left.
- *Through vehicle at intersection.* The pedestrian was struck at a signalized or unsignalized intersection by a vehicle that was traveling straight ahead.
- *Walking along roadway.* The pedestrian was walking or running along the roadway and was struck from the front or from behind by a vehicle.

- *Working/playing in the road.* A vehicle struck a pedestrian who was (1) standing or walking near a disabled vehicle, (2) riding a play vehicle that was not a bicycle, (3) playing in the road, or (4) working in the road.

- *Not in road.* The pedestrian was standing or walking near the roadway edge, on the sidewalk, in a driveway or alley, or in a parking lot when struck by a vehicle.

- *Backing vehicle.* The pedestrian was struck by a backing vehicle on a street, in a driveway, on a sidewalk, in a parking lot, or at another location.

- *Crossing an expressway.* The pedestrian was struck while crossing a limited-access expressway or expressway ramp.

Volume 10 of the NCHRP 500 series presents strategies to address four emphasis areas with respect to pedestrians:

1. Reduce pedestrian exposure to vehicular traffic.
2. Improve sight distance and/or visibility between motor vehicles and pedestrians.
3. Reduce vehicle speeds.
4. Improve pedestrian and motorist safety awareness and behavior.

16.4.2 Non-Motor-Vehicle-Involved Incidents

The number of "same-level" falls is unknown. As Hyde et al. (2002) point out, falls that do not cause injury are not recorded. Therefore, researchers are limited to studying the incidence of such falls as reported by hospitals and similar institutions. Even with a significant number of injury and noninjury incidents going unreported, clearly the magnitude of the fall problem is significant. Data show that falls are the leading cause of injury reported to an emergency room. According to *The Injury Fact Book* (Baker et al. 1992), accidental falls are the second leading cause of unintentional death, the second leading cause of both spinal cord and brain injury, and the most common cause of hospital admission for trauma. While data exist on reported falls, that is, the more serious falls, there is not good information about the circumstances surrounding the fall; that is, did the individual slip on ice, trip on a raised section of sidewalk, or fall on a stairway? In this regard, the author (Eck and Simpson 1996) attempted to use hospital records to obtain detailed information about falls to assist in countermeasure development. Results pointed out the importance of surface conditions to pedestrian safety. Two general types of surface condition problems were identified: slippery surfaces owing to accumulation of ice and snow and holes or openings in the surface itself (e.g., missing, ill-fitting, or defective grates). While the former problem is one of maintenance, the latter issue has facility design, construction, and maintenance implications.

16.4.3 Tort Liability

As discussed earlier, planners and roadway designers and engineers must consider the needs of pedestrians. Design, construction, maintenance, and operation of roads and streets, bridges, and surface conditions must recognize pedestrians. Roadway and recreational facilities that fail to incorporate the needs of all users increase the likelihood of claims against facility owners/managers.

Liability is an important issue for both public agencies and private entities. Implementing an aggressive and well-publicized risk-management program can help to head off these problems. It is useful to look at some of the design, construction, maintenance, and operational problems that are cited commonly in claims or lawsuits involving pedestrians.

- Open drainage grates in travelway
- Inadequate utility box covers (raised/depressed, poor skid resistance, or structural problems)

- Paths that suddenly end at "bad" locations with no transition or escape route provided
- Long-term, severe surface irregularities (e.g., broken pavement, potholes)
- Foreign substances on travel surface (e.g., water, sealants, oil/grease, loose stones or gravel)
- Vertical elevation differences in walkways
- Bridges that are hazardous to pedestrians
- Design errors (e.g., at curb ramps)
- Wheelstops in parking lots
- Lack of railings between path and adjacent slope (and conditions at bottom)

Figure 16.3 shows wheelstops intruding into a walkway, creating a tripping hazard for pedestrians.

16.4.4   Risk-Management Strategies

Outlined below are elements of an effective risk-management strategy. Conscientiously implementing these elements should result in safer pedestrian facilities with fewer injuries, and, therefore, fewer claims, and should increase the likelihood of a successful defense for the claims that proceed to trial.

Perhaps the overarching strategy is to follow commonsense principles for a defensible program. This means providing immediate response to the risks identified, including signing and warning for conditions that cannot be changed immediately and funding spot improvements for those which can be changed. Listen to the public. If a parent calls, an editorial is written, or any other input from a "customer" suggests action, the situation should be evaluated and action taken as quickly and as intelligently as possible. All agencies, departments, and other parties whose duties are affected need to be involved. The risk-reduction effort takes cooperation and ongoing coordination.

Other specific actions that can be taken include

- Incorporate accepted standards and guidelines.
- Use established engineering, planning, and design principles.

FIGURE 16.3   Wheelstops from a parking lot intruding onto and near a sidewalk, creating a tripping hazard for pedestrians.

- Consider all potential users.
- Do it right!
- Promote community involvement and awareness.

16.5 PEDESTRIAN FACILITY DESIGN

16.5.1 Walkways, Sidewalks, and Public Spaces

A successful urban sidewalk should have the following characteristics:

- Adequate width
- Buffer from travel lane
- Gentle cross-slope (2 percent or less)
- Buffer to private properties
- Adequate sight distances around corners and at driveways
- Shy distances to walls and other structures
- Continuity
- Clear path of travel free of street furniture
- Well-maintained condition
- Ramps at corners and flat areas across driveways
- Sufficient storage capacity at corners

Sidewalks require a minimum width of 5 feet if set back from the curb or 6 feet if at the curb face. Walking is a social activity. For two people to walk together, 5 feet of space is the bare minimum needed. In some areas, such as near schools, sporting complexes, some parks, and many shopping districts, the minimum width for a sidewalk is 8 feet.

The desirable width for a sidewalk is often much greater. Some shopping districts require 12, 20, 30, or even 40 feet of width to handle the volume of pedestrian traffic they encounter. Pennsylvania Avenue in Washington, DC, has 30-foot sidewalk sections to handle tour bus operations. The *Highway Capacity Manual* (Transportation Research Board 2000) covers the topics of sidewalk width and pedestrian level of service.

It is important to determine the commercial need for outdoor cafes, kiosks, corner gathering spots, and other social needs for a sidewalk. In commercial areas, designers should consult property owners, chambers of commerce, downtown merchants associations, and landscape architects to ascertain if the desired width is realistic. Corner or midblock bulb-outs can be used to advantage for creating both storage space and for roadway crossing and social space. Figure 16.4 shows a corner bulb-out on an arterial street.

Most sidewalks are made of concrete owing to its long life, distinct pattern, and lighter color. In some cases, asphalt can provide a useful surface.

Paver stones also can be used. These colorful brick, stone, or ceramic tiles are often used to define corners or crosswalks (as illustrated in Figure 16.4), to create a mood for a block or commercial district, or to help those with visual impairments. The blocks need to be set on a concrete pad for maximum life and stability.

Certain cautions are in order relative to paver stones, bricks, and similar materials. The FHWA publication (2001) on best practices in designing trails for access points out that decorative surfaces may create a vibrating, bumpy ride that can be uncomfortable and painful for those in wheelchairs. Pavers or bricks can settle or buckle, creating changes in level. This creates a tripping hazard for ambulatory pedestrians with visibility impairments. Finally, decorative surface materials can make identification more difficult. Decorative crosswalks generally are more difficult for drivers to identify

FIGURE 16.4   Corner bulb-out on an arterial street.

in the roadway. In addition, it may be more difficult for pedestrians with vision impairments to identify detectable warnings that provide critical information about the transition from sidewalk to street. For these reasons, paver, brick, or cobblestone sidewalks and crosswalks are not recommended. Concrete sidewalks with paver or brick trim preserve the decorative qualities but present an easier surface to negotiate.

Desirably, a border area should be provided along streets for the safety of motorists and pedestrians, as well as for aesthetic reasons. The border area between the roadway and the right-of-way line should be wide enough to serve multiple purposes, including provision of buffer space between pedestrians and vehicular traffic, snow storage, an area for placement of underground utilities, and an area for maintainable landscaping. The border may be a minimum of 5 feet, but desirably it should be 10 feet.

Nature strips, particularly in downtown areas, are a good location to use paver stones for easy and affordable access to underground utilities. In downtown areas, nature strips are also a convenient location for the swing width of a door and for placement of parking meters, hydrants, lampposts, and furniture.

On-street parking has two distinct advantages for pedestrians. First, it creates the desired physical separation from motor vehicle traffic. Second, on-street parking has been shown to reduce motorist travel speeds, thus creating a safer environment for street crossings.

On the back side of sidewalks, a minimum-width buffer of 1 to 3 feet is essential. Without such a buffer, vegetation, walls, buildings, and other objects encroach on the usable sidewalk space.

Pedestrians require a shy distance from fixed objects, such as walls, fences, shrubs, buildings, parked cars, and other features. The desired shy distance for a pedestrian is 2 feet. Allowance for this shy distance must be made in determining the functional width of a sidewalk.

The literature points out that attractive windows in shopping districts cause curious pedestrians to stop momentarily. This is a desired element of a successful street. These window watchers take up about 18 to 24 inches of space.

Because of its relatively slow pace, walking is more detail oriented than driving. Walkers are attracted to locations with amenities, interesting storefronts, and outdoor cafés. On the other hand, pedestrians will avoid sidewalks lined with walls or that lack interesting details. Compare Figure 16.5, with its amenities and interesting details, with the wall effect shown in Figure 16.6. Where would you rather walk?

FIGURE 16.5   Sidewalk café and other amenities create an inviting sidewalk.

Newspaper racks, mail boxes, and other street furniture should not encroach into walking space. The items can be placed in a nature strip, a separate storage area behind the sidewalk, or a corner or midblock bulb-out. These items should be bolted in place.

AASHTO's pedestrian guide (2004) presents information on ambience, shade, and other sidewalk enhancements to make the sidewalk environment more comfortable. Landscaping should be arranged to permit sufficiently wide, clear, and safe pedestrian walkways. Combinations of turf, shrubs, and trees are desirable in border areas along roadways. However, care must be taken to ensure that sight distances and clearances to obstructions are preserved, particularly at intersections.

FIGURE 16.6   Empty wall (owing to drawn blinds) and lack of amenities do not attract pedestrians.

Landscaping also can be used to partially or fully control pedestrian crossing points. Low shrubs in commercial areas and near schools are often desirable to channel pedestrians to crosswalks or crossing areas.

Management of corner space is critical to the success of a commercial street. This small public space enhances the corner sight triangle, permits underground piping of drainage so that street water can be captured on both sides of the crossing, provides a resting place, stores pedestrians waiting to cross the roadway, and offers a location for pedestrian amenities. Well-designed corners, particularly in a downtown or village-like shopping district, can become a focal point for an area. Benches, telephones, newspaper racks, mailboxes, bike racks, and other features help to enliven this area. Corners are often the most secure places on a street.

Parking structures in commercial districts ideally should be placed away from popular walking streets. If this is not possible, driveway and curb radii should be kept tight to maximize safety and minimize discomfort to pedestrians. Adhering to driveway access-management principles is a key element of pedestrian-friendly design generally.

If possible, grades should be kept to no more than 5 percent; terrain permitting, avoid grades steeper than 8 percent. Where this is not possible, railings and other aids should be considered to assist older adults. The Americans with Disabilities Act (ADA) does not require designers to change topography but only to work within its limitations and constraints. Do not create a constructed grade that exceeds 8 percent.

Stairs should be avoided, where possible, because they are a barrier to accessibility, and falls are common on poorly designed stairs. It is critical that stairs be well constructed and well maintained, easily detectable, and slip-resistant. The following principles apply (consult local building codes for additional details): Minimum stairway width is 42 inches to allow two people to pass. Stairs require railings on at least one side. Railings must be graspable and extend 18 inches beyond the top and bottom stair. For wide stairs, such as might be present at passenger transportation terminals, there should be railings on both sides and one or more in midstair areas. Open risers should be avoided. The stair should have a uniform grade with constant tread and rise along the stair. For exterior stairs, the tread should have a forward slope of 1 percent to drain water. Stairs should be illuminated at night.

Sidewalks are recommended on both sides of all urban arterials, collectors, and most local roadways. Codes should require sidewalks for new construction. Lack of sidewalks on a road or street means that conflicts with vehicles are maximized. Children, older adults, and people with disabilities may not have mobility under these circumstances. When prioritizing missing sidewalks, the following factors should be considered: schools, transit stops, parks, shopping districts and commercial areas, medical complexes and hospitals, retirement homes, and public buildings.

Experience has shown that the features summarized below are desirable to achieve robust commercial activity and to encourage added walking versus single-occupant motor vehicle trips. Sucher's excellent book (1995) on "city comforts" is recommended for more detailed information.

Trees.   The FHWA text (undated) states that the most charming streets are those with trees gracing both sides of a walkway. The canopy effect is attractive to pedestrians. Trees should be set back 4 feet from the curb.

Awnings.   Retail shops should be encouraged to provide protective awnings to create shade, provide protection from rain and snow, and add color and attractiveness to the street. Awnings are especially important in warm climates on the sunny side of the street.

Outdoor Cafés.   Careful regulation of street vendors, outdoor cafés, and other commercial activity helps enliven a place—the more activity, the better. One successful outdoor café helps to create more activity, and in time, an entire area can be helped back to life. When outdoor cafés are present, it is essential to maintain a reasonable walking passageway. Elimination of two or three parking spaces in the street and addition of a bulb-out area often can provide the necessary extra space when café seating is needed.

Alleys and Narrow Streets.   Alleys can be cleaned up and made attractive for walking. Properly planned and lit, they can be secure and inviting. Some communities have covered over alleys and made them into access points for a number of shops. Alleys can become attractive places for outdoor cafés, kiosks, and small shops.

Gateways.   Gateways identify a place by defining boundaries. They create a sense of welcome and transition.

Kiosks.   Small tourist centers, navigational kiosks, and attractive outlets for other information can be handled through small- or large-scale kiosks. Well-located interpretive kiosks, plaques, and other instructional or historic place markers are essential to visitors. These areas also can serve as safe places for people to meet and generally can help with navigation.

Fountains, Play Areas, and Public Art.   Public play areas and interactive art can enliven a corner or central plaza. Project for Public Spaces, Inc. (2001), points out that for such amenities to work, they must respond to the needs of a location, to the activities that take place there, and to people's patterns of use.

Pedestrian Streets, Transit Streets, and Pedestrian Malls.   Many cities throughout the world have successfully converted streets to transit and pedestrian streets. These conversions need to be made with a master plan so that traffic flow and pedestrian movements are fully provided for.

16.5.2 Pedestrian Plazas

Pedestrian plazas are defined as places of abundant vegetation, artwork, seating, and perhaps fountains that are intended not only as quiet spots for rest and contemplation but also as centers where communities can come together to socialize and take part in a variety of activities (Project for Public Spaces, Inc. 2001). Unfortunately, many recently constructed plazas serve more to enhance the image of the building on the lot in that they are too large and uncomfortable for pedestrians. Problems with plazas include that some are windswept, others are on the shady side of buildings, some break the continuity of shopping streets, and some are inaccessible because of grade changes. Most are without benches, planters, cover, shops, or other pedestrian comforts. To be comfortable, large spaces should be divided into smaller ones. Landscaping, benches, and wind and rain protection should be provided, and shopping and eating should be made accessible. Encourage the use of bandstands, public display areas, outdoor dining spaces, skating rinks, and other features that attract crowds.

The FHWA text (undated) indicates that no extra room should be provided. It is usually better to be a bit crowded than too open and to provide many smaller spaces instead of a few large ones. It is better to have places to sit, planters, and other conveniences for pedestrians than to have clean, simple, and architectural space. It is better to have windows for browsing and stores adjacent to the plaza space with cross-circulation between different uses than to have the plaza serve one use. It is better to have retailers rather than offices border the plaza. It is better for the plaza to be part of the sidewalk instead of separated from the sidewalk by walls. The popular downtown plaza in Montreal, shown in Figure 16.7, has most of these attributes.

Ideally, plazas should be located to provide good sun exposure and little wind exposure in places that are protected from traffic noise and in areas that are easily accessible from streets and shops. Planners should inventory the area for spaces that can be used for plazas, especially small ones. Appropriate spaces include locations where buildings may be demolished and new ones constructed, vacant land, or streets that may be closed to traffic or may connect to parking. The Project for Public Spaces, Inc. (2000), handbook on creating successful public spaces is an excellent resource on this topic.

16.5.3 Intersections

Intersections are locations where the paths of vehicles and pedestrians come together. They can be the most challenging part of negotiating the pedestrian network. If pedestrians cannot cross the street safely, then mobility is severely limited, access is denied, and walking as a mode of travel is

FIGURE 16.7   Popular plaza in downtown Montreal.

discouraged. In designing and operating intersections that are attentive to the needs of pedestrians, the following considerations should be addressed:

- Enhancing visibility of pedestrians through painted crosswalks, moving pedestrians out from behind parked vehicles by using bulb-outs, and increasing sight distances by removing obstructions such as vegetation and street furniture
- Minimizing the time and distance pedestrians need to cross the roadway
- Making pedestrian movements more predictable through the use of crosswalks and signalization
- Using curb ramps to provide transition from walkway to street

The following features of intersections should be designed from the pedestrian as well as motor vehicle standpoint:

Crosswalks.   One way to shorten the crossing distance for pedestrians on streets where parking is permitted is to install curb bulbs or curb extensions. As shown in Figure 16.4, curb bulbs project into the street usually for a distance equal to the depth of a typical parallel parking space, thereby making it easier for pedestrians to see approaching traffic and giving motorists a better view of pedestrians. When designing curb bulbs at intersections where there is low truck traffic, the corner radius should be as small as possible to have the effect of slowing down right-turning traffic.

Signal Timing, Indications, and Detection.   Pedestrians are often confused by pedestrian phase signal timing and pushbuttons because these seem to vary not only from place to place but also from intersection to intersection. The timing of WALK and DON'T WALK phases appears arbitrary. Many pedestrians do not know that the flashing DON'T WALK is intentionally displayed before the average pedestrian can get completely across the street. Or the signal timing may be too fast for slow walkers such as older pedestrians or people with disabilities.

At least two of the amendments to the 2003 edition of the *MUTCD* addressed these problems. First, to provide enhanced pedestrian safety, the existing option of using pedestrian countdown displays has been changed to a requirement for all new installations of pedestrian signals. As shown in Figure 16.8, the countdown clock, located directly under the "walking person" and "upraised hand," indicates the number of seconds remaining until the DON'T WALK signal indication begins.

FIGURE 16.8 Pedestrian countdown display.

Second, signal timing should be calculated, as discussed earlier in this chapter. This will accommodate older pedestrians and those with mobility impairments.

The detectors should be placed at the top of and as near as possible to the curb ramp and clearly in line with the direction of travel. This is especially important for pedestrians with low vision. The pushbutton box should provide a visible acknowledgment that the crossing request has been received.

Refuge Islands. Pedestrian refuge islands are the areas within an intersection or between lanes of travel where pedestrians may wait safely until vehicular traffic clears, allowing them to cross a street. Such islands are commonly found on wide, multilane streets, where adequate pedestrian crossing times cannot be provided without adversely affecting traffic flow. They also provide a resting place for pedestrians (e.g., elderly, wheelchair-bound, or others) unable to completely cross the intersection within the allotted time.

Pedestrian refuge islands may be installed at intersections or midblock locations as determined by engineering studies. They must be designed in accordance with the AASHTO roadway design policy (2004), the AASHTO pedestrian guide (2004), and *MUTCD* (FHWA 2009) requirements. Raised-curb islands need cut-through ramps at pavement level or curb ramps for wheelchair users. The island should be at least 6 feet wide from face of curb to face of curb (minimum width should not be less than 4 feet). The island should not be less than 12 feet long or the width of the crosswalk, whichever is greater. Minimum island size should be 50 square feet. There should be no obstructions to visibility (e.g., barriers, vegetation, or benches).

16.5.4 Midblock Crossings

Designers often assume that pedestrians will cross roadways at established intersections. However, pedestrians routinely cross the street at midblock locations. Pedestrians rarely will go out of their way

FIGURE 16.9   Midblock crossing connecting parking structure with downtown retail mall.

to cross at an intersection unless they are rewarded with a much improved crossing. As noted earlier, pedestrians will take the most direct route, even it means crossing several lanes of high-speed traffic. Pedestrians crossing at random and unpredictable locations create confusion and increase risks to both the pedestrians and drivers. Well-designed and properly located midblock crossings actually can provide safety benefits for pedestrians. Two primary ways to facilitate nonintersection crossings are medians and midblock crossings. Figure 16.9 shows a midblock crossing that incorporates a median and traffic-calming elements. The crossing connects a parking structure with a retail mall.

A median or refuge island is a raised longitudinal space separating opposing traffic directions. A pedestrian faced with crossing one or more lanes in each direction must determine a safe gap in two, four, or six lanes at a time—a complex task. Younger and older pedestrians have reduced gap-assessment skills compared with pedestrians in other age groups. Pedestrian gap-assessment skills are particularly poor at night. A median allows pedestrians to separate the crossing into two tasks, negotiating one direction of traffic at a time. Crossing times are reduced, making the walk across the street much safer.

Desirably, a median should be at least 8 feet wide to allow a pedestrian to wait comfortably in the center, 4 feet from moving traffic. Normally, there will be an open flat cut rather than a ramp owing to the short width.

Midblock crossings are located and placed according to a number of factors, including roadway width, traffic volume, traffic speed and type, desired lines for pedestrian movement, and adjacent land use. Owing to their low speed and volume, local roads generally do not have median treatments. However, there may be exceptions, particularly around schools or hospitals, where traffic calming is desired.

The design of midblock crossings uses warrants similar to those used for conventional intersections (e.g., stopping sight distance, effects of grade, need for lighting, and other factors). The designer should recognize that pedestrians have a strong desire to continue their intended path of travel. Natural patterns should be identified. For example, a parking lot on one side of the street connecting a large office complex or shopping center on another establishes the desired crossing location. Grade-separated midblock crossings have been effective in a few isolated locations. Owing to their cost and potentially low use, engineering studies should be conducted. Given a choice, on most roadways, pedestrians generally prefer to cross at grade.

As noted in the AASHTO pedestrian guide (2004), midblock crossings should be identifiable to pedestrians with vision impairments. If the location is signalized, a locator tone at the pedestrian pushbutton may be sufficient. Some agencies use a tactile strip across the width of the sidewalk to alert pedestrians to the presence of the crossing.

16.5.5 Accessibility

According to the FHWA publication (1999) on designing facilities for access, there are 49 million people in the United States with disabilities. At one time or another, virtually all of them are pedestrians. Anyone can experience a temporary or permanent disability at any time owing to age, illness, or injury. It is estimated that 85 percent of Americans living to their full life expectancy will suffer a permanent disability. Good design is important so that these individuals are not restricted in their mobility.

The ADA was enacted in 1990 to ensure that a disabled person have access to all public facilities in the United States. The law has specific requirements for pedestrian facilities on public and private property. This section provides an overview of basic accessibility requirements that are relevant to designing pedestrian facilities. The complete set of standards can be found in the *Americans with Disabilities Act Accessibility Guidelines* (*ADAAG*) developed by the Access Board. Note that the rules are updated from time to time. The Access Board Web site (www.access-board.gov) should be consulted for the current version of the rules.

Sidewalks. Wheelchairs require a 3-foot minimum width for continuous passage. Therefore, sidewalks should have a minimum clearance width of 5 feet. Sidewalks should be surfaced with a smooth, durable, and slip-resistant material. They should be kept in good condition, free from debris, cracks, and rough surfaces. Sidewalks should have the minimum cross-slope necessary for proper drainage. The maximum cross-slope is 2 percent (1:50). A person using crutches or a wheelchair has to exert significantly more effort to maintain a straight course on a sloped surface than on a level surface. Driveway slopes should not encroach on the sidewalk.

Ramps. Ramps are locations where the grade exceeds 5 percent along an accessible path. Longitudinal grades on sidewalks should be limited to 5 percent but may be a maximum of 1:12 if necessary. Long, steep grades should have level areas every 30 feet because traversing a steep slope with crutches, artificial limbs, or in a wheelchair is difficult, and level areas are needed for the pedestrian to stop and rest.

Street Furniture. Street furniture includes such things as benches, newspaper boxes, trash receptacles, and bus shelters. To accommodate the disabled, street furniture should be out of the normal travel path as much as possible. For greater conspicuity, high-contrast colors such as red, yellow, and black are preferable. Guidelines for street furniture include the following:

- No protruding object should reduce the clear width of a sidewalk or walkway path to less than 3 feet.
- No object mounted on a wall or post or free standing should have a clear open area under it higher than 2.3 feet off the ground.
- No object higher than 2.3 feet attached to a wall should protrude from that wall more than 4 inches.

Pedestrian Signals. Some individuals have difficulty operating conventional pedestrian pushbuttons. There may be a need to install a larger pushbutton or to change the placement of the pushbutton. Pedestrian pushbuttons always should be easily accessible to individuals in wheelchairs and should be no more than 42 inches above the sidewalk. The force required to activate the button should be no greater than 5 pounds.

Accessible pedestrian signals provide audible and/or vibrotactile information coinciding with visual pedestrian signals to let blind or low-vision pedestrians know precisely when the WALK interval begins. Bentzen (1998) identifies intersections that may require evaluation for accessible pedestrian signal installation. These include very wide crossings, secondary streets having little traffic, nonorthogonal or skewed crossings, T-intersections, high volumes of turning vehicles, split-phase signal timing, and noisy locations. Where these conditions occur, it may be impossible for a pedestrian who is blind to determine the onset of parallel traffic or to obtain usable orientation and directional information about the crossing from the cues that are available.

FIGURE 16.10   Truncated dome detectable warning at base of curb ramp.

Curb Ramps.   The single most important design consideration for persons in wheelchairs is to provide curb ramps. New and rebuilt streets with sidewalks should have curb ramps at all crosswalks. It is desirable to provide two curb ramps per corner (single ramps located in the center of a corner are less desirable). Separate ramps provide greater information to visually impaired pedestrians in street crossings, especially if the ramp is designed to be parallel with the crosswalk. These also benefit others with mobility limitations, elderly pedestrians, and persons pushing strollers or carts.

A curb ramp in new construction should be at least 48 inches wide, not including the width of the flared sides. The slope of a curb ramp should not exceed 8.33 percent, and the cross-slope should not exceed 2 percent. If the landing (level area of the sidewalk at the top of a curb ramp) width is less than 48 inches, then the slope of the flares at the curb face should not exceed 8.33 percent. If the landing width is greater than 48 inches, a 10 percent slope is acceptable. Detectable truncated dome warnings (illustrated in Figure 16.10) that are 2 feet wide must be provided for the full width of the ramp.

Ramps should be checked periodically to make sure that large gaps do not develop between the gutter and street surface. Drainage is very important with curb cuts. Standing water can obscure a drop-off or pothole at the base of a ramp and makes the crossing messy. Storm drain inlets should be clear of the crosswalk.

16.5.6   Multiuse Trails

Multiuse or off-road trails provide environments for walking and other nonmotorized users that, as shown in Figure 16.11, are separate from motor vehicle traffic. Such trails are often extremely popular facilities that are in high demand among in-line skaters, bicyclists, joggers, people walking dogs, and a variety of other users. These different trail users have different objectives. The resulting mix of objectives and volume of nonmotorized traffic can create problems that should be anticipated during design by understanding the needs of these users and accommodating expected levels and types of use.

There are many types of trails, including

- Urban trails and pathways
- Rail trails
- Trails in greenways
- Interpretive trails

FIGURE 16.11   Multiuse riverfront trail in Reno, Nevada.

- Historical/heritage trails
- Primitive trails

All these can be designed for use by pedestrians (including joggers, casual strollers, hikers, in-line skaters, and others), people with disabilities, and bicyclists. What distinguishes one type of trail from another is its context.

The NHI workbook (1996) cautions that design alone cannot generally restrict use. For example, if bicycles are to be prohibited on a path, education and enforcement programs should be implemented to back up the prohibition. Design can passively discourage certain users from traveling on a given trail but only to a limited extent. For example, road bicyclists are not likely to ride on an unpaved surface.

It may be easier to separate types of trail users. Several different separation techniques are available, including

- *Parallel pathways*—one for "wheels" and one for "heels" (as shown in Figure 16.12)
- *Striped lanes*—pedestrians and bikes in separate lanes
- *Directional separation*—traffic in one direction stays to the right; traffic in the other direction stays to the left.
- *Signing and enforcement*—For example, "EQUESTRIAN USE ONLY"

National guidelines for the design of multiuse trails are provide by AASHTO's (1999) *Guide for the Development of Bicycle Facilities*. There are a number of other guides as well covering a variety of trail types (e.g., Steinholtz and Vachowski 2001). Schwarz (1993) presents a thorough discussion of greenway trail design, including types of trail layouts. Figure 16.13 shows a stone-surface multiuse trail in steep terrain that incorporates a timber safety railing.

The minimum width for two-directional trails is 10 feet, although 12- to 14-foot widths are preferred where heavy traffic is expected. Centerline stripes should be considered for paths that generate substantial amounts of pedestrian traffic. Appropriate speed limit and warning signs should be posted. Trail etiquette signs should clearly state that bicycles should give an audible warning before passing other trail users.

FIGURE 16.12   Separate pathways for pedestrians and bicycles.

The surfacing material on a trail significantly affects which user groups will be capable of negotiating the path. Soft surfaces such as sand and gravel are more difficult for all users to negotiate. They present special difficulties for those using wheeled devices such as road bicycles, strollers, and wheelchairs.

The FHWA guide to access (1999) notes that local conditions determine the choice of trail surface. Recreational trail surfaces are often composed of naturally occurring soil. Surfaces ranging from concrete to wood chips may be used depending on the designated user types, the expected volume of traffic, the climate, and the surrounding environment.

Trail/roadway intersections can become areas of conflict if not designed carefully. For at-grade intersections, the following characteristics should be included:

FIGURE 16.13   Multiuse trail in steep terrain.

1. Position the crossing at a logical and visible location.

2. Warn motorists of the approaching crossing. Warning signs and pavement markings used to alert motorists of trail crossings should be used in accordance with the *MUTCD* (FHWA 2009).

3. Maintain visibility between trail users and motorists. Vegetation, signs, and other objects in the right-of-way should be removed or relocated so that trail users can observe traffic conditions and motorists can see approaching trail users.

4. Inform trail users of the upcoming intersection. Signs and pavement markings on the trail can provide advance warning of the intersection, especially in areas where the intersection is not clearly visible.

The need for parking should be anticipated during the trail planning process. Adequate parking at trailheads is necessary so that trail users do not park on the shoulder of the road near intersections, blocking sightlines of both motorists and trail users.

Unauthorized motor vehicle access is a problem at some trail/roadway intersections. Trail bollards is the most effective method of restricting motor vehicle traffic. However, care must be taken in their use because they present an obstacle when located in the travel path of pedestrians and bicycles.

Bollards should be painted a bright color and be permanently reflectorized to maintain their visibility. They should be 3 feet tall and can be constructed of a variety of materials. Commercial manufacturers offer bollards that can be unlocked and removed to allow emergency vehicle and maintenance access.

The NHI workbook (1996) points out that trails can be used to provide connections between transportation modes. Intermodal linkage possibilities include (1) to shopping, schools, work, and transit, (2) between parking and transit drop-off points, (3) to ports, rivers, and scenic areas, and (4) to ferry or bridge connections.

Access to trails is a key consideration. A *trailhead*, such as the one shown in Figure 16.14, is a location where people can access a trail. The NHI workbook (1996) indicates that a trailhead can

FIGURE 16.14   Example of trailhead facility, Olympia, Washington.

be as simple as a trail marker and a few parking stalls or a virtual visitor's center with vending machines, snack bar, and interactive trail guide information.

Good trails attract people who need places to park, rest, get trail information, use rest rooms, dispose of trash, and get a drink of water. These are the basic considerations. Trailhead design should consider

- Easy access from public streets
- Provision of adequate parking
- Location relative to transit facilities
- Potential joint use (e.g., also can serve as picnic area)
- Rest rooms and trash disposal
- Weather protection
- Potential for interpretation of area's historic, cultural, and natural features
- Location relative to on-site concessions and ancillary facilities such as bike rentals

16.6 OPERATIONS AND MAINTENANCE

16.6.1 Pedestrian Signs and Pavement Markings

Traffic engineers use a wide variety of signs and pavements markings relative to pedestrians. Some are used to alert motorists to pedestrian activity and others to direct pedestrians to defined crossings.

Signing is governed by the *MUTCD* (FHWA 2009), which provides guidelines for the design and placement of traffic control devices installed on facilities open to the public. It must be noted that signs are often ineffective in modifying driver behavior and overuse of signs breeds disrespect.

Colors for signs and markings should conform to the color schedule recommended by the *MUTCD* to promote uniformity and understanding from jurisdiction to jurisdiction. The Millennium Edition of the *MUTCD* included a new color (in addition to standard yellow) for pedestrian warning signs—fluorescent yellow-green. However, the *MUTCD* cautions that the mixing of standard yellow and fluorescent yellow-green backgrounds within a selected site area should be avoided.

The discussion that follows presents an overview of key issues related to traffic control devices for pedestrians. The discussion is organized by type of device.

Regulatory Signs.   The NO TURN ON RED sign may be used in some situations to facilitate pedestrian movements. Owing to conflicting research results, there has been considerable controversy regarding pedestrian safety and right turn on red. Use of NO TURN ON RED signs at intersections should be evaluated on a case-by-case basis using engineering judgment. Some traffic engineers feel that prohibition of turns on red should be considered only after the need has been fully established and less restrictive methods have been reviewed or tried. Part-time prohibitions should be discouraged; however, they are preferable to full-time prohibitions when the actual need occurs for only short periods of time. A supplemental plate reading WHEN PEDESTRIANS ARE PRESENT may aid the pedestrian without unduly restricting vehicular traffic flow. Education and enforcement play important roles in the benefits and safety of right turn on red. Enforcement is important relative to turns being made only after stopping and yielding to other road users and to the observation of necessary prohibitions.

Other signs include the pedestrian pushbutton signs or other signs at signals directing pedestrians to cross only on the green light or WALK signal. Pedestrian pushbutton signs should be used at all pedestrian-actuated signals. This is helpful to provide guidance to indicate which street the button is for. The signs should be located adjacent to the pushbutton, and pushbuttons should be accessible to pedestrians with disabilities.

Warning Signs. The pedestrian crossing sign (W11-2, as shown in Figure 16.15*a*) may be used to alert road users to locations where unexpected entries into the roadway by pedestrians might occur. The crossing sign should be used adjacent to the crossing location. If the crossing location is not delineated by crosswalk pavement markings, the crossing sign should be supplemented with a diagonal downward pointing arrow plaque showing the location of the crossing. If the crossing location is delineated by crosswalk pavement markings, the diagonal downward pointing arrow plaque is not required. To avoid information overload, this sign should not be mounted with another warning or regulatory sign.

It is important to note that overuse of warning signs breeds disrespect and should be avoided. Care should be taken in sign placement in relation to other signs to avoid sign clutter and to allow adequate motorist response.

The playground sign (W15-1, as shown in Figure 16.15*b*) may be used to give advance warning of a designated children's playground that is located adjacent to the road. This sign is not intended for use on local or residential streets, where children are expected. CAUTION CHILDREN AT PLAY or SLOW CHILDREN signs should not be used. They may encourage children to play in the street and parents to be less vigilant. Such signs provide no guidance to motorists in terms of safe speed, and the signs have no legal basis for determining what a motorist should do. Furthermore, motorists should expect children to be "at play" in all residential areas; the lack of signing on some streets may indicate otherwise. Use of these nonstandard signs also may imply that the involved jurisdiction approves of streets as playgrounds, which may result in the agency being vulnerable to tort liability.

(*a*)

(*b*)

FIGURE 16.15 (*a*) Pedestrian crossing sign (W11-2). (*b*) Playground sign (W15-1).

Informational Signs. Guide or directional signs for pedestrians are intended to assist unfamiliar pedestrians or those who may not know the most direct route to a destination by foot. Use distances meaningful to pedestrians, such as the number of blocks or average walking time.

Crosswalk Markings. Crosswalk markings provide guidance for pedestrians who are crossing roadways by defining and delineating paths on approaches to and within signalized intersections and on approaches to other intersections where traffic stops. Crosswalk markings also serve to alert road users of a pedestrian crossing point across roadways not controlled by traffic signals or STOP signs. At nonintersection locations, crosswalk markings legally establish the crosswalk. The standard in the *MUTCD* is that when crosswalk lines are used, they shall consist of solid white lines that mark the crosswalk. Crosswalk lines should not be used indiscriminately. An engineering study should be performed before they are installed at locations away from traffic signals or STOP signs. Typical types of crosswalk markings are shown in Figure 16.16. Lalani and the ITE Pedestrian and Bicycle Task Force (2001) present an excellent summary of various treatments used by local agencies in the United States, Canada, Europe, New Zealand, and Australia to improve crossing safety for pedestrians, including midblock locations and intersections.

16.6.2 Pedestrians and Work Zones

When construction or maintenance activities take place on or near sidewalks, crosswalks, or multiuse paths, pedestrians may be exposed to a variety of hazards, including detours that are difficult to navigate or that force pedestrians into the street, uneven walking surfaces, walking surfaces contaminated with

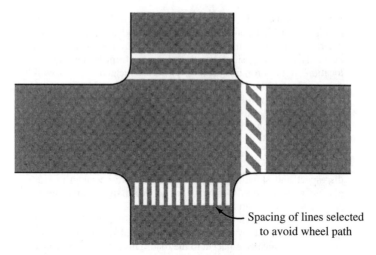

Spacing of lines selected to avoid wheel path

FIGURE 16.16   Crosswalk markings as presented in the *MUTCD*.

foreign substances, restricted sight distances, and conflicts with vehicles and equipment. In addition, the mobility of persons with disabilities may be adversely affected. It is important to develop and implement temporary traffic control zone policies that minimize these problems. All parties involved should be made aware of the needs of pedestrians and made responsible for providing safe and continuous passage.

Developing a workable policy for pedestrian access through work areas requires the cooperation of traffic engineers, construction inspectors, crew chiefs, contractors, and advocates. The policy should apply whenever construction or maintenance work affects pedestrian access, whether the work is done by private firms or city, county, or state personnel.

Permits required for street construction or construction projects that encroach on sidewalks or crosswalks should be contingent on meeting pedestrian access policies. Contractors should be given copies of the standards when they apply for a permit. Crew chiefs and crews should be trained so that they understand and follow the policy. The *MUTCD* (FHWA 2009) provides specific guidance on pedestrian access around work areas.

16.6.3   Facility Maintenance

Walkways are subject to debris accumulation and surface deterioration and require maintenance to function safely and efficiently. Poorly maintained facilities become unusable and a liability because users risk injury.

As noted earlier, while walking, a person typically looks ahead and around without noticing cracks and other discontinuities in the walking surface. A smooth, level surface is critical for young, elderly, and disabled pedestrians. Pedestrians also depend on motorists respecting traffic signs and signals. These also must be maintained properly for pedestrian safety.

A walkway maintenance program is necessary to ensure adequate maintenance of facilities. The program should establish maintenance standards and a schedule for regular maintenance and inspection as outlined below. Recommended maintenance practices include

1. *Sweeping.* Loose gravel, snow/ice control abrasives, broken glass, and other debris are not only unattractive but also can cause slip and fall hazards for pedestrians, particularly when the walkway is on a gradient. A periodic inspection and maintenance program should be implemented so that such loose materials are regularly picked up or swept. Debris from the roadway should not be swept onto sidewalks.

2. *Surface repairs.* A smooth walkway surface free of cracks, potholes, bumps, and other physical problems should be provided and maintained. Surfaces should be inspected regularly, and potentially hazardous conditions should be repaired as soon as possible.

3. *Vegetation.* Vegetation encroaching onto walkways is both a nuisance and a problem. Roots should be controlled to prevent breakup of the surface. Raised sidewalk slabs are a significant cause of trip and fall accidents, particularly for the elderly. It is also important that adequate clearances and sight distances be maintained at driveways and intersections. Pedestrians must be visible to approaching motorists and not hidden by overgrown shrubs or low-hanging branches. Local ordinances should allow authorities to control vegetation that originates from private property.

4. *Traffic control devices.* New pedestrian-related signs and pavement markings (e.g., crosswalks) are highly visible, but over time, they weather and become harder to see, especially at night. Retroreflectivity of signs and markings should be inspected at night and defective devices replaced.

5. *Drainage.* New drainage devices function well, but they deteriorate over time. Repair or relocate faulty drainage at intersections where water backs up onto the curb cut or into the crosswalk.

6. *Utility cuts.* These can leave a rough surface if not constructed carefully. Sidewalk cuts should be finished as smooth as a new sidewalk.

7. *Snow removal.* Snow should be cleared from publicly owned sidewalks. Sidewalks are not appropriate for snow storage. Ordinances regarding removal of snow from private sidewalks should be publicized and enforced.

REFERENCES

AASHTO Task Force on Geometric Design. 1999. *Guide for the Development of Bicycle Facilities.* American Association of State Highway and Transportation Officials, Washington, DC.

American Association of State Highway and Transportation Officials. 2004. *A Policy on Geometric Design of Highways and Streets.* AASHTO, Washington, DC.

———. 2004. *Guide for the Planning, Design, and Operation of Pedestrian Facilities.* AASHTO, Washington, DC.

Architectural and Transportation Barriers Compliance Board (Access Board). 2010. *Americans with Disabilities Act Accessibility Guidelines (ADAAG).* U.S. Access Board, Washington, DC; current edition, available at www.access-board.gov.

Baker, S. P., B. O'Neill, M. J. Ginsburg, and G. Li. 1992. *The Injury Fact Book,* 2nd ed. Oxford University Press, New York, 1992.

Bentzen, B. L. 1998. *Accessible Pedestrian Signals.* U.S. Access Board, Washington, DC, August 4.

Berkovitz, A. 2001. "The Marriage of Safety and Land-Use Planning: A Fresh Look at Local Roadways." *Public Roads* (September–October), 7–19.

Burden, D. 1999. "Pennsylvania Pedestrian and Bicyclist Safety and Accommodation." Assembled for Pennsylvania Department of Transportation.

Eck, R. W., and E. D. Simpson. 1996. "Using Medical Records in Non-Motor-Vehicle Pedestrian Accident Identification and Countermeasure Development." *Transportation Research Record* 1538, 54–60.

Federal Highway Administration (FHWA). 1999. *Designing Sidewalks and Trails for Access,* Part I: *Review of Existing Guidelines and Practices.* FHWA, Washington, DC, July; available at www.fhwa.dot.gov/environment/sidewalk.

———. 2001. *Designing Sidewalks and Trails for Access,* Part II: *Best Practices Design Guide.* FHWA, Washington, DC; available at www.fhwa.dot.gov/environment/sidewalk2.

———. Undated. *FHWA Course on Bicycle and Pedestrian Transportation: Student Workbook.* FHWA, Washington, DC.

———. 2009. *Manual on Uniform Traffic Control Devices.* FHWA, Washington, DC.

———. 1992. *The National Bicycling and Walking Study Case Study No. 1: Reasons Why Bicycling and Walking Are and Are Not Being Used Extensively as Travel Modes.* FHWA-PD-93-041. FHWA, Washington, DC.

Hunter, W. W., J. C. Stutts, and W. E. Pein. 1997. *Pedestrian Crash Types: A 1990's Informational Guide.* FHWA-RD-96-163. FHWA, McLean, Virginia, April.

Hyde, A. H., G. M. Bakken, J. R. Abele, H. H. Cohen, and C. A. La Rue. 2002. *Falls and Related Injuries: Slips, Trips, Missteps and Their Consequences.* Lawyers & Judges Publishing Co., Tucson, AZ.

Lalani, N., and ITE Pedestrian and Bicycle Task Force. 2001. "Alternative Treatments for At-Grade Pedestrian Crossings." Informational Report. Institute of Transportation Engineers, Washington, DC.

Landis, B. W., V. R. Vattikuti, R. M. Ottenberg, D. S. McLeod, and M. Guttenplan. 2001. "Modeling the Roadside Walking Environment—Pedestrian Level of Service." *Transportation Research Record* 1773, 82–88.

National Highway Institute. 1996. *Pedestrian and Bicyclist Safety and Accommodation—Participant Workbook.* FHWA-HI-96-028. FHWA, Washington, DC, May.

Project for Public Spaces, Inc. 2000. "How to Turn a Place Around: A Handbook for Creating Successful Public Spaces." Project for Public Spaces, New York, NY.

Project for Public Spaces, Inc. 2001. "Getting Back to Place: Using Streets to Rebuild Communities." Project for Public Spaces, New York, NY.

Schwarz, L. L., ed. 1993. *Greenways: A Guide to Planning, Design, and Development.* Island Press, Washington, DC.

Steinholtz, R. T., and B. Vachowski. 2001. "Wetland Trail Design and Construction." Technology and Development Program, USDA Forest Service, Missoula, MT, September.

Sucher, D. 1995. *City Comforts—How to Build An Urban Village.* City Comforts Press, Seattle, WA.

Transportation Research Board. 2000. *Highway Capacity Manual.* Special Report 209. National Research Council, Washington, DC.

———. 2004. *National Cooperative Highway Research Program Report 500,* Vol. 10: *A Guide for Reducing Collisions Involving Pedestrians.* National Research Council, Washington, DC.

CHAPTER 17
BICYCLE TRANSPORTATION*

Lisa Aultman-Hall
Department of Civil and Environmental Engineering
University of Connecticut
Storrs, Connecticut

17.1 ARE BICYCLES REALLY TRANSPORTATION IN AMERICA TODAY?

In an automobile-dominated society, it is often difficult for transportation professionals, as well as private citizens to remember that alternative modes of transportation exist and that it is in everyone's interest to promote their use. A diversified transportation system allows for flexibility and choices. The bicycle represents a mode of transportation that is relatively inexpensive, space-efficient, and accessible to almost all members of society. There are many virtuous reasons for promoting the bicycle, including the reduction of air pollution impacts, parking needs, and user costs. But benefits related to physical health, economic development, and tourism have also been shown to increase when more people in a community bicycle or when bicyclists are attracted to a community for bicycling.

Moreover, from a purely traffic engineering point of view, the use of bicycles for more utilitarian or purposeful trips (as opposed to purely recreation) also represents one marginal or incremental solution to the traffic-congestion problems plaguing our transportation systems. For example, while at low traffic volume conditions removing one automobile trip from the system might not be very beneficial, at higher volumes every trip removed represents travel time savings to all other motorized vehicles in the system. The overall widespread benefits of small or marginal reductions in demand on the system when it is operating near or above capacity are what traffic engineers are striving to accomplish today. Changes in traffic signal timings, small increases in ridesharing, and promotion of telecommuting are small efforts to accomplish marginal decreases in demand near or above capacity. It might be tempting to assume that the bicycle is a relatively specialized marginal mode of transportation and that because relatively few people use it, it offers relatively little in terms of benefits to the overall system. However, like traffic signaling timings or other smaller management solutions, a small number of purposeful bicycle trips at congested times represents important travel time savings to all users of the system. For this reason, the bicycle should not be considered an unimportant mode even if few people use it, but rather a mode that contributes to important marginal decreases in travel demand near or above capacity. It is one of many important incremental solutions to our current traffic congestion crisis. Bicycles are worth promoting even if they will not be used by most travelers.

The bicycle has the further benefit of offering relatively independent transportation to many who do not have other options. In 1995, 8 million households did not own an automobile (BTS 1995). Up to 37 percent of a state's population is too young to drive an automobile or is elderly or disabled and may not be able to drive (FHWA 1994). However, despite this potential demand for bicycle use, in

---

*Reprinted from the First Edition.

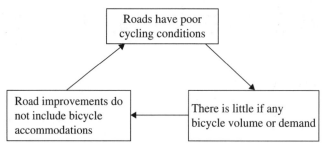

FIGURE 17.1 A transportation planning cycle driven by existing demand.

many places promoting bicycling has become lost in a self-defeating cycle. Often, as planners and engineers, we look at a portion of the system and see no one biking and therefore we assume there is no need to design for or accommodate cyclists on that route. However, the route may well be a major traffic artery with no place for bicycles that people perceive as safe or pleasant. Therefore, no one bicycles and with no existing bicycle demand, a transportation planning system that primarily builds to meet current or future demand does not make accommodations to create the bicycle demand. This cycle is shown in Figure 17.1. One can argue that we need transportation planning that is smarter than simply building to meet a demand. We need transportation planning that designs and builds systems that create the type of demand we want; travel demand that produces more optimal use of space, better communities, and fewer environmental impacts. Instead of building to meet the demand, we need to design the demand with the land use patterns and transportation infrastructure we provide. To help accomplish this goal, this chapter covers many of the aspects of how bicycles should operate safely within our transportation system, as well as an overview of the specific design knowledge that is available today for designing bicycle infrastructure. This chapter is intended to challenge engineers and planners to become informed about the latest information and standards in bicycle transportation in order to become part of educating cyclists, engineers, planners, and the public to make the best use of the bicycle as one of many solutions to our transportation dilemmas.

17.2 STANDARDS AND REGULATIONS FOR BICYCLE TRANSPORTATION

Many standards and regulations at the federal level, and in most states, support the reality that bicycles are vehicles in our transportation system. They are officially considered such in the Uniform Vehicle Code. Both the Intermodal Surface Transportation Equity Act (ISTEA) of 1991 and the Transportation Efficiency Act for the 21st Century (TEA-21) in 1998 contained programs and policies that include the bicycle as a mode of transportation in the multimodal and intermodal transportation system envisioned by these two federal acts. One of the most influential requirements of these acts was that for a bicycle and pedestrian coordinator in each state department of transportation. In some states, this position consists of only a portion of one employee's job. Yet in many other states, several professionals work within a bicycle and pedestrian program. In 2002, the American Association of State Highway and Transportation Officials (AASHTO) established a task force composed of 16 of the state bicycle pedestrian coordinators as a permanent part of its highway design committee. This is one more indication that accommodation of bicycles within the transportation system is no longer just a factor in "special" bicycle-friendly communities but a factor throughout the United States. As a result of ISTEA and TEA-21, federal spending on bicycles and pedestrians increased from $0.49 million in 1988 to $416 million in 2002.

In 2000, the United States Department of Transportation adopted a policy statement *Accommodating Bicycle and Pedestrian Travel: A Recommended Approach* (FHWA 2000) in response to the requirements in TEA-21. The policy requires that "bicycling and walking facilities will be incorporated into all transportation projects unless exceptional circumstances exist." The document, which outlines suggested approaches and guidance to accommodating bicycles in different circumstances, was the product of a national task force on which many groups were represented. However, the implementation of the policy was left to individual states themselves. Some states already had design manuals and regulations in place to specifically address bicycles and pedestrians. Others, such as Kentucky, formed their own task force, drafted design guidance, and adopted state-specific design guidelines in response to the FHWA guidance. Still other states have not yet responded or changed their design process in response to the 2000 FHWA document.

From the above discussion, it is clear that our laws and transportation agencies regard bicycling as a form of transportation. However, one controversial item remains: is there a difference between bicycle transportation and bicycle recreation? Or in other words, which types of bicycling are transportation? The most recent laws, especially TEA-21, have clearly targeted most programs and efforts toward utilitarian or purposeful bicycle trips. These trips are those that might reasonably replace a motorized vehicle trip. This argument suggests that those who bicycle for recreation, sightseeing, or exercise are not undertaking a utilitarian trip and are therefore not transportation. Others argue that all forms of movement between two points within our urban or rural areas are transportation, no matter what mode is used to accomplish this movement. Proponents of this argument cite those who use their cars to travel to the park, purchase ice cream on a hot summer day, or tour through decorative lights in December, and they question whether or not these motorized vehicle trips are transportation. In these cases we still strive as transportation professionals to provide safe travel and good levels of service on our transportation system. By extension, one can argue that all bicycle trips are transportation no matter what purpose motivates an individual to bicycle between two points. Similarly, within this chapter all bicycle trips are assumed to be transportation and the design guidance discussed applies to bicycle facilities for cyclists of all skills and ages and with different travel purposes.

The majority of the discussion of bicycle facility design guidance found in this chapter is based on the 1999 AASHTO *Guide to the Design of Bicycle Facilities,* but other references to design guides are also provided. The *Manual on Uniform Traffic Control Devices (MUTCD)* provides standard signs and lane markings that are used for both dedicated and shared bicycle facilities. The Americans with Disabilities Act is the basis for the *Designing Sidewalks and Trails for Access Guide* (Beneficial Designs Inc. 1999 and 2001), which contains important considerations that are not necessarily well known within the wider transportation design community. Each state and many cities may have a bicycle plan or bicycle design guide that deviates from national recommendations and should be consulted for projects within a given jurisdiction.

17.3 TYPES OF BICYCLE FACILITIES

Oftentimes we think of off-road *shared-use paths*[*] as being the only type of bicycle facility. A shared-use path is an important facility dedicated to nonmotorized transportation because it appeals to new cyclists who may not be able or willing to bicycle with heavy motor vehicle traffic. Even those who are able to bike with traffic sometimes prefer the quieter slower pace along a shared-use path. In addition to providing recreational transportation, shared-use paths often form important backbones of regional bicycle networks, sometimes allowing bypassing of major traffic arterials. Many older shared-use path facilities are not designed according to the solid design parameters that were documented throughout the 1990s. These paths might require upgrading or retrofitting. This is particularly important given the safety issues involved when bicycles, pedestrians, and other nonmotorized modes such as bladers are mixed together on relatively narrow facilities. The different operating characteristics of these users, especially speed, make safe design

---

[*]Note that these facilities are not called *bike paths,* the term sometimes used by popular media as well as engineers. The term *trail* has recently seen less use and is used for less formal, narrow, often dirt paths such as one might see through a wooden area. Paths as described in this chapter are also not those used in less developed areas for mountain biking. Readers are also cautioned that the term *path* has a much different meaning in Europe than it does in the United States and Canada.

and operation of shared-use paths particularly challenging. Research has shown that bicyclists on paths have similar crash and injury rates to those found on roadways (Doherty, Aultman-Hall, and Swaynos 1997).

Bicycle lanes along roadways are a dedicated bicycle facility that has been growing in mileage in the United States. The design parameters for bicycle lanes have also been well documented, but the relative safety rates along bicycle lanes versus roads, as well as their impact on ridership levels in terms of attracting new riders, have not been conclusively determined at this time. Preliminary findings suggest bicycle lanes have improved bicycle safety (Clarke and Tracy 1995; Moritz 1998).

Many jurisdictions have been opting for *wide curb lanes* as a bicycle facility. These facilities allow the bicycle more room to travel with motorized traffic than is available in a standard-width travel lane, but do not require as much space as is needed for a bicycle lane. Some believe these facilities avoid some confusion on the part of drivers and cyclists about the use of bicycle lanes because it is clear the cyclists should act and be treated as a vehicle in the right-hand lane.

In most cases in our cities, towns and rural areas a bicycle facility consists of a *shared roadway*. Shared roadways are simply roads where cycling is not prohibited but dedicated facilities have not been provided. The quality of cycling and the safety of cycling on different shared roadways varies, and not all routes are realistically acceptable for all levels of cyclists.

A *signed bicycle route* is a combination of shared roadways, routes with wide curb lanes or bicycle lanes, and shared-use paths. It is delineated by green bicycle route signs (Figure 17.2) and leads from an origin to a destination. The 1999 AASHTO guide states that a signed bicycle route should offer some advantage over the alternatives. In some states, the bicycle routes are numbered or named in the same ways highways are. In the past, bicycle routes were sometimes established simply along roads where conditions were thought to be good or ideal for cycling. These routes started and ended at points that were not necessarily destinations but rather the location where the ideal bicycle conditions ended. In recent years, the signed bicycle route has been taken very seriously as part of a continuous system or bicycle network connecting areas of a city together. These routes might consist of several types of roads and facilities, but the new focus has been on ensuring continuity and directness. Many studies have shown that cyclists are not willing to travel significant extra distance between points even if the bicycling conditions are perceived to be much better on the longer route. While local and collector roads might seem ideal for inclusion in a bicycle route system, the transportation planner should note that the curvy and disconnected patterns of these streets, particularly in newer suburban areas, often leave cyclists lost, and direction/route signs are most desperately needed in these areas.

FIGURE 17.2   Bicycle route sign.

Readers should specifically note that *sidewalks* are not on the list of bicycle facilities provided here. Numerous studies (Aultman-Hall and Kaltenecker 1999; Aultman-Hall and Hall 1998; Moritz 1997; and Wachtel and Lewiston 1994) have shown that sidewalks are a more dangerous place for bicycles when compared to either roads or shared-use paths. Sidewalks are often narrow, have proximate objects in the clear zone, intersect numerous driveways, and have uneven surfaces. These characteristics make them dangerous for cycling. Child cyclists are often going slower speeds, similar to pedestrians, and therefore they might be the only appropriate cyclists for sidewalks. In some locations special ramps onto a sidewalk have been provided to allow cyclists to use the sidewalk on bridges, especially long bridges or causeways. These sidewalks should have low pedestrian volumes, good bicycle-friendly railings (at least 42 inches) and sufficient width. Care is required to ensure safe access to the sidewalk at both ends without requiring the cyclist to break traffic rules, such as by wrong-way riding.

Selecting the right combination of facilities to create a comprehensive and connective network within a town or city is not straightforward. The process requires significant public input, especially from cyclists who use the routes. Specific guidance on where to use which type of facility has been documented and is available to engineers, planners, and the public (Wilkinson et al. 1994).

17.4   THE BICYCLE AS A DESIGN VEHICLE—
DEFINING PARAMETERS

Four main characteristics define the bicycle as a design vehicle: physical dimensions, speed, stopping distance, and climbing ability. The uninformed designer might unknowingly make incorrect assumptions about the operating characteristics of the bicycle. For example, it is common for bicycles to be assumed slower than they can actually travel or narrower than they actually are. *The Guide for the Design of Bicycle Facilities* (AASHTO 1999) delineates these four operating characteristics for bicycles.

Although the tires of bicycles are very narrow, the effective width of the vehicle plus rider and the width required for safe operation of a bicycle are much greater. Including a 5 inch (0.125 meter) buffer on both the right and the left, the overall vehicle (bicycle plus rider) is 40 inches (1 meter) wide. Given that bicycles cannot be driven, even by the most experienced cyclist, in a perfectly straight line and that some room for lateral shifts is needed, more than 1 meter of space is required for operation of a bicycle. AASHTO further recommends that a design speed of 30 km/hr or 20 mph be used for bicycles on paved surfaces, while 25 km/hr or 15 mph be used for unpaved surfaces. This translates into up to 220 feet of required stopping sight distance on a level, paved surface. Climbing ability varies greatly with bicycle type and the individual bicyclist; however, grades of less than 5 percent are recommended and AASHTO provides recommended grades based on grade length for shared-use paths. Designers should also note that on downgrades speeds are often very high and that ice, debris, or rumble strips are particularly dangerous in these situations.

17.5   HOW BICYCLES SHOULD OPERATE AS VEHICLES

There is a great deal of controversy over how bicycles should operate in the transportation system. Even some experienced cyclists disagree on how best to make left-hand turns or whether they ride with or against traffic. The engineer or planner working with bicycle transportation projects who is not a cyclist is handicapped in meaningfully participating in this debate. Furthermore, the noncyclist is in danger of making incorrect assumptions about how bicycles should operate safely as vehicles. This section of the chapter describes the current philosophy among bicycle safety education professionals as to how bicycles should operate in hopes that designers will implement designs that *promote* this type of riding and that transportation professionals might promote safe cycling.

The League of American Bicyclists promotes an education program entitled *Effective Cycling*. A main tenet of this program is the idea that "Bicyclists fare best when they act and are treated as drivers of vehicles." Acting and being treated as vehicles might also be considered simply following the rules of the road. We have rules and hierarchies of right-of-way within our transportation system in order to increase predictability and minimize ambiguity. This creates safety because users know what to expect from other users in different circumstances. Accomplishing this requires not only that cyclists follow rules and that designers provide appropriate facilities, but also that law enforcement is informed and enforces laws. In most but not all states, the bicycle is granted all the rights and responsibilities of other vehicles.

In 1994, the Federal Highway Administration established a system of categorizing cyclists by their skills and needs. Type A cyclists or advanced cyclists are confident and experienced in most traffic conditions. Type B cyclists or basic cyclists are casual, new, or less confident riders. A child cyclist or type C is self-explanatory. Clearly not all cyclists can operate as vehicles on all types of roads. A type B or C cyclist could not operate in the curb lane of an urban multilane arterial roadway. In these situations, a cyclist might elect to ride on the sidewalk or the wrong way against traffic. Safety research has shown that some of these tactics are simply not safe. A good rule of thumb is that if a cyclist does not feel comfortable riding with the traffic on a given road, then he or she should not be on that road. This creates a challenge for engineers and planners as they seek to provide a comprehensive bicycle network within a region: if all types of cyclists are to be served, bypasses around major arteries must be sought.

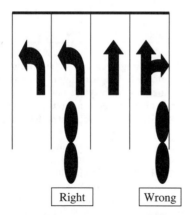

FIGURE 17.3   Bicycle positioning for left turns.

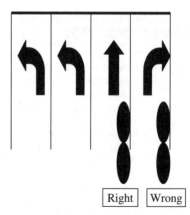

FIGURE 17.4   Bicycle positioning for through movement with a right-turn lane.

Lane positioning is an important consideration for the cyclist who is riding with traffic. Oftentimes we assume the cyclist will be very narrow and as close to the right hand curb as possible. This is not recommended. The edge of the roadway contains many hazards: the curb face, drainage grates, potholes, or debris. Educators recommend between 1 and 3 feet be placed between the cyclist and the road edge or curb. Furthermore, if parallel parked cars are present, the cyclist should consider traveling far enough from the cars so that an open door would not strike him or her. This is perfectly within the cyclist's rights.

It is not appropriate for the cyclists always to be in the rightmost lane, as is often assumed. The League of American Bicyclists promotes the slogan that a cyclist should be "in the rightmost lane that leads to their destination." So, for example, a cyclist making a left-hand turn would not do so from the rightmost lane but rather from the center of the leftmost lane (or the center of the rightmost left-turn lane if multiple left-turn lanes exist). Staying on the absolute right edge would create a conflict point between the bicycle and the through motor vehicle movements, as shown in Figure 17.3. This should make sense to traffic engineers because with all intersection control strategies we seek to minimize conflicting traffic streams in our attempts to maximize safety. Similarly, if a cyclist approaches an intersection where there is a right-hand turn lane and is going straight through, he or she should merge into the through lane and not stay on the rightmost edge of the rightmost lane. Staying in the right lane would create a conflict between the cyclist and any right-turning vehicles, as shown in Figure 17.4. In a shared right-turn through lane, the cyclists should merge to the center of that lane to proceed through the intersection to avoid conflicts with the right-turning vehicles.

These guidelines for lane usage simply follow the premise that bicycles should operate as vehicles. However, they also illustrate why not all cyclists may be able to operate on the roadway as vehicles. These cyclists may choose to use lower-volume single-lane roadways or off-road shared-use paths. It is not recommended that the cyclists improperly use the roadways where they are not comfortable acting as vehicles. Designers should assume all bicycles will operate as vehicles as illustrated in these examples.

17.6   DESIGNING BICYCLE LANES

Although bicycle lanes are relatively new in terms of widespread use, the AASHTO Design Guide provides explicit guidance for lane placement, dimensions, and signage based on decades of use in California and other places. AASHTO defines a bicycle lane as "a portion of the roadway which has been designated by striping, signing and pavement markings for the preferential or exclusive use of bicycles." In special or unique circumstances, alternative solutions and designs are still being tried by various jurisdictions. Readers are directed to use the Web or the state bicycle pedestrian coordinators' network to ensure the most up-to-date options are considered. Bicycle lane designs continue to change and be improved. Innovative design strategies for bicycle lanes and other bicycle facilities have been documented by various groups, including the Institute for Transportation Engineers (ITE) (Nabti and Ridgway 2002).

The typical bicycle lane configuration is shown in Figure 17.5. If a roadway does not have a curb and gutter, a minimum of 4 feet is recommended by AASHTO for a bicycle lane. If a gutter is present, 5 feet including the gutter is recommended and no more than 2 of the 5 feet can be within the gutter. Caution should be used if areas wider than 7 feet are to be used as a bicycle lane. Motor vehicles can fit in this space and will be tempted to use the lanes to bypass traffic, particular for making right turns. This is a very undesirable situation. A 6-inch solid white stripe is used to delineate the bicycle lane

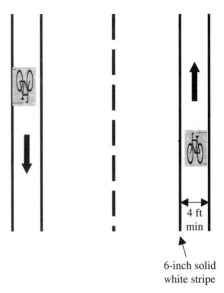

FIGURE 17.5 Typical bicycle lane configuration.

from the adjacent travel lane. If a rumble strip[*] is present, the bicycle lane dimensions do not include the rumble strip, but there is not widespread agreement on whether the bicycle lane should be inside or outside the rumble strip. When parallel parking is present on the roadway, a minimum of 5 feet is recommended for the bicycle lane and a 6-inch solid white strip is placed on both sides of the bicycle lane: one between the bicycle lane and the parked vehicles and one between the bicycle lane and the right-most travel lane. Bicycle lanes with diagonal parking are not recommended. Note that when the bicycle lane approaches an intersection, a transition to a broken white line indicating that turning vehicles and bicycles may be crossing paths is required.

When an existing right-of-way is being restriped for bicycle lanes, a reduction in the width of motorized vehicular lanes may be considered. The AASHTO guide recommends that restriping can reduce lanes to 10 or 10.5 feet if travel speeds are less than or equal to 25 mph. When speeds are between 30 and 40 mph, 11-foot lanes are suggested, with any two-way left-turn lanes being 12 feet wide. If speed is above 45 mph, then standard 12-foot lanes should be provided with the restriping. Sometimes the extra space needed for bicycle lanes can be found by reducing the parking lanes to 7 feet. The city of Chicago installed 75 miles of bicycle lanes between 2000 and 2002. Their detailed design drawings for bicycle lane configurations on streets with different curb-to-curb widths ranging from 44 to 60 feet are contained in a new *Bike Lane Design Guide* (City of Chicago 2002). The drawings contain configurations that include both one-way and two-way streets, bus stops, and parking. These excellent drawings are useful and also illustrate a departure from AASHTO standards: the city of Chicago uses 5 feet as the minimum width for a bicycle lane instead of 4 feet.

The *MUTCD* provides the templates for black-and-white signs indicating the start and end of bicycle lanes. Note that these signs are different than the green signs that indicate bicycle routes and also different from yellow warning signs targeted at motorists to advise caution in certain areas (such as where a shared-use path crosses a roadway).

Two-way bicycle lanes are no longer recommended and are rarely used today. They contradict the notion described in the previous section that bicycles fare best when they act and are treated

---

[*]Readers are cautioned that rumble strips, while shown to improve run-off-the-road motor vehicle safety, are dangerous to cyclists, particularly on downgrades, where cyclists can lose control. Several states have tested and adopted more bicycle-friendly rumble strips that include gaps every 60 to 100 feet to allow cyclists to exit the shoulder for movements such as left turns. The type and size of the rumble strip is also critical for bicycle safety.

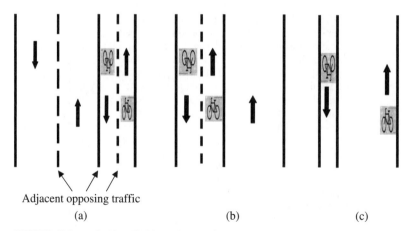

Adjacent opposing traffic

(a)   (b)   (c)

FIGURE 17.6 (*a*) Problematic bicycle lane configuration; (*b*) One-way street; (*c*) Contra-flow bicycle lane.

as vehicles. In a two-way bicycle lane scenario (such as shown in Figure 17.6*a*) the number of opposing traffic steams that are adjacent to each other increases. The predictability and visibility of the wrong-way rider is compromised at driveways and intersections. In some cases two-way bicycle lanes are provided on one-way streets, as shown in Figure 17.6*b*. This type of design can provide an important link in the bicycle network. In cases where space on the one-way street will only allow for one bicycle lane, a contra flow lane can be provided for bicycles as shown in Figure 17.6*c*. In this case, the bicycles traveling the same way as the motorized vehicular traffic use the traffic lane. This is confusing to some inexperienced cyclists who try to use the bicycle lane in the wrong direction. Arrows (templates provided in the *MUTCD*) should be used to indicate where bicycles should travel. In some cases, bicycle lanes may be placed on the left edge of a one-way street if the right edge is being used for bus stops or another conflicting activity.

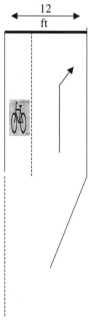

FIGURE 17.7 Shared right-turn and bicycle lane.

In addition to signage and direction arrows, a painted bicycle logo is recommended for placement in bicycle lanes. In previous years, different jurisdictions have had various ways to indicate bicycle lanes, including the use of the words "bike lane." There is now widespread support for the bicycle template as a common nationwide symbol.

The space at intersections is often limited, and therefore striping of bicycle lanes on intersection approaches and through intersections is not straightforward. Some jurisdictions elect to discontinue the bicycle lane at or before intersections. Others stripe the lane through using broken stripes to indicate where right-turning vehicles can cross the bicycle lane. In some places, a shared right-turn bicycle lane (shown in Figure 17.7) has been tested (Hunter 2001), but this design is not yet included in the AASHTO guide.

The road diet transformation illustrated in Figure 17.8 is increasing in popularity throughout the country. In many cases, it is the bicyclists or planners who put forth this idea when a repaving project is considered or when left-turn traffic volumes become problematic. When at least 48 feet of curb-to-curb distance is available on the originally four-lane road, the option of five narrow motorized traffic lanes is also considered (four

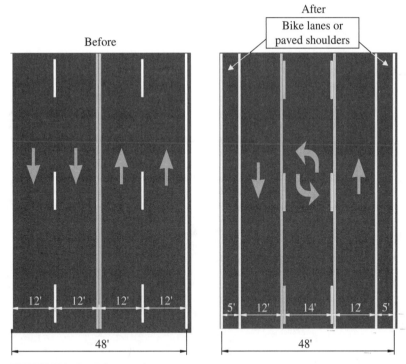

FIGURE 17.8 Road diet example.

through lanes and one two-way center left-turn lane). Instead, the road diet consists of replacing the four lanes with three standard-width travel lanes with two bicycle lanes. To the public and media eye, the transformation often seems to consist of only removing a traffic lane and adding bicycle lanes to the street. But many benefits beyond simply providing good standard bicycle lanes come with the road diet. It removes motorized traffic from proximate fixed objects on the edges of the roadway, provides standard wide travel lanes for trucks and buses, and increases turning radii at intersections and driveways. It also provides a buffer for pedestrians if the sidewalk is immediately behind the curb. Admittedly, the road diet also has both the pros and cons of adding a two-way left-turn lane. The road diet does not necessarily reduce the capacity of the roadway. Capacity changes depend on the volume of turns, particularly left turns. In some cases, these transformed roadways carry over 20,000 vehicles per day.

17.7 DESIGNING SHARED ROADWAYS

Bicycle lanes are, at the current time, widely considered the desirable means to accommodate cyclists on roadways. They make bicycles a visible part of the transportation system even when bicycles are not present due to the lane markings and signage. Bicycle lanes fit better with the rules and conventions of the road network, avoiding the complications often encountered when connecting shared-use paths to the road network. Preliminary safety findings (Clarke and Tracy 1995; Moritz 1998) are positive. However, it is unreasonable to expect that all roads can or should have bicycle lanes, particularly in existing developments, older cities, or completely rural areas. This section of the chapter discusses the use of shoulder bikeways and wide curb lanes as a compromise between true bicycle lanes and no special treatment on roads. The subsequent subsection describes road features or attributes that can make a road good or bad for bicycling even when no special facilities are placed on a given road. The final subsection describes bicycle suitability measures that can gauge quality of service for a bicycle on a given road.

17.7.1   Shoulder Bikeways

Shoulder bikeways are particularly useful in rural areas where shoulders often exist and can be used for biking. Obviously a cyclist might choose to use a shoulder at any time, but designating a shoulder as bikeway requires that signage or another designation be provided. Care needs to be used when rumble strips are present, as noted in the previous section of this chapter. A higher level of responsibility is required for sweeping and maintaining a shoulder bikeway as the debris from the road is often swept by traffic air currents onto the shoulder. Ideally at least 4 feet is recommended by the AASHTO guide for a shoulder bikeway. This width should be increased to at least 5 feet if a guardrail* or other roadside barrier is present. When the shoulder area decreases in width due to bridges or other barriers, warning signs for the cyclists should be considered. A shoulder bikeway must be paved. However, even when it is paved, designers should note that some cyclists prefer to travel on the edge of the travel lane rather than the shoulder and are within their rights to do so. Debris in the shoulder can be a safety hazard, and on some racing or road bikes with smaller tires the risk of tire damage is great.

Shoulder bikeways are even used in some states on freeways or interstates. This requires appropriate consideration of rumble strips and interchanges but has been particularly successful in providing access in places were no other route exists for cyclists such as over rivers. In some cases, access to the freeway shoulder is provided immediately before and after the bridge, avoiding the need for cyclists to negotiate an interchange.

17.7.2   Wide Curb Lanes

Sometimes when restriping is considered for bicycle lanes there is simply not sufficient right-of-way to provide the required space for the bicycle lane but there is space to widen the curb lane. This wide curb lane could also be accomplished in combination with reducing the width of other travel lanes on a multilane roadway. At least 14 feet, and ideally 15 feet, is recommended for a wide curb lane (not including the gutter pan). This bicycle facility is often useful only to experienced cyclists who ride with traffic, but it is consistent with the philosophy of accommodating bicycles as vehicles within the road network and is therefore considered very positive by many bicycle advocates.

17.7.3   Road Attributes Requiring Special Attention for Bicycles

Whether shoulder bikeways or wide curb lanes are provided or not, certain road attributes require special attention on shared roadways to ensure safe and quality service is provided to cyclists. These issues are particularly important where a road is designated as part of a signed bicycle route. Many of these seemingly small details have been shown important to accommodating cyclists. A planner would want to consider these factors carefully for bicycle route networks that include shared roadways as one of the types of facilities connecting origins and destinations throughout an area.

Drainage grates are hazardous to bicycle tires and can cause damage or even crashes. While most jurisdictions have moved to replace grates, some have not, and one can argue that bicycle routes should have priority for replacement. Openings should not only be small but should be crossways to the direction of travel. The potholes and pavement problems that occur at drainage grates should also be considered when bicycle traffic is expected or encouraged.

---

*Designers should consider that guard rails or other barriers can present a hazard to cyclists if they crash into them. Cyclists have a high center of gravity and can flip over guardrails, meaning the conditions on the other side of the barrier should be considered.

Similar to drainage grates, railway crossings represent a hazard to cyclists because of the potential for a tire to be caught in the rail, causing the cyclist to fall or crash. Special textured mats exist to reduce the size of gaps where tires might get caught, but these are often only useful when the bicycle crosses the rails at a 90° angle. When non-90° crossings are required, special pavement widening should be considered as shown in Figure 17.9. Otherwise the cyclist may stray into vehicular traffic in order to try to cross the rails at 90°. Similar consideration should be made when shared-use paths cross rail lines.

The pavement surface quality and edge condition of the roadway are particularly important for cyclists. Uneven surfaces slow cyclists, damage bicycles, and can cause crashes. Cyclists are known to divert to alternative routes based on paving quality and surface conditions. If the edge of the roadway has a sharp drop-off, the effective area for cyclists is greatly reduced. Similar hazards are present on bridge decks and at bridge joints.

Actuated traffic signals can be a great frustration for cyclists who in many cases cannot trigger the signal and may be motivated to ignore it. Detection levels can be altered along bicycle routes or special bicycle detectors can be used. In many cities special paint markings indicate the location of extra wire coils in the inductive loops that can detect cyclists. Many new video and radar detectors do not have these limitations as long as they are aimed at the right edge of the road, where a cyclist is likely to be stopped.

One of the best ways to determine the safety hazards along roads or routes within an area is to hold public input meetings. The cyclists who use a route are the best source of information on needed changes. The designer or planner can then apply solid design principles to address the problems. In some cases, field inspections (Figure 17.10) can point planners to locations where routes are needed.

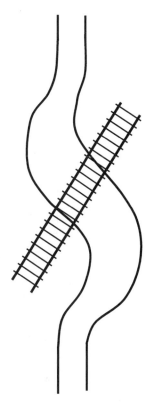

FIGURE 17.9 Widening for angled railway crossings.

FIGURE 17.10 Finding route locations by field inspections.

17.7.4   Bikeability Measures

One of the ways to evaluate the suitability of a road for bicycling is to use one of the formal evaluation tools developed over the last decade by researchers who were motivated by the need to move beyond pure subjective measurement of shared roadways for bicycle traffic. These objective measurement tools are important because planning processes and project prioritization often require solid grounds as a basis for decision making.

The bicycle level of service (BLOS) (Landis, Vattikuti, and Brannick 1997) was developed by having cyclists bike real-world routes and evaluate them. The characteristics or attributes of the route sections were recorded and used as potential explanatory variables. The resulting regression models include motor vehicle traffic volume per lane, motor vehicle speed, traffic mix, cross-traffic levels, pavement surface, and width for cycling. Given the prevalence of city geographic information systems (GIS) containing much of this information on a block-by-block basis, it has become common in some cities to evaluate the whole arterial and collector road network. Sometimes volunteers are used to fill data gaps. The results are used to plot color-coded maps of the suitability of roads for biking based on the BLOS and can be useful in planning connective routes. The BLOS is an interesting contrast to the LOS measures in the *Highway Capacity Manual* (TRB 2000), which are used for motorized traffic on all types of roads. These are typically based on road capacity and travel time or delay. The BLOS reflects the reality that the quality of a ride for a cyclist is based on perceived safety and comfort level; in most places bicycle capacity is not yet an issue.

A second objective measure is the bicycle compatibility index (BCI). This method was originally proposed by Sorton and Walsh (1994) and has been expanded using a larger sample of cyclists (Harkey, Reinfurt, and Knuiman 1998). In this case cyclists view road segments on videotape and evaluate the comfort level. The resultant model bases compatibility on the presence of a bicycle lane, its width, curb lane width, curb lane traffic volume, other lane traffic volume, traffic speed, parking, type of roadside development and adjustments for trucks, parking turnover, and right-turn volume. This measure can be used in the same way discussed above to label and plot block-by-block bicycle compatibility. Both models have been programmed into straightforward Excel-format spreadsheets.

17.8   DESIGNING SHARED-USE PATHS

Shared-use paths are important and popular facilities in a community not only for cyclists but also for other users, including pedestrians and bladers. Paths are particularly important in that they are a low-stress traffic environment, a place for children and novice riders, and can fill missing links in the bicycle network. The mix of users often requires establishment and promotion of trail user rules. However, as Morris (2002) summarizes, greenway paths or trails have been documented to offer much more to the community: healthy lifestyle promotion, economic development, historic preservation, utility corridors, and social interaction.

The basic segments of shared-use path design are straightforward, but the intersections, relationship to roadways, use of structures, and maintenance procedures can create challenges for the transportation or park agency. Many times a path is to be located along an old rail corridor[*] that affords adequate lateral space, limited grades, and existing infrastructure.

17.8.1   Basic Features

A basic shared-use path segment should be a minimum of 10 feet wide with 2 feet of graded shoulder (usually turf) and a minimum of 3 feet of clear zone to proximate objects such as signs, rocks, trees, or fences (AASHTO). A 2 percent minimum cross-slope for drainage is also recommended.

---

[*]Transportation planners might consider doing a background check on the legal requirements for rail banking or maintaining rail corridors for possible future use. In many jurisdictions the ownership and laws regarding abandoned rail lines complicate this process. The grass roots organization the Rails to Trail Conservancy can be of assistance in many cases.

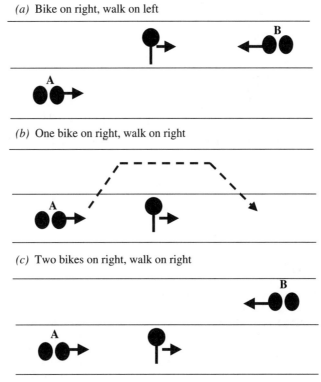

(a) Bike on right, walk on left

(b) One bike on right, walk on right

(c) Two bikes on right, walk on right

FIGURE 17.11   Demonstration of path/trail yielding scenarios.

Use of a striped centerline varies from region to region and is sometimes only used on horizontal curves, where it helps keep cyclists on the right side of the path. If possible, a center stripe should be applied to help with user rules (such as walk and ride on the right, warn before you pass). AASHTO recommends broken stripes where passing sight distance is sufficient; however, this author finds that the labor required to determine passing sight distances for the range of user speeds is not worthwhile. Regardless of whether a stripe is present, pedestrians should walk on the right to ensure the appropriate user is in the correct position to yield the right of way when meeting. Figure 17.11 illustrates this principle: when walkers facing traffic (bicycles riding on the right) see a bicycle approaching, their tendency is to step to the other side of the trail and get out of the way of the bicycle they *see* approaching (bicycle B) (Figure 17.11*a*). Unfortunately, if they do not look behind them, they may step into the path of another cyclist (A) (or other higher-speed user such as a skater). All trail users are quiet, especially on a paved trail, and this situation is likely. The reality that addresses this situation is for all users to be on the right-hand side of the trail in their direction of travel as in the second and third pictures of Figure 17.11. In these cases, the bicycle behind the pedestrian sees and must yield to the oncoming bicycle.

The horizontal curve design recommended by AASHTO for paths is based on a 15° lean angle for the bicycle. Bicycles can lean inward to counter the centrifugal force that pulls the bicycle or any vehicle outwards on a curve. To some extent this means bicycles can turn sharper turns than other vehicles, but the extent of leaning does depend on the skill level of the cyclist. Ultimately the limiting factor is pedal height. For a 20-mph design speed (as one might expect on a paved path) the minimum curve radius is 100 feet (27 meters). Note that path speed limits might be posted substantially lower, between 10 and 16 mph. Superelevation can be used in the same way it is on highways to reduce the required radius at the same design speed. However, designers should note that ADA requires that superelevation cannot exceed 2 to 3 percent. This can reduce the radius at 20 mph to 30 feet.

Grade or vertical curvature must be minimized on paths whenever possible in order to accommodate cyclists of all skill levels. Dismounting and weaving at slow speeds can cause path operational problems. The length of grade affects the maximum percent grade that can be used in the same way it affects highway grades for trucks. Grades of less than 5 percent can be used without length limitation and are recommended as the maximum for paved paths. A grade of 5 percent can only be used for a maximum of 800 feet and 10 percent for 100 feet (AASHTO). In many cases, the downgrade and stopping distance is a more important consideration, particularly if horizontal curves or intersections are located at the bottom of the grade. Cyclist dismount signs can be used in some cases. Stopping sight distance may still be a controlling criterion for the length of vertical curves, as can the lateral clearance on horizontal curves. These factors must be checked in the same manner as for highways. On a level surface the stopping sight distance can be up to 225 feet (75 meters).

17.8.2 Paths and Roads

Inevitably paths have to start at or cross roads, and design of these points is not straightforward. These points often require innovative design treatments as documented in several reports (e.g., Ridgway and Nabti 2002). There are simply too many different scenarios for standard design guidance to exist for all cases, and even then, new treatments are always being tested and evaluated. The basic rule of thumb is to design the path/road intersections in such a way that (1) cyclists are encouraged to follow traffic rules, (2) both cyclists and motorists are warned and anticipate the intersection, and (3) the unambiguous and consistent right-of-way is clearly communicated using signs and pavement markings. When it is appropriate to combine traffic calming into a project, raised path crossings are sometimes considered. The path crossing is typically striped across the roadway. In some places, modeling after European efforts, blue bicycle crossing areas have been tried. Whenever possible, paths should cross roads at a 90° angle. Sometimes curves can be added to the path before the road to ensure this design standard. Midblock crossings should be away from adjacent major intersections. A refuge area between opposing traffic direction is often included, especially on busier roads, but there must be enough space for several queued bicycles. Plantings and vegetation should be selected to ensure sight distance is not obstructed. Ideally crossings will be away from grades. Special care should be used to ensure cyclists are not required to stop too quickly following a downgrade or that they do not start on a significant upgrade. Traffic control, stop signs, or traffic signals are used for both the path and the road, as appropriate.

Paths that are parallel to roadways are particularly challenging, especially when driveways are involved. AASHTO recommends at least 5 feet between the adjacent road edge and the path, but this value is far less than is ideal for a two-way path. On a two-way path the wrong-way bicycle traffic is sometimes trapped between oncoming bicycles on the path and oncoming motor vehicles on the road, increasing the potential for head-on crashes. Furthermore, drivers exiting driveways or on intersecting streets are not expecting faster-moving traffic such as bicycles on their right (see Figure 17.12). Two-way paths adjacent to roads should only be used when driveways and intersecting roads are at a minimum. If cross-streets have even moderate traffic, the path can be pulled back a significant distance from the intersection (Figure 17.13). This simplifies bicycle turning movements and allows cross-bicycle traffic to move without interacting with intersection vehicular traffic. The path should either be close enough to the roadway intersection to be controlled and operated in an integral way or set back far enough that it can operate independently. Additional operational concerns arise when the entrance or start of these paths is located at intersections. The design is often difficult and sometimes requires odd or complex turns on the part of the cyclist (see Figure 17.14).

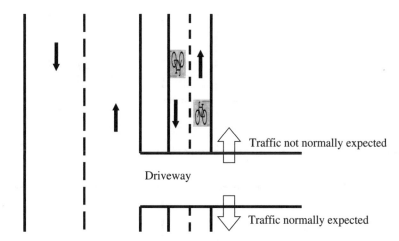

FIGURE 17.12 Entering driver expectations for intersecting traffic.

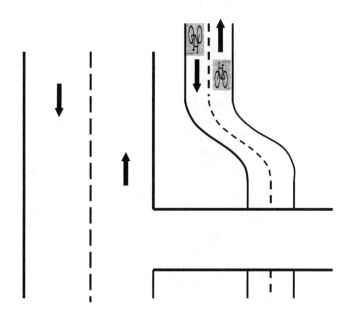

FIGURE 17.13 Shifting of shared-use path away from intersection.

17.8.3 Maintenance, Surface, and Structures

The paving system and structures for a shared-use path must be designed for an emergency vehicle and maintenance trucks. Failure to do so can mean lack of access or premature pavement failure as shown in Figure 17.15. Similarly, any barriers placed at the entrance to paths must be removable for emergency or maintenance access. Posts are often used with a minimum separation of 5 feet for

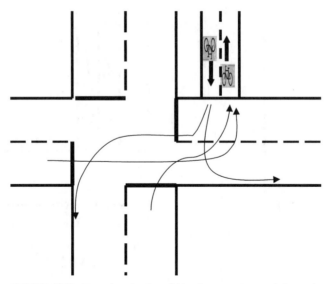

FIGURE 17.14 Examples of awkward bicycle movements needed at path-street junctions.

FIGURE 17.15 Rutting from maintenance vehicles on a shared-use path with substandard paving structure.

bicycle safety and a center post that is locked but removable. Caution should be used to ensure the posts do not constitute a fixed-object danger for cyclists. Some plastic options reduce this risk. Paint or thermal plastic warnings can be used. The separation used must be sufficient for bicycle trailers.

Bicycles are particularly sensitive to maintenance due to the lack of suspension systems and the high-pressure tires. Removal of the glass and debris that creates the risk of tire damage can be a frequent and labor-intensive task. The remote location of some trails makes them attractive for parties or loitering that lead to increased litter. Proper trash receptacles can help alleviate the problem. A well-populated path also helps keep "eyes" on the trail to discourage littering.

Another maintenance threat is creeping vegetation on the path edges that can narrow the path over time and affect safety. Many jurisdictions have programs to allow cyclists to report maintenance issues or other hazards along routes, including both paths and roads.

The decision whether to pave a path requires maintenance considerations as well. Unpaved paths have higher maintenance costs, including drainage and erosion problems. They do come with the added benefit of slowing bicycle traffic. Bladers are usually not found on unpaved paths, meaning that a surface treatment choice can actually impact demand and operating characteristics. Mopeds and horses, which make a poor mix with bicycles, might be found in different numbers on paved versus unpaved paths.

17.9 BICYCLE PARKING AND INTERMODALISM

Studies have shown that parking is important for encouraging bicycle use, especially for utilitarian or transportation trips (Goldsmith 1992). Parking areas must be lighted and visible to counter theft and promote safety for lone cyclists at night. Indoor parking is very desirable for the same reasons it is for motor vehicles: vehicles are protected from the elements and to some degree theft. In some locations, security is provided via video camera or personal attendant. The "Cadillac" of individual bicycle parking is the bicycle locker, which provides personal guaranteed parking for an individual at a given location and consists of a bicycle-sized garage-like shelter. In many areas these lockers have proven to be moneymakers.

Bicycle racks are the most simple and common form of bicycle parking. Many undesirable types of parking racks have been used over the years. The Association of Pedestrian and Bicycle Professionals has recently published a guide for bicycle parking (APBP 2002). They recommend invert U or "post-and-loop" racks and discourage older-style racks where only a portion of one wheel can be locked and holds the bike up. The APBP recommends that bicycles be considered effectively 6 feet in length and that 4 feet be provided between rows of parked bicycles (not the racks). The APBP further recommends that bicycle parking be placed within 50 feet or 30 seconds of a building entrance. In the past, many bicycle parking areas were located so remotely that they were not known to cyclists and therefore not used. Care must be used when placing bicycle parking along sidewalks. Not only does this situation encourage sidewalk riding for access, it often creates negative space issues and blockage on the sidewalk for pedestrians.

Provision of adequate and secure bicycle parking at transit stations has been shown effective in promoting bicycle intermodalism: the use of the bicycle in conjunction with other modes for a complete trip. Formal bike-and-ride programs are becoming more widespread. The use of bicycle racks on buses and the relaxing of rules regarding bicycles on trains and subways has resulted in the bicycle being a common part of intermodal passenger transportation. In some cities insufficient capacity for bicycles on trains or bus racks has become a problem. More people want to bike and ride than can be accommodated.

17.10 CONCLUSION

In summary, the bicycle is a flexible mode of transportation that can be used for a variety of trips within the intermodal transportation system by a wide range of people. Creative designs have been required in many cases to retrofit our auto-dominated road system to promote and encourage bicycling. However, over the last decade strong progress has been made throughout the United States to reintroduce well-engineered bicycle transportation facilities into cities, towns, and rural areas. A wide range of guidance is available for designers who may not be cyclists or bicycle facility designers to ensure safe, appropriate, and consistent bicycle routes, signs, and operations can continue to spread throughout the country.

REFERENCES

American Association of State Highway and Transportation Officials (AASHTO). 1999. *Guide for the Development of Bicycle Facilities.* AASHTO Task Force on Geometric Design, Washington, DC.

Association of Pedestrian and Bicycle Professionals (APBP). 2002. *Bicycle Parking Guidelines.* Washington DC: APBP.

Aultman-Hall, L., and F. L. Hall. 1998. "Ottawa-Carleton Commuter Cyclist On- and Off-Road Incident Rates." *Accident Analysis and Prevention* 30(1):29–43.

Aultman-Hall, L., and K. G. Kaltenecker. 1999. "Toronto Bicycle Commuter Safety Rates." *Accident Analysis and Prevention* 31:675–686.

Beneficial Designs Inc. 1999. *Designing Sidewalks and Trails for Access: Part 1 of 2 Review of Existing Guidelines and Practices.* Washington, DC: U.S. Department of Transportation.

———. 2001. *Designing Sidewalks and Trails for Access: Part 2 of 2 Best Practices Design Guide.* Washington, DC: U.S. Department of Transportation.

Bureau of Transportation Statistics (BTS). 1995. *Our Nation's Travel: 1995 NPTS Early Results Report.* U.S. Department of Transportation, BTS, Washington, DC.

City of Chicago. 2002. *Bike Lane Design Guide.* Pedestrian and Bicycle Information Center, City of Chicago, Chicagol and Bicycle Federation and Association of Pedestrian and Bicycle Professionals.

Clarke, A., and L. Tracy. 1995. *Bicycle Safety-Related Research Synthesis.* Report 94-062, U.S. Department of Transportation, Federal Highway Administration, Washington, DC, April.

Doherty, S., L. Aultman-Hall, and J. Swaynos. 2000. "Commuter Cyclist Accident Patterns in Toronto and Ottawa, Canada." *Journal of Transportation Engineering* 126(1):26–27.

Federal Highway Administration (FHWA). 1994. *The National Bicycling and Walking Study.* FHWA-PD-94-023, U.S. Department of Transportation, FHWA, Washington, DC.

———. 1995. *Bicycle Safety-Related Research.* Synthesis, U.S. Department of Transportation, FHWA, Washington, DC.

———. 2000. *Manual of Uniform Traffic Control Devices, Millennium Edition.* U.S. Department of Transportation, FHWA, Washington, DC.

Goldsmith, S. A. 1992. *Reasons Why Bicycling and Walking Are Not Being Used More Extensively as Travel Modes.* Case Study Number 1, National Bicycling and Walking Study, U.S. Department of Transportation, Federal Highway Administration, Washington, DC.

Harkey, D., D. Reinfurt, and M. Knuiman. 1998. *Development of the Bicycle Compatibility Index: A Level of Service Concept.* FHWA-RD-98-072, U.S. Department of Transportation, Federal Highway Administration, Washington, DC.

Hunter, W. 2002. "Evaluation of a Combined Bicycle Lane/Right-Turn Lane in Eugene, Oregon." In *CD Proceedings of the Transportation Research Board Annual Meeting.* Washington, DC: National Academy of Science, January.

Landis, B. W., V. R. Vattikuti, and M. T. Brannick. 1997. "Real-Time Human Perceptions: Toward a Bicycle Level of Service." *Transportation Research Record* 1578, 119–126.

Moritz, W. E. 1997. "A Survey of North American Bicycle Commuters: Design and Aggregate Results." *Transportation Research Record* 1578, 91–101.

Morris, H. 2002. *Trails and Greenways: Advancing the Smart Growth Agenda.* Washington, DC: Rails-to-Trails Conservancy.

Ridgway, M., and J. Nabti. 2002. *Innovative Bicycle Treatments.* Washington, DC: Institute for Transportation Engineering.

Sorton, A., and T. Walsh. 1994. "Bicycle Stress Level as a Tool to Evaluate Urban and Suburban Bicycle Compatibility." *Transportation Research Record* 1438, 17–24.

Transportation Research Board (TRB). 2000. *Highway Capacity Manual.* National Research Council, TRB, Washington, DC.

Wachtel, A., and D. Lewiston. "Risk Factors for Bicycle-Motor Vehicle Collisions at Intersections." *ITE Journal* (September): 30–35.

Wilkinson, W. C., A. Clarke, B. Epperson, and R. Knoblauch. 1994. *Selecting Roadway Design Treatments to Accommodate Bicycles.* Report No. FHWA-RD-92-073, U.S. Department of Transportation, Federal Highway Administration, Washington, DC.

CHAPTER 18
THE SPECTRUM OF AUTOMATED GUIDEWAY TRANSIT (AGT) AND ITS APPLICATIONS

Rongfang (Rachel) Liu
New Jersey Institute of Technology
Newark, New Jersey

18.1 INTRODUCTION

Automated guideway transit (AGT) is no stranger to the transportation community or to anyone who has recently traveled through large airports, visited cities with downtown people movers (DPMs), or vacationed at amusement parks where monorail trains shuttle visitors around the sprawling resorts. At the turn of the 21st century, while modern communication technology has brought consumers WiFi, Bluetooth, and other high-tech gadgets, the transportation community has its aspirations on the driverless transit: a spectrum of AGT systems, such as automated people movers (APMs), downtown people movers (DPMs), group rapid transit (GRT), driverless metros (DLMs), and personal rapid transit (PRT).

According to the latest tally (Fabian 2010), there are 150 applications along the various spectra of AGT technologies around the world. After more than four decades of emerging and developing processes, the AGT technology is no longer limited to airport use as shuttles or circulators. It has expanded to downtown and metropolitan areas as major activity center circulation and public transit systems. Another surging presence of AGT applications was observed in various leisure and recreation facilities and private and public institutions. As shown in Figure 18.1, overall, AGT applications are almost equally distributed among airports, urban centers, and institutions.

While the general public may be familiar with automated people movers (APMS) and label anything that moves people, including elevators and moving walkways, as people movers, the term preferred by transit professionals is *automated guideway transit* (AGT). Updating the original AGT definition (US Congress, Office of Technology Assessment 1975), AGT is defined as a class of transportation systems in which fully automated vehicles operate along dedicated guideways. The capacities of the AGT vehicles range from 3 or 4 up to 100 passengers. Vehicles are made of single-unit cars or multiple-unit trains. The operating speeds are from 10 to 35 miles per hour (mph), and headways may vary from a few seconds to a few minutes. The guideway system may be made of a single trunk route, multiple branches, or interconnected networks.

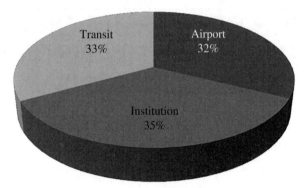

FIGURE 18.1 Distributions of APM applications. (*Source:* Based on data by Fabian 2010.)

18.2 DEFINITIONS

If the preceding definition is a simple, straightforward, and no-frills label for the AGT family, each individual member of the family deserves certain elaboration. Depending on the vehicle size, capacity, and other operating characteristics, AGT may be categorized into various subgroups, such as automated people movers (APMs), group rapid transit (GRT), and personal rapid transit (PRT). Different operating environments often give AGT applications generic names, such as airport circulators, downtown people movers, driverless transit (DLT), or driverless metros (DLM). Diversified track configurations, propulsion powers, and other technological features impart to AGT other names, such as monorail, duorail, and maglev, among others.

AGT systems are different from traditional heavy, light, and commuter rail transit in that they are operated via a central control system without drivers, conductors, or station attendants. Usually AGT systems use narrower right-of-ways, lighter tracks, if any, and smaller vehicles than traditional transit applications. The improved communication and control technology has enabled fully automated, driverless, fail-safe operations of modern AGT to satisfy wider ranges of capacity, spatial coverage, and temporal span of transit services.

As the first effort to provide a consensus definition for the spectrum of AGT technologies in the 21st century, this chapter will provide a brief review and some definitions for each subcategory of the AGT family. The main content of this chapter will focus on APMs and PRT, the two brightest stars of the AGT family. The APM category is important because of its multiple worldwide applications, and the ever-increasing PRT promises that it may be able to combine the advantages of both private automobiles and public transit.

18.2.1 Automated People Movers (APMs)

According to the General Accounting Office (GAO 1980), APMs are driverless vehicles operating on a fixed guideway. Vehicle capacities range up to 100 passengers, and vehicles may be operated as single units or as trains at up to 30 mph. *Headway*, the time interval between vehicles moving along a main route, varies from 15 seconds to 1 minute (Liu and Lau 2008). The realistic operating headway may be limited to 1 minute or more; even headways as short as 1 second were reported and tested in various pilot runs. The system is automated in that there are no drivers on board the vehicles or trains. The system is controlled or monitored by operators from a remote central control facility. Typically, the electromechanical design and physical characteristics of an APM are unique and proprietary to each manufacturer (Elliott and Norton 1999).

The first installation of an APM system at a major U.S. airport was at Tampa International Airport in 1971 (Lin and Trani 2000). Today, close to 50 APM applications can be seen at various airport facilities worldwide, carrying more than 1.6 million passengers daily. Figure 18.2 shows the APM

FIGURE 18.2   An example of airport APM, Newark Liberty International Airport, New Jersey.

system at Newark Liberty International Airport, one of the early APM applications at the airport, recently renewed and expanded.

18.2.2   Personal Rapid Transit (PRT)

If APMs occupy the larger-vehicle spectrum of AGT technology, we can easily place PRT at the other end of the spectrum—very small vehicles with a capacity of three to five persons per car or "pod." Based on definitions in various studies by several authors (Schneider 2008; Muller 2007; Koskinen, Luttinen, and Kosonen 2007; Cottrell 2006; Advanced Transit Association 1988), PRT can be a subcategory of AGT systems that offers on-demand, nonstop transportation using small, automated vehicles on a network of dedicated guideways with off-line stations.

Similar to APM applications, PRT operates automated vehicles along dedicated guideways. In contrast to APMs, PRT vehicles are designed for a single individual or a small group traveling together by choice on a network of guideways, and the trip is nonstop with no transfer. PRT stations are often off-line or bypasses of main lines, so vehicles stop only at their riders' final destination station. PRT trips typically are on-demand, and PRT vehicles or pod cars are supposed to wait at stations prior to the arrival of passengers.

There was strong interest in PRT in the United States during the 1970s (U.S. Congress 1975) when higher gasoline prices and congestion called for more efficient transportation solutions. A PRT application was promoted as the best solution to widespread urban problems in U.S. metropolitan areas. With a much narrower footprint in the right-of-way, smaller vehicles, lighter tracks, and tighter headways, PRT promised to provide a higher level of services with less expensive infrastructure. However, there are still no applications in the United States after its conceptual introduction more than four decades ago.

A number of factors, including political, economic, and technical, jeopardized the initial objectives of the demonstration project. The intended PRT demonstration project in Morgantown, West Virginia, turned out to be a group rapid transit (GRT) application with much higher cost and less

FIGURE 18.3 An example of PRT: ULTRa at Heathrow International Airport, London, United Kingdom. (*Source:* www.atsltd.co.uk, 2010.)

applicability in other places. Starting at the turn of the 21st century, some renewed interest in PRT has spurred a series of feasibility studies and a demonstration project in Heathrow Airport in London, United Kingdom, as shown in Figure 18.3.

18.2.3 Group Rapid Transit

After defining both ends of the AGT spectrum, APM and PRT, it is much easier to envision the middle child—*group rapid transit* (GRT)—which is similar to PRT but with higher-occupancy vehicles and grouping of passengers with potentially different origin–destination pairs. As noted in an early study (U.S. Congress 1975), the starting capacity for GRT is 6 passengers per car, whereas the upper limit is around 16 or 18; there are no clear distinctions between GRT and APMs in terms of vehicle capacities.

As the capacity difference blurs between APMs and GRT, it is possible for a GRT system to have a range of vehicle sizes to accommodate different passenger loading requirements. For example, at different times of the day or on routes with less or more average traffic, a GRT system may constitute an "optimal" surface transportation routing solution in terms of balancing trip time and convenience with resource efficiency. On the other hand, the dynamic coupling may bring complications to the operation processes that have to be evaluated for tradeoffs based on individual entities when such demand arises.

The Morgantown application in West Virginia should be correctly classified as GRT, and it is the only GRT application in the world, even though it is often mislabeled as PRT or an APM. As shown in Figure 18.4, the Morgantown GRT vehicle has seats for 8 people and some room for standees. The cars run on rubber tires in a U-shaped concrete guideway that has power and signal rails along the inner walls. The system is fully automated and does not require human drivers. There are three intermediate stations. Each station has several platforms and also "express tracks" that bypass the station completely.

The Morgantown GRT does not meet the qualifications of PRT because it does not provide nonstop services for a small group of passengers. It does not belong to the APM group, in a more strict definition, owing to its smaller vehicles with limited capacities. Whereas PRT provides nonstop service, GRT carries a larger group of people and stops at multiple requested destinations. In a perfectly parallel world, PRT can be compared with a taxi and an APM with a bus, which leaves no option but to put GRT into the *paratransit* group (Panayotova 2009).

FIGURE 18.4 An example of GRT, Morgantown, West Virginia. (*Source:* West Virginia University 2009.)

18.3 HISTORICAL DEVELOPMENT

Despite the many versions of how AGT systems began, the widely accepted origin of modern AGT has been documented definitively by Fichter (1964). After a brief review of "metropolis centers" and their associated circulation challenges, Fichter introduced the concept of "individualized automated transit," that is, an automated "small car" operating along "small exclusive trafficways" within street right-of-ways. With these descriptions and elaborate vehicle control and network layout, a vivid idea of PRT was born in the United States in the 1960s.

Another group of AGT applications, such as Skybus by Westinghouse Electric Corporation and Peoplemover by Goodyear Tire and Rubber Company, was the baby step taken by the private sector toward modern types of APM applications. AGT technology gained momentum during the 1970s when the Urban Mass Transportation Administration (UMTA), the predecessor of the Federal Transit Administration (FTA) today, signed a contract with West Virginia University to construct the first automated guideway transit (AGT) in the United States (Schneider 1999). Since then, the true markets for driverless metros emerged and grew overseas, whereas airport applications blossomed in the United States.

The following section provides an overview of the historical development of DPMs, APMs, and PRT. While there are a large number of institutional AGT applications around the world, there is little information on the historical background owing to the small scale and private nature of the projects; some of the historical development was intertwined with the general development of AGT applications.

18.3.1 Downtown People Movers (DPMs)

Burdened by increasing transit operating deficits, traffic congestion, and associated air pollution problems, the UMTA turned its hope to the emerging technology, AGT, a future transportation promise. In 1971, the UMTA funded four companies at $1.5 million each to demonstrate its AGT development at a transportation exposition, TRANSPO 72, held at Dulles International Airport near Washington, DC. As a direct result of TRANSPO 72, a few AGT applications were acquired for airports and zoos.

FIGURE 18.5   Downtown people mover (DPM) in Jacksonville, FL.

In 1975, the UMTA announced its Downtown People Mover (DPM) Program and sponsored a nationwide competition among cities. The UMTA DPM Program offered federal funds for the planning, design, and building of AGT systems as part of the demonstration program. Motivated by the "free" money, the response was almost overwhelming. In 1976, after receiving and reviewing 68 letters of interest and 35 full proposals and making on-site inspections of the top 15 cities, the UMTA selected Los Angeles, St. Paul, Cleveland, and Houston as candidates to develop DPM applications (General Accounting Office 1980). As second-tier backup candidates, Miami, Detroit, and Baltimore were selected to develop DPMs if they could do so with existing grant commitments. Pressured by the House of Representatives and the Senate Appropriations Conference Committee, the UMTA included Indianapolis, Jacksonville, and St. Louis on the backup candidate list.

After many rounds of debates and discussions, most of the DPM selectees later withdrew from the program, but Miami, Detroit, and Jacksonville stayed the course and eventually built DPMs. Figure 18.5 shows the DPM network in Jacksonville, Florida. In retrospect, few would regard the UMTA's DPM program as a "success." Among all the three cities that implemented DPMs, Miami was often criticized for its higher initial unit costs. However, a recent examination (Cottrell 2010) indicated that its ridership and costs closely match the original forecast, especially after the network was expanded to connect with other transit systems as originally planned but implemented at a later stage.

18.3.2   Driverless Metros (DLMs)

While DPM and PRT applications have been riding the roller coaster of novelty thrills, government support, and disappointing implementations in the United States, DLM applications have quietly gained momentum overseas. The initial concept of a fully automated, integrated transit system in Lille, France, was conceived in 1971, almost at the same time that the UMTA initiated its DPM Program. The construction for the Lille Metro started in 1978, and the first line was inaugurated in 1983 (Net Resources International 2010). When the entire 13.5 kilometers of Lille Metro Line 1 was opened in 1985, the driverless transit linked 18 stations and operated between 5 a.m. and midnight with 1.5- to 4-minute headways. Today, the DLM in France covers an impressive 60 stations,

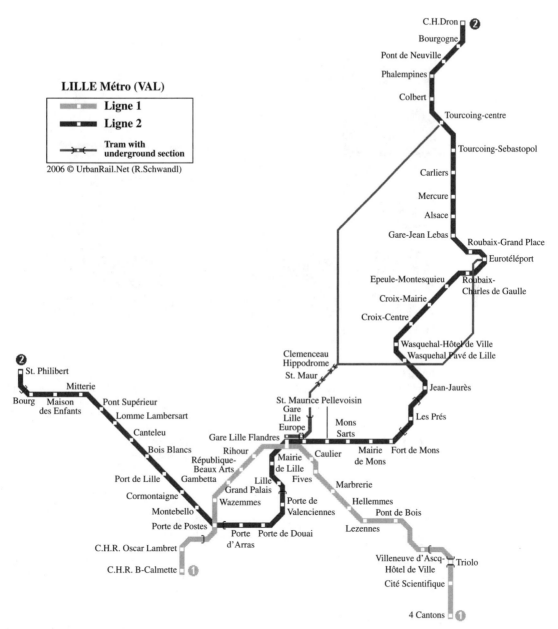

FIGURE 18.6   The Lille Metro Network. (*Source:* Schwandl 2006.)

expanding from Lille north toward the border of Belgium, as shown in Figure 18.6. Surprisingly, Lille's VAL system has run at a profit since 1989, and despite vandalism and concerns over personal safety, ridership figures remain healthy.

Not coincidentally, a fully automated GTS was initiated by our northern neighbor, in Vancouver, Canada, in the mid-1980s. The SkyTrain in Vancouver has three branch lines as of the end of 2009:

FIGURE 18.7   SkyTrain in Vancouver, British Columbia, Canada. (*Source:* Jun Suk 2007.)

Expo, Millennium, and Canada Lines. The Expo Line opened in late 1985 in time for the Expo 86 World's Fair (Economic Expert 2010); the Millennium Line opened in 2002; and the newest kid on the block, the Canada Line, opened in 2009 in time for the 2010 Winter Olympics. Together, the three branches of SkyTrain cover almost 60 miles of track that connects almost 50 stations. It provides easy and convenient access to Vancouver International Airport and two international border crossings. Although most of the system is elevated, hence the name SkyTrain, it runs as a subway through downtown Vancouver, as shown in Figure 18.7, and a short stretch in New Westminster.

18.3.3   Airport APMs

Although the birth of APM systems occurred in the 1970s, the past two decades may be labeled as the blossom period, when a large number of airport APM applications in North America were established. As shown in Figure 18.8, after long experience with only a couple of APM systems during the 1970s and 1980s, an increasing number of them have been installed since 1990 and the new millennium. A quick scan of projects under construction or in the planning stages in 2009 revealed that 10 new systems in the United States will have their inaugural runs in the next three years (Liu and Huang 2010).

Given the various sizes and diversified functions of APM services, the scale of airport APMs spans a wide range. As shown in Figure 18.9, the number of stations ranges from 2 to 16, and the length of each system stretches from 2 to 8.1 miles. Despite the turbulence in the airport market in the past decade, airport APMs have become an inherent part of airport expansion plans worldwide (Liu, Gambla, and Huang 2009).

Similar to transit operations, a number of APM systems at airports are operated by the contractors. Documented in a recent survey (Liu, Gambla, and Huang 2009), approximately two-thirds of APM systems at airports in North America are operated by the system suppliers or contractors, such as Otis Elevator Company, DCC Doppelmayer, and Bombardier Transportation, Inc. Only a third of the APMs at airports are operated by the owners of the systems.

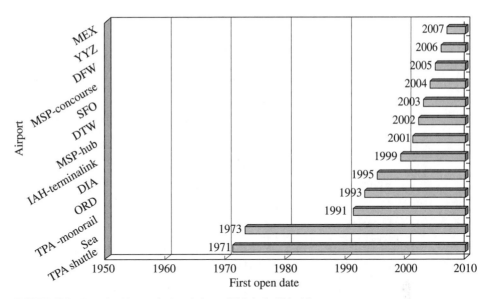

FIGURE 18.8   Operating history of selected airport APMs in the United States.

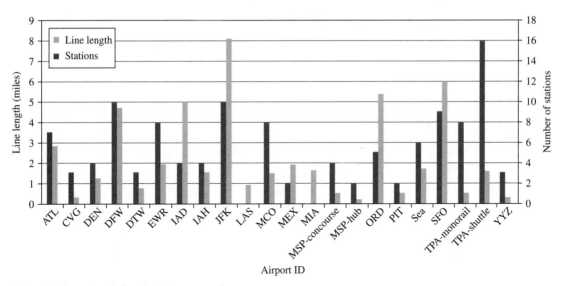

FIGURE 18.9   Scales of selected APM systems at airports.

18.3.4 Personal Rapid Transit

Personal Rapid Transit (PRT) has a long conceptual development history starting approximately in the 1950s. One of the early innovators was Donn Fichter, whose Veyar system was a very small one-person vehicle on a very lightweight elevated infrastructure. Its costs were low enough to enable more miles of network to be installed to make the system accessible to many using only captive single-mode cars. This system was not developed and was criticized for low capacity and an inability to handle group travel such as parents with children.

In 1953, a system called Monocab was conceived as a six-passenger vehicle system with the vehicle suspended below a monorail. This system went through development and sale from Vero, Inc., to Rohr Corporation and finally to Boeing. Monocab visualized a small guideway, but since vehicles were designed to be suspended, the guideway structure had to be high and required a cantilevered beam to displace the vehicles away from the vertical supports. The system later evolved into a magnetic levitation design with linear motor propulsion. Boeing developed this system further under an UMTA program that was initiated in 1964 and terminated in the mid-1980s.

Between 1968 and 1976, the Aerospace Corporation, a nonprofit federally funded research and development (R&D) center, studied the network layout, propulsion and control systems, safety, and traffic management issues and then progressed to experimental work in propulsion and control and ultimately to one-tenth-scale modeling of PRT (Irving 1978). The Aerospace system later was used as the basis for the Taxi 2000 system.

Efforts in Japan contributed to a computer-controlled vehicle system (CVS) that achieved 1-second headway and a 2,000-pound, four-passenger PRT concept. A German system developed by Messerschmitt-Bolkow-Blohm (MBB) and DEMAG called Cabintaxi supported one vehicle above and one vehicle below the guideway. The three-person vehicles ran on rubber tires and were propelled by linear induction motors. This system was licensed by Raytheon, and studies for its application in Indianapolis were conducted. Additional international development efforts were conducted in Canada, Australia, Sweden, and Great Britain (Anderson 1996).

As indicated in Table 18.1, numerous other early innovators, including General Motors, Raytheon, General Research Corporation, IBM, MITRE Corporation, Parsons Company, LTV Aerospace Corporation, Honeywell, Renault Engineering, Bendix, Ford Motor Company, and Otis Elevator Company, contributed to the development of PRT concepts. Many universities and research institutes also were engaged, including Massachusetts Institute of Technology, Johns Hopkins, Ohio State, University of Minnesota, San Diego State, Battelle Columbus Laboratories, Aerospace Corporation, Jet Propulsion Lab, and Booz-Allen Applied Research. Many of the early concepts were collected in proceedings of PRT conferences (Anderson and Romig 1974), but none found its

TABLE 18.1   Historical Development of Personal Rapid Transit

Year	Event	Major Players
1964	Book published: *Individualized Automated Transit in the City*, conceiving the idea of PRT.	Don Fichter
1967	Aramis Project started in Paris	MATRA
1968–1976	Book published: *Fundamentals of PRT*, documenting theoretical and system analyses of PRT	Aerospace Corporation
1972	Exposition of PRT systems, "Transport 72," held to show the benefits and features of PRT systems	Federal Department of Transportation
1972, 1973, and 1975	Three major international conferences held resulting in published proceedings	ASCE Automatic People Mover Committee
1974–1975	Morgantown GRT construction	Boeing
Middle of 1970s	Cabintaxi Project initiated in Germany	Mannesmann Demag
Early 1980s	Design of TAXI 2000 started and proposed to be implemented in Minnesota	Dr. Edward Anderson
1993–1999	Designed, developed, and tested PRT 2000	Raytheon Company
2002	Demonstration project launched for ULTra PRT system	Advanced Transport Systems
2003	Proposed PRISM program	Ford
2004	Feasibility study of PRT in Heathrow Airport	British Airports Authority
2010	Pilot operations in Heathrow Airport	Advanced Transport Systems

Source: Liu and Deng (2006).

place in a real-world commercial application until the installation of ULTra PRT at Heathrow Airport and its operational readiness trials in 2010 (Lowson 2010).

Will PRT be the next dominating transportation mode of the century? While quite a few dominant "authority figures" were quick to dismiss the PRT idea as "inherently unsound," the idea resurges every two decades or so, and there are currently more than one-half million entries on the Internet that are directly related to PRT. With an open mind and out-of-the-box vision, some transportation professionals believe that for PRT to become a reality, it may require a revolution in the way we live and travel. That is, PRT may not be feasible if highways and private automobiles continue to be our anchor mode of transportation in the near future. On the other hand, since our society has already spent billions of dollars and built millions of miles of roads and bridges in the past century and has not complained about the expenses but proudly claims them as civilization and engineering wonders, it may just be possible to layer PRT guideways on top of the existing roadway networks and replace private automobiles with automated PRT pods.

Others who seek more progressive solutions believe that PRT is capable of adapting to existing patterns of living and working, whereas line-haul transit is only efficient in corridor developments. In a large number of metropolitan areas around the world, urban roads are already congested, and land availability and cost forbid any road expansions. With a much smaller footprint and a fraction of life-cycle costs of conventional transit such as LRT, subway, and commuter rail, PRT may be able to combine the benefits of both private automobiles and public transit by providing a no-wait, well-connected, origin-to-destination one-seat ride for most urban dwellers.

Practical engineers and rational planners understand that a single mode does not solve all the urban transportation problems; every mode has a place in the mobility spectrum. The applicability is influenced by a variety of factors, such as changing technology, economic conditions, development patterns, and social acceptance at particular times. Any entity that is contemplating the idea of PRT (or any other form of emerging technology) must undertake systematic research of the PRT technology itself and its advantages and disadvantages. A comparison must be made with other modes, such as GRT, APMs, LRT, or automobiles, as well as the costs and benefits to users and society at large.

An appropriate viability evaluation, however, should not be confined to technology alone. Market analysis, rider preferences such as mode split, given all the travel choices, and cost-benefit analysis also should be part of the viability analysis. Another important aspect to nurture a technology into fruition is the policy framework that will facilitate its implementation. Potential applications of the technology, engineering specifications, procedural implications, and marketing segmentation all should be examined.

18.4 TECHNOLOGY SPECIFICATIONS

Given the long and capricious development process of AGT technologies, it is critical to sort through the volumes of materials to extract accurate and reliable information for potential users. Development and deployment of AGT systems have been the goal of a number of investigators for over 40 years. Their design has been driven by the need to find a way to relieve urban congestion while reducing air pollution, minimizing dependence on oil, and reducing or eliminating the need for transit subsidies.

To reduce congestion, it is necessary to set aside characteristics of conventional transit to find such a solution and, without prejudice, seek to discover transit-system characteristics that would fulfill the desired needs. The new system has to be designed to minimize costs while maximizing ridership and meeting required levels of capacity, safety, reliability, security, and comfort with minimum energy use, less pollution, and integrated land use. A new system has to complement conventional transit systems and make them more effective.

Unlike traditional railroad and conventional transit, which are regulated or overseen by the Federal Railroad Administration (FRA) and the Transportation Safety Board (TSB), respectively, AGT applications do not have a clear jurisdiction in terms of safety oversight and enforcement. However, since AGTs are inherently complex systems that involve multiple interacting subsystems, new technology, and public safety, it is essential to establish minimum standards for their design, construction, operation, and maintenance.

Realizing the benefits of standardization to organizations that specify and procure APM systems, such as regulatory authorities, system suppliers, system operators, system users, and the general public, the American Society of Civil Engineers (ASCE) has taken the lead in developing "Automated People Mover Standards" (Committee of Automated People Mover Standards 2006). The ASCE standards include minimum requirements for design, construction, operation, and maintenance of APM systems, especially on the subject of the physical operating environment, system dependability, automatic train control, and audio and visual communications. The ASCE standards have no legal authority in their own right and have not been adopted by any authority that has jurisdiction over AGT applications; nevertheless, they serve as a general guideline for transportation professionals to plan, build, and manage AGT systems in the years to come.

18.4.1 Vehicles

AGT vehicles are fully automated, driverless, and either self-propelled or propelled by cables. The vehicle speed, capacity, and maximum train size are usually decided by the types of technologies selected. The typical airport APMs or DLMs have a capacity of 50 to 75 passengers depending on sitting arrangements and luggage characteristics. On the two ends of the spectrum, PRT or pod cars usually hold 3 or 4 people, whereas the airside four-car APM trains in the Hartsfield-Jackson Atlanta International Airport are capable of carrying 300 passengers. Figure 18.10 shows the range of vehicle capacity in the airport applications in the United States (Liu, Gambla, and Huang 2009).

The self-propelled APM vehicles are electrically powered by either direct current (dc) or alternating current (ac) from a power distribution subsystem, and small vehicles, such as PRT vehicle or pod cars, can be powered by an onboard battery, which may be charged along the guideway or traveling route. Cable-propelled vehicles are attached to cables and pulled along the guideway. Some cable systems have vehicles permanently attached, whereas the latest applications allow vehicles to be attached or detached depending on the operation needs.

AGT vehicles are usually equipped with thermostatically controlled ventilation and air-conditioning systems, automatically controlled passenger doors, a public address (PA) subsystem, passenger intercom devices, a preprogrammed audio and video message display unit, fire detection

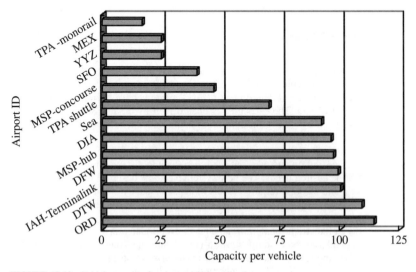

FIGURE 18.10   Vehicle capacity for airport APM applications.

and suppression equipment, seats, and passenger handholds. Some APM vehicles, especially in airport applications, are designed to accommodate luggage and luggage carts.

AGT vehicles can be supported by rubber tires, steel wheels, air levitation, or magnetic levitation. Detailed information on each system can be found in many engineering and technical papers (Anderson 1996; Vuchic 2007). Vehicle steering and guidance mechanics vary by technology. In general, steering inputs are provided to vehicle bogies through lateral guidance wheels or similar devices that travel in continuous contact with the guideway-mounted guide beams or rails. The steering inputs cause the bogie, usually located at both ends of each vehicle, to rotate so that vehicle tires do not "scrub" as they move through horizontal curves. Both central and side guidance mechanisms have been used by different manufacturers, and each type has its own unique characteristics.

18.4.2 Guideway

The guideway is another critical element of AGT applications. It is composed of the track or other running surface, including the supporting structure, that supports, contains, and physically guides AGT vehicles to travel exclusively on the guideway. Exclusive guideways can be constructed at grade, underground, or in an elevated structure. Depending on the selected technology, the supplier, and other considerations, the guideway for AGT may be constructed of steel or reinforced concrete. Figure 18.11 shows the elevated guideway for the APM in Pearson International Airport in Toronto, Canada. The size of the guideway structure varies with span length, train loads, and any applicable seismic requirements. Span length typically ranges from 50 to 120 feet.

The individual components of an AGT guideway usually include running surfaces, guidance and/or running rails, power-distribution rails, signal rails or antennas, communications rails or antennas, and switches. For technologies that employ linear induction motors for propulsion, guideway equipment also may include either reaction rails, also called *rotors*, or the powered element of the motor, also called the *stator*.

Another important segment of guideways is called *crossovers*, and they provide the means for trains to move between guideway lanes. A crossover for most rubber-tired AGT systems is generally composed of two switches, one on each guideway lane, connected by a short length of special track

FIGURE 18.11 Double guideway for the APM system at Toronto Pearson International Airport.

work. Crossover requirements vary significantly among AGT system suppliers, and each supplier's switch and crossover requirements are discrete in that their geometric and other requirements are largely inflexible.

Many APM guideway configurations have guideway switches that allow trains to switch between parallel guideway lanes or between different routes on a system (Vuchic 2007). Different AGT technologies have different types of switches, such as rail-like, side beam replacement, and rotary switches. Steel wheel/steel AGTs use rail switches, and the Siemens VAL systems use a slot-follower switch that is similar to a traditional rail crossover switch. On the other hand, PRT vehicles typically use onboard switching with no moving parts in the guideway.

18.4.3 Propulsion and System Power

Electric power is generally required to propel vehicles and energize system equipment. Propulsion and system power typically are configured such that system operating power will be supplied by power substations spaced along the guideway. The substations house transformers, rectifiers, and the primary and secondary switchgear power-conditioning equipment. The majority of AGT applications are either self- or cable-propelled.

Self-propelled. Self-propelled AGT systems may use electric traction motors or linear induction motors (LIMs). Self-propelled APMs are electrically powered by onboard ac or dc motors using either 750 or 1,300 V dc or 480 or 600 V ac wayside rail-based power-distribution subsystems. Self-propelled AGTs are not limited in guideway length. These technologies can be used for shuttle, loop, pinched-loop, and network guideway configurations.

Cable-propelled. Cable-propelled AGTs use a steel cable or "rope" to pull vehicles along the guideway. The cable is driven from a fixed electric motor driver located along the guideway. These technologies are usually used for shorter shuttle systems, typically for distances up to 4,000 feet. Onboard equipment power is usually provided by 480-V ac wayside power. There has been recent interest among airport owners/operators in reducing power requirements for APM systems. Heightened cost awareness, the variable price of energy, and a focus on sustainability have created a strong interest in lowering the power requirements around the airport, including the APM system.

One of the examples of cable-propelled APM applications is the Toronto Pearson International APM system. The cable wheel drive in the tensioning tower weighs about 75 tons and is powered by two 1,500-kW motors. As shown in Figure 18.12, the eight-strand galvanized "rope" runs in one continuous loop into the guideway. A fixed grip assembly forms the mechanical connection between the train and the cable, which is accelerated, decelerated, and stopped by a stationary machine-drive system. Should there be a power failure, an emergency diesel kicks in to provide power for communication and for heating, ventilating, and air conditioning.

18.4.4 Communications and Control

AGT applications are marked with automated central control and communication systems, which are different from conventional guideway transit, such as commuter rail and subways. All AGT applications use command, control, and communications equipment to operate the driverless vehicles. Each AGT supplier, based on its unique requirements, provides different components to house the automatic train control (ATC) equipment. ATC functions are accomplished by automatic train protection (ATP), automatic train operation (ATO), and automatic train supervision (ATS) equipment.

ATP equipment functions to ensure absolute enforcement of safety criteria and constraints. ATO equipment performs basic operating functions within the safety constraints imposed by the ATP. ATS equipment provides for automatic system supervision by central control computers and permits manual interventions/overrides by central control operators using control interfaces.

The AGT system includes a communications network monitored and supervised by the central control facility (CCF). This network typically includes a station PA system, operation and maintenance

FIGURE 18.12   Cable propeller for the APM at Toronto Pearson International Airport.

(O&M) radio systems, emergency telephone, and closed-circuit television. The basis for many of these communications requirements is emergency egress codes such as National Fire Protection Association Code 130 (NFPA 2010). As shown in Figure 18.13, the CCF is the focal point of the control system and can vary in size from a simple room with one or two operator positions and a

FIGURE 18.13   Control and communications facility at Toronto Pearson International Airport.

minimum of computer and CCTV monitor screens to a large room with multiple operator and supervisor positions and a large array of screens and other information devices.

18.4.5 Stations and Platforms

Similar to conventional transit stations, AGT stations are located along the guideway to provide passenger access to the vehicles. AGT stations generally are equipped with automatic platform doors and dynamic passenger information signs, in contrast to conventional transit. The AGT stations typically have station equipment rooms to house command, control, and communications equipment. In addition to the automatic train doors, the station has doors that align with a stopped train, and the two-door systems operate in tandem. The automatic station platform doors provide a barrier between the passengers and the trains operating on the guideway. These doors are integrated into a platform edge wall, as shown in Figure 18.14. The barrier wall, door sets, and passenger circulation/queuing area within the AGT station and adjacent to the AGT berthing position are commonly referred to as the *platform*. A single station can have multiple platforms. The types of platforms used depend on the type of APM configuration, physical space constraints, and any passenger-separation requirements.

An examination of the roles each platform type is needed to determine the best configuration to suit a particular application in an airport environment. The station platform doors provide protection and insulation from the guideway, noise, heat, and exposed power sources in the guideway. The interface between the station platform and the AGT guideway is defined by the platform edge wall and automated station doors. Dimensions defining the minimum width of AGT platforms and stations are developed based on analyses that take into account train lengths of the design vehicle, reasonable allowance for passenger circulation and queuing at the platform doors and escalators, passenger queuing and circulation requirements based on ridership flow assumptions, and reasonable spatial proportions and other good design practices. Stations for airport APMs typically are "on-line" with all trains stopping at all stations.

FIGURE 18.14 APM station at DFW airport. (*Courtesy of Marilyn Armardi, SkyLink.*)

Spatial, temporal, and institutional issues associated with intermodal connections are important for any transit services but are even more predominant in AGT stations. Dynamic passenger information signs typically are installed above the platform doors and/or suspended from the ceiling at the center of the station to assist passengers using the system. These dynamic signs provide information regarding train destinations, door status, and other operations.

18.4.6 Maintenance and Storage Facilities

Another key element of the AGT system is the maintenance and storage facility (MSF), which provides a location for vehicle maintenance, storage, and administrative offices. The maintenance functions include vehicle maintenance, cleaning, and washing; shipping, receiving, and storage of parts, tools, and spare equipment; fabrication of parts; and storage of spare vehicles.

The MSF typically is located away from the operating alignment in the larger AGT systems, such as airport circulation systems, DPMs, or DLMs. Vehicle testing and test track functions generally are performed on the guideway approaching the MSF when the facility is separate from the operating guideway. Simple, smaller shuttle systems often have the MSF located "under" one of the system stations or toward the end of the operating alignment. Figure 18.15 is an example of an MSF located at the end of the operating alignment at Pearson International Airport in Toronto, Canada.

FIGURE 18.15 Maintenance and storage facility at Pearson International Airport in Toronto, Canada.

18.5 AGT APPLICATIONS

The current status of AGT and its applications are critical to understanding the market need for AGT technologies. A number of feasibility studies of PRT and various stages of APM planning, design, and operating projects may serve as theoretical and practical laboratories to examine various aspects of AGT technologies and their respective successes and failures in meeting the particular travel needs of various markets. A few selected case studies are presented in this section.

18.5.1   Dallas–Fort Worth Airport APM

The Dallas–Fort Worth International Airport (DFW), located between the cities of Dallas and Fort Worth, is the world's fifth busiest airport in terms of passenger enplanements. More than 65 percent of airport usage comes from connecting passengers who usually need to travel between terminals to make their transfers. In order to accommodate the needs of growing international travel demand and to limit transfer time to 30 minutes or less for en-route travelers, DFW chose an APM as an ideal connecting mode among all terminals of DFW International Airport (Corey 2005).

At a price tag of $847 million, SkyLink, a fully automated people mover (APM) system, was constructed at the DFW International Airport and opened in the spring of 2005. As one of the largest high-speed airport train systems in the world, SkyLink features a 4.7-mile double-loop bidirectional guideway that connects six terminals, including a future one, within 8 minutes. With average travel speed in the range of 35 to 37 mph, SkyLink carries more than 3,000 passengers each hour in each direction. Figure 18.16 shows the SkyLink network layout.

18.5.2   Air Train at JFK Airport

The reason for the inclusion of an Air Train at JFK International Airport in this chapter is its unique combination of airport APM and urban metro in one AGT technology application. The John F. Kennedy (JFK) International Airport, located in New York City, is the busiest international air passenger gateway to the United States (Bureau of Transportation Statistics 2006). The Air Train at JFK

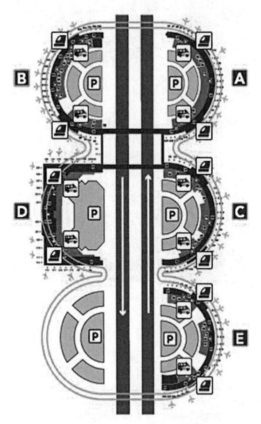

FIGURE 18.16   SkyLink at DFW International Airport.
(*Source:* Dallas–Fort Worth International Airport 2008.)

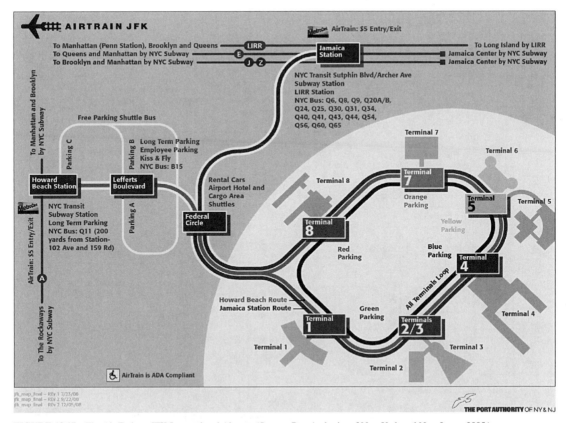

FIGURE 18.17   The Air Train at JFK International Airport. (*Source:* Port Authority of New York and New Jersey 2005.)

International Airport is a completely automated guideway transit system that connects JFK to its adjacent cities. The 8.1-mile-long APM, which cost $1.9 billion, began construction in 1998 and eventually opened in December 2003. It has 10 stations with a 1.8-mile airport circulator loop and two extensions to urban transit systems that equal 6.3 miles.

The Air Train uses AGT technology from Bombardier, and the capacity of the trains ranges from one to four cars with 75 to 78 passengers per car. The headway of the train is approximately 10 minutes, taking about 2 minutes between terminals. The Air Train is comprised of three main routes: All Terminals Route, Howard Beach Route, and Jamaica Station Route. As shown in Figure 18.17, the All Terminals Route is a circle route that connects all six terminal stations. The Howard Beach Route and the Jamaica Station Route connect the terminals and the regional mass-transit hubs, such as the New York urban subway and the Long Island Railroad (LIRR) stations.

18.5.3 Detroit Downtown People Mover

The Detroit DPM opened its service in 1987 with a fully automated guideway transit system operating on an elevated 2.9-mile single-track loop connecting 13 stations through the central business district of downtown Detroit. As shown in Figure 18.18, eight of the 13 DPM stations in Detroit are connected by preexisting structures, over 9 million square feet of commercial and office space, such as the Renaissance Center that houses General Motors Corporation's headquarters. The DPM enables office workers, shoppers, and visitors to travel in the downtown area with great ease. With headway of 3 or 4 minutes, the Detroit DPM takes 15 minutes to traverse the entire loop (Sullivan et al. 2005).

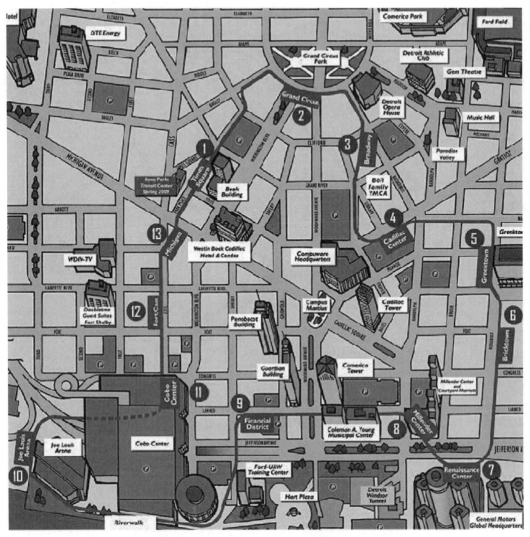

FIGURE 18.18   Detroit downtown people mover (DPM) network. (*Source:* Detroit Transportation Corporation 2010.)

18.5.4   Kuala Lumpur DLM in Malaysia

If there were a competition for AGT applications, Kuala Lumpur easily would win the crown because it has two separate AGT applications, Kelana Jaya and KL Monorail, in the same urban area. Kelana Jaya is ranked as the world's third longest fully automated DLM system at 18 miles, after the SkyTrain in Vancouver, Canada, and the Lille Metro VAL in Lille, France. Catching the expansion wave of DLMs in the world, Kelana Jaya will be extended 17 kilometers (10.6 miles) with 13 new stations by the end of 2010 (Bavani 2009).

Before 2010, there were only two-car trains operating on the Kelena Jaya Line. To increase the capacity during morning peak hours, the trains are currently running in a mixed fleet of two- and four-car train sets. The average frequency of the trains in the daytime is 4 to 7 minutes and 14 minutes after 10:00 p.m. A four-car train is able to carry up to 800 passengers at a time, with each car having a capacity of about 200 passengers. Kelana Jaya uses the newest version of Bombardier Transportation's Advanced Rapid Transit Mark II driverless train, and the maximum operating speed of such a train is about 50 mph (80 kmph) (RapidKL 2009).

In addition to the Kelang Jaya Line, the KL Monorail is the other automated transit system in the Kuala Lumpur Rail Transit System. The KL Monorail is 8.6 kilometers (5 miles) long with 11 stations. The KL Monorail had its maiden voyage in August 2003 with two parallel elevated tracks. It connects the Kuala Lumpur central transport hub with the "Golden Triangle." Two of the 11 stations run on a single track, and four of those serve as interchanges to enable passengers from the Ampang Line or the Kelana Jaya Line to transfer freely. Figure 18.19 shows both AGT trains in Kuala Lumpur.

FIGURE 18.19 Automated guideway transit in Kuala Lumpur, Malaysia. (*Sources:* Yosri 2005; Teo 2009.)

18.5.5 The Las Vegas Monorail

The Las Vegas Monorail is located on the Las Vegas strip, with a total length of 3.9 miles. Running along an elevated guideway with an average height of 30 feet, the Las Vegas Monorail connects Sahara Station in the north end with MGM Grand Station in the south (Stone, Banchik, and Kimmel 2001). The system supplier is Bombardier Transportation, and the alignment is based on an existing monorail between the MGM Grand and Bally's. Opened in July 2004, the seven stations along the monorail provide easy access to several world-class resorts, hotels, and the Las Vegas Convention Center, as shown in Figure 18.20. As the nation's first fully automated urban monorail transit system, the Las Vega Monorail bears a price tag of $650 million, which was completely funded by private entities (Snyder 2005).

Besides the Las Vegas Monorail, there are three more tram or people mover applications along the Las Vegas strip alone. As shown in Figure 18.20, the Mirage–Treasure Island Tram shuttles between the two namesake hotels between 7 and 2 a.m. The Bellagio–City Center–Monte Carlo Tram connects a few more hotels in the north-south direction on the west side of the strip. The Mandalay Bay–Excalibur Tram is completely indoors, connecting the two main hotel resorts via Luxor, another large resort along the Las Vegas strip. With great variations of technology and disconnected alignment, the Las Vegas Monorail and Tram trains may serve as a showcase of AGT applications at best. It would be ideal if some coordination or integration was carried out during the development processes so that the AGT applications could form an integrated transit system with coordination and connection.

The piecemeal development of AGT in Las Vegas underlines a very important issue with AGT development: its coordination and interaction with other modes. As documented by previous studies (Liu, Pendyala, and Polzin 1997; Liu 1996), time loss and frustration associated with transfers between modes or even between vehicles within the same mode are a major impetus that discourages transit use. A true DLM can only establish its market when the transfer impetus is minimized, and travel time and reliability are superior to that of private automobiles.

18.5.6 ULTra PRT at Heathrow Airport

As the main international airport in the United Kingdom and one of the busiest in the world, London Heathrow Airport was committed to the world's first PRT to provide key connectivity for the airport in 2005. As a pilot scheme, the initial application of ULTra PRT in Heathrow was designed to connect

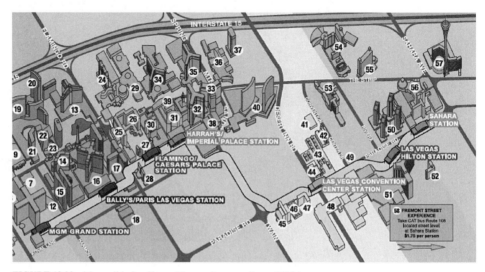

FIGURE 18.20 Monorail in Las Vegas. (*Source:* lvmonorail.com 2010.)

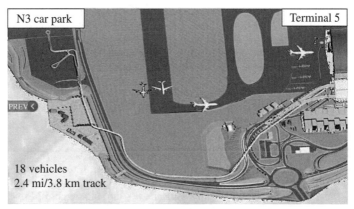

An early design of the 3-station Heathrow pilot system.

FIGURE 18.21 ULTRa PRT at Heathrow International Airport, London, United Kingdom. (*Source:* www.atsltd.co.uk 2010.)

the Terminal 5 building with a commercial parking lot to exploit the opportunities PRT may offer. The PRT service is designed to dramatically reduce the time that passengers need to move from their parked car to check-in counters.

Starting at a small testing scale, the initial ULTra PRT system has a 3.9-kilometer (2.4-mile) single guideway that connects three stations. The ULTRa PRT fleet is made up of 21 vehicles, and total travel time between the two terminals is about 5 minutes. The small footprint of PRT applications is well suited in this particular location because the current alignment traverses two rivers, seven roads, and green-belt land, not to mention negotiating aircraft surfaces and bridge in-ground services while conforming to the Terminal 5 architecture and appearance styles, as shown in Figure 18.21. If it proves successful, the owner of the Heathrow Airport has plans to expand the PRT application into a full network across the north side of the airport and into a newly developed Central Terminal Area (Lowson 2010).

The ULTra pods are battery powered and can hold four adults and two children including luggage, with each pod controlled by an onboard computer with sensor systems. From the perspective of energy efficiency, it is supposed to save 70 percent of the energy compared with cars and 50 percent compared with traditional buses (Rodgers 2007). After completion of construction and installation, the ULTra PRT application in Heathrow Airport has gone through a series of safety and reliability tests, including

- Basic testing at the Cardiff test track
- System integration testing
- Single-vehicle testing
- Multivehicle testing
- Operational readiness

The last testing stage, operational readiness, was conducted at the beginning of 2010. Full-scale operation is expected in the early part of summer 2010, assuming that the current test goes well.

18.6 CURRENT STATES OF AGT DEVELOPMENT

Serving as critical links in many large airports, dense downtown areas, and major activity centers, AGT applications around the world have been performing the vital function of connecting passengers to and from their origins to their destinations every day. However, since most of the AGT applications are short in length, ranging from a few hundred feet to a few miles, generally confined to the environs of airports, and owned by private operators, their importance or vitality is often ignored or taken for granted. The state of research by the AGT community does not help either. As of today, there is no substantial research or performance measures to outline concrete benefit and costs of each AGT system, not to mention their significant impact on surrounding communities.

As stated in the GAO (1980) report almost three decades ago, which is still valid today, "better justifications" and "concrete performance measures" are needed for AGT to move forward as a viable transportation mode. This section presents a brief summary of the current status of AGT development and points out the directions it may take in order to evolve and persist as a viable and sustainable transportation mode.

18.6.1   Rapid Expansion in Airport Applications

As noted by many airline passengers, larger, higher-capacity APM applications have become the normal mode to serve busy and growing airports and "airport cities," which is a recent concept consisting of a number of logically combined elements that reinforce each other not only to guide travelers easily through the airport process but also to meet the individual needs of travelers to the extent possible. APMs are no longer relegated to the peak-hour ridership of a few thousands but a normal presence for airport systems that must carry 9,000 to 10,000 passengers per hour per direction (PPHPD) during peak hours (Lindsey 2001). Atlanta Hartsfield International Airport, Washington Dulles International Airport, and Dallas–Fort Worth International Airport are all operating along the high-capacity range.

Besides the widespread airport APM applications in the United States, as shown in Figure 18.22, more airport applications are springing up in many international airports around the world. For example, Beijing Capital International Airport opened its APM system in time for the 2008 Olympic Games. Mexico City International Airport, Charles De Gaulle International Airport in Paris, and Toronto Pearson International Airport all have just opened their APM systems within the past three years.

18.6.2   Renewed Interest in PRT Technologies

The die-hard ideas for developing PRT systems may be direct offspring of the need to explore sustainable alternatives to private automobiles. Impelled by modern communication and control technologies and "*Star Trek*" quality images, PRT applications that promises to reduce congestion and air pollution and provide minimum trip time point-to-point and nonstop service at any time of the day become increasingly real and appealing. With very short or zero wait times—the vehicles would wait for people rather than requiring people to wait for vehicles—the quality of service will mollify any stubborn opposition. However, the basic requirements for safety and reliability standards can only be tested and validated when a real application is put into place. Not many entities are brave enough to make the commitment before a concrete or comfortable cost range and related ridership estimates are provided, which again will be possible only with adequate scale and real-world applications.

A quick scan of the existing literature and ongoing PRT studies reveals that the specifications of technology and assessment of costs may be relatively straightforward, but quantifying benefits associated with the implementation of a transportation project and evaluating the market conditions are complex. There are a number of analytical tools to assign a dollar value to benefits; however, some impacts, such as congestion relief, safety improvements, and air quality improvements, are often difficult to quantify financially. Other qualities, such as aesthetic appearance, may not even be quantifiable. Environmental and societal impacts are often referred to as *external* effects of transportation activities because they are not reflected directly in monetary costs and benefits of project implementation. By externalizing these factors, cost-benefit analyses often do not capture the full value of beneficial impacts.

The most difficult task so far to convince decision makers or the public of the feasibility of PRT applications is the estimation of ridership in the absence of *revealed preference* (RP) data. While many studies may use *stated preference* (SP) data before a real-world application may take place, the biases associated with SP data are well known, and the discrepancies between the two are difficult to estimate. Further complicating PRT ridership estimates is the intermodal transfer penalties associated with its initial applications. When starting from a short segment, an individual corridor,

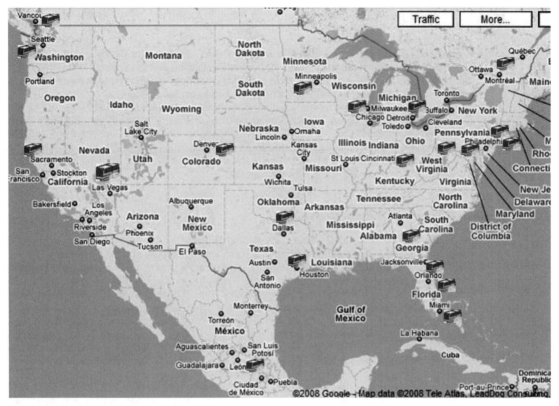

FIGURE 18.22   Widespread airport APM applications in America.

or some other type of limited scale, which is no different from the initial segments of any other transportation project, PRT may suffer from lower ridership owing to its limited network coverage and heavy intermodal penalties.

18.6.3 Evolving Performance Measures

As mentioned earlier, most of the APM systems are short in length, generally confined to the environs of airports, and owned by private operators of the various facilities. There is very limited research on APM systems, not to mention performance measures of such systems. The reason for this lack of linkage between performance measures of APM systems and demand is manifold. One of the key causes may be that the performance of APM systems is narrowly measured for existing systems. In many cases, performance measures of APM systems are specified in contract documents as a means of verifying the operator's compliance with contractual requirements.

Specified performance measures for airport systems typically are based on scheduled and actual operating data. The typical parameters generally are reflected by the following:

- Service mode availability
- Fleet availability
- Station platform door availability
- System service availability

Calculation of the first three items is simply based on the difference between the total operation periods minus the downtime period of each category. The last measure, system service availability, is the final product of the first three measures.

These availability measures are in wide usage in airport APMs worldwide. They provide the most appropriate level of accountability for the system and its elements that most directly affect the quality and level of service experienced by passengers. What is lacking in current performance measures is efficiency, balance between supply and demand, and customer satisfaction.

Given the shortcomings of the existing performance measures of APM systems, Airport Cooperative Research Program (ACRP) Project 03-07 has been charged with exploring and proposing a comprehensive performance measuring system for APM service at airports. After surveying the APM operators in North America and participating in extensive dialog with various stakeholders, the research team has developed a new set of performance measures to reflect a comprehensive evaluation of APM services in airports (Liu and Huang 2010).

18.6.4 Need for Diversified Business Models

As concluded in a recent PRT study (Carnegie and Hoffman 2007), AGT possesses the virtue of sustainability owing to its small footprint, lower cost, and lower impact on the environment. On the other hand, its small size and low-key profile have fostered a large number of applications worldwide without garnering any major headlines, which may, however, suppress its potential as a unique solution to urban circulation and congestion problems. Since the spectrum of AGT technology stems from various geographic, operational, and institutional settings, several business models should be developed and promoted to foster further development of the technology and to expedite the application processes (Liu, Nelson, and Lu 2009).

A set of business models will establish the core operational requirements essential to operate AGT systems safely and profitably. A business model can address the spatial, temporal, and institutional structures associated with implementation of AGT technology. It also will highlight strategies to deal with incremental risks and liabilities associated with new or expanded operations of AGT systems. Business models will identify potential funding sources, market-analysis procedures, and ridership shares when compared with other modes based on domestic and international experience.

ACKNOWLEDGMENTS

I would like to express my sincere appreciation to many people who contributed directly and indirectly to the fruition of this chapter. First, the discussion and dialog with members and friends of Transportation Research Board Committee (AP040), Major Activity Center Circulation Systems, when I first stepped in as the chair, have provided a rich background to shape up my general view of AGT family. Then I would like to express my gratitude to three people who are involved directly in two APM-related Airport Cooperate Research Program (ACRP) Projects: Lawrence Goldstein, senior program manager, Transportation Research Board, David Little, and Christopher Gambla, Lea + Elliot, project manager for ACRP 03-06, Guidebook for Planning and Implementing Automated People Mover Systems at Airports, and ACRP 03-07, A Guidebook for Measuring Performance of Automated People Mover Systems at Airports, respectively. The research projects, both near completion as of the writing of this chapter, undoubtedly will fill large voids in the current literature. They also directly benefited this chapter because I was a project member for ACRP 03-07 and maintained continuous communications with Larry and David on ACRP 03-06. Next, my appreciation goes to the few selected colleagues and friends who have painstakingly reviewed the manuscript and provided valuable suggestions and improvements: Ingmar Andreasson, Royal Institute of Technology, Sweden; Wayne Cottrell, associate of Advanced Transit; Lawrence Fabian, Trans21; and Shannon Sanders, Independent Architecture. Last but not least, I wish to thank my able assistant, Zhaodong (Tony) Huang, who has executed every task, from initial literature search to data analysis, from graph production to final formatting, with great care and patience. This chapter would not be possible without Tony's devotion and highest standard of working ethics. I have accumulated a large number of photographs, tables, and figures based on the information collected and research experience and have tried to provide appropriate credit to the maximum extent possible. I regret any errors or oversights in crediting any material, if any. Of course, any other errors, omissions, and oversights are my responsibility and will be corrected in the near future.

GLOSSARY

ac	alternating current
ACRP	Airport Cooperative Research Program
AGT	automated guideway transit
APM	automated people mover
ASCE	American Society of Civil Engineers
CCF	control and communications facility
CCTV	closed-circuit television
dc	direct current
DFW	Dallas–Fort Worth International Airport
DIA	Denver International Airport
DLM	driverless metros
DLT	driverless transit
DPM	downtown people mover
DTW	Detroit Metropolitan Wayne County Airport
FRA	Federal Railroad Administration
FTA	Federal Transit Administration
GAO	General Accounting Office
GRT	group rapid transit
IAH	Bush Intercontinental Airport
JFK	John F. Kennedy International Airport
kmph	kilometers per hour
LIM	linear induction motor
MEX	Mexico City International Airport
mph	miles per hour
MSP	Minneapolis–St. Paul International Airport
NFPA	National Fire Protection Association
ORD	O'Hare International Airport
PA	public address
PRT	personal rapid transit
SEA	Seattle/Tacoma International Airport (SEA-TAC Airport)
SFO	San Francisco International Airport
SP	stated preference
TPA	Tampa International Airport

TSB   Transportation Surface Board

ULTra   urban light transport

UMTA   Urban Mass Transportation Administration

YYZ   Toronto Pearson International Airport

REFERENCES

Advanced Transit Association. 1988. "Personal Rapid Transit (PRT): Another Option for Urban Transit?" *Journal of Advanced Transportation* 22:192–314.

Anderson, J. Edward. 1996. "Some Lessons from the History of Personal Rapid Transit." *Conference on PRT and Other Emerging Transit Technology*, Minneapolis, MN.

Anderson, J., and S. Romig. 1974. *Personal Rapid Transit II*. Minneapolis, MN: Audio-Visual Extension, University of Minnesota.

Bavani, M. 2009. "Thousand to be Benefit from LRT Extension"; available at http://thestar.com; accessed January 2010.

Bureau of Transportation Statistics, U.S. Department of Transportation. 2006. "U.S.-International Travel and Transportation Trends"; available at www.bts.gov; accessed January 2010.

Carnegie, J., and P. Hoffman. 2007. "Viability of Personal Rapid Transit in New Jersey." Project Report submitted to New Jersey Department of Transportation/New Jersey Transit. Committee of Automated People Mover Standards. 2006. "Automated People Mover Standards," PART 3. American Society of Civil Engineers, New York.

Corey, K. 2005. "DFW APM: Innovative Solutions to Success," in *Proceedings of the 10th International Conference on Automated People Movers*. American Society of Civil Engineers, New York, p. 14.

Cottrell, W. 2010. "CBD Circulators in Cities that Competed for Downtown People Mover Program Funding," in *Proceedings of the 89th Annual Conference of Transportation Research Board*. January.

Cottrell, W. 2006. "Moving Driverless Transit into the Mainstream: Research Issues and Challenges." Transportation Research Record No. 1955. *Journal of the Transportation Research Board*, pp. 69–76.

Dallas–Fort Worth International Airport. 2008. "SkyLink"; available at www.dfwairport.com/connect/index.php; accessed January 2010.

Detroit Transportation Corporation. 2010. " Interactive Map"; available at www.thepeoplemover.com; accessed January 2010.

Economic Expert. 2010. "Vancouver SkyTrain"; available at www.economicexpert.com/a/Vancouver:SkyTrain .html; accessed January 2010.

Elliott, D., and J. Norton. 1999. "An Introduction to Airport APM Systems." *ITE Journal* 33:35–50.

Fabian, L. 2010. "APM Statistics"; available at www.airfront.us/apmguide2008/data/; accessed January 2010.

Fichter, D. 1964. *Individualized Automated Transit and the City*. Chicago: B. H. Sikes.

General Accounting Office. 1980. "Better Justification Needed for Automated People Mover Demonstration Projects." U.S. General Accounting Office, Washington.

Irving, J. 1978. *Fundamentals of Personal Rapid Transit*. Lexington, MA: Lexington Books, 1978.

Jun Suk. 2007. "Sky Train"; available at www.panoramio.com/photo/6626821; accessed January 2010.

Koskinen, K., R. Luttinen, and L. Kosonen. 2007. "Developing a Microscopic Simulator for Personal Rapid Transit Systems," in *Proceedings of 86th Annual Conference of Transportation Research Board,* Washington.

Las Vegas Monorail Company. 2010. "About the Monorail"; available at www.lvmonorail.com/; accessed January 2010.

Lin, Y., and A. Trani. 2000. "Airport Automated People Mover System: Analysis with a Hybrid Computer Simulation Model." *Transportation Research Record* 1703, pp. 45–57.

Lindsey, H., and D. Little. 2001. "Driverless Rapid Transit Systems Take Hold." American Public Transportation Association Rail Transit Conference, Miami, FL.

Liu, Rongfang (Rachel), and Zhaodong (Tony) Huang. 2010. "System Efficiency: Improving Performance Measures of Automated People Mover Systems at Airports" (10-0835), in *Proceedings of Transportation Research Board (TRB) 89th Annual Meeting*, Washington.

Liu, Rongfang (Rachel), and Choikwan (Shirley) Lau. 2008. "Downtown APM Circulator: A Potential Stimulator for Economic Development in Newark, New Jersey," in *Proceedings of First International Symposium on Transportation and Development Innovative Best Practices (TDIBD) by America Society of Civil Engineers (ASCE)*, Transportation and Development Institute (T&DI), Beijing, China, April.

Liu, Rongfang (Rachel), and Y. Deng. 2006. "Research Need for Personal Rapid Transit (PRT) and Its Potential Applications," in *Proceedings of the 85th Annual Conference of Transportation Research Board*, January.

Liu, Rongfang, R. Pendyala, and S. Polzin. 1997. "An Assessment of Intermodal Transfer Penalties Using Stated Preference Data." *Transportation Research Record* 1607, pp. 74–80.

Liu, Rongfang. 1996. "Assess Intermodal Transfer Penalties." University of South Florida, Tampa.

Liu, Rongfang (Rachel), David O. Nelson, and Alexander Lu. 2009. "Business Model for Shared Operations of Freight and Passenger Services" (09-2547). *Transportation Research Record* 2547, pp. 86–92.

Liu, Rongfang (Rachel), Christopher Gambla, and Zhaodong (Tony) Huang. 2009. "Development of Performance Measures for Automated People Mover Systems at Airports" (P09-0442)." Transportation Research Board (TRB) 88th Annual Meeting, Washington.

Lowson, M. 2010. "Preparing for PRT Operations at Heathrow Airport" (10-3267), in *Proceedings of Transportation Research Board (TRB) 89th Annual Meeting*, Washington, January.

Muller, P. 2007. "A Personal Rapid Transit/Airport Automated People Mover Comparison," in *Proceedings of the 29th International Air Transport Conference*, American Society of Civil Engineers.

National Fire Protection Association. 2010. "NFPA 130: Standard for Fixed Guideway Transit and Passenger Rail Systems"; available at www.nfpa.org/aboutthecodes/AboutTheCodes.asp?DocNum=130; accessed January 2010.

NetResourcesInternational.2010."LilleVALAutomatedUrbanMetro,France";availableatwww.railway-technology .com/projects/lille_val/; accessed January 2010.

Panayotova. T. 2003. "People Movers: Systems and Case Studies"; available at www.facilities.ufl.edu/cp/pdf/ PeopleMovers.pdf; accessed October 2009.

Pickering, B. 1998. "Skyway Monorail, Jacksonville, FL"; available at http://world.nycsubway.org/us/jacksonville/; accessed January 2010.

Rapid, K. L. 2010. "Kelana Jaya Line"; available at www.rapidkl.com.my/network/rail/; accessed January 2010.

Rodgers, Lucy. 2007. "Are Driverless Pods the Future?" available at http://news.bbc.co.uk/; accessed January 2010.

Schneider, J. 1999. "A Brief History of UMTA's Downtown People Mover Program"; available at http://faculty. washington.edu/jbs/itrans/dpmhist.htm; accessed January 2009.

Schneider, J. 2008. "Personal Rapid Transit: Is This the Mode of the Future?" *Metro Magazine* 104(4):66–70.

Schwandl. R. 2006. "Lille Metro"; available at www.urbanrail.net/eu/lil/lille.htm; accessed January 2010.

Snyder, T. 2005. "Las Vegas Monorail Innovations," in *Proceedings of the 10th International Conference on Automated People Movers*, Orlando, FL.

Solis, P., et al. 2005. "Zero to Sixty: Managing the Design, Construction and Implementation of the World's Largest Airport Automated People Mover." Automated People Movers 2005, Moving to the Mainstream, 10th International Conference on Automated People Movers, American Society of Civil Engineers.

Stone, T., C. Banchik, and J. Kimmel. 2001. "The Las Vegas Monorail: A Unique Rapid Transit Project for a Unique City," in *Proceedings of the Automated People Movers: Moving Through the Millennium*, San Francisco, CA.

Sullivan, A., et al. 2005. "Detroit People Mover: Automatic Train Control Upgrade (ATCU) Project," in *Proceedings of the Automated People Movers: Moving to the Mainstream*, American Society of Civil Engineers.

Tadi, R., and U. Dutta. 1997. "Detroit Downtown People Mover: Ten Years After," in *Proceedings of the Sixth International Conference on Automated People Movers (APMs)*, American Society of Civil Engineers, pp. 134–142.

Teo, C. 2006. "KL Monorail"; available at http://en.wikipedia.org/wiki/; accessed January 2010.

The Port Authority of New York and New Jersey. 2005. available at www.panynj.gov/airports/jfk-airport-map .html; accessed in January 2010.

"Transportation: Monorails"; available at Lvmonorail.com; accessed January 2010.

"ULTra at London Heathrow Airport"; available at www.atsltd.co.uk; accessed January 2010.

U.S. Congress, Office of Technology Assessment. 1975. "Automated Guideway Transit: An Assessment PRT and Other New Systems," June.

Vuchic, Vukan R. 2007. *Urban Transit Systems and Technology.* Hoboken, NJ: Wiley.

West Virginia University. 2010. "W. Virginia University PRT System Back in Service"; available at www.metro-magazine.com; accessed January 2010.

Yosri. 2005. "Kelana Jaya Line"; available at http://en.wikipedia.org/wiki/; accessed January 201

CHAPTER 19
RAILWAY VEHICLE ENGINEERING*

Keith L. Hawthorne
Transportation Technology Center, Inc.
Pueblo, Colorado

V. Terrey Hawthorne
Newtowne Square, Pennsylvania

(In collaboration with E. Thomas Harley, Charles M. Smith, and Robert B. Watson)

19.1 DIESEL-ELECTRIC LOCOMOTIVES

Diesel-electric locomotives and electric locomotives are classified by wheel arrangement; letters represent the number of adjacent driving axles in a rigid truck (A for one axle, B for two axles, C for three axles, etc.). Idler axles between drivers are designated by numerals. A plus sign indicates articulated trucks or motive power units. A minus sign indicates separate nonarticulated trucks. This nomenclature is fully explained in RP-5523, issued by the Association of American Railroads (AAR). Virtually all modern locomotives are of either B-B or C-C configuration.

The high efficiency of the diesel engine is an important factor in its selection as a prime mover for traction. This efficiency at full or partial load makes it ideally suited to the variable service requirements of routine railroad operations. The diesel engine is a constant-torque machine that cannot be started under load and hence requires a variably coupled transmission arrangement. The electric transmission system allows it to make use of its full rated power output at low track speeds for starting as well as for efficient hauling of heavy trains at all speeds. Examples of the most common diesel-electric locomotive types in service are shown in Table 19.1. A typical diesel-electric locomotive is shown in Figure 19.1.

Most diesel-electric locomotives have a dc generator or rectified alternator coupled directly to the diesel engine crankshaft. The generator/alternator is electrically connected to dc series traction motors having nose suspension mountings. Many recent locomotives utilize gate turn-off inverters and ac traction motors to obtain the benefits of increased adhesion and higher tractive effort. The gear ratio for the axle-mounted bull gears to the motor pinions which they engage is determined by the locomotive speed range, which is related to the type of service. A high ratio is used for freight service where high tractive effort and low speeds are common, whereas high-speed passenger locomotives have a lower ratio.

---

*Reprinted from the First Edition.

TABLE 19.1 Locomotives in Service in North America

Builder	Model	Service	Arrangement	Weight min./max. (1,000 lb)	Tractive effort Starting[b] for min./max. weight (1,000 lb)	Tractive effort At continuous[c] speed (mph)	Number of cylinders	Horsepower rating (r/min)
Bombardier	HR-412	General purpose	B-B	240/280	60/70	60,400 (10.5)	12	2,700 (1,050)
Bombardier	HR-616	General purpose	C-C	380/420	95/105	90,600 (10.0)	16	3,450 (1,000)
Bombardier	LRC	Passenger	B-B	252 nominal	63	19,200 (42.5)	16	3,725 (1,050)
EMD	SW-1001	Switching	B-B	230/240	58/60	41,700 (6.7)	8	1,000 (900)
EMD	MP15	Multipurpose	B-B	248/278	62/69	46,822 (9.3)	12	1,500 (900)
EMD	GP40-2	General purpose	B-B	256/278	64/69	54,700 (11.1)	16	3,000 (900)
EMD	SD40-2	General purpose	C-C	368/420	92/105	82,100 (11.0)	16	3,000 (900)
EMD	GP50	General purpose	B-B	260/278	65/69	64,200 (9.8)	16	3,500 (950)
EMD	SD50	General purpose	C-C	368/420	92/105	96,300 (9.8)	16	3,500 (950)
EMD	F40PH-2	Passenger	B-B	260 nominal	65	38,240 (16.1)	16	3,000 (900)
GE	B18-7	General purpose	B-B	231/268	58/67	61,000 (8.4)	8	1,800 (1,050)
GE	B30-7A	General purpose	B-B	253/280	63/70	64,600 (12.0)	12	3,000 (1,050)
GE	C30-7A	General purpose	C-C	359/420	90/105	96,900 (8.8)	12	3,000 (1,050)
GE	B36-7	General purpose	B-B	260/280	65/70	64,600 (12.0)	16	3,600 (1,050)
GE	C36-7	General purpose	C-C	367/420	92/105	96,900 (11.0)	16	3,600 (1,050)
EMD	AEM-7	Passenger	B-B	201 nominal	50	33,500 (10.0)	NA[d]	7,000[e]
EMD	GM6C	Freight	C-C	365 nominal	91	88,000 (11.0)	NA	6,000[e]
EMD	GM10B	Freight	B-B-B	390 nominal	97	100,000 (5.0)	NA	10,000[e]
GE	E60C	General purpose	C-C	364 nominal	91	82,000 (22.0)	NA	6,000[e]
GE	E60CP	Passenger	C-C	366 nominal	91	34,000 (55.0)	NA	6,000[e]
GE	E25B	Freight	B-B	280 nominal	70	55,000 (15.0)	NA	2,500[e]

[a]Engines: Bombardier—model 251, 4-cycle, V-type, 9 × 10½ in. cylinders.
EMD—model 645E, 2-cycle, V-type, 9 1/16 × 10 in. cylinders.
GE—model 7FDL, 4-cycle, V-type, 9 × 10½ in. cylinders.
[b]Starting tractive effort at 25% adhesion.
[c]Continuous tractive effort for smallest pinion (maximum).
[d]Electric locomotive horsepower expressed as diesel-electric equivalent (input to generator).
[e]Not applicable.

19.2

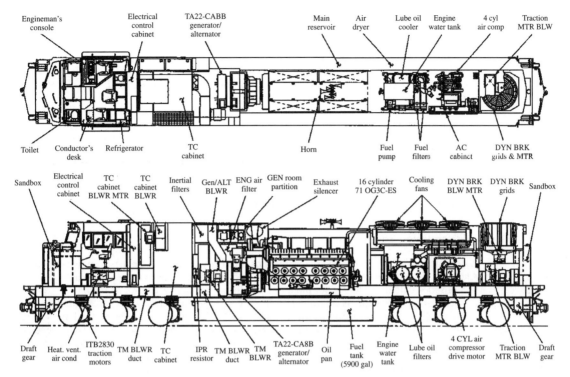

FIGURE 19.1 A typical diesel-electric locomotive. (*Source:* Electro-Motive Division, General Motors Corp.)

19.1.1 Diesel Engines

Most new diesel-electric locomotives are equipped with either V-type, two strokes per cycle, or V-type, four strokes per cycle, engines. Engines range from 8 to 20 cylinders each. Output power ranges from 1,000 hp to over 6,000 hp for a single engine application. These medium-speed diesel engines range from 560 to over 1,000 cubic inches per cylinder.

Two-cycle engines are aspirated by either a gear-driven blower or a turbocharger. Because these engines lack an intake stroke for natural aspiration, the turbocharger is gear-driven at low engine speeds. At higher engine speeds, when the exhaust gases contain enough energy to drive the turbocharger, an overriding clutch disengages the gear train. Free-running turbochargers are used on four-cycle engines, as at lower speeds the engines are aspirated by the intake stroke.

The engine control governor is an electro-hydraulic or electronic device used to regulate the speed and power of the diesel engine.

Electro-hydraulic governors are self-contained units mounted on the engine and driven from one of the engine camshafts. They have integral oil supplies and pressure pumps. They utilize four solenoids, which are actuated individually or in combination from the 74-V auxiliary generator/battery supply by a series of switches actuated by the engineer's throttle. There are eight power positions of the throttle, each corresponding to a specific value of engine speed and horsepower. The governor maintains the predetermined engine speed through a mechanical linkage to the engine fuel racks, which control the amount of fuel metered to the cylinders.

Computer engine control systems utilize electronic sensors to monitor the engine's vital functions, providing both electronic engine speed control and fuel management. The locomotive throttle is interfaced with an on-board computer that sends corresponding commands to the engine speed control system. These commands are compared to input from timing and engine speed sensors. The pulse width and timing of the engine's fuel injectors are adjusted to attain the desired engine speed and optimize engine performance.

One or more centrifugal pumps, gear-driven from the engine crankshaft, force water through passages in the cylinder heads and liners to provide cooling for the engine. The water temperature is automatically controlled by regulating shutter and fan operation, which in turn controls the passage of air through the cooling radiators, or by bypassing the water around the radiators. The fans (one to four per engine) may be motor-driven or mechanically driven by the engine crankshaft. If mechanical drive is used, current practice is to drive the fans through a clutch, since it is wasteful of energy to operate the fans when cooling is not required.

The lubricating oil system supplies clean oil at the proper temperature and pressure to the various bearing surfaces of the engine, such as the crankshaft, camshaft, wrist pins, and cylinder walls. It also provides oil internally to the heads of the pistons to remove excess heat. One or more gear-type pumps driven from the crankshaft are used to move the oil from the crankcase through filters and strainers to the bearings and the piston cooling passages, after which the oil flows by gravity back to the crankcase. A heat exchanger is part of the system; all of the oil passes through it at some time during a complete cycle. The oil is cooled by engine-cooling water on the other side of the exchanger. Paper element cartridge filters in series with the oil flow remove fine impurities in the oil before it enters the engine.

19.1.2 Electric Transmission Equipment

A dc main generator or three-phase ac alternator is directly coupled to the diesel engine crankshaft. Alternator output is rectified through a full-wave bridge to keep ripple to a level acceptable for operation of series field dc motors or for input to a solid-state inverter system for ac motors.

Generator power output is controlled by (1) varying engine speed through movement of the engineer's controller and (2) controlling the flow of current in its battery field or in the field of a separate exciter generator. Shunt and differential fields (if used) are designed to maintain constant generator power output for a given engine speed as the load and voltage vary. The fields do not completely accomplish this, thus the battery field or separately excited field must be controlled by a load regulator to provide the final adjustment in excitation to load the engine properly. This field can also be automatically deenergized to reduce or remove the load, when certain undesirable conditions occur, to prevent damage to the power plant or other traction equipment.

The engine main generator or alternator power plant functions at any throttle setting as a constant-horsepower source of energy. Therefore, the main generator voltage must be controlled to provide constant power output for each specific throttle position under the varying conditions of train speed, train resistance, atmospheric pressure, and quality of fuel. The load regulator, which is an integral part of the governor, accomplishes this within the maximum safe values of main-generator voltage and current. For example, when the locomotive experiences an increase in track gradient with a consequent reduction in speed, traction-motor counter-emf decreases, causing a change in traction-motor and main-generator current. Because this alters the load demand on the engine, the speed of the engine tends to change to compensate. As the speed changes, the governor begins to reposition the fuel racks, but at the same time a pilot valve in the governor directs hydraulic pressure into a load regulator vane motor, which changes the resistance value of the load regulator rheostat in series with the main-generator excitation circuit. This alters the main-generator excitation current and consequently main-generator voltage and returns the value of main-generator power output to normal. Engine fuel racks return to normal, consistent with constant values of engine speed.

The load regulator is effective within maximum and minimum limit values of the rheostat. Beyond these limits the power output of the engine is reduced. However, protective devices in the main generator excitation circuit limit the voltage output to ensure that values of current and voltage in the traction motor circuits are within safe limits.

19.1.3 Auxiliary Generating Apparatus

DC power for battery charging, lighting, control, and cab heaters is provided by a separate generator, geared to the main engine. Voltage output is regulated within 1 percent of 74 V over the full range of engine speeds. Auxiliary alternators with full-wave bridge rectifiers are also utilized for this application. Modern locomotives with on-board computer systems also have power supplies to provide "clean" dc for sensors and computer systems. These are typically 24-V systems.

Traction motor blowers are mounted above the locomotive underframe. Air from the centrifugal blower housings is carried through the underframe and into the motor housings through flexible ducts. Other designs have been developed to vary the air output with cooling requirements to conserve parasitic energy demands. The main generator/alternators are cooled in a similar manner.

Traction motors are nose-suspended from the truck frame and bearing-suspended from the axle (Figure 19.2). The traction motors employ series-exciting (main) and commutating field poles. The current in the series field is reversed to change locomotive direction and may be partially shunted through resistors to reduce counter-emf as locomotive speed is increased. Newer locomotives dispense with field shunting to improve commutation. Early locomotive designs also required motor connection changes (series, series/parallel, parallel) referred to as transition, to maintain motor current as speed increased.

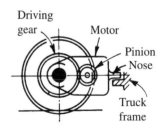

FIGURE 19.2 Axle-hung traction motor for diesel-electric and electric locomotives.

AC traction motors employ a variable-frequency supply derived from a computer-controlled solid-state inverter system, fed from the rectified alternator output. Locomotive direction is controlled by reversing the sequence of the three-phase supply.

The development of the modern traction alternator with its high output current has resulted in a trend toward dc traction motors that are permanently connected in parallel. DC motor armature shafts are equipped with grease-lubricated roller bearings, while ac traction-motor shafts have grease-lubricated roller bearings at the free end and an oil-lubricated bearing at the pinion drive end. Traction-motor support bearings are usually of the plain sleeve type with lubricant wells and spring-loaded felt wicks that maintain constant contact with the axle surface. However, many new passenger locomotives have roller support bearings.

19.1.4 Electrical Controls

In the conventional dc propulsion system, electropneumatic or electromagnetic contactors are employed to make and break the circuits between the traction motors and the main generator. They are equipped with interlocks for various control-circuit functions. Similar contactors are used for other power and excitation circuits of lower power (current). An electropneumatic or electric-motor-operated cam switch, consisting of a two-position drum with copper segments moving between spring-loaded fingers, is generally used to reverse traction-motor field current ("reverser") or to set up the circuits for dynamic braking. This switch is not designed to operate under load. On some dc locomotives these functions have been accomplished with a system of contactors. In the ac propulsion systems, the power-control contactors are totally eliminated since their function is performed by the solid-state switching devices in the inverters.

Locomotives with ac traction motors utilize inverters to provide phase-controlled electrical power to their traction motors. Two basic arrangements of inverter control for ac traction motors have emerged. One arrangement uses an individual inverter for each axle. The other uses one inverter per axle truck. Inverters are typically GTO (gate turn off) or IGBT (insulated gate bipolar transistor) devices.

19.1.5 Cabs

In order to promote uniformity and safety, the AAR has issued standards for many locomotive cab features, RP-5104. Locomotive cab noise standards are prescribed by CFR 49 § 219.121.

Propulsion control circuits transmit the engineer's movements of the throttle lever, reverse lever, and transition or dynamic-brake control lever in the controlling unit to the power-producing equipment of each unit operating in multiple in the locomotive consist. Before power is applied, all reversers must move to provide the proper motor connections for the direction of movement desired. Power contactors complete the circuits between generators and traction motors. For dc propulsion systems excitation circuits then function to provide the proper main-generator field current while the engine speed increases to correspond to the engineer's throttle position. In ac propulsion systems, all power circuits are controlled by computerized switching of the inverter.

To provide for multiple-unit operation, the control circuits of each locomotive unit are connected by jumper cables. The AAR has issued Standard S-512 covering standard dimensions and contact identification for 27-point control jumpers used between diesel-electric locomotive units.

Wheel slip is detected by sensing equipment connected either electrically to the motor circuits or mechanically to the axles. When slipping occurs on some units, relays automatically reduce main generator excitation until slipping ceases, whereupon power is gradually reapplied. On newer units, an electronic system senses small changes in motor current and reduces motor current before a slip occurs. An advanced system recently introduced adjusts wheel creep to maximize wheel-to-rail adhesion. Wheel speed is compared to ground speed, which is accurately measured by radar. A warning light and/or buzzer in the operating cab alerts the engineer, who must notch back on the throttle if the slip condition persists.

19.1.6 Batteries

Lead-acid storage batteries of 280 or 420 ampere-hour capacity are usually used for starting the diesel engine. Thirty-two cells on each locomotive unit are used to provide 64 V to the system. (See also Avallone and Baumeister 1996, Sec. 15.) The batteries are charged from the 74-V power supply.

19.1.7 Air Brake System

The "independent" brake valve handle at the engineer's position controls air pressure supplied from the locomotive reservoirs to the brake cylinders on only the locomotive itself. The "automatic" brake valve handle controls the air pressure in the brake pipe to the train (Figure 19.3). On more recent locomotives, purely pneumatic braking systems have been supplanted by electro-pneumatic systems. These systems allow for more uniform brake applications, enhancing brake pipe pressure control, enabling better train handling. The AAR has issued Standard S-5529, *Multiple Unit Pneumatic Brake Equipment for Locomotives.*

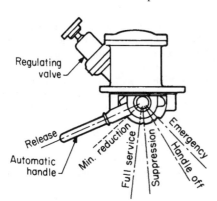

FIGURE 19.3 Automatic-brake valve-handle positions for 26-L brake equipment.

Compressed air for braking and for various pneumatic controls on the locomotive is usually supplied by a two-stage three-cylinder compressor, usually connected directly or through a clutch to the engine crankshaft. An unloader or the clutch is activated to maintain a pressure of approximately 130 to 140 psi (896 to 965 kPa) in the main reservoirs. When charging an empty trainline with the locomotive at standstill (maximum compressor demand), the engineer may increase engine (and compressor) speed without loading the traction motors.

19.1.8 Dynamic Braking

On most locomotives, dynamic brakes supplement the air brake system. The traction motors are used as generators to convert the kinetic energy of the locomotive and train into electrical energy, which is dissipated through resistance grids located near the locomotive roof. Motor-driven blowers, designed to utilize some of this braking energy, force cooler outside air over the grids and out through roof hatches. By directing a generous and evenly distributed air stream over the grids, their physical size is reduced in keeping with the relatively small space available in the locomotive. On some locomotives, resistor-grid cooling is accomplished by an engine-driven radiator/braking fan, but energy conservation is causing this arrangement to be replaced by motor-driven fans, which can be energized in response to need using the parasitic power generated by dynamic braking itself.

By means of a cam-switch reverser, the traction motors are connected to the resistance grids. The motor fields are usually connected in series across the main generator to supply the necessary high excitation current. The magnitude of the braking force is set by controlling the traction motor excitation and the resistance of the grids. Conventional dynamic braking is not usually effective below 10 mph (16 km/hr), but it is very useful at 20 to 30 mph (32 to 48 km/hr). Some locomotives are equipped with "extended range" dynamic braking which enables dynamic braking to be used at speeds as low as 3 mph (5 km/hr) by shunting out grid resistance (both conventional and extended range are shown on Figure 19.4). Dynamic braking is now controlled according to the "tapered" system, although the "flat" system has been used in the past. Dynamic braking control requirements are specified by AAR Standard S-5018. Dynamic braking is especially advantageous on long grades where accelerated brake shoe wear and the potential for thermal damage to wheels could otherwise be problems. The other advantages are smoother control of train speed and less concern for keeping the pneumatic trainline charged. Dynamic brake grids can also be used for a self-contained load-test feature which permits a standing locomotive to be tested for power output. On locomotives equipped with ac traction motors a constant dynamic braking (flat-top) force can be achieved from the horsepower limit down to 2 mph (3 km/hr).

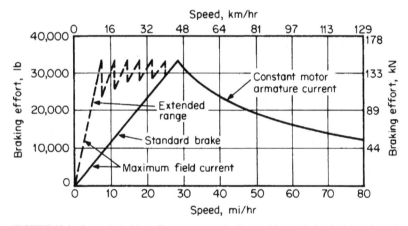

FIGURE 19.4   Dynamic-braking effort versus speed. (*Source:* Electro-Motive Division, General Motors Corp.)

19.1.9   Performance

Engine-Indicated Horsepower.   The power delivered at the diesel locomotive drawbar is the end result of a series of subtractions from the original indicated horsepower of the engine, which take into account the efficiency of transmission equipment and the losses due to the power requirements of various auxiliaries. The formula for the engine's indicated horsepower (ihp) is

$$\text{ihp} = PLAN/33{,}000 \qquad (19.1)$$

where P = mean effective pressure in the cylinder (psi)
   L = length of piston stroke (ft)
   A = piston area (in.^{2})
   N = total number of cycles completed per minute

Factor P is governed by the overall condition of the engine, quality of fuel, rate of fuel injection, completeness of combustion, compression ratio, etc. Factors L and A are fixed with design of engine. Factor N is a function of engine speed, number of working chambers, and strokes needed to complete a cycle.

Engine Brake Horsepower.   In order to calculate the horsepower delivered by the crankshaft coupling to the main generator, frictional losses in bearings and gears must be subtracted from the *indicated horsepower* (ihp). Some power is also used to drive lubricating-oil pumps, governor, water pump, scavenging blower, and other auxiliary devices. The resultant horsepower at the coupling is *brake horsepower* (bhp).

Rail Horsepower.   A portion of the engine bhp is transmitted mechanically via couplings or gears to operate the traction motor blowers, air compressor, auxiliary generator, and radiator cooling fan generator or alternator. Part of the auxiliary generator electrical output is used to run some of the auxiliaries. The remainder of the engine bhp transmitted to the main generator or main alternator for traction purposes must be multiplied by generator efficiency (usually about 91 percent), and the result again multiplied by the efficiency of the traction motors (including power circuits) and gearing to develop rail horsepower. Power output of the main generator for traction may be expressed as

$$\text{Watts}_{\text{traction}} = E_g \times I_m \tag{19.2}$$

where E_g is the main-generator voltage and I_m is the traction motor current in amperes, multiplied by the number of parallel paths or the dc link current in the case of an ac traction system.

Rail horsepower may be expressed as

$$\text{hp}_{\text{rail}} = V \times \text{TE}/375 \tag{19.3}$$

where V = velocity (mph)
TE = tractive effort at the rail (lb)

Thermal Efficiency.   The thermal efficiency of the diesel engine at the crankshaft, or the ratio of bhp output to the rate at which energy of the fuel is delivered to the engine, is about 33 percent. Thermal efficiency at the rail is about 26 percent.

Drawbar Horsepower.   The drawbar horsepower represents power available at the rear of the locomotive to move the cars and may be expressed as

$$\text{hp}_{\text{drawbar}} = \text{hp}_{\text{rail}} - \text{locomotive running resistance} \times V/375 \tag{19.4}$$

where V is the speed in mph. Train resistance calculations are discussed in section 19.5. Theoretically, therefore, drawbar horsepower available is power output of the diesel engine less the parasitic losses and losses described above.

Speed-Tractive Effort.   At full throttle the losses vary somewhat at different values of speed and tractive effort, but a curve of tractive effort plotted against speed is nearly hyperbolic. Figure 19.5 is a typical speed-tractive effort curve for a 3,500 hp (2,600 kW) freight locomotive. The diesel-electric locomotive has full horsepower available over the entire speed range (within the limits of adhesion described below). The reduction in power as continuous speed is approached is known as "power matching." This allows multiple operation of locomotives of different ratings at the same continuous speed.

Adhesion.   In Figure 19.6 the maximum value of tractive effort represents the level usually achievable just before the wheels slip under average rail conditions. Adhesion is usually expressed as a percentage of vehicle weight on drivers, with the nominal level being 25 percent. This means that a force equal to 25 percent of the total locomotive weight on drivers is available as tractive effort.

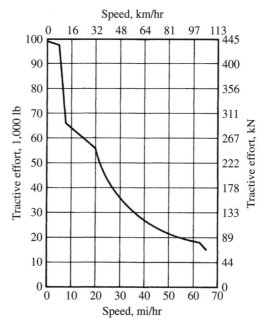

FIGURE 19.5   Tractive effort versus speed.

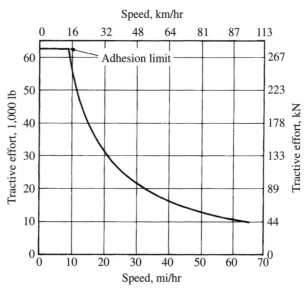

FIGURE 19.6   Typical tractive effort versus speed characteristics. (*Source:* Electro-Motive Division, General Motors Corp.)

Actually, at the point of wheel slip, adhesion will vary widely with rail conditions, from as low as 5 percent to as high as 35 percent or more. Adhesion is severely reduced by lubricants, which spread as thin films in the presence of moisture on running surfaces. Adhesion can be increased with sand applied to the rails from the locomotive sanding system. More recent wheel slip systems permit wheel creep (very slow controlled slip) to achieve greater levels of tractive effort. Even higher adhesion levels are available from ac traction motors; for example, 45 percent at start-up and low speed with a nominal value of 35 percent.

Traction Motor Characteristics.   Motor torque is a function of armature current and field flux (which is a function of field current). Since the traction motors are series connected, armature and field current are the same (except when field shunting circuits are introduced), and therefore tractive effort is solely a function of motor current. Figure 19.7 presents a group of traction motor characteristic curves with tractive effort, speed, and efficiency plotted against motor current for full field (FF) and at 35 (FS1) and 55 (FS2) percent field shunting. Wheel diameter and gear ratio must be specified when plotting torque in terms of tractive effort. (See also Avallone and Baumeister 1996, Sec. 15.)

Traction motors are usually rated in terms of their maximum continuous current. This represents the current at which the heating due to electrical losses in the armature and field windings is sufficient to raise the temperature of the motor to its maximum safe limit when cooling air at maximum expected ambient temperature is forced through it at the prescribed rate by the blowers. Continuous operation at this current level ideally allows the motor to operate at its maximum safe power level, with waste heat generated equal to heat dissipated. The tractive effort corresponding to this current is usually somewhat lower than that allowed by adhesion at very low speeds. Higher current values may be permitted for short periods of time (as when starting). These ratings are specified in time intervals of time (minutes) and are posted on or near the load meter (ammeter) in the cab.

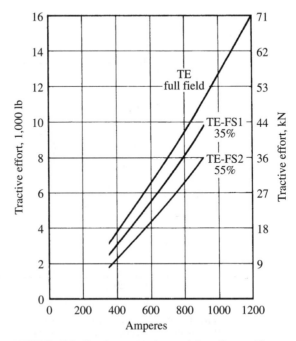

FIGURE 19.7 Traction-motor characteristics. (*Source:* Electro-Motive Division, General Motors Corp.)

Maximum Speed.   Traction motors are also rated in terms of their maximum safe speed in r/min, which in turn limits locomotive speed. The gear ratio and wheel diameter are directly related to speed as well as the maximum tractive effort and the minimum speed at which full horsepower can be developed at the continuous rating of the motors. Maximum locomotive speed may be expressed as follows:

$$(\text{mph})_{max} = \frac{\text{wheel diameter in.} \times \text{maximum motor r/min}}{\text{gear ratio} \times 336} \tag{19.5}$$

where the gear ratio is the number of teeth on the gear mounted on the axle divided by the number of teeth on the pinion mounted on the armature shaft.

Locomotive Compatibility.   The AAR has developed two standards in an effort to improve compatibility between locomotives of different model, manufacture, and ownership: a standard 27-point control system (Standard S-512, Table 19.2) and a standard control stand (RP5132). The control stand has been supplanted by a control console in many road locomotives (Figure 19.8).

TABLE 19.2   Standard Dimensions and Contact Identification of 27-Point Control Plug and Receptacle for Diesel-Electric Locomotives

Receptacle point	Function	Code	Wire size AWG
1	Power reduction setup, if used	(PRS)	14
2	Alarm signal	SG	14
3	Engine speed	DV	14
4*	Negative	N	14 or 10
5	Emergency sanding	ES	14
6	Generator field	GF	12
7	Engine speed	CV	14
8	Forward	FO	12
9	Reverse	RE	12
10	Wheel slip	WS	14
11	Spare		14
12	Engine speed	BV	14
13	Positive control	PC	12
14	Spare		14
15	Engine speed	AV	14
16	Engine run	ER	14
17	Dynamic brake	B	14
18	Unit selector circuit	US	12
19	2d negative, if used	(NN)	12
20	Brake warning light	BW	14
21	Dynamic brake	BG	14
22	Compressor	CC	14
23	Sanding	SA	14
24	Brake control/power reduction control	BC/PRC	14
25	Headlight	HL	12
26	Separator blowdown/remote reset	SV/RR	14
27	Boiler shutdown	BS	14

*Receptacle point 4—AWG wire size 12 is "standard" and AWG wire size 10 is "Alternate standard" at customer's request. A dab of white paint in the cover latch cavity must be added for ready identification of a no. 10 wire present in a no. 4 cavity. From *Mark's Standard Handbook for Mechanical Engineers,* 10th ed.

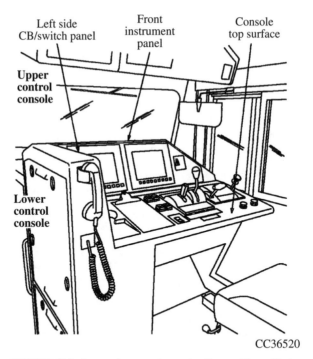

FIGURE 19.8 Locomotive control console. (*Source:* Electro-Motive Division, General Motors Corp.)

Energy Conservation. Efforts to improve efficiency and fuel economy have resulted in major changes in the prime movers, including more efficient turbocharging, fuel injection, and combustion. The auxiliary (parasitic) power demands have also been reduced with improvements including fans and blowers, that only move the air required by the immediate demand, air compressors that declutch when unloaded, and a selective low-speed engine idle. Fuel-saver switches permit dropping trailing locomotive units off the line when less than maximum power is required, while allowing the remaining units to operate at maximum efficiency.

Emissions. The U.S. Environmental Protection Agency has promulgated regulations aimed at reducing diesel locomotive emissions, especially oxides of nitrogen (NO_x). These standards also include emissions reductions for hydrocarbons (HC), carbon monoxide (CO), particulate matter (PM), and smoke. These standards are executed in three tiers. Tier 0 goes into effect in 2000, tier 1 in 2002, and tier 3 in 2005. EPA locomotive emissions standards are published under CFR 40 §§ 85, 89, and 92.

19.2 ELECTRIC LOCOMOTIVES

Electric locomotives are presently in very limited use in North America. Freight locomotives are in dedicated service primarily for coal or mineral hauling. Electric passenger locomotives are used in high-density service in the northeastern United States.

Electric locomotives draw power from overhead catenary or third-rail systems. While earlier systems used either direct current up to 3,000 V or single-phase alternating current at 11,000 V, 25 Hz,

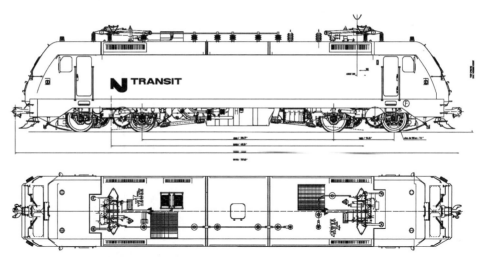

FIGURE 19.9   A modern electric high-speed locomotive. (*Courtesy of* Bombardier Transportation.)

the newer systems in North America use 25,000 or 50,000 V at 60 Hz. The higher voltage levels can only be used where clearances permit. A three-phase power supply was tried briefly in this country and overseas many years ago, but was abandoned because of the complexity of the required double catenary.

While the older dc locomotives used resistance control, ac locomotives have used a variety of systems, including Scott-connected transformers; series ac motors; motor generators and dc motors; ignitrons and dc motors; silicon thyristors and dc motors; and, more recently, chopper control. Examples of the various electric locomotives in service are shown in Table 19.1.

An electric locomotive used in high-speed passenger service is shown in Figure 19.9. High short-time ratings (Figures 19.10 and 19.11) render electric locomotives suitable for passenger service where high acceleration rates and high speeds are combined to meet demanding schedules.

The modern electric locomotive in Figure 19.9 obtains power for the main circuit and motor control from the catenary through a pantograph. A motor-operated switch provides for the transformer change from series to parallel connection of the primary windings to give a constant secondary voltage with either 25-kV or 12.5-kV primary supply. The converters for armature current consist of two asymmetric type bridges for each motor. The traction-motor fields are each separately fed from a one-way connected-field converter.

Identical control modules separate the control of motors on each truck. Motor sets are therefore connected to the same transformer winding. Wheel slip correction is also modularized, utilizing one module for each two-motor truck set. This correction is made with a complementary wheel-slip detection and correction system. A magnetic pickup speed signal is used for the basic wheel slip. Correction is enhanced by a magnetoelastic transducer used to measure force swings in the traction-motor reaction rods. This system provides the final limit correction.

To optimize the utilization of available adhesion, the wheel-slip control modules operate independently to allow the motor modules to receive different current references depending on their respective adhesion conditions.

All auxiliary machines, air compressor, traction motor blower, cooling fans, etc., are driven by three-phase 400-V, 60-Hz induction motors powered by a static inverter that has a rating of 175 kVA at a 0.8 power factor. When cooling requirements are reduced, the control system automatically reduces the voltage and frequency supplied to the blower motors to the required level. As a backup, the system can be powered by the static converter used for the head-end power requirements of the passenger cars. This converter has a 500-kW, 480-V, three-phase, 60-Hz output capacity and has a built-in overload capacity of 10 percent for half of any 1-hour period.

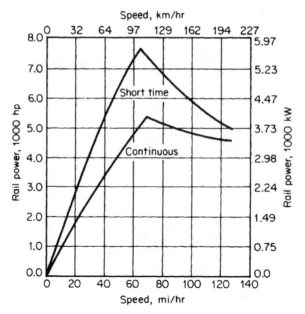

FIGURE 19.10 Speed-horsepower characteristics (half-worn wheels). [Gear ratio = 85/36; wheel diameter = 50 in (1,270 mm); ambient temperature = 60°F (15.5°C).] (*Courtesy of* M. Ephraim, ASME, Rail Transportation Division.)

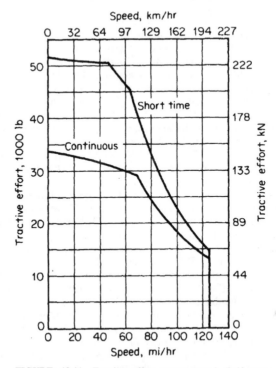

FIGURE 19.11 Tractive effort versus speed (half-worn wheels). [Voltage = 11.0 kV at 25 Hz; gear ratio = 85/36; wheel diameter = 50 in (1,270 mm).] (*Courtesy of* M. Ephraim, ASME, Rail Transportation Division.)

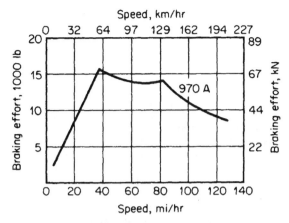

FIGURE 19.12   Dynamic-brake performance. (*Courtesy of* M. Ephraim, ASME, Rail Transportation Division.)

Dynamic brake resistors are roof-mounted and are cooled by ambient airflow induced by locomotive motion. The dynamic brake capacity relative to train speed is shown in Figure 19.12. Regenerative braking can be utilized with electric locomotives by returning braking energy to the distribution system.

Many electric locomotives utilize traction motors identical to those used on diesel-electric locomotives. Some, however, use frame suspended motors with quill-drive systems (Figure 19.13). In these systems, torque is transmitted from the traction motor by a splined coupling to a quill shaft and rubber coupling to the gear unit. This transmission allows for greater relative movement between the traction motor (truck frame) and gear (wheel axle) and reduces unsprung weight.

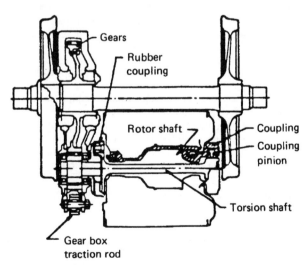

FIGURE 19.13   Frame-suspended locomotive transmission. (*Courtesy of* M. Ephraim, ASME, Rail Transportation Division.)

19.3 FREIGHT CARS

19.3.1 Freight Car Types

Freight cars are designed and constructed for either general service or specific ladings. The Association of American Railroads (AAR) has established design and maintenance criteria to assure the interchangeability of freight cars on all North American railroads. Freight cars that do not conform to AAR standards have been placed in service by special agreement of the railroads over which the cars operate. The AAR *Manual of Standards and Recommended Practices* specifies dimensional limits, weights, and other design criteria for cars that may be freely interchanged between North American railroads. This manual is revised annually by the AAR. Many of the standards are reproduced in the *Car and Locomotive Cyclopedia,* which is revised periodically. Safety appliances (such as end ladders, sill steps, and uncoupling levers), braking equipment, and certain car design features and maintenance practices must comply with the Safety Appliances Act, the Power Brake Law, and the Freight Car Safety Standards covered by the Federal Railroad Administration's (FRA) Code of Federal Regulations (CFR Title 49). Maintenance practices are set forth in the *Field Manual* of the AAR Interchange Rules and repair pricing information is provided in the companion *Office Manual.*

The AAR identifies most cars by nominal capacity and type, and in some cases there are restrictions as to type of load (i.e., food, automobile parts, coil steel, etc.). Most modern cars have nominal capacities of either 70 or 100 tons.[*] A 100-ton car is limited to a maximum gross rail load of 263,000 lb (119.3 ton) with 6½ by 12 in. (165 by 305 mm) axle journals, four pairs of 36-in. (915-mm) wheels, and a 70-in. (1,778 mm) rigid truck wheelbase. Some 100-ton cars are rated at 286,000 lb and are handled by individual railroads on specific routes or by mutual agreement between the handing railroads. A 70-ton car is limited to a maximum gross rail load of 220,000 lb (99.8 ton) with 6 by 11 in. (152 by 280 mm) axle journals, four pairs of 33-in (838-mm) wheels with trucks having a 66-in. (1,676-mm) rigid wheelbase.

On some special cars where height limitations are critical, 28-in. (711-mm) wheels are used with 70-ton axles and bearings. In these cases wheel loads are restricted to 22,400 lb (10.2 ton). Some special cars are equipped with two 125-ton two axle trucks with 38-in. (965-mm) wheels in four wheel trucks having 7 by 14 in. (178 by 356 mm) axle journals and 72-in. (1,829-mm) truck wheelbase. This application is prevalent in articulated double-stack (two-high) container cars. Interchange of these very heavy cars is by mutual agreement between the operating railroads involved.

The following are the most common car types in service in North America. Dimensions given are for typical cars; actual measurements may vary.

Boxcars. There are six popular boxcar types (Figure 19.14*a*):

1. Standard boxcars may have either sliding or plug doors. Plug doors provide a tight seal from weather and a smooth interior. Unequipped box cars are usually of 70-ton capacity. These cars have tongue-and-groove or plywood lining on the interior sides and ends with nailable floors (either wood or steel with special grooves for locking nails). The cars carry typical general merchandise lading: packaged, canned, or bottled foodstuffs; finished lumber; bagged or boxed bulk commodities; or, in the past when equipped with temporary door fillers, bulk commodities such as grain.[†]

2. Specially equipped boxcars usually have the same dimensions as standard cars but include special interior devices to protect lading from impacts and over-the-road vibrations. Specially equipped cars may have hydraulic-cushion units to dampen longitudinal shock at the couplers.

3. Insulated boxcars have plug doors and special insulation. These cars can carry foodstuffs such as unpasteurized beer, produce, and dairy products. These cars may be precooled by the shipper

---

[*]1 ton = 1 short ton = 2,000 lb; 1 ton = 1 metric ton = 1,000 kg = 2,205 lb
[†]70 ton: L = 50 ft 6 in. (15.4 m), H = 11 ft 0 in. (3.4 m), W = 9 ft 6 in. (2.9 m), truck centers = 40 ft 10 in. (12.4 m).

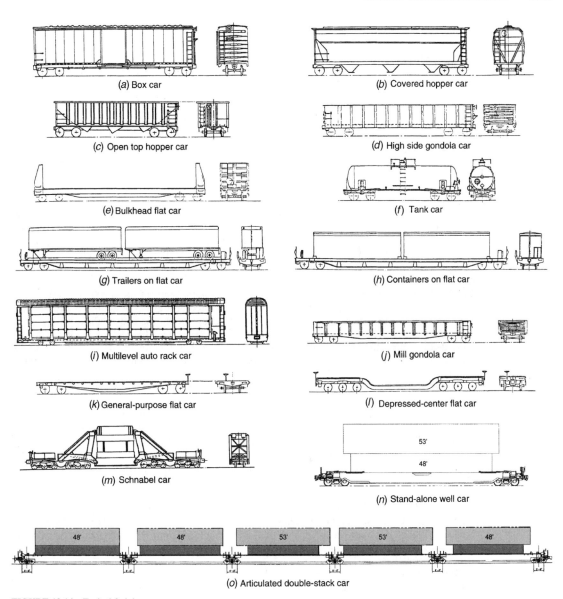

(a) Box car

(b) Covered hopper car

(c) Open top hopper car

(d) High side gondola car

(e) Bulkhead flat car

(f) Tank car

(g) Trailers on flat car

(h) Containers on flat car

(i) Multilevel auto rack car

(j) Mill gondola car

(k) General-purpose flat car

(l) Depressed-center flat car

(m) Schnabel car

(n) Stand-alone well car

(o) Articulated double-stack car

FIGURE 19.14 Typical freight cars.

and maintain a heat loss rate equivalent to 1°F (0.55°C) per day. They can also protect loads from freezing when operating at low ambient temperatures.

4. Refrigerated boxcars are used where transit times are longer. These cars are equipped with diesel-powered refrigeration units and are primarily used to carry fresh produce and meat. They are often 100-ton cars.[*]

---

[*]L = 52 ft 6 in. (16.0 m), H = 10 ft 6 in. (3.2 m), truck centers = 42 ft 11 in. (13.1 m)

5. "All-door" boxcars have doors that open the full length of the car for loading package lumber products such as plywood and gypsum board.

6. High-cubic-capacity boxcars with an inside volume of 10,000 ft³(283 m³) have been designed for light-density lading, such as some automobile parts and low-density paper products.

Boxcar door widths vary from 6 to 10 ft for single-door cars and 16 to 20 ft (4.9 to 6.1 m) for double-door cars. All-door cars have clear doorway openings in excess of 25 ft (7.6 m). The floor height above rail for an empty uninsulated boxcar is approximately 44 in. (1,120 mm) and for an empty insulated boxcar approximately 48 in. (1,220 mm). The floor height of a loaded car can be as much as 3 in. (76 mm) lower than the empty car.

Covered hopper cars (Figure 19.14*b*) are used to haul bulk commodities that must be protected from the environment. Modern covered hopper cars are typically 100-ton cars with roof hatches for loading and from two to six bottom outlets for discharge. Cars used for dense commodities, such as fertilizer or cement, have two bottom outlets, round roof hatches, and volumes of 3,000 to 4,000 ft³ (84.9 to 113.2 m³).* Cars used for grain service (corn, wheat, rye, etc.) have three or four bottom outlets, longitudinal trough roof hatches, and volumes from 4,000 to 5,000 ft³ (113 to 142 m³). Cars used for hauling plastic pellets have four to six bottom outlets (for pneumatic unloading with a vacuum system), round roof hatches, and volumes of from 5,000 to 6,000 ft³** (142 to 170 m³).

Open-top hopper cars (Figure 19.14*c*) are used for hauling bulk commodities such as coal, ore, or wood chips. A typical 100-ton coal hopper car will vary in volume depending on the light weight of the car and density of the coal to be hauled. Volumes range from 3,900 to 4,800 ft³ (110 to 136 m³). Cars may have three or four manually operated or a group of automatically operated bottom doors. Some cars are equipped with rotating couplers on one end to allow rotary dumping without uncoupling.† Cars intended for aggregate or ore service have smaller volumes for the more dense commodity. These cars typically have two manual bottom outlets.‡ Hopper cars used for woodchip service are configured for low-density loads. Volumes range from 6,500 to 7,500 ft³ (184 to 212 m³).

High-side gondola cars (Figure 19.14*d*) are open-top cars typically used to haul coal or wood chips. These cars are similar to open-top hopper cars in volume but require a rotary coupler on one end for rotary dumping to discharge lading since they do not have bottom outlets. Rotary-dump coal gondolas are usually used in dedicated, unit-train service between a coal mine and an electric power plant. The length over coupler pulling faces is approximately 53 ft 1 in. (16.2 m) to suit the standard coal dumper.§ Woodchip cars are used to haul chips from sawmills to paper mills or particle-board manufacturers. These high-volume cars are either rotary-dumped or, when equipped with end doors, end-dumped.¶ Rotary-dump aggregate or ore cars, called "ore jennies," have smaller volumes for the high-density load.

Bulkhead flat cars (Figure 19.14*e*) are used for hauling such commodities as packaged finished lumber, pipe, or, with special inward canted floors, pulpwood. Both 70- and 100-ton bulkhead flats are used. Typical deck heights are approximately 50 in. (1,270 mm).¶¶ Special, center beam, bulkhead flat cars designed for pulpwood and lumber service have a full-height, longitudinal divider from bulkhead to bulkhead.

Tank cars (Figure 19.14*f*) are used for liquids, compressed gases, and other ladings, such as sulfur, that can be loaded and unloaded in a molten state. Nonhazardous liquids such as corn syrup, crude oil, and mineral spring water are carried in nonpressure cars. Cars used to haul hazardous substances such as liquefied petroleum gas (LPG), vinyl chloride, and anhydrous ammonia are regulated by the U.S. Department of Transportation. Newer and earlier-built retrofitted cars equipped

---

*100 ton: L = 39 ft 3 in. (12.0 m). H = 14 ft 10 in. (4.5 m), truck centers = 26 ft 2 in. (8.0 m).
**100 ton: L = 65 ft 7 in. (20.0 m). H = 15 ft 5 in. (4.7 m), truck centers = 54 ft 0 in. (16.5 m).
†100 ton: L = 53 ft ½ in. (16.2 m), H = 12 ft 8½ in. (3.9 m), truck centers = 40 ft 6 in. (12.3 m).
‡100 ton: L = 40 ft 8 in. (12.4 m), H = 11 ft 10⅞ in. (3.6 m), truck centers = 29 ft 9 in. (9.1 m).
§100 ton: L = 50 ft 5½ in. (15.4 m), H = 11 ft 9 in. (3.6 m), W = 10 ft 5⅜ in. (3.2 m), truck centers = 50 ft 4 in. (15.3 m).
¶100 ton: L = 62 ft 3 in. (19.0 m), H = 12 ft 7 in. (3.8 m), W = 10 ft 5⅜ in. (3.2 m), truck centers = 50 ft 4 in. (15.3 m).
¶¶100 ton: L = 61 ft ¾ in (18.6 m), W = 10 ft 1 in. (3.1 m), H = 11 ft 0 in. (3.4 m), truck centers = 55 ft 0 in. (16.8 m).

for hazardous commodities have safety features including safety valves, specially designed top and bottom "shelf" couplers, which increase the interlocking effect between couplers and decrease the danger of disengagement due to derailment, head shields on the ends of the tank to prevent puncturing, bottom-outlet protection if bottom outlets are used, and thermal insulation and jackets to reduce the risk of rupturing in a fire. These features resulted from industry-government studies in the RPI-AAR Tank Car Safety Research and Test Program.

Cars for asphalt, sulfur, and other viscous-liquid service have heating coils on the shell so that steam may be used to liquefy the lading for discharge.*

Intermodal Cars. Conventional 89-ft 4-in. (27.2-m) intermodal flat cars are equipped to haul one 45-ft (13.7-m) and one 40-ft (12.2-m) trailer with or without end-mounted refrigeration units, two 45-ft (13.7-m) trailers (dry vans), or combinations of containers from 20 to 40 ft (6.1 to 12.2 m) in length. Hitches to support trailer fifth wheels may be fixed for trailer-only cars or retractable for conversion to haul containers or to facilitate driving trailers onto the cars in the rare event where "circus" loading is still required. Trailer hauling service (Figure 19.14*g*) is called TOFC (trailer on flat car). Container service (Figure 19.14*h*) is called COFC (container on flat car).**

Introduction of larger trailers in highway service has led to the development of alternative TOFC and COFC cars. One technique involves articulation of skeletonized or well-type units into multiunit cars for lift-on loading and lift-off unloading (standalone well car, Figure 19.14*n*, and articulated well car, Figure 19.14*o*). These cars are typically composed of from 3 to 10 units. Well cars consist of a center well for double-stacked containers.[†]

Another approach to hauling larger trailers is the two-axle skeletonized car. These cars, used either singly or in multiple combinations, can haul a single trailer from 40 to 48 ft long (12.2 to 14.6 m) with nose-mounted refrigeration unit and 36- or 42-in. (914-to 1,070-mm) spacing between the kingpin and the front of the trailer.

Bilevel and trilevel *auto rack cars* (Figure 19.14*i*) are used to haul finished automobiles and other vehicles. Most recent designs of these cars feature fully enclosed racks to provide security against theft and vandalism.[‡]

Mill gondolas (Figure 19.14*j*) are 70-ton or 100-ton open-top cars principally used to haul pipe, structural steel, scrap metal, and, when specially equipped, coils of aluminum or tinplate and other steel materials.[§]

General-purpose or *machinery flat cars* (Figure 19.14*k*) are 70-ton or 100-ton cars used to haul machinery such as farm equipment and highway tractors. These cars usually have wood decks for nailing lading-restraint dunnage. Some heavy-duty six-axle cars are used for hauling off-highway vehicles such as army tanks and mining machinery.[¶]

Depressed-center flat cars (Figure 19.14*l*) are used for hauling transformers and other heavy, large materials that require special clearance considerations. Depressed-center flat cars may have four-, six-, or dual four-axle trucks with span bolsters, depending on weight requirements.

Schnabel cars (Figure 19.14*m*) are special cars for transformers and nuclear power plant components. With these cars the load itself provides the center section of the car structure during shipment. Some Schnabel cars are equipped with hydraulic controls to lower the load for height restrictions and shift the load laterally for wayside restrictions.[¶¶] Schnabel cars must be operated in special trains.

---

*Pressure cars, 100 ton: volume = 20,000 gal (75.7 m³), L = 59 ft 11¾ in (18.3 m), truck centers = 49 ft 0¼ in. (14.9 m). Nonpressure, 100 ton, volume = 21,000 gal (79.5 m³), L = 51 ft 3¼ in. (15.6 m), truck centers = 38 ft 11¼ in. (11.9 m).

**70 ton: L = 89 ft 4 in. (27.1 m), W = 10 ft 3 in. (3.1 m), truck centers = 64 ft 0 in. (19.5 m).

†10-unit, skeletonized car: L = 46 ft 6⅜ in. (14.2 m) per end unit, L = 465 ft 3½ in. (141.8 m).

‡70 ton: L = 89 ft 4 in. (27.2 m), H = 18 ft 11 in. (5.8 m), W = 10 ft 7 in. (3.2 m), truck centers = 64 ft 0 in. (19.5 m).

§100 ton: L = 52 ft 6 in. (16.0 m), W = 9 ft 6 in. (2.9 m), H = 4 ft 6 in. (1.4 m), truck centers = 43 ft 6 in. (13.3 m).

¶100 ton, four axle: L = 60 ft 0 in. (18.3 m), H = 3 ft 9 in. (1.1 m), truck centers = 42 ft 6 in. (13.0 m). 200 ton, 8 axle: L = 44 ft 4 in. (13.5 m), H = 4 ft 0 in. (1.2 m), truck centers = 33 ft 9 in. (10.3 m).

¶¶472 ton: L = 22 ft 10 in. to 37 ft 10 in. (7.0 to 11.5 m), truck centers = 55 ft 6 in. to 70 ft 6 in. (16.9 to 21.5 m).

19.3.2 Freight Car Design

The AAR provides specifications to cover minimum requirements for design and construction of new freight cars. Experience has demonstrated that the AAR Specifications alone do not ensure an adequate car design for all service conditions. The designer must be familiar with the specific service and increase the design criteria for the particular car above the minimum criteria provided by the AAR. The AAR requirements include stress calculations for the load-carrying members of the car and physical tests that may be required at the option of the AAR committee approving the car design. In some cases, it is advisable to operate an instrumented prototype car in service to detect problems that might result from unexpected track or train-handling input forces. The car design must comply with width and height restrictions shown in AAR clearance plates furnished in the specifications (Figure 19.15). In addition, there are limitations on the height of the center of gravity of the loaded car and on the vertical and horizontal curving capability allowed by the clearance provided at the coupler. The AAR provides a method of calculating the minimum radius curve which the car design can negotiate when coupled to another car of the same type or to a standard AAR base car. In the case of horizontal curves, the requirements are based on the length of the car over the pulling faces of the couplers.

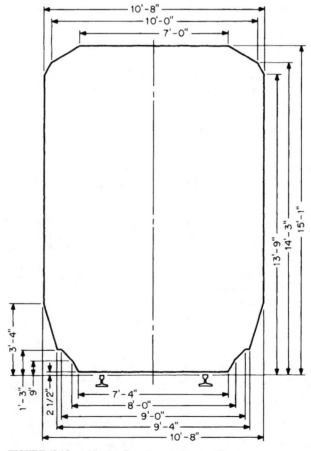

FIGURE 19.15 AAR plate B equipment-clearance diagram.

In the application for approval of a new or untried type of car, the Equipment Engineering Committee of the AAR may require either additional calculations or tests to assess the design's ability to meet the AAR minimum requirements. These tests might consist of a static compression test of 1,000,000 lb (4.4 MN), a static vertical test applied at the coupler, and impact tests simulating yard impact conditions.

Freight cars are designed to withstand single-ended impact or coupling loads based upon the type of cushioning provided in the car design. Conventional friction, elastomer, or combination draft gears or short-travel hydraulic cushion units that provide less than 6 in (152 mm) of travel require a structure capable of withstanding a 1,250,000-lb (5.56-MN) impact load. For cars with hydraulic units that provide greater than 14 in. (356 mm) of travel, the required design impact load is 600,000 lb (2.7 MN). In all cases, the structural connections to the car must be capable of withstanding a static compressive (squeeze) end load of 1,000,000 lb (4.44 MN) or a dynamic (impact) compressive load of 1,250,000 lb (5.56 MN).

The AAR has adopted requirements for unit trains of high-utilization cars to be designed for 3,000,000 mi (4.8 Gm) of service based upon fatigue life estimates. General-interchange cars that accumulate less mileage in their life should be designed for 1,000,000 mi (1.6 Gm) of service. Road environment spectra for various locations within the car are being developed for different car designs for use in this analysis. The fatigue strengths of various welded connections are provided in the AAR *Manual of Standards and Recommended Practices,* Sec. C, Part II.

Many of the design equations and procedures are available from the AAR. Important information on car design and approval testing is contained in AAR *Manual of Standards and Recommended Practices,* Sec. C-II M-1001, Chap. XI.

19.3.3 Freight Car Suspension

Most freight cars are equipped with standard three-piece trucks (Figure 19.16) consisting of two side-frame castings and one bolster casting. Side-frame and bolster designs are subjected to both static and fatigue test requirements specified by the AAR. The bolster casting is equipped with a female centerplate bowl upon which the car body rests and with side bearings located [generally 25 in. (635 mm)] each side of the centerline. In most cases, the side bearings have clearance to the car body and are equipped with either flat sliding plates or rollers. In some cases, constant-contact side bearings provide a resilient material between the car body and the truck bolster.

The centerplate arrangement consists of various styles of wear plates or fiction materials and a vertical loose or locked pin between the truck centerplate and the car body.

Truck springs nested into the bottom of the side-frame opening support the end of the truck bolster. Requirements for spring designs and the grouping of springs are generally specified by the AAR. Historically, the damping provided within the spring group has utilized a combination of springs and friction wedges. In addition to friction wedges, in more recent years some cars have been equipped with hydraulic damping devices that parallel the spring group.

A few trucks have a "steering" feature, which includes an interconnection between axles to increase the lateral interaxle stiffness and decrease the interaxle yaw stiffness. Increased lateral stiffness improves the lateral stability and decreased yaw stiffness improves the curving characteristics.

19.3.4 Freight Car Wheel-Set Design

A freight car wheel set consists of wheels, axle, and bearings. Cast- and wrought-steel wheels are used on freight cars in North America (AAR *Manual of Standards and Recommended Practices,* Sec. G). Freight car wheels are subjected to thermal loads from braking, as well as mechanical loads at the wheel-rail interface. Experience with thermal damage to wheels has led to the introduction of "low-stress" or curved plate wheels (Figure 19.17). These wheels are less susceptible to the development of circumferential residual tensile stresses, which render the wheel vulnerable to sudden failure if a flange or rim crack occurs. New wheel designs introduced for interchange service must be evaluated using a finite-element technique employing both thermal and mechanical loads (AAR S-660).

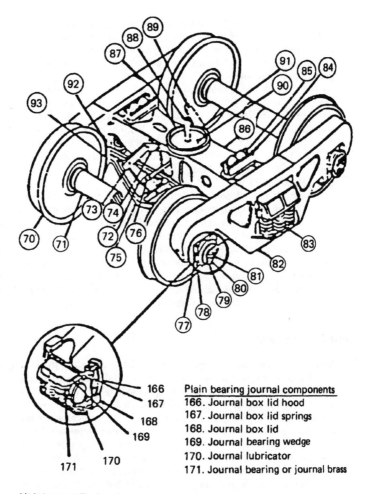

Plain bearing journal components

166. Journal box lid hood
167. Journal box lid springs
168. Journal box lid
169. Journal bearing wedge
170. Journal lubricator
171. Journal bearing or journal brass

Unit beam roller bearing truck components

70. Wheel
71. Axle
72. Truck dead lever
73. Dead lever fulcrum
74. Dead lever fulcrum bracket
75. Brake beam
76. Bottom rod
77. Roller bearing adapter
78. Roller bearing assembly
79. End cap
80. End cap retaining bolt
81. Locking plate
82. Truck side frame
83. Truck springs
84. Truck side bearing
85. Side bearing roller
86. Truck bolster
87. Truck center plate cast
88. Truck live lever
89. Center pin
90 Horizontal wear plate
91. Vertical wear plate
92. Brake shoe key
93. Brake shoe

FIGURE 19.16 Unit-beam roller-bearing truck with inset showing plain-bearing journal. (*Source:* AAR Research and Test Department.)

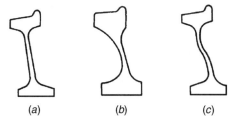

FIGURE 19.17 Wheel-plate designs. (*a*) Flat plate; (*b*) parabolic plate; (*c*) S-curved plate.

TABLE 19.3 Wheel and Journal Sizes of Eight-Wheel Cars

Nominal car capacity, ton	Maximum gross weight, lb	Journal (bearing) size, in.	Wheel diameter, in.
50	177,000	5½ × 10	33
a	179,200	6 × 11	28
70	220,000	6 × 11	33
100	263,000[c]	6½ × 12	36
125[b]	315,000	7 × 12	38

[a]Limited by wheel rating.
[b]Not approved for free interchange.
[c]286,000 in special cases

Freight car wheels range in diameter from 28 to 38 in. (711 to 965 mm) depending on car weight (Table 19.3). The old AAR standard tread profile (Figure 19.18*a*) has been replaced with the AAR-1B (Figure 19.18*c*) profile, which represents a worn profile to minimize early tread loss due to wear and provides a stable profile over the life of the tread. Several variant tread profiles, including the AAR-1B, were developed from the basic Heumann design (Figure 19.18*b*). One of these, for application in Canada, provided increasing conicity into the throat of the flange, similiar to the Heumann profile. This reduces curving resistance and extends wheel life.

Wheels are also specified by chemistry and heat treatment. Low-stress wheel designs of classes B and C are required for freight cars. Class B wheels have a carbon content of 0.57 to 0.67 percent and are rim-quenched. Class C wheels have a carbon content of 0.67 to 0.77 percent and are also rim-quenched. Rim-quenching provides a hardened running surface for a long wear life. Lower carbon levels than those in Class B may be used where thermal cracking is experienced, but freight car equipment generally does not require their use.

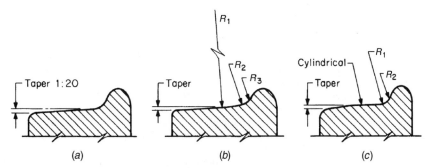

FIGURE 19.18 Wheel-tread designs: (*a*) Obsolete standard AAR; (*b*) Heumann; (*c*) new AAR-1B.

Axles used in interchange service are solid steel forgings with raised wheel seats. Axles are specified by journal size for different car capacities (Table 19.3).

Most freight car journal bearings are grease-lubricated, tapered-roller bearings (see Avallone and Baumeister 1996, Sec. 8). Current bearing designs eliminate the need for periodic field lubrication.

Wheels are mounted and secured on axles with an interference fit. Bearings are mounted with an interference fit and retained by an end cap bolted to the end of the axle. Wheels and bearings for cars in interchange service must be mounted by an AAR-inspected and approved facility.

19.3.5   Special Features

Many components are available to enhance the usefulness of freight cars. In most cases, the design or performance of the component is specified by the AAR.

Coupler Cushioning.   Switching of cars in a classification yard can result in relatively high coupler forces at the time of the impact between the moving and standing cars. Nominal coupling speeds of 4 mph (6.4 km/hr) or less are sometimes exceeded, with lading damage a possible result. Conventional cars are equipped with an AAR-approved draft gear, usually a friction-spring energy-absorbing device, mounted between the coupler and the car body. The rated capacity of draft gears ranges between 20,000 ft-lb (27.1 kJ) for earlier units to over 65,000 ft-lb (88.1 kJ) for later designs. Impact forces of 1,250,000 lb (5.56 MN) can be expected when a moving 100-ton car strikes a string of standing cars at 8 to 10 mph (12.8 to 16 km/hr). Hydraulic cushioning devices are available to reduce the impact force to 500,000 lb (2.22 MN) at impact speeds of 12 to 14 mph (19 to 22 km/hr). These devices may be mounted either at each end of the car (end-of-car devices) or in a long beam that extends from coupler to coupler (sliding centersill devices).

Lading Restraint.   Many forms of lading restraint are available, from tie-down chains for automobiles on rack cars to movable bulkheads for boxcars. Most load-restraining devices are specified by the AAR *Manual of Standards and Recommended Practices* and approved car loading arrangements are specified in the AAR *Loading Rules,* a multivolume publication for enclosed, open-top, and TOFC and COFC cars.

Covered Hopper Car Discharge Gates.   The majority of covered hopper cars are equipped with rack-and-pinion-operated sliding gates that allow the lading to discharge by gravity between the rails. These gates can be operated manually, with a simple bar or a torque-multiplying wrench, or mechanically with an impact or hydraulic wrench. Many special covered hopper cars have discharge gates with nozzles and metering devices for vacuum or pneumatic unloading.

Coupling Systems.   The majority of freight cars are connected with AAR standard couplers. A specification has been developed to permit the use of alternative coupling systems such as articulated connectors, drawbars, and rotary-dump couplers.

19.3.6   Freight Train Braking

The retarding forces acting on a railway train are rolling and mechanical resistance, aerodynamic drag, curvature, and grade, plus that force resulting from friction of the brake shoes rubbing the wheel treads. On locomotives so equipped, dynamic or rheostatic brakes using the traction motors as generators can provide all or a portion of the retarding force to control train speed.

Quick-action automatic air brakes of the type specified by the AAR are the common standard in North America. With the automatic air brake system, the brake pipe extends through every vehicle in the train, connected by hoses between each locomotive unit and car. The front and rear end brake pipe angle cocks are closed.

Air pressure is provided by compressors on the locomotive units to the main reservoirs, usually at 130 to 150 psi (900 to 965 kPa). (Pressure values are gage pressures.) The engineer's automatic brake valve, in "release" position, provides air to the brake pipe on freight trains at reduced pressure, usually at 75, 80, 85, or 90 psi (520, 550, 585, or 620 kPa) depending on the type of service, train weight, grades, and speeds at which a train will operate. In passenger service, brake pipe pressure is usually 90 or 110 psi (620 to 836 kPa).

When brake pipe pressure is increased, the control valve allows the reservoir capacity on each car and locomotive to be charged and at the same time connects the brake cylinders to exhaust. Brake pipe pressure is reduced when the engineer's brake valve is placed in a "service" position and the control valve cuts off the charging function and allows the reservoir air on each car to flow into the brake cylinder. This moves the piston and, through a system of levers and rods, pushes the brake shoes against the wheel treads.

When the engineer's automatic brake valve is placed in the emergency position, the brake pipe pressure (BP) is reduced very rapidly. The control valves on each car move to the emergency-application position and rapidly open a large vent valve, exhausting brake pipe pressure to atmosphere. This will serially propagate the emergency application through the train at from 900 to 950 ft/sec (280 to 290 m/sec). With the control valve in the emergency position both auxiliary- and emergency-reservoir volumes (pressures) equalize with the brake cylinder and higher brake cylinder pressure (BCP) results, building up at a faster rate than in service applications.

The foregoing briefly describes the functions of the fundamental automatic air brake based on the functions of the control valve. AAR-approved brake equipment is required on all freight cars used in interchange service. The functions of the control valve have been refined to permit the handling of longer trains by more uniform brake performance. Important improvements in this design have been (1) reduction of the time required to apply the brakes on the last car of a train, (2) more uniform and faster release of the brakes, and (3) availability of emergency application with brake pipe pressure greater than 40 psi (275 kPa).

The braking ratio of a car is defined as the ratio of brake shoe (normal) force to the car's rated gross weight. Two types of brake shoes, high-friction composition and high-phosphorus cast iron, are used in interchange service. Because these shoes have very different friction characteristics, different braking ratios are required to ensure uniform train braking performance (Table 19.4). Actual

TABLE 19.4 Braking Ratios, AAR Standard S-401

Type of brake rigging and shoes	With 50 lb/in² brake cylinder pressure			Hand brake[a]
	Percent of gross rail load		Maximum percent of light weight	Minimum percent of gross rail load
	Min.	Max.		
Conventional body-mounted brake rigging or truck-mounted brake rigging using levers to transmit brake cylinder force to the brake shoes				
Cars equipped with cast-iron brake shoes	13	20	53	13
Cars equipped with high-friction composition brake shoes	6.5	10	30	11
Direct-acting brake cylinders not using levers to transmit brake cylinder force to the brake shoes				
Cars equipped with cast-iron brake shoes				
Cars equipped with high-friction composition brake shoes	6.5	10	33	11
Cabooses[b]				
Cabooses equipped with cast-iron brake shoes			35–45	
Cabooses equipped with high-friction composition brake shoes			18–23	

[a]Hand brake force applied at the horizontal hand brake chain with AAR certified or AAR approved hand brake.

[b]Effective for cabooses ordered new after July 1, 1982, hand brake ratios for cabooses to the same as lightweight ratios for cabooses.

Note: Above braking ratios also apply to cars equipped with empty and load brake equipment.

From *Marks' Standard Handbook for Mechanical Engineers,* 10th ed.

or net shoe forces are measured with calibrated devices. The calculated braking ratio R (nominal) is determined from the equation

$$R = PLANE \times 100/W \qquad (19.6)$$

where P = brake cylinder pressure, 50 psi gauge
L = mechanical ratio of brake levers
A = brake cylinder area (in^2)
N = number of brake cylinders
E = brake rigging efficiency = $E_r \times E_b \times E_c$
W = car weight (lb)

To estimate rigging efficiency, consider each pinned joint and horizontal sliding joint as a 0.01 loss of efficiency; that is, in a system with 20 pinned and horizontal sliding joints, $E_r = 0.80$. For unit-type (hangerless) brake beams $E_b = 0.90$, and for the brake cylinder $E_c = 0.95$, giving the overall efficiency of 0.684 or 68.4 percent.

The total retarding force in pounds per ton may be taken as

$$F = (PLef/W) = F_g G \qquad (19.7)$$

where P = total brake-cylinder piston force (lbf)
L = multiplying ratio of the leverage between cylinder pistons and wheel treads
ef = product of the coefficient of brake shoe friction and brake rigging efficiency
W = loaded weight of vehicle (tons)
F_g = force of gravity, 20 lb/ton/percent grade
G = ascending grade (%)

Stopping distance can be found by adding the distance covered during the time the brakes are fully applied to the distance covered during the equivalent instantaneous application time.

$$S = \frac{0.0334V_2^2}{\left[\dfrac{W_n B_n (p_a/p_n)ef}{W_a} \right] + \left(\dfrac{R}{2000} \right) \pm (G)} + 1.467 t_1 \left[V_1 - \left(\frac{R + 2000G}{91.1} \right) \frac{t_1}{2} \right] \qquad (19.8)$$

where S = stopping distance (ft)
V_1 = initial speed when brake applied (mph)
V_2 = speed at time t_1
W_n = weight on which braking ratio B_n is based (lb) (see the table below for values of W_n for freight cars); (for passenger cars and locomotives, W_n is based on empty or ready-to-run weight)
B_n = braking ratio (total brake shoe force at stated brake cylinder (psi), divided by W_n)
P_n = brake cylinder pressure on which B_n is based, usually 50 psi
P_a = full brake cylinder pressure, t_1 to stop
e = overall rigging and cylinder efficiency, decimal
f = typical friction of brake shoes (see below)
R = total resistance, mechanical plus aerodynamic and curve resistance (lb/ton)
G = grade in decimal, + upgrade, – downgrade
t_1 = equivalent instantaneous application time, s

Capacity, ton	W_n, 1,000 lb
50	177
70	220
100	263
125	315

Equivalent instantaneous application time is that time on a curve of average brake cylinder buildup versus time for a train or car where the area above the buildup curve is equal to the area below the curve. A straight-line buildup curve starting at zero time would have a t_1 of half the total buildup time.

The friction coefficient f varies with the speed; it is usually lower at high speed. To a lesser extent, it varies with brake shoe force and with the material of the wheel and shoe. For stops below 60 mph (97 km/hr), a conservative figure for a high-friction composition brake shoe on steel wheels is approximately

$$cf = 0.30 \qquad\qquad (19.9)$$

In the case of high-phosphorus iron shoes, this figure must be reduced by approximately 50 percent.

P_n is based on 50 psi (345 kPa) air pressure in the cylinder; 80 psi is a typical value for the brake pipe pressure of a fully charged right train. This will give a 50-psi (345 kPa) brake cylinder pressure during a full-service application on AB equipment and a 60-psi (kPa) brake cylinder pressure with an emergency application.

To prevent wheel sliding, $F_R \leq \phi W$, where F_R = retarding force at wheel rims resisting rotation of any pair of connected wheels (lb), ϕ = coefficient of wheel-rail adhesion or friction (a decimal), and W = weight upon a pair of wheels (lb). Actual or adhesive weight on wheels when the vehicle is in motion is affected by weight transfer (force transmitted to the trucks and axles by the inertia of the car body through the truck center plates), center of gravity, and vertical oscillation of body weight upon truck springs. The value of ϕ varies with speed as shown in Figure 19.19.

The relationship between the required coefficient ϕ of wheel-rail adhesion to prevent wheel sliding and rate of retardation A in mph/sec may be expressed by $A = 21.95\phi_1$.

There has been some encouraging work to develop an electric brake that may eventually make obsolete the present pneumatic brake systems.

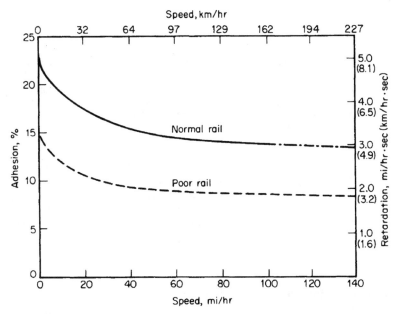

FIGURE 19.19 Typical wheel-rail adhesion. Track has jointed rails. (*Source:* Air Brake Association.)

Test Devices.   Special devices have been developed for testing brake components and cars on a repair track.

End-of-Train Devices.   To eliminate the requirement for a caboose crew car at the end of the train, special electronic devices have been developed to transmit the end-of-train brake-pipe pressure to the locomotive operator by telemetry.

19.4 PASSENGER EQUIPMENT

During the past two decades most main-line or long-haul passenger service in North America has become a function of government agencies, i.e., Amtrak in the United States and Via Rail in Canada. Equipment for intraurban service is divided into three major categories: commuter rail, heavy rail rapid transit, and light rail transit, depending upon the characteristics of the service. Commuter rail equipment operates on conventional railroad rights-of-way, usually intermixed with other long-haul passenger and freight traffic. Heavy rail rapid transit operates on a dedicated right-of-way, which is commonly in subways or on elevated structures. Light rail transit (LRT) utilizing light rail vehicles (LRV) has evolved from the older trolley or streetcar concepts and may operate in any combination of surface, subway, or elevated dedicated rights-of-way, semireserved surface rights-of-way with grade crossing, or intermixed with other traffic on surface streets. In a few cases LRT shares the trackage with freight operations, which are separated to comply with FRA regulations.

Since main-line and commuter rail equipment operates over conventional railroad rights-of-way, the structural design is heavy to provide the FRA-required crashworthiness of vehicles in the event of accidents (collisions) with other trains or at grade-crossings with automotive vehicles. Although HRT vehicles are designed to stringent structural criteria, the requirements are somewhat less severe since the operation is separate from freight equipment and there are usually no highway grade crossings. Minimum weight is particularly important for transit vehicles so that demanding schedules over lines with close station spacing can be met with minimum energy consumption.

19.4.1 Main-Line Passenger Equipment

There are four primary passenger train sets in operation across the world; diesel locomotive hauled, diesel multiple unit (DMU), electric locomotive hauled, and electric MU (EMU). In recent years the design of main-line passenger equipment has been controlled by specifications provided by APTA and the operating authority. Most of the newer cars provided for Amtrak have had stainless steel structural components. These cars have been designed to be locomotive-hauled and to use a separate 480-V three-phase power supply for heating, ventilation, air conditioning, food car services, and other control and auxiliary power requirements. Figure 19.20 shows an Amtrak coach car of the Superliner class.

Trucks for passenger equipment are designed to provide a superior ride as compared to freight-car trucks. As a result, passenger trucks include a form of "primary" suspension to isolate the wheel set from the frame of the truck. A softer "secondary" suspension is provided to isolate the truck from the car body. In most cases, the primary suspension uses either coil springs, elliptical springs, or elastomeric components. The secondary suspension generally utilizes either large coil rings or pneumatic springs with special leveling valves to control the height of the car body. Hydraulic dampers are also applied to improve the vertical and lateral ride quality.

19.4.2 Commuter Rail Passenger Equipment

Commuter rail equipment can be either locomotive-hauled or self-propelled (Figure 19.21). Some locomotive-hauled equipment is arranged for "push-pull" service. This configuration permits the

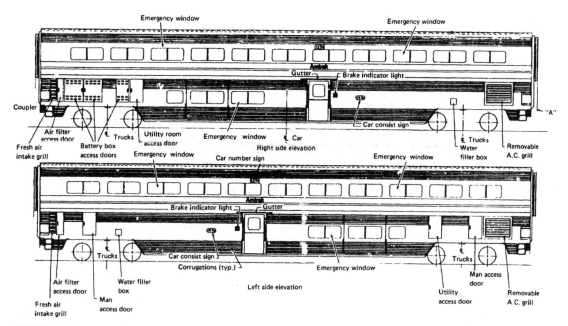

FIGURE 19.20 Main-line passenger car.

train to be operated with the locomotive either pushing or pulling the train. For push-pull service some of the passenger cars must be equipped with cabs to allow the engineer to operate the train from the end opposite the locomotive during the push operation. All cars must have control trainlines to connect the lead (cab) car to the trailing locomotive. Locomotive-hauled commuter rail cars use AAR-type H couplers. Most self-propelled (single-or multiple-unit) cars use other automatic coupler designs that can be mechanically coupled to an AAR coupler with an adaptor.

19.4.3 Heavy Rail Rapid Transit Equipment

This equipment is used on traditional subway/elevated properties in such cities as Boston, New York, Philadelphia (Figure 19.22), and Chicago in a semiautomatic mode of operation over dedicated rights-of-way that are constrained by limiting civil features. State-of-the-art subway-elevated properties include such cities as Washington, Atlanta, Miami, and San Francisco, where the equipment provides highly automated modes of operation on rights-of-way with generous civil alignments.

The cars can operate bidirectionally in multiple with as many as 12 or more cars controlled from the leading cab. They are electrically propelled, usually from a dc third rail which makes contact with a shoe insulated from and supported by the frame of the truck. Occasionally, roof-mounted pantographs are used. Voltages range from 600 to 1,500 V dc.

The cars range from 48 to 75 ft (14.6 to 22.9 m) over the anticlimbers, the longer cars being used on the newer properties. Passenger seating varies from 40 to 80 seats per car, depending upon length and the local policy (or preference) regarding seated to standee ratio. Older properties require negotiation of curves as sharp as 50 ft (15.2 m) minimum radius with speeds up to only 50 mph (80 km/hr) on tangent track, while newer properties usually have no less than 125 ft (38.1 m) minimum radius curves with speeds up to 75 mph (120 km/hr) on tangent track.

All North American properties operate on standard-gage track, with the exception of the 5 ft 6 in. (1.7 m) San Francisco Bay Area Rapid Transit (BART), the 4 ft 10⅞ in. (1.5 m) Toronto Transit Subways, and the 5 ft 2½ in. (1.6 m) Philadelphia Southeastern Pennsylvania Transportation

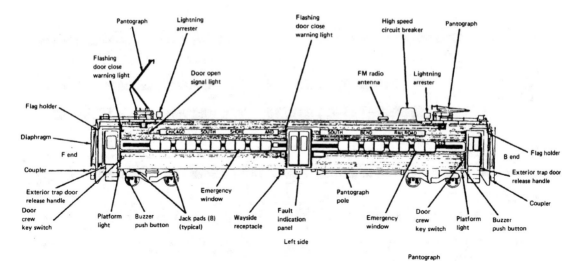

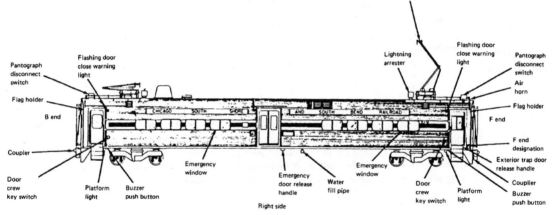

FIGURE 19.21   Commuter rail car.

Authority (SEPTA) Market-Frankford line. Grades seldom exceed 3 percent, and 1.5 to 2.0 percent is the desired maximum.

Typically, newer properties require maximum acceleration rates of between 2.5 and 3.0 mph/sec (4.0 to 4.8 km/hr/sec) as nearly independent of passenger loads as possible from 0 to approximately 20 mph (32 km/hr). Depending upon the selection of motors and gearing, this rate falls off as speed is increased, but rates can generally be controlled at a variety of levels between zero and maximum.

Deceleration is typically accomplished by a blended dynamic and electro-pneumatic friction tread or disk brake, although a few properties use an electrically controlled hydraulic friction brake. Either of these systems usually provides a maximum braking rate of between 3.0 and 3.5 mph/sec (4.8 to 5.6 km/hr/sec) and is made as independent of passenger loads as possible by a load-weighing system that adjusts braking effort to suit passenger loads. Some employ regenerative braking to supplement the other brake systems.

Dynamic braking is generally used as the primary stopping mode and is effective from maximum speed down to 10 mph (16 km/hr) with friction braking supplementation as the characteristic dynamic fade occurs. The friction brakes provide the final stopping forces. Emergency braking rates depend upon line constraints, car subsystems, and other factors, but generally rely on the maximum retardation force that can be provided by the dynamic and friction brakes within the limits of available wheel-to-rail adhesion.

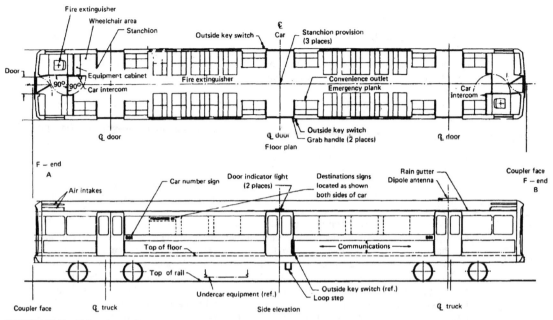

FIGURE 19.22 Full-scale transit car.

Acceleration and braking on modern properties are usually controlled by a single master control-ler handle that has power positions in one direction and a coasting or neutral center position and braking positions in the opposite direction. A few properties use foot-pedal control with a "deadman" pedal operated by the left foot, a brake pedal operated by the right foot, and an accelerator (power) pedal also operated by the right foot. In either case, the number of positions depends upon property policy and control subsystems on the car. Control elements include motor-current sensors and some or all of the following: speed sensors, rate sensors, and load-weighing sensors. Signals from these sensors are processed by an electronic control unit (ECU), which provides control functions to the propulsion and braking systems. The propulsion systems currently include pilot-motor-operated cams, which actuate switches or electronically controlled unit switches to control resistance steps, or chopper or inverter systems, which electronically provide the desired voltages and currents at the motors.

In some applications, dynamic braking utilizes the traction motors as generators to dissipate energy through the on-board resistors also used in acceleration. It is expected that regenerative braking will become more common since energy can be returned to the line for use by other cars. Theoretically, 35 to 50 percent of the energy can be returned, but the present state of the art is limited to a practical 20 percent on properties with large numbers of cars on close headways.

Car bodies are made of welded stainless steel or low-alloy high-tensile (LAHT) steel of a design that carries structural loads. Earlier problems with aluminum, primarily electrolytic action among dissimilar metals and welding techniques, have been resolved and aluminum is also used on a sig-nificant number of new cars.

Trucks may be cast or fabricated steel with frames and journal bearings either inside or outside the wheels. Axles are carried in roller-bearing journals connected to the frames so as to be able to move vertically against a variety of types of primary spring restraint. Metal springs or airbags are used as the secondary suspension between the trucks and the car body. Most wheels are solid-cast or wrought steel. Resilient wheels have been tested but are not in general service on heavy rail transit equipment.

All heavy rail rapid transit systems use high-level loading platforms to speed passenger flow. This adds to the civil construction costs but is necessary to achieve the required level of service.

19.4.4   Light Rail Transit Equipment

The cars are called light rail vehicles (LRVs; Figure 19.23 shows typical LRVs) and are used on a few remaining streetcar systems such as those in Boston, Philadelphia, and Toronto on city streets and on state-of-the-art subway, surface, and elevated systems such as those in Edmonton, Calgary, San Diego, and Portland, Oregon, in semiautomated modes over partially or wholly reserved rights-of-way.

As a practical matter, LRVs' track, signal systems, and power systems utilize the same subsystems as heavy rail rapid transit and are not lighter in a physical sense.

LRVs are designed to operate bidirectionally in multiple with up to four cars controlled from the leading cab. LRVs are electrically propelled from an overhead contact wire, often of catenary design, which makes contact with a pantograph or pole on the roof. For reasons of wayside safety, third-rail pickup is not used. Voltages range from 550 to 750 V dc.

The cars range from 60 to 65 ft (18.3 to 19.8 m) over the anticlimbers for single cars and from 70 to 90 ft (21.3 to 27.4 m) for articulated cars. The choice is determined by passenger volumes and civil constraints.

Articulated cars have been used in railroad and transit applications since the 1920s. They have found favor in light rail applications because of their ability to increase passenger loads in a longer car that can negotiate relatively tight-radius curves. These cars require the additional mechanical complexity of the articulated connection and a third truck between two car-body sections.

Passenger seating varies from 50 to 80 or more seats per car, depending upon length, placement of seats in the articulation, and the policy regarding seated/standee ratio. Older systems require negotiation of curves down to 30 ft (9.1 m) minimum radius with speeds up to only 40 mph (65 km/hr) on tangent track, while newer systems usually have no less than 75 ft (22.9 m) minimum radius curves with speeds up to 65 mph (104 km/hr) on tangent track.

The newer properties all use standard-gage track; however, existing older systems include 4 ft 10⅞ in. (1,495 mm) and 5 ft 2½ in. (1,587 mm) gauges. Grades have reached 12 percent, but 6 percent is now considered the maximum and 5 percent is preferred.

Typically, newer properties require maximum acceleration rates of between 3.0 to 3.5 mph/sec (4.8 to 5.6 km/hr/sec) as nearly independent of passenger load as possible from 0 to approximately

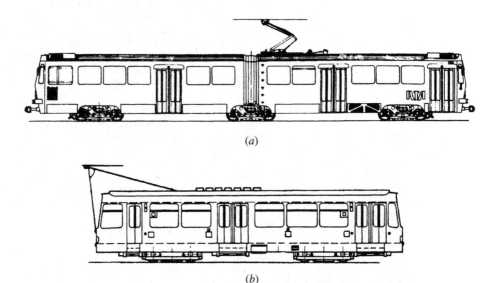

(a)

(b)

FIGURE 19.23   Light-rail vehicle (LRV): (a) Articulated; (b) Nonarticulated.

20 mph (32 km/hr). Depending upon the selection of motors and gearing, this rate falls off as speed is increased, but the rates can generally be controlled at a variety of levels.

Unlike most heavy rail rapid transit cars, LRVs incorporate three braking modes: dynamic, friction, and track brake, which typically provide maximum service braking at between 3.0 and 3.5 mph/sec (4.8 to 5.6 km/hr/sec) and 6.0 mph/sec (9.6 km/hr/sec) maximum emergency braking rates. The dynamic and friction brakes are usually blended, but a variety of techniques exist. The track brake is intended to be used primarily for emergency conditions and may or may not be controlled with the other braking systems.

The friction brakes are almost exclusively disc brakes since LRVs use resilient wheels that can be damaged by tread-brake heat buildup. No single consistent pattern exists for the actuation mechanism. All-electric, all-pneumatic, electro-pneumatic, electro-hydraulic, and electro-pneumatic over hydraulic are in common use.

Dynamic braking is generally used as the primary braking mode and is effective from maximum speed down to about 5 mph (8 km/hr) with friction braking supplementation as the characteristic dynamic fade occurs.

As with heavy rail rapid transit, the emergency braking rates depend upon line constraints, car-control subsystems selected, and other factors, but the use of track brakes means that higher braking rates can be achieved because the wheel-to-rail adhesion is not the limiting factor.

Acceleration and braking on modern properties are usually controlled by a single master controller handle that has power positions in one direction and a coasting or neutral center position and braking positions in the opposite direction. A few properties use foot-pedal control with a deadman pedal operated by the left foot, a brake pedal operated by the right foot, and an accelerator (power) pedal also operated by the right foot. In either case, the number of positions depends upon property policy and control subsystems on the car.

Control elements include motor-current sensors and some or all of the following: speed sensors, rate sensors, and load-weighing sensors. Signals from these sensors are processed by an electronic control unit (ECU) that provides control functions to the propulsion and braking systems.

The propulsion systems currently include pilot-motor-operated cams that actuate switches or electronically controlled unit switches to control resistance steps. Chopper and inverter systems that electronically provide the desired voltages and currents at the motors are also used.

Most modern LRVs are equipped with two powered trucks. In two-section articulated designs (Figure 19.23a), the third (center) truck may be left unpowered but usually has friction and track brake capability. Some European designs use three powered trucks, but the additional cost and complexity have not been found necessary in North America.

Unlike heavy rail rapid transit, there are three major dc-motor configurations in use: the traditional series-wound motors used in bimotor trucks, the European-derived monomotor, and a hybrid monomotor with a separately excited field—the last in chopper-control version only.

The bimotor designs are rated between 100 and 125 shaft hp per motor at between 300 and 750 V dc, depending upon line voltage and series or series-parallel control schemes (electronic or electromechanical control). The monomotor designs are rated between 225 and 250 shaft hp per motor at between 300 and 750 V dc (electronic or electromechanical control).

The motors, gear units (right angle or parallel), and axles are joined variously through flexible couplings. In the case of the monomotor, it is supported in the center of the truck, and right-angle gearboxes are mounted on either end of the motor. Commonly, the axle goes through the gearbox and connection is made with a flexible coupling arrangement. Electronic inverter control drives with ac motors have been applied in recent conversions and new equipment.

Dynamic braking is achieved in the same manner as with heavy rail rapid transit.

Unlike heavy rail rapid transit, LRV bodies are usually made only of welded LAHT steel and are of a load-bearing design. Because of the semireserved right-of-way, the risk of collision damage with automotive vehicles is greater than with heavy rail rapid transit and the LAHT steel has been found to be easier to repair than stainless steel or aluminum. Although LAHT steel requires painting, this can be an asset since the painting can be performed in a highly decorative manner pleasing to the public and appropriate to themes desired by the cities.

Trucks may be cast or fabricated steel with either inside or outside frames. Axles are carried in roller-bearing journals, which are usually resiliently coupled to the frames with elastomeric springs as a primary suspension. Both vertical and a limited amount of horizontal movement occur. Since tight curve radii are common, the frames are usually connected to concentric circular ball-bearing rings, which in turn are connected to the car body. Air bags, solid elastomeric springs, or metal springs are used as a secondary suspension. Resilient wheels are used on virtually all LRVs.

Newer LRVs have low-level loading doors and steps, which minimize station platform costs.

19.5 VEHICLE-TRACK INTERACTION

19.5.1 Train Resistance

The resistance to a train in motion along the track is of prime interest, as it is reflected directly in locomotive energy requirements. This resistance is expressed in terms of pounds per ton of train weight. *Gross train resistance* is that force that must be overcome by the locomotives at the driving-wheel-rail interface. *Trailing train resistance* must be overcome at the rear drawbar of the locomotive.

There are two classes of resistance that must be overcome: *inherent* and *incidental.* Inherent resistance includes the rolling resistance of bearings and wheels and aerodynamic resistance due to motion through still air. It may be considered equal to the force necessary to maintain motion at constant speed on level tangent track in still air. Incidental resistance includes resistance due to grade, curvature, wind, and vehicle dynamics.

19.5.2 Inherent Resistance

Of the elements of inherent resistance, at low speeds rolling resistance is dominant but at high speeds aerodynamic resistance is the predominant factor. Attempts to differentiate and evaluate the various elements through the speed range are a continuing part of industry research programs to reduce train resistance. At very high speeds, the effect of air resistance can be approximated as an aid to studies in its reduction by means of cowling and fairing. The residence of a car moving in still air on straight, level track increases parabolically with speed. Because the aerodynamic resistance is independent of car weight, the resistance in pounds per ton decreases as the weight of the car increases. The total resistance in pounds per ton of a 100-ton car is much less than twice as great as that of a 50-ton car under similar conditions. With known conditions of speed and car weight, inherent resistance can be predicted with reasonable accuracy. Knowledge of track conditions will permit further refining of the estimate, but for very rough track or extremely cold ambient temperatures, generous allowances must be made. Under such conditions, normal resistance may be doubled. A formula proposed by Davis (1926) and revised by Tuthill (1948) has been used extensively for inherent freight-train resistances at speeds up to 40 mph:

$$R = 1.3W + 29_n + 0.045WV + 0.0005AV^2 \qquad (19.10)$$

where R = train resistance (lb/car)
  W = weight per car (tons)
  V = speed (mph)
  n = total number of axles
  A = cross-sectional area (ft^2)

With freight-train speeds of 50 to 70 mph (80 to 112 km/hr), it has been found that actual resistance values fall considerably below calculations based on the above formula. Several modifications of the Davis equation have been developed for more specific applications. All of these equations apply to cars trailing locomotives.

1. Davis equation as modified by Tuthill (1948):

$$R = 1.3W + 29_n + 0.045WV + 0.045V^2 \tag{19.11}$$

Note: In the Totten modification, the equation is augmented by a matrix of coefficients when the velocity exceeds 40 mph.

2. Davis equation as modified by the Canadian National Railway:

$$R = 0.6W + 20_n + 0.01WV + 0.07V^2 \tag{19.12}$$

3. Davis equation as modified by the Canadian National Railway and Erie-Lackawanna Railroad for trailers and containers on flat cars:

$$R = 0.6W + 20_n + 0.01WV + 0.2V^2 \tag{19.13}$$

Other modifications of the Davis equation have been developed for passenger cars by Totten (1937). These formulas are for passenger cars pulled by a locomotive and do not include head-end air resistance.

1. Davis equations modified by Totten for streamlined passenger cars:

$$R = 1.3W + 29_n + 0.045WV + [0.00005 + 0.060725(L/100)^{0.88}]V^2 \tag{19.14}$$

2. Davis equations modified by Totten for non-streamlined passenger cars

$$R = 1.3W + 29_n + 0.045WV + [0.00005 + 0.1085(L/100)^{0.7}]V^2 \tag{19.15}$$

where L = car length in ft

19.5.3 Aerodynamic and Wind Resistance

Wind-tunnel testing has indicated a significant effect on freight train resistance resulting from vehicle spacing, open tops of hopper and gondola cars, open boxcar doors, vertical side reinforcements on railway cars and intermodal trailers, and protruding appurtenances on cars. These effects can cause significant increases in train resistance at higher speeds. For example, the spacing of intermodal trailers or containers greater than approximately 6 ft can result in a new frontal area to be considered in determining train resistance. Frontal or cornering ambient wind conditions can also have an adverse effect on train resistance which is increased with discontinuities along the length of the train.

19.5.4 Curve Resistance

Train resistance due to track curvature varies with speed and degree of curvature. The behavior of rail vehicles in curve negotiation is the subject of several ongoing AAR studies. Lubrication of the rail gage face or wheel flanges has become common practice for reducing friction and the resulting wheel and rail wear. Recent studies indicate that flange and/or gage face lubrication can significantly reduce train resistance on tangent track as well (Allen, "Conference on the Economics and Performance of Freight Car Trucks," October 1983). In addition, a variety of special trucks (wheel assemblies) that reduce curve resistance by allowing axles to steer toward a radial position in curves have been developed. For general estimates of car resistance and locomotive hauling capacity on dry (unlubricated) rail with conventional trucks, speed and gage relief may be ignored and a figure of 0.8 lb/ton per degree of curvature used.

19.5.5 Grade Resistance

Grade resistance depends only on the angle of ascent or descent and relates only to the gravitational forces acting on the vehicle. It equates to 20 lb/ton for each "percent of grade" or 0.379 lb/ton for each foot per mile rise.

19.5.6 Acceleration Resistance

The force (tractive effort) required to accelerate the train is the sum of the forces required for linear acceleration and that required for rotational acceleration of the wheels about their axle centers. A linear acceleration of 1 mph/sec (km/hr/sec) is produced by a force of 91.1 lb/ton. The rotary acceleration requirement adds 6 to 12 percent, so that the total is nearly 100 lb/ton (the figure commonly used) for each mile per hour per second. If greater accuracy is required, the following expression is used:

$$R_a = A(91.05W + 36.36n) \tag{19.16}$$

where R_a = the total accelerating force (lb)
A = acceleration (mph/sec)
W = weight of train (tons)
n = number of axles

19.5.7 Acceleration and Distance

If in a distance of S ft the speed of a car or train changes from V_1 to V_2 mph, the force required to produce acceleration (or deceleration if the speed is reduced) is

$$R_a = 74(V^2/2 - V^2/1)/S \tag{19.17}$$

The coefficient, 74, corresponds to the use of 100 lb/ton. This formula is useful in the calculation of the energy required to climb a grade with the assistance of stored energy. In any train-resistance calculation or analysis, assumptions with regard to acceleration will generally submerge all other variables; for example, an acceleration of 0.1 mph/sec (0.16 km/hr/sec) requires more tractive force than that required to overcome inherent resistance for any car at moderate speeds.

19.5.8 Starting Resistance

Most railway cars are equipped with roller bearings requiring a starting force of 5 or 6 lb/ton.

19.5.9 Vehicle Suspension Design

The primary consideration in the design of the vehicle suspension system is to isolate track input forces from the vehicle car body and lading. In addition, there are a few specific areas of instability that railway suspension systems must address. See AAR *Manual of Standards and Recommended Practice,* Sec. C-II-M-1001, Chap. XI.

Harmonic roll is the tendency of a freight car with a high center of gravity to rotate about its longitudinal axis (parallel to the track). This instability is excited by passing over staggered low rail joints at a speed that causes the frequency of the input for each joint to match the natural roll frequency of the car. Unfortunately, in many car designs this occurs for loaded cars at 12 to 18 mph (19.2 to 28.8 km/hr), a common speed for trains moving in yards or on branch lines where tracks are not well maintained. Many freight operations avoid continuous operation in this speed range.

This adverse behavior is more noticeable in cars with truck centers approximately the same as the rail length. The effect of harmonic roll can be mitigated by improved track surface and by damping in the truck suspension.

Pitch and *bounce* are the tendencies of the vehicle to either translate vertically up and down (bounce), or rotate (pitch) about a horizontal axis perpendicular to the centerline of track. This response is also excited by low track joints and can be relieved by increased truck damping.

Yaw is the tendency of the car to rotate about its axis vertical to the centerline of track. Yaw responses are usually related to *Truck hunting*. Truck hunting is an instability inherent in the design of the truck and dependent on the stiffness parameters of the truck and on wheel conicity (tread profile). The instability is observed as a "parallelogramming" of the truck components at a frequency of 2 to 3 Hz, causing the car body to yaw or translate laterally. This response is excited by the effect of the natural frequency of the gravitational stiffness of the wheel set when the speed of the vehicle approaches the kinematic velocity of the wheel set. This problem is discussed in analytic work available from the AAR Research and Test Department.

19.5.10   Superelevation

As a train passes around a curve, there is a tendency for the cars to tip toward the outside of the curve in response to centrifugal force acting on the center of gravity of the car body (Figure 19.24*a*). To compensate for this effect, the outside rail is superelevated, or raised, relative to the inside rail (Figure 19.24*b*). The amount of superelevation for a particular curve is based upon the radius of the curve and the operating speed of the train. The balance or equilibrium speed for a given curve is that speed at which the centrifugal force on the car matches the component of gravity force resulting from the superelevation between the amount required for high-speed trains and the amount required for slower-operating trains. The FRA allows a railroad to operate with 3 in. of unbalance, or at the speed at which equilibrium would exist if the superelevation were 3 in. greater. The maximum superelevation is usually 6 in. but may be lower if freight operation is used exclusively.

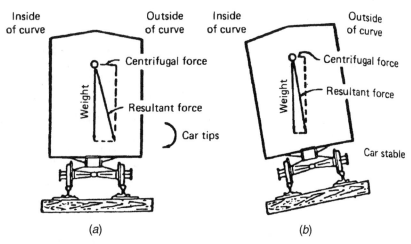

FIGURE 19.24   Effect of super relevation on center of gravity of car body.

19.5.11   Longitudinal Train Action

Longitudinal train (slack) action is a term associated with the dynamic action between individual cars in a train. An example would be the effect of starting a long train in which the couplers between each car had been compressed (i.e., bunched up). As the locomotive begins to pull the train, the slack between the locomotive and the first car must be traversed before the first car begins to accelerate. Next the slack between the first and second car must be traversed before the second car begins to accelerate, and so on. Before the last car in a long train begins to move, the locomotive and the moving cars may be traveling at a rate of several miles per hour. This effect can result in coupler forces sufficient to cause the train to break in two.

Longitudinal train action is also induced by serial braking, undulating grades, or braking on varying grades. The Track-Train Dynamics Program has published guidelines titled *Track-Train Dynamics to Improve Freight Train Performance* that explain the causes of undesirable train action and how to minimize the effects. Analysis of the forces developed by longitudinal train action requires the application of the Davis equation to represent the resistance of each vehicle based upon its velocity and location on a grade or curve. Also, the longitudinal stiffness of each car and the tractive effort of the locomotive must be considered in equations that model the kinematic response of each vehicle in the train. Computer programs are available from the AAR to assist in the analysis of longitudinal train action.

REFERENCES*

Allen, R. A. 1983. In Conference on the Economics and Performance of Freight Car Trucks, October.

American Public Transportation Association (APTA). *Standard for the Design and Construction of PassengerRailroad Rolling Stock.* APTA, Washington, DC. www.apta.com.

American Railway Engineering and Maintenance-of-Way Association (AREMA). *Manual for Railway Engineering.* Landover, MD: AREMA. www.arema.com.

American Society of Mechanical Engineers (ASME). *Proceedings.* New York: ASME. www.asme.org.

Association of American Railroads (AAR). *Field Manual of the AAR Interchange Rules.* AAR, Washington, DC. www.aar.com.

———. *Manual of Recommended Standards and Recommended Practices.* Mechanical Division, AAR, Washington, DC. www.aar.com.

———. *Office Manual of the AAR Interchange Rules.* AAR, Washington, DC. www.aar.com.

———. Multiple Unit Pneumatic Brake Equipment for Locomotives. Standard S-5529. AAR, Washington, DC. www.aar.com.

Avallone, E. A., and T. Baumeister III, eds. 1996. *Marks' Standard Handbook for Mechanical Engineers,* 10th ed. New York: McGraw-Hill.

Davis, W. J. 1926. "Tractive Resistance of Electric Locomotive and Cars." *General Electric Review* 29: 685–708.

Kratville, W. M. 1997. *The Car and Locomotive Cyclopedia of American Practice.* Omaha: Simmons-Boardman.

The Official Railway Equipment Register—Freight Connections and Freight Cars Operated by Railroads and Private Car Companies of North America. East Windsor, NJ: Commonwealth Business Media.

Railway Line Clearances. East Windsor, NJ: Commonwealth Business Media.

Totten, A. I. 1937. "Resistance of Light Weight Passenger Trains." *Railway Age* 103 (July).

Tuthill, J. K. 1948. "High Speed Freight Train Resistance." *University of Illinois Engineering Bulletin* 376.

*In addition to the works cited here, see also the publications of the Association of American Railroads (AAR) Research and Test Department.

CHAPTER 20
RAILWAY TRACK DESIGN*

Ernest T. Selig
Department of Civil Engineering
University of Massachusetts, and
Ernest T. Selig, Inc.
Hadley, Massachusetts

20.1   INTRODUCTION

Railway track as it is considered in this chapter consists of a superstructure and a substructure. The superstructure is composed of steel rails fastened to crossties. The rails are designed to support and guide flanged steel wheels through their prescribed position in space. The superstructure is placed on a substructure. The substructure is composed of a layered system of materials known as ballast, subballast and subgrade. These track components are illustrated in Figure 20.1 (Selig and Waters 1994).

Special track components are added to perform needed functions. These include switches to divert trains from one track to another, crossing diamonds to permit one track to cross another, level grade crossings to permit roads to cross over the train track at the same elevation, types of warning devices such as hot bearing detectors and dragging equipment detectors. The last example is rail attached directly to a reinforced concrete slab in a tunnel or to a bridge structure. The substructure incorporates a drainage system to remove water from the track.

The track design needs to consider soil and rock conditions, weather conditions (precipitation, temperature), traffic requirements (wheel loads, total annual tonnage), and maintenance costs for the designed track.

This chapter will provide a listing of design functions, a description of design methods, and references to sources of information on design details.

20.2   FUNCTIONS OF TRACK COMPONENTS

For each of the main track components the functions are the following (Selig and Waters 1994; Agarwal 1998; Hay 1982).

20.2.1   Rails

1. Guide the flanged wheels in the vertical, lateral, and longitudinal directions.
2. Provide a smooth running surface.
3. Transfer wheel loads to spaced ties without large deflection.

---

*Reprinted from the First Edition.

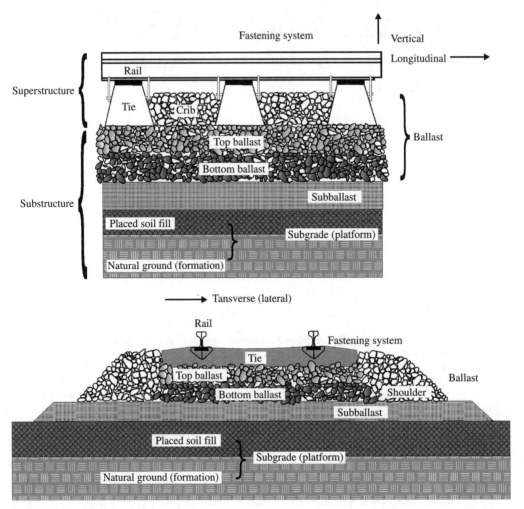

FIGURE 20.1 Track structure components.

4. Resist tension failure from longitudinal tensile force caused by rail temperature reduction.
5. Help resist buckling from longitudinal compression force caused by rail temperature increase.
6. Resist fatigue cracking from repeated wheel loads.
7. Provide strong bolted or welded joints.
8. Limit rail impact by maintaining track geometry and truing wheels to limit "false flange" wear on wheel and rail and reduce wheel defects such as engine burns, corrugations, and flat spots.
9. Permit tracks to cross over each other and permit trains to switch from one track to another.

20.2.2 Fastening Systems

1. Restrain the rail in the vertical, longitudinal, and transverse directions.
2. Resist overturning of rail from lateral wheel force.

3. Connect sections of rail to permit safe and smooth train operation.
4. Create a canted (inclined) surface to provide proper wheel/rail contact—wood ties.
5. Spread the rail seat force over a larger part of the tie surface to reduce tie damage—wood ties.
6. Provide resiliency under the vertical wheel load—concrete ties.
7. Reduce tie abrasion at rail seat—concrete ties.
8. Provide damping of the high frequency wheel-induced vibrations—concrete ties.

20.2.3 Crossties

1. Transfer the vertical wheel load from the rail through the rail seat to the bottom of the ties to provide an acceptable level or stress for the ties and ballast.
2. Hold the fastening system so that it can restrain the rails at the proper vertical, lateral, and longitudinal position and maintain the required gage.
3. Provide a canted (inclined) surface for proper wheel/rail contact—concrete ties.

20.2.4 Ballast

1. Restrain the ties against vertical, lateral, and longitudinal forces from the rails.
2. Reduce the pressure from the tie-bearing area to a level that is acceptable for the under-lying materials.
3. Provide the ability to adjust track geometry by rearranging the ballast particles by tamping and lining.
4. Assist in drainage of water from the track.
5. Provide sufficient voids between particles to allow an efficient migration of unwanted fine particles from the ballast section.
6. Provide some resiliency to the track to decrease rail, rail component, and wheel wear.

20.2.5 Subballast

1. Maintain separation between the ballast and subgrade particles.
2. Prevent attrition of the hard subgrade surface by the ballast.
3. Reduce pressure from the ballast to values that can be sustained by the subgrade without adverse effects.
4. Intercept water from the ballast and direct it to the track drainage system.
5. Provide drainage of water flowing upward from the subgrade.
6. Provide some insulation to the subgrade to prevent freezing.
7. Provide some resiliency to the track.

20.2.6 Subgrade

1. Provide a stable platform on which to construct the track.
2. Limit progressive settlement from repeated traffic loading.
3. Limit consolidation settlement.
4. Prevent massive slope failure.
5. Restrict swelling or shrinking from water content change.

20.2.7 Drainage

Drainage is the single most important factor governing the performance of track substructure. A properly functioning drainage system provides the following:

1. Intersects the water seeping up from the subgrade
2. Diverts the surface water flowing toward the track
3. Removes water falling onto the track
4. Carry off stone dust, sand, and other debris that otherwise could foul the track.

20.3 TRACK FORCES

The forces applied to the track are vertical, lateral (parallel to the ties), and longitudinal (parallel to the rails). These forces are affected by train travel speed. An important point to recognize is that track design involves many force repetitions, not just one load, as in building foundation design. Thus, allowable forces must be considerably smaller than the failure forces in a single load test in order to perform satisfactorily over a period of time. This has long been recognized in the field of material fatigue.

20.3.1 Vertical

The main vertical force is the repetitive downward action of the wheel load. In addition, this wheel/rail interaction produces a corresponding lift-up force on the ties away from the wheel load points.

The nominal vertical wheel force, also called the static force, is equal to the gross weight of the railway car divided by the number of wheels. This force ranges from about 12,000 lb (53 kN) for light rail passenger cars to 39,000 lb (174 kN) for heavy freight cars. However, due to the inertial effects of the moving train traveling over varying geometry on a track with defects, the vertical load can be much greater or much smaller than the nominal value. The largest forces are produced from the impact of the wheel on the rail, which is accompanied by vibration that often can be felt at considerable distance from the track. The passage of trains over the track causes the initial track geometry to deform. The inertia of the moving train causes the vertical wheel force to vary above and below the nominal wheel load. The vertical impact dynamic load has two components, a short-duration larger force and a longer-duration smaller force. The first is expected to be more harmful to the rails and ties, while the second does more damage to the ballast and track geometry.

The major factors affecting the magnitude of the dynamic vertical forces are

- Nominal wheel load
- Train speed
- Wheel diameter
- Vehicle unsprung mass
- Smoothness of the rail and wheel surfaces
- Track geometry
- Track modulus or vertical track stiffness

The traditional approach for representing the geometry-driven dynamic wheel load is to multiply the nominal wheel load by an impact factor that is greater than 1.

The impact factor recommended by AREMA (2003, chap. 16) is a function of the train travel speed and the wheel diameter. The actual maximum dynamic force on any track may be much different from that obtained from this approach.

These dynamic wheel forces increase the rate of track component deterioration. For example, studies in Europe (Esveld 1989) have indicated that the maintenance cost ratio is represented by the force ratio to the power n. Values of n from European work are 1 for rail fatigue, 3 for track geometry deterioration, and 3.5 for rail surface defects.

20.3.2 Lateral

One type of lateral force applied to the rail is the wheel force transmitted through friction between the wheel and top of the rail and by the wheel flange acting against the inside face of the rail head, particularly on curves. Another lateral force is the rail buckling resistance force.

The design lateral wheel force depends upon a number of factors, including

- Vehicle speed
- Track geometry
- Elevation difference between the two rails at the same cross section
- Transverse hunting movement due to the train-track dynamics

As the train speed increases, the lateral force outward on the outside rail of curves increases, and simultaneously the lateral force on the inside rail decreases.

When the field joints in the track are removed and the rails are welded, long lengths of track result, which are subjected to considerable changes in longitudinal stress due to rail temperature changes. Temperature decrease relative to the temperature at the time of welding causes tensile force parallel to the rail, which can result in rail pull-apart, while temperature increase causes compressive force, which can result in track buckling.

The wheels on a railway vehicle are tapered so that the diameter decreases from the inside to the outside. This helps center the wheels on straight track and compensates in part for the greater distance that the outer wheels travel on a curve. Because the wheels are fixed to the axle, both wheels must turn together. Thus, wheel slip is required to the extent that the circumference of the wheels does not compensate for the difference in the inside and outside rail length in a curve. The vehicle wheels on a fixed axle may take a longer time than the curve spiral allows to become oriented to the curvature of the rail. This causes additional stress on the gage or wheel climb on the high rail, and rollover possibilities on the low rail.

A flange on the inside face of each wheel limits the lateral movement of the wheels to the distance between the wheel flange and the inside face (gauge) of the rail. The combination of the wheel and railhead shapes, the inclination of the rails (cant), and the difference in elevation between the inside and outside rail in a curve serve to guide the train wheels along the intended alignment.

A spiral is a transition between the tangent track (straight) and the full radius curve. In a curve the outer rail is at a higher elevation than the inner rail so that the resultant of the weight of the train, the load balance in the car, and the centrifugal force is designed to be perpendicular to the track. This is a function of train speed, so if the train is operating above the design speed there will be a transverse force causing the flanges of the wheel to move against the outer rail. In cases of lower than design speed, the transverse force will move against the inner rail.

To achieve the desired alignment, the geometric components of a track are tangents, spirals (transition between straight track and constant radius curves), and constant radius curves in the horizontal plane, and gradients and vertical curves in the vertical plane.

20.3.3 Longitudinal

Sources of longitudinal rail forces are

- Speed
- Locomotive traction

- Locomotive and car braking
- Expansion and contraction of the rails from temperature change
- Track grade
- Special track, that is, turnouts, at grade crossings, rail crossings, dragging equipment, hot bearing detectors

The ratio of lateral to vertical force (L/V) is also important because it can cause loss of alignment and even track buckling.

20.4   TRACK SYSTEM CHARACTERISTICS

Track system performance is a function of the composite response of the track components under the action of the train loads. Two response characteristics are important to consider in track design: vertical track stiffness and lateral track stability.

20.4.1   Stiffness

The vertical response model is illustrated in Figure 20.2. The vertical track stiffness k is the vertical load on one rail divided by the vertical deflection at the loaded point. The track modulus u is the composite vertical support stiffness of the rails consisting of the fasteners, ties, ballast, subballast, and subgrade. Track modulus cannot be measured directly, but is calculated from track stiffness using the bending stiffness of the rail (Selig and Waters 1994).

Comparable models for horizontal (longitudinal and lateral) track response are not available.

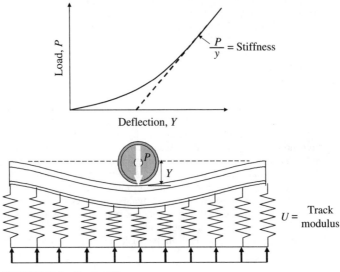

FIGURE 20.2   Track stiffness.

The subgrade is the component that has the greatest influence on the track stiffness. It is also the component with the most variation and the most uncertainty about its property values. The track should be designed to have a stiffness that is neither too high nor too low. Both extremes will shorten the life of the components.

20.4.2 Lateral Stability

Track buckling is a result of increasing longitudinal rail force from increasing temperature. Buckling occurs in the lateral direction. Because this is the least stable direction, the lateral resistance is greatest directly under the wheel loads because the weight of the train increases the lateral restraint provided by the ties. Buckling is most likely to occur on track disturbed by maintenance, when high temperatures develop in the rails, or where rail has been changed at a low temperature so that a large temperature increase can occur and is not adjusted when possible to the laying temperature. Additional information is available from Kish and Samavedam (1991).

Increasing resistance to buckling can be achieved by such means as increasing the ballast shoulder width. Tests have shown that little or no benefit is derived beyond 18 inches (460 millimeters), and most railroads only go to 12 to 16 inches (300 to 400 millimeters). In addition, increasing buckling resistance can be achieved through: (1) dynamic compacting of the crib and shoulder ballast after maintenance or out-of-face tamping, (2) using concrete ties, which are heavier than timber ties, and (3) avoiding tamping in extremely hot weather. The problem is that tamping involves lifting the track and rearranging the ballast to fill the space below the ties. This puts the track in its least stable condition until dynamic compaction is completed and/or further traffic, at decreased speeds, stabilizes the ballast.

Compaction and development of residual stresses in the ballast around the ties from train traffic greatly increases the lateral track stability. When the geometry deteriorates to the point when the track has to be lifted and tamped, the lateral track stability diminishes considerably. Consequently, it is common practice to run trains slowly at first after tamping until the lateral stability has increased from the train traffic. *Dynamic stabilizing* is used to reduce speed restrictions to only the first train over a stabilized area, whereas 10 heavy trains are commonly used for track that has not been stabilized.

20.5 RAILS

The goal of rail cross-section design should be to select the shape, size, material, and rail hardness to provide the most economical rail with the required strength and ductility for wear durability (Agarwal 1998; Marich, Mutton, and Tew 1991a). The rails are produced in the factory by hot-rolling the steel, and then cooled and straightened to form finished lengths of 80 feet (24 meters). Fixed welding plants, where the customer can send rail to be welded, are located all over North America. Mobile welding plants will come to a railroad, and in-track-flash butt welding machines will go on their track and weld jointed rails together. Most railroads still use thermite welds for rail change-outs and other short jobs in the field. These lengths can be electric flash-butt-welded (EFW) to form longer lengths, typically 1,440 feet (439 meters), that can be transported to the field for installation.

The rail may be heat-treated at the mill for increased hardness to increase the wearing resistance. Varying the steel composition can also give increased hardness with a slight loss of ductility. Overall rail life increases as the rail weight increases largely due to the ability to maintain profile over the thicker head with increased maintenance grinding. Rail properties to consider when choosing the size and type of rail are

1. Wearability
2. Hardness
3. Ductility
4. Manufacture defects in the rail material
5. Rail straightness

The rail sections are connected in the field by either bolted joints or welding. The locations of the bolted joints are high-maintenance areas because of the impact of the wheels passing the rail end gap at the center of the joints. The combination of the impact load and reduced rail stiffness of the supporting joint bars causes greater stress on the fasteners, ties, ballast, and subgrade. This in turn causes fastener looseness, plate cutting (wood ties), pad deterioration and concrete tie seat abrasion (concrete ties) and more rapid track settlement and geometry deterioration. This problem can be reduced by eliminating the joints by field welding. This approach is preferred on high-speed and heavy axle load (HAL) lines.

The spacing of the rails is standardized at a value termed *gauge*. The gauge is the distance between the inside faces of the rail at $\frac{5}{8}$ in. (14 mm) below the top of the railhead. Gauge limitations and excesses are defined by the U.S. Federal Railroad Administration (FRA) Track Safety Standards (FRA 1998) based on the class of track. For example, in North America the gage is 4 ft 8½ in. (1435 mm) with various tolerances as defined by the FRA. A range of standardized rail cross-sections is available for the designer to choose from, for example, AREMA Manual of Engineering (AREMA 2003, chap. 4). Properties for a light rail and a heavy rail are given in Table 20.1 as an indication of representative values.

TABLE 20.1   Example of Rail and Tie Properties

Rail weight	Size	Mass lb/yd (kg/m)	Area in.^{2} (cm^2)	Moment of inertia horizontal axis in.^{4} (cm^4)
Light	115 RE	115 (56.9)	11 (72.6)	66 (2730)
Heavy	136 RE	136 (67.6)	13 (86.1)	95 (3950)
Tie material	Base width in. (mm)	Mass lb (kg)	Length ft (mm)	Spacing in. (mm)
Wood	9 (229)	200 (91)	8.5 (2590)	19.5 (495)
Concrete	11 (286)	800 (360)	8.6 (2629)	14 (610)

20.6 TIES

Concrete ties are both prestressed and reinforced. AREMA (2003, chap. 10) recommends that the average ballast pressure at the base of concrete ties not exceed 85 psi (590 kPa) for high-quality abrasion-resistant ballast. AREMA (2003, chap. 16) recommends a limit of 65 psi (450 kPa). The pressure would be reduced for lower quality ballast. The limits should also consider the durability of the tie bottoms, but this is not a part of the AREMA consideration for the maximum pressure. The reason it should be considered is because the abrasion resistance of the cement in concrete is less than the resistance of much of the rock currently used for ballast.

Timber ties are both hardwood and softwood. Natural wood used for timber ties will have defects such as knots, splits, checks, and shakes. Specifications exist (for example, see AREMA (2003, chap. 3)) for the maximum size of allowed defects. Wood ties are treated with a preservative for protection against deterioration from bacteria, insects, and fire. The performance of the track can help project the need for maintenance. The upper curve represents a low-quality track because of the rapid increase in roughness with time.

Timber is the most common material used for the manufacture of crossties in North America. Next most common is prestressed/reinforced concrete. A small percentage of crossties are manufactured from other materials such as steel and cast iron. Some new materials are being introduced such as glued wood laminates and recycled plastic. Representative values of concrete and timber tie properties are given in Table 20.1. Concrete ties, at approximately 800 pounds (360 kilograms), are heavier than timber ties, at approximately 200 pounds (91 kilograms), so concrete resists track buckling better but timber ties are easier to handle. Concrete ties generally have more secure fastening systems than timber, so concrete holds the rails better. Timber ties have natural resiliency, whereas concrete ties require compressible pads for some resiliency.

One design consideration is the bending stresses in the ties caused by the wheel loads moving over the tie. These bending stresses are significantly affected by the pressure distribution of the ballast along the bottom of the ties. When the track is lifted and tamped to smooth the geometry, a gap is produced under the middle of the ties to cause the tie-bearing area to be limited to the tamp zone on both sides of the rail. With traffic the track will settle, eventually bringing the center of the tie into contact with the ballast. This condition, called center binding, will greatly increase the bending stresses in the ties. Because it is not possible to predict the exact pressure distribution along the bottom of the tie, some simplified assumptions are commonly used (Marich, Mutton, and Tew 1991b).

Analysis of the pressure distribution using the vertical track model shows that the distribution is dependent upon the flexibility of the tie, the contact-bearing area between the ballast and the tie, the compactness of the ballast under the tie, and stiffness of the subballast, ballast, and subgrade. The peak values of pressure distribution are also a function of the tie base dimensions and the center-to-center tie spacing.

The vertical track model, if available, can determine the maximum rail seat loads. The rail seat loads can also be estimated using the beam on elastic foundation model, which requires an estimate of the track modulus and the bending characteristics of the rail (Selig and Waters 1994; Hay 1982).

The maximum tie-bending moments depend on all the same factors as the maximum contact pressure at the base of the tie. The maximum bending moments are at the rail seat and the center of the tie length. Because of the difficulty of accurately predicting the maximum bending moments and bending stresses as well as contact pressures, it is quite common to select the ties based on experience in track, in the environment, and under the loading for the design conditions. In this regard, both the maximum magnitude and the number of repetitions must characterize the load. The latter affects the durability requirement (e.g., ballast crushing, tie abrasion, and fatigue life).

To complete the design based on flexural considerations, the maximum allowable bending stress needs to be determined. This can be calculated from the maximum bending moment. The maximum bending stress depends upon the material from which the tie is constructed. The use of tie plates between the rails and the ties will spread the rail seat load and therefore further reduce the bending moment.

AREMA (2003, chap. 3) indicates that an estimate of the maximum allowable bending stress in the timber ties under repeated wheel loading could be taken as 28 percent of the modulus of rupture in bending test to failure. Accordingly, values were reported as 1 ksi (7 MPa) for softwood and 1.3 ksi (9 MPa) for hardwood. Similar methods have been developed for reinforced concrete and steel ties. The designer must ensure that the appropriate values of the maximum bending stress under repeated loading are obtained for the ties being considered.

A less conservative assumption that the ballast bearing pressure under concrete ties is uniformly distributed may be appropriate for these materials because their properties are better controlled and the ties are more expensive.

20.7 FASTENING SYSTEMS

A rail joint is desired to be as stiff and strong as the rail itself. Welded joints approach this condition, but bolted joints do not (Talbot 1933). The bolted joints have bars that fit within the railhead and base fillets and against the web of the rail. Holes are drilled through the rail concentric with the holes in the joint bars. Insulation can be placed between the bars and the rail to electrically isolate the signal circuits.

For timber ties, steel tie plates are fit to the rail base with a $1/4$-inch (6.4-millimeter) shoulder on either side of the base for line-spiking and up to 18 inches (460 millimeters) in length secured to the timber tie with 6-inch (150-millimeter) cut spikes or screw spikes. The plates are placed between the rail and the tie surface to spread the rail seat load. The plates work together with a variety of rail anchors or other elastic fasteners for horizontal and lateral restraint of the track. The tie plates provide the cant to the rail. Tie plates are available in a variety of sizes (AREMA 2003, chap. 5).

Tie plates come with four holes. Cut spikes are used in a variety of patterns depending on the geometry of the track, tangent, or curve. At least one spike is driven through the hole immediately outside the shoulder on each side of the rail for line stability. Again, at least one spike is driven through the opposing corner at each end of the plate to hold the plate in position on the tie. The primary function of the spike is to hold the plate to the tie and provide line stability for the rail as it fits within the shoulders of the tie plate. The heads of the spikes are driven down to a $\frac{1}{8}$-inch (3-millimeter) height above the top of the rail base, but through time and the natural plate cut that occurs from the flexing and uplift of the rail, spikes are lifted up somewhat while still providing stable line. Specifications and proper spike driving patterns are given in AREMA (2003, chap. 5).

For concrete ties, spring clips, known as elastic fasteners, are connected to the top of tie and press down on the top of the rail base and against the web of the rail. These same fasteners come secured within the concrete tie pour, which makes fewer parts and more stable holding ability. There are several other elastic fastener designs on the market, all of which are designed to secure the rail in vertical, lateral, and longitudinal directions. A pad is placed between the bottom of the tie and the rail seat to provide resiliency and insulation for signal conductivity and help prevent rail seat and tie abrasion due to the L/V forces resulting from the load.

20.8 BALLAST

The ballast component of track shown in Figure 20.1 is subdivided into four zones:

1. Crib—material between the ties
2. Shoulder—material beyond the tie ends down to the bottom of the ballast layer
3. Top ballast—upper portion of supporting ballast layer that is disturbed by tamping
4. Bottom ballast—lower portion of supporting ballast layer, which is not disturbed by tamping and generally is the more fouled

The mechanical properties of the ballast layer result from a combination of the physical properties of the individual particles and the degree of fouling together with the in-place density of the assembly of particles. Fouling refers to the small particles that infiltrate the space between the ballast particles. The main factors producing the density are tamping, and train traffic. Tamping involves the insertion of tools into the ballast to rearrange the particles to fill the space under the ties resulting from track lift. This leaves the ballast in a relatively loose state. The many load cycles from the trains produce most of the compaction. Most of the major freight and passenger railroads use a combination of measured dynamic stabilizing and restricted speed over disturbed track.

20.8.1 Ballast Particle Requirements

Index tests have been established for characterizing the ballast properties. These cover mechanical strength, shape, water absorption, specific gravity, surface texture, particle size, and breakdown from cycles of freezing and thawing. Each railroad has a set of ballast specifications that stipulates limits for the values from the index tests. These specifications are known to be insufficient for ensuring satisfactory performance. One major limitation is that no correlation exists between index tests so that trade-offs can be established between two ballast materials, which differ in the values of the individual index properties. Petrographic analysis of the parent rock is a valuable aid in assessing ballast suitability. This information should be supplemented by observations of performance in track.

For ballast to perform its intended functions (section 20.2.4), it should consist of the following characteristics:

1. Most particles in the 0.8- to 2.5-inch (19- to 64-millimeter) size range
2. Produced by crushing hard, durable rock
3. Planar fractured faces intersecting at sharp corners (to give angularity)

4. Particles with a maximum ratio of 3:1 for largest to smallest dimensions

5. Rough surface texture preferred

6. Low water absorption

The relatively small range of particle size limits segregation when the particles are rearranged during tamping and also minimizes the loosening effect of the tamping process. The large size of particles creates large void spaces to permit migration and holding of fine particles while delaying the time when the ballast performance is significantly degraded by accumulation of the fine particles. The condition also permits rapid flow of water through the ballast layer. The fractured faces, with rough texture and high angularity together with restrictions on the amount of flat and elongated particles, provide high strength and stability for the assembly of ballast particles. Hard, durable rock is needed to reduce the particle breakage caused by the repeated train loading and from the tamping action during maintenance to smooth the track geometry. The low water absorption indicates stronger particles and reduces breakdown from water expansion during freezing temperatures. The stress-reduction function depends on the above characteristics and the layer thickness.

The optimum choice of particle characteristics depends on the magnitude of axle load and number of repetitions, together with the cost to deliver the ballast. Lower-quality ballast can be more cost-effective than higher-quality ballast on low-traffic lines, especially when the lower-quality ballast is closer.

20.8.2 Ballast Fouling

Over a period of time in track the ballast gradation typically becomes broader and finer than the initial condition because the larger ballast particles will break into smaller particles and additional smaller particles from a variety of sources will infiltrate the voids between the ballast particles. This process is known as fouling. Five categories of fouling material have been identified:

1. Particles entering from the surface such as wind-blown sand or coal fines falling out of cars

2. Products of wood or concrete tie wear

3. Breakage and abrasion of the ballast particles by train loading

4. Particles migrating upward from the granular layer underlying the ballast

5. Migration of particles from the subgrade

The main causes of ballast fouling should be identified so that proper steps can be taken to reduce the rate of fouling. The most frequent cause of ballast fouling is ballast breakdown, but there are individual situations in which each one of the other categories dominates. Geotextiles (filter fabrics) generally have not been found to be useful in solving ballast fouling problems (Selig and Waters 1994). A proper subballast layer is the best cure for fouling from the underlying granular layer and from the subgrade. When subgrade is the source of fouling material one of two main mechanisms usually is present: (1) abrasion of the subgrade surface by ballast particles in contact with the subgrade, or (2) crack pumping resulting from hydraulic erosion of water-filled cracks in the subgrade subjected to repeated train loading.

Most commonly observed fouling problems are restrictions of drainage and interference with track maintenance. However, as the voids become completely filled with fines, the ballast begins to take on the characteristics of the fines, with the ballast particles acting as filler. Soaked fines represent mud and hence the ballast becomes soft and deformable. When wet fouled ballast becomes frozen the resiliency is lost. When the fines become dry (but still moist), they act as a stiff binding agent for the crushed rock particles. This also causes loss of resiliency. All of these conditions prevent proper track surfacing.

The term *cemented ballast* is frequently used in the railroad industry to represent a condition in which the ballast particles are bound together. Although this term has not been officially defined, in most cases it appears to be used to represent dried fouled ballast. However the word *cemented* has led to the notion that a chemical bonding is involved, such as in the case of portland cement, a derivative of limestone rock. This is one of the reasons given by the railroad industry for preferring not to use limestone ballast.

A thorough examination of cemented ballast conditions is needed to determine the cause. Such a study could very well show that chemical bonding as in cement is not the main bonding mechanism in cemented ballast because it is not normally the type of bonding in dried fouled ballast.

20.8.3 Petrographic Analysis

The value of petrographic analysis as a means of assessing and/or predicting behavior of an aggregate has been long recognized by the concrete industry. Techniques for evaluating aggregate for use in concrete and for examining hardened concrete have been established by ASTM in standards C295 for aggregate and C856 for hardened concrete. The purposes of this petrographic examination are

1. To determine the physical and chemical properties of the material that will have a bearing on the quality of the material for the intended purpose
2. To describe and classify the constituents of the sample
3. To determine the relative amounts of the constituents of the sample, which is essential for the proper evaluation of the sample, especially where the properties of the constituents vary significantly

The value of the petrographic analysis depends to a large extent on the ability of the petrographer to correlate data provided on the source and proposed use of the material with the findings of the petrographic examination.

Petrographic analysis is very helpful in the selection of a suitable quarry for ballast and also for prediction of the shape and character of the components of future ballast breakdown (i.e., the fines generated by breakage and abrasion of the ballast). An experienced petrographer can estimate the relative mechanical properties, including hardness, shape, type of fracture, and durability in track.

20.8.4 Ballast Compaction

At the time when surfacing is required to correct track geometry irregularities, the ballast is in a dense state, particularly beneath the tie-bearing areas. When the rail and tie are raised to the desired elevation, tamping tines are inserted in the crib next to the rail to displace the ballast into the voids under the tie that were created by the raise. This tamping process disturbs the compact state of the ballast and leaves it loosened (Figure 20.3). The more fouled the ballast the greater the raise, the looser the ballast is after tamping.

The loosened ballast beneath the tie results in renewed settlement as the traffic, or track equipment made for this purpose, stabilizes the ballast. The loosened crib ballast results in a significant reduction in lateral buckling resistance of the rail in the unloaded state. Crib surface vibratory compactors can be used to compact the crib ballast immediately after tamping, but not the ballast under the tie.

Traffic is the most effective means of compacting ballast under the tie, but this takes time and results in nonuniform track settlement. Traffic also causes crib ballast to stabilize.

In addition to increasing density, there is evidence that both the traffic and the crib compactor produce residual horizontal stresses in the ballast. These residual stresses may be one of the most important factors influencing ballast performance in track. Fouled ballast in the crib will reduce densification of crib ballast by traffic after tamping and hence diminish any tendency for the development of lateral residual stress against the sides of the ties.

At present no adequate correlation exists between ballast index tests as a group and ballast performance in track. What is needed to select ballast is a method that takes into account the effect

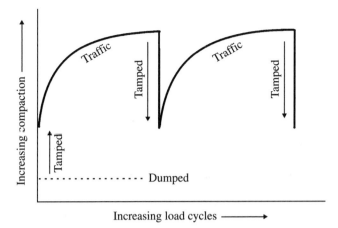

FIGURE 20.3   Effect of tamping and traffic on ballast compaction.

of differences in ballast gradation and particle composition and, in addition, simulates field service conditions such as ballast depth, subgrade characteristics, traffic loading, and track parameters.

20.8.5 Ballast Layer Thickness

The thickness of the ballast layer beneath the ties should generally be the minimum that is required for the ballast to perform its intended functions. Typically this will be 9 to 18 inches (230 to 460 millimeters). The minimum subballast layer thickness beneath the ballast should be 6 inches (150 millimeters). Thus, the minimum granular layer thickness beneath the ties would be 15 to 24 inches (380 to 600 millimeters). A check then needs to be made to determine whether this is enough granular layer thickness to prevent overstressing the subgrade (Li and Selig 1998). If not, then the granular layer thickness needs to be increased. Increasing the subballast thickness, not the ballast thickness, should do this. A thick ballast layer, when unconfined, can be prone to settlement from particle movement, particularly in the presence of vibration.

20.8.6 Track Stiffness

Clean ballast provides only a small amount of the total track stiffness, unless the subgrade is rock or a thick concrete slab.

20.9 SUBBALLAST

Subballast is a very important but not adequately recognized track substructure component. It serves some of the same functions as ballast, but it also has some unique functions. Like ballast, it provides thermal insulation and resiliency. Also, like ballast, it is a structural material that further reduces pressure on the subgrade. Water falling on the track will enter the ballast and flow downward to the subballast. Generally the water will enter the subballast, and this will be diverted to the sides of the track by the subgrade surface unless the subgrade is similar to the subballast in permeability. If the subgrade surface is not properly sloped for drainage, then some of the water will be retained under the track, where it will weaken the substructure. The remainder of the water will be shed by the subballast. The subballast also allows water flowing up from the subgrade to discharge without eroding the subgrade soil, which ballast cannot do because ballast is too coarse.

One unique function of subballast is to prevent the fine subgrade particles from migrating into the ballast voids, whether from repeated train loading or from flow of water. Finally, a particularly important function of subballast is to prevent the ballast particles from coming into contact with the subgrade soil, where they abrade or grind away the subgrade surface (subgrade attrition). The fine soil particles produced then mix with water and form mud that squeezes into the ballast voids. This is mainly a problem with hard subgrade. Inserting a 6-inch (150-millimeter) layer of properly graded and durable subballast between the ballast and the subgrade solves the problem.

Crushing durable rock to form sand-and gravel-sized particles forms subballast. Suitable subballast materials are commonly found in natural deposits. The aggregate must be resistant to breakdown from cycles of freezing and from repeated cycles of train loading. However, the durability requirements are not as severe as for ballast because the subballast particles are smaller and the stresses are lower. The finest particles less than 0.003 inch (0.075 millimeter) must be nonplastic. Depending on the permeability requirements for drainage, the fine particles must not exceed 5 to 10 percent by weight and may be less than 0 to 2 percent in some cases. Subballast materials satisfying these requirements, when placed and compacted, will satisfy the structural requirements of pressure reduction to the subgrade and resiliency.

Subballast must be well drained so that it is not saturated during repeated train loading, particularly dynamic loading from impact forces. Saturated and undrained subballast materials can deform significantly during train loading and even liquefy.

To provide separation between the ballast and subgrade particles, the subballast gradation must satisfy the requirement in Figure 20.4. This provides that the finest subballast particles are smaller than the largest subgrade particles, and correspondingly the largest subballast particles must be larger than smallest ballast particles. There must also not be gaps in the gradation of the subballast. Subballast satisfying these requirements will also be satisfactory for preventing attrition on the hard subgrade surface by the ballast.

The subballast must also be permeable enough to serve the drainage functions discussed more fully in section 20.11.

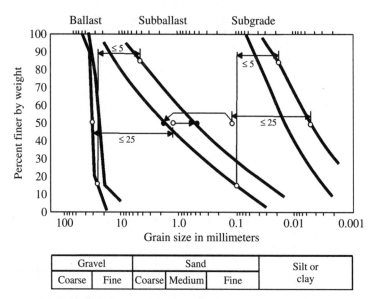

FIGURE 20.4   Subballast satisfying filter criteria.

To provide freezing protection of the subgrade and not contribute to frost heave-thaw softening problems, the subballast must be well drained and contain less than 5 percent fines (silt and clay-sized particles).

20.10 SUBGRADE

The subgrade is the platform upon which the track structure is constructed. Its main function is to provide a stable foundation for the subballast and ballast layers. The influence of the traffic-induced stresses extends downward as much as 5 meters below the bottom of the ties. This is considerably beyond the depth of the ballast and subballast. Hence, the subgrade is a very important substructure component that has a significant influence on track performance and maintenance. For example, subgrade is a major component of the superstructure support resiliency and hence contributes substantially to the elastic deflection of the rail under wheel loading. In addition, the subgrade stiffness magnitude is believed to influence ballast, rail, and sleeper deterioration. Subgrade also is a source of rail differential settlement due to movement of the subgrade from various causes.

The various types of subgrade problems are listed in Table 20.2 together with their causes and features.

The subgrade may be divided into two categories (Figure 20.1): (1) natural ground (formation) and (2) placed soil (fill). Anything other than soils existing locally are generally uneconomical to use for the subgrade. Existing ground should be used without disturbance as much as possible. However, techniques are available to improve soil formations in place if they are inadequate. Often some of the formation must be removed to construct the track at its required elevation, which is below the existing ground surface. This puts the track in a cut with the ground surface sloping downward toward the track. If the excavation intercepts the water table, slope erosion or failure can occur, carrying soil onto the track. Placed fill is used either to replace the upper portion of unsuitable existing ground or to raise the subgrade surface to the required elevation for the superstructure and the remainder of the substructure.

The subgrade is often the weakest substructure layer. Thus, a combined ballast and subballast thickness is required that will reduce the pressure on the subgrade to a level that produces an acceptably small deformation from the repeated train loading for the desired design life. The design method must consider the type and strength of the subgrade soil, the distribution of dynamic wheel loads and number of repetitions, and the substructure layer resilient moduli. The various levels of wheel loads and their corresponding numbers of cycles are converted to a single representative design load and equivalent number of cycles.

The design method is described in detail in Li and Selig (1998). Two analyses are performed:

- Limiting the cumulative plastic strain on the subgrade surface accompanying progressive shear to an acceptably small value over the life of the track to restrict the subgrade squeeze
- Limiting the cumulative plastic settlement of the compressible subgrade to prevent forming a "bathtub" depression in the subgrade that traps water

20.10.1 Limiting Strain Method

The steps in the method are

1. Select the allowable strain limit based on design life desired.
2. Determine the equivalent number of design load cycles.
3. Estimate the static compressive strength of the subgrade soil.
4. From Figure 20.5, calculate the allowable cyclic stress.
5. Calculate the strain influence factor.
6. From Figure 20.6, determine the required minimum granular layer thickness.

TABLE 20.2   Major Subgrade Problems and Features

Type	Causes	Features
(1) Progressive shear failure	• Repeated over stressing subgrade • Fine-grained soils • High water content	• Squeezing of subgrade into ballast shoulder • Heaves in crib and/or shoulder depression under ties trapping water
(2) Excessive plastic deformation (ballast pocket)	• Repeated loading of subgrade • Soft or loose soils	• Differential subgrade settlement ballast pockets
(3) Subgrade attrition with mud pumping	• Repeated loading of subgrade stiff hard soil • Contact between ballast and subgrade • Clay-rich rocks or soils • Water presence	• Muddy ballast • Inadequate subballast
(4) Softening subgrade surface under subballast	• Dispersive clay • Water accumulation at soil Surface • Repeated train loading	• Reduces sliding resistance of subgrade soil surface
(5) Liquefaction	• Repeated dynamic loading • Saturated silt and fine sand • Loose state	• Large track settlement • More severe with vibration • Can happen in subballast
(6) Massive shear failure (slope stability)	• Weight of train, track, and subgrade • Inadequate soil strength	• Steep embankment and cut slope • Often triggered by increase in water content
(7) Consolidation settlement	• Embankment weight • Saturated fine-grained soils	• Increased static soil stress as from weight of newly constructed embankment • Fill settles over time
(8) Frost action (heave and softening)	• Periodic freezing temperature • Free water • Frost-susceptible soils	• Occurs in winter/spring period • Heave from ice lens formation • Weakens from excess water content on thawing • Rough track surface
(9) Swelling/shrinkage	• Highly plastic or expansive soils • Changing moisture content	• Rough track surface • Soil expands as water content increases • Soil changes as water content decreases
(10) Slope erosion	• Surface and subsurface water movement • Wind	• Soil washed or blown away • Flow onto track fouls ballast • Flows away from track can undermine track
(11) Slope collapse	• Water inundation of very loose soil deposits	• Ground settlement
(12) Sliding of side hill fills	• Fills placed across hillsides • Inadequate sliding resistance • Water seeping out of hill or down slope is major factor	• Transverse movement of track

20.10.2   Limiting Deformation Method

The steps in the method are

1. Select the allowable deformation limit.
2. Determine the equivalent number of design load cycles (same as method 1).
3. Estimate the static compressive strength of the subgrade soil (same as method 1).
4. Calculate the deformation influence factor.
5. From a figure similar to Figure 20.6, determine the required minimum granular layer thickness.

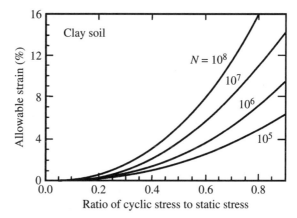

FIGURE 20.5 Allowable deviator stress.

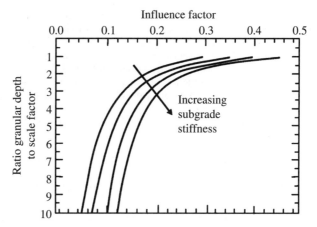

FIGURE 20.6 Granular layer thickness chart.

Sections 20.8 and 20.9 indicate the minimum ballast and subballast thickness. The design granular layer thickness is the greater of (1) the combined ballast and subballast layer minimum thickness and (2) the thickness required for the subgrade protection.

Some alternatives are available to reduce the settlement in cases where the subgrade is overstressed. Reducing the wheel load and the total annual traffic million gross tons are assumed to be unacceptable alternatives in most cases. Two general categories may be designated: (1) with the track in place and (2) with the track removed (this would include new construction).

20.10.3 Track in Place

With the track remaining in place, there are several options for improving the subgrade performance:

1. Improve drainage.
2. Increase the granular layer thickness.

3. Add tensile reinforcement in the subballast (such as geogrid, geoweb).

4. Use special on-track machines that can renew substructure conditions while working beneath the track.

20.10.4 Track Removed

With the track removed or not yet placed, additional options become available:

1. Install proper drainage.

2. Remove soft soils and replace with compacted suitable soils.

3. Place impermeable membrane to prevent water from coming into contact with the soil.

4. Lime or cement stabilization of soils by mechanical mixing.

5. Insert hot mix asphalt concrete layer on subgrade.

Clearly, designing and installing the substructure to meet the track needs without the track in place is easier and more effective. Obviously many reasons exist why this is not done.

AREMA (2003, chap. 16) recommends a method for determining ballast depth to limit wheel load-induced stress on top of subgrade so that the subgrade will not fail. The method involves determining the depth for a given track modulus and wheel load that results in an allowable pressure of 25 psi. This value applies to all soils. The number of wheel load repetitions is not considered in the AREMA method. This is another major deficiency in this manual. The correct allowable stress at top of subgrade is not constant but depends on the soil conditions and number of wheel load repetitions. The allowable stress on the top of subgrade for good track performance is determined by cumulative deformation (settlement) rather than by bearing capacity. For a mix of traffic the heaviest loads mainly cause the deformation.

The following are a few examples of subgrade remedial treatment methods to fix the problems in Table 20.2:

1. *Grouting:* Some grouts penetrate the voids of the soils and strengthen them or reduce water seepage. Other grouts compact and reinforce the soils to strengthen them or displace the soils to compensate for settlement. Jet grouting mixes cement with soil to form columns of strengthened soil.

2. *Soil mixing:* This is a process in which soil is mixed with augers and paddles to create a mixture of soil and cement based grout. Soil mixing creates a column of strengthened soil for compression and shear reinforcement.

3. *Modification of clay properties with lime:* There are several alternatives: quick lime is placed in boreholes to strengthen the soil; lime is mechanically mixed with soil to form columns of material with increased strength; lime and water mixed to form slurry is injected into clay soil under pressure with the expectation of improving the clay properties. This last is a common but not usually effective treatment with undesirable side effects. It fractures the clay instead of penetrating the voids and also solidifies ballast.

4. *Reconstruction:* Compaction of existing soils in layers at proper water content or substitutions of better soils will give improved subgrade. Chemicals such as cement or quick hydrated lime, mechanically mixed with the soils in layers before compaction, will form a stronger or less reactive soil after compaction. The chemistry of the soils should be checked or tests performed to verify the effectiveness of the treatment, because some combinations can be harmful. All of these methods generally require removal of the track.

5. *Reinforcement:* Various plastic grids, metal strips, or cellular materials placed in the soils give tensile reinforcement. Alternatively, steel reinforcing can be installed in grout-filled boreholes.

6. *Stress reduction:* Increasing the thickness of the ballast and subballast will reduce the pressure on the weaker subgrade caused by the train loading. Contrary to the AREMA engineering manual,

the allowable pressure is not constant but varies widely and must be determined in each case for correct design. The correct strength considers the magnitude of the repeated loading from the trains and the number of repetitions. For a given axle load, a high-tonnage line would have a much lower apparent strength than a low-tonnage line. Thus, the high-tonnage line needs a greater ballast/subballast thickness for the same subgrade properties.

20.11   TRACK DRAINAGE

Drainage of railway tracks is essential to achieve acceptable track performance. Water in the track substructure originates from three potential sources (Figure 20.7):

1. Precipitation onto the track

2. Surface flow from areas adjacent to the track

3. Groundwater flow

A complete drainage system must include provisions for handling water from all three sources (Heyns 2000).

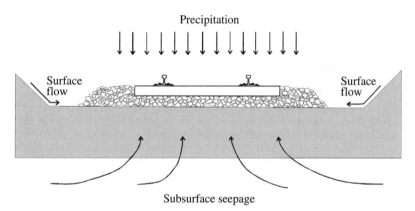

FIGURE 20.7   Sources of water entering the track.

20.11.1   Drainage of Precipitation Falling on the Track

Precipitation onto the track will enter the ballast, unless the ballast is highly fouled. The water will then flow laterally out of the ballast into the trackside drainage system or enter the subballast. The water entering the subballast will either drain laterally out of the subballast or continue downward onto the subgrade. The ability of water to drain laterally requires that the drainage paths at the edge of the ballast and subballast layers not be blocked. Two conditions need to be met to achieve this requirement: (1) the ballast shoulder and the edge of the subballast must be free-draining and (2) discharged water must be able to flow away from the track. The lowest point in the granular layer system (bottom of subballast) is most critical, assuming low subgrade permeability. Surface ditch drains can collect water from the ballast and subballast. The type of rainstorm event, as well as the permeability of the materials, will affect the amount of water entering the subballast.

The degree of fouling generally controls the form of seepage through the ballast. Under clean ballast condition, the ballast will have a high void ratio and water will drain freely. As fouling of the ballast progresses, voids are filled with fouling material and the permeability and flow velocities are reduced. Due to the reduced flow velocities, the subballast is exposed for a longer time to the precipitation draining out of the ballast, resulting in a higher subballast infiltration rate. When the permeability of the subballast is low, there is less infiltration and more water discharge laterally out of the ballast. Also, the greater the lateral slope of the subballast toward the side, the less subballast infiltration occurs.

In a similar manner, for a given ballast degree of fouling and a given subballast permeability, low-intensity storms of long duration cause more seepage into the subballast than high-intensity storms of short duration. Storms of high intensity and short duration cause a higher water level within the ballast and hence a high-energy head, and so water is discharged quickly out of the ballast. The subballast is therefore exposed to the precipitation for a relatively short period of time. During a storm of low intensity but long duration, however, water build-up in the ballast is less and so the energy head is low. The subballast is therefore exposed to the precipitation for a longer time period.

Particularly difficult is drainage of water from tracks surrounded on one or both sides by other tracks. Not only is the drainage path to the side longer than for a single track, but a suitable drainage path is more difficult to maintain. Cross-drains under the outer tracks or longitudinal drains between tracks may be needed.

20.11.2   Surface Drainage Systems

A proper surface drainage system is necessary to remove surface water in the right-of-way. Sources of this water are seepage out of the ballast and subballast, runoff from the cut slope, and surface runoff from areas adjacent to the track. Open ditches parallel to the track are the most common component of a surface drainage system. In cuts, ditches parallel to the track usually drain water discharged from the cut face as well as lateral discharge out of the tracks.

Ditches on top of cuts should intercept water from drainage basins adjacent to the cut before it reaches the cut slopes and divert it to a drainage inlet structure or to a natural watercourse nearby. This reduces slope erosion problems and also reduces the required capacity of the trackside ditches. Because the cutoff ditches are placed on top of cuts and are usually not visible from the tracks, they are often overlooked during regular track maintenance.

To predict the quantity of surface runoff and lateral discharge out of the tracks that the system will need to handle, the rainfall at the site being evaluated needs to be characterized. This is done in terms of frequency of occurrence of a storm of a particular duration and intensity. Intensity-duration-frequency curves can be developed from available meteorological records for the site. Then an appropriate storm return period needs to be selected. As the return period becomes longer, the maximum storm intensity likely to be encountered will increase. Then the larger the selected return period for design, the smaller the risk of the ditch overflowing and causing damage to the tracks, but the higher the cost of the system to accommodate the larger quantity of water. A design return period of 5 to 10 years is typically appropriate for ditch design.

20.11.3   Considerations for Ditch Design

Ditches parallel to the track are usually unlined and require a high level of maintenance to remove vegetation and sedimentation and to restore the ditch side-slopes where they have eroded. For the ditches to remain functional and to avoid deposition of sedimentation, a longitudinal minimum slope of 0.5 percent is recommended (Heyns 2000). Ditches placed in long cuts or on flat terrain with the track profile less than the recommended minimum ditch slope therefore become deep (and hence far bigger than the required flow capacity) as they drain towards the outlet. As an alternative, the ditch may be lined to allow placement at a grade shallower than 0.5 percent. For example, smooth-lined concrete ditches can typically be placed at a minimum grade of 0.25 percent.

Stabilization against erosion of ditches is necessary for severe hydraulic conditions. Stabilization measures include rigid linings, such as concrete, or flexible linings, such as vegetation or riprap.

Rigid linings are impermeable and are useful in flow zones where high shear stress or nonuniform flow conditions exist, such as at a cut-to-fill transition where the ditch outlets onto a high fill. Although rigid linings are nonerodible, they are susceptible to structural failure. The major causes of such failures are underlying soil movement from freeze-thaw cycles, swelling-shrinking cycles, and undermining by water.

20.11.4 Subsurface Drainage Systems

Groundwater flow into the subballast from the subgrade is a problem only in cuts where water can enter the ground from higher elevations. This water needs to be intercepted by subsurface drains to prevent it from weakening the subgrade. However, groundwater also can be a problem in level ground when the water table in the track subgrade is within the zone of influence of the train loading 13 feet (4 meters). Subsurface drains can be used to lower this water table and may also be needed to help drain the subballast. An exploration should be made to estimate the extent and nature of the groundwater in order to design the subsurface drainage system properly.

A subsurface drainage system is an underground means of collecting gravitational or free water from the track substructure. Provided that a proper surface drainage system is present, gravitational or free water in the track substructure comes from both precipitation onto the track and groundwater flow. If a proper surface drainage system is not present, surface flow also could be a source of subsurface water and must be considered in the design of the drainage system.

The flow rate of water out of the subballast to a drainage ditch may not be fast enough to keep the subballast from saturating. Factors causing this include long seepage distance, low permeability of the subballast, and settlement of the subgrade surface causing a depression. In these cases, a subsurface drainage system is appropriate for removing the water trapped within the subballast.

Subsurface drains that run laterally across the track are classified as transverse drains and are commonly located at right angles to the track centerline. If the ground water flow tends to be parallel to the track, transverse drains can be more effective than longitudinal drains in intercepting and/or drawing down the water table. Also, where ballast pockets exist in the subgrade, transverse drains can be an effective way to drain water from the low location in the ballast pocket. Transverse systems usually connect to the longitudinal subsurface system or the surface drainage system, such as a ditch.

In a multitrack system the tracks in the center should have lateral subballast slopes that match (or are higher than) the lateral slopes of the outside tracks to allow water to discharge under the outside tracks. Where the center tracks are lower than the outside tracks, the granular layer thickness of the outside tracks can be increased to allow continuous lateral drainage or a longitudinal drainage system should be placed between the tracks. In either case the seepage distance is long, resulting in slow drainage. An alternative would be to install transverse drains to carry water from the inside tracks under the outside tracks to a discharge point.

In a multitrack system the path for water to flow to the surface drainage system may become very long, even where a proper subballast lateral slope exists. For example, in a four-track system a water particle falling between the center tracks has to drain two track widths; thus, drainage may be inadequate. Therefore, it may always be desirable practice to drain water with a proper drainage system between the tracks.

20.12 MAINTENANCE IMPLICATIONS

The decisions made during design and construction of new track have a major effect on the cost of track maintenance. Special attention should be paid to subballast and subgrade drain-age from under the track because they are very difficult to fix after the track is in service. Cutting construction costs on these important components may result in large maintenance cost for years afterwards to compensate for the construction shortcomings.

ACKNOWLEDGMENTS

Vincent R. Terrill is acknowledged for sharing his extensive railway experience with the writer over many years, and in particular for his willingness to review the manuscript of this chapter and provide many valuable suggestions.

REFERENCES

Agarwal, M. M. 1998. *Indian Railway Track*, 12th ed. New Delhi: Prabha & Co.

American Railway Engineering and Maintenance-of-Way Association (AREMA). 2003. *Manual for Railway Engineering.* Landover, MD: AREMA.

ASTM C295. "Standard Practice for Petrographic Examination of Aggregates for Concrete." *ASTM Annual Book of Standards,* Section 4, *Construction,* vol. 04.02, *Concrete Mineral Aggregates.*

ASTM C856. "Standard Recommended Practice for Petrographic Examination of Hardened Concrete." *ASTM Annual Book of Standards,* Section 4, *Construction,* vol. 04.02, *Concrete and Mineral Aggregates.*

Esveld, C. 1989. *Modern Railway Track.* Duisburg: MRT-Productions.

Federal Railroad Administration (FRA). 1998. *Track Safety Standards,* Part 213, Subpart A to F, Class of Track 1 to 5 and Subpart G for Class of Track 6 and higher. U.S. Department of Transportation, FRA, Washington, DC.

Hay, W. W. 1982. *Railroad Engineering,* 2nd ed. New York: John Wiley & Sons.

Heyns, F. J. 2000. "Railway Track Drainage Design Techniques." Ph.D. dissertation, University of Massachusetts, Department of Civil and Environmental Engineering, May.

Kish, A., and G. Samavedam. 1991. "Dynamic Buckling of Continuous Welded Rail Track: Theory, Tests, and Safety Concepts." Rail-Lateral Track Stability, 1991—*Transportation Research Record* 1289.

Li, D., and E. T. Selig. 1998. "Method for Railroad Track Foundation Design: Development" and "Method for Railroad Track Foundation Design: Applications." *Journal of Geotechnical and Geoenvironmental Engineering, ASCE* 124(4):316–22 and 323–29.

Marich, S., P. J. Mutton, and G. P. Tew. 1991a. *A Review of Track Design Procedures,* vol. 1, *Rails.* Melbourne: BHP Research-Melbourne Labs.

———. 1991b. *A Review of Track Design Procedures,* vol. 2, *Sleepers and Ballast.* Melbourne: BHP Research-Melbourne Labs.

Selig, E. T., and J. M. Waters. 1994. *Track Geotechnology and Substructure Management.* London: Thomas Telford.

Talbot, A. N. 1933. "Sixth Progress Report of the Special Committee on Stresses in Track." *Bulletin* 358.

CHAPTER 21
IMPROVEMENT OF RAILROAD YARD OPERATIONS*

Sudhir Kumar
Tranergy Corporation
Bensenville, Illinois

21.1 INTRODUCTION

21.1.1 The Importance of Railroad Yards

Railroad yards play an important role in railroad operations. Yards consist of a large number of tracks grouped for the purpose of disassembling, sorting, and assembling cars in a train. Trains are brought into a receiving yard and sent on to a classification yard where cars are sorted and assembled into new trains, which are then dispatched to their new destinations. Railroads assemble a new train of cars based on the delivery requirements of their customers. (For a general discussion of yards and terminals see Petracek et al. 1997; *Railway Age* 2001; Wong et al. 1978; Christianson et al. 1979; AREMA 2001.)

Figure 21.1 shows a simplified diagram of car movement of which the classification yard process is a major component. Cars arrive in a yard in one of three ways:

1. From over the road trains
2. From an interchange yard where the cars of different railroad companies are interchanged and dispatched as regrouped cars
3. Loaded by a shipper and moved by a switch engine from an industry siding or yard to be dispatched to a destination specified by the shipper

After arriving in a receiving yard, these cars are moved into an adjacent yard called the classification yard. Here they are classified by being moved into different tracks and regrouped to make up various trains. These trains are finally moved into a departure yard and become one of the three types of train—over the road train or an interchange yard or delivered to an industry yard or siding via a switch engine as the final destination of that car.

For a railroad, the yard process is an essential but non-revenue-producing component. It has been estimated that nearly one-fourth of a railroad's expense is yard-related. In a DOT study (Petracek et al. 1997) it was reported that a typical freight car spends an average of 62 percent of its time in either terminal yards or intermediate yards. By comparison, the car spends only 14 percent of its time in line haul operations and, of this, only 6.6 percent in revenue producing service. This means that in that time a car spends only one in 15 days in revenue producing operation and more than 9 days in different

---

*Reprinted from the First Edition.

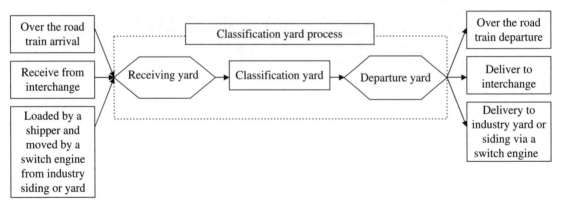

FIGURE 21.1 A simplified car movement diagram showing the classification yard process.

yard processes. The importance of yard operations was noted in an interview with Hunter Harrison (*Railway Age* 2001) regarding the seven steps he proposes for superior railroad service and a better bottom line. Of the seven steps, four concerned yard operations are

1. Minimize car dwell time in yards.
2. Minimize classifications in yards.
3. Use multiple traffic outlets between yards to keep traffic moving.
4. Space trains to support a steady workload flow through various yard processes.

21.1.2 Types of Railroad Yards

The present chapter is focused largely on railroad classification yards. There are several other types of yards, including industrial yards, passenger terminals, TOFC/COFC yards, storage yards, and interchange yards.

Classification Yards.   As stated earlier, classification yards are used primarily to make new trains from the cars of arriving trains.

Industrial Yards.   Industrial yards are used to collect and distribute freight cars to local industries. Cars are received from a classification yard to a local industrial yard for final delivery. In these yards, a resorting of cars takes place and local or industrial switch trains are made up, which are then sent to industrial sidings using switch engines.

Passenger Terminals.   Passenger terminals can be thought of as yards in which the passengers travel to different trains on different tracks. The trains are not sorted and stay on one track as an integral unit.

Storage Yards.   As the name implies, storage yards are used to store rolling stock, which often consists of empty cars. Depending on their demand, surplus cars are kept at such locations.

Interchange Yards.   These yards are used as points of interchange for cars between connecting railroads. Some yards are used exclusively for interchange movements and operations.

TOFC/COFC Yards.   These yards handle both the trailer-on-flat-car (TOFC) or container-on-flat-car (COFC) loadings and unloadings. Their design can vary considerably based on the movement of cars and the loading and unloading processes. They also generally carry high-speed cars, which are classified as intermodal traffic cars.

21.1.3 Classification Yards

All classification yards are designed to sort and reassemble cars into new trains. The type of classification yard used depends on the volume of cars to be handled per day. Other factors that influence the type of yard suitable are the character of traffic to be handled and train schedules. An economic study is therefore necessary to determine how to minimize the costs of classifying each car.

Most classification yards provide several supporting railroad operations other than classification. These include locomotive and freight car service and repair, inspection of cars for mechanical defects, and weighing of cars for revenue purposes. The inclusion of all these activities makes the classification yard a major center of railroad operations.

There is no fixed number of cars to be handled by each type of yard. However, the following numbers give an idea of the type of yard to be used, which is based on, among other things, the volume of car traffic that it handles.

1. Hump yards or gravity yards handle the largest volumes of cars, with most of them humping over 1,000 cars per day and some more than 2,000 cars per day.

2. Mini-hump yards and ladder track yards, as the name implies, are smaller versions of regular hump yards and handle 600 to 1,500 cars per day.

3. Double flat yards handle between 300 and 600 cars per day.

4. Single flat yards handle the smallest volumes, up to 300 cars per day.

Hump Classification Yards.   Hump yards, or gravity yards, are the largest car volume carriers of the railroads. There are presently over 50 hump yards in the United States. On average, each hump yard classifies about 1,500 cars a day. This translates to a nationwide total of 75,000 cars per day and 27,000,000 cars per year classified in these yards.

Figure 21.2 shows the general arrangement of a hump classification yard with inspection, maintenance, and servicing facilities. Inbound trains enter the receiving yard, and when the hump yard is ready to receive them the cars are moved on to the main hump. One or more cars are cut from the train and released on a hilltop whose incline varies based on the speed required to take the car(s) down to the classification track. On the way to their final classification, cars may pass through master, intermediate, and group retarders. Only three groups of classification tracks are shown in Figure 21.2. There may be up to eight groups in a yard. There may also be intermediate retarders (not shown) before the cars reach the group retarder. Once all the cars for a particular track have been classified, they are moved to the outbound or departure yard. Here they are coupled with one or more locomotives and the final train leaves the departure yard for the main line. Depending on the number of groups and tracks in a yard, there are generally two or more towers. One tower controls and oversees the classification process while another oversees the trim and departure process. Controllers in the towers have a wide view of the yard and control the cars through retardation and switches to classify them and make up a train. There may also be an inspection and running repair shop, a light car repair shop, a heavy-car repair track, locomotive servicing shop, and inspection and fuel tracks. The figure shows a nearly full service yard for a railroad.

Figure 21.3 is a condensed view of the profile of a typical hump yard. As the name implies, the cars are pushed to a hump by a locomotive, a point that is usually 10 feet or more than the average elevation of the yard. Most hump yards are shaped like a bowl, with cars entering the yard with a downhill (negative) grade and leaving the yard in the trim area with a slight uphill or positive grade. The hump crest follows the hump lead. At the crest the resistance pin is pulled from the coupler of the cars, releasing one or more cars on a downhill slope track at about 3 percent grade. While in motion, the car is weighed by an electronic weighing scale. The car enters the master retarder at a speed of 9.9 to 10.4 mph (the grade of the master retarder is often over 4 percent) and exits at 6 to 12 mph. The car(s) then enter the group retarder at 6 to 11 mph and exit at about 5 mph. Using this method, a car enters the classification track and couples at a speed lower than the exit speed from the group retarder. The ideal coupling speed of approximately 4 mph minimizes damage to the car and the contents inside. Railroads prefer lowering exit speeds from the group retarder to make cars roll and couple consistently at about 4 mph. This is often

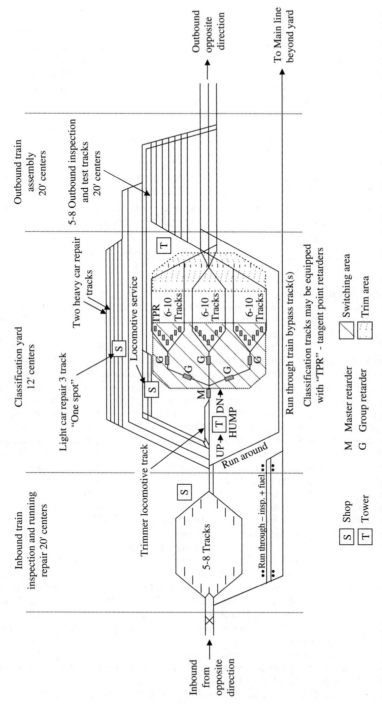

FIGURE 21.2 General arrangement of a classification yard with inspection, maintenance, and servicing facilities. (Based on a drawing by the late David G. Blaine.)

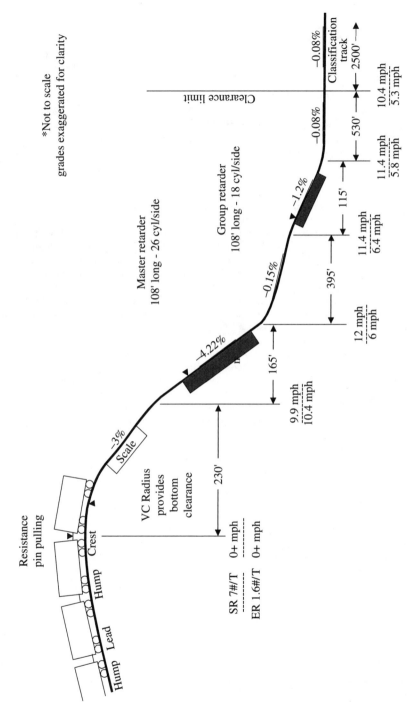

FIGURE 21.3 Profile of a typical hump yard. (Based on a drawing by the late David G. Blaine.)

not possible largely due to continually changing weather-dependent wheel-rail rolling friction. Many yard controllers therefore allow cars to exit the group retarder at higher speeds so that even the slowest rolling car reaches its destination and couples with the car ahead. As a result, most cars couple at speeds much higher than 4 mph, and some collide and derail. This can also cause the track to become damaged. The reasons for this and a solution to this problem will be discussed later.

Group retarders primarily control car speed. However, techniques developed recently combine different types of retarders for more accurate car speed control. The distances shown in Figure 21.3 of the different grades and slopes in the yards are for a small hump yard. These distances can be considerably larger for a bigger hump yard. The number of classification tracks shown in Figure 21.3 is only between 20 and 30. In modern larger hump yards, the number of tracks can exceed 60. Large-classification hump yards are expensive to operate and maintain, causing many railroads and industries with their own railroads to use a smaller yard, commonly known as a mini-hump yard.

Mini-Hump Yards. The mini-hump yard is a variation of the conventional large hump yard. It can handle a reasonably large traffic volume of 500 to 1,500 cars per day ("Pint Sized Gravity Yards" 1975). It has a smaller hump crest 3 to 10 feet in height and may not use any retarders. The distance between the crest and clear point is also less, about 500 feet, and it has fewer classification tracks, generally less than 20. Some of these yards have only one master retarder, and some may have group retarders as well. Some of these yards use hydraulic retarders called Dowty retarders on tangent track (to be discussed below). The control of these retarders may be fully automatic, semi-automatic, or manual. Because of their smaller size, these yards cost less than conventional hump yards and often do not exceed the cost of flat yards. They can achieve humping rates comparable to conventional yards, which are about 2 to 3 cars per minute. The flat yards discussed below generally have much smaller rates of classification.

Flat Yards. Most of the railroad yards in the United States are flat yards. These yards are designed to handle a smaller volume of cars. A good classification rate for such yards is 200 to 300 cars per day. Figure 21.4 shows a typical flat yard for switching from either end. A train approaches the yard backwards (the locomotive pushes the cars into the yard). As the cars to be uncoupled and classified approach the entry point, the locomotive accelerates the train to between 4 and 10 mph and then brakes. With the resistance pin at the coupling of the car pulled, the cars become uncoupled by inertia and continue to

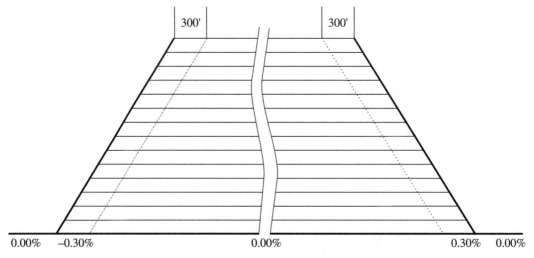

FIGURE 21.4 Typical flat yard for switching at both ends.

travel at the speed at which they become uncoupled. The rest of the train slows to a stop due to the applied brakes. The uncoupled car(s) enter the flat yard and go on track AB or CD depending on the side that the train enters the flat yard. The switch of the desired track in the yard is closed so that the car or group of cars can then roll into the classification track, where they are intended to make up the new train. The point of entry A may have flat track as shown in the diagram or may be a minor hump of a few feet. The mini-hump used in certain yards has the advantage of providing the potential energy needed by the cars to keep rolling on AB or CD. In any case, a typical yard has a 300-feet-wide strip at a grade of approximately 0.3 percent downhill so that cars keep moving to enter the switch without slowing down or getting hung up. If a car stalls anywhere along the track and fails to reach its destination, the entire train enters the yard at point A and along track AB or CD to push the car into place. This "trimming" operation can become time-consuming if many cars stall during classification. Stalls occur in spite of train engineer experience because cars roll differently based on the weather and time of day. The inconsistent rollability and high energy consumption in going through a switch are serious problems that have been solved only recently; they will be discussed later. In a well-designed flat yard, cars may enter from either side, thus allowing a train to kick cars into the yard from point A or point C. The central section of the yard EFGH has zero slope to ensure that car rollability is similar for cars entering from either direction.

Ladder Track Yards.   Mini-hump yards with ladder track are capable of handling larger volumes of cars than flat yards. The ladder track or mini-hump yards take advantage of gravity and occasionally, retarders. Both the clasp-type friction retarders and the hydraulic Dowty retarders are used. This allows the train to back into the mini-hump and release cars one by one or in a group so that they reach the ladder track and eventually the individual track in which they have to be switched. Figure 21.5 is an example of a mini-hump yard with ladder tracks. The hump may only be a few feet above the level of the yard. If it is located at position O in the figure, the car accelerates for a

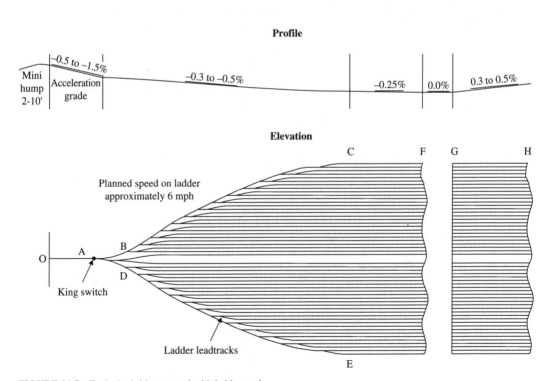

FIGURE 21.5   Typical mini-hump yard with ladder track.

couple of hundred feet. This part of the track, known as the acceleration grade, can be at a grade of anywhere between –1.5 and 0.5 percent. Before reaching the king switch, a clasp-type friction retarder may be located at point A. The ladder lead tracks, which essentially carry cars to individual tracks, are spread from B to C and D to E. The grade for these ladder lead tracks can be between –0.3 and –0.5 percent to keep cars rolling at a speed of approximately 6 mph. Dowty retarders may be located along the ladder lead track to slow faster rolling cars to the desired speed. The ladder lead tracks may stretch for several hundred feet, eventually coming to a steady grade tangent track of a few hundred feet length in the region C–F and E–I. The center of the yard at F–G and I–J is relatively flat, almost 0 percent grade. Toward the end of the yard there is an upward grade between 0.3 and 0.5 percent that slows cars if they are traveling at higher speeds. This section can also be a few hundred feet in length.

21.1.4 Retarders and Car Control

Cars move through yards due either to gravity (hump yard) or a locomotive kick (flat yards). That is, cars move with their own energy, derived either through potential or kinetic means. It is therefore necessary to have a good method to control car motion, particularly speed. Devices known as retarders are used to control car speed. Cars are coupled through impact at low speed. Railroads try to maintain that speed at approximately 4 mph. In practice, good coupling speeds are between 4 and 6 mph, preferably closer to 4 mph. Retarders are used to help achieve this coupling speed. Another reason that retarders are necessary to control the speed of free-rolling freight cars is to maintain a certain separation in cars as they are either humped or kicked. This separation is necessary to throw the different switches along the way in time for the track to receive the car. In practice, cars are switched in a 6- to 10-second interval and a minimum distance of 50 feet must be maintained between cars sent down the hump. The majority of retarders are clasp-type friction-based retarders. The Dowty system, which is a series of hydraulic cylinders bolted to the gage side of the rail, is also in use.

Friction-Type Retarders.   These retarders use a series of clasps with braking shoes that apply a predetermined force to the rims of car wheels a short distance above the rail to develop braking friction. The magnitude of braking force applied is based on the speed desired for the car. The braking shoes may be a composite material or a long beam that pushes against the wheels in their motion through the retarder. Retarder brake shoes should be kept free of grease, oil, and other lubricants in order to avoid compromising their retarding ability. When the wheels are clasped, an ear-piercing squeal radiates from the shoe-wheel contact. Some railroads set up 8-foot tall sound barriers on both sides of the retarder to mitigate this sound. The most common retarder mechanism for wheel braking is electropneumatic air cylinders. Other retarders use electrically actuated hydraulic power or electrically actuated spring power for the same purpose. Some other retarders are all-electric models suitable for heavy-duty jobs and amenable to automatic control. Finally, some retarders use the weight of the car wheel to determine the retardation force. These are often hydraulically, and sometimes mechanically, actuated. The degree of retardation in this case is determined by the speed and weight of the car, which needs to be slowed to a desired speed. Friction is the most utilized method of speed retardation because it is the most economical and efficient way to dissipate car energy.

Dowty Retarders.   These retarders are made up of a series of relatively small hydraulic cylinders mounted to the gage side of both rails for a certain length. Figure 21.6 shows a cross-section of one such Dowty cylinder unit (Petracek et al. 1997; Bick 1984; Melhuish 1983). As a car wheel flange rolls over the cylinder, the sliding piston is depressed by the weight of the car and oil or hydraulic fluid is moved from the chamber below the piston to the chamber above the speed-control valve. This motion of hydraulic fluid through an orifice dissipates car energy. A hydraulic oil pressure cylinder is mounted sequentially along the rail, and as the car moves forward it depresses each successive cylinder. Car speeds are maintained at a relatively low range of 4 to 6 mph in this process. Dowty cylinders function moderately well at low speeds but are prone to frequent failure. When car speeds are high, they do not respond well, resulting in car wheels riding on the cylinder and wheel tread

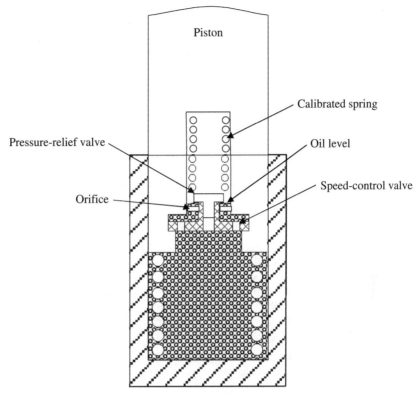

FIGURE 21.6 Dowty cylinder unit. (*Source:* Petracek et al. 1997.)

temporarily losing contact with the head of the rail. Heavy, faster cars cause a larger degree of damage to the cylinders. These oil pressure units have an advantage in that they do not produce wheel "squeal" but produce instead a "ringing bell" type of sound as each cylinder is depressed. They are also not affected by wheel contamination that may result from leakage of freight car contents such as oil, grease, molasses, etc.

Other types of retarders, including electrodynamic, pure hydraulic, and linear induction types, have been experimented with, but none have been effective or efficient enough to warrant their cost and be adopted by the industry.

Skates. Skates are devices installed on top of the railhead to block the movement of a wheel in a given direction. They consist of a steel frame with a blocking head on top of the railhead. They stop a car from moving unless the lateral force becomes excessive, at which point they slide. The friction force between the skates and the rail produces retardation. Although still used by some yards, skates are considered obsolete and have been replaced by spring-loaded retarders and other devices.

21.2 RECENT IMPROVEMENTS IN YARD EQUIPMENT AND OPERATIONS

Yards have traditionally been manpower-intensive. In the last several decades, therefore, the most important development has been computer control of the retarders, switches, and hump engine speed. Nearly all hump yards in the United States use computers for such operations. One or more manned hump towers are used to observe and control yard operations. The tower, or an adjacent building, houses computers and offices for the operation of the yard. The computers control the degree of retardation and switches in the classification section of the yard. Switches are operated by an electrically powered switch machine. Computers also control hump engine speed. A desired speed is maintained during humping to facilitate higher hump rates and complete humping of a train in one operation called a "cut." This is somewhat simpler than using variable humping speeds, which can also be computed and utilized to maximize the output of the yard. Thus, the computer control extends to switch operation, retarder control and operation, hump engine speed control, and overall traffic process control in the yard. The theory of operations advanced significantly in the 20th century, particularly during and after World War II. Yard operations have benefited from this theory as well, including processes such as multistage switching, sorting by block sequence, initial sorting by outbound trains, triangular switching, geometrical switching, and preblocking (Landow 1972; Daganzo, Dowling, and Hall 1982; Kraft 2000; Kubala and Raney 1983; Mundy, Heide, and Tubman 1992; Finian 1994). No attempt will be made here to explain any detail of these operations. There have been many additions to the yard operation that have made yard classification more efficient. Electronic devices mounted on cars called transponders identify the cars as they pass by. These are passive devices that respond to a radiated signal and bounce back the identification codes of the cars. In addition, there are devices like hotbox detectors that detect bad bearings on a car, dragging equipment detectors, broken wheel flange detectors, and loose wheel detectors. All these devices are used to identify bad cars, which are then diverted to a car maintenance shop in the yard to correct the defects. Each car is weighed in motion, within a certain accuracy, when it rolls over rail that is mounted on a scale.

Car computer control utilizes an important input that is measured by a device called a distance-to-couple measurement system. This measurement is derived from track circuits installed on the classification tracks. One method is to measure the impedance of a classification track from the clearance point to the nearest axle of a car present on the track; the axle acts as a shunt across the track. Because the impedance of the rails is proportionate to the distance from the circuit origin to the nearest shunt, it becomes possible to correlate impedance with the distance to couple. Other variations and methods also exist to measure the distance to couple. This is an important measurement needed to determine proper car coupling at a defined speed.

Rail lubrication, by wayside greasers on the gage side of the rail or through a hole in the railhead, is another addition that has developed to reduce car rolling resistance on sharp curves. Unfortunately, these greasers create a terrible mess in their surroundings. A long soaking pad between the rails absorbs excess grease and must occasionally be replaced—a process that can be filthy, expensive, and manpower-intensive.

21.3 THEORETICAL CONSIDERATIONS FOR CLASSIFICATION YARD DESIGN AND OPERATION

Railroad classification yards are designed to disassemble and sort cars without the use of locomotives. In hump yards, cars travel with the help of gravity and switches to a desired classification track and couple with the car ahead at speed of 4 to 6 mph. In flat yards, cars are kicked at a certain speed to a desired track and couple with the car ahead at speeds of 4–6 mph. In both cases, cars roll to their destination on their own, generally without additional motive power. The majority of their energy is used in overcoming wheel rail friction through sharp curves, switches, and tangent track. In this section, therefore, wheel rail friction will be discussed before discussing the mechanics of a yard.

21.3.1 Wheel Rail Rolling Friction

Wheel rail rolling friction coefficient μ, also known as adhesion coefficient, is defined as

$$\mu = \frac{\text{tangential force in wheel rail contact}}{\text{normal force on the wheel}} = \frac{\mu N}{N} \qquad (21.1)$$

where N is the normal force.

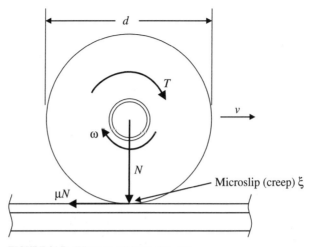

FIGURE 21.7 Wheel on rail in tractive mode.

In Figure 21.7, T is the torque applied to the wheel, V is the forward velocity, ω is the angular velocity, and d is the wheel diameter.

$$T = (\mu N d)/2 \tag{21.2}$$

Rolling friction is different from Coulomb friction in that there is only a microslip in the wheel rail contact called creep (ξ), which is not visible. Coulomb friction involves a macroslip. The microslip increases as torque T increases with μ until it reaches a maximum value, μ_{max}, for a certain microslip. If more power is applied, T or μ will not increase. Only wheel slip or creep will increase, and for dry rail μ will actually decrease.

The wheel microslip or creep ξ is defined as

$$\xi = \frac{\text{microslip velocity}}{\text{forward velocity}} \tag{21.3}$$

or

$$\%\xi = \frac{(\omega D/2) - V}{V} \times 100 \tag{21.4}$$

For a freely rolling wheel on tangent track, creep in the rolling direction is a small fraction of 1 percent. On curves, this value increases because the rolling radii of the two conical wheels mounted on a rigid axle are different, forcing a microslip of one or both wheels. Thus, the friction in the rolling direction (longitudinal adhesion) and the microslip (longitudinal creep) increase significantly on sharp curves. Other factors that also come into play on curves will be discussed below. (Considerable detail on wheel-rail friction adhesion and creep characteristics is provided by Alzoubi 1998; Kumar 1995; TCRP 1997; Kumar and Mangasahayam 1980; Kumar, Rajkumar, and Sciammarella 1980; Kumar, Krishnamoorthy, and Rao 1986.)

Figure 21.8 shows two typical adhesion creep plots of wheel and rail. The highest friction is produced under normal clean dry rail conditions, with μ reaching values of 0.6 or higher for steel wheel and rail. This benefits locomotive wheels, which are designed to produce traction through good friction levels, but is not so good for rolling car wheels, for which the rolling resistance is increased. It is particularly not beneficial for car wheels in a yard where the rolling of a car to its destination depends on a fixed amount of potential energy stored from a hump or kinetic energy transferred from a locomotive push. Adhesion levels on the rail, varying anywhere from 0.6 to 0.1 depending on rail contamination, the weather, and time of day, further exacerbate the problem. This makes it difficult for computer controls to function efficiently without continuous manual adjustment and leads to car collisions and/ or stalls before they reach their destination.

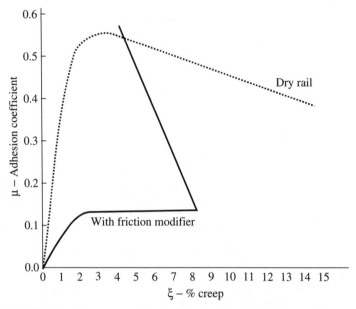

FIGURE 21.8   Adhesion-creep plots for dry, clean rail and with friction modifier.

In addition to friction in the rolling or longitudinal direction, wheel sets experience a much larger friction force in the lateral direction while negotiating a sharp curve, switch, or frog. Figure 21.9 shows a truck about to navigate a turnout. The wheels are forced to change direction by an angle θ, called the angle of attack. This results in a lateral slip of the wheel on the tread contact.

$$\text{Lateral creep } \xi_L = \frac{\text{lateral sliding velocity}}{\text{forward velocity}}$$

$$\xi_L = \frac{V\sin\theta}{V} = \sin\theta \cong \theta \tag{21.5}$$

or

$$\xi_L = \theta \text{ (radians)}$$

With the sign convention, a positive angle of attack is associated with positive force on the wheel.

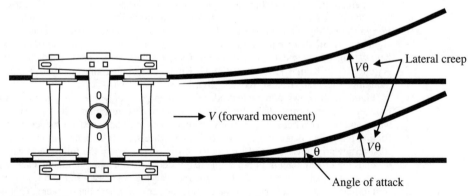

FIGURE 21.9   Lateral creep for a truck going through a switch or curve.

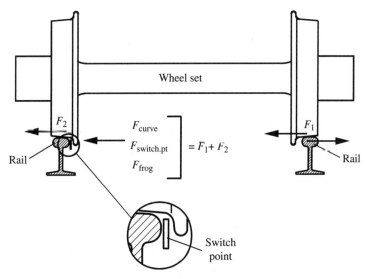

FIGURE 21.10   Lateral friction forces on a single axle negotiating a curve, switch or frog.

Figure 21.10 shows a wheel set sitting on rail through the region of contact, going over a curve, switch point, or frog. The wheel flange on the wheel opposite the turning direction hits the rail gage corner with a large force (up to 20,000 lb). This is responsible for wheel flange wear and the large gage-side wear of the high rail. It can sometimes result in derailments due to tipped rail or wheel climb on the rail. What is generally not recognized is that this large flange force is produced by lateral creep wheel friction F_1 and F_2 on top of both rails, which is produced by the lateral creep-related friction.

$$\mu_L = \frac{\text{lateral creep force}}{\text{normal force}}$$

$$\mu_L = \frac{\xi_L \cdot k \cdot n}{N} = k\xi_L \tag{21.6}$$

for the linear part of the μ, ξ relation. In equation (21.6), k is a proportionality constant that changes with rail surface contamination. It is highest in value for dry, clean rail and lowest for lubricated top of rail. As lateral creep increases, so does the lateral force.

Lateral creep is unavoidable on a curve because it is geometry-related, but the high lateral force associated with it is avoidable. This is achieved by using a suitable friction modifier on the wheel tread and the top of the rail, as discussed below.

In addition to the longitudinal and lateral creep, spin creep is experienced by the throat of the wheel flange and tread when this part of the wheel contacts the rail gage corner, such as in a conformal contact. The effect of this is not significant on the factors being discussed here and is therefore not further elaborated.

21.3.2 Mechanics of Yard Operation and Design

A car entering a classification yard has a given amount of energy that it can use to get to its destination. In the case of hump yards it has a certain amount of potential energy or energy head, and in the case of flat yards it has a certain amount of kinetic energy or velocity head, determined by the speed at which it is kicked. With this available energy it needs to overcome

- Rolling friction (resistance) on tangent track
- Rolling friction (resistance) on curved track
- Rolling friction on turnout, switch, and frog
- Air and wind resistance

Yard classification tracks are designed with downward or negative grades such that the speed of a car is generally maintained constant. This does not always happen, because the rolling resistance (lb/ton) of lighter axle load cars such as empty cars is considerably higher than the rolling resistance of heavy axle load cars. Thus, heavy cars tend to roll much faster and collide with cars ahead while light cars may stop short on the curve and need to be trimmed. This is further complicated by the change of rail friction coefficient with time of day, weather, etc., which will be discussed below.

Figure 21.11 shows the profile and grades of a hump yard track without retarders.

Let V_0, V_1, V_2, V_3 be car speeds at 0, 1, 2, and 3 in ft/sec

m be the mass of the car in lb

h_1 h_2 be car heights between 0 and 1

D_1, D_2, D_3 be distances in ft from 0 to 1, 1 to 2, and 2 to 3

G_1, G_2 be gradient coefficients between 0, 1 and 1, 2

$$G_1 = h_1/D_1 \qquad G_2 = h_2/D_2$$

R_T be total resistance in lb per lb in car motion

H_0, H_1, H_2 be velocity heads at 0, 1, and 2

g be acceleration due to gravity

w be car weight (lb)

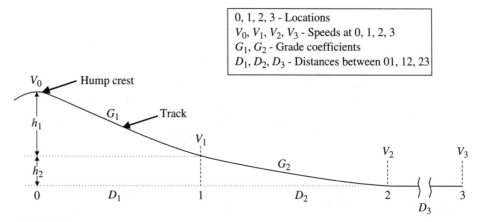

FIGURE 21.11   Profile of track grade in a hump yard.

For a car traveling from 0 to 1,

Potential energy   its kinetic   energy used to overcome
(based on energy $= mgh_1 =$ energy $+$ all resistance (work done)
head of h_1 in ft) $\frac{1}{2}mV_1^2$ $R_T \times D_1 \times W/2{,}000$

or
$$mgh_1 = \tfrac{1}{2}mV_1^2 + [(R_T \cdot mg)/2{,}000] \times D_1$$

or
$$2gh_1 = V_1^2 + [(R_T \times D_1 \times g)/1{,}000] \tag{21.7}$$

Similar relations can be written when the car travels from 1 to 2 and 2 to 3:

$$V_1 = \sqrt{[2gh_1 - (R_T D_1 g/1{,}000)]} \tag{21.8}$$

Rollability and Rollability Coefficient. Cars have good rollability if they have low rolling resistance. Rolling resistance is expressed as pound force per ton needed to move the car at constant speed on level tangent track. A rollability coefficient C is defined as the resistance force in lb per lb weight of the car:

$$C = \text{resistance force (lb)/lb weight of car} \tag{21.9}$$

$$\text{Resistance force} = (R_T \text{ lb/ton}) \times [W(\text{lb})/2{,}000 \text{ lb/ton}] = (R_T W/2{,}000) \text{ lb}$$

and so

$$C = (R_T W/2{,}000)/W = R_T/2{,}000 \tag{21.10}$$

If, for example, the force resisting the motion of the car is 5 lb/ton

$$C = 5 \text{ lb}/2{,}000 \text{ lb} = 0.0025 \text{ lb force/lb weight} = 0.25\%$$

This car will travel at constant speed on a -0.25% grade tangent track. The higher the value of the rollability coefficient, the harder it is for that car to roll.

Velocity Head Consideration. Velocity head at 0

$$H_0 = 0$$

Coming down to point 1 has a drop of height h_1 and

$$h_1 = D_1 G_1$$

Then the velocity head at 1 is

$$H_1 = D_1 G_1 - \text{loss due to rolling resistance}$$
$$= D_1 G_1 - D_1 C_1$$

Similarly, velocity head at point 2 is

$$H_2 = D_1 G_1 - D_1 C_1 - D_2 C_2, \text{ and so on}$$

The car resistance on one track consists of

  Resistance on tangent track (lb/ton)
  Resistance on curved track (lb/ton/degree of curve)
  Aerodynamic and wind resistance (lb/ton at a given car speed)

Accordingly, the total rollability coefficient C_T is a sum of four elements.

$$C_T = C_0 + C_D + C_A + C_W \qquad (21.11)$$

where typical values of these coefficients are given below:

  C_0 = rollability coefficient on tangent track = 0.0005 to 0.006 lb/lb
  C_D = rollability coefficient on curve of degree D = 0.00045 lb/lb/degree
  C_A = rollability coefficient due to aerodynamics = 0.00016 lb/lb/ft/sec car speed
  C_W = rollability coefficient due to wind = 0.0001 lb/lb/ft/sec of wind velocity

The rollability of a car is the inverse of its rollability coefficient. In other words, the smaller the rollability coefficient of a car, the better its rollability and vice versa.

The values of rollability coefficients given above are approximate and can be estimated in yards where cars operate at generally low speeds. In fact, these values change significantly with axle load, car shape, weather, and speed.

Example

  Rolling coefficient of a car:

$$\text{Car weight} = 100 \text{ tons} = 200,000 \text{ lb}$$

$$\text{On a good tangent track} \quad C_0 \cong 0.001$$

If the car is operating on a 15° curve in the yard, then

$$C_D = 0.00045 \times 15 = 0.00675$$

Assuming a car speed of 6 mph or 8.8 ft/sec

$$C_A = 0.00016 \times 8.8 = 0.00141$$

If there is no wind, $C_W = 0$
Thus, the total rollability coefficient is

$$C_A = 0.001 + 0.00675 + 0.00141 = 0.00916 \text{ lb/lb}$$

It should be noted that in this example the curve rollability coefficient of the car is by far the largest. In addition, there are large energy losses when a car negotiates a switch, which are not included in the above calculation.

There is no good estimate available for rollability coefficient on a switch. It is a well-known fact that the cars in a yard lose considerable speed when they go through a switch. This is a major problem that has been solved only recently.

The rolling resistance of cars on curves and tangent track is quite high and varies with time and weather. For efficient car control, it is necessary to reduce it and keep it consistent. This will be discussed in the next section.

Rolling Resistance and Loss in Speed with Wheel Tread and Top of Rail Friction Modification. It has been well accepted in the rail industry that top of rail friction modification (TOR-FM) reduces the rolling resistance and the lateral forces produced on a wheel on a curve (Kumar 1999; Davis, Kumar, and Sedelmeier 2000; FRA 2000; Clapper et al. 2001; Kumar, Yu, and Witte 1995; Reiff, Gage, and Robeda 2001; Reiff and Davis 2003; Reiff 2000). Yards are good candidates for the application of this phenomenon. This section presents a brief analysis of such an application. Technical details of achieving this will be discussed later.

The rolling resistance of a car can be expressed as

$$R(\text{lb/ton}) = a + bV + cV^2$$

where a, b, and c are constants and V is car speed in mph. In a yard where car speeds are small (≤ 10 mph), the cV^2 term is small enough to be neglected.

So car resistance in a yard can be expressed as

$$R = a + bV \tag{21.12}$$

Using this expression and starting with a car arriving at point 1 (Figure 21.11), let us use two scenarios for analysis. There are two conditions of operation, A and B. For condition A, the rail head and wheel treads are dry with high friction, and for condition B the top of rail and wheel treads are coated with a friction modifier. The friction modifier used acts as a lubricant for rolling car wheels but does not compromise the traction and braking of locomotive wheels.

$$\text{Loss of speed from 1 to 2 in dry condition} = \Delta V_A$$

$$\text{Loss of speed from 1 to 2 in friction modified condition} = \Delta V_B$$

$$\Delta V_A = (V_1 - V_2)_A$$
$$\Delta V_B = (V_1 - V_2)_B$$

As a good first approximation, it can be assumed that the loss of speed by a car is affected only by the force of its rolling resistance. The grade and curve are the same for both conditions.

Kinetic energy of the car at location 1 is

$$\tfrac{1}{2}mV_1^2$$

While rolling to location 2, some of this energy is used up in doing work against the rolling resistance R. Kinetic energy of the car at location 2 is therefore

$$\tfrac{1}{2}mV_2^2 = \tfrac{1}{2}mV_1^2 - mRD_2$$

or $\qquad \tfrac{1}{2}\left(V_1^2 - V_2^2\right) = RD \qquad$ and $\qquad \Delta V = V_1 - V_2 = 2RD/(V_1 + V_2)$

so $$\Delta V_A = \frac{2R_A D_2}{V_1 + (V_2)_A} \tag{21.13}$$

and $$\Delta V_B = \frac{2R_B D_2}{V_1 + (V_2)\,B} \tag{21.14}$$

Percentage change in speed as a result of friction modification is

$$= 100 \times \frac{\Delta V_B - \Delta V_A}{\Delta V_B}$$

$$= 100 \times \frac{2D_2 \{R_B / [V_1 + (V_2)_B] - R_A / [V_1 + (V_2)_A]\}}{2D_2 \{R_B / [V_1 + (V_2)_B]\}}$$

$$= 100 \times \left[1 - \frac{R_A}{R_B} \times \frac{V_1 + (V_2)_B}{V_1 + (V_2)_A} \right]$$

For most practical values of speed, the term $V_1 + (V_2)_B / V_1 + (V_2)_A$ is very close to 1 and therefore, % change in loss of speed from 1 to 2

$$= 100 \, (1 - R_A / R_B) \tag{21.15}$$

due to the friction modifier. This is the same as percentage change in rolling resistance due to friction modification. Therefore, we conclude that

% change in loss of speed = % change in rolling resistance

This means that the % change in loss of velocity is a very meaningful number that correlates directly to the % change in rolling resistance of the car. It is also a measure of the improvement in rollability.

Example

Car exit speed at 1 = 8 mph

For case A—dry rail

Speed at point 2 = 6 mph

(Loss of speed)$_A$ = 2 mph

For case B—with friction modifier

Speed at point 2 = 7 mph

(Loss of speed)$_B$ = 1 mph

$$\text{\% change in speed due to use of friction modifier} = \frac{2-1}{2} \times 100$$

$$= 50\%$$

So improvement in rollability = 50%

21.4 YARD FRICTION MODIFIERS SOLVE MANY PROBLEMS

In spite of considerable modernization and automation, yards continue to face serious problems in the areas of safety, lost productivity, loss and damage, increased track maintenance, and environmental concerns.

21.4.1 Problems in Yard Operations

Yard Safety Issues.   Railroad yards continue to face the hazards of car collisions and derailments. In addition to being safety concerns, these events can cost thousands of dollars in damage and lost productivity. Some of the problems associated with car derailments are tipped rail, crossed couplers, short stops, worn-out rail and switch points, and damaged and misaligned track.

Car collisions are not uncommon in a yard. They are caused mainly by rollout, insufficient speed control, inefficient skates due to excessive grease on the rails, and worn out of maintenance retarders.

Loss and Damage.   The number of derailments and collisions that take place in a yard is a significant portion of the total derailments or collisions that occur in a railroad. Some of these are due to human error, but a significantly large percentage are due to insufficient control of car speeds. When collisions or derailments take place, there is significant damage to the car and the freight inside, as well as the track, which in turn increases the track maintenance cost.

Loss of Productivity.   Collisions, derailments, and cars stalled before their destination cause a loss of productivity in a yard because classification is suspended while these problems are remedied. In theory, 10 or more cars can be humped every minute and one or more cars can be kicked in a flat yard in the same time. In reality, however, output is considerably less, about one-tenth of the maximum. One of the factors that contribute to this is stalled cars that have to be trimmed by locomotives.

Main Problem.   At present it is not possible to achieve consistent rolling speeds in hump yards or flat yards. In hump yards, computer-controlled retarders cannot be set once for ever-changing friction conditions. In flat yards, engineers have to make intelligent guesses on prevailing friction conditions and then estimate kicking speed. Both scenarios often result in either car collisions or short stops. Sharply curved track sometimes produces such high lateral forces that it tips the rail and causes derailments. High lateral forces combined with cross-coupler impact at high speeds are a recipe for derailments.

Wayside Greasers.   Until recently, many yards used wayside greasers to reduce the rolling resistance of cars on curves. Greasers are not generally effective in solving the problem and in some cases make things worse. In order to make rolling easier, it is a common practice to set greasers at high rates of grease application. Excessive grease makes the surroundings messy by forming grease pools under the applicator. A coating of grease also forms between the rails for a stretch of track beyond the application site, making the area a safety hazard. Thus, the effective overall cost of these greasers is quite high considering the large amounts of grease used and the manpower required to maintain them and clean up excess grease.

21.4.2 The New Yard Friction Modifiers

The newly developed yard friction modifiers solve many of the problems stated above. Many yard personnel use the abbreviated form "modifiers" to describe these units. This system sprays a clean, environmentally safe friction modifier fluid on the wheels of a car as it enters its classification track. The application is in the form of microburst lubricant jets, which hit both wheels of the lead axle of an approaching truck of a car. Figure 21.12 shows how the modifier works. It is located just before the main switch of a group of tracks into which the classification is intended. In the case of a hump yard it is located right after the retarder. There are two sensors mounted on the rail PS and FS as shown in the figure. There are also two nozzle holder units marked N on each of the rails. As the car wheelset trips sensor PS, a signal is sent to start a pump in the main unit C to develop a desired pressure in the system. When the axle trips sensor FS, a predetermined small quantity of friction modifier fluid is fired on the wheels by both nozzle units. The shot is fired in such a way that both the flange and tread of the wheel receive some of the friction modifier (FM). The box C contains a microprocessor and software that controls the entire process. Each modifier unit is set to apply a specified amount of FM to each group of tracks based on its requirements. The system also does not fire when it detects locomotive wheels. The unit can also be set to skip a desired number of cars before firing

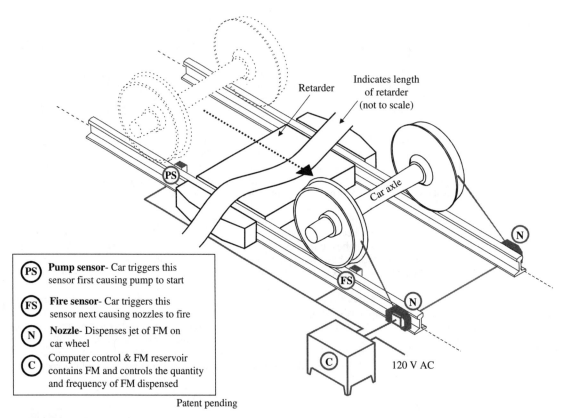

Retarder

Indicates length
of retarder
(not to scale)

Car axle

120 V AC

Patent pending

PS	**Pump sensor**- Car triggers this sensor first causing pump to start
FS	**Fire sensor**- Car triggers this sensor next causing nozzles to fire
N	**Nozzle**- Dispenses jet of FM on car wheel
C	Computer control & FM reservoir contains FM and controls the quantity and frequency of FM dispensed

FIGURE 21.12   How a yard friction modifier works.

again if so desired. The amount of FM dispensed for each shot can also be adjusted within certain limits. Such systems have replaced greasers on a large number of hump yards and some flat yards in the United States. The modifiers offer a clean and cost-effective solution to yard problems and lead to increased productivity and improved performance. They are currently produced by Tranergy Corporation in the United States.

21.4.3   Friction Modifier Fluid

Friction modifier fluid is a thin, synthetic polymer liquid, and not a grease. It is safe to handle and biodegradable. It contains no solids such as graphite or molybdenum disulphide and dissipates as trains pass over it, leaving little or no buildup on rails or wheels. It is effective for a wide range of temperatures, from −20°F to 150°F, ensuring smooth delivery flow throughout the year. In the United States, it is currently produced for Tranergy Corporation by Shell Oil Co.

21.4.4   How Many Modifier Units Does a Yard Need?

The number of units needed by a yard is determined by the groups of tracks present in the yard. Figure 21.13 shows the layout of a hump yard equipped with modifiers. It has five groups of tracks, five corresponding retarders, and one primary or master retarder. The modifier units needed are shown in the diagram. In this example there are five modifiers present, corresponding to the five

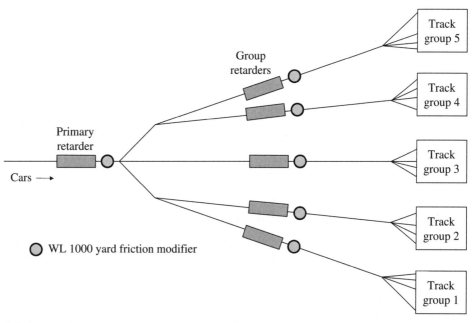

FIGURE 21.13   Layout of friction modifiers required in a yard with five groups.

groups, and a possible sixth unit placed right after the master retarder. At the minimum, a yard needs one modifier unit after the final retarder of each group of tracks. The friction developed on each group of tracks is different and hence each unit is tuned to its group. The unit after the master retarder provides consistent rolling and entry speeds to the group retarders further down the track. It also reduces wear and tear on the track that follows it.

21.4.5   Proven Improvements in Car Rollability and Reduction of Short Stops

A considerable amount of initial testing was conducted with the modifiers. It is generally accepted in the United States that in addition to being very clean, modifiers improve car rollability and reduce short stops in a yard. Data were first gathered by the Union Pacific Railroad Neff Yard in Kansas City, Kansas. Testing was conducted over a period of one month in the year 2000. The yard classified about 950 cars a day before the installation of the modifier units. After the installation of the modifier units and computer control adjustment of the retarder, the yard classified over 1,200 cars per day. While the modifiers were not solely responsible for this increase, they helped to improve consistent car rolling, which enabled the yard to be more productive. The test data in Figure 21.14 shows that car rollability improved anywhere between 30 and 57 percent for cars of various weights.

The Burlington Northern and Santa Fe Railroad also gathered data in their yard in Gales-burg, Illinois. The rollability test data (not presented here) were similar to data gathered in the Union Pacific Yard. The Galesburg yard also gathered data for short stops in the yard for a day on one group of tracks (Figure 21.15). The data clearly indicate that the number of short stops was reduced by over 60 percent. To date, the use of modifiers in over a dozen yards in the United States has shown the following:

- Reduced derailments
- Reduced short stops and needed locomotive trim
- Reduced exit speeds from group retarders
- Reduced car-coupling speeds

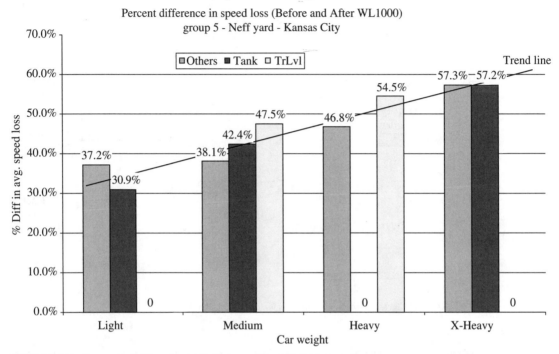

FIGURE 21.14 Improved rollability of cars with friction modifier (WL1000) opposed to greasers.

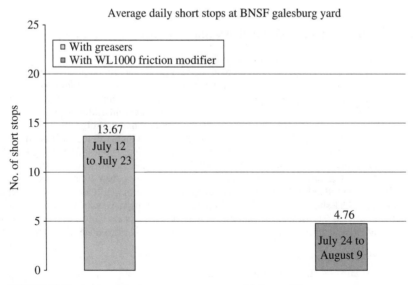

FIGURE 21.15 Average daily short stops: greasers versus friction modifiers.

- Eliminated tipped rail
- Consistent car-rolling speeds
- Reduced lateral forces on sharp curves
- Reduced rollouts and collisions
- Reduced track maintenance
- Increased yard output per day

21.5 NEW RAIL SWITCH ENHANCER IMPROVES YARD OPERATION AND SAFETY

A new electro-hydraulic system called the rail switch enhancer (RSE) has been developed to enhance yard operations. It automatically lubricates rail switches and the top of switch rail using computer-controlled quantities of lubricant to improve switch life, performance, and safety. For the benefit of the general reader unfamiliar with rail switches, a brief description of switches is presented before RSE is discussed.

21.5.1 General Background

Switches, found extensively in railroad industrial yards, are used in turnouts and crossovers to divert trains to other tracks. A typical hump yard may have 200 to 400 switches. Mini-hump yards and flat yards have 50 to 150 switches. Switches are a high-maintenance item for a railroad engineering department. It is estimated that nearly 30 percent of the engineering maintenance budget is used for switches and frogs. Enhancements in switch performance and life can thus lead to major savings and profitability for the railroad. Figure 21.16 shows the essential elements of a rail switch. The stock

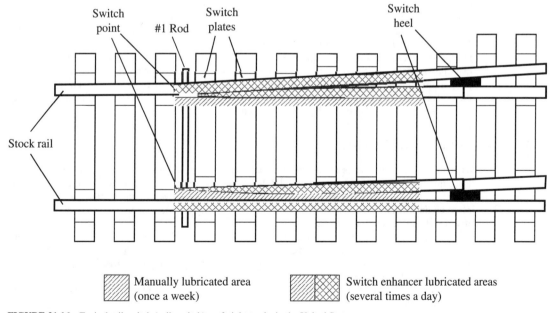

FIGURE 21.16 Typical rail switch (split switch) on freight tracks in the United States.

rail remains straight while turnout rail turns away from the main track. The switch in the diagram is in a position that would make a car travel straight. The switch point shown is moved with the help of switch rods and when moved into the turnout position causes the wheels to go on the turnout. The switch point, a rail with a sharp knife-like end, pivots about the switch heel. Only one rod is shown in the diagram, but there may be as many as five, depending on the length of the switch point. The switch point slides over switch plates in order to get into one of two positions—one on stock rail and the other on the turnout rail. Rail switches are used in conjunction with frogs and sometimes guardrails. All such railroad track components experience serious impact and wear depending on the sharpness of the turns in which they are located. It is therefore necessary to maintain these components regularly. In spite of maintenance, however, engineering departments find it necessary to replace switch rails and other moving components much more frequently than other rail. Switches are also a major factor in derailments. Studies by the U.S. Federal Railroad Administration have shown that the majority of derailments take place within 200 to 300 yards of a switch. As a car enters a switch, a sudden change in direction results in a lateral impact force on the wheels of the car (Figure 21.10). This force is produced by a sudden change in the lateral creep force on top of the switch rail and the other two rails. This contributes to the dynamic instability of the car, which can lead to derailments under certain conditions. In a yard, rolling cars often stall at or near a switch, indicating that considerable car energy is taken away by the switch, which is also due to the lateral creep forces mentioned above. It has been reported by certain rail transit systems that a large percentage of rail fractures occur near switches. It is theorized that these fractures are also related to the lateral creep impact force mentioned above. The current practice of maintaining switches involves using a lubricant or grease to lubricate the sliding plates of the switch. This reduces friction and makes the switch move more easily, but it does not provide any reduction of lateral force on top of the rails. The same is true for rail frogs, which are present at all rail turnouts and crossovers along with switches. Frogs, like switches, are affected by the creep force impact. At present there is no consistent protection or performance enhancement available for either rail switches or frogs. The current lubrication practice is manpower-intensive and leads to irregular maintenance of the switch, especially in remote areas. Switches are operated either manually or by a powered switch machine. When switches become difficult to throw, the life of the switch machine is reduced. If it is a manual switch, then an injury hazard develops for the person throwing the switch. The present practice of maintaining and operating a switch involves high costs. The highest among these is derailment cost. Another large cost component is disruption of traffic. In a yard, several principal switches are essential to yard operation. The failure of these switches can lead to severe disruption of traffic. These switches are prime candidates for performance enhancement and automatic maintenance. On the main line, disruption of traffic costs can be even higher. The maintenance and replacement of points is also a major cost. Other costs include wear on the power machine and/or injury to the person throwing the switch. Thus, an avoidance of all these costs and a new approach to improving the performance, operation, and safety of rail switches has been badly needed. Such a new approach is now available and is discussed in the next section.

21.5.2 The New Rail Switch Enhancer

The rail switch enhancer is an electro-hydraulic system that automatically lubricates rail switches and enhances their performance. Figure 21.17 shows a RSE placed on the switch of a rail turnout. The stock rail turnout rails along with frog and guardrails are shown in the diagram. The switch rails pivot on the switch heels. The switch enhancer components include a set of check valves and nozzles in a nozzle holder on brackets attached to each rail. These nozzles are supplied with lubricant under pressure by a hose or pipe from a pressurized tank containing the lubricant. The controller, located in a box, is mounted on or near the tank. The tank and the box may be enclosed in another box for security. The nozzle spray occurs at a specified frequency, dispensing a specified amount of lubricant to cover the switch, plates, switch rods, base of the stock rails, and top of the switch and the stock rails, providing protection and enhancing switch performance. On main line switches with heavy traffic, it applies a lubricant shot prior to every train going through the turnout. This is triggered by the signal from the switch controller that opens the switch for the turnout. The lubricant sprayed on

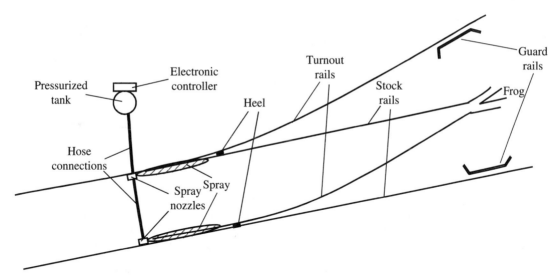

FIGURE 21.17   Schematic of a switch enhancer installed on a switch.

top of the stock and switch rails is carried forward by the train wheels and lubricates the top of all four rails, reducing the lateral creep impact forces and thus protecting the frog and the guard rails as well. Figure 21.18 shows a cross section of the switch point rail and the stock rail. Figure 21.19 shows a cross-sectional frontal view of the stock rail and the switch rail. The stock rail is mounted on a tie plate. The switch rail is supported on its own switch plate. The current occasional practice of manual lubrication provides lubricant coverage on the upper surface of the switch plate and

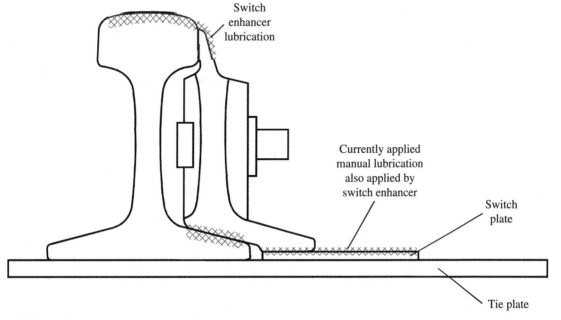

FIGURE 21.18   Cross-sectional view of a switch point and surfaces needing lubrication.

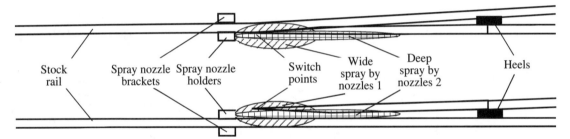

FIGURE 21.19   Plan view of a switch enhancer showing dual nozzle sprays covering the switch and the stock rail (crosshatched).

the stock rail base. Lubrication is actually required for the upper surfaces of the rails in addition to the lubrication of the switch plate and the stock rail base. These are shown as crosshatched surfaces that are lubricated by the switch enhancer. Figure 21.19 shows a plan view of the switch enhancer spraying the lubricant jets. The spray nozzle brackets and the nozzle holders are mounted on the stock rail. The nozzle holders each have two separate orifices. In order to increase the coverage area laterally, a nozzle with a wider spray, nozzle 1, is used in addition to another nozzle with a long range of spray application, nozzle 2. This nozzle spray may reach the heel, if so desired. This arrangement provides lubrication to all the surfaces discussed earlier as well as the top of the rails of the switch point and stock rail.

In this way the rail switch enhancer continues to make an automatic lubricant application several times a day to switch components and the top of the rails at a defined time and quantity with one lubricant charge in the tank. The regularity of a small lubricant application reduces switch wear and improves train rolling on the switch, thereby enhancing switch operation.

ACKNOWLEDGMENTS

The author is most thankful to Mr. Suneet Cherian of Tranergy for his great support throughout the article, in transcribing the text, creating all figures, and editing the manuscript through each cycle of revision. The author also thanks Mr. Nikul Patel and Mr. Naved Tirmizi of Tranergy for transcribing the first version of the theoretical section of the text. Finally, he also thanks Mr. Saud Al Dajah, his Ph.D. student at I.I.T. Chicago, for participating in the Adhesion-Creep subsection.

REFERENCES

Alzoubi, M. 1998. "Adhesion-Creepage Characteristics of Wheel-Rail System under Dry and Contaminated Rail Surfaces." Ph.D. thesis, Illinois Institute of Technology, Chicago, IL.

American Railway Engineering and Maintenance-of-Way Association (AREMA). 2001. *Manual for Railway Engineering,* vol. 3, ch. 14, "Yards and Terminals." Landover, MD: AREMA.

Bick, D. E. 1984. "A History of Dowty Marshalling Yard Wagon Control System." *Proceedings of the Institute of Mechanical Engineers* 198B(2):19–26.

Christianson, H. B., et al. 1979. "Committee 14, Yards and Terminals,Report on Assignment #7 Yard System Design for Two-Stage Switching." *American Railway Engineering Association Bulletin* 81(11):145–55.

Clapper, J., V. Dyavanapalli, S. Kumar, and E. Wolf. 2001. "Top of Rail Lubrication Pilot Project Results on Wheeling & Lake Erie Railway." In *Proceedings of 2001 ASME International Mechanical Engineering Congress and Exposition,* New York, November 11–16.

Daganzo, C. F., R. G. Dowling, and R. W. Hall. 1982. "Railroad Classification Yard Throughput: The Case of Multistage Triangular Sorting." *Transportation Research A* 17A(2):95–106.

Davis, K., S. Kumar, and G. Sedelmeier. 2000. "Further Developments in Top of Rail Lubrication Testing." In *Proceedings of the 62nd Annual Meeting, Locomotive Maintenance Officers Association,* September 18–20.

Federal Railroad Administration (FRA). 2000. *Evaluation of a Top of Rail Lubrication System.* Research results, RR00-01, U.S. Department of Transportation, FRA, Washington, DC.

Finian, D. H. 1994. *Current Methods for Optimizing Rail Marshalling Yard Operations.* NTIS#ADA289371, Kansas State University, Manhattan, KS.

Kraft, E. R. 2000. "A Hump Sequencing Algorithm for Real Time Management of Train Connection Reliability." *Journal of Transportation Research Forum* 39(4):95–115, published jointly with *Transportation Quarterly* 54(4)

Kubala, R., and D. Raney. 1983. "A Modular Approach to Classification Yard Control." *Transportation Research Record* 927, 62–67.

Kumar, S. 1995. "Wheel/Rail Adhesion Wear Investigation Using a Quarter Scale Laboratory Testing Facility." Paper delivered at 1995 ASME International Mechanical Engineering Congress and Exposition, San Francisco, CA, November 15.

Kumar, S. 1999. "Top of Rail Lubrication System for Energy Reduction in Freight Transportation by Rail." Society of Automotive Engineers Technical Paper Series, 1999-01-2236.

Kumar, S., and R. Mangasahayam. 1980. "A Parametric and Experimental Analysis of Friction, Creep and Wear of Wheel and Rail on Tangent Track." *Proceedings of ASME Symposium on the General Problem of Rolling Contact,* AMD 40:139–55.

Kumar, S., P. K. Krishnamoorthy, and D. L. P. Rao. 1986. "Influence of Car Tonnage and Wheel Adhesion on Rail and Wheel Wear: A Laboratory Study." *ASME Transactions, Journal of Engineering for Industry* 108(1):48–58.

Kumar, S., B. R. Rajkumar, and C. Sciammarella. 1980. "Experimental Investigation of Dry Frictional Behavior in Rolling Contact under Traction and Braking Conditions." Proceedings of 7th Leeds-Lyon Symposium on Tribology, Leeds, UK, Sept. 1980, *Friction and Traction.* Mechanical Engineering Publications, September, 207–19.

Kumar, S., G. Yu, and A. C. Witte. 1995. "Wheel-Rail Resistance and Energy Consumption Analysis of Cars on Tangent track with Different Lubrication Strategies." Paper delivered at IEEE/ASME Joint Railroad Conference, Baltimore, April 4–6.

Landow, H. T. 1972. "Yard Switching with Multiple Pass Logic." *Railway Management Review Quarterly* 72(1):11–23.

Melhuish, A. W. 1983. "Developments in the Application of the Dowty Continuous-Control Method." *Transportation Research Record* 927, 32–38.

Mundy, R. A., R. Heide, and C. Tubman. 1992. "Applying Statistical Process Control Methods in Railroad Freight Classification Yards." *Transportation Research Record* 1341, 53–62.

Petracek, S. J., A. E. Moon, R. L. Kiang, M. W. Siddiquee. 1997. "Railroad Classification Yard Technology: A Survey and Assessment." FRA/ORD 76/304, Stanford Research Institute, Menlo Park, CA, January.

"Pint Sized Gravity Yards Keep the Traffic Rolling." 1975. *Southern Pacific Magazine* (January/February): 10–11.

Railway Age. 2001. Interview with Hunter Harrison. (May): 28.

Reiff, R. P. 2000. *Top of Rail Lubrication Implementation Issues.* Report# RS-00-001, Transportation Technology Center Inc., July.

Reiff, R. P., and K. Davis. 2003. "Implementing Locomotive-Based Top-of-Rail Friction Control for Improving Network Efficiency." Paper delivered at International Heavy Haul Association, 2003 Specialist Technical Session, Dallas, 5.87, 5.95, May 5–9.

Reiff, R. P., S. E. Gage, and J. A. Robeda. 2001. "Alternative Methods for Rail/Wheel Friction Control." In *Proceedings of 7th International Heavy Haul Conference,* Brisbane, Australia, June, 559–62.

Transit Cooperative Research Program (TCRP). 1997. "Improved Methods for Increasing Wheel/Rail Adhesion in the Presence of Natural Contaminants." *Research Results Digest* 17.

Wong, P. J., C. V. Elliot, R. L. Kiang, M. Sakasita, and W. A. Stock. 1978. Railroad Classification Yard Technology, Design Methodology Study. Federal Railroad Administration, Washington, DC, September.

CHAPTER 22
MODERN AIRCRAFT DESIGN TECHNIQUES*

William H. Mason

Department of Aerospace and Ocean Engineering
Virginia Polytechnic Institute and State University
Blacksburg, Virginia

22.1 INTRODUCTION TO AIRCRAFT DESIGN

This chapter describes transport aircraft design. We discuss the key issues facing aircraft designers, followed by a review of the physical principles underlying aircraft design. Next we discuss some of the considerations and requirements that designers must satisfy and the configuration options available to the designer. Finally, we describe the airplane design process in some detail and illustrate the process with some examples. The modern commercial transport airplane is a highly integrated system. Thus the designer has to have an understanding of a number of aspects of engineering, the economics of air transport, and the regulatory issues.

Large transports are currently manufactured by two fiercely competitive companies: Boeing in the United States and Airbus in Europe. Smaller "regional jets" are manufactured by several companies, with the key manufacturers being Bombardier of Canada and Embraer of Brazil. Any new airplane designs must offer an advantage over the products currently produced by these manufacturers (known as "airframers"). Key characteristics of current designs can be found in the annual issue of *Aviation Week and Space Technology,* the *Source-book.* The other standard reference is *Jane's All the World's Aircraft* (Jackson 2002). An electronic appendix to Jenkinson, Simpkin, and Rhodes (1999) provides an especially complete summary. Also, essentially all new transport aircraft use turbofan engines for propulsion, although there are a number of smaller turboprop airplanes currently in service.

In picking the basis for a new aircraft design, the manufacturer defines the airplane in terms of range, payload, cruise speed, and takeoff and landing distance. These are selected based on marketing studies and in consultation with potential customers. Two examples of decisions that need to be made are aircraft size and speed. The air traffic system operates near saturation. The hub-and-spoke system means that many passengers take several flights to get to their destination. Often this involves traveling on a regional jet carrying from 50 to 70 passengers to a major hub, and then taking a much larger airplane to their destination. They may even have to transfer once again to a small airplane to get to their final destination. From an airport operations standpoint, this is inefficient. Compounding the problem, the small regional jets require the same airspace resources as a large plane carrying perhaps 10 times the number of passengers. Thus, the designer needs to decide what size is best for both large and small passenger transports. At one time United Airlines operated two sections of a flight between Denver and Washington's Dulles Airport using wide-body aircraft. Frequently,

---

*Reprinted from the First Edition.

both aircraft were completely full. Thus, there is a need for even larger aircraft from the airspace operations viewpoint even though it may not be desirable on the basis of the operation of a single plane. Based on the system demands, Airbus has chosen to develop a very large airplane, the A380. Alternatively, Boeing predicts that the current hub-and-spoke system will be partially replaced by more point-to-point operations, leading to the major market for new aircraft being for about the size of B767. They reached this conclusion based on the experience of the North Atlantic routes, where B747s have been replaced by more frequent flights using the smaller B767s.

Another consideration is speed. Designers select the cruise speed in terms of the Mach number M, the speed of the plane relative to the speed of sound. Because the drag of the airplane rises rapidly as shock waves start to emerge in the flow over the airplane, the economical speed for a particular configuration is limited by the extra drag produced by these shock waves. The speed where the drag starts to increase rapidly is known as the drag divergence Mach number M_{DD}. Depending on the configuration shape, the drag divergence Mach number may occur between $M = 0.76$ and $M = 0.88$. It is extremely difficult to design an airplane to fly economically at faster speeds, as evidenced by the decision to withdraw the Concorde from service. Numerous supersonic transport design studies since the introduction of the Concorde have failed to produce a viable successor. In addition to the aerodynamic penalties, the sonic boom restriction for supersonic flight over land and the difficulty of achieving low-enough noise around airports (so-called community noise) makes the challenge especially severe.

The choice of design characteristics in terms of size and speed is at least as important as the detailed execution of the design. Selecting the right combination of performance and payload characteristics is known as the "you bet your company" decision. The small number of manufacturers building commercial transports today provides the proof of this statement.

The starting point for any vehicle system design work is to have information about current systems. In this chapter, we will use 15 recent transports as examples of current designs. We have divided them into three categories: narrow-body transports, which have a single aisle; wide-body aircraft, which have two aisles; and regional jets, which are small narrow-body aircraft. Table 22.1 provides a summary of the key characteristics of these airplanes. The values shown are the design

TABLE 22.1   Key Current Transport Aircraft

Aircraft	TOGW (lb)	Empty weight (lb)	Wingspan (ft)	Number of passengers	Range (nm)	Cruise (mach)	Takeoff distance (ft)	Landing distance (ft)
				Narrow body				
A320-200	169,800	92,000	111.8	150	3,500	0.78	5,900	4,800
B717-200	121,000	68,500	93.3	106	2,371	0.76	5,750	5,000
B737-600	143,500	81,000	112.6	110	3,511	0.782	5,900	4,400
B757-300	273,000	141,690	124.8	243	3,908	0.80	8,650	5,750
				Wide body				
A330-300	513,670	274,650	197.8	440	6,450	0.82	8,700	5,873
A340-500	811,300	376,800	208.2	375	9,960	0.83	10,450	6,601
A380-800	1,234,600	611,000	261.8	555	9,200	0.85	9,350	6,200
B747-400	875,000	398,800	211.4	416	8,356	0.85	9,950	7,150
B747-400ER	911,000	406,900	211.4	416	8,828	0.85	10,900	7,150
B767-300	345,000	196,000	156.1	218	5,450	0.80	7,550	5,200
B777-300	660,000	342,900	199.9	368	6,854	0.84	12,150	6,050
B777-300ER	750,000	372,800	212.6	365	8,258	0.84	10,700	6,300
				Regional jets				
CRJ200(ER)	51,000	30,500	69.7	50	1,895	0.74	5,800	4,850
CRJ700(ER)	75,000	43,500	76.3	70	2,284	0.78	5,500	4,850
ERJ135ER	41,888	25,069	65.8	37	1,530	0.76	5,052	4,363
ERJ145ER	54,415	26,270	65.8	50	1,220	0.76	5,839	4,495

TOGW, takeoff gross weight.

values, and the range and payload and associated takeoff and landing distances can vary significantly. Detailed performance data can be found for Boeing airplanes on their website: www.boeing.com. Other airframers may provide similar information.

While this section provides an overview, numerous books have been written on airplane design. Two that emphasize commercial transport design are Jenkinson, Simpkin, and Rhodes (1999) in the United Kingdom and Schaufele (2000) in the United States. Paul Simpkin had a long career at Rolls Royce, and the book he coauthored includes excellent insight into propulsion system considerations. Roger Schaufele was involved in numerous Douglas Aircraft Company transport programs. Two other key design books are by Raymer (1999) and Roskam (1987–1990) (an eight-volume set).

22.2 ESSENTIAL PHYSICS AND TECHNOLOGY OF AIRCRAFT FLIGHT

Aircraft fly by exploiting the laws of nature. Essentially, lift produced by the wing has to equal the weight of the airplane, and the thrust of the engines must counter the drag. The goal is to use principles of physics to achieve efficient flight. A successful design requires the careful integration of a number of different disciplines. To understand the basic issues, we need to establish the terminology and fundamentals associated with the key flight disciplines. These include

- Aerodynamics
- Propulsion
- Control and stability
- Structures/materials
- Avionics and systems

Shevell (1989) describes these disciplines as related to airplane design, together with methods used to compute airplane performance.

To understand how to balance these technologies, designers use weight. The lightest airplane that does the job is considered the best. The real metric should be some form of cost, but this is difficult to estimate. Traditionally, designers have used weight as a surrogate for cost. For designs using similar technology and sophistication, the lightest airplane costs least. One airplane designer has said that airplanes are like hamburger, you buy them by the pound. A study carried out at Boeing (Jensen, Rettie, and Barber 1981) showed that an airplane designed to do a given mission at minimum takeoff weight was a good design for a wide range of operating conditions compared to an airplane designed for minimum fuel use or minimum empty weight.

We can break the weight of the airplane up into various components. For our purposes, we will consider the weight to be

$$W_{TO} = W_{empty} + W_{fuel} + W_{payload} \qquad (22.1)$$

where W_{TO} = takeoff weight
   W_{empty} = empty weight, mainly the structure and the propulsion system
   W_{fuel} = fuel weight
   $W_{payload}$ = payload weight, which for commercial transports is passengers and freight

Very crudely, W_{empty} is related to the cost to build the airplane and W_{fuel} is the cost to operate the airplane. The benefit of a new technology is assessed by examining its effect on weight.

Example.   Weight is critically important in aircraft design. This example illustrates why. If

$$W_{TO} = W_{struct} + W_{prop} + W_{fuel} + \underbrace{W_{playload} + W_{systems}}_{W_{fixed}}$$

$$= W_{TO}\left(\frac{W_{struct}}{W_{TO}} + \frac{W_{prop}}{W_{TO}} + \frac{W_{fuel}}{W_{TO}}\right) + W_{fixed}$$

or
$$\left[1 - \left(\frac{W_{struct}}{W_{TO}} + \frac{W_{prop}}{W_{TO}} + \frac{W_{fuel}}{W_{TO}}\right)\right] W_{TO} = W_{fixed}$$

and

$$W_{TO} = \frac{W_{fixed}}{\left[1 - \left(\dfrac{W_{struct}}{W_{TO}} + \dfrac{W_{prop}}{W_{TO}} + \dfrac{W_{fuel}}{W_{TO}}\right)\right]}$$

Using weight fractions, which is a typical way to view the design, the stuctural fraction could be 0.25, the propulsion fraction 0.1, and the fuel fraction 0.40. Thus

$$W_{TO} = \frac{W_{fixed}}{(1 - 0.75)} = 4 \cdot W_{fixed}$$

Here 4 is the *growth factor*, so for each pound of increased fixed weight, the airplane weight increases by 4 pounds to fly the same distance. Also, note that the denominator can approach zero if the problem is too difficult. This is an essential issue for aerospace systems. Weight control and accurate estimation in design are very important.

The weight will be found for the airplane carrying the design payload over the design range. To connect the range and payload to the weight, we use the equation for the range R, known as the Brequet range equation:

$$R = \frac{V(L/D)}{sfc} \ln\left(\frac{W_i}{W_f}\right) \tag{22.2}$$

where R = range of the airplane (usually given in the design requirement)
V = airplane speed
L = lift of the airplane (assumed equal to the weight of the airplane, W)
D = drag

Since the weight of the plane varies as fuel is used, the values inside the log term correspond to the initial weight W_i and the final weight W_f. The specific fuel consumption (sfc) is the fuel used per pound of thrust per hour. The aerodynamic efficiency is measured by the lift to drag ratio L/D, the propulsive efficiency is given by sfc, and the structural efficiency is given by the empty weight of the plane as a fraction of the takeoff weight.

22.2.1 Aerodynamics

The airplane must generate enough lift to support its weight, with a low drag so that the L/D ratio is high. For a long-range transport this ratio should approach 20. The lift also has to be distributed around the center of gravity so that the longitudinal (pitching) moment about the center of gravity can be set to zero through the use of controls without causing extra drag. This requirement is referred to as *trim*. Extra drag arising from this requirement is *trim drag*.

Drag arises from several sources. The viscosity of the air causes friction on the surface exposed to the airstream. This "wetted" area should be held to a minimum. To account for other drag associated with surface irregularities, the drag includes contributions from the various antennas, fairings, and manufacturing gaps. Taken together, this drag is generally known as *parasite drag*. The other major contribution to drag arises from the physics of the generation of lift and is thus known as *drag due to lift*. When the wing generates lift, the flowfield is deflected down, causing an induced angle over the wing. This induced angle leads to an induced drag. The size of the induced angle depends on the span loading of the wing and can be reduced if the span of the wing is large. The other contribution to drag arises due to the presence of shock waves. Shock waves start to appear as the plane's speed approaches the speed of sound, and the sudden increase in drag once caused an engineer to describe this "drag rise" as a "sound barrier."

To quantify the aerodynamic characteristics, designers present the aerodynamic characteristics in coefficient form, removing most of the size effects and making the speed effects more clear. Typical coefficients are the lift, drag, and pitching moment coefficients, which are

$$C_L = \frac{L}{\frac{1}{2}\rho V_\infty^2 S_{ref}} \qquad C_D = \frac{D}{\frac{1}{2}\rho V_\infty^2 S_{ref}} \qquad C_m = \frac{M}{\frac{1}{2}\rho V_\infty^{2} c S_{ref}} \tag{22.3}$$

where L, D, and M are the lift, drag, and pitching moment, respectively. These values are normalized by the dynamic pressure q and a reference area S_{ref} and length scale c, as appropriate. The dynamic pressure is defined as $q = 1/2 \rho V^2$. Here ρ is the atmospheric density. The subscript infinity refers to the freestream values. One other nondimensional quantity also frequently arises, known as the aspect ratio, $\mathrm{AR} = b^2/S_{ref}$.

In particular, the drag coefficient is given approximately as a function of lift coefficient by the relation

$$C_D = C_{D_0} + \frac{C_L^2}{\pi \mathrm{AR} E} \tag{22.4}$$

where C_{D_0} is the parasite drag and the second term is the drag due to lift term mentioned above. E is the airplane efficiency factor, usually around 0.9. Many variations on this formula are available, and in particular, when the airplane starts to approach the speed of sound and wave drag starts to arise, the formula needs to include an extra term, $C_{D_{wave}} (M, C_L)$. Assuming that wave drag is small and that the airplane is designed to avoid flow separation at its maximum efficiency, the drag relation given above can be used to find the maximum value of L/D (which occurs when the parasite and induced drag are equal) and the corresponding C_L:

$$\left(\frac{L}{D}\right)_{max} = \frac{1}{2}\sqrt{\frac{\pi \mathrm{AR} E}{C_{D_0}}} \tag{22.5}$$

and

$$(C_L)_{L/D_{max}} = \sqrt{\pi \mathrm{AR} E C_{D_0}} \tag{22.6}$$

These relations show the importance of streamlining to achieve a low C_{D_0}. They also seem to suggest that the aspect ratio should be large. However, the coefficient form is misleading here, and the way to reduce the induced drag D_i is actually best shown by the dimensional form:

$$D_i = \frac{1}{q \pi E}\left(\frac{W}{b}\right)^2 \tag{22.7}$$

where b is the span of the wing.

Finally, to delay the onset of drag arising from the presence of shock waves, the wings are swept. We will present a table below containing the values of wing sweep for current transports.

The other critical aspect of aerodynamic design is the ability to generate a high enough lift coefficient to be able to land at an acceptable speed. This is characterized by the value of $C_{L_{max}}$ for a particular configuration. The so-called stalling speed (V_{stall}) of the airplane is the slowest possible speed at which the airplane can sustain level flight, and can be found using the definition of the lift coefficient as

$$V_{stall} = \sqrt{\frac{2(W/S)}{\rho C_{L_{max}}}} \qquad (22.8)$$

and to achieve a low stall speed we need either a low wing loading, W/S, or a high $C_{L_{max}}$. Typically, an efficient wing loading for cruise leads to a requirement for a high value of $C_{L_{max}}$, meaning that high-lift systems are required. High-lift systems consist of leading and trailing edge devices such as single-, double-, and even triple-slotted flaps on the rear of the wing, and possibly slats on the leading edge of the wing. The higher the lift requirement, the more complicated and costly the high lift system has to be. In any event, mechanical high-lift systems have a $C_{L_{max}}$ limit of about 3. Table 22.2 provides an example of the $C_{L_{max}}$ values for various Boeing airplanes. These values are cited by Brune and McMasters (1990). A good recent survey of high-lift systems and design methodology is van Dam (2002).

TABLE 22.2   Values of $C_{L_{max}}$ for Some Boeing Airplanes

Model	$C_{L_{max}}$	Device type
B-47/B-52	1.8	Single-slotted Fowler flap
367-80/KC-135	1.78	Double-slotted flap
707-320/E-3A	2.2	Double-slotted flap and Kreuger leading edge flap
727	2.79	Variable camber Kreuger and triple-slotted flap
747/E-4A	2.45	Variable camber Kreuger and triple-slotted flap
767	2.45	Slot and single-slotted flap

22.2.2 Propulsion

Virtually all modern transport aircraft use high-bypass-ratio turbofan engines. These engines are much quieter and more fuel efficient than the original turbojet engines. The turbofan engine has a core flow that passes through a compressor and then enters the combustor and drives a turbine. This is known as the hot airstream. The turbine also drives a compressor that accelerates a large mass of air that does not pass through the combustor and is known as the cold flow. The ratio of the cold air to the hot air is the bypass ratio. From an airplane design standpoint, the key considerations are the engine weight per pound of thrust and the fuel consumption.

$$W_{eng} = \frac{T}{(T/W)_{eng}} \qquad (22.9)$$

where the engine thrust is given by T. Typical values of the T/W of a high-bypass-ratio engine are around 6–7. The fuel flow is given as

$$sfc = \frac{\dot{w}_f}{T} \qquad (22.10)$$

where $\dot{w}_f$ is the fuel flow in lb/hr and the thrust is given in pounds. Thus, the units for sfc are per hour. There can be some confusion in units because the sfc is sometimes described as a mass flow. But in the United States the quoted values of sfc are as a weight flow. Table 22.3 provides the characteristics of the engines used in the aircraft listed in Table 22.1.

TABLE 22.3   Engines for Current Transport Aircraft

Aircraft	Engine	Thrust (lb)	Weight (lb)	sfc	T/W_{eng}
		Narrow body			
A320-200	IAE V2527-A5	26,500	5,230	0.36	5.1
B717-200	RR BR 715	21,000	4,597	0.37	4.6
B737-600	CFM56-7B	20,600	5,234	0.36	3.9
B757-300	PW 2040	41,700	7,300	0.345	5.7
		Wide body			
A330-300	Trent 768	71,100	10,467	0.56	6.8
A340-500	Trent 553	53,000	10,660	0.54	5.0
A380-800	Trent 970	70,000	—	0.51	—
B747-400	GE CF6-80C2	58,000	9,790	0.323	5.9
B747-400ER	GE CF6-80C2	58,000	9,790	0.323	5.9
B767-300	GE CF6-80C2	58,100	9,790	0.317	5.9
B777-300	RR Trent 892	95,000	13,100	0.56	7.25
B777-300ER	GE90-115	115,000	18,260	—	6.3
		Regional jets			
CRJ200(ER)	GE CF34-3B1	9,220	1,670	0.346	5.5
CRJ700(ER)	GE CF34-8C1	13,790	2,350	0.37	5.9
ERJ135ER	AE3007-A3	8,917	1,586	0.63	5.6
ERJ145ER	AE3007-A1/1	8,917	1,586	0.63	5.6

Values for thrust and fuel flow of an engine are quoted for sea-level static conditions. Both the maximum thrust and fuel flow vary with speed and altitude. In general, the thrust decreases with altitude, and with speed at sea level, but remains roughly constant with speed at altitude. The sfc increases with speed and decreases with altitude. Examples of the variations can be found in Appendix E of Raymer (1999). More details on engines related to airplane design can be found in Cumpsty (1998).

22.2.3 Control and Stability

Safety plays a key role in defining the requirements for ensuring that the airplane is controllable in all flight conditions. Stability of motion is obtained either through the basic airframe stability characteristics or by the use of an electronic control system providing apparent stability to the pilot or autopilot. Originally airplane controls used simple cable systems to move the surfaces. When airplanes became large and fast, the control forces using these types of controls became too large for the pilots to be able to move surfaces and hydraulic systems were incorporated. Now some airplanes are using electric actuation. Traditionally, controls are required to pitch, roll, and yaw the airplane. Pitch stability is provided by the horizontal stabilizer, which has an elevator for control. Similarly, directional stability is provided by the vertical stabilizer, which incorporates a rudder for directional control. Roll control is provided by ailerons, which are located on the wing of the airplane. In some cases one control surface may be required to perform several functions, and in some cases multiple surfaces are used simultaneously to achieve the desired control. A good reference for control and stability is Nelson (1997).

Critical situations defining the size of the required controls include engine-out conditions, cross-wind takeoff and landing, and roll response. Longitudinal control requirements are dictated by the ability to rotate the airplane nose up at takeoff and generate enough lift when the airplane slows down to land. These conditions have to be met under all flight and center-of-gravity location conditions.

22.2.4 Structures/Materials

Aluminum has been the primary material used in commercial transports. However, composite materials have now reached a stage of development that allows them to be widely used, providing the required strength at a much lighter weight. The structure is designed for an extremely wide range of loads, including taxiing and ground handling (bump, touchdown, etc.) and flight loads for both sustained maneuvers and gusts.

Typically, transport aircraft consists of a constant cross-section pressurized fuselage that is essentially round and a wing that is essentially a cantilever beam. The constant cross-section of the fuselage allows the airplane to be stretched to various sizes by adding additional frames, some in front of and some behind the wing, to allow the plane to be properly balanced. However, if the airplane becomes too long, the tail will scrape the ground when the airplane rotates for takeoff. The wing typically consists of spars running along the length of the wing and ribs running between the front and back of the wing. The wing is designed so that fuel is carried between the front and rear spars. Fuel is also carried in the fuselage, where the wing carry-through structure is located. Carrying fuel in the wing as well as the wing support of pylon-mounted engines helps reduce the structural weight required by counteracting the load due to the wing lift. Because the wing is a type of cantilever beam, the wing weight is reduced by increasing the depth of the beam, which increases the so-called thickness-to-chord ratio (t/c). This increases the aerodynamic drag. Thus, the proper choice of t/c requires a system-level trade off. An excellent book illustrating the structural design of transport aircraft is by Niu (1998).

22.2.5 Avionics and Systems

Modern aircraft incorporate many sophisticated systems to allow them to operate efficiently and safely. The electronic systems are constantly changing, and current periodicals such as *Aviation Week* should be read to find out about the latest trends. The survey by Kayton (2003) provides an excellent overview of the electronics systems used on transports. Advances in the various systems allowed modern transports airplanes to use two-man crews. Fielding (1999) has a good summary of the systems use on transport aircraft. The basic systems are

Avionics Systems

Communications

Navigation

Radar

Auto pilot

Flight control system

Other Systems

Air conditioning and pressurization

Anti-icing

Electrical power system

Hydraulic system

Fuel system

Auxiliary power unit (APU)

Landing gear

Each of these systems, listed in a single line, is associated with entire companies dedicated to providing safe, economical components for the aircraft industry.

22.3 TRANSPORT AIRCRAFT DESIGN CONSIDERATIONS AND REQUIREMENTS

22.3.1 The Current Environment and Key Issues for Aircraft Designers

In addition to the overall selection of the number of passengers and design range, described above, the designer has to consider a number of other issues. One key issue has been the selection of the seat width and distance between seats, the pitch. The seating arrangements are closely associated with the choice of the fuselage diameter. This has been a key design issue since the selection of the fuselage diameter for the DC-8 and B-707s, the first modern jet transports. This can be a key selling point of the aircraft. For example, currently Boeing uses the same fuselage diameter for its 737 and 757 transports: 148 inches. The comparable Airbus product, the A320, uses a fuselage diameter of 155 inches. Because of the details of the interior arrangements, both companies argue that they have superior passenger comfort. Typically, in economy class the aisles are 18 inches wide and the seats are approximately 17.5 to 19 inches wide, depending on how they are measured (whether the armrest is considered). In general, the wider the aircraft, the more options are available, and the airlines can select the seating arrangement.

The distance between rows, known as the pitch, can be selected by the airline and is not as critical to the design process. Airplanes can be lengthened or shortened relatively cheaply. The fuselage diameter essentially cannot be changed once the airplane goes into production. Typically, the pitch for economy class is 30 inches, increasing for business and first class seating.

Emergency exits (which are dictated by regulatory agencies), overhead bins, and lavatories are also key considerations. In addition, access for service vehicles has to be considered. In some cases, enough ground clearance must be included that carts can pass underneath the airplane.

In addition to passengers, transport airlines depend on freight for a significant portion of their revenue. Thus, the room for baggage and freight also requires attention. There are a number of standardized shipping containers, and the fuselage must be designed to accommodate them. The most common container, known as an LD-3, can fit two abreasts in a B777. The LD-3 is 64 inches high and 60.4 inches deep. The cross-section is 79 inches wide at the top and 61.5 inches wide at the bottom, the edge being clipped off at approximately a 45° angle to allow it to fit efficiently within the near circular fuselage cross-section. This container has a volume of 158 cubic feet and can carry up to 2,830 pounds.

The modern transports turn out to have about the right volume available as a natural consequence of the near-circular cylindrical fuselage and the single passenger deck seating. Regional jets, which have smaller fuselage diameters, frequently cannot fit all the passenger luggage in the plane. When you are told that "The baggage didn't make the flight," it probably actually means it didn't fit on the plane. A similar problem exists with large double-deck transports, where some of the main deck may be required to be used for baggage and freight.

Details of passenger cabin layout are generally available from the manufacturer's website. Boeing and Embraer are particularly good. Texts such as Jenkinson, Simpkin, and Rhodes (1999) provide more details.

22.3.2 Regulatory Requirements

The aircraft designer has to accommodate numerous requirements. Safety is of paramount importance and is associated with numerous regulatory considerations. Environmental considerations are also important, with noise and emissions becoming increasingly critical, especially in Europe. In addition, security has become an important consideration. These requirements arise independently of the aircraft economics, passenger comfort, and performance characteristics of the introduction of a successful new airplane.

In the United States the Federal Aviation Administration (FAA) must certify aircraft. The requirements are given in Federal Airworthiness Regulations (FARs). In Europe the regulations are Joint Airworthiness Requirements (JARs). Commercial aircraft are generally governed by

- Regulatory design requirements:
  - FAR Pt 25: the design of the aircraft
  - FAR Pt 121: the operation of the aircraft
  - FAR Pt 36: noise requirements
  - Security
  - Airport requirements
  - Icing
  - Extended-range twin-engine operations (ETOPS)

An airplane design has to be consistent with the airports it is expected to use. Details of airport design for different size airplanes can be found in Ashford and Wright (1992). Table 22.4 defines the basic characteristics. The FAA sets standards and defines airplanes within six categories, related to the airplane wingspan. A key consideration for new large airplanes is the maximum wingspan on the class VI airport of 262 feet, the so-called 80-meter gatebox limit. The new Airbus A380, listed in Table 22.1, is constrained in span to meet this requirement. Because we have shown that the wingspan is a key to low induced drag, it is clear that the A380 will be sacrificing aerodynamic efficiency to meet this requirement.

TABLE 22.4 FAA Airplane Design Groups for Geometric Design of Airports

Airplane design group	Wingspan (ft)	Runway width (ft)	Runway centerline to taxiway centerline (ft)
I	up to 49	100	400
II	49–79	100	400
III	79–118	100	400
IV	118–171	150	400
V	171–197	150	Varies
VI	197–262	200	600

Another issue for airplane designers is the thickness of the runway required. If too much weight is placed on a tire, the runway may be damaged. Thus you see fuselage-mounted gears on a B747, and the B777 has a six-wheel bogey instead of the usual four-wheel bogey. This general area is known as flotation analysis. Because of the weight concentrated on each tire, the pavement thickness requirements can be considerable. The DC-10 makes the greatest demands on pavements. Typically, it might require asphalt pavements to be around 30 inches thick and concrete pavements to be 13 inches thick. An overview of landing gear design issues is available in Chai and Mason (1996).

22.4 VEHICLE OPTIONS: DRIVING CONCEPTS—WHAT DOES IT LOOK LIKE?

22.4.1 The Basic Configuration Arrangement

The current typical external configuration of both large and small commercial transport airplanes is similar, having evolved from the configuration originally chosen by Boeing for the Boeing B-47 medium-range bomber shortly after World War II. This configuration arose following the development of the jet engine by Frank Whittle in Britain and Hans von Ohain in Germany, which allowed for a significant increase in speed (Gunston 1995). The discovery of the German aerodynamics development work on swept wings to delay the rapid increase in drag with speed during World War II was incorporated into several new jet engine designs, such as the B-47, immediately after the war. Finally, Boeing engineers found that jet engines could be placed on pylons below the wing without excessive drag. This defined the classic commercial transport configuration. The technical evolution of the commercial transport has been described by Cook (1991), who was an active participant. A broader view of the development, including business, financial, and political aspects of commercial transports, has been given by Irving (1993). The other key source of insight into the development of these configurations is by Loftin (1980).

So where do we start when considering the layout of an airplane? In general, form follows function. We decide on candidate configurations based on what the airplane is supposed to do. Generally, this starts with a decision on the type of payload and the mission the airplane is supposed to carry out with this payload. This is expressed generally in terms of

- What does it carry?
- How far does it go?
- How fast is it supposed to fly?
- What are the field requirements? (How short is the runway?)
- Are there any maneuvering and/or acceleration requirements?

Another consideration is the specific safety-related requirements that must be satisfied. As described above, for commercial aircraft this means satisfying the Federal Air Regulations (FARs) and JARs for Europe. Satisfying these requirements defines the takeoff and landing distances, engine-out performance requirements, noise limits, icing performance, and emergency evacuation, among many others.

With this start, the designer develops a concept architecture and shape that responds to the mission. At the outset, the following list describes the considerations associated with defining a configuration concept. At this stage we begin to see that configuration design resembles putting a puzzle together. These components all have to be completely integrated.

- Configuration concept:
  - Lifting surface arrangement
  - Control surface(s) location
  - Propulsion system selection
  - Payload
  - Landing gear

The components listed above must be coordinated in such a fashion that the airplane satisfies the requirements given in the following list. The configuration designer works to satisfy these requirements with input from the various team members. To be successful, the following criteria must be met:

- Good aircraft:
  - Aerodynamically efficient, including propulsion integration (streamlining)
  - Must balance near stability level for minimum drag
  - Landing gear must be located relative to cg to allow rotation at takeoff
  - Adequate control authority must be available throughout the flight envelope
  - Design to build easily (cheaply) and have low maintenance costs
  - Today, commercial airplanes must be quiet and nonpolluting

Two books do an especially good job of covering the aerodynamic layout issues: by Whitford (1987) and Abzug and Larrabee (1997). The titles of both these works are slightly misleading. Further discussion of configuration options can be found in Raymer (1999) and Roskam (1987–1990).

We can translate these desirable properties into specific aerodynamic characteristics. Essentially, they can be given as

- *Design for performance*
  - Reduce minimum drag:

    Minimize the wetted area to reduce skin friction
    Streamline to reduce flow separation (pressure drag)
    Distribute area smoothly, especially for supersonic aircraft (area ruling)
    Consider laminar flow
    Emphasize clean design/manufacture with few protuberances, steps or gaps
  - Reduce drag due to lift:

    Maximize span (must be traded against wing weight)
    Tailor spanload to get good span e (twist)
    Distribute lifting load longitudinally to reduce wave drag due to lift (a supersonic requirement, note R. T. Jones's oblique wing idea)
    Camber as well as twist to integrate airfoil, maintain good two-dimensional characteristics
  - Key constraints:

    At cruise: buffet and overspeed constraints on the wing
    Adequate high lift for field performance (simpler is cheaper)
    Alpha tailscrape, $C_{L\alpha}$ goes down with sweep
- *Design for handling qualities*
  - Adequate control power is essential:

    Nose-up pitching moment for stable vehicles
    Nose-down pitching moment for unstable vehicles
    Yawing moment, especially for flying wings and fighters at high angle of attack
    Consider the full range of *cg*'s.
    Implies: Must balance the configuration around the *cg* properly
- *FAA and military requirements*
  - Safety: For the aerodynamic configuration this means safe flying qualities:

    FAR Part 25 and some of Part 121 for commercial transports
    MIL STD 1797 for military airplanes
    Noise: Community noise, FAR Part 36, no sonic booms over land (high *L/D* in the takeoff configuration reduces thrust requirements, makes plane quieter)

To start considering the various configuration concepts, we use the successful transonic commercial transport as a starting point. This configuration is mature. New commercial transports have almost uniformly adopted this configuration, and variations are minor. An interesting comparison of two different transport configuration development philosophies is available in the papers describing the development of the original Douglas DC-9 (Shevell and Schaufele 1966) and Boeing 737 (Olason and Norton 1966) designs. Advances in performance and reduction in cost are currently obtained by improvements in the contributing technologies. After we establish the baseline, we will examine other configuration component concepts that are often considered. We give a summary of the major options. Many, many other innovations have been tried, and we make no attempt to be comprehensive.

The Boeing 747 layout is shown in Figure 22.1. It meets the criteria cited above. The cylindrical fuselage carries the passengers and freight. The payload is distributed around the *cg*. Longitudinal stability and control power comes from the horizontal tail and elevator, which has a very useful moment arm. The vertical tail provides directional stability, using the rudder

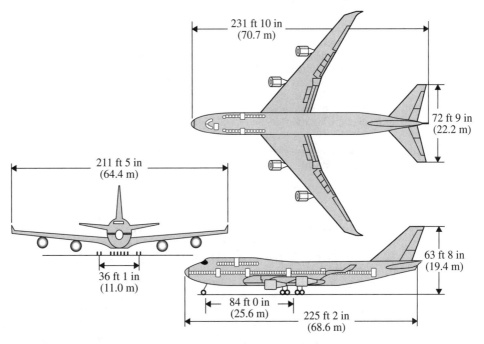

FIGURE 22.1   The classic commercial transport, the Boeing 747 (www.boeing.com).

for directional control. The swept wing/fuselage/landing gear setup allows the wing to provide its lift near the center of gravity and positions the landing gear so that the airplane can rotate at takeoff speed and also provides for adequate rotation without scraping the tail (approximately 10 percent of the weight is carried by the nose gear). The wing has a number of high-lift devices. This arrangement also results in low trimmed drag. The engines are located on pylons below the wing. This arrangement allows the engine weight to counteract the wing lift, reducing the wing root bending moment, resulting in a lighter wing. This engine location can also be designed so that there is essentially no adverse aerodynamic interference.

22.4.2 Configuration Architecture Options

Many another arrangements are possible, and here we list a few typical examples. All require attention to detail to achieve the claimed benefits.

- *Forward swept wings:* reduced drag for severe transonic maneuvering conditions
- *Canards:* possibly safety, also possibly reduced trim drag, and supersonic flight
- *Flying wings:* elimination of wetted area by eliminating fuselage and tail surfaces
- *Three-surface configurations:* trim over wide *cg* range
- *Slender wings:* supersonic flight
- *Variable sweep wings:* good low speed, low altitude penetration, and supersonic flight
- *Winglets:* reduced induced drag without span increase

Improvements to current designs can occur in two ways. One way is to retain the classic configuration and improve the component technologies. This has been the recent choice for new aircraft, which are mainly derivatives of existing aircraft, using refined technology, for example, improved aerodynamics, propulsion, and materials. The other possibility for improved designs is to look for another arrangement.

Because of the long evolution of the current transport configuration, the hope is that it is possible to obtain significantly improved aircraft through new configuration concepts. Studies looking at other configurations as a means of obtaining an aircraft that costs less to build and operate are being conducted. Two concepts have received attention recently. One integrates the wing and the fuselage into a blended wing body concept (Liebeck 2002), and a second uses strut bracing to allow for increased wingspan without increasing the wing weight (Gundlach et al. 2000).

22.4.3 The Blended Wing Body

The blended wing body concept (BWB) combines the fuselage and wing into a concept that offers the potential of obtaining the aerodynamic advantages of the flying wing while providing the volume required for commercial transportation. Figure 22.2 shows the concept. This configuration offers the potential for a large increase in L/D and an associated large reduction in fuel use and maximum take-off gross weight (TOGW). The major overview is given by Liebeck (2002), who predicted that the BWB would have an 18 percent reduction in TOGW and 32 percent in fuel burn per seat compared to the proposed A380-700.

Because the BWB does not have large moment arms for generating control moments, and also requires a nontraditional passenger compartment, the design is more difficult than traditional designs and requires the use of multidisciplinary design optimization methods to obtain the predicted benefits (Wakayama 1998). Recently the concept has been shown to be able to provide a significant speed advantage over current commercial transports (Roman, Gilmore, and Wakayama 2003). Because of the advantages of this concept, it has been studied by other design groups.

22.4.4 The Strut-Braced Wing

Werner Pfenninger suggested the strut-braced wing concept around 1954. His motivation was actually associated with the need to reduce the induced drag to balance his work in reducing parasite drag

FIGURE 22.2   The blended wing body concept. (*Courtesy* Boeing.)

FIGURE 22.3 The strut-braced wing concept.

by using active laminar flow control to maintain laminar flow and reduce skin friction drag. Since the maximum L/D occurs when the induced and parasite drag are equal, the induced drag had to be reduced also. The key issues are

- Once again, the tight coupling between structures and aerodynamics requires the use of MDO (multidisciplinary design optimization) (see section 22.7) to make it work.
- The strut allows a thinner wing without a weight penalty and also a higher aspect ratio and less induced drag.
- Reduced t/c allows less sweep without a wave drag penalty.
- Reduced sweep leads to *even lower* wing weight.
- Reduced sweep allows for some natural laminar flow and thus reduced skin friction drag.

The benefits of this concept are similar to the benefits cited above for the BWB configuration. The advantage of this concept is that it does not have to be used on a large airplane. The key issue is the need to provide a mechanism to relieve the compression load on the strut under negative g loads. Work on this concept was done at Virginia Tech (Grasmeyer et al. 1998; Gundlach et al. 2000). Figure 22.3 shows the result of a joint Virginia Tech–Lockheed Martin study.

There are numerous options for the shape of the aircraft. Other possibilities exist, and there is plenty of room for imagination. See Whitford (1987) for further discussion of configuration options.

22.5 VEHICLE SIZING—HOW BIG IS IT?

Once a specific concept is selected, the next task is to determine how big the airplane is, which essentially means how much it weighs. Typically, for a given set of technologies the maximum takeoff gross weight is used as a surrogate for cost. The lighter the airplane, the less it costs, both to buy and operate. Some procedures are available to estimate the size of the airplane. This provides a starting point for more detailed design and sizing and is a critical element of the design. The initial "back-of-the-envelope" sizing is done using a database of existing aircraft and developing an airplane that can carry the required fuel and passengers to do the desired mission. This usually means acquiring data similar to the data presented in Tables 22.1 and 22.2 and doing some preliminary analysis to obtain an idea of the wing area required in terms of the wing loading W/S and the thrust to weight ratio T/W, as shown in Table 22.5.

TABLE 22.5 Derived Characteristics of Current Transport Aircraft

Aircraft	TOGW (lb)	Empty weight (lb)	Wing area (ft²)	Sweep (quarter chord)	Aspect ratio	W/S	W/b	T/W
Narrow body								
A320-200	169,800	92,000	1,320	25.0	9.47	129	1,519	0.312
B717-200	121,000	68,500	1,001	24.5	8.70	121	1,297	0.347
B737-600	143,500	81,000	1,341	25.0	9.45	107	1,274	0.287
B757-300	273,000	141,690	1,951	25.0	7.98	140	2,188	0.305
Wide body								
A330-300	513,670	274,650	3,890	30.0	10.06	132	2,597	0.272
A340-500	811,300	376,800	4,707	30.0	9.21	172	3,897	0.261
A380-800	1,234,600	611,000	9,095	33.5	7.54	136	4,716	0.227
B747-400	875,000	398,800	5,650	37.5	7.91	155	4,139	0.265
B747-400ER	911,000	406,900	5,650	37.5	7.91	161	4,309	0.255
B767-300	345,000	196,000	3,050	31.5	7.99	113	2,210	0.337
B777-300	660,000	342,900	4,605	31.6	8.68	143	3,302	0.278
B777-300ER	750,000	372,800	4,694	31.6	9.63	160	3,528	0.307
Regional jets								
CRJ200(ER)	51,000	30,500	520	26.0	9.34	98	732	0.36
CRJ700(ER)	75,000	43,500	739	26.8	7.88	102	983	0.37
ERJ135ER	41,888	25,069	551	20.3	7.86	76	637	0.43
ERJ145ER	54,415	26,270	551	20.3	7.86	82	690	0.39

Following Nicolai (1975), consider the TOGW, called here W_{TO}, to be

$$W_{TO} = W_{fuel} + W_{fixed} + W_{empty} \tag{22.11}$$

where the fixed weight includes a nonexpendable part, which consists of the crew and equipment, and an expendable part, which consists of the passengers and baggage or freight. W_{empty} includes all weights except the fixed weight and the fuel. The question becomes: For a given (assumed) TOGW, is the weight left enough to build an airplane when we subtract the fuel and payload? We state this question in mathematical terms by equating the available and required empty weight:

$$W_{Empty\ Avail} = W_{Empty\ Reqd} \tag{22.12}$$

where $W_{Empty\ Reqd}$ comes from the following relation:

$$W_{Empty\ Reqd} = KS \cdot A \cdot TOGW^{B} \tag{22.13}$$

and KS is a structural technology factor and A and B come from the data gathered from information in Table 22.5. Now, the difference between the takeoff and landing weight is due to the fuel used (the mission fuel). Figure 22.4 shows how this relation is found from the data. Note that KS is very powerful and should not be much less than 1 without a very good reason.

Next we define the mission in terms of segments and compute the fuel used for each segment. Figure 22.5 defines the segments used in a typical sizing program. Note that the mission is often defined in terms of a radius (an obvious military heritage). Transport designers simply use one-half of the desired range as the radius. At this level of sizing, reserve fuel is included as an additional range, often taken to be 500 nm. To use the least fuel, the airplane should be operated at its best cruise Mach number (BCM), and its best cruise altitude (BCA). Often, air traffic control or weather conditions may prevent being able to fly at these conditions in actual operation.

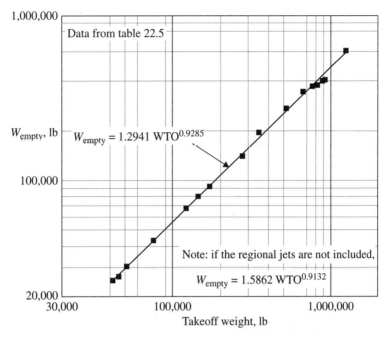

FIGURE 22.4   Relationship between empty weight and takeoff weight for the airplanes in Table 22.5.

Mission segment definitions for Figure 22.5:

1–2   engine start and takeoff
2–3   accelerate to subsonic cruise velocity and altitude
3–4   subsonic cruise out

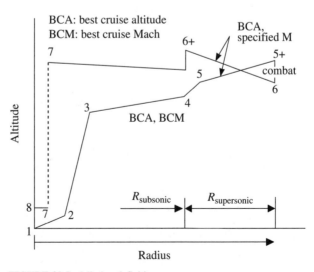

FIGURE 22.5   Mission definition.

4–5	accel to high speed (supersonic) dash/cruise
5–5+	supersonic cruise out
	combat (use fuel, expend weapons)
6–6+	supersonic cruise back
6+–7	subsonic cruise back
7–8	loiter
8	land

To get the empty weight available, compute the fuel fraction for each mission segment. For the fuel fraction required for the range, invert the Brequet range equation given above:

$$\frac{W_{i+1}}{W_i} = e^{-R \cdot \mathrm{sfc}/(V \cdot L/D)} \tag{22.14}$$

and for loiter:

$$\frac{W_{i+1}}{W_i} = e^{-R \cdot \mathrm{sfc}/(L/D)} \tag{22.15}$$

The values of the cruise L/D and sfc have to be estimated, and the velocity for best range also has to be estimated, so it takes some experience to obtain these values. Note that this approach can also be used to establish the values of L/D and sfc required to perform a desired mission at a desired weight. Values for takeoff and climb are typically estimated and can be computed for more accuracy. However, for a transport aircraft the range requirement tends to dominate the fuel fraction calculation, with the rest of the fuel fractions being near unity. Therefore, we compute the mission weight fraction as

$$\frac{W_{\text{final}}}{W_{\text{TO}}} = \frac{W_8}{W_1} = \underbrace{\frac{W_2}{W_1} \cdot \frac{W_3}{W_2} \cdot \frac{W_4}{W_3} \cdots \frac{W_8}{W_7}}_{\text{fuel fraction for each segment}} \tag{22.16}$$

and solve for the fuel weight in equation (22.16) as

$$W_{\text{fuel}} = \left(1 + \frac{W_{\substack{\text{reserve} \\ \text{fuel}}}}{W_{\text{TO}}} + \frac{W_{\substack{\text{trapped} \\ \text{fuel}}}}{W_{\text{TO}}}\right)\left(1 + \frac{W_8}{W_1}\right)W_{\text{TO}}$$

$$= \left(1 + \frac{W_{\substack{\text{reserve} \\ \text{fuel}}}}{W_{\text{TO}}} + \frac{W_{\substack{\text{trapped} \\ \text{fuel}}}}{W_{\text{TO}}}\right)(W_{\text{TO}} - W_{\text{landing}}) \tag{22.17}$$

so that we can compute $W_{\text{Empty Avail}}$ from

$$W_{\text{Empty Avail}} = W_{\text{TO}} - W_{\text{fuel}} - W_{\text{fixed}} \tag{22.18}$$

The value of W_{TO} that solves the problem is the one for which $W_{\text{Empty Avail}}$ is equal to the value of $W_{\text{Empty Reqd}}$, which comes from the statistical representation for this class of aircraft. An iterative procedure is often used to find this value. The results of this estimate are used as a starting point for the design using more detailed analysis. A small program that makes these calculations is available on the Web (Mason n.d.).

We illustrate this approach with an example, also from Nicolai (1975). The example is for a C-5. In this case we pick

Range: 6000 nm

Payload: 100,000 lb

sfc: 0.60 @ M = 0.8

h = 36,000 ft altitude

L/D = 17

Figure 22.6 shows how the empty and available weight relations intersect, defining the weight of the airplane, which is in reasonable agreement with a C-5A. Note that as the requirements become more severe, the lines will start to become parallel, the intersection weight will increase, and the uncertainty will increase because of the shallow intersection.

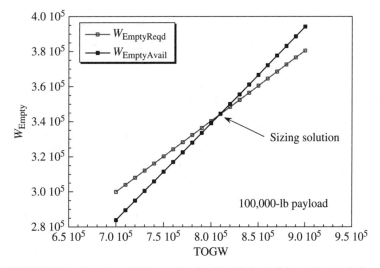

FIGURE 22.6   Illustration of sizing results using Nicolai's back-of-the-envelope method.

Note that the author's students have used this approach to model many commercial transport aircraft and it has worked well, establishing a baseline size very nearly equal to existing aircraft and providing a means of studying the impact of advanced technology on the aircraft size.

Once the weight is estimated, the engine size and wing size are picked considering constraints on the design. Typical constraints include the takeoff and landing distances and the cruise condition. Takeoff and landing distance include allowances for problems. The takeoff distance is computed such that in case of engine failure at the decision speed the airplane can either stop or continue the takeoff safely, and includes the distance required to clear a 35-foot obstacle. The landing distance is quoted including a 50-foot obstacle, and it includes an additional runway distance. Other constraints that may affect the design include the missed approach condition, the second segment climb (the ability to climb if an engine fails at a prescribed rate between 35 and 400 feet altitude), and the top-of-climb rate of climb. Jenkinson, Simpkins, and Rhodes (1999) have an excellent discussion of these constraints for transport aircraft. Figure 22.7 shows a notional constraint diagram for T/W and W/S. Typically, the engines of long-range airplanes are sized by the top-of-climb requirement and the engines of twin-engine airplanes are sized by the second-segment-climb requirement (Jenkinson, Simpkin, and Rhodes 1999).

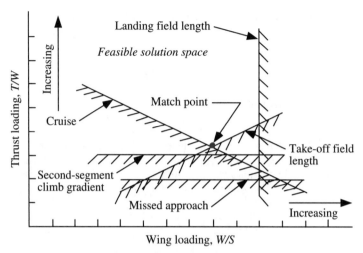

FIGURE 22.7 Typical constraint diagram (after Loftin 1980).

22.6 CURRENT TYPICAL DESIGN PROCESS

The airplane design process is fairly well established. It starts with a conceptual stage, where a few engineers use the sizing approaches slightly more elaborate than those described above to investigate new concepts. Engine manufacturers also provide information on new engine possibilities or respond to requests from the airframer. If the design looks promising, it progresses to the next stage: preliminary design. At this point the characteristics of the airplane are defined and offered to customers. Since the manufacturer cannot afford to build the airplane without a customer, various performance guarantees are made, even though the airplane has not been built yet. This is risky. If the guarantees are too conservative, you may lose the sale to the competition. If the guarantees are too optimistic, a heavy penalty will be incurred.

If the airplane is actually going to be built, it progresses to detail design. The following, from John McMasters of Boeing, describes the progression.

- *Conceptual design (1 percent of the people)*:
  - Competing concepts evaluated   What drives the design?
  - Performance goals established   Will it work/meet requirement?
  - Preferred concept selected   What does it look like?

- *Preliminary design (9 percent of the people)*:
  - Refined sizing of preferred concept   Start using big codes
  - Design examined/establish confidence   Do some wind tunnel tests
  - Some changes allowed   Make actual cost estimate (you bet your company)

- *Detail design (90 percent of the people)*:
  - Final detail design   Certification process
  - Drawings released   Component/systems tests
  - Detailed performance   Manufacturing (earlier now)
  - Only "tweaking" of design allowed   Flight control system design

22.7   MDO—THE MODERN COMPUTATIONAL DESIGN APPROACH

With the increase of computer power, new methods for carrying out the design of the aircraft have been developed. In particular, the interest is in using high-fidelity computational simulations of the various disciplines at the very early stages of the design process. The desire is to use the high-fidelity analyses with numerical optimization tools to produce better designs. Here the high-quality analysis and optimization can have an important effect on the airplane design early in the design cycle. Currently, high-fidelity analyses are used only after the configuration shape has been frozen. At that point it is extremely difficult to make significant changes. If the best tools can be used early, risk will be reduced, as will the design time. Recent efforts have also focused on means of using large-scale parallel processing to reduce the design cycle time. These various elements, taken all together, are generally known as multidisciplinary design optimization (MDO). One collection of papers has been published on the subject (Alexandrov and Hussaini 1997), and there is a major conference on MDO every other year sponsored by the AIAA, ISSMO, and other societies. Perhaps the best survey of our view of MDO for aircraft design is Giunta et al. (1996). We will outline the MDO process and issues based on these and other recent publications.

Our current view of MDO is that high-fidelity codes cannot be directly coupled into one major program. There are several reasons. Even with advanced computing, the computer resources required are too large to perform an optimization with a large number of design variables. For 30 or so design variables, with perhaps 100 constraints, hundreds of thousands of analyses of the high-fidelity codes are required. In addition, the results of the analyses are invariably noisy (Giunta et al. 1994), so that gradient-based optimizers have difficulty in producing meaningful results. In addition to the artificial noise causing trouble, the design space is nonconvex and many local optima exist (Baker 2002). Finally, the software integration issues are complex, and it is unlikely that major computational aerodynamic and structures codes can be combined. Thus, innovative methods are required to incorporate MDO into the early stages of airplane design.

Instead of a brute-force approach, MDO should be performed using surrogates for the high-fidelity analyses. This means that for each design problem, a design space should be constructed that uses a parametric model of the airplane in terms of design variables such as wingspan and chords, etc. The ranges of values of these design variable are defined, and a database of analyses for combinations of the design variables should be constructed. Because the number of combinations will quickly become extremely large, design of experiments theory will need to be used to reduce the number of cases that need to be computed. Because these cases can be evaluated independent of each other, this process can exploit coarse-grain parallel computing to speed the process. Once the database is constructed, it must be interpolated. In statistical jargon, this means constructing a response surface approximation. Typically, second-order polynomials are used. This process automatically filters out the noise from the analyses of the different designs. These polynomials are then used in the optimization process in place of the actual high-fidelity codes. This allows for repeated investigations of the design space with an affordable computational cost. A more thorough explanation of how to use advanced aerodynamics methods in MDO, including examples of trades between aerodynamics and structures, has been presented by Mason et al. (1998).

Current issues of interest in MDO also include the consideration of the effects of uncertainty of computed results and efficient geometric representation of aircraft. MDO is an active research area and will be a key to improving future aircraft design.

REFERENCES

Abzug, M. A., and E. E. Larrabee. 1997. *Airplane Stability and Control.* Cambridge: Cambridge University Press.

Alexandrov, N. M., and M. Y. Hussaini. 1997. *Multidisciplinary Design Optimization.* Philadelphia: SIAM.

Ashford, N., and P. H. Wright. 1992. *Airport Engineering*, 3rd ed. New York: John Wiley & Sons. *Aviation Week and Space Technology.* 2003. *Sourcebook.* New York: McGraw-Hill, January 13.

Baker, C. A., B. Grossman, R. T. Haftka, W. H. Mason, and L. T. Watson. 2002. "High-Speed Civil Transport Design Space Exploration Using Aerodynamic Response Surface Approximations." *Journal of Aircraft* 39(2):215–20.

Brune, G. W., and J. H. McMasters. 1990. "Computational Aerodynamics Applied to High Lift Systems." In *Applied Computational Aerodynamics*, ed. P. Henne. Progress in Astronautics and Aeronautics 125. Washington, DC: AIAA, 389–413.

Chai, S. T., and W. H. Mason. 1996. *Landing Gear Integration in Aircraft Conceptual Design.* MAD Center Report MAD 96-09-01, September 1996, revised March 1997. http://www.aoe.vt.edu/ mason/~Mason_f/ MRNR96.html.

Cook, W. H. 1991. *The Road to the 707.* Bellevue: TYC.

Cumpsty, N. A. 1998. *Jet Propulsion: A Simple Guide to the Aerodynamic and Thermodynamic Design and Performance of Jet Engines.* Cambridge: Cambridge University Press.

Fielding, J. P. 1999. *Introduction to Aircraft Design.* Cambridge: Cambridge University Press.

Giunta, A. A., J. M. Dudley, B. Grossman, R. T. Haftka, W. H. Mason, and L. T. Watson. 1994. "Noisy Aerodynamic Response and Smooth Approximations in HSCT Design." AIAA Paper 94-4376, Panama City, FL, September.

Giunta, A. A., O. Golovidov, D. L. Knill, B. Grossman, W. H. Mason, L. T. Watson, and R. T. Haftka. 1996. "Multidisciplinary Design Optimization of Advanced Aircraft Configurations." In *Proceedings of the 15th International Conference on Numerical Methods in Fluid Dynamics,* ed. P. Kutler, J. Flores, and J.-J. Chattot. Lecture Notes in Physics 490. Berlin: Springer-Verlag, 14–34.

Grasmeyer, J. M., A. Naghshineh, P.-A. Tetrault, B. Grossman, R. T. Haftka, R. K. Kapania, W. H. Mason, and J. A. Schetz. 1998. *Multidisciplinary Design Optimization of a Strut-Braced Wing Aircraft with Tip-Mounted Engines.* MAD Center Report MAD 98-01-01, January.

Gundlach, J. F., IV, P.-A. Tetrault, F. H. Gern, A. H. Naghshineh-Pour, A. Ko, J. A. Schetz, W. H. Mason, R. K. Kapania, B. Grossman, and R. T. Haftka. 2000. "Conceptual Design Studies of a Strut-Braced Wing Transonic Transport." *Journal of Aircraft* 37(6):976–83.

Gunston, B. 1995. *The Development of Jet and Turbine Aero Engines.* Sparkford: Patrick Stevens.

Irving, C. 1993. *Wide-Body: The Triumph of the 747.* New York: William Morrow & Co.

Jackson, P., ed. 2002–2003. *Jane's All the World's Aircraft.* Surrey: Jane's Information Group.

Jenkinson, L. R., P. Simpkin, and D. Rhodes. 1999. *Civil Jet Aircraft Design.* London: Arnold; Washington, DC: AIAA. Electronic Appendix at http://www.bh.com/companions/034074152C/appendices/ data-a/default.htm.

Jensen, S. C., I. H. Rettie, and E. A. Barber. 1981. "Role of Figures of Merit in Design Optimization and Technology Assessment." *Journal of Aircraft* 18(2):76–81.

Kayton, M. 2003. "One Hundred Years of Aircraft Electronics." *Journal of Guidance, Control, and Dynamics* 26(2):193–213.

Liebeck, R. H. 2002. "The Design of the Blended-Wing-Body Subsonic Transport." AIAA Paper 20020002, January.

Loftin, L. K. 1980. *Subsonic Aircraft: Evolution and the Matching of Size to Performance.* NASA Reference Publication 1060, August.

Mason, W. H. ACsize, Software for Aerodynamics and Aircraft Design. http://www.aoe.vt.edu/~mason/Mason_f/ MRsoft.html#Nicolai.

Mason, W. H., D. L. Knill, A. A. Giunta, B. Grossman, R. T. Haftka, and L. T. Watson. 1998. "Getting the Full Benefits of CFD in Conceptual Design." Paper delivered at AIAA 16th Applied Aerodynamics Conference, Albuquerque, NM, June. AIAA Paper 98-2513.

Nelson, R. C. 1997. *Flight Stability and Automatic Control,* 2nd ed. New York: McGraw-Hill.

Nicolai, L. M. 1975. *Fundamentals of Aircraft Design.* San José, CA: METS.

Niu, M. C. Y. 1999. *Airframe Structural Design,* 2nd ed. Hong Kong: Conmilit Press. [Available through ADASO/ADASTRA Engineering Center, P.O. Box 3552, Granada Hills, CA, 91394, Fax: (888) 735-8859, http://www.adairframe.com/~adairframe/.]

Olason, M. L., and D. A. Norton. "Aerodynamic Design Philosophy of the Boeing 737." *Journal of Aircraft* 3(6):524–28.

Raymer, D. P. 1999. *Aircraft Design: A Conceptual Approach,* 3rd ed. Washington, DC: AIAA.

Roman, D., R. Gilmore, and S. Wakayama. 2003. "Aerodynamics of High-Subsonic Blended-Wing-Body Configurations." AIAA Paper 2003-0554, January.

Roskam, J. *Airplane Design.* 1987–1990. 8 vols. Lawrence, KS: DARcorporation. http://www. darcorp.com/Textbooks/textbook.htm.

Schaufele, R. D. 2000. *The Elements of Aircraft Preliminary Design.* Santa Ana: Aries.

Shevell, R. S. 1989. *Fundamentals of Flight,* 2nd ed. Upper Saddle River, NJ: Prentice Hall.

Shevell, R. S., and R. D. Schaufele. 1966. "Aerodynamic Design Features of the DC-9." *Journal of Aircraft* 3(6):515–23.

Van Dam, C. P. 2002. "The Aerodynamic Design of Multi-element High-Lift Systems for Transport Airplanes" *Progress in Aerospace Sciences* 38:101–44.

Wakayama, S. 1998. "Multidisciplinary Design Optimization of the Blended-Wing-Body." AIAA Paper 98-4938, September.

Whitford, R. 1987. *Design for Air Combat.* Surrey: Jane's Information Group.

CHAPTER 23
AIRPORT DESIGN*

William R. Graves
School of Aeronautics
Florida Institute of Technology
Melbourne, Florida

Ballard M. Barker
School of Aeronautics
Florida Institute of Technology
Melbourne, Florida

23.1 INTRODUCTION

Transportation engineering is a broad and somewhat synthetic field, involving a myriad of public interests and multiple professional disciplines. The majority of engineers working on transportation issues were probably educated as civil engineers and are graduates of the many ABET-accredited programs in the United States that carefully prescribe the essentials of engineering civil works, which must be imparted in a four-year curriculum. Graduate degrees in civil engineering, as in most disciplines, tend to develop more depth than scope of expertise in the field. As a consequence, many competent and experienced engineers newly employed on transportation engineering works have little formal training or direct experience with transportation matters. Airport planning and design, a subset of transportation engineering, gets even less attention in civil engineering education. The majority of engineers probably receive their first education in airport matters in on-the-job training after being employed by a firm doing airport projects. Indeed, a persistent lament of many airport planning firms and architectural and engineering firms involved in airport projects has been that engineers generally lack necessary education and knowledge of airport-related matters.

It is the goal of this chapter to make the reader aware of the scope of airport-specific planning and design issues and to point toward the modest body of literature that addresses routine elements of airport planning and design, as well as issues, challenges, and approaches to help the transportation engineer address community needs. The central focus, however, is on those elements of airport design typically performed by transportation engineers, rather than those commonly performed by aviation planners, architects, or narrower engineering specialties such as electrical or mechanical.

23.2 AIRPORT PLANNING

Airport *planning* is a prerequisite of proper airport *design* activities. The need for thorough airport planning in the context of addressing a region's transportation, socioeconomic, and environmental

---

*Reprinted from the First Edition.

concerns is globally recognized. The International Civil Aviation Organization (ICAO), an operating agency of the United Nations with nearly 190 contracting states, addresses the need for guidance material to assist states with planning for construction and expansion of international airports in its Annex 14 to the convention on civil aviation. ICAO publishes and periodically revises its Airport Planning Manual (ICAO Doc. 9184-AN/902), which details a model airport planning process for airside and landside development, operations and support facilities, and land use and environmental controls, and provides generic guidance for the selection of airport consultants and construction services.

In the United States, there is a Federal Aviation Administration (FAA)–driven, four-tiered bureaucratic hierarchy of products dealing with airport planning in one way or another. It is FAA-driven in the sense that since implementation of the Airport Development Aid Program in 1970, any of the nearly 4,000 national interest airports in the United States that wish to apply for federal airport development grants under the Airport Improvement Program, or receive FAA approval of local Passenger Facility Charges for airport development, must participate in the FAA-steered process.

At the local level, airport operators develop *airport master plans*. The airport master plan is a substantial document designed to reflect the airport's, and hopefully the community's, realistic vision of the airport's missions, roles, and facilities 5, 10, and 20 years into the future. It lays out phased airport development based upon forecast needs, alternative assessments, and financial planning for accomplishment. The airport master plan is the one planning document commonly used by transportation engineers as a basis for gross design guidance. The form and process of the airport master plan will be discussed in more detail in a later section.

It is common for major metropolitan areas and self-defined regions within or between states to develop general plans to address regional airport and other transportation needs. These *regional/ metropolitan airport systems plans* incorporate the visions of local airport master plans and seek to harmonize them while accommodating present and future transportation needs in the subject area. Adequacy of air service to the region and coordination of the roles of the various airports in the area are typical elements. Each state transportation agency also compiles a *statewide integrated airport system plan* that is similar in level of detail and scope of concerns to the regional/metropolitan plans. Statewide plans are typically designed to identify and coordinate airport development needs for enhancement of state economies, and sometimes assist in prioritization of state planning and development aid to airports and communities.

The *National Plan of Integrated Airport Systems* (NPIAS) incorporates much of the information on future airport development needs from the states' plans and complements it with FAA forecasts of airport terminal area and national aviation activity. The NPIAS is required by the Airport and Airways Development Act of 1970 to be published biennially, ostensibly to aid in the appropriation and programming of AIP funds by the federal government. It is quite literally nothing more than a compiled summary of the estimated cost of desired airport development projects at each of the nation's national interest airports for running 10-year periods, and is of little practical value for airport-to-airport planners and engineers.

23.2.1 The Airport Master Plan

As stated earlier, the airport master plan is the airport's concept for future development of the airport to meet community needs and concerns. It is the planning document with the most relevance and utility to engineers working on airport projects. Its preparation is the airport's responsibility, but at most airports it is developed by aviation consulting firms working with key members of airport staff. Broad guidance on the process and content of master plans for U.S. airports is presented in FAA Advisory Circular 150/5070-6A, *Airport Master Plans*. Some states, such as Florida, also publish excellent guidelines for airport master planning. These state guides are typically intended to give practical guidance for the development of useful and cost-effective planning products for general aviation and small commercial service airports.

The airport master plan has several objectives in addition to presenting a technically, economically, and environmentally sound development concept. Primary is development of a meaningful *capital improvement program* (CIP) that addresses at least the subsequent 5-year period in detail. The CIP should be accompanied by a corresponding conceptual financial plan that will allow for the phased implementation of the CIP as projected development needs become actual needs. The airport community's actual and forecast aeronautical demands are documented in association with the community's role and priorities for airport development. The airport master plan process should also provide ample opportunity for public involvement in plan development, to include review by regional and state agencies. Last, the master plan should become the foundation and framework for responsive and effective evolution of the airport to meet future needs of the community.

The elements of airport master planning vary with each airport's particular situation and may include

1. An inventory of all relevant airport infrastructure, land uses, operational activities, and development issues.

2. Forecasts of aeronautical demand on airport infrastructure for subsequent 5-, 10-, and 20-year periods. This is arguably the most critical step of the planning process because of its effect on all subsequent planning actions. This step should also include realistic predictions of future aircraft types that are likely to serve the airport.

3. Determination of future airport requirements and alternative concepts for demand satisfaction. This step involves comparison of present capabilities with forecast demand at 5-year intervals, and the development of concepts to best satisfy different infrastructure needs. Various simulations may be employed to assist analysis in complex situations.

4. Site location and situation of new, relocated, or expanded infrastructure.

5. Environmental analyses and procedural compliance. These may include environmental assessments, archeological studies, and documentation of the situation.

6. Airport plan drawings. A series of large-scale drawings of overall airport layout (airport layout plan), terminal area plan, land use plan, surface transportation access plan, and the airspace plan. The airport layout plan (ALP) is the single most important and commonly used drawing because it includes all minimally required information and is an FAA prerequisite for any federally assisted airport development.

General guidelines and useful references for completing each of the several airport master plan elements are provided in FAA Advisory Circular 150/5070-6A, *Airport Master Plans,* and will not be repeated in this text.

23.3 AIRPORT DESIGN

23.3.1 Airport Layout Design

Once master planning has determined the desired roles of the airport, the needed capacity, and the critical aircraft for which the airport should be designed, airport layout (also referred to as airport geometry) can be determined using the appropriate ICAO or FAA standards. Airport geometry refers to the two-dimensional horizontal layout of airport surfaces as they are depicted on the airport layout plan drawing contained in the master plan. A principal goal of the airport layout should be to allow for future growth and change in the aircraft movement area and other airside facilities.

The airport needs to be designed to the standards associated with the most demanding airplane(s) to be accommodated during the planning window. The most demanding airplane for runway length may not be the same airplane used for determining the appropriate airplane design group or for pavement strength. If the airport is to have two or more runways and associated movement areas, it is highly desirable to design all airport elements to the standards associated with the most demanding airplane(s). However, it may be more cost-effective for some airport elements, for example, a secondary runway and its associated taxiway, to be designed to standards associated with less demanding airplanes that will use the airport.

Runway Location and Orientation.   Runway location and orientation are paramount to aviation safety, airport efficiency, airline operating economies, and environmental compatibility. The weight given to each of the following runway location and orientation factors depends, in part, on (a) the characteristics of airplanes expected to use each runway; (b) the volume of air traffic expected on each runway; (c) the meteorological conditions to be accommodated; and (d) the nature of the airport environment.

Runway location and orientation factors include

1. *Area winds.* A wind analysis must be performed to determine the optimum runway orientation for purposes of wind coverage and to determine the necessity for a crosswind runway. Appendix 1 of AC 150/5300-13, *Airport Design,* provides information on wind data analysis for airport planning and design.

2. *Airspace availability.* Existing and planned instrument approach procedures, missed approach procedures, departure procedures, control zones, special-use airspace, restricted airspace, and other traffic patterns should be carefully considered in the development of new airport layouts and locations.

3. *Environmental factors.* Environmental studies should be made to ensure that runway development will be as compatible as possible with the airport environs. These studies should include analyses of the impact upon air and water quality, wildlife, existing and proposed land use, and historical/ archeological factors.

4. *Obstructions to air navigation.* An obstruction survey should be conducted to identify those objects that might affect airplane operations. Approaches free of obstructions are desirable and encouraged, but, as a minimum, runways require a location that will ensure that the approach areas associated with the ultimate development of the airport can be maintained clear of airport hazards.

5. *Land consideration.* The location and size of the site, with respect to the airport's geometry, should be such that all of the planned airport elements, including the runway clear zone, are located on airport property. If it is anticipated that the airport will need to purchase additional land to accommodate development, AC 150/5100-17, *Land Acquisition and Relocation Assistance for Airport Improvement Program Assisted Projects,* should be referenced.

6. *Topography.* Topography affects the amount of grading and drainage work required to construct a runway and its associated movement areas. In determining runway orientation, the costs of both the initial work and ultimate airport development should be considered. For guidance see AC 150/5320-5, *Airport Drainage.*

7. *Airport facilities.* The relative position of a runway to associated facilities such as other runways, the terminal, hangar areas, taxiways, aprons, fire stations, navigational aids, and the airport traffic control tower, will affect the safety and efficiency of operations at the airport. A general overview of the siting requirements for navigational aids located on, or in close proximity to, the airport, including references to other appropriate technical publications, is presented in AC 150/5300-2, *Airport Design Standards—Site Requirements for Terminal Navigational Facilities.*

Additional Runways.   Many airports may have conditions that indicate one or more additional runways are necessary to accommodate the local circumstances such as

1. *Wind conditions.* When a single runway or set of parallel runways cannot be oriented to provide 95 percent wind coverage, an additional runway (or runways), oriented in a manner to raise coverage to at least that value, should be provided. The 95 percent wind coverage is computed on the basis of the crosswind not exceeding 10.5 knots for Airport Reference Codes A-I and B-I, 13 knots for Airport Reference Codes A-II and B-II, 16 knots for Codes A-III, B-III, and C-I through D-III, and 20 knots for Codes A-IV through D-VI (AC 150/5300-13, App. 1).

2. *Environmental considerations.* New runways may be justified to reduce adverse impacts on areas around the airport during certain seasons or hours of the day or for certain types of aircraft operations. An example might be an additional runway that would be preferential for diverting night air cargo operations from overflight of dense residential areas.

3. *Operational demands.* An additional runway or runways of the configurations listed below may be warranted when the traffic volume exceeds the existing runway capacity. With rare exceptions, capacity-justified runways should be oriented parallel to the primary runway. In addition,

runways will have different centerline separation and threshold offset distances depending on the type of operations to be accommodated.

Additional runways may have any of the following combinations of intended use and configuration: *Parallel Runways Designed for Simultaneous Use During Instrument Flight Rule (IFR) Operations* (AC 150/5300-13). When more than one of the listed conditions applies, the largest separation is required as the minimum. When centerline spacing of less than 2,500 feet (750 meters) is involved, wake turbulence avoidance procedures must be observed by aircraft and air traffic managers. Additionally, runways may be separated farther than the minimum distance to allow for placement of terminal facilities between runways in order to minimize taxi times and runway crossings. Types of simultaneous operations are

1. *Simultaneous approaches.* For operations under instrument meteorological conditions (IMC), specific electronic navigational aids and monitoring equipment, air traffic control, and approach procedures are required. Simultaneous precision approaches for parallel runways require centerline separation of at least 4,300 feet.

2. *Simultaneous departures—nonradar environment.* May be conducted from parallel runways whose centerlines are separated by at least 3,500 feet (1,000 meters).

3. *Simultaneous departures—radar environment.* Departures may be conducted from parallel runways whose centerlines are separated by at least 2,500 feet (750 meters).

4. *Simultaneous approach and departure—radar environment.* Simultaneous, radar-controlled approach and departure may be conducted on parallel runways whose centerlines are separated as follows:

   - Thresholds are not staggered—a separation distance between runway centerlines of 2,500 feet (750 meters) is required.

   - Thresholds are staggered and the approach is to the nearer threshold—the 2,500 feet (750 meters) separation may be reduced by 100 feet (30 meters) for each 500 feet (150 meters) of threshold stagger to a minimum limiting separation of 1,000 feet (300 meters). For Airplane Design Group V, however, a minimum centerline separation of 1,200 feet (360 meters) is recommended.

   - When the thresholds are staggered and the approach is to the threshold—the 2,500 feet (750 meters) separation must be increased by 100 feet (30 meters) for every 500 feet (150 meters) of threshold stagger.

Parallel Runway Separation—Simultaneous VFR Operations. For simultaneous landings and takeoffs using visual flight rules (VFR), the standard minimum separation between centerlines of parallel runways is 700 feet (210 meters). The minimum runway centerline separation recommended for Airplane Design Group V is 1,200 feet (360 meters). Wake turbulence separation standards apply for runways separated by less than 2,500 feet. Separation standards are presented in Tables 23.1 and 23.2 as extracted from AC 150/5300-13.

Aircraft Weight Limitations

1. Maximum takeoff weight (MTW) and maximum landing weight (MLW) are structural limitations established by the manufacturer.

2. Maximum allowable takeoff weight (MATW) and maximum allowable landing weight (MALW) are climb performance limitations.

3. Desired takeoff weight (DTW) and desired landing weight (DLW) are operational requirements.

23.3.2 Runway Length Design

Runway design length is computed using as its basis the takeoff and landing requirements of the most demanding aircraft that will use the runway. FAA worksheets such as the one used below are typical of the type used to determine full-strength pavement requirements. Landing distance is defined as

TABLE 23.1 Runway Separation Standards for Aircraft Approach Categories C & D

Item	DIM[a]	Airplane Design Group					
		I	II	II	IV	V	VI
Nonprecision instrument and visual runway centerline to:							
Parallel runway centerline	H	Refer to discussion in section 23.2.1 on simultaneous operations					
Hold line[b]		250 ft	250 ft	250 ft	250 ft	250 ft	250 ft
		75 m	75 m	75 m	75 m	75 m	75 m
Taxiway/taxilane centerline[b]	D	300 ft	300 ft	400 ft	400 ft	c	600 ft
		90 m	90 m	120 m	120 m	c	180 m
Aircraft parking area	G	400 ft	400 ft	500 ft	500 ft	500 ft	500 ft
		120 m	120 m	150 m	150 m	150 m	150 m
Helicopter touchdown		Refer to Advisory Circular 50/5390					
Precision instrument runway centerline to:							
Parallel runway centerline	H	Refer to section 23.2.1					
Hold line[b]		250 ft	250 ft	250 ft	250 ft	280 ft	325 ft
		75 m	75 m	75 m	75 m	85 m	98 m
Taxiway/taxilane centerline[b]	D	400 ft	400 ft	400 ft	400 ft	c	600 ft
		120 m	120 m	120 m	120 m	c	180 m
Aircraft parking area	G	500 ft	500 ft	500 ft	500 ft	500 ft	500 ft
		150 m	150 m	150 m	150 m	150 m	150 m
Helicopter touchdown pad		Refer to Advisory Circular 150/5390-2					

[a]Letters correspond to the dimensions on Figure 2-1 (AC 150/5300-13 change 4), page 12.
[b]The separation distance satisfies the requirement that no part of an aircraft (tail tip, wing tip) at the holding location or on a taxiway centerline is within the runway safety area or penetrates the obstacle free zone (OFZ). Accordingly, at higher elevations, an increase to these separation distances may be needed to achieve this result.
[c]For Airplane Design Group V, the standard runway centerline to parallel taxiway centerline separation distance is 400 feet (120 meters) for airports at or below an elevation of 1,345 feet (410 meters); 450 feet (135 meters) for airports between elevations of 1,345 feet (410 meters) and 6,560 feet (2,000 meters); and 500 feet (150 meters) for airports above an elevation of 6,560 feet (2,000 meters).

the horizontal distance necessary to land and come to a complete stop from a point 50 feet above the landing surface.

Runway Length Calculation Sheet

Design Conditions

Airplane: Boeing 737-200 C (JT8D-15 Eng.) 1 + 15 reserve

Mean daily maximum temperature (°F):	50°F
Airport elevation (ft):	1,500 ft
Effective runway gradient (percent):	1.0%
Length of haul (statute miles):	400
Payload (lb):	31,930

Landing Weight Limitations. Landing gross weight must not exceed the lowest maximum weights allowed for:

TABLE 23.2 Taxiway and Taxilane Separation Standards

ITEM	DIM[a]	Airplane Design Group					
		I	II	III	IV	V	VI
Taxiway centerline to:							
Parallel taxiway/taxilane	J	69 ft	105 ft	152 ft	215 ft	267 ft	324 ft
centerline		21 m	32 m	46.5 m	65.5 m	81 m	99 m
Fixed or movable object[b,c]	K	44.5 ft	65.5 ft	93 ft	129.5 ft	160 ft	193 ft
		13.5 m	20 m	28.5 m	39.5 m	39.5 m	59 m
Taxilane centerline to:							
Parallel taxilane centerline		64 ft	97 ft	140 ft	198 ft	245 ft	298 ft
		19.5 m	29.5 m	42.5 m	60 m	74.5 m	91 m
Fixed or movable object[b,c]		39.5 ft	57.5 ft	81 ft	112.5 ft	138 ft	167 ft
		12 m	17.5 m	24.5 m	34 m	42 m	51 m

[a]Letters correspond to the dimensions on RPZ designations.
[b]This value also applies to the edge of service and maintenance roads.
[c]Consideration of the engine exhaust wake impacted from turning aircraft should be given to objects located near runway/taxiway/taxi-lane intersections.
[d]The values obtained from the following equations are acceptable in lieu of the standard dimensions shown in "this table":
  • Taxiway centerline to parallel taxiway/taxilane centerline equals 1.2 times airplane wingspan plus 10 feet.
  • Taxiway centerline to fixed or movable object equals 0.7 times airplane wingspan plus 10 feet.
  • Taxilane centerline to parallel taxilane centerline equals 1.1 times airplane wingspan plus 10 feet.
  • Taxilane centerline to fixed or movable object equals 0.6 times airplane wingspan plus 10 feet (3 meters).

1. *Compliance with runway length requirements (landing distance must be equal to or less than 0.6 times the landing distance available).* The landing distance set forth in the *Airplane Flight Manual* may not exceed 60 percent of the landing distance available.

2. *Compliance with approach requirements (landing weight ≤ MALW).* An aircraft on approach to landing must be able to execute a missed approach under the following conditions. Maximum allowable landing weight (MALW) is the highest weight at which the aircraft can meet this requirement with existing temperature and pressure altitude.

Approach Climb Requirement. In the approach configuration corresponding to the normal all-engine-operating procedure in which V_S for this configuration does not exceed 110 percent of the V_S for the related landing configuration. The steady gradient of climb may not be less than 2.1 percent for two-engine airplanes, 2.4 percent for three-engine airplanes, and 2.7 percent for four-engine air planes, with

1. The critical engine inoperative, the remaining engines at the available takeoff power or thrust

2. The maximum allowable landing weight

3. A climb speed established in connection with normal landing procedures, but not exceeding 1.5 V_S

Landing Climb Requirement. In the landing configuration, the steady gradient of climb may not be less than 3.2 percent, with

1. All engines operating

2. The engines at the power or thrust that is available 8 seconds after initiation of movement of the power or thrust controls from the minimum flight idle to the takeoff position

3. A climb speed of not more than 1.3 V_S

4. Structural limit of the airplane (landing weight ≤ MLW)

<u>Desired Landing Weight</u>

Typical operating empty weight plus reserve fuel:	70,138
Payload:	+31,930
Desired landing weight:	102,068

<u>Landing Runway Length</u> (flaps 40°, Table 23.3)

Temperature:	50°F
Airport elevation:	1,500 ft
Maximum landing weight: MLW = MALW = 103.0 > DLW	103,000
Landing runway length:	5,544 ft
Landing runway length:	5,544 ft

Using Table 23.3, runway length versus weight and pressure altitude, find 5.33 thousand feet at 100,000 pounds and 1,000 feet elevation, and 5.47 at 2,000 feet elevation. The maximum allowable weight for landing is 103,000 pounds, so we must interpolate between 100,000 and 105,000 pounds and 1,000 and 2,000 feet elevation. The numbers below are selected from Table 23.3 as described below.

The answer of 5.544 below is multiplied by 1,000 to determine the runway length required for landing.

$$
\begin{array}{ccccccc}
 & & 1{,}000 & & 1{,}500 & & 2{,}000 \\
\begin{bmatrix} 100 \\ \\ 103 \\ 5 \\ 105 \end{bmatrix} & 3 & x\begin{bmatrix} 5.33 \\ 0.144 \\ 5.474 \\ \\ 5.57 \end{bmatrix}0.24 & & y\begin{bmatrix} 5.47 \\ 0.144 \\ 5.614 \\ \\ 5.71 \end{bmatrix}0.24 & & \begin{array}{r} 5.614 \\ 5.474 \\ \hline 11.088 \div 2 = 5.544 \times 1{,}000 = 5{,}544 \end{array}
\end{array}
$$

$3/5 = x/0.24$ $3/5 = y/0.24$

$x = (0.24)(0.6) = 0.144$ $y = (0.25)(0.60) = 0.144$

<u>Desired Takeoff Weight</u>

Length of haul:		400
Average fuel consumption:	×	15
Haul fuel:		6,000
Typical operating empty weight plus reserve fuel:	+	70,138
Weight, no payload:		76,138
Payload:	+	31,930
Desired takeoff weight:		108,068

Takeoff Weight Limitations.   Takeoff gross weight must not exceed the lowest the maximum weights allowed for:

1. Compliance with runway length requirements. [takeoff distance (TOD) ≤ takeoff distance available (TODA), accelerate stop distance (ASD) ≤ accelerate stop distance available (ASDA)]

In determining the allowable gross weight for takeoff for any given runway, the performance of the airplane must be related to the dimensions of the airport; that is, the required takeoff distance and accelerate-stop distance for the gross weight must not exceed the runway length available.

2. (Takeoff weight ≤ MATW)

TABLE 23.3   Aircraft Performance, Landing (Boeing 737-200 Series) JT8D-15 Engine, 40° Flaps, Maximum Allowable Landing Weight (1,000 lb)

Temp °F	Airport elevation (ft)								
	0	1,000	2,000	3,000	4,000	5,000	6,000	7,000	8,000
50	103.0	103.0	103.0	103.0	103.0	103.0	102.7	98.8	95.0
55	103.0	103.0	103.0	103.0	103.0	103.0	102.7	98.8	95.0
60	103.0	103.0	103.0	103.0	103.0	103.0	102.7	98.8	95.0
65	103.0	103.0	103.0	103.0	103.0	103.0	102.7	98.8	95.0
70	103.0	103.0	103.0	103.0	103.0	103.0	101.9	98.0	94.0
75	103.0	103.0	103.0	103.0	103.0	103.0	100.8	97.1	93.2
80	103.0	103.0	103.0	103.0	103.0	103.0	99.6	95.9	92.2
85	103.0	103.0	103.0	103.0	103.0	101.8	98.1	94.5	91.0
90	103.0	103.0	103.0	103.0	103.0	100.1	96.5	93.0	89.5
95	103.0	103.0	103.0	103.0	101.9	98.2	94.7	91.2	87.8
100	103.0	103.0	103.0	103.0	99.8	96.2	92.7	89.2	85.9
105	103.0	103.0	103.0	101.1	97.5	93.9	90.5	87.1	83.8
110	103.0	103.0	102.1	98.5	94.9	91.5	88.1	84.7	81.5

****

Weight 1,000 lb	Runway length (1,000 ft) Airport elevation (ft)								
	0	1,000	2,000	3,000	4,000	5,000	6,000	7,000	8,000
70	3.95	4.05	4.14	4.23	4.31	4.40	4.50	4.59	4.70
75	4.15	4.25	4.35	4.44	4.54	4.64	4.75	4.86	4.98
80	4.35	4.46	4.56	4.66	4.77	4.88	5.00	5.12	5.25
85	4.56	4.67	4.78	4.89	5.01	5.13	5.25	5.39	5.53
90	4.77	4.88	5.00	5.12	5.25	5.38	5.51	5.65	5.80
95	4.98	5.11	5.23	5.36	5.49	5.63	5.77	5.92	6.07
100	5.20	5.33	5.47	5.60	5.74	5.88	6.03	6.18	6.34
105	5.42	5.57	5.71	5.85	5.99	6.14	6.29	6.44	6.61

****

Airplane characteristics	Measure	Unit of advanced options	
		200	200C
Maximum takeoff weight	lb	109,000	115,500
Maximum landing weight			
Flaps 30°	lb	98,000	103,000
Flaps 40°	lb	89,700	103,000
Typical operating empty	lb	67,238	70,138[a]
Weight plus reserve fuel	lb	71,480	74,380[b]
Average fuel consumption	lb/mile	15	15
Typical maximum passenger			
Load @200 lb/passenger	lb	26,000	26,000
Maximum structural payload	lb	34,830	31,930

[a]Based on 1.25 hours of reserve fuel.
[b]Based on 2.00 hours of reserve fuel.

3. Compliance with en route performance requirements.

4. Compliance with maximum landing weight, considering normal fuel burnout en route.

5. Structural limit of the airplane (takeoff weight ≤ MTW)

<u>Takeoff Runway Length</u> (Flaps 5°, Table 23.4)

Temperature:	50°F
Airport evaluation:	<u>1,500</u>
Maximum takeoff weight: MATW $= \dfrac{115.5+114.1}{2} = 114.8 > \text{DTW}$	<u>114,800</u>
Reference factor "R":	<u>51.25</u>
Limiting weight: $103,000 + 6,000 > \text{DTW} = 108,068$	<u>108,068</u>
Runway length:	<u>5,967</u>
<u>Gradient correction: 5967 (R/WL) $\times 0.10 \times$ 1.0 (ERG) =</u>	<u>597</u>
Corrected runway length:	<u>6,564</u>

<u>Summary</u>

Landing weight:	<u>103,000</u>
Takeoff weight:	<u>108,068</u>
Landing runway length:	<u>5,544</u>
Takeoff runway length:	<u>6,564</u>
Design runway length:	<u>6,600</u>

$$3.068/5 = x/0.54$$
$$x = (0.61)(0.54) = 0.331$$

$$\text{WT} \qquad\qquad \underline{50} \quad \underline{51.25} \quad \underline{60}$$

$$\begin{bmatrix} 105 \\ {}_{1}\!\begin{matrix}108.068\end{matrix} \\ {}_{5} \\ 110 \end{bmatrix}\!{}^{2}3.068\ {}^{3}\ x\begin{bmatrix} 5.48 \\ .331 \\ 5.811 \\ 6.02 \end{bmatrix}\!{}_{0.54}\ {}^{4}\ y\begin{bmatrix} 6.64 \\ .417 \\ 7.057 \\ 7.32 \end{bmatrix}\!{}_{0.68}\ {}^{5}$$

$$1.25/10 = z/1.246$$

$$3.068/5 = y/0.68$$
$$y = (0.61)(0.68) = 0.417$$

$$z = (0.125)(1.246) = 0.156 + 5.811 = 5.967 \times 1,000 = 5,967'$$

Runway Threshold Placement.   The landing threshold identifies (by markings and lighting) the beginning of that portion of the full-strength runway surface that is available for landing. The threshold should normally be located at the beginning of the full-strength runway surface, but it may be displaced down the runway length when an object that obstructs the airspace required for landing airplanes is beyond the airport authority's power to remove, relocate, or lower. The new location is the *displaced threshold* and is treated as the new end of the runway for landing on that end. The runway pavement preceding the displaced thresh old is available for takeoff in either direction and for landing rollout from the opposite direction. The following alternatives should be considered if an object penetrates an approach surface as defined in the preceding section:

1. Remove or lower the object so that it will not penetrate the applicable surface.

2. Apply a less demanding surface; for example, convert the runway from a precision ornon-precision approach to a VFR-only approach.

TABLE 23.4 Aircraft Performance, Takeoff (Boeing 737-200 Series) JT8D-15 Engine, 5° Flaps, Maximum Allowable Takeoff Weight (1,000 lb)

Temp °F	Airport Elevation (ft)								
	0	1,000	2,000	3,000	4,000	5,000	6,000	7,000	8,000
50	115.5	115.5	114.1	110.1	106.0	102.0	98.1	94.4	91.0
55	115.5	115.5	114.1	110.1	106.0	102.0	98.1	94.4	91.0
60	115.5	115.5	114.1	110.1	106.0	102.0	98.1	94.4	91.0
65	115.5	115.5	114.1	110.1	106.0	102.0	98.1	94.4	91.0
70	115.5	115.5	113.5	109.6	105.6	101.5	97.4	93.6	90.0
75	115.5	115.1	111.6	107.7	103.8	99.8	95.8	92.1	88.6
80	115.5	113.3	109.7	105.8	101.9	98.0	94.2	90.6	87.2
85	114.8	111.4	107.7	104.0	100.1	96.3	92.6	89.1	85.8
90	113.1	109.5	105.8	102.1	98.3	94.6	91.0	87.5	84.3
95	111.3	107.6	103.9	100.2	96.5	92.9	89.4	86.0	82.9
100	109.5	105.7	102.0	98.3	94.7	91.1	87.3	84.5	81.5
105	107.8	103.9	100.1	96.4	92.8	89.4	86.1	83.0	80.0
110	106.0	102.0	98.1	94.5	91.0	87.7	84.5	81.5	78.6

****

Temp °F	Reference factor "R" Airport elevation (ft)								
	0	1,000	2,000	3,000	4,000	5,000	6,000	7,000	8,000
50	48.0	49.5	53.0	57.5	62.2	67.1	72.4	78.1	84.5
55	48.1	50.1	53.3	57.8	62.4	67.4	72.7	78.5	84.8
60	48.2	50.8	53.8	58.3	62.9	67.9	73.2	79.1	85.5
65	48.3	51.4	54.5	59.0	63.6	68.7	74.1	80.0	86.5
70	48.5	51.3	55.5	59.9	64.6	69.7	75.2	81.3	87.8
75	48.6	52.4	56.6	61.0	65.8	71.0	76.7	82.8	89.3
80	50.0	53.7	57.9	62.4	67.3	72.6	78.4	84.6	91.4
85	51.4	55.2	59.4	63.9	69.0	74.4	80.4	86.8	93.7
90	53.0	56.8	61.0	65.7	70.9	76.5	82.6	89.2	96.3
95	54.7	58.6	62.9	67.7	73.1	78.9	85.2	91.9	99.2
100	56.5	60.5	65.0	70.0	75.5	81.5	88.0	95.0	102.5
105	58.4	62.6	67.2	72.4	78.1	84.4	91.1	98.3	106.1
110	60.4	64.8	69.7	75.1	81.0	87.5	94.5	102.0	110.0

****

Weight 1,000 lb	Runway length (1,000 ft) Reference factor "R"						
	50	60	70	80	90	100	110
70	2.61	3.12	3.56	3.96	4.36	4.78	5.25
75	2.93	3.49	4.00	4.50	5.01	5.57	6.21
80	3.28	3.90	4.49	5.08	5.70	6.39	7.16
85	3.67	4.36	5.04	5.72	6.44	7.23	8.10
90	4.06	4.86	5.64	6.42	7.23	8.09	9.04
95	4.51	5.41	6.29	7.17	8.06	8.89	9.96
100	4.98	6.00	7.00	7.98	8.94	9.91	10.87
105	5.48	6.64	7.76	8.84	9.87	10.85	11.78
110	6.02	7.32	8.57	9.75	10.84	11.82	12.67
115	6.59	8.05	9.44	10.77	11.86	12.82	

TABLE 23.5 Minimum Positive Climb Gradient—
One Engine Inoperative

1st segment	Two-engine aircraft	Positive
	Three-engine aircraft	0.3%
	Four-engine aircraft	0.5%
2nd segment	Two-engine aircraft	2.4%
	Three-engine aircraft	2.7%
	Four-engine aircraft	3.0%
3rd segment	Two-engine aircraft	1.2%
	Three-engine aircraft	1.5%
	Four-engine aircraft	1.7%
	Net takeoff flight path	
Takeoff flight path reduced by:		
	Two-engine aircraft	0.8%
	Three-engine aircraft	0.9%
	Four-engine aircraft	1.0%

3. Displace the threshold so that the object will not penetrate the applicable surface, and accept a shorter landing surface. Relevant factors to be evaluated include

- Types of airplanes that will use the runway and their performance characteristics.
- Operational disadvantages associated with accepting higher landing minima.
- Cost of removing, relocating, or lowering the object.
- Effect of the reduced available landing length when the runway surface is adversely affected by precipitation or snow and ice accumulations.
- Cost of extending the runway if insufficient runway would remain as a result of displacing the threshold. The environmental and public acceptance aspects of a runway extension must also be evaluated under this consideration.
- Cost and feasibility of relocating visual and electronic approach aids such as threshold lights, visual approach slope indicator, runway end identification lights, localizer, glide slope (to provide a threshold crossing height of not more then 60 feet), approach lighting system, and runway markings.
- Effect of the threshold change on aircraft noise-abatement procedures.

Displacing the Threshold. If, after consideration of alternatives, the decision is made to displace the threshold, the required displacement distance can be determined in a three-step process. Given (a) type of approach to that runway end, (b) obstacle location, and (c) obstacle height above the ground; one calculates (a) approach surface height at the obstacle, (b) difference in height between obstacle and approach surface, and (c) obstacle displacement distance required.

Clearways and Stopways. There are certain runways where there is inadequate full-strength runway length to accommodate the full takeoff distance (TOD) of certain airplanes. Federal Aviation Regulations permit the use of a *clearway* to provide part of the takeoff distance required for turbine-powered airplanes. The clearway is defined as a plane, above which no object protrudes, extending from the end of the runway with an upward slope not exceeding 1.25 percent, not less than 500 feet wide, centrally located about the extended centerline of the runway, and under the control of the airport authorities. Threshold lights may protrude above the clearway plane, however, if their height above the end of the runway is 26 inches or less and if they are located to each side of the runway. Although the use of a clearway is a technique that permits higher allowable operating weights without an increase in runway length, the runway length recommended without use of a clearway (or stopway—see paragraph below) for the most demanding airplane should be provided.

A runway is normally designed with a critical aircraft and the maximum performance mission requirement in mind. The clearway should serve only as a means of accommodating the takeoff distance requirements for that occasional heavy operation requiring a greater takeoff distance than the most demanding airplane for which the runway length is designed. When the frequency of this "occasional" operation increases to a certain point, a new "most demanding" airplane for runway design length exists, and additional runway length should be provided. An airport owner interested in providing a clearway should be aware of the requirement that the clearway be under his control, although not necessarily by direct ownership. The purpose of such control is to ensure that no takeoff operation intending to use a clearway is initiated unless it has been absolutely determined that no fixed or movable object will penetrate the clearway plane during that operation.

A *stopway* is an area beyond the runway used for takeoff that is designated by the airport authority for use in decelerating an airplane during an aborted takeoff. A stopway is at least as wide as the runway it serves and is centered on the extended centerline of the runway. It should be able to support an airplane during an aborted takeoff without causing structural damage to the airplane.

Declared Distances. Introduction of stopways and clearways and the use of displaced thresholds on runways have created a need for accurate information concerning the distances available and suitable for the landing and takeoff of airplanes. There are also situations when additional runway length is needed but the construction of additional full-strength runway is not feasible or practical. The concept of *declared distances* has been developed to accommodate certain operations in those circumstances. The declared distances that must be calculated for each runway direction are (a) takeoff run available (TORA), (b) takeoff distance available (TODA), (c) accelerate stop distance available (ASDA), and (d) landing distance available (LDA). When a runway is not provided with either a stopway or clearway, and the threshold is located at the extremity of the runway, the four declared distances should normally be equal to the length of the runway, as shown in Figure 23.1*a*. When a runway is provided with a clearway, then the TODA will include the length of clearway, as shown in Figure 23.1*b*. Where a runway is provided with a stopway, the ASDA will include the length of stopway, as shown in Figure 23.1*c*. The LDA will be reduced by the length of the threshold displacement when a runway has a displaced threshold as shown in Figure 23.1*c*. A displaced threshold affects only the LDA for approaches made to that threshold; declared distances for landings in the opposite direction and takeoffs in either direction are unaffected. Figures 23.1*b* through 23.1*d* illustrate a runway provided with a clearway or a stopway or having a displaced threshold. More than one of the declared distances will be modified when more than one of these features exists, but the modification will follow the same principle illustrated. An example of situation where all these features exist is shown in Figure 23.1*d*. If a runway direction cannot be used for takeoff or landing or both because it is operationally forbidden, then this should be declared and the words "not usable" or the abbreviation "NU" entered.

23.3.3 Airport Approach and Departure Design

The preceding sections on airport layout design and runway length design focused on development of safe and efficient airport surfaces containing aircraft movement areas and the balance of the airport operating area. This section will address three-dimensional design standards for the protection of aircraft when operating in the immediate airport airspace, such as during approaches and departures. These standards are prescribed in three regulations: FAR Part 77, *Civil Airport Imaginary Surfaces;* FAR Part 25, *Airworthiness Standards: Transport Category Airplanes;* and FAR Part 121, *Operating Requirements: Domestic, Flag, and Supplemental Operators.*

FAR Part 77 was developed to help protect airspace in the vicinity of airports by defining imaginary surfaces that objects should not penetrate and thus constitute hazards or obstructions to air traffic. A *hazard* to air navigation is defined as a fixed or mobile object of a greater height than any of the heights or surfaces presented in Subpart C of FAR Part 77 that has not been properly charted

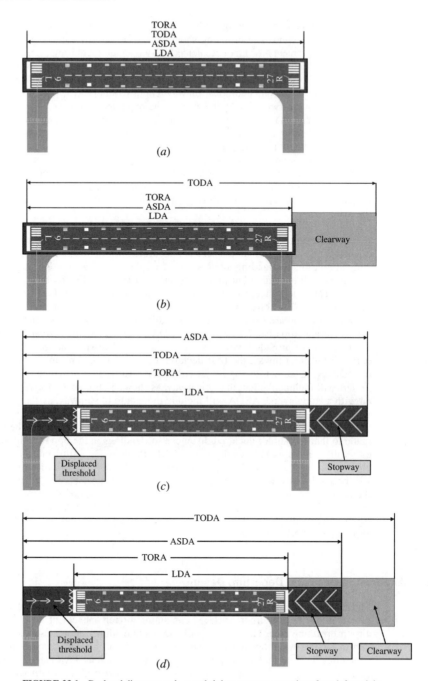

FIGURE 23.1   Declared distances *a*, *b*, *c*, and *d* show runway operations from left to right.

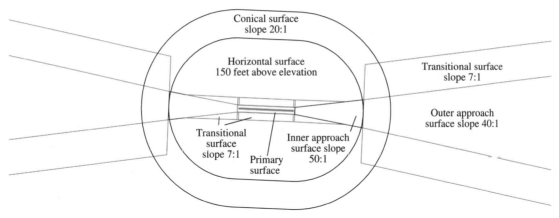

FIGURE 23.2   FAR Part 77 imaginary surfaces.

and marked or lighted. An *obstruction* is an equivalent object that has been so marked, lighted, and charted. One of the surface sets referred to in Subpart C is the set of civil airport imaginary surfaces that is defined in FAR Part 77.25. The various imaginary surfaces affect operational procedures and threshold placement at all runways. The following civil airport imaginary surfaces (see Figure 23.2) are established with relation to the airport and to each runway. The size of each such imaginary surface is based on the category of each runway according to the type of approach available or planned for that runway. The slope and dimensions of the approach surface applied to each end of a runway are determined by the most precise approach existing or planned for that runway end.

Horizontal Surface.   A horizontal plane 150 feet above the established airport elevation, the perimeter of which is constructed by swinging arcs of specified radii form the center of each end of the primary surface of each runway and connecting the adjacent arcs by lines tangent to those arcs. The radius of each arc is

- 5,000 feet for all runways designated as utility or visual
- 10,000 feet for all other runways

The radius of the arc specified for each end of a runway will have the same arithmetical value. That value will be the highest determined for either end of the runway. When a 5,000-foot arc is encompassed by tangents connecting two adjacent 10,000-foot arcs, the 5,000-foot arc shall be disregarded on the construction of the perimeter of the horizontal surface.

Conical Surface.   A conical surface is extended outward and upward from the periphery of the horizontal surface at a slope of 20:1 for a horizontal distance of 4,000 feet.

Primary Surface.   A surface longitudinally centered on a runway. When the runway has a specially prepared hard surface, the primary surface extends 200 feet beyond each end of that runway; but when the runway has no specially prepared hard surface or planned hard surface, the primary surface ends at each end of that runway. The elevation of any point on the primary surface is the same as the elevation of the nearest point on the runway centerline. The width of a primary surface will be the width prescribed for the most precise existing or planned approach to either end of that runway:

- 250 feet for utility runways having only visual approaches
- 500 feet for utility runways having nonprecision instrument approaches

- For other than utility runways:
  - 500 feet for visual runways having only visual approaches
  - 500 feet for nonprecision instrument runways having visibility minima greater than three-fourths statute mile
  - 1,000 feet for a nonprecision instrument runway having visibility minima as low as three-fourths of a statute mile, and for all precision instrument runways

Approach Surface.   A surface longitudinally centered on the extended runway centerline and extending outward and upward from each end of the primary surface. An approach surface is applied to each end of each runway based upon the type of approach available or planned for that runway end.

- The inner edge of the approach surface is the same width as the primary surface, and it expands uniformly to a width of
  - 1,250 feet for that end of a utility runway with only visual approaches
  - 1,500 feet of a runway other than a utility runway with visual approaches
  - 2,000 feet for that end of a utility runway with a nonprecision approach
  - 3,500 feet for that end of a nonprecision instrument runway other than utility, having visibility minimums greater than three-fourths of a statute mile
  - 4,000 feet for that end of a nonprecision instrument runway, other than utility, having a nonprecision instrument approach with visibility minimums as low as three-fourths statute mile
  - 16,000 feet for all precision instrument runways
- The approach surface extends for a horizontal distance of
  - 5,000 feet at a slope of 20:1 for all utility and visual runways
  - 10,000 feet at a slope of 34:1 for all nonprecision instrument runways other than utility
  - 10,000 feet at a slope of 50:1 with an additional 40,000 feet at a slope of 40:1 for all precision instrument runways
  - The outer width of an approach surface to an end of a runway will be that width prescribed in this subsection for the most precise approach existing or planned for that runway end

Transitional Surface.   These surfaces extend outward and upward at right angles to the runway centerline and the runway centerline extended at a slope of 7:1 from the sides of the primary surface and from the sides of the approach surfaces. Transitional surfaces for those portions of the precision approach surface that project through and beyond the limits of the conical surface extend a distance of 5,000 feet measured horizontally from the edge of the approach surface and at right angles to the runway centerline.

Takeoff Path Requirements.   FAR Part 25 prescribes airworthiness standards for certification of transport category airplanes, and FAR Part 121 prescribes rules governing air carrier flight operations. The following excerpts from those regulations demonstrate the concept of climb gradient and obstacle clearance that must be accommodated in airport design. Major portions of the regulations have been omitted or edited in the interest of brevity and clarity. For example, because of the inherent differences in the performance characteristics of reciprocating and turbine-powered engines, the certification and operating standards for these airplane types differ. In this text, only the regulations applicable to turbine-powered airplanes are summarized; however, if one understands the concepts presented in this text, there should be no difficulty addressing situations involving reciprocating engine airplanes.

Takeoff Path (FAR Part 25.111)

1. The takeoff path extends from a standing start to a point in the takeoff at which the airplane is 1,500 feet above the takeoff surface, or at which the transition from the aircraft's takeoff to the enroute configuration is completed, whichever point is higher. In addition:

a. The airplane must be accelerated on the ground to V_{EF}, at which point the critical engine must be made inoperative and remain inoperative for the rest of the takeoff

b. After reaching V_{EF}, the airplane must be accelerated to V_2

2. During the acceleration to speed V_2, the nose gear may be raised off the ground at a speed V_R. Landing gear retraction begins once airborne.

3. During the takeoff path determination:

a. The slope of the airborne part of the takeoff path must be positive at each point

b. The airplane must reach V_2 before it is 35 feet above the takeoff surface and must continue at a speed as close as practical to, but not less than V_2, until it is 400 feet above the takeoff surface

c. At each point along the takeoff path, starting at the point at which the airplane reaches 400 feet above the take-off surface, the available gradient of climb may not be less than

i. 1.2 percent for two-engine airplanes

ii. 1.5 percent for three-engine airplanes

iii. 1.7 percent for four-engine airplanes

d. Except for gear retraction and propeller feathering, the airplane configuration may not be changed, and no change in power or thrust that requires action by the pilot may be made, until the airplane is 400 feet above the takeoff surface.

Takeoff Distance and Takeoff Run (FAR Part 25.113)

1. Takeoff distance is the greater of

a. the horizontal distance along the takeoff path from the start of the takeoff to the point at which the airplane is 35 feet above the takeoff surface, as determined under Part 25.111, or

b. 115 percent of the horizontal distance along the takeoff path, with all engines operating, from the start of the takeoff to the point at which the airplane is 35 feet above the takeoff surface, as determined by a procedure consistent with Part 25.111.

Takeoff Flight Path (FAR Part 25.115)

1. The takeoff flight path begins 35 feet above the takeoff surface at the end of the takeoff distance determined in accordance with 25.113 (a).

2. The net takeoff flight path data must be determined so that they represent the actual takeoff flight paths (determined in accordance with 25.111 and with paragraph (1) of this section) reduced at each point by a gradient of climb equal to

a. 0.8 percent for two-engine airplanes;

b. 0.9 percent for three-engine airplanes; and

c. 1.0 percent for four-engine airplanes.

Climb: One-Engine-Inoperative (FAR Part 25.121)

1. *Takeoff with landing gear extended.* Critical takeoff configuration exists along the flight path between the points at which the airplane reaches V_{LOF} and the point at which the landing gear is fully retracted. The configuration used in 25.111, but without ground effect, the steady gradient of climb must be positive for two-engine airplanes, and not less than 0.3 percent for three-engine airplanes of 0.5 percent for four-engine airplanes, at V_{LOF} and with

a. the critical engine inoperative and the remaining engines at the power or thrust available when retraction of the landing gear is begun in accordance with Part 25.111 unless there is a more critical power operating condition existing later along the flight path but before the point at which the landing gear is fully retracted; and

b. the weight equal to the weight existing when retraction of the landing gear is begun, determined under Part 25.111.

2. *Takeoff with landing gear retracted.* The takeoff configuration exists at the point of the flight path at which the landing gear is fully retracted. The configuration used in Part 25.111 but without ground

effect, the steady gradient of climb may not be less than 2.4 percent for two-engine airplanes, 2.7 percent for three-engine airplanes, and 3.0 percent for four-engine airplanes, at V_2 and with

a. the critical engine inoperative, the remaining engines at the takeoff power or thrust available at the time the landing gear is fully retracted, determined under 25.111 unless there is a more critical power operating condition existing later along the flight path but before the point where the airplane reaches a height of 400 feet above the takeoff surface; and

b. the weight equal to the weight existing when the airplane's landing gear is fully retracted determined under Part 25.111.

Transport Category Airplanes: Turbine Engine Powered Takeoff Limitations (FAR Part 121.189)

An airplane certified after September 30, 1958, is allowed a net takeoff flight path that clears all obstacles either by a height of at least 35 feet vertically, or by at least 200 feet horizontally within the airport boundaries and by at least 300 feet horizontally after passing airport boundaries. The following information is provided to further clarify the climb gradient and obstacle clearance requirements

1. In order to achieve compliance with this regulation, the takeoff gross weight for any given flight must not exceed the lowest of the maximum weights allowed for:
   a. Compliance with runway length and obstacle clearance requirements
   b. Compliance with takeoff climb requirements
   c. Compliance with en route performance requirements
   d. Compliance with maximum landing weight taking into account normal fuel burnout en route
   e. Structural limit of the airplane

2. *Takeoff climb requirement.* The maximum weight for takeoff may not exceed that weight that will allow airplane performance equal to the climb gradients specified in Table 3. The takeoff path may be considered as the trajectory, or elevation profile made good on a takeoff with an engine failure occurring at V_1 speed. The path is considered as extending from the standing start to a point in the takeoff where a height of 1,500' above the takeoff surface is reached, or to a point in the takeoff where transition from takeoff to en route configuration is complete, whichever is higher. The airborne part of the takeoff is comprised of the following parts:
   a. *First Segment:* Starts at liftoff and ends when gear retraction is complete.
   b. *Second Segment:* Starts at gear retraction and ends when at 400 feet above the runway.
   c. *Third Segment:* Starts at 400 feet above the runway, and continues to 1,500 feet above the takeoff surface, or until transition to the aircraft's enroute configuration is completed, which ever occurs last.
   d. *Net Takeoff Flight Path:* A profile starting at the end of the takeoff distance, having a gradient specified in Table 3, below the takeoff flight path. The "net" flight path must clear all obstacles by 35 feet vertically or 300 feet horizontally. As the actual flight path altitude is greater than the "net" flight path, the airplane will have clearance above obstacles in the flight path.

A climb gradient profile is depicted in Figure 23.3. The climb segments are located relative to the departure end of the runway. Adherence to the manufacturer's limits on maximum allowable

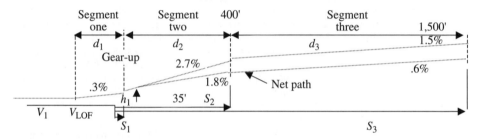

FIGURE 23.3 Climb gradient profile.

takeoff weight and to the appropriate speed schedules will ensure compliance with takeoff climb requirements. During our classroom sessions we will be locating the climb segments based on typical takeoff data (liftoff distance, gear retraction distance and takeoff distance). Once the climb segments are located, the net takeoff flight path can be plotted to determine compliance with obstacle clearance requirements.

23.3.4 Airport Pavement Design

Airport traffic forecasts for airports encompass a variety of aircraft having different landing gear geometry and weights, the effects of which must be identified in terms of the critical aircraft for that design feature. Each aircraft must be converted to an equivalency of the landing gear type represented by the design aircraft, as discussed in AC 150/5320-6D. The factors below should be used to make that conversion:

Aircraft in question landing gear wheel configuration	Design aircraft landing gear wheel configuration	Multiply departures by
Single	Dual	0.8
Single	Dual tandem	0.5
Dual	Dual tandem	0.6
Double dual tandem	Dual tandem	1.0
Dual tandem	Single	2.0
Dual tandem	Dual	1.7
Dual	Single	1.3
Double dual tandem	Dual	1.7

The conversion to equivalent annual departures of the design aircraft continues using the following formula:

$$R_1 = R_2 \sqrt{W_2 / W_1}$$

where R_1 = equivalent annual departures by the design aircraft
R_2 = annual departures expressed in design aircraft landing gear
W_1 = wheel load of the design aircraft
W_2 = wheel load of the aircraft in question

Ninety-five percent of the gross weight of an aircraft rests on the main landing gear. Wide-body aircraft have radically different landing gear assemblies from narrow-bodied airplanes, and therefore special consideration is given to maintain the relative effects. Each wide-body is treated as if it were a 300,000 pound dual-tandem wheel aircraft when computing equivalent annual departures. Remember, the factor multiplier is based on the actual gear configuration for wide-body aircraft, but the $R1$, the equivalent annual departures, are determined using the modification described above. The number of equivalent annual departures for each aircraft is computed and the aircraft population is added to define the total equivalent annual departures (TEAD).

23.4 CONCLUSION

This chapter has introduced and explained just a few of the key concepts in airport design. Airport design involves a broad array of airside and landside elements and often requires a multidisciplinary approach. It definitely requires familiarization with, and reference to, the sizable body of ICAO and FAA documents as a foundation. There is also a growing body of literature in technical and scholarly journals dealing with airport design issues. The most important thing in the end is that the transportation engineer contribute to airport developments that are effective, economical to build and operate, flexible in design, and capable of expansion and change without undue stress on the host community and airport users.

REFERENCES

Ashford, N., and P. Wright. 1991. *Airport Engineering*, 3d ed. New York: John Wiley & Sons.

Caves, R., and G. Gosling. 1999. *Strategic Airport Planning*. London: Pergamon.

De Neufville, R., and A. Odoni. 2003. *Airport Systems: Planning, Design, and Management*. New York: McGraw-Hill.

Federal Aviation Administration (FAA). 1970. *Planning the Metropolitan Airport System*. Advisory Circular 150/5070-5, FAA, Washington, DC.

———. 1975. *The Continuous Airport System Planning Process*. Advisory Circular 150/5050-5, FAA, Washington, DC.

———. 1983a. *Airport Capacity and Delay*. Advisory Circular 150/5060-5, FAA, Washington, DC.

———. 1983b. *Noise Control and Compatibility Planning for Airports*. Advisory Circular 150/5020-1, FAA, Washington, DC.

———. 1985. *Airport Master Plans*. Advisory Circular 150/5070-6A, FAA, Washington, DC.

———. 1987. *A Model Zoning Ordinance to Limit Height of Objects Around Airports*. Advisory Circular 150/5190-4A, FAA, Washington, DC.

———. 1988. *Planning and Design Guidelines for Airport Terminal Facilities*. Advisory Circular 150/ 5360-13, FAA, Washington, DC.

———. 1989a. *Planning the State Aviation System*. Advisory Circular 150/5050-3B, FAA, Washington, DC.

———. 1989b. *Standards for Specifying Construction Standards on Airports*. Advisory Circular 150/ 5370-10A, FAA, Washington, DC.

———. 1990. *Runway Length Requirements for Airport Design*. Advisory Circular 150/5325-4A, FAA, Washington, DC.

———. 1994. *Architectural, Engineering, and Planning Consultant Services for Airport Grant Projects*. Advisory Circular 150/5100-14C, FAA, Washington, DC.

———. 1996a. *Airport Pavement Design and Evaluation*. Advisory Circular 150 / 5320-6D, FAA, Washington, DC.

———. 1996b. *Land Acquisition and Relocation Assistance for Airport Improvement Program Assisted Projects*. Advisory Circular 150/5100-17, FAA, Washington, DC.

———. 1996c. *Proposed Construction or Alteration of Objects That May Affect the Navigable Airspace*. Advisory Circular 70/7460-2J, FAA, Washington, DC.

———. 1999. *The National Plan of Integrated Airport Systems (NPIAS) 1998–2002*. FAA, Washington, DC.

———. 2002a. *Airport Design* (with Change 7). Advisory Circular 150/5300-13, FAA, Washington, DC.

———. 2002b. *Operational Safety on Airports During Construction*. Advisory Circular 150/5370-2D, FAA, Washington, DC.

———. *Airport Environmental Handbook*. Order 5050.4A, FAA, Washington, DC.

———. Federal Aviation Regulations Part 25, *Airworthiness Standards: Transport Category Airplanes*. FAA, Washington, DC.

———. Federal Aviation Regulations Part 77, *Objects Affecting Navigable Airspace*. FAA, Washington, DC.

———. Federal Aviation Regulations Part 121, *Operating Requirements: Domestic, Flag, and Supplemental Operations*. FAA, Washington, DC.

———. Federal Aviation Regulations Part 150, *Airport Noise Compatibility Planning*. FAA, Washington, DC.

———. Federal Aviation Regulations Part 157, *Notice of Construction, Alteration, Activation and Deactivation of Airports*. FAA, Washington, DC.

Florida Department of Transportation Aviation Office. *Guidebook for Airport Master Planning*. Florida Department of Transportation, Tallahassee, FL.

Horonjeff, R., and F. McKelvey. 1994. *Planning and Design of Airports*, 4th ed. New York: McGraw-Hill.

International Air Transport Association (IATA). 1995. *Airport Development Reference Manual*, 8th ed. Montreal: IATA.

International Civil Aviation Organization (ICAO). 1983. *Airport Planning Manual, Part 3: Guidelines for Consultant/Construction Services*, 1st ed. (Doc 9184-AN / 902). Montreal: ICAO.

———. 1985. *Airport Planning Manual, Part 2: Land Use and Environmental Control*, 2d ed. (Doc 9184-AN/902). Montreal: ICAO.

———. 1987. *Airport Planning Manual, Part 1: Master Planning*, 2d ed. (Doc 9184-AN/902). Montreal: ICAO.

———. 1988. *Annex 16 to the Convention on International Civil Aviation, Environmental Protection*, vol. 1, *Aircraft Noise*. Montreal: ICAO.

———. 1997a. *Aerodrome Design Manual, Part 2: Taxiways, Aprons, and Holding Bays*, 3d ed., Corr. 1 (Doc 9157). Montreal: ICAO.

———. 1997b. *Aerodrome Design Manual, Part 3: Pavements*, 2d ed., Amdts. 1 and 2 (Doc 9157). Montreal: ICAO.

———. 1999. *Annex 14 to the Convention on International Civil Aviation, Aerodromes*, vol. 1, *Aerodrome Design and Operations*, 3d ed. Montreal: ICAO.

———. 2000. *Aerodrome Design Manual, Part 1: Runways*, 2d ed., Amdt. 1 (Doc 9157). Montreal: ICAO.

Transportation Research Board (TRB). 2002. *Aviation Demand Forecasting: A Survey of Methodologies*. Transportation Research Circular E-C040, National Research Council, TRB, Washington, DC.

CHAPTER 24
AIR TRAFFIC CONTROL SYSTEM DESIGN*

Robert Britcher
Montgomery Village, Maryland

Aviation is a recent form of travel. During the 17th and 18th centuries, humans tried to build airplanes with wings that flapped like birds. The planes were called ornithopters. They did not succeed. In the 19th century, armies used lighter-than-air balloons to reconnoiter, while a few pioneers dreamed of and experimented with heavier-than-air machines. In 1804, George Cayley, often called the father of aviation, designed, built, and flew a small model glider. It was the first modern configuration airplane in history, with a fixed wing and a horizontal and vertical tail that could be adjusted. Powered flight awaited the German aviator Otto Lilienthal. Some 90 years after Cayley, Lilienthal constructed a glider with flapping wing tips that was to be powered by a small motor using compressed carbonic gas. His efforts were truncated. On August 9, 1896, he was killed when he stalled and crashed to the ground while gliding.

In 1903, the Wright brothers designed and built a flying craft that could be controlled while in the air. Every successful aircraft built since has had controls to roll the wings right or left, pitch the nose up or down, and yaw the nose from side to side. These three controls, roll, pitch, and yaw, let a pilot navigate an airplane in all three dimensions, making it possible to fly from place to place. The entire aerospace business, the largest industry in the world, depends on this simple but brilliant idea. More important, the Wright brothers changed the way we view our world. Before flight became commonplace, people traveled in two dimensions, crossing the borders that separate town from town and nation from nation. Seen from above, the artificial boundaries that divide us disappear. Distances shrink, the horizon stretches. The world seems grander and more interconnected.

Airplanes proved their worth in World War I. Soon after, commercial flying burgeoned. At first, civil aviation was limited to exploration and hauling cargo. But by the late 1920s, planes were carrying passengers. In 2000, U.S. commercial airlines flew over 600 million passengers a year.

In the United States, air traffic control began with the Air Commerce Act of 1926. An aeronautics branch of the Department of Commerce was established. It chartered safety standards, licensing, certification, rule-making, and the management of airways and navigational aids. These services remain the linchpin of civil aviation authorities around the world, including today's U.S. Federal Aviation Administration (FAA), now under the Department of Transportation.

This chapter emphasizes the design of systems to control commercial and general aviation in the United States. Our military and aviators of other nations also require air traffic control (ATC) systems. But, in the main, they use the same technology and techniques as the U.S. civil system.

One of the first design principles to shape domestic air traffic control was the creation of airways. Early flights traversed airspace at the whim of the pilots. There were no airways. In 1926, that changed. Like the corners and turns of highways, airways would be marked. Navigational aids became known as fixes, and the leg between fixes as an *airway*. (Navigational aids, or *navaids,* consist

---

*Reprinted from the First Edition.

FIGURE 24.1 Controller airspace.

of landmarks, lights, and radio signals—even bonfires in the early days. We can now add satellites to the list.) Planes would fly point to point, from a runway to a fix at a planned altitude, to the next fix, and so on until the plane landed. This procedural approach inherently constrains air travel, as do roads and railways on the ground. (Figure 24.1 shows how sector boundaries and airways are mapped onto the controller's display.)

Point-to-point control and radios enabled planes to be tracked from the ground. The nation's first air route traffic control center (ARTCC) was created in Newark, NJ, in 1935. (Figure 24.2 shows the 20 ARTCCs that monitor traffic across the continental United States.) The first controllers included Glen Gilbert and J. V. Tighe, who designed shrimp boats to track flights on a table map. Gilbert is often credited as the father of air traffic control.

Gilbert and others not only monitored flights, they issued voice clearances. A clearance gives pilots the go-ahead to depart from, arrive at, or hold at a *fix,* flight level (altitude), or runway. Ground control of airplanes soon became an FAA-regulated practice for flights flying instrument flight rules (IFR), such as commercial airliners. Flights using visual flight rules (VFR) are not considered "controlled"; the pilot is on his or her own: see and avoid. For controlled flights, the pilot must give priority to the controllers' directions, regardless of what he or she may see. The exception is weather. The pilot may use his or her judgment in negotiating weather.

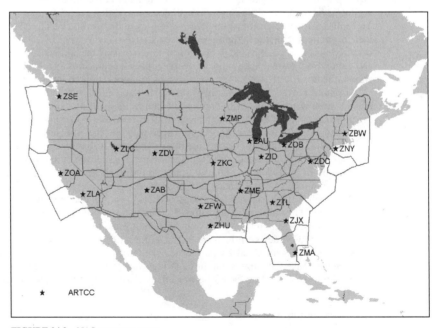

FIGURE 24.2 NAS en route centers.

Shortly before World War II a system coalesced. Regulations, maps, navigational aids, ground facilities, and communications began to interact in the pursuit of what the FAA calls separation assurance. Despite improvements in technology, the step-by-step, fix-by-fix, clearance-by-clearance approach, often called the board game, still dominates the design of air traffic control. All that is needed for the game to work is a flight plan (flight plans identifies an aircraft's call sign, intended speed, departure fix, proposed en route altitude, and route of flight, fix by fix, to the arrival fix; see Figure 24.6) and a calculator to determine the arrival time at each fix. Start the game. Planes take off. For every controlled flight, shortly before it reaches the next fix, a flight progress strip is printed that shows the time of arrival at that fix (see Figure 24.3).

World War II brought forth a plethora of technology and techniques. Radar and encryption are perhaps the most obvious. But the list includes computing, communications theory, information theory, game theory, and operations research. All would become important to modern living.

Radar proved crucial to ground control of aircraft. It gives controllers a view of the airspace situation in near real-time. After point-to-point routing, the situation display is the second great hallmark of air traffic control design.

In concept, radar is simple. A rotating signal bounces off airborne targets, planes among them. The time and angle of the return signal allows the location of the target to be calculated. One reply gives the position in terms of range and azimuth. From the second and succeeding replies the heading

FIGURE 24.3   En route controller flight strips.

and velocity can be deduced. By the 1960s, many airplanes were equipped with transponders. The transponder intercepts the radar signal and adds a four-digit octal beacon code to the reply. This secondary return more clearly identifies the aircraft. What remains is to feed the return, with its position, velocity, heading, and maybe a beacon code, to a ground facility and the controllers' displays. This is done by land lines similar to telephone lines.

Radar helped shape the topology of the air traffic control system, that is, how it is laid out. Long-range radars monitor wide-area en route airspace. In the United States, their reports feed the 20 ARTCCs. Figure 24.4 shows the "R" controller's sector display of weather and traffic, supported by the latest 20- × 20-inch console.

At airports, short-range radars feed terminal radar (approach) control facilities (TRACONs), which control arrivals and departures in concert with airport towers. In the United States, there are slightly fewer than 200 TRACONs. They are placed near major airports, where they control departing and arriving traffic within a radius of about 60 nautical miles. There are over a thousand FAA-supported towers, and many more municipal and private airport towers.

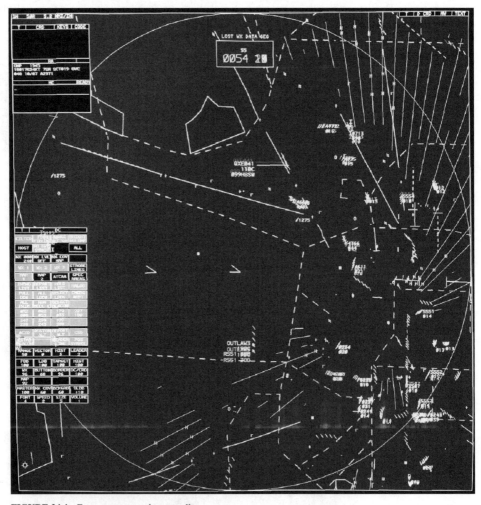

FIGURE 24.4   En route center radar controller.

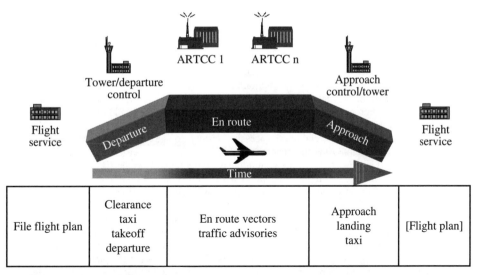

ARTCC 1   ARTCC n

Tower/departure
control

Approach
control/tower

Flight
service

Flight
service

En route

Departure

Approach

Time

| File flight plan | Clearance
taxi
takeoff
departure | En route vectors
traffic advisories | Approach
landing
taxi | [Flight plan] |

FIGURE 24.5 Gate-to-gate ATC.

A typical flight leaves the gate under the control of an airport tower. It takes off and ascends under TRACON control. About 60 nautical miles out, the flight transitions to en route airspace. A transcontinental flight will fly under the jurisdiction of several en route ARTCCs. It arrives as it left, but in reverse. The plane descends through approach control airspace, under control of a TRACON, and lands, guided by airport tower controllers. Along the way, controllers at each facility track the plane with radar. (See Figure 24.5.)

Controllers at TRACONs and ARTCCs manage only a sector of the approach or en route airspace. Therefore, control must be handed off from sector to sector, from controller to controller. The Chicago en route ARTCC watches over 60 sectors of airspace. A plane crossing Illinois will thus be handed off several times. As it is leaving the Chicago center's airspace, which spans a few hundred miles around the city's compass, control will be handed off to the next ARTCC. If the flight is eastbound, for example, that will be the Indianapolis ARTCC.

The first computer used in air traffic control was located in Indianapolis. In the late 1950s, it stored flight plan data on a drum. The drum was connected to a printer that printed the flight progress strips. The computer calculated the times of arrival at each fix along the route of each flight, much like estimated times at rail stops along a train route.

The first automated system was developed by Burroughs and UNIVAC for the TRACONs. Unlike the en route ARTCCs, which are concerned with airway traffic and thus longer planning horizons, the TRACONs objective is to help guide airplanes through the tender process of departing and arriving. In this they rely mostly on radar and visual tracking, a more immediate, near-real-time job. The TRACON system is a radar tracking system. It was and is still called the Automated Radar Terminal System, or ARTS.

The ARTS computers, originally UNIVAC computers, are located in the TRACONs and provide (originally analog, now digital) displays of the departing and arriving traffic to the dozen or fewer approach controllers, who monitor the airspace around the airport. The ARTS also transmits the digital display data to the airport towers, which contain little in the way of computers, except for output devices like the displays attached to the remote ARTS TRACON, and flight strip printers, connected to the en route centers.

So much has changed since 1965, the year that the first ARTS sites went operational: technologies, social trends, space exploration, the geopolitical landscape. But not air traffic control automation. The ARTS equipment has been improved over the years, but its functions and algorithms have changed little.

The same can also be said for en route automation. IBM developed the en route automation system for the 20 ARTCCs. It was originally called the 9020 NAS En Route Stage A, for the 9020 multiprocessor IBM built for the FAA. The computer, twice upgraded, is now referred to simply as "the Host."

Over the decades in which automation has aided the controllers, the FAA has put more energy into safe and dependable automation than into rich automation. The 9020 multiprocessor is an excellent example of what the FAA calls its "failsafe" system. In the 1960s, the ARTS and the 9020 system used multiprocessors, so that if one computer failed, another would resume its processing. These machines were pioneering. Both UNIVAC and IBM built upon the concepts developed for the FAA to produce the first commercial multiprocessing, virtual storage computers. IBM's 9020 became the prototype for its System/370 virtual machines, which automatically mask storage, instruction fetch, cache memory, and channel faults while continuously computing. (That the IBM Enterprise System/390 today has a mean-time-to-failure of about 25 years is due largely to the objectives set by the FAA.)

The en route system combines radar tracking and flight planning, using algorithms developed by MIT, based on what had been learned at the early Indianapolis center. Thus, en route control is divided into two parts that the automation must reconcile, but that favor two different jobs: the "R" (radar) controller and the flight or "D" (data) controller. Each sector of airspace is under the control of a radar controller for near-real-time actions, and a flight controller, who is engaged in more long-term planning. (Figure 24.6 shows the fields of a filed flight plan.)

The ARTS required little more than 2 years to develop; the en route system took 9. Two aspects of the en route system design contributed to this: the complexities and size of en route airspace and the tricky business of pairing real-time radar tracking with intermediate- and long-term flight plans and their frequently amended altitudes and routes of flight. The en route software is about 10 times larger than the ARTS. Not the least of the automation challenges is automatically resynchronizing tracking and flight data during and after computer failures.

U.S. department of transportation federal aviation administration **Flight plan**	(FAA use only)	☐ Pilot briefing ☐ Stopover	☐ VNR	Time started	Specialist initials

1. Type VFR IFR DVPR	2. Aircraft identification	3. Aircraft type / special equipment	4. True airspeed KTS	5. Departure point	6. Departure time	7. Cruising altitude
					Proposed (Z) / Actual (Z)	

8. Route of flight

9. Destination (Name of airport and city)	10. EST. time enroute		11. Remarks
	Hours	Minutes	

12. Fuel on board		13. Alternate airport(s)	14. Pilot's name, address & telephone number & aircraft home base	15. Number aboard
Hours	Minutes		17. Destination contact/telephone (optional)	

16. Color of aircraft	Civil aircraft pilots, FAR Part 91 requires you file an IFR flight plan to operate under instrument flight rules in controlled airspace. Failure to file could result in a civil penalty not to exceed $1,000 for each violation (Section 901 of the Federal Aviation Act of 1958, as amended). Filing of a VFR flight plan is recommended as a good operating practice. See also Part 99 for requirements concerning DVFR flight plans.

FAA form 7233-1 (8-82)   Close VFR flight plan with_____ FSS on arrival

FIGURE 24.6 Flight plan.

Flight planning is still the Achilles' heel of air traffic control. No flight or collection of flights can be planned perfectly. The exigencies of weather, dense traffic, and limited runways disrupt even careful planning. No airplane flies according to its projected plan, not for an hour, not for 10 minutes. The flight route estimation algorithms have not been precise or accurate enough to keep up with the shifts in a flight's trajectory. In addition, individual flight strips are still the norm. As a flight progresses along its route, strips are printed at each fix and stacked along the side of the radar display. Several flights converging on a fix at various times and altitudes are manually sorted so that the controller can read and envision any conflicts. Needless to say, this nongraphical approach challenges even the most experienced en route flight controller, who must internalize the aircrafts' positions and velocities, convert hardcopy integer data into spatial relationships, and *imagine* the airspace picture.

In the early 1970s, the FAA added to the TRACON and en route automation altitude tracking and metering, which provides arrival controllers in both the centers and the TRACONs a list of flights arriving at destination fixes, organized by time of arrival and approach altitude. Before 1980, FAA had added to the ARTS and the 9020 en route systems two automation functions to enhance safety: conflict alert, which evaluates the position, velocity, and heading of tracks within a geographical are of interest and alerts the controller if separation standards will be violated; and minimum safe altitude warning (MSAW), which supports separation assurance between aircraft and terrain, such as tall buildings or mountains.

In 2001, Lockheed Martin, FAA's primary automation supplier, implemented an algorithm developed by MITRE to assist the flight ("D") controller in projecting aircraft routes. The tool has two advantages: it uses a more precise and accurate route estimation algorithm—called a trajectory modeler—and it shows the airplanes' trajectories graphically, aligned with airspace maps. The algorithm, packaged under the name of the user request evaluation tool (URET), where the user is the "D" controller and ultimately the pilot, also allows the controller to probe a flight beyond its current position, in order to evaluate where it and nearby flights will be in, say, 20 minutes. URET also allows the controller to proffer trial flight plans to investigate how conflicts might be avoided. Figure 24.7, shows, with red lines, two flights potentially converging as they approach the northwest geographic area of interest.

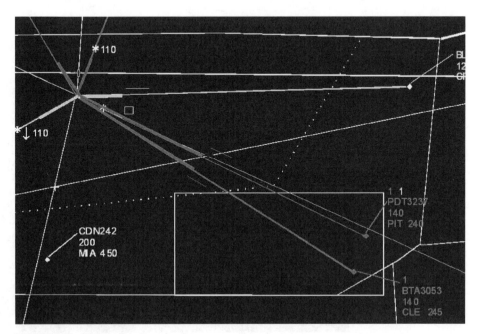

FIGURE 24.7   Flight trajectories and probe.

To recap, the air traffic control domains under FAA jurisdiction include the airport towers, with dominion over departures and arrivals, TRACONs, which guide a flight for about 60 miles in the approach areas, and the 20 national en route enters. The Department of Defense also owns and operates military airports and towers, as well as TRACONs. Military airspace is considered separate from civilian airspace. Thus, the civilian system under the FAA is marked by airspace that is not contiguous, with military and other special-use airspace fragmenting the national picture. There are automation links between the FAA and the military—mostly they exchange flight plans and voice alerts, but the two systems are run separately. This arrangement was sorely tested during the attack on America in 2001. That the FAA was able to land hundreds of flights within minutes of the attack on the World Trade Center is a testament to FAA-DoD coordination and air traffic controller proficiency.

Critical partners in commercial aviation are the airlines. Since deregulation, the airlines have been more active in using automation and information in concert with the FAA. In addition to the regional and local airline operation centers, major airlines operate ramp towers at our largest airports. The ramp towers manage traffic between and among the aprons, concourse, runways, ramps, and gates. (Some municipal airport authorities also use ramp towers airport-wide as part of their airport management.) It stands to reason that the FAA and the airlines would want to share what they know about flight conditions, including runway availability, ground delays, weather, special-use airspace restrictions, and flow control. To that end, the airlines and FAA have recently instituted collaborative decision making (CDM). The practice was immediately effective. CDM is largely carried out by voice. Fully automatic sharing of data between airlines and the FAA remains an objective.

One of the areas of air traffic control that has emerged as a result of burgeoning traffic and crowded skies and runways is called traffic flow management (TFM). The FAA considers TFM an "essential," not a "critical," service. (Critical services involve the ground controller's actions to ensure aircraft-aircraft and aircraft-airspace separation.) Many believe that in the future long-range airspace analysis and flow control will be the most important function on the ground; that with on-board systems such as GPS, collision avoidance, conflict detection, advanced weather sensors, and automatic data link (automatically exchanging data computer to computer between cockpit and ground facilities), separation assurance will fall largely to the pilot, using the automation on the flight deck. (See Figure 24.8.)

FIGURE 24.8 Flight deck circa 2003.

Certainly the commercial airlines would benefit from improved TFM. The FAA can satisfy safety constraints simply by increasing flight separation standards and holding planes on the ground and, en route, at fixes until the competing traffic or weather dissipates. But this relies on air traffic controller intervention using rule-making and voice procedures more than automation. A more automated and collaborative TFM system would allow commercial and general aviators more flexibility in choosing routes and thus eliminate one-at-a-time delays.

The traffic flow management system is run by the FAA. It overlies the tactical control systems of the towers, TRACONs and ARTCCs. Each of these facilities send proposed and active flight plans, second-order amendments, and tracking data to the Air Traffic Control System Command Center (ATCSCC) in Herdon, VA. The ATCSCC receives the data rather circuitously. The control facilities send their flight and tracking data to the FAA's support facility, the W. J. Hughes Technical Center, in Pomona, NJ, and from there to the Volpe National Transportation System Center in Cambridge, MA. The two support facilities transform the data from a local facility view to a regional and national view and integrate weather information from the National Weather Service. The picture the ATCSCC has to work with is then conducive to evaluating large volumes of airspace.

ARTSCC staff use traffic volume algorithms to predict and analyze traffic patterns throughout the United States. The TFM hub shares the results with traffic planners at the control facilities as well as the airlines, which can better plan for possible delays.

Oceanic air traffic is under the control of centers located in Hawaii, Alaska, off the west coast of the United States at the Oakland ARTCCs, and off the east coast of the United States at the New York ARTCC. The continental sites at Oakland and New York maintain oceanic control rooms separate from their en route counterparts. Virtually the same automation as used for en route control is used to ensure separation over the ocean: tracking and flight planning and flight progress strips. However, over the ocean, instead of radar and land-based navigational aids being used, position, velocity, and heading are provided to the ground control facilities by radio, the coastal navigational system LORAN, inertial navigation, and GPS, which, like radar, uses signal distance and time—but from space—to derive position, velocity, and heading.

In 2003, Lockheed Martin tailored a New Zealand oceanic system that automates pilot reports, ground controller clearances, and supports automatic dependent surveillance (automatic position and velocity reports from the onboard GPS) over data link, allowing the FAA to reduce separation standards from 60 to 30 nautical miles over the sea.

The design of the U.S. air traffic control system has evolved conservatively. The FAA places safety first. United States air travel is the safest in the world. In operating one of most complex systems ever realized, the FAA has chosen to introduce changes deliberately. The system is largely procedural; humans are constantly in the loop. New technology is assimilated at a pace far slower than in the commercial sector and the military. This is the case in computing, telecommunications, navigation, graphical systems, and meteorology.

The FAA (see Figure 24.9) is organized to ensure that no advanced functions are introduced before they are deemed absolutely safe. Designs are ultimately concerned with reconciling the "new artificial" with the "old," both natural and artificial. As such, how people organize to design is part of the design. Two aspects of the FAA's organization are decisive in that regard.

The first is that the air traffic controllers, as well as the operations and maintenance staff and administrative employees, belong to unions. The unions throttle abrupt changes in technology, especially as it might disrupt the well-worn and effective habits of the air traffic controllers, without suffocating it. For technologists and those who wish to fly without any personal sacrifice, this arrangement must seem atavistic. But it provides a built-in governor.

The second aspect of the FAA organization also ensures against unwelcome change. It is the W. J. Hughes Technical Center (WJHTC), where FAA systems are tested and maintained. Regardless of the intensity of testing done by suppliers, all systems must pass a strict set of tests at the WJHTC, which reports directly to the administrator for research and acquisition, independent of systems development. As such, the WJHTC is the third-party quality assurance arm of the FAA.

The future of air traffic control design is certain: it will improve, but in specific areas and incrementally. In terms of major changes, economics and security concerns have reduced commercial air travel and the amount of money available for system upgrades. In the near term, improvements will be modest and directed.

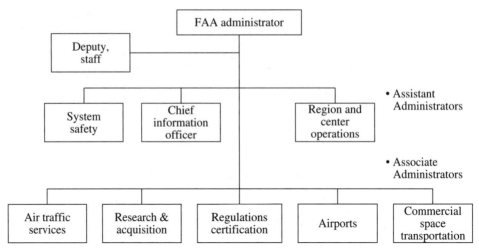

FIGURE 24.9 FAA organization.

In 1998, Mrs. Jane Garvey, the current FA Administrator, started the Free Flight initiative. The objectives for Free Flight Phase 1 were to use CDM, deploy URET, implement automated data link, and improve arrival and runway flow into the major airports. These objectives were met.

In 2003, the FAA is undertaking a new plan, the Operational Evolution Plan (OEP). The OEP increases the FAA's commitment to a modernization strategy that encapsulates the entire aviation community, including the airlines, airports, cargo carriers, the DOD, and NASA. No single objective or technique characterizes the OEP. It is a collection of specific objectives that will increase capacity and efficiency at airports, relieve en route congestion, and improve weather prediction. Airport efficiency, for example, will soon benefit from GPS and air navigation to help guide landings, augmenting the long-time instrument landing systems in use worldwide. Traffic management advisories will be used to smooth the arrival of many flights to few airports and runways, helping controllers find the optimum path from en route airspace to the final approach and runway, and ultimately the gate. To relieve en route congestion and reduce voice communications, automated data link—computer-to-computer—will add more messages to the core set.

Perhaps the most safety-enhancing element of the OEP falls in the weather domain. A quote from the OEP summarizes one such initiative: Cockpit surface movement maps have shown promise in improving crew situational awareness in low visibility. These tools supplement the pilot's out-the-window assessment of aircraft position, direction and speed. When coupled with positive identification of other surface traffic, procedures can be changed to direct one aircraft to follow another without visual references outside the cockpit. These changes may enhance pilot confidence and efficiency in moving about the airport surface. The key to success for this initiative as an OEP capacity enhancement is the ability to go beyond improvement in situational awareness to improved efficiency in surface movement.

Progress is not so much technological as geographical. The U.S. air traffic control design is being inserted into the developing nations around the world. Flight data, radar, oceanic control, voice clearances, digital maps, data link, GPS, runway management, flow control—all have their roots in the U.S. system, and all are being made available overseas, thanks to companies like Lockheed Martin. Already, the fundamentals of air traffic control design are at work in the United Kingdom, Scotland, Argentina, New Zealand, Taiwan, China, and Korea, and soon they will be in Africa and most of the developing nations in Europe and Asia.

To repeat from the introduction, the world is becoming smaller and therefore friendlier: Before flight became commonplace, people traveled in two dimensions, crossing the borders that separate town from town and nation from nation. Seen from above, the artificial boundaries that divide us disappear. Distances shrink, the horizon stretches. The world seems grander and more interconnected.

CHAPTER 25
MARITIME TRANSPORT AND SHIP DESIGN

Apostolos Papanikolaou
Director of the Ship Design Laboratory
National Technical University of Athens
Athens, Greece

25.1 FOREWORD

This chapter deals with ship design in the frame of maritime transport. It aims at introducing interested readers systematically into the basic concepts of ship design and the preliminary sizing of a ship, fulfilling the requirements of a hypothetical transportation scenario.

Section 25.2 introduces the reader into conventional and innovative ship concepts. Section 25.3 outlines the framework of maritime transport, elaborates on the efficiency of conventional ships and advanced marine vehicles compared with other modes of transport, and concludes with an examination of the impact of maritime operations on the marine and atmospheric environment. Section 25.4 focuses on methodologies of ship design, introducing the main phases of ship design and the governing objectives, commenting on the design procedure and on typical transportation requirements (shipowner's statement of work), detailing the preliminary ship design procedure for the main ship categories and types, and finally, outlining preliminary ship sizing. The material presented is supported by an appendix of diagrams deduced by regression analysis of main technical data of built ships and allowing quick preliminary estimations of a ship's main dimensions and of other main ship properties.

The material presented herein to a great extent is based on parts of my lecture notes on methods of preliminary ship design, taught in the seventh semester of the 5-year diploma engineering program of the School of Naval Architecture and Marine Engineering at the National Technical University of Athens, Greece. These lecture notes are published in the textbook *Ship Design (Methodologies of Preliminary Design)* by the author.[28] For the needs of this chapter, this lecture material has been enhanced with additional material from relevant recent research work by the author.

25.2 INTRODUCTION: CONVENTIONAL AND ADVANCED MARINE VEHICLES

Human beings traveled for thousands of years throughout the oceans without first knowing how or why this was possible. Archeological findings indicate that the first shiplike floating devices were operating in the Aegean Sea in 7,000 BC. The Phoenicians and Egyptians appear to have been the leaders in the art of early shipbuilding, followed by the Greeks of the Cycladic and Crete islands

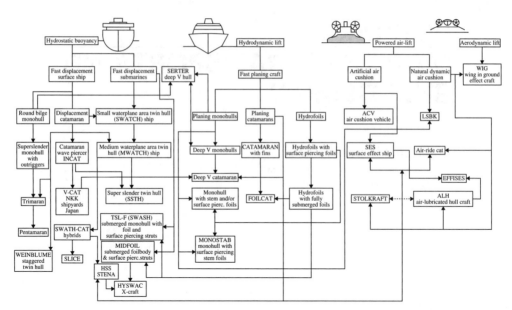

Comments on Chart of Advanced Marine Vehicles and Explanation of Used Acronyms

1. **ACV: Air Cushion Vehicle** - Hovercraft, excellent calm water and acceptable seakeeping (limiting wave height), limited payload capacity.
2. **ALH: Air Lubricated Hull**, various developed concepts and patents, see type STOLKRAFT.
3. **Deep V**: ships with **Deep V** sections of semidisplacement type acc. to E. Serter (USA) or of more planing type, excellent calm water and payload characteristics, acceptable to good seakeeping, various concepts AQUASTRADA (RODRIQUEZ, Italy), PEGASUS (FINCANTIERI, Italy), MESTRAL (former BAZAN, Spain), CORSAIR (LEROUX & LOTZ, France).
4. **EFFISES**: Hybrid ALH twin hull with powered lift, patented by SES Europe A.S. (Norway)
5. **FOILCAT**: Twin hull (catamaran) hydrofoil craft of KVAERNER (Norway), likewise MITSUBISHI (Japan), excellent seakeeping (but limiting wave height) and calm water characteristics, limited payload.
6. **HYSWAC – X-Craft**: Hybrid SWATH with midfoil, prototypes currently tested by US Navy
7. **LWC: Low Wash Catamaran**, twin hull superslender semidisplacement catamaran with low wave-wash signature of FBM Marine Ltd. (United Kingdom), employed for river and closed harbour traffic.
8. **LSBK: Längs Stufen- Bodenkanalboot- Konzept**, optimized air-lubricated twin hull with stepped planing demihulls, separated by tunnel, aerodynamically generated cushion, patented in Germany.
9. **MIDFOIL**: Submerged Foil-body and surface piercing twin struts of NAVATEK-LOCKHEED (USA).
10. **MONOSTAB**: Semiplanning monohull with fully submerged stern fins of RODRIQUEZ (Italy).
11. **MWATH: Medium Waterplane Area Twin Hull Ship**, as type SWATH, however with larger waterplane area, increased payload capacity and reduced sensitivity to weight changes, worse seakeeping.
12. **PENTAMARAN**: Long slender monohull with four outriggers, designs by Nigel Gee and former IZAR (Spain).
13. **SES: Surface Effect Ship**, Air Cushion Catamaran Ship, similar to ACV type concept, however w/o side skirts, improved seakeeping and payload characteristics.
14. **SLICE**: Staggered quadruple demihulls with twin struts on each side, acc. to NAVATEK-LOCKHEED (USA), currently tested as prototype.
15. **SSTH: Superslender Twin Hull**, semidisplacement catamaran with very slender long demihulls of IHI shipyard (Japan), similar to type WAVEPIERCER.
16. **STOLKRAFT**: Optimized air-lubricated V-section shape catamaran, with central body, reduced frictional resistance characteristics, limited payload, questionable seakeeping in open seas, patented by STOLKRAFT (Australia).
17. **Superslender Monohull with Outriggers**: Long monohull with two small outriggers in the stern part, EUROEXPRESS concept of KVAERNER-MASA Yards (Finland), excellent calm water performance and payload characteristics, good seakeeping in head seas.
18. **SWATH Hybrids**: SWATH type bow section part and planing catamaran astern section (STENA's HSS of Finyards, Finland, AUSTAL hybrids, Australia), derived from original type SWATH & MWATH concepts.
19. **SWATH: Small Waterplane Area Twin Hull Ship**, synonym to SSC (Semi-Submerged Catamaran of MITSUI Ltd.), ships with excellent seakeeping characteristics, especially in short period seas, reduced payload capacity, appreciable calm water performance.
20. **TRICAT**: Twin hull semidisplacement catamaran with middle body above SWL of FBM Marine Ltd. (United Kingdom).
21. **TRIMARAN**: Long slender monohull with small outriggers at the center, introduced by Prof. D. Andrews - UCL London (United Kingdom), currently tested as large prototype by the UK Royal Navy (TRITON), similarities to the Superslender Monohull with outriggers concept of KVAERNER-MASA.
22. **TSL-F - SWASH**: Techno-Superliner Foil version developed in Japan by shipyard consortium, submerged monohull with foils and surface piercing struts.
23. **V-CAT**: Semidisplacement catamaran with V section shaped demihulls of NKK shipyard (Japan), as type WAVEPIERCER.
24. **WAVEPIERCER**: Semidisplacement catamaran of INCAT Ltd. (Australia), good seakeeping characteristics in long period seas (swells), good calm water performance and payload characteristics.
25. **WEINBLUME**: Displacement catamaran with staggered demihulls, introduced by Prof. H. Söding (IfS-Hamburg-Germany), very good wave resistance characteristics, acceptable seakeeping and payload, name to the honor of late Prof. G. Weinblum.
26. **WFK: Wave Forming Keel High Speed Catamaran Craft**, employment of stepped planning demihulls, like type LSBK, but additionally introduction of air to the planning surfaces to form lubricating film of microbubbles or sea foam with the effect of reduction of frictional resistance, patented by A. Jones (USA)
27. **WIG: Wing In Ground Effect Craft**, various developed concepts and patents, passenger/cargo carrying and naval ship applications, excellent calm water performance, limited payload capacity, limited operational wave height, most prominent representatives the ECRANOPLANS of former USSR.

FIGURE 25.1 Development of basic types and hybrids of advanced marine vehicles. (*Source:* Papanikolaou.[24])

(Minoan period, 1,700 to 1,450 BC). However, it was the work of the great Archimedes in 300 BC that explained a ship's floatability and stability; even this work remained practically unexploited until relatively modern times (18th century AD).[19]

Having in mind the Archimedean principle of carrying a ship's weight by hydrostatic forces, the various types of modern ship concepts, ranging from conventional ships to unconventional, innovative ship concepts [which we call *advanced marine* vehicles, (AMVs)], may be illustrated through a comprehensive ship development chart (Figure 25.1). This chart is based on a categorization of the various marine vehicles by considering the main physical concepts leading to the forces balancing the weight of the ship, namely, *hydrostatic buoyancy force, hydrodynamic lift force, powered fan-lift force,* and *aerodynamic lift force.* In the chart we may distinguish in the first row the fundamental ship concepts. The *derivatives* of these basic concepts (so-called hybrids) are filed column-wise according to the *major* physical force balancing the ship's weight, notwithstanding the fact that during operation, forces derived from other physical concepts might as well contribute to their weight balance. For example, the weight of a planning craft is not entirely carried by the hydrodynamic lift force but to a certain degree, depending on the speed of operation, also by the hydrostatic, buoyancy force, according to the displaced water volume. Historically, technological developments are understood to have taken place from the upper left corner (Archimedean principle) toward the right and then down.[24]

25.3   MARITIME TRANSPORT: INNOVATIVE CONCEPTS, ENERGY EFFICIENCY, AND ENVIRONMENTAL IMPACT

Ships are built for covering needs of society through the provision of specific services. These services may be on a commercial or noncommercial basis. Whereas in the first case (commercial ships), the objective is to generate profit for the ship owner, the latter case is related to a public service of some kind, the cost of which is in general carried by a governmental authority. The main bulk of commercial ships are cargo ships, which carry all types of cargo (solid and liquid cargo or passengers) and provide in fact the largest [by volume of cargo and transport distance (ton-miles)] worldwide transportation work compared with other modes of transport. The categorization of ships will be addressed in section 25.4.6.

The transport efficiency of ships and marine vehicles in general may be defined in various ways, and many researchers have addressed this in the past. In particular, when introducing efficiency indicators (efficiency indices or metrics), we need to ensure as wide as possible applicability of the introduced *performance indices* (or *merit functions*) on a "fair" basis when assessing sometimes competing alternative transport concepts (and modes of transport). In the following, a brief review of related past work is conducted and complemented by more recent work.

The *transport efficiency* may be defined as a function of the vessel's deadweight W_d ($\equiv$ DWT), service speed V_S (kn), and total installed power P(kW). That is,

$$E_1 = \frac{W_d \cdot V_S}{P} \tag{25.1}$$

Noting the difference between the deadweight and payload,[*] the transport efficiency also may be expressed in terms of the vessel's payload W_p instead of deadweight:

$$E_2 = \frac{W_p \cdot V_s}{P} \tag{25.2}$$

When comparing the transport efficiency of marine vehicles with that of alternative modes of transport (land and airborne), it is very useful to employ the well-known *Karman-Gabrielli transport efficiency diagram.* Akagi[1] has replotted the original Karman-Gabrielli diagram in terms of

---

[*]Deadweight = payload + fuel + lub oil + water + crew and other effects + water ballast.

the *reciprocal* transport efficiency as a function of the total installed power P (PS), displacement W (tons), and maximum speed V (km/h):

$$\frac{1}{E_3} = \frac{P}{W \cdot V} \tag{25.3}$$

Akagi-Morishita[2] added more recent developments of various transport vehicles. Figure 25.2 presents reciprocal transport efficiency once more updated by sample data of the Ship Design

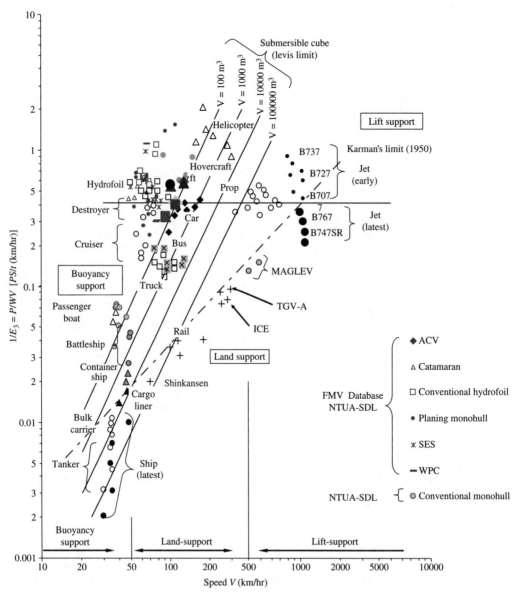

FIGURE 25.2 Reciprocal transport efficiency of alternative modes of transport. (*Source:* Akagi[1]; data supplemented by NTUA-SDL.[25])

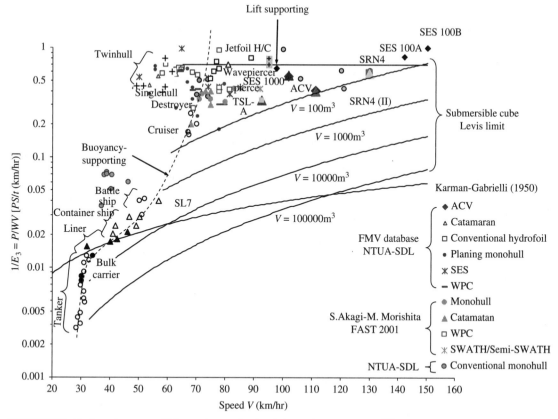

FIGURE 25.3 Reciprocal transport efficiency of conventional and advanced marine vehicles.[2,25]

Laboratory–National Technical University of Athens (NTUA-SDL) database, whereas Figure 25.3 focuses on the performance of marine vehicles only.

The reciprocal transport efficiency (specific power) also may be based on payload W_p:

$$\frac{1}{E_4} = \frac{P}{W_p \cdot V^2} \tag{25.4}$$

and is presented in Figure 25.4.

When comparing alternative modes of transport with respect to speed, it makes sense to plot the *payload ratio* (W_p/W) against maximum speed (in km/h; Figure 25.5), because the earnings and likely profit are directly related to payload.

Kennel[13] introduced a different *transport factor*, namely,

$$TF = \frac{K_2 \cdot W}{SHP_{TI}/(K_1 \cdot V_K)} \tag{25.5}$$

where K_2 is a constant ($K_2 = 2,240$ lb/long ton), W is ship's displacement in long tons, SHP_{TI} is the total installed power in horsepower, K_1 is a constant ($K_1 = 1.6878/550$ hp/lb-kn), and V_k is the design speed in knots. Figure 25.6 presents Kennel's transport factor versus speed updated with the relevant NTUA-SDL database data.

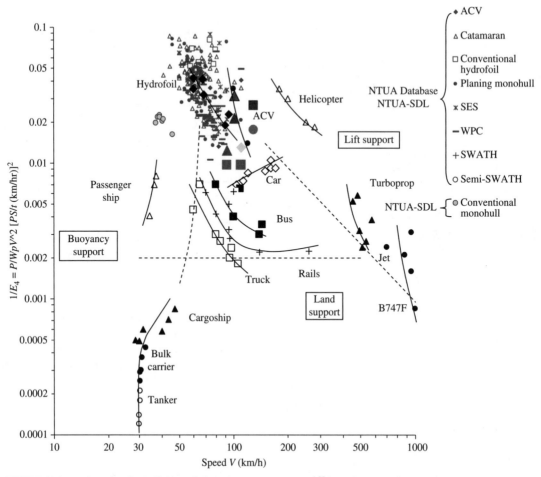

FIGURE 25.4   Reciprocal payload efficiency of alternative transport systems.[1,25]

Following Kennel's approach, the displacement and transport factors may be decomposed as follows:

$$W = W_{ship} + W_{cargo} + W_{fuel} \tag{25.6}$$

$$TF = TF_{ship} + TF_{cargo} + TF_{fuel} \tag{25.7}$$

where W_{ship}, W_{cargo}, and W_{fuel} are the lightship, cargo, and fuel oil weights, respectively (in long tons) and TF_{ship}, TF_{cargo}, and TF_{fuel} are the transport factors calculated for each weight group, respectively. W_{ship} and W_{fuel} are obtained from the following equations:

$$W_{ship} = W - W_{cargo} - W_{fuel} \tag{25.8}$$

$$W_{fuel} = SFC_{avg} \cdot K_{SHP} \cdot SHP_{TI} \frac{R}{K_S \cdot V_K} \tag{25.9}$$

where SFC_{avg} is the average effective fuel consumption rate, K_{SHP} is the endurance-power-to-design-power ratio, R the range (nautical miles), and K_S the endurance speed to design speed ratio.

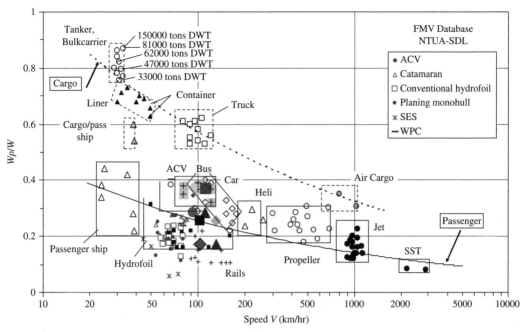

FIGURE 25.5 Payload ratio of alternative transport systems.[1,25]

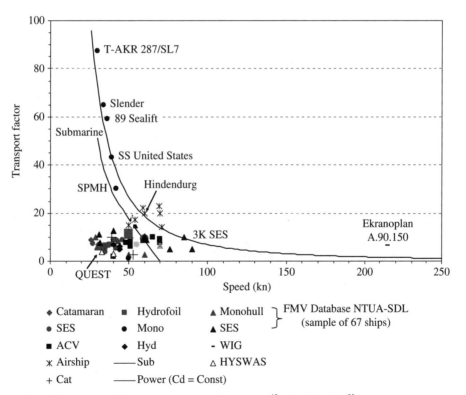

FIGURE 25.6 Transport factor versus speed according to Kenell[13] and NTUA-SDL.[25]

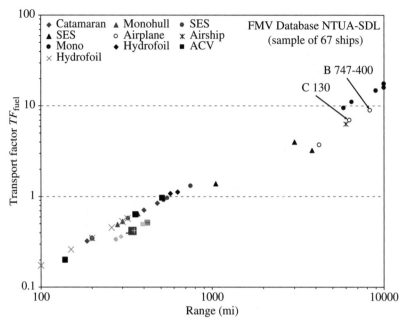

FIGURE 25.7 Fuel transport factor versus range according to Kenell[13] and NTUA-SDL.[25]

Figures 25.7 through 25.9 show the fuel transport factor versus range and the trends of transport factors and various fractions thereof as plotted by Kennell and updated by the NTUA-SDL database ships.

For some more general data regarding the *fuel efficiency* of transport of cargo and passengers by alternative modes of transport, Tables 25.1 and 25.2 may be consulted.

From the preceding, the high efficiency of waterborne transport is evidenced, followed by rail transport.* However, comparing waterborne with other modes of transport (land and air-borne), the speed of transport also needs to be taken into account, especially when dealing with the transport of so-called just-in-time (JIT) products and passengers, for which the *value of time* and the demand for high speed are of high importance, so higher fuel and transport cost might be accepted.[1,24,25]

Regarding the impact of shipping operations on the marine and atmospheric environment, there are two major factors to consider, namely, the likely pollution of the marine environment by crude and other oil products when transported by tankers and the toxic gas emissions of marine engines to the atmosphere. Both factors are strictly regulated by international authorities (International Maritime Organisation, www.imo.org) and have a significant impact on ship design, outfitting, and operation.

The likely pollution of the marine environment is regulated by MARPOL 73/78 (International Convention for the Prevention of Pollution from Ships), which is one of the major International Maritime Organization (IMO) conventions; over the course of the years, after its introduction in 1973, MARPOL underwent several amendments and improvements that contributed to today's quite satisfactory state of affairs in terms of tanker accidents and environmental consequences

---

*In this comparison, the high investments for the building and maintaining of rail infrastructure compared with relevant costs for port infrastructure are not considered.

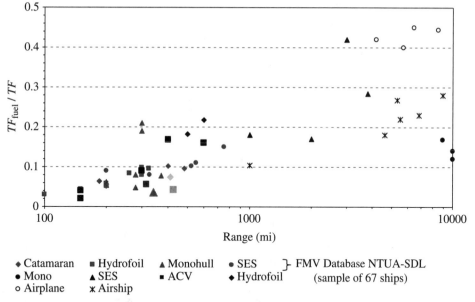

◆ Catamaran   ■ Hydrofoil   ▲ Monohull   ● SES   }- FMV Database NTUA-SDL
● Mono   ▲ SES   ■ ACV   ◆ Hydrofoil   (sample of 67 ships)
○ Airplane   ✕ Airship

FIGURE 25.8   Fuel transport factor ratio versus range according to Kenell[13] and NTUA-SDL.[25]

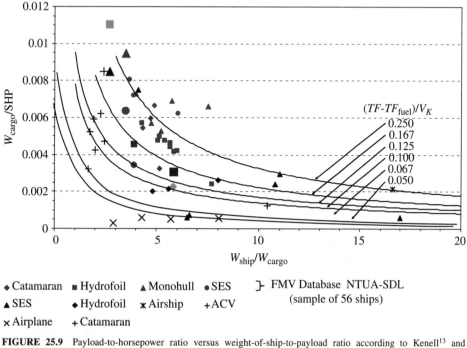

◆ Catamaran   ■ Hydrofoil   ▲ Monohull   ● SES   }- FMV Database NTUA-SDL
▲ SES   ◆ Hydrofoil   ✕ Airship   + ACV   (sample of 56 ships)
✕ Airplane   + Catamaran

FIGURE 25.9   Payload-to-horsepower ratio versus weight-of-ship-to-payload ratio according to Kenell[13] and NTUA-SDL.[25]

TABLE 25.1 Specific Fuel Consumption for Break Bulk Cargo Transport

Ship	0.4 kg/(ton 100 km)
Truck	1.1/1.6 kg/(ton 100 km)
Rail	0.7/1.6 kg/(ton 100 km)
Airplane	6/8 kg/(ton 100 km)
	This refers to tons payload and includes the weight of fuel.
	11/14 kg/(ton 100 km) This refers to tons payload and includes the weight of fuel for *transatlantic flights.*

TABLE 25.2 Specific Fuel Consumption for Passenger Transport (Assuming Fully Loaded Vehicles)

Private car, only driver	About 8 kg/(person 100 km)
Bus (55 passengers, 100 km/h)	1 kg/(person 100 km)
Train type IC (10 cars of 60 seats, 160 km/h)	3 kg/(person 100 km)
Train type D (14 cars of 72 seats, 140 km/h)	1.5 kg/(person 100 km)
Airplane in transatlantic flight (including other cargo)	17 kg/(person 100 km)
Airplane in European flight (without other cargo)	3.6/6 kg/(person 100 km)
Air-cushion high-speed vehicle (600 passengers)	5 kg/(person 100 km)
Modern large cruise ship (500 to 1000 passengers)	16/18 kg/(person 100 km)
Ro-Ro passenger ferry with deck passengers (1500 passengers)	5/6 kg/(person 100 km)
Small riverboat, with deck passengers	1.5 kg/(person 100 km)
Large rivership, with deck passengers	0.5 kg/(person 100 km)

(Figure 25.10).[*] Following a series of catastrophic single-hull tanker accidents, current MARPOL regulations (and long before U.S. Oil Pollution Act 1990) recognize double-hull tanker designs as the only acceptable solution for the safe carriage of oil in tanker ships. According to current MARPOL regulations, the tank arrangement of the cargo block of an oil tanker should be designed properly to provide adequate protection against accidental oil outflow, as expressed by the so-called mean outflow parameter. Further improvements of MARPOL may be expected in the future.

Finally, it is today well established that human activities have a significant impact on the levels of greenhouse gases in the atmosphere, that is, the gases that absorb and emit radiation within the thermal infrared range. The gases with the most important release to the atmosphere are, in descending order, water vapor, carbon dioxide (CO_2), methane, and ozone. The Intergovernmental Panel on Climate Change (IPCC) recently released a report stating that "most of the observed increase in global average temperatures since the mid-20th century is very likely due to the observed increase in anthropogenic greenhouse gas concentrations."[11] One of the main contributors of emissions of greenhouse gases owing to human activity is the burning of fossil fuels. The total CO_2 emissions from shipping (domestic and international) amount to about 3.3 percent of the global emissions from fuel consumption, according to International Energy Agency (IEA).[6]

---

[*]The statistics presented cover the period 1978 to 2003; it is noted that the very low accidental rates, as of year 2003, are confirmed by more recent statistical studies (Papanikolaou et al., "Assessment of Safety of Crude Oil Transport by Tankers," *Proceedings Annual Meeting of the German Society of Naval Architects—STG*, Berlin, 2009).

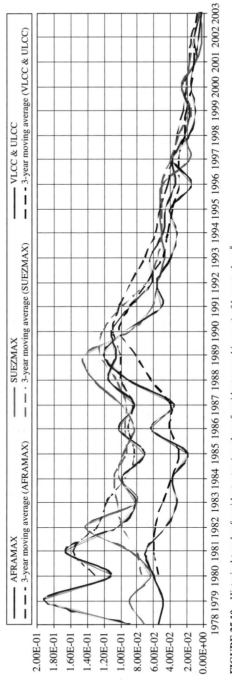

FIGURE 25.10 Historical trends of accident rates (number of accidents per ship-year) of large tankers.[8]

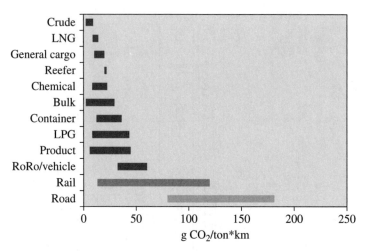

FIGURE 25.11   Typical ranges of CO_2 efficiencies of ships compared with rail and road transport.[6]

Climate stabilization will require significant reductions of CO_2 emissions by 2050, and the international shipping industry needs to participate in this process. Independently, of the fact that maritime transport is the most efficient mode of transport (ton-km) and least pollutant in terms of greenhouse gas emissions, present discussions and expected regulatory measures suggest the collaboration of all major stakeholders of shipbuilding and ship operations to address this complex technoeconomical and highly political problem efficiently and call eventually for the development of proper design and operational knowledge and assessment tools for the energy-efficient design and operation of ships.[4]

Typical design and outfitting measures for reducing CO_2 emissions are related to hull form optimization for least powering (and fuel consumption), improved diesel engine combustion, improved fuel technology, and last but not least, a drastic operational measure for reducing CO_2 emissions is reduction of service speed, with a major impact on a ship's competitiveness and economy, especially when the ship is in liner service (e.g., for container and passenger ships).

Finally, societal concerns about the safety of human lives and protection of the environment recently have led the maritime industry to increased efforts in the design and operation of ships for enhanced safety. Applications of risk-based approaches in the maritime industry actually started in the early 1960s with introduction of the concept of probabilistic ship's damage stability. In the following years, such approaches were widely applied within the offshore sector and are now being adapted and used within the ship technology and shipping sector. The main motivation to use risk-based approaches is twofold: Implement a novel ship design that is considered to be safe but—for some formal reason—cannot be approved today and/or rationally optimize an existing design with respect to safety without compromising on efficiency and performance.[29]

25.4   SHIP DESIGN

25.4.1   Introduction: Main Phases of Ship Design

Ship design in the past was more of an art than a science, highly dependent on experienced naval architects with good backgrounds in various fundamental and specialized scientific and engineering subjects, next to practical experience. The design space (multitude of solutions to the design problem)

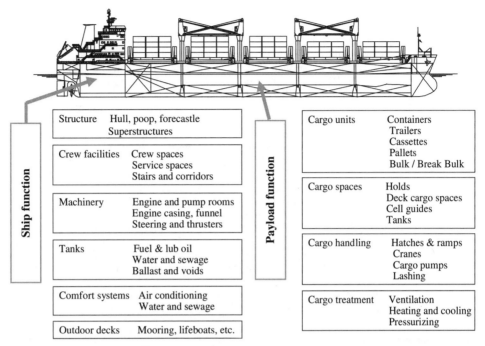

Ship function		
Structure	Hull, poop, forecastle Superstructures	
Crew facilities	Crew spaces Service spaces Stairs and corridors	
Machinery	Engine and pump rooms Engine casing, funnel Steering and thrusters	
Tanks	Fuel & lub oil Water and sewage Ballast and voids	
Comfort systems	Air conditioning Water and sewage	
Outdoor decks	Mooring, lifeboats, etc.	

Payload function		
Cargo units	Containers Trailers Cassettes Pallets Bulk / Break Bulk	
Cargo spaces	Holds Deck cargo spaces Cell guides Tanks	
Cargo handling	Hatches & ramps Cranes Cargo pumps Lashing	
Cargo treatment	Ventilation Heating and cooling Pressurizing	

FIGURE 25.12   Ship functions.[16]

was practically explored using heuristic methods, namely, methods deriving from a process of trial and error often over the course of decades. Gradually, trial-and-error methods were more and more replaced by gained knowledge, which eventually formed a knowledge base, namely, semiempirical methods and statistical data on existing ships and successful designs.

A modern, systems-based approach to ship design may consider the ship as a complex system integrating a variety of subsystems and their components, for example, subsystems for cargo storage and handling, energy/power generation and ship propulsion, accommodation of crew/passengers, ship navigation, etc. They serve well-defined *ship functions*. Ship functions may be divided into two main categories, namely, *payload functions* and *inherent ship functions* (Figure 25.12). For cargo ships, the payload functions are related to the provision of cargo spaces, cargo handling, and cargo treatment equipment. Inherent ship functions are those related to the carriage of payload, at specified speed and safely from port to port.

Considering that ship design actually should address the whole ship's *life cycle*, we may consider ship design composed of various stages, namely, besides the traditional concept-preliminary-contractual and detailed design, the stages of ship construction and fabrication, ship operation for her economic life, and scrapping/recycling. It is evident that the optimal ship with respect to her whole life cycle is the outcome of a *holistic** optimization of the entire complex ship system for its entire life cycle.[26]

Mathematically, every constituent of the defined life-cycle ship system and design stage forms a complex nonlinear optimization problem of the design variables, with a variety of constraints and criteria/objective functions to be jointly optimized. Even the simplest component of the ship design process, namely, the traditional first loop (conceptual–preliminary design), is complex enough to be simplified (*reduced*) in practice. Also, inherent to ship design optimization are the conflicting requirements resulting from the design constraints and optimization criteria (merit or objective

---

*Principle of holism according to Aristotle (*Metaphysics*): "The whole is more than the sum of the parts."

functions), reflecting the interests of the various ship design stakeholders: ship owners/operators, shipbuilders, classification society/coast guard, regulators, insurers, cargo owners/forwarders, port operators, etc. Assuming a specific set of requirements (usually the ship owner's requirements for merchant ships or mission statement for naval ships), a ship needs to be optimized for the lowest construction cost, the highest carrying capacity and operational efficiency or the lowest required freight rate (RFR), the highest safety and comfort of passengers/crew, the satisfactory protection of cargo and the ship herself as hardware, and last but not the least, the minimum environmental impact, particularly for oil carriers with respect to marine pollution in case of accidents, for high-speed vessels with respect to wave wash, and recently, for all ships with respect to engine emissions and air pollution. Many of these requirements are clearly conflicting, and a decision regarding the optimal ship design for a set of design requirements needs to be made rationally.

To make things more complex, but coming closer to reality, even the specification of a set of design requirements with respect to ship type, cargo capacity, speed, range, etc. is complex enough to require another optimization (or decision-making) procedure that satisfactorily considers the interests of all shareholders of the ship as an industrial product servicing the needs of international markets or others. Actually, the initial set of ship design requirements is the outcome of a compromise of intensive discussions between highly experienced decision makers, mainly the shipbuilders and end users (ship owners), who attempt to promote their interests, while accepting some tradeoffs during contract negotiations. A way to undertake and consolidate this kind of discussions in a rational manner has been advanced by the European Union (EU)–funded project LOGBASED (LOGistics-BAsed ship design) (2004–2007).[5]

Modern approaches to ship design are reviewed by Andrews et al.[3] and Papanikolaou et al.[27] on behalf of expert committees of the International Marine Design Conference (www.imdc.cc).

25.4.2 Main Phases of Ship Design

Traditionally, ship design has been decomposed into four main phases, namely:

1. Concept design, feasibility study
2. Preliminary design
3. Contract design
4. Detailed design

This chapter deals with the first two phases of ship design, which are also known as *basic design*; they are often merged into the more general definition of preliminary design.[*] Figure 25.13 sketches the course of the design of a ship, which is designed to service specific requirements or a mission (mission), disposing certain functional (function), form, space, weight (form), technical performance (performance), and economic characteristics (economics).[†]

Preliminary ship design is the early stage of design in which based on a ship owner's or mission requirements and specifications, the main technical and economic ship characteristics are determined by optimization, particularly the ship characteristics that decisively affect the cost of shipbuilding (and indirectly the cost of acquisition) and the economy of operation.

25.4.3 Objectives of Preliminary Design

Preliminary ship design encompasses the following more detailed objectives:

- Selection of main ship dimensions
- Development of ship's hull form (wet and above-water part)

---

[*]The last two phases are briefly commented on in section 12.4.4.

[†]See also A. Papanikolaou (coordinator), P. Andersen, H.-O. Kristensen, et al., "State of the Art on Design for X," in *Proc. 10th International Marine Design Conference*, IMDC09, Trondheim, May 2009.[27]

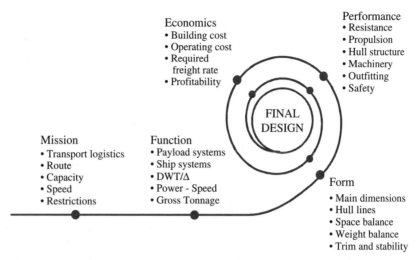

FIGURE 25.13 Ship design according to K. Levander.[16]

- Specification of main machinery and propulsion system type and size (powering)
- Estimation of auxiliary machinery type and powering
- Design of general arrangement of main and auxiliary spaces (cargo spaces, machinery spaces, accommodations)
- Specification of cargo handling equipment
- Design of main structural elements for longitudinal and transverse strength
- Control of floatability, stability, trim, and freeboard (stability and load line regulations)
- Tonnage measurement (Gross Register Tons)

It is understood that determination of all these elements of ship design is subject to compliance with various national and international maritime rules and regulations, which are enforced by national and international authorities (flag and port states and the International Maritime Organization) or by an internationally recognized classification society. In cases of lack of regulatory specifications, it is understood that the ship as designed and built will correspond to the modern state of the art of shipbuilding science and technology.

Preliminary ship design is a technoeconomic feasibility study of the subsystem *ship* as one of the most important *earning* elements in the global maritime transportation system (or maritime network chain) of transport services and of the maritime operation (shipping) industry; trivially, a ship is also a high-investment product of the shipbuilding/maritime technology industry. Taking into account the most recent developments of shipbuilding and marine technology, the physical and technical constraints, the technoeconomic specifications of the ship owner, and the national and international regulations and conventions regarding the building and safety of operation of ships, preliminary design aims at consolidating the various parties' conflicting requirements and determining the most economic design solution for the highest return on investment.

The main difficulties of ship design are a result of the complexities of the various technoeconomic requirements, which are partly contradictory to each other, and the in force and foreseeable future maze of safety requirements of national and international regulations. In terms of fundamental fluid mechanics, the unique operation of the ship on the free sea surface, which represents an irregular boundary surface between two fluids of substantially different density (namely, water and air and so defines the surface profile of the sea waves, which is a priori unknown) and results in a time-varying (dynamic) loading on the ship's structure and rigid-body ship motions in six degrees of freedom, the complex flow around the ship's hull and a variety of other problems of ship hydrodynamics and dynamic ship loading, forms a series of unique scientific problems and theoretical as well as technological solutions. To address these difficulties and provide proper solutions to particular problems requires the collaboration of scientists, designers, and engineers of various disciplines, particularly in the development of new buildings (prototypes) without having empirical data of sister ships in hand.

The design of a ship crosses the strict boundaries of technology and science in many instances of development, coming closer to the disciplines of the arts. Here we understand beyond the aesthetics and architectural elements of ship design, which greatly affect the design of specific ship types (e.g., passenger/cruise ships, yachts, etc.), the many smaller and larger problems arising in ship design and construction that are addressed more by the intuition (mastering) of the naval architect, following the tradition of small shipbuilders, rather than deciding rationally through the use of modern decision support tools and systems. The reasons for this approach in practice involve first the lack of time for an exhaustive investigation of all parameters of the set design problem, whereas a decision is due immediately, and second, the complexity of some problems, with manifold possible solutions, without having the certainty of a rationally optimal solution with respect to technology and economy. In this respect, the experience of the ship designer, shipbuilder, or production manager complements the lack of design data that would be obtained only after tedious theoretical elaborations. Nevertheless, in recent years, information technology (IT) has been widely introduced to all phase of ship design, production, and operation, closing more and more the gaps resulting from the current often lack of experienced ship designers and engineers in many parts of the world.

25.4.4   Design Procedure: Design Spiral

The design procedure sketched in the last section and the main phases of ship design may be illustrated by the well-known design spiral, originally introduced by J. H. Evans[32] (Figure 25.14). The spiral effectively illustrates the sequential course of ship design through the various steps, the repeating, iterative procedure for determination of ship dimensions and of other properties and, finally, the gradual approach to the final stage of detailed ship design. In the figure, some indicative effort in man-days for completion of each stage of ship design is given, pertaining to the design of a large merchant ship in the late 1950s. The ship design procedure also may be illustrated by other, more modern and comprehensive approaches, encompassing, besides the design, the manufacturing procedure as well, as illustrated in Figures 25.15 and 25.16.

Commenting on the iterative ship design procedure illustrated by the design spiral (Figure 25.14), following points are noted:

Concept Design–Feasibility Study: First Iteration.   In this design stage, the mission or owner's requirements are transferred to some equivalent technical ship characteristics. Preliminary estimations of the basic ship dimensions, that is, length L, beam B, side depth D, draft T, block coefficient C_B, and powering P_B, complying with the mission requirements are conducted. Alternative designs with respect to more economical solutions are investigated, however, without necessarily going into details. Nevertheless, the feasibility of the set design problem is clarified by determining one or more *feasible* design solutions. According to R. K. Kiss,[32] the corresponding effort for this stage of design of a large merchant ships has been in the late 1950s about 20 man-days. However, with the development of computers and software, this effort has been reduced today to about 1/20th. Thus today the feasibility study may be accomplished in one day (or even less) by a naval architect, assuming a well-organized design office with proper software infrastructure.

Preliminary Design: Second to Fourth Iteration.   This stage of design is a more comprehensive elaboration of the various ship design steps partly addressed in the first phase. In other words, it involves exact determination of the ship's main characteristics and dimensions, namely, ship length L, beam B, side depth D, draft T, block (and other) hull form coefficients, the ship hull form and powering, and the preliminary structural design, while fulfilling the owner's requirements and national, international, and class society regulations. The design solution achieved after (commonly) two to

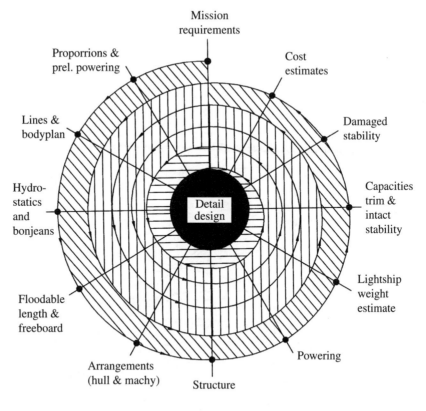

FIGURE 25.14   Design spiral. (After J. H. Evans, 1959.[32])

four iterations corresponds to an *optimal* solution with respect to a set economic criterion.* The outcome of the preliminary design forms the basis for compilation of the shipbuilding contract between the ship owner and the shipbuilder. Typically, this effort is about 15 times larger than the estimated effort for the first phase. The combined first two phases are also known as *basic design*.

---

*This economic criterion may be the shipbuilding cost or the required freight rate (RFR); nowadays, multiple criteria may be considered by use of multiobjective optimization procedures such as genetic algorithms.[26]

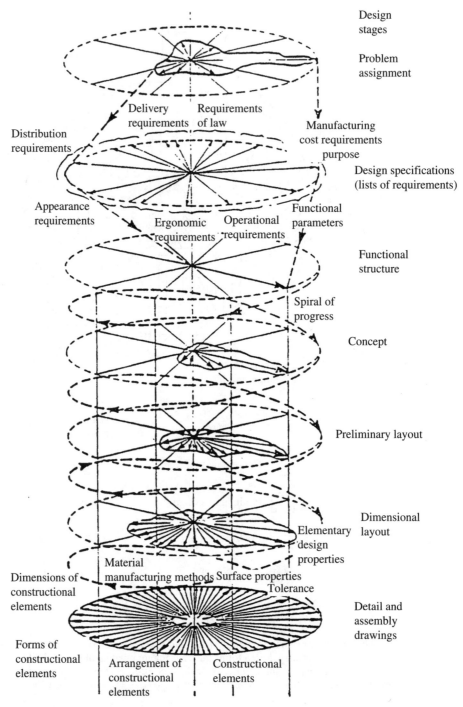

FIGURE 25.15   3D design spiral, according to IMDC (www.imdc.cc).

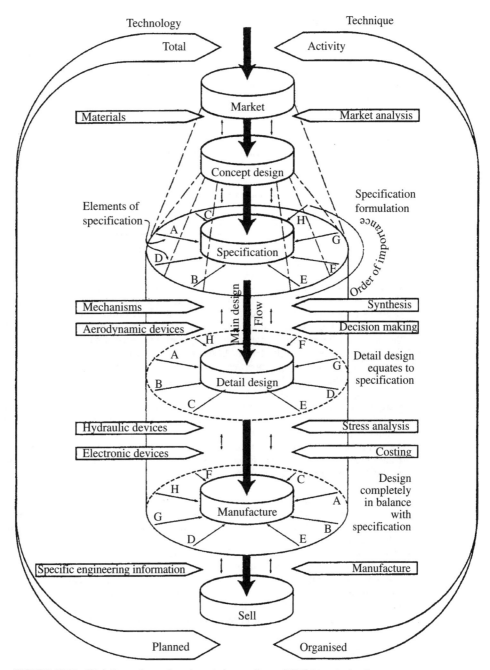

FIGURE 25.16   3D design and manufacturing spiral, according to IMDC (www.imdc.cc).

TABLE 25.3   List of Required Typical Naval Architectural Plans/Drawings/Studies to be Developed during the Contract Design of Merchant Ships

Outboard profile, general arrangement	Power and lighting system—one line diagram
Inboard profile, general arrangement	Fire control diagram by decks and profile
General arrangement of all decks and holds	Ventilation and air-conditioning diagram
Arrangement of crew quarters	Diagrammatic arrangements of all piping systems
Arrangement of commissary spaces	Heat balance and steam flow diagram—normal power at normal operating conditions
Lines	Electric load analysis
Midship section	Capacity plan
Steel scantling plan	Curves of form
Arrangement of machinery—plan views	Floodable length curves
Arrangement of machinery—elevations	Preliminary trim and stability booklet
Arrangement of machinery—sections	Preliminary damage and stability calculations
Arrangement of main shafting	

Source: From ref. 32.

Contract Design: Fifth Iteration.   The objective of this stage of ship design is completion of the necessary studies, calculations, and naval architectural drawings, as well as the shipbuilding specifications document, which all form an indispensable part of the shipbuilding contract between the ship owner and the contract-awarded shipyard. This design phase includes a detailed description of the ship's hull form through a faired ship lines plan, an exact estimation of the powering requirement for achieving a specified speed on the basis of model experimental data of a towing tank,* an analysis of the ship's maneuvering properties,† consideration of alternative propulsion systems (propeller-machinery system), details of ship's structural design,‡ the development of the main and auxiliary ship's supply networks (electric, piping, etc.), and finally, more exact estimation of the various ship weight components or of ship's total weight (and displacement) and of corresponding centers.

It is estimated that the third phase of ship design is about 17 times more demanding than the second phase, which corresponded to about 5,000 man-days for a large merchant ship in the late 1950s (according to Taggart[32]), whereas today it is estimated to be a small fraction of the preceding value. The drawings/plans developed, the studies, and the building specifications determined during contract design are detailed in Tables 25.3 and 25.4.

Detailed Design.   In the last phase of design, a detailed design of the ship's structural elements is created, along with the setup of the construction and fitting specifications for the shipyard's production workshops and the external suppliers of mechanical machinery and other outfitting. It is characteristic of this phase of design that whereas the prepared drawings/plans and specifications are the result of efforts of expert engineers (naval architects and marine engineers), the subsequent implementation of the ship's design in practice depends solely on the capabilities of the shipyard's production units (and their productivity) in terms of both hardware infrastructure and human resources (foremen and technicians of the yards). According to Kiss,[32] this stage of design corresponds to the tremendous effort of 60,000 man-days in the late 1950s, whereas today is small fraction depending on the degree of introduction of IT technology in the yard's design and production departments.

---

*The conduct of seakeeping model experiments for assessing the behavior of the ship in seaways is uncommon for merchant ships (except for cruise and ro-ro passenger ships, containerships, and some others on a case basis); they are conducted commonly for all naval ships and offshore supply vessels; independently, the seakeeping assessment of a ship may be conducted by use of modern theoretical/numerical methods with quite satisfactory results and reliable conclusions about ship's behavior in various seaway and weather conditions.

†On the basis of theoretical/numerical or more seldom model experimental studies.

‡Without going into the detailed design of structural elements and to the production drawings.

TABLE 25.4 List of Required Typical Main Technical Specifications to be Developed during the Contract Design of Merchant Ships

General	Forced draft system
Structural hull	Steam and exhaust systems
Houses and interior bulkheads	Machinery space ventilation
Sideports, doors, hatches, manholes	Air-conditioning, refrigeration equipment
Hull fittings	Ship's service refrigeration
Deck coverings	Cargo refrigeration—direct expansion system
Insulation, lining, and battens	Liquid cargo system
Kingposts, booms, masts, davits, rigging, and lines	Cargo hold dehumidification system
Ground tackle	Pollution abatement systems and equipment
Piping—hull systems	Tank level indicators
Air conditioning, heating, and ventilation	Compressed air systems
Fire detection and extinguishing	Pumps
Painting and cementing	General requirements for machinery pressure piping systems
Navigating equipment	Insulation—lagging for piping and machinery
Life-saving equipment	Emergency generator engine
Commissary spaces	Auxiliary turbines
Utility spaces and workshops	Tanks—miscellaneous
Furniture and furnishings	Ladders, gratings, floor plates, platforms, and walkways in machinery spaces
Plumbing fixtures and accessories	Engineers' and electricians' workshop, stores, and repair equipment
Hardware	Hull machinery
Protection covers	Instruments and miscellaneous cage boards—mechanical
Miscellaneous equipment and storage	Spares—engineering
Name plates, notices, and markings	Electrical systems, general
Joiner work and interior decoration	Generators
Stabilization systems	Switchboards
Container stowage and handling	Electrical distribution
Main and auxiliary machinery	Auxiliary motors and controls
Main turbines	Lighting
Reduction gears—main propulsion	Radio equipment
Main shafting, bearings, and propeller	Navigation equipment
Vacuum equipment	Interior communications
Distilling plant	Storage batteries
Fuel oil system	Test equipment, electrical
Lubricating oil system	Centralized engine room and bridge control
Seawater system	Planning and scheduling, plans, instructions, books, etc.
Freshwater system	Tests and trials
Feed and condensate systems	Deck, engine, and stewards' equipment and tools, portable
Steam generating plant	

Source: From ref. 32.

Reviewing all phases of ship design, it may be concluded that on the basis of the results of the basic design, the main technical and economic ship characteristics can be reliably estimated. Thus the yard and the interested ship owner may proceed to the conclusion of a contract on the basis of a yard's tender, which will be developed using the outcome of the first two phases. In the case of acceptance of a yard's tender, the more detailed and demanding third and fourth design phases are completed.

25.4.5   Main Owner's Requirements: Statement of Work

The main requirements of a ship owner with respect to the design and construction of a merchant ship, as laid down in the shipbuilding contract and technoeconomic ship specifications, are driven by a variety of factors that are all related to the attractiveness of a shipping business in terms of return on investment and that might lead to a shipbuilding contract. A sample of these factors is listed below:

1. *Replacement or conversion of aged or less competitive ships.* These are ships with unsatisfactory payload, speed, and/or operational cost characteristics or ships not complying with newly introduced safety regulations pertaining to the safety of life at sea (SOLAS) or protection of the marine environment (MARPOL, OPA 90).
2. Extension or change of activities of a shipping company in an already serviced market (increase of competitiveness).
3. Development of new services in other geographic areas (geographic extension of business activities).
4. Transport of new types of cargo in an existing market (increase of share in local trade).
5. Introduction of advanced marine technology in terms of
   a. Advanced marine vehicles: high-speed and innovative design vessels
   b. Cargo loading-unloading systems:
      • Modern and innovative loading-unloading systems
      • Transport of high-value cargo in standardized transport units (containers, pallets, etc.)
   c. Intermodal transport systems: integrated sea-land-river transport systems, etc.
6. Development of special types of ships supporting shipping, offshore, and ocean surveillance activities: tugboats, icebreakers, pilot boats, offshore supply vessels, search and rescue (SAR) vessels, hydrographic/research vessels, etc.

Typical main requirements of a ship owner, interested in awarding a shipbuilding contract are listed and briefly commented on below:

1. *Transport capacity*, expressed by ship's *deadweight,*[*] capacity of cargo spaces (volume), the number of transported containers, number and type of transported vehicles and passengers, as applicable
2. *Speed in trial conditions*, at 100 percent maximum continuous rating (MCR) of engine power
3. *Range* (expressed in seamiles or days of operation without refueling) for a specified routing scenario at service speed and with indication of ports for refueling and replenishment
4. *Classification:* Society and class

These requirements are supplemented by national and international safety regulations (International Maritime Organization, www.imo.org), which pertain to ship's stability and floatability in the case of loss of the ship's watertight integrity, to fire safety, to the ship's navigational equipment, to lifesaving equipment, and to the ship's evacuation procedures (SOLAS), as well as determination of the ship's load line and freeboard (International Convention on Load Lines), the ship's tonnage measurement (International Convention on Tonnage Measurement of Ships), protection of the marine environment from oil pollution (MARPOL, International Convention for the Prevention of Pollution from Ships), the number and type of crew (according to flag state regulations), the manner of transport of dangerous goods, etc.

---

[*]A ship's *deadweight* is the sum of the weights that are added to the *light ship weight*, such as weights of cargo (payload), engine fuel and lubrication oil, passengers and crew including effects, various types of carried water (fresh, drinking, engine cooling, steam boiler, ballast), and varying outfitting. It is noted that the *light ship weight* is the sum of the weights of ship's structure, machinery installation, accommodation and outfitting, which corresponds to ship's state when delivered by the yard, namely, completely outfitted and ready for operation, without cargo and any provisions/supplies.

The extent and detailing of these requirements depend on the degree of organization/preparation of the shipping company's services and may vary between some brief requirements, as stated earlier (in the case of small shipping companies), and up to a comprehensive technical specification of the ship's construction in the case of large shipping companies with field expertise.

The contract award procedure of shipbuilding to a yard is begun by exploration of solicited tenders of competing yards for the shipping company's contract. The tenders are developed in accordance with the ship owner's requirements and consist (at least) of an estimation of the ship's main dimensions and of the other main design characteristics, a preliminary general arrangement (GA) on main spaces and equipment, and a preliminary estimation of costs, along with a time schedule for completion of the ship from day 1 after conclusion of the award/contract up to the day of ship's delivery (*tender design*). In general, the ship owner awards the contract to the yard with the "best" offer (lowest building cost or best value for money), considering, however, the offered financial terms by the yard (extent of downpayment, support in securing competitive loans from banks, etc.). It should be noted that if the tendering yards are likewise reliable in terms of offered technology and quality of production, the yard offer associated with the lowest-cost ship, complying with the set requirements, indeed will be the most attractive for the ship owner.

The preceding list of main requirements of an interested ship owner are briefly commented on below:

1. *Specification of transport capacity or deadweight (DWT)*. The cargo-carrying capacity of a cargo ship is expressed by her deadweight because the contracted tons DWT may be easily controlled during (or shortly after) the ship's delivery by the difference in the ship's displacement in the fully loaded and light ship conditions (this is practically established by the readings of draft marks/ ahmings at the ship's bow, stern, and amidships).

The ship owner may, to a certain degree, adapt the amount of carried payload to the actual market needs by corresponding changes in the amount of carried fuel without changing ship's deadweight. Regarding the degree of achievement of the contracted deadweight capacity, in general, the following (or similar) provisions are set in the shipbuilding contract:

- Difference of specified and finally achieved deadweight less than about 2 percent: No penalty provisions

- Difference up to about 5 percent: Proportional reduction of shipbuilding cost

- Difference more than 5 percent: Significant reduction of payments or rejection of ship by the ship owner at full return of payments

It is evident that for special types of ships with their main mission beyond cargo or passengers transportation, the deadweight capacity as a main requirement is replaced by other ship characteristics, representing the actual value of the vessel; for example, for tugboats, a main requirement is the towing and propulsion power; for icebreakers, an additional factor is the maximum ice thickness in which the ship may operate at a certain speed.

Specification of the cargo hold capacity (volume) is set for covering the stowage needs of certain cargo. Thus, for tanker ships, the crude oil tank volume is specified; for "reefer" cargo ships, the net hold volume for refrigerated cargo is specified; for general cargo ships, the tank volume for the additional carriage of olive oil or other fluids is specified.

The reference to unitized cargo, namely, the number of containers (TEU: twenty feet equivalent unit, cross-section 8×8 feet, length 20 feet), number of roll-on/roll-off vehicles (private cars, trucks, trailers, etc.), or the lengths of vehicle lanes on car decks may be related to specialized ships for the carriage of this type of cargos [e.g., cellular type containerships or roll-on/roll-off (ro-ro) ships] but also to multipurpose cargo ships carrying this type of cargo from time to time. The same refers to the number of passengers, beyond the crew, that may apply to combined ro-ro passenger ships (ferries or RoPAX, ships with length over about 60 m) rather than pure passenger ships. Pure passenger ships today are encountered as cruise ships only, except for small-boat passenger-ship transport services. It should be noted that today the combined cargo-passenger ship (thus a cargo ship carrying more than 12 passengers beyond the crew) has practically disappeared as a ship type, with small exceptions in specific routes and in some modern ro-ro cargo ships (carrying truck drivers).

2. *Specification of trial speed.* The requirement for a minimum speed in trial conditions (thus wind force up to a maximum 2 to 3 Beauforts, calm and deep water, without currents, clean ship hull surface) and a specified draft results from the easy control of ship's propulsive efficiency (performance of main machinery and propulsion system) in relation with ship's hull form and displacement on the basis of the speed achieved. During delivery, the speed is commonly measured by the time to pass 1 nautical sea (nautical) mile (1,852 meters) as specified geographically near the shipyard and commonly agreed. The trial procedures are detailed in a separate document attached to the shipbuilding contract.

The *specified* speed refers in general to the design draft (and ship's displacement) and to the MCR of the main machinery (MCR-*maximum continuous rating* at 100 percent engine rpms). Because during the trials, except for tankers and some passenger ships, the design draft and displacement cannot be achieved, the trial speed may be measured for a reduced draft (e.g., at the ballast condition), and this speed may be transferred to the full draft value by an agreed calculation procedure. Regarding the possible deviations between the contracted speed and the speed at delivery, the following penalties (rarely premiums) apply in general:

- Deviation up to about 2 percent (or about 1/2 knot at 20 knots speed): No or small penalty.
- Deviation up to 5 percent (or about 1 knot less speed): Significant reduction of shipbuilding price.
- Deviation by more than 5 percent: Ship owner reserves the right to reject the ship delivery.
- In case of achieving a higher speed than contracted, a premium might be paid to the yard. The same often applies for early ship delivery, if so agreed in the contract.

3. *Specification of range.* The specification of the ship's range determines the amount (weight) and required volume of fuel and other liquid tanks as necessary for the ship's operation (lubrication oil, fresh and drinking water, etc.).

4. *Specification of classification society and class.* This must be an internationally recognized classification society that is entitled to award the ship a specific "class" and as necessary for the various authorities to allow the ship's operation. The award of a "class" corresponds to the issuance of a series of safety certificates ensuring the integrity of the ship's structure and of the ship's vital equipment and outfitting that affect ship's safety (ship's machinery, propulsion, and steering system, including auxiliary devices). Selection of the class society is in general a matter of the ship owner.

The internationally most important[*] class societies are members of the International Association of Classification Societies (IACS; www.iacs.org.uk) and are as follows:

- *Germany:* Germanischer Lloyd (GL; www.gl-group.com)
- *United Kingdom:* Lloyds Register of Shipping (LR; www.lr.org)
- *Norway:* Det Norske Veritas (DNV; www.dnv.com)
- *Unites States:* American Bureau of Shipping (ABS; www.eagle.com)
- *France:* Bureau Veritas (BV; www.bureauveritas.com)
- *Italy:* Registro Italiano Navale (RINA; www.rina.org)
- *Japan:* Nippon Kaiji Kyokai (NKK; www.classnk.or.jp)
- *PR China:* China Classification Society (CCS; www.ccs.org.cn)
- *Korea:* Korean Register of Shipping (KR; www.krs.co.kr)
- *Russia:* Russian Maritime Register of Shipping (RS; www.rusregister.ru)

In the common ship owner's requirements and specifications, the following additional items may be included: the type of propulsion system and number of propellers; the type and manufacturer of

---

[*]Mainly in terms of volume of activities (total fleet tonnage under class).

main machinery (for diesel engines according to engine listings, however, *without* specifying the machinery powering, which will be an essential result of ship design); the type, number, and arrangement of cargo handling equipment (for tankers, pump power); and the quality of crew accommodations and especially passenger cabins and public spaces if the ship is a cruise ship or RoPAX ferry.

Regarding the ship's main dimensions, there are, in general, no specifications or boundary limits set by the ship owner, except for navigational constraints (passing through canals and narrow streets: limits on maximum draft and beam, seldom on length; approaching harbors: limits mainly on draft, seldom on length). Also, there are in general no specifications regarding a ship's stability properties in the various loading conditions, except for the initial stability (minimum $\overline{GM}$) for fishing vessels and sometimes for ro-ro passenger ships, containerships, and "reefer" ships; clearly, the built ship is assumed to fulfill relevant national and international safety regulations, including those for intact and damage stability and floatability.

The preceding safety regulations specify in detailed form the requirements (criteria) pertaining to the safety of the global system "ship" (vessel, crew, passengers, and cargo) and "marine environment" (marine biology and coastal areas) in normal and extreme ship operating conditions (dangerous weather conditions, collision with other ships, grounding, flooding, explosion, and fire).

Finally, where the regulatory framework and the ship owner's specifications do not literally prescribe a specific ship performance measure or property, it is tacitly understood that the ship needs to perform in the frame of contemporary shipbuilding technology and science. Especially regarding the operability of the ship, the following are expected (without literally specifying them):

1. Good seakeeping performance (seaworthiness)
2. Good maneuvering properties: stability of course keeping, turning diameter, stopping distance
3. Good arrangement of cargo spaces: ease of cargo stowage and access to holds and lower decks
4. Good arrangement of functional spaces: access to and ergonomic arrangement of equipment and arrangement of machinery space and navigational bridge
5. Good arrangement of accommodations and public spaces and access ways: design of simple access ways to spaces, corridors, etc., especially in passenger ships; optimization of pathways of crew from their cabins to working areas; and comfortable accommodations for passengers and crew

Finally, the design and construction of naval ships are governed by other types of criteria, namely, those referring to the fulfillment of a mission under specific operational (and weather) conditions in the frame of needs of the department of defense of a country. The main factors affecting the fighting capability and main requirements for design of a naval ship are as follows:

1. Type of naval ship and mission (corvette, frigate, cruiser, destroyer, aircraft carrier, surveillance vessel, etc.)
2. Type and extent of armament and electronic/operational outfitting
3. Number of crew and accommodation requirements
4. Structural enforcements
5. Floatability and stability after damage, damage control
6. Sustained speed in calm waters and in specified seaways (top and cruise speeds at specific engine ratings)
7. Specification of seakeeping and maneuvering capabilities
8. Range

Commonly, and despite some recent developments, the design and construction of naval ships are governed mainly by technological and physical performance criteria because they result from latest developments of science and technology and, to a lesser degree, economic considerations. The history of shipbuilding (as in other branches of technology) is rich in examples of innovative technological solutions applied first to naval ships and later successfully adapted to merchant ships, for example, the use of new construction materials: higher tensile steels, aluminum alloys, and synthetic materials; the use gas turbines as main machinery; and the introduction of electronic control systems, onboard computers, etc.

25.5   INTRODUCTION TO PRELIMINARY SHIP DESIGN

25.5.1   General

The traditional approach to preliminary ship design includes the following main steps:

1. Critical *assessment* of the owner's requirements (statement of work) and especially of those affecting decisively the selection of the ship's main dimensions and other main design characteristics.

2. *Collection* of data (ship type, size, deadweight, speed, installed horsepower) of similar ships by search in specific bibliographies, including databases of built ships, e.g., Lloyd's Register Fairplay database, www.lrfairplay.com: data on more than 45,000 merchant ships of all types, with deadweight > 1,000 tons, database of the Ship Design Laboratory of NTUA, www.naval.ntua.gr/sdl: data on more than 700 European passenger ships, ro-ro passenger ships, and ro-ro cargo ships (with tonnage > 1,000 GRT). For shipyards and design offices: use of data from their own design data files.

3. *Inventory and study* of relevant rules and safety regulations for ship design and construction: national regulatory provisions, national and international maritime regulations, rules of specified classification society, technical ship specifications.

25.5.2   Ship Types

Before presenting an outline of a generalized approach to ship design, it is rational to proceed to a categorization of the various ship types into some main ship categories that may be characterized by common design procedures. These categories, referring to common design features of various ship types, are as follows:

1. *Deadweight carriers*, with a decisive design characteristic—their deadweight capacity. These are ships that carry relatively heavy cargos with a *stowage factor*[*] that is less than about 1.3 m³/t (e.g., ores, cement, coal, grain, oil, etc.). Typical representatives of this ship category are the bulkcarriers (bulk/ore carriers) and tankers (crude oil carriers); also included are the general cargo ships on charter trade (tramp ships), transporting dry cargo with relatively low stowage factor in bulk or as break cargo. The common design characteristic of this type of ship is that there is available space in the cargo holds to accept even more cargo; however, the allowable maximum draft, according to the provisions of the Load Line Convention (minimum freeboard), restricts further loading. The ship's *capacity factor*[†] is relatively low and generally less than about 1.5 m³/t DWT.

2. *Volume carriers*, with the most significant design characteristic being their hold volume capacity. These are ships that carry relatively light cargos with a stowage factor of more than about 2.0 m³/t (e.g., cotton, tobacco, fruits, high-value industrial goods, electronic and electric equipment, private cars and trucks, etc.). Typical representatives of this ship category are the ro-ro cargo ships, car carriers in general (PCC [pure car carrier], PCTC [pure car and truck carrier]), ro-ro passenger ships (ferries), containerships, "reefer" ships, general cargo ships in liner service (liners), and passenger/cruise ships; they dispose in general at least one continuous deck *above* the freeboard deck (bulkhead deck), whereas they do not fully exploit, in general, the maximum allowable draft, as it results from the provisions of the Load Line Convention (they dispose excessive freeboard,

---

[*]Expresses the required volume for the stowage of 1 ton of cargo; it is a property of the cargo.
[†]Is the ratio between the ship's cargo hold volume and ship's deadweight; it is a property of the ship.

because there is a lack of available hold volume to accept more cargo); they dispose a relatively high *capacity factor* of more than about 2.5 m³/t DWT. Ships carrying intermediately heavy cargos (*stowage factor* between about 1.3 and 2.0 m³/t) or alternative cargos of strongly varying stowage factor may be designed as deadweight or volume carriers.

3. *Linear dimension ships* are ships with one linear dimension (length, beam, draft, or side depth) restricted by physical external boundaries or constraints set by the cargo carried. These are ships with restrictions because of passing major canals, such as the canals of the St. Lawrence Seaway (Lake Ontario, Great Lakes bulkcarriers), with a maximum allowable beam of 22.85 m; the Panama canal, with a maximum overall length of 294.13 m (965 ft), beam of 32.31 m (106 ft), and draft of 12.04 m (39 ft, 6 in), the so-called PANMAX* ships, or operating near the mouth of important rivers, for example, La Plata River (South America), of importance to "reefer" bananas ships, with a maximum draft of 8.2 m. Also, ships carrying standardized cargo units, such as containers (i.e., cellular-type containerships), have a well-defined beam (and side depth height) that is determined by the number of stowed containers in the transverse (and vertical) direction, considering that the beam (and height) of the containers is standardized (cross-section: 8 × 8 ft, 8 ft = 2.438 m).† The same applies to other box-type cargo ships, such as ships carrying floating barges of standardized dimensions, LASH (lighter aboard ship), and SEABEE, ships carrying vehicles of standard size (ro-ro cargo and ro-ro passenger ships, rail ships, etc.). Common characteristic of all these ship types is the stepwise (discontinuous) change of their beam and the relatively increased length, especially if the beam happens to be restricted (e.g., PANMAX ships), thus in general these are ships for which the relationship between main dimensions and displacement is less optimal.

4. *Special-purpose ships.* These are ships that cannot be categorized in the preceding main categories owing to specific conditions of their design and operational profile (e.g., tugboats, icebreakers, fishing vessels, and offshore support vessels). Likewise, all unconventional ships are inherently special-purpose ships, and their design greatly depends on specific type and size (high-speed craft in general, advanced marine vehicles, twin- and multihull vessels: catamarans, trimarans, pentamarans, air-cushion vehicles, submarines, etc.).[15,24,25]

5. Other methods or criteria of categorization of ship types are as follows:

- Mission profile
  - Merchant ships
  - Naval and coast guard ships
  - Research/hydrographic vessels
  - Sport vessels
  - Tug boats
  - Ice breakers
  - Dredgers
  - Supply vessels; drilling ships and platforms; floating production, storage, and offloading (FPSO) vessels; crane ships; etc.
  - Pilot boats
  - Cable ships
- Operational area
  - Open/deep-water ships
  - Inland ships—river and lake boats

---

*An expansion of the Panama Canal is under way (completion in year 2014), in the way to allow the passing of ships with maximum lengths of up to 426.72 m (1,400 ft), beam up to 54.86 m (180 ft), and draft up to 18.29 m (60 ft). These dimensions correspond to the size of the new generation of MEGA (JUMBO)-containerships, with a carrying capacity of up to about 15,000 TEU.

†Some containers may be 8.5 ft high.

- Floatability
  - Surface ships
  - Underwater vehicles
    - With forward speed (submarines)
    - Without or with very small forward speed (bathyscaphs)
- Propulsive power
  - Main machinery/engine type
    - Steam engines
    - Turbines
      - Steam-powered
      - Gas-powered
    - Diesel engines
    - Otto gas engines
    - Diesel/electric engines
    - Combined diesel and gas turbines (CODAG, etc.)
    - Nuclear steam powered turbines
    - "Green" environmentally friendly prime or auxiliary energy sources
      - Wind energy (airfoil sails and kites)
      - Solar energy systems
      - Fuel cells
      - LNG diesel engines
      - NYK super eco ship 2030
  - Wind sails
  - Oars
- Propulsion type
  - Paddle
  - Propeller
    - Fixed pitch
    - Controllable pitch
    - Voith-Schneider
  - Water jets
  - Azipods
- Construction material
  - Steel
  - Aluminum alloys
  - Wood
  - Synthetic materials
  - Marine concrete
- Type of transported cargo
  - General cargo ships
  - Bulkcarriers
  - Tankers
  - Gas carriers
    - Liquefied petroleum gas (LPG) tankers
    - Liquefied natural gas (LNG) carriers
  - Break bulk carriers
    - Break bulk cargo ships
    - Containerships

- Floating-barge carriers
  - LASH (lighter aboard ship)
  - SEABEE
  - BACO (barge-container carrier)
- Vehicles carriers
  - PCC (pure car carriers) and PCTC (pure car truck carriers)
  - Ro-ro cargo ships
  - Passenger/ro-ro-roPAX
  - Rail and combined ro-ro rail ships
- Heavy-lift transport ships
- Multipurpose cargo ships
- Passenger ships
  - Cruise ships
    - Day cruise ships
    - Overnight cruise ships
  - Short sea passenger transport ships
    - Day ships
    - Overnight ships
  - Excursion boats

Table 25.5 presents a breakdown of the world's fleet by basic ship types for the year 2008.

TABLE 25.5   World Cargo Ships Fleet and Breakdown of New Building Orders for Year 2008 According to Ship Type

Total Cargo Ship Fleet (million tons dead weight)	Year End				October 1, 2008		Order book		
	2004	2005	2006	2007	No.	Million tons dead weight	No.	Million tons dead weight	% Fleet
Oil tankers > 10,000 tons dead weight	299.5	320.2	337.2	356.3	3,515	363.3	1,321	170.3	46.9 %
Oil tankers < 10,000 tons dead weight	11.0	11.1	11.3	11.5	5,111	11.6	133	0.8	6.6 %
Chemical tankers	23.9	26.7	29.7	33.7	3,137	37.3	1,023	18.9	50.7 %
Other tankers	3.2	3.3	3.3	3.3	644	3.4	44	0.4	11.6 %
Bulkers	322.7	345.3	368.7	392.8	6,958	413.9	3,404	295.7	71.5 %
Combos	10.2	9.4	8.9	8.2	58	8.2	9	2.8	34.9 %
LPG carriers	11.7	11.8	12.5	13.4	1,113	13.4	205	3.4	25.5 %
LNG carriers	13.9	15.4	17.5	20.6	284	20.6	104	9.1	44.3 %
Containerships	99.6	111.5	128.0	144.0	4,657	157.4	1,285	75.1	47.7 %
Multipurpose	22.4	22.9	23.7	24.7	2,817	25.5	684	8.6	33.9 %
General cargo	37.8	38.0	38.1	38.6	15,114	38.7	220	1.5	3.8 %
Ro-ro	10.2	10.2	10.4	10.5	3,539	10.5	123	1.5	14.1 %
Car carriers	7.0	7.6	8.2	9.0	676	9.8	217	3.6	37.1 %
"Reefers"	7.9	7.9	7.8	7.6	1,992	7.6	20	0.3	3.5 %
Offshore (AHTS/PSV)	3.9	4.2	4.6	5.0	3,912	5.2	803	2.0	38.7 %
Other cargo	8.9	9.0	9.1	9.0	1,456	9.0	38	0.5	5.5 %
World cargo fleet	894.0	954.4	1,019.0	1,088.2	55,010	1,135.3	9,633	594.5	52.4 %

Source: "Clarkson Research Studies," *Shipping Intelligence Weekly* 841, October 10, 2008.

25.5.3   Estimation of Main Dimensions and Form Coefficients

There are two basic methods in ship design for preliminary estimation of the main dimensions and basic form characteristics, namely, the *relational* or *empirical method* and the *parametric method* or *method of independent parameters.*

The Relational or Empirical Method.   The estimation of main dimensions and main form characteristics (form coefficients) is based on comparative data from similar built ships, with the data stemming from open bibliographies, commercial and internal databases, and data files. The data then are deduced by interpolation of comparative data of similar ships. In the frame of this empirical method, we may consider the use of empirical design formulas, of relevant statistical diagrams, or of properly defined design coefficients, with the help of which the sought data (e.g., main dimensions, weight components, and powering) are related to the initially given, specified ship data (e.g., ship's deadweight or, indirectly, ship's displacement). For successful application of the empirical method, it is assumed that the available comparative data or empirical relationships are sufficient and reliable for the type (and often size) of the ship under investigation. Of course, it is additionally assumed that the comparative built ships represent economically competitive and reliable design solutions and that the relationship between the main design parameters and optimization criteria is quite flat (of small gradient) in the region of actual design parameters (i.e., a small change in a design parameter does not lead to a significant change in an optimization criterion).

The Parametric Method.   When comparative data from similar ships are lacking (e.g., when designing innovative ships or ships whose absolute ship size exceeds common limits or, independently, when looking for further improved design solutions), it is necessary to conduct a parametric study from scratch, in which the best combination of main dimensions and main design characteristics is sought for optimizing some selected design criteria. In particular, a mathematical model (algorithm and corresponding software) is used to optimize an economic criterion, such as the ship's building cost or the required freight rate for 1 ton of transported cargo (RFR-*required freight rate*[*]), or return on investment. As a result, the optimal set of design parameters is identified, minimizing or maximizing a set criterion.[†]

The setup of a sufficient mathematical model, in which the ship's main design parameters are rationally related to the ship's performance (physical and economic characteristics), is a very demanding task and obviously strongly related to the specific conditions of a ship type. The model may be (and often is) supported by empirical data [e.g., when estimating hydrodynamic performance, systematic model experimental data (of model series) are embedded in the mathematical modeling, and therefore, they support the validity of the overall modeling]. Identification of the optimal ship design solution is one of the fundamental tasks of computer-aided ship design (CASD) and, mathematically, a typical nonlinear multiparametric optimization problem with multiple constraints.[20]

A classic example of systematic parametric optimization for identifying the "least-cost ship" is given in Figure 25.17. It refers to the optimization procedure of a cargo ship on the basis of main requirements of a hypothetical ship owner for speed (V: velocity), payload (WC: weight of cargo), stowage factor (SFR: stowage factor required), and range (R: range) according to Murphy et al.[32] It should be noted that this approach was developed in the early 1960s by use of the very limited computer hardware and software available at that time. Today, the availability of modern optimization methods and of strong computer infrastructure enables the consideration of many more design parameters, objective functions, and constraints; identification of the optimal solution (or the Pareto front of best solutions in multiobjective optimization) is achieved with a minimum number of parametric iterations compared with the "brute force" parametric optimization used in the initial stages of

---

[*]Definition of RFR = (annual expenses + annual depreciation of ship's value)/annual transport volume [$/ton]. This definition holds for uniform annual cash flow; clearly, ships with lower RFRs are more competitive than others.
[†]Modern ship design optimization methods consider multiobjective optimization procedures, optimizing simultaneously a series of partly contradicting criteria and identifying the so-called Pareto front of the best design solution.[26]

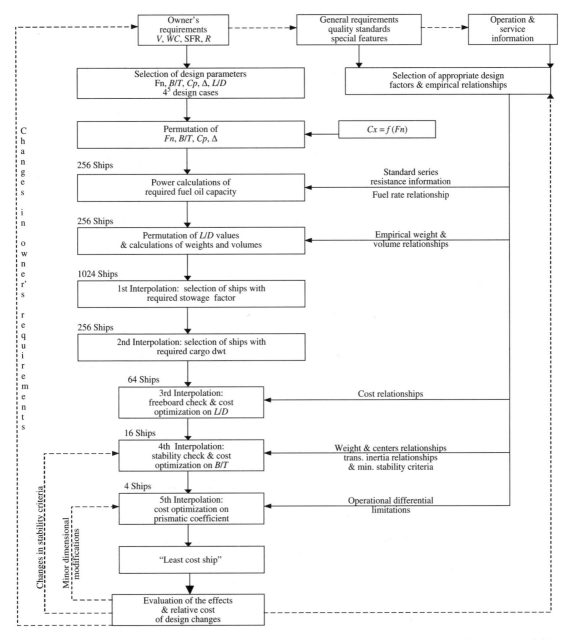

Notes: Owner's requirements: *V* (velocity), *WC* (weight of cargo), *SFR* (stowage factor required), Δ (displacement weight) *R* (range), *Fn* = *V*/$\sqrt{gL}$, Froude number

FIGURE 25.17 Parametric optimization by permutation of main design parameters for a "least-cost cargo ship," according to Murphy et al.[32]

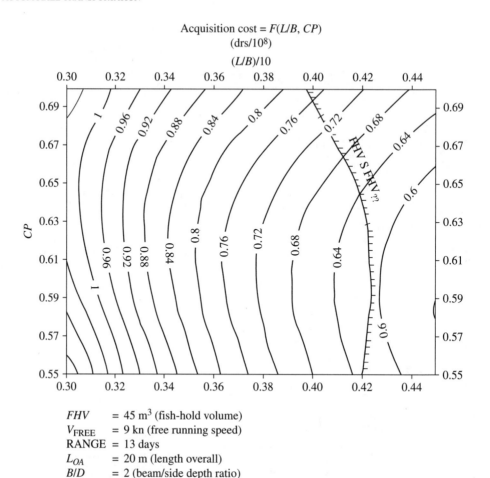

Acquisition cost = $F(L/B, CP)$
$(drs/10^8)$

$(L/B)/10$

FHV   = 45 m³ (fish-hold volume)
V_{FREE} = 9 kn (free running speed)
RANGE = 13 days
L_{OA} = 20 m (length overall)
B/D   = 2 (beam/side depth ratio)

FIGURE 25.18   Optimization of medium fishery vessel (stern trawler) with respect to shipbuilding/acquisition cost.[21]

CASD (see mathematical optimization methods and nonlinear programming problems[20] and modern ship design optimization with genetic algorithms.[26]

Some examples of modern ship design optimizations conducted in the frame of research work of the Ship Design Laboratory of NTUA are outlined below:

1. *Single-objective optimization of a fishing vessel with respect to building/acquisition cost.*[21] Figure 25.18 shows the dependence of building cost (represented by isolines of 10^8 Greek currency units in the early 1990s) on the ship's prismatic, form coefficient, and length-to-beam ratio.

2. Multi-objective optimization of ro-ro passenger ship with respect to structural weight, payload (as expressed by the length of lanes of carried vehicles), and the attained subdivision index A^* by use of genetic algorithms[33] (Figures 25.19 through 25.22).

---

*which is a measure of ship's survivability after a collision damage; it is the conditional probability that a ship survives collision damage caused by another ship

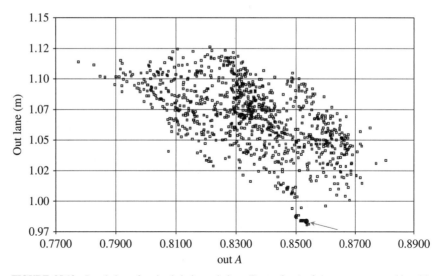

FIGURE 25.19 Population of optimal design solutions (Pareto front) of a ro-ro passenger ship with respect to the attained subdivision index A and the achieved length of vehicle lanes (initial design indicated by arrow). (Ship Design Laboratory–NTUA.[33])

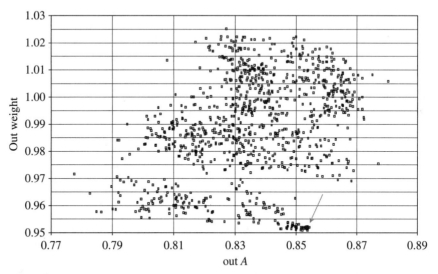

FIGURE 25.20 Population of optimal design solutions (Pareto front) of a ro-ro passenger ship with respect to the attained subdivision index A and the ship's structural weight (initial design indicated by arrow). (Ship Design Laboratory–NTUA.[33])

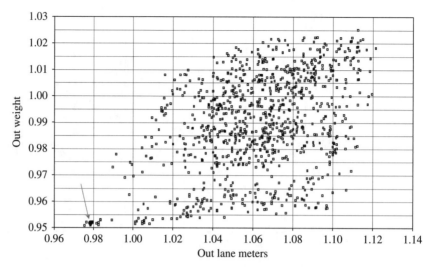

FIGURE 25.21   Population of optimal design solutions (Pareto front) of a ro-ro passenger ship with respect to the achieved length of vehicle lanes versus ship's structural weight (initial design indicated by arrow). (Ship Design Laboratory–NTUA.[33])

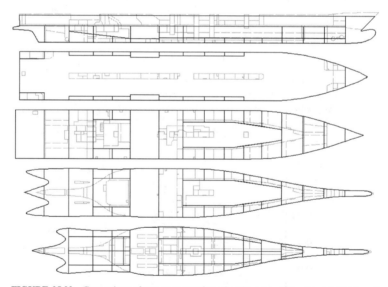

FIGURE 25.22   Comparison of compartmentation of optimal (*dark lines*) and initial (*gray*) ro-ro passenger ship design. (Ship Design Laboratory–NTUA.[33])

In the following, the ship design problem is formulated as a *decision process* in the frame of systems theory, and its optimization is achieved by nonlinear programming methods[20] (Figure 25.23):

Ship Design = Decision Progress

$E_1(I)$: Input = initial, given data = shipowner's requirements (deadweight, speed, range, operational conditions) and other initial conditions

Ship design = decision process

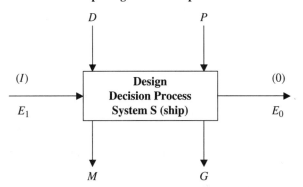

FIGURE 25.23   Systems approach to ship design as a decision process.

$E_0(0)$: Output = sought data—design; technoeconomic ship characteristics of an optimized ship with respect to one (or more) assessment (decision) criteria

D: Decision—design-free variables = all variables that can be varied freely by the designer (independent and dependent variables, e.g., main ship dimensions, length, beam, draft, dimensional ratios, etc.)

P: Restriction—parameters, constraints = all values that cannot be influenced (controlled) by the designer (decision maker, e.g., physical constraints, limiting dimensions of canals and ports, state of shipping market, weather conditions, etc.)

M: Evaluation criterion(a)—merit function(s) = $M_i(D, P)$ = formulation of one (or more) assessment criterion(a) in terms of an objective function, which will be relating the design and constraint parameters

G: Constraint functions—constraints = $G_j(D, P)$ = formulation of constraint functions relating the design and constraint parameters by linear of nonlinear algebraic equalities and nonequalities, for example, implementation of stability regulations (required minimum GM value), structural rules (requirement for minimum structural moment of inertia amidships), load line convention (required minimum freeboard), etc.

S: Design of system ship = decision process = mathematical model relating the input variables and parameters I, D, and P with the output data O, $M(D, P)$, and $G(D, P)$.

In a *life-cycle approach* to ship design, the entire life of a ship from concept design up to demolition and recycling needs to be considered and optimized (Figure 25.24).

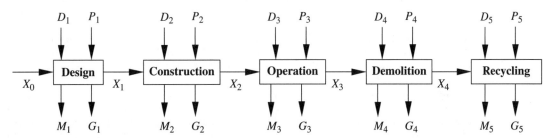

FIGURE 25.24   Life-cycle approach to ship design.

25.5.4   Comments on the Implementation of Design Methods

Regarding the practical application of the preceding basic methods in practice, we may note the following:

Basic Principles.

1. *Theory and practice (theoretical and empirical methods).* Only when considering both approaches may we arrive at good or truly optimal design solutions.

2. *Exploitation of data of prototypes.* The use of empirical data from similar ships greatly reduces the design work effort and serves also as validation of obtained design data.

Selection of Sample Ships and Use of Comparative Data from Similar Ships (Prototypes).
Typical Comparative Data from Main Ship Types.

Cargo ships: Deadweight, speed, trade type (tramp, charter, or liner), main machinery powering

Tankers and bulk carriers: Deadweight, speed, powering, passing limits through canals and narrow straits

"Reefer" ships: Deadweight, speed, powering, refrigerated cargo hold volume (net and net net).

Containerships: Deadweight, speed, powering, number of containers (above and below deck, dry and "reefer," TEU and FEU), passing limits through canals

Ro-ro passenger ships: Speed, powering, number of passengers (with and without cabins), number of vehicles (private cars and lorries), lane meters, extent and quality of accommodations, type of service (day and overnight trip)

Cruise ships: Speed, powering, number of passengers, extent and quality of accommodations and public spaces, type of service (day and overnight trips), passing limits through canals

Tugboats: Operational area (open sea and harbor area), speed and powering, towing power

Fishing vessels: Free-running and fishing (net towing) speeds, powering of main machinery, towing power, fish-hold volume, extent of accommodations, range, type of fishing vessel and fisheries (trawler, purse seiner, factory mother ship, coastal, oceanic, etc.)

When *even one* of these characteristics differs substantially from the comparative ship, then the direct use of the empirical data in hand is problematic and requires great caution. There are, however, methods such as the so-called method of Normand,[28] according to which, after using some transfer functions, the comparative data may be used (when better data are available), assuming that the differences in main parameters are small (up to a maximum of 10 percent, exceptionally up to 20 percent).

Use of Design Data.   Assessment and exploitation of as much as possible comparative data. The *interpolation* between comparative data in hand is in general seamless; however, *extrapolation* on basis of comparative data is often very problematic, unless for small violations of boundary limits.

Use of Design Diagrams.   The ship design bibliography offers a plethora of design diagrams in which typical design data for various types of ships are presented and main ship characteristics (such as dimensions, etc.) are depicted as a function of a typical ship owner's requirements; for example, for cargo ships, main dimensions versus deadweight; for containerships, versus the number of TEU; for fishing vessels, versus the fish-hold volume; etc. These types of diagrams should be used only in the initial conceptual design stage and should be avoided in later design stages, except as a way of checking/validating the design data obtained (see Appendix for some sample diagrams).

Use of Design Constants and Coefficients.   A basic tool of traditional ship design is the use of various empirical and semiempirical design constants and coefficients that are properly defined constant values (may vary with vessel size) that account for the impact of the variation in design parameters on certain design properties, such as weight components and engine power. Well-defined design constants and coefficients are those which do not change significantly when the underlying

design parameters vary. Design constants and coefficients may be dimensional or dimensionless, and care should be taken to consider the method of nondimensionalization. In the case of dimensional coefficients, the dimensional units used need to be observed; design coefficients may be used in the initial design stage for early and quick estimations.

Examples.

1. *Admiralty constant*:

$$C_N = \frac{\Delta^{2/3} V^3}{P}$$

where Δ is displacement weight (tons), V is speed (knots), and P is horsepower (typically installed horsepower, hp or kW); it is a *dimensional* design coefficient allowing the quick assessment of the powering of a ship; that is, the required horsepower may be estimated on the basis of the initially estimated displacement and the specified speed.

Assuming that the C_N constant is known from data for similar ships, it can be used for estimating the required horsepower (for given Δ and V) or the anticipated speed for the *same* ship when changing her loading condition (change of displacement). The constant is due to the British Admiralty and has a long history as a very effective way to quickly estimate speed/powering values for a given ship displacement; care should be taken when determining the value of the constant, besides taking care of proper units, to consider data for ships for similar *absolute length* because of the underlying physics and similitude law of frictional resistance components (effect of Reynolds number).

2. *Structural weight coefficient*:

$$P_{ST} = \frac{W_{ST}}{LBD}$$

where W_{ST} is the ship's structural weight (tons, kp, or kN), L is length, B is beam, and D is side depth (all in meters). This is a dimensional coefficient (weight unit/volume unit) that also may be defined for other ship components, such as light ship weight and weight of outfitting.

3. *Dimensionless form coefficients:*

A ship's (wetted) hull form is described globally by a series of dimensionless form coefficients that are the most significant form parameters when designing a ship (Figure 25.25). These coefficients include

- *Block coefficient* C_B, which corresponds to the ratio of the ship's displaced volume to the volume of an orthogonal parallelpiped with the same main dimensions L (mainly length between perpendiculars L_{PP}), B, and T.[*]
- The *prismatic coefficient* C_P expresses the ratio of ship's displaced volume to the volume of a prism with the same base area as the ship's midship section and a height equal to her length L.
- The *midship section* coefficient C_M expresses the ratio of the ship's midship section area to the square area $B \times T$.
- It may be shown easily that the block coefficient C_B may be expressed by the so-called prismatic coefficient C_P and the midship section coefficient C_M, namely:

$$C_B = C_P \cdot C_M$$

where $C_B = \dfrac{\nabla}{LBT}$ and $C_P = \dfrac{\nabla}{A_M L}$

---

[*]Obviously, a large C_B means a relatively full ship hull form, whereas a small C_B is related to a slender hull form.

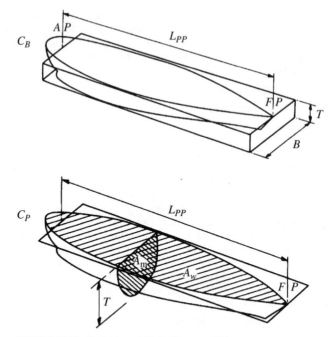

FIGURE 25.25 Definition of ship hull form coefficients.

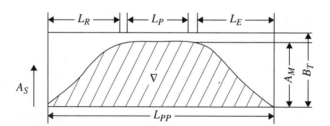

FIGURE 25.26 Definition of sectional area curve—lengthwise distribution of displacement.

For large merchant ships, the midship section coefficient is close to 1.0. Thus the block and prismatic coefficients dispose similar values and have the same physical importance. For smaller ships and boats, C_M decreases to values close to 0.6, and C_B and C_P differ significantly. The prismatic coefficient describes more effectively the slenderness of a hull, with small values corresponding to hull forms with displacement concentrating amidships and larger values to hulls with more evenly distributed displacement lengthwise and a large, parallel hull body (Figure 25.26). Typical form coefficients for various types of merchant ships are shown in Table 25.7.

Ship Design Equation. The so-called ship design equation is deduced from the Archimedean principle, namely, that the weight of the ship is equal to the weight of the displaced water. Methods related to the ship design equation for the initial estimation of a ship's main dimensions are based on an analysis of both sides of the equation by expressing them through empirical coefficients and dimensional ratios and deducing a final algebraic equation for a main dimension, such as the ship's beam or length. This is elaborated below:

$$\Delta = \rho_{SW}\, g\, \nabla^*$$

where ρ_{SW} = seawater density

∇^* = displacement volume (includes ship's shell thickness) = $C_B L B T k_A$

k_A = correction coefficient accounting for an average thickness of a hull's shell and of appendices, which are not included in the molded displacement volume ∇

Introducing the ratios L/B and B/T, which may be estimated from similar ships, the design equation may be written in the form

$$\Delta = \rho_{SW} g (L/B) B^2 [B/(B/T)] C_B k_A$$

or

$$\Delta = \rho_{SW} g C_B [(L/B)/(B/T)] B^3 k_A$$

Thus we obtain for the beam

$$B = \left[\frac{\Delta(B/T)}{\rho_{SW} g\, C_B (L/B) k_A} \right]^{1/3}$$

Likewise, assuming that the ratios L/B and L/T are known, we obtain for the length

$$L = \left[\frac{\Delta(L/B)(L/T)}{\rho_{SW} g\, C_B k_A} \right]^{1/3}$$

It is noted that a ship's displacement Δ may be estimated easily from ship's deadweight via the DWT/Δ ratio from similar types of ships.

Computer-Aided Ship Design (CASD).   Beyond parametric and mathematical ship design optimization outlined in the preceding section, a number of ship design–specific software programs are employed nowadays in the various stages of ship design. Some typical examples of specialized software applications, covering the calculatory needs of ship design are listed below:

- *Hydrostatic calculations:* Hydrostatic sheets and diagrams, parametric stability/Bonjean curves, floodable length design curves, stability booklets, probabilistic stability calculations, and stability criteria in intact and damage conditions, etc.)
- *Resistance and propulsion calculations:* Selection of main machinery and propulsion
- *Calculations of load line convention:* Determination of freeboard and allowable draft
- *Weight component calculations:* Structural weight, weight of machinery and outfitting
- *Structural strength calculations:* Analysis of static and dynamic ship strength control of classification society rules, strength assessment by first principles methods—finite-element methods
- *Assessment of seakeeping:* Determination of wave-induced motions
- *Assessment of ship's maneuverability*
- *Analysis of vibrations of structure, machinery,* and *propeller*

Further typical software applications in ship design, beyond the pure calculatory tasks, include

- Ship design optimization with respect to various criteria, for example, minimization of resistance and seakeeping (hydrodynamic optimization), minimization of required freight rate, minimization of structural weight, and maximization of survivability in the case of hull damage as single- or multiple-criteria optimization

- Development of ship hull lines from initial integral form characteristics, from existing hull form lines by distortion, or from systematic model series (Figure 25.27)
- Fairing of ship lines and of hull surfaces (skinning) (Figure 25.28)
- Development of general arrangement of spaces and outfitting (conventional 2D and 3D graphic presentation) (Figure 25.29)
- Simulation of ship evacuation (Figure 25.30)
  - EVI, www.safety-at-sea.co.uk/evi
  - EXODUS, www.fseg.gre.ac.uk/exodus
  - AENEAS, www.gl-group.com/maritime
- Simulation of the ship's behavior in waves and of dynamic intact and damage stability by use of the software CAPSIM (Figure 25.31)

Modern *integrated* naval architectural software packages and software platforms, which are able to support partly or entirely the various phases of ship design, are listed below:

- NAPA, www.napa.fi
- TRIBON, www.aveva.com
- FORAN, www.foransystem.com
- GHS, www.ghsport.com
- AUTOSHIP, www.autoship.com
- RHINOS 3D, www.rhino3D.com

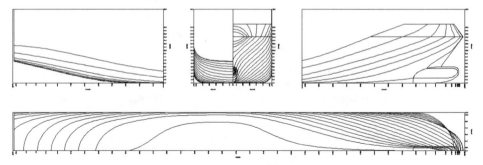

FIGURE 25.27   Development of ship hull lines for a ro-ro passenger ship by use of the software package NAPA. (Ship Design Laboratory, NTUA.)

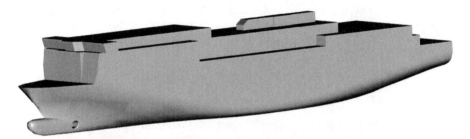

FIGURE 25.28   Development of faired 3D hull surface (skinning) for a ro-ro passenger ship by use of the software package NAPA. (Ship Design Laboratory, NTUA.)

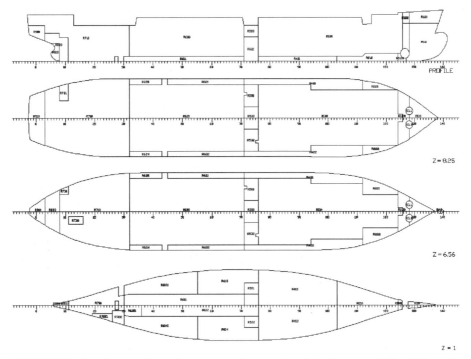

FIGURE 25.29 Development of general arrangement of spaces and outfitting (conventional 2D and 3D graphic presentation).

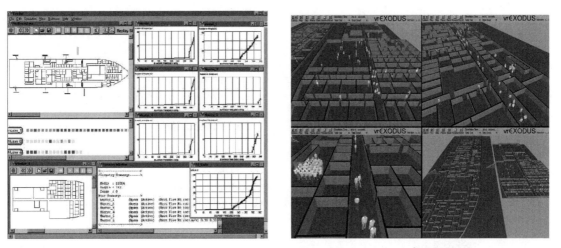

FIGURE 25.30 Simulation of ship evacuation by software tools: EVI (www.safety-at-sea.co.uk/evi) and EXODUS (www.fseg.gre.ac.uk/exodus).

FIGURE 25.31   Simulation of a ship's behavior in waves and of dynamic intact and damage stability by use of software CAPSIM. (Ship Design Laboratory, NTUA, www. naval.ntua.gr/sdl.)

25.6 BASIC DESIGN PROCEDURES FOR MAIN SHIP CATEGORIES

Following the preparatory steps outlined in the preceding two sections, the designer may proceed to a gradual estimation of the ship's main characteristics in a well-defined, sequential order, assuming the same order of steps for the main ship categories, namely, deadweight carriers, volume carriers, and linear dimension ships.

25.6.1 Deadweight Carriers

1. Estimation of displacement Δ on the basis of the specified (given) deadweight DWT (see Table 25.6, or use of regression data from the Appendix[28])

2. Estimation of main dimensions and form coefficients in the order outlined in Table 25.7, steps 1 through 6, and use of regression data from the Appendix[28] or Table 25.8.

3. Preliminary estimation of powering (by use of the Admiralty's constant or Figures 25.32 through 25.34)

4. Development of a *sketch* of the ships lines and of the general arrangement (preliminary estimation of displaced volume)

5. Control of balance of the sum of the ship's weight components (displacement weight) and weight of displaced water on the basis of the sketched ship lines (balance of geometric displacement and displacement weight)

6. Estimation of cargo hold volume

7. Preliminary estimation and control of minimum freeboard

8. Control of stability and trim

9. Preliminary estimation of building cost

10. Review and summary of results

TABLE 25.6 Typical Percentages of Weight Groups for Main Merchant Ship Types

	1	2	3	4	5	6
	Range		DWT/Δ	W_{ST}/W_L	W_{OT}/W_L	W_M/W_L
Ship type	Low	High	(%)	(%)	(%)	(%)
General cargo	5,000 tdw	15,000 tdw	65–80	55–64	19–33	11–22
Cargo coasters	499 GRT	999 GRT	70–75	57–62	30–33	9–12
Bulk carriers	20,000 tdw	50,000 tdw	74–80	68–79	10–17	12–16
	50,000 tdw	150,000 tdw	80–87	78–85	6–13	8–14
Crude oil tankers	25,000 tdw	120,000 tdw	68–83	73–83	5–12	11–16
	200,000 tdw	Max	83–88	75–88	9–13	9–16
Containerships	10,000 tdw	15,000 tdw	60–76	58–71	15–20	9–22
	15,000 tdw	50,000 tdw	60–70	62–72	14–20	15–18
Ro-ro	$L \cong 80$ m	16,000 tdw	50–60	68–78	12–19	10–20
"Reefers" (net ref. volume)	300,000 (ft³)	500,000 (ft³)	45–55	51–62	21–28	15–26
Small ferries	$L \cong 85$ m	120 m	16–33	56–66	23–28	11–18
Large passenger ships	$L \cong 200$ m	Max	23–34	52–56	30–34	15 – 20
Small passenger ships	$L \cong 50$ m	$L \cong 120$ m	15–25	50–52	28–31	20–29
Fishing vessels—stern trawlers	$L \cong 44$ m	82 m	30–58	42–46	36–40	15–20
Tugboats (tow power)	$P_B \cong 500$ kW	3000 kW	20–40	42–56	17–21	38–43
River ships (self-propelled)	$L \cong 80$ m	$L \cong 110$ m	78–79	69–75	11–13	13–19

Notes: Displacement weight: $\Delta = W_L + \text{DWT}$, where DWT = deadweight; W_L = light ship weight; $W_L = W_{ST} + W_{OT} + W_M$; W_{ST} = weight of steel structure; W_{OT} = weight of outfitting; W_M = weight of machinery installation; tdw = tons deadweight; GRT = gross register tons (tonnage capacity).
Sources: From refs. 28 and 31.

After completion of the last step in this procedure for estimation of the main dimensions and form coefficients, a more detailed *reassessment* of the aforementioned quantities is initiated; in particular, the more complex design studies related to steps 3 to 8 are conducted in the frame of the ship's preliminary design. In particular, in the second iteration, confirmation of the initially estimated absolute values of main dimensions is necessary, and they need to correspond with technical design solutions and fulfill the technical requirements and criteria set up in the statement of work by the ship owner. In addition, they must correspond to the extent possible to economically optimal solutions.

In the following, the preliminary main naval architectural plans are developed, namely, the ship lines plan, the general arrangement plan and sectional areas, and the lengthwise volume distribution plan, enabling estimation of the available cargo hold volume. The technical part of the preliminary ship design study is completed by the International Load Line Convention (ILLC) Study, leading to the minimum freeboard and allowable maximum ship draft, the analysis and control of stability and trim of the *intact* ship in various main loading conditions (i.e., departure, ship fully loaded at design draft, arrival at port, fuel tanks partly empty, etc.). Finally, the ship's stability and floatability in damage condition are assessed, which is related to the ship's internal watertight compartmentation and weight distribution. Following the most recent international regulations, this damage stability analysis is of a probabilistic nature; namely, it is assumed that the ship suffers a random side collision damage, and the likelihood (probability) that the ship survives this damage is calculated, which needs to be larger than a required survivability level set by the regulation (SOLAS-IMO). After the end of the technical design study, a preliminary calculation of the shipbuilding cost is conducted, along with a critical review and concise presentation of the results (steps 9 and 10).

The preceding study steps are repeated in a *trial-and-error iterative procedure* until, after about the third iteration, this approach to the ship's various characteristics converges to the final values. Step-by-step descriptions of the preliminary design of various types of ships may be found in references 9, 15, 23, 28, 30, 31, and 32.

TABLE 25.7 Typical Form Coefficients and Main Dimensional Ratios of Various Merchant Ships

Ship type	Form coefficients				Dimensional ratios				Notes
	C_P	C_M	C_B	C_{WP}	L/B	B/T	$L_{PP}/\nabla^{1/3}$	L_{PP}/D	
Oceangoing cargo ships (fast)	0.57–0.65	0.97–0.98	0.56–0.64	0.68–0.74	6.5–7.1[1]	2.2–2.6	5.6–5.9	11.0–12.0	[1]$L/B > 7.0$ seldom
Oceangoing cargo ships (slow)	0.66–0.74	0.97–0.995	0.65–0.73	0.80–0.86	6.3–7.2[1]	2.1–2.3	5.2–5.4		
Coasters	0.69–0.73	Up to 0.985	0.58–0.72	0.78–0.83	4.5–5.5	2.5–2.7	4.2–4.8	10.0–12.01 (7.5)[2]	[1]Closed-type tonnage [2]Open-type tonnage
Transatlantic passenger liners (old)	0.56–0.58	0.94–0.97	0.54–0.56	0.67–0.70	8.2–9.0	2.8–3.2	(7.6)[1] 7.0–7.3	10.4–11.8	[1]Former transatlantic liner FRANCE: $L_{PP}/\nabla^{1/3} = 7.6$
Transatlantic cruise ships	0.58–0.635	0.93–0.97	0.56–0.59	0.71–0.76	6.3–7.0	2.8–3.4	6.2–6.6	8.0–10.0	
Small RoPAX ferries	0.61–0.63	0.82–0.85	0.51–0.53	0.65–0.70	5.8–6.5	3.3–3.9	6.3–6.6	10.4–11.6	
Ferries	0.53–0.62	0.91–0.98	0.50–0.60	0.69–0.81	5.9–6.2[1] 5.2–5.4[2]	3.7–4.0	6.2–6.9[1] 5.7–5.9[2]	8.6–10.3	[1]For $L > 100$ m [2]For $L = 80$–95 m
Fishing vessels	0.61–0.63	0.87–0.90	0.53–0.56	0.76–0.79	5.1–6.1	2.3–2.6	5.0–5.4	8.2–9.0	
Tug boats	0.61–0.68	0.75–0.85	0.50–0.58	0.79–0.84	3.8–4.5	2.4–2.6	4.0–4.6	7.7–10.0	
Bulk carriers	0.79–0.84	0.990–0.997	0.78–0.83	0.88–0.92	7.2–7.6[1] 5.9–6.5[2]	2.2–6[1] 2.5–2.7[2]	5.3–5.5[1] 4.9–5.2[2]	11.5[2]–13.5[1]	[1]For limited B [2]For unlimited B
Tankers, $F_n = 0.15$	0.835–0.855	0.992–0.996	0.83–0.85	0.88–0.94	6.8–7.1[1] 6.0–6.5[2]	2.4–2.8	5.3–5.5[1] 5.0–5.2[2]	12.7–14.0[1] 12.0–13.0[2]	[1]For limited T [2]For unlimited T
Tankers, $F_n = 0.16$–0.18	0.79–0.83	0.992–0.996	0.79–0.82	0.88–0.92					Dimensional ratios like bulk carriers
Oceangoing "reefer" ships (fast)	(0.55)[1] 0.59–0.62	0.96–0.985	(0.53)[1] 0.57–0.59	0.68–0.72	6.7–7.2	2.8–3.0	6.1–6.5	–11.0	[1]Seldom: $C_P\, C_B < 0.57$

TABLE 25.8 Order of Estimation of Main Dimensions and Form Coefficients for deadweight carriers

Quantity	Basis for calculation
1. Length L	Slenderness ratio:
	$L/\nabla^{1/3}$, where ∇ = displacement volume
2. Block coefficient C_B	Length L, dimensionless Froude number $F_n = V/\sqrt{gL}$,
	where V = given speed, g = gravitational acceleration
3. Beam B	Ratios L/B, B/T
4. Draft T	Ratios B/T, L/T
5. Side depth D	Required cargo hold volume, ratio L/D
6. Remaining hull form coefficients: midship section coefficient C_M, prismatic coefficient C_P, and waterplane area coefficient C_{WP}	C_B or through Froude number F_n

Source: From ref. 28.

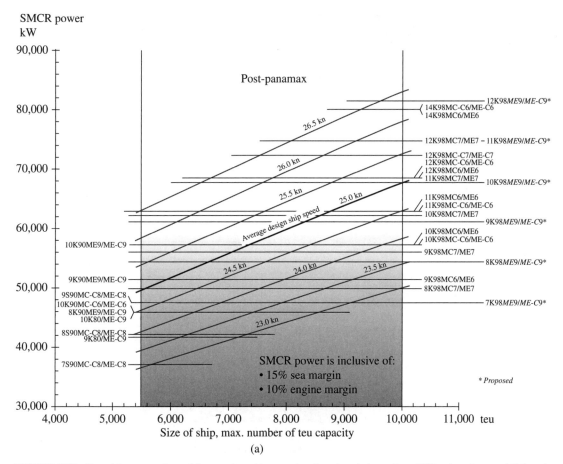

FIGURE 25.32 Propulsion power demand for container ships as a function of carried containers TEU and service speed V (knots), according to MAN Diesel SE (www.manbw.com).

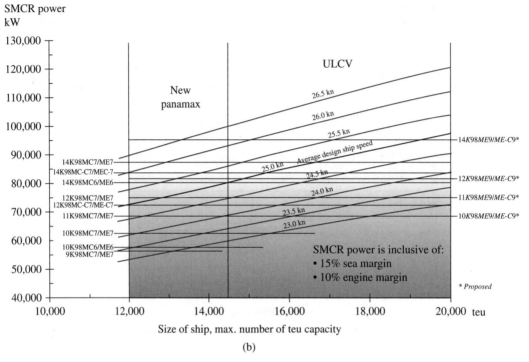

FIGURE 25.32 (*Continued*)

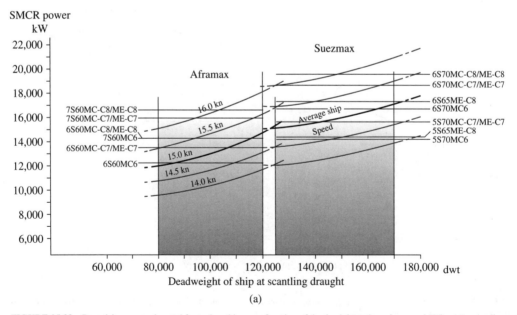

FIGURE 25.33 Propulsion power demand for tanker ships as a function of deadweight and service speed *V* (knots) according to MAN Diesel SE (www.manbw.com).

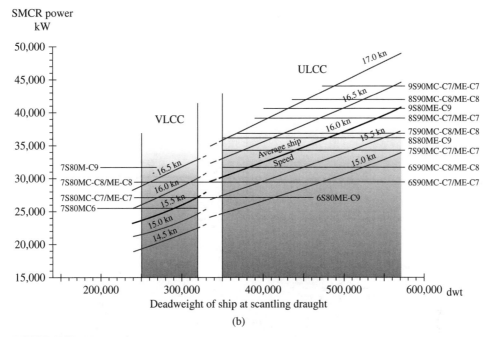

FIGURE 25.33 (*Continued*)

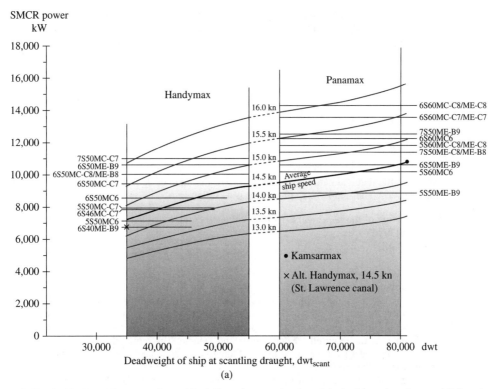

FIGURE 25.34 Propulsion power demand for bulk carriers as a function of deadweight and service speed *V* (knots), according to MAN Diesel SE (www.manbw.com).

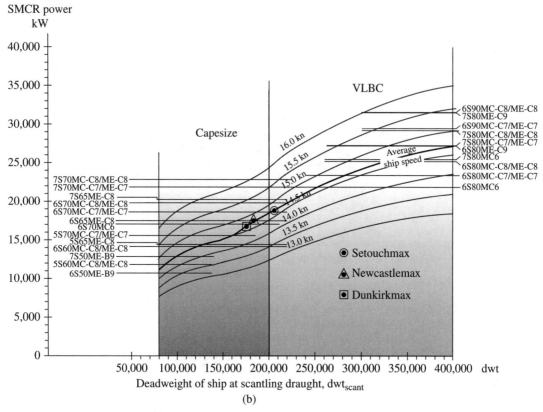

FIGURE 25.34   (*Continued*).

25.6.2   Volume Carriers

Compared with deadweight carriers, the procedure for the volume carriers commences with an estimation of the required cargo hold volume below the main deck on the basis of the required overall hold capacity. The step-by-step procedure is as follows:

1. Estimation of the *required cargo hold volume below the main deck* on the basis of the specified overall hold capacity
2. Estimation of the main dimensions and form coefficients in the sequence order outlined in Table 25.9

The following procedure is the same with that for deadweight carriers. Some special provisions need to be taken for the design of ro-ro passenger ships and ro-ro ships in general: The required volumes for the accommodation of passengers, crew, and public spaces; of machinery room; and of cargo hold spaces (for ro-ro cargo; space for carried vehicles) are estimated by use of empirical coefficients for the required area/passenger and vehicle. The required vehicles' areas and volumes in general are specified in terms of the total length of vehicle lanes, considering the standard size of private cars and trucks/lorries.[23]

The allocation of space above and below the main deck of passenger ships is determined by aspects of intact stability (as to the vertical position of the ship's center of mass), the beam-to-deck height

TABLE 25.9 Order of Estimation of Main Dimensions and Form Coefficients for Volume Carriers

Quantity	Basis for calculation
1. Length L	Hold capacity ∇_{BALE} (dry-cargo ships)
	Refrigerated cargo capacity ∇_{NET} ("reefers")
	Number of containers n_{TEU} (containerships)
2. Block coefficient C_B	L, F_n
3. Beam B	Ratio L/B or on the same basis as for the calculation of L
4. Depth D	Ratio B/D or on the same basis as for the calculation of L, or by use of the empirical coefficient (hold capacity/$L \cdot B \cdot D$)
5. Light ship weight W_L	Empirical coefficient $W_L/L \cdot B \cdot D$ from tables, diagrams, or similar ships, correcting for difference in C_B
6. Deadweight DWT	Considering weight of cargo/payload, fuel, freshwater, effects, etc.
7. Displacement Δ	$W_L + $ DWT
8. Draft T	Δ, L, B, C_B
9. Remaining form coefficients C_M, C_P, and C_{WP}	C_B, F_n (See Table 25.8, step 6)

Hold capacity BALE = required volume for bale cargo; Hold capacity NET = required net volume for refrigerated cargo; number of standard containers TEU ($8 \times 8 \times 20$ feet) below deck (considering, however, also the number of above-deck containers).
Source: From ref. 28.

ratio B/D, and the extent/height of superstructures, and these greatly affect ship's stability. Finally, as to the floatability and stability of passenger ships after damage (loss of watertight integrity), care should be taken at the preliminary design stage to ensure survivability according to safety rules (SOLAS) by internal watertight subdivision in transverse, longitudinal, and vertical direction (transverse and longitudinal bulkheads, and horizontal decks).

A more generic preliminary design procedure that pertains mainly to volume carriers is outlined in Figure 25.35.

25.6.3 Linear Dimension Ships

With at least one main dimension being specified, in terms of maximum values, for example, the beam B through limits of canals or by the dimensions of box-type cargo, the following preliminary design procedure of linear dimension ships does not differ from the ones outlined before for deadweight and volume carriers. Care should be taken, however, when using comparative data for similar ships because of the noncontinuous change of main dimensions and their impact in the other ship characteristics.

25.6.4 Special-Purpose Ships and Advanced Marine Vehicles

The individual character of these ships does not permit generalized design methods. However, we may note that if an initial estimation of the ship's displacement is possible (e.g., for tug boats, through empirical relationships to towing power; for ice breakers, likewise, through empirical relationships to installed horsepower) or of required hold volume (e.g., for fishing vessels, through the refrigerated fish hold volume), then the procedure will be like those given for deadweight and volume carriers.

Comprehensive data for the design of a variety of special-purpose ships can be found in the book *Ship Design and Construction* (T. Lamb ed., SNAME, 2003).

Finally, the design of advanced marine vehicles, which are ships of individual character and of generally very high speed, of light weight structure, and high powering, is greatly determined by hydrodynamic design aspects and requires first-principles approaches to ship design. In reference 34, some computerized design optimization procedures for high-speed ships are presented.

Assumptions	Steps in process	Sources

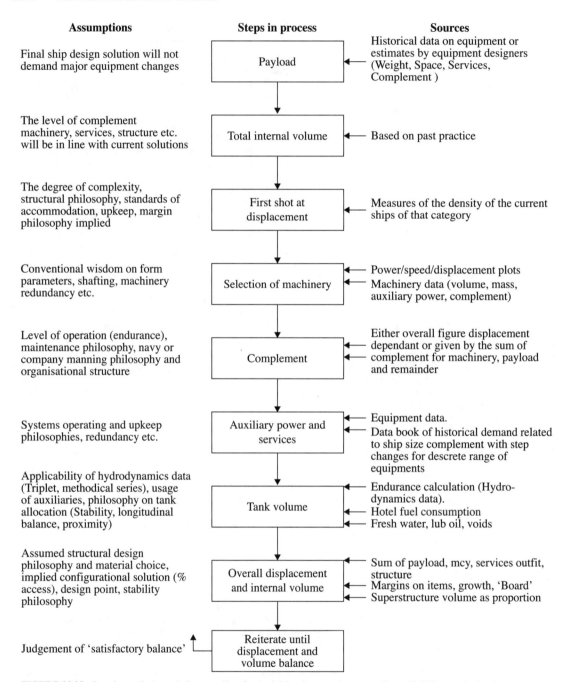

FIGURE 25.35   Iterative preliminary design procedure for (mainly) volume carriers, according to IMDC (www.imdc.cc).

REFERENCES

1. Akagi, S., "Synthetic Aspects of Transport Economy and Transport Vehicle Performance with Reference to High Speed Marine Vehicles," *Proc. FAST 91* (1991).

2. Akagi, S., and Morishita, M., "Transport Economy–Based Evaluation and Assessment of the Use of Fast Ships in Passenger–Car Ferry and Freighter Systems," *Proc. FAST 2001* (2001).

3. Andrews, D. (coordinator), Papanikolaou, A, Erichsen, S., and Vasudevan, S., "IMDC2009 State of the Art Report on Design Methodology," *Proc. 10th International Marine Design Conference*, IMDC09, Trondheim, May 2009.

4. Boulougouris, E., and Papanikolaou, A., "Energy Efficiency Parametric Design Tool in the Frame of Holistic Ship Design Optimization," *Proc. 10th International Marine Design Conference*, IMDC09, Trondheim, May 2009.

5. Brett, P. O., Boulougouris, E., Horgen, R., et al., "A Methodology for Logistics-Based Ship Design," *Proc. Ninth International Marine Design Conference (IMDC)*, ed. M. G. Parsons. Ann Arbor, MI: Department of Naval Architecture and Marine Engineering, University of Michigan, May 2006, pp. 123–146.

6. Buhaug, Ø., Corbett, J. J., Endresen, Ø., et al., *Updated Study on Greenhouse Gas Emissions from Ships: Phase I Report*. London: International Maritime Organization (IMO), September 1, 2008.

7. Buxton, I. L., *Engineering Economics and Ship Design*, 2d. BSRA Report, 1976.

8. Eliopoulou, E., and Papanikolaou, A., "Casualty Analysis of Large Tankers," *Journal of Marine Science and Technology* 12, 240–250, 2007.

9. Friis, A. M., Andersen, P., and Jensen, J. J., *Ship Design* (Parts I and II). Section of Maritime Engineering, Department of Mechanical Engineering, Technical University of Denmark, 2002.

10. International Marine Design Conference (IMDC, www.imdc.cc), *Proceedings of Series of Conferences,* 1982–2009.

11. IPCC, "Summary for Policymakers." In: *Climate Change 2007: The Physical Science Basis. Contribution of Working Group I to the Fourth Assessment Report of the Intergovernmental Panel on Climate Change,* Solomon, S., Qin, D., Manning, M., et al., eds. Cambridge, UK: Cambridge University Press, 2007.

12. *ITTC Symbols and Terminology List*, Version 2008, International Towing Tank Conference, http://ittc.sname.org.

13. Kennel, C., "Design Trends in High-Speed Transport," *Journal of Marine Technology* 35(3), 1998.

14. JANE's Information Group's Catalogues, Sea Transport, http://catalog.janes.com.

15. Lamb, T., ed., *Ship Design and Construction*, Vols. I and II. Jersey City, NJ: SNAME, 2003–2004.

16. Levander, K., "Innovative Ship Design: Can Innovative Ships be Designed in a Methodological Way?" *Proc. of 8th International Marine Design Conference.* IMDC03, Athens, Greece, May 2003.

17. Lewis, E. V., ed., *Principles of Naval Architecture*, Vols. I to III. Jersey City, NJ: SNAME, 1988.

18. MARPOL 73/78: Marine Environment Protection Committee, Resolution MEPC.117(52), Amendments to the Annex of the Protocol of 1978 Relating to the International Convention for the Prevention of Pollution from Ships, 1973. International Maritime Organization. Adopted on October 15, 2004.

19. Nowacki, H., and Ferreiro, L. D., "Historical Roots of the Theory of Hydrostatic Stability of Ships," *Proc. of 8th International Conference on the Stability of Ships and Ocean Vehicles*, STAB2003, Madrid, Spain, 2003.

20. Nowacki, H., "Developments of Marine Design Methodology: Roots, Results and Future Trends," *Proc. of 10th International Marine Design Conference,* IMDC09, Trondheim, May 2009.

21. Papanikolaou, A., and Kariambas, E., "Optimization of the Preliminary Design and Cost Evaluation of Fishing Vessels," *Journal Ship Technology Research, Schiffstechnik,* 41, 1994.

22. Papanikolaou, A., and Boulougouris, E., "Design Aspects of Survivability of Surface Naval and Merchant Ships." In: *Contemporary Ideas on Ship Stability.* New York: Elsevier, September 1999.

23. Papanikolaou, A., "Design and Safety of Ro-Ro Ferries" (in German). In: *Handbuch der Werften*, Vol. XXVI. Hamburg, Germany: Schiffahrts-Verlag "HANSA" C. Schroedter & Co., 2002.

24. Papanikolaou, A., "Developments and Potential of Advanced Marine Vehicles Concepts," *Bulletin of the KANSAI Society of Naval Architects* 55, 50–54, 2002.

25. Papanikolaou, A., "Review of Advanced Marine Vehicles Concepts," *Proc. of 7th International High Speed Marine Vehicles Conference* (HSMV05), Naples, Italy, September 2005.

26. Papanikolaou A., "Holistic Ship Design Optimization," *Journal of Computer-Aided Design,* 2009.

27. Papanikolaou, A. (coordinator), Andersen, P., Kristensen, H.-O., et al., "State of the Art on Design for X," *Proc. of 10th International Marine Design Conference,* IMDC09, Trondheim, May 2009.

28. Papanikolaou, A., *Ship Design: Methodologies of Preliminary Design,* Vols. 1 and 2 (in Greek). Athens: SYMEON 2009.

29. Papanikolaou, A., ed., *Risk-Based Ship Design: Methods, Tools and Applications.* Berlin: Springer, 2009.

30. Schneekluth, H., and Bertram, V., *Ship Design for Efficiency and Economy,* 2d ed. Boston: Butterworth-Heinemann, 1998.

31. Schneekluth, H., *Ship Design* (in German). Herford: Koehler Verl., 1985.

32. Taggart, R., ed., *Ship Design and Construction.* New York: SNAME, 1980.

33. Zaraphonitis, G, Boulougouris, E., and Papanikolaou, A., "An Integrated Optimisation Procedure for the Design of Ro-Ro Passenger Ships of Enhanced Safety and Efficiency," *Proc. of 8th International Marine Design Conference,* IMDC03, Athens, May 2003.

34. Zaraphonitis, G., Skoupas, S., and Papanikolaou, A., "Parametric Design and Optimization of High-Speed, Twin-Hull Ro-Ro Passenger Vessels," *Proc. of 10th International Marine Design Conference*, IMDC09, Trondheim, May 2009.

APPENDIX 25.A

REGRESSION ANALYSIS OF MAIN TECHNICAL SHIP DATA

BULK CARRIERS[28]

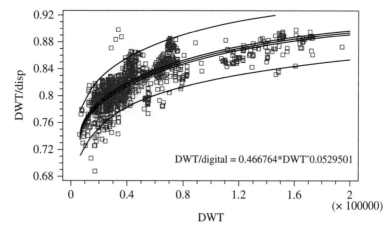

FIGURE 25A.1   Regression analysis of (DWT/Δ) versus DWT for bulk carriers.

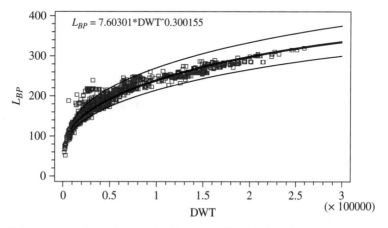

FIGURE 25A.2   Regression analysis of L_{BP} versus DWT for bulk carriers.

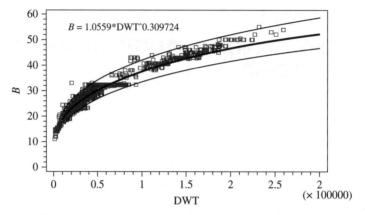

FIGURE 25A.3   Regression analysis of *B* versus DWT for bulk carriers.

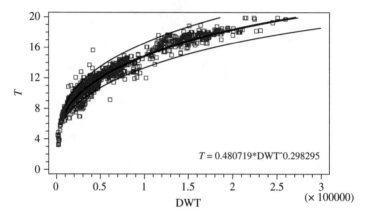

FIGURE 25A.4   Regression analysis of *T* versus DWT for bulk carriers.

TANKERS[28]

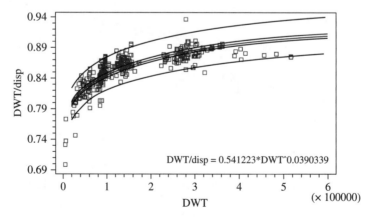

FIGURE 25A.5   Regression analysis of (DWT/Δ) versus DWT for tankers.

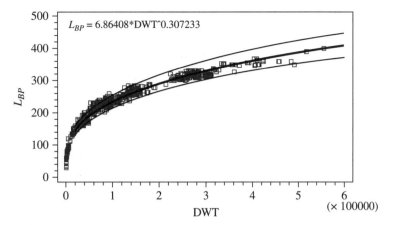

FIGURE 25A.6   Regression analysis of L_{BP} versus DWT for tankers.

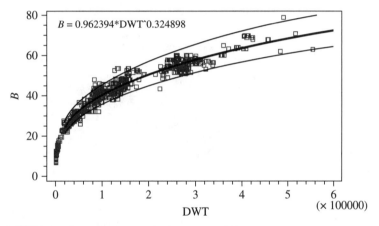

FIGURE 25A.7   Regression analysis of B versus DWT for tankers.

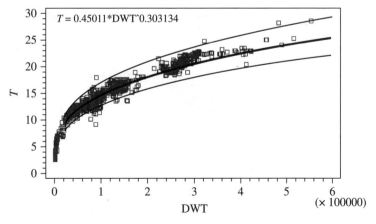

FIGURE 25A.8   Regression analysis of T versus DWT for tankers.

CONTAINER SHIPS[28]

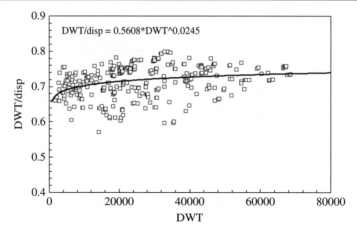

FIGURE 25A.9 Regression analysis of (DWT/Δ) versus DWT for container ships.

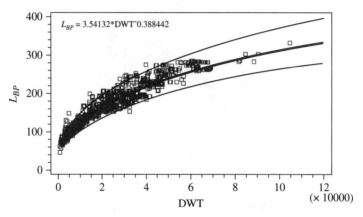

FIGURE 25A.10 Regression Analysis of L_{BP} versus DWT for container ships.

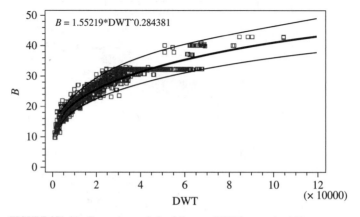

FIGURE 25A.11 Regression analysis of B versus DWT for container ships.

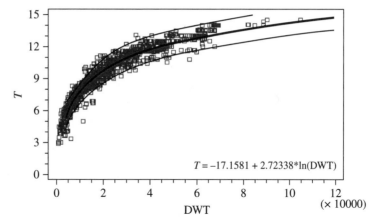

FIGURE 25A.12 Regression analysis of T versus DWT for container ships.

REEFERS[28]

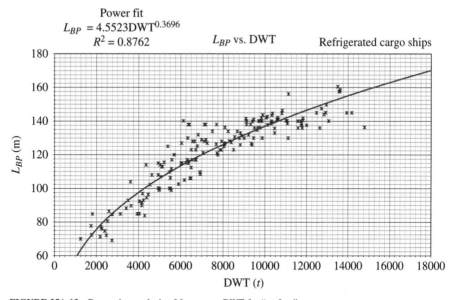

FIGURE 25A.13 Regression analysis of L_{BP} versus DWT for "reefers."

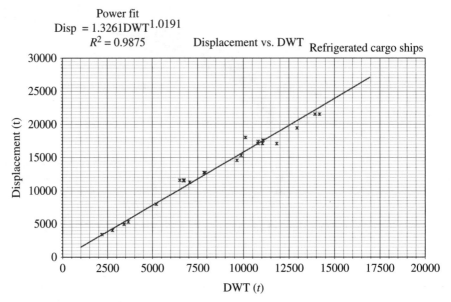

FIGURE 25A.14   Regression analysis of Δ versus DWT for "reefers."

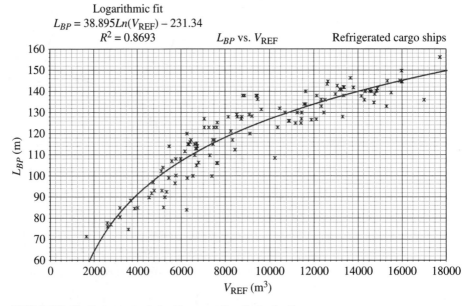

FIGURE 25A.15   Regression analysis of L_{BP} versus V_{REF} for "reefers."

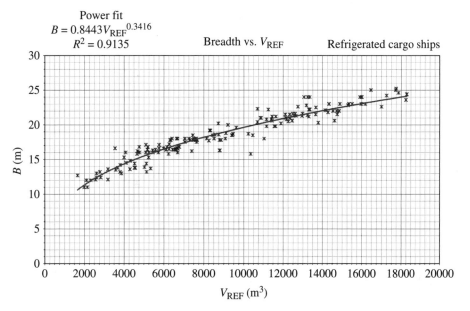

FIGURE 25A.16 Regression analysis of B versus V_{REF} for "reefers."

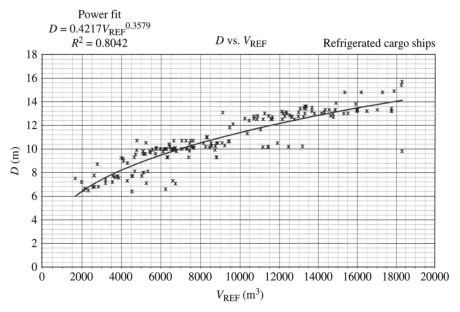

FIGURE 25A.17 Regression analysis of D versus V_{REF} for "reefers."

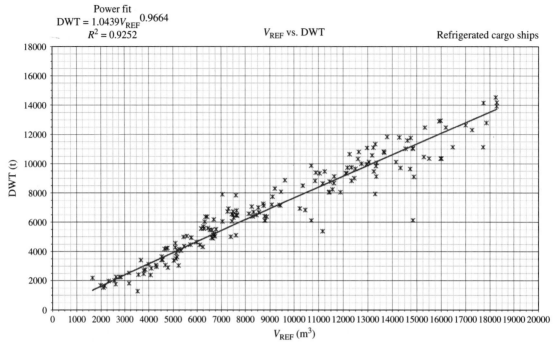

FIGURE 25A.18   Regression analysis of DWT versus V_{REF} for "reefers."

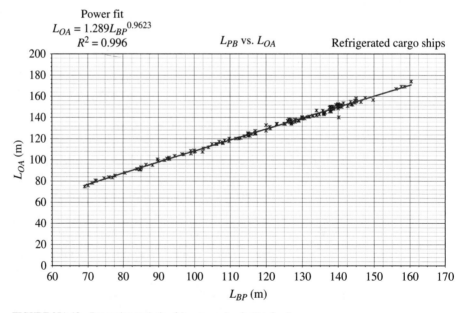

FIGURE 25A.19   Regression analysis of L_{OA} versus L_{BP} for "reefers."

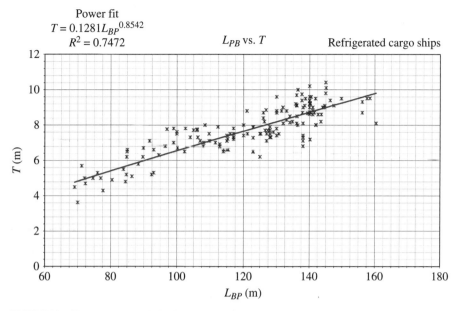

FIGURE 25A.20 Regression analysis of T versus L_{BP} for "reefers."

RO-RO PASSENGER SHIPS[23]

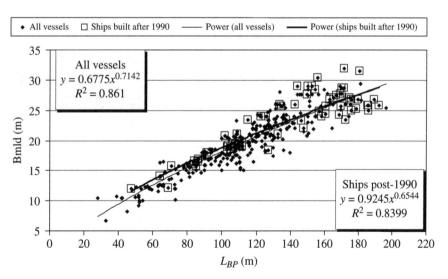

FIGURE 25A.21 Regression analysis of B versus L_{BP} for ro-ro passenger ships.

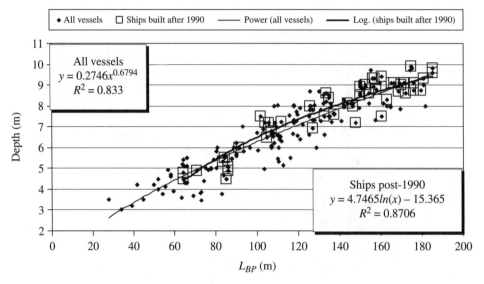

FIGURE 25A.22 Regression analysis of D versus L_{BP} for ro-ro passenger ships.

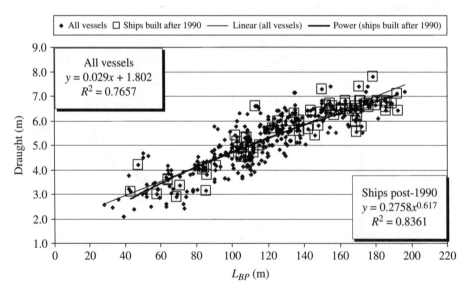

FIGURE 25A.23 Regression analysis of T versus L_{BP} for ro-ro passenger ships.

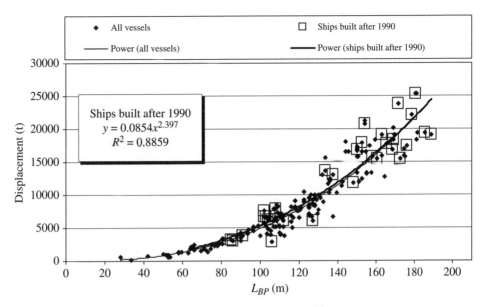

FIGURE 25A.24 Regression analysis of Δ versus L_{BP} for ro-ro passenger ships.

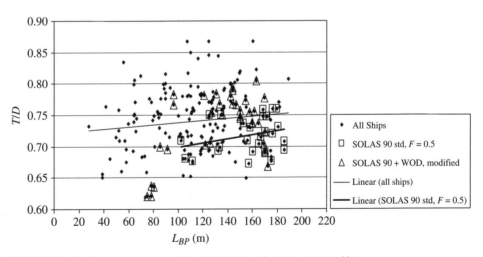

FIGURE 25A.25 Regression analysis of T/D ratio versus L_{BP} for ro-ro passenger ships.

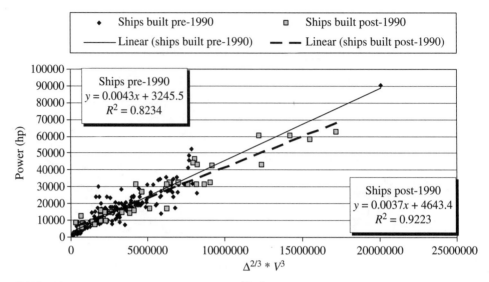

FIGURE 25A.26 Regression analysis of power versus $\Delta^{2/3}*V^3$ for ro-ro passenger ships.

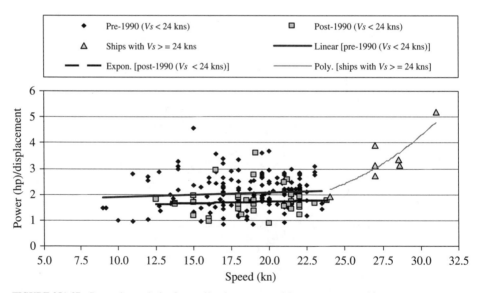

FIGURE 25A.27 Regression analysis of power/Δ ratio versus speed for ro-ro passenger ships.

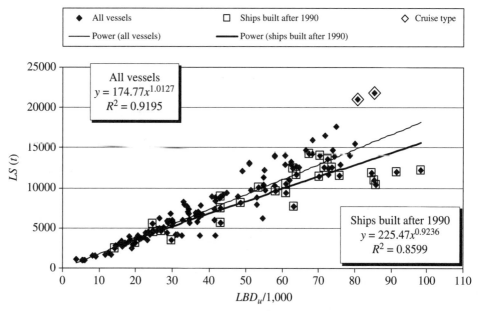

FIGURE 25A.28 Regression analysis of *LS* versus $LBD_u/1{,}000$ for ro-ro passenger ships.

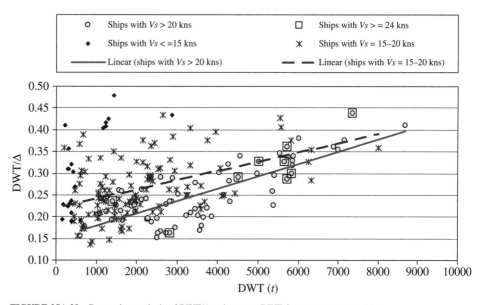

FIGURE 25A.29 Regression analysis of DWT/Δ ratio versus DWT for ro-ro passenger ships.

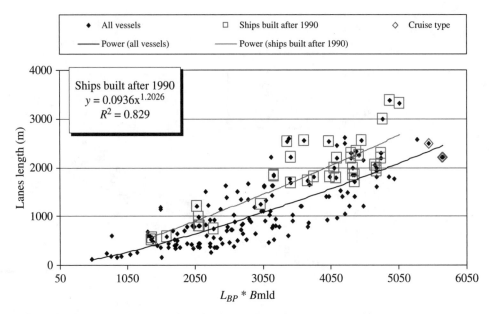

FIGURE 25A.30 Regression analysis of lane length versus $L_{BP}*B$ for ro-ro passenger ships.

CHAPTER 26
OIL AND GAS PIPELINE ENGINEERING

Thomas Miesner
Pipeline Knowledge & Development
Katy, Texas

David Vanderpool
Vanderpool Pipeline Engineers Inc.
Littleton, Colorado

26.1   INTRODUCTION

Using hollow tubes to direct the movement of fluids has been around for many years. Some say the first use dates back to ancient Chinese salt works. Legend has it that hollowed-out bamboos logs were sealed together with mud and used to transport natural gas, produced along with salty brine, short distances, where it fueled fires to evaporate water from the brine, leaving salt. Others cite the bath works of ancient Rome. Regardless of when pipelines were first employed, they came into extensive use during the 20th century.

Pipelines are critical to modern society. Millions of miles crisscross cities, towns, states, countries, and even continents and carry essential products for modern lifestyles: water, wastewater, liquid fuels, gaseous fuels, and chemical feedstocks. Regardless of nationality, location, or fluids moved, all these pipelines obey the same laws of physics.

But the practices and precautions involved when transporting natural gas, crude oil, refined products, natural gas liquids, chemical feed stocks, CO_2, and other hazardous and flammable fluids are quite different from those involved with moving clean water, liquid wastes, slurries, and many other fluids. This chapter leaves behind other pipelines to focus on oil and gas transmission (long-distance) pipelines and specifically onshore rather than offshore pipelines. Before moving on to the details of engineering, however, a brief introduction to oil and gas pipelines is in order.

26.2   OIL AND GAS PIPELINE FUNCTIONS AND CLASSIFICATIONS

Oil and gas pipelines are categorized according to the fluids they transport, the pressures involved, where they are located (onshore or offshore), and their function. Oil pipelines primarily move crude oil and refined products (gasoline, diesel, jet, and other fuels and chemical feedstocks). Gas pipelines move natural gas, composed of methane (CH_4) with smaller concentrations of ethane (C_2H_6), propane (C_3H_8), and perhaps other light hydrocarbons.

Functionally, lines consist of the following:

- *Gathering.* Generally low pressure (<500 psi), used to aggregate oil or gas production from wells and producing facilities and deliver it to crude oil gathering stations and terminals, gas plants, or sometimes directly into transmission lines
- *Transmission* (also called *trunk lines* or *main line*). Generally high pressure (up to 2,500 psi), receives fluids from gathering lines, storage facilities, crude oil and refined-products terminals, other transmission lines, refineries, and gas plants and delivers them to storage facilities, crude oil or refined-products terminals, other transmission lines, or natural gas city gates
- *Distribution.* Generally low pressure (<500 psi), receives fluids from transmission lines, storage facilities, refined-products terminals, and city gates and delivers them to final destinations, including individual homes and businesses

Crude oil and gas gathering lines, natural gas transmission, crude oil and refined main or trunk lines, and natural gas distribution lines are each shown in Figure 26.1.

Some gathering lines operate in two-phase flow, moving both liquids and gases. All refined-products main or trunk lines are designed to operate primarily in the liquid phase. Although both the Trans Alaska pipeline and the Baku—Tbilisi—Ceyhan pipeline have a portion designed for 2-phase flow. Different types of crude oil sometimes are moved in batches on the same pipeline, as are grades of refined products. For example, 87, 89, and 92 octane gasoline are moved on a pipeline in batches, and moved end to end on the same pipeline along with diesel and jet fuel (Figure 26.2). Physics—turbulent flow—keeps the batches from mixing with those at their beginning and end.

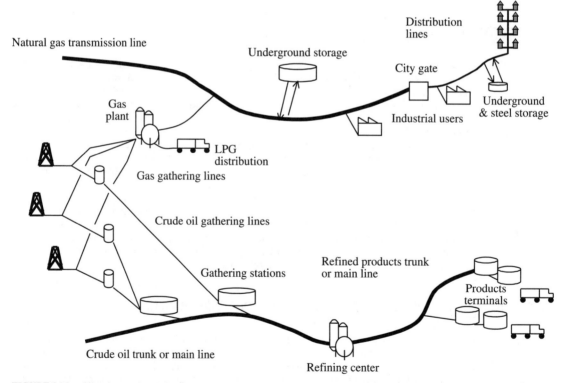

FIGURE 26.1. Oil and gas value chain. (*Source: Introduction to the Oil and Gas Pipeline Business for Executives.* Courtesy of Pipeline Knowledge & Development.)

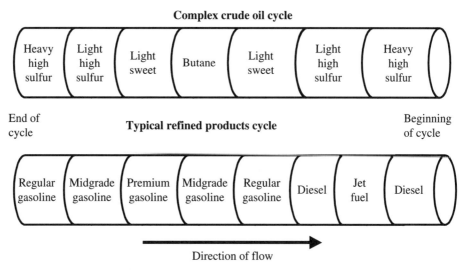

FIGURE 26.2 Typical refined-products and crude oil pipeline batching sequences. (*Source:* Miesner and Leffler, *Oil and Gas Pipelines in Nontechnical Language*, Pennwell Corp., 2006.)

Natural gas transmission and distribution lines, on the other hand, move natural gas as a relatively homogeneous commodity. Batching for oil lines and compressibility for gas lines introduce some significant difference in how the two types of pipelines are designed and operated.

26.3 HYDRAULICS: THE PHYSICS OF FLUIDS

The great theoretical physicists of the 17th and 18th centuries, Robert Boyle, Isaac Newton, Daniel Bernoulli, and others, set the stage for the likes of Henry Darcy, Julius Weisbach, James Joule, and Osborne Reynolds, who furthered the study of fluid mechanics during the 19th century. The concepts and empirical formulas developed in the 19th century still form the basis of today's hydraulics. Pipeline engineers use most of the same USC and SI measures as other engineers (feet, meters, miles, kilometers, seconds, hours, °F, °C), but two—pressure and volume—merit pipeline-specific amplification.

26.3.1 Pressure Measures and Units

Pressure, the key measure of energy in pipeline operations, is expressed in pounds-force per square inch (psi) in the USC system, and pascals (Pa) and kilopascals (kPa) in the SI system. Pipeline engineers working with SI units also frequently use the bar as a unit of pressure, which is equal to 100 kPa and also almost equal to 1 atmosphere. As with all pressure measurement, differentiating between gauge and absolute is important.

Oil pipeline engineers use still another unit of measurement—*pressure head*, also known as *static pressure head*, or *static head*. Expressed in feet or meters of a specified liquid, it represents the height to which a column (piezometric head) of that particular fluid would rise if subject to the applicable pressure. (The challenge of attempting to express force in terms of a column of compressible fluid such as gas with its attendance changes in density means that gas engineers seldom use static head.)

Pressure and static head are forms of measurement of force. Therefore, *pressure* and *head* are synonymous with *force* and are simply different units of measurement. One cubic foot of water weighs approximately 62.43 pounds at 5°C. It exerts this weight on 144 square inches, so 1 foot of head of water at 5°C equals a little more than 0.433 psi. Liquid density varies by temperature and pressure, so both must be understood when using head. Because pipelines traverses great distances over variable ground elevation, the energy of position, or elevation, is critical in the evaluation of energy requirements; thus expressing pressure energy in units of elevation or static head enables the engineer to evaluate energy requirements at every point along the pipeline. More on this important concept will be discussed later.

26.3.2 Volume Measures and Units

One barrel (abbreviated bbl), equal to 42 U.S. gallons, is the USC standard oil measure, as is the cubic meter (m^3) in the SI system. The tonne (t), although normally a commercial measure rather than an operating measure, is not an official SI term but is sometimes used to express volume. It is, however, actually a measure of mass, so it must be converted to volume based on the density of the fluid. One tonne of gasoline occupies more volume than 1 tonne of diesel fuel because gasoline is less dense than diesel.

Natural gas volume and mass are expressed in standard cubic feet (SCF) and either standard or normal cubic meters (m^3 or cm) as volume measurements, but introduce the British thermal unit (Btu), therm (equal to 100,000 Btu), and the dekatherm (equal to 1,000,000 Btu) when it comes to commercial transactions in some countries. All volume measurements must be at an agreed-on or specified temperature and pressure owing to the compressibility of these valuable products.

26.3.3 Fluid Properties

Petroleum products and crude oil mixtures of complex chemical molecules, and the characteristics and interactions of these molecules and their thermodynamic properties can involve substantial study. This discussion will focus on the few selected properties of particular importance to pipeline transportation.

Density. Specific gravity, with water as the oil reference and air as the gas reference, is the normal density expression for oil and gas pipeline engineering and business transactions. But another density measurement, °API is an important concept to understand because of its historical significance. Degree API is not a different measurement, only a different expression of density without any true physical significance. Conversions between specific gravity and API gravity are based on the following equation:

$$\text{Specific gravity at } 60°F = 141.5/(API \text{ gravity} + 131.5)$$

The API gravity of water is 10. Some equations are designed to use API gravity and others specific gravity. Commercial transactions specify which gravity measure is being employed.

Viscosity. Overcoming the intermolecular forces between the molecules moving through a pipeline and the frictional force between the molecules and the pipe wall requires energy. This energy, measured as pressure, is added to the pipeline system by compressors, pumps, and the force of gravity. From a technical perspective, *viscosity* is defined as the shearing stress necessary to induce a unit velocity shear gradient in a substance.[*] From a practical standpoint, viscosity is the internal resistance to flow. Higher-viscosity fluids require more energy to move than do lower-viscosity fluids, all other things being equal. Fluid viscosity, then, is one of the key determinants of pipeline energy consumption and a key variable in pipeline flow equations. Liquid viscosity is highly sensitive to temperature in crude oils.

Most pipeline equations use kinematic viscosity, which is commonly expressed in terms of centistokes (cSt) (1 cSt = 1 milllimeter^2/second), pascal-seconds (Pa-s), or directly as milllimeter^2/second (mm^2/s). Absolute viscosity is measured with one of the several types of viscometers, which measure the amount of time required for a given amount of fluid to flow through the device.

---

[*]Lester, C. B., *Hydraulics for Pipeliners*, Vol. 1, 2d ed., Houston: Gulf Publishing Company, p. 54.

Kinematic viscosity is the expression of absolute viscosity relative to the density of the fluid and aids the convenience of hydraulic calculations.

$$\text{Kinematic viscosity} = \text{Absolute viscosity/density}$$

For oils, viscosity generally is inversely related to temperature, increasing as temperature decreases, whereas gas viscosity increases with increasing temperatures. Oil viscosity generally is independent of pressure, within the pressure ranges of pipeline operations, but gas viscosity increases with increasing pressure which is important within the pressure ranges of gas pipeline operations. Viscosity is an important variable, and pipeline engineers must understand the temperature and pressure ranges of the pipeline system under design and the viscosities of the fluid within the expected operating ranges.

Compressibility.   Ideal gases, by definition, follow the ideal gas law. Essentially, there is a predictable relationship between pressure, volume, temperature, molecular weight, and a universal gas constant, applicable generally at pressures near atmospheric pressure. Elements of the ideal gas law are given by

- Boyle's law $(V_1/V_2 = P_2/P_1)$—volume and pressure are inversely related.
- Charles' law $(V_1/V_2 = T_2/T_2)$—volume and temperature are inversely related.
- Gay-Lussac's law $(P_1/P_2 = T_1/T_2)$—pressure and temperature are directly related.

For these laws, the factors are

V = absolute volume

P = absolute pressure

T = absolute temperature expressed in degrees Rankin (°R) or degrees Kelvin (K)

The factor not included in the equation must be kept constant.
   From the three previous laws arose the ideal gas law:

$$PV = nRT$$

where, in addition to the factors defined previously, n is the amount of gas measured in moles and R is the universal gas constant (10.73 psia ft^3/lb mol °R or 8.314472 J/mol K). Since the ideal gas law only predicts gas behavior fairly accurately at low pressures, a compressibility factor Z was developed, resulting in the real gas law:

$$PV = ZnRT$$

The compressibility factor Z depends on the temperature and pressure of the gas. Over the years, graphs were developed allowing engineers to quickly choose the appropriate Z factor. Now, Z factors are built into commercially available gas hydraulic packages.
   For pipeline capacity and friction-loss calculations, the compressibility of petroleum liquids can be ignored for simplicity. There are two important exceptions, however, where the compressibility of the liquid must be considered:

- Surge calculations
- Leak-detection modeling and evaluation

These two topics are beyond the scope of this text but nevertheless are important and critical in an analysis of sudden pipeline shutdown or sophisticated leak-detection models.

26.4 HYDRAULICS

The need to safely and efficiently transport natural gas and oil from producing areas to population centers created the need to turn the scholarly topic of fluid mechanics into the practical one of hydraulics.

26.4.1 The First Law of Thermodynamics

The first law of thermodynamics asserts that energy is neither created nor destroyed. It simply changes form. Compressors and pumps convert mechanical, chemical, and electrical energy into pressure energy applied to pipeline fluids to move them from one location to another. Friction converts part of that energy to heat energy as the fluids move through the pipe. When fluids run from the top of a hill to the bottom, the force of gravity helps to pull them down. Keeping track of these energy flows allows pipeline engineers to design and operate pipelines.

26.4.2 Bernoulli's Principle

In fluid dynamics, Bernoulli's principle states that for an inviscid flow, an increase in the speed of the fluid occurs simultaneously with a decrease in pressure or a decrease in the fluid's potential energy. Bernoulli's principle is named after the Dutch-Swiss mathematician Daniel Bernoulli, who published this principle in his book *Hydrodynamica* in 1738. For application by a pipeline engineer, Bernoulli's principle is reduced to an understanding of the pressure difference between two points and is explained by the difference in elevation head and friction losses of the moving fluid. Simply,

$$P_1 - P_2 = H_2 - H_1 + \text{losses}$$

Figure 26.3 demonstrates Bernoulli's principle. The lower velocity of flow in sections A and C result in lower dynamic pressure than in section B, so the static pressure (as measured by the liquid in the vertical section of pipe) is higher in sections A and C than in section B. Note that the velocities are different but the flow rates in mass per unit time are the same in each section (Figure 26.3).

Based on this principle, Bernoulli developed several equations to explain the behavior and flow of ideal fluids in ideal situations. These equations became the basis of more practical algorithms developed empirically over time.

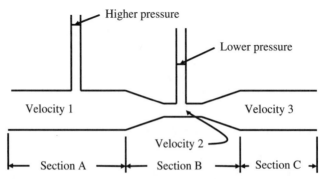

FIGURE 26.3 Demonstration of Benoulli's principle. (*Source:* Oil and Gas Pipeline Fundamentals Course, Courtesy Pipeline Knowledge & Development.)

26.4.3 Flow Characteristics

As flow is initiated from a container into a pipeline, both liquid and gas molecules move in generally straight lines. As flow velocity increases, the randomness of molecular flow increases, a phenomena known as *turbulence*. Flow through pipelines, therefore, is characterized as laminar or turbulent. There is a flow velocity (function of fluid properties) at which the flow is fully turbulent, and higher velocities do not increase turbulence. Therefore, we can say that flow in an oil pipeline is laminar, transitional turbulent, or fully turbulent. *Laminar* flow is so named because the molecules move in layers over each other, with those near the center of the pipe moving faster and those near the pipe wall moving slower (Figure 26.4).

In turbulent flow, the molecules move laterally as well as longitudinally along the pipe (Figure 26.5).

The lateral movements are caused by the molecules impacting the sides of the pipe and each other. Transitional turbulent flow demonstrates some characteristics of both laminar and turbulent flow.

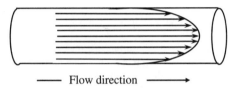

—— Flow direction ——→

FIGURE 26.4 Laminar flow velocity profile. (*Source:* Oil and Gas Pipeline Fundamentals Course. Courtesy Pipeline Knowledge & Development.)

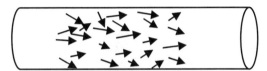

FIGURE 26.5 Turbulent flow velocity profile. (*Source:* Oil and Gas Pipeline Fundamentals course. Courtesy Pipeline Knowledge & Development.)

26.4.4 Reynolds Number

Sir Osborn Reynolds is widely credited with characterizing and classifying the different types of liquid (not gas) flow ("regimes") around 1880. He injected dye into a stream of fluid and found that at lower flow rates, the flow remained streamlined. At higher velocities, the flow went from streamlined to turbulent. In 1882 he published a paper regarding this topic and introduced an empirical concept to characterize flow, now known as the *Reynolds number*.

The Reynolds number is a dimensionless number that depends on viscosity, velocity, diameter, temperature, and pressure and is used to predict whether flow is laminar, turbulent, or transitional turbulent. For flow in a pipeline, the Reynolds number can be calculated as

$$Re = dVD/u$$

where Re = the Reynolds number
d = density
V = mean velocity
D = internal diameter of the pipe
u = dynamic viscosity

Recalling that kinematic viscosity is equal to dynamic viscosity divided by density yields

$$Re = VD/v$$

where v is kinematic viscosity, and the other factors are as defined for the previous equation.

This simplified equation is generally acceptable for oil lines where density and velocity do not change (much) over the length, but it is not appropriate for long-distance gas lines, where viscosity, velocity, and density all change as gas travels along the line, losing pressure to friction and accordingly losing density and gaining velocity. As with all equations, engineers must keep in mind the units in use and which of the empirically derived equations is most appropriate for the particular pipeline. Generally, the higher the Reynolds number, and therefore turbulence, the greater is the frictional energy loss through the pipeline.

26.4.5   Friction Loss

For long-distance pipelines, friction loss is the major factor requiring pump or compressor stations to add energy to the system. Friction loss is dictated by

- Viscosity
- Density
- Velocity (a function of pipe diameter and flow rate)
- Pipe length
- Roughness of the inside of the pipe

Friction loss is normally expressed in terms of pressure loss per unit length, psi per mile, or kpa per kilometer at a given flow rate for a given fluid.

Most long-distance pipelines operate in turbulent flow. Given that turbulent flow cannot be modeled rigorously, however, empirical formulas have been developed (a recurring theme in this chapter) to predict pressure loss owing to friction. A critical component in these equations is the friction factor f. Fortunately, the Reynolds number can be calculated and the friction factor can be related to the Reynolds number through the Moody diagram (Figure 26.6).

Prior to the advent of computers, engineers calculated the Reynolds number, assumed the pipeline roughness, and read the corresponding friction factor from the Moody diagram. These factors are now built in to hydraulic programs or can be programmed easily into a spreadsheet.

The implicit friction factor derived from the Moody diagram is commonly called the *Darcy friction factor*, presumably because it was originally developed for use with the Darcy-Weisbach pressure-loss formula. The Moody diagram is the result of trial and error in the laboratory to derive the implicit friction factor under various flow conditions. The Fanning friction factor, equal to one-quarter the Darcy friction factor, is sometimes used instead. Much experimentation has followed to develop explicit expressions of the friction factor because the Darcy friction factor is an implicit equation requiring an iterative solution. A popular explicit friction factor is the Jain equation.[*] For turbulent flow with Reynolds' numbers ranging from 5,000 to 1×10^8, it will produce an explicit friction factor within about 1 percent of the Moody diagram:

$$\frac{1}{\sqrt{f}} = 1.14 - 2 \log\left(\frac{e}{d} + \frac{21.25}{R^{0.9}}\right)$$

where f = friction factor
$\quad e$ = absolute pipeline roughness
$\quad d$ = pipe inside diameter
$\quad R$ = Reynolds number

26.4.6   Elevation Loss (or Gain)

One more important force works on pipelines—earth's gravitation pull, gravity. Gravity causes pressure loss as fluids flow uphill and gains as it flows downhill. Thus, pipelines lose pressure owing to

---

[*] Swamee, P.K., and A.K. Jain, *Journal of the Hydraulics Division*, ASCE, 1976.

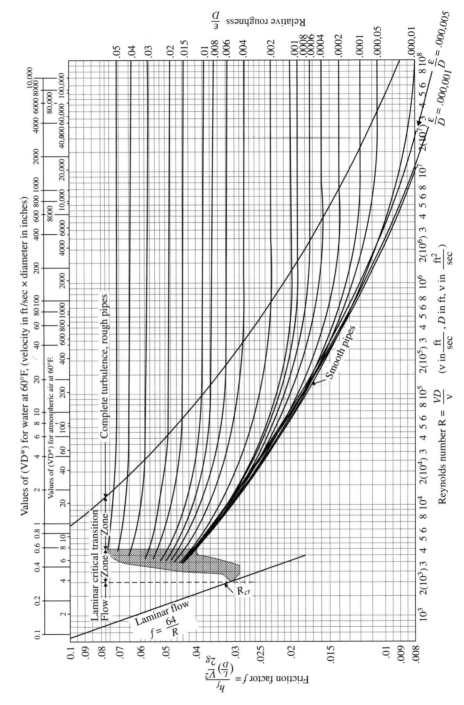

FIGURE 26.6 Moody diagram. (Courtesy ASME.)

friction loss and gain or lose pressure owing to elevation changes. Accordingly, pressure at any point along a segment of pipeline is equal to

$$P = P \text{ at origin} - P \text{ loss owing to friction} \pm P \text{ change owing to elevation}$$

26.4.7 Flow Rates and Capacities[*]

The flow rate of a pipeline is simply the volume that can or does flow past a given point in a set amount of time. The generic flow rate equation is

$$\text{Mass} = \text{density} \times \text{area} \times \text{velocity} \times \text{time}$$

The amount of mass going past a point in a given amount of time is equal to the density of the material times the cross-sectional area times the velocity times the unit of time. There are many different ways flow is measured, but each is a form of this equation. Mass is often converted to volume.

Since liquids are essentially incompressible, over a reasonable time, the flow into an oil line should equal the flow out. If not, there is a problem. Not so for gas. A gas line can be filled with gas at low pressure, and more can be forced into it, up to its maximum operating pressure, with no gas coming out. The amount it takes to "fill up" a gas line is called *line pack*.

The volume of fluid over time through a pipeline is its flow rate, usually a mass flow rate. Of common interest for pipelines is the current flow rate and the maximum flow rate, or capacity. Natural gas lines servicing cities must have sufficient capacity to serve the town on the coldest day of winter and the local gas-fired power plants on the hottest day of summer. They normally flow well below their capacity in the off-season.

Generally, the input into gas lines is more uniform than the output. Gas wells produce into gas lines at a relatively constant rate. At the other end, consumers generally use more gas during the day than at night. Over a 24-hour period, gas pipelines can go from trough to peak demand. Gas storage and line pack are used to help level off this imbalance. Week-to-week and season-to-season demands can vary by more than 50 percent, and storage and compression capacity usually handle these variations.

Oil lines operate at more constant flow rates because oil is relatively easy to store in inventory in tanks at each end and along the way.

26.4.8 Friction-Loss Equations

The Frenchmen Gaspard de Prony and Henry Darcy and the German Julies Weisbach in the late 18th and early 19th centuries generally are credited with developing the equations on which modern pipeline designers base their pressure-loss calculations. Prony developed and Darcy and Weisbach refined the equations that finally became

$$h = f(L/D)(V^2/2g)$$

where h = pressure loss owing to friction
  f = Darcy friction factor
  L = pipe length
  D = pipe internal diameter
  G = acceleration owing to gravity

---

[*]This section was excerpted in its entirety from T. O. Miesner, and W. L. Leffler, *Oil and Gas Pipelines in Nontechnical Language*. Tulsa, PennWell Corporation, used by permission of the authors.

This equation, called the *Darcy-Weisbach formula*, forms the basis for numerous other equations developed by engineers to determine pressure loss in specific applications. The text *Gas Pipeline Hydraulics*, by Shasi Menon, for example, lists 11 different equations used for calculating gas hydraulics. Each equation has the same underlying factors but uses different dimensions and unique friction factors for its particular application. Some of the equations are more reliable for smaller-diameter pipe; others, for larger-diameter pipe. Fluid type makes a difference as well.

For liquid pipelines, the Darcy equation can be reduced to a more convenient form:

$$p = \frac{34.87 \, fQ^2 sg}{d^5}$$

where P = frictional energy loss, psi per mile
  f = friction factor
  Q = volumetric flow rate, barrels per hour
  d = pipe inside diameter
  sg = fluid density, specific gravity

Prior to selecting friction loss or other empirically derived equations, engineers must understand which is appropriate for the particular project.

26.4.9 Hydraulic Tools

Using the appropriate flow equations, pressure loss per mile at various flow rates can be calculated for the design fluids across the range of specified operating conditions. The results of these calculations can be plotted to create two tools.

Pipeline system head curves.   A parabolic curve, this graph shows the pressures required to produce a range of flow rates in a given system for a given fluid at a given temperature (Figure 26.7). Systems curves are particularly useful for sizing compressors and pumps and for understanding how pipeline operating variables react to pressure and fluid property changes.

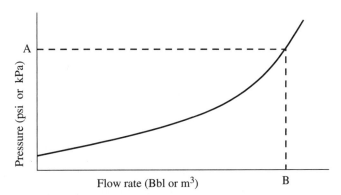

FIGURE 26.7 System curve. Point *A* is the pressure required to produce flow rate *B*. (*Source:* Oil and Gas Pipeline Fundamentals. Courtesy Pipeline Knowledge & Development.)

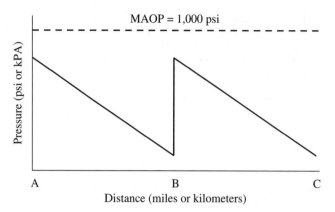

FIGURE 26.8 Oil hydraulic gradient 1. Point *A* is the beginning of the pipeline. Point *B* is an intermediate pump station, and point *C* is the end of the line. The MAOP of the line, in this case 1,000 psi, is included as a check to ensure that pressures do not exceed allowable operating pressure. The slope of the line indicates the pressure loss per mile. Since the line is straight, the pressure loss per mile is uniform, indicating one type of oil with no receipt and delivery points along the line. (*Source:* Hydraulics for Control Room Operators Course, Courtesy Pipeline Knowledge & Development.)

Hydraulic Pressure Gradient. The slope of the pressure-gradient line indicates the pressure loss per mile along the pipeline. Oil pipeline hydraulic gradients are straight lines, indicating uniform pressure loss per mile (Figure 26.8).

If the pressure loss per mile changes owing to fluid or flow-rate changes, the slope of the line changes as well (Figure 26.9).

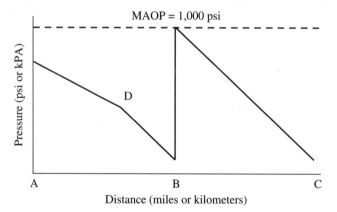

FIGURE 26.9 Oil hydraulic gradient 2. Points *A*, *B*, and *C* are the same as in Figure 26.8. Note that at point *D* the slope changes, indicating increased friction loss per unit. This increased friction loss could have been caused by an increased flow rate at *D* owing to an injection into the line or by a change in the type of oil being pumped. In either event, the higher friction loss means the discharge pressure at pump station *B* must increase to pump the same amount of oil from point *A* as in Figure 6. (*Source:* Hydraulics for Control Room Operators Course. Courtesy Pipeline Knowledge & Development.)

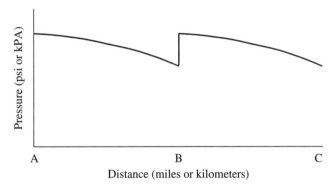

FIGURE 26.10 Natural gas hydraulic gradient. Point *A* is the beginning of the pipeline. Point *B* is an intermediate compressor station, and point *C* is the end of the line. The slope of the line droops, indicating pressure loss per unit increases as the pressure in the pipeline decreases and the gas expands. Since the gas is expanding and, accordingly, less dense, the molecules must travel at higher velocity to maintain the same quantity flow rate. Higher velocity means higher pressure loss per mile. (*Source:* Hydraulics for Control Room Operators Course. Courtesy Pipeline Knowledge & Development.)

Gas pipeline hydraulic gradients are curved downward, indicating increased pressure loss per mile, as the gas decompresses under reduced pipeline pressure (Figure 26.10).

Pipeline Profiles. Showing the elevation of the pipeline along its route and the maximum allowable operating head (MAOH), line profiles are important tools for liquid pipeline (but not gas) design and operations (Figure 26.11).

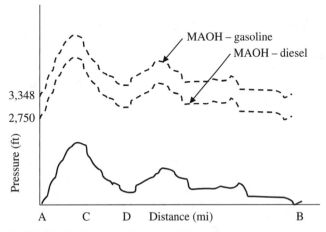

FIGURE 26.11 Pipeline profile with MAOH. Point *A* is the beginning of the pipeline, and point *B* is the end of the line. MAOH at points *A*, *B*, *C*, and *D*, 2,750 ft of 0.84 sg diesel fuel or 0.69 sg gasoline are equal to the MAOP of 1,000 psi. (*Source:* Hydraulics for Control Room Operators. Courtesy Pipeline Knowledge & Development.)

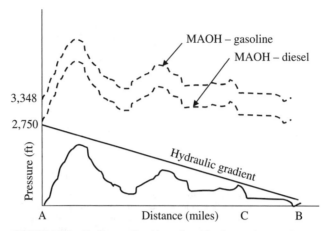

FIGURE 26.12 Pipeline profile with gradient. The distance between the profile and the gradient is the pressure head in the line at that point. The distance between the hydraulic gradient and the MAOH line is how far below MAOH the line is operating at that point. If the line is moving diesel, points *A* and *C* are both control points. If the pressure increases at *A*, the line will exceed MAOH. If the pressure drops at *C*, the pressure will fall to zero. If the line is moving gasoline, *A* is not a control point because the line is well below MAOH. (*Source:* Hydraulics for Oil Control Room Operators Course. Courtesy Pipeline Knowledge & Development.)

Combining pipeline profiles with hydraulic gradients allows liquid pipeline engineers and operators to understand whether the pipeline will operate in the case of design, or is operating in the case of operations below MAOH but above the vapor pressure of the oil inside the line (Figure 26.12).

Hydraulics introduced the concepts and provided an insight into tools. Equipment and components explains the building blocks of pipeline systems, and then engineering and design apply the tools to put the equipment and components together into optimal transportation systems.

26.5 EQUIPMENT AND COMPONENTS

26.5.1 Line Pipe

Metallurgy of the parent steel and the manner in which the steel is made into pipe are the two main ways line pipe is characterized. API 5L, *Specification for Line Pipe*, and ISO 3183, *Petroleum and Natural Gas Industries—Steel Pipe for Pipeline Transportation Systems*, are the two key standards.

Strength of Materials. Specified minimum yield stress (SMYS) is one of the key steel properties because it, along with wall thickness and pipe diameter, directly affects the maximum allowable operating pressure (MAOP) of the pipeline. Table 26.1 shows the SMYS for common grades of line pipe. In addition to strength, other mechanical properties, such as toughness, ductility, and weldability, must be considered in the chemical formulation of the steel, which is defined in API Specification 5L.

Manufacturing Methods. Line pipe is manufactured in the following four ways:

- *Longitudinally welded.* Flat plates of steel called *skelp* are formed into a tube. The longitudinal seam then is joined with filler material deposited by submerged arc welding. Commonly,

TABLE 26.1 Common Pipe Grades and Their SMYS

Pipe Grade	Minimum Yield Strength	
	(KSI)	(MPa)
X-42	42	290
X-46	46	320
X-52	52	360
X-56	56	390
X-60	60	415
X-65	65	450
X-70	70	485
X-80	80	550

FIGURE 26.13 Submerge arc-welded pipe. The weld seam runs longitudinal to the pipe. (Courtesy Pipeline Knowledge & Development.)

pipe produced by this method is called *SAW,* for submerged arc-welded, or *DSAW,* for double-submerged arc-welded (Figure 26.13).

- *Spirally welded.* Flat plates of steel are twisted into a spiral, and the spiral seam is welded together with filler materials deposited from submerged arc welding. The key advantage to spiral welded over longitudinally welded pipe is that larger-diameter pipe can be made with narrower skelp (Figure 26.14).

- *Seamless.* An ingot of steel is heated and pierced with a mandrel or plug depending on the process, and the pierced ingot is formed into its diameter and wall thickness through a series of rollers.

- *Electric resistance welded (ERW) and high-frequency induction (HFI).* As with SAW and DSAW pipe, the skelp is formed into a tube. Unlike SAW and DSAW pipe, no filler material is deposited; rather, the edges are forced together under pressure and heated, causing the seam to fuse. HFI has largely taken the place of ERW to heat the seam.

FIGURE 26.14   Spiral-wound pipe. The weld seam spirals around the pipe. (Courtesy Pipeline Knowledge & Development.)

From a quality standpoint, good specification writing and modern manufacturing methods mean they all produce essentially the same quality pipe, so pipe selections are driven primarily by economics. Spirally welded pipe is often less expensive in larger diameters than SAW pipe, for example.

26.5.2 Coatings

Coatings are used to protect many structures from corrosion, not just pipelines. Early pipeline coatings consisted of pouring a coal tar over the pipe and spreading it with a rag (Figure 26.15). The two primary coatings used for new oil and gas pipelines are

- *Fusion-bonded epoxy (FBE)*. Usually applied in a plant setting, the material in powder form is sprayed onto heated pipe. The powder liquefies and flows around the pipe, forming a protective layer several mils thick. Most people agree that FBE is the premier coating product for pipeline service.
- *Coal tar epoxy*. For many years the standard, this is normally installed embedded in a fiberglass mat that is then wrapped with craft paper.
- *Extruded plastic* is also used as an external coating in some cases.

FBE, coat tar, and extruded plastic are applied in plant settings and then transported to the site where the joints of pipe are welded together. After welding, a small gap remains between the two ends of plant coated pipe. This small section is abrasive blasted prior to field applying a coating to cover the area (Figure 26.16).

FIGURE 26.15 Pipeline coating technology circa 1930s. (Courtesy Association of Oil Pipelines.)

FIGURE 26.16 Girth weld seam abrasively blasted and ready for coating. Note that the areas on either side of the weld seam are factory coated with FBE. (Courtesy Pipeline Knowledge & Development.)

FIGURE 26.17   Concrete-coated pipe. This pipe will be welded together and pulled through a direction drill across a body of water. (Courtesy Pipeline Knowledge & Development.)

Pipe is sometimes coated with concrete or other extremely rugged, thick coating to protect the pipe as it is being pulled through pipeline bores or for buoyancy purposes (Figure 26.17).

26.5.3 Fittings

For gentle undulations in the terrain or slight turns, the pipe may be carefully cold bent in the field to fit, but for more complex turns and connections, prefabricated fittings, with descriptive names such as *tees*, *elbows*, and *reducers,* are used (Figure 26.18).

FIGURE 26.18   An assortment of pipeline fittings, some of which are already wended together, for use in a pump station construction project. (Courtesy Pipeline Knowledge & Development.)

FIGURE 26.19   Thread-O-lets welded to the pipe to provide for sensor connections. (Courtesy Pipeline Knowledge & Development.)

Fittings come in various pressure ratings and must be compatible with the rest of the installation. ASME/ANSI B16.9, *Factory-Made Wrought Steel Buttwelding Fittings,* and ISO 15590, *Petroleum and Natural Gas Industries—Induction Bends, Fittings and Flanges for Pipeline Transportation Systems—Part 2: Fittings,* are two standards applying to fittings. ASME/ANSI B16.11, *Forged Steel Fittings, Socket-Welding and Threaded,* applies to a smaller type of fitting that is used for tapping into the pipe (Figure 26.19).

26.5.4 Flanges

In addition to welding, pipes and components are often joined with flanges. One of the most used flanges for pipeline station work is the raised-face-weld neck flange (Figure 26.20).

FIGURE 26.20   Raised-face weld-neck flange. (Courtesy Pipeline Knowledge & Development.)

Gaskets are placed between the raised faces to seal and prevent leaks. ASME/ANSI B16.5, *Pipe Flanges and Flanged Fittings*, and ISO 15590, *Petroleum and Natural Gas Industries—Induction Bends, Fittings and Flanges for Pipeline Transportation Systems—Part 3: Flanges,* are two standards that apply to flanges. In addition to raised-face mating surfaces, flat-faced and ring joints are sometimes used. Other neck configurations include slip on, lap joint, boss, and socket weld. A special type of flange, a blind, is used to close up an opening and block flow.

26.5.5 Valves

Block, control, check, and relief—each one is a valve function. The primary pipeline valve types are the gate, ball, plug, and globe valve. API 6D, *Specifications for Pipeline Valves,* and ISO 14313, *Petroleum and Natural Gas Industries—Pipeline Transportation Systems—Pipeline Valves,* have been harmonized with each other and are the prevailing pipeline valve standards.

Gate Valves.   Used extensively to block flow, the gate valve consists of a rounded or rectangular gate or wedge that is positioned across an opening to stop flow or removed from the opening to allow flow. The slab or wedge is moved up or down by a screw mechanism driven by a hand wheel or actuator. In the closed position, it comes to rest against seats, which, assisted by differential pressure across the valve, provides the seal (Figure 26.21). Gate valves are quite prevalent in liquid service to block flow.

Ball Valves.   Rather than a slab or wedge, ball valves consist of a ball with a hole through it. This quarter-turn valve is open when the longitudinal axis of the opening is parallel to flow and closed when the longitudinal axis is perpendicular to flow (Figure 26.22).

Ball valves are used extensively in natural gas service to block flow. In both gas and oil, they are also used to control, that is, modulate, flow rates. As the valve closes, it creates more friction loss, thereby slowing flow. As it opens, less friction is generated, so the flow increases.

FIGURE 26.21   Partially open slab gate valve. When the valve is fully open, the hole in the slab is exactly lined up with the internal diameter of the pipeline. (Courtesy Pipeline Knowledge & Development.)

FIGURE 26.22   Partially open ball valve. Ball valves rotate to open and close. (Courtesy Pipeline Knowledge & Development.)

Plug Valves.   Plug valves are also quarter turn valves. Resembling ball valves in design, they have a tapered plug rather than a ball to stop flow. Rather than spring-loaded seats or a pressure differential providing the sealing force, however, the tapered plug can be forced downward, providing a tight mechanical seal between the plug and seats. After the plug is mechanically engaged, a cavity exits inside the plug and perpendicular to the flow. This cavity can be vented to the outside to verify that no flow is leaking past. The forced mechanical seal with the ability to verify sealing makes plug valves a frequent choice for manifold segregation and custody transfer service. The reduced port feature of plug valves means that they are not used when unrestricted flow through the pipe is desired to allow passage of the internal line inspection devices required (Figure 26.23).

Check Valves.   These unidirectional valves normally have a flapper that opens to allow flow in one direction but swings shut if flow tries to reverse (Figure 26.24).

Check valves are installed in pump, compressor, and meter stations to allow flow in one direction while stopping it in the other. They are also sometimes installed on the downstream pipeline flow side of river and stream crossings to prevent fluid from backflowing into the river in the event of an unintentional release, upstream of the valve.

Globe Valves.   This valve takes its name from its sphere- or globe-shaped mechanism. Since the opening between the seat and globe varies linearly as the globe travels up and down, globe valves are used extensively as control valves (Figure 26.25). These valves are produced with various globe and seat configurations to achieve the desired flow modulation pattern.

Pressure-Relief Valves.   Installed to relieve pressure in closed systems caused by temperature rise or to relieve pressure surges caused by operations, pressure-relief valves are of several types. During normal operations, relief valves are held closed by springs, line pressure, pressure from an outside

FIGURE 26.23 Fully open plug valve. Like ball valves, plug valves rotate to open. (Courtesy Pipeline Knowledge & Development.)

FIGURE 26.24 Partially open swing check valve. Flow in this picture would be into the page. (Courtesy Pipeline Knowledge & Development.)

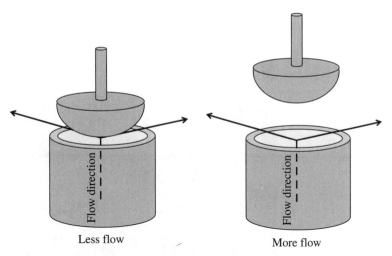

Less flow   More flow

FIGURE 26.25   Drawing of a globe valve. (*Source:* Oil and Gas Pipeline Line Fundamentals Course. Courtesy Pipeline Knowledge & Development.)

source, or a combination of springs and pressure. When system pressures overcome the forces holding them closed, relief valves momentarily open, allowing flow into a tank or other vessel, thereby preventing an overpressure situation. Technicians carefully calibrate these valves to ensure that they open within the anticipated pressure ranges.

26.5.6 Actuators

Manually opening valves requires hand wheels, cranks, or handles. Actuators (sometimes called *operators*) are installed on valves to automate opening and closing. The three basic types of actuators are motor, gas, or hydraulic. Motor operators are simply electric motors mounted onto the valve. One direction opens the valve, the other closes it (Figure 26.26).

Gas-driven operators use gas pressure from the line itself or station air in compressor stations (Figure 26.27). Gas pipelines frequently use gas-driven actuators, and oil pipelines use primarily electric motor operators. Hydraulic actuators use a piston much like gas actuators, and can be used for either gas or oil pipelines.

26.5.7 Pumps and Compressors

Of the various types of pumps and compressors available, oil and gas pipelines use primarily positive-displacement (PD) piston and rotating centrifugal units. While centrifugal compressors are gaining in popularity in the natural gas industry, PD compressors are still used more widely in the natural gas transmission industry than centrifugal compressors. The liquids pipeline industry uses more centrifugal than PD pumps. Compressor and pump details are addressed thoroughly in other texts, so here we focus on pump and compressor operating characteristics.

Positive Displacement (PD) Compressors and Pumps.   Their name says it all—positive displacement—with every stroke of the piston through the cylinder, they displace the same volume. API 674, *Positive Displacement Pumps—Reciprocating*; ISO 13710, *Petroleum and Natural Gas Industries—Reciprocating Positive Displacement Pumps*; ISO 13631, *Petroleum*

FIGURE 26.26   Motor operator. Mounted on a plug valve, the motor, located in the box behind the hand wheel, turns to open or close the valve. The valve also can be operated manually with the hand wheel in case of power loss. (Courtesy Pipeline Knowledge & Development.)

and Natural Gas Industries—Packaged Reciprocating Gas Compressors; and ISO 13707, *Petroleum and Natural Gas Industries—Reciprocating Compressors*, are several of the standards that apply.

Since liquids are essentially noncompressible, PD piston pumps move the same amount of liquid with each stroke. The amount of gas displaced with each stroke of the PD piston compressor depends on the gas density in the cylinder prior to the beginning of the compression stroke and the pressure of the gas in the discharge piping. PD piston compressors and pumps are either single- or double-acting depending on whether they compress gas or pump oil in both directions (Figure 26.28).

Volumes and pressures are controlled by either varying the pump or compressor speed (strokes per minute) or installing a bypass line to essentially recirculate from the discharge to the suction side of the unit. Gas compressors add a third control element, pockets, which vary the volume of the cylinder (Figure 26.29).

PD compressors and pumps seek to overcome system backpressure regardless of what is causing the backpressure. Consequently, closing off downstream flow without providing another flow path can result in mechanical damage or failure.

FIGURE 26.27   Gas valve operator. Mounted on a ball valve, line pressure moves a piston to open or close the valve. (Courtesy Pipeline Knowledge & Development.)

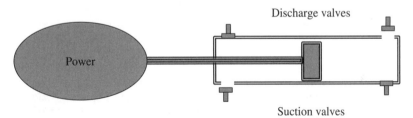

FIGURE 26.28   Schematic of a double-acting piston pump. Valves are held closed by springs and are opened by differential pressure. (Courtesy Pipeline Knowledge & Development.)

For liquid service, PD pumps normally are installed in parallel. That is, the oil only goes through one pump, and the flow capacity of each pump is additive to achieve total system flow rate. Total pressure is determined by the total resistance of the system. Whether or not PD compressors are installed in parallel or series for natural gas service depends on the necessary compression ratios (outlet pressure/inlet pressure). Higher compression ratios are associated with higher force requirements on compressor shafts and other components and a higher temperature rise during compression. For natural gas PD compressors, the allowable temperature rise, and not the component forces, normally provides the compression ratio limit. Normal compression ratios for transmission PD compressors generally are less than 2 except for stations feeding into pipelines where larger pressure increases are usually needed.[*] Installing two PD compressors in series rather than parallel essentially reduces the compression ratio by 50 percent.

---

[*]Mokhatab, S., Poe, W., and Speight J., *Handbook of Natural Gas Transmission and Processing*, Gulf Professional Publishing, 2006, p. 309.

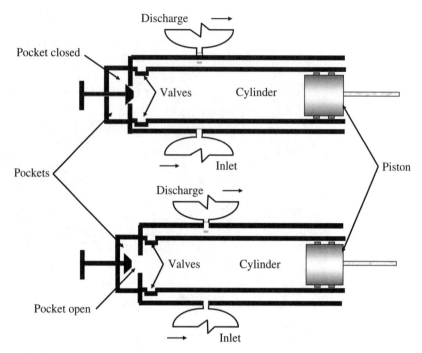

FIGURE 26.29 Schematic of compressor pockets. When the pocket is open (*bottom*), the cylinder has more volume than when the pocket is closed. The piston, however, displaces the same volume with each stroke. This means lower compression ratios with the open pockets than the closed pockets. (*Source:* Hydraulics for Control Room Operators Course. Courtesy Pipeline Knowledge & Development.)

Centrifugal Pumps and Compressors.   API 617, *Centrifugal Compressors*; ISO 10439, *Petroleum, Chemical and Gas Service Industries—Centrifugal Compressors*; API 610, *Centrifugal Pumps for Petroleum, Petrochemical and Natural Gas Industries*; and ISO 13709, *Petroleum and Natural Gas Industries—Centrifugal Pumps,* are some of the standards that apply.

Imparting kinetic energy and then converting it back to potential energy (pressure) is how these pumps and compressors generate pressure. Fluid enters the eye of an impeller located along a shaft and accelerates in a radial or "centrifugal" manner along the impeller vane toward the outside of the pump case (Figure 26.30).

On exciting the impeller at high velocity, the fluid enters the volute, a spiral-shaped passageway which is designed to decelerate the fluid, thereby converting kinetic to potential energy. The high-pressure fluid exits the case via a discharge nozzle. This conversion of kinetic to potential energy gives rise to the characteristic centrifugal curve (Figure 26.31).

Note that the vertical axis in Figure 26.31 shows units in feet of head. Centrifugal pumps discharge the same pressure measured in feet of head at the same flow rates. Said another way, they produce different pressure measured in psi or kPa based on density of the fluid moving through the pump. The same is true of centrifugal compressors, but since density does not vary much between types of natural gas, this phenomenon does not have much impact on gas compressors.

Centrifugal pumps and compressors can have more than one impeller. Each impeller and its associated housing are referred to as a *stage*. Each stage generally produces the same amount of pressure differential, so doubling the number of stages essentially doubles the pressure differential produced at the same flow rate.

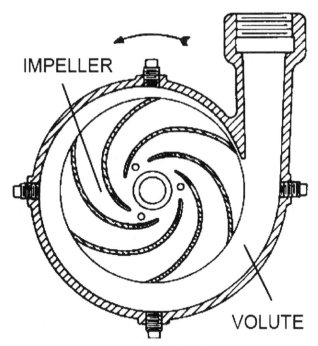

FIGURE 26.30   Schematic of a single stage centrifugal pump. Fluid enters the center of the stage and is discharged out the top. (*Source:* Oil and Gas Pipeline Fundamentals Course. Courtesy Pipeline Knowledge & Development.)

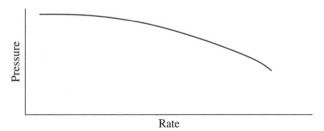

FIGURE 26.31   Typical centrifugal pump discharge curve. (*Source:* Hydraulics for Control Room Operators Course. Courtesy Pipeline Knowledge & Development.)

Centrifugal pump and compressor performance obeys a set of physical laws called the *affinity laws*, where

S_1 = speed 1

S_2 = speed 2

D_1 = diameter 1

D_2 = diameter 2

Q = quantity

dH = Change in head

P = power required

At constant impeller diameter, as the speed changes, quantity moved varies linearly with speed:

$$S_1/S_2 = Q_1/Q_2$$

pressure change varies as the square of the speed change:

$$(S_1/S_2)^2 = dH_1/dH_2$$

and power required varies as the cube of the speed change:

$$(S_1/S_2)^3 = P_1/P_2$$

Similarly, at constant speed, as impeller diameters change, quantity moved varies linearly with diameter:

$$D_1/D_2 = Q_1/Q_2$$

pressure change varies as the square of the diameter change:

$$(D_1/D_2)^2 = dH_1/dH_2$$

and power required varies as the cube of the diameter change:

$$(D_1/D_2)^3 = P_1/P_2$$

Centrifugal pumps and compressors are controlled with variable-speed drives or control valves that move the pump curve or throttle pressure, respectively, to meet system needs. When volumes or other system design features change, impellers sometimes are trimmed to optimize efficiencies.

Two more important notes regarding centrifugal pumps and compressors: As liquids enter the eye of the impeller and begin to accelerate, the pressure is reduced. If this pressure falls below the vapor pressure of the liquid, it turns to a gas. As the fluid moves through the impeller and into the volute, the pressure rises, recompressing the gas into a liquid with an attendant release of energy. This sudden release of energy is called *cavitation*, and it can damage the impeller and housing. Consequently, pumps are produced with a net positive suction head (NPSH) required. Upstream conditions must be designed and controlled so that the NPSH available at the pump suction always exceeds the NPSH required.

Compressors do not experience cavitation because the fluid always remains in gaseous form. They can, however, experience several types of instability, the two most common being surging and stall. These are rather complex phenomena, the treatment of which is beyond the scope of this text, but design engineers should be aware of and design systems and select compressors with these two in mind.

26.5.8 Prime Movers

Engines powered by natural gas drawn from the transmission line, motors powered by electricity, or turbines powered by natural gas, diesel, or kerosene are the primary drivers used to power pumps and compressors. Sometimes the driver is coupled directly to the pump or compressor, sometimes it is connected through a gear box, and sometimes it is connected through a variable-speed drive. In most locations, natural gas provides power for gas transmission lines, and electricity provides the power for oil pipelines. In some remote locations, however, more exotic fuels such as crude oil withdrawn from the pipeline or diesel refined on site at the pump station is used to fuel engines or turbines.

Small-diameter liquid lines may use drivers in the range of 500 (373 kW) to 1,000 horsepower (746 kW). In a 2006 EIA study, gas pipeline compressor stations averaged 4 compressor units per station, with an average horsepower per unit of 3,590 (2,657 kW). The average pressure added per station was approximately 250 psi (1,724 kPa), with discharge pressures in the range of 1,500 psi (10,342 kPa) to 1,750 psi (12,066 kPa) into 36- to 42-inch-diameter lines.[*] Given this average, there are many individual compressors that are much larger.

The engines, motors, and turbines driving compressors and pumps are identical to other large units operating in hazardous environments. Pump and compressor station design is covered later in this chapter.

26.5.9 Variable-Speed Drives

Varying compressor or pump strokes per minute in the case of reciprocating units or revolutions per minute in the case of centrifugal units is an often-used control strategy. The most common types of variable-speed drives (VFDs) are fluid drives and variable-frequency drives.

Fluid Drives.   The basic design of fluid drives involves an impeller mounted on a shaft connected to the shaft of the prime mover and an opposing impeller mounted on a shaft connected to the pump or compressor shaft. Both the impellers are contained in an enclosure containing fluid—normally oil. The prime mover turns one impeller at a constant speed. To increase the speed of the pump or compressor, more fluid is added inside the enclosure by a small pump located on the drive, thereby reducing the slip between the two impellers.

Variable-Frequency Drives.   Electric motors contain windings. The number of these windings, in conjunction with the frequency of the alternating current powering the motor, determines the motor's rotational speed. Variable-frequency drives (VFDs) vary the frequency of the electricity to vary the speed of the VFD output shaft.

26.5.10 Meters[†]

Measuring the amount of fluid entering and leaving the pipeline is important for monetary as well operating and leak-detection purposes, making accurate meters and metering critical. Most people assume that meters measure the amount of fluid moving through them, but, except for PD meters, they don't. Instead, they measure the velocity and density of the fluid moving through them. Knowing these two variables and the cross-sectional area across which the fluid moves, the amount is calculated.

Positive-Displacement Meters.   While there are several styles of PD meters, rotary-vane PD meters are the type used most in the pipeline business. The cavity inside a cylindrical case is divided into sections by vanes connected to a rotating inner element. The amount of volume between the vanes is known, so a known amount of fluid is displaced with each rotation. Revolutions multiplied by volume per revolution equal volume. The API *Manual of Petroleum Measurement Standards,* Chapter 5 "Metering," contains sections dealing with the various types of meters for liquid service.

PD meters handle liquids well but don't work well with gases owing to compressibility issues. PD meters on oil lines require enough pressure to keep the oil above its vapor pressure as it passes

---

[*]Tobin, J., Natural Gas Compressor Stations on the Interstate Pipeline Network: Developments Since 1996. U.S. Energy Information Administration, p. 2.

[†]The metering and proving sections are taken primarily from Miesner, T. O., and Leffler, W. L., *Oil and Gas Pipelines in Nontechnical Language*. Pennwell Publishing, 2006, Chapter 11, pp. 261–267. Edits have been made and new figures added as appropriate for the purpose and scope of this text.

FIGURE 26.32   PD meter. Flow is from left to right. The counter on the top records meter revolutions and sends the counts to a flow computer for processing. (*Source:* Equipment and Components Training Module. Courtesy Pipeline Knowledge & Development.)

through the meter. Sometimes the friction loss or elevation past the meter is sufficient to maintain this pressure. Sometimes a backpressure valve is added to ensure that the oil does not vaporize. PD meters have been the standard for oil pipeline custody transfer for many years and are still used extensively (Figure 26.32).

Turbine Meters.   Turbine meters are inference meters that began to receive recognition in the 1960s and have become well accepted as their technology has developed. American Gas Association Report No. 7, *Measurement of Natural Gas by Turbine Meters*, and ISO 9951, *Measurement of Gas Flow in Closed Conduits—Turbine Meters*, are two of the key natural gas turbine meter standards. Turbine meters determine flow rate by measuring the speed of a bladed rotor suspended in the flow stream (Figure 26.33).

The rotational speed of the rotor is proportional to the flow rate. For oil, the volume flow rate is equal to the cross-sectional area times the rotors velocity adjusted by the meter factor and corrected for temperature and pressure. The natural gas calculation is the same except that compressibility is also taken into account.

Turbine meters do best when the flow rate is constant and the properties of the fluid are consistent, that is, when density and viscosity do not vary. Turbine meters also like smooth, straight flow. This calls for a certain length of straight pipe before the oil or natural gas enters a meter and after it leaves. Straightening vanes, essentially small pipes inside the larger-diameter pipeline, can be installed if there is not enough room to install the straight pipe runs.

FIGURE 26.33 PD meter. Flow can be bidirectional. The pickup coils on the top and bottom record meter revolutions and send the counts to a flow computer for processing. (*Source:* Equipment and Components Training Module. Courtesy Pipeline Knowledge & Development.)

Orifice Meters. For many years the standard for measuring gas flow, orifice meters are based on Bernoulli's principle. American Gas Association Report No. 3, *Orifice Metering of Natural Gas*, and ISO 5167, *Measurement of Fluid Flow by Means of Pressure Differential Devices Inserted in Circular Cross-Section Conduits Running Full—Part 2: Orifice Plates*, are two of the key orifice meter standards. Orifice meters consist of a plate with a circular opening of known size and configuration contained within a case. Sensors measure line temperature, line pressure, and differential pressure from the front of the orifice plate to the back (Figure 26.34).

From the pressure changes, velocity can be calculated. Since the cross-sectional area is known and the temperature is measured, only density remains to be determined before volume can be calculated. Density can be measured directly via a densitometer, measured after the fact through sampling, or agreed on between the parties. Formerly, temperature and pressure were collected on paper roll charters, estimated, and averaged, but modern flow computer techniques have rendered roll charts essentially obsolete.

Ultrasonic Meters. Like orifice meters, ultrasonic meters calculate flow rate from stream velocity, the cross-sectional area of the meter, and fluid density. American Gas Association Report No. 9, *Measurement of Gas by Multi-path Ultrasonic Meters*; ISO 12242, *Measurement of Fluid Flow in Closed Conduits—Ultra Sonic Meters for Liquid*; and ISO 17089, *Measurement of Fluid Flow in Closed Conduits—Ultrasonic Meters for Gas,* are three of the key ultrasonic meter standards.

Using the same technology as used in medical ultrasounds, they transmit high-frequency sound pulses or waves from a transmitter on one side of the pipe to a receiver on the other side. The

FIGURE 26.34   Orifice meter. Flow is from right to left for this small gathering system orifice meter. The flow computer located above the meter calculates flow quantities. (*Source:* Equipment and Components Training Module. Courtesy Pipeline Knowledge & Development.)

pulses can also be bounced from the opposite wall to a receiver on the same side as the transmitter. The velocity of the flowing stream affects how long it takes for the pulse to travel across the pipe. Ultrasonic meters have several transmitters and receivers located around the meter. Each transmitter sends pulses frequently, and computers average the reading to get accurate results (Figure 26.35). In addition to measuring velocity, ultrasonic meters also can measure density. Based on velocity and cross-sectional area, along with temperature, volumes are calculated.

Coriolis Meters.   Gaspard-Gustave de Coriolis (1792–1843) discovered Coriolis force, the principle on which these meters operate. This same force influences weather patterns, sea currents, and which way water turns as it flows down a toilet or drain. The force has to do with angular momentum, making it a bit difficult to explain. For this text, suffice it to say that the mass, not the flow rate or velocity, of the fluid moving through Coriolis meters is measured as it acts on a vibrating tube,

FIGURE 26.35 Ultrasonic meter. Flow can be bidirectional on this meter located at a gas interchange facility. (*Source:* Equipment and Components Training Module. Courtesy Pipeline Knowledge & Development.)

deflecting the tube. The amount of deflection is proportional to the amount of mass acting on the tube. Coriolis meters work for both gas and liquids. In pipeline applications, where flow rate rather than mass is needed, density is used to convert the mass to volume. American Gas Association Report No. 11, *Measurement of Natural Gas by Coriolis Meters*, and ISO 10790, *Measurement of Fluid Flow in Closed Conduits—Guidance to the Selection, Installation and Use of Coriolis Meters* (mass flow, density and volume flow measurements), are two of the key Coriolis meter standards.

Meter Selection. Each meter type has advantages and disadvantages.

26.5.11 Provers

To ensure accuracy, meters are calibrated on initial setup and are recalibrated periodically through the use of meter factors. As an example, if gasoline were pumped through a prover, and the meter said that it pumped 1,001 gallons but the prover said that it pumped 1,000 gallons, the meter factor is 1000/1001 = 0.9990. The API *Manual of Petroleum Measurement Standards,* Chapter 4, "Proving Systems," Section 1, "Introduction," serves as a good start to learn more about oil meter provers, and the *American Gas Association Gas Measurement Manual*, Part 12, "Meter Proving," covers natural gas provers.

26.5.12 Storage

Pipelines operate most efficiently when flow is less variable. But demand, and to a lesser extent supply, are variable. Residential natural gas demand, for example, peaks in the winter when the weather is coldest at over four times the summertime residential demand. Storage serves to balance supply and demand.

Oil Storage. Crude oil and refined products are stored mainly in above-ground welded steel storage tanks ranging in size from about 10,000 barrels (1,590 m³) to over 1 million barrels (158,990 m³). API 620, *Design and Construction of Large, Welded, Low Pressure Storage Tanks*, and API 650, *Welded Tanks for Oil Storage,* are two key new tank design and construction standards, and API 653, *Tank Inspection, Repair, Alteration, and Reconstruction*, addresses maintenance of these tanks.

The weight of the tank steel, combined with the weight of the fluid in the tank, means that tanks must be constructed on specially prepared pads. Depending on soil load-bearing properties, a concrete ring wall may be built to distribute the weight of the tank wall. Flat plates of steel are placed on the tank pad and welded together, forming the floor. Rolled sheets of steel welded together comprise the sides. Roofs are of two main types. Cone roofs are just that, steel cones constructed of steel plate. Steel columns support the roof. The desire to prevent evaporative losses and protect the environment (since cone roof tanks filled with the vapors between the fluid surface and the tank roof had vents to allow release of the vapors to prevent overpressuring the tank) leads to the introduction of external floating-roof tanks (Figure 26.36).

External floating-roofs are normally made of steel. The three primary types are pan, pontoon, or double-deck design. Pan roofs are the simplest, but even one small leak in a pan roof can cause the entire roof to sink over time. Pontoon and double-deck roofs essentially compartmentalize floatation, mitigating sinking potential. The area between the tank shell and the roof has a seal to prevent vapor losses, and depending on size and rainfall levels, one of more roof drains are included to remove water from the roof.

Emerging environmental regulations and the need to dispose of rain and snow water, as well as to keep snow off the roof, lead to the installation of geodesic domes over external floating roof tanks, especially in areas of high rain or snow fall (Figure 26.37).

Given the success of external floating roofs in preventing vapor losses, most cone roof tanks have been retrofitted with internal floating roofs. Internal floating roofs do not have to carry water or snow weight, so roofs constructed of aluminum sheets buoyed by aluminum pontoons are the norm for internal roofs. The area between the tank shell and roof is sealed as in external floating roof tanks, and so are the columns holding up a cone roof.

Natural Gas Storage. Natural gas is stored under pressure, primarily underground in depleted natural gas or oil reservoirs, salt caverns, or naturally occurring aquifers. API 1114, *Recommended*

FIGURE 26.36 External floating-roof tanks. These tanks are open at the top with a steel roof that floats on top of the oil. (*Source: Equipment and Components Training Module. Courtesy Pipeline Knowledge & Development.*)

FIGURE 26.37 Geodesic dome. This external floating-roof tank has a geodesic dome to keep out rain water and snow. (*Source:* Equipment and Components Training Module. Courtesy Pipeline Knowledge & Development.)

Practice for the Design of Solution-Mined Underground Storage Facilities, and API 1115, *Recommended Practice on the Operation of Solution-Mined Underground Storage*, deal with salt domes. Additionally, a small but important storage source is line pack, the amount of gas contained in the line. Since gas is compressible, the amount of gas contained in any line at any time depends on the pressure. Natural gas control room operators make sure that the line is fully packed as a cold morning dawns. During the day, the line unpacks, and the controller supplements from storage. At night, the opposite happens. A small but growing amount of natural gas is also stored as liquid natural gas (LNG) in above-ground steel refrigerated tanks.

Steel storage tanks used for liquids are rather precise; each one is a certain measurable size and holds a specific amount of liquid. It can all be drained out without affecting the tank, and if too much is put in, it runs over. Natural gas storage capacity and injection and delivery rates vary for each location depending primarily on the underlying geology and the amount and pressure of gas in storage. Storage capacity is expressed in cubic feet or cubic meters. Natural gas storage terms and their definition include the following:

- *Total gas storage capacity* is the maximum volume of gas that can be stored in an underground storage facility in accordance with its design, which consists of the physical characteristics of the reservoir, installed equipment, and operating procedures particular to the site.

- *Total gas in storage* is the volume of storage in the underground facility at a particular time.

- *Base gas (or cushion gas)* is the volume of gas intended as permanent inventory in a storage reservoir to maintain adequate pressure and deliverability rates throughout the withdrawal season.

- *Working gas capacity* refers to total gas storage capacity minus base gas.

- *Working gas* is the volume of gas in the reservoir above the level of base gas. Working gas then is the amount of gas available to the marketplace.

- *Deliverability* is most often expressed as a measure of the amount of gas that can be delivered (withdrawn) from a storage facility on a daily basis. The deliverability of a given storage facility is variable and depends on factors such as the amount of gas in the reservoir at any particular time, the pressure within the reservoir, compression capability available to the reservoir, the configuration and capabilities of surface facilities associated with the reservoir, and other factors.

• *Injection capacity (or rate)* is the complement of the deliverability or withdrawal rate. It is the amount of gas that can be injected into a storage facility on a daily basis. The injection rate varies inversely with the total amount of gas in storage, it is at its lowest when the reservoir is most full and increases as working gas is withdrawn.[*]

Depleted reservoirs generally have larger capacities than do salt-dome caverns but require a larger amount of cushion gas to maintain reservoir integrity (make sure that it does not collapse). Salt domes generally have higher injection and delivery rates.

26.5.13 Instruments and Electrical Equipment

Motor starters, switches, temperature and pressure sensors and transmitters, densitometers, and other electrical equipment are the same as or very similar to those used in any hazardous industrial setting and are governed by various electrical codes and standards. Consequently, these components are not included in this text.

26.6 ENGINEERING AND DESIGN

Engineering and design are the processes of applying the principles of physics to achieve the desired output from the constructed work and the process of creating drawings and details to communicate to suppliers and constructors exactly how the facility should be constructed or assembled to achieve the desired results. Engineering and design are complicated processes that brings together a vast array of components and elements to create a constructed work or utility. This text specifically separates the engineering process from the design process although they are often performed by the same person. Engineering is science. Design is art.

26.6.1 Standards, Best Practices, and Codes[†]

Since the first use of manufactured gas for lighting over 200 years ago, engineers and operators have merged the laws of physics with practical experience to produce engineering standards and accrediting organizations for those standards, all in the interest of assisting and advancing the industry. The American Society of Mechanical Engineers formed in 1880 did much to further pipeline standards. Since that time the name has been changed to ASME International as a way to downplay the fact that this association for much of its history had an American base. ASME International defines standards as follows:

> A standard can be defined as a set of technical definitions and guidelines, "how to" instructions for designers, manufacturers and users. Standards promote safety, reliability, productivity and efficiency in almost every industry that relies on engineering components or equipment. Standards can run from a few paragraphs to hundreds of pages, and are written by experts with knowledge and expertise in a particular field who sit on many committees.[‡]

When a standard is adopted by a governmental body, it assumes the force of law and achieves *code* status. The distinction between standards and codes is important and so merits a bit more

---

[*]Definitions are taken from the U.S. Energy Information Administration Website, extracted February 19, 2010, from www.eia.doe.gov/pub/oil_gas/natural_gas/analysis_publications/storagebasics/storagebasics.html.

[†]Standards are based largely on an educational module, from *Oil and Gas Pipeline Fundamentals*, a class developed and taught by T. O., Miesner, *Pipeline Knowledge & Development,* piplineknowledge.com; used by permission.

[‡]Extracted November 30, 2009, from http://asme.org/Codes/About/FAQs/Codes_Standards.cfm.

discussion. Standards are based on universal laws of physics. For example, the determination of maximum allowable pressure in a pipeline is the application of a hoop stress analysis, the same the world over. This is a standard. When the hoop stress calculation is applied to determine the maximum allowable operating pressure, however, different regulatory bodies will set different factors of safety. This is a code.

Some of the standards-issuing organizations include the following:

- American National Standards Institute (ANSI)
- American Petroleum Institute (API)
- American Society for Testing and Materials (ASTM)
- ASME International (ASME)
- ASTM International (ASTM)
- British Standards Institution (BSI)
- European Committee for Standardization (CEN)
- Gas Technology Institute (GTI)
- International Organization for Standardization (ISO)
- Manufacturers Standardization Society of the Valve and Fittings Industry, Inc. (MSS)
- NACE International (NACE)
- National Fire Protection Association (NFPA)
- Pipeline Research Council International (PRCI)
- Plastics Pipe Institute, Inc. (PPI)
- Russian Industry Standards (SNIP)

Providing a complete listing of standards is beyond the scope of this text, but two widely used standards for the pipeline industry are ASME B31.8, *Gas Transmission and Distribution Piping Systems*, and ASME B31.4, *Pipeline Transportation Systems for Liquid Hydrocarbons and Other Liquids*. These standards are certainly the longest standing, and referring to them will cover the vast majority of designs anywhere in the world.

Codes often incorporate standards by reference, but codes—rules and regulations—vary depending on the country involved. The primary code governing natural gas transmission and distribution safety in the United States is contained in CFR Title 49, Part 192—"Transportation of Natural and Other Gas by Pipeline," enforced by the Department of Transportation. The U.S regulations for oil are contained in CFR Title 49, Part 195—"Transportation of Hazardous Liquids by Pipeline." In the United Kingdom, the regulations are Statutory Instrument 1996 No. 825, "The Pipeline Regulations of 1996," enforced by the Health and Safety Executive. Engineers are advised to determine the particular codes applicable to the country in which the pipeline they are designing will operate.

One final note, standards and codes are not design manuals and do not take the place of experience and operating knowledge. Engineers cannot just "take the standards off the shelf" and design an effective system, but they may be able to properly engineer a system using only standards. Standards and codes define the constraints or boundaries on design, and only a competent engineer can in consultation with knowledgeable operators determine the most effective design elements.

26.6.2 Design and Engineering Process

As shown in Figure 26.38, project concepts typically are developed by other groups. Those groups usually lead the concept development and feasibility study phases, asking the engineering department for assistance as needed.

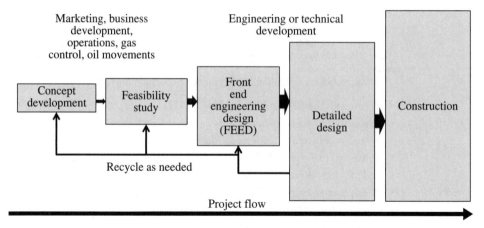

FIGURE 26.38   Project flow diagram. (*Source: Introduction to the Oil and Gas Pipeline Business for Executives.* Courtesy Pipeline Knowledge & Development.)

Front-End Engineering and Design (FEED).   Put the key words *front-end engineering and design* into any search engine, and hundreds of hits show up. The process details vary, but in general, these front-end processes seek to accomplish the same end: Solidify the design basis and key project variables early in the project's development while the ability to influence design changes is relatively high and the cost to make these changes is relatively low, and identify key project risks so that they can be managed effectively. The design decisions made during FEED typically are captured and communicated through the use of a design-basis document and a preliminary process-flow diagram (PFD).

Design Basis.   While there is no universal standard as to what must be contained in a design-basis document, most pipeline operating companies and engineering design firms have their own requirements and format. At a minimum, the following parameters normally are captured and communicated in the design basis

- *Design capacity*—a point value to which the design is optimized; for pipelines, pump stations, or truck loading or unloading facilities, the design capacity is normally quantified as a volumetric flow rate
- *Operating range*—the range of operating flow rates anticipated by the design and for which equipment is to be sized to accommodate
- *Identification of fluid properties* (density, viscosity, specific heat)
- *Operating temperature range*
- *Applicable codes and standards*
- *Process control philosophy*
- *Preliminary route*, including topography, and receipt and delivery locations with volumes
- *Sensitivities* regarding customers, regulators, public opinion, geologic, environmental, or other considerations that would affect line design or routing
- *Project economics and economic sensitivities*
- *Special operating considerations*

Process-Flow Diagram.   Prior to starting detailed design, engineers prepare a process-flow diagram, a schematic representation of the major equipment components, flow direction, and receipt and delivery locations (Figure 26.39). The amount of detail contained on the process-flow diagram

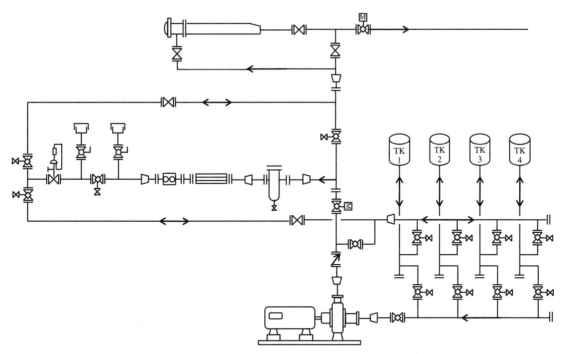

FIGURE 26.39 Process-flow diagram. This process-flow diagram of a receipt station provides the piping designer with detailed information regarding the intent of the facility. (Courtesy Vanderpool Pipeline Engineers Inc.)

varies between companies, but whatever the level of detail, it must be sufficient to convey to the design engineer what is expected.

With the overall flow and specific parameters defined, it is time to get down to the business of detailed design and engineering, an often iterative process. Engineering design tasks often are divided between design of the cross-country pipeline and design of the stations—pump, compressor, and receipt and delivery.

26.6.3 Detailed Pipeline Design and Engineering

The two largest tasks associated with cross-country pipeline design are selecting the route and choosing the pipe diameter, wall thickness, and materials strength.

Route Selection. Engineers start with a straight line between the specified beginning and ending points and then begin looking for geographic, topographic, ecologic, social, and political obstacles. These obstacles include water bodies, population centers, severe topographic features (e.g., mountains or canyons), commercial or industrial development, and sensitive environmental and cultural locations. Generally, the route is selected based on limiting the total pipeline length, offset by constructability and permitting.

Cross-country pipeline construction is most often contracted on the price per foot of pipe installed, with adders for river, stream, road, and other crossings, as well as nontypical construction. Accordingly, after several preliminary routes are identified, the cost of the project is evaluated by estimating the different types of installations required and the length and cost of each. For example, constructing across farm pasture or fields, constructing up a steep slope, crossing waterways or highways, and constructing in congested suburban or urban areas are increasingly more costly on a per-foot basis. Classifying and identifying the degree of construction difficulty, as measured by estimated cost, allows the engineer to compare different route alternatives on a total-installed-cost basis and to select the most cost-effective route for further analysis.

With several routes preliminarily selected, the engineer works with land specialists, identifying and classifying sections of the pipeline along the route as

- Rural undeveloped (agricultural, grazing lands, crop lands, etc.)
- Residential developed
- Residential undeveloped
- Industrial
- Parks or recreational
- Reservoirs and water storage
- Forested lands

Together, the engineer and land-acquisition specialist develop the project implications associated with each of these land classes; difficulty and cost of acquiring easements, public sentiment, constructability, and often other factors are considered.

Advantageous features for pipeline routing include

- Existing utility corridors
- Existing roads that may accept pipeline installation
- Access for transport of workers and materials for construction and maintenance personnel for later operations
- Level or near-level terrain

Features to avoid or mitigate include

- Major freeways
- Residential areas
- Parks and open spaces
- Earthquake fault lines
- Areas subject to flooding
- Other unstable geologic areas
- Highly rugged terrain and side slopes
- Highly environmental sensitive areas

With all these factors considered, final route selection is made on the basis of project economics, construction availability, and environmental and permitting analysis.

Permitting.   *Permits* are authorization by a governmental agency to cross or otherwise impact country, state, or locally managed resources or utilities. Securing permits generally is separate and apart (but closely connected to) securing rights-of-ways (ROW), way leaves, easements, and private access rights required for pipeline installation.

In many cases, there is no single permitting process. Each country, state, or local government has one (or more) permitting processes. Major projects usually require extensive studies to identify their environmental, cultural, social, and economic impact. In the United States, for example, when projects will cross federal lands or otherwise require federal action for implementation (such as a FERC permit), they require extensive environmental review as provided by the Environmental Protection Act of 1970. The review process may require one of two levels of environmental analysis: environmental impact statement or environmental analysis. Other counties have similar processes. Countries that do not have their own process may follow a World Bank process and use World Bank guidelines.

Another common permitting requirement for pipeline construction and installation in the United States is the Clean Water Act, Section 404 permit, administered by the U.S. Army Corps of Engineers. This permit is required any time a water body of the United States is crossed, and this includes most water bodies, rivers, and streams, including some that are intermittent. Additionally, if sensitive wildlife or aquatic environments are impacted, the U.S. Fish and Wildlife Service may be involved based on the prospective impact of the project.

Other permit requirements may include archeological sites, usually administered by a state historical administration, stream and river crossing state permits (Department of Environmental Quality), railroad crossing permits, permits for crossing other utilities, highway and road crossing permits, and local building or access permits. A pipeline is a very long facility that crosses many miles, and the associated permitting is usually complex and involved.

Sizing the Pipe.   Selecting the optimal line diameter and wall thickness begins with the flow-rate design basis and fluid properties. Receipts into the line, deliveries out of it, and any other factors that define the quantity and characteristics of fluids are all considered as the design engineer, using hydraulic software, models line performance to determine the appropriate pipe size in combination with compressor (or pump) unit sizing and station location.

The single most important variable in pipeline sizing, friction loss owing to fluid movement, is directly related in an exponential fashion to the diameter of the pipe. (Recalling that the velocity component in the Darcy-Weishbach formula is squared explains the exponential relationship.) Selecting the optimal pipe size requires evaluating:

- Pipeline hydraulics to determine total system pressure requirements
- Maximum allowable operating pressure (MAOP) to ensure safe operating conditions
- Pump or compressor station sizing to determine horsepower requirements
- Station locations and number to determine total energy requirements

These operations are all performed through computer modeling, but conceptually (and what was done prior to the advent of computers), the system resistance curve from Figure 26.7 is being overlaid with the pump curve from Figure 26.31 to establish the operating point for the pipeline (Figure 26.40). The system resistance curve changes as pipe diameter and fluid properties change. Compressor or pump curves are matched to the resulting systems resistance curves to establish the best range of operating points to meet the established design parameters.

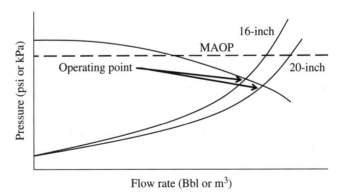

FIGURE 26.40 System resistance curves for 16- and 20-inch pipelines overlaid with a centrifugal pump curve. (*Source:* Oil and Gas Pipeline Fundamentals Course, Courtesy Pipeline Knowledge & Development.)

The MAOP line shown in Figure 26.40 is established through the use of Barlow's hoop stress formula:

$$\text{MAOP} = (2t \times \text{SMYS})SF/D$$

where t = wall thickness
     SMYS = specified minimum yield strength of the pipe parent metal
          D = outside diameter of the pipe
          SF = safety factor established through standards or codes

Based on various combinations of line sizes and properties, as well as energy requirements, cost estimates demonstrating the trade offs between line pipe diameter, wall thickness and steel strength, and number of pumping or compressor stations required to achieve the design basis flow rate are prepared (Table 26.2).

TABLE 26.2   Unit Cost Comparison Pipe Size versus Station Trade-Off

	12-inch	16-inch	20-inch
Total pressure required	6,000 psi	2,000 psi	700 psi
Number of stations required	5	2	1
Pressure per station (balanced system)	1,200 psi	1,000 psi	700 psi
Pipeline investment (materials and construction)	$12,000,000	$16,000,000	$22,400,000
Station investment (materials and construction)	$5,550,000	$2,100,000	$1,000,000
Total investment	**$17,550,000**	**$18,100,000**	**$23,400,000**
Annual operating costs			
• Salaries, wages, station operation, power, maintenance, etc.	$686,000	$353,000	$244,000
Depreciation (annual cost of the investment)			
• Pipeline @3%	$360,000	$480,000	$672,000
• Stations @4%	$220,000	$84,000	$40,000
• Property taxes @1%	$175,000	$181,000	$234,000
Total annual costs (present value)	**$1,441,000**	**$1,098,000**	**$1,190,000**
Annual cost per unit of capacity	$0.079	$0.060	$0.065

Source: Adapted from "Hydraulics for Pipeline Engineers," unpublished. Courtesy of Vanderpool Pipeline Engineers Inc.

The capital cost of line pipe is compared with the operating costs and capital costs of the stations. As line-pipe size increases, the cost of the pipeline material and construction can increase significantly. As line-pipe size increases, however, the friction losses decrease greatly, as do the requirements for compressor or pump stations. The *economical line size*, where costs are minimized considering both construction costs and operating costs, is chosen as the solution for continued engineering design. As shown by Table 26.2, pipeline diameter is critically important in the flow-resistance calculations. It also has a large impact on the flow-rate capacity.

Wall Thickness and Material Strength.   From the MAOP calculations, it is clear that pipe wall thickness and steel strength are critical elements in the structural capability of the pipe to hold pressure. The pipe behaves as a thin-walled pressure vessel, with the pressure inside the pipe trying to push the pipe outward in every direction, thereby stressing the steel. The strength of the material resists the tensile force applied by the pressure, and the thickness of the material provides the structure to resist the applied pressure. Table 26.3 shows the MAOP range for a 20-inch-diameter, 0.375-inch-wall thickness for various steel grades. Table 26.4 shows the MAOP range for a 20-inch-diameter X-60 pipe for various standard wall thicknesses.

TABLE 26.3 Comparison of Grade and MAOP

Grade	MAOP (psi)	MAOP (kpa)
X-42	1,134	7,819
X-52	1,404	9,680
X-56	1,512	10,425
X-60	1,620	11,170
X-70	1,890	13,031

Source: Adapted from Oil and Gas Pipeline Fundamentals Course. Courtesy Pipeline Knowledge & Development.

TABLE 26.4 Comparison of Wall Thickness and MAOP

Wall Thickness (in)	MAOP (psi)	MAOP (kpa)
0.218	942	6,495
0.250	1080	7,446
0.375	1620	11,170
0.500	2160	14,893

Source: Adopted from Oil and Gas Pipeline Fundamentals Course. Courtesy Pipeline Knowledge & Development.

Pipe generally is priced by the ton, so it seems the best solution is always to buy the highest-strength (and thinnest walled) pipe. But higher-strength pipes generally are less ductile and require different welding process than lower-strength pipes. Additionally, if the diameter-to-wall-thickness ratio (D/t) is too large (> 80 is a general rule of thumb), the pipe may not have the structural integrity to maintain its roundness during handling and many not adequately support the weight of backfill materials. Two other useful rules of thumb: Always use a wall thickness greater than 0.250 inch (6.35 mm) for buried pipe and 0.375 inch (9.53 mm) for above-ground pipe to mitigate the risk of outside force damage.

Optimization: Balancing the Factors. Oil and gas pipeline design and engineering require balancing many critical factors. Pipelines are expensive and permanent. Making wise choices regarding the proper line sizing requires the best possible balancing of foreseeable economics and creating options for an unpredictable future.

Generally, if there is any possibility of expansion requirements in the future beyond current foreseeable needs, it is wise to install or construct the largest possible diameter pipeline that can be justified. Because of the physics of friction loss, flow-rate capacity is gained exponentially with incremental increases in pipe diameter. This has the impact of increasing flow rate exponentially while increasing construction costs only linearly, so an incremental dollar spent on larger pipe diameter will have a greater than unity effect on flow-rate capacity. This best positions the pipeline for expansion years into the future because the reasonably expected life of a steel pipeline system that is maintained property can be 50 to 100 years.

Looping and Future Expansions. When the flow-rate capacity of an existing pipeline becomes a limiting factor in the business of the operating company, the capacity can be expanded in several ways. One is *looping,* or adding a second pipeline parallel to the first. From an engineering perspective, this is analogous to an electrical circuit problem, when looping the conductor allows the same current flow at half the resistance, or twice the flow at the same total resistance. Natural gas is a largely homogeneous mixture, so looping is a common expansion solution. However, looping of batched liquids lines is seldom undertaken because loops allow increased mixing at the interface between the various grades of refined products or types of crude oil. The first expansion of liquid lines is commonly accomplished through adding chemical drag-reducing agents (DRAs), which reduce the friction loss per mile by reducing turbulence within the fluid.

Oil Pump Station Engineering and Layout.   Oil pipelines require significant amounts of energy to overcome friction loss and push oil up and over elevation changes. Pump stations, consisting of one or more pump units, provide this energy. For cross-country pipelines, the pumping units are usually multistage centrifugal pumps driven by electric motors. Where a stable and substantial electrical grid is not available, the pumps may be driven by natural gas turbines or even by engines fueled with diesel, crude oil, or natural gas.

The first pump station on the pipeline is referred to as the *origination station*, and subsequent stations are called *booster stations*. Origination pump stations normally are part of a larger receipt station, including meter stations that measure the oil before it enters the pump units. Some stations are designed to either originate or boost depending on the need (Figure 26.41).

FIGURE 26.41   Oil pump station. One grade of oil is received into one of the tanks from local crude oil gathering, whereas another grade of crude oil originating upstream is boosted by this station. When the local gathering tank is full, flow from upstream is diverted into the other tank, and this station originates the local oil into the pipeline. (*Source:* Oil and Gas Pipeline Fundamentals Course. Courtesy Pipeline Knowledge & Development.)

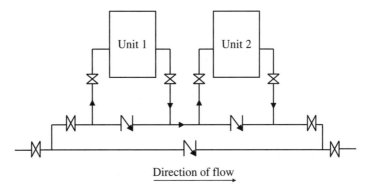

FIGURE 26.42 Schematic of a pump station. Either pump can be used individually, or both pumps can be used in series. (*Source:* Oil and Gas Pipeline Fundamentals Course. Courtesy Pipeline Knowledge & Development.)

Almost all liquid pump stations employ centrifugal pumps arranged in series; that is, the discharge of one connects to the suction of the next. Piping is arranged so that either or both pumps can be used (Figure 26.42). In the case of series operations, the flow rates through each pump are the same, and pressures are additive.

The system rate depends on the discharge pressure from the pump station, which is controlled with either a control valve or a variable-speed controller. Pump station control loops monitor the overall station as well as individual unit, suction, and discharge pressures. Programmable logic controllers (PLCs) compare these pressures to the setpoints programmed into the PLC for each pressure and make adjustment to the control valve or variable-speed controller as required, thereby maintaining station pressures within the setpoints.

Practical pipe wall thickness economics dictate that pump stations typically are designed to discharge between 1,400 and 2,500 psi (9,653 and 17,240 KPa), and typical pump station spacing ranges between 30 and 100 miles. When pipelines are forecast to gain increased volume over their life, the originating and then every other station may be constructed initially. Midpoint stations are added later between the original stations as pipeline volumes increase. As a rule of thumb, adding midpoint station increases capacity by the square root of 2 or about 40 percent.

Pump station mechanical designers consider flow and pressure requirements as they select the optimal pumps and drivers. Then they lay out station piping for efficiency and ease of maintenance, all the while working with the other engineers on the team, including instrumentation, controls, and electrical power engineers, to achieve station and pipeline control.

Natural Gas Compressor Station Engineering and Layout. Analogous to liquid pipeline pump stations, natural gas compressor stations are installed on gas pipelines to boost pressures. But the compressible nature of natural gas means that in addition to overcoming friction loss, compressor stations also compress the gas, increasing its density so that it can be moved more efficiently. Natural gas at 1,200 psi (8,273 kPa) and 60°F, for example, is 100 times more dense than at standard temperature and pressure. Compression requires power and generates heat, both of which are considered as engineers select equipment and layout stations.

The compressible nature of natural gas introduces the compression ratio, a concept not found in pump selection or pump station design. *Compression ratio* is the ratio of inlet to outlet pressure. A compressor with 100 psi (689 kPa) inlet and 300 psi (2,068 kPa) outlet pressures has a compression ratio of 3, for example. Higher compression ratios mean higher heat generation and higher forces on compressor parts. Compression ratios of between 1.5 and 2 are common, but origination stations generally have higher compression ratios because they compress gas from gathering up to transmission pressures. Compressors are arranged either in parallel or series depending primarily on the compression ratio needed.

Positive-displacement compressors driven by gas engines for years have been the compression strategy of choice. Centrifugal compressors, driven by gas turbines or electric motors, however, are gaining popularity owing to their generally lower initial costs.

Use of electric motors have the added attraction that the emissions they generate are released at power plants and not at the station site. Of course, using electric motors rather than gas engines means total emissions are increased owing to electrical generation and transmission inefficiencies. When selecting compressors, design engineers should consider

- Flow rate
- Gas composition
- Inlet pressure and temperature
- Outlet pressure
- Train arrangement
- Number or units[*]

Reciprocating positive-displacement compressor advantages include

- Ideal for low-volume flow and high pressure ratios
- High efficiency at high pressure ratios
- Relatively low capital cost in small units (less than 3,000 hp)
- Less sensitive to changes in composition and density[†]

Centrifugal compressors advantages include

- Good efficiency over a wide range of operating conditions
- Maximum flexibility of configuration
- Low life-cycle cost
- Acceptable capital cost
- High availability[‡]

Friction loss as the gas moves along means pressures drop, and the gas decompresses. As it decompresses, its velocity rises, and because friction loss varies as the square of the velocity, friction loss per mile increases. Consequently, the amount of pressure loss between stations is limited. Gas pipelines might operate with compressor station discharges at 1,400 psi (9,653 kPa), with suction limits set at 1,000 psi (6,895 kPa). Liquids pipelines, in contrast, operate with suction pressures as low as 50 psi (345 kPa) and discharge pressures up to 1,400 psi (9,653 kPa). From a practical standpoint, discharge pressures are often limited to 1,440 psi owing to the need to significantly increase flange and component pressure rotings from 600 ANSI to 900 ANSI above 1,440 psi, although higher discharge is possible.

Safety and Control. Protective instrumentation and controls are essential for modern high-pressure compressor and pump stations. Safety is of paramount importance, so all possible upset scenarios must be identified and addressed during the design phase. This is often accomplished through a hazards analysis process, a tedious table top simulation process where experts in various disciplines (process, mechanical, electrical, instrumentation, safety engineering, rotational equipment specialists) laboriously identify upset scenarios and design

---

[*]Mokhatab, S., Poe, W., and Speight J., *Handbook of Natural Gas Transmission and Processing*, Jersey city, NJ: Gulf Professional Publishing, 2006, p. 300.
[†]*Ibid*, p. 299.
[‡]*Ibid*, p. 300.

safety systems to address each one or develop engineered solutions to alleviate the possibility of the scenarios.

Safety identification starts at the electrical incoming equipment by ensuring that breakers and fault situations are properly addressed. As the safety identification process works its way through the stations, some of the issues addressed include

- What are the electrical requirements of the power company?
- Where should high-voltage transformers be placed?
- How are high-voltage terminations made in the pump station control building?
- How many protective devices are advisable to facilitate maintenance?

These are just a sampling of incoming power issues.

Examination of operational parameters begins with developing an operations procedures manual from which the "brains" of the pump station are programmed into the PLC. The PLC must be designed to monitor each and every electrical, pressure, flow, and equipment monitoring parameter and to constantly check those parameters against preprogrammed limits. The PLC is programmed either to send signals to local station equipment for immediate involuntary shutdown or to the control center when parameters approach limits so that intervention may be taken before involuntary shutdown is initiated.

Some examples of primary station parameters that are monitored for safety and station control include

- Station suction pressure
- Station discharge pressure
- Unit case pressures
- Station incoming and outgoing temperatures and temperatures on flow at each unit
- Station incoming voltage and voltage fed to each unit
- Amperage draws of each unit
- Flow rates through the station (if equipped with flowmeters)
- Sump-level alarm
- Hazardous gas release

Equipment is normally heavily instrumented to monitor safety and performance parameters. Some examples of unit parameters that are monitored include

- Motor windings temperatures (two for each voltage phase)
- Motor bearing temperature (inboard and outboard)
- Unit bearing temperature (inboard and outboard)
- Motor vibration (inboard and outboard)
- Pump or compressor vibration (shaft inboard and outboard)
- Seal pressure differential (inboard and outboard)
- Lube pump vibration, flow, and temperature

While ANSI/ASME B31.4 sets forth many engineering limits and topics that must be respected in pump station design, the code does not cover how to lay out the station piping to ensure efficient flow through the station, ease of maintenance, and respect of surrounding facilities or neighbors. Nor does ANSI/ASME B31.4 cover how to lay out drain and miscellaneous piping for positive drain, evacuation, safety from vibration or other loading, or convenience of maintenance. Accordingly, most pipeline operating and engineering design firms have their own, more detailed design standards and specifications based on years of experience.

FIGURE 26.43 Refined-products receipt station. The incoming scraper receiver is in the foreground, with the sample shed immediately behind it. The tanks in the background will receive the refined products. (Courtesy Pipeline Knowledge & Development.)

Oil Pipeline Receipt and Delivery Stations. As with any transportation system, pipelines stage and account for cargo as it enters and exits the system. Receipt stations accept volumes into the pipeline. Delivery stations handle exiting volumes. When one pipeline delivers to another, the station involved can be both a receipt and a delivery station. Receipt and delivery stations both contain, at a minimum, metering and pressure-control equipment along with the needed instrumentation and controls. Additionally, they may contain facilities to filter, sample, and temporarily store the oil (Figure 26.43).

Designing these stations involves understanding incoming and outgoing flows, pressures, specific shipper needs, and company metering and control strategies and then laying out the optimal assembly of meters, valves, fittings, piping, instrumentation, and other components and equipment to accommodate the flows and pressures that meet shipper needs and comply with regulations. Figure 26.44 is the piping drawing for a rather straightforward delivery station.

Natural Gas Receipt and Delivery Stations. Reservoir pressure pushes gas to the surface, and it stays under pressure its entire trip until finally reaching the burner tip where it is consumed. Along the way, natural gas is delivered into (received by) and out of pipelines. Sometimes it goes into storage, but often the molecule makes its entire trip without ever entering storage, expect perhaps for the fact that line pack is used as temporary storage. Orifice meters have long been the standard for gas measurement (Figure 26.45).

Interconnections facilitate competition because the same gas can access multiple markets. As a consequence, pipelines in the United States following FERC Order 636 and other FERC Orders leading up to and subsequent to 636 and pipelines in Europe seeking to establish a Pan-European market are installing interconnections. Often these new connections employ ultrasonic meters (Figure 26.46).

Natural gas delivery stations to local distribution companies (LDCs) are called *town* or *city gates*. Their purpose is to receive the gas, measure it, ensure that it is clean and free of liquids, ensure the proper energy content, and in some cases add the odorant so that it has the classic natural gas smell (Figure 26.47).

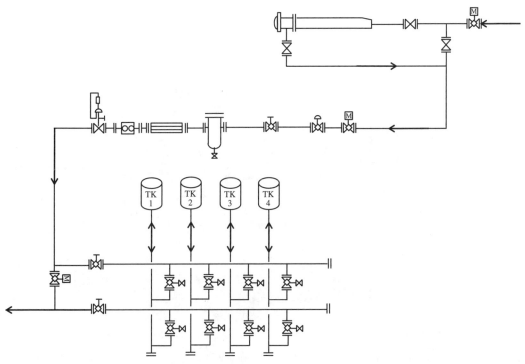

FIGURE 26.44 Typical oil receipt facility line drawing. Oil enters through the scrapper trap at the top, passes through the strainer and meters, and is directed to one of the tanks. (*Source:* Liquid Control Room Operations Teaching Module. Courtesy Pipeline Knowledge & Development.)

FIGURE 26.45 Orifice meter. Two taps shown at the right center of the meter sense pressures upstream and downstream of the orifice plate. These pressures are sent to the flow computer, which calculates quantities. (Courtesy Pipeline Knowledge & Development.)

FIGURE 26.46 Gas interconnection facility. Two natural gas transmission pipelines transfer gas across this facility through an ultrasonic meter located just below the crossover. Valves with gas-operated actuators are located behind the crossover and direct flow. A control building is shown in the background. (*Source:* Natural Gas Pipeline Economics Teaching Module. Courtesy Pipeline Knowledge & Development.)

FIGURE 26.47 City gate station. Gas enters the station through the large pipe in the foreground. Pressure is reduced, and the gas is measured prior to oderant injection. (*Source:* Natural Gas Pipeline Economics Teaching Module. Courtesy Pipeline Knowledge & Development.)

26.6.4 Station Design and Layout Rules of Thumb

Designing and laying out any of the stations discussed in this chapter are rather more of an art based on common sense and learned through experience than a science involving calculations. The following rules of thumb are from the Vanderpool Pipeline Engineers Inc., Design Standards and are offered as an aid to those new to station design and layout.[*]

- Develop process-flow diagrams (PFDs) and piping and instrumentation drawings (PIDs) prior to beginning detailed design.
- Conduct thorough reviews of these documents with operations personnel early in the design process, and involve an instrumentation technician with field experience as part of the design team.
- Develop operating procedures and cause-and-effect matrices to facilitate integrating safety and upset management into station design.
- Standardize equipment between stations to facilitate management of replacement and spare parts.
- Minimize turns and bends and maintain pipe, valve, and fitting diameters to minimize vibration within the piping system.
- Install sufficiently long straight-pipe runs prior to meters, and install straightening vanes to condition flow and increase metering accuracy.
- Minimize the length of pipe between valves in common manifolds to minimize "dead-leg" and contamination.
- Provide pressure relief for thermal pressure caused by increases in temperature. Station piping is small volume, so small changes in temperature can greatly increase pressure if the station is idle and flow is blocked in.
- Understand and install safety devices and emergency shutdowns to provide protection.
- Ensure sufficient clearance between pipe runs and equipment to allow easy access and maintenance.
- Think about stepovers and head-bump dangers.
- Provide escape routes in case of emergency.
- Provide sufficiently large drain sumps to handle the drain down from several pieces of equipment at a time.
- Lay out auxiliary piping and electrical conduits to allow ease of access and safety.
- Consider an auxiliary bypass (small-diameter) to facilitate maintenance and/or future station modification while continuing to allow the pipeline to flow, rather than requiring long lengths of drain-up or blow-down.
- Add a few "extra" ports (threaded-O-rings) at strategic locations to facilitate access to the flow stream for future needs.
- Involve experienced operators in the design and review process to ensure that operating needs are understood and met.
- Provide security as needed to prevent unauthorized access to the site.
- Design piping and supports to eliminate stress on pump flanges.
- Complete a piping stress analysis to understand and provide for stresses.
- Be sensitive to the ambient temperature during fabrication and installation; ambient temperature changes cause great amounts of piping stress.
- Provide careful consideration for the amounts of above-ground and underground piping.
- For safety and access, above-ground piping should be maximized, but turns, bends, and below-ground piping should be installed to manage piping stress.
- Provide lighting for security and maintenance purposes. Consider supplemental lighting controlled by switch for late-night call-out maintenance.

---

[*]Extracted from Vanderpool Pipeline Engineers Inc., Standards, by permission from Vanderpool Pipeline Engineers Inc.

- Provide access to control building just inside gate and well separated from equipment in case of fire or other malfunction.
- Consider adding sufficient select fill to bring the station up beyond the 500-year flood level because pump station flooding can be expensive and disruptive.
- Consider installation of remote cameras, along with "fire eyes" and gas detectors, to provide visual checks from remote-control center.
- Provide station data recorders in the control building for all electrical, pressure, and temperature recordings for efficiency diagnosis and other troubleshooting.
- Consider noise management of equipment for neighbors.
- Consider visual impacts (perhaps landscaping for noise and visual management).
- Evaluate electrical power requirements for immediate as well as future requirements. Voltage available and design of incoming power can be a significant limitation on future expansion and affect station constraints for decades.

Clean, efficient stations with easy access have a significant impact on cost, safety, quality, and worker morale.

Block Valves.   Valves are installed at locations along a pipeline for emergency purposes and to isolate the line for maintenance. Block valves are a function, not a type of valve. They shut off or block flow on the pipeline mainline. Often block valves on oil pipelines are gate valves, and block valves on gas pipelines are ball valves. Natural gas block valves commonly are installed with a smaller-diameter bypass to allow pressure equalization across the valve prior to operating the valve (Figure 26.48).

FIGURE 26.48   Buried ball valve used as a block valve in a natural gas pipeline. This picture was taken before the security fencing was installed around the valve. Note the locks on the hand wheel and the two plug valves on the bypass. (Courtesy Pipeline Knowledge & Development.)

FIGURE 26.49   Above-ground gate valve used as a block valve on an oil line. An electric motor operator is located on the valve, and the satellite dish can receive signals from the central control room, which can operate the valve remotely. (Courtesy Pipeline Knowledge & Development.)

Block valves sometimes are buried in line with the pipe flow, as in Figure 26.48, and sometimes are brought above ground, as in Figure 26.49. Whether the valve is buried or above ground seems to have more to do with normal company practice than with any specific law, regulation, or standard.

Regulations specify the minimum placement of block valves to protect natural resources and population centers, but pipeline operators often install additional block valves to section the pipeline for maintenance. Block valves may be operated manually, which slows response time in the event of emergency, or they can be actuated and controlled remotely.

Crossings.   Intersections of the pipeline with another utility, road, waterway, railroad, or some other obstacle requires careful design to avoid damaging the object or feature being crossed while still allowing maintenance of the pipe. Typically, the wall thickness of the pipe is increased at crossings to provide additional strength of the pipe as a structure. The additional strength is required to withstand increased overburden of the soil, loads from trains and vehicles, and additional margins of safety for corrosion because the pipe installed under obstacles often has reduced accessibility in the future for maintenance purposes. There are three basic types of crossing: open trench, boring, and horizontal directional drill. Typically, design engineers specify which of these techniques to use, but sometimes the choice is left to the construction contractor.

26.6.5 Control System Design[*]

Many transportation systems are monitored and dispatched extensively. Air traffic controllers, for example, monitor the movement of airplanes and communicate vectors to pilots, depending on the pilots to actually control the airplane. Likewise, railroad dispatchers tell engineers how fast to travel

[*]Pipeline controls are extracted from Pipeline Knowledge & Development's Introduction to SCADA and Controls Training Module and has been edited and used by permission.

and when to stop and start, but the engineers perform the control movements required to meet the dispatch orders. Pipelines go beyond dispatching, adding an element of direct control by central control room operators.

Control can be with respect to an individual component (e.g., motor, engine, pump, compressor, meter, or valve), a particular group of components (e.g., compressor station, pump station, or metering station), or an entire pipeline network. Control systems provide operators with the ability to ensure that all components are working together safely and efficiently to receive, transport, and deliver the desired quantities at the specific rates and pressures, all the while meeting product specifications. Control requires understanding the status of individual units and stations as well as overall system status and provides the ability to influence (start, stop, open, or close) units, stations, and networks. Supervisory control and data-acquisition (SCADA) systems provide those functions at both the local station and the centralized system levels (Figure 26.50).

At the local station (shown on the right side in Figure 26.50), pressures, temperatures, equipment status, and other information are aggregated and processed by one or more intelligent devices. The station intelligent device decides if it should send a control command to a device or simply report the information to a central SCADA server. The central server may decide to send a control signal or simply report the information to a human operator via the HMI. Most process industries employ SCADA or distributed controls, so the details of SCADA control are left to other texts. But some of the pipeline design strategy issues are addressed here.

Control Strategy.   Since the point of SCADA systems is controlling the pipeline, the first two design strategy questions are

- Which control decisions will be made at the local station and which will be made at a centralized location?
- Which control decisions will be made by machines and which will be made by humans?

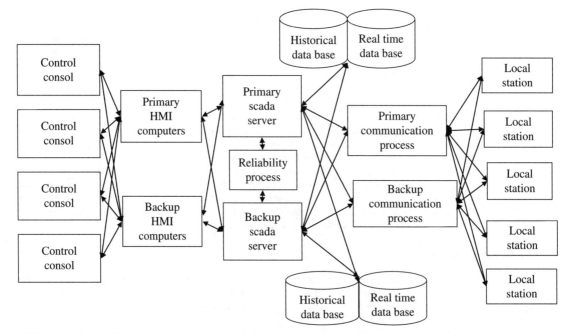

FIGURE 26.50   SCADA schematic. Local stations are shown on the right. The central control room is on the left. Communications processes connect the two. (*Source:* Control Room Operations Training Module. Courtesy Pipeline Knowledge & Development.)

Local decision making is milliseconds faster than remote decision making, but the primary advantage to local decision making is removing long-distance communications as a potential point of failure. Local decision making, however, implies station staffing or, more likely, additional computing capabilities at the local station. Common practice includes making routine unit–related commands such as compressor selection or device shutdown when preset parameters are exceeded locally. Additional control often is transferred to the local station if communication is lost for a predetermined amount of time.

How much of the decision making should be done automatically without human intervention and how much should be left to human controllers also must be agreed. Machines can make decisions quickly and consistently based on the information given to them. They do not "think" broadly about the information. Humans can introduce intuition, judgment, and feelings.

"Routine" decisions commonly are left to the control system. But what is *routine* and what is *abnormal* must be decided in advance so that the control system can be built and programmed accordingly. Industry standards have not been developed to address these questions. Some recommended practices include

- ISA18.2, *Instrument Signals and Alarms*
- API Recommended Practice 1113, *Developing a Pipeline Supervisory Control Center*
- API Recommended Practice 1149, *Pipeline Variable Uncertainties and Their Effects on Leak Detectability*
- API 1165, *Recommended Practice for Pipeline SCADA Displays*
- API Recommended Practice 1162, *Pipeline Control Room Management*
- API Recommended Practice 1130, *Computational Pipeline Modeling for Liquids*

Data Strategy.   Data acquisition, manipulation, and storage are key SCADA functions. Data design considerations include

- What data should be acquired?
- On what frequency should these data be acquired?
- For how long should these data be retained?
- How should these data be acquired?
- What error checking should be performed, and how should suspect data be managed?

Key pipeline process data include pressures, flow rates, temperatures, and device status, but a wealth of other data can be acquired as well. Acquiring more data may or may not result in more information and requires more bandwidth or causes slower polling, processing, and response times.

Communication Strategy.   The nature of pipelines means that data are collected from and commands are sent to locations along the pipeline that are remote from the central control room. Both local station and station-to-control-center communication strategies must be developed. Figure 26.51 shows one long-distance communication strategy.

At the local station level, analog and digital signals must move from pressure and temperature sensors, meters, and other devices to the station intelligent devices. Likewise, controls must be sent from the intelligent devices to actuaters, starters, and other devices. The available strategies for these movements, as well as the station intelligent device to SCADA host server, are the same as those available in other industrial applications.

Alarm Strategy.   When process variables rise above or fall below preset limits, the SCADs system takes one or more actions. Depending on the control strategy, these actions may be issuing commands, alerts, or alarms or issuing a combination of the three. Alerts are less significant than alarms. They inform the operator of a situation that may or may not need immediate action. Alarms are reserved for situations that do need immediate action. The alarm filtering strategy involves deciding

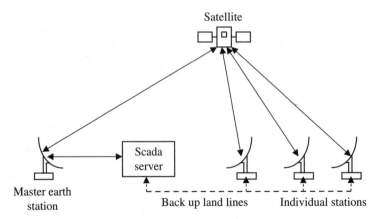

FIGURE 26.51 Communication Schematic. Primary communications are provided through satellite systems. Backup is provided through phone lines with modems located at stations along the pipeline. (*Source:* Control Room Operations Training Module. Courtesy Pipeline Knowledge & Development.)

which events should be signaled to the operator, which of those signals should be alerts and alarms, how the alert or alarms should be announced, and the operating action required for a particular alert or alarm.

Applications Strategy. Basic SCADA packages acquire data and control the pipeline system, but features can be added for applications such as

- Measurement and scheduling
- Batch tracking
- Real-time operations support
- Leak detection

Which, if any, additional applications will be included and the requirements for those applications are addressed based on decisions made prior to the start of detailed engineering.

Leak-Detection Systems. Pipeline leak-detection systems generally are classified as internal or external. Internal systems use data generated by the SCADA system, either monitoring the data versus preset limits or using the data to model line performance and compare actual to modeled operation. External systems are applied external to the pipeline as a means to discover escaping hydrocarbons. The key leak detection criteria are the following:

- *Sensitivity.* The ability to detect a given size leak in a given time period. This is normally measured in percent of flow rather than a finite number.
- *Accuracy.* The ability to determine the location of a leak, its rate, and the volume lost.
- *Reliability.* Whether or not a leak actually exists when the systems says it does.
- *Robustness.* The ability to work under less than ideal conditions, that is, the minimum level of sensitivity, accuracy, and reliability the system can sustain under specified conditions determines robustness.

These four aspects must be considered together. More sensitive systems produce higher rates of false alarms, for example. Accurate systems under ideal conditions may lose their accuracy under less than ideal systems. One more important note: Leak-detection systems are only as good as the data feeding them. More sensitive systems require more careful instrument calibration than as less sensitive systems.

26.7 PIPELINE CONSTRUCTION

Design engineers produce drawings depicting how the facilities should be installed and specifications for materials, equipments, and construction. These form the basis of material and equipment procurement and project construction bid documents. Construction project managers, engineers, and supervisors interface with design engineers during this transition to ensure that care is taken during design and is reflected in the final work. After construction contracts have been awarded, and as materials and equipment begin to arrive at staging yards along the route, the shift to construction begins in earnest.

The linear nature of pipeline projects creates the requirement for constant movement of workers and equipment, often over wide varieties of terrain. High equipment and personnel utilization is the key to managing construction costs, and each day of reduced productivity is expensive. Communication along the project between engineers, projects managers, project inspectors, and the contractor's crews can be challenging because of the distances and sometimes because of remoteness.

26.7.1 Construction Staffing, Planning, and Project Management

Project size and complexity determine project staffing. Large projects such as the BTC crude oil line from Baku, Azerbaijan, to Ceyhan, Turkey, or the REX gas line from Rio Blanco County, Colorado, to Monroe County, Ohio, require a large project management staff. Small projects may be managed by one person. No matter the size, the roles of managing and directing a pipeline project are complex and varied. The roles may be filled by employees of the owner company or by trusted and talented consultants dedicated to the project. Roles include

- *Program manager.* Executive-level individual with extensive talents in planning, communications, and content knowledge of pipelines.
- *Project manager, pipeline.* Executive-level individual with extensive talents in planning, communications, and content knowledge of pipelines. This role is for management of the cross-country pipeline.
- *Project manager, stations.* Executive-level individual with extensive talents in planning, communications, and content knowledge of pipelines. This role is for management of the pump stations or compressor stations and receipt and delivery facilities.
- *Technical manager or project engineer.* Team leader to direct all technical activities of the project, including hydraulics, pump/compressor sizing and specifications, material specifications, surveying, detailed civil, mechanical, and electrical design and construction specification writing.
- *Controls engineer.* Team leader responsible for project scheduling and cost controls and reporting.
- *Contracts administration.* Procurement specialist with appropriate legal support in the development of contracting and procurement strategies.
- *Environmental manager.* Team leader to direct all environmental studies and procurement of government permitting, including federal and state.
- *Right-of-way manager.* Team leader to direct all activities related to easement procurement over the pipeline route and additional land requirements for construction and staging of equipment and materials.
- *Materials manager.* Team leader to work with technical personnel on quoting, purchasing, transporting, and warehousing of all project required materials.

- *Regulatory manager.* Team leader to ensure compliance with safety and environmental regulations during construction and establishing documentation for operations.
- *Safety manager.* Team leader for establishment and enforcement of project safety procedures and incident reporting.
- *Document manager.* Team leader responsible for organization, management, and distribution of project critical documents, including drawings, specifications, permits, easements, correspondence, as-constructed records, weld maps, weld inspections, and other important project documentation.
- *Construction manager.* Team leader to plan and direct all quality control associated with construction and installation of the pipeline.
- *Craft inspection team.* Boots on the ground representing the owner with each contractor's crew to provide quality-control inspection of construction and installation of the pipeline.
- *Public relations or community manager.* Pipelines often require significant public communications, sometimes as part of the permitting process and sometimes it's just good business.
- *Engineers.* Technical specialists responsible for technical analysis, calculations, design, drafting, and design document management.

The contractor also requires a diverse and broad set of talents to execute the project. Each task of the construction project requires various crews consisting of the following roles and crafts:

- *Project manager.* Onsite direct communication facility between owner and contractor.
- *Superintendent.* Construction specialist responsible for overseeing the foreman of each crew.
- *Foreman.* One foreman for each of the contractor's crews: right-of-way clearing and grubbing, stringing, ditching, bending, alignment and stringer, welding, coating, lowering, backfill, cleanup, and restoration.
- *Time keeper.* Team leader responsible for knowing and tracking each individual and each piece of equipment working each day on each crew.
- *Personnel manager.* Construction crews often have some level of turnover during a project, so the contractor requires someone as a point contact for recruiting crafts as required for the project on a daily basis.
- *Surveying.* Once the right-of-way is established by the owner, the contractor is usually responsible for maintaining ditch alignment.
- *Safety director.* Ensuring that each crew member follows company-established safety policies and OSHA compliance. Also ensuring that each piece of equipment is safe and working properly.

26.7.2   Route Selection

Route selection has a significant impact on construction costs and schedule because it determines the difficulty and cost of easement acquisition; the difficulty and cost of construction; the number of river crossings, road crossings, railroad crossings, utility crossings, and environmental permitting requirements associated with publically owned land. Sometimes the difficulty of many of these factors may not be well understood until the process begins and the route must be altered slightly to meet unanticipated circumstances during the early part of construction.

26.7.3   Right-of-Way and Land Acquisition

Usually, the right-of-way (ROW), way leave, or easement procurement and land-acquisition processes begins with the ROW managers personally contacting land owners to discuss the potential of purchasing options first rather than actually purchasing the ROW at that time. Procuring an option costs less than an easement and gives the owner the right, within a reasonable period of time, to acquire the easement at a known cost. Early ROW and land-procurement activities afford project management information about the difficulty and costs of acquiring the requisite easements. With this information, routes may be altered or the project reevaluated and reestimated for costs and timing.

Another critical step is determining access to the construction ROW. Rather than simply starting at one end and proceeding directly along the ROW, access to the ROW at various locations along the construction path is usually desirable. These perpendicular access points require additional permissions and access and sometime road building, but they are important considerations for contractor bidding or pricing of the job. Planning and land or ROW acquisition for material staging and contractor's equipment are also required.

26.7.4 Material Procurement

Procurement of major piece of equipment with long lead times such as compressors, pumps, meters, and even the pipe normally starts well in advance of construction to ensure that the equipment is available when needed. Simultaneous with land and rights acquisition, the remainder of the materials and equipment is purchased.

26.7.5 Mainline Construction

The notion of a *moving assembly line* with the different elements of the process organized in a sequential manner characterizes construction of the mainline (Figure 26.52). Clearing vegetation from the construction corridor, leveling and grading the corridor, stringing pipe along the corridor,

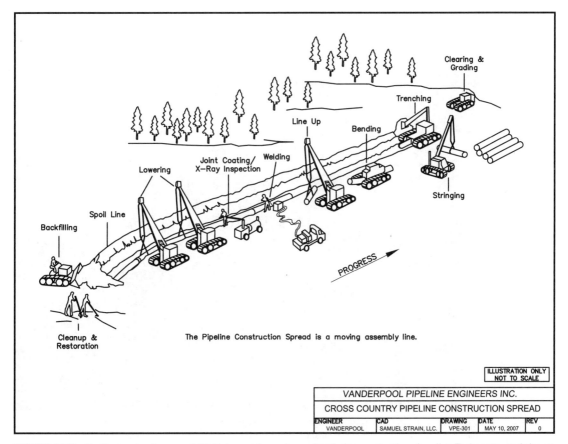

FIGURE 26.52 Pipeline construction spread. Multiple spreads may be employed when constructing a long line. Each constructs their part, and the parts are tied together. (*Source:* Vanderpool Pipeline Engineers Inc., Standards. Courtesy Vanderpool Pipeline Engineers Inc.)

welding the pipe, digging the ditch, lowering the pipe into the ditch, backfilling the ditch, final grading of the corridor, installing erosion control, and revegetating the corridor all proceed in order, interspersed with inspection and quality assurance as required.

Surveying and Staking the Centerline.   With the pipeline route established and ROW purchased, surveyors move along the route marking the pipeline centerline with stakes at each point of intersection (PI) or more often, if needed, to maintain the route. This activity may require some clearing to allow a line of sight between stakes. Once final clearing and grading are completed, additional stakes are added to guide the ditching operation.

ROW Clearing and Preparation.   Construction corridor width is an important variable in the pipeline construction process. A 60-foot-wide construction corridor is common for a 12-inch-diameter pipeline and is sufficient to enable the contractor to install the pipeline and move equipment up and down the corridor in a parallel manner without disturbing the efficiency of the process. Larger-diameter pipelines may require a greater width, and smaller-diameter pipelines may require less, but 60 feet in cleared width is a good approximation.

Required ROW width for construction varies with topography, congestion, and population density. Narrow widths are workable, but with reduced efficiency and increased costs because equipment is forced to "take turns" occupying the same space (Figure 26.53). Whatever the width, the construction corridor must be cleared of obstructions so that equipment and personnel can move about safely and efficiently.

FIGURE 26.53   Congested construction location. Sandwiched between two apartment complexes, the pipe from the directional drill emerges from the ground. Careful coordination of equipment is required in this close space. (*Source:* Introduction to Pipeline Construction Training Module. Courtesy Pipeline Knowledge & Development.)

In addition to allowing room for equipment and personnel, a primary goal of preparing the ROW is smoothing the route sufficiently that the line pipe fits the contour of the ditch and is continuously supported by it. Special considerations must be made for crossing obstacles, including waterways, sensitive habitat, and roads. Toward the end of the clearing process, the top-soil layer normally is stripped off and stockpiled so that it can be spread during the ROW restoration process.

Ditching.   Accomplished with a ditching machine, back hoe, or by hand, if necessary, ditching usually occurs prior to stringing the pipe. But ditching may be delayed until after the pipe joints has been strung, welded together, and tested if an open ditch would present a dangerous situation. The ditch is normally offset to the "left" side of the crew progress, establishing a working side (Figure 26.54).

Stringing.   Joints of pipe are strung along the ROW sequentially so that they can be lined up and welded together. Rather than being laid directly on the ground, they are normally cribbed up on skids (Figure 26.55).

Welder Testing.   API 1104, *Welding of Pipelines and Related Facilities,* is the most widely used pipeline welding standard. Prior to welding on the line, each welder, and the welding process itself, must be tested and certified. Testing involves performing test welds from which straps are cut and tested to ensure strength and ductility at least equivalent to that of the pipe. Welders expected to perform complex welds such as branch connections perform their test welds on more complex shapes (Figure 26.56).

FIGURE 26.54   Pipeline ditch. This ditch was dug by a backhoe. Note the top soil stockpiled on the right. This topsoil will be spread over the ditch after the lesser-quality material on the left is retuned to the ditch. (*Source:* Introduction to Pipeline Construction Training Module. Courtesy Pipeline Knowledge & Development.)

FIGURE 26.55 Joints of pipe strung along the ditch. Note the ropes around the pipe joints placed to protect the coating during transport. (*Source:* Introduction to Pipeline Construction Training Module. Courtesy Pipeline Knowledge & Development.)

FIGURE 26.56 Welder test piece. Straps were cut from this piece and tested to certify the welder who welded it. (*Source:* Introduction to Pipeline Construction Training Module. Courtesy Pipeline Knowledge & Development.)

Lining Up, Stabbing, and Root Pass. The process of welding individual joints of pipe begins with carefully aligning two joints of pipe and making the initial weld, called the *root pass*. The ends of each joint are beveled to facilitate penetration to the inside of the pipe. Internal alignment tools are used to ensure that the insides are exactly aligned (Figure 26.57). During this process, the pipe is stacked on wooden skids in preparation for the firing, line welding crew.

FIGURE 26.57 Internal alignment tool. The shoes inside the pipe have been expanded, centering the alignment tool. After the next joint is slipped into place, the other set of shoes will be aligned so that the inside pipe wall meets squarely with the mating pipe joint. (*Source:* Introduction to Pipeline Construction Training Module. Courtesy Pipeline Knowledge & Development.)

Firing Line.   The next crew consists primarily of welders who make two or more additional welding passes, depending on pipe wall thickness, around the pipe to fully strengthen the connection between the joints. These welds can be made by hand or with automated welding machines. Between each weld bead, the weld is cleaned to remove any slag or other impurities created when welding the pervious bead.

Inspection and Coating.   Immediately on completion of the cap weld pass, the welds are inspected by radiography or ultrasonically to ensure quality (Figure 26.58). When the weld passes inspection, it is cleaned to white metal and then coated for corrosion protection. If the weld fails radiographic inspection, it must be repaired or cutout and rewelded until it passes inspection.

Lowering.   After the joints are welded together and tested, the ditch is cleaned of any large stones that could damage the coating, and the entire pipe is lowered into the ditch. As it is lowered, the pipe coating is inspected one more time with a holiday detector, commonly called a *jeep* after the fictional jeep character in the Popeye cartoon strip. In rocky soil, the ditch may be padded with sand to further protect the coating.

Backfill.   Finally, the pipe is backfilled with soil suitable for backfill and deemed safe for the pipe corrosion-protection coating. The last step is to replace the top soil removed earlier to promote land cover (crop, grasses, etc.) restoration to prevent erosion.

FIGURE 26.58   Radiographic girth weld testing. Films and secured around each girth weld and exposed to a radioactive source. After exposure they are developed in the mobile dark room located on the back of the truck. Technicians read the films to verify weld integrity. (*Source:* Introduction to Pipeline Construction Training Module. Courtesy Pipeline Knowledge & Development.)

Crossings, Bending, and Tie-ins.   Specialist crews, different from the production crews described previously, typically handle crossings of streams, roads, and other obstacles and perhaps places where bends are required in the otherwise straight or gently undulating pipeline.

Open-trench crossing is the simplest crossing method and requires the least design. As the name implies, the crossing is simply ditched through with a backhoe, the pipe is buried, and the trench is filled. Open-trench crossing installation across two-lane rural or local roadways is common, as is open-trench crossing of small streams or rivers during times of low water flow. Open-trench crossing is usually the least expensive method of crossing and requires only heavier-wall pipe typically buried with 5 or 6 feet of cover to provide additional safety from traffic. For stream and river crossings, concrete-coated pipe is often used for two reasons; (1) negative buoyancy under flood conditions and (2) physical armoring of the pipe from stones and rocks that get washed downstream. In addition to burying the pipe across the stream at least 5 feet below scour depth, after covering the pipe with a safe material to prevent damage to the pipe, the trench is often backfilled with large rip-rap material or commercial bank armoring or stabilization products to preclude erosion along the disturbed trench line.

A second crossing method is the *direct bore* or *tunnel under an obstacle*. Boring is usually applied to short crossings when surface disturbance is unacceptable, such as roadways and railroads. Boring or tunneling consists of excavating a deep trench on both sides of the obstacle. Then a boring machine is placed in the ditch on one side and aligned, and the heavy-wall carrier pipe or a casing pipe is installed just behind the boring drill bit and the pipe is continuously pushed under the obstacle and provides its own structure to maintain the hole until the bore is complete to the other side. Finally, the pipe under the road is welded to the main section of pipe (Figure 26.59).

From a design standpoint, understanding the medium through which the auger will bore is important. Since the bit is not guided, it can be deflected by rocks or boulders surfacing in surprising places.

Finally, a horizontal directional drill (HDD) uses drilling technology developed for the oil and gas industry. An HDD installation is usually a long installation with a radius of bend less than the normal radius of curvature of the pipe. The HDD driller controls the drill bit direction to first install a pilot along the desired route (Figure 26.60).

FIGURE 26.59 Road bore. A tunnel is bored under the road with an auger, and a pipe is inserted. Then the balance of the pipeline is welded to it. (*Source:* Introduction to Pipeline Construction Training Module. Courtesy Pipeline Knowledge & Development.)

FIGURE 26.60 HDD machine. The drill pipe enters the ground midcenter of the picture. Note that the pipe to the left is already welded together and ready to install after the hole is reamed to size. (*Source:* Introduction to Pipeline Construction Training Module. Courtesy Pipeline Knowledge & Development.)

After the pilot hole is installed and the location is verified, a reaming process is started to ream or clean the hole to the required diameter. Then the pipe is pulled into the hole. The pipeline design engineer normally works with the HDD engineer to establish the route, depth, and radius of the hole. Depths on the order of 50 to 75 feet under the feature being crossed are common, and HDDs of several thousand feet are common. Lengths well over a mile have been installed.

When the bend radius is less than can be accomplished with normal pipe flexibility, the contractor employs a skilled surveyor called a *bending engineer*. Special equipment, bending shoes, are applied to generate small, long-radius cold-worked bends so that the pipe will fit the ditch. It is important that the ditch be cut prior to stringing so that the bending engineer can take his or her measurements and bend the pipe as required to fit the contour of the ditch or a horizontal turn in the route.

Tie-in is the process of connecting the various lengths of pipes together at bends and crossings to finally make the pipeline contiguous along its entire length (Figure 26.61).

Hydrostatic Testing.   Between the time the joints of pipe are welded together and prior to putting the pipeline into service, it is tested hydrostatically. That is, it is filled with water, pressured well above its operating range, and held at that pressure for a given time as one last check to ensure that it can safely handle the allowed pressures. Following testing, the water is evacuated, the pipeline is dried, and final connections are made.

Reclamation and Reseeding.   The final step is reseeding of the land to restore to its preconstruction condition to the extent practical. Often the owner's contract will require that the contractor make a follow-up inspection 1 year after construction to correct any erosion issues or repair anyplace where restoration was not complete or acceptable.

FIGURE 26.61 Tieing in. Welders making the final weld to complete this replacement project. Note the pipe on the left, which is being replaced. (*Source:* Introduction to Pipeline Construction Training Module. Courtesy Pipeline Knowledge & Development.)

26.7.6 Station Construction

Unlike pipeline construction, stations are located on a site rather than along a route. Station construction involves mechanical, civil, and electrical components and is quite similar to other types of plant construction projects. Accordingly it is not covered in detail in this text.

26.8 SUMMARY

Pipelines serve a critical energy transportation role and will continue to do so for the foreseeable future. They are safer than any other form of transportation, but when they do fail, the event is usually spectacular, drawing media and public attention. They are efficient—the entire revenue for energy pipelines in the United States is less than $30 billion annually, according to the annual survey published by the *Oil and Gas Journal*. But, as with other modes of transportation, understandably, most people do not want them in their backyards. The largest challenge facing gas and oil pipelines today, then, is balancing the needs (and wants) of the industry's myriad of stakeholders in a safe and environmentally responsible manner, which still operating reliably and efficiently.

PART · V

SAFETY, NOISE, AND AIR QUALITY

CHAPTER 27
TRAFFIC SAFETY*

Rune Elvik

Institute of Transport Economics
Norwegian Center for Transportation Research
Oslo, Norway

27.1 INTRODUCTION

This chapter presents essential elements of traffic safety analysis. The topics covered in this chapter include

1. Basic concepts of road accident statistics
2. Safety performance functions and the empirical Bayes method for estimating safety
3. Identifying and analyzing hazardous road locations
4. The effects on road safety of some common traffic engineering measures
5. How to assess the quality of road safety evaluation studies
6. Formal techniques for setting priorities for safety treatments

27.2 BASIC CONCEPTS OF ROAD ACCIDENT STATISTICS

The basic concepts of road accident statistics were developed by the French mathematician Simeon Denis Poisson more than 150 years ago. Poisson investigated the properties of binomial trials. A binomial trial is an experiment that has two possible outcomes: success or failure. The probability of success is the same at each trial. The outcome of a trial is independent of other trials. Repeated tosses of a coin are an example of a sequence of binomial trials. If a coin is tossed four times, the outcomes can be zero heads, heads once, heads twice, heads three times, or heads all four times. At each trial, the probability of heads (p) is 50 percent. When a coin is tossed four times, the probability of getting heads n times ($n = 0, 1, \ldots, 4$) is

Heads 0 times:	0.0625
Heads 1 time:	0.2500
Heads 2 times:	0.3750
Heads 3 times:	0.2500
Heads 4 times:	0.0625

---

*Reprinted from the First Edition.

The expected number of heads in N trials is $N \cdot p$. Since $p = 0.5$, the expected number of heads when $N = 4$ is 2. Poisson studied what happened to the binomial probability distribution when the number of trials N became very large while at the same time the probability of failure p became very low. Denote the expected value in N trials by λ. Poisson found that the probability of x failures in N trials could be adequately described by the following probability function, which bears his name:

$$P(X = x) = \frac{\lambda^x e^{-\lambda}}{x!} \tag{27.1}$$

The parameter lambda (λ) indicates the expected value of the random variable X, x is a specific value of this variable, e is the base of the natural logarithms ($e = 2.71828$), and $x!$ is the number of permutations of x. If, for example, $x = 3$, then $x! = 1 \cdot 2 \cdot 3$. If $x = 0$, then $x! = 1$. $\lambda = N \cdot p$, when N is very large and p is very small. A random variable is a variable that represents the possible outcomes of a chance process. Translating these abstract terms to a language more familiar to traffic engineers gives

Expected number of accidents (λ) = exposure (N) · accident rate (p)

Accident rate is traditionally defined as the number of accidents per unit of exposure:

$$\text{Accident rate} = \frac{\text{number of accidents}}{\text{units of exposure}}$$

Exposure denotes the number of trials, and a commonly used unit of exposure in traffic engineering is one kilometer of travel (Hakkert and Braimaister 2002). The idea, deeply rooted in probability theory, that the expected number of accidents depends on exposure and accident rate, is perhaps the source of the assumption traditionally made in traffic engineering that one can account for the effects of traffic volume on accidents by using accident rates. However, as will be discussed in the next section, this assumption is no longer tenable.

Definitions of some key concepts are given in Figure 27.1.

Expected number of accidents

The mean number of accidents expected to occur in the long run for a given combination of exposure and accident rate.

Random variation in the number of accidents—regression-to-the-mean

Variation in the recorded number of accidents around a given expected number of accidents. Regression-to-the-mean is the return of an abnormally high or low recorded number of accidents to figures closer to the expected number.

Systematic variation in the number of accidents

Variation in the expected number of accidents (across time, space, modes of transportation, etc.).

Exposure

The volume of an activity exposed to risk. Risk is the product of the probability of an unwanted event (in our case an accident) and the consequences of that event (in our case accident severity).

FIGURE 27.1 Definition of key concepts of accident statistics.

TABLE 27.1   Accident Data for Highway-Railroad Grade Crossings in Norway 1959–1968

| | Highway-railroad grade crossings by type of protective device | | | |
| | Signals only | | Automatic gates | |
Count of accidents	Actual distribution	Poisson distribution	Actual distribution	Poisson distribution
0	108	105	71	68
1	48	54	13	16
2	17	14	1	2
3	3	3	2	1
N	176	176	87	87
Mean	0.517		0.241	
Variance	0.545		0.344	
Chi-square		1.395		2.194
Degrees of freedom		2		2
Exact P-value		0.498		0.334

Source: Amundsen and Christensen (1973).

It is essential to keep in mind that accidents are subject to random variation. The variation observed in counts of accidents will nearly always be a mixture of pure random variation and systematic variation. Two questions immediately come to mind:

1. How can we know if the variation in the count of accidents in a set of study units is not just random?

2. If accidents are found to occur at random, is it still possible to reduce the number of accidents?

Table 27.1 presents data that can be used to answer both of these questions. The table shows the distribution of highway-railroad grade crossings on public roads in Norway by the number of accidents recorded during the 10 years, 1959–1968 (Amundsen and Christensen 1973). There are two categories of grade crossings: those protected by signals only, and those protected by automatic gates. For each category, the first column shows the actual distribution of crossings by number of accidents. There were, for example, 48 crossings protected by signals that had 1 accident during the years 1959–1968. The mean number of accidents per crossing was 0.515, and the variance was 0.545. Next to the actual distribution of accidents is shown the distribution that would be expected if accidents occurred entirely at random. This distribution, the Poisson distribution, was estimated by means of equation (27.1) by inserting the value 0.515 for the mean (λ). It can be seen that the Poisson distribution is very similar to the actual distribution of accidents. Whether the two distributions really differ can be tested by means of a chi-square test. This statistical test is described in any elementary textbook in statistics. For grade crossings protected by signals, the value of chi-square is 1.395. There are two degrees of freedom (the number of categories (4) minus 1, minus the number of parameters of the Poisson distribution). The exact P value is 0.498, which means that the actual distribution of accidents does not differ significantly from the Poisson distribution. An identical analysis for grade crossings protected by automatic gates leads to the same conclusion.

It is fair to conclude on the basis of these tests that accidents occurred at random in grade crossings protected by signals only. Yet had all these grade crossings been protected by automatic gates, the mean number of accidents would have been reduced by more than 50 percent. The mean number of accidents for grade crossings protected by automatic gates was 0.241, which is less than half the mean number of accidents for grade crossings protected by signals only (0.515). A random distribution of accidents in a sample of study units does not necessarily mean that it is impossible to reduce the mean number of accidents. It just means that it is impossible to identify one of the study units as having a higher expected number of accidents than another.

In the Poisson distribution, the variance is equal to the mean. Looking at Table 27.1, the variance for grade crossings protected by signals only was 0.545, which is close to the mean number of accidents (0.515). For the grade crossings protected by automatic gates, however, the variance (0.344) is somewhat larger than the mean (0.241). Since, by definition, the size of the purely random variation in the count of accidents equals the mean number of accidents, the total variation in the count of accidents found in a sample of study units can be decomposed into random variation and systematic variation (Hauer 1997):

$$\text{Total variation} = \text{random variation} + \text{systematic variation} \qquad (27.2)$$

There is systematic variation in number of accidents whenever the variance exceeds the mean. This is usually referred to as overdispersion. The amount of overdispersion found in a data set can be described in terms of the overdispersion parameter, which is defined as follows:

$$\text{Var}(x) = \lambda \cdot (1 + \mu\lambda) \qquad (27.3)$$

Solving this with respect to the overdispersion parameter gives

$$\mu = \frac{\dfrac{\text{var}(x)}{\lambda} - 1}{\lambda} \qquad (27.4)$$

For the distributions listed in Table 27.1, the overdispersion parameter can be estimated to 0.105 for grade crossings protected by signals and 1.773 for grade crossings protected by gates. For grade crossings protected by signals only, the total variance consists of 95 percent random variation and 5 percent systematic variation. For grade crossings protected by gates, there is 70 percent random variation in accidents, 30 percent systematic variation.

A stepwise approach to the statistical analysis of accident occurrence within a system for which a traffic engineer is professionally responsible is suggested in Figure 27.2. The first step of analysis is to define sets of study units. These sets are often categories of roadway elements, such as road sections, intersections, or curves. In a city there may, for example, be a few thousand intersections. It is important that the study units be identically defined and can be counted.

Step 1: Define suitable sets of study units

Examples of study units used in traffic engineering studies include road sections of a given length, intersections, driveways, horizontal curves, highway-railroad grade crossings, bridges, tunnels.

Step 2: Analyze distribution of accidents in each set of study units

For each set of study units defined, the distribution of accidents should be analyzed with respect to the mean number of accidents and the variance.

Step 3: Identify the safety performance function in each set of study units

A safety performance function is an equation that describes the sources of systematic variation in accidents, fitted by means of appropriate multivariate techniques of analysis.

Step 4: Estimate safety for each study unit using the empirical Bayes method

The empirical Bayes method combines information from two clues to safety and can be used to estimate the expected number of accidents for each study unit.

Step 5: Define hazardous road locations and identify them statistically

A hazardous road location is any study unit for which the expected number of accidents is abnormally high.

FIGURE 27.2   Stepwise approach to statistical analysis of accidents.

For each set of study units, the distribution of accidents should be analyzed as illustrated above for grade crossings (step 2 in Figure 27.2). The count of accidents should represent at least a few years of data. The objective of the analysis is to determine the amount of systematic variation in the number of accidents. If there is very little systematic variation, as was the case for grade crossings protected by signals only, there is little point in continuing to the next steps of analysis indicated in Figure 27.2. If, on the other hand, the number of accidents is found to contain significant systematic variation, the next step of analysis is to identify sources of systematic variation (step 3 in Figure 27.2). The effects of the variables that explain systematic variation in the number of accidents are usually summarized in terms of a safety performance function, whose general form will be discussed in the next section. Once a safety performance function has been estimated, the information that this function gives about safety can be combined with the accident records for each study unit to estimate the expected number of accidents for each unit (step 4 in Figure 27.2). Combining these two sources of information about safety is the essential feature of the empirical Bayes method to the estimation of road safety, which is illustrated in the next section.

When the expected number of accident has been estimated for each study unit, a distribution of study units according to the expected number of accidents can be formed and hazardous road locations can be identified (step 5 in Figure 27.2). Several definitions of hazardous road locations can be imagined; common to all definitions is that they are intended to help identify statistically road locations with a high expected number of accidents.

27.3   SAFETY PERFORMANCE FUNCTIONS: THE EMPIRICAL BAYES METHOD FOR ESTIMATING ROAD SAFETY

A safety performance function is any mathematical function that relates the normal, expected number of accidents to a set of explanatory variables. The following form for the safety performance function is widely applied:

$$E(\lambda) = \alpha Q^\beta \, e^{\Sigma \gamma_i x_i} \tag{27.5}$$

The estimated expected number of accidents $E(\lambda)$ is a function of traffic volume Q and a set of risk factors $X_i(i = 1, 2, 3, \ldots n)$. The effect of traffic volume on accidents is modeled in terms of an elasticity, that is, a power β to which traffic volume is raised (Hauer 1995). This elasticity shows the percentage change of the expected number of accidents, which is associated with a 1 percent change in traffic volume. If the value of β is 1.0, the number of accidents is proportional to traffic volume, as traditionally assumed when using accident rates in road safety analysis. If the value of β is less than 1, the number of accidents increases by a smaller percentage than traffic volume. If the value of β is greater than 1, the number of accidents increases by a greater percentage than traffic volume.

A number of recent studies (Fridstrøm et al. 1995; Persaud and Mucsi 1995; Mountain, Fawaz, and Jarrett 1996; Fridstrøm 1999) have estimated safety performance functions for total accidents, fatal and injury accidents, and specific types of accidents (such as single-vehicle and multiple-vehicle). It is typically found that the value of β is less than 1 for fatal accidents, close to 1 for injury accidents, and in some cases greater than 1 for property damage-only accidents. Figure 27.3 shows functions based on a value for β of 0.7 for fatal accidents, 0.9 for injury accidents, and 1.1 for property-damage-only accidents.

The effects of various risk factors that influence the probability of accidents, given exposure, are generally modeled as an exponential function, that is, as e (the base of natural logarithms, see section 27.2) raised to a sum of the product of coefficients γ_i and values of the variables x_i, denoting risk factors. A model of the form shown in equation (27.5) can be fitted with several commercially available computer software packages. When estimating a safety performance function, it is important to specify the distribution of the residual terms correctly. The residual term of a model is the part of systematic variation in accident counts, which is not explained by the model. If a model explains all the systematic variation in accident counts there is in a data set, the

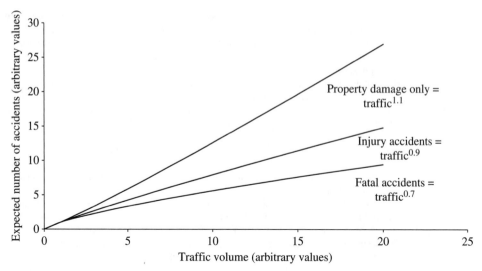

FIGURE 27.3   Typical relationships between traffic volume and the expected number of accidents.

residuals will by definition contain random variation only and can be specified as Poisson distributed. Usually, however, a model will not be able to explain all systematic variation in accident counts. The residuals will then contain some overdispersion, which can usually be adequately described by the negative binomial distribution.

Two questions need to be answered with respect to the use of safety performance functions:

1. How should the variables to be included in a model be selected?
2. How can one determine the success of a model in explaining accidents?

The choice of variables to be included in a safety performance function will often be dictated by the availability of data. In traffic engineering analyses, variables referring to highway design and traffic control would normally be included. Such variables include number of lanes, lane width, number of intersections or driveways per kilometer of road, type of roadside terrain, horizontal curvature, and speed limit. The success of a model in explaining accidents can be evaluated by comparing the overdispersion parameter of a fitted model to the overdispersion parameter in the original data set. Table 27.2 presents a data set for national highways in Norway, showing the number of fatalities per kilometer of road during 1993–2000 (Ragnøy, Christensen, and Elvik 2002). The mean number of road accident fatalities per kilometer of road was 0.0646, and variance was 0.0976. The overdispersion parameter can be estimated to 7.91. A multivariate accident model was fitted to the data, assuming a negative binomial distribution for the residuals. The coefficients estimated for this model are listed in Table 27.2. The overdispersion parameter for the model was 2.39. Inserting this into equation (27.3) gives an estimated variance of 0.0745. The contributions of various factors to the observed variance in accident counts can be determined as follows:

Random variation = 0.0646/0.0976 = 0.662 = 66.2% of all variance

Systematic variation = (0.0976 − 0.0646)/0.0976 = 0.338 = 33.8%

Systematic variation explained by safety performance function = (0.0976 − 0.0745)/0.0976 = 0.237 = 23.7%

Remaining systematic variation not explained by safety performance function = (0.0745 − 0.0646)/0.0976 = 0.101 = 10.1%

TABLE 27.2 Number of Road Accident Fatalities per Kilometer of Road, National Highways, Norway 1993–2000

Number of fatalities	Distribution of road sections by number of fatalities			Safety performance function	
	Actual	Poisson	Negative binomial	Explanatory variables	Coefficient
0	19,957	19,728	19,974	Constant	−7.154
1	895	1,274	854	Ln(AADT)	0.842
2	135	41	163	Speed limit 50 km/hr	Reference category
3	43	1	39	Speed limit 60 km/hr	−0.020
4	9	0	10	Speed limit 70 km/hr	0.385
5	3	0	3	Speed limit 80 km/hr	0.172
6	1	0	1	Speed limit 90 km/hr, rural road	0.090
7	0	0	0	Speed limit 90 km/hr, class B road	0.610
8	1	0	0	Speed limit 90 km/hr, class A road	0.879
N	21,044	21,044	21,044	Number of lanes	−1.967
				Number of intersections/km	0.082
				Dummy for trunk road	0.255
Mean	0.0646				
Variance	0.0976			Estimated variance	0.0745
Overdispersion parameter	7.91			Overdispersion parameter	2.39

Source: Ragnøy, Christensen, and Elvik (2002).

The safety performance function explains $0.237/0.338 = 0.701 = 70.1\%$ of all systematic variation found in this data set.

The coefficients given in the safety performance functions show the effects of each variable, controlling for the other variables included in the model. By inserting these coefficients into equation (27.5), the normal expected number of fatalities during eight years for a road section of 1 kilometer can be estimated for any combination of values observed for the explanatory variables.

What is the best estimate of the long-term expected number of accidents or accident victims for a given roadway element, given the fact that we know some but not all of the factors affecting accident occurrence? According to the empirical Bayes method (Hauer 1997), the best estimate of safety is obtained by combining two sources of information: (1) the accident record for a given site, and (2) a safety performance function, showing how various factors affect accident occurrence. Denote by R the recorded number of accidents and by A the normal, expected number of accidents as estimated by a safety performance function. The best estimate of the expected number of accidents for a given site is then

$$E(\lambda/r) = \alpha \cdot \lambda + (1 - \alpha) \cdot r \qquad (27.6)$$

The parameter α determines the weight given to the estimated normal number of accidents for similar sites when combining it with the recorded number of accidents in order to estimate the expected number of accidents for a particular site. The best estimate of α is

$$\alpha = \frac{1}{1 + \dfrac{\lambda}{k}} \qquad (27.7)$$

λ is the normal expected number of accidents for this site, estimated by means of a safety performance function, and k is the inverse value of the overdispersion parameter of this function, that is, $1/\mu$. To illustrate the use of the empirical Bayes method, suppose that the normal expected number of accidents for a 1-kilometer road section during a period of eight years has been estimated by means of a safety performance function to be 3.73. The over-dispersion parameter for the safety performance function is 0.3345; hence k is 2.99. The weight to be given to the estimate based on the safety performance function thus becomes $1/[1 + (3.73/2.99)] = 0.445$. Seven accidents were recorded. The long-term expected number of accidents is estimated as

$$E(\lambda/r) = 0.445 \cdot 3.73 + (1 - 0.445) \cdot 7 = 5.54$$

The interpretation of the three different estimates of safety can be explained as follows: 3.73 is the number of accidents one would normally expect to occur at a similar site, that is, one that has the same traffic volume, the same speed limit, the same number of lanes, etc., as the site we are considering. Seven accidents were recorded. Part of the difference between the recorded and normal number of accidents for this type of site is due to random variation. An abnormally high number of accidents due to chance cannot be expected to continue; a certain regression-to-the-mean must be expected. In the example given above, the regression-to-the-mean expected to occur in a subsequent 8-year period is $(7 - 5.54)/7 = 0.209 = 20.9$ percent. The difference between the site-specific expected number of accidents (5.54) and the normal, expected number of accidents for similar sites (3.73) can be interpreted as an effect of local risk factors for the site, causing it to have a higher expected number of accidents than similar sites.

Applying the empirical Bayes method to a site that had 0 recorded accidents, but was otherwise identical to the site used as an example above, gives a site-specific expected number of accidents during 8 years of 1.66 (the normal, expected number of accidents for similar sites was 3.73). For this site, the difference between the site-specific expected number of accidents and the normal, expected number of accidents for similar sites can be interpreted as the effect of local safety factors, that is, factors causing the site to be safer than otherwise similar sites.

27.4   IDENTIFYING AND ANALYZING HAZARDOUS ROAD LOCATIONS

No standard definition exists of a hazardous road location. The following definition conforms to common usage of the term (Elvik 1988): A hazardous road location is any site at which the site-specific expected number of accidents is higher than for similar sites, due to local risk factors present at the site. Hazardous road locations should be identified in terms of the expected number of accidents, not the recorded number of accidents. An abnormally high recorded number of accidents could to a large extent be the result of random variation and does not necessarily mean that a site has a high expected number of accidents.

The empirical Bayes method illustrated in section 27.3 offers an ideal way of identifying hazardous road locations. For each road location, one would estimate the normal, expected number of accidents by means of a safety performance function. This would then be combined with the accident record for each site, yielding an estimate of the site-specific expected number of accidents. A road location would be considered hazardous if the site-specific expected number of accidents was (substantially) higher than the normal, expected number of accidents for similar sites. This approach to the identification of hazardous road locations conforms to the definition of the concept given above and utilizes all information that gives clues to safety.

Once hazardous road locations have been identified statistically, the task of finding remedies to reduce the hazards can be started. The traditional approach to this task has been to perform a detailed analysis of accidents, producing collision diagrams, analyzing the distribution of accidents by time of the day, road surface condition, type of accident, and so on. In addition to analyzing accidents, elements of the traditional approach to black spot treatment include site visits, detailed traffic counts, filming the site, and driving through it from all directions. Although this kind of detailed examination can no doubt give valuable information, there are inherent limitations of the approach that render it less than perfectly reliable as a means to test hypotheses about why there is a concentration of accidents at a particular site.

Accident analysis at hazardous road locations amounts to proposing hypotheses based on an analysis of known data, which means that the data that generated the hypotheses cannot also be used for testing them. Take the analysis presented in Table 27.3 as an example. Eight accidents were recorded at a hazardous road location. The analysis shows that five of the eight accidents were pedestrian accidents, whereas one would normally expect one in eight accidents to involve pedestrians. If the distribution of accidents by type is modeled as a binomial trial (each accident is either a pedestrian accident or nonpedestrian accident), it is found that recording five pedestrian accidents in a total of eight accidents is a highly unlikely outcome, given that the initial probability of a pedestrian accidents is 0.125 (the probability of a nonpedestrian accident is 0.875).

On the whole, the predominant accident pattern found in Table 27.3, pedestrian accidents occurring at night on a wet road surface, suggests that local risk factors related to the amount of pedestrian traffic, road surface friction, and visual obstructions may be present at the location. Yet a more careful investigation would be needed in order to determine whether the factors suggested are actually responsible for the abnormally high number of pedestrian accidents at this particular location.

Convincing evidence showing that current methods for analyzing accidents at hazardous road locations are able to separate the true black spots from the false ones does not exist. In a very interesting paper, Jarrett, Abbess, and Wright (1988) estimated the regression-to-the-mean effect for two samples of road accident black spots. These black spots had been identified according to the recorded number of accidents only, not the expected number of accidents. One of the samples consisted of black spots that had been selected for treatment but for which the treatment had, for various reasons, been deferred or given up. Accident analyses had been made for these candidate sites, with a view to identifying the dominant accident pattern for the purpose of selecting an appropriate treatment. If these analyses were really able to tell the true black spots from the false ones, one would expect the regression-to-the-mean effect to be smaller for the candidate sites than for black spots in general. But Jarrett, Abbess, and Wright found no evidence for this. The number of accidents regressed to a lower mean value by almost the same percentage at the candidate sites as for black spots in general. In other words, accident analyses did not correctly identify those sites that had a higher long-term expected number of accidents.

These results are supported indirectly in a study reported by Elvik (1997). This study investigated the effects attributed to black spot treatment depending on which confounding factors evaluation studies controlled for. It was found that the effect of black spot treatment was zero in the best-controlled studies. Belief in the effectiveness of black spot treatment seems at least in part to have rested on an uncritical acceptance of poor evaluation studies whose findings may to a great extent reflect the effects of uncontrolled confounding factors.

A better way to analyze hazardous road locations for the purpose of identifying contributing risk factors and propose treatment is by means of a blinded matched-pair comparison. The essential elements of this approach can be stated as follows:

1. For each hazardous road location, find a safer-than-average comparison location, matched as closely to the hazardous road location as possible with respect to variables included in the safety performance function.

2. For each matched pair of sites, search for local risk factors or safety factors from a list of factors drawn up on the basis of the analysis of accidents at the hazardous road location.

3. Blind analysts to accident records. Analysts should not know which site was hazardous and which site was safer than average.

The use of this approach is shown in Table 27.4. Hazardous and safe sites are matched in pairs according to the values observed for the variables included in the safety performance function. Two matched pairs are shown in Table 27.4. Once the pairs have been formed, each site is inspected and data collected regarding local risk factors. A sample of such data, not necessarily exhaustive, is shown in Table 27.4.

In the first pair of sites it was found that road surface friction was significantly worse, there were more pedestrians crossing the road, and there were more sources of visual obstruction at the hazardous site than at the safe site. This information confirms the hypotheses regarding contributing factors proposed on the basis of the analysis of accidents. The analysis has therefore successfully identified local risk factors. Keep in mind that the analysts identifying risk factors should be blinded to accident records to prevent their knowledge of accident records from biasing their observations.

TABLE 27.3 Hypothetical Results of Accident Analysis at a Hazardous Road Location

Accident number	Type of accident	Time of day	Road surface	Vehicles involved	Alcohol involved	Excessive speed	Failure to see
1	Pedestrian	11 PM	Wet	Car	Yes, pedestrian	Yes	Yes
2	Rear-end	10 AM	Wet	Truck	No	No	No
3	Rear-end	5 PM	Dry	Car	No	No	No
4	Pedestrian	8 PM	Dry	Car	No	Yes	No
5	Pedestrian	9 PM	Wet	Car	Yes, pedestrian	No	Yes
6	Pedestrian	11 AM	Wet	Car	No	Yes	Yes
7	Overturning	1 PM	Dry	Motorcycle	No	Yes	No
8	Pedestrian	11 PM	Wet	Truck	Yes, pedestrian	No	Yes
Key finding	5 pedestrian	4 in evening	5 on wet road	Nothing abnormal	3 with alcohol	4 speeders	4 did not see
Normal value	1 pedestrian	2 in evening	2 on wet road		1 with alcohol	3 speeders	2 did not see
P value (binomial)	0.0011	0.0865	0.0231	Not tested	0.0561	0.2112	0.0865
Predominant accident pattern	The predominant type of accident is a pedestrian accident at night on a wet road surface, in which the parties did not see each other. Some over-involvement of alcohol among pedestrians.						

TABLE 27.4 Verification of Traditional Accident Analysis by Identification of Risk Factors Contributing to Accidents

	Case 1: Local risk factors successfully identified		Case 2: Local risk factors not identified	
	Hazardous	Safe	Hazardous	Safe
Matching variables	AADT, speed limit, number of lanes, number of intersections, trunk road status = nearly same values observed for both sites		AADT, speed limit, number of lanes, number of intersections, trunk road status = nearly same values observed for both sites	
Local risk factors (sample):				
Road surface friction—dry	0.70	0.82	0.78	0.77
Road surface friction—wet	0.25	0.48	0.47	0.49
Pedestrians crossing per day	2,500	1,000	1,200	1,250
Sources of visual obstruction	5	2	3	3
Minimum sight distance (m)	100	155	110	115
Driveways per km of road	2	2	0	0
Public bar nearby	Yes	No	No	No
Accident records:	8 accidents in total, of which 5 involving pedestrians on a wet road surface	0 accidents	7 accidents in total, but no clear pattern	0 accidents

The other case shown in Table 27.4 was less successful. It turned out that there were no differences between the hazardous and the safe site with respect to the risk factors surveyed. Hence, accidents must be attributed to other factors, for example, a widespread violation of speed limits or other traffic control devices.

The implications for safety treatment of a successful analysis of local risk factors at a hazardous road location are rarely obvious. For the successful case presented in Table 27.4, one can imagine many treatments that might be effective, including lowering the speed limit to allow pedestrians and cars more time for observation, removing obstacles to vision, improving road surface friction, providing or upgrading road lighting, upgrading pedestrian crossing facilities, conducting random breath testing (there was a bar nearby), or stepping up police enforcement in general. It is not obvious which of these measures would be the most cost-effective; choice of treatment should rely on a detailed analysis of the cost-effectiveness of alternative treatments.

27.5 EFFECTS ON ROAD SAFETY OF SELECTED TRAFFIC ENGINEERING MEASURES

A very large body of information exists concerning the effects on road safety of traffic engineering measures. Unfortunately, as pointed out by Elvik (2002b) and Hauer (2002), not all of this information can be trusted; in fact, some of it deserves to be labeled disinformation. The next section of this chapter attempts to explain why this is so and provides some advice to traffic engineers about how to read evaluation studies critically and how to extract information from such studies. Information is given on the effects of selected highway traffic engineering and traffic control measures (Table 27.5) for which meta-analyses summarizing evidence from several studies have been reported.

Amundsen and Elvik (2002) summarized evidence on the road safety effects of constructing or upgrading urban arterial roads. All the studies included in their review controlled for regression-to-the-mean and long-term trends in the number of accidents. It was found that the construction of new urban arterial roads tends to induce new traffic, which offsets the reduction of the accident rate, resulting in very small changes in the number of accidents. Upgrading urban arterial roads by means of lane additions and a median is associated with a large reduction of the number of accidents.

TABLE 27.5 Best Estimates of the Effects on Accidents of Selected Highway and Traffic Engineering Measures

Measure	Accidents affected	Fatal accidents		Injury accidents		Property-damage-only accidents	
		Best estimate	95% CI	Best estimate	95% CI	Best estimate	95% CI
New urban arterial roads	All accidents	-14	(-50, +50)	-1	(-9, +8)	Not available	Not available
Lane additions and median	All accidents	-71	(-91, +35)	-51	(-65, -32)	+15	(+8, +22)
Bypass roads	All accidents	-17	(-56, +57)	-25	(-33, -16)	-27	(-38, -13)
Roundabouts (yield on approaches)	At intersections	-81	(-93, -50)	-47	(-56, -34)	+1	(-19, +25)
Road lighting (previously unlit)	In darkness	-64	(-74, -50)	-30	(-35, -25)	-18	(-22, -13)
Upgrading road lighting	In darkness	-50	(-79, +15)	-32	(-39, -25)	-17	(-22, -11)
Guard rails along roadside	Run-off-road	-44	(-56, -32)	-47	(-52, -41)	-7	(-35, +33)
Median guard rails	Median crossings	-43	(-53, -31)	-30	(-36, -23)	+24	(+21, +27)
Area wide urban traffic calming	All accidents	-30	(-78, +128)	-11	(-21, -1)	-5	(-19, +11)
Treatment of hazardous curves	In curves	Not available	Not available	-16	(-35, +9)	-18	(-44, +21)
Speed limit (mean speed -5%)	All accidents	-6	(-11, -2)	-10	(-11, -9)	Not available	Not available
Speed limit (mean speed -10%)	All accidents	-24	(-32, -15)	-17	(-19, -15)	Not available	Not available
Speed limit (mean speed +5%)	All accidents	+17	(+13, +22)	+19	(+16, +21)	Not available	Not available
Speed limit (mean speed +10%)	All accidents	Not available	Not available	+8	(-2, +20)	Not available	Not available

Source: See text, section 27.5

Bypass roads around small towns have been found to reduce the number of accidents by about 25 percent (Elvik, Amundsen, and Hofset 2001).

The conversion of intersections to roundabouts (Elvik 2003), the installation of road lighting (Elvik 1995a), and the provision of guardrails (Elvik 1995b) reduce both the number and severity of accidents. The effects of area-wide urban traffic calming schemes (Bunn et al. 2003) and of treatment of horizontal curves (Elvik 2002a) are somewhat more uncertain. The effects of speed limits can be appraised in terms of the effects on accidents of certain percentage changes in the mean speed of traffic. On the average, a 10 km/hr change in the speed limit can be expected to result in a change of 3 to 4 km/hr in the mean speed of traffic (Elvik, Mysen, and Vaa 1997). Nearly all studies evaluating the effects on accidents of changes in speed have found that increases in speed are associated with increases in the number of fatal and injury accidents, whereas reductions in speed are associated with reductions in the number of fatal and injury accidents. The effect of changes in speed on property-damage-only accidents is less well known.

27.6 HOW TO ASSESS THE QUALITY OF ROAD SAFETY EVALUATION STUDIES

As noted above, an enormous number of road safety evaluation studies exist. How can users of this research assess whether or not to trust the findings reported by these studies? The first thing a reader should look for is a description of study design. If a study employed an experimental design involving the random assignment of safety treatment to study units, it is in most cases safe to assume that the study shows the effects of the safety treatment only and has, by way of randomization, controlled for all confounding factors. Most road safety evaluation studies do not employ an experimental study design. Various versions of before-and-after designs are common. Hauer (1997) gives a comprehensive guide to the critical assessment of before-and-after studies.

One may assess the credibility of before-and-after studies by checking whether the studies have controlled for important confounding factors such as (Elvik 2002a)

1. Regression-to-the-mean

2. Long-term trends affecting the number of accidents or injured road users (during several before periods)

3. General changes of the number of accidents from before to after the road safety measure is introduced (from one before period to one after period)

4. Changes in traffic volume

5. Any other specific events introduced at the same time as the road safety measure

A confounding variable is any exogenous (i.e., not influenced by the road safety measure itself) variable affecting the number of accidents or injuries whose effects, if not estimated, can be mixed up with effects of the measure being evaluated. Two main approaches can be taken to control these variables: (1) estimate the effects of a confounding factor statistically, and (2) use a comparison group. In many cases, both approaches will be used in the same study. Regression-to-the-mean is usually controlled for by means of statistical estimation. The effects on accidents of changes in traffic volume are often also estimated statistically. A comparison group is used to control for all confounding factors whose effects cannot be estimated statistically.

The more confounding factors an evaluation study has controlled for, the less likely are the results to have been caused by confounding factors not controlled for by the study. The results of before-and-after studies of road safety measures can be greatly influenced by the approach taken to controlling for confounding factors in such studies. There is a tendency for studies that do not control for any confounding factors to exaggerate the effects of the road safety measure being evaluated. In some cases this exaggeration can be substantial, amounting to a difference of some 20 to 30 percent in the size of the effect attributed to the road safety measure.

Readers of observational before-and-after studies of road safety measures should always pay very careful attention to what these studies say about control of confounding factors. If a study does not state explicitly that it controlled for a certain confounding factor, one is almost always right in believing that the study did not control for that factor. Simple before-and-after studies, which do not control for any confounding factors, should never be trusted.

Provided a study has controlled adequately, but perhaps not perfectly, for important confounding factors, the most important aspect of study quality is the size and diversity of the accident sample. The larger the accident sample, the more precise are the estimates of the effects of the safety measure. Moreover, a large accident sample may permit estimates to be made of the effect of a measure for different types of accidents, different levels of accident severity, and different types of traffic environment. A common problem in many studies evaluating traffic engineering safety treatments is that little or no information is given concerning how the sample of sites was obtained. This is regrettable, as it is then difficult to know the generality of study findings.

27.7   FORMAL TECHNIQUES FOR PRIORITY SETTING OF ROAD SAFETY MEASURES

Broadly speaking, three formal techniques have been developed to help policymakers choose the most effective road safety measures from a set of potentially effective measures:

1. Cost-effectiveness analysis, which seeks to identify those road safety measures that give the largest reductions in accidents or injuries per dollar spent to implement the measures.
2. Cost-benefit analysis, which seeks to identify those road safety measures for which the benefits are greater than the costs. All benefits are converted into monetary terms.
3. Multiattribute utility analysis, which seeks to maximize the overall attainment of a set of goals, for each of which the utility function of the responsible decision makers is determined.

The simplest of these techniques is cost-effectiveness analysis. It does not require a monetary valuation of accidents or injuries, nor of any other relevant policy objective. A cost-effectiveness analysis simply estimates the cost-effectiveness ratio, which can be defined as

$$\text{Cost-effectiveness ratio} = \frac{\text{number of accidents prevented}}{\text{cost of measure}}$$

The number of accidents prevented forms the numerator, consistent with the idea that one wants to maximize the cost-effectiveness ratio. Cost refers to the direct costs of implementing the measure. There are three limitations of cost-effectiveness analysis:

1. The concept of cost-effectiveness becomes a problem if accidents of different severities are to be considered. It may then be necessary to estimate a cost-effectiveness ratio for each level of accident severity and then compare ratios across levels of severity.
2. Cost-effectiveness analysis does not include a criterion stating when a certain measure should be regarded as cost-ineffective, that is, as giving too small safety benefits compared to the costs of the measure. Cost-effectiveness analysis can only be used to rank order measures by cost-effectiveness.
3. Cost-effectiveness analysis cannot be used to make tradeoffs against other policy objectives. It seeks to maximize a single objective only, that of preventing accidents or injuries.

Cost-benefit analysis seeks to overcome these limitations of cost-effectiveness analysis. Accidents or injuries of different severities are made comparable by estimating the benefits to society, stated in monetary terms, of preventing them. Measures are rejected as inefficient if benefits are smaller than costs. Tradeoffs against other policy objectives are made possible by converting all policy objectives to monetary terms. As far as road safety policy is concerned, the most important potentially conflicting policy objectives are those related to travel time, costs of transport (vehicle operating costs), and quality of the environment (noise, air pollution).

Estimates of road accident costs to be used in cost-benefit analysis have been made in most highly motorized countries. For the United States, the National Safety Council recommends the following comprehensive costs of traffic injury for use in cost-benefit analyses (National Safety Council 2002; U.S. dollars 2000 prices):

Injury severity	Cost per injured person (US$ 2000)
Fatal injury	3,214,290
Incapacitating injury	159,449
Nonincapacitating evident injury	41,027
Possible injury	19,528
No injury	1,861

These cost estimates refer to the so-called KABCO scale for injury severity, used by the police in reporting accidents to state and federal highway agencies.

Multiattribute utility analysis is rarely applied in formal analyses of road safety measures. It is a complex technique, conceptually closely related to cost-benefit analysis but differing from it by not requiring relevant policy impacts to be converted to monetary terms. Each policy impact is expressed in "natural" units (number of accidents, hours or minutes of travel time, vehicle operating costs in dollars, noise in decibels, and so on), but a preference function (utility function) is defined for each impact and for the pairwise trade-offs between sets of impacts. Readers will find an excellent introduction to the technique in a book by Keeney and Raiffa (1976).

Regardless of which of the three formal techniques for assessing the effectiveness of road safety measures a policy-maker wants to rely on, the first step of any formal analysis is to survey potentially effective road safety measures. Figure 27.4 lists the essential steps of the application to road safety of formal techniques of priority setting.

Step 1: Survey potentially effective road safety measures

A measure is potentially effective if (1) credible evaluation studies have found that it improves safety, or (2) it favorably affects one or more risk factors known to contribute to accidents or injuries, and (3) it has not yet been fully implemented.

Step 2: Estimate costs and effects of each potentially effective measure

Assess the scale of use conceivable for each potentially effective road safety measure and estimate the attendant costs and likely effects on safety.

Step 3: Choose a formal criterion of efficiency for ranking measures

There are three possible criteria: (1) cost-effectiveness, (2) benefit-cost ratio, and (3) multiattributive utility value.

Step 4: Do a marginal analysis of efficiency for each measure

The objective of a marginal analysis is to determine the marginal costs and marginal benefits of each measure.

Step 5: Choose those measures that maximize overall efficiency

Determine the mix of measures that maximizes overall efficiency according to the criterion chosen (cost-effectiveness, benefit-cost ratio, or overall utility).

FIGURE 27.4 Steps in the application of formal techniques of priority setting for road safety measures.

It is essential to conduct a broad survey of potentially effective road safety measures, to make sure that one does not miss the most efficient measures (step 1 in Figure 27.4). Once a list of potentially effective road safety measures has been made, the scale of use of each measure should be assessed (step 2 in Figure 27.4). By scale of use is meant the number of locations where it is, in principle, possible to introduce the measure. This could refer to the number of intersections that can be converted to roundabouts, the length of road for which public lighting can be provided, and so on. Based on an assessment of the scale of use of each measure, it is possible to develop alternatives for the use of each measure, such as: convert 10 intersections to roundabouts, convert 20, convert 30, and so on. A formal criterion of efficiency by which to compare all measures should be chosen (step 3 in Figure 27.4). The most frequently used criterion is probably the benefit-cost ratio.

A marginal analysis of each road safety measure is essential (step 4 in Figure 27.4). By a marginal analysis is meant an analysis of the additional costs and additional benefits of increasing the use of a measure by one unit. This refers, for example, to the additional costs and benefits of the eleventh conversion of an intersection to a roundabout, when the first 10 conversions are ranked in order of declining benefit-cost ratio. A marginal analysis is needed if one wants to put together an optimal mix of safety measures, that is, a mix that gives the largest benefits for a given cost. Once the marginal benefits and marginal costs of all measures are known, an optimal mix is obtained by using each measure up to the point where marginal benefits equal marginal costs (step 5 in Figure 27.4).

REFERENCES

Amundsen, A. H., and R. Elvik. 2002. *Evaluering av hovedvegomlegginger i Oslo.* TØI Report 553, Institute of Transport Economics, Oslo.

Amundsen, F. H., and P. Christensen. 1973. *Statistisk opplegg og bearbeiding av trafikktekniske effektma°linger.* TØI Report, Institute of Transport Economics, Oslo.

Bunn, F., T. Collier, C. Frost, K. Ker, I. Roberts, and R. Wentz. 2003. "Area-Wide Traffic Calming for Preventing Traffic Related Injuries." Cochrane Collaboration systematic review. *The Cochrane Library* 3.

Elvik, R. 1988. "Ambiguities in the Definition and Identification of Accident Black Spots." In *Proceedings of International Symposium on Traffic Safety Theory and Research Methods,* Amsterdam, April 26–28, Session 1, Context and scope of traffic-safety theory.

———. 1995a. "A Meta-Analysis of Evaluations of Public Lighting as an Accident Countermeasure." *Transportation Research Record* 1485, 112–23.

———. 1995b. "The Safety Value of Guardrails and Crash Cushions: A Meta-analysis of Evidence from Evaluation Studies." *Accident Analysis and Prevention* 27:523–49.

———. 1997. "Evaluations of Road Accident Blackspot Treatment: A Case of the Iron Law of Evaluation Studies?" *Accident Analysis and Prevention* 29:191–99.

———. 2002a. "The Importance of Confounding in Observational Before-and-After Studies of Road Safety Measures. *Accident Analysis and Prevention* 34:631–35.

———. 2002b. "Measuring the Quality of Road Safety Evaluation Studies: Mission Impossible?" Paper presented at Transportation Research Board annual meeting, Washington, DC, January, special session 539. Available on request.

———. 2003. "Effects on Road Safety of Converting Intersections to Roundabouts: A Review of Evidence from Non-US studies." Paper 03-2106, Transportation Research Board, annual meeting, Washington, DC, January.

Elvik, R., F. H. Amundsen, and F. Hofset. 2001. "Road Safety Effects of Bypasses." *Transportation Research Record* 1758, 13–20.

Elvik, R., A. B. Mysen, and T. Vaa. 1997. *Trafikksikkerhetshåndbok,* 3rd ed. Oslo: Institute of Transport Economics.

Fridstrøm, L. 1999. *Econometric Models of Road Use, Accidents, and Road Investment Decisions,* vol. 2. Report 457, Institute of Transport Economics, Oslo.

Fridstrøm, L., J. Ifver, S. Ingebrigtsen, R. Kulmala, and L. Krogsgård Thomsen. 1995. "Meauring the Contribution of Randomness, Exposure, Weather, and Daylight to the Variation in Road Accident Counts." *Accident Analysis and Prevention* 27:1–20.

Hakkert, A. S., and L. Braimaister. 2002. *The Uses of Exposure and Risk in Road Safety Studies.* Report R-2002-12, SWOV Institute for Road Safety Research, Leidschendam, The Netherlands.

Hauer, E. 1995. "On Exposure and Accident Rate." *Traffic Engineering and Control* 36:134–38.

———. 1997. *Observational Before-After Studies in Road Safety.* Oxford: Pergamon Press.

———. 2002. "Fishing for Safety Information in the Murky Waters of Research Reports." Paper presented at Transportation Research Board annual meeting, Washington, DC, January, special session 539. Available at www.roadsafetyresearch.com.

Jarrett, D. F., C. R. Abbess, and C. C. Wright. 1988. "Empirical Estimation of the Regression-to-Mean Effect Associated with Road Accident Remedial Treatment." In *Proceedings of International Symposium on Traffic Safety Theory and Research Methods,* Amsterdam, April 26–28, Session 2, Models for evaluation.

Keeney, R. L., and H. Raiffa. 1976. *Decisions with Multiple Objectives: Preferences and Value Tradeoffs.* New York: John Wiley & Sons.

Mountain, L., B. Fawaz, and D. Jarrett. 1996. "Accident Prediction Models for Roads with Minor Junctions." *Accident Analysis and Prevention* 28:695–707.

National Safety Council. 2000. "Estimating the Cost of Unintentional Injuries. Paper accessible at www.nsc.org. Accessed November 7, 2002.

Persaud, B. N., and K. Mucsi. 1995. "Microscopic Accident Prediction Models for Two-Lane Rural Roads." *Transportation Research Record* 1485, 134–39.

Ragnøy, A., P. Christensen, and R. Elvik. 2002. *Skadegradstetthet. Et nytt mål på hvor Yarlig en vegstrekning er.* TØI-report 618, Institute of Transport Economics, Oslo.

CHAPTER 28
TRANSPORTATION HAZARDS*

Thomas J. Cova
Center for Natural and Technological Hazards
Department of Geography, University of Utah
Salt Lake City, Utah

Steven M. Conger
Center for Natural and Technological Hazards
Department of Geography, University of Utah
Salt Lake City, Utah

28.1 INTRODUCTION

Transportation systems are designed to move people, goods, and services efficiently, economically, and safely from one point on the earth's surface to another. Despite this broad goal, there are many environmental hazards that commonly disrupt or damage these systems at a variety of spatial and temporal scales. Whereas road-curve geometry and other engineered hazards can be addressed through design (Persaud, Retting, and Lyon 2000), hazards such as extreme weather, landslides, and earthquakes are much more difficult to predict, manage, and mitigate. These adverse events can dramatically reduce network serviceability, increase costs, and decrease safety. The economic livelihood of many individuals, firms, and nations depends on efficient transportation, and this is embodied in 20th-century innovations like just-in-time manufacturing and overnight shipping. As the movement of people, goods, and services increases at all scales due to population growth, technological innovation, and globalization (Janelle and Beuthe 1997), the systematic study of these events becomes increasingly important.

Research in the area of transportation hazards aids governments in allocating scarce resources to the four phases of emergency management: mitigation, preparedness, response, and recovery. New fields of study are emerging to address this need, as in the case of *Highway Meteorology*, which focuses on the adverse effects of extreme weather on transportation systems (Perry and Symons 1991). The growing importance of this particular field in the United States can be seen in the recent publication of *Weather Information for Surface Transportation—National Needs Assessment Report* (OFCM 2002). Some transportation agencies organize special teams to manage and mitigate the effects of one or more of these hazards. Recurrence intervals for an event span from daily to centuries, while the associated consequences range from inconvenient to catastrophic. In some cases one event may cause another–torrential rain can trigger a landslide that blocks a road. Some occur unexpectedly, while others arrive with significant warning, but all are amenable to some level of prediction and mitigation.

---

*Reprinted from the First Edition.

Transportation systems also create hazards. Accelerated movement comes with risks, and the corresponding accidents that occur disrupt lives and transportation systems daily. Vehicles collide, trains derail, boats capsize, and airplanes crash often enough to keep emergency managers and news reporters busy. The transportation of hazardous materials (HazMat) is a controversial example in this regard because it places substantial involuntary risks on proximal people and the environment. From the *Lusitania* to the World Trade Center, we are occasionally reminded that transportation disasters can be intentional acts. Lesser-known transportation hazards include elevated irrigation canals, gas pipelines, and electrical transmission lines. Intramodal risks are present in many transportation systems, as in wake turbulence behind large aircraft (Gerz, Holzapfel, and Darracq 2002; Harris et al. 2002), but intermodal risks are also a significant factor—a train might collide with a truck at an at-grade crossing (Austin and Carson 2002; Panchaanathan and Faghri 1995), or a river barge might bump a bridge, leading to the derailment of a train.

Transportation systems that are disrupted by a hazardous event also play a critical role in emergency management. Transportation lifelines are generally considered the most important in an emergency because of their vital role in the restoration of all other lifelines. Emergency managers must route personnel to an accident site, restore lifelines, relocate threatened populations, and provide relief, all of which rely on transportation. Research in this area is increasing, and there are many methods and tools to aid in addressing problems in this domain. The 2000 Cerro Grande Fire in Los Alamos, New Mexico, is a case where a low-capacity transportation network was partially disabled yet successfully used to manage a large fire and safely relocate more than 10 thousand residents.

This chapter reviews recent research and practice in three areas related to transportation and hazards: environmental hazards to transportation systems, transportation risks to proximal people and resources, and the role of transportation in emergency management.

28.2 HAZARD, VULNERABILITY, AND RISK

The study of adverse transportation events can be broadly divided into transportation *hazard analysis, vulnerability analysis,* and *risk analysis.* The focus in hazard analysis is identifying threats to a transportation system, its users, and surrounding people and resources. This is also referred to as *hazard identification.* The term *hazard* is often used to refer to environmental threats like fog, wind, and floods, but transportation hazards exist at all scales, from a sidewalk curb that might trip a pedestrian to the potential for sea-level rise to flood a coastal highway. In the most general sense, a hazard is simply a threat to people and things they value. Vulnerability analysis focuses on variation in the susceptibility to loss from hazardous events. Vulnerability can be viewed as the inverse of resilience, as resiliency implies less susceptibility to shocks. Risk analysis incorporates the likelihood of an event and its consequences, where an event can range from a minor road accident to a dam break that inundates an urban area. For example, identifying the lifelines in a given area that might be compromised by a landslide would be transportation hazard analysis. The loss of a lifeline to a landslide, or a reduction in its service, will have varying consequences depending on the design of the lifeline, its importance in the system, and the spatial economic consequences to the region. Analyzing this variation would constitute vulnerability analysis. In risk analysis, the likelihood of a landslide and its associated consequences would both be incorporated, often with the goal of identifying potential landslides that represent an "unacceptable" risk. The following sections review these three areas in greater depth.

28.2.1 Hazard Analysis

There are many questions that drive transportation hazard analysis. In the simplest case, we could assemble a list of the potential hazards that might affect transportation systems in a region. This could be accomplished by creating a hazard matrix (hazard against travel mode) that indicates whether a given hazard threatens a mode. The next level would be to identify where and when these events might occur. This is typically approached from two perspectives. In one case, we might map the potential for each hazard in a region and overlay areas of high hazard with road, rail, pipeline, and transmission networks to identify points where the two coincide. In the second case, we could select a link and inventory its potential hazards. The first approach requires a method for hazard mapping. This can be further divided into deductive and inductive modeling approaches to hazards mapping (Wadge, Wislocki, and Pearson 1993). In a deductive approach, an analyst builds a physical process model using governing equations. For example, if landslides are the hazard in question, one could use slope instability equations to determine landslide hazard along a road. In an inductive approach to landslide hazard mapping, an empirical study is undertaken to map past events to determine the conditions that lead to their occurrence. Areas with similar characteristics are then identified, often with techniques in map overlay, because they may also be hazardous. The line between inductive and deductive approaches should not be drawn too sharply, because most hazard analyses rely on both. For example, past events may be studied to help build a deductive process model.

There are a number of important dimensions in transportation hazard analysis, most notably the spatial and temporal scales. The spatial scale includes both the extent of the study and the resolution or detail. The spatial extent might be global, national, regional, local, or an individual link in a network. Detail and spatial extent are correlated, but as computer storage continues to increase, this is weakening, and we may soon see national (or larger) studies with very fine spatial and temporal detail. The temporal extent and resolution are also important. A central question is the time-horizon of the study, which can range from a single time period (cross-sectional) to any duration (longitudinal). Time is also important because of the many cycles that affect the potential for hazards. Road icing is most common at night in the winter, and thus it varies seasonally and diurnally. Landslides occur more often during the rainy season, avalanches occur in the winter, and fires occur during the dry season. Figure 28.1 depicts the changing likelihood of hazardous events over time, and this becomes more important in risk analysis.

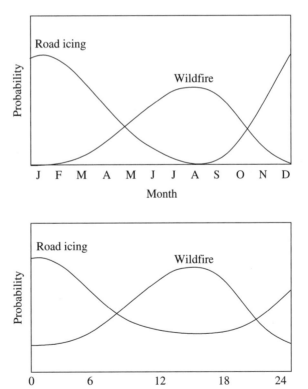

FIGURE 28.1 Seasonal and diurnal variation in the probability for two North American transportation hazards.

28.2.2   Vulnerability Analysis

Vulnerability is an increasing focus in researching threats to transportation systems (Berdica 2002; Lleras-Echeverri and Sanchez-Silva 2001; Menoni et al. 2002). There are many definitions of hazard vulnerability in the research literature (Cutter 1996). As noted, vulnerability in a transportation context recognizes that susceptibility is not uniform across people, vehicles, traffic flow, infrastructure, and the environment. Vulnerability can refer to the physical vulnerability of the users or the potential for an incident to decrease the serviceability of the transportation system. Vulnerability in a transportation context can also be approached from the point of view of network *reliability,* as a reliable network is less vulnerable, and Berdica (2002) links these two concepts. For an example of differing road network vulnerability, a road accident in a two-way tunnel may temporarily cripple a regional transportation system, leading to significant delays, but a system with a separate tunnel in each direction would be less vulnerable to an incident halting traffic in both directions. People and environmental resources in proximity to a transportation corridor are also vulnerable to adverse events. For example, in transporting hazardous materials along a populated corridor, vulnerability along the corridor may vary significantly from point to point, and two potential incidents a few miles apart can have very different outcomes. There are also regional economic vulnerabilities because adverse events can disrupt commerce. Individuals can miss meetings, retail outlets can lose customers, commodities can be delayed, and tourism can be adversely impacted, all of which have economic consequences.

28.2.3   Risk Analysis

The most common definition of risk incorporates both the likelihood of an event and its consequences. It is not possible to avoid all risks, only to choose from risk-benefit tradeoffs (Starr 1969). Kaplan and Garrick (1981) define risk as a set of triplets:

$$(s, p, c) \tag{28.1}$$

where s is a scenario, p its probability, and c its consequences. Risk analysis can be viewed as the process of enumerating all triplets of interest within a spatial and temporal envelope. The probability of a scenario varies inversely with its consequences, which is embodied in the concept of a risk curve (Figure 28.2). In Kaplan and Garrick's framework, the definition of a scenario can be arbitrarily precise. For example, one scenario might be an intoxicated driver crashing on a wet road at night, while

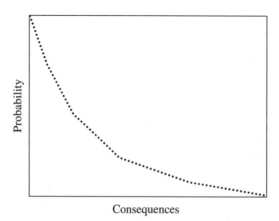

FIGURE 28.2   An example risk curve.

another might be an earthquake-induced landslide above a town. The concept of vulnerability enters the triplet through the consequence term, which varies as a function of the unique vulnerabilities of the scenario elements. In accident analysis, the consequence term can be held constant for comparison purposes, as in a *road casualty*. This effectively removes the *c* term, which allows an analyst to focus on estimating *p* for different scenarios and levels of risk exposure (Thorpe 1964; Chapman 1973; Wolfe 1982). An example would be comparing the probability of a daytime versus nighttime road casualty. It is difficult to estimating *p* for extreme events with little historical data, and Bier et al. (1999) provide an excellent survey of current methods to address this problem.

A thought experiment might help convey the related concepts of hazard, vulnerability, and risk in a transportation context. Imagine two motorcyclists riding in adjacent lanes with the hazard in question being a crash. All characteristics of the drivers, vehicles, and the environment are equal. We would say that the two face the same hazard, vulnerability, and risk because the likelihood and consequences of either motorcyclist crashing are equal. To understand vulnerability, place a helmet on one rider. The likelihood of a crash has not been altered, but both the vulnerability and the risk of the rider with the helmet have decreased. Now imagine that both riders are wearing a helmet, but the surface of one lane is wet. The vulnerability of both drivers is equal, but the likelihood of a crash is higher (as is the risk) for the rider in the wet lane. To make it tricky, imagine that the rider in the wet lane is wearing a helmet but the rider in the dry lane is not. One has a greater likelihood of a crash and the other a greater vulnerability to a crash, but which rider is at greater risk? An empirical approach to this problem would be to compare the casualty rate for motorcyclists wearing a helmet in rainy conditions with the rate for riders without a helmet in dry conditions, attempting to control for all other variables.

Despite the challenges presented by quantitative risk assessment and its many assumptions, risk analysis has many benefits that outweigh the drawbacks. Evans (1997) reviews risk assessment practices by transport organizations for accidents and notes that the benefits of quantitative risk assessment include

1. It makes possible the prioritization of safety measures when resources are scarce, or where there are different approaches to achieving the same end.

2. It makes possible the design of systems (engineering or management) aimed at achieving specified safety targets or tolerability limits.

3. It facilitates proactive rather than just reactive safety regulation.

4. It provides a basis for arguing against safety measures whose benefits are small compared with their costs, and for justifying such decisions on a rational basis.

An overarching goal in quantitative risk assessment is to determine if a given transportation risk is "acceptable." If it is not, mitigation actions are in order. One approach to this problem is to compare the given risk with commonly accepted risks. Thus, a rock fall study along a highway might compare the results with other risks like air travel, drowning, lightning, or structural failure to determine if the risk of a rock fall fatality is significantly greater than other risks (Bunce, Cruden, and Morgenstern 1997). Another approach is to compare the risk of two scenarios to compute their relative risk using a risk ratio. For example, if there were 10 road accidents on rainy weekends on average and 5 on dry weekends, then the risk ratio of rainy-day weekend driving to fair-weather weekend driving would be $10/5 = 2$, or twice as risky, assuming that the amount of driving (aggregate exposure) was roughly the same from weekend to weekend.

28.3 HAZARDS TO TRANSPORTATION SYSTEMS

There are many environmental hazards that may damage or disrupt transportation systems, and we review only the more common ones here. For example, Figure 28.3 depicts familiar road hazards grouped by their principal effect along with some of their causal relationships. In general, road hazards can: (1) compromise the quality of the surface, (2) block or damage infrastructure, (3) compromise

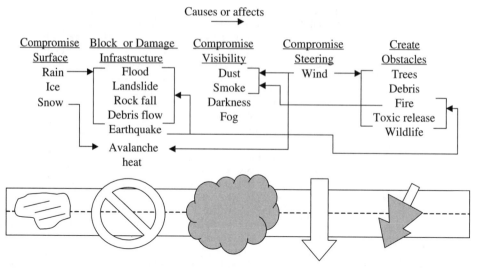

FIGURE 28.3   Example road hazards grouped by their principle effect, including some of their causal relationships.

user visibility, (4) compromise steering, (5) create a temporary obstacle, or (6) some combination of the prior five. From the figure, it is clear that rain, wind, and earthquakes have causal links with many other hazards. Rain and earthquakes can both induce a flood, landslide, rock fall, or debris flow. Earthquakes can also start a fire or result in a toxic release. Extreme wind can kick up dust, start a fire, drive smoke from a fire, blow trees and debris into the roadway, or redeposit snow, leading to an avalanche. This is only a sample of the many hazards and relationships that might exist. Hazards can also coincide, as in a nighttime earthquake in severe rain. This section reviews recent research in the analysis of many of these hazards, but it should not be considered comprehensive. The review is multimodal and driven primarily by these questions:

- What is the hazard?
- What has been done to address the hazard in research and practice?
- What travel modes does the hazard affect?
- How well can we predict the hazard in space and time?
- What are the consequences of the hazard and how are they defined and measured?
- What mitigation actions exist and what might be developed?

28.3.1   Avalanches

An avalanche is a sudden transfer of potential energy inherent in a snow pack into kinetic energy. The principal contributing factors include snow, topographic effects, and wind which can redeposit snow. *Snow structure* refers to the composition of its vertical profile, which can become unstable as new layers are added. An avalanche occurs when the strength of the snowpack no longer exceeds the internal and external stresses. Avalanches are typically divided into dry or wet and loose or slab avalanches. Dry slab avalanches accelerate rapidly and can reach speeds in excess of 120 mph, but wet avalanches move much more slowly.

The systematic study of avalanches in North America dates back to the 1950s in Alta, Utah. Figure 28.4 depicts the most active and damaging slide paths in Alta. The most useful, general reference is McClung and Schaerer's (1993) avalanche handbook. Avalanches typically reduce the

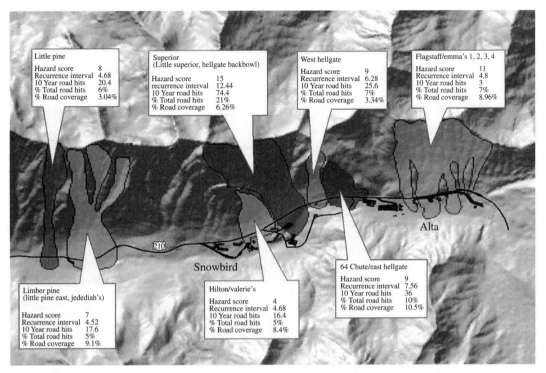

FIGURE 28.4 The most active avalanche slide paths in Little Cottonwood Canyon, Utah shown with their mean recurrence interval (years), the number of road hits in the last 10 years, the percentage of times they hit the road, and the percentage of the length of the road they covered. (*Source:* William Naisbitt, Alex Hogle, and Wendy Bates.)

serviceability of a road, but they can also damage infrastructure and cause injury or death. Other modes affected include rail, pipelines, and transmission lines. The science of predicting the timing of avalanches is called *forecasting* (Schweizer, Jamieson, and Skjonsberg 1998), and it has improved significantly over the last 50 years. Snow pits, weather instrumentation, field observation, and remote sensing are combined to forecast avalanches. The corridors that receive the greatest attention are those with high traffic volume and a documented avalanche history. Avalanche path identification using terrain and vegetation is also a common task in areas where historical records may not be available.

Three challenges that transportation agencies face in avalanche control are (1) selecting paths where mitigation would be most beneficial, (2) evaluating mitigation measures, and (3) comparing the risks of different roads. The avalanche-hazard index (Schaerer 1989) combines forecasting with traffic flow volumes to address these needs. The index includes the likelihood of vehicles being impacted by an avalanche along a road as well as the potential consequences. It also incorporates the observation that loss of life can occur when a neighboring avalanche overcomes traffic halted by another slide. The composite avalanche-hazard index for a road is

$$I = \sum_i \sum_j w_j (P_{mij} + P_{wij}) \tag{28.2}$$

where P_{mij} = the likelihood that *moving* traffic might be hit by an avalanche of class j at path i
P_{wij} = the likelihood that *waiting* traffic might be hit by an avalanche of class j at path i
w_j = the consequence of an avalanche of class j

The index can also be calculated separately for each avalanche path along a road to determine where mitigation would make the largest contribution to overall hazard reduction.

A number of avalanche risk case studies for transportation corridors have been performed, including at Glacier National Park (Schweizer, Jamieson, and Skjonsberg 1998), the Colorado Front Range (Rayback 1998), and the Himalayas (De Scally and Gardner 1994). Avalanche-mitigation options and increase in cost, include explosives, snow sheds, and deflection dams. Rice et al. (2000) provide an example of system for automatically detecting avalanches on rural roads.

28.3.2 Earthquakes

The study of earthquakes and seismic risk spans many fields in the sciences and social sciences. They are widely researched by transportation engineers from a variety of perspectives because they can severely damage and disrupt transportation systems. A devastating earthquake epitomizes a low-probability, high-consequence event in risk analysis. The recurrence interval for a large earthquake in a region can be centuries, varying inversely with magnitude, yet devastating earthquakes occur almost every year somewhere in the world. For many populated areas without a history of severe earthquake loss, the likelihood of facing an earthquake that damages transportation lifelines is a near certainty because the geologic record reveals past large earthquakes (Clague 2002). No major transport mode is exempt from the adverse affects of an earthquake. Roadways, railways, pipelines, transmission lines, and airports and seaports can all be damaged, with tremendous economic costs (Cho, Shinozuka, and Chang 2001). Earthquakes can also start fires, trigger landslides (Refice and Capolongo 2002), release toxic chemicals (Lindell and Perry 1996), cause dam failures, and create sudden earthen dams via landslides leading to inevitable flooding (Schuster 1986).

Preimpact earthquake research in transportation engineering focuses on vulnerable structures like bridges (Malik 2000), tunnels (Hashash et al. 2001), and water-delivery systems (Chang, Svekla, and Shinozuka 2002). The central problem is estimating the response characteristics of these structures to ground shaking and liquefaction (Price et al. 2000; Selcuk and Yuceman 2000; Romero, Rix, and French 2000). Werner (1997) notes that earthquake losses to highway systems depend not only on the response characteristics of the highway components, but also on the nature of the overall highway system's configuration, redundancy, capacity, and traffic demand (see also Basoz and Kiremidjian 1996). For example, two bridges may be equally susceptible to ground shaking, but one may be much more important in serving the daily travel demand to an important destination. Retrofitting is typically in high order when a bridge highly susceptible to the effects of an earthquake is also essential in serving a large volume of travel demand.

Postimpact earthquake research focuses on immediate damage assessment (Park, Cudney, and Inman 2001), the performance of the transportation system (Chang 2000), and the lifeline restoration process (Isumi, Nomura, and Shibuya 1985; Opricovic and Tzeng 2002). Chang (2000) examines postearthquake port performance following the Kobe quake in 1995 and frames the economic loss (and thus vulnerability) in terms of three types of traffic:

1. Cargo originating from or destined to the immediate hinterland
2. Cargo from/to the rest of Japan
3. Foreign transshipment cargo

By examining the pre- and postconditions of these cargo types, Chang concludes that 2 and 3 suffered the most, resulting in both short-term loss of revenue and long-term loss of competitive position. Economic impacts may last beyond the point where the infrastructure has been repaired. Kobe demonstrates that 3 is especially important, and the central port vulnerability question can be framed as the percentage of a port's revenue tied to transshipment cargo.

28.3.3 Floods and Dam Breaks

Floods cause the greatest loss in many countries because they occur frequently and their severity is compounded by dense development along many rivers. The National Weather Service (NWS) in the United States estimates that greater than half of all flood-related deaths occur in vehicles at low-water crossings. Flood damage to transportation systems represents one of the largest losses in the public sector. Intense rainfall is the chief cause of floods, but hurricanes also hold the potential to cause a significant amount of storm surge inundation. Dam breaks are included here as a special type of technologically induced flood. This includes earthen dam breaks caused by earthquake-induced landslides (Schuster 1986). The modeling of dam breaks has increased in recent years because agencies such as the U.S. Bureau of Reclamation (USBR) are required to submit a report and associated inundation animations of potential dam breaks to local emergency managers downstream from all dams for emergency planning purposes.

Figure 28.5 depicts an example of modeling flooding across a transportation network. The depth of the flood is shown in meters, with the direction and velocity of the flood depicted using a vector field. This example is output from the MIKE 21 flood simulation system for modeling two-dimensional free surface flows. The system can model many conditions that occur in a floodplain, including flooding and drainage of floodplains, embankment overtopping, flow through hydraulic structures, tidal forces, and storm surge. MIKE 21 is an excellent example of a deductive process-oriented hazard mapping approach because the system solves nonlinear equations of continuity and conservation of momentum for flooding.

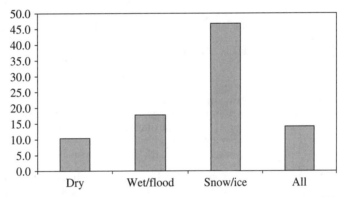

FIGURE 28.5 Example skidding rates (the percentage of accidents where skidding is a factor) for Great Britain (adapted from Perry and Symons 1991).

Flooding is a serious problem in many areas because of its ability to degage the serviceability of a transportation network at various points rapidly. Ferrante, Napolitano, and Ubertini (2000) combine a numerical model for flood propagation in urban areas with a network path-finding algorithm to identify least-flood-risk paths for rescuing people as well as providing relief. They use Dijkstra's (1959) shortest-path algorithm, but calculate cost of a link in a very novel manner using the flood flow depth and velocity across the road. In this way, the "cost" of traversing a link is a function of both the length of the road and its flood characteristics:

$$c_{ij} = \frac{L_{ij}}{\alpha_h \alpha_v} \qquad (28.3)$$

where c_{ij} = the cost/risk of traversing the link
L_{ij} = the length of the link
α_h = a parameter (0–1) related to flood height
α_v = a parameter related to water velocity (0–1)

Each alpha parameter *decreases* as flood height or velocity increases, respectively, until the maximum allowable flood height (e.g., 0.3 meters) or velocity (e.g., 1 m/sec) is reached, whereby they become 0. At this point, link cost is infinite and is no longer traversable. Thus, the travel cost of a link without flooding is its length, but as flood height and velocity across the link increase, its cost and risk quickly increase. This example links a hazard process model with a network algorithm, which points to a valuable opportunity for analysts, as many hazards reduce the serviceability of network links. Real-time path finding in a network degraded by a hazard is a very valuable application. The challenge is to develop a means for acquiring accurate, timely information on the hazard as well as to manage and convey the uncertainty in the results.

28.3.4 Fog, Dust, Smoke, Sunlight, and Darkness

Fog, dust, smoke, sunlight, and darkness are transportation hazards that compromise the visibility of system users. This hazard category does not apply to pipeline networks, transmission lines, and other networks where visibility is not an issue. From a roadway perspective, Perry and Symons (1991) provide an excellent source on these hazards. Musk (1991) thoroughly covers the fog hazard, and Brazel (1991) describes a dust storm case study for Arizona. Although smoke from wildfires routinely disrupts roadways and inhibits operations at airports each summer, it appears to be an underresearched topic in transportation hazards. Darkness also has an understandably adverse effect on road safety, especially when combined with fog, smoke, or dust.

Fog can cause spectacular road accidents involving hundreds of vehicles on a roadway. Musk (1991) describes the fog potential index (FPI), which expresses the susceptibility of a location p on a road to thick radiation fog on a scale from 0 to 100. The values of two locations are comparative in that a value of 30 at location A and a value 20 at location B means that location A should experience 50 percent more hours of thick radiation fog than location B. The index is of the form:

$$I_p = 10d + 10t + 2s + 3e \tag{28.4}$$

where d = the distance of the location p from standing surface water
t = a function of the local topography at p (e.g., hill or valley)
s = a function of the road site topography (e.g., bridge or embankment)
e = the incorporation of other environmental features likely to affect the formation of radiation fog (e.g., proximity to power station cooling)

The index coefficients are weights that affect the relative importance of the variables. This index can be applied at any linear resolution, but 1 kilometer is common. The index can then be tested against in situ observations of visibility.

28.3.5 Rain, Snow, and Ice

Rain, snow, and ice are common hazards that compromise visibility and the quality of a road, rail, or airport surface (Benedetto 2002; Andrey 1990). All road users are familiar with road signs like "slippery when wet" or "bridges may be icy" (Carson and Mannering 2001). Ice is also a hazard for aircraft because of its effect on lift, as well as sea travel because it creates obstacles (Tangborn and Post 1998). In a road network context, skidding is the most common explanation for accidents that occur in the context of these hazards. The *skidding rate* is the statistic used to quantify this factor, which is the percentage of accidents where one or more vehicles are reported to have skidded (Perry and Symons 1991). Example skidding rates for cars are given in Figure 28.6 for Great Britain in 1987. This figure shows that rain roughly doubles the percentage of accidents where skidding is a factor over dry conditions, and snow and ice quadruple the rate over dry conditions. The overall skidding rate for cars for all road conditions is about 14 percent.

The question of how rain, snow, and ice affect the total number of road accidents is not straightforward. Palutikof (1983) found that people drive more carefully in snow or simply postpone or cancel journeys. This leads to reduction in the total number of accidents over that which would be expected. Rain does not seem to have the same effect on travel decision making, and Brodsky and Hakkert (1988) found that the number of accidents increases in wet conditions. Al Hassan and Barker (1999) found a slightly greater drop in traffic activity owed to inclement weather on the weekend (>4 percent) than on weekdays (<3 percent). In a case study in Chicago, Bertness (1980)

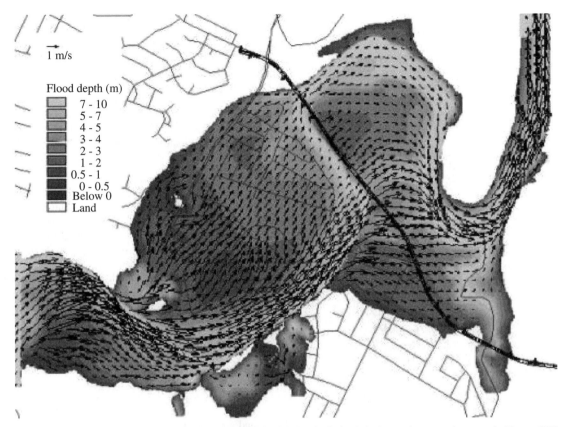

FIGURE 28.6   A floodplain inundation map depicting flood depth (m) and velocity (m/sec) over a transportation network. (*Source:* DHI Water & Environment and Camp Dresser & McKee, Inc. ©.)

found that rain roughly doubled the number of road accidents, with the greatest effect in rural areas. It is important to keep in mind that rain, snow, and ice studies tend to underestimate the risk because road accidents are typically underreported.

Hazards that affect the road surface represent the most costly maintenance function for many cities, counties, and state transportation departments. Salt is the most common road deicer, with about 10,000,000 tons applied each year in the United States (Perry and Symons 1991). This is expensive and comes with environmental side effects. Eriksson and Norman (2001) note that road weather information systems have a very high benefit–cost ratio in reducing weather-related risk. The widespread adoption of Doppler radar has greatly improved the reporting of precipitation, and some systems can now report rain intensity to levels as detailed as an individual street segment. There is much work in developing and installing in situ road sensors to automatically detect poor road conditions. This can greatly improve road maintenance procedures because managers can apply mitigation measures like salt where it is most needed.

28.3.6   Landslides, Rock Fall, and Debris Flow

Many miles of roads, rail, and pipeline travel through areas with rock faces and steep slopes in mountainous terrain. Geomorphic hazards that commonly affect transportation corridors include landslides, rock fall, and debris flow. A debris flow is essentially a fast-moving landslide. These hazards can damage or reduce the serviceability of infrastructure, crush or bury vehicles, and result in death. In some cases they occur without little or no warning, but they are typically preceded by intense rain (Al Homoud, Prior, and Awad 1999). They can also be volcanically or earthquake-induced (Dalziell and Nicholson 2001) and create sudden earthen dams that lead to flooding (Schuster, Wieczorek, and Hope 1998). An excellent general source on landslides and debris flows is the Transportation Research Board (TRB) report on landslides edited by Turner and Schuster (1996). In terms of case studies, Marchi, Arattano, and Deganutti (2002) examine 10 years of debris flows in the Italian Alps, Evans and Savigny (1994) examine landslides in Canada; He, Ma, and Cui (2002) looked at debris flows along the China-Nepal Highway; Budetta (2002) conducted a risk assessment for a 1-kilometer stretch of road subject to debris flows in Italy; and Petley (1998) examined geomorphic road hazards along a stretch of road in Taiwan. Fish and Lane (2002) discuss a rock-cut management system, and Franklin and Senior (1997) describe a rock fall hazard rating system.

Bunce, Cruden, and Morgenstern (1996) provide an excellent example method for assessing the risk of loss of life from rock-fall along a highway. They used rock-fall impact-mark mapping supplemented by documented rock-fall records to establish a rock-fall frequency for the Argillite Cut on Highway 99 in British Columbia. The method relies on separate calculations for the risk of a rock hitting a stationary vehicle versus a moving vehicle, as well as a moving vehicle hitting a rock on the road. The probability that one or more vehicles will be hit is given as

$$P(S) = 1 - (1 - P(S \mid H))^N \tag{28.5}$$

where $P(S \mid H)$ is the probability that a vehicle occupies the portion of the road affected by a rockfall and N is the number of rocks that fall. This equation states that the probability of a vehicle being hit is one minus the probability that a vehicle will not be hit. With a series of assumptions, they estimate the risk of death due to rockfall for a one-time road user and daily commuter at 0.00000006 and 0.00003 per year, respectively.

28.3.7 Wind, Tornadoes, and Hurricanes

Wind is a significant hazard to road, rail, sea, and air transport (Perry and Symons 1994). Gusts, eddies, lulls, and changes in wind direction are often greatest near the ground in extreme wind episodes. In these episodes, the majority of fatalities are generally transport-related. It is difficult to summarize the effects of wind on road and rail transport because little data exists, although it is generally viewed as less of a hazard than ice, snow, and rain. Figure 28.7 depicts wind that is blowing smoke across an interstate and blocking traffic. Perry and Symons (1994) divide the wind hazard into three categories: direct interference with a vehicle, obstructions, and indirect effects. Direct interference includes its effects on vehicle steering, which may push one vehicle into another or run a vehicle off the road. Extreme winds can overturn high-profile trucks and trains when the wind vector is orthogonal to the direction of travel because the force of the wind is proportional to the vehicle area presented (Baker 1988). Wind can impede transport by blowing dust or smoke across a road, which can reduce visibility. It can also blow trees and other debris onto a road or railway and create temporary obstacles. Indirect effects include the redeposition of snow leading to an avalanche, as well as its adverse effect on bridges and air and sea-based termini. Overall, wind can impede transport operation or damage vehicles and infrastructure, all of which can result in economic impacts, injuries, and fatalities.

Air transport faces the greatest hazard from wind. A violent downdraft from a thunderstorm (microburst) on takeoff or landing is one example, but any exceptionally large local wind gradient (wind shear) can affect lift adversely at low altitudes (Vorobtsov 2002; Goh and Wiegmann 2002). In many air disasters, wind is considered the primary contributing factor. Small aircrafts are much more vulnerable to in-flight storms and are often warned to avoid storms completely. Measures to reduce wind hazard include permanent wind breaks, warnings, road closures, and low-level wind shear alert systems. An airport wind-warning system generally consists of a set of anemometers that are analyzed by computer. A warning is issued when levels differ by some threshold. Automated

FIGURE 28.7   A wildfire adjacent to an interstate blocking traffic. (*Source:* http://www.commanderchuck.com).

wind-warning systems for individual roads may appear soon because of advances in weather instrumentation. The finest level that wind warnings are commonly issued is at a county scale. Improved weather forecasting is generally viewed as the principal means for reducing the hazard (Perry and Symons 1994).

Hurricanes and tornadoes represent special cases of extreme winds. Due to satellite, radar, and other in situ sensor networks, their prediction has greatly increased in recent years. Much of the transportation research in this area focuses on evacuation. Wolshon (2001) reviews the problems and prospects for contraflow freeway operations to reduce the vulnerability of coastal communities by reversing lanes to increase freeway capacities in directions favorable for evacuation. This problem is simple conceptually but represents a significant challenge for both traffic engineers and emergency managers.

28.3.8 Wildlife

Wildlife is a familiar hazard to most drivers because of the many warning signs along roadways. Wildlife accidents typically result in vehicle damage, but they can also result in injury or death. Two common examples of wildlife hazards include the threat that ungulates such as moose (Joyce and Mahoney 2001) present to vehicles and the threat that birds present to aircraft. The number of these collisions is staggering, and it is estimated that in 1991 greater than half a million deer were killed by vehicles in the United States (Romin and Bissonette 1996). Lehnert and Bissonette (1997) review research on deer-vehicle collisions and describe a field experiment on the effectiveness of highway crosswalk-structures as a means of mitigation. The crosswalk system evaluated forces deer to cross at specific areas that are well marked for motorists. Although deer fatalities decreased by 42 percent following the installation of the crosswalks, they were unable to attribute this reduction to the crosswalks because there was an 11 percent probability that it might have occurred by chance.

Bird hazards to aircraft are also a significant concern, and Lovell and Dolbeer (1999) provide a recent review with a study to validate the results of the U.S. Air Force (USAF) bird avoidance model (BAM). BAM provides information to pilots regarding elevated bird activity based on refuge surveys, migration dates, and routes. Lovell and Dolbeer note that since 1986 birds have caused 33 fatalities and almost $500 million in damage to USAF aircraft alone. On average, USAF aircraft incur 2,500 bird strikes a year, with most occurring in the fall and spring migration. Waterfowl and raptors account for 69 percent of the damaging strikes to low-level flying military aircraft. Lovell and Dolbeer found that BAM predicted significantly higher hazard for routes where bird strikes have occurred in the past and thus can assist in minimizing strikes.

28.4   TRANSPORTATION AS HAZARD

In addition to the many environmental hazards that threaten transportation systems, transportation itself presents hazards to people, property, and the environment. Road traffic accidents are the most common example, accounting for the majority of transportation casualties in most countries. The contributing factors for road accidents are typically classified into those associated with the driver, vehicle, and the environment. Contributing factors associated with the driver include error, speeding, experience, and blood-alcohol level. Factors associated with the vehicle include its type, condition, and center of gravity. Environmental factors include the quality of the infrastructure, weather, and obstacles. The majority of road accidents are attributed to driver factors (Evans 1991), and this holds for many other modes such as boats (Bob-Manuel 2002), bicycles (Cherington 2000), snowmobiles (Osterom and Eriksson 2002), and all-terrain vehicles (Rogers 1993). Taken together, this implies that most transportation casualties in the world are road accidents chiefly attributed to the driver. Not surprisingly, research on driver factors represents the largest area of transportation hazards research (see the journal *Accident Analysis and Prevention*). Transportation accidents have severe effects on those directly involved, as well as side effects on others. Other effects might include severe traffic delays leading to missed meetings, lost sales to businesses, delayed commodity shipments, and increased insurance costs. Research in accident analysis spans all modes and typically focuses on assessing the role of various driver, vehicle, and environmental factors as well as methods for mitigating accidents. (See chapter 18 on incident management.)

In addition to common traffic incidents, there are also low-probability, high-consequence transportation events that place risks on people and environmental resources in proximity to transportation corridors and ports. Rail, road, pipeline, and marine HazMat transport is the prime example in this area because it places considerable involuntary risks on people (and resources) who do not perceive much benefit to the transport of hazardous materials. HazMat has been studied from a number of perspectives for many materials and modes, so there are numerous frameworks for analysis (Bonvicini, Leonelli, and Spadoni 1998; Cassini 1998; Erkut and Ingolfsson 2000; Fabiano et al. 2002; Helander and Melachrinoudis 1997; Jacobs and Warmerdam 1994; Klein 1991). Aldrich (2002) provides an historic perspective on rail HazMat shipments from 1833 to 1930, and Cutter (1997) reviews recent trends in hazardous material spills. Caputo and Pelagagge (2002) present a system for monitoring pipeline HazMat shipments. Singh, Shonhardt, and Terezopoulos (2002) examine spontaneous coal combustion in sea transport. Raj and Pritchard (2000) present a risk-analysis tool used by the Federal Railway Administration. Hwang et al. (2001) present a comprehensive risk analysis approach for all modes that includes 90 percent of the dangerous chemicals. Abkowitz, Cheng, and Lepofsky (1990) describe a method for evaluating the economic consequences of HazMat trucking. Verter and Kara (2001) review HazMat truck routing in Canada. Dobbins and Abkowitz (2002) look at inland marine HazMat shipments. Saccommano and Haastrup (2002) focus on HazMat risks in tunnels. Marianov, ReVelle, and Shih (2002) propose that proximal communities receive a tax reduction to offset the risk of proximal HazMat shipments.

Following the events of September 11, 2001, *transportation security* has become a national research priority led by the Transportation Security Administration (TSA), recently reorganized under the Department of Homeland Security. Transportation terrorism has not been a focus of transportation hazards researchers in the past, so there is little to review at this point. However, reports and proposals are beginning to surface that indicate that this will be one of the largest areas of transportation hazards research for many years.

28.5 *TRANSPORTATION IN EMERGENCY MANAGEMENT*

Transportation lifelines are vital during an emergency and play an important role in all four phases of emergency management: mitigation, preparedness, response, and recovery. The concern in the mitigation phase is reducing the likelihood of an event, its consequences, or both. The focus of the preparedness phase is improving operational capabilities to respond to an emergency such as training emergency personnel, installing notification systems, and redeploying resources to maximize readiness (Sorensen 2001). The mitigation and preparedness phases both help reduce the impact of hazardous events. The response phase begins immediately following an event, and this is when plans devised in the preparedness phase as well on-the-fly plans are activated. Common concerns include evacuating and sheltering victims, providing medical care, containing the hazard, and protecting property and the environment. The recovery phase addresses longer-term projects like damage assessment and rebuilding, which feeds back into the mitigation phase because this phase presents an opportunity to rethink hazardous areas.

Mitigation strategies for specific hazards and assets were discussed in the prior section on hazards to transportation systems. The overarching challenge in the mitigation phase is identifying and prioritizing mitigation projects in a region and allocating scarce resources to their completion. Benefit-cost analysis is a valuable method in this regard, but it must be preceded by risk assessments for all potential hazards. The effectiveness of the mitigation strategy is also important, and this can be considered part of the benefit.

Research in the preparedness and response phase has been fueled by new technologies. Enhanced 911 (E-911) is a significant relatively recent innovation, and this is covered in chapter 18 on incident management. Relevant topics that are actively researched in this phase include optimally locating emergency teams (List and Turnquist 1998), locating and stocking road maintenance stations, finding optimal fire station location for urban areas and airports (Revelle 1991; Tzeng and Chen 1999), and installing hazard-specific warning systems. Evacuation planning in this phase focuses on delimiting emergency planning zones (Sorensen, Carnes, and Rogers 1992), designing and simulating evacuations (Sinuany-Stern and Stern 1993; Southworth 1991; Cova and Johnson 2002), developing and testing evacuation routing schemes (Dunn 1993; Yamada 1996; Cova and Johnson 2003), and identifying potential evacuation bottlenecks (Cova and Church 1997). Reverse 911 systems that allow police to call evacuees are becoming increasingly important in dealing with notification. State-of-the-art systems allow emergency managers to send custom messages with departure timing and routing instructions to zones defined on the fly with a mouse. Other research in the preparedness and response phase includes methods for keeping roadways open following an earthquake or landslide (Santi, Neuner, and Anderson 2002).

One problem that complicates emergency planning by transportation agencies is the increasing amount of development in many hazardous areas. This is nearly universal as populations increase in floodplains, coastal areas subject to hurricanes, fire-prone wildlands, areas near toxic facilities (Johnson and Zeigler 1986), regions at risk to seismic activity, and so on. This presents a problem because in many of these areas (and at many scales) the transportation system is not being improved to deal with these increasing populations. This means that evacuating threatened populations is becoming increasingly difficulty at all scales as new development occurs. In other words, vulnerability to environmental hazards is continually increasing due to the fact that populations in hazardous areas are increasing at the same time that the ability of emergency managers to invoke protective actions such as evacuation are decreasing.

Figure 28.8 is two maps from Cova and Johnson (2002) that show the effect of a new road on household evacuation times for a community at risk to wildfire near Salt Lake City. Before the construction of the new road, homes in the back of the canyon had the greatest evacuation times, as the sole road out of the canyon would get congested. In this scenario, the average vehicle departure-time

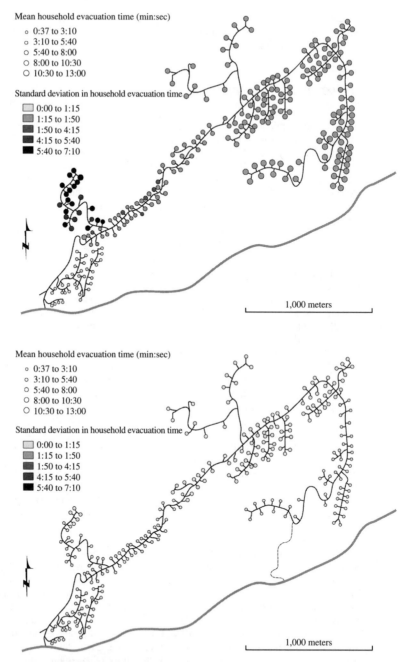

FIGURE 28.8 The effect of the construction of a second access road (dashed line) on household evacuation times in a fire-prone canyon east of Salt Lake City, Utah.

following notification to evacuate was 10 minutes and the average number of vehicles per household was 2.5, so it can be considered a reasonably urgent evacuation when most residents are at home. Note that houses in the back of the canyon stand to gain much more from the construction of the new exit because their evacuation times decrease much more than those for homes near the original exit from this community. Also, all evacuation times become more consistent because the second exit reduces the delay caused by everyone using one exit. Viewed another way, the new exit reduces the number of households per exiting road from 250/1 = 250 to 250/2 = 125.

28.6 NEW TECHNOLOGIES

There are many new technologies that hold promise to aid transportation agencies in reducing the effects of transportation hazards. Weather instrumentation is prime example that is improving both in terms of breadth of measurement, as well as the number of installed road weather stations. The suite of geospatial technologies including the global positioning system (GPS), geographic information systems (GIS), and remote sensing also hold much promise to improve the amount of information available to transportation users, planners, and emergency responders. The recent formation of the National Consortia for Remote Sensing in Transportation (NCRST) is dedicated to this task (Gomez 2002). The consortia are divided into four themes: hazards, environment, infrastructure, and flow. NCRST Hazards (NCRSTH) is the most relevant in the context of applying geospatial technologies to monitoring and mitigating transportation hazards.

A simple benefit of GPS in accident analysis is that it is an inexpensive means for greatly improving the locational component of crash data (Graettinger 2001). Remote sensing can be used to detect and monitor fires, volcanoes (Oppenheimer 1998), landslides, avalanches, and many other hazards. One technology that is having a significant effect on the study of transportation hazards is GIS. For a comprehensive review of GIS in transportation (GIS-T) see Miller and Shaw (2001). GIS is being used in transportation applications such as mapping collision data (Arthur and Waters 1997; Austin, Tight, and Kirby 1997), routing HazMat shipments (Brainard, Lovett, and Parfitt 1996; Lepofsky, Abkowitz, and Cheng 1993), identifying hazardous highway locations (Spring and Hummer 1995), and modeling the vulnerability of populations to toxic spills (Chakraborty and Armstrong, 1996), among many other transportation hazard applications.

28.7 CONCLUSION

Transportation and hazards is a growing field in terms of research and practice. New methods for predicting and mitigating hazards are continually being developed, and researchers are linking these methods and models to tasks in transportation planning and mitigation. Globalization is increasing our dependence on transportation systems at all scales. For this reason, disruptions will only become more costly and important to mitigate. New information technologies are converging that promise to change this field drastically in the coming years. These technologies are emerging during a shifting emphasis toward transportation security. This will bring information-based research on reducing the effects of transportation terrorism to the forefront of this research area. Finally, development in hazardous areas is increasing along with corresponding increases in traffic volumes along many lifelines at risk to hazards. This will present substantial challenges to transportation researchers and analysts for the foreseeable future.

ACKNOWLEDGMENTS

The authors would like to thank Justin P. Johnson for his time in creating figures. The lead author was supported through the U.S. DOT Research and Special Programs Administration (RSPA) as part of the National Center for Remote Sensing in Transportation Hazards Consortium (NCRST-H) administered as Separate Transactions Agreement #DTRS56-00-T-003.

REFERENCES

Abkowitz, M., P. D. M. Cheng, and M. Lepofsky. 1990. "Use of Geographic Information Systems in Managing Hazardous Materials Shipments." *Transportation Research Record* 1261, 35–43.

Aldrich, M. 2002. "Regulating Transportation of Hazardous Substances: Railroads and Reform, 1883–1930." *Business History Review* 76(2):267–97.

Al Hassan, Y., and D. J. Barker. 1999. "The Impact of Unseasonable or Extreme Weather on Traffic Activity Within Lothian Region, Scotland." *Journal of Transport Geography* 7(3):209–13.

Al Homoud, A. S., G. Prior, and A. Awad. 1999. "Modeling the Effect of Rainfall on Instabilities of Slopes along Highways." *Environmental Geology* 37:317–25.

Andrey, J. 1990. "Relationships Between Weather and Traffic Safety: Past, Present, and Future Directions." *Climatological Bulletin* 24:124–36.

Arthur, R., and N. Waters. 1997. "Formal Scientific Research of Traffic Collision Data Utilizing a GIS." *Transportation Planning and Technology* 21:121–37.

Austin, K., M. Tight, and H. Kirby. 1997. "The Use of Geographical Information Systems to Enhance Road Safety Analysis." *Transportation Planning and Technology* 20:249–66.

Austin, R. D., and J. L. Carson. 2002. "An Alternative Accident Prediction Model for Highway-Rail Interfaces." *Accident Analysis and Prevention* 34:31–42.

Baker, C. J. 1988. "High-Sided Articulated Lorries in Strong Cross Winds." *Journal of Wind Energy and Industrial Aerodynamics* 31:67–85.

Basoz, N., and A. Kiremidjian. 1996. *Risk Assessment for Highway Transportation Systems.* J. A. Blume Earthquake Engineering Center, Department of Civil Engineering, Stanford University, Report No. 118.

Benedetto, A. 2002. "A Decision Support System for the Safety of Airport Runways: The Case of Heavy Rainstorms." *Transportation Research A* 8:665–82.

Berdica, K. 2002. "An Introduction to Road Vulnerability: What Has Been Done, Is Done and Should Be Done." *Transport Policy* 9:117–27.

Bertness, J. 1980. "Rain Related Impacts on Selected Transportation Activities and Utility Services in the Chicago Area." *Journal of Applied Meteorology* 19:545–56.

Bier, V. M., R. Zimmerman, Y. Y. Haimes, J. H. Lambert, and N. C. Matalas. 1999. "A Survey of Approaches for Assessing and Managing the Risk of Extremes." *Risk Analysis* 19:83–94.

Bob-Manuel, K. D. H. 2002. "Probabilistic Prediction of Capsize Applied to Small High-Speed Craft." *Ocean Engineering* 29:1841–51.

Bonvicini, S., P. Leonelli, and G. Spadoni. 1998. "Risk Analysis of Hazardous Materials Transportation: Evaluating Uncertainty by Means of Fuzzy Logic." *Journal of Hazardous Materials* 62:59–74.

Brainard, J., A. Lovett, and J. Parfitt. 1996. "Assessing Hazardous Waste Transport Using a GIS." *International Journal of Geographical Information Science* 10:831–49.

Brazel, A. I. 1991. "Blowing Dust and Highways: The Case of Arizona, USA." In *Highway Meteorology,* ed. A. H. Perry and L. J. Symons. New York: E. & F. N. Spon, 131–61.

Brodsky, H., and A. S. Hakkert. 1988. "Risk of a Road Accident in Rainy Weather." *Accident Analysis and Prevention* 17:147–54.

Budetta, P. 2002. "Risk Assessment from Debris Flows in Pyroclastic Deposits along a Motorway, Italy." *Bulletin of Engineering Geology and the Environment* 61:293–301.

Bunce, C. M., D. M. Cruden, and N. R. Morgenstern. 1997. "Assessment of the Hazard from Rock Fall on a Highway." *Canadian Geotech Journal* 34:344–56.

Caputo, A. C., and P. M. Pelagagge. 2002. "An Inverse Approach for Piping Networks Monitoring." *Journal of Loss Prevention in the Process Industries* 15:497–505.

Carson, J., and F. Mannering. 2001. "The Effect of Ice Warning Signs on Ice-Accident Frequencies and Severities." *Accident Analysis and Prevention* 33:99–109.

Cassini, P. 1998. "Road Transportation of Dangerous Goods: Quantitative Risk Assessment and Route Comparison." *Journal of Hazardous Materials* 61:133–38.

Chakraborty, J., M. P. Armstrong. 1996. "Using Geographic Plume Analysis to Assess Community Vulnerability to Hazardous Accidents." *Computers, Environment, and Urban Systems* 19:341–56.

Chang, S. E. 2000. "Disasters and Transport Systems: Loss, Recovery and Competition at the Port of Kobe after the 1995 Earthquake." *Journal of Transport Geography* 8:53–65.

Chang, S. E., W. D. Svekla, and M. Shinozuka. 2002. "Linking Infrastructure and Urban Economy: Simulation of Water-Disruption Impacts in Earthquakes." *Environment and Planning B* 29(2):281–301.

Changnon, S. A. 1999. "Record Flood-Producing Rainstorms of 17–18 July 1996 in the Chicago Metropolitan Area. Part III: Impacts and Responses to the Flash Flooding." *Journal of Applied Meteorology* 38:273–80.

Chapman, R. 1973. "The Concept of Exposure." *Accident Analysis and Prevention* 5:95–110.

Cherington, M. 2000. "Hazards of Bicycling: From Handlebars to Lightning." Seminars in Neurology 20(2): 247–53.

Cho, S., M. Shinozuka, and S. Chang. 2001. "Integrating Transportation Network and Regional Economic Models to Estimate the Costs of a Large Urban Earthquake." *Journal of Regional Science* 41: 39–65.

Clague, J. J. 2002. "The Earthquake Threat in Southwestern British Columbia: A Geologic Perspective." *Natural Hazards* 26:7–34.

Collins-Garcia, H., M. Tia, R. Roque, and B. Choubane. 2000. "Alternative Solvent for Reducing Health and Environmental Hazards in Extracting Asphalt—An Evaluation." Construction 2000—*Transportation Research Record* 1712, 79–85.

Cowen, D. J., J. R. Jensen, C. Hendrix, M. E. Hodgson, and S. R. Schill. 2000. "A GIS-Assisted Rail Construction Econometric Model That Incorporates LIDAR Data." *Photogrammetric Engineering and Remote Sensing* 66:1323–28.

Cova, T. J. 1999. "GIS in Emergency Management." In *Geographical Information Systems: Principles, Techniques, Applications, and Management,* ed. P. A. Longley, M. F. Goodchild, D. J. Maguire, and D. W. Rhind. New York: John Wiley & Sons, 845–58.

Cova, T. J., and R. L. Church. 1997. "Modelling Community Evacuation Vulnerability Using GIS." *International Journal of Geographical Information Science* 11(8):763–84.

Cova, T. J., and J. P. Johnson. 2002. "Microsimulation of Neighborhood Evacuations in the Urban-Wildland Interface." *Environment and Planning A* 34(12):2211–29.

———. 2003. "A Network Flow Model for Lane-Based Evacuation Routing." *Transportation Research A,* 37:579–604.

Cutter, S. 1996. "Vulnerability to Environmental Hazards." *Progress in Human Geography* 20:529–39

———. 1997. "Trends in U.S. Hazardous Materials Spills." *Professional Geographer* 49:318–31.

Dalziell, E., and A. Nicholson. 2001. "Risk and Impact of Natural Hazards on a Road Network." *Journal of Transportation Engineering* 127:159–66.

De Scally, F. A., and J. S. Gardner. 1994. "Characteristics and Mitigation of the Snow Avalanche Hazard in Kaghan Valley, Pakistan Himalaya." *Natural Hazards* 9:197–213.

Dobbins, J. P., and M. D. Abkowitz. 2002. "Development of an Inland Marine Transportation Risk Management Information System." Marine Transportation and Port Operations—*Transportation Research Record* 1782, 31–39.

Dunn, C. E., and D. Newton. 1992. "Optimal Routes in GIS and Emergency Planning Applications." *Area* 24:259–67.

Eriksson, M., and J. Norrman. 2001. "Analysis of Station Locations in a Road Weather Information System." *Meteorological Applications* 8:437–48.

Erkut, E., and A. Ingolfsson. 2000. "Catastrophe Avoidance Models for Hazardous Materials Route Planning." *Transportation Science* 34:165–79.

Evans, S. G., and K. W. Savigny. 1994. "Landslides in the Vancouver-Fraser Valley-Whistler Region." *Bulletin— Geological Survey of Canada* 481:251–86.

Evans, A. W. 1997. "Risk Assessment by Highway Organizations." *Transport Reviews* 2:145–63.

Evans, L. 1991. *Traffic Safety and the Driver.* New York: Van Nostrand Reinhold.

Fabiano, B., F. Curro, E. Palazzi, and R. Pastorino. 2002. "A Framework for Risk Assessment and Decision-Making Strategies in Dangerous Good Transportation." *Journal of Hazardous Materials* 93: 1–15.

Ferrante, M., F. Napolitano, and L. Ubertini. 2000. "Optimization of Transportation Networks during Urban Flooding." *Journal of the American Water Resources Association* 36:1115–20.

Fish, M., and R. Lane. 2002. "Linking New Hampshire's Rock Cut Management System with a Geographic Information System." *Transportation Research Record* 1786, 51–59.

Franklin, J. A., and S. A. Senior. 1997. "The Ontario Rockfall Hazard Rating System." In *Proceedings of the International Association of Engineering Geologists Conference, Engineering Geology and the Environment,* vol. 1, 647–56.

Fridstrom, L., J. Ifver, S. Ingebrigtsen, R. Kulmala, and L. K. Thomsen. 1995. "Measuring the Contribution of Randomness, Exposure, Weather, and Daylight to the Variation in Road Accident Counts." *Accident Analysis and Prevention* 27:1–20.

Gerz, T., F. Holzapfel, and D. Darracq. 2002. "Commercial Aircraft Wake Vortices." *Progress in Aerospace Sciences* 38(3):181–208.

Goh, J., and D. Wiegmann. 2002. "Human Factors Analysis of Accidents Involving Visual Flight Rules Flight into Adverse Weather." *Aviation Space and Environmental Medicine* 73:817–22.

Gomez, R. B. 2002. "Hyperspectral Imaging: A Useful Technology for Transportation Analysis. *Optical Engineering* 41:2137–43.

Graettinger, A. J., T. W. Rushing, and J. McFadden. 2001. "Evaluation of Inexpensive Global Positioning System Units to Improve Crash Location Data." Highway Safety: Modelling, Analysis, Management, Statistical Methods, and Crash Location—*Transportation Research Record* 1746:94–101.

Harris, M., R. I. Young, F. Kopp, A. Dolfi, and J. P. Cariou. 2002. "Wake Vortex Detection and Monitoring. *Aerospace Science and Technology* 6:325–31.

Hashash, Y. M. A., J. J. Hook, B. Schmidt, and J. I. C. Yao. 2001. "Seismic Design and Analysis of Underground Structures." *Tunnelling and Underground Space Technology* 16(4):247–93.

He, Y., D. Ma, and P. Cui. 2002. "Debris Flows along the China-Nepal Highway." *Acta Geographica Sinica* 57:275–83.

Helander, M. E., and E. Melachrinoudis. 1997. "Facility Location and Reliable Route Planning in Hazardous Material Transportation." *Transportation Science* 31:216–26.

Hwang, S. T., D. F. Brown, J. K. O'Steen, A. J. Policastro, and W. E. Dunn. 2001. "Risk Assessment for National Transportation of Selected Hazardous Materials." Multimodal and Marine Freight Transportation Issues, *Transportation Research Record* 1763, 114–24.

Isumi, N., N. Nomura, and T. Shibuya. 1985. "Simulation of Post-Earthquake Restoration for Lifeline Systems." *International Journal of Mass Emergencies and Disasters* 87–105.

Jacobs, T. L., and J. M. Warmerdam. 1994. "Simultaneous Routing and Siting for Hazardous Waste Operations." *Journal of Urban Planning and Development* 120:115–31.

Janelle, D. G., and M. Beuthe. 1997. "Globalization and Research Issues in Transportation." *Journal of Transport Geography* 5:199–206.

Johnson, J. H., and D. J. Zeigler. 1986. "Evacuation Planning for Technological Hazards: An Emerging Imperative." *Cities* (May):148–56.

Joyce, T. L., and S. P. Mahoney. 2001. "Spatial and Temporal Distributions of Moose-Vehicle Collisions in Newfoundland." *Wildlife Society Bulletin* 29:281–91.

Kaplan, S., and B. J. Garrick. 1981. "On the Quantitative Definition of Risk." *Risk Analysis* 1:11–27.

Kim, K., and N. Levine. 1996. "Using GIS to Improve Highway Safety." *Computers, Environment, and Urban Systems* 20:289–302.

Klein, C. M. 1991. "A Model for the Transportation of Hazardous Waste." *Decision Sciences* 22:1091–1108.

Lehnert, M. E., and J. A. Bissonette. 1997. "Effectiveness of Highway Crosswalk Structures at Reducing Deer-Vehicle Collisions." *Wildlife Society Bulletin* 25:809–18.

Lepofsky, M., M. Abkowitz, and P. Cheng. 1993. "Transportation Hazard Analysis in Integrated GIS Environment." *Journal of Transportation Engineering* 119:239–54.

Levine, N., and K. E. Kim. 1998. "The Location of Motor Vehicle Crashes in Honolulu: A Methodology for Geocoding Intersections." *Computers, Environment, and Urban Systems* 22:557–76.

Lindell, M. K., and R. W. Perry. 1996. "Addressing Gaps in Environmental Emergency Planning: Hazardous Materials Releases During Earthquakes." *Journal of Environmental Planning and Management* 39:529–43.

List, G. F., and M. A. Turnquist. 1998. "Routing and Emergency-Response-Team Siting for High Level Radioactive Waste Shipments." *IEEE Transactions on Engineering Management* 45:141–52.

Lleras-Echeverri, G., and M. Sanchez-Silva. 2001. "Vulnerability Analysis of Highway Networks: Methodology and Case Study." *Proceedings of the Institution of Civil Engineers—Transport* 147:223–30.

Lovell, C. D., and R. A. Dolbeer. 1999. "Validation of the United States Air Force Bird Avoidance Model." *Wildlife Society Bulletin* 27:167–71.

Malik, A. H. 2000. "Seismic Hazard Study for New York City Area Bridges." Fifth International Bridge Engineering Conference, vols. 1 and 2. *Transportation Research Record* 1696, 224–28.

Marchi, L., M. Arattano, and A. M. Deganutti. 2002. "Ten Years of Debris-Flow Monitoring in the Moscardo Torrent (Italian Alps)." *Geomorphology* 46:1–17.

Marianov, V., C. ReVelle, and S. Shih. 2002. "Anticoverage Models for Obnoxious Material Transportation." *Environment and Planning B* 29:141–50.

McClung, D., and P. Schaerer. 1993. *The Avalanche Handbook*. Seattle: The Mountaineers.

Menoni, S., V. Petrini, F. Pergalani, and M. P. Boni. 2002. "Lifelines Earthquake Vulnerability Assessment: A Systemic Approach." *Soil Dynamics and Earthquake Engineering* 22:1199–1208.

Miller, H. J., and S. L. Shaw. 2001. *Geographic Information Systems for Transportation: Principles and Applications*. New York: Oxford University Press.

Musk, L. F. 1991. "Climate as a Factor in the Planning and Design of New Roads and Motorways." In *Highway Meteorology,* ed. A. H. Perry and L. J. Symons. New York: E. & F. N. Spon, 1–25.

Office of the Federal Coordinator for Meteorological Services and Supporting Research (OFCM). 2002. *Weather Information for Surface Transportation*. National Needs Assessment Report, FCM-R18-2002, U.S. Department of Commerce, National Oceanic and Atmospheric Administration (NOAA).

Oppenheimer, C. 1998. "Volcanological Applications of Meterological Satellites." *International Journal of Remote Sensing* 19:2829–64.

Opricovic, S., and G. H. Tzeng. 2002. "Multicriteria Planning of Post-Earthquake Sustainable Reconstruction." *Computer-Aided Civil and Infrastructure Engineering* 17:211–20.

Osterom, M., and A. Eriksson. 2002. "Snow Mobile Fatalities: Aspects on Preventive Measures from a 25-Year Review." *Accident Analysis and Prevention* 34:563–68.

Panchaanathan, S., and A. Faghri. 1995. "Knowledge-Based Geographic Information System for Safety Analysis at Rail-Highway Grade Crossings." *Transportation Research Record* 1497, 91–100.

Park, G., H. H. Cudney, and D. J. Inman. 2001. "Feasibility of Using Impedance-based Damage Assessment for Pipeline Structures." *Earthquake Engineering and Structural Dynamics* 30:1463–74.

Perry, A. H., and L. J. Symons. 1991. *Highway Meteorology*. New York: E. & F. N. Spon.

————. 1994. "The Wind Hazard in the British Isles and Its Effects on Transportation." *Journal of Transport Geography* 2:122–30.

Persaud, B., R. A. Retting, and C. Lyon. 2000. "Guidelines for Identification of Hazardous Highway Curves, Highway and Traffic Safety: Crash Data, Analysis Tools, and Statistical Methods." *Transportation Research Record* 1717, 14–18.

Petley, D. N. 1998. "Geomorphological Mapping for Hazard Assessment in a Neotectonic Terrain." *Geographical Review* 164:183–201.

Price, B. E., M. Stilson, M. Hansen, J. Bischoff, and T. L. Youd. 2000. "Liquefaction and Lateral Spread Evaluation and Mitigation for Highway Overpass Structure—Cherry Hill Interchange, Davis County, Utah." Soil Mechanics 2000—*Transportation Research Record* 1736:119–26.

Raj, P. K., and E. W. Pritchard. 2000. "Hazardous Materials Transportation on US Railroads: Application of Risk Analysis Methods to Decision Making in Development of Regulations." *Transportation Research Record* 1707:22–26.

Rayback, S. A. 1998. "A Dendrogeomorphological Analysis of Snow Avalanches in the Colorado Front Range." *Physical Geography* 19:502–15.

Refice, A., and D. Capolongo. 2002. "Probabilistic Modeling of Uncertainties in Earthquake-Induced Landslide Hazard Assessment." *Computers and Geosciences* 28:735–49.

ReVelle, C. 1991. "Siting Ambulances and Fire Companies: New Tools for Planners." *Journal of the American Planning Association* 57:471–84.

Rice, R., R. Decker, N. Jensen, R. Patterson, and S. Singer. 2000. "Rural Intelligent Transportation System for Snow Avalanche Detection and Warning." *Transportation Research Record* 1700:17–23.

Rogers, G. B. 1993. "All Terrain Vehicle Injury Risks and the Effects of Regulation." *Accident Analysis and Prevention* 25:335–46.

Romero, S., G. J. Rix, and S. P. French. 2000. "Identification of Transportation Routes in Soils Susceptible to Ground Motion Amplification in the New Madrid Seismic Zone." *Transportation Research Record* 1736, 127–33.

Romin, L. A., and J. A. Bissonette. 1996. "Deer-Vehicle Collisions: Status of State Monitoring Activities and Mitigation Efforts." *Wildlife Society Bulletin* 24:276–83.

Saccomanno, F., and P. Haastrup. 2002. "Influence of Safety Measures on the Risks of Transporting Dangerous Goods Through Road Tunnels." *Risk Analysis* 22:1059–69.

Santi, P. M., E. J. Neuner, and N. L. Anderson. 2002. "Preliminary Evaluation of Seismic Hazards for Emergency Rescue Route, U.S. 60, Missouri." *Environmental and Engineering Geoscience* 8:261–77.

Schaerer, P. 1989. "The Avalanche-Hazard Index." *Annals of Glaciology* 13:241–47.

Schuster, R. L., G. F. Wieczorek, and D. G. Hope II. 1998. "Landslide Dams in Santa Cruz County, California Resulting from the Earthquake." U.S. Geological Survey Professional Paper 1551-C, 51–70.

Schweizer, J., B. Jamieson, and D. Skjonsberg. 1998. "Avalanche Forecasting for Transportation Corridor and Backcountry in Glacier National Park (BC, Canada)." In *Proceedings of 25 Years of Snow Avalanche Research,* ed. E. Hestnes. Publication 203, Norwegian Geotechnical Institute, 238–44.

Selcuk, A. S., and M. S. Yucemen. 2000. "Reliability of Lifeline Networks with Multiple Sources under Seismic Hazard." *Natural Hazards* 21:1–18.

Singh, R. N., J. A. Shonhardt, and N. Terezopoulos. 2002. "A New Dimension to Studies of Spontaneous Combustion of Coal." *Mineral Resource Engineering* 11:147–63.

Sinuany-Stern, Z., and E. Stern. 1993. "Simulating the Evacuation of a Small City: The Effects of Traffic Factors." *Socio-Economic Planning Sciences* 27:97–108.

Sorensen, P. A. 2001. "Locating Resources for the Provision of Emergency Medical Services." Ph.D. dissertation, Department of Geography, University of California Santa Barbara.

Sorensen, J., S. Carnes, and G. Rogers. 1992. "An Approach for Deriving Emergency Planning Zones for Chemical Stockpile Emergencies." *Journal of Hazardous Materials* 30:223–42.

Southworth, F. 1991. *Regional Evacuation Modeling: A State-of-the-Art Review.* ORNL/TM-11740, Oak Ridge National Laboratory.

Spring, G. S., and J. Hummer. 1995. "Identification of Hazardous Highway Locations Using Knowledged-Based GIS: A Case Study." *Transportation Research Record* 1497, 83–90.

Starr, C. 1969. "Societal Benefit versus Technological Risk." *Science* 165:1232–38.

Tangborn, W., and A. Post. 1998. "Iceberg Prediction Model to Reduce Navigation Hazards." In *Ice in Surface Waters,* ed. H. T. Shen. Rotterdam: A. A. Balkema, 231–36.

Thorpe, J. 1964. "Calculating Relative Involvements Rates in Accidents without Determining Exposure." *Australian Road Research* 2:25–36.

Turner, A. K., and R. L. Schuster. 1996. *Landslides: Investigation and Mitigation.* Transportation Research Board Special Report 247, National Academy Press, Washington, DC.

Tzeng, G.-H., and Y.-W. Chen. 1999. "The Optimal Location of Airport Fire Stations: A Fuzzy Multi-objective Programming and Revised Genetic Algorithm Approach." *Transportation Planning and Technology* 23:37–55.

Ullman, G. L. 2000. "Special Flashing Warning Lights for Construction, Maintenance, and Service Vehicles—Are Amber Beacons Always Enough?" Work Zone Safety; Pavement Marking Retroreflectivity—*Transportation Research Record* 1715, 43–50.

Verter, V., and B. Y. Kara. 2001. "A GIS-Based Framework for Hazardous Materials Transport Risk Assessment." *Risk Analysis* 21:1109–20.

Vorobtsov, S. N. 2002. "Estimation of the Hazard of Aircraft Flight in Conditions of Shears of a Three-Dimensional Wind." *Journal of Computer and Systems Science International* 41:703–15.

Wadge, G., A. Wislocki, and E. J. Pearson. 1993. "Spatial Analysis in GIS for Natural Hazard Assessment." In *Environmental Modelling with GIS,* ed. M. F. Goodchild, B. O. Parks, and L. T. Steyaert. Oxford: Oxford University Press, 332–38.

Werner, S., C. E. Taylor, and J. E. Moore. 1997. "Loss Estimation Due to Seismic Risks to Highway Systems." *Earthquake Spectra* 13:585–604.

Wolfe, A. 1982. "The Concept of Exposure and the Risk of a Road Traffic Accident and an Overview of Exposure Data Collection Methods." *Accident Analysis and Prevention* 14:337–40.

Wolshon, B. 2001. "One-Way-Out: Contraflow Freeway Operation for Hurricane Evacuation." *Natural Hazards Review* 2(3):105–112.

Yamada, T. 1996. "A Network Flow Approach to a City Emergency Evacuation Planning." *International Journal of Systems Science* 27:931–36.

CHAPTER 29
HAZARDOUS MATERIALS TRANSPORTATION

Linda R. Taylor
North Carolina State University
Raleigh, North Carolina

29.1 INTRODUCTION

The use of hazardous materials is prevalent throughout society. Hazardous materials can be found in thousands of consumer goods such as cleaning solutions for our homes and fertilizers and pesticides for our gardens, as well as in the processes used and products manufactured by manufacturing and industry. Some hazardous materials are manufactured and used at the same location. However, many hazardous materials are transported throughout the United States. Hazardous materials or dangerous goods include flammable liquids or solids; explosives; poisonous, flammable, or compressed gases; oxidizers; infectious substances; and toxics, radioactive materials, and corrosives.

The U.S. Department of Commerce collects data on the shipment of commercial goods, including hazardous materials, and estimates that hundreds of thousands of hazardous materials shipments occur every day and are an integral part of the U.S. economy and our daily lives. Each year in the United States, companies ship over 2 billion tons of hazardous materials by truck, rail, vessel, or aircraft valued at over $1.4 trillion (Table 29.1). If the hazardous materials are handled improperly or released from their shipping containers, they may cause injury to individuals, damage to property, and/or harm to the environment owing to the risk of fire, explosion, toxicity, or other harmful nature of the material.

29.2 HAZARDOUS MATERIALS LEGISLATION

According to the U.S. Department of Transportation (USDOT), a *hazardous material* is defined as a substance or material that presents an unreasonable risk to health, safety, and property and therefore is regulated during transportation. The Hazardous Materials Transportation Act (HMTA), originally passed in 1974, is the basis for the regulations administered by the USDOT. The HMTA was amended substantially in 1990 with the passage of the Hazardous Materials Transportation Uniform Safety Act (HMTUSA). HMTUSA provided the opportunity to harmonize U.S. requirements with international regulations, which are based on the UN Recommendations on the Transport of Dangerous Goods, including the International Civil Aviation Organization's Technical Instructions for the Safe Transport of Dangerous Goods and the International Maritime Organization's International Maritime Dangerous Goods Code.

TABLE 29.1 Shipment of Hazardous Materials by Mode, 2007

Mode description	Value ($ millions)	Value (percent of total)	Tons (thousands)	Tons (percent of total)	Ton-miles (millions)
All modes	1,448,218	100.0	2,231,133	100.0	323,457
Single modes	1,370,615	94.6	2,111,622	94.6	279,105
Truck	837,074	57.8	1,202,825	53.9	103,997
For-hire truck	358,792	24.8	495,077	22.2	63,288
Private truck	478,282	33.0	707,748	31.7	40,709
Rail	69,213	4.8	129,743	5.8	92,169
Water	69,186	4.8	149,794	6.7	37,064
Shallow draft	57,022	3.9	124,396	5.6	22,411
Great Lakes	S	S	S	S	S
Deep draft	11,626	0.8	24,181	1.1	13,767
Air (incl. truck and air)	1,735	0.1	S	S	S
Pipeline	393,408	27.2	628,905	28.2	S
Multiple modes	71,069	4.9	111,022	5.0	42,886
Parcel, USPS, or courier	7,675	0.5	236	Z	151
Truck and rail	7,052	0.5	11,706	0.5	10,120
Truck and water	23,451	1.6	36,588	1.6	12,380
Rail and water	5,153	0.4	5,742	0.3	2,937
Other multiple modes	27,739	1.9	56,750	2.5	17,297
Other and unknown modes	6,534	0.5	8,489	0.4	1,466

S = withheld because estimated did not meet publication standards; Z = zero
Source: U.S. Bureau of the Census, 2007, Commodity Flow Survey.

29.3 *HAZARDOUS MATERIALS REGULATIONS*

The Hazardous Materials Regulations (HMR), found in Title 49 of the *Code of Federal Regulations*, parts 100 to 180, govern the safety aspects of transportation. They apply to the transportation of hazardous materials in interstate, intrastate, and foreign commerce by aircraft, railcar, vessel, and motor vehicle. The USDOT enforces the regulations through the various modal-specific administrations within the USDOT. The Federal Motor Carrier Safety Administration (FMCSA) enforces the requirements for motor carriers, the Federal Aviation Administration (FAA) oversees air carrier regulations, the Federal Railroad Administration (FRA) regulates shipments by rail, and the U.S. Coast Guard (USCG) enforces the maritime shipments of hazardous materials. The regulations apply to the following entities who

- Transport hazardous materials in commerce (common, contract, and private carriers)
- Offer hazardous materials for interstate, foreign, and intrastate transportation in commerce (offerers, sometimes called *shippers*)
- Design, manufacture, fabricate, inspect, mark, maintain, recondition, repair, or test a package, container, or packaging component that is represented, marked, certified, or sold as qualified for use in transporting hazardous material in commerce
- Prepare or accept hazardous materials for transportation in commerce
- Are responsible for the safety of transporting hazardous materials in commerce
- Certify compliance with any requirement under the federal hazmat law

29.3.1 Civil and Criminal Penalties

The USDOT has the ability to investigate and apply sanctions to those failing to comply with the regulations. If an entity violates the law, the USDOT may issue a civil penalty or refer a criminal case to the U.S. court system. Civil penalties can range from a minimum of $250 per violation to a maximum assessment of $50,000 per violation per day. The maximum penalty may be raised to $100,000 if the violation results in death, serious or severe injury to any person, or substantial destruction of property. Criminal penalties may be imposed on individuals for willful or reckless violations. For individuals, the criminal sanctions may be up to $250,000 and five years imprisonment and for corporations, up to $500,000 and five years imprisonment. Criminal penalties of up to 10 years imprisonment may be imposed where the violations were willful and resulted in death or serious bodily injury.

29.3.2 Hazardous Materials Registration

The HMRs require the registration of "persons who offer for transportation or transport certain hazardous materials in intrastate, interstate or foreign commerce." More specifically, those shipping any of the following categories of hazardous materials (including a hazardous wastes) must register with the USDOT:

- A highway route controlled quantity of a class 7 radioactive material
- More than 25 kilograms (55 pounds) of a division 1.1, 1.2, or 1.3 substance (explosive materials in a motor vehicle, railcar, or freight container)
- More than 1 liter (1.06 quarts) per package of a "material extremely toxic by inhalation"
- A hazardous material (including hazardous wastes) in bulk packaging having a capacity equal to or greater than 13,248 liters (3,500 gallons) for liquids or gases or more than 13.24 cubic meters (468 cubic feet) for solids
- A shipment in other than bulk packaging of 2,268 kilograms (5,000 pounds) gross weight or more of one class of hazardous materials (including hazardous wastes) for which placarding of a vehicle, railcar, or freight container is required for that class
- A quantity of hazardous material that requires placarding

Some groups are specifically exempted from the registration requirements, including local, state, and federal agencies; Indian tribes; and most farmers. *Note:* More information about the registration process and the forms used to register can be found on the USDOT Pipeline and Hazardous Material Safety Administration (PHMSA) Web site at www.phmsa.dot.gov/hazmat.

The purposes of the registration program are to gather information about the transportation of hazardous materials and to fund the Hazardous Materials Emergency Preparedness (HMEP) grants program and additional related activities. The registration fees range from a minimum of $250 to a maximum of $3,000 per entity. The HMEP grants provide financial and technical assistance to states and Indian tribes to (1) develop, improve, and carry out emergency plans, (2) train public-sector hazardous materials emergency-response employees to respond to accidents and incidents involving hazardous materials, (3) determine flow patterns of hazardous materials with a state and between states, and (4) determine the need within a state for regional hazardous materials emergency-response teams.

29.3.3 Emergency Response and Incident Reporting

When a shipment of hazardous materials is involved in an incident (e.g., a spill or release), the carrier must immediately contact the USCG National Response Center (NRC) (by telephone at 800-424-8802 or online at www.nrc.uscg.mil) or, for infectious substances, the Centers for Disease Control and

Prevention (CDC) (at 800-232-0124) in place of the NRC. Reporting must include information such as contact information for the person reporting the incident; the date, time, and location of the incident; the name, hazard class, and quantity of the material spilled, if known; and any other emergency information about the spill (e.g., extent of injuries, disposition of the material, emergency actions taken, and quantity material remaining at the scene). Additionally, a more detailed follow-up hazardous materials incident report (USDOT Form F5800.1) is required to be submitted within 30 days of the incident to either the Pipeline and Hazardous Materials Safety Administration or the FAA for incidents involving transportation by aircraft.

The number of hazardous materials incidents varies from year to year and across the different modes of transportation. Highway transportation continues to be the most dangerous mode, with typically over 14,000 hazardous materials incidents per year. Figures 29.1 and 29.2 show that for 2009 the largest risk for both incidents and serious incidents was due to the transport of hazardous materials by highway. The USDOT defines a *serious incident* as an incident with a fatality or major injury involving the evacuation of 25 or more people, the closure of a major transportation artery, alteration of an aircraft flight plan or operation, the release of radioactive materials, the release of over 11.9 gallons or 88.2 pounds of a severe marine pollutant, or the release of a bulk quantity (>119 gallons or 882 pounds) of a hazardous material.

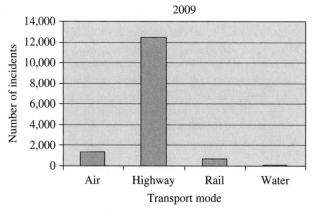

FIGURE 29.1   Hazardous materials incident by mode (2009).

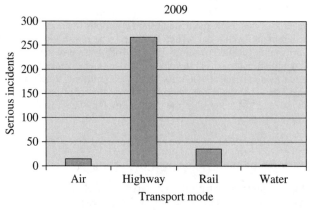

FIGURE 29.2   Serious hazardous materials incidents by mode (2009).

Table 29.2 summarizes the number of incidents and damages associated with incidents from 2009 and shows the greatest amount of damages owing to highway and rail transport. Although the total number of rail shipments and incidents by rail are low, the rail incidents tend to be larger in scope than highway incidents owing to the quantity of hazardous materials carried on adjacent railcars. Hazardous materials highway incidents typically involve single-transport vehicles carrying smaller quantities of hazardous materials.

Tables 29.3 and 29.4 show the cumulative number of incidents and serious incidents by mode for the last 10 years. On a positive note, although the total number of incidents per year has remained somewhat steady over that past 10 years, the number of serious incidents has been declining.

TABLE 29.2 Hazardous Materials Incidents by Mode, Including Damages, 2009

	Number of incidents	Damage ($)
Air	1,330	690,939
Highway	12,493	28,204,435
Rail	636	17,209,268
Water	84	11,360
Total	14,543	46,116,002

TABLE 29.3 Hazardous Materials Incidents by Mode by Year

	2000	2001	2002	2003	2004	2005	2006	2007	2008	2009
Air	1,419	1,083	732	750	933	1,655	2,408	1,556	1,278	1,330
Highway	15,063	15,804	13,502	13,594	13,071	13,460	17,153	16,905	14,784	12,493
Rail	1,058	899	870	802	765	745	702	750	748	636
Water	17	6	10	10	17	69	68	61	98	84
Total	17,557	17,792	15,114	15,156	14,846	15,929	20,331	19,272	16,908	14,543

Source: U.S. Department of Transportation, Hazardous Materials Safety, Hazardous Materials Incident Database.

TABLE 29.4 Serious Hazardous Materials Incidents by Mode by Year

	2000	2001	2002	2003	2004	2005	2006	2007	2008	2009
Air	35	37	15	13	6	18	17	13	10	15
Highway	449	488	377	399	400	422	398	394	360	267
Rail	83	63	71	58	82	85	75	82	54	36
Water	2	1	3	2	4	2	1	1	3	2
Total	569	589	466	472	492	527	491	490	427	320

Source: U.S. Department of Transportation, Hazardous Materials Safety, Hazardous Materials Incident Database.

Responders to incidents involving hazardous materials have access to emergency-response information provided with the shipping papers accompanying the material. Shipping papers must include the following items found in the HMRs:

- Proper shipping name
- Hazard class/division
- UN/NA identification number
- Packing group

This information may be used in conjunction with the *North American Emergency Response Guidebook* to help responders quickly identify the material involved in the incident and protect themselves and the general public during the initial response phase of the incident. Each guide page includes information on the potential hazards, public safety information, and emergency-response steps to be taken by the first responder arriving at the scene. Figure 29.3 shows a page (guide 128) out of the *North American Emergency Response Guidebook* (*NAERG*). The guidebook (*NAERG*) was developed

GUIDE 128 — FLAMMABLE LIQUIDS (NON-POLAR/WATER-IMMISCIBLE) — ERG2008

POTENTIAL HAZARDS

FIRE OR EXPLOSION

- HIGHLY FLAMMABLE: Will be easily ignited by heat, sparks or flames.
- Vapors may form explosive mixtures with air.
- Vapors may travel to source of ignition and flash back.
- Most vapors are heavier than air. They will spread along ground and collect in low or confined areas (sewers, basements, tanks).
- Vapor explosion hazard indoors, outdoors or in sewers.
- Those substances designated with a "P" may polymerize explosively when heated or involved in a fire.
- Runoff to sewer may create fire or explosion hazard.
- Containers may explode when heated.
- Many liquids are lighter than water.
- Substance may be transported hot.
- If molten aluminum is involved, refer to GUIDE 169.

HEALTH

- Inhalation or contact with material may irritate or burn skin and eyes.
- Fire may produce irritating, corrosive and/or toxic gases.
- Vapors may cause dizziness or suffocation.
- Runoff from fire control or dilution water may cause pollution.

PUBLIC SAFETY

- CALL Emergency Response Telephone Number on Shipping Paper first. If Shipping Paper not available or no answer, refer to appropriate telephone number listed on the inside back cover.
- As an immediate precautionary measure, isolate spill or leak area for at least 50 meters (150 feet) in all directions.
- Keep unauthorized personnel away.
- Stay upwind.
- Keep out of low areas.
- Ventilate closed spaces before entering.

PROTECTIVE CLOTHING

- Wear positive pressure self-contained breathing apparatus (SCBA).
- Structural firefighters' protective clothing will only provide limited protection.

EVACUATION

Large Spill
- Consider initial downwind evacuation for at least 300 meters (1000 feet).

Fire
- If tank, rail car or tank truck is involved in a fire, ISOLATE for 800 meters (1/2 mile) in all directions; also, consider initial evacuation for 800 meters (1/2 mile) in all directions.

GUIDE 128 — FLAMMABLE LIQUIDS (NON-POLAR/WATER-IMMISCIBLE)

EMERGENCY RESPONSE

FIRE

CAUTION: All these products have a very low flash point: Use of water spray when fighting fire may be inefficient.

CAUTION: For mixtures containing alcohol or polar solvent, alcohol-resistant foam may be more effective.

Small Fire
- Dry chemical, CO_2, water spray or regular foam.

Large Fire
- Water spray, fog or regular foam.
- Use water spray or fog; do not use straight streams.
- Move containers from fire area if you can do it without risk.

Fire involving Tanks or Car/Trailer Loads
- Fight fire from maximum distance or use unmanned hose holders or monitor nozzles.
- Cool containers with flooding quantities of water until well after fire is out.
- Withdraw immediately in case of rising sound from venting safety devices or discoloration of tank.
- ALWAYS stay away from tanks engulfed in fire.
- For massive fire, use unmanned hose holders or monitor nozzles; if this is impossible, withdraw from area and let fire burn.

SPILL OR LEAK

- ELIMINATE all ignition sources (no smoking, flares, sparks or flames in immediate area).
- All equipment used when handling the product must be grounded.
- Do not touch or walk through spilled material. • Stop leak if you can do it without risk.
- Prevent entry into waterways, sewers, basements or confined areas.
- A vapor suppressing foam may be used to reduce vapors.
- Absorb or cover with dry earth, sand or other non-combustible material and transfer to containers. • Use clean non-sparking tools to collect absorbed material.

Large Spill
- Dike far ahead of liquid spill for later disposal.
- Water spray may reduce vapor; but may not prevent ignition in closed spaces.

FIRST AID

- Move victim to fresh air. • Call 911 or emergency medical service.
- Give artificial respiration if victim is not breathing.
- Administer oxygen if breathing is difficult.
- Remove and isolate contaminated clothing and shoes.
- In case of contact with substance, immediately flush skin or eyes with running water for at least 20 minutes.
- Wash skin with soap and water.
- In case of burns, immediately cool affected skin for as long as possible with cold water. Do not remove clothing if adhering to skin. • Keep victim warm and quiet.
- Ensure that medical personnel are aware of the material(s) involved and take precautions to protect themselves.

FIGURE 29.3 *North American Emergency Response Guidebook* (2008).

jointly by the USDOT, Transport Canada, and the Secretariat of Communications and Transportation of Mexico (SCT) for use by firefighters, police, and other emergency services personnel who may be the first to arrive at the scene of a transportation incident involving a hazardous material.

The *NAERG* is updated every three to four years to accommodate new products and technology. The current version was updated in 2008. The next version is scheduled for 2012. The USDOT's goal is to place one *NAERG* in each emergency-service vehicle nationwide through distribution to state and local public safety authorities. To date, nearly 11 million copies have been distributed without charge to the emergency-response community. Copies are made available free of charge to public emergency responders through state coordinators. The *NAERG* is also available for download in English or Spanish by the public at the PHMSA Web site. The USDOT and Health and Human Services (HHS) National Library of Medicine have recently announced that they will make the *NAERG* available through electronic access for laptops and personal digital assistants (PDAs) under a new software application called the *Wireless Information System for Emergency Responders* (http://wiser.nlm.nih.gov/).

29.4 REGULATORY STRUCTURE

The major sections of the HMRs are listed in Table 29.5. Subchapter C (Parts 171 to 180) occupies the largest portion of the regulations. Part 171 includes critical definitions, reporting requirements, and other procedural requirements. Part 172 contains a listing of hazardous materials in the Hazardous Materials Table (49 CFR 172.101) and various communications requirements of the shipping paper descriptions, marking and labeling of packages, placarding of vehicles and bulk packagings, emergency-response communication, and security awareness and planning. Part 173 contains the various hazard class definitions for classifying materials and lists the packages authorized for use during transportation. Parts 174 to 177 contain requirements applicable to specific transport modes: part 174 for railcar, part 175 for aircraft, part 176 for vessels, and part 177 for motor vehicles. Part 178 contains the performance-oriented specification packaging information necessary for manufacturers to develop, manufacture, and test both bulk and nonbulk hazardous materials packagings (including carboys, drums, barrels, boxes, cases, drums, tubes, bags, and cylinders). Part 179 provides the packaging specifications for tank cars, and finally, part 180 contains the requirements for the continuing qualification and maintenance of packaging.

TABLE 29.5 Hazardous Materials Regulations (49 CFR 171–180)

Part	Heading
171	General Information, Regulations and Definitions
172	Hazardous Materials Table, Special Provisions, Hazardous Materials Communications
173	Shippers General Requirements for Shipments and Packagings
174	Carriage by Rail
175	Carriage by Aircraft
176	Carriage by Vessel
177	Carriage by Public Highway
178	Specifications for Packagings
179	Specifications for Tank Cars
180	Continuing Qualification and Maintenance of Packagings

29.4.1 Hazardous Material Classification

The development of the HMRs has been an evolutionary process. Regulations originally addressed only the most acute safety hazards, such as the risks of explosives and flammable materials. As new materials presenting different risks entered transportation systems, new hazard classes were added. Currently, the HMRs divide the hazardous materials into nine major hazard classes according to their

TABLE 29.6 USDOT Hazard Classes

Hazard class	Division	Definition	49 CFR Part
1	1.2	Explosives (with a mass projection hazard)	173.50
1	1.2	Explosives (with a projection hazard)	173.50
1	1.3	Explosives (predominately a fire hazard)	173.50
1	1.4	Explosives (with a minor explosion hazard)	173.50
1	1.5	Explosives (insensitive explosives)	173.50
1	1.6	Explosives (very insensitive detonating articles)	173.50
2	2.1	Flammable gas	173.115
2	2.2	Non flammable, nonpoisonous compressed gas	173.115
2	2.3	Gas (poisonous by inhalation)	173.115
3	—	Flammable liquids	173.120
4	4.1	Flammable solid	173.124
4	4.2	Spontaneously combustible material	173.124
4	4.3	Dangerous-when-wet materials	173.124
5	5.1	Oxidizer	173.127
5	5.2	Organic peroxides	173.128
6	6.1	Poisonous material	173.132
6	6.2	Infectious substances	173.134
7	—	Radioactive materials	173.403
8	—	Corrosive material	173.136
9	—	Miscellaneous hazardous materials	173.140

physical, chemical, and nuclear properties. Table 29.6 lists the nine hazard classes, the divisions within the hazard classes, the definition of each hazard class, and a reference to the part of the regulations with the full description of the hazard class.

29.4.2 Hazardous Materials Table

The Hazardous Materials Table (49 CFR 172.101) provides a list of over 3,000 chemicals and items classified into one of the nine hazard classes and is an essential step in compliance with the regulations. The table contains 10 columns of information vital to the proper shipping of hazardous materials. Figure 29.4 provides a look at a portion of the Hazardous Materials Table.

§ 172.101 Hazardous Materials Table									
Symbols	Hazardous materials descriptions and proper shipping names	Hazard class or division	Identification Numbers	Packing Group	Label(s) required (if not excepted)	Special provisions	(8) Packaging authorizations (§173.***)		
							Exceptions	Nonbulk packaging	Bulk packaging
(1)	(2)	(3)	(4)	(5)	(6)	(7)	(8A)	(8B)	(8C)
———	Poisonous solids, self heating, n.o.s. ...	6.1	UN3124	I	Poison, spontaneously combustible	A5_____	None	211	241

FIGURE 29.4 Hazardous Materials Table (49 CFR 172.101).

Column 1 contains one of six symbols or is blank. The six symbols (+, A, D, G, I, or W) provide information on which modes of transport the material may be regulated under (A for air, W for water, D for domestic shipments only, and I for international shipments only) or indicates that the material requires the technical name in addition to the proper shipping name (G for generic), or finally, fixes the proper shipping name, hazard class, and packing group for that entry (+).

Proper shipping names are listed in column 2 of the Hazardous Materials Table (HMT). The proper shipping name may be a unique chemical, such as acetone, or it may refer to a group of materials that possess the same hazard but are not specifically named. These materials are identified with a phrase such as *n.o.s.* (not otherwise specified). Column 3 shows the materials hazard class or the word *Forbidden*. Materials that are forbidden may not be offered for transportation in their current state. In order to transport a forbidden material, it must be diluted, stabilized, incorporated into a device, or modified in some manner to remove or reduce the original hazard of the material. Column 4 lists the four-digit United Nations or North American Identification Number, which can be easily referenced by emergency responders using the *NAERG*. Column 5 indicates one or more of three roman numerals (I, II, or III), which indicate the materials packing group. The packing group has implications related to the materials packaging requirements. A roman numeral I indicates major danger, II indicates medium danger, and III indicates minor danger. The final columns of the Hazardous Materials Table provide information on labeling (column 6), special provisions (column 7), packaging (columns 8a, b, and c), quantity limitations for air transport (columns 9a and b), and vessel stowage requirements (column 10).

29.4.3 Packaging, Marking and Labeling, and Shipping Papers

In addition to the requirements for classification of materials, the HMRs include regulations relating to packaging (including manufacture, continuing qualification, and maintenance), hazard communication (i.e., package marking, labeling, placarding, and shipping documentation), transportation and handling, incident reporting, and the safety and security of all hazardous materials shipments.

According to the HMRs, anyone who offers or accepts a hazardous materials shipment and those who manufacture packaging used for the shipment of hazardous materials are regulated. The packaging used must conform to the performance-oriented packaging (POP) standards developed by the United Nations. The POP standards provide manufacturers with a set of specific tests that packing must pass to be used for hazardous materials shipments. The specifications include requirements for drop, leakproofness, hydrostatic pressure, and stacking. Package manufacturers are required to mark containers with the UN specification marks to communicate to shippers that the containers are acceptable for hazardous materials shipments. The UN symbol is a circle with the lowercase letters *un* on top of each other. Figure 29.5 shows a typical UN mark. Table 29.7 includes a list of

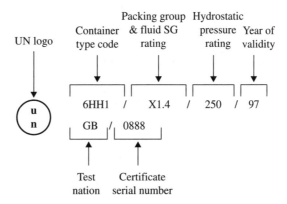

FIGURE 29.5 Example of UN performance-oriented packaging mark.

TABLE 29.7 UN Performance-Oriented Packaging Codes

Container type	Material of manufacture	Other
1. Drum	A. Steel	1. Closed head
2. Barrel	B. Aluminum	2. Open head
3. Jerrican	C. Natural wood	
4. Box	D. Plywood	
5. Bag	F. Reconstituted wood	
6. Composite packaging	G. Fiber	
7. Pressure receptacle	H. Plastic	
	L. Textile	
	M. Paper	
	N. Metal (other than steel or aluminum)	
	P. Glass or porcelain	

the container-type marks and what type of material they designate. After the UN symbol and the container type, the mark lists either an uppercase *X*, *Y*, or *Z*, which is the performance level to which the package is manufactured. *X* meets the requirements for packing groups I, II, and III. *Y* meets the requirements for packing groups II and III, and *Z* meets the requirements for packing group III only. The code also lists the specific gravity or gross mass for which the package has been tested, followed by the hydrostatic test pressure. The year the package was manufactured, the country of manufacture, and a registration number for the manufacture are listed last.

It is the responsibility of the shipper to place the hazardous materials in the proper container for shipping. Column 8 of the Hazardous Materials Table for each proper shipping name provides a reference to the list of acceptable packagings for each hazardous material. The numbers found in column 8 indicate the location in 49 CFR Part 173 where the shipper may find the list of acceptable packaging.

Once a proper package is selected, it must be marked and labeled in accordance with the regulations. Marks and labels provide visual and written information on the hazards contained in each hazardous materials package leaving a shipper's control. Individuals handling the packages must know what is inside in order to best protect themselves and the public from the hazards associated with each material.

Marking refers to placing the required information on the outer packaging containing the hazardous material. The HMRs state in 49 CFR 172.300 that "each person who offers a hazardous material for transportation shall mark each package, freight container, and transport vehicle containing the hazardous material in the manner required. Marking means a descriptive name, identification number, . . . or UN marks, or combinations thereof, required . . . on outer packagings of hazardous materials."

Packages in transportation get handled many times on their way to their destination. Markings are very important. They help emergency-response personnel to identify the exact nature of the material in case of a spill. Appropriate remedial action can be taken when the markings describe the material fully. Therefore, accurate and consistent descriptions for the various hazardous materials transported must be available at all times.

Marking requirements are provided for nonbulk packaging as well as for bulk packaging. Considering the size of bulk packaging, that is, liquid containers with capacity over 119 gallons or solid containers with capacity over 400 kilograms, it is especially important to identify these materials because they represent the potential for significant release in transit.

Some of the required markings for packaging are listed below. Not all of them will be required for every hazardous materials package because the marks required will depend on what is being shipped and what packaging is used. In general, for nonbulk packaging, the following marks are required:

1. The proper shipping name
2. UN/NA identification number
3. Consignee's or consignor's name and address

Other markings that may be required (depending on the hazardous material) include

4. Orientation arrows for liquids
5. The letters *RQ* for reportable quantity for those materials also meeting the definition of a hazardous substance
6. The words *HAZARDOUS WASTE* for Resource Conservation and Recovery Act (RCRA) wastes
7. The words *Inhalation Hazard* for packages that meet the criteria for poison-inhalation hazards or toxic inhalation hazard

The markings may be applied in several ways; they simply may be printed on the package, stenciled on, or applied with an adhesive-backed sticker. The USDOT requires all markings to be durable, in English, and printed on or affixed to the surface of a package or on a label, tag, or sign. They must be displayed on a background of sharply contrasting color and unobscured by other labels or attachments that could reduce their effectiveness substantially. Markings must be located near the USDOT hazard label. Other markings may be required depending on the specific hazard of the material.

Labeling is required for all nonbulk packaging containing hazardous materials. The required labels for each proper shipping name are specified in column 6 of the Hazardous Materials Table. The hazard warning labels are intended to accurately represent the hazard in the package. Some materials possess more than one hazard (more than one label code found in column 6) and must be labeled accordingly. Multiple labels must be displayed next to each other. The labels must conform to the specifications provided by the USDOT on the color, symbol, number, wording, and size. The USDOT prints a chart (Chart 13) that provides a picture of each of the labels required.

Placards are similar to labels but larger in size and are required for bulk packages and transport vehicles such as railcars, tank trucks, and freight containers holding hazardous materials. In general, bulk packaging, freight containers, or railcars containing greater than 1,000 pounds of hazardous materials must be placarded. Also in general, explosives and radioactive materials require placarding regardless of the quantity being shipped. There are additional placarding requirements for each mode of transportation, and they can be found in subpart F of the HMRs.

Shipments of hazardous materials also must be accompanied by written documentation communicating the hazards of each hazardous material. Types of shipping papers include bills of lading, air waybills, hazardous waste manifests, and shippers' declaration for dangerous goods by air. In most cases, no specific shipping paper is required, only specific information that must be included on the shipping paper. The shipping papers must include the following information:

- UN identification number
- Proper shipping name
- Hazard class
- Packing group
- The number of packages and total quantity of material being shipped

The shipper's name and address, as well as the consignee's name and address, are necessary as well. Emergency-response information is also required. A 24-hour emergency-response number must be included, as well as written information containing basic safety information, including

- Immediate dangers to health
- Fire and explosion risks
- Precautions to take in the event of a spill or release of the hazardous material
- Initial first-aid measures that may be implemented after exposure to the hazardous material

This information may be included in written form or shippers may include guide pages from the *NAERG* that contain the required information. The driver of the hazardous materials vehicle must keep the shipping papers on the seat next to him or her or in the door pocket, such that the information is readily and easily available to authorities in the event of a release or inspection.

Shipments of hazardous waste, designated by the U.S. Environmental Protection Agency (USEPA), must be accompanied by the Uniform Hazardous Waste Manifest (Figure 29.6) as the form of its shipping paper. Air shipments must use the "Shippers Declaration for Dangerous Goods" (Figure 29.7).

All shipping papers must be signed and have the shipper's certification included. One of the following statements must be included:

> This is to certify that the above-named materials are properly classified, described, packaged, marked, and labeled and are in proper condition for transportation according to the applicable regulations of the Department of Transportation. *Or*

> I hereby declare that the contents of this consignment are fully and accurately described above by the proper shipping name, and are classified, packaged, marked, and labeled/placarded, and are in all respects in proper condition for transport according to applicable international and national governmental regulations.

For air shipments, the following statement also must be included:

> I hereby declare that all the applicable air transport requirements have been met.

29.5 TRAINING

When HMTUSA was enacted, the USDOT required employers to provide training for their hazmat employees. The purpose of the training is to increase a hazmat employee's safety awareness and be an essential element in reducing hazmat incidents. The regulations established several categories of required training:

- *General awareness.* Ability to recognize and identify hazardous materials.
- *Function-specific training.* Specific to the hazardous material function each employee may conduct (e.g., marking, labeling, documentation, packaging, loading and unloading, separation, and segregation).
- *Safety.* Training on emergency-response information and how to protect themselves from the hazards of exposure.
- *Security awareness.* Training on the security risks and threats associated with the transport of hazardous materials.
- *Drivers' training.* Training on general awareness and function-specific topics as well as the safe operation of a motorized vehicle under the Federal Motor Carrier Safety Regulations.

Employers are required to provide training to employees within 90 days of the initial assignment to a hazardous material function and recurrent training at least once every 3 years. Employers must include a testing component to their training to ensure compliance with the regulations and must keep records of the training for each employee. The record must include

- Hazmat employee's name
- Completion date of most recent training
- Training materials (copy, description, or location)
- Name and address of hazmat trainer
- Certification that the hazmat employee has been trained and tested

Please print or type. (Form designed for use on elite (12-pitch) typewriter.)   Form Approved. OMB No. 2050-0039

UNIFORM HAZARDOUS WASTE MANIFEST	1. Generator ID Number	2. Page 1 of	3. Emergency Response Phone	4. Manifest Tracking Number

5. Generator's Name and Mailing Address   Generator's Site Address (if different than mailing address)

Generator's Phone:

6. Transporter 1 Company Name	U.S. EPA ID Number

7. Transporter 2 Company Name	U.S. EPA ID Number

8. Designated Facility Name and Site Address	U.S. EPA ID Number

Facility's Phone:

9a. HM	9b. U.S. DOT Description (including Proper Shipping Name, Hazard Class, ID Number, and Packing Group (if any))	10. Containers No.	Type	11. Total Quantity	12. Unit Wt./Vol.	13. Waste Codes
	1.					
	2.					
	3.					
	4.					

14. Special Handling Instructions and Additional Information

15. GENERATOR'S/OFFEROR'S CERTIFICATION: I hereby declare that the contents of this consignment are fully and accurately described above by the proper shipping name, and are classified, packaged, marked and labeled/placarded, and are in all respects in proper condition for transport according to applicable international and national governmental regulations. If export shipment and I am the Primary Exporter, I certify that the contents of this consignment conform to the terms of the attached EPA Acknowledgment of Consent. I certify that the waste minimization statement identified in 40 CFR 262.27(a) (if I am a large quantity generator) or (b) (if I am a small quantity generator) is true.

Generator's/Offeror's Printed/Typed Name	Signature	Month	Day	Year

16. International Shipments	☐ Import to U.S.	☐ Export from U.S.	Port of entry/exit: _____
Transporter signature (for exports only):			Date leaving U.S.:

17. Transporter Acknowledgment of Receipt of Materials

Transporter 1 Printed/Typed Name	Signature	Month	Day	Year
Transporter 2 Printed/Typed Name	Signature	Month	Day	Year

18. Discrepancy

18a. Discrepancy Indication Space	☐ Quantity	☐ Type	☐ Residue	☐ Partial Rejection	☐ Full Rejection

Manifest Reference Number:

18b. Alternate Facility (or Generator)	U.S. EPA ID Number

Facility's Phone:

18c. Signature of Alternate Facility (or Generator)	Month	Day	Year

19. Hazardous Waste Report Management Method Codes (i.e., codes for hazardous waste treatment, disposal, and recycling systems)

1.	2.	3.	4.

20. Designated Facility Owner or Operator: Certification of receipt of hazardous materials covered by the manifest except as noted in Item 18a

Printed/Typed Name	Signature	Month	Day	Year

EPA Form 8700-22 (Rev. 3-05) Previous editions are obsolete.   **DESIGNATED FACILITY TO DESTINATION STATE (IF REQUIRED)**

GENERATOR | TRANSPORTER INT'L | DESIGNATED FACILITY

FIGURE 29.6 Uniform hazardous waste manifest.

SHIPPER'S DECLARATION FOR DANGEROUS GOODS

Shipper	Air Waybill No. Page of Pages Shipper's Reference Number *(optional)*
Consignee	*For optional use* *For* *company logo* *name and address*

Two completed and signed copies of this Declaration must be handed to the operator.	**WARNING**
TRANSPORT DETAILS	Failure to comply in all respects with the applicable Dangerous Goods Regulations may be in breach of the applicable law, subject to legal penalties.

This shipment is within the limitations prescribed for: *(delete non-applicable)*	Airport of Departure:
PASSENGER AND CARGO AIRCRAFT / **CARGO AIRCRAFT ONLY**	

Airport of Destination:	Shipment type: *(delete non-applicable)* NON-RADIOACTIVE | RADIOACTIVE

NATURE AND QUANTITY OF DANGEROUS GOODS
UN Number or Identification Number, proper shipping name, class or division (subsidiary risk), packing group (if required), and all other required information.

Additional Handling Information

I hereby declare that the contents of this consignment are fully and accurately described above by the proper shipping name, and are classified, packaged, marked and labelled/placarded, and are in all respects in proper condition for transport according to applicable international and national governmental regulations. I declare that all of the applicable air transport requirements have been met.	Name/Title of Signatory Place and Date Signature *(see warning above)*

FIGURE 29.7 Shippers declaration for dangerous goods.

29.5.1 Drivers' Training

Hazardous materials shippers are not the only regulated entity within the hazardous materials transportation arena. The USDOT, through the collection of data on transportation incidents, has come to realize that highway transportation is a significant contributor to the overall number of hazardous materials incidents and emergencies. To increase the safety of the highway transportation of hazardous materials, the USDOT requires drivers of vehicles carrying hazardous materials to be trained. Drivers must be knowledgeable in the identification and recognition of hazardous materials, in loading and unloading of hazardous materials, and in hazardous materials separation and segregation requirements, as well as in the safe operation of a motor vehicle. The Federal Motor Carrier Safety Regulations (FMCSRs) are contained in a different part of the *Code of Federal Regulations* (49 CFR Parts 390–397).

The FMCSRs require drivers to ensure the safe operation of their vehicles through thorough pretrip safety inspections, including the inspection of braking mechanisms, steering devices, lights and reflectors, horns, windshield wipers, mirrors, and other important vehicle operations. Drivers must be trained on emergency actions, including the operation of critical equipment such as fire extinguishers, fuses, and warning devices (e.g., flares and reflective triangles). In addition, drivers must be tested on the proper operation of the vehicle, including turning, backing, and parking, and the dangers of unsafe conditions related to poor weather, speed, or other instability conditions.

Drivers must know how to properly load and unload hazardous materials to ensure the safety of the shipment throughout the transportation process. Materials must be loaded securely so that they do not shift during transportation. Incompatible hazardous materials must be segregated according to the guidelines provided by the USDOT. Proper operation of the vehicle prior to loading and unloading must be followed, including the requirements to ensure that engines are turned off, handbrakes are set, and wheels are properly chocked against movement.

The FMCSRs contain other provisions for drivers, including the requirement to be with or near the hazardous material shipment at all times. Drivers must be with 100 feet of the vehicle even during breaks with an unobstructed view, and drivers may not park on the roadway unless vehicle or road conditions prevent the safe parking of the vehicle more than 5 feet from the roadway. Smoking is not permitted by drivers carrying explosives, flammable materials, or oxidizing substances.

Drivers are also required to understand important routing limitations and incident-reporting requirements for hazardous materials transportation. Hazardous materials transportation routes should avoid dangerous conditions such as congested areas, tunnels, or other heavily populated centers. Some roads prohibit the transportation of hazardous materials altogether. Drivers are responsible for knowledge of these routing limitations. Also, if drivers of hazardous materials are involved in incidents that result in the spill or release of a hazardous material or the injury of people or the environment, they are required to provide a report to the National Response Center with critical emergency-response information.

Additional requirements for highway transportation of hazardous materials include the requirement for drivers of hazardous materials to possess a commercial driver's license (CDL) with a hazmat endorsement and the testing of drivers for controlled substance and alcohol use. Drivers about to perform or immediately available to perform hazardous materials functions may be subject to random testing. Employers are also allowed to conduct testing when there is a "reasonable suspicion" that the driver has violated the alcohol provisions of the regulations. Significant conditions are in place to prevent the misuse of this provision by employers and employees with the overarching goal of the safe transportation of hazardous materials.

There are additional modal-specific requirements within the HMRs for transportation by air, rail, and water. For example, some air carriers are designated as "will not carry," indicating that they will not transport hazardous materials by air. They must train their employees on how to properly identify hazardous materials and reject the shipment. Air carriers choosing to accept the shipment of hazardous materials must understand the more stringent regulations that apply to air shipments. Some hazardous materials are forbidden for transport by air, others are restricted to cargo aircraft only, and still others are limited in the quantity per package allowed on passenger and cargo aircraft.

Each of the modal-specific requirements are found in 49 CFR Parts 174–177 and must be complied with fully by the carriers of hazardous materials.

The transportation of hazardous materials is a complicated process requiring the completion of many crucial steps. However, the safe transport of these essential materials to their destination is of paramount concern owing to their dangerous properties. The USDOT and other national and international agencies continue to enhance the regulations to minimize the present and future costs associated with hazardous materials transportation.

CHAPTER 30
INCIDENT MANAGEMENT

Ahmed Abdel-Rahim
Department of Civil Engineering
University of Idaho
Moscow, Idaho

30.1 INTRODUCTION

Incident management has become an important component of the activities of departments of transportation nationwide. With much of the nation's roadway system operating very close to capacity under the best of conditions, the need to reduce the impact of incident-related congestion has become critical. Traffic incidents are causing thousands of hours of congestion and delay annually. Incident-related delay accounts for between 50 and 60 percent of total congestion delay. In smaller urban areas, it can account for an even larger proportion (FHWA 2000b). Incidents also increase the risk of secondary crashes and pose safety risks for incident responders on the incident scene and elsewhere. While incidents on roadways cannot be predicted or prevented entirely, the implementation of an effective incident management system can mitigate the impacts of the incidents.

The success of an incident management program depends largely on coordination and collaboration between different agencies involved in the incident management operations. The emergence of intelligent transportation system (ITS) technologies in the 1990s has brought opportunities to increase the timeliness, effectiveness, and efficiency of incident management operations. The evolution of ITS promises not only new tools for real-time communication, but also tools for incident detection and verification, integrated network-wide management, and the provision of up-to-date traveler information.

This chapter contains a review of the latest developments in incident management programs, with the intention of providing an incident management practical reference and guide to transportation professionals and also to other professionals who might be involved in the incident management process. The topics in this chapter are not covered in great detail, as the breadth of the traffic incident management activities is too wide for this handbook. More extensive references can be consulted (FHWA 2000a,b; Koehne, Mannering, and Hallenbeck 1995; Carvell et al. 1997).

30.2 CHARACTERISTICS OF TRAFFIC INCIDENTS

30.2.1 Incident Definition

Traffic incidents can be defined as nonrecurring events that cause congestion and delay by restricting normal traffic flow. Traffic incidents can result from either a reduction in the roadway capacity or an increase in the traffic demand. Based on this definition, incidents can be classified into two groups: planned and random events. Planned events include

- Highway maintenance and reconstruction projects
- Special nonemergency events (e.g., sports activities, concerts, or any other event that significantly affects roadway operations)

Random events include

- Traffic crashes
- Disabled vehicles on the road
- Spilled cargo
- Natural or man-made disasters

The incident management activities covered in this chapter are focused primarily on random events incidents.

30.2.2   Incident Types

Incident types and frequencies are location-dependent. Differences are due to factors such as road geometry, level of traffic demand, weather, grade, and shoulder availability. The reported relative frequencies of incident types are quite diverse. However, the majority of the reported incidents were found to involve minor events, such as disabled vehicles on the shoulder and other incidents that have little impact on the roadway capacity. An example of a typical composite profile of reported freeway incidents is presented in Figure 30.1, which illustrates the distribution of freeway incidents by incident type, duration, and incident-related delay (ATA and Cambridge Systematics, Inc. 1997).

30.3   INCIDENT IMPACTS

The magnitude of incident-related problems is severe. Traffic incidents pose three primary concerns:

1. Reduction in the operational efficiency of the transportation network, causing significant delay to motorists, including delay to emergency responders
2. Increased risk of secondary crashes
3. Safety risks to incident responders due to increased danger at the incident scene

Problems associated with incidents are not limited to these primary areas; they also include (FHWA 2000b)

- Increased response time by police, fire, and emergency medical services due to roadway delay and reduction in manpower
- Lost time and a reduction in productivity and increased cost of goods and services
- Increased fuel consumption and reduced air quality and other adverse environmental impacts
- Negative public image of public agencies involved in incident management activities
- Traveler frustration and road rage

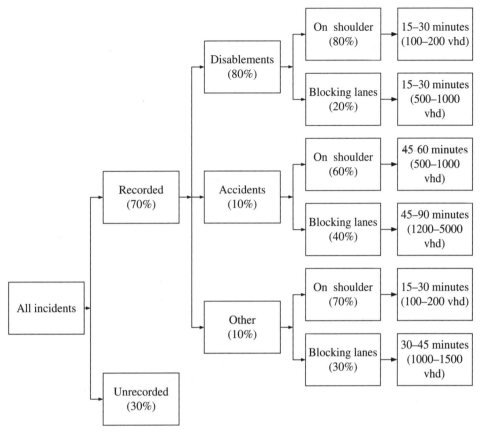

FIGURE 30.1   Incident types. (*Source:* Cambridge Systematics, Inc.)

30.3.1 Incident-Related Congestion

Incidents are estimated to cause somewhere between 52 and 58 percent of total delay experienced by motorists in all urban area population groups. The amount of incident-related congestion for the 435 urban areas in the United States in 2007 ranged from 1,280,000 to 2,031,000 person-hours of delay with an annual cost of approximately $48.2 billion in wasted time and fuel consumption (TTI 2009). The incident-related delay is caused by the reduction of roadway capacity that results from the incident. The *Highway Capacity Manual* (*HCM 2000*) (TRB 2000) provides guidelines on the expected reduction in the proportion of freeway segment capacity as a function of number of freeway lanes and the type of incident. Incident types included in the guidelines range from shoulder disablement to full lane closure, as shown in Table 30.1.

Figure 30.2 is a graphical representation of traffic behavior upstream and downstream of an incident location when the incident partially or totally blocks the roadway. Upstream of the incident, the characteristics represent traffic conditions moving at normal speeds and normal density. The area located immediately upstream of the incident represents the high-density congested area where vehicles are queuing and traveling at low speeds. The region immediately downstream of the incident reflects traffic flowing at a metered rate with low density and slightly higher speed than

TABLE 30.1   Proportion of Freeway Segment Capacity Available under Incident Conditions

Number of freeway lanes by direction	Shoulder disablement	Shoulder accident	One lane blocked	Two lanes blocked	Three lanes blocked
2	0.95	0.81	0.35	0.00	N/A
3	0.99	0.83	0.49	0.71	0.00
4	0.99	0.85	0.58	0.25	0.13
5	0.99	0.87	0.65	0.40	0.20
6	0.99	0.89	0.71	0.50	0.26
7	0.99	0.91	0.75	0.57	0.36
8	0.99	0.93	0.78	0.63	0.41

Source: Highway Capacity Manual 2000.

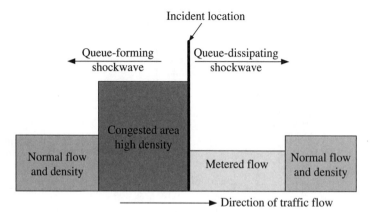

FIGURE 30.2   Behavior of traffic under an incident.

normal flow. The far downstream area from the incident represents traffic flow similar to the area far upstream of the incident, with normal traffic flow and normal density. This behavior continues until the incident is completely removed and the queued vehicles upstream of the incident are completely discharged. The length of the congested area, and thus the total incident-related delay, will depend on the incident severity (number of lanes blocked) and the incident duration. Figure 30.3 illustrates incident-based delay represented by cumulative arrivals and departures diagrams throughout the incident duration. Figure 30.4 illustrates possible reduction in incident-based delay with an incident management system. Reducing incident detection, verification, response, and clearance times and diverting some of the freeway traffic to alternate routes can significantly reduce the incident-based delay and queues, as illustrated in the figure.

30.3.2   Increased Risk of Secondary Crashes

The severity of secondary crashes is often greater than that of the original incident. A study conducted in Minnesota found that 13 percent of all peak-period crashes were secondary crashes (Minnesota DOT 1982). Similar results were found in a study by the Washington State Department of Transportation. The study found that 3,165 shoulder crashes had occurred on interstate, limited access, or other state highways during a seven-year period. These collisions caused a total of 40 deaths and 1,774 injuries. The study reported that 41 percent of all shoulder collisions involved

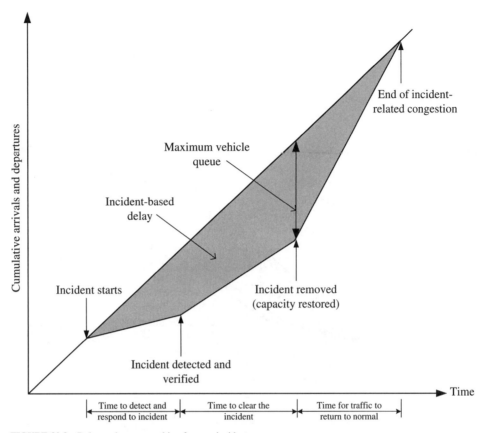

FIGURE 30.3 Delay and queue resulting from an incident.

injuries (IACP 1996). An analysis of accident statistics in California revealed that secondary crashes represent an increase in collision risk of over 600 percent (Volpe 1995).

30.3.3 Safety Risks to Incident Responders

Hazards at the incident scene put emergency personnel who respond to the incident at risk of being struck by passing vehicles. In 1997, nearly 40 percent of all law enforcement officers who died in the line of duty died in traffic (*The Police Chief* 1998). In addition to the hazards of the incident scene, emergency responders traveling to and from the incident scene are also at risk because of incidents. In 1998, there were 143 fatalities in the United States involving emergency vehicles, 77 of which occurred when the vehicle was responding to an emergency (NHTSA 1998).

30.4 *THE INCIDENT MANAGEMENT PROCESS*

Incident management consists of a series of activities involving personnel from a variety of response agencies and organizations. These activities are not necessarily performed sequentially. The activities include incident detection, incident verification, incident response, site management and

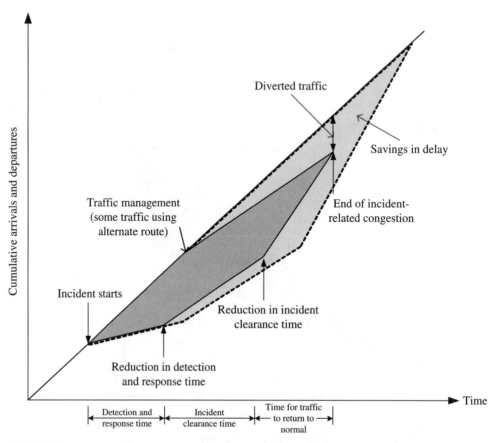

FIGURE 30.4   Possible reduction in delay and queue with an incident management system.

clearance, and motorist information, as shown in Figure 30.5. Table 30.2 lists the roles and responsibilities of different agencies involved in the incident management activities. The following sections discuss each of these components of the incident management process.

30.4.1   Incident Detection

Incident detection is the process by which an incident is identified and brought to the attention of the responsible agency. Some of the methods commonly used to detect incidents include (FHWA 2000b)

- Cellular telephone calls from motorists
- Closed-circuit television (CCTV) cameras viewed by operators in traffic management centers
- Automatic incident detection algorithms (AID)
- Motorist aid telephones or call boxes
- Police and service patrol vehicles
- Fleet vehicles (transit and trucking)

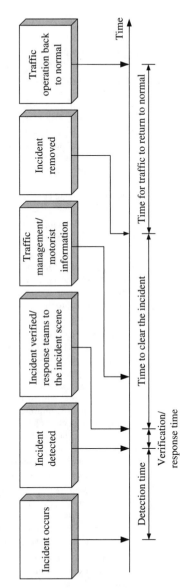

FIGURE 30.5 The incident management process.

TABLE 30.2   Role and Responsibilities of Different Agencies Involved in the Incident Management Process

Agency	Role and Responsibility
Police department	• Assist in incident detection • Secure the incident scene • Assist disabled motorists • Provide emergency medical aid until help arrives • Direct traffic • Conduct accident investigations • Serve as incident commander • Safeguard personal property • Supervise scene clearance
Fire and rescue	• Protect the incident scene • Provide traffic control until police or DOT arrival • Provide emergency medical care • Provide initial hazmat response and containment • Perform fire suppression • Rescue crash victims from wrecked vehicles • Arrange transportation for the injured • Serve as incident commander • Assist in incident clearance
Hazardous materials (hazmat)	• Rescue crash victims from contaminated environments • Clean up and dispose of toxic or hazardous materials.
Emergency medical service (EMS)	• Provide advanced emergency medical care • Determine of destination and transportation requirements for the injured • Coordinate evacuation with fire, police, and ambulance or airlift • Serve as incident commander for medical emergencies • Determine approximate cause of injuries for the trauma center • Remove medical waste from incident scene
Transportation agencies	• Assist in incident detection and verification • Initiate traffic management strategies on incident impacted facilities • Protect the incident scene • Initiate emergency medical assistance until help arrives • Provide traffic control • Assist motorists with disabled vehicles • Provide motorist information • Determine incident clearance and roadway repair needs • Establish and operate alternate routes • Coordinate clearance and repair resources • Serve as incident commander for clearance and repair functions • Repair transportation infrastructure
Towing and recovery	• Remove vehicles from incident scene • Protect victims' property and vehicles • Remove debris from the roadway • Provide transportation for uninjured vehicle occupants • Serve as incident commander for recovery operations
Media and information service providers	• Report traffic incidents • Broadcast information on delays • Provide alternate route information • Update incident status frequently • Provide video or photography services

With the widespread use of wireless phones, most incidents are being reported by multiple wireless calls to 911 emergency numbers or 511 travel numbers (Figure 30.6). The percentages of incidents detected by cell phone callers to 911 or other dedicated phone numbers are as high as 90 percent, with most incidents being reported within 2 minutes of their occurrence. When multiple calls are received for the same incident, it is important to identify accurately the location of the callers to determine the exact location of the incident. Some agencies use 0.1 milepost to aid callers in identifying their locations. Cellular phones with GIS-based systems will allow agencies to accurately identify the caller's location and hence expedite the incident detection and verification process.

Automated incident detection (AID) algorithms have been the focus of much research and have been implemented and tested in many systems throughout North America. AIDs can be divided into four major categories:

FIGURE 30.6 911 emergency cellular phone calls help expedite incident detection.

- Comparative (or pattern recognition) algorithms that compare traffic parameters at a single detector station or between two detector stations against thresholds that define the incident conditions. This category includes 10 modified California algorithms, as well as the pattern recognition (PATREG) algorithm.

- Statistical algorithms that use statistical techniques to determine whether observed detectors' data differ statistically from historical or defined conditions.

- Time-series/smoothing algorithms that compare short-term predictions of traffic conditions to measured traffic conditions.

- Modeling algorithms that use standard traffic flow theories to model expected traffic conditions on the basis of current traffic measurements. Dynamic model and McMaster algorithms fall into this category.

The effectiveness of incident detection algorithms depends mainly on the detection system configuration and the prevailing traffic flow profiles. Successful implementation of incident detection algorithms require a dense network of detectors capable of transmitting real-time speed and occupancy data. The performance of incident detection algorithms can be measured using the following three factors:

- Percentage of incidents detected
- Time required to detect an incident
- False alarm rate

Algorithms with high detection sensitivity and short detection time will typically yield high false alarms. Agencies must decide what is an acceptable balance between detection sensitivity and false alarm rates for their detection system. False alarms can be tolerated in order to achieve a higher detection sensitivity. Research indicates that the best performing algorithms detect 70 to 85 percent of all incidents, with a false alarm rate of 1 percent or less (Hellinga and Knapp 2000; Mahamassani et al. 1999; Chang 1998; Barton-Aschman Associates, Inc. 1993; FHWA 1997.) The state-of-the-practice is to use AID algorithms to flag operators to any change in traffic flow patterns, then use CCTV cameras to verify the nature of problem.

30.4.2 Incident Verification

Incident verification is the process of confirming that an incident has occurred, determining its exact location, and obtaining as many relevant details about the incident as possible before dispatching responders to the incident scene. Verification includes gathering enough information to dispatch the proper initial response. Methods of incident verification include the following:

- Operators viewing CCTV cameras
- Dispatching field units such as police or service patrols to the incident site
- Combining information from multiple cellular phone calls

CCTV cameras can be used to provide the traffic operation centers with valuable information about the specific location, nature, and extent of incidents. If an incident has occurred within the CCTV coverage area, operators can monitor the incident from the operation center. Agencies with CCTV cameras covering more than 90 percent of their freeway system can verify incidents reported by cell phones in less than 2 minutes from the time they are reported. This can reduce the incident response and clearance time significantly.

30.4.3 Incident Response

Incident response includes dispatching the appropriate personnel and equipment to the incident scene. It also includes activating the preestablished communication links between different agencies and the motorist information system. The response should start as soon as the incident is verified. Effective incident response requires preparedness on the part of each responding agency. This is achieved through planning and personnel training, both on an individual basis and collectively with other response agencies.

Two ITS technologies can facilitate better incident response: computer-aided dispatch (CAD) and traffic signal preemption. CAD combines computer and communications technologies to manage communications better among emergency responders and each group's dispatch center. Computer-aided dispatch systems are in place in thousands of fire, police, and other emergency service agencies throughout North America. CAD functions assist dispatchers in tracking the status of field units and in assigning units to respond to an incident based on the severity of the incident, the unit's location, and the equipment required at the incident scene. They also provide police and fire personnel with access to multiple databases containing information on issues such as hazardous materials identification and handling procedures. In addition, CAD systems archive incident response activities, which can be used to conduct postincident review and evaluations.

Traffic signal preemption provides emergency vehicles with green lights, accelerating their move through signalized intersections en route to the incident scene. Signal preemption systems consist of an emitter-equipped vehicle, a detector, and a signal phase selector. Each emitter's communication signal is encoded with a unique identifier and user authorization code. The detector receives the emitter's message and transfers it to the signal controller. The phase selector validates the signal and requests priority control from the traffic controller. Command priority results in the system providing a green light to authorized vehicles after appropriate intersection timing is complete, usually within a few seconds. However, emergency vehicle preemption benefit comes with considerable disutility cost. Every emergency vehicle preemption causes disruption to the corridor operation and may cause considerable delay to motorists.

30.4.4 Site Management

Site management is defined as the process of managing the incident site and coordinating on-scene resources. The primary objective of site management is to ensure the safety of response personnel, incident victims, and other motorists on or near the incident scene (Figure 30.7). When multiple agencies respond, the unified command structure concept has helped solve some of the coordination and communication problems. The unified command structure provides a management tool to facilitate cooperative participation by representatives from many agencies and/or jurisdictions. It encourages the incident commanders of the major agencies to work together in a unified command post. Communication to the unified command post from the traffic management center is of key importance.

FIGURE 30.7 Site management for an incident on a multilane freeway section—safety and operational challenges.

30.4.5 Incident Clearance

Incident clearance is the process of removing vehicles, wreckage, debris, spilled material, and other items to return roadway capacity to normal levels. Clearance is the most critical step in managing major incidents, due to the length of time required to restore traffic flow. The objectives of incident clearance are to

- Restore the roadway to its preincident capacity as quickly and safely as possible to minimize motorist delays.
- Make effective use of all clearance resources.
- Enhance the safety of responders and motorists.
- Protect the roadway system and private property from unnecessary damage during the removal process.

Planning for efficient incident clearance involves scheduling and deploying of response personnel and selecting the appropriate methods to remove all types of incidents. It also involves identifying the resources available and obtaining agreements for the use of resources. Training for response personnel and postincident reviews can significantly improve the incident-clearance operation.

Tow trucks, typically operated by private companies, are the most common resource used in incident-clearance activities. Before dispatching tow trucks to the incident scene, tow companies need to be provided with specific information on the incident type so they can dispatch the appropriate equipment to the incident location. For example, heavy-duty wreckers are used for removal or recovery of large trucks, and the availability of heavy recovery equipment is the primary factor in the clearance of major truck-related incidents (Figure 30.8).

30.4.6 Accident Investigations

Accident investigations are part of any incident and play a significant role in incident clearance time. Evidence preservation and determining collision dynamics require specific procedures and take a significant amount of time. When collision scenes become crime scenes, priority is given to thorough investigations. Incident scenes are thus closed and access is limited to authorized personnel only. The priorities for thorough investigation often conflict with the desire of transportation agencies

FIGURE 30.8   Dispatching the right equipment to the incident scene reduces incident clearance time.

to reopen the road as quickly as possible. Investigators can often prioritize their work to facilitate movement of traffic. They may complete the portions that allow partial opening of the roadway first, provided there is adequate traffic control available to maintain their safety while finishing other investigative tasks. Shoulder investigation for minor accidents minimizes traffic disruption and reduces incident-based delay (Figure 30.9).

Accident investigations require precise and detailed measurements of the crash scene. The survey needed to take these measurements often lengthens the closure time. New techniques and technology advances, such as total station surveying systems and photogrammetry, have significantly reduced the time required for the on-site surveys. Total station surveying systems expedite the on-scene measurements and can require fewer lane closures. Photogrammetry is quickly emerging as an alternative to total station measuring. Digital photographs are taken at different angles and locations. A software program has been designed to determine accurate measurements from the photographs. The Oregon, California, Arizona, Washington, and Utah State Police are now using this system (FHWA 2000b).

FIGURE 30.9   Shoulder accident investigation reduces traffic disruption and improves network operation.

30.5   *INCIDENT TRAFFIC MANAGEMENT*

Traffic management, in the context of an incident, is the implementation of temporary traffic control measures at the incident site and on facilities affected by the incident to minimize traffic disruption while maintaining a safe workplace for responders. Traffic control measures can be categorized into two groups: measures intended to improve traffic flow past the incident scene and measures intended to improve traffic flow in affected areas and alternate routes.

Measures to improve flow past the incident site include

- Establishing point traffic control at the incident scene and deploying appropriate personnel to assist in managing traffic
- Effectively managing the incident scene space and taking necessary measures to block lanes and other areas to ensure the safety of the responders. This also includes staging and parking emergency vehicles and equipment in a way that minimizes the impact on traffic flow and ensures maximum safety

Measures to improve traffic flow in areas affected by the incident and on alternate routes include

- Real-time management of traffic control devices (including ramp meters, lane control signs, and traffic signals) in the areas where traffic flow is affected by the incident
- Designating, developing, and operating alternate routes

30.5.1   Traffic Management and Incident Site Operation

After an incident has occurred, it is critical to establish a safe worksite as quickly as possible. Two primary concepts that should be the focus of traffic management personnel are establishing point traffic control at the scene and managing the roadway space.

Establish Point Traffic Control at the Scene.   In order for traffic to move smoothly and safely past the incident, traffic control needs to be established at the incident scene. If lanes or roadways will be closed, traffic needs to be clearly marked to merge into lanes or shoulders that will remain open to traffic. Portable signs that warn motorists of the lane closure transition should be placed upstream from the incident site. Shoulders and even the oncoming lanes can be utilized to provide a lane of travel around the incident scene (Figure 30.10). Responders who will set up traffic control at the incident scene should be familiar with the requirements of the *Manual on Uniform Traffic Control Devices* (*MUTCD*).

Manage the Roadway Space at the Incident Location.   The primary concepts in managing the roadway space are twofold: (1) to close only those lanes that are absolutely essential for protection of the incident responders and victims, and (2) to minimize the time that those lanes are closed.

The number of lanes that must be closed may change with the stage of the incident management and clearance efforts. This means that traffic control may be established and then changed several times. The incident responders responsible for traffic management will need to stay informed about the planned sequence of work to clear the incident. They should continually assess the impacts of the incident on traffic flows and monitor the extent of the queues. This information should be communicated to the traffic management center staff, who, in turn, can pass the information on to the motorist information functions. Information can also be passed on from the center to those responsible. This information should be used to adjust and modify the site traffic control plan.

Methods for reducing the duration of incidents or the number of lanes closed may include (FHWA 2000b)

- Immediately move any vehicle which can move under its own power.
- Equip responders with push-bumpers to facilitate expedient clearance.

FIGURE 30.10   Point traffic management at the incident scene to ensure responders and motorist safety.

- Open individual lanes as soon as they are cleared. In some cases this opportunity occurs when one of the response vehicles leaves the scene.
- Clear incident scenes from left to right, gradually yet systematically shrinking the scene and moving toward the right shoulder and the nearest freeway exit.
- Encourage the first responder from any agency to remove accident debris or a small portion of a spilled load to clear one or two lanes.
- Pour oil dry or sand onto small spills to restart traffic flow in one lane.
- Require the first arriving wrecker to clear travel lanes first, aggressively and in a safe manner.
- Do not allow elaborate rigging, cargo off-loading, vehicle repairs, or the loading and securing of cars on roll back (flatbed) wreckers until all the travel lanes are cleared.
- Make sure that appropriate type and number of tow units is requested very early in the incident.
- Consider including in towing and recovery agreements the requirement that all wreckers that respond to freeway incidents have push-bumpers installed. This includes heavy-duty units and flatbeds.

A high priority must also be given to the vehicles trapped between the diversion point, usually the last upstream exit and the incident site. In some cases these motorists can be allowed to proceed past the scene by opening a shoulder or even a portion of one lane for a brief period then securing the scene and continuing the incident management activities.

30.5.2   Improve Traffic Flow on Alternate Routes—Integrated Incident Management Systems

Many states and metropolitan areas have already initiated integrated incident management systems to ensure optimal utilization of the existing freeway and arterial system networks. These often consist of an extensive network of incident detection, surveillance, and information-dissemination devices strategically located along the freeway and arterial networks and connected to a local or

FIGURE 30.11 Traffic management center operation (courtesy ADA County High District, Boise, Idaho).

regional traffic management center (Figure 30.11). Careful analysis of diversion strategies, which includes examination of the real-time operational characteristics of both the freeway and alternate routes, can lead to more efficient and effective incident management strategies.

Some studies suggest that the implementation of dynamic traffic assignment algorithms, which provide analytical tools to develop optimal integrated incident management plans, can significantly improve the flow of traffic on the network during incident situation (Messmer, Albert, and Papageorgiou 1995; Al-Deck and Kanafani 1991; Reis, Gartner, and Cohen 1991). The developments in dynamic algorithms and in sensing and communication technologies, however, have not been accompanied by corresponding developments in evaluation and decision-making tools that can take advantage of real-time data to produce more efficient control strategies for managing freeway incidents. The state-of-the-practice is often limited to having diversion plans that cannot effectively account for the stochasticity of demand, incident severity, and system disturbances such as drivers' compliance with the provided control.

Alternate Route Planning.   When a major incident occurs that severely limits roadway capacity, motorists will naturally find ways to divert around the incident. Some regions have chosen to formally establish alternate routes to direct traffic to the routes that are best suited to handle this increased traffic demand. If a region chooses to implement alternate routes, it is critical that all agencies affected by the implementation and operation of alternate routes be involved in every step of the planning and operation of those diversion plans. Diversion practices are also discussed in detail in the National Cooperative Highway Research Program Synthesis 279, "Roadway Incident Division Practices" (NCHRP 2000).

Alternate Route Operations and Real-Time Control.   When an incident occurs that results in traffic diverting to other routes, whether as a result of using a planned diversion route or because motorists chose to continue their trip on a less congested route, measures should be taken to manage the increased traffic on alternate routes.

FIGURE 30.12   Example of alternate route plan (courtesy ADA County High District, Boise, Idaho).

The existing traffic control plan should be adjusted to accommodate the diverted traffic. This may include the following (Figure 30.12):

- Use variable message signs and highway advisory radio to disseminate information on the incident or to direct traffic to the most appropriate alternate route.
- Adjust ramp metering rates to account for changes in traffic flows.
- Adjust signal timing to account for changing traffic conditions.

Since freeways and arterial streets are typically operated by different agencies, personnel from the agencies responsible for operating these traffic control devices will need to implement the changes. Close cooperation and collaboration in these situations is usually facilitated by early planning efforts and ongoing interaction.

The operational procedures of implementing alternate routes must be developed to meet the needs of individual regions. The typical chronology of establishing alternate routes include the following steps (FHWA 2000b):

- Determination that an alternate route should be implemented
- Identification of most applicable preplanned alternate route
- Notification of noninvolved agencies that will be affected by the alternate routes
- Modification of traffic signal timings on alternate routes
- Activation of traffic control devices and traveler information sources
- Monitoring of the alternate routes

- Communication with noninvolved agencies affected by alternate routes that the use is about to be terminated
- Termination of the use of the alternate routes

Uncertainties Associated with Alternate Route Planning and Operation.   There are several uncertainties associated with incidents that affect the incident management and alternate route planning and operations, including

- The actual duration of the incident cannot be accurately predicted.
- There is no effective way in practice to achieve a theoretical optimum diversion from an open freeway. When drivers are allowed to make their own choice based on the advisory messages disseminated to them through traveler information systems (uncontrolled and elective), diversion transportation professionals have very little control over how many drivers actually divert.
- The routing of vehicles that are diverted cannot be accurately predicted. Specifically, not all diverted vehicles will return to the freeway, as some drivers will choose to complete their trip on the arterial network once they have been diverted from the freeway. Many of the drivers who are willing to divert will pursue routes other than the designated diversion route to complete their trip.

30.6 MOTORIST INFORMATION

Motorist information involves activating various means of disseminating incident-related information to affected motorists. Media used to disseminate motorist information include

- Commercial radio broadcasts
- Highway advisory radio (HAR)
- Dynamic message signs (DMS)
- Telephone information systems
- In-vehicle or personal data assistant information or route guidance systems
- Commercial and public television traffic reports
- Internet/on-line services
- A variety of dissemination mechanisms provided by information service providers

Motorist information includes the dissemination of incident-related information to two different motorist groups. The first group are motorists who are at or approaching the scene of the incident and receive motorist information en route. The second group are motorists who plan to travel using a route that passes through the incident location and receive pretrip motorist information.

Motorist information should be provided as early in the incident management process as possible and should continue until the incident has been cleared completely and the traffic queues resulting from the incident have completely dissipated. Motorist information reduces incident-based delay and has been shown to help incident response and clearance activities. It does this by reducing traffic demand at and approaching the incident scene, reducing secondary incidents, and improving responder safety on-scene.

30.6.1 En Route Motorist Information

Dynamic message signs (DMS), also known as variable message signs (VMS) or changeable message signs (CMS), can be used to disseminate travel information on a real-time basis. Fixed-location and portable signs can be used to support incident management functions. DMSs are most often used to

- Inform motorists of varying traffic, roadway, and environmental conditions (including variable speed limits in adverse conditions).
- Provide specific information regarding the location and expected duration of incident-related delays.
- Suggest alternate routes because of construction or a roadway closure.
- Redirect diverted drivers back onto the freeway.
- Inform drivers when passage along the shoulder is permissible (e.g., in the event of a major incident, where this can help restore the traffic flow safely).
- Display amber alerts.
- Provide information on special events that may affect freeway traffic (concerts, football games, etc.).
- Provide specific travel time information to freeway users regarding expected travel time to specific destination.

Highway Advisory Radio (HAR) information is communicated to drivers via their vehicles' AM radio receivers (Figure 30.13). With signs and flashing lights placed upstream of the incident scene, drivers are instructed to tune into a specific frequency to hear a message about upcoming congestion. Message transmission can be controlled either on-site or from a remote location. Information disseminated by HAR is similar to that provided by VMS. An advantage of HAR, according to research, is that since the information is audible, the driver experiences less sensory overload than when having to read and process a written message, which also requires taking his or her eyes off the road. Another advantage of HAR over VMS is that longer, more complex messages can be communicated.

HAR does have disadvantages: drivers who do not have functioning radios cannot access the information and drivers who tune in midway through the message have to listen through another cycle. It also requires specific action on the part of the driver: that he or she pay attention to the advance signal and then tune in to the specified station.

FIGURE 30.13 Highway advisory radio (HAR) effectively disseminate information to motorists at and near the incident scene.

With the spread of cellular telephone usage, transportation agencies have a new and quite power-ful way to communicate with travelers who are already on the road. Wireless phone hotlines can be used by drivers to access construction, congestion, and incident-related travel infor mation on the route or routes they are interested in traveling, via touch-tone menus. Commercial radio has long provided traffic information to commuters and travelers. Its clear advantage is that it is well under-stood and utilized by millions of drivers.

30.6.2 Pre-Trip Motorist Information

The increasing number of Internet users at home, in the workplace, and in schools has made traf-fic websites relatively inexpensive and effective tools to disseminate pre-trip traffic information to motorists. Many agencies maintain websites with easy-to-read real-time traffic information such as system maps that are color-coded to indicate travel speeds and congestion on given links.

Commercial television and radio stations in most major cities provide traffic reports to indicate incident locations, other congestion, malfunctioning traffic signals, and other traffic-related informa-tion. To get around the limited time that commercial TV will typically devote to traffic information, another possibility is through public access TV. Travel and traffic information stations, which may take the form of video monitors mounted in a wall or on a countertop at fixed public locations, can also be used to provide incident-related information to travelers such as expected travel times, road-way closures, and alternate routes.

Alphanumeric pagers can be used to provide real-time traffic information for specified routes. This technology has already been tested as part of ITS programs in Seattle, Minneapolis, and New Jersey. A limitation of this technology is the small number of characters that can be broadcast in a given message. Personal data assistants (PDAs) allow users to interact directly with travel informa-tion systems. This interaction allows users to obtain route-planning assistance, traffic information broadcasts, and other information.

30.7 ASSESSING INCIDENT MANAGEMENT PROGRAM BENEFITS

The greatest benefits of an effective incident management program are achieved through the reduc-tion of incident duration. Reducing the duration of an incident can be achieved by

- Reducing the time to detect incidents
- Reducing incident clearance time, defined as the time between awareness of an incident and removal of all evidence of the incident, including debris or remaining assets, from shoulders
- Reducing roadway clearance time, defined as the time between awareness of an incident and restoration of lanes to full operational status
- Reducing the number of secondary incidents

Substantial reductions in response and clearance of incidents can be achieved through the imple-mentation of policies and procedures that are understood and agreed on by each player in the incident management process. Benefits resulting from an effective incident management program can be characterized as both quantitative and qualitative.

30.7.1 Quantitative Benefits

No consistent standard has been identified that can be uniformly applied to evaluate the quantifiable benefits of an effective incident management program. In part, this results from the relatively diverse structure and operations of incident management programs (FHWA 2000b). Each program is devel-oped to meet the unique identified needs of the given region. Incident management programs are

TABLE 30.3 Traffic Incident Management Objectives and Performance Measures (FHWA 2009)

Objective	Performance measure(s)
1. *Reduce incident notification time* (defined as the time between the first agency's awareness of an incident and the time to notify needed response agencies).	a. The time between the first agency's awareness of an incident and the time to notify needed response agencies
2. *Reduce roadway clearance time* (defined as the time between awareness of an incident and restoration of lanes to full operational status).	a. Time between first recordable awareness (detection/notification/verification) of incident by a responsible agency and first confirmation that all lanes are available for traffic flow
3. *Reduce incident clearance time* (defined as the time between awareness of an incident and removal of all evidence of the incident, including debris or remaining assets, from shoulders).	a. Time between first recordable awareness (detection/notification/verification) of incident by a responsible agency and time at which all evidence of incident is removed (including debris cleared from the shoulder) b. Time between first recordable awareness and time at which the last responder has left the scene
4. *Reduce "recovery" time* [defined as time between awareness of an incident and restoration of affected roadway(s) to "normal" conditions].	a. Time between awareness of an incident and restoration of impacted roadway(s) to "normal" conditions (*Note:* Participants noted that "normal" conditions could be difficult to define.)
5. *Reduce time for needed responders to arrive on-scene after notification.*	a. Time between notification and arrival of first qualified response person to arrive on incident scene
6. *Reduce number of secondary incidents and severity of primary and secondary incidents.*	a. Number of total incidents (regardless of primary or secondary) and severity of primary incidents [National Highway Transportation Safety Administration (NHTSA) classification] b. Number of secondary of incidents and severity (NHTSA classification) c. Number fatalities
7. *Develop and ensure familiarity with regional, multi-disciplinary TIM goals and objectives and supporting procedures by all stakeholders.*	a. Existence/availability of program-level plan for implementing traffic control devices and/or procedures b. Existence of/participation in multiagency/jurisdictional training programs on the effective use of traffic control/staging devices and procedures c. Percent of workforce trained on National Incident Management System as well as local/regional/program-level procedures d. Percent of agencies with active, up-to-date memoranda of understanding (MOUs) for program-level TIM e. Number of certified courses taken f. Number of attendees at various courses
8. *Improve communication between responders and managers regarding the status of an incident throughout the incident.*	a. Number or percent of agencies with a need to communicate who are able to communicate (sharing information or communications systems) within an incident
9. *Provide timely, accurate, and useful traveler information to the motoring public on regular basis during incident.*	a. Comparison of information provided at any given time to what information could have been provided b. Customer perceptions on usefulness of information provided c. Time of updates to various sources d. Number of minutes it takes to disseminate informational updates to the public (after something changes regarding incident status) e. Number of sources of information to the public f. Number of system miles that are covered/density of coverage by traveler information systems (seek to increase these)

(Continued)

Objective	Performance measure(s)

10. Regularly evaluate and use customer (road user) feedback to improve TIM program assets and practices.	a. Percent incidents managed in accordance with program-level procedures
	b. Percent of incidents for which multiagency reviews occur
	c. Perceived effectiveness (by involved stakeholders) of use of traffic control devices to achieve incident management goals developed for each incident
	d. Correlation of use of program-level traffic control devices by incident type
	e. Number of instances of sending the needed equipment (presumes that needed quantities and types of equipment are defined) for the incident
	f. Frequency of dissemination of multiagency/program-level and customer feedback back to partners
	g. Measures of customer feedback:
	  • Number of Web site feedback
	  • Number of surveys conducted/focus groups
	  • Number of complaint logs
	  • Number of service patrol comment cards
	  • Number of 1-800 feedback system calls
	  • Number of media/government outlets providing information
	  • Number of 511 calls

also generally developed to fit within the existing institutional framework. In addition, baseline data against which to measure a new program's benefits (e.g., incident response times) are rarely available. The Federal Highway Administration Traffic Incident Management (TIM) Self-Assessment Guide (FHWA 2002) provides general guidelines that can help agencies assess the effectiveness of their traffic incident management programs.

Quantifiable benefits generally associated with an effective incident management programs include

- Increased survival rates of crash victims
- Reduced delay
- Improved response time
- Improved air quality
- Reduced occurrence of secondary incidents
- Improved safety of responders, crash victims, and other motorists

30.7.2 Qualitative Benefits

Qualitative benefits generally associated with an effective incident management program include

- Improved public perception of agency operations
- Reduced driver frustration
- Reduced travel time and improved air quality
- Improved coordination and cooperation of response agencies

Table 30.3 lists incident management program objectives and their performance measures (FHWA 2009).

30.8 INCIDENT MANAGEMENT WITHIN NATIONAL ITS ARCHITECTURE

One of the market packages defined in the National ITS Architecture (USDOT 1998, 2001) is the Incident Management market package (ATMS08—Incident Management System), which includes incident-detection capabilities through roadside surveillance devices and regional coordination with other traffic management centers. Information from these diverse sources is collected and correlated by this market package to detect and verify incidents and implement an appropriate response.

ATMS08 also supports traffic operations personnel in developing an appropriate regional response in coordination with emergency and other incident response personnel to confirmed incidents. Incident response also includes dissemination of information to affected motorists using the Traffic Information Dissemination market package (ATMS06) and dissemination of incident information to travelers through the Broadcast Traveler Information (ATIS1) or Interactive Traveler Information (ATIS2) market packages. Other market packages specific to incident management are hazmat management, emergency response, and emergency routing.

30.9 PLANNING AN EFFECTIVE INCIDENT MANAGEMENT PROGRAM

30.9.1 Define Mission, Goals, and Objectives

The process of defining mission, goals, and objectives and identifying existing problems or limitations helps determine incident management program needs. The incident management program mission is ultimately to create a safe and reliable transportation system. Goals are the desired effects of an effort. They provide ways of defining the mission in terms of specific achievements. Common goals of traffic incident management are to reduce delay and congestion caused by traffic incidents on freeways and reduce the number and severity of secondary crashes. Visible outcomes help define opportunities for system improvement and specific results to be achieved. A clear set of quantitative objectives must be defined at the early stages of the planning incident management planning process. An example of an objective might be a 50 percent reduction in average detection and clearance time for minor traffic incidents. Another important component of any effective incident management planning process is the identification of problems that limit the ability to meet stated goals and objectives.

30.9.2 Compare Alternatives and Define Performance Measures

Through each step of the incident management process, many alternatives that address specific objectives and deliver results should be considered. This generally includes operational and procedural alternatives combine d in a comprehensive system to support the program mission, goals and objectives. Performance measures are used to evaluate how well various alternatives meet program objectives. Performance measures are most clearly applied in terms of quantifying an objective, but they can be measured in less quantitative ways. Responder observation and public feedback have been used as qualitative performance measures by many agencies.

30.10 BEST PRACTICE IN INCIDENT MANAGEMENT PROGRAMS

The following represent the best practice in incident management programs (FHWA 2000b):

30.10.1 Transportation Agencies

- Develop predesigned response plans for freeway closures, which include diversion routes and traffic control. These response plans have to be developed and reviewed in coordination with police, fire, and other local officials.
- Deploy service patrol vehicles to remove debris from travel lanes and assist motorists broken down on the freeway shoulder or in travel lanes. These service vehicles are critically important during peak periods and must be equipped with arrow boards to assist with traffic control for incidents.
- Create video links from traffic management centers to share with law enforcement and fire/rescue agencies.
- Participate in the incident command system on the incident scene and communicate with fire and police agencies for the prompt clearance of the scene.
- Set up safe traffic control around the crash scene, divert traffic upstream of an incident through the use of changeable message signs, and provide traffic information to the media and general public.
- Install reference markers at 2/10th-mile increments, which will allow cellular phone callers to report incident locations accurately.

30.10.2 Law Enforcement Agencies

- Coordinate with transportation, fire, and transportation agencies to develop incident response plans.
- Within the unified incident command system at the incident scene, communicate with transportation agencies to establish traffic management plans/detours and direct a partial or complete reopening of the roadway as quickly as possible.
- For accident investigations, efficiently collect evidence and survey scene using advanced tools such as total station equipment or aerial surveying.
- For minor noninjury crashes that involve property damage only, have dispatchers provide guidance to drivers on local policy for moving vehicles from travel lanes, and exchanging information as per state law.

30.10.3 Fire and Emergency Medical Agencies

- Dispatch the minimum amount of equipment necessary, to reduce the exposure of personnel at the scene.
- Provide for effective training in the identification of hazardous materials, to avoid lengthy lane closures for material that does not pose a threat to people or the environment.
- Provide effective training in temporary traffic control around incidents.
- Set up an effective communication system as part of the incident command system, so that partner response agencies are aware of progress in rescue efforts and can make correct decisions regarding traffic management, and provide traveler information to local media.

30.10.4 Towing and Recovery

- Prequalify towing companies so the towing company called to the incident scene has the capability needed to handle the vehicles involved.
- Train law enforcement personnel to ensure that responders can request the correct equipment be dispatched to the incident.

- Weigh the cost-benefit of calling in third-party recovery teams, and consider whether their distance/time of travel will have excessive impact on the amount of time lanes remain closed.

- Move commercial vehicles or trailers to the roadside or shoulder to restore as many travel lanes as possible, as soon as possible, then perform any necessary salvage operations after the peak hour.

REFERENCES

Al-Deek, H., and A. Kanafani. 1991. "Incident Management with Advanced Traveler Information Systems." In *Proceedings of the Vehicle Navigation and Information Systems Conference,* Warrendale, PA, 563–76.

American Trucking Association (ATA) and Cambridge Systematics, Inc. 1997. *Incident Management: Challenges, Strategies and Solutions for Advancing Safety and Roadway Efficiency.* Final Technical Report, ATA and Cambridge Systematics, Inc., February.

Barton-Aschman Associates, Inc. 1993. *Incident Detection and Response System in North Dallas County.* Prepared for the Texas Department of Transportation, October.

Carvell, J. D., K. Balke, J. Ullman, K. Fitzpatrick, L. Nowlin, and C. Brehmer. 1997. *Freeway Man-agement Handbook.* U.S. Department of Transportation, Federal Highway Administration, August.

Chang, E. C.-P. 1998. "Operational Sensitivity Evaluation of a Speed-Based Incident Detection Algorithm." In *Transportation Technology for Tomorrow: Conference Proceedings,* ITS America 8th Meeting, Detroit.

Federal Highway Administration (FHWA). 1997. *Development and Testing of Operational Incident Detection Algorithms.* Executive Summary, September.

——. 2000a. *Incident Management Successful Practices—A Cross-Cutting Study: Improving Mobility and Saving Lives.* Publication No. FHWA-JPO-99-018, U.S. Department of Transportation, FHWA, ITS Joint Program Office, April.

——. 2000b. *Traffic Incident Management Handbook.* Publication No. DOT-T-01-01, U.S. Department of Transportation, FHWA, Office of Traffic Management. Prepared by PB Farradyne, November.

——. 2002. *Traffic Incident Management (TIM)* Self-Assessment Guide. U.S. Department of Transportation, FHWA Office of Operation, November.

——. 2009. *Federal Highway Administration Focus States Initiative: Traffic Incident Management Performance Measures. Final Report.* Publication No. FHWA-HOP-10-010, U.S. Department of Transportation, FHWA, Washington, DC.

Hellinga, B., and G. Knapp. 2000. "Automatic Vehicle Identification Technology-Based Freeway Incident Detection." *Transportation Research Record* 1727, 142–53.

International Association of Chiefs of Police (IACP). 1996. *The Highway Safety Desk Book.* IACP, April.

Koehne, J., F. L. Mannering, and M. E. Hallenbeck. 1995. *Framework for Developing Incident Man-agement Systems.* U.S. Department of Transportation, Federal Highway Administration, October.

Lomax, T., S. Turner, H. L. Levinson, and R. H. Pratt. 1996. *Quantifying Congestion.* Phase III Final Report, National Research Council, Transportation Research Board, National Cooperative Highway Research Program, September.

Mahamassani, H. S., C. Haas, J. Peterman, and S. Zhou. 1999. *Evaluation of Incident Detection Meth-odologies.* Report FHWA/TX-00-1795-S, Project Summary, Federal Highway Administration/Texas Department of Transportation / University of Texas, Austin, October.

Messmer, A., and M. Papageorgiou. 1995. "Route Diversion Control in Motorway Networks via Nonlinear Optimization." *IEEE Transactions on Control Systems Technology* 3(1):144–54.

Minnesota Department of Transportation. 1982. *I-35 Incident Management and the Impact of Incidents on Freeway Operation,* January.

National Highway Traffic Safety Administration (NHTSA). 1998. *Fatality Analysis Reporting System.* U.S. Department of Transportation, NHTSA.

Police Chief, The. 1998. "National Police Week Observed." International Association of Chiefs of Police, May.

National Cooperative Highway Research Program (NCHRP). 2000. *Roadway Incident Division Practices.* Synthesis 279, National Research Council, Transportation Research Board, NCHRP, Washington, DC.

Reis, R. A., N. H. Gartner, and S. L. Cohen. 1991. "Dynamic Control and Traffic Performance in a Freeway Corridor: A Simulation Study." *Transportation Research A* 25A(5):267–76.

Transportation Research Board (TRB). 2000. *Highway Capacity Manual.* National Research Council, TRB, Washington, DC.

U.S. Department of Transportation: Texas Transportation Institute (TTI). 2009. "Urban Mobility Report 2009." Texas A&M University, College Station, TX, 2009; available at http://mobility.tamu.edu/ums.

U.S. Department of Transportation (USDOT). 1998. *Developing Freeway and Incident Management Systems Using the National ITS Architecture.* USDOT, Intelligent Transportation System Joint Program Office, August.

——. 2001. Version 4.0 of the National ITS Architecture.

Volpe National Transportation Systems Center. 1995. *Intelligent Transportation Systems Impact Assessment Framework.* Final Report, September.

CHAPTER 31

SECURITY AND SURVIVABILITY OF SURFACE TRANSPORTATION NETWORKS

Ahmed Abdel-Rahim
Civil Engineering Department
University of Idaho
Moscow, Idaho

Paul W. Oman
Department of Computer Science
University of Idaho
Moscow, Idaho

31.1 INTRODUCTION

As transportation systems continue to be stressed by increasing traffic needs, traffic engineers have incorporated intricate control networks to optimally signalize intersections and gather information in real time. Such "intelligent" transportation systems (ITS) require extensive interaction with other critical infrastructures such as communications and power (Peerenboom 2001). Rapid incremental design of these systems has focused almost exclusively on safety and efficiency while typically downplaying the equally important need of survivability (Sheldon et al. 2004). Supporting infrastructures are improvised to meet transportation needs while ignoring survivable design. As a result, the delicate interaction of these systems is often compromised at multiple single points of failure, resulting in failures of one infrastructure that profoundly affect others (Krings and Oman 2003). The reluctance of system designers to account for survivability is understandable—it is costly to incorporate redundant designs because backup components provide diminishing returns on continued maintenance of the system (Smith and Sielken 1999). Therefore, when seeking to improve system survivability, an effort must be made to identify components that are both essential to the needs of the system and sufficiently vulnerable to justify the cost of design improvements or redundant backups (Mead et al. 2000).

Several approaches to assessing transportation systems for their ability to maintain service have been considered. An approach typically consists of a measure of evaluation and an analytical strategy. Common measures of evaluation are reliability, vulnerability, and survivability. Typical analytical strategies are connectivity estimation, game theory, microsimulation, Monte Carlo estimation, probabilistic risk assessment, and survivable systems analysis. These measures and strategies are described in this chapter, and an important relationship among some of them is established.

31.2   MEASURES OF EVALUATION

The first major step in analyzing a system is to choose a measure for evaluation—in other words, to identify the priorities on which the analysis is based. A wide variety of measures can be chosen, and this section summarizes the most common measures used in transportation and network analysis: reliability, vulnerability, risk, and survivability.

31.2.1   Reliability

Most strategies for assessing transportation systems for their ability to maintain performance have focused on *reliability*, which, in general, is defined as "the probability of a device performing its purpose adequately for the period of time intended under the operating conditions encountered" (Wakabayashi and Iida 1992). Such a definition covers many possible approaches to defining *adequate* performance. To define adequate performance involves adopting a performance measure, and many different performance measures have been considered in the past, the most prominent of which are described here.

Connectivity, or *terminal reliability*, is often defined as "the probability that nodes are connected, such that it is possible to reach a destination from a given source" (Nicholson et al. 2003). While this is a simple measure, it is also limited, only considering the binary case that links are either available or blocked. Therefore, either there exists a path between a source and destination, or no such path exists. In cases where network state fluctuates or is unknown, the probability that a path exists can be expressed in terms of the probability that each link is available or blocked (Wakabayashi and Iida 1992).

While connectivity is a valid measure for some networked systems (e.g., power systems or sparsely used communications networks), it does not adequately reflect capacity constraints of links and therefore overestimates reliability because a connected node pair may be connected by paths for which traffic demand exceeds service capacity. A more accurate measure in this sense is *capacity reliability*, defined as "the probability that the network can accommodate a specific demand level," where demand level typically is expressed as an origin-destination matrix describing the demand for traversing given source-destination node pairs (Chen et al. 1999).

Although capacity reliability reflects the capacity of links and therefore accounts for the ability of a system to satisfy demand through adequate service, it does not exclusively indicate how well demand is satisfied. *Travel time reliability* or, more generally, *cost reliability* is defined as "the probability that a trip can be successfully finished within a specified time interval (or cost, respectively)" (Schmöcker and Bell, 2001).

While it may be claimed that by emphasizing travel time or cost, the focus on capacity thus is diminished, it is important to point out that the two measures are intricately related. A failure to satisfy capacity results in an increase in travel time or cost (Yang et al. 2000). However, there exists a paradox that owing to the nature of how traffic adapts to changing conditions given insufficient information, an increase in capacity potentially can result in an increase in travel time or cost (Sheffi, 1985).

31.2.2   Vulnerability

Reliability serves as an adequate measure for assigning probabilities of satisfying a fixed level of performance based on the presence of benign factors, but when accounting for malicious acts, it becomes important to assess the vulnerability (weakness) of components in a system and to measure the degree of performance satisfaction.

Vulnerability, in the transportation literature, typically takes one of two definitions. According to (Nicholson and Du 1994), *vulnerability* is defined as the susceptibility to incidents that result in route closures and increases as probability and/or detrimental consequences of failure increase. In this sense, vulnerability can be viewed as a product of probability and consequence. D'Este and Taylor (2001), on the other hand, define *vulnerability* as the likelihood of severe adverse consequences given the degradation of a small number of links, distinguishing between connective vulnerability (considering the cost of travel between a pair of nodes) and access vulnerability (considering a single node and quality of access to it). Therefore, this definition ignores probability and focuses exclusively on the consequences of degrading components.

While both measures successfully reflect a potential change in performance of a system, it is important to distinguish between these two definitions. This has been done in the network analysis literature, as shown in section 31.2.4.

31.2.3 Risk

In the transportation literature, *risk* is regarded as the product of the probability of an event occurring and the cost of such an event (Dalziell and Nicholson 2001). This is similar to the first definition of vulnerability (Nicholson and Du, 1994) given above.

31.2.4 Survivability

Survivability, while not commonly addressed by name in the transportation literature, forms a prominent area of active research in networked systems. The CERT Center at Carnegie-Mellon University's Software Engineering Institute devised an analytical approach for networked systems called *survivable systems analysis* (Mead et al. 2000) that defines the following:

- *Essential* components provide services or satisfy properties that must be maintained during an attack.
- *Compromisable* components can be penetrated and damaged by intrusion (attack).
- *Critical*, or *softspot*, components are both essential and compromisable.

While these are qualitative definitions, it is clear that these measures are closely related to the definitions of vulnerability given above. In fact, the second definition of vulnerability given in D'Este and Taylor (2001) can be viewed as a quantification of essentiality, whereas the first definition given in Nicholson and Du (1994) can be viewed as a quantification of criticality.

Accordingly, *survivability* can be viewed as an umbrella term incorporating security, reliability, fault tolerance, dependability, and other aspects of a system that contribute to the maintained fulfillment of the system's mission. A further discussion of survivable systems analysis appears later in this chapter.

31.3 ANALYTICAL STRATEGIES

The chosen evaluative measure must be matched with an appropriate analytical technique. In both transportation and network systems analysis, many particular techniques have been introduced. This section summarizes these techniques, defining the measures being calculated and describing the models or tools used for each approach.

31.3.1 Connectivity Estimation

To precisely calculate the probability of connectivity of a node pair, given corresponding probabilities for individual nodes and links, is an NP-hard problem because it requires evaluating the connectivity of each acyclic path between the given nodes and computing the corresponding probability that at least one path is connected, which makes precise calculation infeasible for large networks. To counter this, Bell and Iida (2001) adapted the Floyd-Warshal all-routes algorithm (Brassard and Bratley 1996) not to return the shortest path or even the probability of connectivity but to return bounds on the probability that given pairs of nodes are connected.

31.3.2 Game Theory

To account for malicious behavior, transportation analysts sometimes have chosen to use game theory as a means for describing an equilibrium between which noncooperative users and attackers interact in a "game" where individual users each attempt to minimize their costs through appropriate route choice, whereas attackers attempt to maximize them through component degradation. Users are given information about degraded links, in the form of link cost estimation, so that they have an opportunity to adjust their routes. This, in turn, causes a change in the malicious users' strategy for degrading components to increase user cost as much as possible. Thus a cycle between users' and attackers' reactions to each other's behavior is established. This cycle reaches a *Nash equilibrium* (Nash 1950), at which no player (user or attacker) can expect to benefit by changing his or her strategy while others' strategy remains unchanged. This equilibrium yields link failure probabilities that then are applied to reliability calculations (Bell 2000).

31.3.3 Microsimulation

The preceding two techniques are based mostly on equilibrium analysis to measure the effect of long-term degradations. However, when any incident occurs, traffic adapts in the short term, before a new equilibrium is reached in the case of long-term degradations or before the degradation is repaired and the original equilibrium is restored in the case of short-term degradations. Equilibrium modeling therefore cannot account for these immediate effects.

Microscopic simulation (abbreviated to *microsimulation*) is the process of modeling individual vehicles in a simulated environment (PTV 2003). This can be used to estimate the effects of incidents during equilibrium transitions, which often have more magnitude than the net effects after equilibrium is reached (Berdica et al. 2001). Microsimulation also can be used to simulate equilibrium itself, either in isolation or to verify mathematical models, although this is uncommon in earlier research owing to the computing time required to conduct the simulations.

31.3.4 Monte Carlo Estimation

In many cases, it is infeasible to exhaustively account for all possible states of a system. *Monte Carlo estimation* approximates properties of a population by evaluating a sample of the population. In the assessment of traffic system reliability or survivability, this typically involves determining the likelihood of a set of events, choosing a sampling of those sets of events, and evaluating the system, given the presence of each sampled set of events in order to statistically derive an overall evaluation, calculating an error based on the properties of the sample (Vose 2000). This technique was applied to transportation systems by Dalziell and Nicholson (2001).

31.3.5 Probabilistic Risk Assessment

Probabilistic risk assessment (PRA) is a popular analytical strategy that uses reliability indicators, including fault trees, reliability block diagrams (RBDs), event trees, and a subsidiary technique, either failure modes and effects analysis (FMEA) or failure modes and effects criticality analysis (FMECA) (Atwood et al. 2003). FMEA analyzes a design by identifying individual components, determining potential failure modes of those components, determining the effects, and then determining the cost of those effects. FMECA adds the additional step of determining the probability of these failures to assess criticality. Typically, both these techniques have been quantitative in terms of determining failure probabilities but qualitative when defining the impact of the failures on overall system performance.

31.3.6 Survivable Systems Analysis

Survivable systems analysis (SSA), formerly known as *survivable network systems analysis* or *survivable network analysis* (SNA), has become increasingly common since its inception in 1997 at Carnegie-Mellon University's Software Engineering Institute's CERT Coordination Center. The canonical SSA method, as defined by CERT, consists of four steps that are combined and repeated as necessary. After identifying softspot components, which are both essential and compromisable, these components are targeted for improvements with respect to survivability by focusing on three (or four) stages of threat analysis—recognition, resistance, and recovery (as well as adaptation, where applicable) (Mead et al. 2000).

31.4   RELATIONSHIP OF MEASURES AND APPROACHES

We clearly see similarities and differences in the preceding evaluative measures and analytical strategies. Of particular interest are the qualitative definitions related to survivability assessment that have analogous quantitative definitions related to vulnerability assessment. These similarities and relationships are described further in this section.

Survivability is defined qualitatively as the capability of a system to perform its mission even in the presence of faults. *Reliability* is defined quantitatively as the probability that a system continues to maintain a given minimum level of performance, given probabilities of individual faults. *Essentiality* is defined qualitatively as the importance of a component with respect to fulfilling the mission of a system (Mead et al. 2000), whereas *vulnerability* is defined quantitatively as the consequence on the system (in terms of cost) resulting from component failure (D'Este and Taylor 2001).

In a parallel manner, *criticality* is defined qualitatively as the conjunction of essentiality and compromisability (Mead et al. 2000), whereas *risk* is defined quantitatively as the product of vulnerability and failure probability (Dalziell and Nicholson 2001; Nicholson and Du 1994).

Hence a correspondence between qualitative and quantitative definitions of performance analysis can be demonstrated, as shown in Table 31.1. Furthermore, it is clear that an initial qualitative assessment can be complemented and validated with a quantitative approach.

TABLE 31.1   Comparison of Qualitative and Quantitative Definitions of Performance Maintenance

Measure	Qualitative definition	Quantitative definition
Survivability	The capability of a system to perform its mission, even in the presence of faults	
Reliability		The probability that a system continues to maintain a given minimum level of performance given probabilities of individual faults
Essentiality	The importance of a component with respect to fulfilling the mission of a system	
Vulnerability		The consequence on the system (in terms of cost) resulting from component failure
Criticality	The conjunction of essentiality and compromisability	
Risk		The product of vulnerability and failure probability

Term	Type	Definition
Survivability	Qualitative	Essentiality = (compromisability ∩ criticality)
Reliability	Quantitative	Vulnerability = (failure probability × risk)

31.5   QUALITATIVE SURVIVABLE SYSTEMS FOR TRANSPORTATION NETWORKS

31.5.1   Survivable System Analysis

The combined security and survivability analysis process presented here is based on the CERT SSA process but expanded into seven steps, as illustrated in Figure 31.1. The following subsections further describe each of these stages.

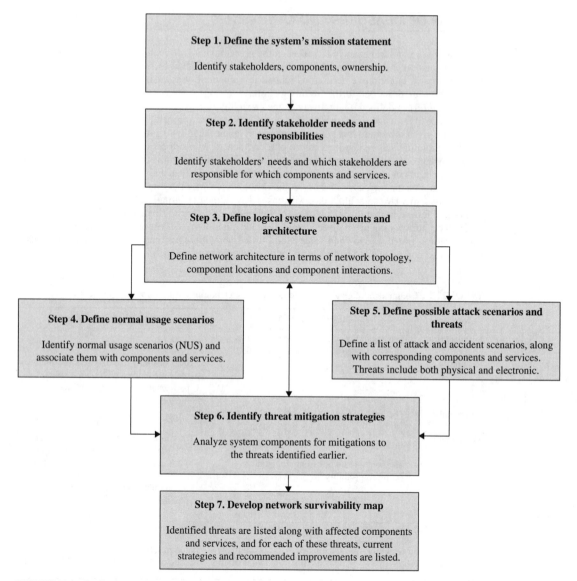

FIGURE 31.1   Qualitative survivable system analysis.

31.5.2   Mission Statement

The first stage of the modified SSA is performing a review of the mission statement of the project (if one exists and is available). This first stage is important for understanding the scope of the project. The mission statement generally will contain information that helps in further stages of the project. Attention should be focused on identifying project scope, stakeholders, different project components, time frame, ownership, and other details that may be relevant to later stages in the modified SSA analysis.

31.5.3 Stakeholder Needs, Responsibility, and Access

The second stage of the modified SSA involves identifying stakeholders' needs and which stakeholders are responsible for what portions of the project. It also involves identifying who will have access to the system once it is complete. Identifying stakeholder needs can be a difficult process and may require interviewing representatives from various stakeholder groups. At this stage, ownership and responsibility for components are also discussed. Later, as the project components are better defined, a more specific mapping of responsibility may be accomplished. In some cases, the owner of a component may not be the entity who is required to operate and maintain it. Access control to the various components of the system is discussed and mapped at this stage. Some components of the system may need to be accessed by several stakeholders as well as nonstakeholders such as private contractors. This has implications for the security and survivability of the system and needs to be addressed and documented at this stage of the analysis.

31.5.4 Logical System Components and Architecture

The third stage of our modified SSA is a more complete definition of system components and architectures. This stage may be revisited several times as the project evolves and implementation decisions are made. In this phase, the basic system components are reviewed and documented. These components are used later in the process to identify softspots and critical points and components. Once a listing of all the components is completed, the architecture of the system is reviewed. This includes network routing topologies, server location and connection, and physical layout of wire or wireless systems.

31.5.5 Normal Usage Scenarios (NUS)

The mission objectives of the system are accomplished through the normal usage scenarios that are elicited from each of the stakeholders. Once identified, a stakeholder–normal usage scenarios matrix is constructed to convey overlapping scenarios and needs. The normal usage scenarios are important because they are mapped to essential services that need to maintain availability.

31.5.6 Attack Scenarios and Threats

Using knowledge of the potential access points and of the softspot components obtained in previous stages of the process, a list of possible attack and failure scenarios then is conceived and ranked in order of likelihood. Components and services affected by the failure are listed underneath each scenario. Accidents, natural disasters, and vandalism are also possibilities that should be taken into account in every survivability analysis. A list of critical components and threats to the components are cataloged, and then a threat component matrix that displays the threats and the corresponding compromised components is generated. For surface transportation networks, the threats are stratified into two categories, physical threats and electronic threats.

31.5.7 Threat Mitigation Strategies

During this phase, all system components are analyzed for potential threats and for mitigation techniques to minimize the effect of each of the threats. Threats are either physical (affecting the structure of a component) or electronic (affecting the operation of a component without affecting its overall structure). Threats should include natural damage, intentional damage, accidental damage, design or structural failures, and electronic failures and attacks, among others. Mitigation techniques are strategies that minimize the frequency or severity of anticipated threats. Once a sufficient list of general threats is completed for each component, analysts often will find that many threats are common to multiple components and that the same mitigation techniques can be applied to address the same threat for different components. By grouping components within a single threat category, analysts greatly simplify the process of establishing a reasonably complete list of mitigation strategies.

This step also provides a preliminary overview of how various components can be protected to develop a survivable system. Subsequent stages of this analysis will identify components that are critical to the operation of a system as well as components that are most vulnerable to threats. Once these components have been identified, mitigation strategies corresponding to every anticipated threat can be evaluated for their effectiveness and feasibility.

31.5.8   Survivability Map

In this step, the three components of survivability—recognition, resistance, and recovery—are mapped with recommended mitigation strategies along with components and services affected. The derivation of the survivability map (or table) results from an analysis of attack scenarios and the threat component matrix conducted in earlier steps.

31.6   QUANTITATIVE SURVIVABLE SYSTEM ANALYSIS FOR ITS NETWORK: MULTILAYER NETWORK ANALYSIS

31.6.1   Case Study: Traffic Management Center Priorities

We now showcase our modified SSA process with an actual study. The objective of the analysis was to identify typical normal usage scenarios and possible attack scenarios and their impact on traffic management center (TMC) operations. Data were collected through phone interviews with 17 TMC operators throughout North America and applied to the Ada County Highway District (ACHD) ITS system surrounding Boise, Idaho. A stylized example of different components of the ACHD ITS system is presented in Figure 31.2, showing a typical ITS architecture with several project components and communication vectors. Additionally, a theoretical multilayer representation of a typical ITS network is presented in Figure 31.3 which shows three different infrastructure layers within the system: (1) electric power network layer, (2) communication network layer, and (3) physical (roadway) network layer.

Our survey of TMC operators identified three primary and core functions for the TMCs: (1) provide optimal traffic management and control, (2) provide surveillance and monitoring of the network operations, and (3) disseminate travel information to the public. These three functions are clearly interdependent. As part of the phone interview, TMC operators were asked to rank the effect of loosing the functionality of one or more of the field devices in the network. Excerpts of the results are presented in Tables 31.2 and 31.3.

Table 31.2 contains an excerpt from a larger table showing the effect of communication network failure on the functionality of different field devices. Based on the survey results, loss of communication to/from a local controller will have a low/moderate effect on the network operations as controllers will implement the time-of-day default control plans stored in the local controllers. However, loss of communication to a closed-circuit TV camera (CCTV) or a changeable message sign (CMS) can have a major (high) effect on network operations, especially during peak travel periods or during incident situations.

Table 31.3 contains an excerpt from a larger table showing the effect of electric power grid failure on the functionality of different field devices. Loss of power at a local controller will have a moderate/high effect on network operations. Intersection control in such case will be done manually, which requires dispatching personnel to the site, or through an all-way-stop-control mode, which could cause excessive network delay, especially at major intersections. Effective traffic management plans for critical intersections during power outages was identified by TMC operators as the

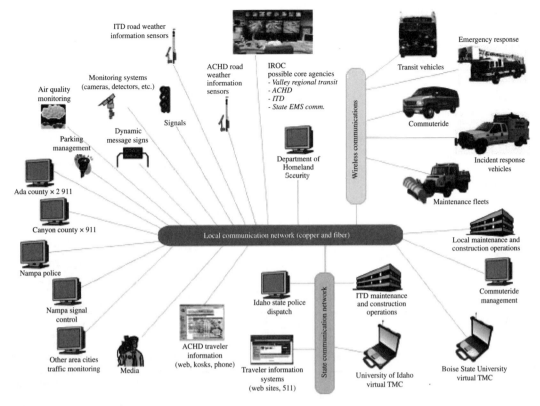

FIGURE 31.2 Typical components of an intelligent transportation system.

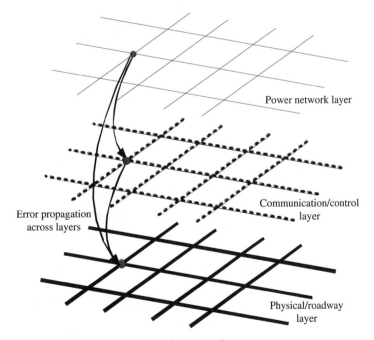

Power network layer

Communication/control layer

Error propagation across layers

Physical/roadway layer

FIGURE 31.3 ITS multilayer network representation.

TABLE 31.2   Effects of Communication Network Failure on the Functionality of ITS Field Devices (an Excerpt)

Field device	Functionality	Failure effect	Criticality level of the failure
Controllers/cabinets/ ramp meters	Local control	Loss of real-time communication with the TMC; controllers will use default local control plans stored in the controllers	Low/moderate
Central control system	Network-wide optimal control	100 percent loss of communication with all field devices; loss of major functionality	Very high
CCTVs	Network surveillance	100 percent loss of functionality	High, especially for CCTV at critical points such as major intersections and interchanges
CMSs	User information	100 percent loss of functionality	Moderate/high, especially during incident situations

TABLE 31.3   Effects of Power Grid Failure on the Functionality of ITS Field Devices (an Excerpt)

Field device	Functionality	Failure effect	Criticality level of the failure
Controllers/cabinets/ ramp meters	Local control	Control will be done manually or will revert to all-way stop control operations	Moderate/high, especially at major intersections in the network
Central control system	Network-wide optimal control	100 percent loss of functionality	Extremely high
CCTVs	Network surveillance	100 percent loss of functionality	High, especially for CCTV at critical points such as major intersections and interchanges
CMSs	User information	100 percent loss of functionality	Moderate/high, especially during incident situations

key mitigation strategy for such a case. Findings from the analysis also showed that a loss of power supply or communication at the TMC could have a paralyzing effect on the ability of the network to provide several of its essential services. An operational-ready alternate TMC and a redundant communication system to the TMC were identified as possible mitigation strategies for these threats. Table 31.4 shows some examples of physical and electronic threats mitigation strategies, also excerpted from a larger table constructed during our analysis.

31.6.2   Case Study: Quantifying Component Essentiality

The main goal of the ACHD multilayered quantitative graph-based analysis was to compute the essentiality of each component with respect to each infrastructure layer and rank-order these essentialities to determine the most essential components with respect to that layer and other layers on which the components depend. The stylized multilayer representation of ITS networks is presented

TABLE 31.4 Examples of Physical and Electronic Threat Mitigation Strategies (an Excerpt)

	Threat	Component	Mitigations
Physical threats	Vehicles	Fiberoptics	Height, barriers for poles, pole location, and periodic automated testing
		Fiber cabinets	Cabinet structure, color, signage, location
		Signal heads	Height, warning signs, chains, sag mitigation, color
	Malicious cutting	Fiberoptics	Shielding, location, height, signage, periodic automated testing, climbing, safeguards, burying
		CCTV/video detectors	Conduit, height, location, climbing, safeguards
	Vandalism	CCTV/video detectors	Height, location, shielding, signage, periodic manual testing
	Break-ins	Fiber cabinets	Location, shielding, tactile deterrent, lock mechanisms, signage, clean junctions, perimeter fencing
		Signal cabinets	Same as fiber cabinets
	Flooding	Communication switchgear	Waterproof shielding, location elevated rack mounting
		Loop detectors	Waterproof shielding
		Fiber splices	Waterproof shielding, elevated rack mounting
		Fiber and signal cabinet	Complete junctions, waterproof shielding
		Servers and wireless	Elevated rack mounting
	Power outage	Switchgear	Battery backup
		CCTV/video detectors	Multiple power feeds
		Signal controller/signal heads	UPS
		Data archive	UPS
		IT	UPS
		Servers	UPS
		Wireless	Battery backup
Electronic threats	Denial of service	Switchgear	IP filtering, access restrictions, programmable switch
		CCTV/video detectors	Port restrictions, IP restrictions, periodic self-test
		Signal controllers	Same as CCTV/video detectors
		IT	IP filtering, access restrictions, port restrictions, intrusion detection system, firewall, drive partitioning, redundant IT servers, formal periodic OS patch procedures
		Servers	Same as IT
		Wireless	Defensive sniffing, encryption, port restrictions, IP restrictions
	Setting changes	Switchgear	Set/reset procedures, initial testing, overburdened test
		Signal controllers	Same as switchgear
		Conflict monitor	Same as switchgear
	Data storm	Switchgear	Self-test, failover switch with isolation logic Remote test/resets
		Signal controllers	Remote test/reset procedures, self-test Failover controller with isolation logic
	Signal degradation	Fiberoptics	Periodic automated testing
		Fiber splices	Periodic automated testing
		Signal controllers	Periodic automated testing
	Unauthorized access	Switchgear	Password protection IP filtering
		CCTV/video detectors	Same as switchgear
		Signal controllers	Password protection, IP filtering, audit logging
		Archive	Audit logging, intrusion detection system firewall, system log monitoring, backup and restore procedures, password protection IP filtered, defensive sniffing
		IT	Same as archive (above)

in Figure 31.3 and shows error and failure propagation across network layers. For example, a power layer failure will affect both the communication and control layers and therefore affect what can transit on the physical layer. A failure in the communication layer will affect the control and hence the physical layer, but a failure in the physical (roadway) layer will affect only that layer. The essentiality of each network component with respect to its layer can be determined using the following relationship:

$$e_L(c) = p_L - p_L(c) \tag{31.1}$$

where $e_L(c)$ denotes the essentiality of component c with respect to layer L, p_L denotes the performance of layer L during normal operations (when no components have been damaged or disabled), and $p_L(c)$ denotes the performance of layer L when component c is disabled.

Given these essentialities, we can define a weighting scheme that apportions weights $w_L(c)$ to components proportional to their essentiality. One convenient way is to compute the total of essentialities of all components under consideration, denoted t_L, and assign

$$w_L(c) = \frac{e_L(c)}{t_L} \tag{31.2}$$

This assignment has the property that the sum of the weights is 1, and thus $w_L(c)$ describes relative importance in terms of a percentage relative to all components under consideration. Given this notation and the performance measures defined earlier, we can compute performance, essentiality, and weight for each component with respect to each layer. Then, given a suitable choice of weights when measuring performance, we can evaluate $p_L(c)$, $e_L(c)$, and $w_L(c)$ for each layer L and component c. However, since previous analysis showed that communication failures were less essential than power component failures (Abdel-Rahim et al. 2004), we concentrate on the analysis of component essentialities with respect to the relationship between electric power and vehicle traffic on the physical/control layer. We show how full or partial loss of electric power affects vehicle traffic and what areas are affected and how, and then we briefly discuss mitigation schemes to enhance survivability.

31.6.3 Essentiality of Network Components: Graph-Based Measures

The electric power layer represents the power supply for all essential network components that provide network control (traffic controllers, signals, and ramp meters), network monitoring and surveillance (CCTVs, loops, and video detectors), and traveler information dissemination via CMS. In large-scale ITS networks, the power network typically consists of multiple high-voltage transmission power lines and distribution substations. Mapping the power network components and different field devices they serve to a graph is nontrivial. Figure 31.4a and b shows simplified versions of the graph representation for the ACHD physical (roadway) and signalized intersections network.

The essentiality of each of these components was assessed using two different methods following equation (31.1). The first method determined the essentiality based on the number of field devices (signalized intersections) affected by a loss of service or failure of the power network component. The second method determined the essentiality based on the total number of vehicles affected by a loss-of-service failure of the power network component. Average hourly volumes during the afternoon peak period at each signalized intersection were used in this analysis.

The essentiality values obtained from the graph analysis using either of the two proposed methods enable decision makers to assess the damage expected from failure in any of the power network components. For example, a loss of service in the Boise power substation would affect approximately 28 percent of the signalized intersections, and if this happens during afternoon peak periods, this will affect approximately 25 percent of the vehicles using the network. These data can be used by the decision makers to take policy decisions regarding possible failure mitigation strategies (Benke et al. 2006).

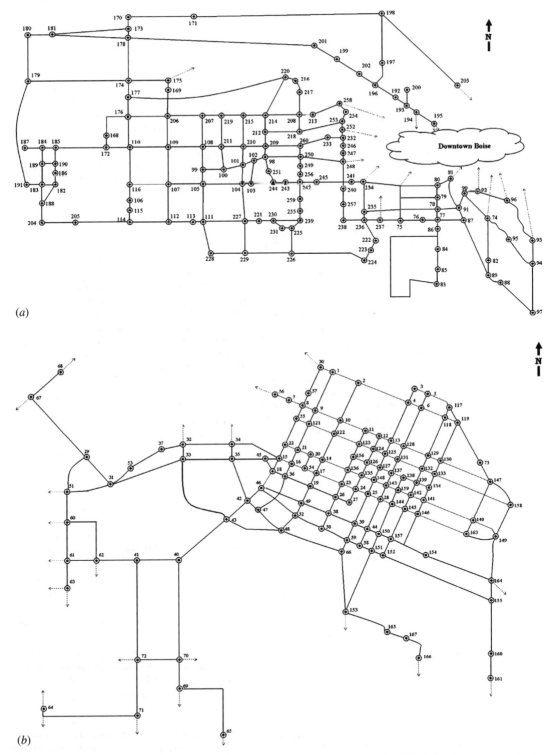

(a)

(b)

FIGURE 31.4 (a) Example of physical and control graph for the Boise area. (b) Example of physical and control graph for the downtown Boise area.

31.6.4 Essentiality of Network Components: Network-Modeling-Based Measures

The graph-based essentiality values presented in the preceding subsection provide a general and broad quantitative measure of the essentiality of each network component. To be able to better quantify the essentiality of each network component and hence to assess the survivability of the network in absence of one or more network components, essentiality values that are based on a more detailed modeling of network operations should be used. For medium to large urban ITS networks, similar to the ACHD ITS network, the actual run time for a single run of a microscopic model often exceeds several hours, especially if dynamic route-assignment models requiring several iterations are employed. Accordingly, planning-level macroscopic models are more appropriate for this type of analysis. Several of these models are available such as Transim, Dynasmart-P, Cube, and VISUM; the latter was used in this analysis.

VISUM is a macrosimulation model that uses a dynamic stochastic assignment procedure to determine the shortest route between each origin-destination pair (PTV 2003). For each simulation time slice, the shortest route for departure in the time slice is determined using a time-dependent best-path algorithm. The model reports detailed output including zone-to-zone travel time as well as network-wide travel time as vehicle-hour-travel.

Our macroscopic VISUM network simulation consists of 76 zones and includes 260 signalized intersections. Traffic volumes, based on the 2004 traffic counts at different locations throughout the network, were used to calibrate and validate the model. Analysis of variance (ANOVA) was used to test the null hypothesis that there is no difference between VISUM output and actual traffic counts at the 0.05 level of significance level. The F value is 0.14, smaller than the F-critical value of 2.1, which means that the statistical result fails to reject the null hypothesis of equality of the mean values for these two traffic count values. Furthermore, the distribution of VISUM and 2004 average daily traffic (ADT) data is close to a 45° line with no outliers, and it has the R^2 value 0.9482, which further validates the output of the simulation model.

Several performance measures can be used to describe the quality of operations in the network. In this analysis, total network travel time (TTT) was used as the primary measure. The TTT during the normal operation scenario was used as the benchmark against which the TTTs recorded during event-based (aka failure) scenarios with a given component c disabled are compared. An event-based network measure was calculated using the formula

$$p_P(c) = \frac{\text{TTT}_{\text{normal}}}{\text{TTT}_{c\_\text{disabled}}} \tag{31.3}$$

The event-based scenarios tested in the analysis are based on the following assumptions:

1. Failure in a power substation will lead to a power outage at all devices served by this substation. The average duration of the power outage is 1 hour, representing the average time needed to manually switch the devices to a functioning substation.

2. Failure at the power transformer supplying local field devices will result in a 1-hour power outage at this device, representing the time needed to restore the power by replacing the transformer or switching to another transformer.

3. Power outage at a signalized intersection will switch the control at that intersection to all-way-stop-control (AWSC).

4. Loss of communication to/from a signalized intersection will switch the control to the default control plan. This automatic switch to local control causes relatively minor disruption to overall network operations, especially if the default control plans are continuously updated.

5. Power outage or communication loss at CCTVs or CMS locations will lead to a complete loss of functionality of those devices.

The final part of our analysis focused on power failure events at signalized intersections and power substations. A total of 38 failure events were modeled representing a 1-hour failure at each of the 13 power substations and a 1-hour power failure at 25 major signalized intersections. The

25 major intersections were chosen based on the total traffic using the intersection during the afternoon peak period. Each failure event was modeled using the VISUM model. The simulation duration for all runs was 4 hours, with the 1-hour power failure occurring 30 minutes after the start of the simulation. In addition, the simulation model was run with no failure events to determine the normal total travel time on the network. Traffic volumes used in the macroscopic simulation analysis represented the afternoon peak hour traffic volume in the network (Abdel-Rahim et al. 2007).

31.7 CONCLUSION

In this chapter, two different survivability assessment techniques for surface transportation networks were briefly introduced. Full details of these analysis and implementation procedures can be found in several of the citations listed in the References. The first analytical approach was based on CMU's SSA as modified for critical infrastructures. The qualitative analysis is lengthy and relatively subjective, but it successfully identifies which components are essential and compromisable (i.e., softspots). However, SSA does not provide a quantification of how essential or compromisable the component is and offers no indication of the degree to which suggested mitigations actually improve the system. The modified SSA presented in this chapter and elsewhere does, however, provide a systematic means of including multiple stakeholders with individual goals and responsibilities, classifying the wide variety of components found in transportation systems, and combining existing knowledge with modern contributions to efficiently identify physical and cyber threats to each component and mitigations for each threat.

The second analytical approach described in this chapter applies a multilayered graph-based method to provide a quantitative assessment of network survivability. For most ITS the multilayered approach would incorporate three layers: (1) electric power, (2) communications, and (3) physical roads and traffic controls. In this chapter we focused our analysis on the interrelationships between electric power and traffic controls in order to assess the effect electric power outages have on vehicular traffic on the physical transportation network. Macroscopic modeling was used to investigate the effect of different component failures on the network's operational characteristics. The analyses presented here provide the means of assessing the importance of different components in critical infrastructure layers, thus allowing decision makers to better understand failure events and prioritize threat mitigation alternatives.

REFERENCES

Abdel-Rahim, A., P. Oman, J. Waite, M. Benke, and A. Krings. 2004. "Integrating Network Survivability Analysis in Traffic Systems Design." Intelligent Transportation Systems Safety and Security Conference, Paper SS-46, Miami, FL, March 24–25, 2004.

Abdel-Rahim, A., P. Oman, B. Johnson, and L. W. Tung. 2007. "Survivability Analysis of Large Scale Intelligent Transportation System Networks," Transportation Research Record: Journal of the Transportation Research Board, No. 2022, TRB of the National Academies, Washington, D.C., pp. 9–20.

Atwood, C. L., J. L. Lachance, H. F. Martz, D. J. Anderson, M. Englehardt, D. Whitehead & T. Wheeler. 2003. "Handbook of Parameter Estimation for Probabilistic Risk Assessment." Technical Report, Office of Nuclear Regulatory Research, U.S. Nuclear Regulatory Commission, Washington.

Bell, M. G. H. 2000. "A Game Theory Approach to Measuring the Performance Reliability of Transportation Networks." *Transportation Research* 34B:533–549.

Bell, M. G. H., and Y. Iida. 2001. "Estimating the Terminal Reliability of Degradable Transportation Networks." Triennial Symposium on Transportation Analysis IV (TRISTAN), São Miguel, Azores Islands, Portugal, June 13–19.

Benke, M., A. Abdel-Rahim, P. Oman, and B. Johnson. 2006. "Case Study of a Survivability Analysis of a Small Intelligent Transportation System," Transportation Research Record: Journal of the Transportation Research Board, No. 1944, TRB of the National Academies, Washington, D.C., pp. 98–106.

Berdica, K., Z. Andjic, and A. J. Nicholson. 2001. "Simulating Road Traffic Interruptions: Does it matter what model we use?" In *The Network Reliability of Transport*. New York: Pergamon Press, pp. 353–368.

Brassard, G., and P. Bratley. 1996. *Fundamentals of Algorithmics*. Englewood Cliffs, NJ: Prentice-Hall.

Chen, A., H. Yang, H. K. Lo, and W. H. Tang. 1999. "A Capacity Related Reliability for Transportation Networks." *Journal of Advanced Transportation* 33(2):183–200.

D'Este, G. M., and M. A. P. Taylor. 2001. "Network Vulnerability: An Approach to Reliability Analysis at the Level of National Strategic Transport Networks," in *The Network Reliability of Transport*. New York: Pergamon Press, pp. 23–44.

Dalziell, E. P., and A. J. Nicholson. 2001. "Risk and Impact of Natural Hazards on a Road Network." *Journal of Transportation Engineering* 127(2):159–166.

Krings, A., and P. Oman. 2003. "A Simple GSPN for Modeling Common Mode Failures in Critical Infrastructures." 36th Annual Hawaii International Conference on System Sciences, Minitrack on Secure and Survivable Software Systems, Paper STSSS02, Waikola, HI, January 6–9.

Mead, N. R., R. J. Ellison, R. C. Linger, R. C. Longstaff, T. Longstaff, and J. McHugh. 2000. "A Survivable Network Analysis Method." Technical Report, Software Engineering Institute, Carnegie Mellon University.

Nash, J. F. 1950. "Non-cooperative Games," Dissertation, Department of Mathematics, Princeton University.

Nicholson, A. J., and Z. P. Du. 1994. "Improving Network Reliability: A Framework." 17th Australian Road Research Board Conference, Gold Coast, Australia, August 14–19.

Nicholson, A. J., J.-D. Schmöcker, M. G. H. Bell, and Y. Iida. 2003. "Assessing Transport Reliability: Malevolence and User Knowledge," in *The Network Reliability of Transport*. New York: Pergamon Press, pp. 1–22.

Peerenboom, J. 2001. "Infrastructure Interdependencies: Overview of Concepts and Terminology," NSF Workshop, Washington, D.C. June 14–15.

Planung transport Verkehr AG (PTV). 2003. *VISSIM User Manual,* Version 3.70. Planung Transport Verkehr AG, Karlsruhe, Germany.

Schmöcker, J.-D., and M. G. H. Bell. 2001. "The PFE as a Tool for Robust Multi-Modal Network Planning." *Traffic Engineering and Control* 44(3):108–115.

Sheffi, Y. 1985. *Urban Transportation Networks*. Englewood Cliffs, NJ: Prentice-Hall.

Sheldon, F., T. Potok, A. Loebl, A. Krings, and P. Oman. 2004. "Management of Secure and Survivable Critical Infrastructures Toward Avoiding Vulnerabilities." Eighth IEEE International Symposium on High Assurance Systems Engineering, Tampa, FL, March 25–26.

Smith, B. L., and R. S. Sielken. 1999. "Survivability of Intelligent Transportation Systems." Technical Report, Virginia Transportation Research Council.

Vose, D. 2000. *Risk Analysis: A Quantitative Guide*. Hoboken, NJ: Wiley.

Wakabayashi, H., and Y. Iida. 1992. "Upper and Lower Bounds of Terminal Reliability of Road Networks: An Efficient Method with Boolean Algebra." *Journal of Natural Disaster Science* 14:29–44.

Yang, H., H. K. Lo, and W. H. Tang. 2000. "Travel Time versus Capacity Reliability of a Road Network," in *Reliability of Transport Networks*. London: Research Studies Press, Ltd., pp. 119–138.

CHAPTER 32
OPTIMIZATION OF EMERGENCY EVACUATION PLANS

Hossam Abdelgawad
Department of Civil Engineering
University of Toronto
Toronto, Ontario, Canada

Baher Abdulhai
Department of Civil Engineering
University of Toronto
Toronto, Ontario, Canada

32.1 EMERGENCY EVACUATION: BRIEF OVERVIEW

Emergency evacuation is the collective movement of people using multiple modes of transport from a hazard area [emergency protection zone (EPZ)] to safe destinations (shelters) via specific routes. Emergency evacuation plans are necessary both in the case of human-made disasters (e.g., nuclear reactor failures or leaks or terrorist attacks) and natural disasters (e.g., hurricanes, floods, tsunamis, earthquakes, or tornados).

Transportation networks in cities evolve over long spans of time in tandem with population growth and evolution of travel patterns. In cases of emergency, travel demand and travel patterns change drastically from everyday regular volumes and patterns. Given that most U.S. and Canadian cities are already congested and operating near capacity during peak periods, network performance can deteriorate severely if such drastic changes in origin-destination (O-D) demand patterns occur during or after a disaster (Tuydes and Ziliaskopoulos 2004). Loss of capacity owing to the disaster and associated incidents can complicate the matter further. The primary goal when a disaster or a hazardous event occurs is to coordinate, control, and possibly optimize use of the existing transportation network capacity. Emergency operation management centers face multifaceted challenges in anticipating evacuation flows and providing proactive actions to guide and coordinate the public toward safe shelters (Chiu et al. 2006).

Designing a transportation network for evacuation demand patterns is financially infeasible. One option is to better use the available network capacity by reallocating it more efficiently, possibly in the form of "contraflow" operation (i.e., reversing the opposite direction capacity into the direction of evacuation). Another alternative is staging the evacuation demand such that the population is advised to evacuate according to an announced schedule rather than simultaneously. In general, the evacuation problem can be modeled as an optimization problem to achieve certain objectives (e.g., minimize the evacuation time and the network clearance time). The control variables could combine capacity allocation, evacuation scheduling, destination (shelter) choice, traffic routing, and traffic control, to name a few.

Emergency evacuation has been investigated extensively in the last two decades owing to frequent natural and human-made disasters. Numerous studies explored the emergency evacuation problem, some of which are robust and have addressed pressing issues quantitatively; others presented qualitative measures of emergency evacuation. This chapter reviews and highlights the challenges in emergency evacuation planning and modeling.

32.1.1 Contraflow Operation in Emergency Evacuation

Contraflow operations refer to reversing the direction flow on one or more lanes of a road segment, thereby increasing the capacity in the direction with heavier flow without constructing additional lanes (Urbina 2001). The contraflow concept has been considered in many cities in the United States as a way to improve the roadway capacity during routine rush traffic flow hours. Contraflow also has been implemented during special events (e.g., concerts and football games) to accommodate the outbound traffic at the end of the event.

The concept of contraflow was first proposed in the 1980s by the Federal Emergency Management Agency (FEMA) for use as a last resort during emergencies and was planned for use originally during potential nuclear missile attacks. Considering the increased frequency of hurricanes in heavily populated areas, contraflow operations have become a valuable option for moving people out of threatened areas (Brian Wolshon 2001).

Although contraflow evacuation has been widely implemented, there is little comprehensive research on the costs and benefits of its use. FEMA (Post 2000) investigated the cost/benefit of capacity improvements resulting from the implementation of contraflow operations. In this study, traffic volumes collected by Florida, Georgia, and South Carolina during Hurricane Floyd were considered as a base for computing planning-level roadway capacities for evacuating traffic under different conditions. According to this report, depending on the configuration of contraflow lanes, a 30 to 70 percent increase in capacity over conventional operations has been achieved by the use of contraflow lanes. Table 32.1 shows an example of such values.

TABLE 32.1   Contraflow Rates for Four-Lane Freeways

Strategies	Estimated Average Total Outbound Capacity (veh/hr)
Normal Two Way Operations	3,000
Three Lane (one contraflow lane)	3,900
Three Lane (using outside shoulder)	4,200
All lanes reversed for evacuation (no shoulder lanes)	5,000

Source: Post, 2000.

Recently, there has been strong interest among researchers and planners in adapting contraflow measures during emergency evacuation to optimize the use of existing infrastructure. Although most of these studies (Theodoulou and Wolshon 2004; Tuydes and Ziliaskopoulos 2004; Lim and Wolshon 2005; Kim et al. 2008; Meng et al. 2008) show operational improvements, safety and practical operation continue to be an issue, especially with the increase in unfamiliar drivers with the contraflow operation. Brian Wolshon (2001) pointed out that reverse-flow scenarios are not without significant problems. These problems include the safety risks associated with reverse flow on interstate freeways, the fact that traffic control devices and safety appurtenances are not designed to accommodate contraflow, and potential problematic operation near access and termination points. Despite a wide acceptance of contraflow operations in practice, limited research has been published regarding how to choose the links or lanes to be reversed for contraflow operations in an optimal manner that maximizes its effectiveness under resource limitations.

Most of the research conducted on contraflow operations is based on the cell transmission model (CTM). The CTM is based on fundamental traffic flow theory diagrams (e.g., trapezium shape) (Muñoz et al. 2006). The CTM discretizes the time period of interest into small intervals and then divides every link of the transportation network into homogeneous segments called *cells* such that the length of each cell is traveled in one time interval while moving at free-flow speed. For each cell, the hydrodynamic flow equations result in two sets of equations, cell mass conservation and flow propagation. However, this approach models the contraflow reversibility capacity as a continuous variable, which may not be implementable practically (Bell and Iida 1997). Furthermore, the absence of the dynamics incorporated in the traffic assignment process is a major limitation of such models. Theodoulou and Wolshon (2004) assessed a microscopic simulation model, CORSIM, to model contraflow operations. However, CORSIM, like most traffic simulators, does not support the creation of reversible-flow freeway segments or the behavioral characteristics of evacuation drivers.

32.1.2 Evacuation Scheduling

Scheduled/staged evacuation is another widely used control strategy to guide evacuation flows. Unlike changing the network geometry, as in contraflow design, or enforcing route choice restrictions, evacuation scheduling aims to better distribute/manage the evacuation demand over an evacuation horizon. In simultaneous evacuations, evacuees are advised to evacuate immediately to their destination, whereas in staged evacuation, evacuees are advised when to evacuate so as to achieve certain objective (e.g., minimize network clearance time) (Sbayti and Mahmassani 2006). Managing the evacuation surge by holding some evacuees at their origins, scheduled evacuation can effectively reduce overall network congestion and, more important, mitigate potential casualties, stress levels, and chaos caused by evacuees being blocked in hazard areas (Liu 2007).

In staged evacuation, the most critical decision is the time to issue the evacuation orders for the evacuation zones. Once an evacuation order is announced, the evacuees' responses will determine the demand generation, which requires continuous monitoring to track network conditions and the evolution of the evacuation process. Obtaining such starting times during a staged evacuation is the subject of many studies with varying degrees of comprehensiveness. Chen and Zhan (2006) investigated the effectiveness of simultaneous (concurrent) and staged evacuation strategies in three road network structures using Paramics as a microscopic simulator. The study concluded that staging the evacuation process is essential in communities where the street networks have "Manhattan structure" and the population density is high.

Using a small hypothetical network, Mitchell and Radwan (2006) proposed a heuristic prioritization of emergency evacuation to reduce the network clearance time. Zonal parameters that might affect the staging decision are defined; these include population density, road exit locations/capacity, and major evacuation routes. Chiu et al. (2006) presented a system optimal dynamic traffic modeling technique for solving the evacuation destination-route-flow-staging problem for the nonnotice events. The algorithm was based on the CTM (Daganzo1994) and the LP formulation proposed by Ziliaskopoulos (2000). The framework was applied on a simple hypothetical evacuation event in which cell flows at each time interval are reported. The study concluded that the optimal solution of the presented LP depicts the optimal joint evacuation destination-flow-staging decision in an effective manner.

Sbayti and Mahmassani (2006) introduced an optimal evacuation scheduling approach with two assumptions to relax the destination selection and evacuees' compliance in which they assumed that evacuees would adhere to the evacuation guidance information and not switch to different departure times, destinations, or paths. It is also assumed that a controller will provide pretrip variable message signs and en-route information for nonevacuees who are on their way to the affected zone for necessary detours and trip changes. The final output of the study is an optimal loading curve that minimizes the total network clearance time. Abdelgawad and Abdulhai (2009) proposed an optimal spatiotemporal evacuation (OSTE) strategy that optimizes the scheduling and destination-choice problems simultaneously. The OSTE platform is built on the interaction between dynamic traffic assignment (DTA) and evolutionary algorithms (EAs). The output of OSTE is guidance to evacuees as to when to leave their origins, where to go (optimal destination), and how to get there (which route) in the quickest possible way.

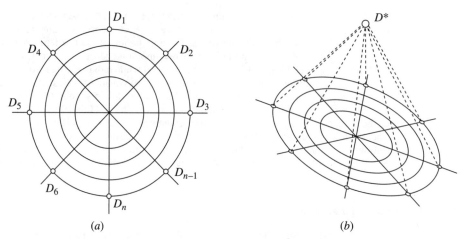

FIGURE 32.1 Original multiple-destination (*nD*) and modified one-destination (*1D*) networks (Yuan *et al.*, 2006).

32.1.3 The Concept of One Destination

In conventional evacuation planning models, evacuees are assigned to predetermined destinations that are based primarily on the geographic context and their daily activities. The traditional method of assigning evacuees to a prespecified fixed O-D table might result in suboptimal system performance owing to congestion, road blockage, chaos, incidents, hazards associated with the emergency situation, and limited destination/shelter capacity. One promising concept to address this problem is to relax the constraint of assigning evacuees to fixed destinations. In other words, instead of assigning the demand to fixed destinations, evacuees will be directed to the destinations that they can reach in minimal time. From a modeling perspective, this is achievable by adding one augmented dummy destination beyond all destinations, as shown in Figure 32.1.

Chiu et al. (2006) and Yuan et al. (2006) proposed the concept of a one-destination (*1D*) evacuation model in which the traditional road network with *m* origins to *n* destinations (as shown in Figure 32.1*a*) has been transformed to an *m* origins to 1 destination network (as shown in Figure 32.1*b*). The modified network is augmented with dummy edges that link each real-world destination to the one common dummy destination (*D**). The added dummy links are assumed to have unlimited capacity and zero costs. Yuan et al. (2006) reported that around 60 percent reduction in the overall evacuation time can be achieved when testing a case of regional evacuation owing to a nuclear power plant mishap and 80 percent reduction in the overall evacuation time when modeling traffic routing and en-route information accompanied with the *1D* framework.

32.1.4 Traffic Signal Control

As an efficient control strategy, traffic signal control has been widely accepted to improve arterial road capacity and reduce congestion during daily traffic and, more important, during emergency evacuation. Sisiopiku et al. (2004) used SYNCHRO (a software for optimizing traffic signal timing) to set up the optimal signal-timing plans for a small area in Birmingham, Alabama. CORSIM then was used to test different evacuation scenarios and evaluate the impacts of signal-timing optimization on selected measures of effectiveness. The study concluded that signal optimization during evacuation could reduce average vehicle delays significantly and improve network clearance time. Chen et al. (2007) applied CORSIM for two evacuation corridors in Washington, DC, and examined four different signal-timing plans: red flash, yellow flash, minimal green, and ordinary peak-hour plan.

Although this study offered some insights into the effect of various timing plans, its analysis of plan selection under various evacuation scenarios is mostly qualitative.

Another area of interest in emergency signal control is emergency vehicle preemption (EVP), in which the emergency vehicles override all other traffic movements and thus may affect the evacuation traffic. An impact analysis of EVP using CORSIM was conducted by McHale and Collura (2003). Among other series of studies to evaluate EVP impacts on traffic conditions, Louisell et al. (2004) proposed a conflict-point-analysis approach to evaluate the potential safety benefits of EVP. Furthermore, they developed a worksheet method to assess the crash-reduction benefits of EVP on a given intersection during a preemption signal phase. However, to the best of our knowledge and based on the literature, no published research has been reported in the area of emergency vehicle preemption during evacuation scenarios.

32.1.5 Traffic Routing and Control Strategies in Emergency Evacuation

Traffic routing, as one of the main traffic control strategies, aims to identify the best set of routing decisions so as to fully use the available capacity of a transportation network and assign traffic to those routes accordingly. User equilibrium (UE) assignment means that drivers follow time-dependent least-travel-time paths, whereas system-optimal (SO) assignment results from drivers following time-dependent least-marginal-travel-time paths. Under UE, all used routes between a given O-D pair have the same travel time, and hence the network reaches a stable equilibrium state. Under SO assignment, on the other hand, some users may be assigned to routes that are longer, whereas others are assigned to shorter routes. SO traffic patterns therefore are neither stable nor equitable. Simulation-based approaches have become flexible enough to perform UE, SO, or multiple-user-class assignment, although the equilibrium conditions are only approximated heuristically (Peeta and Ziliaskopoulos 2001).

From this perspective, there are two schools of thoughts in dynamic traffic assignment (DTA) during evacuation. One group argues that in emergency evacuations, the major concern of planners is overall system performance; therefore, it is more plausible to use system-optimized traffic assignment models. Among these studies is the work conducted by Sbayti and Mahmassani (2006) on evacuation scheduling and by Han and Yuan (2005) on destination optimization for emergency evacuation assignment using SO DTA. Conversely, the other group argues that aiming for SO assignment in actual evacuation operations is neither practical nor desirable from an equity perspective. The SO traffic assignment procedure assumes full adherence of evacuees to the evacuation guidance information, departure times, destinations, and paths. However, during large evacuations, it is expected that selfish behavior will increase among evacuees and lead toward user-optimal decisions, a concern that motivated the work done by Chiu and Mirchandani (2008) on dynamic traffic management for emergency evacuation. They quantified system performance in case of emergency evacuation for a small hypothetical network. The study compared the results from the route-choice decisions made by evacuees and the SO pretrip route guidance scheme. A controller was designed to influence the system performance toward an optimal level. The results showed suboptimal system performance owing to driver deviation from optimal paths.

32.1.6 Role of Intelligent Transportation Systems in Emergency Evacuation

Real-time traffic information is crucial during emergency evacuations. Several states/cities plan to incorporate ITS technologies into the evacuation process. ITS allow traffic information to be collected remotely from the field and disseminated through a traffic management center. This information can help emergency managers monitor the status of contraflow operations, evacuation scheduling, congested traffic routes, and incidents and thus proactively deal with the evacuation situation. A wide matrix of tools can be used in case of emergency evacuation. Closed-circuit television cameras (CCTVs) can provide detailed information to transportation officials and emergency managers. CCTV cameras can provide real-time traffic data for incident detection and verification that an incident has been cleared. This information then can be disseminated to evacuees or used by the traffic management center (TMC) to divert evacuees to less congested routes (Urbina 2001).

In addition, variable message signs (VMSs) and highway advisory radio (HAR) (FHWA 2010) have been shown to be useful tools for disseminating evacuation information and controlling traffic routing. VMSs can be fixed or mobile and typically are installed before bifurcation points to reroute traffic during emergency evacuation. Under routine conditions, HAR is used for traffic information broadcasting over small geographic areas. During emergency evacuation, evacuees can use these information technology (IT) systems to obtain alternative evacuation route information, congestion locations, incident information, and shelter locations. In summary, numerous ITS technologies and communication techniques should be considered to improve emergency evacuation and potentially reduce evacuation time.

32.1.7 Travel Behavior in Emergency Evacuation

Travel behavior and the compliance of evacuees with an evacuation order are important factors that should be modeled or realistically assumed during evacuation operation. Travel behavior under emergency conditions is expected to differ considerably from day-to-day travel patterns. Fu and Wilmot (2004) proposed a sequential logit model to simulate evacuee behavior in the case of hurricane evacuation. However, little research has been conducted in the area of driver stress and aggression that certainly would increase in such conditions. Zhi et al (2010) investigated the driver perception-reaction times (PRTs) under emergency evacuation situations. Using a driving simulator to model emergency situations and a survey to validate the driving simulator environment, the study concluded that the value of PRTs in normal situations is greater than that under emergency situations. It also has been hypothesized that higher levels of confusion might result from the unfamiliarity of driving in contraflow conditions. Moreover, it is anticipated that more incidents will occur under emergency evacuations; however, incidents during evacuation are rarely modeled in the literature except the sensitivity analysis investigated by Chen et al (2007). They randomly assumed three incident levels (minor, medium, and major) in different locations on a small tested network to capture incident effect.

Alsnih and Stopher (2004) acknowledged that there is a major gap in modeling drivers' behavior. It is a challenging issue to resolve because it involves identifying how evacuees would perceive an evacuation order whether it is mandatory or only recommended. Alsnih and Stopher (2004) also concluded that households may not follow an evacuation order for numerous reasons, such as preferring to stay to protect their property, not seeing neighbors evacuate, and obviously, the inconvenience associated with the evacuation process. How to model and capture this behavior is still a big concern that needs further investigation. Southworth and Chin (1987) posed the question of whether evacuees seek the safest, the nearest, or the farthest destination from the hazard location. The study showed that evacuees' choice of exit (destination) can be one of the following possibilities:

- Exit to the nearest shelter
- Exit depends on location of friends and relatives and the travel speed of the approaching hazard
- Exit toward prespecified destinations depending on the evacuation plan in operation
- Exit based on underlying traffic conditions of the network at the time of evacuation (allows for myopic evacuee behavior)

32.1.8 Summary

Despite the numerous approaches that significantly contributed to improving evacuation strategies, an integrated optimal evacuation strategy still needs further research. More effort is needed to synergistically combine all or some of the promising strategies to further improve the efficiency of the evacuation process. Traffic routing, evacuation staging, and destination optimization can be potentially combined into a comprehensive portfolio of solutions. However, optimizing the evacuation problem necessitates extensive modeling, design, and analyses that should capture the dynamic interaction among the analytical part of the evacuation process, the operation side of the transportation network, and the behavior of evacuees. Some analytical challenges that are facing the development of such an integration include

- Lack of a hazard prediction method that potentially could define the evacuation area and the duration and risk level associated with the hazard
- Lack of a travel demand estimation method or model for predicting the population to be evacuated within the hazard area and its flow patterns [This demand model should incorporate, identify, and predict demand by mode (transit, driving, etc.) under the unusual evacuation circumstances. The model also should capture the effect of evacuees compliance with an evacuation order.]
- Lack of understanding of how to best issue evacuation orders that affect the desired scheduling or staging of the evacuation demand
- Lack of destination selection and shelter capacity estimation procedures, which can affect the success or the failure of the evacuation process
- Lack of methods to dynamically assign traffic to the transportation network while capturing the spatiotemporal characteristics of travel in such chaotic situations

    Along with the analytical challenges, there are many pressing operational issues as well, such as

- Lack of traffic management tools that propose optimal control strategies while capturing the operational constraints embedded in implementing each control strategy
- Lack of comprehensive traveler information dissemination systems to update the system state and inform travelers of any disruptions during the evacuation process
- Lack of understanding of how to integrate the various control strategies in field, such as contra-flow, staging, traffic routing, signal control, and the use of ITS.

    While the aforementioned issues are separate modeling, analysis, or operational tasks, they are closely interrelated. Each is indispensable for the design and implementation of an effective emergency evacuation plan.

32.2 OPTIMIZATION OF AUTOMOBILE EVACUATION

This section discusses the conceptualization and development of the optimal spatiotemporal evacuation (OSTE) approach to the evacuation problem. It provides a high-level description of the framework components. It then introduces the time structures incorporated in emergency evacuation and its linkage to achieving certain objective in emergency situations. It also provides a mathematical formulation for the approach. Discussion of the solution algorithm is presented while emphasizing the multidimensional and nondeterministic nature of the problem. Application of the approach to a large-scale network is presented in section 32.6.

32.2.1 Scheduling and Destination Choice

In emergency situations that require population evacuation, particularly in dense urban areas, evacuees may react chaotically, causing severe congestion, gridlock, and excessive delays. Such uncontrolled evacuation also may expose evacuees to further harm, especially in cases of a high time pressure to evacuate. Therefore, the challenge of prompt evacuation of dense urban areas has made evacuation demand management an essential priority in planning for emergency situations. In addition, given the fact that the distribution of evacuation demand has unique spatiotemporal characteristics that are different from common everyday demand patterns, studying the evacuation demand distribution in both time and space is crucial. Concurrent temporal and spatial management of evacuation demand is central for successful emergency evacuation planning. Optimal spatiotemporal evacuation (OSTE) can be achieved through optimizing the evacuation scheduling problem and the destination choice problem simultaneously. OSTE, a comprehensive and extensible tool that is capable of generating optimal demand management plans for realistic-size networks, is founded on the interaction between DTA, evacuation demand scheduling, and destination choice.

Conceptually, the evacuation process follows two stages: mobilizing evacuees into the transportation network and evacuating the hazard area toward safe shelter. The two stages are represented by two cumulative flow curves, as shown in Figure 32.2. The loading (mobilization) curve $L(t)$ represents the demand entering the system, and the evacuation curve $E(t)$ represents the arrivals to safety, that is, exit flows. In the first stage, evacuee loading can be managed through temporally staging demand in an optimal manner that fulfills predefined objective function. Objectives can include, but are not limited to, maximization of the number of evacuees reaching safe destinations over a predefined evacuation horizon, minimization of the number of evacuees en route, or minimization of waiting time.

Destination choice refers to guiding evacuees to the nearest safe refuge and not necessarily their homes or routine daily destinations. Choice of destinations to which evacuees seek refuge can improve the efficiency of the evacuation process significantly by reducing unnecessary longer trips (Chiu et al. 2006; Yuan et al. 2006). Unlike fixed, predefined destinations, as in the conventional transportation planning models, optimal destination choice offers more flexibility in both destination and route choice. This may prevent overloading critical routes given the fact that some routes to destinations may be blocked or damaged by the disastrous event. Therefore, it may be better to dynamically direct traffic to the nearest safe zone; that is, optimize destination choice to fulfill the desired objective. This can be achieved by amalgamating all destinations into one hypothetical superzone or safe destination. Evacuees would seek this destination in the fastest possible manner, that is, reaching a safe zone quickly that is not necessarily their home.

For illustration purposes, three sets of curves are sketched conceptually in Figure 32.2 in which each set characterizes a certain level of demand control: simultaneous evacuation (SE), optimal temporal evacuation (OTE), and optimal spatiotemporal evacuation (OSTE). The case of SE mimics an evacuation scenario where evacuees decide to leave the area instantaneously, that is, no control. In the case of OTE, evacuees receive guidance on when to start the evacuation but without specifying a destination; that is, evacuees are expected to head "home." In the case of OSTE, evacuees receive guidance on when to evacuate (departure time) and where to go (optimal destination), as well as how to get there (optimal route). As depicted in the figure, the horizontal distance between the loading and evacuation curves is the travel time experienced by evacuees, and the vertical distance is the evacuees queued in the system, that is, en route $Q(t)$. Therefore the ultimate goal is to push the loading and evacuation curves upward and to the left, as shown in the figure, while minimizing the area in between them and hence minimizing both the travel time and vehicles en route. It is clear that the optimal evacuation curve and the optimal loading curve are interdependent.

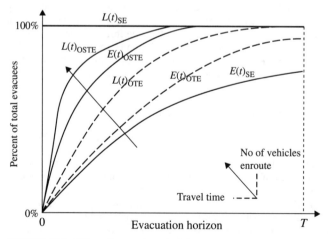

FIGURE 32.2 Effect of evacuation scheduling and destination choice on the loading and evacuation curves.

The concept of evacuees seeking a safe superzone destination in the case of emergency evacuation can be modeled using any plausible simulation tool that has the capability of modeling complex network structures. With a modified network representation, adding hypothetical zero-travel-time links from all destination zones to the superzone, drivers seek and reach the superzone destination in the simulation model in the shortest or fastest way. Therefore, the simulation model produces the optimal "exit" point for each vehicle, as will be illustrated further in the case study. To deploy the evacuation plan, as in our case, evacuees are instructed to head for the corresponding exit points/destinations obtained from the analysis. Communicating such information to evacuees is possible through numerous emerging ITS technologies and pervasive mobile communication and computing platforms.

32.2.2 Time Structures in Evacuation

The evacuation problem consists of three interconnected time structures: warning time, mobilization time, and evacuating time (Sorensen and Sorensen 2006). The warning time is announced based on the nature of the event; in case of no-notice evacuation, the warning time is zero, whereas in case of expected evacuation scenarios such as hurricanes, jurisdictions typically announce a warning to the public according to a predefined plan. The mobilization time/loading time (how groups of people are evacuated over time) is highly variable, depending on the nature of the event and the level of urgency. The mobilization or loading pattern does affect the overall evacuation process significantly. Following the mobilization time is the evacuation time, which is the time evacuees spend traveling through the transportation network seeking safe destinations. Figure 32.2 illustrates the temporal patterns of mobilization and evacuation, that is, the mobilization/loading curve $L(t)$ and evacuation curve $E(t)$. The time at which all evacuees reach a safe destination is defined as the network clearance time T. It is important to note that both curves are interacting dynamically in emergency evacuation. In other words, the loading curve can be optimized rather than being assumed as an input, as in the case of most conventional hurricane evacuation studies. Also, the evacuation curve can be optimized by shifting it as close as possible to the loading curve in order to minimize the evacuees entrapped en route.

32.2.3 Typical Objective Functions in Emergency Evacuation

The evacuation problem typically has been solved as an optimization problem that minimizes/maximizes certain objective functions subject to supply, demand, and time constraints. Numerous objective functions have been formulated in the literature with the goal of expediting the evacuation process. The most common objective functions addressed in the emergency evacuation literature can be summarized according to the following taxonomy (see Figure 32.3):

- *Minimize the evacuation travel time* (area A_1 in Figure 32.3) (Hobeika and Kim 1998; Tuydes and Ziliaskopoulos 2004; Chiu et al. 2006; Tuydes and Ziliaskopoulos 2006; Yuan et al. 2006; Liu et al. 2007).

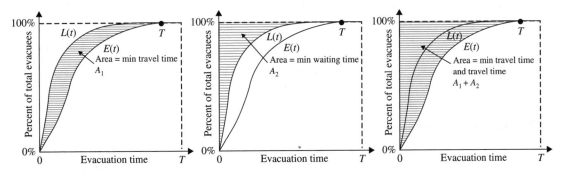

FIGURE 32.3 Multiple objectives in evacuation.

- *Minimize network clearance time* (point T in Figure 32.3) (Sattayhatewa and Ran 2000; Sbayti and Mahmassani 2006).

- *Maximum throughput for a specified clearance time* (Liu 2007; Abdelgawad and Abdulhai 2009).

- *Minimize total travel time and total waiting (mobilization) time* (area $A_1 + A_2$ in Figure 32.3) (Sbayti and Mahmassani 2006; Abdelgawad et al. 2010).

As can been seen from this taxonomy, various objectives have been investigated in the literature. Although some objective functions seem similar, some key differences rest on how the objective function is formulated and solved. For instance, some studies show that minimization of the total system travel time is equivalent to providing system optimal information for evacuees through a traffic management center with 100 percent compliance. System optimal traffic assignment is an appealing system-wide target, but it implies that some evacuees will experience travel times longer than their best attainable individual travel time. It seems unrealistic to expect evacuees to offer such sacrifice during evacuation in order for the system to be optimized, not to tort liability of the system operator. In addition, minimizing travel (en route) times ignores the wait time, that is, how long evacuees need to be held at the origin before being allowed to begin evacuation, an important factor in nonnotice evacuation events. Although the network clearance time is essential for evacuation planning, solely considering the network clearance time (endpoint T in Figure 32.3) might be misleading. This is due to the fact that two evacuation curves might result in the same network end/clearance time, but the network conditions at intermediate times and the system stability for both cases might differ significantly (an upward or downward curve at intermediate points) (Han et al. 2007). Alternatively, another measure of the effectiveness of the evacuation process is the time at which multiple percentages of evacuees have reached a safe destination; for example, Yuan et al. (2006) measured the times at which 25, 50, 75, 95, and 100 percent of the population have evacuated the hazard area.

Very few studies explicitly considered minimizing the wait time of evacuees at origins as well as the travel time of evacuees in the transportation network. A good evacuation model is one that accounts for minimizing total system evacuation time, including both wait time and travel time; however, both objectives might be conflicting, and the tradeoff between the two is challenging. As can be seen from Figure 32.3, minimizing the waiting time for evacuees implies evacuating all the population instantly, that is, simultaneous evacuation, which may lead to longer travel times in the system and a longer total evacuation time. Simultaneous evacuation typically results in early gridlock and underutilization of the available infrastructure. Also, minimizing the travel time implies delaying evacuees at the origin (increase in wait time) so that network conditions remain stable and evacuees can reach their destinations in the least possible travel time. Although Sbayti and Mahmassani (2006) considered both the waiting time and the travel time of evacuees in the optimal scheduling problem, the waiting time was not explicitly optimized. One way to achieve this compromise is defined by Abdelgawad et al. (2010), who solved the problem of minimize the waiting time of evacuees at the origin as well as the en-route travel time of evacuees.

32.2.4 Optimal Scheduling and Destination Choice Problem Formulation

The decision variable in the following formulation is the staging percentages $\mu = (\ldots, \mu_i, \ldots)$ with a modified superdestination network representation to account for the optimal destination. The problem is formulated as a multiobjective optimization problem in which wait time and travel time is minimized as opposed to the typical minimization of vehicle travel time. The formulation is presented in Table 32.2.

32.2.5 The Optimization Approach

Typical Solution Algorithms versus Evolutionary Algorithm Approach. The evacuation planning problem typically has been formulated as an iterative bilevel optimization problem wherein the upper level the objective function is optimized while keeping the traffic assignment parameters fixed

TABLE 32.2 OSTE Formulation

Sets	
T	evacuation planning horizon divided into equal evacuation (departure) intervals o time t, where $t = 1, \ldots, T$
I	set of evacuation zones (origins); $i \in I$
J	set of safe shelters (destinations); $i \in J$

Parameters	
D_{ij}	evacuation demand from origin i to destination j;

Decision Variables	
μ	demand scheduling/staging vector for evacuation: $\mu = (\ldots, \mu_t, \ldots)$ where μ_t is the percentage of demand released at time interval t

Objective Function

- **Min Vehicle Travel Time**

$$Min \sum_{t \in T} \sum_{i \in I} \sum_{j \in J} Y_{ijt} TT_{ijt}$$

Where TT_{ijt} is the evacuation travel time between origin i to destination j at time t. It should be noted that travel times are dynamically changing according to the scheduling vector and the DTA model.

- **Min Waiting Time**

$$Min \sum_{t \in T} \sum_{i \in I} \sum_{j \in J} Y_{ijt} WT_{ijt}$$

Where WT_{ijt} is the waiting time for evacuees traveling from origin i to destination j at time t. It should be noted that waiting times are also dynamically changing according to the scheduling vector.

- **Min Waiting Time and Vehicle Travel Time**

$$Min \sum_{t \in T} \sum_{i \in I} \sum_{j \in J} Y_{ijt} (TT_{ijt} + WT_{ijt})$$

Where,
Y_{ijt} = number of evacuees leaving evacuation zone i to destination j at time interval t
$Y_{ijt} = \mu_t D_{ij}$

Constraints	
Flow Conservation Constraint	$\sum_{t \in T} \sum_{i \in I} \sum_{j \in J} Y_{ijt} = D_{ij} \forall i \in I, j \in J$
Scheduling Range	$1 \geq \mu_t \geq 0$
	$\sum_{t \in T} \mu_t = 1$

(route flow patterns), whereas in the lower level the traffic assignment problem is solved while keeping the upper-level optimization parameters fixed (Chen and Zhan 2006). The conventional solution algorithms for such bilevel optimization problems iterate between the two optimization levels (Sbayti and Mahmassani 2006; Meng et al. 2008; Xie et al. 2009). To the best of our knowledge and based on the literature, the optimization problem is solved at the upper level with deterministic approaches; these include a hill-climbing search method, the Simplex method, the Frank-Wolfe algorithm, and a gradient-descent approach. These traditional optimization methods have been found to be problematic when applied to nonlinear or highly dimensional problems in which the objective function cannot be represented analytically in a closed form (Kruchten 2003). Furthermore, these approaches inherently search in the vicinity of the starting point and thus create a high dependency on the initial solution, which can be an issue in very large search spaces. Moreover, traditional deterministic approaches follow a single path in the search space and may get stuck in local minima. This is to be contrasted with methods that rely on the evolution of multiple solutions. In addition, large-scale applications such as evacuation of a large city may require ample computer processing power or even parallel processing. Gradient-based algorithms are not amenable to parallelization (Bethke 1976). Therefore, the existing bilevel deterministic iterative optimization approaches may not be the best to tackle large-scale evacuation problems, particularly when a significant difference exists between the optimal solution and the initial starting point and when the search process may get stuck in local minima.

Global optimization offers a myriad matrix of potential heuristic approaches that can be used to optimize the evacuation problem. Among the most promising global optimization approaches are evolutionary algorithms (EAs). Simply, EAs are methods of searching in multidimensional space while satisfying certain criteria (Bäck 1996). In recent years, EAs have emerged as one of the leading methodologies for powerful search and optimization of problems in high-dimensional and nondifferentiable search spaces. Examples of EAs are genetic algorithms (GAs) and evolution strategies (ESs). OSTE uses GAs instead of the more traditional approaches in the literature. GAs sidestep all the problems associated with the traditional deterministic optimization methods; they start the search from a population of solutions and not from a single point. Therefore, the odds of finding the global optima without getting stuck in local minima are more than with most conventional approaches, and they don't require differentiation of the objective function. Moreover, GAs are inherently parallelizable, allowing for harnessing the power of several computers or CPUs and for future use of high-performance computing (HPC) clusters (see below). The number of processors in a typical HPC can be in the hundreds; that is, extensible computing power can be used as the size of the evacuation problem grows, thereby reducing the computation time almost linearly with the number of available processors (Kruchten et al. 2004). In GAs, a population of artificial solutions (chromosomes) is created. A *chromosome* is one feasible point or a candidate solution that carries the encoded values of the decision variables in the form of a string of *genes*. Each candidate solution from the population then is *evaluated* to give some measure of its *fitness* (Garey et al. 1979). Evaluation can be as simple as substituting variable values in a mathematical function, or it can be a full simulation experiment such as in this study. After evaluating an initial population by calculating the fitness of each candidate solution, *selection* and a series of *genetic operators* work on the population to *reproduce* a sequence of populations (*children*) through methods of *crossover* and *mutation* with increasingly enhanced solutions. This cycle of evaluation, fitness assignment, selection, and reproduction continues through a number of *generations* (iterations) until a certain stopping criterion is met (Figure 32.4). It is clear that GAs are inspired from the process of natural selection and evolution and mimic the biodiversity of the world we live in.

An important property of EAs is that they can be parallelized (Bäck 1996). In general, three types of EAs exist: *panmictic* EAs, *diffusion-style parallel* EAs, and *island model parallel* EAs. In *panmictic* EAs, reproduction can be conducted between any two chromosomes in the population, whereas in *diffusion-style parallel* EAs, the chromosomes are spatially distributed (e.g., two-dimensional grid), and only neighboring chromosomes can be recombined. In *island model parallel* EAs, semi-independent subpopulations, *demes*, evolve independently with periodic exchange of some chromosomes through a *migration* process (Tomassini 1999). This type of EA exhibits even more correspondence with species evolution theory, where thousands of subpopulations or demes exist and coevolve in parallel in the same continuous geography.

It is important to note that the *parallel* in parallel EAs is not to be confused with the *parallel* in parallel computing. Parallel computing *distributes* the computation across multiple processors simultaneously, in which chromosomes are farmed out to multiple processors for evaluation (i.e., in parallel). On the other hand, in a parallel EA, *parallel* typically refers to the spatial structure of the population. Therefore, a parallel EA can be either executed sequentially on a single processor (in which subpopulations or demes evolve independently and migration takes places across multiple demes) or across multiple processors. In this chapter, the term *distributed* is used to refer to computation instead of parallel, and the term *parallel* will only refer to the EA's population structure (Figure 32.5).

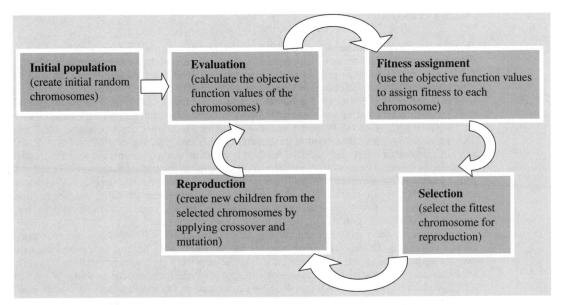

FIGURE 32.4   The basic cycle of EAs.

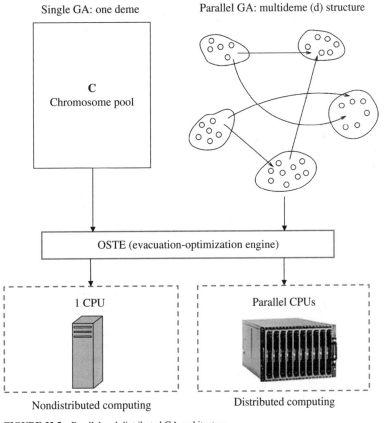

FIGURE 32.5   Parallel and distributed GA architecture.

Parallel Distributed Genetic Algorithm (PDGA) for OSTE.   A noniterative formulation for the optimization of the spatiotemporal evacuation pattern, where a global search optimization model that we call OSTE, is developed. A noniterative approach means that the simulated flows and the scheduling strategy are computed simultaneously rather than iteratively for a given scheduling vector. Therefore, OSTE works in a noniterative fashion, invoking a simulation model only for evaluating a given scheduling vector in a superzone destination representation. OSTE uses and expands on GENOTRANS (Generic Parallel Genetic Algorithms Framework for Optimizing Intelligent Transportation Systems), developed at the University of Toronto (Mohamed 2007). GENOTRANS is a GA platform in which the objective function is evaluated and constraints are satisfied through a simulation model. The simulation model is used to replicate the transportation network and perform dynamic traffic assignment as well as representing the superdestination approach. Vehicles are loaded onto the superdestination network according to their optimized schedules, and they navigate through the network toward their destination(s) while dynamically updating their travel paths until they reach a safe destination.

In addition to the parallelized nature of GAs, OSTE and GENOTRANS also use distributed genetic algorithms (DGAs). This distribution is very efficient when dealing with computationally demanding problems (either large network size or the need for very short computation time). The implementation of parallel DGAs is believed to provide better performance in terms of solution quality and convergence speed (Tomassini 1999). A multideme DGA is designed to simultaneously optimize the evacuation schedule and the destination choice for large-scale evacuation problems. As clearly shown in Figure 32.5, in OSTE and GENOTRANS, the core GA engine is both distributed and parallelized.

An HPC cluster is used to deploy the parallel DGA. The cluster has 64 processing nodes, 48 with 4 GB of memory and 16 with 8 GB of memory, all with two processors, XEON 5150 2.66-GHz dual-core Woodcrests, for a total of four processing cores per node (i.e., 265 processors), a 36-GB 15,000 rpm SAS hard disk, Dual Gig Ethernet, one public port, and one dedicated cluster port. The cluster is used to deploy the parallel DGA optimization platform to solve large-scale evacuation optimization problems.

In OSTE, the chromosome is encoded as the row vector μ encoding the evacuation percentages over discrete time intervals. Each gene in the decision variable vector is encoded as a real number with a range between 0 and 1. The chromosome size (number of genes) represents the number of departure time intervals. The GA fitness function is the same as the objective function in Table 32.2, which requires the output of the traffic simulator each time an evaluation is performed. Figure 32.6 describes the general structure of the solution algorithm. Figure 32.6a and b depict the single GA (SGA) and parallel DGA structures, respectively, in which GENOTRANS interacts with a simulation model of the transportation network to be evacuated.

As shown in Figure 32.6b, a master process generates the initial population, divides it into the specified number of demes (islands), and manages each deme's evolution and migration process. Each chromosome is evaluated via a complete simulation run covering the entire evacuation period. Based on the literature and pilot experimentation, the multideme structure is selected as follows:

1. Deme topology: Fully connected multiple-demes topology.
2. Number of demes (d) and deme/subpopulation size: Scenario-dependent.
3. Migration policy: Good migrants replace bad individuals.
4. Migration rate (the number of individuals to migrate): 15 percent of the population.

The design of a GA for a particular application involves the choice of methods and parameters for the population size, initial population, selection, crossover, mutation, and stopping criterion. Significant testing is undertaken to find a combination of parameters to refine the choice of the GA parameters. The preceding values and the following GA parameters are used to illustrate application of the algorithm itself rather than the sensitivity of its parameters. The solution algorithm iterates through the following steps until termination:

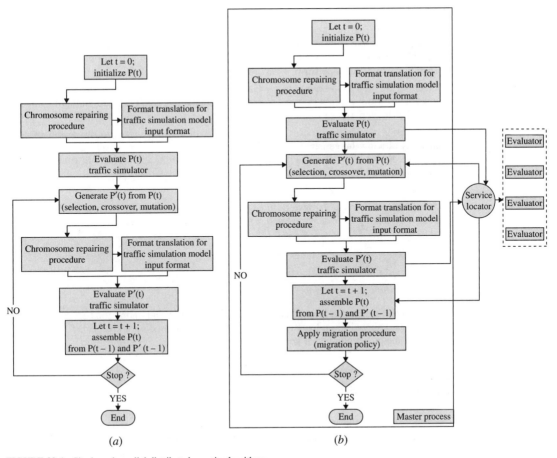

FIGURE 32.6 Single and parallel distributed genetic algorithms.

- *Step 0: Initialization:*
  - Set the user-specific GA design parameters (probability of crossover = 0.9 and mutation = 0.05), range for each decision variable = 0 to 1.
  - Set the generation counter $t = 0$.
- *Step 1: Chromosome representation:*
  - Code the decision variable μ to form a chromosome C of real numbers.
  - Adjust the chromosome using a scaling algorithm to sum up the staging vector to 100 percent of the demand being released at each generation.
- *Step 2: Evaluate objective function:*
  - Using the evacuation schedule encoded in each chromosome, a simulation-based traffic assignment is performed to evaluate the chromosome's fitness. In the simulation model, traffic seeks an amalgamated superzone, and hence the optimal exit points (actual destination) are ultimately produced. The process is repeated for all the chromosomes in population $P(t)$. Based on the output of each simulation run, the objective function is evaluated.

- *Step 3: Apply genetic operators:*
  - Based on the fitness of each solution, the selection process, which resembles the "survival of the fittest" principle, picks a number of candidate solutions for "breeding" to produce the next generation. Better individuals have higher probabilities of being selected to produce the next generation. Breeding is performed via crossover and mutation.
  - The crossover step, also known in the GA literature as *recombination*, is the process of combining genes from parents in order to produce better offspring. This step mainly exploits the regional characteristics of parents.
  - The mutation step is the process responsible for exploring new areas in the search space by randomly altering the genetic structure of some chromosomes. The offspring is similar to the parent except for a few changes in the parental genetic information. This will guarantee diversity in the search space.
- *Step 4: Chromosome repair:*
  - After the genetic operators have been executed, and if the offspring is an infeasible solution, repair it by scaling the evacuation percentages to add up to 100 percent.
- *Step 5: Fitness assignment:*
  - Use the objective function values to determine fitness values of the new population.
- *Step 6: Apply the migration policy in case of PGA.*
- *Step 7: Test termination criterion.*

If the termination criterion is satisfied, then terminate the GA and output the best solutions from the last iteration. Otherwise, go back to step 3. Termination is based on a prespecified convergence threshold or if the preset number of generations is exhausted.

32.3 OPTIMIZATION OF TRANSIT EVACUATION

This section discusses the conceptualization and development of the transit shuttling approach during evacuation. It provides a general description of the framework components and the linkage to the typical vehicle routing problem (VRP). It then introduces an analogy between the VRP and the mass-transit evacuation problem. It also provides a mathematical formulation for the approach. Application of the approach to a large-scale network is presented in section 32.6.

32.3.1 Vehicle Routing Problem and the Transit Evacuation Problem

The VRP is investigated extensively in the literature; numerous techniques are offered, with the common goal of modeling and solving the VRP. However, the solution efficiency differs significantly according to the exactness of the solution and the problem size. Table 32.3 provides a comparative literature survey for the VRP and its extensions. The survey is organized to highlight the most relevant studies according to the following categories: exact, approximate, and metaheuristic approaches. The rows represent the method/approach used to solve the problem, and the columns represent the criteria for each study in light of relevance to the transit evacuation problem.

32.3.2 VRP and Emergency Evacuation: Background

Sayyady (2007) investigated the use of public transit systems in nonnotice evacuation of urban areas. The author formulated the public transit routing plan (PTRP) problem as a mixed integer linear program in which a network flow problem is solved with the objective of evacuating as many people

TABLE 32.3 Comparative Literature Survey

Method	Solution Procedure	Objective	Time Window	Multiple Depot	Pick-up and Delivery location	Problem Size and Computation Time
Exact (Fisher, 1994)	Solve degree-constraint k-tree problem	Min Total Routing Cost	—	—	—	50–199 customers, CPU Time (6–680 min)
Approximate (Clarke and Wright, 1964)	Assign a vehicle to each customer and then improve the routing using saving algorithm	Min Tour Length	—	—	—	Small size problems
(Fisher and Jaikumar, 1981)	Generalized assignment procedure with routes starting and ending at depots	Min Total Routing Cost using the TSP Heuristic	—	—	—	50–199 customers, CPU Time (9–25 min)
(Kindervater and Savelsbergh, 1997)	K-Exchange algorithm to improve and extend the TSP solution	Min Total Cost of Travel	Yes	—	Yes	10–30 customers, CPU Time (0.2–0.5 sec)
Meta-Heuristic (Rochat and Taillard, 1995)	Probabilistic Diversification and Intensification to overcome local minima and improve computation time	Min Total Tour Length	Yes	—	—	50–385 customers, CPU Time (0.2–3000 min) to get 1% improvement over the best known problems
(Shaw, 1998)	Constraint Programming using Large Neighborhood Search	Min Total Distance Travelled by Vehicles	Yes	—	—	Perform better than benchmark problems (C, R, RC)
(Xu and Kelly, 1996)	Tabu Search to Solve Network Flow Model	Min Total Travel Cost	Yes	—	—	50–199 customers, CPU (4.89–207.8 min)
(Toth and Vigo, 2003)	Granular Tabu Search	Min Total Routing Length	Yes	—	—	50–199 customers, CPU (0.8–3.18 min)
(Gambardella et al., 1999)	Multiple Ant Colony Optimization	Min Total Travel Cost and Fleet Size Cost	Yes	—	—	50–199 customers, results are reported after stopping criterion (100, 200, ... 1800 sec)

as possible from a set of source nodes (transit stations) to a set of exit nodes (shelters) in a given time frame without violating the capacity constraints. However, the bus schedules were predefined with a limited number of tasks to be performed. The evacuation demand was assumed to follow a normal distribution that forms the demand at each visit. The study provided a sensitivity analysis to the available fleet size and percentage of people captive to transit. The computation times of two solution algorithms, CPLEX and Tabu Search, are evaluated and found to be a bottleneck. CPLEX runs out of memory after running for 3 days without reporting the optimal solution. The running time to evacuate 300 to 2,000 people ranges from 166 to 167 minutes using CPLEX and 4 to 16 minutes using Tabu Search, respectively.

Murray-Tuite and Mahmassani (2003) provided two linear integer programs to express household behavior during evacuation conditions. The first formulation determines the meeting location for household members, whereas the second considers a modified version of the VRP in which the sequence of family member pickup is determined. The fleet of vehicles available to households is assumed to be heterogeneous, depending on the available car capacity. In addition, vehicles are distributed depending on the location of their drivers at the onset of the evacuation.

Pagès et al. (2006) introduced the mass-transport vehicle routing problem (MTVRP) in which a fleet of vehicles (of given capacity) is routed to pick up and deliver passengers. The problem is solved iteratively between two levels: the transit problem (TP) and the passenger problem (PP). The TP is solved once to generate an initial solution, and then the PP works on improving the first solution by assigning passengers to routes. The authors compared the computation time of CPLEX with the PP for different network sizes (range from 5 to 56 links). The benefits of using the proposed algorithm are found to be greater in the case of large networks. However, network size and the problem dimensions are relatively small compared with real-life evacuation scenarios.

Song et al. (2009) proposed an optimal transit routing for emergency situations. The problem is formulated as a location routing problem in which the decision variable is routing of vehicles and choice of shelter points so as to minimize vehicle routing times. The problem is solved using hybrid GAs in which 15 constraints have to be satisfied in the initialization and reproduction processes. The problem has a special network structure that is different from the actual transportation network, which necessitates building a set of different link categories for streets, intersections, and U-turn links. The proposed algorithm was tested on a small network with hypothetical evacuation parameters to test the feasibility of the proposed hybrid GA algorithm compared with the traditional GA, and the former turned out to perform better.

32.3.3 Vehicle Routing and Mass Transit Evacuation: An Analogy

The VRP is a generic class of problems in which sets of customers are visited by vehicles. The goal is to solve the routing problem by assigning vehicles to traverse the transportation network so as to visit customers, satisfy given sets of constraints, and optimize certain objective functions (Figure 32.7). The VRP consists of several interacting elements that are summarized as follows with the key attributes of each element:

- *Customers:* Demand, time constraints, pickup and delivery locations, priority
- *Vehicles:* Capacity, cost, time window of vehicle availability
- *Depot:* Number, location, capacity
- *Network:* Time, distance, geographic representation

To effectively use the transit fleet in emergency evacuation situations, the traditional VRP can be extended to include (1) *multiple depots* to account for the dispersed transit vehicles in the transportation

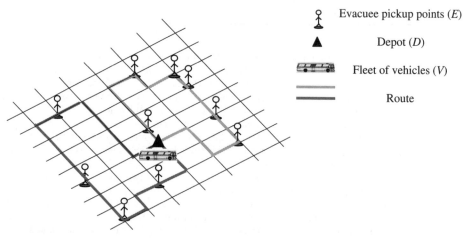

FIGURE 32.7 Traditional vehicle routing problem representation.

network, (2) *time constraints* to account for the desired evacuation time window, and (3) *pickup and delivery locations* for evacuees to allow for picking up evacuees from dispersed stops to avoid excessive walk distances. The multidepot time constrained pickup and delivery VRP (MDTCPD-VRP) is known to be an NP-hard problem (Garey et al. 1979). The following points highlight the analogy and the correspondence between the traditional VRP and the presented emergency evacuation MDTCPD-VRP (Figure 32.8), where customers become evacuees, pickup points are hazard areas to be evacuated, delivery points are safe shelters, and vehicles are transit shuttle buses:

- Pickup and delivery problem (PDP) where evacuees (customers) are picked up from hazard zones and delivered to shelters (safe destinations) and where the vehicles (shuttle buses) head back empty to pick up more evacuees until all the customers are delivered to shelters. Shuttling is not necessarily confined to two fixed ends (pickup and delivery points) but is rather optimized.

- Multiple-depot VRP (MDVRP) where the transit fleet is stored at multiple locations as opposed to the traditional, fixed, one-depot location. This is particularly convenient for cases where buses are initially scattered at different hubs, more so in the case of multiple bus companies such as public transit buses, commuter service buses, school buses, and so on. In cases of emergency, buses can be assembled from different providers in accordance with prior agreements and globally managed from a unified emergency command center.

- Time-constrained problem (known as *VRP with time windows*, VRPTW). Time-window problems occur in many business sectors. The greater the time constraints introduced to the problem, the more challenging is the routing plan. The severity and nature of the emergency situation may dictate certain evacuation time horizons to minimize evacuee exposure to risk. Also, if evacuees wait for service for excessively long times, they may turn impatient and seek other (uncalculated) alternatives. Therefore, the following time windows are modeled in the evacuation problem:

  - Evacuees have to be picked up before a time threshold (T_{min}).
  - Evacuees have to be delivered to a safe destination before a time threshold (T_{max}).

- Alternative delivery sites. Shuttle buses will seek the optimal safe destination. An optimal destination procedure is designed to search for the destination that forms the quickest exit for each vehicle (Chiu et al. 2006; Yuan et al. 2006).

The MDTCPD-VRP is formulated by defining the sets, parameters, decision variables, objective functions, and constraints as shown in Table 32.4.

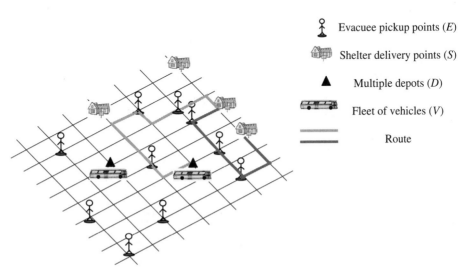

 Evacuee pickup points (E)

 Shelter delivery points (S)

 Multiple depots (D)

 Fleet of vehicles (V)

 Route

FIGURE 32.8   MDTCPD-VRP problem representation.

TABLE 32.4 MDTCPD-VRP Formulation

Sets	
N	$= \{1, 2, \ldots, n\}$ All modeled nodes that construct the network (graph)
E	Evacuee pick-up points, a subset of N
s	Shelter delivery point
A	All arcs (i, j) linking connected nodes $i, j \in N$
D	$= \{1, 2, \ldots, d\}$ Set of depots nodes where vehicle v is located
V_d	Set of all vehicles assigned to depot d
V	$= \{1, 2, \ldots, v\}$ Available fleet of vehicles, $V = \displaystyle\sum_{d=1}^{D} V_d$

Parameters	
CAP_V	Capacity of vehicle v
$COST_V$	Cost of using vehicle v
DEM_e	The demand of evacuees located at node e
C_{ij}	The routing cost (travel time) on link (i, j)
T_{Max}	Time global threshold before which evacuees have to be delivered (can also be visit specific)

Decision Variables	
X_{vije}	$= z$ if vehicle v travels along are (i, j) z times to pickup evacuess from point e; $z \in z$

Objective Function

Min Routing Cost

$$\text{MIN} \sum_{e \in E} \sum_{(i,j) \in A} \sum_{v \in V} X_{\text{vije}} C_{ij}$$

Min Routing Cost and Waiting Time

$$\text{MIN} \left[W_{RC} \sum_{e \in E} \sum_{(i,j) \in A} \sum_{v \in V} X_{\text{vije}} C_{ij} + W_{WT} \sum_{e \in E} \text{DEM}_\in \sum_{v \in V} \sum_{(i,j) \in A_{VE}} C_{ij} \right]$$

Min Routing Cost and Vehicle Cost

$$\text{MIN} \left[W_{RC} \sum_{e \in E} \sum_{(i,j) \in A} \sum_{v \in V} X_{\text{vije}} C_{ij} + W_{VC} \sum_{v \in V} X_v \, \text{COST}_v \right]$$

Min Routing Cost, Vehicle Cost, and Waiting Time

$$\text{MIN} \left[W_{RC} \sum_{e \in E} \sum_{(i,j) \in A} \sum_{v \in V} X_{\text{vije}} C_{ij} + W_{VC} \sum_{v \in V} X_v \, \text{COST}_v + W_{WT} \sum_{e \in E} \text{DEM}_\in \sum_{v \in V} \sum_{(i,j) \in A_{VE}} C_{ij} \right]$$

Where

$$X_v = \begin{cases} 1 & \text{if vehicle } v \text{ dispatched (i.e. if } X_{\text{vije}} \geq 1 \text{ for any } e \in E, i, j \in N) \\ 0 & \text{otherwise} \end{cases}$$

A'_{ve} = set of arcs (i, j) linking connected nodes $i, j \in N'_{ve}$

N'_{ve} = set of nodes visited by vehicle v before reaching pickup point e

(Continued)

TABLE 32.4   MDTCPD-VRP Formulation (*Continued*)

Constraints
Vehicle Capacity and Tour Constraints

Vehicle Capacity and Tour Constraints — All evacuees must be picked up (no evacuee storage in the evacuation area) while each time a vehicle v picks up at maximum CAP_v evacuees from point e

$$\sum_{v \in V} X_{ve} CAP_v \geq DEM_e \qquad \forall e \in E$$

Where

$$X_{ve} = \begin{cases} z \text{ if vehicle } v \text{ assigned } z \text{ times to pickup evacuaees at point } e \text{ (i.e. if } \max_{i,j \in N} (X_{vije}) \\ 0 \qquad \text{otherwise} \end{cases}$$

All evacuees must arrive at the safe destination (s)

$$\sum_{i \in N} X_{vise} = X_{ve} \qquad \forall v \in V, e \in E$$

All vehicles do not return back to depot d (open vehicle routing problem).

$$\sum_{v \in V} \sum_{i \in N} \sum_{e \in E} X_{vije} = 0 \qquad \forall j \in D$$

Time Windows

$$\sum_{v \in V} \sum_{(i,j) \in A} C_{ij} X_{vije} \leq T_{Max} \qquad \forall_{e \in E}$$

32.3.4 Solution Algorithm: A Constraint-Programming Approach

Constraint programming (CP) aims to simultaneously solve a constraint-satisfaction problem (CSP) and an optimization problem. CP uses multiple algorithms to find feasible solutions to constraint-satisfaction and optimization problems. In CP problems, search strategies are defined to model how the decision variables change with iterations to satisfy the constraints. Numerous algorithms were found in the literature; among the most relevant to the constrained VRP in emergency situations are domain reduction and constraint propagation.

Each decision variable in the CSP has its domain. The domain-reduction algorithm works to remove the values from the domain of variables that are apart from any feasible solution (i.e., that violate the set of constraints). This process is performed for all the variables in each constraint. At some point, the algorithm might discover that eliminating some values from the variable domain results in an unfeasible solution; at this stage, the previous domain value is retrieved. CP determines how the domain reduction technique is conducted among several constraints.

CP appears to be a good approach for tackling real-world VRPs because of its ability to address problem-specific constraints effectively. CP produces initial solutions that satisfy the constraints set; however, an improvement mechanism must be in place to enhance the first solution (Shaw et al. 2002). Local search techniques fit well in the evacuation problem to improve the initial routing plan by exploring neighborhoods. Integrating CP and local search techniques would synergize their potential utility in more challenging real-world VRPs. Such integration is particularly appealing in the case of emergency evacuation because side constraints, such as bus capacities, evacuation time window, and pickup and delivery, play a paramount role in the success of the evacuation plan.

The emergency evacuation routing and scheduling problem is solved in two stages. First, the model of the area to be evacuated is built, and then the routing and scheduling problem is solved/optimized.

Building the Evacuation Model.   Building the model consists of four basic objects: nodes, visits, vehicles, and costs. These objects are used to construct the network and model the evacuation problem in a CP environment.

- *Nodes.* Any transportation network is composed of a set of intersections or nodes with specific coordinates and a set of roads or arcs connecting the nodes. ILOG Dispatcher represents these intersections as nodes N. These nodes then are used to compute distances and times (and subsequently cost) between node pairs (i, j). A set of arcs A (links) then is defined to connect these nodes and subsequently construct the graph (network).

- *Visits.* A visit represents an activity that the vehicle has to perform within the specified time window $(T_{min} - T_{max})$. A visit is located at a single node and is performed by only one vehicle v. Visit locations in emergency evacuation are the evacuee pickup points E. Multiple visits might be created at the same node in case the demand for pickups is greater than vehicle capacity CAP_v. For example, a specific pickup point e along the graph might be visited more than once if the demand located at this point is greater than the capacity of the vehicle assigned to perform this visit. Visits have quantities, which can be weight, volume, or numbers of objects. In the evacuation problem, these quantities are the number of evacuees boarding and alighting at the pickup and delivery points, respectively.

- *Vehicles.* Vehicles represent the supply that serves the demand of visits. A vehicle has a start and end visit and can have variable start and end times associated with each visit. In the evacuation problem, visits are the pickup points from hazard areas E and the safe shelter (destination) points S. Vehicles have limited capacities CAP_v. These capacities represent the total number of people the vehicle can carry at any point along the route, which is user-defined. Therefore, one vehicle might have multiple runs given an assigned (or optimized) schedule.

- *Cost.* Cost attributes are objects closely associated with visits and vehicles. The most common costs are time and distance. Cost is used to model side constraints such as capacity, time windows, etc. For example, costs could be attributed to the distance between nodes and projected travel time along arcs. In addition, waiting time of evacuees at the stop can be added as necessary to the cost function. Cost is also used to calculate the objective function.

Solving the Evacuation Routing and Scheduling Problem. Solving the evacuation bus routing and scheduling problem consists of finding a value for each decision variable while simultaneously satisfying the constraints and optimizing the objective function. As described earlier, given the network to be evacuated, the evacuation demand, and the available buses, the routing and scheduling of transit vehicles are solved to minimize the total cost (travel time and/or waiting time) while satisfying bus capacity, time window constraints, and pickup and delivery constraints. Search strategies and constraint propagation are used to solve the problem. Two types of constraint propagation are used: initial constraint propagation and constraint propagation during search. The initial constraint propagation removes all values from domains that will not take part in any solution. After initial constraint propagation, the search space is greatly reduced, and search strategies are used to explore the search tree (the remaining part of the search space) for feasible solutions that satisfy the constraints and further improve the objective function.

The search strategy consists of two steps. First, an initial solution is found using route construction techniques. A known routing heuristic, the insertion algorithm (Jaw et al. 1986), is used to provide the initial solution. Once the initial solution is obtained, the second step is to apply a local search procedure to improve the solution obtained by introducing small changes (called *neighborhoods*) to the current solution. The new solution is tested again for constraint feasibility using *constraint propagation during search*, and its cost is computed. If the new solution is feasible and has reduced the cost, it is accepted as the new solution; otherwise, the algorithm backtracks along the search tree and tries a different value for the variables. Backtracking offers the flexibility to retract the search moves that may turn out to be wrong. Alternatives can be tried, and if they do not succeed, they can be reversed. In this way, only moves to feasible and improved solutions are accepted. Solutions that do not improve the cost or that violate the problem constraints are rejected. This is known as a *greedy improvement algorithm* because it only makes changes to the solutions that improve the cost. This process is repeated to the point at which a certain stopping criterion is met. The stopping criterion is met when no further changes can be found. The solution procedure is illustrated in Figure 32.9.

Route Construction Heuristic to Find Initial Solution. Route construction heuristics select visits sequentially until a feasible solution is found. Sequential methods construct one route at a time, whereas parallel methods build several routes simultaneously. The insertion algorithm (Jaw et al. 1986), one of the most commonly used route construction heuristics in vehicle routing problems, is used to route emergency evacuation buses.

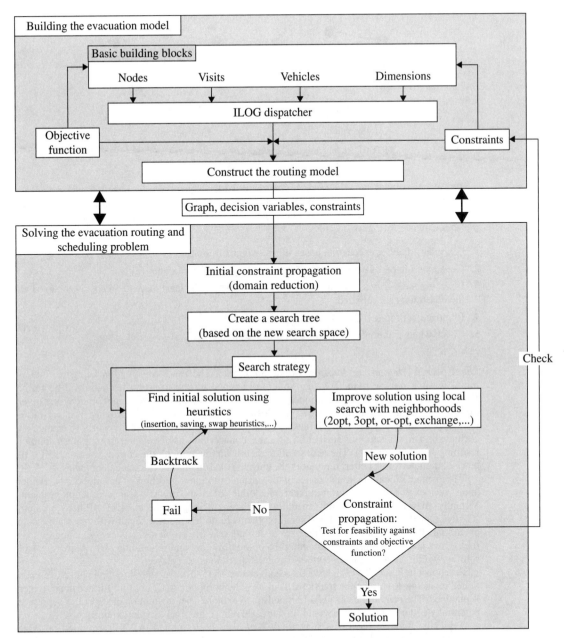

FIGURE 32.9 Framework for integrating constraint programming and local search techniques for MDTCPD-VRP emergency evacuation.

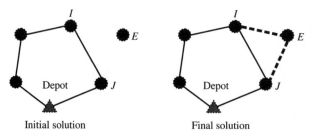

FIGURE 32.10 Insertion heuristic.

INSERTION HEURISTIC. The insertion heuristic works sequentially by inserting evacuees E at a time into the schedule of the available vehicles V. The insertion is conducted according to the order in which the visits were created at the best possible place in terms of cost. The heuristic comprises the following steps (Figure 32.10):

1. Consider a fleet of evacuation buses to be assigned to routes.
2. Construct a list L of unassigned pickups and drop-offs for evacuees (E, S).
3. Insert evacuees E in a route at a feasible position where the least increase in cost is achieved and the constraints are satisfied.
4. Remove the drop-off visit of the same trip S from L.
5. Check whether there are any unrouted evacuees in the list L; if yes, go to step 3.

Local Search Methods to Improve Initial Solution. The route-construction heuristic finds an initial feasible routing plan. The first solution obtained by the insertion heuristic can be further improved using local search techniques. Local search methods explore neighborhoods iteratively to improve the initial solution. To design a local search algorithm, the improving mechanism and the stopping criterion have to be defined. The improving mechanism typically works to change one attribute or a combination of attributes (e.g., arcs connecting sets of customers) of a given solution, resulting in a new solution. The new solution then is compared with the current solution, and if the new solution performs better, it replaces the current solution, and so the search continues.

The computational effort associated with various solution algorithms is an important consideration when exploring optimal routing and scheduling solutions for large-scale evacuation problems. Virtually all vehicle routing and scheduling problems belong to the class of NP-hard problems (Garey et al. 1979) in which solving small problems to optimality is difficult with reasonable computational effort. Consequently, in the case of solving real-life problems such as evacuation by mass transit, one should not insist on obtaining the optimal routing and scheduling plan. A good feasible solution within a reasonable amount of computation time is acceptable.

As shown in Figure 32.11, a local search algorithm starts from an initial routing plan R_0 and continues replacing R with better solutions from its neighborhoods $N(R)$ until no further improvement is obtained. The central idea of neighborhoods is to explore a set of solution changes that represents potentially better solutions that would, in our case, reduce the total evacuation time.

Typically, iterative local search methods that have been applied to VRP are based on edge-exchange algorithms* (Braysy and Gendreau 2005). Logically, the difference between neighborhood algorithms depends on how many routes are involved in the improvement process and how many arcs (edges) within each route are exchanged. The improvement mechanism assumes that the cost is proportional to the route cost (e.g., length, travel time or combination of both) and that, therefore, a shorter routing plan is less costly (e.g., a tour that minimizes the total evacuation time), and consequently,

---

*Edge-exchange neighborhoods: set of tours that can be obtained from an initial tour by replacing a set of k of its edges by another set of k edges (Braysy and Gendreau 2005).

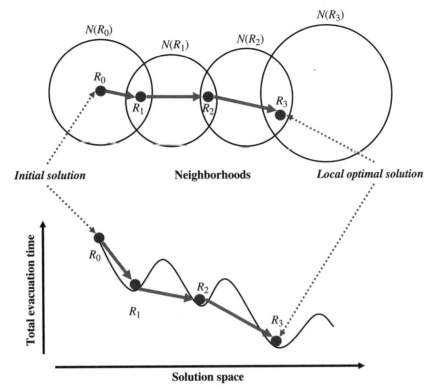

FIGURE 32.11 Local search concept.

the initial solution is improved. The local search methods proposed by Braysy and Gendreau (2005) are used in this research; this includes 2-opt neighborhood, or-opt neighborhood, relocate neighborhood, cross neighborhood, and exchange neighborhood.

32.4 MULTIMODAL EVACUATION FRAMEWORK

After the terrorist attacks of September 11, 2001, and the massive distractions of the 2004 and 2005 hurricane seasons, U.S. attention has focused on harnessing the capacity of transportation systems to efficiently respond to emergencies and to evacuate the population in a timely and safe manner. Lately, it has been recognized that public transportation systems play a significant role in emergency evacuation. The history of disasters in Canada is not comparable with that of the United States; however, the latter has contributed to shaping the emergency management systems in Canada and worldwide. Despite the fact that each disaster or emergency event has its sole circumstances, lessons learned from major disasters have shaped the emergency management systems in Canada and the United States (Litman 2006; Lindsay 2009). Numerous lessons were learned after studying the consequences of each disaster; therefore, a special committee was formed by the Transportation Research Board (TRB) to examine how the potentially critical role of public transit can be fulfilled in an emergency evacuation, and the committee produced a special report entitled, "The Role of Transit in Emergency Evacuation." The committee reviewed the literature and analyzed the emergency-response and evacuation plans of the 38 largest urbanized areas in the United States (TRB 2008).

This section summarizes the relevant lessons and findings of this report to build on these lessons while integrating multiple modes in the presented framework. It then highlights the modes that potentially can be incorporated into emergency evacuation plans. It ends with a description of a multimodal evacuation framework that integrates the two optimization platforms discussed in Sections 32.2 and 32.3.

32.4.1   Role of Transit in Emergency Evacuation: Lessons Learned

Numerous factors affect transit systems in emergency evacuation; these factors include internal factors (e.g., transit system, emergency nature, and evacuee behavior) and external factors (e.g., urban area characteristics). These factors are charted in Figure 32.12, and the interaction between them is highlighted. As shown in the figure, these factors are interrelated; for example, in *dense urban areas* that have *connected* and *large transit systems*, people typically are more *trained* and *knowledgeable* of emergency evacuation plans, resulting in less chaos and disruption to the system in cases of emergency, as opposed to *limited transit systems* in *small urban areas* that can serve small portion of the population, leaving more transit-dependent people under risk.

Considering the preceding structure from external and internal factors that shapes the role of public transit in emergency evacuation, the examination and review of the plans of the 38 largest urbanized areas in the United States reveal the following conclusions:

- Most evacuation plans were found to be inadequate to manage catastrophic events.
- Lack of incorporating all available modes of transportation, including transit, in evacuation plans is found to be a major concern.
- Lack of identifying and accommodating special-needs populations and those who are transit-dependent is a major weakness of most evacuation plans.
- Capacity and congestion issues owing to the demand surge in automobile traffic precluded the travel of transit vehicles on urban areas highways.

Therefore, the transportation infrastructure (including transit infrastructure and automobile infrastructure) may come to a halt in the case of improper planning for emergency evacuation.

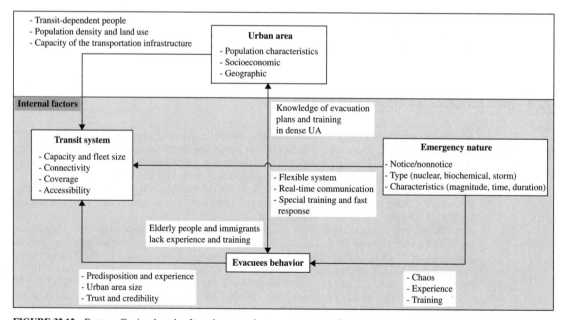

FIGURE 32.12   Factors affecting the role of transit systems in emergency evacuation.

34.4.2 Modes in Evacuation

Automobiles have been the dominant mode of emergency evacuation in most cities and urban areas. In the United States, only a small number of states have incorporated multiple modes of transportation into their emergency evacuation plans (TRB 2008). Also, the lack of coordination between transit agencies and traffic operators may further hinder the potential for integrating multiple modes in emergency evacuation.

Optimized multimodal evacuation is still largely missing in most emergency evacuation plans. However, using the readily available transit capacity can improve the evacuation process significantly and lessen evacuation casualties. During evacuation events, available buses (e.g., public transit, commuter service, and school buses) in the area can be used and routed to shuttle evacuees to safety, a process that can be explicitly optimized, as presented in section 32.3. A single bus-only highway lane can carry up to six times the passengers as a passenger-car-only highway lane (Litman 2006). In addition, standard buses, LRT, and rapid transit (subway or metro) can carry up to 5,400, 28,800, and 72,000 spaces/hour, respectively (Vuchic 2005). Therefore, transit services afford huge capacity that can reduce the clearance time significantly in the case of evacuation. In addition, many evacuees, when a disaster hits, do not have access to their vehicles and hence must be evacuated by transit. Transit has played a vital role in the nonnotice evacuation of both New York City and Washington, DC, that followed the events of the September 11, 2001.

32.4.3 Multimodal Evacuation Plan

The system presented attempts to optimize the use of multiple modes during emergency evacuation. Figure 32.13 illustrates the steps toward achieving this goal by using both OSTE and MDTDPD-VRP

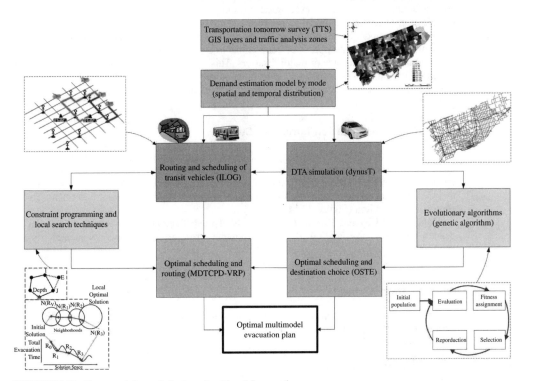

FIGURE 32.13 Framework for optimization of multimodal evacuation.

in one platform. The system starts by estimating the evacuation demand using a regional demand survey (e.g., TTS) and a representation for the traffic analysis zones. The output of this is a representation of the spatial and temporal distribution of the population and its modes of travel. OSTE plans then are generated for the vehicular demand using genetic algorithms as a global optimization technique and a dynamic traffic assignment tool. OSTE generates an optimal evacuation schedule, optimal destination choices (if requested), and optimal routes to destinations. It also produces link travel times that are used as input for the optimal routing and scheduling of transit vehicles. The routing and scheduling of transit vehicles then are solved using constraint programming. The auto-OSTE plan and the transit optimal routing and scheduling plan are finally combined for dissemination to evacuees. It is to be noted that the current platform does not loop back from the transit-assignment component to the traffic-assignment component, a potentially worthy step that is deferred to future research. However, while extracting the travel times from the DTA model to form the input to the MDTCPD-VRP, the most congested travel times are used as a worst case for buses while traveling through the network. Although it might overestimate the travel times for buses, it may compensate for the uncertainty of travel times for such heavily used transit vehicles; that is, if the process errs, it does so on the conservative side. The next section describes the demand estimation model from a regional travel survey.

32.5   EMERGENCY EVACUATION DEMAND ESTIMATION FROM REGIONAL TRAVEL SURVEY

An accurate description of the spatial distribution of the population, by time of day and mode of travel, is essential to realistically model major population evacuation. Unlike day-to-day travel patterns, planning for emergency evacuation has unique demand distribution that should be carefully examined in order for the model to produce accurate evacuation performance measures. Typically, travel demand modeling in evacuation is based on postsurvey data after disasters or based on trip generation and participation rates of geographic areas. However, many cities lack such information owing to the rare occurrence of major disasters. This section presents a demand estimation model in which not only the value of the evacuation demand per traffic analysis zone (TAZ) is determined but also the spatiotemporal distribution of demand is estimated.

32.5.1   Data Source: The Transportation Tomorrow Survey

The Transportation Tomorrow Survey (TTS) is the largest and most comprehensive travel survey in Canada and is conducted once every 5 years. The TTS covers 5 percent of all households in the greater Toronto area (GTA) and surrounding areas selected at random. The demand estimation model includes the entire GTA, which is divided into six regions, namely, Toronto, Durham, York, Peel, Halton, and Hamilton (Figure 32.14).

Data reported by the TTS include two sets of data: demographic characteristics such as age, gender, household size, and dwelling type, to name a few, and travel patterns such as trip purpose and mode of travel.

32.5.2   Demand Estimation Method

The detailed records of each person in each household are tracked during the course of a 24-hour period. The following attributes are used to construct a query to extract the demand data each half hour for the entire day.

- Household sample number
- Person number within the household

Greater Toronto Area

FIGURE 32.14 GTA regions for year 2001 (*Source:* http://www.jpint .utoronto.ca/gta01/GTA.html).

- Start time of the trip (24-hour clock) 400–2800 (4 a.m. on the trip day to 4 a.m. the next day)
- Primary mode of the trip*
- GTA zone of trip destination (1996 GTA zones)
- GTA zone of the household (1996 GTA zones)

The estimation process includes the following steps for each time interval:

- Group people according to the start time of their trip.
- Identify people who drive, and identify their home location (zone). This results in an O-D matrix in which origins are the current location of people who drive, and their default destinations in cases of evacuation are their homes (unless another safe destination is suggested by the destination-choice module).
- Identify people who do not drive, and identify their home location (zone). This results in an O-D matrix in which origins are the current location of nonauto people, and their default destinations in cases of evacuation are their homes.
- Identify people who returned home.
- Identify people who have not yet made a trip.
- Identify people who are at their homes by combining the preceding two steps.

---

*In this application, modes are categorized to *Drive* (auto driver) and *NonDrive* modes (auto passenger), local transit, GO train, walk and cycle, and other.

The output of the demand estimation process is the spatial and temporal characteristics of the trips that are made by each of the three classes of people (travelers using auto, travelers using other modes, and travelers who are at home still or returned home). For those who are traveling when the crisis hits, their home locations are known and assumed to be their default destinations in absence of a better destination choice. Ultimately, this method identifies where people are located by time of day and mode of travel.

32.5.3   City of Toronto Demand Estimation Results

The output of the demand estimation method is illustrated in Figure 32.15, where the temporal distribution of the total number of people present in the city of Toronto is plotted for a 24-hour clock (400–2800) that starts at 4 a.m. on the trip day and goes to 4 a.m. the next day. As shown in the figure, three groups of people are considered in case of emergency evacuation: people who commute with the *Drive* mode, people who commute with the *NonDrive* mode, and people *Resident* at home. The plots show the people by mode within the geographic bounds of the city of Toronto at any instant in time regardless of whether they are traveling within the city, heading out of the city, heading into the city, or simply present at home. In total, the number of people in Toronto peaks at 108 percent of the city's population (residents at 4:00 a.m.). This increase is attributed to the high concentration of economic activities in the city of Toronto and particularly the business and financial district. It is also interesting to show a wide peak-activity period that starts from 7:00 a.m. and ends at 6:00 p.m. The peak demand is found to be 2.56 million people and occurs at the 11:30 a.m. to noon interval, which constitutes the worst-case scenario for evacuating the city of Toronto. It is important to note that total trips in the GTA as processed by the demand estimation process sum up to the total population in the GTA of 5.368 million, 2.56 million of which are present in Toronto around noon, and the rest are outside the bounds of the city (DMG 2003). An example for the spatiotemporal distribution of evacuation demand in the city of Toronto is shown in Figure 32.16 for the *Drive* mode. Only two time intervals were selected for the sake of illustration—6:00 a.m. and 12:00 noon.

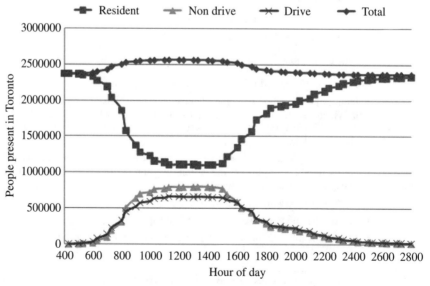

FIGURE 32.15   Temporal distribution of evacuees in the city of Toronto.

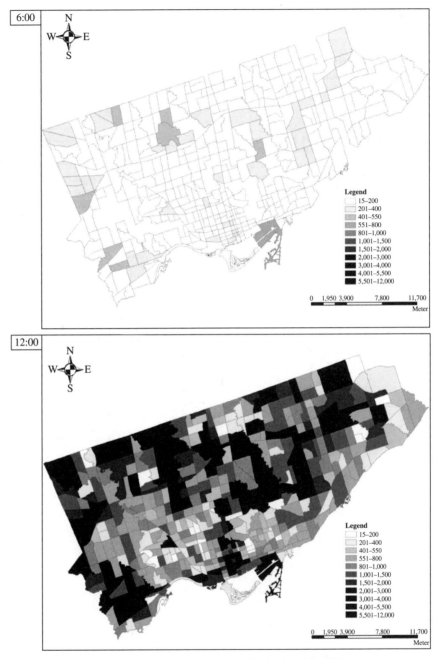

FIGURE 32.16   Drive population spatial and temporal distribution in the city of Toronto.

32.6   LARGE-SCALE APPLICATION: EVACUATION OF THE CITY OF TORONTO

This section documents the effort of applying the OSTE model presented in section 32.2 and the transit shuttling model presented in section 32.3. The methodologies outlined in preceding sections are applied to develop a large-scale evacuation of the city of Toronto (Figure 32.17).

32.6.1   Application Context and Analysis Scope

The city of Toronto is located in the center of the greater Toronto area (GTA). The city of Toronto is bounded on the south by Lake Ontario, on the west by the city of Mississauga and the city of Brampton, on the east by the Durham Region, and on the north by the York Region. Toronto is a unique city, and more so is Toronto's waterfront. It is the oldest, densest, most diverse area in the region. It contains one of the highest concentrations of economic activity in the country. The key economic activity in downtown Toronto is the high-value-added office sector, particularly in financial and business services.

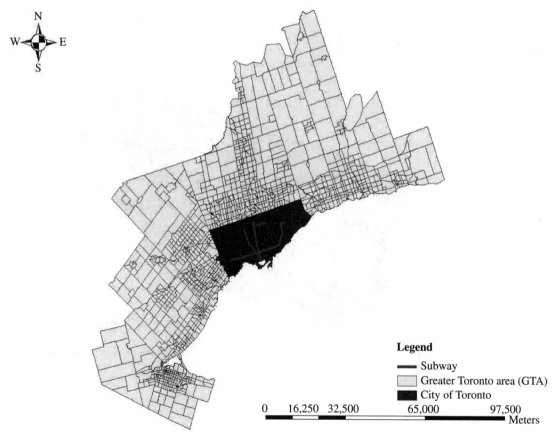

FIGURE 32.17   The GTA traffic analysis zones.

It is clear that the frequency and impact of natural/human-made disasters are increasing worldwide, and Canada (and specifically the city of Toronto) is not immune to this trend. Among the most frequent disasters are earthquakes, hurricanes, tsunamis, forest fires, tornados, ice storms, and severe rain storms. In 1998, the largest Canadian disaster, an ice storm, struck Quebec and Ontario, and more than 5 million people were affected by at least one power outage.

In 2005, Toronto and the surrounding area were hit with a severe rainstorm and tornadoes that led to the second-largest insurance payout in Canada's history (Insurance Bureau of Canada 2007[*]). Massive numbers of homes were damaged in the areas of Kitchener, Guelph, and possibly Toronto. The damage spread from Stratford (20 kilometers west of Kitchener) to Peterborough and along the Georgian Bay near Collingwood.

32.6.2 Supply Modeling

The GTA and City of Toronto Roadway Networks.   In this effort, the GTA and the city of Toronto road networks are developed in DynusT (Dynamic Urban System in Transportation), a mesoscopic DTA model that is well suited for dynamic traffic simulation and assignment on a regional scale. Mesoscopic models simulate the movement of individual vehicles in the transportation network but move them in groups according to fundamental traffic theory diagrams. Mesoscopic models achieve the compromise between microscopic and macroscopic models; they don't suffer from the curse of dimensionality with an increase in problem/population size, as microscopic models do, and they also provide more insightful analysis and detailed results than macroscopic models.

In this implementation, data were obtained from the EMME/2 planning model. Although this is a rich and comprehensive planning model, some essential data to develop the GTA and city of Toronto models at the mesoscopic level were not available. Therefore, the following tasks were carried out to develop the simulation model.

1. Traffic analysis zones (TAZs) were created using the information available from ArcMap. The vertices of each zone in the GTA were exported from the TAZ layer, and then a series of programs was created to automate the importation process. The process resulted in 160,324 TAZ vertices imported into the simulation model. Special attention was given to keep the mapping consistent between the TAZ zoning system and the zone numbers created in the simulation model.

2. Unlike the centroid-connector method of generating traffic planning models, existing roads (links) have to be defined to generate traffic in the mesoscopic simulation model. To the best of our knowledge and familiarity with the network and based on the available data, generation links were created to release traffic from parking lots, parking garages, residential areas, etc. Special attention was given to model generational links at intersections to avoid unrealistic congestion at the beginning or the end of generation links. In addition, the traffic generated from these links is distributed across the road segments proportionally to the capacity and length of these road segments.

3. The freeway system across the GTA is carefully modeled to not include any generation links or destination nodes. On- and off-ramps link categories are created to realistically model the movements at the entrances and exits of the freeways.

4. Traffic flow models have to be identified in the mesoscopic model, as described previously. A typical Greensheild traffic flow model is used while constructing the speed-density function for different road segments according to their category (e.g., highway, freeway, collector, on- and off-ramps, etc.). The uncongested portion of the traffic fundamental diagrams was set to approximately match the typical link-volume delay function for different road segments provided by EMME/2.

5. Traffic signals in the model are designed as fully actuated to automatically handle the anticipated traffic fluctuations in case of emergency evacuation. In reality, approximately 25 percent of traffic

---

[*]Source: www.ibc.ca/en/Natural_Disasters/.

light in Toronto use adaptive control (SCOOT). The rest are either actuated or pretimed. The pretimed signals would require evacuation-specific timing plans in reality, which was approximated by fully actuated operation in the model.

As discussed in section 32.6.1, the evacuation area includes the city of Toronto. Therefore, a replica of the GTA model has been developed to include only the city of Toronto. The GTA model was developed to estimate the noncomplaint traffic in emergency evacuation and to account for intercity trips made to/from the city of Toronto, as discussed in section 32.6.3. The GTA and the city of Toronto simulation models are illustrated in Figures 32.18 and 32.19, respectively.

Transit Infrastructure: The Toronto Transit Commission (TTS) System Fleet. The TTC fleet consists of bus, light-rail transit (streetcar), and rapid-rail transit services. The fleet includes about 1,500 vehicles (1,320 buses + 180 streetcars) during the a.m. peak period (TTC 2001). Rapid transit service includes 4 subway (metro) lines (Bloor–Danforth, Yonge–University Spadina, Sheppard, and Scarborough RT) that cover a large area of the city of Toronto. The locations of subway stops are extracted from the transportation network and placed on the subway lines, as illustrated in Figure 32.20. The rapid transit fleet characteristics are attached to the figure (TTC 2001). Subway lines are coded as double track lines in which each station is represented by two platforms, one for each direction. Also, additional constraints are added to model first in, first out (FIFO) operation of subway trains while departing and arriving at subway stations. It is assumed that in cases of emergency, the entire bus fleet will be available for our system to reschedule and reroute based on the needs of the evacuation process. Regular transit services will be no longer in effect. The whole bus fleet would operate as shuttles to the nearest safe zone. Streetcars are not used in this application.

FIGURE 32.18 The Greater Toronto area simulation model.

FIGURE 32.19   The city of Toronto simulation model.

32.6.3 Demand Modeling

Unlike the typical day-to-day demand patterns, in this application, the evacuation demand has the following unique characteristics:

- As discussed in section 32.5.3, the worst time for a crisis to hit is around noon. Trip start times were considered to extract the demand data. Based on the TTS, it is reported that the average auto in-vehicle travel time in the GTA is 12 minutes, and the average transit in-vehicle travel time is 25 minutes (Roorda et al. 2006). Therefore, in this analysis, the trip start times were discretized to 30-minute intervals.
- The demand estimation process resulted in three main categories of trips that are modeled in evacuating the city of Toronto. Table 32.5 shows the breakdown of trips according to the location of evacuees and their mode of transport at the onset of the evacuation event, that is, at noon.

It is interesting to show that the total number of evacuation trips at noon (2.56 million) is greater than the population of the city of Toronto (2.36 million). This increase (8.2 percent) is attributed to the high concentration of economic activities in the city of Toronto, particularly in the waterfront area, where key financial, office sector, and business services are located. It is also important to note that this increase is the net difference between internal-external and external-internal trips made from/to the city of Toronto.

In the case of evacuation, it may be harder to persuade drivers to abandon their cars and take any other mode. Also, transit users and nondrivers are captive to transit modes because their choices are

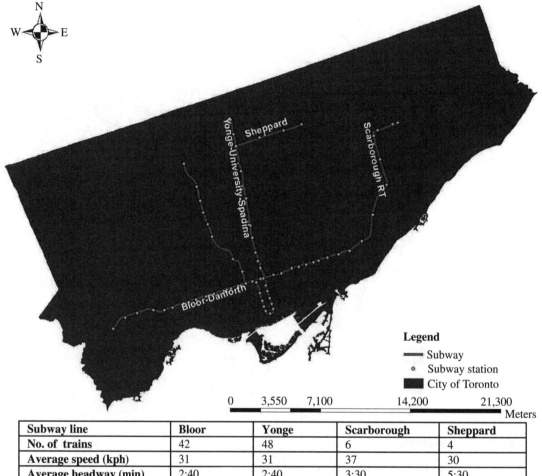

Subway line	Bloor	Yonge	Scarborough	Sheppard
No. of trains	42	48	6	4
Average speed (kph)	31	31	37	30
Average headway (min)	2:40	2:40	3:30	5:30
Average runtime (min)	110	118	21	22
Line distance (km)	52.5	60.5	13	11
No. of cars/train	6	6	4	4
No. of stations/line	31	32	6	5

FIGURE 32.20   The TTC rapid transit lines, stops, and fleet characteristics.

limited. Therefore, it seems practical to assume that evacuees will use the same mode of transport while commuting to city of Toronto; that is, the not-at-home drivers will evacuate using their cars and the not-at-home nondrivers will evacuate using transit modes. Available transit modes in this implementation include the rapid transit system (subway lines) and surface street buses used as shuttles. For at-home evacuees, trips are assigned to modes based on the mode split reported by the TTS for the trips made by residents of the city of Toronto (DMG 2003). The overall results are 1,216,886 evacuation trips by auto and 1,344,942 evacuation trips by transit.

Representation of Noncompliant Traffic.   In emergency evacuation in general and hurricane evacuations in particular, considerable percentages of evacuees sought their homes and relatives first. In a

TABLE 32.5 Evacuation Demand in Each Zone in City of Toronto by Mode at Noon

	Trip	No. of Trips	Description	Spatial Distribution of Trips
At Home Evacuees		1,107,353	Number of people located at their homes in the subject zone at the onset of the evacuation if occurs at noon. These trips include trips that are not yet started and trips that started and ended at home at the time of the evacuation event.	
Not-At-Tome Evacuees	Transit	791,264	Number of people starting from any zone other than their homes at the onset of the evacuation and ending at the subject zone in the 11:30-noon interval using *NonDrive* mode. These trips include trips made by the following modes: passenger, transit, walking, cycling.	
	Auto	663,209	Number of people starting from any zone other than their homes at the onset of the evacuation and ending at the subject zone in the 11:30-noon interval using *Drive* mode. These trips include trips made by auto-drivers.	

comparison of trip-generation models in hurricane evacuations, Wilmot and Mei (2004) reported that 50 to 70 percent of the destinations were found to be relatives and friends. In our implementation, evacuees are advised to go to the optimal shelter; however, deviation from the evacuation "plan" is unavoidable. Therefore, certain percentages of evacuees are assumed to not comply (noncompliant evacuees) with the optimal plan and seek their homes. The percentage of noncompliant evacuees is treated as a sensitivity parameter, that is, exogenously supplied by the modeler. This study makes no attempt to estimate this percentage on the basis of evacuee behavior, which is beyond our scope. During the demand estimation process, some of the evacuees who are not at home are directed toward their homes based on the specified percentage.

Two components form the total demand in the evacuation planning model: (1) the compliant evacuation demand (superzone demand) and (2) noncompliant demand—as follows:

$$\text{Total demand } D = \text{evacuation demand (superzone demand)} + \text{noncompliant demand} \qquad (32.1)$$

To the best of our knowledge and based on the literature, evacuee compliance to guidance is rarely modeled or reported. Although it is challenging to model evacuee behavior, stated-preference surveys and postanalysis surveys for certain evacuation scenarios may be plausible avenues to model such behavior. In the absence of past evacuation surveys in Ontario and Toronto, it is assumed that 25 percent of evacuees will not comply with the provided guidance and will seek their homes first. This assumed percentage has no scientific evidence. It is only for the purpose of analysis and illustration until a better behavioral approach is available.

To represent the noncompliant demand, a dynamic user equilibrium traffic assignment is executed at the GTA level in which origins are the locations of people at the onset of the evacuation and destinations are their homes. And then an analysis is made to capture these trips at the gateways of the city of Toronto. The output of this analysis is an O-D matrix that identifies the demand pattern of people who might return to their homes instead of the advised destination through the optimized evacuation plan. Finally, a certain percentage of this demand is assumed to not comply with the plan, whereas the destination of the rest of the population is optimized.

Demand on Subway Stations.   An access distance buffer zone (1,000 m) is defined around subway lines, and evacuees within the accessible buffer zone are assumed to be carried by the subway to safe destinations. The spatial distribution of evacuees within each TAZ is important to estimate the demand within the buffer zone. In this study, a randomly uniform distribution of evacuees within each TAZ is assumed for lack of better information. Safe destinations for subway travelers are the subway terminals (e.g., Kipling, Finch subway stations). The demand from the buffer zones of the subway system is estimated to be 615,434 evacuees, which form a large portion of the transit-dependent population (54 percent). This is not surprising given the nature of the city of Toronto and the high density and transit-orientated development throughout the city, especially along the Yonge-Spadina subway line (Miller and Shalaby 2003). Evacuees then are spatially assigned to the nearest subway station. Figure 32.21 shows the distribution of evacuees at each subway station. The average walking distance to the closest station is found to be 460 meters, with the maximum demand located at King and Union subway stations.

Demand on Bus Network Stops.   Main bus stops are extracted from the EMME/2 planning model, and the TTC bus routes are provided by the University of Toronto Map Library.* Then evacuees are spatially assigned to the nearest bus stop, which resulted in an average walking distance to the nearest bus stop of 332 meters. The demand distribution at each bus stop is shown in Figure 32.22.

Estimation of TTC Initial Conditions.   In this implementation, special attention is given to estimate the initial conditions for operating the TTC system in case of emergency evacuation. As discussed in section 32.6.2, the primary modes to evacuate transit-dependent population are the rapid transit system (subway) and shuttle buses. The initial conditions for both systems are estimated based on the available data and the operational characteristics of each system.

---

*The University of Toronto Map Library (online source www.main.library.utoronto.ca).

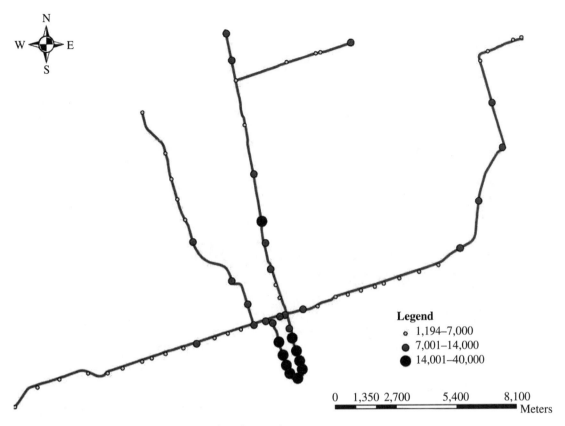

FIGURE 32.21 Distribution of evacuation demand at subway stations.

According to the reported average speed and average headway for each subway line, the locations of subway trains are identified and used as the initial condition while applying the scheduling algorithm for subway trains. In this implementation, each subway line is modeled separately, and transfers between subway lines are barred. This decision is made to minimize the chaos associated with subway operations in case of emergency evacuation. It is also worth noting that in the case of subway operation, the routing is predetermined owing to the nature of the subway tracks and the determinant backtrack locations. The problem is rather a scheduling problem that is concerned with the sequencing of subway runs.

The spatial distribution of transit shuttle buses in the TTC at the onset of evacuation is crucial to realistically model startup conditions for buses. To the best of our knowledge and the available data, the location of buses within the transportation network is not readily available. However, these data can be extracted using the output of a recent research effort at the University of Toronto, where a dynamic transit assignment model (namely, MILATRAS) has been developed. MILATRAS is capable of simulating passengers and bus schedules at transit stations (Wahba 2009).* A postanalysis method has been developed to obtain the locations of buses at any point of time through the extracted transit route records from MILATRAS. Although MILATRAS results were reported for the a.m. peak only (6:00 to 9:00 a.m.), the model was run until noon, which coincides with the onset of the worst-case

---

*Permission has been obtained from Professors Moahmed Wahba and Amer Shalaby to access MILATRAS data.

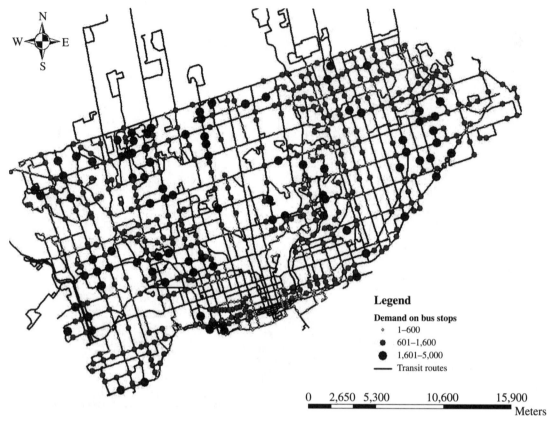

Legend

Demand on bus stops
- ∘   1–600
- ●   601–1,600
- ⬤   1,601–5,000
- —— Transit routes

0   2,650   5,300   10,600   15,900

Meters

FIGURE 32.22   Distribution of evacuation demand at bus stations.

evacuation scenario of interest. Therefore, a series of programs was developed to extract the bus location (e.g., at transit stops) around noon for all the TTS bus routes (inbound and outbound bus routes). Figure 32.23 shows the spatial distribution of the TTC fleet in the city of Toronto at noon at the onset of the evacuation scenario. It is found that the maximum number of transit vehicles per route is along the 504 King streetcar route. This is not surprising because the 504 King streetcar route is the most congested transit route in the TTC.

32.6.4   Results and Discussion

OSTE of the City of Toronto.

Genetic Optimization Process.   The output of the genetic-based optimization process is a scheduling vector μ that minimizes the prespecified objective function (see section 32.2.5) with a modified network representation to model the destination-choice problem (superzone). The OSTE framework, integrated with the GENOTRANS engine, outputs the optimized scheduling vector and the detailed routing plan for the evacuation scenario.

Designing genetic-based optimization is problem-specific and requires extensive testing to select among the many variables, as discussed in section 32.2.5. Table 32.6 shows the GA methods and parameter values chosen based on these tests.

FIGURE 32.23   Spatial distribution of the TTC fleet at the onset of the evacuation scenario.

As discussed previously, genetic optimization approaches require careful identification of the set of parameters that best replicate the specific problem at hand. A good design for the GA parameters is the one that captures the fundamental properties of genetic optimization, which are the survival of the fittest and evolution of the population from one generation (iteration) to the next. Ultimately, the *average* fitness of a population of chromosomes will approach the *optimal* fitness function as the number of generations goes to infinity. While it is practically infeasible to continue the generic optimization procedure indefinitely, typically, a stopping criterion is specified to bring the optimization process to an end. In this implementation, the optimization process is terminated if the value of the fitness function does not change more than 1 percent for 25 generations. Figure 32.24 illustrates the evolution of the *minimum* fitness function value with the number of generation across multiple demes.

In general, the unconstrained genetic-based optimization procedure is generic because it does not restrict the scheduling vector values except for the feasibility requirements. This implies that the relationship between the optimization parameters (genes of each chromosome) is not determined a priori but rather by an output of the optimization process. In this application, in order to achieve the feasibility requirements, two constraints must be met: (1) the nonnegativity constraint for all the

TABLE 32.6 Genetic Algorithm Methods and Parameter Values

GA Design Element	Method or Chosen Value	Description and Chosen Values
Population Size	Population Size ≥ Chromosome Size	75
Initial Population	Random	Each gene in the decision variable vector is encoded as a real number with a range between 0 and 1. The chromosome size (number of genes) represents the number of departure time intervals.
Selection	Linear Ranking Selection	The individuals in a population of n chromosomes are ranked in descending order of fitness, with a rank of n points given to the best individual and a rank of 1 given to the worst individual.
Recombination	Real Blend Crossover	Exchange the genetic information between the population individuals; it acts on two parents in the intermediate population by combining their traits to form two new offsprings. This is not applied to all chromosomes but depends on the probability Pc defined by the crossover rate. Pc = 0.9 (90 percent of the time the crossover operator is applied).
Mutation	Real Gaussian Mutation	Randomly chosen genes are mutated with a range of ± 5%
Stopping Criteria	Number of generation with no decrease in the fitness function value	25
Parallel GA	Fully Connected Topology	Number of demes: 3 Migration policy: good migrants replace bad individuals Migration rate (the number of individuals to migrate): 15% of the population

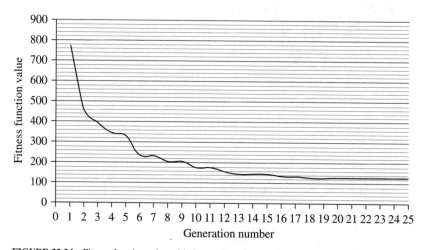

FIGURE 32.24 Fitness function value with the number of generations.

parameters (gene values) and (2) the summation of all the parameters must be 1 (this is to ensure that 100 percent of evacuees are released in the transportation network).

An advantage of using a simulation-optimization approach is the explicit representation of network congestion and capacity constraints, an essential matter in emergency evacuation. Therefore, relatively poor scheduling and destination optimization parameters will lead to more congestion and result in poor fitness function values. This is clearly captured in Figure 32.25, where the best and

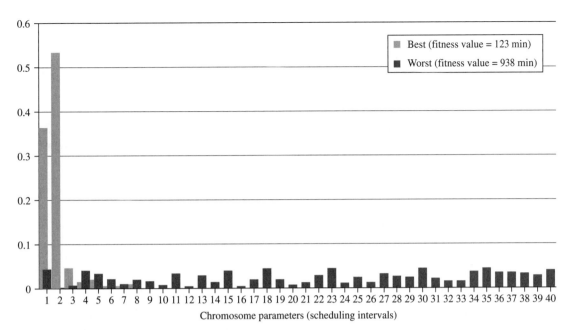

FIGURE 32.25   Best vs worst chromosome fitness values.

worst chromosomes of the GA evolution are plotted with the associated fitness function value. A significant difference in the scheduling vector parameter values is shown with corresponding significant values in the fitness function (123 versus 938 minutes). When the scheduling vector is spread across the whole evacuation horizon, it results in the worst fitness value owing to the fact that the waiting time of evacuees increases significantly. On the other hand, the set of parameters with the best fitness (lowest total system evacuation time) demonstrates a good compromise between in-vehicle travel time and waiting times at the origins.

Traffic Assignment Outputs.   The output of the genetic optimization process (the optimal scheduling vector corresponding to the best-chosen chromosome) is evaluated using the traffic assignment model to produce the detailed routing plan and the measures of effectiveness. The mesoscopic representation of the traffic assignment simulation model provides sufficient details for the analysis of departure time, destination choice (shelters), and routing plan for each vehicle. An important factor in planning for emergency evacuations is the status of the evacuees in/out of the network with the evolution of the evacuation plan. In this implementation, special attention is paid to study the dynamic interaction between the loading and evacuation curves and the area between them. Three scenarios are evaluated: OSTE, OTE, and SE. Each scenario examines certain levels of integration for different evacuation scenarios. While OSTE integrates evacuation scheduling and destination choice optimization and OTE examines the effectiveness of evacuation scheduling only, SE, on the other hand, presents the do-nothing scenario, where evacuees are rushed to the transportation network without a preannounced schedule. In fact, the SE evacuation scenario is evaluated twice, once with destination choice optimization (SE-DC) and once without destination choice optimization (SE). The latter replicates the worst-case scenario, where evacuees immediately seek their preferred destination, which is not necessarily the optimal one.

The optimal loading and evacuation curves are shown in Figure 32.26 for the four scenarios (OSTE, OTE, SE, and SE-DC). The following measures of effectiveness are extracted to judge the efficiency of evacuation strategies when synergized compared with the base-case scenario: average waiting time, average travel time, average total system evacuation time, average trip distance, network clearance time, and average stop time.

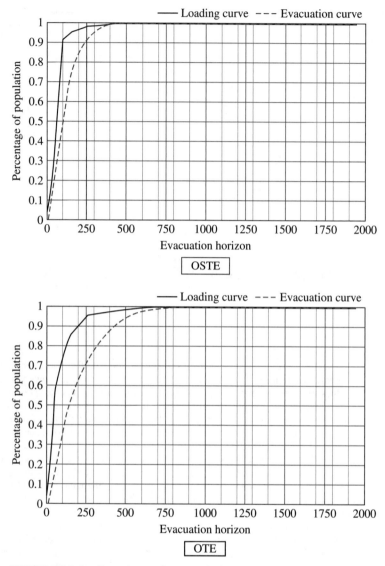

FIGURE 32.26 Loading and evacuation curves for four strategies.

It is clearly shown in Table 32.7 that the SE strategy performs the worst because it results in the longest network clearance time (NCT, end-of-evacuation curve) and the most congested travel times (area between the loading and evacuation curves). This is not surprising given the demand surge in emergency evacuation. Simultaneous evacuation with destination optimization can reduce in-vehicle travel time but not to the level that any scheduling strategy can achieve; that is, OTE and OSTE always result in less in-vehicle travel times compared with SE and SE-DC. The average stop time (time where vehicles are caught in congestion) is the largest in the cases of SE and SE-DC. In terms of NCT, OSTE performs the best; however, SE-DC performs slightly better than OTE, given the fact the OTE explores the optimal scheduling curve so as to minimize the total system evacuation time, which typically results in holding evacuees from being rushed to the transportation network (in this

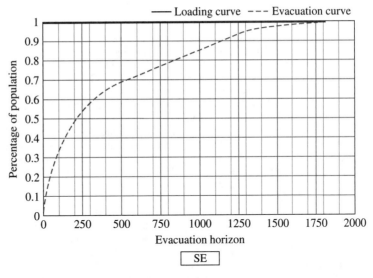

SE

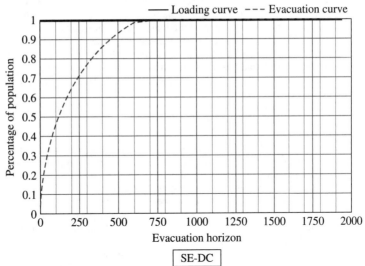

SE-DC

FIGURE 32.26 (*Continued*)

TABLE 32.7 Comparative Analysis of Evacuation Strategies

Scenario	Average Waiting Time (min)	Average In-Vehicle Travel Time (min)	Average Total System Evacuation Time (min)	Average Trip Distance (km)	Network Clearance Time (min)	Average Stop Time (min)
SE	0	412	412	11	1815	380
SE-DC	0	175	175	11	800	148
OTE	82	112	194	14	940	88
OSTE	65	50	115	12	445	33

case up to 82 minutes). Also, SE-DC cuts down travel distances significantly owing to selecting more accessible destinations. It is to be noted, on the other hand, that network stability is another performance factor to consider. SE and SE-DC may result in a network (infrastructure) that has no further wiggle room and potentially can come to gridlock in the case of further panic and/or secondary events (i.e., possibly less stable). Stability, however, is not explicitly tested in this research. This conclusion is hypothesized based on observing that SE and SE-DC have the highest stop time; that is, traffic is more in a stop-and-go condition.

It is also obvious that OSTE outperforms all the other strategies because it synergizes the scheduling and destination choice in one-shot optimization. The improvements are remarkably vivid, especially in the NCT and average stop time. This reflects the essence of a *wise* evacuation strategy that holds the evacuees up to a point that if they are released in the network and sought the dynamic optimal destination, they would clear the network promptly with the minimum stopping time and encounter less congestion, and this also would result in less overall network clearance time. An order-of-magnitude savings in NCT (75 percent), total system evacuation time (72 percent), average stop time (92 percent), and in-vehicle travel time (89 percent) are reported in comparison with the do-nothing case. This means that in OSTE, the network can be cleared *four* times faster than in the do-nothing strategy, evacuees travel *eight* times less than in the do-nothing strategy, evacuees stop *eleven* times less than in the do-nothing strategy, and finally, the total system evacuation time can be reduced to one-fourth.

Optimal Routing and Scheduling of Transit Vehicles.   In general, transit services are designed based on the seating and standing capacity of transit vehicles; however, in emergency evacuation, evacuees are expected to tolerate more crowding and make the best use of any available space in the transit vehicle. Therefore, it is plausible to consider the crush load/capacity for a transit vehicle when planning for emergency situations.[*] For a transit bus, it is assumed that one bus can carry at most 90 passengers;[†] for a subway vehicle, it is assumed to carry at most 330 passengers; that is, a subway train with six vehicle can carry up to 2,000 passengers. Dwell times that reflect the physical and operating characteristics of transit units are calculated at transit stops (Vuchic 2005). It is to be noted that dwell times are mode-dependent; therefore, dwell times for buses are differed from dwell times for subway trains.

The output of the MDTCPD-VRP is the optimal routing and scheduling of each transit vehicle. In addition to optimized routing plans, the model provides extensive vehicle-by-vehicle detailed output for each scenario, which is beneficial in many ways. The analyst can examine the scheduling of transit vehicles through the reported data at the transit stop level, individual bus level, or route level. For instance, at the stop level, the analyst can identify stop ID, bus arrival times, departure times, onboard passengers on arrival, onboard passengers on departure, alighting passengers, and boarding passengers. At the vehicle level, the analyst can observe and assess the fleet capacity requirements by examining vehicle ID, total number of passengers transported by that vehicle, travel distance, and travel time. In addition, at the route level, the analyst can construct the route that a given transit vehicle takes by examining vehicle ID, first-stop ID, sequence of nodes that construct a route, pickup points along that route, and destination stop ID.

Subway.   As discussed in section 32.6.3, subway trains are scheduled to transport evacuees who are within the buffer zone to safe shelters (subway terminals). The model generates the optimal scheduling and timetable for each train. It should be noted that a train does not necessarily stop at each station; it rather picks up evacuees from selected stations to achieve certain objective functions until its capacity is reached. Typically, the demand at stations is larger than the capacity of a train; therefore, trains travel in cycles until all the demand is exhausted.

---

[*]Crush load (maximum people per vehicle) typically loads above 150 percent of a bus's seating capacity. Such loads are unacceptable to regular day-to-day passengers. In typical day-to-day operation of buses in rush hour, crush loads might preclude circulation of passengers at intermediate stops and therefore result in delay and reduce the overall vehicle and system capacity (TCRP 2003). However, in the case of bus shuttling in an emergency evacuation situation, buses are not allowed to stop at intermediate stops; rather, typically, buses are loaded to capacity (owing to the huge surge in demand that exceeds the supply of buses) and stop only at safe shelters.

[†]The characteristics of common bus transit vehicles for the United States and Canada are reported in TCRP (2003). A typical low-floor 40-foot transit bus has passenger capacity that ranges from 55 to 70; therefore, it is plausible to assume a slightly higher crush capacity in the case of emergency situations, that is, a maximum of 90 passengers per transit bus.

Each subway line operates independently according to the operation characteristics defined above in section 32.6.2; that is, no transfers are allowed between lines and evacuees are assumed to ride the subway from one station and stay until they reach safe destinations. This assumption is made to minimize the service disruptions and the chaos associated with emergency evacuation.

The demand assignment process described in section 32.6.3 resulted in evacuees being carried by the closest subway line. The output is illustrated in Figure 32.21 where each subway station has a certain demand of evacuees that need to be transported to safe destinations. The Yonge and Bloor subway lines are the major lines that cover large geographic areas within the city of Toronto and pass through the dense core of the city; therefore, the majority of the demand is attracted to these lines. It is not surprising to find that the Yonge line has the highest number of evacuees; it carries around 373,360 evacuees, which is almost double the demand assigned to the Bloor line (176,189). This is so because of the denser land use around the Yonge line and the transit-oriented development within that area. The Scarborough and Sheppard lines carry 20,006 and 9,692 evacuees, respectively.

Two scenarios that represent two objective functions are evaluated in this implementation; the first objective minimizes the travel time (routing time) for transit vehicles (TT), whereas the second minimizes the travel time of transit vehicles and the waiting time of evacuees at the stops (TT.WT) simultaneously, thus minimizing the total system evacuation time. Each subway line is evaluated based on a matrix of measures; this includes average number of runs per transit vehicle, average in-vehicle travel time, average total in-vehicle travel time, average travel distance, and average waiting time. Table 32.8 demonstrates the Measure of Effectiveness (MOEs) for each subway line for each scenario. For illustration, Figure 32.27 demonstrates the MOE for both scenarios for the Yonge line.

Analysis of the results leads to the following conclusions:

- The Yonge line is the busiest line, where each transit unit makes, on average, four runs to transport evacuees to safe destinations. Including the waiting time in the objective function (as shown in the TT.WT scenario) has evened out the average in-vehicle travel time. This is clearly shown by the significant drop (70 percent) in the standard deviation of the average total in-vehicle travel time across transit units (see Figure 32.27). Although the average in-vehicle travel time appears to be the same in both scenarios (TT and TT.WT), the pattern across transit units is significantly

TABLE 32.8 Measures of Effectiveness of Rapid Transit Lines[*]

Line/Mode MOE	Bloor Line		Yonge Line		Scarborough Line		Sheppard Line		Shuttle Buses	
	TT	TT.WT	TT	TT.WT	TT	TT.WT	TT	TT.WT	TT	TT.WT
Average No of Runs[†]	2.1 (1.2)	2.1 (0.8)	4 (3)	4 (0.6)	1.6 (0.9)	1.6 (0.4)	1.2 (0.5)	1.2 (0.5)	5.59 (30)	5.59 (3)
Average Total In-Vehicle Travel Time (min)	96 (70)	90 (35)	228 (117)	216 (35)	47 (29)	47 (10)	19 (8)	19 (8)	114 (425)	122 (44)
Average In-Vehicle Travel Time (min)[‡]	42	43	55	56	27	29	16	16	23	24
Average Travel Distance (km)	20.5	21	27.4	27.8	15.17	16.61	6.9	6.6	21	21.2
Average Waiting Time (min)	72 (63)	40.6 (36)	178 (157)	98 (70)	27 (27)	23 (15)	12 (9)	9 (8)	914 (1844)	53 (43)

[*]Numbers between parentheses represent the standard deviation of the MOE across the number of vehicles.
[†]Average Number of Runs: The average number of runs each transit vehicle travels to serve all the evacuees.
[‡]In-vehicle travel time: the total in-vehicle travel time divided by the number of runs per vehicle.

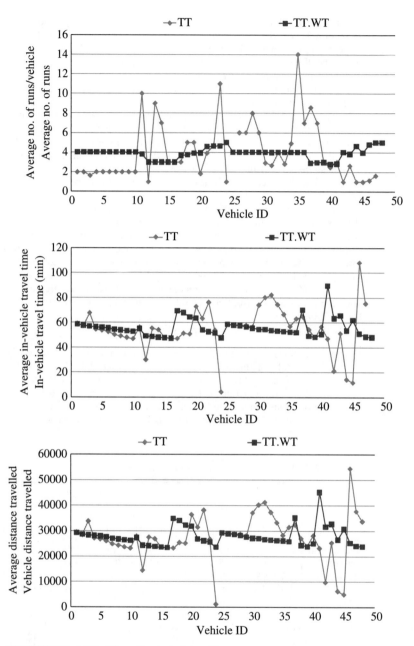

FIGURE 32.27   MOE of Yonge line for both scenarios.

different. This observation confirms balancing the workloads assigned to each transit unit/driver because it is clearly shown by the consistent pattern in travel times for the TT.WT scenario in Figure 32.27. Including the waiting time in the objective function has increased the in-vehicle travel time and the total travel distance. The average in-vehicle travel time increased by only 2 percent and the average travel distance by only 2 percent, yet the average waiting time of evacuees dropped by 45 percent. Furthermore, the standard deviation of the average waiting time for evacuees dropped by 55 percent when including the waiting time in the objective function. This allows us to conclude that the TT.WT scenario provides a more reliable service for evacuees in emergency situations. This is contrary, however, to what people and transit authorities may be inclined to expect, which is that all trains stop at all stations.

- The Bloor line is the second busiest line, where each transit unit makes, on average, two runs to transport evacuees to safe destinations. Including the waiting time in the objective function (as shown in the TT.WT scenario) has evened out the average in-vehicle travel time. This is clearly shown by the significant drop (50 percent) in the standard deviation of the average total in-vehicle travel time across transit units. Although the total in-vehicle travel time appears to be the same in both scenarios (TT and TT.WT), the pattern across transit units is different. Including the waiting time in the objective function has increased the in-vehicle travel time and the total travel distance. However, it is worth noting that the average in-vehicle travel time increased by only 2.5 percent and the average travel distance by only 2.5 percent, yet the average waiting time of evacuees dropped by 43 percent and the standard deviation of the average waiting times dropped by 43 percent.

- Unlike the Yonge and Bloor lines, Sheppard and Scarborough lines are not heavily used. However, the same conclusions are drawn when comparing scenario T.T with WT.TT by examining Table 32.8. For the Scarborough line, including the waiting time resulted in average in-vehicle travel time increasing by only 7.4 percent and average travel distance increasing by only 9.5 percent, yet the average waiting time of evacuees dropped by 15 percent. For the Sheppard line, including the waiting time resulted in the same average in-vehicle travel time and the average travel distance, yet the average waiting time of evacuees dropped by 25 percent.

Shuttle Buses. The following results demonstrate the optimal routing and scheduling for the shuttle buses for each scenario. The MDTCPD-VRP is modeled and solved using constraint propagation and optimization techniques, as discussed in section 32.3.4. ILOG Dispatcher and Solver are used to model and solve the problem, respectively (ILOG 2008). It is be noted that the VRP and its extensions (e.g., MDTCPD-VRP) are NP-hard problems; that is, solving small problems to optimality is difficult with reasonable computational effort (Garey et al. 1979). Consequently, in the case of solving large-scale problems such as evacuation by mass transit of the city of Toronto, it might take days for the algorithm to obtain a reasonable "near optimal" solution.

The model generates the optimal routing and timetable for each bus as it shuttles between pickup points and shelters. It is to be noted that buses are initially assigned to pickup points according to their location at the onset of the evacuation, as described in section 32.6.3. After the initial pickup, buses shuttle to the nearest shelter and then back into the system, but not necessarily to the same pickup points. This means that each bus seeks the *best* pickup point in order to achieve a certain *objective function*. Shuttle buses loop between the *optimal* pickup points and *optimal* safe shelters until all the demand is exhausted and finally head back to safe shelters. Owing to the unprecedented sheer size of the problem and the huge search space, a feasible initial solution is attainable with a few hours of CPU time; however, the improvement process might take up to a few days of CPU time depending on the objective function being optimized. For example, to minimize the in-vehicle travel time for buses, on a quad core machine with 8 GB of RAM, it took 5 hours to obtain an initial feasible solution and 3 days to improve the solution and completely solve the optimization problem.

Similar to the rapid transit scenarios, two scenarios that represent two objective functions are examined: minimizing total travel time (TT) and minimizing total travel time and waiting time (TT.WT). Examination of the results leads to the following conclusions:

- On average, transit shuttle buses make around six runs to transport evacuees to safe destinations. Including the waiting time in the objective function (as shown in the TT.WT scenario) has evened out the average in-vehicle travel time and the average number of runs/vehicle. This is clearly shown by the significant drop (44 versus 425) in the standard deviation of the average total in-vehicle travel time across transit units (see Figure 32.28).

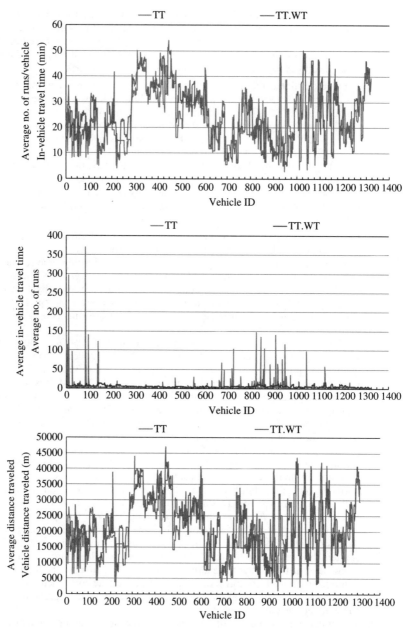

FIGURE 32.28   MOE of shuttle buses for both scenarios.

- The average in-vehicle travel time is found to be 23 minutes in the TT scenario and 24 minutes in the TT.WT scenario; although close to each other, the pattern across transit units is significantly different including the waiting time in the objective function has increased the in-vehicle travel time and the total travel distance. However, it is worth noting that the average total in-vehicle travel time increased by only 7 percent, yet the average waiting time of evacuees dropped by an order of magnitude (53 versus 914).

- Furthermore, the standard deviation of the average waiting time for evacuees dropped by an order of magnitude (43 versus 1,844) when including the waiting time in the objective function. This allows us to conclude that the TT.WT scenario provides a more reliable service for evacuees in emergency situations. It is to be noted that a careful examination of the absolute average of in-vehicle travel times and travel distance for the TT and TT.WT scenarios is required. Because these averages are calculated based on the average number of runs, which depends on the objective function (TT versus TT.WT), they already take into the account the variability in the load assigned to each vehicle.

For the sake of conciseness, a detailed routing and scheduling for one bus is shown in Figure 32.29 for scenario TT.WT. As shown in the figure, the arrival and departure times of the bus at each pickup point (visit) are extracted, and the associated travel time and distance traveled are illustrated. The optimal route is a sequence of nodes starting from a pickup point to a shelter to the next pickup point. For example, one bus (vehicle 401) is scheduled to start from the depot (location of the bus at the onset of the evacuation) at time 0, picks up 90 passengers (visit 6075), and travels along the optimal route, as shown in Figure 32.29. The vehicle then drops off the evacuees at Shelter1 and continues to pick up evacuees located at Visit 5826 and drops them off at Shelter 2. The routing and scheduling plan continues until all vehicles accomplish the assigned tasks. At the end of the evacuation, all vehicles return to the shelter (terminal), as in the case of bus 401 shown in the last row of Figure 32.29.

32.7 TOWARD A COMPLETE EVACUATION DEMAND AND SUPPLY MODELING AND MANAGEMENT PROCESS

Day-to-day travel patterns typically are modeled using conventional urban transportation planning models. These models, in a variety of ways, assess trip generation, trip distribution, mode split, and trip assignment either sequentially or concurrently (e.g., combining destination and mode choice). Under emergency evacuation scenarios, the behavior of the transportation system is vastly different from day-to-day travel patterns. Such emergency situations are characterized by sudden sheer non-discretionary demand, vulnerable supply (infrastructure), and poor system performance in the form of longer travel times, chaos, severe congestion, uncertainty, and destination vulnerability, to name a few. Moreover, travelers themselves may act differently in emergency situations compared with their regular daily travel. For instance, evacuees may more likely follow directions from officials as to which route to use instead of their habitual routes (Fu and Wilmot 2004). Also, in daily travel, trip makers decide on their trip start times to maximize their utility of travel; however, in the case of emergency, evacuees may be urged to follow an evacuation schedule or get directed to safe shelters that are not necessarily their preplanned destination choices.

Despite these unique characteristics of travel under emergency situations, a complete set of integrated tools for assessing demand generation, distribution, mode choice, destination choice, and route choice is still lacking. For instance, a tremendous body of recent literature on evacuation planning, although extremely useful, assumes that demand is known or given or focuses on one mode of evacuation (predominantly cars) with little attention to multimodal evacuation using both cars and mass transit (Sbayti and Mahmassani 2006). Although widely used transportation planning approaches comprehensively cover all aspects of travel starting from the generation of demand all the way to dynamic travel assignment and link travel times, this is not the case yet in emergency evacuation planning, which motivates us to close this gap.

32.7.1 Evacuation Planning versus Transportation Planning Models

In this section we attempt to compare evacuation planning models with the more mature and well-established transportation planning models. In a variety of ways, traditional planning models capture

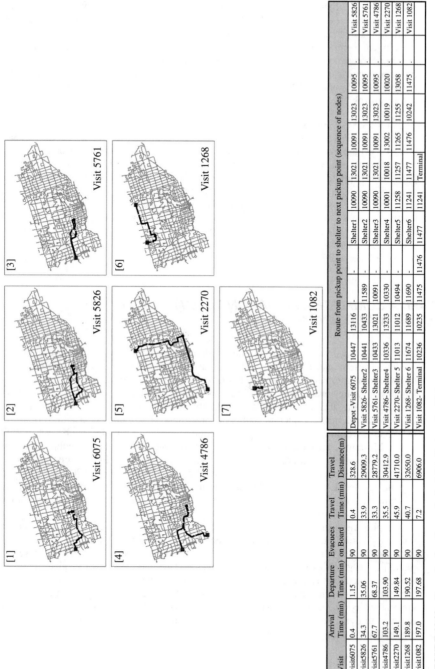

Routing maps: [1] Visit 6075, [2] Visit 5826, [3] Visit 5761, [4] Visit 4786, [5] Visit 2270, [6] Visit 1268, [7] Visit 1082.

	Route from pickup point to shelter to next pickup point (sequence of nodes)										
Depot -Visit 6075	10447	13116	-	Shelter1	10090	13021	10091	13023	10095	Visit 5826	-
Visit 5826- Shelter2	10441	10433	11589	Shelter2	10090	13021	10091	13023	10095	Visit 5761	-
Visit 5761- Shelter3	10433	13021	10091	Shelter3	10090	13021	10091	13023	10095	Visit 4786	-
Visit 4786- Shelter4	10336	13233	10330	Shelter4	10001	10018	13002	10019	10020	Visit 2270	-
Visit 2270- Shelter 5	11013	11012	10494	Shelter5	11258	11257	11265	11255	13058	Visit 1268	-
Visit 1268- Shelter 6	11674	11689	11690	Shelter6	11241	11477	11476	10242	11475	Visit 1082	-
Visit 1082- Terminal	10236	10235	11475	11477	11241	Terminal					

Visit	Arrival Time (min)	Departure Time (min)	Evacuees on Board	Travel Time (min)	Travel Distance(m)
visit6075	0.4	1.15	90	0.4	328.6
visit5826	34.3	35.06	90	33.9	29009.3
visit5761	67.7	68.37	90	33.3	28779.2
visit4786	103.2	103.90	90	35.5	30412.9
visit2270	149.1	149.84	90	45.9	41710.0
visit1268	189.8	190.52	90	40.7	32650.0
visit1082	197.0	197.68	90	7.2	6906.0

FIGURE 32.29 Example of routing and scheduling for transit vehicle.

Trip generation (trip rates, cross classification, logit models)
Trip generation (gravity models)
Mode split (utility and logit models)
Traffic assignment (UE, SO, deterministic, stochastic)

FIGURE 32.30   State-of-the-art four staging model.

four main processes: trip generation, trip distribution, mode split, and trip assignment, as shown in Figure 32.30. These planning models reasonably capture the typical daily origin-destination patterns; however, they are not applicable to emergency evacuation modeling owing to the vastly different spatiotemporal travel patterns.

Figure 32.31 illustrates our summary of the five-stage evacuation modeling process in a manner analogous to the well-known four-stage transportation planning process. A fifth layer is added to the process to account for the departure pattern during evacuation, that is, evacuation schedule.

Evacuation trip generation (historical data, hurricane post survey data, evacuees rate, day-to-day OD matrices) (Mie 2000; Mei and Wilmot 2004)
Departure pattern (assume/optimize mobilization distribution) (corps of engineers 2000; Radwan et al. 1985; Tweedie el al. 1986; Chiu Yi 2006; Sbyati and Mahmassani 2006)
Trip distribution (historical data, post survey) % friends/relatives > % hotels/motels > % shelters (Southworth 1991; Mei 2002)
Mode split (automobile evacuation is dominant)
Traffic assignment System optimal is primary used in traffic assignment (Liu et al. 2006; Chiu et al. 2006; Yuan et al. 2006; Sattayhatewa and Ran 1999; Sbayti and Mahmassani 2006; and Liu 2007)

FIGURE 32.31   State-of-the-practice evacuation planning models.

In emergency situations, the mobilization pattern of evacuees plays a paramount role in the performance of the system and in the success or failure of the evacuation process. Despite this importance, mobilization curves typically are assumed. Only in recent years have efforts emerged that focus on optimizing the mobilization pattern for evacuees so as to minimize or maximize an objective (Sbayti and Mahmassani 2006). Trip distribution typically is assumed on the basis of past emergency events. Recent research started to address the potential of optimizing evacuee destination (Chiu et al. 2006; Yuan et al. 2006). Most evacuation modeling studies focus on automobile-based evacuation. Therefore, mode split is rarely modeled or even realistically assumed and is certainly not optimized (TRB 2008).

32.7.2 Evacuation Planning Modeling Process

Despite the traditional transportation planning models and the state-of-the practice evacuation models that significantly contributed to improving the evacuation process, an integrated evacuation model is still largely missing. Most evacuation planning models deal with each layer (stage) separately, whereas evacuee decision-making stages are closely interrelated. For example, the departure time of evacuees may influence their destination choice, and their destination choice may be affected by congestion on the routes to the chosen destination. This is after assuming that the mode choice is known a priori, that is, how many evacuees own and/or have access to cars and how many are transit captives. It is indeed clear that more effort is needed to synergistically integrate some or all of these decision elements to further improve the efficiency of the evacuation process. Our approach combines evacuation scheduling (departure curve), destination choice (trip distribution), and route choice (trip assignment) into a single comprehensive solution.

In addition, an accurate representation of the spatial distribution of population, by time of day and mode of travel, is essential to realistically address major population evacuation. Unlike day-to-day travel patterns, emergency evacuation has unique nonrecurrent demand distribution that depends on the time an emergency strikes and how the population is distributed at that time. Our approach attempts to carefully assess evacuation demand based on knowledge of people's likely location at different times of the day, which is important for the model to produce accurate evacuation management measures.

Furthermore, automobile evacuation has received the most attention; consequently, multimodal evacuation is still largely missing in most emergency evacuation studies. A significant portion of the population in cities such as Toronto use public transit particularly within, toward, and out of the downtown core. This portion of the population does not have access to automobiles during the day regardless whether the own one. Using the readily available transit capacity therefore is essential not only to shuttle the transit captives to safety but also to expedite the overall evacuation process and reduce network clearance time by moving people in masses. Therefore, our approach explicitly optimizes mass-transit-based evacuation.

In summary, our approach considers the following elements to be essential to realistically plan for emergency evacuation:

- Accurate estimation of the spatial and temporal distribution of the population (*trip generation*)
- Accurate identification of available modes and captive population to certain modes (*mode split*)
- Integrated framework that accounts for various evacuation strategies such as evacuee scheduling, destination choice, and route choice simultaneously (*departure curve, trip distribution,* and *trip assignment*)
- Multimodal evacuation strategies that synergize the effect of multiple modes
- Robust and extensible optimization and solution algorithms that can tackle such multidimensional nondeterministic problem

The following paragraphs highlight the components of our system. As shown in Figure 32.32, the platform uses two optimization modules: an optimal spatiotemporal evacuation (OSTE) module

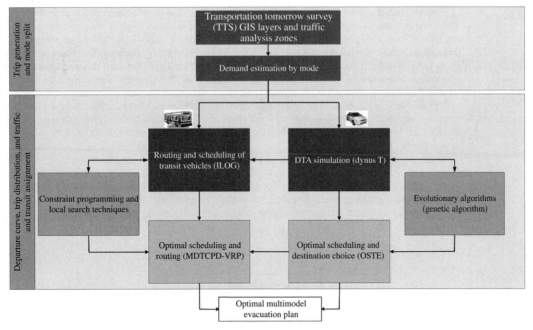

FIGURE 32.32   Emergency evacuation planning model.

for optimizing the auto-evacuation side of the problem and a multiple-depot time-constrained pickup and delivery vehicle routing module (MDTCPD-VRP) to optimize the transit-evacuation side of the problem (Abdelgawad et al. 2010) with travel time estimates from the first module. The process starts by estimating the evacuation demand using a regional demand survey and a representation for the traffic analysis zones (implementation in Toronto is discussed later in this chapter). The output of this is a representation of the spatial and temporal distribution of the population and their modes of travel, which represents the *trip generation* and *mode split* stages simultaneously.

32.7.3   Open-Loop versus Closed-Loop Evacuation Management: Future Directions

Although planning for emergency evacuation is paramount to ensure public safety from human-made and nature disasters, prediction of evacuation scenarios is challenging owing to the highly dynamic and stochastic nature of emergency situations. No matter how well an evacuation plan is scrutinized or optimized, actual evacuation patterns almost definitely will deviate from plans. Therefore, actual system behavior will need to be monitored and managed in real time, and the evacuation plan will need to be reoptimized in accordance with the measured state of the system in a rolling-horizon closed-loop control fashion. Very few studies have addressed real-time traffic management in emergency evacuation in real time (Liu et al. 2007). Numerous factors can contribute to considerable deviation of evacuation evolution from the *offline optimized* evacuation plans. For instance, potential chaotic behavior of evacuees, transportation network vulnerability to any disruption (e.g., incidents), and potential evacuee noncompliance with announced plans all can cause such deviations. Therefore, there is a strong need for closing the evacuation control loop via feeding back actual system conditions measured in real time into the online evacuation optimization engine. A closed-loop evacuation control system can guide and drive the transportation network toward optimal performance despite any unexpected disturbances or deviations from original plans.

The envisioned closed-loop evacuation control system is illustrated schematically in Figure 32.33. The system starts with disseminating an offline optimized plan from a system such as OSTE. An online monitoring system reports real-time information about the status of evacuees and the current road network conditions. The monitoring system also updates the status of the transit system in a form of current location of transit vehicles and number of passengers already evacuated. Numerous ITS technologies can be used in the process, including, for instance, Global Positioning Systems (GPS), mobile devices, automatic passenger counters, etc. Once the real-time status of the system is updated, a new time-dependent O-D matrix can be estimated on the basis on measured flows in the network (Kattan and Abdulhai 2006). The estimated demand matrix then is input into the two optimization engines (OSTE and MDTCPD-VRP) to generate new plans for the next horizon.

In large-scale applications such as evacuating a large city like Toronto, the real-time implementation of such a closed-loop approach can be challenging and will require ample computer processing power or even parallel processing. Our ongoing research is using a recently acquired high performance computing cluster at the Department of Civil Engineering of the University of Toronto that has 256 processors in an attempt to achieve evacuation optimization in a few minutes.

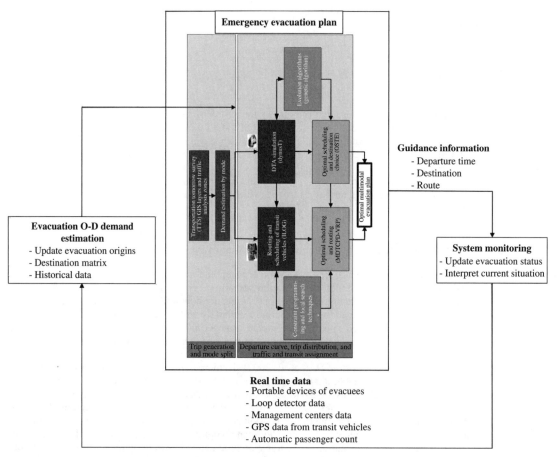

FIGURE 32.33   Closed loop evacuation control syste.

ACKNOWLEDGMENTS

We gratefully acknowledge the financial support of the Connaught Fellowship of the University of Toronto, the International Transport Forum, NSERC, OGSST Scholarship, and the Canada Research Chairs Program. Travel demand data were provided by the Data Management Group (DMG) of the Urban Transportation Research Advancement Centre (UTRAC). This research was enabled by the Toronto Intelligent Transportation Systems Centre. The methodology and results presented in this chapter reflect the views of the authors only.

REFERENCES

Abdelgawad, H., and B. Abdulhai. 2009. "Optimal Spatio-Temporal Evacuation Demand Management: Methodology and Case Study in Toronto," in *Proceedings of the 88th Annual Meeting of the Transportation Research Board*, Washington, DC.

Abdelgawad, H., B. Abdulhai, and M. Wahba. 2010. "Multi-Objective Optimization of Multimodal Evacuation." *Journal of Transportation Research Record* (forthcoming).

Abdelgawad, H., B. Abdulhai, and M. Wahba. 2010. "Optimizing Mass Transit Utilization in Emergency Evacuation of Congested Urban Areas," in *Proceedings of the 89th Annual Meeting of The Transportation Research Board*, Washington, DC.

Alsnih, R., and P. Stopher. 2004. "Review of Procedures Associated with Devising Emergency Evacuation Plans." *Transportation Research Record* 1865(1):89–97.

Bäck, T. 1996. *Evolutionary Algorithms in Theory and Practice: Evolution Strategies, Evolutionary Programming, Genetic Algorithms.* New York: Oxford University Press.

Bell, M., and Y. Iida. 1997. *Transportation Network Analysis.* Hoboken, NJ: Wiley.

Bethke, A. D. 1976. "Comparison of Genetic Algorithms and Gradient-Based Optimizers on Parallel Processors: Efficiency of Use of Processing Capacity." Technical Report 197, Logic Computing Group, University Michigan, Ann Arbor.

Braysy, O., and M. Gendreau. 2005. "Vehicle Routing Problem with Time Windows, Part I: Route Construction and Local Search Algorithms." *Transportation Science* 39(1):104–118.

Brian Wolshon, P. E. 2001. "'One Way Out': Contraflow Freeway Operation for Hurricane Evacuation." *Natural Hazards Review* 2(3):105–112.

Chen, M., L. Chen, and E. Miller-Hooks. 2007. "Traffic Signal Timing for Urban Evacuation." *Journal of Urban Planning and Development* 133(1):30–42.

Chen, X., and F. Zhan. 2006. "Agent-Based Modeling and Simulation of Urban Evacuation: Relative Effectiveness of Simultaneous and Staged Evacuation Strategies." *Journal of the Operational Research Society* 59(1):25–33.

Chiu, Y., and P. Mirchandani. 2008. "Online Behavior: Robust Feedback Information Routing Strategy for Mass Evacuation." *IEEE Transactions on Intelligent Transportation Systems* 9(2):264–274.

Chiu, Y., J. Villalobos, B. Gautam, and H. Zheng. 2006. "Modeling and Solving the Optimal Evacuation Destination-Route-Flow-Staging Problem for No-Notice Extreme Events," in *Proceedings of the 85th Transportation Research Board,* TRB, National Research Council, Washington, DC.

Clarke, G., and J. Wright. 1964. "Scheduling of Vehicles from a Central Depot to a Number of Delivery Points." *Operations Research* 12:568–581.

Daganzo, C. F. 1994. "The Cell Transmission Model: A Dynamic Representation of Highway Traffic Consistent with the Hydrodynamic Theory." *Transportation Research Part B Methodological* 28:269–269.

Data Management Group (DMG). 2003. *Transportation Tomorrow Survey Summaries 2001.* Data Management Group, University of Toronto, Joint Program in Transportation.

Federal Highway Administration (FHWA). 2010. "Using Highways for Non-Notice Evacuations: Routes to Effective Evacuation Planning Primer Series." Available at http://ops.fhwa.dot.gov/publications/evac_primer_nn/primer.pdf; accessed February 8, 2010.

Fisher, M. 1994. "Optimal Solution of Vehicle Routing Problems Using Minimum k-Trees." *Operations Research* 626–642.

Fisher, M., and R. Jaikumar. 1981. "A Generalized Assignment Heuristic for Vehicle Routing." *Networks* 11(2):109–124.

Fu, H., and C. Wilmot. 2004. "Sequential Logit Dynamic Travel Demand Model for Hurricane Evacuation." *Transportation Research Record* 1882(1):19–26.

Gambardella, L., É. Taillard, and G. Agazzi. 1999. "MACS-VRPTW: A Multiple Ant Colony System for Vehicle Routing Problems with Time Windows." *McGraw-Hill's Advanced Topics in Computer Science Series.* New York: McGraw-Hill, pp. 63–76.

Garey, M., D. Johnson, R. Backhouse, G. von Bochmann, D. Harel, C. van Rijsbergen, J. Hopcroft, J. Ullman, A. Marshall, and I. Olkin. 1979. *Computers and Intractability: A Guide to the Theory of NP-Completeness.* Berlin: Springer.

Han, L., and F. Yuan. 2005. "Evacuation Modeling and Operations Using Dynamic Traffic Assignment and Most Desirable Destination Approaches," in *Proceedings of the 84th Transportation Research Board*, TRB, National Research Council, Washington, DC.

Han, L. D., F. Yuan, and T. Urbanik. 2007. "What Is an Effective Evacuation Operation?" *Journal of Urban Planning and Development* 133(1):3–8.

ILOG. 2008. *ILOG Dispatcher and Solver User's Manual and Reference Manuals.*

Jaw, J., A. Odoni, H. Psaraftis, and N. Wilson. 1986. "Heuristic Algorithm for the Multi-Vehicle Advance Request Dial-A-Ride Problem with Time Windows." *Transportation Research Part B Methodological* 20(3):243–257.

Kattan, L., and B. Abdulhai. 2006. "Noniterative Approach to Dynamic Traffic Origin-Destination Estimation with Parallel Evolutionary Algorithms." *Transportation Research Record* 1964(1):201–210.

Kim, S., S. Shekhar, and M. Min. 2008. "Contraflow Transportation Network Reconfiguration for Evacuation Route Planning." *IEEE Transactions on Knowledge and Data Engineering* 20(8):1115–1129.

Kindervater, G., and M. Savelsbergh. 1997. "Vehicle Routing: Handling Edge Exchanges." *Local Search in Combinatorial Optimization* 337–360.

Kruchten, N. 2003) "Galapagos: A Distributed Parallel Evolutionary Algorithm Development Platform." Faculty of Applied Science and Engineering, University of Toronto, Canada.

Kruchten, N., B. Abdulhai, L. Kattan, and D. de Koning. 2004. "Galapagos: A Generic Distributed Parallel Genetic Algorithm Development Platform for Computationally Demanding ITS Optimization Problems," in *Proceedings of the 83rd Transportation Research Board*, TRB, National Research Council, Washington, DC.

Lim, E., and B. Wolshon. 2005. "Modeling and Performance Assessment of Contraflow Evacuation Termination Points." *Transportation Research Record* 1922(1):118–128.

Lindsay, J. 2009. "Emergency Management in Canada: Near Misses and Moving Targets." Available at www.training.fema.gov/EMIWeb/edu/ARRPT/March percent2010, percent202009.doc; accessed January 2010.

Litman, T. 2006. "Lessons from Katrina and Rita: What Major Disasters Can Teach Transportation Planners." *Journal of Transportation Engineering* 132(1):11–18.

Liu, H., J. Ban, W. Ma, and P. Mirchandani. 2007. "Model Reference Adaptive Control Framework for Real-Time Traffic Management under Emergency Evacuation." *Journal of Urban Planning and Development* 133:43.

Liu, Y. 2007. "An Integrated Optimal Control System for Emergency Evacuation." Ph.D. dissertation, Department of Civil and Environmental Engineering, University of Maryland, College Park, MD.

Louisell, C., J. Collura, D. Teodorovic, and S. Tignor. 2004. "Simple Worksheet Method to Evaluate Emergency Vehicle Preemption and Its Impacts on Safety." *Transportation Research Record* 1867(1):151–162.

McHale, G., and J. Collura. 2003. "Improving Emergency Vehicle Traffic Signal Priority System Assessment Methodologies."

Meng, Q., H. L. Khoo, and R. L. Cheu. 2008. "Microscopic Traffic Simulation Model-Based Optimization Approach for the Contraflow Lane Configuration Problem." *Journal of Transportation Engineering* 134(1):41–49.

Miller, E. J., and A. Shalaby. 2003. "Evolution of Personal Travel in Toronto Area and Policy Implications." *Journal of Urban Planning and Development* 129:1.

Mitchell, S., and E. Radwan. 2006. "Heuristic Prioritization of Emergency Evacuation Staging to Reduce Clearance Time," in *Proceedings of the 85th Transportation Research Board*, TRB, National Research Council, Washington, DC.

Mohamed, M. 2007. "Generic Parallel Genetic Algorithms Framework for Optimizing Intelligent Transportation Systems (GENOTRANS)." Master's thesis, University of Toronto.

Muñoz, L., X. Sun, R. Horowitz, and L. Alvarez. 2006. "Piecewise-Linearized Cell Transmission Model and Parameter Calibration Methodology." *Transportation Research Record* 1965(1):183–191.

Murray-Tuite, P., and H. Mahmassani. 2003. "Model of Household Trip-Chain Sequencing in Emergency Evacuation." *Transportation Research Record* 1831(1):21–29.

Pagès, L., R. Jayakrishnan, and C. Cortés. 2006. "Real-Time Mass Passenger Transport Network Optimization Problems." *Transportation Research Record* 1964(1):229–237.

Peeta, S., and A. Ziliaskopoulos. 2001. "Foundations of Dynamic Traffic Assignment: The Past, The Present And The Future." *Networks and Spatial Economics* 1(3):233–265.

Post, B. 2000. "Reverse Lane Standards and ITS Strategies Southeast United States Hurricane Study." Technical Memorandum 3.

Rochat, Y., and É. Taillard. 1995. "Probabilistic Diversification and Intensification in Local Search for Vehicle Routing." *Journal of Heuristics* 1(1):147–167.

Roorda, M. J., E. J. Miller, and N. Kruchten. 2006. "Incorporating Within-Household Interactions into a Mode Choice Model Using a Genetic Algorithm for Parameter Estimation," in *Proceedings of the 83rd Transportation Research Board*, TRB, National Research Council, Washington, DC.

Sattayhatewa, P., and B. Ran. 2000. "Developing a Dynamic Traffic Management Model for Nuclear Power Plant Evacuation." Preprint CD.

Sayyady, F. 2007. "Optimizing the Use of Public Transit System in Non-Notice Evacuations in Urban Areas." Master's thesis, Mississippi State University Department of Industrial and Systems Engineering.

Sbayti, H., and H. Mahmassani. 2006. "Optimal Scheduling of Evacuation Operations." *Transportation Research Record* 1964(1):238–246.

Shaw, P. 1998. "Using Constraint Programming and Local Search Methods to Solve Vehicle Routing Problems." *Lecture Notes in Computer Science* 417–431.

Shaw, P., V. Furnon, and B. Backer. 2002. "A Constraint Programming Toolkit for Local Search," in S. Voss, ed., *Optimization Software Class Libraries*. Boston: Kluwer Academic Publisher, 1520:219–262.

Sisiopiku, V., S. Jones, A. Sullivan, S. Patharkar, and X. Tang. 2004. "Regional Traffic Simulation for Emergency Preparedness." UTCA Report 03226, Final Report.

Song, R., S. He, and L. Zhang. 2009. "Optimal Transit Routing Problem for Emergency Evacuations." *Transportation Research* 9(7):154–160.

Sorensen, J., and B. Sorensen. 2006. "Interactive Emergency Evacuation Guidebook." Available at http://emc .ornl.gov/CSEPPweb/evac_files/files; accessed December 13, 2007.

Southworth, F., and S. Chin. 1987. "Network Evacuation Modeling for Flooding as a Result of Dam Failure." *Environment and Planning A* 19:1543–1558.

TCRP. 2003. *Transit Capacity and Quality of Service Manual* (Part 4: "Bus Transit Capacity"), 2nd ed. Washington, DC: Transportation Research Board, Transit Cooperative Research Program, pp. 1–116.

Theodoulou, G., and B. Wolshon. 2004. "Modeling and Analyses of Freeway Contraflow to Improve Future Evacuations," in *Proceedings of the 83rd Transportation Research Board*, TRB, National Research Council, Washington, DC.

Tomassini, M. 1999. "Parallel and Distributed Evolutionary Algorithms: A Review." *Evolutionary Algorithms in Engineering and Computer Science: Recent Advances in Genetic Algorithms, Evolution Strategies, Evolutionary Programming, Genetic Programming, and Industrial Applications,* 113–131.

Toth, P., and D. Vigo. 2003. "The Granular Tabu Search and Its Application to the Vehicle-Routing Problem." *INFORMS Journal on Computing* 15(4):333.

TRB. 2008. "The Role of Transit in Emergency Evacuation." Special Report 294.

TTC. 2001. *Service Summary Report*. Service Planning Department of the Toronto Transit Commission (unpublished).

Tuydes, H., and A. Ziliaskopoulos (2004). "Network Re-design to Optimize Evacuation Contraflow," in *Proceedings of the 83rd Transportation Research Board*, TRB, National Research Council, Washington, DC.

Urbina, E. A. 2001. "A State-of-the-Practice Review of Hurricane Evacuation Plans and Policies." Ph.D. dissertation, Louisiana State University.

Vuchic, V. 2005. *Urban Transit: Operations, Planning and Economics.* Hoboken, NJ: Wiley.

Wahba, M. 2009. "MILATRAS MIcrosimulation Learning-based Approach to TRansit ASsignment." Thesis, University of Toronto.

Wilmot, C. G., and B. Mei. 2004. "Comparison of Alternative Trip Generation Models for Hurricane Evacuation." *Natural Hazards Review* 5:170.

Xie, C., D. Y. Lin, and S. Travis Waller. 2009. "A Dynamic Evacuation Network Optimization Problem with Lane Reversal and Crossing Elimination Strategies." *Transportation Research, Part E: Logistics and Transportation Review* 46(3):295–316.

Xu, J., and J. Kelly. 1996. "A Network Flow-Based Tabu Search Heuristic for the Vehicle Routing Problem." *Transportation Science* 30(4):379.

Yuan, F., L. Han, S. Chin, and H. Hwang. 2006. "Proposed Framework for Simultaneous Optimization of Evacuation Traffic Destination and Route Assignment." *Transportation Research Record* 1964(1):50–58.

Zhi, X., H. Guan, X. Yang, X. Zhao, and L. Lingjie. 2010. "Exploration of Driver Perception-Reaction Times under Emergency Evacuation Situations," in *Proceedings, the 89th Annual Meeting of the Transportation Research Board*, Washington, DC.

Ziliaskopoulos, A. K. 2000. "A Linear Programming Model for the Single Destination System Optimum Dynamic Traffic Assignment Problem." *Transportation Science* 34(1):37–49.

CHAPTER 33
TRANSPORTATION NOISE ISSUES

Judith L. Rochat
U.S. Department of Transportation/RITA/Volpe Center
Cambridge, Massachusetts

33.1 INTRODUCTION

Noise is a serious issue that should be considered in all stages of transportation system projects from original design and construction to modifications. Transportation-related noise affects millions of people and in many cases requires local, state, and federal governments to provide noise abatement to help improve or restore quality of life. The impact of noise on the quality of life can be substantial, especially with expanding transportation systems. This is why there is an increasing need for noise control and why the field of transportation-related noise is thriving.

There are three modes of transportation where noise issues typically are addressed: highway traffic, aircraft, and rail. Although less common, other transportation-related noise concerns also can warrant noise control or at least consideration; these include construction noise, noise inside vehicles, recreational vehicle noise, and underwater noise.

This chapter provides an overview of noise issues, followed by discussions of the specific issues relating to different modes of transportation, including noise sources, noise prediction, noise metrics, noise control, and relevant resources.

33.2 SOUND, NOISE, AND ITS EFFECTS

33.2.1 Sound

Sound is a vibratory disturbance created by a moving or vibrating source. Examples of transportation sound sources include steady traffic on a highway, construction equipment used to build a highway or bridge, and a jet flying overhead. Each sound source can be described by its associated spectrum, amplitude, and time history. The *spectrum* reveals the frequency content of a sound. Figures 33.1 and 33.2 show two examples of typical spectra corresponding to transportation noise sources. Figure 33.1 shows an example spectrum of steady highway traffic; as can be seen, the most dominant frequency range is between 200 and 2000 Hz. Figure 33.2 shows an example of a propeller-driven aircraft in flight; for this spectrum, it is seen that there are some tonal components (perceived as a pitch by humans) in the lower frequencies, a noticeable tone being at 100 Hz. Both these sound sources are broadband in nature—they are complex sounds containing multiple frequencies.

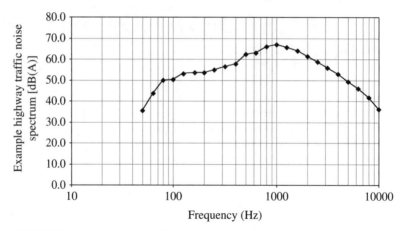

FIGURE 33.1   Example spectrum of highway traffic noise.

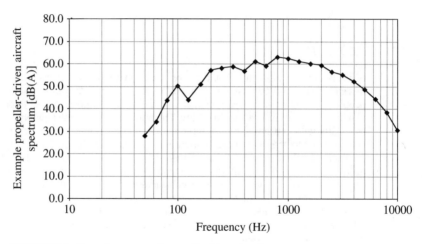

FIGURE 33.2   Example spectrum of a propeller-driven aircraft.

The *amplitude* of a sound indicates its volume; in general, higher amplitudes indicate louder sounds. The amplitude of a sound can be measured as a sound pressure level with a microphone/ sound level meter or spectrum analyzer system; for guidance on instrumentation, please refer to section 33.3. Time-averaged sound levels or maximum sound levels are metrics often used to quantify sound, the associated quantities presented in units of decibels (dB). A decibel level is a logarithmic representation of sound energy and is calculated using a medium-dependent (e.g., air or water) reference energy level. The reference level is based on human perception such that the sound energy associated with the threshold of unimpaired human hearing will correspond to 0 dB. In general, a sound that is 10 dB higher than another sound with the same spectral characteristics is said to be twice as loud; please refer to Table 33.1 for different perceptions of loudness. The previously presented spectral data (Figures 33.1 and 33.2) show the amplitude for each third-octave frequency band; the band levels logarithmically added together equate to the overall amplitude.

The *time history* of a sound reveals its amplitude variation over time. The perception of a sound (to be discussed further in the noise section that follows) also depends on its time-varying characteristics.

TABLE 33.1   Perceptions of Loudness.

Sound-level change	Descriptive change in perception
+20 dB	Four times as loud
+10 dB	Twice as loud
+5 dB	Readily perceptible increase
+3 dB	Barely perceptible increase
0 dB	Reference
−3 dB	Barely perceptible reduction
−5 dB	Readily perceptible reduction
−10 dB	Half as loud
−20 dB	One quarter as loud

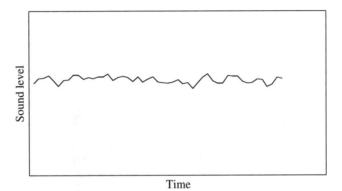

FIGURE 33.3   Example time history of continuous highway traffic.

Sound with little variation over time is a continuous sound; sound that exists only for a brief time is a transient sound. Figures 33.3 and 33.4 show two examples of typical time histories. Figure 33.3 shows the time history for steady highway traffic, a relatively continuous sound, and Figure 33.4 shows the time history for a jet flying overhead, a transient sound. The onset of the transient sound in Figure 33.4 is relatively gradual; in general, a commercial jet can be heard approaching, flying overhead, and then departing. Many transient sounds have more rapid onset times (sometimes referred to as *impulsive sounds*), where the amplitude increases substantially within a fraction of a second.

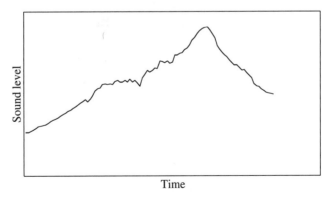

FIGURE 33.4   Example time history of a jet flyover.

33.2.2 Noise

Certain sounds are considered to be pleasant and therefore found to be acceptable to a listener; some sounds, on the other hand, are unwanted and therefore are considered to be noise. Noise is defined in different ways depending on the listener; it depends on the spectrum, amplitude, and time history of the sound. However, some types of noise are understood to be objectionable. Many types of transportation noise sources are often included in this category. Figure 33.5 shows differing sound levels for typical noise sources, including transportation noise sources. As can be seen, a train horn 200 feet (60 meters) away from the source reaches a level of about 105 dB(A); the train horn noise is a loud burst that is tonal and startling. A jet flying overhead at an altitude of 1,000 feet (305 meters) reaches a maximum level on the ground of about 90 dB(A); the jet noise has a longer onset time than the train horn and is less startling with broader frequency content. Highway traffic noise at a distance of about 300 feet (90 meters) from the highway has an almost continuous noise level of about 60 dB(A); often highway traffic noise is a continuous broadband sound.

It is straightforward to measure sound pressure levels in order to quantify sound, but describing noise involves quantifying its perception. Quantifying perception accounts for the hearing abilities of the listener. Humans do not hear equally well at all frequencies; humans ideally hear frequencies from

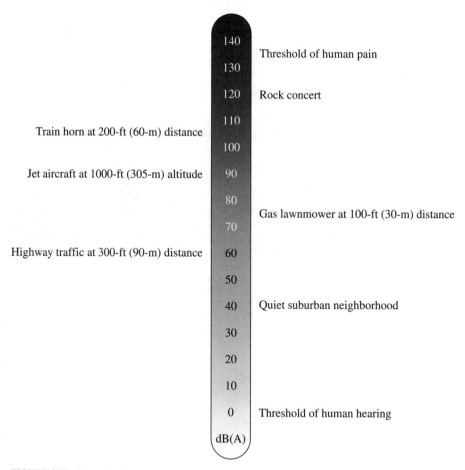

FIGURE 33.5   Example noise sources.

20 to 20,000 Hz (although sensitivity to higher frequencies decreases with age), where the hearing is most sensitive from about 1,000 to 6,300 Hz. To describe sound levels in a manner that closely approximates normal human hearing, the actual sound levels typically are modified by applying *A-weighting*. This is a response function that spans the audible frequency range, emphasizing frequencies in the most sensitive range and deemphasizing frequencies out of the sensitive range. Although the A-weighted sound level is the most widely used measure of environmental noise and is accepted internationally, there are other frequency-weighting functions that may be more suitable to specific noise analyses.

33.2.3 Effects of Noise

The effects of noise on people and wildlife can be substantial; as a result, serious consideration should be given to transportation noise impacts. Effects of noise on people range from annoyance to stress-related illnesses. Noise can interfere with communication and concentration, an example being that noise diminishes the ability of students to learn in a classroom environment (Seep 2000). Noise can cause sleep deprivation and stress, contributing to high blood pressure and heart disease. For wildlife, there are serious concerns about noise interfering with communication, migration, and reproduction. For a review on noise effects, see Sandberg (2002), FRA (2005), Lee (2001), and Griefahn (2009), and for specifics on marine life, refer to Richardson (1995), Rochat (1998), and Caltrans (2009).

33.2.4 Noise Metrics

The metric applied to quantify noise depends on the noise source and purpose for the noise measurement. Hundreds of metrics are defined (see Shultz 1982). Table 33.2 lists noise metrics that are used

TABLE 33.2 Summary of Commonly Used Transportation Noise Metrics.

Metric name	Metric symbol	Metric abbreviation	Metric description
A-weighted equivalent sound level	$L_{\mathrm{Aeq}T}$ (T = time increment)	LAEQ	Sound level associated with the sound energy averaged over a specified time period
Day-night average sound level	L_{dn}	DNL	L_{Aeq24h} with a 10 dB(A) penalty between the hours of 10 p.m. and 7 a.m. (sleeping hours)
Community noise equivalent level	L_{den}	CNEL	L_{Aeq24h} with a 10 dB(A) penalty between the hours of 10 p.m. and 7 a.m. (sleeping hours) and a 5-dB(A) penalty between the hours of 7 p.m. and 10 p.m. (relaxation, conversation hours)
A-weighted maximum sound level	L_{Amax}	LAMAX	A-weighted maximum sound level during a noise event or specified time period
Percent exceeded sound level	L_X (e.g., L_{10})	—	The sound level exceeded X percent of the time during a specified time period
A-weighted sound exposure level	L_{AE}	SEL	The time integral of sound level over the course of a single event
Effective Perceived noise level	L_{EPN}	EPNL	A sound level based on human perception and accounting for tonal components and event duration, used primarily for assessing aircraft noise
Time above/percent time above	TA, %TA	TA, %TA	Time or percentage of time that the A-weighted noise level is above a user-specified sound level during a specified time period
Detectability level	$D'L$	$D'L$	A measure of the ability to detect a particular sound in the presence of other noise; a function of the signal-to-noise ratio

commonly in transportation noise engineering. Some of these metrics require additional calculations beyond sound level measurements, but noise-prediction software usually calculates the desired metric for the user. With each of these metrics, different frequency weightings, as mentioned previously, also can be applied. See the appropriate noise policy to find out which metrics are acceptable in a particular situation. In later sections of this chapter, when addressing specific modes of transportation, typical noise metrics will be listed.

33.2.5 Vibratory Disturbances

Vibratory disturbances other than sound are also generated by transportation systems. Acoustic waves traveling through the ground or air cause structures, such as windows, walls, and floors, to vibrate. These vibrations can adversely affect building inhabitants because of structural noise, uncomfortable sensations, or interference with delicate procedures where vibration isolation is mandatory (e.g., surgical procedures).

Vibrations also can adversely affect the structures themselves, potentially causing structural damage (e.g., cracks). Example sources of such vibrations are trains and aircraft. Vibrations from trains extend beyond the track, the waves traveling through the ground into nearby communities. Low-altitude or fast-flying aircraft can generate loud sounds that cause structures to vibrate, sometimes violently. (Sonic booms from supersonic aircraft have, on rare occasion, shattered windows in nearby buildings.)

For guidance on vibration vocabulary, please refer to ISO 2041. For guidance on the measurement and evaluation of vibration in buildings, please refer to ISO 4866. For guidance on methods for analysis and presentation of vibrational data, refer to ANSI S2.10. References for the appropriate mode of transportation will provide guidance to sources of vibration, measuring vibration, vibratory analysis, vibration control, and related policy issues.

33.2.6 General Noise Resources

References are listed by topic. See the References at the end of this chapter for full citations.

Overall: Beranek (1992), Harris (1991), ANSI S1.1 (terminology)

Transportation noise: Fleming (2000)

Noise measurements: Beranek (1988)

Noise metrics: Shultz (1982)

Noise effects: Seep (2000), Sandberg (2002), FRA (2005), Lee (2001), Griefahn (2009), Richardson (1995), Rochat (1998)

33.3 INSTRUMENTATION FOR MEASURING NOISE

It is common to measure noise using a microphone/sound level meter or spectrum analyzer system. The choice of field instrumentation and settings depends greatly on the type of noise being measured and the purpose of the measurements. For specific guidance on instrumentation requirements, please refer to the references cited in this section.

33.3.1 Microphones

Microphone choice is based largely on sensitivity, frequency response, directivity, and environmental performance. A few examples will be given for the four elements listed:

Microphone Sensitivity. Larger-diameter microphones are used for measuring quieter sounds and smaller-diameter microphones for louder sounds; for typical community noise measurements, ½-inch-diameter microphones are used most commonly.

Frequency Response. Smaller microphones generally exhibit better high-frequency response; for community noise measurements, microphones with good frequency response over the range of 50 to 10,000 Hz will convey the information required for most transportation noise measurements.

Microphone Directivity. Directional microphones give best results when pointed directly at the sound source (normal incidence), whereas pressure-response microphones give uniform responses to sound sources in a plane perpendicular to the axis of the microphone (grazing incidence) (Figure 33.6). Microphone orientation requirements are determined by regulatory and practical concerns.

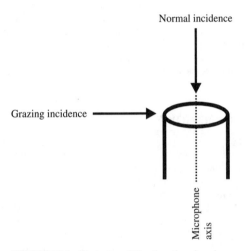

FIGURE 33.6 Illustration of directions for microphone orientation.

Environmental Performance. Electret condenser microphones provide improved reliability in a humid environment over standard condenser microphones, which combine a flat frequency response and high sensitivity.

For more information on microphones, please refer to Beranek (1988) and microphone manufacturers' documentation.

33.3.2 Sound Level Instrumentation

Sound level instrumentation choice depends on the required noise metric (spectrum analyzer when frequency content is required, integrating capability for time-averaged sound levels, desired frequency-weighting capabilities, etc.). It is also important to consider requirements for portability and environmental reliability/durability for a particular set of measurements. In addition, regulations and standards for particular noise measurements recommend or require specific types of sound level instrumentation. Sound level meters meeting the class 1 specifications of IEC 61672 are usually recommended/required for transportation noise measurements.

For resources on sound level instrumentation, refer to IEC 61672, IEC 61260, and Beranek (1988).

For the latest on available instrumentation, refer to current issues of *Sound and Vibration Magazine* or *Physics Today* magazine. Also, many vendors attend noise-related conferences [such as Inter-Noise (I-INCE, Web), Noise-Con (INCE, Web), Acoustical Society of America (ASA, Web), and Transportation Research Board (TRB, Web)], where product literature and demonstrations are available.

33.4   PREDICTION OF TRANSPORTATION NOISE

There are several computer programs available for the prediction of a specific mode of transportation noise. These models help to determine which residences/communities are affected by transportation noise and allow the user to design some noise-abatement strategies. Most noise evaluations are conducted for a single mode of transportation, so the computer programs will be listed under the corresponding mode of transportation. Multimodal noise prediction is less common, although steps are being taken to further develop this capability. The IMAGINE project (IMAGINE) provides guidance and calculation methodologies for the European Union, and an Airport Cooperative Research Program project (ACRP) will provide guidance for multimodal predictions for U.S. applications. (Some commercial software packages allow prediction of noise from multiple modes of transportation; see the CandaA, SoundPLAN, LimA, and IMMI references.)

33.5   HIGHWAY TRAFFIC NOISE

Highway traffic noise is something many of us encounter on a daily basis, although those most affected are people who live, work, or attend school next to a busy highway or roadway. Highways dominate our ground transportation systems. In the United States, the road network is over 3.9 million miles (6.3 million kilometers) in length. Millions of people drive on the highways, including those driving automobiles commuting to work (where a single automobile is typically the quietest vehicle type on the road) and those driving tractor-trailers hauling freight across the country (where a single heavy truck is typically the loudest vehicle type on the road). All the vehicles on the highway contribute to the noise that affects adjacent communities.

33.5.1   Highway Traffic Noise Sources

The source of highway noise is caused by tire/pavement interaction, aerodynamic sources (turbulent airflow around and partly through the vehicle), and the vehicle itself (the power-unit noise created by the engine, exhaust, transmission, etc.). At highway speeds, tire/pavement interaction generally is the most dominant source. Figure 33.7 shows a generalized plot of highway noise levels as a function of speed for five vehicle types and "average" pavement. (The sound levels are averaged over different types of pavement.)

The interaction between tires and pavement is quite complex, the generated noise level being highly dependent on the road surface and the tire tread pattern and construction. A brief overview of different pavement types will be described here; see Sandberg (2002) for more details.

Two broad categories of pavements are asphalt and cement concrete. Asphaltic concrete (AC) can vary in chipping or stone size (the aggregate), filler, and binder, the final formula and construction affecting its acoustical properties. Dense-graded asphaltic concrete (DGAC or DAC) is less porous, and open-graded asphaltic concrete (OGAC or PAC for porous AC) is more porous, the latter allowing some noise to be absorbed. In general, OGAC is "quieter" than DGAC; also in general, increasing the maximum aggregate size increases the noise. Portland cement concrete (PCC) is paved with a smooth surface, after which a surface treatment is applied for reasons of safety. An example of a PCC surface treatment is tining. Tining (small grooves in the pavement) is done longitudinally

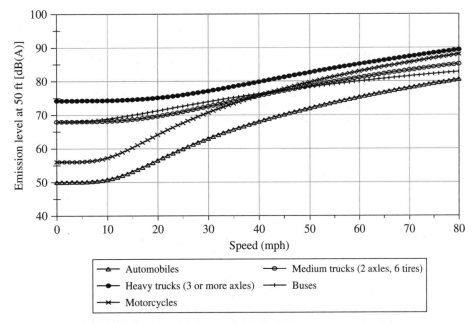

FIGURE 33.7 Emission level data as a function of speed for five vehicle types.

(in the direction of traffic flow) or transversely (perpendicular to the direction of traffic flow); longitudinal tining is generally "quieter" than transverse tining, although variation occurs depending on tine spacing and depth. In general, DGAC and OGAC surfaces are "quieter" than PCC surfaces; however, some PCC surfaces, such as diamond grinding, can be as "quiet" as asphalt pavements.

33.5.2 Highway Traffic Noise Prediction

Several computer software packages are available for obtaining highway traffic noise predictions. These include

1. The Federal Highway Administration's Traffic Noise Model (TNM) (Anderson 1998; Menge 1998)
2. The University of Florida's Community Noise Model (CNM) (Wayson 1998)

   TNM is the official U.S. federal government model, and it must be used in order for state departments of transportation to receive federal funding for highway noise-abatement projects.

33.5.3 Highway Traffic Noise Metrics

There are some generally accepted practices for quantifying highway traffic noise. For quantifying highway traffic noise in communities, the A-weighted equivalent sound level (L_{AeqT} or LAEQ), the day-night average sound level (DNL or L_{dn}), the community noise equivalent level (L_{den} or CNEL), and the percent exceeded sound level (L_X, e.g., L_{10} for 10 percent) are commonly applied. For vehicle emission levels and analyzing pavements, the A-weighted maximum sound level (L_{Amax} or LAMAX) is applied; it is also useful to look at spectral information. For a single number to characterize pavement, several methods have been or are being developed. For wayside measurements, the statistical pass-by method (SPB; ISO 11819-1) is used most commonly; some developing methods include the statistical isolated pass-by method (SIP; Rochat 2009) and the continuous-flow traffic time-integrated method (CTIM; Rochat 2009). For measuring the tire-pavement interaction noise near the source, there are two developing methods: the close-proximity method (CPX; ISO 11819-2) and the on-board sound intensity method (OBSI; AASHTO TP076). The metric $L_{A\text{max}}$ is measured in SPB and SIP, the metric $L_{A\text{eqT}}$ is measured in CTIM and CPX, and the sound intensity level (A-weighted, time-averaged) is measured in OBSI. For most highway noise measurements, an A-weighted metric is usually applied. Refer to Table 33.2 for metric descriptions. Also, see the appropriate federal, state, or local noise policy to find out which metrics are acceptable for highway noise analysis in a particular area.

33.5.4   Highway Traffic Noise Control

Communities with residences, schools, parks, etc. that meet the criteria for noise abatement are eligible to receive some form of noise control. The qualifications for noise abatement can be found by accessing the appropriate noise policy; refer to Knauer (2000) for U.S. policies by state, or contact the state noise representative (see state departments of transportation in the References for the Web site listing the contacts).

The most typical highway noise-abatement tool is a noise wall or wall-type noise barrier; many are designed to meet a goal of 10 dB(A) reduction. These walls are constructed alongside a highway, blocking the line of sight between the vehicles and people in the community. Blocking the line of sight typically reduces the sound by 5 dB; extending the wall above the line of sight helps to reduce the noise further. There are many variations in noise wall material and construction, where considerations are made for the available space, community acceptance, and durability, among other items. Refer to Knauer (2000) for more details.

In some cases, noise walls are not feasible or effective, and buildings next to a highway may need to be altered (e.g., double-paned windows installed). A "quiet" pavement on the highway, such as one with an acoustically absorptive surface, also may be used to reduce highway noise; such pavement would require periodic examination to determine the potential need for repaving, retreating the surface, or cleaning.

33.5.5   Highway Traffic Noise Resources

References are listed by topic. See the References at the end of this chapter for full citations.

Overview: Sandberg (2002), Rochat (1999), FHWA (2006), Hendriks (1998)

Measurement: ANSI S12.8, ANSI S1.13, Lee (1996), Sandberg (2002), ISO 11819-1, ISO 11819-2, AASHTO TP076, Rochat (2009)

Prediction software: Anderson (1998), Menge (1998), Wayson (1998)

Noise-abatement policy: Knauer (2000), state DOTs, FHWA (1995), 23 CFR Part 772

Tires/roadways: Sandberg (2002), Rasmussen (2007)

Noise barriers: Knauer (2000)

33.6   AIRCRAFT NOISE

There are over 14,000 airports in the United States, some of the major airports being the busiest in the world. In the United States and other countries around the world, noise from aircraft is affecting millions of people daily, people who live, work, go to school, etc. near the airports or helipads and under the flight paths.

33.6.1 Aircraft Noise Sources

The noise generated from aircraft is attributed to the engines and aerodynamic noise sources. In general, each type of aircraft generates a distinctive type of sound that depends on its construction, its mode of propulsion (e.g., jet or propeller), and its state of operation (e.g., departure, approach, cruise, etc.). For fixed-wing aircraft, jets in general can be heard on the ground as a broadband noise source, with some having higher-frequency tones (>1,000 Hz); the sound of propeller-driven aircraft in general is more biased toward lower frequencies, usually with a prominent low-frequency tone (< 500 Hz). For rotorcraft, helicopters can be heard on the ground sometimes as noise similar to that of jets and sometimes similar to that of propeller-driven aircraft, often with a repetitive impulsive quality; tiltrotor aircraft (a multimodal aircraft that uses both wings and rotors to generate lift) can sound like helicopters or propeller-driven aircraft depending on the mode of flight.

For each aircraft, the noise generated is affected by the aircraft's operations. Three operational states are (1) taking off from a runway or helipad, (2) approaching a runway or helipad, and (3) en-route flight (cruise). As part of the operations, the physical aerodynamic configuration of the aircraft (flaps, landing gear, etc.), the engine settings (thrust, propeller speed, etc.), and the relative motion of the aircraft (airspeed, attitude, etc.) all can affect the noise. The specific design of the aircraft (shape, size, structural configuration, etc.) also affects the noise. Lastly, the total weight of the aircraft (including fuel, passengers, and cargo) affects the noise; this variable depends on both the design limitations and the aircraft's operations.

Some aircraft operations can cause structural vibrations. A sonic boom, for example, created by an aircraft flying faster than the speed of sound, is impulsive and can be very loud. The received noise can reveal itself as window and wall vibrations (sometimes causing structural damage).

As a note, aircraft noise received on the ground can be substantially affected by the atmosphere because the sound is traveling over great distances when the aircraft is in flight. Wind, turbulence, and temperature gradients can affect the received sound level and the spectral content.

33.6.2 Aircraft Noise Prediction

Several computer software packages are available for obtaining aircraft noise predictions. These include

1. The Federal Aviation Administration's Integrated Noise Model (INM) (He 2007; Boeker 2008)
2. The Federal Aviation Administration's Area Equivalent Method (AEM) (FAA 2008)
3. The Federal Aviation Administration's Noise Integrated Routing System (NIRS) and NIRS Screening Tool (NST) (FAA NIRS)
4. The U.S. Air Force's NOISEMAP (Galloway 1974)
5. NASA Langley Research Center's Rotorcraft Noise Model (RNM) (Page 2002)

[Newer Federal Aviation Administration and Department of Defense models are under development: Aviation Environmental Design Tool (AEDT) and Advanced Acoustic Model (AAM), respectively.]

INM is the official U.S. federal government model, and it must be used to receive federal funding for aircraft noise-abatement projects.

33.6.3 Aircraft Noise Metrics

There are some generally accepted practices for quantifying aircraft noise. For aircraft noise in communities, the day-night average sound level (DNL or L_{dn}) is often applied. For noise certification of aircraft, the effective perceived noise level (L_{EPN} or EPNL) and A-weighted sound exposure level (L_{AE} or SEL) are applied, the metric depending on the type of aircraft; the noise certification process is overseen by the Federal Aviation Administration (FAA) and is necessary to certify that new aircraft or new configurations of aircraft generate acceptable noise levels (CFR 14 Part 36). For noise in national parks, possible supplementary metrics that can be applied are time above/percent time above (TA/%TA) and detectability level ($D'L$) (NPS 1994); the study of aircraft noise in wilderness areas is important to preserving the natural quiet environment. Refer to Table 33.2 for metric descriptions. Newer metrics are being discussed, such as one associated with the number of nighttime awakenings (ANSI S12.9, Part 6), to help assess the effect of aircraft noise. See the appropriate noise policy to find out which metrics are acceptable for aircraft noise analysis in a particular area.

33.6.4   Aircraft Noise Control

For those who meet the criteria to receive noise abatement, some measures can be taken to help reduce the noise. Noise walls can be constructed to help reduce idle and takeoff noise in the nearby communities. Where there is no option to block the line of sight between the source and receiver, buildings can be better insulated to help the reduction; an example of this includes the installation of double-paned windows or the purchase of air-conditioning units to allow the closure of windows.

33.6.5   Aircraft Noise Resources

References are listed by topic. See the References at the end of this chapter for full citations.

Overview: Hubbard (1995), FAA (Web)

Measurement: ANSI S1.13, Fleming (1996), SAE 4721_1, SAE 4721_2

Prediction software: He (2007), Boeker (2008), FAA (2008), FAA NIRS, Galloway (1974), Page (2002)

Noise-abatement policy: 14 CFR Part 36, 14 CFR Part 150, FAA (Web), FAA Order 1050, FAA (2000)

Sonic booms: Plotkin (1990), Rochat (1998), Hubbard (1995)

33.7   RAIL NOISE

The total length of the rail network in the United States extends over 145,000 miles (225,000 kilometers). The noise and/or vibrations from high-speed trains, lower-speed trains, and train horns all affect communities near the tracks or guideways.

33.7.1   Rail Noise Sources

The sources creating the noise from rail operations are (1) engine noise (propulsion machinery), (2) mechanical noise (wheel-rail interactions or guidance vibrations), (3) aerodynamic noise (for speeds greater than 160 mph), and (4) warning-device noise (horns). For safety purposes, the train horns are sounded at a minimum of 96 dB measured at a distance of 100 feet (30 meters) away.

33.7.2   Rail Noise Predictions

Most rail noise predictions are accomplished through mathematical calculations rather than software packages. The Federal Transit Administration (FTA) provides the Noise Impact Assessment Spreadsheet (FTA, Web) to allow practitioners to predict rail noise. The Federal Railroad Administration (FRA) provides CREATE, based on the FTA spreadsheet, HSRNOISE, for high-speed rail, and the horn model.

33.7.3 Rail Noise Metrics

There are some generally accepted practices for quantifying rail noise. For single events, such as a train passing through a community, the A-weighted maximum sound level (L_{Amax} or LAMAX) or the sound exposure level (L_{AE} or SEL) is applied. For the cumulation of multiple events, the A-weighted equivalent sound level (L_{AeqT} or LAEQ) or the day-night average sound level (DNL or L_{dn}) is applied. For train horns, L_{Amax} is applied. Refer to Table 33.2 for metric descriptions. Also see the appropriate noise policy to find out what metrics are acceptable for rail noise analysis in a particular area.

33.7.4 Rail Noise Control

Abatement of rail noise can be addressed in multiple ways. First, the noise source can be modified (modifications to the cars, wheels, or rails). To help reduce the noise before it reaches the community, noise walls can be installed, and the buffer zone between the rail line and the community can be increased. Also, improved sound insulation in buildings will help to reduce the noise level.

Because train horns are an important element of highway-rail grade crossing safety, their sounds often go unmitigated. There are, however, some noise-control measures that have been implemented or are proposed. These include (1) removing the safety hazard at the crossing to remove the need for the horn (e.g., by constructing a bridge over the tracks, by introducing retractable solid barriers to block vehicles from entering the track area during train operations, etc.), (2) limiting the maximum sound level, and (3) limiting the direction of the sound.

33.7.5 Rail Noise Resources

References are listed by topic. See the References at the end of this chapter for full citations.

Overview: FRA (2003) (horns), FRA (2005) (high-speed rail), Stusnick (1982) (rail), Hanson (2006) (rail)

Measurement: FRA (2003) (horns), FRA (2005) (high-speed rail), Stusnick (1982) (rail), Hanson (2006) (rail)

Prediction software: FTA (Web), CREATE, HSRNOISE, horn model

Noise-abatement policy: 40 CFR Part 201, 49 CFR Part 210, FRA (2003) (horns), FRA (2005) (high-speed rail), Stusnick (1982) (rail), Hanson (2006) (rail)

33.8 OTHER TRANSPORTATION-RELATED NOISE

Although the three modes of transportation discussed in sections 33.5, 33.6, and 33.7 are those most commonly referred to in the noise-abatement literature, other types of transportation-related noise also should be mentioned.

The construction of our transportation systems also causes noise. The noise from machinery used to build our roadways, for example, varies in frequency content, amplitude, and time history—the noise generated from jackhammers to bulldozers to backup alarms. Although the machinery itself has noise limitations, an entire construction site as a whole can be quite disturbing to its neighbors. Some noise abatement is applied temporarily in these situations; usually noise barriers are placed between the construction site and the community, and restricted hours of operation are enforced. The Federal Highway Administration (FHWA) developed a handbook that provides an overview of highway construction noise, including the following topics: effects, measurement, criteria and descriptors, prediction, project coordination, and policy (Knauer 2006). In addition, the FHWA Roadway Construction Noise Model (RCNM; Reherman 2006) is available to predict noise in the vicinity of a construction site, and the Highway Construction Noise Computer Program (HICNOM;

Bowlby 1982) is available for a more complex prediction analysis. Another resource for construction noise is Schexnayder (1999); some of the references for specific modes of transportation also discuss construction noise (e.g., FRA 2005).

In addition to the communities affected by transportation vehicle noise, it also should be noted that the people operating the vehicles are also affected. The noise in the cab of a train, for example, can be quite loud; strict noise limits must be adhered to, and in some cases, hearing protection is essential. See the preceding section on "rail noise" for resources.

Recreational vehicle noise is also a source of concern. Open wilderness areas, such as national parks attract vehicles such as snowmobiles and all-terrain vehicles (ATVs), contributing to noise in areas trying to maintain a natural quiet environment, as well as to noise in neighboring communities. For example, the noise from snowmobiles is an issue at Yellowstone National Park. For resources on recreational vehicle noise, refer to the specific area or park in question.

Underwater noise can be created by several transportation sources: shipping traffic, bridge construction, underwater naval operations, and aircraft. All must be considered in order to quantify the underwater sound levels. In some cases, noise control is warranted in order to reduce adverse effects on marine life. Resources for underwater noise are Richardson (1995), Rochat (1998), and Caltrans (2009).

33.9 OTHER NOISE INFORMATION

The elements required for a successful noise study include

1. Proper identification of the noise source(s)
2. Proper noise metric selection
3. Proper instrumentation and field measurements
4. Proper noise prediction
5. Proper data analysis and interpretation

Additionally, for successful noise-control projects, other items include

1. Adequate public participation
2. Proper application of federal, state, and local requirements and limitations

ACKNOWLEDGMENTS

I wish to thank Dave Read for his valuable contributions. I also wish to thank other members of the John A. Volpe National Transportation Systems Center, Acoustics Facility, particularly Gregg Fleming, Christopher Roof, Amanda Rapoza, and Eric Boeker, for their input.

REFERENCES

Acoustical Society of America (ASA) Web site: http://asa.aip.org/.

Aircraft Cooperative Research Program. 2009. "A Comprehensive Development Plan for a Multi-Modal Noise and Emissions Model." ACRP 02-09; available at 144.171.11.40/cmsfeed/TRBNetProjectDisplay.asp?ProjectID=2102.

American Association of State Highway and Transportation Officials. 2008. *Standard Test Method for Measurement of Tire/Pavement Noise Using the On-Board Sound Intensity Method.* AASHTO TP076, Washington, DC.

American National Standards Institute and the Acoustical Society of America Standards. 1971. *Methods for Analysis and Presentation of Shock and Vibration Data.* ANSI S2.10-1971 (R2001), Acoustical Society of America, New York.

American National Standards Institute and the Acoustical Society of America Standards. 1994. *Acoustical Terminology.* ANSI S1.1-1994 (R2004), Acoustical Society of America, New York.

American National Standards Institute and the Acoustical Society of America Standards. 1995. *Measurement of Sound Pressure Levels in Air.* ANSI S1.13-1995 (R2005), Acoustical Society of America, New York.

American National Standards Institute and the Acoustical Society of America Standards. 1998. *Methods for Determining the Insertion Loss of Outdoor Noise Barriers.* ANSI S12.8-1998 (R2008), Acoustical Society of America, New York.

American National Standards Institute and the Acoustical Society of America Standards. 2008. *Quantities and Procedures for the Description and Measurement of Environmental Sound, Part 6: Methods for Estimation of Awakenings Associated with Outdoor Noise Events Heard in Homes.* ANSI S12.9/Part 6-2008, Acoustical Society of America, New York.

Anderson, Grant S., Cynthia S. Y. Lee, Gregg G. Fleming, and Christopher W. Menge. 1998. *FHWA Traffic Noise Model, Version 1.0: User's Guide.* Report Nos. FHWA-PD-96-009 and DOT-VNTSC-FHWA-98-1, U.S. Department of Transportation, Volpe National Transportation Systems Center, Acoustics Facility, Cambridge, MA; available at www.fhwa.dot.gov/environment/noise/tnm/index.htm.

Beranek, Leo L., and István L. Vér, eds. 1992. *Noise and Vibration Control Engineering: Principles and Applications.* Wiley, New York.

Beranek, Leo L. 1998. *Acoustical Measurements.* Acoustical Society of America, Woodbury, NY.

Boeker, Eric, et al. 2008. *Integrated Noise Model (INM), Version 7.0: Technical Manual.* Report No. FAA-AEE-08-01, U.S. Department of Transportation, Federal Aviation Administration, Office of Environment and Energy, Washington, DC.

Bowlby, W., and L.F. Cohn. 1982. *Highway Construction Noise: Environmental Assessment and Abatement,* Volume 1: *Executive Summary and Prediction Methods.* Vanderbilt University, Nashville, TN.

California Department of Transportation (Caltrans), ICF Jones and Stokes, and Illingworth and Rodkin. 2009. *Technical Guidance for Assessment and Mitigation of the Hydroacoustic Effects of Pile Driving on Fish.* Sacramento: California Department of Transportation.

DataKustik. *CandaA (A Software Package).* DataKustik, Greifenberg, Germany.

European Union. *IMAGINE—Improved Methods for the Assessment of Generic Impact of Noise in the Environment;* available at www.imagine-project.org.

Federal Aviation Administration (FAA), *Code of Federal Regulations,* "Noise Standards: Aircraft Type and Airworthiness Certification." 14 CFR Part 36, U.S. Department of Transportation, Federal Aviation Administration, Washington, DC; available at www.gpoaccess.gov/cfr/index.html.

Federal Aviation Administration. *Code of Federal Regulations,* "Airport Noise Compatibility Planning." 14 CFR Part 150, U.S. Department of Transportation, Federal Aviation Administration, Washington, DC; available at www.gpoaccess.gov/cfr/index.html.

Federal Highway Administration. *Code of Federal Regulations,* in *Procedures for Abatement of Highway Traffic Noise and Construction Noise.* 23 CFR Part 772, U.S. Department of Transportation, Federal Highway Administration, Washington, DC; available at www.gpoaccess.gov/cfr/index.html.

Federal Railroad Administration. *Code of Federal Regulations,* in *Noise Emission Standards for Transportation Equipment: Interstate Rail Carriers,* 40 CFR Ch. 1 (7-1-01 edition) Part 201, U.S. Department of Transportation, Federal Railroad Administration, Washington, DC; available at www.gpoaccess.gov/cfr/index.html.

Federal Railroad Administration. *Code of Federal Regulations,* in *Railroad Noise Emission Compliance Regulations,* 49 CFR Ch. 11 (10-1-00 edition) Part 210, U.S. Department of Transportation, Federal Railroad Administration, Washington, DC; available at www.gpoaccess.gov/cfr/index.html.

Federal Aviation Administration (FAA). *Noise Integrated Routing Systems (NIRS) and NIRS Screening Tool (NST).* U.S. Department of Transportation, Federal Aviation Administration, Washington, DC; available at www.faa.gov/about/office_org/headquarters_offices/aep/models/nirs_nst/.

Federal Aviation Administration (FAA), Airport Environmental Program; available at www.faa.gov/airports/environmental/.

Federal Aviation Administration (FAA). *Policies and Procedures for Considering Environmental Impacts.* FAA Order 1050, U.S. Department of Transportation, Federal Aviation Administration, Washington, DC; available at www.faa.gov/regulations_policies/orders_notices/index.cfm/go/document.information/documentID/13975.

Federal Aviation Administration (FAA). 2008. *Area Equivalent Method (AEM) Version 7.0 User's Guide.* U.S. Department of Transportation, Federal Aviation Administration, Office of Environment and Energy, Washington, DC; available at www.faa.gov/about/office_org/headquarters_offices/aep/models/aem_model/.

Federal Aviation Administration (FAA). 2000. "Aviation Noise Abatement Policy 2000," *Federal Register* 65(136); available at www.faa.gov/airports/environmental/policy_guidance/media/fr_vol65_no136.pdf.

Federal Highway Administration (FHWA). 2006. *Highway Traffic Noise in the United States—Problem and Response.* U.S. Department of Transportation, Federal Highway Administration, Office of Human and Natural Environment, Washington, DC; available at www.fhwa.dot.gov/environment/probresp.htm.

Federal Highway Administration (FHWA). 1995. *Highway Traffic Noise Analysis and Abatement Policy and Guidance.* U.S. Department of Transportation, Federal Highway Administration, Office of Human and Natural Environment, Washington, DC; available at www.fhwa.dot.gov/environment/noise/polguide/index.htm.

Federal Railroad Administration (FRA). 2005. *High-Speed Ground Transportation Noise and Vibration Impact Assessment.* U.S. Department of Transportation, Federal Railroad Administration, Office of Railroad Development, Washington, DC; available at www.fra.dot.gov/downloads/RRDev/final_nv.pdf.

Federal Railroad Administration (FRA). 2003. *Final Environmental Impact Statement: Interim Final Rule for the Use of Locomotive Horns at Highway-Rail Grade Crossings.* U.S. Department of Transportation, Federal Railroad Administration, Office of Railroad Development, Washington, DC: available at www.fra.dot.gov/downloads/rrdev/HORNS_FEIS_MASTER.pdf.

Federal Railroad Administration (FRA). "Horn Model." Federal Railroad Administration, Washington, DC, available at www.fra.dot.gov/us/content/254.

Federal Transit Administration (FTA). Internet information and link for download of the FTA noise impact assessment spreadsheet: www.fta.dot.gov/planning/environment/planning_environment_2233.html.

Fleming, Gregg G. 1996. "Aircraft Noise Measurement: Instrumentation and Techniques." Letter Report No. DTS-75-FA653-LR5, U.S. Department of Transportation, Volpe National Transportation Systems Center, Acoustics Facility, Cambridge, MA.

Fleming, Gregg G., et al. 2000. *Transportation-Related Noise in the United States.* TRB A1F04 Committee on Transportation-Related Noise and Vibration, Millennium Publication. Transportation Research Board, National Research Council, Washington, DC.

Galloway, W. J. 1974. "Community Noise Exposure Resulting from Aircraft Operations: Technical Review." Report No. AMRL-TR-73-106, Wright-Patterson Air Force Base, Aerospace Medical Research Laboratory.

Griefahn, Barbara, and Mathias Basner. 2009. "Noise-Induced Sleep Disturbances and Aftereffects on Performance, Well-Being and Health." Inter-Noise, Ottawa, Canada.

Hanson, Carl E., et al. 2006. "Transit Noise and Vibration Impact Assessment." Report No. FTA-VA-90-1003-06, U.S. Department of Transportation, Federal Transit Administration, Washington, DC.

Harris, Cyril M., ed. 1991. *Handbook of Acoustical Measurements and Noise Control.* New York: McGraw-Hill.

Harris, Miller, Miller, and Hanson, Inc. 2006. *Chicago Rail Efficiency and Transportation Efficiency Model (CREATE)* (spreadsheet prediction model); available at www.fra.dot.gov/us/content/253.

Harris, Miller, Miller, and Hanson, Inc. 2005. *High-Speed Rail Initial Noise Evaluation (HSRNOISE)* (spreadsheet prediction model). Federal Railroad Administration, Washington, DC; available at www.fra.dot.gov/us/content/167.

He, H., et al. 2007. "Integrated Noise Model (INM) Version 7.0 User's Guide." Report No. FAA-AEE-07-04, U.S. Department of Transportation, Federal Aviation Administration, Office of Environment and Energy, Washington, DC.

Hendriks, Rudolf W. 1998. *Technical Noise Supplement: A Technical Supplement to the Traffic Noise Analysis Protocol.* California Department of Transportation, Division of Environmental Analysis, Sacramento, CA; available at www.dot.ca.gov/hq/env/noise/index.htm.

Hubbard, Harvey H., ed. 1995. *Aeroacoustics of Flight Vehicles: Theory and Practice*, Vols. 1 and 2. Woodbury, NY: Acoustical Society of America.

International Electrotechnical Commission (IEC). 1995. *Electroacoustics: Octave Band and Fractional Octave Band Filter.* IEC 61260:1995, International Electrotechnical Commission, Geneva, Switzerland.

International Electrotechnical Commission (IEC). 2002. *Electroacoustics: Sound Level Meters, Part 1: Specifications.* IEC 61672-1:2002, International Electrotechnical Commission, Geneva, Switzerland.

Institute of Noise Control Engineering of the USA (INCE): www.inceusa.org.

International Institute of Noise Control Engineering (I-INCE): www.i-ince.org.

International Organization for Standardization (ISO). 2009. *Mechanical Vibration, Shock and Condition Monitoring: Vocabulary.* ISO 2041:2009, International Organization for Standardization, Geneva, Switzerland; available at www.iso.org.

International Organization for Standardization (ISO). 1990. *Mechanical Vibration and Shock—Vibration of Buildings: Guidelines for the Measurement of Vibrations and Evaluation of Their Effects on Buildings.* ISO 4866:1990, International Organization for Standardization, Geneva, Switzerland; available at www.iso.org.

International Organization for Standardization (ISO). 1997. *Acoustics—Measurement of the Influence of Road Surfaces on Traffic Noise, Part 1: Statistical Pass-By Method.* ISO 11819-1:1997, International Organization for Standardization, Geneva, Switzerland; available at www.iso.org.

International Organization for Standardization (ISO). 1997. *Acoustics—Measurement of the Influence of Road Surfaces on Traffic Noise, Part 2: Close-Proximity Method* (draft). ISO 11819-1:1997, International Organization for Standardization, Geneva, Switzerland; available at www.iso.org.

Knauer, Harvey S., et al. 2006. *FHWA Highway Construction Noise Handbook.* Report Nos. FHWA-HEP-06-15, DOT-VNTSC-FHWA-06-02, and NTIS PB2006-109012, U.S. Department of Transportation, Federal Highway Administration, Office of Natural and Human Environment, Washington, DC; available at www.fhwa.dot.gov/environment/noise/handbook/index.htm.

Knauer, Harvey S., Soren Pedersen, Cynthia S. Y. Lee, and Gregg G. Fleming. 2000. *FHWA Highway Noise Barrier Design Handbook.* Report Nos. FHWA-EP-00-05 and DOT-VNTSC-FHWA-00-01, U.S. Department of Transportation, Volpe National Transportation Systems Center, Acoustics Facility, Cambridge, MA.

Lee, Cynthia S. Y., and Gregg G. Fleming. 2001. "General Health Effects of Transportation Noise." Letter Report No. DTS-34-RR297-LR2, U.S. Department of Transportation, Volpe National Transportation Systems Center, Acoustics Facility, Cambridge, MA.

Lee, Cynthia S. Y., and Gregg G. Fleming. 1996. *Measurement of Highway Related Noise.* Report Nos. FHWA-PD-96-046 and DOT-VNTSC-FHWA-96-5, U.S. Department of Transportation, Volpe National Transportation Systems Center, Acoustics Facility, Cambridge, MA.

Menge, Christopher W., Christopher F. Rossano, Grant S. Anderson, and Christopher J. Bajdek. 1998. *FHWA Traffic Noise Model, Version 1.0: Technical Manual.* Report Nos. FHWA-PD-96-010 and DOT-VNTSC-FHWA-98-2, U.S. Department of Transportation, Volpe National Transportation Systems Center, Acoustics Facility, Cambridge, MA; available at www.fhwa.dot.gov/environment/noise/tnm/index.htm.

National Park Service (NPS). 1994. *User's Manual for the National Park Service Overflight Decision Support System.* BBN Report 7984, prepared by BBN Systems and Technologies, U.S. Department of Interior, National Park Service, Denver, CO.

Page, J. A., et al. 2002. *Rotorcraft Noise Model (RNM 3.0) Technical Reference and User Manual.* Report No. WR 02-05, Wyle Laboratories, Arlington, VA.

Plotkin, Kenneth J., and Louis C. Sutherland. 1990. *Sonic Boom: Prediction and Effects.* Tallahassee, FL: AIAA Professional Studies Series.

Rasmussen, Robert Otto, et al. 2007. *Little Book of Quieter Pavements.* Report No. FHWA-IF-08_004, Federal Highway Administration, Office of Pavement Technology, Washington, DC.

Reherman, Clay, et al. 2006. *FHWA Roadway Construction Noise Model Version 1.0 User's Guide.* Report Nos. FHWA-HEP-05-054 and DOT-VNTSC-FHWA-05-01, U.S. Department of Transportation, Federal Highway Administration, Office of Natural and Human Environment, Washington, DC; available at www.fhwa.dot.gov/environment/noise/cnstr_ns.htm.

Richardson, W. John, Charles R. Greene, Jr., Charles I. Malme, and Denis H. Thomson. 1995. *Marine Mammals and Noise.* San Diego, CA: Academic Press.

Rochat, Judith L. 2009. "Developing U.S. Wayside Methods for Measuring the Influence of Road Surfaces on Traffic Noise." Inter-Noise, Ottawa, Canada.

Rochat, Judith L., and Gregg G. Fleming. 1999. *Acoustics and Your Environment—The Basics of Sound and Highway Traffic Noise.* Video Production and Letter Report No. DTS-34-HW966-LR1, U.S. Department of Transportation, Volpe National Transportation Systems Center, Acoustics Facility, Cambridge, MA.

Rochat, Judith L. 1998. "Effects of Realistic Ocean Features on Sonic Boom Noise Penetration into the Ocean: A Computational Analysis." Ph.D. thesis, Pennsylvania State University, University Park, PA.

Sandberg, Ulf, and Jerzy A. Ejsmont. 2002. *Tyre/Road Noise Reference Book.* INFORMEX, Harg, SE-59040 Kisa, Sweden; available at www.informex.info.

Schexnayder, C. J., and J. Ernzen. 1999. *Mitigation of Nighttime Construction Noise, Vibration, and Other Nuisances.* NCHRP Synthesis of Highway Practice 218, Transportation Research Board, National Research Council, Washington, DC.

Schultz, Theodore J. 1982. *Community Noise Rating*, 2nd ed. New York: Applied Science Publishers.

Seep, Benjamin, et al. 2000. *Classroom Acoustics.* Melville, NY: Acoustical Society of America; available at http://asa.aip.org/map_publications.html.

Society of Automotive Engineers (SAE). 2006. *Monitoring Aircraft Noise and Operations in the Vicinity of Airports: System Description, Acquisition, and Operation.* SAE-ARP-4721/1, Society of Automotive Engineers, Philadelphia; available at www.sae.org.

Society of Automotive Engineers (SAE). 2006. *Monitoring Aircraft Noise and Operations in the Vicinity of Airports: System Validation.* SAE-ARP-4721/2, Society of Automotive Engineers, Philadelphia; available at www.sae.org.

SoundPLAN, LLC. *SoundPLAN (A Software Package).* Braunstein + Berndt GmbH, Shelton, WA.

Stapelfeldt Ingenieurgesellschaft mbH. *LimA (A Software Package).* Stapelfeldt Ingenieurgesellschaft mbH, Dortmund, Germany.

State DOTs: www.adc40.org/adc40membershiplist.pdf.

Stusnick, E., M. L. Montroll, K. J. Plotkin, and V. K. Kohli. 1982. *Handbook for the Measurement, Analysis, and Abatement of Railroad Noise.* Report No. DOT/FRA/ORD-82/02H, U.S. Department of Transportation, Federal Railroad Administration, Office of Research and Development, Washington, DC.

Transport noise. U.S. federal government website on transportation noise: http://ostpxweb.dot.gov/policy/safetyenergyenv/noise.htm.

Transportation Research Board (TRB) Web site: www.trb.org.

Wayson, R. L., and J. M. MacDonald. 1998. *The AMAA Community Noise Model, Version 5.0, User Guide.* University of Central Florida, Orlando, FL.

Wolfel. *IMMI (A Software Package).* Wolfel, Hoechberg, Germany.

CHAPTER 34
TRANSPORTATION-RELATED AIR QUALITY

Shauna L. Hallmark
Department of Civil and Construction Engineering
Iowa State University
Ames, Iowa

34.1 INTRODUCTION

Air pollution is the presence of undesirable material in the air in sufficient quantities to pose health risks to humans, damage vegetation, negatively impact ecological systems, reduce visibility, or damage property (de Nevers 2000). A number of naturally occurring processes contribute to air pollution, such as dust, forest fires, and volcanoes. In urban areas, most pollution originates from human activity and includes stationary sources, such as factories or power plants, area sources including facilities such as dry cleaners, and mobile sources (USEPA 2000). Mobile sources include automobiles, trucks, buses, aircraft, trains, marine activity, farming equipment, construction equipment, lawn mowers, etc. Based on most recently available estimates from the U.S. Environmental Protection Agency (USEPA 2005), on- and off-road mobile sources contribute approximately 84 percent of carbon monoxide (CO), 58 percent of nitrogen oxides (NO_x), 44 percent of volatile organic compounds (VOCs), 9 percent of particulate matter that is 2.5 microns in diameter and smaller ($PM_{2.5}$), and 2 percent of PM_{10}. Relative contributions from the transportation sector for CO, VOCs, and NO_x are provided in Figures 34.1 through 34.3.

34.2 NATIONAL AMBIENT AIR QUALITY STANDARDS

The Clean Air Act (CAA) was passed in 1963 and subsequently amended in 1970, 1977, and 1990. The Clean Air Act and Amendments (CAAA) provide the legal basis for air pollution laws in the United States. The role of the U.S. Environmental Protection Agency (USEPA) is to formulate and publish regulations demonstrating how air quality laws should be applied. Regulations set forth by the USEPA are subject to public hearings, approval by the Office of Management and Budget (OMB), and in some cases litigation before they have the force of law (de Nevers 2000).

The CAA set National Ambient Air Quality Standards (NAAQS). Primary standards were instituted at levels to protect the public health of the most sensitive members of the population, which includes children, the elderly, and asthmatics. Secondary standards also were set with the intent to protect the public welfare and the environment, including damage to crops, ecosystems, vegetation, buildings, and decreased visibility. The six criteria pollutants are ozone, carbon monoxide, nitrogen dioxide, lead (Pb), sulfur dioxide (SO_2), and particulate matter (PM).

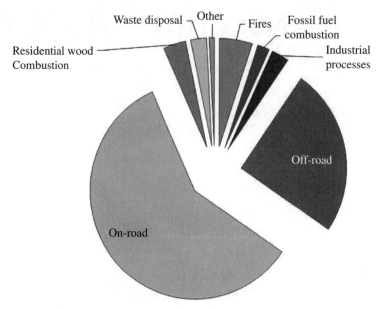

FIGURE 34.1   Major sources of carbon monoxide. (*Data source:* USEPA, 2005.)

The CAAA require nonattainment areas to reduce emissions to meet NAAQS. Areas that are not in compliance with NAAQs for ozone, CO, NO_x, SO_2, and PM must develop statewide implementation plans (SIPs) to demonstrate conformity. Areas in nonattainment for the criteria pollutants are required to show compliance by projecting mobile-source inventories for SIPs and estimating differences between alternatives (including "do nothing") for transportation plans, programs, and

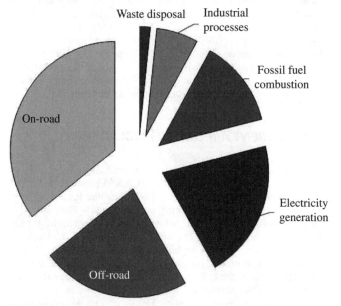

FIGURE 34.2   Major sources of NO_x. (*Data source:* USEPA, 2005.)

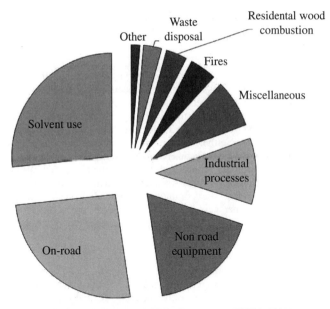

FIGURE 34.3 Major sources of VOCs. (*Data source:* USEPA, 2005.)

projects. Emissions from alternatives must be less than "do nothing" alternatives in order for transportation plans and programs to proceed. *Do nothing* is defined as the current transportation system including future projects that receive environmental approvals under National Environmental Policy Acts (NEPA) (Chatterjee et al. 1997). Table 34.1 shows the primary standards for NAAQS.

NAAQS include two types of pollutants. Primary pollutants are those emitted directly into the atmosphere. Secondary pollutants are those formed indirectly. Primary pollutants include carbon dioxide (CO), sulfur dioxide (SO_2), and lead (PB). Ozone is a secondary pollutant and is a by-product of a photochemical reaction in the atmosphere between volatile organic compounds (VOCs) and nitrogen oxides (NO_x). Airborne PM is a combination of primary and secondary pollutants (TRB 2000). Mobile sources contribute a significant amount of pollution for only four of the six criteria pollutants: VOCs, CO, NO_x, and PM. Only minor proportions of lead and SO_2 are released by mobile sources. Consequently, they are typically not evaluated.

TABLE 34.1 Primary Standards for NAAQS as of May 2010

Pollutant	Type of average	Concentration
Carbon monoxide	8-Hour	9 parts per million (ppm)
	1-Hour	35 ppm
Lead	Maximum quarterly average	1.5 µg/m³
Nitrogen dioxide	Annual	53 parts per billion (ppb)
	1-Hour	100 ppb
Ozone	8-Hour average	0.08 ppm
Particulate matter (PM_{10})	24-Hour	150 µg/m³
Particulate matter ($PM_{2.5}$)	Annual arithmetic mean	15 µg/m³
	24-Hour	35 µg/m³
Sulfur dioxide	Annual arithmetic mean	0.03 ppm
	24-Hour	0.14 ppm

Source: USEPA (2010a).

34.3   TRANSPORTATION-RELATED POLLUTANTS

34.3.1   Carbon Monoxide

Carbon monoxide is a colorless, odorless gas that can be lethal when inhaled at high concentrations. It also causes headaches and fatigue. Epidemiologic studies have indicated that a relationship may exist between elevated levels of CO and cardiovascular problems (Godish 1997). In the body, hemo-globin binds with CO rather than with oxygen, and the body is deprived of oxygen (Wark, Warner, and Davis 1998). As shown in Table 34.1, the 1-hour standard for CO is 35 ppm, and the 8-hour standard is 9 ppm. Neither standard can be exceeded annually. Carbon monoxide typically is evaluated on a localized basis because of its immediate health effects.

The amount of CO produced depends on the air to fuel ratio (A/F) in the combustion process. Under rich combustion (lower A/F), CO production is increased (de Nevers 2000). Fuel-rich conditions occur when cold starts or under engine loading, such as acceleration, high speeds, or operation on steep grades when cars are not tuned properly and at higher altitudes (USEPA 2001). Lean fuel mixtures (higher A/F) occur when cruising or decelerating (Homburger et al. 1996).

34.3.2   Oxides of Nitrogen

Two main oxides of nitrogen are formed during the combustion process. Nitric oxide (NO) is formed during combustion from naturally occurring N_2 and O_2 under high temperatures. Vehicle engines emit a mixture composed of primarily NO with some nitrogen dioxide (NO_2). NO oxidizes to NO_2 in the atmosphere and is a major contributor to photochemical smog. The concentrations that are produced depend on the temperature of combustion and spark ignition timing (Homburger et al. 1996). Oxides of nitrogen are one of the principal components in smog and react with VOCs in the presence of sunlight to form ozone.

34.3.3   Ozone

Ozone is a severe irritant. It occurs naturally in the upper atmosphere and with NO_2 as a major component of smog. Ozone is a secondary pollutant and is a by-product of a photochemical reaction in the atmosphere between VOCs and nitrogen oxides. The USEPA requires reductions of both NO_x and VOCs to control ozone formation but focuses on reduction of VOCs as the most effective strategy. Ozone damages materials, causes eye and respiratory irritations, decreases pulmonary function, and decreases heart rate and oxygen intake. Effects include a sore throat, chest pain, cough, and head-ache. The effects are more adverse for the young and the old. Ozone also affects plant growth and causes crop and vegetation damage (Wark, Warner, and Davis 1998).

34.3.4   Volatile Organic Compounds

VOCs are not a criteria pollutant but react with NO_x to form ozone. Hydrocarbon emissions (HCs), a form of VOCs, are a result of incomplete combustion or fuel evaporation, as shown in Figure 34.4. About half of HC emissions are tailpipe emissions (de Nevers 2000). Hydrocarbons are evaporated during vehicle operation as well as while the vehicle is parked. Evaporative emissions make up a significant portion of HC emissions and can be categorized as (USEPA, 1994)

Diurnal. Diurnal emissions are a result of evaporation that occurs as the fuel tank heats and cools throughout the day owing to changes in ambient temperature.

Running losses. Running losses occur during vehicle operation. Additional evaporation occurs owing to fuel being in contact with a hot engine and exhaust system. Evaporative emissions escape while the vehicle is operating. Emissions depend on ambient temperature, fuel vapor pressure, and driving cycle (TRB 2000).

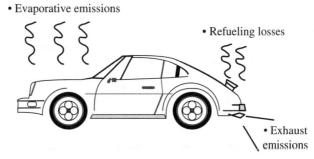

• Evaporative emissions

• Refueling losses

• Exhaust emissions

FIGURE 34.4 Schematic of vehicle emissions. (*Source:* USEPA, 1994.)

Hot soak. Hot soak emissions result from evaporation during the first hour following engine shutdown. While the engine is still hot, fuel is heated above ambient temperatures, and evaporation occurs primarily from the fuel tank and carburetor bowl in carburetor vehicles. Hot soak emissions depend on a number of factors, including ambient temperature, fuel vapor pressure, proportion of fuel in the fuel tank, and vehicle technology (TRB 2000). Most hot-soak emissions occur in the first 10 minutes following engine shutdown (USEPA 1999).

Refueling. During refueling, fuel vapors are forced from the fuel tank, contributing to HC emissions.

34.3.5 Particulates

Particulates are fine particles of material suspended in the air. Particulate matter is produced by industry, natural sources, motor vehicles, agricultural activities, mining and quarrying, and wind erosion. Particles emitted from vehicles consist primarily of lead compounds, motor oil, and carbon particles (Homburger et al. 1996). Particulates are classified by size. PM_{10} refers to all particles with a diameter of 10 microns or less. Sources include road dust, windblown dust, etc. PM_{10} particles settle rapidly, and consequently, their range of influence is limited to the immediate area. $PM_{2.5}$ particles have a diameter of 2.5 microns or less. They are formed from fuel combustion, fireplaces, and wood stoves. $PM_{2.5}$ particles are much smaller than PM_{10} particles and settle less rapidly. Consequently, they are transported greater distances in the atmosphere and are more uniformly dispersed in urban areas (USEPA 2001).

Both PM_{10} and $PM_{2.5}$ particles are inhaled and accumulate in the respiratory system. Coarser particulate matter aggravates respiratory conditions such as asthma, whereas fine particulate matter results in decreased lung function and increased hospital admissions and emergency department visits. Particulate matter is a major cause of reduced visibility, including in national parks. It also affects the acidity balance in land or water systems, and deposition on plants can corrode plants or interfere with plant metabolism (USEPA 2001).

34.4 *ESTIMATING TRANSPORTATION-RELATED EMISSIONS*

The CAAAs and subsequent amendments set forth a number of requirements for nonattainment areas to work toward meeting NAAQs. For transportation activities, estimates of both area-wide and project-level transportation emissions are necessary to meet the various requirements established by the legislation (TRB 1998). Mobile-source emissions (typically in grams) are estimated by

$$\text{Total emissions} = \text{vehicle activity} \times \text{emission rate} \tag{34.1}$$

Vehicle activity estimates vary by agency and type of analysis but frequently are provided in miles of vehicle activity. Emission rates are provided by either the USEPA's MOBILE model or in California, the Emission Factors (EMFAC) model. The MOBILE model is the primary emission factor tool to estimate on-road emissions. It is used by national, state, and local agencies to estimate emissions, make policy decisions, measure environmental impacts, and create national, regional, and urban emission inventories (TRB 2000). In California, the EMFAC model is the primary emission factor model. Estimates of mobile-source emission are used almost exclusively to determine both the current and projected future contributions of mobile sources.

34.4.1   Motor Vehicle Emission Simulator (MOVES)

Calculation of emission factors historically was done using the USEPA's series of MOBILE models, which estimated fleet average, in-use fleet emission rates for hydrocarbons (HCs), carbon monoxide (CO), and oxides of nitrogen (NO_x). The USEPA currently has released its most recent emission modeling system, Motor Vehicle Emission Simulator (MOVES). MOVES is based on the analysis of millions of emission test results, including data from dynamometer tests, vehicle inspection and maintenance (I/M) programs, remote-sensing-device testing, certification testing, and portable emission measurement systems (PEMS). The model also represents advances in USEPA's understanding of motor vehicle emissions (USEPA 2010b). One of the main differences between MOVES and MOBILE is that MOBILE emission rates are based on aggregate driving cycles, whereas MOVES emission rates are based on operating mode. Additionally, MOBILE models relied on certification data for heavy-duty diesel emissions, whereas MOVES has included a chassis dynamometer and on-board emissions measurements (USEPA 2010b).

MOVES can model HCs, VOCs, CO, NO_x, PM, SO_2, energy consumption, greenhouse gas emissions (i.e., CO_2, methane, and NO), and several mobile-source air toxics (MSATs), including benzene, 1, 3-butadiene, formaldehyde, acetaldehyde, acrolein, naphthalene, ethanol, and methyl tertiary butyl ether (MTBE) (USEPA 2009a, 2010b, 2010c).

MOVES is able to calculate either total inventory estimates of emissions (total emissions in units of mass) or emission rates, which represent emissions per unit of distance for running emissions or emission per vehicle for starts, extended idling, and resting evaporative emissions. The model also allows modeling at the national, county, or project-level scale. The national scale uses national defaults and can be used to estimate emissions for the entire country, a state, or a group of counties. The county scale uses local meteorology, fleet, and activity data at the level of a county. The project scale can be used to conduct microscale analysis of emissions for individual roadway links or for hot-spot analysis (USEPA 2010d).

MOVES can model the calendar year 1990 and calendar years 1999 to 2050 and can estimate emissions by month of the year or day of the week. The model includes 13 vehicles types, which are termed *source types*, as shown in Table 34.2, and 4 fuel types, including gasoline, diesel, compressed natural gas, and electricity. Five different road types are included in the model (USEPA 2010d):

- *Off-network locations*, where the predominant activity is vehicle starts, and parking and idling, such as parking lots, truck stops, rest areas, and freight or bus terminals
- *Rural restricted access*, where the rural highways are only accessed by on-ramps
- *Rural unrestricted access*, where rural roads not included in the preceding categories, which include arterials, connectors, and local streets
- *Urban restricted access,* urban roadways that are accessed only by on-ramps
- *Urban unrestricted access*, urban roads not included in the preceding category, which include arterials, connectors, and local streets

Speed is modeled in MOVES as a distribution rather than an average value, as was done for MOBILE (USEPA 2010d). Speed distributions are shown in Table 34.3. Vehicle activity is modeled using a vehicle-specific poser (VSP), which is a measure of the energy a vehicle is using for a

TABLE 34.2 Vehicle Sources in MOVES

Source types in MOVES	HPMS[a] vehicle type
Motorcycle	Motorcycles
Passenger car	Passenger cars
Passenger truck	Other two-axle, four-tire vehicles
Light commercial truck	Other two-axle, four-tire vehicles
Intercity bus	Buses
Transit bus	Buses
School bus	Buses
Refuse truck	Single-unit truck
Single-unit short-haul truck	Single-unit truck
Single-unit long-haul truck	Single-unit truck
Motor home	Single-unit truck
Combination short-haul truck	Combination truck
Combination long-haul truck	Combination truck

[a]Highway performance monitoring system.
Source: Data from USEPA (2010d).

TABLE 34.3 Speed Ranges in MOVES

Average bin speed, mph	Speed range in bin Speeds, mph
2.5	Speed < 2.5
5	$2.5 \leq$ speed < 7.5
10	$7.5 \leq$ speed < 12.5
15	$12.5 \leq$ speed < 17.5
20	$17.5 \leq$ speed < 22.5
25	$22.5 \leq$ speed < 27.5
30	$27.5 \leq$ speed < 32.5
35	$32.5 \leq$ speed < 37.5
40	$37.5 \leq$ speed < 42.5
45	$42.5 \leq$ speed < 47.5
50	$47.5 \leq$ speed < 52.5
55	$52.5 \leq$ speed < 57.5
60	$57.5 \leq$ speed < 62.5
65	$62.5 \leq$ speed < 67.5
70	$67.5 \leq$ speed < 72.5
75	$72.5 \leq$ speed

Source: Data from USEPA (2010d).

particular moment in time. VSP is an operating mode and is modeled in MOVES by speed and VSP. Twenty-three operating mode bins are defined (USEPA 2010c). VSP represents the tractive power required for a vehicle to move itself and accompanying passengers and cargo and is estimated by equation (34.2) for light-duty vehicles (USEPA 2009b):

$$P_{V,t} = \frac{Av_t + Bv_t^2 + Cv_t^3 + mv_t a_t}{m} \tag{34.2}$$

where $P_{V,t}$ = vehicle specific power in kW/tone
A = rolling resistance in kW-sec/m
B = rotational resistance in kW-sec^2/m^2
C = aerodynamic drag in kW-sec^3/m^3
v_t = speed at time t (m/sec)
a_t = acceleration at time t (m/sec^2)
m = weight (mass in tone)

34.5 DISPERSION MODELING

The total amount of pollution produced is important. Health effects, however, are measured in concentrations of pollution, so the NAAQs represent amounts of ambient pollution. Dispersion models are used to model how pollutants are transported and dispersed once they are emitted from vehicles or sources. Dispersion of pollutants depends on a number of factors, including

- Chemical and physical characteristics of the pollutants
- Meteorologic conditions such as atmospheric stability, wind speed and direction, and ambient temperature
- Topography
- Distance from the source

Models predict the concentrations of pollutants present at a specific location. The urban airshed model (UAM) is the most common regional dispersion model. It uses a three-dimensional (3D) photochemical grid to mathematically simulate the atmosphere and calculate pollutant concentrations based on physical and chemical processes of the atmosphere. Typical input requirements include hourly gridded emission for NO$_x$ and VOCs, hourly estimates of the height of the mixed layer, ambient temperature, ambient humidity, solar radiation, cloud cover, atmospheric pressure, and an hourly 3D wind fields. Typical model outputs include hourly average concentrations by grid square for pollutants of interest, instantaneous concentrations of pollutants by grid square at the beginning of the averaging period, calculation of summer ozone levels, concentrations of winter CO, and projections of future emission scenarios (Chatterjee et al. 1997).

Microscale analysis is usually performed to measure whether local violations of the NAAQs for CO may occur. Microscale analysis is used frequently to model concentrations at intersections where vehicles may spend significant amounts of time queuing and idling. Locations of queuing and idling often are characterized by elevated emission rates. CAL3QHC is the microscale dispersion model required by the USEPA. In California, CALINE-4 is the microscale dispersion model used. Both are Gaussian plume models. For microscale analysis, concentrations typically are measured at specific receptor locations and during peak hours.

34.6 CONTROL OF AIR POLLUTION FROM VEHICLES

Most improvements in reducing motor vehicle emissions, with the exception of lead, have been technological improvements to the vehicles themselves. One type of improvement is engine design features such as fuel injectors with computer-controlled fuel and air injection rates that optimize the combustion mixture, chamber design, compression ratio, spark timing, and exhaust gas recirculation. Add-on pollution-control technology also has contributed to emission reduction. The most significant add-on pollution-control improvement was the catalytic converter. Catalytic converters treat engine exhaust by reducing NO$_x$ to N$_2$ and O$_2$ and oxidizing VOCs and CO to H$_2$O and CO$_2$. Another technological improvement was carbon canisters. The canisters absorb evaporative VOC emissions from the hot engine once the vehicle has been turned off. When the vehicle is started again, intake air is circulated through the canister, and absorbed VOCs are sent back into the cylinders for combustion. Another device, positive crankcase ventilation valves, routes air from the engine crankcase back into the cylinders for burning (Cooper and Alley 2002). Reduction of lead from motor vehicles came about almost entirely by removal of lead from gasolines.

34.6.1 Fuels

Gasoline is a mixture of olefinic, aromatic, and paraffinic HC compounds. Composition varies from refinery to refinery and even from one geographic area to another. The composition of fuel is an important factor in the amount of emissions that are produced by both evaporation and combustion (Godish 1997). One improvement in fuels was a reduction in the maximum allowable Reid vapor pressure (RVP) of gasoline, resulting in reductions in VOCs from vehicles as well as in evaporative emissions at gas stations and gasoline storage areas (Cooper 2002).

Oxygenated fuels are another improvement in fuels. Oxygenated compounds include MTBE, ethyl tert-butyl ether (ETBE), tertiary amyl methyl ether, and ethanol. MTBE is the most frequently used oxygenated compound. Oxygenated fuels have one or more oxygen atoms embedded in the fuel molecule. They are commonly added during winter months to reduce CO emissions. A 3 percent oxygen by weight blend is estimated to reduce CO emission by 30 percent. HC emissions are slightly decreased as well, although NO_x emissions may increase slightly (Cooper 2002). Several alternative fuels are in use or have been evaluated as replacements for gasoline. They include compressed natural gas, liquefied petroleum gas, pure methanol, pure ethanol, and alcohol-gasoline blends ("gasohol"). Reductions in emission vary by fuel type. Natural gas and propane produce significantly less CO and lower emissions of VOCs. Evaporative emissions from ethanol and methanol are less reactive.

34.6.2 Transportation Control Measures

Although technological controls on vehicles have led to significant reductions in the amount of pollutants released by each vehicle, increases in the total number of vehicles on the road and increases in the number of vehicle miles traveled annually by each vehicle threaten to outpace technological advances. As a result, methods to reduce overall travel and to reduce amounts of time vehicles spend in activity where emissions are higher such as idling are needed. The CAAAs identify transportation control measures (TCMs) that are expected to decrease motor vehicle use and decrease emissions. TCMs are listed in Section 108(f) of the Clean Air Act Amendments of 1990. They are also shown in Table 34.4.

TCMs are strategies intended to both reduce the total number of vehicle miles traveled (VMTs) and to make that travel more efficient (Cambridge Systematics 1991). Transportation control measures are required to help reduce the amount of pollution released by the transportation sector to improve air quality and meet federal requirements. Although various definitions exist for TCMs, a general description is that they are actions designed to change travel demand or vehicle operating characteristics to reduce motor vehicle emissions, energy consumption, and congestion. TCMs include transportation supply improvement strategies and transportation demand management strategies. Transportation supply improvement strategies either change the physical infrastructure or implement actions for more efficient use of existing facilities to improve traffic flow and decrease stop-and-go movement. Supply improvement strategies take the form of bottleneck relief, construction improvements, improved signal timing, ramp metering, applications of intelligent transportation system technology, and alterations to land-use patterns. Demand management measures attempt to change driver behavior to reduce the frequency and length of automobile trips. Demand management measures include, but are not limited to, no-drive days, employer-based trip-reduction programs, parking management, park-and-ride programs, work-schedule changes, transit-fare subsidies, ride sharing, and public-awareness programs (Guensler 1998).

TABLE 34.4 TCMs Listed in the CAAAs

Public transit
High-occupancy vehicle facilities
Employer-based transportation management plans
Trip-reduction ordinances
Traffic flow improvements
Park-and-ride/fringe parking
Vehicle-use limitations or restrictions in downtown areas during peak periods
Limitations on certain roads for use by pedestrians and nonmotorized vehicles
Ride-share and high-occupancy-vehicle (HOV) programs
Bike lanes and storage facilities
Control of extended vehicle idling
Reduction of cold-starts
Flexible work schedules
Programs to encourage nonautomobile travel and reduce the need for single-occupant vehicle travel (includes provision for special events and major activity centers)
Pedestrian and nonmotorized vehicle paths
Scrappage of older vehicles

REFERENCES

Cambridge Systematics, Inc. 1991. *Transportation Control Measure Information Documents*. Prepared for the U.S. Environmental Protection Agency Office of Mobile Sources, Cambridge, MA.

Chatterjee, Arun, Terry L. Miller, John W. Philpot, Thomas F. Wholley, Jr., Randall Guensler, David Hartgen, Richard A. Margiotta, and Peter R. Stopher. 1997. "NCHRP Report 394: Improving Transportation Data for Mobile Source Emission Estimates." Transportation Research Board, National Research Council, National Academy Press, Washington, DC.

Cooper, David C., and F. C. Alley. 2002. *Air Pollution Control: A Design Approach*, 3rd ed. Prospect Heights, IL: Waveland Press.

de Nevers, Noel. 2000. *Air Pollution Control Engineering*, 2nd ed. New York: McGraw-Hill.

Godish, Thad. 1997. *Air Quality*, 3rd ed. Boca Raton, FL: Lewis Publishers.

Guensler, R. 1998. "Increasing Vehicle Occupancy in the United States," in *L'Avenir Des Deplacements en Ville* (The Future of Urban Travel). Lyon, France: Laboratoire d'Economie des Transports.

Homburger, Wolfgang S., Jerome W. Hall, Roy C. Loutzenheiser, and William R. Reilly. 1996. *Fundamentals of Traffic Engineering*, 14th ed. Institute of Transportation Studies, University of California, Berkeley.

Transportation Research Board (TRB). 1998. *NCHRP Research Results Digest: Number 230*. Washington, DC: Transportation Research Board, National Research Council.

TRB. 2000. *Modeling Mobile-Source Emissions*. Washington, DC: Transportation Research Board, National Research Council, National Academy Press.

U.S. Environmental Protection Agency (USEPA). 1994. *Automobile Emissions: An Overview*. Fact Sheet OMS-5, EPA 400-F-92-007. USEPA, Washington, DC.

USEPA. 1999. *Hot Soak Emissions as a Function of Soak Time*. EPA420-P-98-018, M6.EVP.007, USEPA Office of Mobile Sources, Washington, DC.

USEPA. 2000. *Latest Findings on National Air Quality: 1999 Status and Trends*. EPA-454/F-00-002, USEPA Office of Air Quality Planning and Standards, Research Triangle Park, NC.

USEPA. 2001. *National Air Quality and Emissions Trends Report, 1999*. EPA 454/R-01-004, USEPA Office of Air Quality Planning and Standards, research Triangle Park, NC.

USEPA. 2002. *User's Guide to MOBILE 6.0: Mobile Source Emission Factor Model*. EPA420-R-02-001, USEPA Office of Air and Radiation, Washington, DC.

USEPA. 2005. "Air Source Emissions." Used as data source; available at. www.epa.gov/air/emissions/index.htm; accessed May 2010.

USEPA. 2009a. *Motor Vehicle Emission Simulator (MOVES) 2010: User Guide*. EPA-420-B-09-041, USEPA Office of Transportation and Air Quality, Washington, DC.

USEPA. 2009b. "Development of Emission Rates for Light-Duty Vehicles in the Motor Vehicle Emission Simulator (MOVES2009)," draft report. EPA-420-B-09-002, USEPA Office of Transportation and Air Quality, Washington, DC.

USEPA. 2010a. *National Ambient Air Quality Standards (NAAQS)*. Available at www.epa.gov/air/criteria.html#4; accessed May 2010.

USEPA. 2010b. "MOVES (Motor Vehicle Emission Simulator)." USEPA Office of Transportation and Air Quality; available at www.epa.gov/otaq/models/moves/index.htm; accessed May 2010.

USEPA. 2010c. "An Introduction to MOVES2009." USEPA Office of Transportation and Air Quality; available at www.epa.gov/otaq/models/moves/420b09026 pdf; accessed May 2010.

USEPA. 2010d. Technical Guidance on the Use of MOVES2010 for Emission Inventory Preparation in State Implementation Plans and Transportation Conformity." EPA-420-B-10-023, USEAP, Transportation and Regional Programs Division, Office of Transportation and Air Quality, Washington, DC.

Wark, Kenneth, Cecil F. Warner, and Wayne T. Davis. 1998. *Air Pollution: Its Origin and Control*, 3rd ed. Reading, MA: Addison-Wesley.

CHAPTER 35
CLIMATE CHANGE AND TRANSPORTATION

Zhong-Ren Peng,[*,†] **Suwan Shen,**[*] **Qingchang Lu,**[†,*] **and Sarah Perch**[*]

35.1 INTRODUCTION

Climate change is generally recognized as a reality that must be confronted. Influenced by the past and what is happening in the present, it will be a part of the future and therefore must be studied thoroughly. Climate change very much affects transportation and infrastructure. The demands placed on infrastructure vary and will alter with climate changes. For example, coastal roads and other infrastructure will be more likely to be inundated as sea level rises. There are two ways in which climate changes must be taken into account: mitigation and adaptation. Both are necessary considerations in long-range transportation planning and engineering processes. This chapter explores what climate change is and what it means for the future of transportation planning and engineering.

The International Panel on Climate Change (IPCC) defines *climate change* as "a change in the state of the climate that can be identified . . . by change in the mean and/or the variability of its properties and that persists for an extended period . . ." (Hegerl 2007). Climate is influenced by many factors, but what experts believe to be one of the most influential is greenhouse gas emission. Using climate models, researchers are able to estimate various scenarios of climate change; however, there is much uncertainty in what the future will bring. Although climate change is recognized, many reports are still based on historical data, which might not be appropriate for future predictions. As the Pew Center on Global Climate Change writes, "It is generally understood that impacts felt today are a result of emissions from decades past, and based on our current, and even higher emission levels today, we are already committed to greater warming, precipitation changes, and sea level rise in the future" (Cruce 2009).

The term *climate change* is vast and encompasses many aspects of social, economic, and environmental activities. In respect to the transportation sector, there are a few main aspects of climate change that are more important to study, namely, the potential increase in extreme events, such as more intensive hurricanes, sea level rise, changes in patterns of water availability, rising temperatures, increased heat waves and more very hot days, and demand-induced effects, such as increased usage on some roads and less on those more likely to be inundated. There is great potential for transportation infrastructure to be damaged by extreme heat, that rising sea levels could inundate coastal roads and threaten bridges and ports, that runoff could wash away roads, and that hurricanes also could severely damage these structures. Transportation planners, engineers, and decision makers must address these concerns (Neumann 2009).

---

[*]Department of Urban and Regional Planning, University of Florida, Gainesville, Florida.
[†]School of Transportation Engineering, Tongji University, Shanghai, China.

In summary, climate change and transportation are interrelated. Greenhouse gases from the transportation sector accelerate the pace of climate change, which, in return, affects the way we plan, design, and operate transportation systems. Correspondingly, there are two basic strategic approaches to climate change: mitigation and adaptation, which are defined by the United Nations Environment Program (2010) as "building resilience to climate change" and "facilitating a transition toward low carbon societies." Both are necessary to combat the effects of climate change. The following sections will first provide an introduction to the interactions between climate change and transportation and then summarize the challenges for transportation professionals to deal with climate change.

35.1.1   Interactions Between Climate Change and Transportation

Transportation is inherently linked to climate change both by the nature of the transport sector and transportation infrastructures. Transportation contributes toward and affects climate change, which, in turn, affects transportation infrastructures and costs. The transport sector contributes about one-quarter of global CO_2 emissions, and this number is only rising because urbanization is increasing rapidly throughout the world (UNFCCC 2010). The relationship between transportation and climate change is one that has been, and must continue to be studied in detail so as to develop more efficient and effective mitigation and adaptation strategies.

There are two ways in which transportation and climate change relations must be studied. First, the role transport plays on accelerating climate change must be understood so that practices can be adjusted to curb negative effects. Second, the effects of climate change on transportation infrastructures must be studied, and measures must be taken to mitigate and adapt to these changes.

Sustainable transportation measures must be taken to combat climate change. The cost of not adapting to climate change is to risk increasing damage and loss of land. Adaptations can be technological, policy-based, behavioral, or managerial (McNeil 2009). Along coastal areas, which are especially vulnerable to climate change, there is, for example, the risk of losing land owing to sea level rise (World Bank 2010). Adapting infrastructure and its use is a way in which greenhouse gas (GHG) emissions can be reduced. Potential efforts to curb emissions include

1. Using alternative fuel sources
2. Promoting fuel-efficient vehicles
3. Reducing vehicle miles driven
4. Integrating land use and transportation planning
5. Increasing the use of Intelligent Transportation Systems (ITS) and Telecommuting
6. Coordinating transportation systems
7. Avoiding unnecessary travel (FHWA 2005; UNFCC 2010)

Incorporating these measures will help to reduce potential effects of future climate change.

However, there is also the question of what can be done now. Mitigation is challenging due to the constraints of current infrastructure, policies, and conventional practices. Often it is difficult to modify the present. In addition, mitigation strategies are heavily connected to adaptation strategies, including the need to avoid excess travel, shifting modes of travel, and implementing technological improvements (UNFCCC 2010).

35.1.2   Challenges for Transportation Planning, Engineering, and Decision Making

Climate change is a global concern. It is often thought of in the context of global organizations, institutions, and decision makers (e.g., IPCC, World Bank, and UN Environment Program). However, as these organizations recognize, climate change must be understood and dealt with on many levels, including local, regional, national, and international levels.

There are many challenges to implement mitigation and adaptation strategies throughout the necessary governmental levels. For smaller, local organizations, insufficient access to funding is a frequent barrier. Regional coordination faces problems in unification of smaller localities and the challenge of a region with varying climate change scenarios. On a national level, individual areas face their own climate change concerns and will be affected differently. Globally, climate change must account for the economic, social, and environmental situations of each individual area. Efforts from each level are necessary in order to encourage the use of climate-friendly technology and infrastructure, as well as research on climate change. Partnerships must be formed and cultivated, and agreements need to be made.

Climate change is an increasing concern to planners, engineers, and decision makers. It will potentially affect much in the field of transportation. However, there are still many uncertainties as to how this will happen. Planners, engineers, and decision makers must understand climate change and its potential impacts. In the face of unknown future scenarios, decisions must be made as to the planning, design, and implementation of transportation policy that will be affected by climate change. The purpose of this chapter is to outline some of the potential ways in which climate change can be understood, addressed, and logically approached.

35.2 THE ROLE OF TRANSPORTATION IN CLIMATE CHANGE MITIGATION

Considering the undesirable and even dangerous impacts that climate change will have on the environment, it is necessary to reduce GHG emissions that cause climate change. According to the U.S. Greenhouse Gas Inventory (U.S. Environmental Protection Agency 2009), transportation, electricity generation, and industry sectors are the three major sources of GHG emissions in the United States. Figure 35.1 shows the proportion of GHG emissions by economic sector. As demonstrated by the figure, transportation is the second largest source of GHGs, contributing to almost one-third of all nationwide emissions (National Research Council 2009). In addition, transportation is the fastest-growing sector in emissions, growing at 2 percent annually (Environmental Protection Agency 2009). Among the subsectors within transportation, light-duty vehicles and heavy-duty vehicles constitute more than 80 percent of total CO_2 emissions, whereas passenger rail and buses contribute just 1 percent of total emissions, as seen in Figure 35.2. Considering the large share transportation contributes to GHG emissions, this sector is a crucial consideration in the climate change mitigation process.

By 2008, there were 23 states with GHG emissions Targets (Figure 35.3), and 21 states had adopted GHG reduction plans (EPA n.d.; Pew Center n.d.). State departments of transportation and local metropolitan planning organizations are in the process of developing mitigation policies

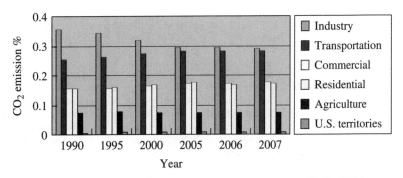

FIGURE 35.1   U.S. CO_2 emissions by economic sectors. (*Data Source:* U.S. EPA 2009.)

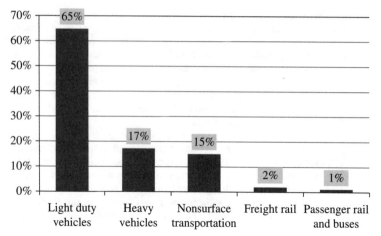

FIGURE 35.2   Estimate CO_2 emissions by mode, U.S. transportation sector, 2006. (*Data Source:* EIA, 2007.)

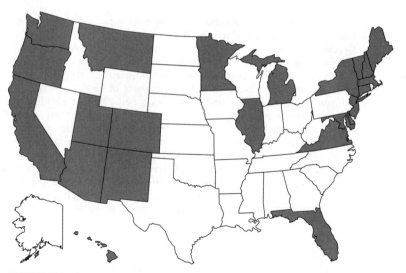

FIGURE 35.3   States with greenhouse gas emissions targets. (*Source:* Pew Center on Global Climate Change. 2009.)

with respect to the transportation sector, including mandates and incentives promoting biofuels and vehicle miles traveled (VMT)–reduction-related policies and incentives.

There is much research focusing on the interaction between transportation and climate change mitigation (U.S. Environmental Protection Agency 2009; Environmental Protection Agency 2000; Chu and Meyer 2009; Yang et al. 2009; Sperling et al. 2009). These research efforts address topics ranging from GHG estimations to the impact of land-use policy on GHG emissions, contributing toward and enriching the knowledge base of climate change mitigation strategies and practices. The following sections will provide an overview of GHG estimation practices and major mitigation strategies.

35.2.1 How to Measure, Inventory, and Forecast GHG Emissions?

There are many mitigation strategies that have been proposed to reduce GHG emissions in the transportation sector, including vehicle efficiency improvements, the use of lower-GHG-content fossil fuels, and the use of intelligent transportation system technologies to improve system efficiencies (Lutsev et al. 2008). It is clear that a GHG inventory and measurement tools are needed to provide a baseline against which the effectiveness of these mitigation strategies can be measured.

Usually, a GHG emission inventory is composed of the following three factors: the chemical identity of pollutants, the geographic area and time period addressed, and the types of contributing factors (Transportation and Climate Change Clearing House n.d.). The pollutants in the inventory often include carbon dioxide (CO_2), methane (CH_4), nitrous oxide (N_2O), fluorinated gases, and so on.

Currently, there are three levels of emission inventories within the United States, namely, those at the national, state, and local levels. At the national level, the U.S. Energy Information Administration and the Environmental Protection Agency provide annual GHG emissions inventory reports, keeping record of the total emissions, as well as emissions by economic sector (U.S. Energy Information Administration 2009; U.S. Environmental Protection Agency 2009). EIA and EPA also provide forecasts of emission trends in different sectors. Since EPA follows the methodologies recommend by the IPCC, most transportation-related emissions fall under the category of energy source (Transportation and Climate Change Clearing House n.d.).

At the state level, the EPA provides state CO_2 emission inventories from fossil fuel combustion using fuel consumption data from the Department of Energy (DOE)/EIA State Energy Data Consumption tables and emission factors from the Inventory of U.S. Greenhouse Gas Emissions and Sinks 1990–2007 (EPA State and Local Climate Change and Energy Program n.d.). They also provide tools (e.g., state inventory and projection tools) and guidelines to assist state governments in creating their own GHG inventory, tracking activities that lead to emissions or removals, and also the methods needed to make GHG estimations. As of 2010, 46 states have completed their own GHG inventories (Figure 35.4).

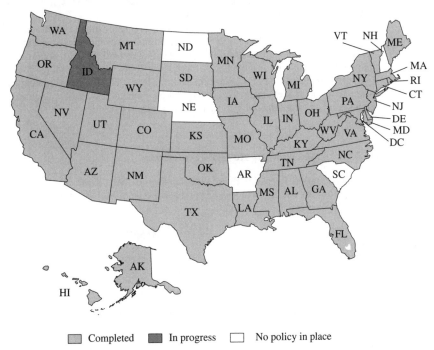

FIGURE 35.4 State GHG inventory. (*Source:* EPA State and Local Climate and Energy Program, n.d., State Planning and Incentive Structures; Available at www.epa.gov/statelocalclimate/state/tracking/state-planning-and-incentive-structures.html#a02.)

At the local level, emission inventories include emissions from all community activities within local jurisdiction boundaries and sectors within the community. These inventories serve as a useful planning tool to help local decision makers develop mitigation action plans (EPA State and Local Climate Change and Energy Program n.d.). The data required to create local GHG emission inventories include basic activity data and emission factors from the U.S. Greenhouse Gas Emissions Inventory (EPA State and Local Climate Change and Energy Program n.d.). There are two methods to collect activity data. One relies on aggregated data collected at the state, national, or global level. The other involves data collected from local users (e.g., utilities) (EPA State and Local Climate Change and Energy Program n.d.). In addition, the EPA provides a number of protocols and guidance documents with default values to help local governments build their own inventory. Local transportation planners have a variety of tools and software available to them [e.g., NONROAD Model, Emission FACtor (EMFAC) model] to help measure and project GHG emissions.

There are various tools available to help with GHG calculations and inventory compilations. NONROAD and EMFAC are direct GHG emission calculation tools that produce emission factors or emission estimates for gases during vehicle use, focusing only on transportation sources (ICF Consulting 2006). In addition to these kinds of tools, there are also multisector inventory tools (e.g., state inventory tool, state inventory projection tool), life-cycle GHG emission calculation tools (e.g., GREET model, MOVES), transportation/emissions strategy analysis tools (e.g., COMMUTER model), and energy/economic forecasting analysis tools (e.g., National Energy Modeling System, VISION) (ICF Consulting 2006). All these estimate GHG emissions and quantify the impacts of proposed transportation alternatives and strategies. Specifically, the multisector inventory tools are used to develop GHG inventories by economic sector (ICF Consulting 2006). The life-cycle GHG calculation tools evaluate the efficiency of advanced vehicle technologies and new transportation fuels and generate emission factors from project-level analyses (ICF Consulting 2006). Transportation/emissions strategy analysis tools are designed to evaluate the travel and emissions impacts of transportation strategies given specified inputs of the strategies (ICF Consulting 2006). The energy/economic forecasting analysis tools use economic factors, such as fuel prices, to forecast energy consumption at the national level (ICF Consulting 2006).

While it is useful to have tools with which one can estimate GHG emissions, they come with limitations. First, most of the tools focus on national-level analysis, providing little insight into local-level concerns (ICF Consulting 2006). Second, it is often difficult to obtain accurate data (e.g., motor vehicle fuel consumption) required for a state-level inventory, especially considering the differences between jurisdictions (ICF Consulting 2006). Third, vehicle operating characteristics generally are ignored in the models (ICF Consulting 2006). Nonetheless, this is an important consideration in developing and implementing mitigation strategies.

35.2.2   General Approaches to Mitigate GHG Emissions

In Transportation Research Board Special Report 290 (2008), it is proposed to use the following equation to characterize the CO_2 emissions from fuel combustion by transportation vehicles (National Research Council 2008):

$$G = A * Si * Ii * F_{ij} \tag{35.1}$$

$G = CO_2$ emissions from fuel combustion by transport
A = total transport activity
Si = modal structure of transport activity
Ii = energy consumption (fuel intensity) of each transport mode
$F_{i,j}$ = GHG emissions characteristics of each transport fuel (i = transport mode, j = fuel type)

$A*Si$ represents "the demand for transport services provided by transport mode i" (National Research Council 2008). $Ii*F_{ij}$ represents "the GHG generated by each unit of transportation service provided by mode i using fuel type j" (National Research Council 2008).

This causal relationship means that GHG reductions in the transportation sector can be achieved by three alterations: increasing fuel efficiency, encouraging alternative modes of transportation with fewer GHG emissions, and reducing travel demand. The primary approaches to achieving these objectives include improving fuels and vehicle technology, surface multimodal transportation planning and land-use integration, and transportation system operation efficiency improvements.

Fuels and Vehicle Technology. The literature provides a summary of the vehicle technologies that have the potential to reduce fuel consumption and GHG emissions (National Research Council 2008). There are five primary vehicle technologies: engine, nonengine, and aircraft technologies, as well as waterborne vessel and railroad engines improvements (Table 35.1).

Certain states have begun to adopt these new technologies in their mitigation plans. As of June 2009, 38 states mandated and incentivized the promotion of biofuels, particularly ethanol production and use. Twelve states have introduced renewable fuels standards (RFS) (Figure 35.5). Nevertheless, there is a debate about whether these new technologies could contribute to GHG reductions if broad issues, such as agriculture land changes or the replacement of conservation land with agricultural land, are taken into consideration (Searchinger et al. 2008).

Surface Multimodal Transportation Planning and Land-Use Integration. GHG mitigation can be incorporated into surface multimodal transportation planning and land-use integration through travel demand management (TDM), transit investment, bicycle/pedestrian projects, and land-use policies such as growth management, transit-oriented development (TOD), and traditional neighborhood development (TND) strategies. Using these methodologies and tools, DOTs and MPOs are able to work toward VMT reduction, which correlate strongly with emission reductions. Pricing and other forms of TDM have been identified as tools to control VMT (Meyer 1999). Recent studies have continued to exploring the relationship between compact development and VMT (National Research

TABLE 35.1 Categories of Fuels and Vehicle Technology

Type of technology	Subtypes	Technology examples
Engine	Spark-ignition internal combustion engines	Variable value control, controlled auto ignition (CAI)
	Compression-ignition engines	Diesel engines
	Hybrid vehicles	Series hybrids, parallel hybrids, full hybrids
	Fuel cell vehicles	Proton-exchange-membrane fuel cells
Nonengine	Transmission technologies	Continuously variable transmission (CVT)
	Technologies to reduce vehicle weight	Composite materials, the replacement of mechanical or hydraulic systems by electrical or electronic systems
	Tire technologies	Energy-efficient tires
	Technology to improve vehicle aerodynamics	
	Technologies to reduce the energy requirements of onboard equipment	Energy-efficient onboard components
	Technologies with the potential to reduce fuel consumption by nonroad vehicle (e.g., aircraft, waterborne vessels, and railroad locomotives)	
Aircraft	Reduce specific fuel consumption, increase aerodynamic efficiency, and improved structural efficiency	Engine efficiency, aerodynamic efficiency, structural efficiency
Waterborne vessel	Diesel engine improvement	
Railroad engines	Electric power, efficiency of electric locomotives	Use of ac power

Source: National Research Council 2008.

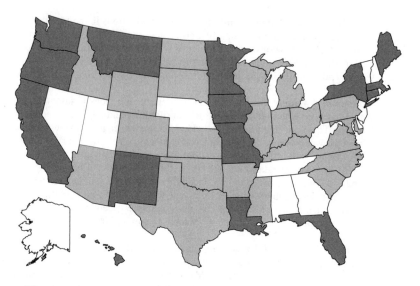

☐ Tax exemptions, credits and/or grants

■ Above + RFS with biofuel mandates

FIGURE 35.5   States with mandates and incentives promoting biofuels. (*Source:* Pew Center on Global Climate Change, n.d., Available at www.pewclimate.org/what_s_being_done/in_the_states/map_ethanol.cfm.)

Council 2009). For example, compact development with higher residential and employment densities are more likely to reduce VMT (National Research Council 2009). Similarly, TOD and TND are able to improve connectivity and therefore reduce external trips and reduce VMT (Johnston et al. 1999). A number of states have developed policies to reduce VMT either directly or though land-use policies, such as smart growth, which supports high-density, compact, mixed-use development (Pew Center on Global Climate Change n.d.). Current modeling techniques are not able to evaluate the effectiveness of multimodal transportation plans and land-use mitigation policies. Research efforts must be put forth to encourage "studies of threshold population and employment densities to support alternatives to automobile travel" and "studies of changing household and travel preferences" (National Research Council 2009).

System Operation Efficiency Improvement.   By reducing vehicle delays and improving traffic flows, transportation system operation strategies can prevent unnecessary emissions. The tools available in this field include incident management, traveler information, transit signal priority management, and freeway management. The *ITS/Operations Resource Guide* (U.S. DOT 2009) provides the principles, technologies, and training necessary for transportation planners and engineers to implement these strategies.

This section summarized the current practice to measure, forecast, and mitigate GHG emissions in the transportation sector. It provided an overview of current models, technologies, and policies available to reduce GHG emissions. However, even if no GHGs will be produced in the future, the existing amount of GHGs are changing climates. To only consider mitigation without adaptation will put society at a high risk of suffering tremendous losses in the future. Therefore, the next sections will address climate change's impact on transportation and corresponding adaptation strategies, which are two important parts of the greater picture.

35.3 *IMPACTS OF CLIMATE CHANGE ON TRANSPORTATION*

Debates regarding the inevitability of climate change and its effects on scientific fields have raised public concerns as to the potential impacts of climate change. Rising sea levels, more frequent precipitations, and storms will significantly challenge current urban infrastructures. In the transportation sector, the National Research Council (2008) announced that the performance of the national transportation system will be profoundly affected by climate change. Even if GHGs do not increase further, the accumulation of GHG emissions over the past centuries is enough to cause climate change that quite likely will increase both the frequency and intensity of extreme weather in the near future (National Research Council 2008; U.S. Environment Protection Agency 2005). Consequently, it is especially important to understand how climate change interacts with the current transportation system and how to minimize the negative impacts of climate change with minimal cost in the long-range transportation planning process. This section will provide an overview of the impacts analysis of several of the most relevant climate change variables on the transportation system.

35.3.1 Most Relevant Climate Change Variables to Transportation

The National Research Council (2008) has identified five types of climate extremes through which climate change may have primary effects on transportation. These include very hot days and heat waves, increases in Arctic temperatures, rising sea levels, intense precipitation, and extreme hurricanes (National Research Council, 2008).

1. *Increases in very hot days and heat waves.* During the 21st century, more intense, longer-lasting, and more frequent heat extremes and heat waves have the possibility of greater than 90 percent to continue occurring (Peterson et al. 2006). These temperature changes potentially will impact pavement and structural design, therefore challenging current materials standards and maintenance practices.

2. *Increases in Arctic temperatures.* According to the IPCC (2007), arctic warming is virtually certain (with a probability greater than 99 percent). Increases in Arctic temperatures will significantly impact permafrost, raising the subsidence risk of transportation infrastructures.

3. *Rising sea levels.* Rising sea levels are virtually certain (with probability of greater than 99 percent) in the 21st century (IPCC 2007). The rising of sea levels will challenge the transportation infrastructures in coastal areas, with more frequent and severe flooding, as well as increasing the frequency of road and bridge interruptions caused by inundation.

4. *Increases in intense precipitation events.* It is highly likely (with probability of greater than 90 percent) that more intense and frequent precipitation events will occur in United States (IPCC 2007). Changing precipitation levels could increase flooding levels and raise the moisture content in soils, as well as affect pavement, foundation, and drainage designs.

5. *Increases in hurricane intensity.* It is estimated that increased tropical storm intensities will become likely (with probability of greater than 66 percent) in the next century (IPCC 2007). Larger and more frequent storm surges and more powerful wave action will increase the possibility of infrastructure failures (e.g., bridge damage and road erosions) and threat the harbor infrastructure.

While the transportation system as a whole is vulnerable to all these impacts, individual vulnerabilities differ by mode of transportation and the geographic location of the infrastructure (National Research Council 2008). Table 35.2 summarizes the impacts of the five climate change variables on various modes of transportation.

Given the clustered nature of populations and the increased exposure to climate change variables, such as sea level rise and storm surges, coastal regions are particularly susceptible to the effects of climate change. For instance, with the increased intensity and frequency of precipitation and storm events with storm surges of 7 meters (23 feet), more than half the Gulf Coast's major highways

TABLE 35.2   Impacts of Climate Change on Different Modes of Transportation

Climate change variables and situations	Land transportation (highways, rail, pipeline)	Marine transportation	Air transportation
Increase in very hot days and heat waves	Pavement and concrete construction Thermal expansion on bridge expansion joints and paved surfaces Limitations of constructions activities Vehicle overheating and tire deterioration Pavement integrity	Impact of warmer water on shipping	Delays and flights canceled due to extreme heat Energy consumption Pavement and concrete facilities
Increases in Arctic temperatures	Subsidence of infrastructure due to thawing of permafrost Short season for ice roads	Longer ocean transport season, possible northern sea route	Undermining runway foundations caused by thawing of permafrost
Rising sea levels	Interruptions on coastal and low-lying infrastructure Flooding and evacuation Inundation and erosion Reduced clearance under bridges Land subsidence	High tides cause change in harbor and port facilities, reduce clearance under bridges, change navigability of channels	Closure and restrictions caused by inundation
Increases in intense precipitation events	Increase of weather-related delays, traffic disruptions, flooding and evacuation, disruption of construction and maintenance operations, affects structure integrity	Increase of weather-related delays, impacts on harbor infrastructure, affects channel depth	Increase of weather-related delays and airport closings, structural integrity of airport facilities, destruction of navigation aid instruments, damage to pavement drainage systems
Increases in hurricane intensity	More debris on roads and rail lines, interrupting travel and shipping, extensive emergency evacuations, greater probability of infrastructure failures, threat to stability of bridge decks, decreases infrastructure lifetimes	Emergency evacuation, challenge to robustness of infrastructure, damage to harbor infrastructure, damage to cranes and other dock and terminal facilities	More frequent interruptions, damage to landside facilities

Source: National Research Council 2008, pp. 135–141; ANNEX 3-1, Potential Climate Changes and Impacts on Transportation.

(64 percent of interstates 57 percent of arterials), almost half of the rail miles, 29 airports, and virtually all the ports are subject to flooding (U.S. Climate Change Program 2008). When considering the mid-Atlantic region, including the District of Columbia, Maryland, North Carolina, and Virginia, sea level rise is generally considered to be the most critical climate issue in the region (ICF International 2007). Extending beyond coastal areas, transportation infrastructures are also vulnerable to climate change. For example, roads, rail lines, runways, and pipelines in Alaska have already suffered from land subsidence issues caused by thawing of the permafrost (National Research Council, 2008). In the Midwest, severe flooding as a result of intense precipitation threatens transportation infrastructures (National Research Council 2008). Nearly all regions in the United States will be affected somehow by climate change.

35.3.2   General Approach for Climate Change Impact Analysis

As indicated in Table 35.2, the greatest impact of climate change for North America's transportation systems is predicted to be flooding-related (National Research Council 2008). While extreme temperatures challenge transportation infrastructure materials, sea level rise, increased precipitation, and more intense storms all impose greater threats to the transportation system because of the associated flood risks (Zimmermanrch 2003).

Studies have been conducted to evaluate the impacts of climate change on transportation infrastructures (Burkett 2002; Titus 2002; Suarez et al. 2005; Jacob et al. 2007; ICF International 2007; U.S. Climate Change Program 2008; U.S. Environmental Protection Agency 2009). These studies have addressed many associated concerns, ranging from local climate change (e.g., sea level rise) predictions to general conclusions about transportation facilities vulnerable to climate change and the associated economic costs, providing valuable information and instruction as to how to adapt to climate change in coastal areas. However, global average climate projections and 10- to 30-meter Digital Elevation Models (DEMs) are often used in studies to identify the number of infrastructures vulnerable to floods. This level of accuracy is too rough to make specific suggestions for decision making.

In order to incorporate climate change into the transportation planning and operation processes, two gaps must be addressed. First, local climate change projections should be developed based on global climate change model outputs in order to estimate climate change's impacts at the local level. Global average projection delivers an overall simple and broadly general idea of climate change. It does not provide an accurate description of climate change at the local level. Detailed local climate change projections should consider factors such as differences in terrain, land use, coastal erosion, and local subsidence. Second, a series of quantitative measures should be developed to provide a comprehensive and detailed delineation of climate change's impacts on the transportation system. Currently, most studies simply use number of roads being inundated, vehicle hours traveled (VHT), and vehicle miles traveled (VMT) as indicators of the impacts climate change on system performance (ICF International 2007; U.S. Climate Change Program 2008; Peterson et al. 2008; Jacob et al. 2007; Suarez et al. 2005). A comprehensive index needs to be developed to assist multimodal system impact assessments and further risk assessments.

Transportation professionals, hydrologists, and climate scientists should work together to overcome the aforementioned gaps. The following describes a general approach that can be followed to estimate the impacts of climate change on transportation:

1. Identify the climate change variables of greatest relevance for the transportation system within the study area.
2. Determine the time frame over which each variable will become an influential factor.
3. Determine the geographic scale at which these climate change variables can be projected with acceptable confidence.
4. Assess each transportation mode's vulnerability to these climate change variables.
5. Estimate each climate variable's direct impacts on transportation infrastructures by transportation mode.
6. Estimate each climate variable's indirect impacts (e.g., social effects, economic costs, increase in operation and maintenance costs).

Climate change should be accounted for and incorporated into the long-range transportation planning process (National Research Council Committee on Climate Change and U.S. Transportation 2008; Meyer 2008; Committee on Climate Change and U.S. Transportation Research Board 2008; ICF 2008). A Federal Highway Administration report, "Integrating Climate Change into the Transportation Planning Practice" (2008), concluded that climate change impact assessments and adaptations could be incorporated into the following steps of long-range transportation planning: establishing goals, defining performance criteria and data needs, evaluating deficiencies, developing alternative plan scenarios, and evaluating alternatives and selecting a preferred alternative. Together with traditional transportation planning methods, this general approach could be implemented through the following procedures (ICF 2008):

1. Define performance measures, indicators, and variables that reflect climate change impact on the transportation system.
2. Estimate system performance under different climate change scenarios.
3. Estimate the risk and economic cost of each scenario.

4. Develop an adaptation plan.

5. Evaluate risk, economic cost, as well as effectiveness and efficiency based on the system performance measures of each plan.

35.3.3 Impacts of Sea Level Rise on Transportation

Sea level rise and associated storm surges will significantly increase the risk of inundation and corruption of coastal transportation infrastructures, threatening the reliability of transportation networks in coastal regions. Increased risk of bridge failure during extreme weather, periodic or permanent inundation of coastal infrastructures, and increased maintenance and protection costs with respect to vulnerable infrastructures are several examples of the many potential impacts of sea level rise on transportation systems. Unless these effects are considered at an early stage, the inevitable maintenance and repair costs could be tremendous.

According to the IPCC (2001), since the peak of the last ice age, the average global sea level has risen more than 120 meters. Moreover, the global average rate of sea level rise in the 20th century was greater than that in the 19th century (IPCC 2007). The IPCC Fourth Assessment Report (IPCC 2007) projected that using conservative estimates, average global sea levels will continue to rise by 0.18 to 0.59 meters before the year 2100. While global average sea level rise indicates a long-term acceleration trend, the rate of sea level rise differs significantly at the local level. For instance, the historic rate of sea level rise in the Gulf of Mexico is much higher than in many other regions of the United States (IPCC 2007).

At the local level, relative sea level is influenced by a variety of factors, including the global average sea level (eustatic sea level), gradual uplift or subsidence of land elevation, abrupt changes owing to a seismic event, gradual erosion, rapid bluff collapse, atmosphere pressure, weather systems such as El Niño and tides (California Coastal Commission 2001). Local-scale sea level is usually referred to as *relative sea level*, different from the global change in sea level and eustatic sea level, and is most often estimated by isolating the changes in water level from tide records (California Coastal Commission 2001).

As sea levels increase, they will provide a greater base from which storm surges can sweep inland. This means that the damage to coastal infrastructures will be significantly more destructive than with previous storm surges. In addition, flooding will affect low-lying areas and cause increased coastal erosion, which is a major challenge to transportation facilities (U.S. Climate Change Science Program 2009). For example, inundation and erosion are especially threatening to the state of Florida, where the highest point is only 53 feet above sea level, and the Florida Keys are all less than 10 feet above sea level (U.S. Global Change Research Program 2003). Studies suggest that a 1-foot rise in sea level will cause land erosion of 100 to 1,000 feet off the coastline (U.S. Global Change Research Program 2003). Losses caused by occasional flooding have already been observed in Dixie County on Florida's Gulf Coast (U.S. Global Change Research Program 2003).

Many studies use global average sea level rise instead of local sea level predictions to estimate the impacts of sea level rise on transportation systems (National Assessment Synthesis Team and U.S. Global Change Research Program 2000; National Research Council 2008; Peterson et al. 2008). Although several studies focus on local sea level projections, they only offer general conclusions about the length of road, area of road segments, and traffic flow that would be affected by generated projected sea level levels (Burkett 2002; Titus 2002; Suarez et al. 2005; U.S. Climate Change Program 2008; DuVair et al. 2002). These analyses usually use 10- to 30-meter DEM data from the USGS National Elevation Dataset (NED) under the assumption that DEM represents true elevation value (Suarez et al. 2005; ICF International 2007; U.S. Climate Change Program 2008). However, few studies examine the elevation data and accuracy issues. Although several papers recognize the issue of horizontal accuracy and use interpolation as a solution (U.S. Environmental Protection Agency 2009), vertical accuracy issues are widely ignored in climate change studies. When considering slight increments of sea level rise, the vertical accuracy matters more than horizontal accuracy. The vertical accuracy of DEM from NED varies from 7 to 15 meters (USGS 2009), whereas the estimation of sea level rise is usually within 1 meter. This implies that the DEM data from NED are not suitable for sea level rise analysis because the sea level increase is well within the error range of the DEM data. Since vertical accuracy is an important factor in impact assessment, more research efforts are needed to help local practitioners understand how the vertical accuracy of terrain data affects impact analysis (Shen et al. 2010).

Moreover, in addition to direct inundation, sea level rise will lead to more destructive storm waves and increased erosion, as well as corresponding social and economic effects, all of which will greatly affect the durability and reliability of the transportation system. Further studies should take these indirect impacts into account.

35.3.4 Impacts of Precipitation on Transportation

According to the Environmental Protection Agency, observations compiled by National Oceanic and Atmospheric Administration's (NOAA) National Climatic Data Center indicate an average increase in annual precipitation at a rate of 6.1 percent per century, with considerable variation by region: "[T]he greatest increases came in the East North Central climate region (11.6 percent per century) and the South (11.1 percent). Hawaii was the only region to show a decrease (−9.25 percent)" (U.S. Environmental Protection Agency, United States Precipitation Changes 2010).

Change in precipitation will have two impacts on transportation. For some regions, the change in rainfall pattern will result in increases in drought conditions. Simultaneously, for other regions, the change in seasonal precipitation and river flow patterns will increase the risk of floods (National Research Council 2008). Specifically, the region around the Great Lakes is estimated to experience drier conditions, resulting in lower water levels, whereas the Houston region will experience an increase in intense precipitation events (National Research Council 2008). Figure 35.6 shows the distribution of precipitation change in United States.

As demonstrated in Figure 35.6, most regions have experienced an increase in precipitation, which has the potential to increase the severity of flooding events. The National Research Council (2008)

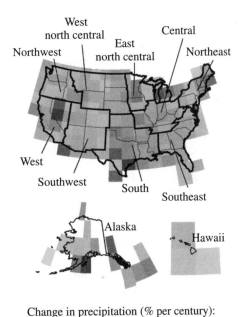

Change in precipitation (% per century):

−30 −20 −10 0 10 20 30

FIGURE 35.6 Change of precipitation in United States, 1901–2005. (*Source:* Environmental Protection Agency, 2010, Precipitation and Storm Changes. Available at www.epa.gov/climatechange/science/recentpsc.html.)

uses the Great Flood of 1993, which occurred in the Mississippi and Missouri River system, to illustrate the potential impacts of such an increase. During this event, a catastrophic flood caused disruptions to the surface transportation system within 500 miles of the river system, affecting rail, truck, and marine traffic. Increases in precipitation also will increase moisture levels in soils, therefore affecting the stability of pavement subgrades. Runoff from increased precipitation also may affect stream flow and sediment and negatively impact bridge foundations.

Traditionally, the frequency, intensity, and duration of extreme precipitation of different return periods have been used by civil engineers to design transportation infrastructures (National Research Council 2008). The *Project Development and Design Manual* proposed by the U.S. Department of Transportation (U.S. Department of Transportation 2008) provides detailed standards, criteria, and recommended methods of dealing with hydrology-related issues in highway and bridge design. There are specific requirements regarding roadway hydraulics with respect to the culverts, ditches, pavement drainage, storm drains, energy dissipaters, and alternative pipe materials. The limitation of these standards is that they are based solely on historical records of flood magnitude and runoff. If the intensity and frequency of precipitation increase in the near future as a result of climate change, it would lead to a higher magnitude of flood and a larger amount of runoff. This requires that the current design specifications be updated.

Estimating the impacts of precipitation on transportation requires multidisciplinary corporation. In order to generate quantified impact assessment of changing precipitation at the local level, the following research gaps need to be overcome:

1. Spatial interpolation of current discharge and rainfall data. Daily discharge and rainfall data are recorded at specific stations, and spatial interpolation is needed to estimate discharge and rainfall at other locations.
2. Establish the relationship between rainfall pattern and riverine discharge values, which require an estimation of regional hydrologic lag time.
3. Based on global climate change model output, estimate the magnitude of future extreme precipitation with a certain return period.
4. Create an accurate hydrologic data inventory for the study area so as to convert the discharge value into a flooding map in GIS software.

35.3.5 Summary

This section provides an overview of the climate variables that are most relevant to transportation, a rationale for why they are important, and the potential impacts of these variables. At the moment, although overall impact assessments are available in various regions, detailed estimations are still insufficient and inaccurate at the local level. In order to help transportation decision makers adapt to climate change, more accurate impact assessments should be conducted at the local level. The following improvements should be incorporated in future to achieve this goal:

1. Use high-resolution elevation data such as LiDAR (Light Detection and Ranging) data.
2. Develop improved climate models that estimate the changes in storm and rainfall frequency and intensity at the local level.
3. Create accurate inventories of major infrastructures, including location and associated economic values.
4. Calculate susceptibility of infrastructures to flood with consideration of their design standards.
5. Develop GIS arithmetic that can calculate system risk under different scenarios.

In order to minimize the risk and costs of climate change, corresponding adaptation strategies should be implemented at the local level. The following section will focus on the climate change adaptations, including topics such as estimating the cost of doing nothing, vulnerability estimation, adaptation approaches, and risk assessment.

35.4   CLIMATE CHANGE ADAPTATIONS

The impacts of global climate change are diverse and are constantly changing the current knowledge base, particularly the effects of climate change on the transportation sector. In order to respond to the challenges posed by climate change, mitigation strategies are needed to eliminate GHG emissions. GHG emissions can remain in the atmosphere for decades once they are emitted, meaning that today's emissions will affect the climate far into the future, no matter what mitigation measures are taken. It is because of the GHG emissions released into the air over the past century that current climate change occurs as it does today. Therefore, adaptation is necessary to address the impacts of global warming (IPCC 2007). Furthermore, the IPCC (2007) defines *adaptation* as the "adjustment in natural or human systems in response to actual or expected climatic stimuli or their effects, which moderates harm or exploits beneficial opportunities" (p. 750). Adaptation strategies should be integrated throughout the transportation planning, designing, and decision-making process to adapt to future climate changes.

Proactive transportation adaptations to climate change can be divided into an assessment of transportation system vulnerability, adaptation strategy alternatives, cost-benefit analysis, and the prioritization of adaptation options.

The first step is vulnerability assessment, which then could be used to identify the transportation segments that are most susceptible to climate change and therefore most vulnerable. These segments have the potential for the most severe consequences on the performance of transportation network. The second step is to develop alternative adaptation strategies in transportation planning, design, construction, operation and maintenance, and the decision-making process. All the possible adaptation strategies should be evaluated based on a cost-benefit analysis of each option. Finally, the adaptation alternatives should be prioritized for decision-making purposes. Methodologies and measures of adaptation should be incorporated into the entire process of transportation decision making. The following sections will give an in-depth review of the steps of climate change adaptation analysis.

35.4.1   The Cost of Doing Nothing

Before adaptation strategies can be put into place, the potential disruptions to the transportation system caused by climate change need to be evaluated. The evaluation will help us to better understand the total damage that may happen to the whole system. Disruptions to the transportation system are the *cost* of what the outcome would be if nothing is done. Furthermore, in addition to cost estimation, risk assessment is another important element to adapt to climate change. The Federal Highway Administration (FHWA 2005) recommends that risk and cost assessments be conducted for infrastructure design over long-term time spans, such as 100- and 500-year designs, specifically considering extreme events, such as water levels rising over structures.

The costs brought to transportation infrastructures by climate change include cost increases in infrastructure investment travel times, GHG emissions, traffic accidents, travel time unreliability, operating costs, and so on as a result of disruption. To make the results of assessment comparable, all these costs should be converted into a uniform value. Certain costs, such as infrastructure damage, are easily calculated using money spent on restoration. On the other hand, however, the impacts on the transportation system, such as travel time and delay increases, as a result of disruptions could not be calculated easily. The general way to generate these figures is to run the traditional four-step transportation planning procedure under network disruption scenario. Here the results, such as increased congestion time, and total VMT and VHT, could be compared with those before the network disruption. The costs of these increases together with the infrastructure damage costs could be converted into monetary values. In addition, the increase in operation and maintenance costs owing to more frequent extreme adverse weather also must be added to the total cost calculation. However, other costs, such as canceled trips because of a lack of available routes, an increase in traffic accidents, and sea water erosion, are not easily captured, complicating the cost calculation process. Although this is difficult to predict, the cost of climate change impacts on transportation infrastructure is very likely to amount to billions of dollars, in addition to other costs associated with population migration and land use changes. For example, according to Cambridge Systematics, Inc. (2009), in the long run, a study estimates that climate change could increase the annual cost of flooding in the United Kingdom almost 15-fold by the 2080s.

Risk assessment is another factor. One of the well-developed methodologies for risk assessment is a probabilistic risk assessment (PRA), which is a comprehensive method of evaluating potential risk. Probabilistic risk assessment is composed of two components: the probability of hazards occurring and the probability of their consequences. According to the Transportation Research Board (TRB Special Report 290, 2008, p. 144), the risk is formulated as follows:

$$\text{Risk} = \sum_{\substack{\text{all assets} \\ \text{all hazards}}} \text{prob(hazard)} \times \text{prob(consequence)}$$

where the first component is the probability of the occurrence of a hazard, and the second component is the cost probability of each consequence. The total risk is the summation of the two results of all the possible hazards.

With respect to climate change and transportation, the *probability of hazards* refers to the occurrence probability of the climate change variables discussed earlier, such as intense precipitation, storms, and sea level rise. The *probability of consequences* refers to the probable impacts of climate change on transportation networks and segments. For instance, the impact of one infrastructure disruption is calculated, as is the direct monetary cost and the cost of network performance reduction. Following this methodology, each consequence is associated with a probability of transportation facility failure and of a value of economic loss. Using this approach, decision makers could effectively manage and prioritize possible risks.

35.4.2 Identify Vulnerable Transportation Systems

In order to respond to climate change effectively, the most vulnerable transportation systems and segments should be clearly identified in order to minimize the disruption costs on the transportation system. Vulnerability analysis should be conducted at both the system level and the individual infrastructure level. The system vulnerability calculates vulnerability from a system point of view, which is usually represented by system performance, whereas the latter deals with vulnerability based on the components of the system and is reflected by the possibility of the component failure. The transportation segment vulnerability should be calculated before calculating the transportation system analysis, which includes both system vulnerability and system reliability. *System reliability* refers to the probability for the system to be able to continue working and fulfilling its normal functions, whereas *system vulnerability* examines the reverse, focusing on the nonfunctioning aspects of the system, especially the consequences to the whole system.

The vulnerability analysis focuses on the systems susceptible to possible climate change scenarios. The vulnerability analysis is challenging because there are many definitions of vulnerability, and no consensus thus far has been reached. One of the most popularly used definitions of *vulnerability* is given by Berdica (2002), who states, "vulnerability in a road transportation system is a susceptibility to incidents that can result in considerable reductions in road network serviceability" (p. 19).

As definitions of vulnerability vary, so do the ways in which it can be measured with respect to the transportation system. There are multiple ways to measure the vulnerability of a transportation system. The most commonly used method is an accessibility-based network vulnerability method calculation, measured by a reduction in the accessibility index. It assumes that road network performance will be degraded by the disruption of any segment of the entire network, which will result in an accessibility reduction to the whole network. The vulnerability of the network is represented by the extent of the accessibility reduction under certain network degradation scenarios.

The accessibility index used by Taylor and D'Este (2006) is a Hansen accessibility index, as shown below:

$$A_i = \frac{\sum_j B_j f(c_{ij})}{\sum_j B_j} \qquad i \neq j \qquad (35.2)$$

where A_i = the accessibility of location (city) i
B_j = the attractiveness of location (city) j, for example, the number of opportunities available at j
$f(c_{ij})$ = the impedance function representing the separation between location (city) i and j

Here, the number of employment opportunities, or the total population of a location, is used to reflect the attractiveness of the location. The impedance function is an inverse function of travel cost. This accessibility index is location-based. It computes the accessibility for each location or city. Other accessibility indices (e.g., damage-based analyses) also may be used in transportation network vulnerability analysis (Sohn 2006; Chen et al. 2007).

In order to measure the road network vulnerability under potential climate change scenarios, a transportation modeling process with, for example, the traditional four-step model, including trip generation, trip distribution, mode choice, and trip assignment, should be conducted in a before-and-after (disfunction) study. The accessibilities for each traffic analysis zone (TAZ) and the whole system are computed under various scenarios. The one with the most system accessibility reductions is considered to be the most challenging scenario for the transportation system.

Once the impact analysis is done, the accessibility index–based methodology can be adopted to evaluate the vulnerability of different transportation systems, such as the road transportation system, the rail transportation system, the marine transportation system, and so on. This process identifies the most vulnerable transportation systems or subsystems to climate change and helps to prioritize transportation systems for retrofitting. The aforementioned procedure also could be employed to identify the most potentially dangerous climate change scenarios. Transportation planners, engineers, and decision makers can use this to prepare for the worst-case scenario or to identify the vulnerability of different transportation systems under other specific climate change scenarios.

Vulnerable transportation system identification presents an overall view of the systems or zones that suffer as a result of climate change. Since budget is often a concern, this system provides a way in which involved parties can make informed decisions based on the projected costs and physical weaknesses of the transportation system in the face of climate change. This methodology also can be used to identify vulnerable transportation systems and mark the *criticality* of the whole system. The accessibility reduction of the whole transportation system is computed by removing a component (e.g., one section of road segment) of the system. The accessibility reduction without this link is compared with the original intact system, showing a representation of the criticality of the one component. Similarly, the accessibility reduction of each component is calculated through this process, which is called a *network scan*. The criticality of each component then is compared, and the most critical components are identified. Using this process, planners, engineers, and decision makers can better understand what measures must be taken to strengthen the components of the transportation system and at what risk.

35.4.3 Adaptation Strategies, Approaches, and Methods

Adaptation strategies may be incorporated in the long-range transportation planning process from transportation planning to maintenance and operations. For example, it can occur in the transportation planning process when locating corridors or selecting transportation routes and facilities. In the short term, adaptation strategies may be incorporated into transportation operations and maintenance processes. These strategies are also useful when considering the reexamination of current facility designs and constructions standards. As climate change becomes an increasing concern, adaptation strategies should be incorporated along the way, in various forms, and with various methods. These strategies can involve other agents, and partnerships should be formed. For example, in operation and maintenance, flood insurance should be considered. However, for the purposes of this chapter, the discussion of adaptation strategies, methodologies, and measures focuses on transportation planning, operation, and maintenance.

Adaptation Planning. One of the most effective adaptation strategies to reduce the negative impacts of climate change is to locate people and infrastructure away from vulnerable areas, such as costal and low-lying areas. Currently, most transportation planning is performed based on the transportation supply and demand interaction without considering the impacts of climate change. This is inadequate given the possible impacts of climate change on the transportation system.

In vulnerable areas such as the coastal areas, the impact of climate change variables, such as sea level rise, land subsidence, and hurricanes, should be considered carefully and measured, and the affected areas should be delineated accurately. Flood insurance maps were created originally for individual residencies and buildings in coastal or low-lying areas that have the potential of flooding risk. These maps are often used by transportation engineers to design transportation infrastructures for the 100- or 500-year flood zones. However, these maps are not of the appropriate precision for transportation infrastructure design in the climate change scenarios discussed earlier in this chapter. More accurate flooding maps with multiple climate change scenarios, especially with transportation infrastructure highlighted, are needed so that transportation planners and engineers are better informed of the risks to locate transportation infrastructure in certain areas.

Considering the uncertainty in climate change projection, scenario planning should be applied to adapt to climate change. Transportation plan alternatives should correspond with varying scenarios of climate changes. The potential risks of planning transportation infrastructures in vulnerable coastal or low-lying areas could be illustrated with the results of an impact analysis that details the levels of risk. In order for this process to work successfully, there must be an inventory of transportation infrastructures, especially critical infrastructures, in vulnerable areas. The inventory should detail the attributes of infrastructures as well as provide a priority list indicating which segments need additional support with respect to transportation planning, risk analysis, and funding prioritization.

Adaptation at the Project Design and Construction Level. As mentioned previously, current transportation facility design standards may be inadequate to adapt to climate change. Therefore, transportation engineers must reconsider whether existing transportation facilities design and construction standards are capable of resisting extreme events caused by climate changes. It is important to examine the basis on which transportation infrastructure design and construction standards are designed and the corresponding logic behind this (Meyer 2008). Transportation engineers must consider how climate change will affect infrastructure within 50 to 100 years in future and propose new measures of design standards to adapt to the new environment.

The most important component in design and construction adaptation is the updating of current design specifications. The risks and uncertainties posed by climate change make modifications to existing design and construction standards especially important. Design and construction standards are based on past experiences and knowledge. However, past experience and knowledge have a potential to change significantly in the coming future. According to Meyer (2008), "In the coastal environment, design practice assumes that flood events would essentially behave in a manner similar to a riverine environment. Bridge failure mechanisms associated with recent storm events have resulted in a reevaluation of these assumptions" (p. 8). For example, hurricane damage to Gulf Coast bridges resulted primarily from the combination of storm surge and wave crests, and the implication of this assumed frequency of storms is that designs do not consider the effect of wave actions on bridges (Meyer 2008). Design standards need to be modified to adapt to climate change. In conclusion, climate change must be understood by transportation planners and engineers, and current design standards must reflect this knowledge.

Adaptation at the Operations and Maintenance Level. While transportation planning and infrastructure design and construction address long-term adaptation strategies to climate change, daily transportation operation and maintenance practices are the most visible components affected by climate change. According to the Transportation Research Board (2008), about 15 percent of total delays on the nation's highways are caused by bad weather such as snow, ice, rain, hurricanes, and fog, and approximately 75 percent of air travel delays are weather-related (TRB Special Report 290 2008). With increased frequency, intensity, and incidence of bad weather as a result of climate change, it becomes increasingly important to have effective adaptation strategies and measures within daily transportation operations and maintenance schedules and practice.

Transportation agencies must address weather-related emergency traffic operations. The use of contraflow operations for emergency evacuation is common in vulnerable areas, such as those prone to hurricanes and snow. Proactive planning and emergency operation management for poor weather conditions are critical to respond to weather-related emergency situations. The capacity of transportation networks for evacuation purposes must be evaluated. Emergency operations could close roadways or lanes, impacting the nonemergency links of the transportation network. Proper traffic management measures are needed to accommodate traffic flows successfully under these emergency operation situations.

Communications among state and local transportation agencies are necessary because climate change has no boundaries. Transportation system operators must work together with weather forecasters and emergency response planners to reduce the response time of emergency transportation incidences and to improve the efficiency and effectiveness of their responses. Emergency management should be considered as a crucial role of DOTs and other transportation agencies in the face of climate change.

Maintenance plays a large role in this process. Inspection frequencies may be increased to ensure that infrastructures remain structurally sound, or culverts may need to be maintained on a more frequent interval to deal with flash-flood stream conditions (TRB Special Report 290 2008). Better maintenance materials such as heat-resistant asphalt and stronger concrete should be selected to increase the strength of transportation infrastructures. This includes using materials that can withstand increased freeze-thaw cycles in northern regions and less frost-susceptible foundation materials (TRB 2008, Haas et al. 2009). Other measures include debris clearing. Investments in road maintenance on older facilities must be considered in the planning, design, and construction process.

35.4.4 Assess and Prioritize Adaptation Strategies

Both short- and long-term adaptation strategies should be adopted. Each possible strategy should be assessed to estimate potential tradeoffs in its adaptation. Alternative adaptations then should be prioritized according to certain criteria and constraints.

A commonly used methodology to assess adaptation strategies is the cost-benefit analysis (CBA). CBA could evaluate the tradeoff between the costs of the adaptation strategy and its benefits. Cost-benefit analysis has been applied for many years to set priorities for road safety measures (Trilling 1978). The goal of this methodology is to maximize the benefits considering each available alternative. Using CBA, the benefit of adapting to climate change will be the cost spent if nothing is done. The effectiveness of adaptation strategies also will be assessed based on the cost spent and benefits received. Adaptation strategies therefore could be prioritized using the value of the net benefits of different adaptation strategies.

There are many other possible methodologies to select strategies. One methodology to consider is the optimization method. Different from CBA, the optimization method seeks to find the best adaptation strategy to achieve certain goals under a given budget or other constraints. For instance, linear programming is a commonly used optimization method. Here, an objective function is formulated so as to maximize the benefits of the adaptation. Limited budget, maximum network disruptions, and other similar factors can be treated as constraints. A simple example of linear program is as follows. The objective function maximizes the performance of the transportation network, such as to maximize the total accessibility and to minimize the total VMT of the whole network. The budget, the free-flow travel time, and so on serve as the model constraints. An optimal solution could be found by solving this linear program.

The method of optimization (Scaparra and Church 2006; Bell et al. 2008) assumes a worst-case scenario of climate change, which potentially will maximize the disruptions of the transportation system. The problem then can be treated as a game between two players—one is climate change, who wants to do the maximum damage to transportation system, and the other is transportation decision makers, who want to minimize this effect by taking adaptation strategies. A bilevel model can be used to solve this problem. The best adaptation strategy with the minimum effect on the transportation system could be found under the worst climate change scenario. However, this methodology is more complicated than CBA.

Other system optimization methods and program evaluation measures could be used in the assessment and prioritization of adaptation strategies. The most important component to these methods is choosing of criteria. The traditional monetary cost, benefit, and transportation system performance must be considerations because the goal of adaptation is to make the transportation system run more efficiently with the least amount of damage.

35.4.5   Response Challenges

Climate change poses challenges to the field of transportation. While much work has been done to study the possible effects of climate change and adaptation strategies, much remains unknown. Further studies must address the following problems.

Uncertainties in Short-Term Climate Change Projections.   There is a mismatch between climate change projections and transportation planning and infrastructure design. Climate experts usually forecast climate changes on a continental or national level, over hundreds or thousands of years. However, climate forecasts on the state or local level over short periods of time is currently limited. As a result, considering the relative short time span of transportation planning and operation, obtaining accurate climate change projection is still a challenge. This mismatch makes it difficult to plan and design transportation infrastructure to adapt to climate change because we know little about how it will change. Researchers must provide detailed and accurate climate change information that relates to transportation over smaller regions and shorter periods of time. Dynamic planning and engineering strategies, as well as the decision-making process, should be modified in such ways as to adapt to newly emerging information.

Institutional Collaborations Among Different Transportation Agencies.   In the United States, transportation facilities are owned and operated by both public agencies and private sectors. Highways, bridges, and public transportation infrastructure are owned and operated by state and local governments; railroads and pipelines are privately owned and operated; ports are joint public-private operations; and many airport capital improvements are funded by the federal government and supplemented by state and local grants and passenger facility charges (TRB Special Report 290 2008). Transportation facilities involve many players, and they are owned and managed in a way that makes coordination, communicating, and decision-making a challenge. Climate change challenges transportation systems across multiple modes and over multiple geographic regions, as illustrated by Hurricane Katrina. Transportation agencies should work together to share information and experiences, promoting efficiency and effectiveness among climate change responders.

Multiresponse to Changes in the Transportation World.   Climate change is challenging transportation in multiple ways. The majority of investments are aimed at retrofitting or upgrading existing transportation systems or providing new capacity at the margins. Existing transportation planning, design, and construction are based on the previous knowledge and cannot accommodate the changing climate. New planning regulations and design specifications take time to be implemented. Transportation professionals in both the public and private sectors should be engaged in operational planning so as to respond to short-term traffic congestion, and disrupted system operation, thereby promoting more effective traffic emergency and evacuation management procedures. This calls for improved capability when dealing with daily traffic operation and maintenance. The responses to climate change in the transportation sector should be in multiple ways so as to build a more robust transportation system not only in the future but also today.

35.5 CONCLUSION

Climate change is a real concern to transportation planners, engineers, and decision makers. Infrastructures must be both mitigated and adapted to accommodate climate changes, although many aspects of which cannot be precisely determined at the moment. This uncertainty is one of the many challenges that climate change presents. Climate change is not only a global concern but also a national and local concern. At each level, the changes will manifest themselves differently. As information of climate change is revealed, the many involved parties must understand their available options and work toward developing solutions that will better prepare our transportation system for the future.

In order to effectively understand the potential impacts of climate change, the solutions also must be clearly understood. Mitigation and adaptation are both necessary components of climate change and transportation practice. First, mitigation efforts should be taken. This involves looking at GHG emissions, transportation practices, and land-use developments, as discussed previously. Fuel standards, multimodal transportation, and smart growth policies are all important components of mitigation.

In addition to mitigation practices, steps to adapt current infrastructure and practices to climate change must be made. Vulnerable links should be identified and understood in the context of the whole network. There are multiple ways to determine which links and systems are most vulnerable. Perhaps the most common approach is cost-benefit analysis. Here, as in other scenarios, the cost of doing nothing always must be a consideration. Mitigation must involve identification, but also short- and long-term interventions, such as those in the maintenance process and design standard changes, so as to holistically address climate change.

As transportation officials consider how to react to climate change, mitigation and adaptation strategies can be effective tools to better understand the problem and potential solutions to address and minimize the costs of climate change on the transportation system and infrastructure.

ACKNOWLEDGMENTS

This research was supported in part by U.S. National Science Foundation Award BCS-0616957, National Natural Science Foundation of China Award 50738004, and China National High Technology Research and Development Program 863 Award 2009AA11Z220. Any opinions, findings, and conclusions or recommendations expressed in this chapter are those of the authors and do not necessarily reflect the views of the sponsors.

REFERENCES

American Association of State Highway and Transportation Officials. 2004a. AASHTO LRFD Bridge Design Specifications. Washington, DC: AASHTO.

Bell, M. G. H., et al. 2008. "Attacker-Defender Models and Road Network Vulnerability." *Mathematical & Physics Science Engineering* 366:1893–1906.

Berdica K. 2002. "An Introduction to Road Vulnerability: What Has Been Done, Is Done and Should Be Done." *Transport Policy* 9:117–127.

Bruton, M. L., and Hicks, M. J. 2005. "Hurricane Katrina: Preliminary Estimates of Commercial and Public Sector Damages." Available at www.marshall.edu/cber/research/katrina/Katrina-Estimates.pdf.

Burkett, V. R. 2002. "Potential Impacts of Climate Change and Variability on Transportation in the Gulf Coast/ Mississippi Delta Region." In: *The Potential Impacts of Climate Change on Transportation: Summary and Discussion Papers*.

California Coastal Commission. 2001. "Overview of Sea Level Rise and Some Implications for Coastal California." Available at www.coastal.ca.gov/climate/SeaLevelRise2001.pdf.

Cambridge Systematics, Inc. 2009. "Transportation Adaptation to Global Climate Change." *National Transportation Policy Project Report.* Available at www.infrastructureusa.org/wp-content/uploads/2009/12/transportationadaptation.pdf.

Chen, A., et al. 2007. "Network-Based Accessibility Measures for Vulnerability Analysis of Degradable Transportation Networks." *Network Spatial Economy* 7:241–256.

Chu H. C. and Meyer, M. D. 2009. "An Approach to Measure CO_2 Emissions of Truckonly Toll Lanes." Presented at 88th Annual Meeting of Transportation Research Board, Washington, DC.

Cruce, T. 2009. "Adaptation Planning: What U.S. States and Localities Are Doing." Pew Center on Global Climate Change. Available at www.pewclimate.org/docUploads/state-adapation-planning-august-2009.pdf.

DuVair, P. et al. 2002. "Climate Change and Potential Implications for California's Transportation System." In: *The Potential Impacts of Climate Change in Transportation: Workshop Summary and Proceedings.* Washington, DC: U.S. Department of Transportation, Center for Climate Change and Environmental Forecasting. Available at http://climate.volpe.dot.gov/workshop1002/.

Energy Information Administration. 2007. "Emissions of Greenhouse Gases in the United States, 2006." Washington, DC: Office of Integrated Analysis and Forecasting. U.S. Department of Energy. Available at ftp://ftp.eia.doe.gov/pub/oiaf/1605/cdrom/pdf/ggrpt/057306.pdf.

Environmental Protection Agency (EPA). 2005. "Emission Facts: Average Carbon Dioxide Emissions Resulting from Gasoline and Diesel Fuel." Available at www.epa.gov/otaq/climate/420f05001.htm.

Environmental Protection Agency (EPA). 2000. "Emission Facts: Average Annual Emissions and Fuel Consumption for Passenger Cars and Light Trucks." Available at www.epa.gov/otaq/consumer/f00013.pdf.

Environmental Protection Agency (EPA). 2010. "Precipitation and Storm Changes." Available at www.epa.gov/climatechange/science/recentpsc.html.

EPA State and Local Climate Change and Energy Program. "Developing a GHG Inventory." Available at www.epa.gov/statelocalclimate/state/activities/ghg-inventory.html.

EPA. "State Climate Action Plans Database." Available at http://yosemite.epa.gov/gw/StatePolicyActions.nsf/webpages/index.html.

Federal Highway Administration. 2005. "Coastal Bridges and Design Storm Frequency." Washington, DC: Office of Bridge Technology.

Haas, R., et al. 2006. "Climate Impacts and Adaptations on Roads in Northern Canada." Transportation Research Board 2006 Annual Meeting, Washington, DC.

Hegerl, G. C., Zwiers, F. W., Braconnot, P. et al., 2007. "Understanding and Attributing Climate Change." In: *Climate Change 2007: The Physical Science Basis. Contribution of Working Group I to the Fourth Assessment Report of the Intergovernmental Panel on Climate Change,* Solomon, S., Qin, D., Manning, M., et al., eds. Cambridge, England: Cambridge University Press.

ICF Consulting. 2006. "Assessment of Greenhouse Gas Analysis Techniques for Transportation Projects." American Association of State Highway and Transportation Officials (AASHTO), NCHRP 25-25(17).

ICF International. 2007. "The Potential Impacts of Global Sea Level Rise on Transportation Infrastructure." Phase 1: Final Report: The District of Columbia, Maryland, North Carolina and Virginia.

ICF International. 2008. "Integrating Climate Change into the Transportation Planning Practice: Final Report." Federal Highway Administration. Available at www.fhwa.dot.gov/hep/climatechange/climatechange.pdf.

IPCC. 2001. *Climate Change. 2001: Impacts, Adaptation, and Vulnerability. Contribution of Working Group II to the Third Assessment Report of the Intergovernmental Panel on Climate Change.* New York: Cambridge University Press. Available at www.ipcc.ch/.

IPCC. 2007. "Summary for Policymakers." In: *Climate Change 2007: Impacts, Adaptation and Vulnerability. Contribution of Working Group II to the Fourth Assessment Report of the Intergovernmental Panel on Climate Change,* M. L. Parry, O. F. Canziani, J. P. Palutikof, P. J. van der Linden, and C. E. Hanson, Eds. Cambridge, England: Cambridge University Press, pp. 7–22.

IPCC. 2007. "Summary for Policymakers." In: *Climate Change 2007: The Physical Science Basis. Contribution of Working Group I to the Fourth Assessment Report of the Intergovernmental Panel on Climate Change,* S. Solomon, D. Qin, M. Manning, Z. Chen, M. Marquis, K. B. Averyt, M. Tignor, and H. L. Miller, eds. Cambridge, England: Cambridge University Press.

Jacob. K. et al. 2007. "Vulnerability of the New York City Metropolitan Area to Coastal Hazards, Including Sea-Level Rise: Inferences for Urban Coastal Risk Management and Adaptation Policies." In *Managing Coastal Vulnerability.* New Yorks: Elsevier, pp. 139–156.

Johnston, R. A. and Rodier, C. J. 1999. "Synergisms Among Land Use, Transit and Travel Pricing Policies." *Transportation Research Record 1670*, 1999, pp. 3–7.

Lutsey, Nicholas P. and Daniel Sperling. 2008. "Transportation and Greenhouse Gas Mitigation." *Climate Action, United Nations Environmental Program* 2007, pp. 191–194.

McNeil, S. 2009. Special Report 299: "Reducing Transportation Greenhouse Gas Emissions and Energy Consumption: A Research Agenda." Prepared for TRB. Available at http://onlinepubs.trb.org/onlinepubs/sr/SR299Adaptation.pdf.

Meyer, M. D. 1999. "Demand Management as an Element of Transportation Policy: Using Carrots and Sticks to Influence Travel Behavior." *Transportation Research Part A: Policy and Practice*. 33, 7–8.

Meyer, M. D. 2008. "Design Standards for U.S. Transportation Infrastructure: The Implications of Climate Change." Transportation Research Board. Available at http://onlinepubs.trb.org/onlinepubs/sr/sr290Meyer.pdf.

National Assessment Synthesis Team and US Global Change Research Program. 2000. "National Assessments of the Potential Consequences of Climate Variability and Change."

National Research Council. 2008. "Potential Impacts of Climate Change on U.S. Transportation." Washington, DC: Transportation Research Board. Available at http://onlinepubs.trb.org/onlinepubs/sr/sr290.pdf.

National Research Council. 2009. "Driving and the Built Environment The Effects of Compact Development on Motorized Travel, Energy Use, and CO_2 Emissions." Washington, DC: Transportation Research Board.

Neumann, J. and Price, J. 2009. "Adapting to Climate Change: The Public Policy Response." RFF Report (Resources for the Future). Available at www.rff.org/rff/documents/RFF-Rpt-Adaptation-NeumannPrice.pdf.

Peterson, T. C. et al. 2008. "Climate Variability and Change with Implications for Transportation." Washington, DC: Transportation Research Board. Available at onlinepubs.trb.org/onlinepubs/sr/sr290Many.pdf.

Peterson, T. C., McGuirk, M. Houston, T. G., et al. 2006. *Climate Variability and Change with Implications for Transportation*. Washington, DC: National Oceanic and Atmospheric Administration and Lawrence Berkeley National Laboratory.

The Pew Center on Global Climate Change. "Greenhouse Gas Emissions Targets." Available at www.pewclimate.org/what_s_being_done/in_the_states/emissionstargets_map.cfm.

Rossiter, L. 2004. "Climate Change Impacts on the State Highway Network: Transit New Zealand's Position." *Transit New Zealand*, Wellington, New Zealand.

Scaparra, M. P., and Church, R. L. 2006. "A Bilevel Mixed-Integer Program for Critical Infrastructure." *Computers & Operations Research* 35:1905–1923.

Searchinger, T., Heimlich, R., Houghton, R. A., et al. 2008. "Use of U.S. Croplands for Biofuels Increases Greenhouse Gases Through Emissions from Land-Use Change." *Science* 319, 5867:1238–1240.

Shen, Suwan, Zhong-Ren Peng, Zun Wang, and Hemant Salokhe. 2010. "Impacts of Vertical Data Accuracy on Estimating the Effects of Sea Level Rise on Local Highway Infrastructure." In: 89th TRB Annual Meeting, Washington, DC, TRB Paper 10-3961.

Shladover. S.E. 2002. "Potential Contributions of Intelligent Vehicle/Highway Systems (IVHS) to Reducing Transportation's Greenhouse Gas Production." *Transportation Research Part A: Policy and Practice* 27, 3: 207–216.

Sohn, J. 2006. "Evaluating the Significance of Highway Network Links Under Flood Damage: An Accessibility Approach." *Transportation Research Part A* 40:491–50.

Sperling, Daniel, and Yeh, Sonia. 2009. "Low Carbon Fuel Standards." *Issues in Science and Technology* 25, 2.

Suarez, P. et al. 2005. "Impacts of Flooding and Climate Change on Urban Transportation: A Systemwide Performance Assessment of the Boston Metro Area." *Transportation Research Part D* 10:231–244.

Titus. J. G. 2002. "Does Sea Level Rise Matter to Transportation Along the Atlantic Coast?" In: *The Potential Impacts of Climate Change on Transportation, Summary and Discussion Papers*. Washington, DC: Federal Research Partnership Workshop, Brookings Institution, Oct. 1–2, pp. 135–150.

Transportation and Climate Change Clearing House. "GHG Inventories, Forecasts, and Transportation Data." Available at http://climate.dot.gov/ghg-inventories-forcasts/index.html.

Transportation Research Board. 2008. "Potential Impacts of Climate Change on U.S. Transportation." *Transportation Research Board Special Report 290*. Washington, DC.

Trilling, D. R. 1978. "A Cost-Effectiveness Evaluation of Highway Safety Countermeasures." *Traffic Quarterly* 32:41–67.

Turner S. M. et al. 1998. "ITS Benefits: Review of Evaluation Methods and Reported Benefits." Report number: FHWA/TX-99/1790-1.

United Nations Environment Program. "Adaptation." Available at www.unep.org/climatechange/UNEPsWork/Adaptation/tabid/241/language/en-US/Default.aspx.

United Nations Envrionment Program. "Climate Change." Available at www.unep.org/climatechange/.

United Nations Framework Convention on Climate Change. "Adaptation." Available at www.unfccc.int/adaptation/items/4159.php.

U.S. Climate Change Program. 2008. "Impacts of Climate Change and Variability on Transportation Systems and Infrastructure: Gulf Coast Study, Phase I."

U.S. Climate Change Science Program. 2009. "Coastal Sensitivity to Sea-Level Rise: A Focus on the Mid-Atlantic Region." Available at www.climatescience.gov/Library/sap/sap4-1/final-report/.

U.S. Department of Transportation, Federal Highway Administration. 2008. *Project Development and Design Manual*. Available at http://flh.fhwa.dot.gov/resources/manuals/pddm/Chapter_07.pdf#7.1.1.

U.S. DOT. 2009. *ITS/Operations Resource Guide*. Available at www.resourceguide.its.dot.gov/default.asp.

U.S. Energy Information Administration. 2009. "Emissions of Greenhouse Gases Report." Available at www.eia.doe.gov/oiaf/1605/ggrpt/carbon.html.

U.S. Environmental Protection Agency. "Climate Change." Available at www.epa.gov/climatechange/.

U.S. Environmental Protection Agency. 2009. "2009 U.S. Greenhouse Gas Inventory Report." Available at www.epa.gov/climatechange/emissions/usinventoryreport09.html.

U.S. Environmental Protection Agency. 2009. "Coastal Sensitivity to Sea Level Rise: A Focus on the Mid-Atlantic Region. Available at www.climatescience.gov/Library/sap/sap4-1/final-report/default.htm.

U.S. Environmental Protection Agency. 2009. *Inventory of U.S. Greenhouse Gas Emissions and Sinks: 1990 to 2007*. Washington, DC: Available at www.epa.gov/climatechange/emissions/usinventoryreport09.html.

US Global Change Research Program. 2003. "US National Assessment of the Potential Consequences of Climate Variability and Change Educational Resources Regional Paper: The Southeast." Available at www.usgcrp.gov/usgcrp/nacc/education/southeast/se-edu-5.htm#Economic.

USGS. 2009. http://edc.usgs.gov/guides/dem.html. Accessed on July 23, 2009.

The World Bank. "Water and Climate Change." Available at http://web.worldbank.org/WBSITE/EXTERNAL/TOPICS/EXTWAT/0,,contentMDK:21723353~pagePK:148956~piPK:216618~theSitePK:4602123,00.html.

The World Bank. 2010. *World Development Report 2010: Development and Climate Change*. Washington, DC. Available at http://econ.worldbank.org/WBSITE/EXTERNAL/EXTDEC/EXTRESEARCH/EXTWDRS/EXTWDR2010/0,,contentMDK:21969137~menuPK:5287816~pagePK:64167689~piPK:64167673~theSitePK:5287741,00.html.

Yang, C., McCollum, D., McCarthy, R., and Leighty, W. 2009. "Meeting an 80% Reduction in Greenhouse Gas Emissions from Transportation by 2050: A Case Study in California." *Transportation Research Part D: Transport and Environment*. 14, 3:147–156.

Zimmerman, R. 2003. "Global Climate Change and Transportation Infrastructure: Lessons from the New York Area." In: *The Potential Impacts of Climate Change on Transportation: Workshop Summary and Proceedings*. Washington, DC: U.S. DOT (Center for Climate Change and Environmental Forecasting) in cooperation with the U.S. EPA, U.S. DOE, U.S.GCRP, pp. 91–101.

INDEX